MICRO-ORGANISMS IN FOODS

6

MICRO-ORGANISMS IN FOODS

6

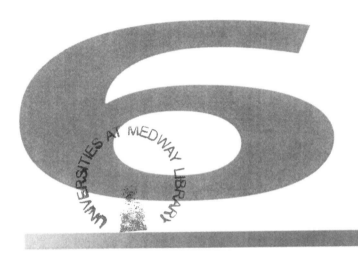

MICROBIAL ECOLOGY OF FOOD COMMODITIES

SECOND EDITION

ICMSF

Kluwer Academic / Plenum Publishers
New York, Boston, Dordrecht, London, Moscow

The author has made every effort to ensure the accuracy of the information herein. However, appropriate information sources should be consulted, especially for new or unfamiliar procedures. It is the responsibility of every practitioner to evaluate the appropriateness of a particular opinion in in the context of actual clinical situations and with due considerations to new developments. The author, editors, and the publisher cannot be held responsible for any typographical or other errors found in this book.

ISBN: 0-306-48675-X

eISBN: 0-306-48676-8

233 Spring Street, New York, New York 10013

(Formerly published by Chapman & Hall, ISBN 0-7514-0430-6)

http://www.wkap.nl

10 9 8 7 6 5 4 3 2 1

A C.I.P. record for this book is available from the Library of Congress.

Printed in the United States of America

Contents

Preface

The second edition of *Microbiology of Foods 6: Microbial Ecology of Food Commodities* was written by the ICMSF, comprising 16 scientists from 11 countries, plus consultants and other contributors to chapters.

The intention of the second edition was to bring the first edition (published in 1996) up to date, taking into account developments in food processing and packaging, new products, and recognition of new pathogens and their control acquired since the first edition.

The overall structure of the chapters has been retained, viz each covers (i) the important properties of the food commodity that affect its microbial content and ecology, (ii) the initial microflora at slaughter or harvest, (iii) the effects of harvesting, transportation, processing, and storage on the microbial content, and (iv) an assessment of the hazards and risks of the food commodities and (v) the processes applied to control the microbial load.

In 1980s, control of food safety was largely by inspection and compliance with hygiene regulations, together with end-product testing. *Microorganisms in Foods 2: Sampling for Microbiological Analysis: Principles and Specific Applications (2nd ed. 1986)* put such testing on a sounder statistical basis through sampling plans, which remain useful when there is no information on the conditions under which a food has been produced or processed, e.g. at port-of-entry. At an early stage, the Commission recognized that no sampling plan can ensure the absence of a pathogen in food. Testing foods at ports of entry, or elsewhere in the food chain, cannot guarantee food safety.

This led the Commission to explore the potential value of HACCP for enhancing food safety, particularly in developing countries. *Microorganisms in Foods 4: Application of the Hazard Analysis Critical Control Point (HACCP) System to Ensure Microbiological Safety and Quality (1988)* illustrated the procedures used to identify the microbiological hazards in a practice or a process, to identify the critical control points at which those hazards could be controlled, and to establish systems by which the effectiveness of control could be monitored. Recommendations are given for the application of HACCP from production/harvest to consumption, together with examples of how HACCP can be applied at each step in the food chain.

Effective implementation of HACCP requires knowledge of the hazardous microorganisms and their response to conditions in foods (e.g. pH, a_w, temperature, preservatives). The Commission concluded that such information was not collected together in a form that could be assessed easily by food industry personnel in quality assurance, technical support, research and development, and by those in food inspection at local, state, regional or national levels. *Microorganisms in Foods 5: Characteristics of Microbial Pathogens (1996)* is a thorough, but concise, review of the literature on growth, survival, and death responses of foodborne pathogens. It is intended as a quick reference manual to assist making judgements on the growth, survival, or death of pathogens in support of HACCP plans and to improve food safety.

The second edition of *Microorganisms in Foods 6: Microbial Ecology of Food Commodities (2004)* is intended for those primarily in applied aspects of food microbiology. For 17 commodity areas, it describes the initial microbial flora and the prevalence of pathogens, the microbiological consequences of processing, typical spoilage patterns, episodes implicating those commodities with foodborne illness, and measures to control pathogens and limit spoilage. Those control measures are presented in a standardized format, and a comprehensive index has been added.

The second edition of *Microorganisms in Foods 6: Microbial Ecology of Food Commodities* has been written following *Microorganisms in Foods 7: Microbiological Testing in Food Safety Management (2002)*. The latter illustrates how systems such as HACCP and GHP provide greater assurance of safety than microbiological testing, but also identifies circumstances where microbiological testing still plays a useful role in systems to manage food safety. It continues to address the Commission's objectives to: (a) assemble, correlate, and evaluate evidence about the microbiological safety and quality of foods; (b) consider whether microbiological criteria would improve and assure the microbiological safety of particular foods; (c) propose, where appropriate, such criteria; (d) recommend methods of sampling and examination; (e) give guidance on appraising and controlling the microbiological safety of foods. It introduces the reader to a structured approach for managing food safety, including sampling and microbiological testing. The text outlines how to meet specific food safety goals for a food or process using Good Hygienic Practice (GHP) and the HACCP system. Control measures as used in GHP and HACCP are structured into three categories: those that influence the initial level of the hazard, those that cause reduction, and those that may prevent increase, i.e. during processing and storage. In *Microorganisms in Foods 6*, a control section following each commodity group uses this structured approach.

Microorganisms in Foods 5, 7, and the second edition of *Microorganisms in Foods 6* (2005) are intended for anyone using microbiological testing and/or engaged in setting Microbiological Criteria, whether for the purpose of Governmental Food Inspection and Control or in Industry. The contents are essential reading for food processors, food microbiologists, food technologists, veterinarians, public health workers and regulatory officials. For students in Food Science and Technology, they offer a wealth of information on Food Microbiology and Food Safety Management, with many references for further study.

Editorial committee

T. A. Roberts (Joint Chairman)
J.-L. Cordier
L. Gram
R. B. Tompkin
J. I. Pitt (Joint Chairman)
L. G. M. Gorris
K. M. J. Swanson

ICMSF Members during preparation of the second edition of *Microbiology of Foods 6: Microbial Ecology of Food Commodities*

Chairman	M. B. Cole	
Secretary	M. van Schothorst (retired 2003)	
	L. Gram (from 2003)	
Treasurer	J. M. Farber	
Members	R. L. Buchanan	J.-L. Cordier
	S. Dahms	R. S. Flowers
	B. D. G. M. Franco	L. G. M. Gorris
	J.-L. Jouve	F. Kasuga
	A. M. Lammerding	Z. Merican
	J. I. Pitt (to 2002)	M. Potter
	K. M. J. Swanson	P. Teufel
	R. B. Tompkin (to 2002)	
Consultants	J. Braeunig (2000)	M. Germini (2003)
	L. G. M. Gorris (2000)	F. Kasuga (2002–03)
	H. Kruse (2000)	X. Lui (2003)
	J. I. Pitt (2003)	M. Potter (2002–03)
	T. A. Roberts (2001–03)	R. Stephan (2003)
	K. M. J. Swanson (2000)	R. B. Tompkin (2003)
	M. Zwietering (2003)	

Contributors and reviewers

	Chapter	*Contributors*	*Reviewers*
1	Meat	J. Greig (Can)	
		T. Nesbakken (Norway)	—
		R. Stephan (Switz)	
2	Poultry	F. Kasuga (Japan)	J. E. L. Corry (UK)
			T. Humphrey (UK)
3	Fish	F. Kasuga (Japan)	Q. L. Yeoh (Malaysia)
4	Feeds	—	B. Veldman (Neth)
			F. Driehuis (Neth)
			C. Jakobsen (Den)
5	Vegetables	M. L. Tortorello (USA)	M. Kundura (USA)
6	Fruit	—	—
7	Spices	—	—
8	Cereals		T. Smith (USA)
			S. Hood (USA)
9	Nuts	—	—
10	Cocoa	—	—
11	Oils & fats	—	R. van Santen (Neth)
			G. Naaktgeboren (Neth)
12	Sugar	L. Eyde (Aus)	
13	Soft drinks	C. Stewart (Aus)	
		K. Deibel (USA)	—
14	Water	—	—
15	Eggs	R. Buchner (USA)	J. E. L. Corry (UK)
			T. Humphrey (UK)
16	Milk	J. Braunig (Ger)	
		P. Hall (USA)	
17	Fermented beverages	A. Lillie (Den)	
		P. Sigsgaard (Den)	—
18	Index	J. Eyles (Aus)	—

1 Meat and meat products

I Introduction

Red meat is derived from a number of animal species (e.g. cattle, sheep, goat, camel, deer, buffalo, horse, and pig). Total world production of red meats and quantities in international trade can be obtained from http://apps.fao.org/page/collections?subset=agriculture, a part of http://www.fao.org.

Red meat has the potential to carry pathogenic organisms to consumers. In the past, the main public health problem was caused by the classical zoonoses, i.e. diseases or pathogens that can be transmitted from animals to human beings, such as bovine tuberculosis, and also produce pathological changes in animals. However, the measures introduced by classical meat inspection (inspection, palpation, and incision) have proved highly effective against them. Thus, tuberculosis shows very typical changes of the lymph nodes (granulomatous lymphadenitis); they can be reliably detected by incision of the nodes during meat inspection. However, today, the main problem is latent zoonoses. These pathogens occur as a reservoir in healthy animals, where they produce no pathological conditions or changes. However, they can contaminate the food chain in meat production, for instance during slaughtering. The slogan "healthy animals, healthy food" is not true from this point of view. Strict maintenance of good practices of slaughter hygiene in meat production is of central importance, because microbiological hazards are not eliminated in the slaughtering process. Bacteria able to cause food-borne disease, and which can constitute a hazard in at least some meat products, include *Salmonella* spp., thermophilic *Campylobacter* spp., enterohemorrhagic *Escherichia coli* (e.g. serogroup O157; EHEC), some serovars of *Yersinia enterocolitica*, *Listeria monocytogenes*, *Clostridium perfringens*, *Staphylococcus aureus*, *Cl. botulinum*, and *Bacillus cereus*. Meats are also subject to microbial spoilage by a range of microorganisms including *Pseudomonas* spp., *Shewanella*, Enterobacteriaceae, *Brochothrix thermosphacta*, lactic acid bacteria (LAB), psychrotrophic clostridia, yeasts, and molds.

In recent years, bovine spongiform encephalopathy (BSE) ("mad cow disease") has attracted public health attention. The first cases of BSE were reported in Great Britain in November 1986. It appears probable that the disease can be transmitted to humans by food. The prions that cause the disease are very resistant to chemical and physical influences, i.e. to heat, UV, and ionizing radiations and disinfectants. Prions are sensitive to certain alkaline substances and moist heat under high pressure. An effective disinfectant measure is steam sterilization at 133°C and 3 bar pressure for 20 min. On the basis of current knowledge, the cause of the BSE epidemic was animal feed (meat- and bone-meal and the like) containing brain, eyes or spinal cord of infected animals, and other tissues that had been inadequately heated during the production process.

To protect human health, the use of certain bovine organs (so-called specified risk materials: brain, eyes, spinal cord, spleen, thymus (sweetbread), bovine intestines of cattle >6 months old, visible lymph and nerve tissue, as well as lymph nodes) is prohibited for manufacturing foodstuffs, gelatine, tallow, drugs or cosmetics. More information and actual data can be obtained from the following web-sites: http://www.oie.int/eng/en_index.htm; http://www.who.int/mediacentre/factsheets/fs113/en/; http://www.defra.gov.uk/animalh/bse/index.html; http://www.aphis.usda.gov/oa/bse/; http://www.tseandfoodsafety.org/; http://www.unizh.ch/pathol/neuropathologie/.

This chapter, however, mainly describes the microorganisms that contaminate red meats and meat products, and factors and operations that increase or decrease the numbers or spread of microorganisms during processing, storage, and distribution. It also contains sections on the microbiology of froglegs and snails as foods.

A *Definitions*

Red meat is primarily the voluntary striated skeletal muscular tissue of "red" meat animals. The muscle is made up of contractile myofibrillar proteins, soluble sarcoplasmic proteins (e.g. glycolytic enzymes and myoglobin) and low molecular weight soluble organic and inorganic compounds. Connective tissue is in intimate association with muscle cells and can constitute up to 30% of total muscle protein. Fat cells occur subcutaneously and both within and surrounding the muscle. Within a muscle, fat cells are located in the perimysial space. Up to one-third of the weight of some muscles may be fat. Muscle tissues also contain 0.5–1% phospholipid.

Meat as legally defined commonly includes various organs ("variety meats" or "offals"). The organs and other parts of the carcass that are regarded as edible vary between countries. The heart has some similarities to skeletal muscle and is composed of striated involuntary muscle, connective tissue, and some lipid. The liver contains uniform liver cells with a network of blood vessels and epithelial-lined sinusoids. In the kidney, there is a meshwork of connective tissue that supports renal tubules, small veins, and arteries.

B *Important properties*

Meat has a high water and protein content, is low in carbohydrates and contains a number of low molecular weight soluble constituents (Table 1.1). The vitamin content (μg/g) of muscle is approximately: thiamine, 1; riboflavin, 2; niacin, 45; folic acid, 0.3; pantothenic acid, 10; B_6, 3; B_{12}, 0.02 and biotin, 0.04 (Schweigert, 1987). The concentrations of vitamins vary with species, age, and muscle. Pork muscle has 5–10 times more thiamine than is found in beef or sheep muscle. Vitamins tend to be higher in organs (e.g. liver and kidney) than in muscle.

Meat is a nutritious substrate with an a_w (0.99) suitable for the growth of most microorganisms. Growth is primarily at the expense of low molecular weight materials (carbohydrates, lactate, and amino acids). Microbial proteolysis of structural proteins occurs at a very late stage of spoilage (Dainty *et al.*, 1975).

Table 1.1 Approximate composition of adult mammalian muscle after *rigor mortis*

Component	% Wet weight
Water	75
Protein	19
Lipid	2.5
Glycogen[a]	0.1
Glucose[a,b] and glycolytic intermediates[a]	0.2
Lactic acid[a]	0.9
Inosine monophosphate[b]	0.3
Creatine[b]	0.6
Amino acids[b]	0.35
Dipeptides (carnosine and anserine)[b]	0.35
pH[a]	(5.5)

Lawrie (1985).
[a]Varies between muscles and animals.
[b]Varies with time after rigor mortis.

During death of the animal when the oxygen supply to the muscle is cut off, anaerobic glycolysis of stored glycogen to lactic acid lowers the pH. Post-mortem glycolysis continues as long as glycogen is available or until a pH is reached which inhibits the glycolytic enzymes. In typical muscles this pH is 5.4–5.5. In some muscles (e.g. beef *sternocephalicus* muscle), glycolysis ceases at a pH near 6 even though considerable glycogen remains. The ultimate pH varies between muscles of the same animal and between animals, and is determined by the glycogen content of the muscle and the accessibility of glycogen to glycolysis. The pH of post-rigor muscle can vary from 5.4–5.5 (lactate content close to 1%) to 7.0 (very little lactate present). The lactate content of muscle is inversely proportional to its pH. On the surfaces of beef and sheep carcasses, the availability of oxygen permits aerobic metabolism to continue, and much of the exposed surface tissue has a pH >6 (Carse and Locker, 1974), which facilitates microbial growth.

In the live animal, the glycogen concentration of muscle averages 1%, but varies considerably. Glycogen in pig muscle is readily depleted by starvation and moderate exercise, whereas glycogen in the muscles of cattle is more resistant to starvation and exercise. In both species, pre-slaughter stress (e.g. excitement and cold) depletes muscle glycogen. Glycogen is more concentrated in liver (2–10%) than in muscle, and its content is also affected by pre-slaughter conditions. A low concentration of glycogen in muscles results in a high ultimate pH, which gives rise to "dark-cutting" beef or dark, firm and dry meat (DFD).

The amount of glucose in post-rigor muscle varies with pH (Newton and Gill, 1978) being virtually absent in muscle of pH > 6.4. In normal-pH (5.5–5.8) muscle, glucose is present at about 100–400 μg/g (Gill, 1976). Liver has a high glucose content (3–6 mg/g), which appears to be independent of pH (Gill, 1988).

By the time the ultimate pH is reached, adenosine triphosphate has largely broken down to inosine monophosphate (IMP). During the storage of meat, IMP and inosine continue to degrade to hypoxanthine, ribose, and ribose phosphate. Ribose, inosine, and IMP can be used as energy sources by a number of fermentative Gram-negative bacteria, and ribose by *Broch. thermosphacta*, and a number of lactic acid bacteria.

Fatty tissue contains less water than muscle, has a pH near neutrality with little lactate, and contains low molecular weight components (glucose and amino acids) from serum (Gill, 1986). Consequently, microbial growth on fat is slower than on the surface of muscle.

C Methods of processing and preservation

Animals are raised on farms where some are grazed and some are raised under intensive or almost industrial conditions. The microflora in the intestinal tract or on the external surfaces of the animals may vary with the systems of animal production (e.g. more fecal material on the hides of feed-lot cattle). Animals may be slaughtered when young (e.g. calves at 3–4 weeks of age), or when 1 or 2, or several, years old (e.g. cattle and sheep). At the abattoir, the skin of cattle and sheep is removed, the skin of pigs is usually scalded (although it is removed in some plants), then the intestinal tract and viscera are removed. The carcass may then be washed, where regulations permit it, or not, and then chilled.

Spoilage organisms grow rapidly on meat, which is a highly perishable commodity. Thus, trade in meat, even at the local level, depends on some degree of preservation that controls the spoilage flora.

The most important means of preservation are chilling or freezing, cooking (includes canning), curing, drying, and packaging. Packaging affords extension of shelf-life. Several procedures to reduce microbial growth are often combined. Chilled temperature storage enables fresh meat to be held for only a limited time before spoilage ensues. However, by vacuum-packaging chilled meat in films of low permeability to gases, or by packaging in modified atmospheres, storage-life may be extended for up to at least 12 weeks.

D Types of meat products

Red meats are traded as chilled or frozen carcasses, large primal pieces or retail size portions, chilled or frozen offals, chilled vacuum-packed meat, dried meats, fermented meat, raw or cooked cured products, cooked uncured meat and cooked canned products.

II Initial microflora

A Ruminants

At birth, the digestive tract of a ruminant is physiologically that of a monogastric. The rumino-reticulum complex develops quickly between 2 and 6 weeks of age when the animals are fed roughage. Initially, large numbers of *E. coli, Cl. perfringens* and streptococci are in the gut and are shed in feces (10^7–10^8 cfu *Cl. perfringens*/g, 10^9 cfu *E. coli*/g). After about 2 weeks, *Cl. perfringens* declines to about 10^4 cfu/g and *E. coli* to ca. 10^6 cfu/g at about 3 months of age. When comparing fecal excretion of coliforms, the mean count for eight calves between 3 and 8 weeks of age was $\log_{10}$ 7.2 cfu/g and for adult cows was $\log_{10}$ 4.9 cfu/g (Howe *et al.*, 1976).

Invasive serotypes of salmonellae, such as *Salmonella* Typhimurium and *S.* Enteritidis, are more difficult to control in the live animal than serovars occasionally found in feed. In the first few days of life, young ruminants are more susceptible to salmonellae. Calves dosed with *S.*Typhimurium prior to 3 days of age were more easily infected, and excreted salmonellae for longer periods and in greater numbers, than calves inoculated at 18 days (Robinson and Loken, 1968). At slaughter, salmonellae were also detected more frequently in mesenteric and cecal lymph nodes from the younger animals. Young calves that are surplus to dairy farm requirements may be sold through markets and dealers to rearing farms. In England, salmonellae have been found in 3.7% of environmental samples taken at calf markets and in 20.6% of swab samples from vehicles used to transport calves (Wray *et al.*, 1991). Salmonellae have also been detected on the walls (7.6% of swabs) and floors (5.3% of swabs) at dealers' premises (Wray *et al.*, 1990). The mixing of young susceptible calves and their subsequent transport to rearing farms disseminates salmonellae. On arrival at rearing farms, the prevalence of salmonellae in calf feces is relatively low but can increase rapidly. When fecal samples were taken from 437 calves within 2 days of arrival at a rearing farm, salmonellae were detected in 5.3% (Hinton *et al.*, 1983). After about 2 weeks on the farm, salmonellae were found in 42.2% of 491 animals sampled. The shedding rate of salmonellae peaked at 2–3 weeks and then declined; this is possibly associated with the development of a more adult-type intestinal flora.

The high concentration of volatile fatty acids and the pH of the fluid in the developed rumen of the well-fed animal provide some protection to infection with salmonellae and verotoxin-producing *E. coli* (often of the serogroup O157; VTEC) (Chambers and Lysons, 1979; Mattila *et al.*, 1988). Viable cells of these organisms disappear from rumen fluid at a rate faster than expected from wash-out. Starved or intermittently fed ruminants are more susceptible to infection as salmonellae and VTEC O157 can then grow in the rumen. This probably influences the percentage of infected animals on farms during periods of low feed intake (e.g. drought, mustering, shearing or dipping and high stocking densities). On farms, the prevalence of salmonellae in the intestinal tract varies (Edel and Kampelmacher, 1971). Outbreaks of clinical bovine salmonellosis tend to show seasonal patterns. In the UK, most incidents of bovine salmonellosis occur in summer–autumn and peak near the end of the grazing season (Williams, 1975). Peaks of clinical salmonellosis in sheep in New Zealand during summer–autumn have been associated with movement and congregation of sheep for shearing and dipping.

In a study of the prevalence of salmonellae in cow–calf operations (Dargatz *et al.*, 2000), of 5 049 fecal samples collected from 187 beef cow–calf operations, salmonellae were recovered from 1 or more fecal samples collected on 11.2% (21 of 187) of the operations. Overall 78 salmonellae representing 22 serotypes were isolated from 1.4% (70 of 5 049) of samples, and multiple serotypes from eight samples from a single operation. The five most common serotypes were *S.* Oranienburg (21.8% of isolates) and *S.* Cerro (21.8%), followed by *S.* Anatum (10.3%), *S.* Bredeney (9.0%) and *S.* Mbandaka (5.1%).

Although it is broadly accepted that human salmonellosis is derived from foods, especially meat and poultry, firm proof is elusive. Sarwari *et al.* (2001) concluded from US data for 1990–1996, that there was a significant mismatch between the distribution of *Salmonella* species isolated from animals at the time of slaughter and that of isolates found in humans. This questions the validity of assumptions that raw animal products are the primary source for human salmonellosis, or whether there are methodological reasons for the difference.

The increased susceptibility to infection resulting from changes in the rumen can also affect the prevalence of salmonellae in cattle and sheep during transport from farm to slaughter, or in long transport from farm to farm when feeding patterns and type of feed are changed. Frost *et al.* (1988) reported a high prevalence of salmonellae in the mesenteric lymph nodes and rumen fluid of adult cattle during the first 18 days of entering a feed-lot from a market. After 80 days in the feed-lot, there was little evidence of salmonellae infection. Some of the deaths of sheep during sea-shipment from Australia to Singapore and the Middle East have been due to salmonellosis, which was associated with empty gastrointestinal tracts, loss of appetite and poor adjustment from grazing green pastures to dry feed.

Although healthy cattle may excrete thermophilic campylobacters in their feces, numbers are generally low (NACMCF, 1995). While thermophilic campylobacters are frequently found in the lower intestinal tract of ruminants (prevalence range 0–54%), it is usually present in numbers <1000/g. The organism occurs more frequently and in higher numbers in the feces of very young calves (<3–4 weeks old). It can be present in small numbers (<100/g) in the rumen, where it is probably only part of the transient flora.

Streams, fields, wild-life and other livestock are all likely to be sources of salmonellae and *C. jejuni*. The opportunity for animal-to-animal spread is increased in intensively reared animals. Salmonellae contaminated feeds can be a source of infection. Jones *et al.* (1982) reported an infection of cattle on three dairy farms that was directly attributable to consumption of a vegetable fat supplement contaminated with *S.* Mbandaka.

L. monocytogenes can exist as a saprophyte in the plant–soil ecosystem, and clinical outbreaks of listeriosis in cattle and sheep have long been linked with feeding silage of inferior quality. *L. monocytogenes* has been reported in the feces of apparently normal cattle in many countries, whether animals were examined on the farm or at slaughter (Table 1.2). On Danish farms, where there was a high occurrence of the organism in dairy cows, it was commonly found in the feed (silage from different crops, and alkalized straw). Silage and decaying vegetable material can contain large numbers of *Listeria* spp. The higher incidence in Danish cattle than in Danish pigs has been associated with feeding wet plant material to cattle and providing dry feed to pigs (Skovgaard and Norrung, 1989).

VTEC is a group of *E. coli* that produces one or more verocytotoxins (VT) also known as Shiga toxins (STX). This group of bacteria has many synonyms. In the United States and to a varying extent in Europe, the notation Shiga-toxin producing *E. coli* (STEC) is used. The term, EHEC was originally used to denote VTEC causing hemorrhagic colitis (HC) in humans; later EHEC has been used as a synonym for VTEC in the medical domain in some European countries (SCVPH, 2003).

VTEC are frequently present in the feces of calves, cattle, buffaloes, sheep and goats (Mohammad *et al.*, 1985; Suthienkul *et al.*, 1990; Beutin *et al.*, 1993; Clarke *et al.*, 1994). These VTEC strains belong to a large number of serotypes. Some (e.g. O5:NM, O8:H9, O26:H11 and O111:NM) may cause diarrhea or dysentery with attaching–effacing lesions in calves (Moxley and Francis, 1986; Schoonderwoerd *et al.*, 1988; Wray *et al.*, 1989). However, VTEC has emerged as a pathogen that can cause food-borne

Table 1.2 *Listeria monocytogenes* in red-meat animals

Species	Sample site	Country	Number	% Positive	Reference
Cattle	Feces	Belgium	25	25	van Renterghem *et al.*, 1991
		Denmark	75	52	Skovgaard and Morgen, 1988
		New Zealand	15	0	Lowry and Tiong, 1988
		Yugoslavia	52	19	Buncic, 1991
	I.R.P.N.[a]	Yugoslavia	52	29	Buncic, 1991
Cattle	Feces		30	11	Johnson *et al.*, 1990
Cattle (dairy)	Feces	Finland	3 878	6.7	Husu, 1990
	Feces	USA	40	3	Siragusa *et al.*, 1993
Cattle (beef)	Lymph nodes	B&H	8	0	Loncarevic *et al.*, 1994
Cattle	Feces	Germany	138	33.3	Weber *et al.*, 1995
Cattle	Content of large intestine	Japan	9 539	2	Iida *et al.*, 1998
Cattle (dairy)	Feces	Sweden	102	6	Unnerstad *et al.*, 2000
Cattle	Feces	Scotland	29	31	Fenlon *et al.*, 1996
Sheep	Feces	New Zealand	20	0	Lowry and Tiong, 1988
Sheep	Feces	Germany	100	8	Adesiyun and Krishnan, 1995
Pig	Feces	Belgium	25	20	van Renterghem *et al.*, 1991
		Denmark	172	1.7	Skovgaard and Norrung, 1989
		Hungary		25.6	Ralovich, 1984.
		Yugoslavia	97	3	Buncic, 1991
	Tonsils	Yugoslavia	103	45	Buncic, 1991
Pig	Lymph nodes	B&H	21	5	Loncarevic *et al.*, 1994
Pig	Rectal swabs	Trinidad	139	5	Adesiyun and Krishnan, 1995
Pig	Feces	Germany	34	5.9	Weber *et al.*, 1995
Pig	Content of large intestine	Japan	5 975	0.8	Iida *et al.*, 1998
Pig	Tonsils	Finland	50	12	Autio *et al.*, 2000
Horse	Feces	Germany	400	4.8	Weber *et al.*, 1995

[a]I.R.P.N., Internal retropharyngeal nodes.

infections and severe and potentially fatal illness in humans. VTEC are the cause of human gastroenteritis that may be complicated by hemorrhagic colitis (HC) or hemolytic-uremic syndrome (HUS).

VTEC strains causing human infections belong to a large, still increasing number of O:H serotypes. A review of the world literature on isolation of non-O157 VTEC (by K.A. Bettelheim) is available on the MicroBioNet website (http://www.sciencenet.com.au). Most outbreaks and sporadic cases of HC and HUS have been attributed to O157:H7 VTEC strains. However, especially in Europe, infections with non-O157 strains, such as O26:H11 or O26:H−, O91:H−, O103:H2, O111:H−, O113:H21, O117:H7, O118:H16, O121:H19, O128:H2 or O128:H−, O145:H−, and O146:H21 are frequently associated with severe illness in humans.

Pathogenicity of VTEC is associated with several virulence factors. The main factor is the ability to form different types of exotoxins (verotoxins). They can be subdivided into a Verotoxin 1 group (Stx1) and a Verotoxin 2 group (Stx2). Characterization of the *stx1* and *stx2* genes revealed the existence of different variants in both Stx groups. At present, three *stx1* subtypes (*stx1*, *stx1c*, and *stx1d*) and several *stx2* gene variants have been described (e.g. *stx2*, *stx2c*, *stx2d*, *stx2e* and *stx2f*). Apart from the capability to produce verotoxins, these pathogroups may possess accessory virulence factors such as intimin (*eae*), VTEC auto-agglutinating adhesin (*saa*) or enterohemolysin (*ehxA*). Characterization of *eae* genes revealed the existence of different *eae* variants. At present, 11 genetic variants of the *eae* gene have been identified and are designated with letters of the Greek alphabet. It is believed that different intimins may be responsible for different host- and tissue cell tropism.

E. coli O157 is found in the feces of cattle and sheep (Table 1.3) and of water buffalo (Dorn and Angrick, 1991). It has been isolated from healthy cattle, from dairy and beef cattle and from pasture-fed and feed-lot cattle (Tables 1.3 and 1.4). In some studies, the highest prevalence appears to occur in young calves shortly after weaning (Meng *et al.*, 1994). Although individual animal infection with

Table 1.3 *Escherichia coli* O157:H7 in the feces of cattle and sheep

Country	Animal	No. samples	% Positive	Reference
Germany	Dairy cow	47	<2	Montenegro *et al.*, 1990
Germany	Bull	212	0.9	Montenegro *et al.*, 1990
Scotland	Cattle[a]	1 247	0.4[b]	Synge and Hopkins, 1992
Scotland	Sheep	450	<0.2	Synge and Hopkins, 1992
Spain	Diarrhoeic calf	78	1.3	Blanco *et al.*, 1988
Spain	Cattle	328	0.3	Blanco *et al.*, 1996a,b
USA	Adult dairy			
USA	Cow[c]	662	0.15	Wells *et al.*, 1991
USA	Dairy heifer[c]	394	3.0[d]	Wells *et al.*, 1991
USA	Dairy Calf [c]	210	2.3[d]	Wells *et al.*, 1991
USA	Cattle	10 832	1.0[e]	Hancock *et al.*, 1997
USA	Cattle	205	3.4[e]	Rice *et al.*, 1997
USA	Cattle	1 091	4.9	Besser *et al.*, 1997
USA	Cattle	327	28[e]	Elder *et al.*, 2000
USA	Cattle	1 668	1.3	McDonough *et al.*, 2000
USA	Sheep	70	15.7[f]	Kudva *et al.*, 1997
Thailand	Cattle	55	1.8[e]	Vuddhakul *et al.*, 2000
Czech Rep	Cattle	365	20[e]	Čížek *et al.*, 1999
Japan	Cattle	387	1.8	Miyao *et al.*, 1998
Japan	Cattle	306	1.6	Shinagawa *et al.*, 2000
France	Cattle	471	0.2	Pradel *et al.*, 2000
Italy	Cattle	450	12.9	Bonardi *et al.*, 1999
England	Cattle	48 000	15.8[e]	Chapman *et al.*, 1997
England	Sheep	1 000	2.2[e]	Chapman *et al.*, 1997
Netherlands	Cattle	540	10.6[e]	Heuvelink *et al.*, 1998
Netherlands	Sheep	101	4.0[e] (3.96)	Heuvelink *et al.*, 1998
Australia	Cattle	588	1.9	Cobbold and Desmarchelier, 2000
Canada	Cattle	98	11.2	Power *et al.*, 2000
Canada	Cattle	1 478	0.8	Wilson *et al.*, 1996
Canada	Cattle	1 000	0.4	Schurman *et al.*, 2000
Canada	Cattle	1 247	7.5	Van Donkersgoed *et al.*, 1999
Brazil	Cattle	197	1.5	Cerqueira *et al.*, 1999

B&H, Bosnia and Herzogovina.
[a]Five calves positive.
[b]Not stated if H7 or non-motile (NM).
[c]Some herds implicated in human illness.
[d]One of 17 from the heifer-calf group was *E. coli* O157:NM.
[e]Not stated if H7.
[f]Seventy sheep where tested over a 16-month period and 11 tested positive at least one time.
[g]Herd implicated in human illness.

E. coli O157:H7 appears to be transient, herd infection may be maintained (Wells *et al.*, 1991; Zhao *et al.*, 1995; Faith *et al.*, 1996). Drinking water may be a source of dissemination or maintenance of *E. coli* O157 on farms (Faith *et al.*, 1996). Growth of *E. coli* O157:H7 in rumen fluid is restricted by the pH and volatile fatty acid concentration in well-fed animals, but is not when the animal is fasted for 24–48 h (Rasmussen *et al.*, 1993). The impact of diet on fecal shedding of *E. coli* O157:H7 remains unclear (Tkalcic *et al.*, 2000).

Several outbreaks with life-threatening illness resulted in huge efforts to understand VTEC in ruminants (SCVPH, 2003).

More recent investigations have confirmed that the gastrointestinal tract of ruminants is the main reservoir of *E. coli* O157:H7 (Duffy *et al.*, 2001a), with a prevalence of VTEC O157:H7 in fecal samples of healthy cattle ca. 2–3% (United States) and 7.9% elsewhere. Group prevalence (i.e. a herd with at least one animal shedding) was 22.8% in Scotland. Prevalence as high as 64.1% has been reported in heifers (Conedera *et al.*, 2001).

Table 1.4 Type of cattle and incidence of *Escherichia coli* O157:H7 in feces

Cattle type	Country	No. of samples	% Positive	Reference
Unweaned dairy calf		649	<0.15	Hancock *et al.*, 1994
Unweaned dairy calf		6 894	0.36	NDHEP survey quoted in Zhao *et al.*, 1995
Unweaned dairy calf <10 days		304	<0.3	Martin *et al.*, 1994
Dairy calf <8 weeks old[a]		423	1.4	Garber *et al.*, 1995
Dairy calf >8 weeks old[a]		518	4.8	Garber *et al.*, 1995
Dairy calf 24 h to weaning[b]		570	1.9	Zhao *et al.*, 1995
Weaned dairy calf <4 months[b]		395	5.1	Faith *et al.*, 1996
Weaned dairy calf <4 months		560	1.8	Faith *et al.*, 1996
Cow and heifer[c]		193	4.7	Faith *et al.*, 1996
Bull and steer[c]		51	3.9	Faith *et al.*, 1996
Weaned calf[c]		273	2.9	Faith *et al.*, 1996
Cull dairy cow	USA	1 668	1.3	McDonough *et al.*, 2000
Feed-lot cattle	USA	327	28	Elder *et al.*, 2000
Dairy cows	USA	1 091	4.9	Besser *et al.*, 1997
Cull dairy cow	USA	205	3.4[d]	Rice *et al.*, 1997
Weaned heifers	USA	3 483	1.7[d]	Hancock *et al.*, 1997
Unweaned calves	USA	1 040	1.3[d]	Hancock *et al.*, 1997
Dairy cows	USA	4 762	0.4[d]	Hancock *et al.*, 1997
Bull calves 50–100 kg	Czech Rep	163	5.5[d]	Čížek *et al.*, 1999
Bulls 100–200 kg	Czech Rep	47	59.6[d]	Čížek *et al.*, 1999
Bulls 200–400 kg	Czech Rep	36	44.4[d]	Čížek *et al.*, 1999
Bulls 400–600 kg	Czech Rep	71	22.5[d]	Čížek *et al.*, 1999
Bulls slaughtered	Czech Rep	48	6.2[e]	Čížek *et al.*, 1999
Beef cattle	England	1 840	13.4[e]	Chapman *et al.*, 1997
Dairy cattle	England	1 661	16.1[e]	Chapman *et al.*, 1997
Adult cattle	Netherlands	540	10.6[e]	Heuvelink *et al.*, 1998
Veal calves	Netherlands	397	0,5[e]	Heuvelink *et al.*, 1998
Calves <7 days old	Australia	79	1.3[e]	Cobbold and Desmarchelier, 2000
Weanlings 1–14 weeks old	Australia	109	5.5[e]	Cobbold and Desmarchelier, 2000
Heifers	Australia	106	2.8[e]	Cobbold and Desmarchelier, 2000
Cattle type	Australia	588	1.9	Cobbold and Desmarchelier, 2000
Dairy calf <4 month	Brazil	64	3.1	Cerqueira *et al.*, 1999
Beef cattle >2 years	Brazil	76	1.3	Cerqueira *et al.*, 1999
Veal calves	Italy	90	0	Bonardi *et al.*, 1999
Feedlot cattle	Italy	223	16.6[e]	Bonardi *et al.*, 1999
Dairy cows	Italy	137	16.1[e]	Bonardi *et al.*, 1999
Yearling cattle	Canada	654	12.4	Van Donkersgoed *et al.*, 1999
Cull cows	Canada	593	2	Van Donkersgoed *et al.*, 1999
Calves <4 months	Spain	23	4.3	Blanco *et al.*, 1996 a,b

[a]One quarter of herds sampled were previously positive in the National Dairy Heifer Evaluation Project (NDHEP); three quarters of herds sampled were previously negative for *E. coli* O157. Calves that were older than 8 weeks were up to 4 months old.
[b]Fourteen herds sampled that were previously positive in NDHEP survey; fifty herds sampled that were previously negative for *E. coli* O157.
[c]Five herds sampled that were previously positive in Wisconsin survey; seven herds sampled that were previously negative for *E. coli* O157.
[d]Results from studies in the USA.
[e]At % positive = Not stated if H7.

Chapman *et al.* (1997) reported prevalence in bovines, ovines, and porcines to be 2.8, 6.1 and 4%, respectively, but found none in poultry. In dairy herds, only 1% of samples (113/10 832) were positive, whereas 9 of 15 herds had one or more positive isolate (Hancock *et al.*, 1997). Vågsholm (1999) found similar levels in Sweden: individual prevalence of 1–2% in calves and heifers sampled on the farm, and a herd prevalence of VTEC O157 of 10%, on 249 herds. In Canada, Van Donkersgoed *et al.* (1999) found a prevalence of non-O157 VTEC of 43%, and 8% prevalence of VTEC O157:H7 in fecal samples taken at slaughter, higher prevalence of VTEC in cull cows, and highest VTEC O157:H7 prevalence in calves. A European Community report on trends of zoonoses for 2000 (EC, 2002) reported prevalence of VTEC O157 in cattle herds (10% or more), individual bovines (1–5% or more) and beef or minced meat (0–1%).

In Australia, VTEC prevalence among cattle reared specifically for beef production (6.7%) was lower than that in dairy cattle (14.6%). Seasonal variation in shedding results in most cattle being positive in late summer–early autumn (Chapman *et al.*, 1997; Hancock *et al.*, 1997; De Zutter *et al.*, 1999; Tutenel *et al.*, 2002). Other animals carrying *E. coli* O157 include sheep, goats, wild deer, pigs, and seagulls (Synge, 1999; Chapman, 2000), feral pigeons (Dell'Omo *et al.* 1998), and zebu cattle (Kaddu-Mulindwa *et al.*, 2001).

Laegreid *et al.* (1999) reported the prevalence of *E. coli* O157:H7 in beef calves at weaning, prior to arrival at the feed-lot or mixing with cattle from other sources. Thirteen of 15 herds (87%) yielded one, or more than one, isolation of *E. coli* O157:H7 in fecal samples. All herds had high prevalence of anti-O157 antibodies (63–100% of individuals within herds seropositive) indicating *E. coli* O157:H7 occurrence before weaning and prior to entering feed-lots. Serological evidence suggested that most calves (83%) and all herds (100%) had been exposed to *E. coli* O157.

The site of colonization of EHEC has been identified as the lymphoid follicle-dense mucosa at the terminal rectum (Naylor *et al.*, 2003).

Elder *et al.* (2000) estimated the frequency of *E. coli* O157:H7 or O157:non-motile (EHEC O157) in feces and on hides within groups of cattle from single sources (lots) at meat processing plants. Of 29 lots sampled, 72% had at least one EHEC O157-positive fecal sample and 38% had positive hide samples. Overall, EHEC O157 prevalence in feces and on hides was 28% (91 of 327) and 11% (38 of 355), respectively.

Thran *et al.* (2001) suggested that screening fecal samples should not be limited to *E. coli* O157:H7, and that identification of STEC-positive cattle prior to slaughter should help to reduce the risk of beef contamination.

Other verotoxin-producing serotypes (e.g. O26:H11, O103:H2, O111:NM, O113:H21 and O157:NM) associated with human bloody diarrhea and HUS have also been isolated from sheep, calves, and cattle feces (Dorn *et al.*, 1989; Montenegro *et al.*, 1990; Wells *et al.*, 1991).

Cobbold and Desmarchelier (2000) examined 588 cattle fecal samples and 147 farm environmental samples from three dairy farms in southeast Queensland, Australia. STEC were isolated from 16.7% of cattle fecal samples and 4.1% of environmental samples: 10.2% serotyped as O26:H11 and 11.2% as O157:H7, with prevalences in the cattle samples of 1.7% and 1.9%. Prevalences for STEC and EHEC in dairy cattle feces were similar to those derived in surveys within the northern and southern hemispheres. Calves at weaning were identified as the cattle group most likely to be shedding STEC, *E. coli* O26 or *E. coli* O157. Cattle, particularly 1–14-week-old weaning calves, appear to be the primary reservoir for STEC and EHEC on the dairy farm.

Identifying environmental sources of *E. coli* O157:H7 in two feed-lots in southern Alberta, to identify management factors associated with the prevalence and transmission, Van Donkersgoed *et al.* (2001) isolated *E. coli* O157:H7 in pre-slaughter pens of cattle from feces (0.8%), feedbunks (1.7%), water troughs (12%) and incoming water supplies (4.5%), but not from fresh total mixed rations. Fresh total mixed rations did not support the growth of *E. coli* O157:H7.

Large populations of microorganisms are present on the hide and fleece and are composed of normal resident skin flora (e.g. micrococci, staphylococci and yeasts) and organisms, including *Salmonella* spp. and *L. monocytogenes*, derived from the environment (soil, pasture and feces). *Staph. xylosus* is the major species of that genus found on cattle hide (Kloos, 1980). The skin of sheep and goats carries relatively large populations of *Staph. xylosus* and *Staph. lentus*. *Staph. aureus* can be found in udders, teat canals, and milk, particularly when animals are mastitic. In colder climates, there is a greater proportion of psychrotrophic flora on the hide and fleece than in warmer climates. Growth on wet hide or fleece can change the numbers of some types of microorganisms. The amount of fecal material on the skin under feed-lot conditions can be large (several kg). Consequently, there is considerable variation in the microflora from animal-to-animal and site-to-site on the hide and fleece.

During transport of sheep and cattle from farm to abattoir, salmonellae, and other organisms shed in the feces (e.g. *L. monocytogenes* and *E. coli* O157:H7) will contaminate transport vehicles, markets, and abattoir holding areas. Although thermophilic campylobacters are a relatively fragile organism in the laboratory, they survive well in the environment.

The longer the animals are held before slaughter, the greater the salmonellae contamination of the outside of the animal, and the greater the prevalence of salmonellae in the intestinal tract. Anderson *et al.* (1961) found that 0.5% of calves were infected at the market, 0.6% infected when held for only a few hours before slaughter, but 35.6% when the calves were held in lairage for 2–5 days before slaughter. The prevalence of salmonellae in cattle feces can be 10 times that on the farm (Galton *et al.*, 1954). When sheep awaiting slaughter were held for 7 days, the incidence and numbers of salmonellae on the fleece, in the rumen liquor and in feces increased with time of holding (Grau and Smith, 1974).

In a UK survey (Small *et al.*, 2002), prevalences of *E. coli* O157, *Salmonella* spp. and *Campylobacter* spp. from swabs taken along the unloading-to-slaughter routes of animal movement in lairages of six commercial abattoirs, three for cattle and three for sheep, were 27.2, 6.1 and 1.1%, respectively, in cattle lairages, and 2.2, 1.1 and 5.6%, respectively, in sheep lairages. On cow hides, prevalences of the three pathogens were 28.8, 17.7 and 0%, respectively, and on sheep pelts 5.5, 7.8 and 0%.

Much of the contamination on the carcass is derived from the hide/fleece contaminated with gut contents. Contamination from the exterior of the animal can be reduced by not accepting for slaughter animals that are visibly dirty. Although this is difficult administratively, there is some evidence that it has helped to improve carcass hygiene, e.g. in the UK.

At the laboratory-scale, sub-atmospheric steam applied to bovine hide pieces inoculated with *E. coli* O157:H7 in fecal suspensions (McEvoy *et al.*, 2001), effected some reduction in viable numbers, indicating that steam condensing at $\leq 80 \pm 2°C$ can reduce *E. coli* O157:H7 when it is present on bovine hide, and suggesting a possible means of reducing cross-contamination to the carcass during slaughter and dressing.

B Pigs

In young piglets, the initial microflora of the intestinal tract is composed principally of high populations of *E. coli, Cl. perfringens* and streptococci (Smith, 1961). As the animal grows the numbers of these organisms decline, and non-sporing strict anaerobes become the predominant population in the lower intestine.

Young animals are more susceptible than older animals to infection with salmonellae. Clinical illness was formerly mostly caused by the host-adapted *S.* Cholerae-suis, but control measures have significantly reduced the number of outbreaks due to this serotype to <5% of the salmonella isolations reported from pigs in the UK (Hunter and Izsak, 1990), whereas *S.*Typhimurium and *S.* Derby accounted for 40–50% of isolations.

However, pigs shed a wide range of serotypes, often intermittently or transiently, without any evident symptoms of illness. Animal feed, whether derived from rendered and dried animal material (e.g. meat and bone meal, fish meal and feather meal) or derived from vegetable sources (e.g. grain, cotton seed and peanut meal), are a source of salmonellae. At slaughter, salmonellae were detected in the intestinal tract of 23% of pigs that were fed a mash containing contaminated fish meal, but were found in <2% when the fish meal was not used (Lee *et al.*, 1972). The form in which feed is presented affects the extent of salmonellae excretion. Excretion may be transient only when freshly prepared mash is fed (Linton *et al.*, 1970). However, the small numbers of salmonellae in the dry meal can grow in the mash during holding of bulk mash and in the residues in pipelines and troughs. This growth increases the rate of infection and the duration of excretion. Pelleting reduces the salmonellae contamination of meal with the extent of reduction dependent on the temperature achieved and the duration of exposure to high temperatures. In pigs raised on pelleted meal, salmonellae were detected in only 1 of 6 047 fecal samples, but were detected in almost all fecal samples after pigs were fed the same meal as a mash for 10 weeks (Edel *et al.*, 1967). Reduced infection rates in pigs have been observed in a number of other studies where pelleted meal was used (Edel *et al.*, 1970; Ghosh, 1972).

Elimination of salmonellae from feeds does not ensure the absence of salmonellae from pig-fattening farms. There are a number of other sources of salmonellae including pigs previously in the pens, birds, rats, and other animals. Breeding sows and boars may be infected. Movement of stock and animal attendants may spread salmonellae. Control of salmonellae contamination on the farm requires a multi-pronged approach (e.g. structural changes to the farm, restriction of movement of stock and personnel, disinfection of pens, feed pelleted at high temperature and by stocking with salmonellae-free animals) (Ghosh, 1972; Linton, 1979). It is very difficult to eradicate salmonellae from the environment of pigs in intensive piggeries (Oosterom and Notermans, 1983; Swanenburg *et al.*, 2001).

Nevertheless, in some countries, measures have essentially eradicated salmonellae from pigs and from pork. In five Swedish pig slaughterhouses, each visited six times, with a total of 3 388 samples from pork carcasses and the slaughterhouse environment all cultured negative for salmonellae (Thorberg and Engvall, 2001).

Because salmonellae can cause major economic losses to the swine industry, and the gut is a major reservoir for *Salmonella*, novel strategies to reduce their concentration in pigs immediately before processing have been explored. Respiratory nitrate reductase activity possessed by salmonellae catalyzes the intracellular reduction of chlorate to chlorite, which is lethal to salmonellae. Weaned pigs orally infected with 8×10^7 cfu of an antibiotic-resistant strain of *S.* Typhimurium were treated 8 and 16 h later via oral gavage (10 mL) with 0 or 100 mM sodium chlorate. Chlorate treatment significantly reduced caecal concentrations of salmonellae, the greatest reductions occurring 16 h after receiving the last chlorate treatment, indicating a possible means of reducing numbers of salmonellae before slaughter (Anderson *et al.*, 2001).

Attempts are also being made to identify bacteria that can be used as competitive exclusion cultures to prevent colonization by *S.* Typhimurium in pigs (Hume *et al.*, 2001a).

Thermophilic *Campylobacter* spp. are found at a very high frequency (61–100%) in the lower intestinal tract of pigs, often at counts of 10^3–10^4 cfu/g of feces (Teufel, 1982; Stern and Line, 2000). Presumably, animal-to-animal spread is the major mechanism for this widespread occurrence. The vast majority of strains isolated are *C. coli*.

Permanent colonization of the gut of neonatal pigs appears to be related to constant exposure of the piglets to feces containing campylobacters and is reduced by early removal of the piglets from the sows and rearing in nurseries isolated from sows (Harvey *et al.*, 2000). *Arcobacter* spp. can also be isolated from nursing sows and developing pigs (Hume *et al.*, 2001b).

L. monocytogenes has been detected in the feces and the lymph nodes, and on the tonsils of slaughtered pigs (Table 1.2). Samples (373) from 10 low-capacity slaughterhouses in Finland were examined for *Listeria* spp. (Autio *et al.*, 2000), 50 from carcasses, 250 pluck sets, and 73 from the slaughterhouse environment. Six slaughterhouses and 9% of all samples were positive for *L. monocytogenes*. Of the samples taken from pluck sets, 9% were positive for *L. monocytogenes*, the highest prevalence occurring in tongue (14%) and tonsil samples (12%). Six of 50 (12%) carcasses were contaminated with *L. monocytogenes*. In the slaughterhouse environment, *L. monocytogenes* was detected on two saws, in one drain, on one door and one table. Carcasses were contaminated with *L. monocytogenes* in two slaughterhouses where mechanical saws used for both brisket and back-splitting also tested positive for *L. monocytogenes*. Pulsed-field gel electrophoresis typing indicated that *L. monocytogenes* from the tongue and tonsils can contaminate the slaughtering equipment and in turn spread to carcasses.

Healthy pigs often carry serotypes of *Y. enterocolitica* that appear indistinguishable from human pathogenic strains (Table 1.5). The isolation rates from the throat, tonsils and tongue are often higher than those from the cecum or feces (Schiemann, 1989). The carriage rate varies greatly between herds and in different geographic locations. In one survey in England, although non-pathogenic biotypes were frequently encountered, pathogenic strains were rarely found (Table 1.5). In Danish herds, 82% were shown to contain pigs carrying *Y. enterocolitica*, and no association could be found between carriage rate of the organism and different types of herd management (Andersen *et al.*, 1991).

In Norway, IgG antibodies against *Y. enterocolitica* O:3 were found in sera from 869 (54.1%) of samples from 1 605 slaughter pigs from 321 different herds. In the positive herds, there were significantly fewer combined herds of piglets and fatteners than fattening herds.

In Denmark, Norway, Sweden, Holland, and Belgium, serotype O:3 is commonly found in pigs. Although serotype O:3 is common in pigs in Eastern Canada, it is rare in Western Canada where O:5,27 occurs in the swine population. In the United States, serotypes O:3, O:5,27, and O:8 have been detected on the tongues of pigs (Table 1.5). The appearance of strains of serovars O:3 and O:9 in Europe, Japan in the 1970s and in North America by the end of the 1980s, is an example of a global pandemic (Tauxe, 2002).

Colonization of pigs appears to be from animal contact rather than from environmental sources. Risk factors included: using a farm-owned-vehicle for transport of slaughter pigs to abattoirs and using straw bedding for slaughter pigs. Epidemiological data suggested that the herd prevalence of *Y. enterocolitica* O:3 can be reduced by minimizing contact between infected and non-infected herds (Skjerve *et al.*, 1998). Young pigs become carriers within 1–3 weeks of entering contaminated pens. Within a short time of infection, large numbers (10^6 cfu/g) of *Y. enterocolitica* are excreted in the feces. This may continue for some weeks before the numbers fall to <100/g (Fukushima *et al.*, 1983).

Transport of pigs to slaughter appears to result in increased shedding of salmonellae in feces (Williams and Newell, 1967, 1970). Part of the explanation for this may be that the stress of transport increases the flow of material along the intestinal tract. Salmonellae in the cecum and colon can then more readily appear in the feces. However, real differences in salmonellae prevalence in feces have been observed between pigs killed on the farm and at abattoirs (Kampelmacher *et al.*, 1963). Transport vehicles and lairages in which animals are held at abattoirs become contaminated. Cross-contamination of feet, skin and intestinal tract can then take place.

Holding pigs for long times in lairages at abattoirs has long been known to increase the prevalence of salmonellae in the intestinal tract (Tables 1.6 and 1.7). When pigs from one producer were killed at two abattoirs, salmonellae were isolated from 18.5% of pigs killed on the first day, 24.1% on the second day and 47.7% on the third day after leaving the farm (Morgan *et al.*, 1987b).

Y. enterocolitica can also be transferred between pigs and appear in low numbers in caecal contents when pigs are held for about 20 h in abattoir lairages (Fukushima *et al.*, 1991). The skin can be contaminated by *Yersinia* spp. excreted during transport and into the pens.

The skin of pigs carries a large population of microorganisms composed of the resident flora together with organisms acquired from the environment on the farm, in transport and in lairage at the slaughterhouse. *Staph. hyicus* is quite common in the nares and the hairy cutaneous areas of pigs, and *Staph. aureus* is also carried on pig skin (Devriese, 1990).

Table 1.5 *Yersinia enterocolitica* in pigs

Country	Site	No. samples	% Positive	Reference
Belgium	Tongue	29	97	Wauters *et al.*, 1988
	Tonsil	54	61	Wauters *et al.*, 1988
Canada (Quebec)	Feces	200	12.5	Mafu *et al.*, 1989
Canada (Ontario)	Throat	20	50	Schiemann and Fleming, 1981
	Tonsil	20	20	Schiemann and Fleming, 1981
	Tongue	20	55	Schiemann and Fleming, 1981
Canada (Alberta)	Throat	100	22[a]	Schiemann and Fleming, 1981
Canada (BC)	Throat	98	3[a]	Schiemann and Fleming, 1981
Canada (PEI)	Tonsils	202	28.2	Hariharan *et al.*, 1995
	Tonsils	202	7.4[b]	Hariharan *et al.*, 1995
Canada (Quebec)	Feces	1 010	13.5[c]	Pilon *et al.*, 2000
Canada (Alberta)	Cecal contents	1 420	16.5	Letellier *et al.*, 1999
	Cecal contents	1 420	1.8[d]	Letellier *et al.*, 1999
	Cecal contents	120	0.07[e]	Letellier *et al.*, 1999
	Cecal contents	1 420	0.6[f]	Letellier *et al.*, 1999
	Cecal contents	1 420	0.35[g]	Letellier *et al.*, 1999
Denmark	Feces	1 458	24.7	Andersen, 1988
	Tonsil	2 218	25	Andersen *et al.*, 1991
England	Feces	1 300	0	Hunter *et al.*, 1983
	Tonsil	631	0	Hunter *et al.*, 1983
Finland	Tongue	51	98[h]	Fredriksson-Ahomaa *et al.*, 1999
	Carcass	80	21.3	Fredriksson-Ahomaa *et al.*, 2000
	Liver	13	38.5	Fredriksson-Ahomaa *et al.*, 2000
	Kidney	13	84.6	Fredriksson-Ahomaa *et al.*, 2000
	Heart	8	62.5	Fredriksson-Ahomaa *et al.*, 2000
Japan	Feces	1 200	3.8[i]	Fukushima *et al.*, 1991
Japan	Tonsils	140	24.3	Shiozawa *et al.*, 1991
	Cecal contents	140	24.3	Shiozawa *et al.*, 1991
	Oral cavity swabs	40	85	Shiozawa *et al.*, 1991
	Masseter muscles	25	36	Shiozawa *et al.*, 1991
Norway	Tonsil	461	31.7	Nesbakken and Kapperud, 1985
Netherlands	Feces	100	16	de Boer and Nouws, 1991
		100	1[j]	de Boer and Nouws, 1991
	Tonsil	86	38.4	de Boer and Nouws, 1991
		86	3.5[a]	de Boer and Nouws, 1991
	Tongue	40	15	de Boer and Nouws, 1991
		40	5[j]	de Boer and Nouws, 1991
Switzerland	Tonsils or mesenteric lymph nodes	570	6.7	Offermann *et al.*, 1999
Trinidad	Rectal swabs	141	16.1	Adesiyun and Krishnan, 1995
	Tongue swabs	141	6.4	Adesiyun and Krishnan, 1995
	Tonsils	150	7.3	Adesiyun and Krishnan, 1995
USA	Tongue	31	6.5	Doyle *et al.*, 1981
		31	19.4[k]	Doyle *et al.*, 1981

Percent positive is for serotype O:3 (biotype 4), except as indicated by superscripts.
[a]Serotype O:5,27.
[b]Serotype O:5,27.
[c]93.5% were serotype O:3.
[d]Serotype O:5.
[e]Serotype O:8.
[f]Serotype O:9.
[g]non-typeable.
[h]98% were serotype O:3.
[i]In addition, there were 3.6% serotype O:3 (biotype 3) and a single isolate of O:5,27.
[j]Serotype O:9.
[k]Serotype O:8; serotype O:5,27 was found in 16% of another 49 pig tongues (Doyle and Hugdahl, 1983).

Table 1.6 Salmonellae in animals at slaughter

Country	Species	% Samples positive								Reference
		Feces	MLN	HLN	Liver	Gall bladder	Spleen	Cecal	Rumen	
Canada	Calf	–	–	0.7	–	–	–	–	–	Lammerding *et al.*, 1988
Canada	Cattle	–	1.5	1.2	–	–	–	–	–	Lammerding *et al.*, 1988
Canada	Pig	–	13.6	4.7	–	–	–	–	–	Lammerding *et al.*, 1988
Canada (PEI)	Pig	18	–	–	–	–	–	–	–	Mafu *et al.*, 1989
	Cattle	–	–	–	–	–	–	4.6	–	Abouzeed *et al.*, 2000
Canada	Swine	–	–	–	–	–	–	5.2	–	Letellier *et al.*, 1999
Canada	Cattle	0.08	–	–	–	–	–	–	–	Van Donkersgoed *et al.*, 1999
Canada	Cattle	–	–	–	–	–	–	–	0.3	Van Donkersgoed *et al.*, 1999
Denmark	Pig	3	4	–	–	–	–	–	–	Skovgaard and Nielson, 1972
Denmark	Swine	–	–	–	–	–	–	22	–	Baggesen *et al.*, 1996
England	Calf	1.1	1.3	2.2	1.3	0.6	–	–	–	Nazer and Osborne, 1976
England	Pig	7	6	–	–	–	–	–	–	Skovgaard and Nielson, 1972
England	Pig	5.8	–	–	–	–	–	–	–	Ghosh, 1972
Germany	Swine Swab 3.7%	–	3.3	–	–	–	–	–	–	Käsbohrer *et al.*, 2000
Holland	Calf	4.9	6	3.1	0.7	6.4	1.2	–	–	Guinee *et al.*, 1964
Holland	Cow	–	0.3	–	–	–	–	–	–	Guinee *et al.*, 1964
Holland	Pig	11.7	13.9	6.4	5	10	3.1	–	–	Kampelmacher *et al.*, 1963
India	Goat	2.6	2.3	–	0.7	–	0.3	–	–	Kumur *et al.*, 1973
India	Sheep	2	1.2	–	0.2	–	0.1	–	–	Kumur *et al.*, 1973
Ireland	Pig	16.4	–	–	–	–	–	–	–	Timoney, 1970
NZ	Calf	4.2	2.1	–	1.6	–	2.9	–	–	Nottingham *et al.*, 1972
NZ	Cattle	9.4	7.2	–	2.2	–	1.6	–	–	Nottingham and Urselmann, 1961
NZ	Sheep	1.7	4	–	–	–	0	–	–	Kane, 1979
N Ireland	Cattle	7.6	–	–	–	4.7	–	–	–	McCaughey *et al.*, 1971
Papua-N.G.	Pig	27	54	–	–	11.2	–	–	–	Caley, 1972
Saudi Arabia	Sheep/Goats	4.7	14.7	–	–	–	0.8	–	–	Nabbut and Al-Nakhli, 1982
Switzerland	Swine	0.9[e]	–	–	–	–	–	–	–	Offermann *et al.*, 1999
USA	Cattle	0.6	<0.4	–	–	–	–	–	–	Gay *et al.*, 1994
USA	Horse[a]	15.2	–	–	–	–	–	–	–	Anderson and Lee, 1976
USA	Pig[b]	23.5	–	–	–	–	–	–	–	Hansen *et al.*, 1964
USA	Pig[c]	–	50	–	–	–	–	–	–	Keteran *et al.*, 1982
USA	Cattle	–	–	–	–	–	–	–	0	Galland *et al.*, 2000
USA	Swine	–	–	–	–	–	–	61[d]	–	Harvey *et al.*, 1999
USA	Swine	12	–	–	–	–	–	–	–	Davies *et al.*, 1998
USA	Swine	–	21	–	–	–	–	61	–	Gebreyes *et al.*, 2000

MLN, Mesenteric lymph nodes; HLN, hepatic lymph nodes; PEI, Prince Edward Island.
[a]Cecal samples.
[b]Pigs killed within 3 h = 10% positive; pigs killed after 3 days in abattoir holding pens = 35%.
[c]Sows killed after 10–14 days in abattoir holding pen = 58.2% positive; slaughter hogs killed after 1–3 days = 31.3%.
[d]Lleocolic lymph node or cecal.
[e]Samples from tonsils or MLN (Offerman, 1999).

Table 1.7 Prevalence of *Salmonella* fecal shedding by animal species and class (various NAHMS national studies)

Year	Species and class	No. of samples	% Positive
1991–1992	Dairy calves	6862	2.1
1994	Feedlot cattle	4977	5.5
1995	Swine	6655	6.0
1996	Dairy cows	4299	5.4
1997	Beef cow-calf	5049	1.4
2001	Cull cows at market	2287	14.9

Wells *et al.*, 2001.

III Primary processing

A Ruminants

Pre-mortem inspection should remove from slaughter excessively dirty and obviously diseased animals. However, inspection cannot prevent slaughter of stock carrying human pathogens in the intestinal tract or on the hide or fleece. During slaughter and dressing, hocks, head, hide, or fleece and viscera are removed. These operations are important. The object is to do this with as little contamination as possible of the exposed sterile carcass tissue and of edible offals. The rumen, lower intestinal tract and the hide and fleece all carry a very large microbial population.

Feces may contain up to 10^6 spores of *Cl. perfringens*/g (Smith, 1961) as well as salmonellae (Tables 1.6 and 1.7) at levels of up to 10^8 cfu/g, thermophilic campylobacters and *L. monocytogenes* (Table 1.2). In the feces of healthy bobby calves there can be 10^6 cfu *C. jejuni*/g. In more adult animals, numbers are fewer (Grau, 1988). Rumen fluid may contain salmonellae and *C. jejuni* in low numbers. The hide and fleece can carry considerable numbers of salmonellae. Patterson and Gibbs (1978) found up to 4×10^6 cfu salmonellae/g of cattle hair, and 200 cfu salmonellae/cm^2 have been reported on sheep fleece (Grau and Smith, 1974). *L. monocytogenes* may also be on hide and fleece (Lowry and Tiong, 1988). Yeasts (e.g. *Candida, Cryptococcus* and *Rhodotorula* spp.) usually form only a small percentage of the microflora but can be as high as 12.7% of the microbial load (Dillon and Board, 1991). Hooves usually also carry a large microbial population. Scrapings from cattle hooves have yielded 260 cfu salmonellae/g (Patterson and Gibbs, 1978). Hides and hooves may be heavily contaminated with fecal material, particularly when cattle are intensively raised. Udders may be infected with *Staph. aureus* and other organisms. A significant percentage of *Staph. aureus* strains from mastitic cows, goats, and sheep produced enterotoxin C (Bergdoll, 1989; Gutierrez *et al.*, 1982; Stephan *et al.*, 2001).

The equipment used in the slaughter-dressing operation, and the hands and clothing of personnel can contaminate and spread contamination from animal-to-animal. Unless properly cleaned, saws, steel-mesh gloves, knives, scabbards, and other equipment can carry a high bacterial load and can be sources of salmonellae contamination. Intestinal tract material (rumen and lower intestine) is most likely to be the major source of VTEC (including *E. coli* O157:H7), salmonellae, *C. jejuni, Cl. perfringens,* and other clostridia for carcass and offals. Hide and fleece add most of the mesophilic aerobes (including bacilli) and the psychrotrophs (including psychrotrophic yeasts) to the carcass. The hide and fleece is also a source of staphylococci, *L. monocytogenes* and clostridia.

The extent and nature of contamination of the carcass and offal meat are reflections of the microbial status of the animal as presented for slaughter, and the care and standards of hygiene and sanitation used. Strict maintenance of food practices of slaughter hygiene in meat production is of central importance, because microbiological hazards are not eliminated in the slaughtering process. Chilling of carcasses and offals prevents growth of mesophilic pathogens and reduces the growth rate of psychrotrophic pathogens and spoilage organisms.

Stunning and bleeding. After animals are stunned, they fall to the floor where the hide can pick up fecal contamination.

It is possible for bacteria on the instruments used in slaughtering to contaminate some deep tissues through the blood stream (Mackey and Derrick, 1979). When the end of a captive bolt, heavily coated with bacteria, penetrated the skull of cattle, the organism could be recovered from the spleen but not from muscle. When cattle were pithed with heavily contaminated pithing rods, bacteria were found in the spleen and, at times, in muscles of the neck and flank. Bacteria from inoculated stick-knives (10^8–10^{11} cfu) were isolated from the blood and sometimes from the heart, liver, kidney, spleen, and lung of sheep though rarely from muscles. With relatively modest hygiene, it is unlikely that muscle is often contaminated by either the stick-knife or captive bolt. The relative ease with which sterile muscle can be obtained for experimental purposes from normal animals suggests that muscle is essentially sterile.

The bleeding process should be completed as quickly as possible, although efficiency of bleeding has little effect on microbial growth on meat (Gill, 1991).

The esophagus is cleared from surrounding tissue and tied or clamped close to the rumen ("rodding") to prevent leakage of rumen fluid, which would contaminate the neck and, during evisceration, the pleural region.

Skinning. Most of the microbial load on the carcass is derived from the skin, hide or fleece during skinning. Bacterial contamination includes the normal skin flora as well as organisms from soil and feces, which are on the skin, and includes yeasts, bacilli, micrococci, staphylococci, corynebacteria, moraxella, acinetobacter, flavobacteria, Enterobacteriaceae, *E. coli*, salmonellae and *Listeria* spp. In New Zealand, cattle hide and sheep fleece appear to be the major source of *L. monocytogenes* on carcasses (Lowry and Tiong, 1988). The predominant contamination is mesophilic. The percentage of psychrotrophs varies with season and geographic location, being highest in winter and in colder climates. Sometimes animals are washed before slaughter to remove loose dirt. However, this pre-slaughter washing can have a significant effect on microbial contamination of the carcass.

Hocks are removed and incisions through the skin are made along the inside of the legs, along the neck, sternum and abdomen and around the anus. Knives and the operator's fist are used to separate the skin from the underlying tissue before the rest of the skin is pulled away manually or mechanically. The hands of workmen handling hocks and skin become heavily contaminated, as do their knives, steels, and aprons. Salmonellae can often be found on the hands and equipment of these workers (Stolle, 1981). In one study in Germany, the highest contamination of cattle carcasses with salmonellae was associated with removal of hooves and freeing of the skin around the legs (Stolle, 1981). The incision through the contaminated skin carries microorganisms onto the carcass tissue. The knife blade and handle, and the fist of the operator, as these are used to free the skin, transfer mechanically organisms onto the carcass. Bacterial numbers are highest on regions of the carcass where the initial manual removal of the skin takes place and lowest where the skin is mechanically pulled away (Empey and Scott, 1939; Kelly *et al.*, 1980). The brisket is a site that is usually considered as a "dirty" site in terms of total bacterial contamination (Roberts *et al.*, 1980b). Organisms are also transferred to the carcass when fleece or hide touches exposed tissue, or when exposed tissue is handled by operators.

Cutting the skin around the anus and freeing the anal-sphincter and rectal end of the intestine are major sources of carcass contamination. The hide or fleece around this site and the tail are often contaminated with feces. Care taken during this operation is critical in limiting fecally derived contamination. Samples taken immediately after tissue was exposed during hide removal showed that there was considerably more contamination with *E. coli* and salmonellae of the perianal and rectal channel than of the hind-leg or brisket (Grau, 1979). The rectal end of the lower intestinal tract of beef animals is often enclosed in a plastic bag to limit contamination of the rectal channel and abdominal cavity. During subsequent carcass trimming, some of the contamination on the fatty tissue around the anal opening is removed. In the operation of releasing the anal-sphincter and rectum of sheep, the operator may handle the anus, and with this hand then handle the exposed tissue of the hind-leg. After the anal-sphincter and rectum are cut free, there can be about a 100-fold increase in *E. coli* and a significant increase of salmonellae on sheep carcasses without any detectable increase in the total aerobic viable count (Grau, 1986).

During mechanical hide-pulling on cattle, the intestine may be squeezed occasionally through cuts in the abdomen, made from the initial knife incision, and the intestine may rupture contaminating the abdomen and chest regions.

Elder *et al.* (2000) sampled carcasses at three points during processing: pre-evisceration, post-evisceration before antimicrobial intervention, and post-processing after carcasses entered the cooler. In 30 lots, 87% had at least one EHEC O157-positive pre-evisceration, 57% were positive post-evisceration and 17% were positive post-processing. Prevalence of EHEC O157 in the three post-processing samples was 43% (148 of 341), 18% (59 of 332), and 2% (6 of 330), respectively. Fecal and hide prevalence were significantly correlated with carcass contamination ($P = 0.001$), indicating the importance of strict maintenance of good practices of slaughter hygiene in the slaughtering process.

Chapman *et al.* (1993a,b) reported prevalence of *E. coli* O157:H7 on beef carcasses at abattoir level as 8.0% from rectal swab-negative and 30% from rectal swab-positive cattle. Elsewhere the reported prevalences of *E. coli* O157 in beef and veal carcasses were <1% (Daube, 2001).

Byrne *et al.* (2000) reported that power washing for 3 min significantly reduced *E. coli* O157:H7 counts on contaminated hides, but did not significantly reduce *E. coli* O157:H7 counts when transferred to the carcass. Prohibiting access to slaughter facilities of soiled animals is judged by some to be an important preventive measure in the dissemination of food pathogens including VTEC, but the efficacy of prohibiting entry of animals on the basis of selection of visible soiled cattle, to reduce carcass contamination, has been questioned (Van Donkersgoed *et al.*, 1997; Jordan *et al.*, 1999). In Dutch cattle- and calf-slaughtering establishments Heuvelink *et al.* (2001) obtained a significant reduction of visibly contaminated chilled carcasses (from 22 to 7%) over 4 months by introducing a statutory zero-tolerance policy of visible fecal contamination. Bolton *et al.* (2001) advocated the application of non-intervention HACCP systems, like the Hygiene Assessment Scheme (HAS) in operation in the UK, as an effective tool for reducing the microbial levels on beef carcasses. After implementation of the HAS system bacterial counts of $<2 \log_{10}$ cfu/cm^2 have been obtained. In addition, knife trimming, water wash, and application of steam vacuum are possible means to reduce or eliminate visible fecal contamination from carcasses (Castillo *et al.*, 1998b). Brown (2000) overviewed the implementation of HACCP in the meat industry.

Barkocy-Gallagher *et al.* (2001) studied the implied relationships between shedding of VTEC and carcass contamination. Within lots, 68.2% of post-harvest (carcass) isolates matched pre-harvest (animal) isolates. For individual carcasses, >65% of isolates recovered post-evisceration and in the cooler matched those recovered pre-evisceration, suggesting that most *E. coli* O157 carcass contamination originates from animals within the same lot and not from cross-contamination between lots.

Castillo *et al.* (1998c) using a chemical dehairing process under laboratory conditions, found reductions of *E. coli* O157:H7 counts on artificially contaminated bovine skin ranging from 3.4 to $>4.8 \log_{10}$ cfu/cm^2.

Evisceration. As part of the evisceration process, the brisket is cut, the abdomen is opened and the organs of the thorax and abdomen are removed. Offals are separated from the viscera and inspected. Care is needed to prevent puncture of the rumen during brisket cutting. Similarly, use of the correct style of knife and care by the eviscerator to prevent his knife piercing the rumen or intestine tract is needed. Puncture of the intestine or spillage of its contents can cause massive contamination of the carcass and offals, but this is a rare event. Technological solutions have already been found that allow removal of the rectum without soiling the carcass, e.g. by sealing-off of the rectum with a plastic bag immediately after it has been freed.

Campylobacters can occur in bile (Bryner *et al.*, 1972). The gall bladder and the mesenteric and hepatic lymph nodes can be infected with salmonellae (Tables 1.6 and 1.7). Normally, salmonellae are found in <10% of these lymph nodes. However, in cattle and sheep held for some days in contaminated abattoir environments >50% of jejunal, cecal, and colonic lymph nodes can harbor salmonellae (Samuel *et al.*, 1981), and there can be $>10^3$ cfu salmonellae/g of mesenteric nodes (Samuel *et al.*, 1980a). In cattle where there was a high prevalence and a high count of salmonellae in these nodes draining the lower intestine, infections of the tonsils, retropharyngeal, ruminal, and abomasal nodes were rare (Samuel *et al.*, 1980a, 1981). Salmonellae were also not found in the tracheobronchial, caudal mediastinal, lumbar aortic, medial iliac, superficial inguinal or caudal deep cervical lymph nodes. In spite of a high infection rate of mesenteric lymph nodes, spread beyond is limited. Organisms that spread systemically appear to be localized in the spleen or liver. Following intranasal inoculation of sheep with salmonellae (10^3–10^4 cfu), salmonellae could be isolated from lymph tissue of the head and neck regions of some infected animals (tonsil, suprapharyngeal, mandibular, parotid, and bronchial nodes) (Tannock and Smith, 1971).

Desmarchelier *et al.* (1999) reported that the contamination of beef carcasses with coagulase-positive staphylococci (CPS) at three beef abattoirs ranged from 20% to 68.6% on the hides. Carcass contamination increased after evisceration. Average numbers of CPS on carcasses prior to and after overnight chilling was <50 cfu/cm^2, increasing to ca. 100 cfu/cm^2 after chilling. Of the isolates tested, 71.4% produced staphylococcal enterotoxin and 21% could not be classified phenotypically. At one abattoir, the hands of workers were heavily contaminated and the likely source of CPS contamination.

Vanderlinde *et al.* (1999) reported that, at an Australian abattoir, genotyping patterns of CPS from beef carcasses and workers' hands were indistinguishable. Genotypes from non-evisceration abattoir personnel and clerical staff were distinct from patterns among isolates collected from the slaughter floor personnel. During evisceration, carcasses were handled extensively and the hands of workers were the primary source of staphylococcal contamination of carcasses.

General contamination of the heart, liver, and diaphragm of cattle and sheep has been shown to take place during removal from the carcass cavity, from contact with the evisceration table and from handling during separation of the different organs (Sheridan and Lynch, 1988).

The measures introduced by meat inspection have proved highly effective in controlling classical zoonoses (e.g. bovine tuberculosis) that produce pathological and anatomical changes in animals. However, nowadays this classical system is under discussion. For example, the US Food Safety and Inspection Service (FSIS) published a Federal Register notice informing the public of its intent to change from an inspection system requiring extensive carcass palpation to an inspection system that requires no carcass palpation for lambs because (i) extensive carcass palpation in lambs does not routinely aid the detection of food safety hazards that result in meat-borne illnesses; (ii) hands are capable of spreading or adding contamination to the carcasses and (iii) FSIS inspection systems must reflect science-based decisions as they pertain to meat-borne illnesses consistent with a Pathogen Reduction/Hazard Analysis and Critical Control Point environment (Walker *et al.*, 2000).

Trimming and washing. Trimming and washing are done to improve the appearance of the carcass and to remove blood, bone-dust, hair and soil. Trimming removes some bacterial contamination. Washing removes some bacteria and redistributes some organisms from one site to another. The effectiveness of washing varies with the temperature, pressure and volume of water, the design of the system and the time spent.

Washing with water ≤40–50°C gives relatively small and variable reductions in bacterial contamination. Counts at more highly contaminated sites may be reduced, whereas counts are unchanged at sites with an initial low level of contamination (Kelly *et al.*, 1981). Sheridan (1982) obtained a reduction of 60-fold in the microbial load on sheep carcasses, but the mean initial counts on the sites sampled were high ($\log_{10}$ 4.6–4.9 cfu/cm^2). Raising the temperature of the wash water >80°C gives a greater reduction in carcass contamination, but even then the reduction may be small (Bailey, 1971). When a spray system is used to wash carcasses, there is a marked fall in temperature of the water after it leaves the nozzle. When the temperature of sprayed water at impact on the carcass is 56–63°C, the psychrotrophic population is reduced about 10-fold (Bailey, 1971). At impact temperatures of 65°C, the reduction in the mesophilic bacterial load still tends to be variable ($\log_{10}$ 0.2–0.9; Kelly *et al.*, 1981). Impact temperatures of ≥80°C appear to be needed to give at least a 10-fold reduction in the numbers of mesophilic organisms on carcasses (Kelly *et al.*, 1981).

For hot water to be effective, all surfaces of the carcass need to be contacted for sufficient time. Though immersion in water at 80°C for 10 s seems to achieve this and reduces the coliform and mesophilic count on carcasses by 10–100-fold (Smith and Graham, 1978), this is not a practical procedure. An automatic hot-water wash-cabinet, designed to distribute a continuous stream of water over all surfaces of a beef-side and to prevent heat and vapor loss to the environment, can give >2 $\log_{10}$ reduction in numbers of

E. coli on the surfaces of beef sides (Davey and Smith, 1989; Smith and Davey, 1990). Experiments with this system show that the temperature of the film of water at the carcass surface must be ≥55°C to achieve any significant lethal effect.

The addition of chlorine to wash water appears to have only a small effect on reducing carcass contamination (Bailey, 1971; Anderson *et al.*, 1977; Kelly *et al.*, 1981). Normally, there is not >5-fold reduction in microbial count. Low concentrations of chlorine (20–30 mg/l) give some reduction which is not markedly changed with increasing chlorine concentration. Populations of *E. coli* on beef were not significantly reduced by 800 ppm chlorine (Cutter and Siragusa, 1995).

Both acetic (Anderson *et al.*, 1977) and lactic (Smulders, 1987) acid solutions, when applied to carcass surfaces, reduce bacterial contamination. A 1% solution of lactic acid reduced the mesophilic count on beef, veal, and pork carcasses by between $\log_{10}$ 0.8 and 1.9. Both acetic and lactic acids have a residual effect, reducing the rate of microbial growth on chilled meats. However, acid sprays appear to cause little reduction in *E. coli, E. coli* O157, and salmonellae on meat surfaces (Brackett *et al.*, 1994). The reduction often differs little from that given by water (Cutter and Siragusa, 1994). Particular attention must be paid to the acid tolerance of *E. coli* O157, which allows them to survive.

Normally, washing has only a small effect on the overall microbial load on the carcass. If special systems are used (such as water at high impact temperatures and lactic or acetic solutions), significant reductions in bacterial contamination are possible. In such a case, temperature, time of application, concentration and/or volume will need to be controlled to ensure efficient operation of the system.

Sanitizing treatments, including hot-water sprays (74–80°C), steam pasteurization, organic acid sprays, and other chemicals, e.g. trisodium phosphate (TSP), mixtures of nisin with 50 mM EDTA, and acidified sodium chlorite solutions (ASC) have been used in an attempt to reduce contamination with *E. coli* O157:H7 (Castillo *et al.*, 2002) on cattle carcasses. Hot-water spraying on different hot carcass surface regions reduced numbers of *E. coli* O157:H7 by 3.7 $\log_{10}/\text{cm}^2$ (Castillo *et al.*, 1998a). Steam pasteurization in a chamber operating above atmospheric pressure reduced *E. coli* O157:H7 by 4.4–3.7 $\log_{10}$ cycles on surfaces of freshly slaughtered beef (Phebus *et al.*, 1997).

Spraying organic acids (acetic, citric or lactic acid) at different concentrations failed to reduce *E. coli* O157:H7 on beef sirloin pieces (Brackett *et al.*, 1994), supporting findings by Cutter and Siragusa (1994) and Uyttendaele *et al.* (2001). However, when applied at 55°C, lactic or acetic sprays reduced levels of *E. coli* O157:H7 (Hardin *et al.*, 1995; Castillo *et al.*, 1998b), lactic acid being more effective than acetic acid. Spraying TSP solutions at 55°C resulted in reduction of *E. coli* O157:H7 on lean beef muscle ranging from 0.8 to 1.2 $\log_{10}/\text{cm}^2$ (Dickson *et al.*, 1994). Cutter and Siragusa (1996) sprayed a mixture of nisin+lactate and nisin+EDTA and obtained statistically significant but practically insignificant reductions of *E. coli* O157:H7. Carneiro *et al.* (1998) reported that, in combination with sublethal concentrations of TSP, nisin successfully inhibited Gram-negative cells, including *S.* Enteritidis, *C. jejuni,* and *E. coli* and could therefore be used for decontamination of surface of food products.

Cutter and Rivera-Betancourt (2000) evaluated different interventions used by the meat industry in the United States, on pre-rigor beef surfaces previously inoculated with a bovine fecal slurry containing *S.* Typhimurium and *S.* Typhimurium DT 104, *E. coli* O157:H7 and O111:H8, or *E. coli* O157:H7 and O26:H11, then spray washed with water, hot water (72°C), 2% acetic acid, 2% lactic acid or 10% TSP (15 s, 125 ± 5 psi, 35 ± 2°C). Spray treatments with TSP were the most effective, resulting in pathogen reductions of >3 $\log_{10}$ cfu/cm^2, followed by 2% lactic acid and 2% acetic acid (>2 $\log_{10}$ cfu/cm^2). Interventions reduced all the pathogens tested were effective immediately after treatment and after long-term, refrigerated, vacuum-packaged storage.

Organic acid was applied post-chill as a 30 s lactic acid spray at 55°C (Castillo *et al.*, 2001) to outside rounds contaminated with *E. coli* O157:H7 and *S.* Typhimurium, after pre-chill hot carcass treatments consisting of water wash alone or water wash followed by a 15 s lactic acid spray at 55°C. The pre-chill treatments reduced both pathogens by 3 $\log_{10}$ (water wash alone) to 5 $\log_{10}$ (water wash and lactic acid).

In all cases, the post-chill acid treatment produced an additional reduction in *E. coli* O157:H7 (ca. 2 $\log_{10}$) and ca. 1.6 $\log_{10}$ for *S.* Typhimurium.

Samelis *et al.* (2001) examined the survival of *E. coli* O157:H7, *S.* Typhimurium DT 104 and *L. monocytogenes* in meat decontamination fluids (washings) after spray-washing fresh beef top rounds sprayed with water (10°C or 85°C) or acid solutions (2% lactic or acetic acid, 55°C) during storage of the washings at 4°C or 10°C in air to simulate plant conditions. Inoculated *S.* Typhimurium DT 104 died off in lactate (pH 2.4±0.1) and acetate (pH 3.1±0.2) washings by 2 days at either storage temperature, but *E. coli* O157:H7 and *L. monocytogenes* survived in lactate washings for at least 2 days and in acetate washings for at least 7 and 4 days, respectively; their survival was better in acidic washings stored at 4°C than at 10°C.

Table 1.8 Bacterial numbers on beef carcasses before chilling

	European Union Member State[a]							
	1	2	3	4	5	6	7	
Survey 1								
Mean $\log_{10}$ cfu/cm^2 (30°C)	3.85	2.77	2.29	3.14	2.45	2.75	3.23	
Enterobacteriaceae (% +ve)	74	67	39	79	47	79	61	
Survey 2								
Mean $\log_{10}$ cfu/cm^2 (30°C)	3.78	3.15	2.35	3.50	2.48	3.11	3.33	
	Norway[b]							
	1	2	3	4	5	6	7	8
Mean $\log_{10}$ cfu/cm^2 (20°C)	2.46	2.24	2.87	2.09	2.56	2.42	2.36	1.98
	England[c]							
	1	2	3	4	5	6	7	8
Mean $\log_{10}$ cfu/cm^2 (37°C)	2.68	2.98	1.83	–	–	–	–	–
Mean $\log_{10}$ cfu/cm^2 (20°C/30°C)	3.12[d]	2.95[d]	2.03[d]	3.03[e]	3.06[e]	3.60[e]	3.57[e]	3.76[e]
Mean $\log_{10}$ cfu/cm^2 (1°C)	1.96	1.83	–	–	–	–	–	–
Enterobacteriaceae (% +ve)	19	–	–	66	70	–	–	–
Coliforms (% +ve)	17	–	19	–	–	–	–	–
	Other Countries							
	NZ[f]	In1[g]	In2[g]					
Mean $\log_{10}$ cfu/cm^2 (25°C)	1.85	3.62	4.02	–	–	–	–	–
Mean $\log_{10}$ cfu/cm^2 (0°C)	0.56	–	–	–	–	–	–	–
Coliforms (% >10/cm^2)	–	77.3	85.7	–	–	–	–	–
Salmonella (% positive)	–	15.5	13.6	–	–	–	–	–

[a]Samples taken in seven countries, by wet-dry swabbing 50 cm^2 at each of four sites on a carcass. Aerobic viable count determined by incubation at 30°C. For each survey, means ($\log_{10}$ cfu/cm^2) are the overall means of 4 sites/carcass, 10 carcasses/visit, 3 visits/abattoir, 3 abattoirs/country, excepting country 4 where only two abattoirs were sampled (Roberts *et al.*, 1984; and personal communication). The limit of detection of Enterobacteriaceae was 0.4 cfu/cm^2 and contamination is expressed as the percent of samples from which Enterobacteriaceae were detected.

[b]Samples obtained, at nine abattoirs, by wet-dry swabbing of 50 cm^2 at each site on a carcass. Aerobic viable count determined by incubation at 30°C. For each survey, means ($\log_{10}$ cfu/cm^2) are the overall means of 8 sites/carcass, 10 carcasses/visit, 3–6 visits/abattoir for each abattoir sampled (Johanson *et al.*, 1983).

[c]Samples obtained by wet-dry swabbing of 100 cm^2 at each of 13 sites on 10, 10 and 6 carcasses for abattoirs 1, 2 and 3 (Roberts *et al.*, 1980b). For abattoirs 4 and 5, means ($\log_{10}$ cfu/cm^2) are the overall means of 7 sites/carcass, 10 carcasses/visit, 12 visits/abattoir. Each site swabbed was 100 cm^2. Abattoir 4 was a manual (cradle) system and abattoir 5 an automated (rail) system (Whelehan *et al.*, 1986). The limit of detection of Enterobacteriaceae and coliforms for abattoirs 1–5 was 0.2 cfu/cm^2. For abattoirs 6, 7 and 8, 4 sites (each 50 cm^2) were swabbed from each of 10 carcasses on 6 visits. Abattoirs 6–8 were the same slaughterhouse at different stages of modernization (Hudson *et al.*, 1987).

[d]Aerobic viable count determined by incubation at 20°C.

[e]Aerobic viable count determined by incubation at 30°C.

[f]Samples obtained by wet swabbing of 100 cm^2 at 2 sites on each of 28 carcasses (Newton *et al.*, 1978).

[g]Samples obtained by wet-dry swabbing of 50 cm^2 at each of three sites on 22 carcasses from abattoir 1 (traditional) and 84 carcasses from abattoir 2 (semi-modern; Mukartini *et al.*, 1995).

Cattle and sheep carcasses, at the completion of hygenic slaughter and dressing, can typically have on their surface tissues about 10^2–10^4 cfu mesophiles/cm^2 (Tables 1.8–1.10). Counts consistently above about 10^5 cfu/cm^2 indicate that more care is needed in carcass dressing (Mackey and Roberts, 1993). A large proportion of the organisms are Gram-positive (e.g. micrococci, staphylococci and coryneforms). Enterobacteriaceae and coliforms can be detected (Tables 1.8–1.10). However, these microorganisms are usually present in small numbers (e.g. only 2% of 182 swab samples from beef carcasses examined in the UK had >11 cfu Enterobacteriaceae/cm^2; Mackey and Roberts, 1993). The psychrotrophic population will be a variable percentage (from about 0.2 to 10%) of the mesophilic count (Tables 1.8–1.10; Newton *et al.*, 1978). The proportion of psychrotrophs varies with ambient temperature so that there will be both seasonal and geographic differences. Psychrotrophs that may be found on carcasses include *Pseudomonas*, *Acinetobacter*, *Psychrobacter*, Enterobacteriaceae, *Broch. thermosphacta*, lactic acid bacteria, and yeasts (*Cryptococcus*, *Candida*, and *Rhodotorula* spp.). Siragusa (1995) reviewed the effectiveness of carcass decontamination systems for controlling pathogens on meat animal carcasses.

One survey of dressed camel carcasses (Hamdy, 1981) found counts of 3.6×10^5 mesophiles, 4.8×10^3 Enterobacteriaceae, and 2.2×10^3 *Staph. aureus*/cm^2.

The need for assurance that hygienic measures at slaughter are effective has led to several surveys of the microbiological quality of carcasses (e.g. Murray *et al.*, 2001 on beef carcasses in Ireland; Phillips *et al.*, 2001a on Australian beef; Phillips *et al.*, 2001b on Australian sheep meat; Duffy *et al.*, 2001b on lamb carcasses in the United States; Zweifel and Stephan, 2003 on lamb carcasses in Switzerland and Rose *et al.*, 2002 on contamination of raw meat and poultry products with salmonellae).

For example, Phillips *et al.* (2001a) sampled beef carcasses and frozen boneless beef derived from 1 275 beef carcasses and drilled from 990 cartons of frozen boneless beef. Mean log total viable counts (TVC) were 2.42 cfu/cm^2 for carcasses and 2.5 cfu/g for boneless meat. *E. coli* was detected on 10.3% carcasses and 5.1% of boneless beef samples; CPS on 24.3% of carcasses and 17.5% of boneless beef; salmonellae on 0.2% of carcasses and 0.1% of boneless beef; *E. coli* O157:H7 on 0.1% carcasses and was not detected on 990 boneless beef samples. Results indicated small but significant improvements in several microbiological criteria for carcasses and boneless meat over a similar survey in 1993–1994.

Table 1.9 Bacterial numbers on sheep carcasses before chilling

	Abattoir								
	E1[a]	E2[a]	E3[a]	E4[a]	E5[a]	E6[a]	Irel[b]	NZ[c]	Aust[d]
Mean $\log_{10}$ cfu/cm^2 (37°C)	3.18	3.01	2.54	2.84	3.25	2.76	–	–	–
Mean $\log_{10}$ cfu/cm^2 (30°C)	–	–	–	–	–	–	–	–	–
Mean $\log_{10}$ cfu/cm^2 (20°C/25°C)	3.45	3.13	2.67	2.90	3.32	3.13	3.77[f]	2.78[f]	3.18[f]
Mean $\log_{10}$ cfu/cm^2 (7°C)	–	–	–	–	–	–	–	–	–
Mean $\log_{10}$ cfu/cm^2 (1°C)	2.13	–	–	–	–	–	–	0.77[g]	–
Enterobacteriaceae (% positive)	57	48	–	69	–	–	–	–	–
Coliforms (% positive)	44	28	18	59	–	–	–	–	(0.99)[h]
No. carcasses	18	6	6	12	6	6	7	29	10

[a] Abattoirs in England. Samples obtained by wet-dry swabbing of 100 cm^2 at each of 10 sites on each carcass. Aerobic viable count determined by incubation at 37°C, 20°C and 1°C. Mean $\log_{10}$ number is the mean $\log_{10}$ value for all sites (Roberts *et al.*, 1980b).
[b] Abattoir in Ireland. Samples obtained by wet swabbing of 25 cm^2 at each of 12 sites on each carcass before spray washing (Kelly *et al.*, 1980).
[c] Abattoir in New Zealand. Samples obtained by wet swabbing of 100 cm^2 at two sites on each carcass (Newton *et al.*, 1978).
[d] Abattoir in Australia. Samples obtained by dry swabbing of 100 cm^2 at each of 10 sites on each carcass immediately after spray washing (Grau, 1979).
[e] Abattoir in Spain. Samples obtained by swabbing, and excision of the swabbed and scrapped areas of 45 cm^2 at each of the three sites on each carcass (Prieto *et al.*, 1991).
[f] Aerobic viable count determined by incubation at 25°C.
[g] Aerobic viable count determined by incubation at 0°C.
[h] Mean $\log_{10}$ cfu *E. coli*/cm^2.

Table 1.10 Bacterial numbers on sheep carcasses before chilling

	1NZ^{a,1}	1NZ^{a,2}	1NZ^{a,3}	1NZ^{a,4}	1NZ^{b,1}	1NZ^{b,2}	1NZ^{b,3}	1NZ^{b,4}	2NZ^{a,1}	2NZ^{a,2}
APC	5.02	3.97	5.36	4.54	4.54	4.42	5.16	4.36	4.10	3.98
E. coli	1.49	0.89	2.11	0.54	0.99	0.72	1.32	0.38	0.96	1.05
Ent	–	–	–	–	–	–	–	–	–	–
Lm	–	–	–	–	–	–	–	–	–	–
Colif.	–	–	–	–	–	–	–	–	–	–
No.	25	25	25	25	25	25	25	25		
	Irel^{a,1}	Irel^{b,1}	Irel^{a,2}	Irel^{b,2}	Irel^{a,3}	Irel^{b,3}	Irel^{a,4}	Irel^{b,4}	Can	Scot
APC	4.52	4.88	4.81	4.84	5.01	4.84	4.63	4.63	3.43^{Cana}	–
E. coli	–	–	–	–	–	–	–	–	2.61^{Canb}	–
Ent	2.10	1.67	1.35	1.31	1.40	1.13	0.87	0.66	–	–
Lm	24^{c}	–	–	–	–	–	–	–	–	14.3
Colif.	–	–	–	–	–	–	–	–	2.73^{Canc}	–
No.	5	5	5	5	5	5	5	5	–	7

APC, mean APC $\log_{10}/cm^2$ (30°C); *E. coli*, mean *E. coli* count $\log_{10}/cm^2$ (37°C); Ent, Enterobacteriaceae $\log_{10}/cfu/cm^2$ (30°C); Lm, % of carcasses positive for *Listeria* spp.; Colif., coliforms (35°C).

NZ[a,1] 25 carcasses from slaughterhouse A. Animals were washed 20 h before slaughter and clean/shorn (maximum wool length 2 cm). Clean, faecal staining of the pelt was absent, or localized covering <5% of the surface; collection faecal material around the perineum and only minor contamination of the fleece with dirt and dust. 10 excision samples (5 cm^2 each) were removed aseptically from each carcass (six samples from the forequarters, four from the hind-legs (Bliss and Hatheway, 1996a).

NZ[a,2] 25 carcasses from slaughterhouse A. Animals were not washed but were clean/shorn (Bliss and Hatheway, 1996a).

NZ[a,3] 25 carcasses from slaughterhouse A. Animals were washed 20 h before slaughter and were clean/woolly (minimum wool length 6 cm) (Bliss and Hatheway, 1996a).

NZ[a,4] 25 carcasses from slaughterhouse A. Animals were not washed and were clean/woolly (Bliss and Hatheway, 1996a).

NZ[b,1] 25 carcasses from slaughterhouse B. Animals were washed and clean/shorn (Bliss and Hatheway, 1996a).

NZ[b,2] 25 carcasses from slaughterhouse B. Animals were not washed but were clean/shorn (Bliss and Hatheway, 1996a).

NZ[b,3] 25 carcasses from slaughterhouse B. Animals were washed and clean/woolly (Bliss and Hatheway, 1996a).

NZ[b,4] 25 carcasses from slaughterhouse B. Animals were unwashed and clean/woolly (Bliss and Hatheway, 1996a).

2NZ[a,1] Counts from pre-wash ovine carcasses visibly contaminated with fecal material but sampled from an uncontaminated site (uncontaminated carcass) (Bliss and Hatheway, 1996b).

2NZ[a,2] Counts from pre-wash ovine carcasses visibly contaminated with fecal material but sampled from an uncontaminated site (contaminated carcass) (Bliss and Hatheway, 1996b).

Irel[a,1] In plant 1, 50 cm^2 sternum/abdominal area swabbed after evisceration, using wet and dry gauze swabs, then placed in 50 mL MRD (Sierra *et al*, 1997).

Irel[b,1] In plant 1, 50 cm^2 sternum /abdominal area swabbed after washing, using wet and dry gauze swabs, then placed in 50 mL of MRD (Sierra *et al*, 1997).

Irel[a,2] In plant 2, 50 cm^2 sternum/abdominal area swabbed after evisceration, using wet and dry gauze swabs, then placed in 50 mL of MRD (Sierra *et al*, 1997).

Irel[b,2] In plant 2, 50 cm^2 sternum/abdominal area swabbed after washing, using wet and dry gauze swabs, then placed in 50 mL of MRD (Sierra *et al*, 1997).

Irel[a,3] In plant 3, 50 cm^2 sternum/abdominal area swabbed after evisceration, using wet and dry gauze swabs, then placed in 50 mL of MRD (Sierra *et al*, 1997).

Irel[b,3] In plant 3, 50 cm^2 sternum/abdominal area swabbed after washing, using wet and dry gauze swabs, then placed in 50 mL of MRD (Sierra *et al*, 1997).

Irel[a,4] In plant 4, 50 cm^2 sternum/abdominal area swabbed after evisceration, using wet and dry gauze swabs, then placed in 50 mL of MRD (Sierra *et al*, 1997).

Irel[b,4] In plant 4, 50 cm^2 sternum/abdominal area swabbed after washing, using wet and dry gauze swabs, then placed in 50 mL of MRD (Sierra *et al*, 1997).

[c]Percent of carcasses positive for spp. in four plants during four visits (Sierra *et al*, 1997).

[Cana]Total aerobic counts per cm^2 from a randomly selected site on each of 25 carcasses at trimming (25°C) (Gill and Baker, 1998).

[Canb]*E. coli* counts per 100 cm^2 from a randomly selected site on each of 25 carcasses at trimming (35°C) (Gill and Baker, 1998).

[Canc]Coliforms per 100 cm^2 from a randomly selected site on each of 25 carcasses at trimming (35°C) (Gill and Baker, 1998).

[Scot]Trimmings from the carcass (25 g) taken from the lowest part of the hanging carcass (Fenlon *et al*., 1996).

Duffy *et al.* (2001b) sampled 5 042 lamb carcasses from six major lamb-packing facilities in the United States to develop a microbiological baseline for *Salmonella* spp. and *E. coli* after chilling for 24 h. The incidence of *Salmonella* spp. from chilled carcasses was 1.9% in fall or winter and 1.2% in spring (1.5% combined). Mean counts (log cfu/cm^2) were aerobic plate count (APC) 4.42, total coliform count (TCC) 1.18 and *E. coli* counts (ECC) 0.70 on chilled carcasses. APC were lower ($P<0.05$) in spring than in fall or winter, whereas TCC were higher in spring. There was no difference ($P>0.05$) between ECC in samples in the spring and winter. Only 7 of 2 226 total samples (0.3%) tested positive for *C. jejuni/coli*, across all sampling sites.

Zweifel and Stephan (2003) examined at three Swiss abattoirs, 580 sheep carcasses at 10 sites by the wet–dry double-swab technique. Median APCs (log cfu/cm^2) ranged from 2.5 to 3.8, with the brisket and neck sites showing the most extensive contamination. Enterobacteriaceae were detected on 68.1% of the carcasses and in 15.2% of the samples. The proportion of positive results ranged from 2.6% (for the hind-leg and the flank at abattoir C) to 42.2% (for the perineal area at abattoir A). The percentage of samples testing positive for *stx* genes by polymerase chain reaction was 36.6%. A significant relationship between APC and the detection of STEC was found for abattoirs A and B (depending on the sampling site), whereas a significant relationship between Enterobacteriaceae and STEC detection was confirmed only for abattoir A ($P<0.05$).

Comparisons of hygienic performance within a country and between countries are often not possible, because there is no international agreement on the sites to be sampled, the methods of sampling those sites, or which microbiological tests and methods should be applied. Nevertheless, since 2001, there are directives in the EU countries (directive 2001/47/EG). There is some evidence of improvement in hygiene since the introduction of a more structured approach to slaughter procedures and to hygiene in slaughterhouses.

Offal can also typically have on surface tissue about 10^3–10^5 cfu mesophiles/cm^2 (Tables 1.11 and 1.12). The flora on offal is similar to that on carcasses, being mainly Gram-positive, but Enterobacteriaceae can be a significant percentage of the initial flora (Gardner, 1971; Gill and De Lacy, 1982; Oblinger *et al.*, 1982). Several hundred bacteria per gram can be found in the internal tissue of liver (Gardner, 1971). Internal tissues of the open sinusoidal structure of organs like the liver can be contaminated during organ removal from the carcass (Gill, 1988).

B Pigs

Primary processing consists of stunning and bleeding, scalding, dehairing, singeing, scraping, and polishing, pre-evisceration washing, evisceration and washing. The process differs from ruminants in that the skin is usually not removed, and scalding and singeing steps are used. Pre-mortem inspection removes from slaughter excessively dirty and obviously diseased pigs, but cannot prevent the slaughter of pigs with potential human pathogens in the intestine or on the skin.

Scalding and singeing reduce the microflora on the skin, depending on the time and temperature used in these processes. Organisms from the intestinal tract may contaminate carcasses during dehairing and evisceration. Contamination of offal meats can also occur at the evisceration stage. The dehairing and scraping operations are often sources of major contamination of carcass surfaces. The equipment used can contaminate and spread contamination from carcass to carcass. Special care is needed in the cleaning of dehairing and scraping equipment to prevent a build-up of contaminants. This equipment can be a major source of the final microbial population on carcasses as they leave the slaughtering area and enter the chill rooms.

Potential pathogens from the intestinal tract are *Cl. perfringens, C. coli, L. monocytogenes* (Table 1.2), *Y. enterocolitica* (Table 1.5) and *Salmonella* spp. (Tables 1.6 and 1.7). Mesenteric and hepatic lymph nodes may contain salmonellae. Skin is a source not only of *Staph. aureus* and *Staph.*

Table 1.11 Bacterial counts ($\log_{10}$ cfu/cm^2) on offal before chilling

Species	Liver	Heart	Kidney	Diaphragm	Tongue	Tail	Reference
Beef	4.35[a]	3.72[a]	3.73[a]	6.46[a]	5.99[a]	–	1
	5.51	4.58	–	4.86	4.97	4.45	2
	4.55	4.61	4.72	3.75	4.98	4.86	2
	2.50	2.28	2.77	–	–	–	3
	3.36	–	–	–	3.69	–	4
	3.69	3.92	3.51	–	4.20	–	5a
	3.58	3.85	3.66	–	3.61	–	5b
	2.61	3.13	2.85	–	2.32	–	5c
Sheep	4.62[a]	4.92[a]	–	5.46[a]	6.38[a]	–	1
	4.23	3.05	3.29	–	–	–	3
	3.18	–	–	–	4.28	–	4
Pig	3.28	3.04	3.80	–	–	–	3
	4.04	–	–	–	5.76	–	4
	4.78	–	–	–	–	–	6
	3.28	–	–	–	–	–	5a
	3.27	–	–	–	–	–	5b
	1.69	–	–	–	–	–	5c

[a]$\log_{10}$ cfu/g.
1. Samples excised, incubated at 25°C (Sheridan and Lynch, 1988).
2. Swab samples, incubated at 22°C (Patterson and Gibbs, 1979).
3. Samples excised, incubated at 25°C (Hanna *et al.*, 1982).
4. Samples excised, incubated at 25°C (Vanderzant *et al.*, 1985).
5. Swab samples, incubated at 37°C (5a), at 20°C (5b), at 7°C (5c), (Oblinger *et al.*, 1982).
6. Samples excised, incubated at 22°C (Gardner, 1971).

Table 1.12 Bacterial counts ($\log_{10}$ cfu/cm^2) on offal before chilling

Species	Liver	Diaphragm	Tongue	Tail	Large intestine	Head-meat	Reference
Beef							
–[a]	2.2–4.7	–	3.6–5.5	2.9–5.7			Delmore *et al.*, 2000
–[b]	–	–	–	3.29	–	–	Gill *et al.*, 1999
–[c]	–	–	4.63	–	–	–	Gill *et al.*, 1999
–[a]	2.2–4.7	–	3.6–5.5	2.9–5.7	3.5–4.7	–	Delmore *et al.*, 2000
–[d]	1.9–3.4	–	0.2–3.1	1.5–4.2	1.8–3.0	–	Delmore *et al.*, 2000
–[e]	1.6–3.0	–	0.0–0.7	1.3–2.3	1.3–2.7	–	Delmore *et al.*, 2000
Pork							
	3.60 ± 0.481	–	–	–	–	–	Delmore *et al.*, 2000
–[f]	–	–	–	–	–	4.52 ± 0.26	Moore and Madden, 1998
–[g]	–	–	–	–	–	2.37 ± 0.42	Laubach *et al.*, 1998
–[h]	–	–	–	–	–	2.25 ± 0.42	Laubach *et al.*, 1998

[a]Range of $\log_{10}$ cfu/g aerobic plate counts (Delmore, 2000).
[b]Mean of 25 aerobic counts ($\log_{10}$) from randomly selected tails after skinning.
[c]Mean of 25 aerobic counts ($\log_{10}$) from randomly selected tongues within the mouths of carcasses.
[d]Range of $\log_{10}$ cfu/g coliform counts.
[e]Range of $\log_{10}$ cfu/g *E. coli* counts.
[f]APC $\log_{10}$ cfu/g.
[g]Coliforms $\log_{10}$ cfu/g.
[h]*E. coli* log 10 cfu/g.
[j]*Campylobacter* spp. $\log_{10}$ cfu/g.

hyicus, bacilli, and other mesophiles but also of organisms of fecal origin as a result of contamination acquired on the farm, during transport or in lairages at abattoirs. The throat, tongue, and mainly tonsils of some pigs can carry large numbers of *Y. enterocolitica*, which, during slaughter and dressing, can be spread to the carcass and especially to head-meat.

Chilling prevents growth of mesophilic pathogens and reduces the growth of psychrotrophic pathogens and spoilage organisms.

Stunning and bleeding. After animals are stunned, they are exsanguinated by severing the carotid arteries and veins. The prevalence of potential pathogens on the skin can be increased by cross-contamination and from organisms on surfaces at the stunning-sticking area (e.g. *Listeria* spp.; Gobat and Jemmi, 1991). After bleeding, the skin can carry 10^6–10^7 cfu mesophiles, 10^4–10^5 cfu Gram-negative organisms and 10^3–10^4 cfu Enterobacteriaceae/cm^2 (Snijders and Gerats, 1976b).

Scalding. After bleeding, the carcasses are treated with hot water or steam to aid the removal of hair and scurf. The time and temperature of heat treatment are primarily determined by the need for efficient removal of the bristles by the dehairer. Autumn hair is most difficult to remove, and summer hair the easiest (Snijders and Gerats, 1976a). Temperatures between 58°C and 62°C are normally used for 5–6 min. Temperatures ≥63°C for 4–5 min damage the skin. At some plants, carcasses are pre-cleaned by water-sprays or by pre-scald brushing machines. Though pre-cleaning reduces the load of organic material in the scald system, the use of brushing systems does not appear to result in the production of carcasses with lower final levels of microorganisms (Snijders and Gerats, 1976b; Rahkio *et al.*, 1992).

During scalding, feces may escape from the anus, blood seeps from the sticking wound and material from hide and hooves accumulates in the water. In vat systems, water and colloidal size particles can enter the lungs and through the sticking wound into the heart and aorta (Jones *et al.*, 1984). In combined scalding and dehairing systems (vats or sprayed water), bacilli and micrococci from scald water have been frequently found in the pelvic aorta, but were rarely detected when separate vat or spray scalding systems were used (Troeger and Woltersdorf, 1990).

Whether vat or spray systems are used the microbial content of the water depends on its temperature and pH (Snijders, 1975). During use, the pH is 7.3–8.3 but may be raised to pH values of 10–12 by the addition of alkali. Mesophilic aerobes can vary between about 10^2 and 10^5 cfu/mL and are predominantly spore-formers and thermoduric micrococci. There can be several hundred spores of *Cl. perfringens*/mL. In vat water at 61–62°C, contaminating staphylococci appear to be destroyed and only occasionally can fecal streptococci be detected (Sorqvist and Danielsson-Tham, 1986). Gram-negative bacteria may be found in low numbers when the temperature is not >60°C (Snijders, 1975; Sorqvist and Danielsson-Tham, 1986). The numbers of viable *Campylobacter*, *Salmonella*, and *Yersinia* spp. that contaminate the scald water from the animals are rapidly reduced (Sorqvist and Danielsson-Tham, 1991). Salmonellae have been detected in some samples of scald water. For instance, Chau *et al.* (1977) detected salmonellae in 3 of 20 samples of scald water at 60°C when there was a high prevalence of the organism in the pigs being processed. In general, the numbers of campylobacters, salmonellae, and yersiniae that may survive transiently in the scald water will be low compared with those in fecal material. Even when *Listeria* spp. were detected on about 50% of skin swabs of pigs before scalding, they were not found in scald water at 60–63°C (Gobat and Jemmi, 1991).

Scalding can significantly reduce the microflora on the skin. The numbers of mesophiles, psychrotrophs, bacilli, clostridia, Enterobacteriaceae, and the prevalence of salmonellae and listeriae are reduced (Kampelmacher *et al.*, 1963; Dockerty *et al.*, 1970; Snijders and Gerats, 1976b; Scholefield *et al.*, 1981; Gobat and Jemmi, 1991). Dockerty *et al.* (1970) found a reduction in mesophilic contamination on the skin by $\log_{10}$ 2–2.5 when vat scald-temperatures were from 58.5°C to 60°C. Snijders and Gerats (1976b) reported that, on pigs spray-scalded with water of 60°C, the mesophilic count was reduced from about 10^6 to 2×10^3 cfu/cm^2, the Gram-negative flora was reduced from $>10^4$ to <100, and the numbers of Enterobacteriaceae from about 4×10^3 to <70 cfu/cm^2. Rahkio *et al.* (1992) reported that the extent of the reduction in the number of mesophiles was less when the initial contamination on pigs was low.

Dehairing. The hair loosened by scalding is removed mechanically. Fecal material can escape from the anus and contaminate carcasses and the machinery. Dehairing can be a major source of carcass contamination. Snijders and Gerats (1976b) observed about a 100-fold increase in contamination of the skin with mesophiles and Gram-negative organisms, and a significant increase in Enterobacteriaceae, after dehairing. Dehairing equipment can also become contaminated with *Campylobacter* and *Salmonella* spp. and spread the organism among carcasses (Gill and Bryant, 1993). Bacterial recontamination of carcasses is reduced if hot water (60–62°C) is used in the dehairing machinery to remove hair and other material. If not properly cleaned, growth of organisms on tissue residues in the equipment can be a significant source of contamination.

Singeing. Residual hair on the carcass is removed by singeing. This destroys organisms on the skin exposed to heat. However, heat treatment of surfaces may be uneven, for instance at fore-leg pits, groin, and ears (Roberts, 1980). Scalding at 60–63°C damages some of the epithelial layer and during dehairing some bacteria may be massaged into sub-epithelial tissue where they may be protected. At this stage, with efficient singeing, carcasses have the lowest surface count of aerobic mesophiles and of Enterobacteriaceae. On excised samples of abdominal skin the count of mesophiles was reported to be mostly <100 cfu/cm^2 and Gram-negative organisms were only very rarely detected (Snijders and Gerats, 1976b). After singeing, a mean count was found of only 52 cfu mesophiles/cm^2 with the count varying with the site sampled (e.g. neck samples had about a 10-fold higher count than samples from the abdomen) (Rahkio *et al.*, 1992).

Scraping and polishing. The carcass is scraped manually or mechanically and washed to remove burnt material and remaining hair and scurf. Scraping and cleaning can spread contamination from one site to another. This stage frequently results in significant increases in bacterial contamination of the skin, and can cause contamination with salmonellae. The extent of the contamination varies between abattoirs and with time of day, and can be as high as 1000-fold for mesophiles and Gram-negative bacteria (Snijders and Gerats, 1976b). In inadequately cleaned and sanitized scraping and polishing equipment, microbial growth occurs and results in the machinery acting as a continuous source of contamination (Gerats *et al.*, 1981), and in the pigs slaughtered first being the most contaminated (Snijders and Gerats, 1976b). Bacterial contamination of the carcasses with mesophiles and Enterobacteriaceae can be related to the care taken in cleaning equipment (Gerats, 1987). When equipment is properly cleaned and sanitized, the bacterial load on carcasses approximates that immediately after singeing (Gerats, 1987).

Evisceration. During evisceration, the abdomen is opened and the brisket is cut. Organs in the thorax and abdomen are removed, and edible offal separated from the viscera and inspected. Organisms such as *Salmonella* and *Campylobacter* spp. from the intestinal tract frequently contaminate both carcass and offal. Care taken during evisceration reduces contamination with salmonellae and Enterobacteriaceae (Gerats *et al.*, 1981; Gerats, 1987). Puncture of the intestinal tract can cause massive contamination. However, even without obvious damage to the lower intestinal tract, bacteria from fecal material can be found on carcasses. Freeing the anal-sphincter and rectum from attachment to the carcass is a critical operation. The anus may be handled and organisms transferred to the carcass. Hand-washing and disinfection of knives between each carcass reduces contamination with *E. coli* and salmonellae (Childers *et al.*, 1977). Newer methodologies in freeing the anal-sphincter, including enclosing the rectum in a plastic bag, have also been shown to reduce significantly contamination of the hind-quarter with *Y. enterocolitica* from feces (Andersen, 1988). Data from the Norwegian National Institute of Public Health indicates that the occurrence of human yersiniosis has fallen by about 30% since the introduction of the plastic bag technique (Nesbakken *et al.*, 1994) in about 90% of the pig slaughterhouses in Norway.

When the pluck (the organs from the thorax) is removed together with the tongue, and hung by the tongue, there can be considerable contamination with *Y. enterocolitica* from the tonsils. Removal of the pluck without the tongue or tonsils reduces this contamination (Andersen *et al.*, 1991).

Live animals carrying *Salmonella* spp. are 3–4 times more likely to result in *Salmonella*-positive carcass than *Salmonella*-free animals (Berends *et al.*, 1997). It is estimated that 70% of all carcass contamination is caused by animals carrying the organism, whereas the remaining 30% is caused by cross-contamination from animal carriers slaughtered concurrently. Dirty polishing machines and, especially, errors during evisceration are the most important risk factors. An estimated 5–15% of carcass contamination with *Salmonella* spp. occurs during polishing after singeing; 55–90% from evisceration practices, and 5–35% from dressing and meat inspection procedures.

Salmonellae may occur in the mesenteric and hepatic lymph nodes, and sometimes in the gall bladder and in the liver and spleen (Tables 1.6 and 1.7). Salmonellae may be present in these lymph nodes even when the organisms are not detected in intestinal contents. Salmonellae can also be found in palatine tonsils and mandibular lymph nodes (Wood *et al.*, 1989). Campylobacters have been found in the gall bladder and associated bile ducts (Rosef, 1981).

During evisceration, there frequently is no increase in the number of mesophilic aerobes on the carcass (Rahkio *et al.*, 1992) even though there may be significant increases in Enterobacteriaceae (Gerats, 1987). This is particularly apparent for carcasses with a relatively high number of bacteria on the skin before evisceration when even insanitary evisceration may not result in a noticeable increase in mesophilic bacteria.

Washing. Washing pig sides with water removes some bacteria and re-distributes some organisms from one site to another. Washing appears to have little effect on contamination with salmonellae and *E. coli* (Childers *et al.*, 1973) and, unless prolonged, has little effect on the number of psychrotrophs (Skelley *et al.*, 1985). The addition of sodium hypochlorite (200 mg/l) to the wash water often has little extra effect on reducing bacterial contamination (Skelley *et al.*, 1985). Reductions ($\geq$10-fold) in the number of mesophilic organisms and of Enterobacteriaceae have been obtained when organic acid solutions were applied to the surface tissue of pigs (Reynolds and Carpenter, 1974; Smulders, 1987).

At the completion of slaughter and dressing, the number and types of bacteria on the surfaces of pig sides are determined by the survival of flora from heating during scalding and singeing and by recontamination during dehairing, scraping, evisceration and handling. In general, there are about 10^2–10^4 cfu mesophiles/cm^2 on carcasses (Tables 1.13 and 1.14) with somewhat higher counts on offal (Tables 1.11 and 1.12). Enterobacteriaceae, coliforms and *E. coli* tend to be more common on on pig carcasses than on cattle carcasses (Tables 1.8, 1.13 and 1.14; Mackey and Roberts, 1993). The counts of psychrotrophic organisms on pig carcasses are usually 10^2–10^3 cfu/cm^2 (Scholefield *et al.*, 1981; Skelley *et al.*, 1985; Tables 1.11 and 1.12), and appear mainly derived from equipment. Pathogens that may occur on carcasses include salmonellae, *Y. enterocolitica*, thermophilic campylobacters, *Staph. aureus*, *L. monocytogenes*, *B. cereus*, *Cl. perfringens*, and *Cl. botulinum*.

Chilling. Adequate chilling prevents the growth of mesophiles and mesophilic pathogens (e.g. salmonellae), and limits growth of psychrotrophs.

Carcasses. Carcasses of sheep and young calves and sides of beef and pork are chilled in air. Air temperatures are maintained—except for the period during loading of the chiller—at <7–8°C. The

Table 1.13 Bacterial numbers on pig carcasses before chilling

	Abattoir								
	N1[a]	N2[a]	N3[a]	N4[a]	N5[a]	N6[a]	N7[a]	N8[a]	N9[a]
Mean $\log_{10}$ cfu/cm^2 (20°C)	2.65	3.33	3.90	2.99	2.86	3.29	2.76	2.58	2.72
	E1[b]	E2[b]	E3[b]	E4[b]	E5[b]	A1[c]	A2[c]		
Mean $\log_{10}$ cfu/cm^2 (37°C)	3.65	4.06	3.32	2.20	2.76	3.30	3.72		
Mean $\log_{10}$ cfu/cm^2 (20°C)	4.03	4.22	3.21	2.11	2.75	3.41[d]	3.92[d]		
Mean $\log_{10}$ cfu/cm^2 (1°C)	2.65	2.21	2.03	–	–	–	–		
Enterobacteriaceae (% +ve)	–	–	77	60	68	–	–		
Coliforms (% +ve)	–	–	68	49	51	(1.30)[e]	(1.50)[e]		
No. carcasses	10	14	10	6	6	50	50		

[a]Abattoirs in Norway (N). Samples obtained by wet-dry swabbing of 50 cm^2 at each of six sites on 10 carcasses at each of 3–6 visits. Mean $\log_{10}$ number is the mean $\log_{10}$ value for all sites (Johanson *et al.*, 1983).
[b]Abattoirs in England (E). Samples obtained by wet-dry swabbing of 100 cm^2 at each of 12 sites on each carcass (Roberts *et al.*, 1980b).
[c]Abattoirs in Australia (A). Samples obtained by wet-dry swabbing of 40 cm^2 at each of four sites on 10 carcasses at each of five visits (Morgan *et al.*, 1987a).
[d]Aerobic viable count determined by incubation at 21°C.
[e]Mean log cfu *E. coli*/cm^2. *E. coli* was detected (1 cfu/cm^2) on 96% and salmonellae on 4% of samples from abattoir A1. *E. coli* was detected on 98% and salmonellae on 19.5% of samples from abattoir A2.

Table 1.14 Bacterial numbers on pig carcasses before chilling

	Abattoir									
	C1	C2	C3	C4	C5	C6	C7	C8	C9	S1
Mean $\log_{10}$ (cfu/cm^2) (25°C)	3.48[a]	2.94[a]	2.84[a]	2.64[a]	2.33[a]	1.93[a]	2.01[a]	1.06[a]	1.06[b]	–
Mean $\log_{10}$ (cfu/cm^2) (37°C)	–	–	–	–	–	–	–	–	–	3.66±0.50[c]
Coliform counts (cfu/100 cm^2) (35°C)	2.09[d]	1.14[d]	1.12[d]	1.52[d]	1.01[d]	–	–	–	–	–
Coliform counts (number/cm^2) (35°C)	–	–	–	–	–	–	–	–	0.15[e]	0.14±0.32[c]
Enterobacteriaceae ($\log_{10}$ cfu/cm^2) (37°C)										
E. coli counts (cfu/100 cm^2) (35°C)	1.62[f]	0.63[f]	0.81[f]	–	–	–	–	–	–	0.13±0.34[c]
E. coli counts ($\log_{10}$ cfu/cm^2) (44.5°C)										

Gill *et al.*, 2000a,b; Gill and Jones, 1997; Rivas *et al.*, 2000.
[a]Abattoirs in Canada (C). Mean of 25 total aerobic counts from pig carcasses at 8 plants after dressing and the final wash (Gill *et al.*, 2000a,b).
[b]Abattoirs in Canada (C). Mean of 25 total aerobic and coliform counts from 25 randomly selected pig carcasses entering a chiller at a pig slaughtering plant (Gill, 1997).
[c]Counts for the surface of carcasses of Iberian pigs at the end of the processing line in Spain (Rivas, 2000).
[d]Abattoirs in Canada (C). Mean of 25 total coliform counts from pig carcasses at 8 plants after dressing and the final wash (Gill *et al.*, 2000a,b).
[e]Aerobic viable count determined by incubation at 21°C.
[f]Abattoirs in Canada (C). Mean of 25 *E. coli* counts from pig carcasses at eight plants after dressing and the final wash (Gill *et al.*, 2000a,b).

temperature used depends on the species and subsequent stages of production. If beef or sheep is to be boned immediately after chilling (18–24 h from slaughter), air temperatures <7°C are used. If the meat is for chilled or frozen trade as carcasses, sides or quarters, carcasses are often chilled with air near −1°C to 2°C. Rapid chilling causes cold-shortening and toughening of the meat. This is largely prevented by electrical stimulation after stunning. As pork is much less likely to cold-shorten, fast rates of chilling can be used without affecting tenderness (temperatures of −5 to −12°C at 1.5–2 m/s may be used for the first 1.5–4 h).

Initially, water is lost from surface tissue at a rate predominately dependent on the difference in temperature between surface tissue and the surrounding air stream. As the surface temperature falls and approximates that of the air, the rate of evaporation depends on air speed and relative humidity. In the first stage of chilling, surfaces become dry. Later, as the rate of moisture movement from within the tissue to the surface exceeds the evaporation rate, the surface rehydrates. Fatty tissue tends to remain dry as there is less moisture migration from sub-surface tissue. Cut lean remoistens most. The lowering of surface a_w accompanying air-chilling delays or prevents growth of organisms, and is lethal to some (Scott and Vickery, 1939). Campylobacters are particularly sensitive to this drying of surface tissue (Oosterom *et al.*, 1983; Grau, 1988). Mesophiles are held in check by the lowered a_w till the surface temperature falls <7–8°C when temperature prevents growth (e.g. of salmonellae, *E. coli*). Psychrotrophs are controlled by the lowered a_w for some time but they eventually grow.

The extent of surface drying varies at different sites on the carcass. Drying and microbial control is most readily achieved on the thickest and slowest cooling portions of the carcass (Scott and Vickery, 1939). At such sites there can be significant decreases in the number of some microorganisms (e.g. psychrotrophs and *E. coli*). On thinner regions of the carcass, drying is less and there is a greater tendency for growth of psychrotrophs (e.g. the neck and parts of the abdomen of beef sides). In sheep carcasses, the body cavity and flap surfaces are likely to undergo the least drying during chilling (Gill, 1987). As there are gradients of air movement and temperature within a chill room, there will also be differences in the rate of cooling and the extent of surface drying of carcasses. When there is contact between carcasses, contacting surfaces will remain warm and moist and significant growth of mesophiles can occur. Thus, spacing of carcasses is important for control of microbial growth.

The doors and walls of chillers can be a source of contamination of carcasses with psychrotrophs (Newton *et al.*, 1978) and with *Listeria* spp. (Gobat and Jemmi, 1991).

An extensive USDA-FSIS survey of chilled steer and heifer carcasses indicates the microbiological profile that can be expected under good manufacturing conditions (Tables 1.15 and 1.16) (see also McNamara, 1995).

Bone-taint ("sticky post-mortem ageing") is the development of putrid or sour odors in the deep-seated portions of the carcass. The defect is sporadic, and is associated with slow cooling of the deep tissues and with heavy carcasses that have a thick layer of fat. Bone-taint is detected most commonly in the region of the hip-joint and less commonly in the region of the scapula near the joint with the humerus. Microscopic examination of bone-tainted meat has shown the presence of spore-forming large Gram-positive rods, cocci and small rods in the synovial fluid from the femur joint, in the bone marrow of the femur and/or in the fatty connective tissue surrounding the blood and lymph vessels and lying

Table 1.15 Bacteria on chilled beef carcasses

Organism	No. carcasses examined	Result
Aerobic plate count	2 089	93.1% $<10^4$/cm^2
Coliforms	2 089	96.4% <100/cm^2
E. coli	2 089	95.9% <10/cm^2
Staph. aureus	2 089	4.2% positive[a]
Cl. perfringens	2 079	2.6% positive[b]
L. monocytogenes	2 089	4.1% positive[c]
C. jejuni/coli	2 064	4% positive[c]
Salmonella	2 089	1% positive[c]
E. coli O157:H7	2 081	0.2% positive[c]

USDA Nationwide Beef Microbiological Baseline Data Collection Program: Steers and Heifers (USDA, 1994; McNamara, 1995). Carcasses in chiller for at least 12 h before rump, brisket and flank surface tissue excised. Pooled sample from the three sites examined.
[a]Limit of detection ~0 cfu/cm^2.
[b]Limit of detection ~5 cfu/cm^2.
[c]60 cm^2 of surface tissue tested.

Table 1.16 Bacteria on chilled beef carcasses

Organism	Carcasses examined	Result	Reference
$\log_{10}$ mean APC (25°C)	899	3.13	Vanderlinde *et al.*, 1998
$\log_{10}$ mean APC (5°C)	899	2.2	Vanderlinde *et al.*, 1998
$\log_{10}$ mean TVC (25°C)	1 275	2.42/cm^2	Phillips *et al.*, 2001a
$\log_{10}$ mean TVC (22°C)	420	2.8 ± 0.70 cfu/cm^2	Murray *et al.*, 2001
$\log_{10}$ mean TVC (37°C)	420	2.75 ± 0.64 cfu/cm^2	Murray *et al.*, 2001
E. coli (% +ve)	16[a]	43.8	Ingham and Schmidt, 2000
E. coli (% +ve)	16[b]	31.2	Ingham and Schmidt, 2000
E. coli (% >100 cfu/cm^2)	30[c]	0–6.7	Sofos *et al.*, 1999
E. coli (% >100 cfu/cm^2)	30[d]	0–6.7	Sofos *et al.*, 1999
E. coli (% >100 cfu/cm^2)	30[e]	0–1.7	Sofos *et al.*, 1999
E. coli (% >100 cfu/cm^2)	30[f]	0	Sofos *et al.*, 1999
E. coli (% >100 cfu/cm^2)	30[g]	1.1–5.6	Sofos *et al.*, 1999
E. coli (% >100 cfu/cm^2)	30[h]	0–4.4	Sofos *et al.*, 1999
E. coli (% +ve)	899	21.5	Vanderlinde *et al.*, 1998
Coliforms (% +ve)	899	39.6	Vanderlinde *et al.*, 1998
Enterobacteriaceae (22°C)	420 (15% +ve)	0.41 ± 0.37 cfu/cm^2	Murray *et al.*, 2001
Enterobacteriaceae (37°C)	420 (21% +ve)	0.40 ± 0.30 cfu/cm^2	Murray *et al.*, 2001
Salmonella spp.	899	0.22% positive	Vanderlinde *et al.*, 1998
Campylobacter spp.	629	0.16% positive	Vanderlinde *et al.*, 1998
Listeria spp.	170	0.59% positive	Vanderlinde *et al.*, 1998
E. coli O157:H7	899	0.45% positive	Vanderlinde *et al.*, 1998
E. coli O157:H7	1 275	0.10% positive	Phillips *et al.*, 2001a
Staphylococcus spp.	390	29% positive	Vanderlinde *et al.*, 1998

[a]Presumptive *E. coli* counts on carcasses chilled for 24 h.
[b]Presumptive *E. coli* counts on carcasses chilled for 7 days.
[c]% *E. coli* counts >100 cfu/cm^2, steer/heifer carcasses chilled 24 h, on brisket.
[d]% *E. coli* counts >100 cfu/cm^2, steer/heifer carcasses chilled 24 h, on flank.
[e]*E. coli* counts >100 cfu/cm^2, steer/heifer carcasses chilled 24 h, on rump.
[f]*E. coli* counts >100 cfu/cm^2, cow/bull carcasses chilled 24 h, on brisket.
[g]*E. coli* counts >100 cfu/cm^2, cow/bull carcasses chilled 24 h, on flank.
[h]*E. coli* counts >100 cfu/cm^2, cow/bull carcasses chilled 24 h, on rump.

between the muscles (e.g. adductor, pectineus and vastus intermedius) adjacent to the femur. Strains of *Clostridium* spp. have been isolated from bone-tainted pork and beef. The source of these deep-seated organisms is not clear, but thermometers used for checking temperatures in deep tissue (e.g. beef rounds) is one means of inoculating these sites and transferring the spoilage bacteria among carcasses within a lot and from day to day.

Stephan *et al.* (1997) studied different assays (pH, NH_3, TVB-N, H_2S, R-value, IMP, pyruvate, and various volatile acids such as lactic acid, acetic acid, butyric acid, uric acid, and pyroglutamic acid) for their suitability to serve as parameters to characterize sticky post-mortem ageing in beef muscle samples. A high concentration of butyric acid seemed typical for sticky post-mortem ageing in beef. A distinction from the onset of spoilage is possible owing to the high butyric acid concentration, a low content of ammonia (10.5 mg 100 g^{-1}), and a low pH value. It can be differentiated from meat after normal ageing by sensory, color, smell, consistency, a high butyric and pyroglutamic acids content, and a low IMP concentration.

Offals. Offals often have a higher initial contamination, and are more likely to be contaminated with potential pathogens, than carcass meats.

Larger offal (e.g. beef liver) may be placed on racks and smaller offal on trays and moved to chillers or freezers. However, offal is commonly placed while still warm in temporary containers before transfer to chillers. Warm livers, hearts, kidneys, and ox-tails are often bulk-packed into containers or cartons after inspection and washing, and the containers or cartons are chilled or frozen (Gill, 1988).

If offal is cooled slowly, e.g. on trays, significant increases in bacterial numbers can occur (Sheridan and Lynch, 1988). Contacting surfaces remain moist and may remain >10°C for many hours. In bulk-packed offal, even when placed in a freezer, the temperature may remain >10°C for 6–15 h. In such frozen offal, growth can result in > 100-fold increase in the numbers of *E. coli* (Gill and Harrison, 1985). Measurements of the temperature history during chilling and freezing allow the extent of proliferation of *E. coli* (and hence of salmonellae) to be calculated. Calculated increases agree closely with measured increases (Gill and Harrison, 1985). The lack of surface drying of offals with most cooling systems is likely to be an important factor in the prevalence of thermophilic *Campylobacter* spp. on retail offals being higher than on retail carcass meats. Poor shelf-life of offal is often a result of inadequate chilling.

Carcass storage and transport. Carcasses are stored at abattoirs for further processing, transported to retail stores and exported to other countries. Temperature control during transport and storage is critical in limiting microbial growth.

After the initial chilling phase, the microbial flora on carcasses is a mixture of mesophiles and psychrotrophs. The types of organisms that grow, and their growth rates, are determined by the storage temperature and the surface a_w of the carcass. When maximum storage-life is desired, carcasses are stored and shipped at a temperature of $-1 + 0.5$°C and with some controlled weight loss. Normally, however, storage temperatures are 3–5°C where storage-life is about half that at 0°C.

Under most conditions, pseudomonads are the fastest growing component of the flora. The percentage of *Psychrobacter, Moraxella phenylpyruvica,* and *Acinetobacter* in the flora usually declines during storage (Sauter *et al.*, 1980; Prieto *et al.*, 1992). Psychrotrophic Enterobacteriaceae form only a minor percentage of the developing flora. *Broch. thermosphacta* is the major Gram-positive organism growing on chilled carcass meat. The slower growing micrococci and yeasts are favored by dry conditions limiting the growth of the faster growing flora. These become a significant proportion of the final population when the relative humidity of the air and its speed over the carcass result in a low surface a_w-value. High numbers of *Staph. aureus* have occasionally been reported on pig carcasses delivered to a processing plant (Schraft *et al.*, 1992). Counts $> 10^3$ cfu/cm^2 were found on 2.4% of carcasses, with a few carcasses having counts of 10^6 cfu/cm^2. Growth of *Staph. aureus* might be expected on carcasses if poor chilled storage conditions allowed surface temperatures to be >7–10°C for significant periods while drying of surface tissues repressed growth of the normal spoilage flora. Because of the lower water content of fatty tissue, drying occurs more readily on this tissue. Diffusion of water in lean tends to maintain sufficient moisture for the growth of pseudomonads, though at reduced rates, even with relative humidities as low as 85–90%. When carcasses are moved into surroundings where the dew point is above the surface temperature of the carcass, condensation will occur and cause some increase in surface temperature and a_w-value.

C Spoilage

Under chilled conditions, mesophiles will not grow and the psychrotrophs eventually cause spoilage of the carcass. The rate of spoilage increases with (i) the number of contaminating psychrotrophs on the carcass, (ii) increases in storage temperature, and (iii) increases in the a_w of surface tissue (dependent on air velocity and relative humidity). On the lean of carcasses, off-odors are detected when the population reaches about 10^7 cfu/cm^2. Slime is apparent only when the a_w is near 0.99 and the population about 10^8 cfu/cm^2. At a lower a_w-value, colonies fail to coalesce and remain small and discrete. Spoilage occurs first on moist areas of the carcass (abdominal cavity of sheep, near the diaphragm, in folds between the fore-leg and the chest and on the cut muscles of the neck).

The spoilage flora is dominated by psychrotrophic, aerobic, Gram-negative motile and non-motile rods, identified as *Pseudomonas* spp., *Acinetobacter* and *Psychrobacter immobilis* (Dainty and Mackey, 1992). Pseudomonads normally account for >50% of the spoilage flora with *Ps. fragi* (clusters 1 and 2), *Ps. lundensis* and *Ps. fluorescens* being the most important species. *Broch. thermosphacta* and psychrotrophic Enterobacteriaceae usually form only a small proportion of the spoilage flora but appear to be more prevalent on the fat surfaces of sheep and pork. Storage at 5°C rather than at 0–1°C tends to favor their growth (Dainty and Mackey, 1992). At high ambient storage temperatures (25–30°C), Enterobacteriaceae and *Acinetobacter* spp. dominate the spoilage flora (Gill and Newton, 1980; Rao and Sreenivasamurthy, 1985). When storage has caused the surfaces to remain dry, colonies of micrococci, yeasts, and molds may appear. Growth is uneven over the carcass mostly owing to variations in surface a_w. Limited information on the bacterial flora of horses and goats has been reported (Cantoni, 1977; Sinha and Mandar, 1977).

D Pathogens

Salmonella. The prevalence of salmonellae on beef, sheep, and pig carcasses varies widely. Sometimes salmonellae are rarely found (Biemuller *et al.*, 1973). Sometimes they can be found on close to half of the carcasses (Oosterom and Notermans, 1983), and at other times on all carcasses from a herd (Grau and Smith, 1974). In a large survey in the United States, salmonellae were found on 1% of excised 25 g samples of brisket from 3 075 chilled carcasses of steers, heifers, bulls, and cows and on 5% of samples from 397 calves (Hogue *et al.*, 1993). A more recent US survey (Tables 1.15 and 1.16), also detected salmonellae on 1% of samples excised from chilled steer and heifer carcasses (USDA, 1994). In Canada, salmonellae were detected on 11.2% of 596 pork carcasses, on 4.1% of 267 veal and on 1.7% of 666 beef carcasses when neck muscle samples, excised from carcasses before chilling, were examined (Lammerding *et al.*, 1988). Contamination rates for animals in Eygptian abattoirs were: buffalos 1–2% (Lotfi and Kamel, 1964; El Moula, 1978); camels up to 44% (El Moula, 1978); sheep 3–4% (Lotfi and Kamel, 1964); a wide range of serotypes were found.

The extent of carcass contamination is strongly influenced by the prevalence and concentration of salmonellae in the intestinal tract and, for sheep and cattle, by the contamination of the fleece and hide. It is also influenced by the care taken in slaughter and dressing. The salmonellae status of animals at slaughter is determined by contamination acquired at the farm and by holding conditions before slaughter.

Salmonellae can be found in the internal tissues of liver and spleen from apparently normal animals (Tables 1.6 and 1.7).

Normally, only small numbers of salmonellae are on carcass or offal meats. However, inadequate chilling, storage or transport, at temperatures above about 7°C, can permit growth.

Outbreaks of salmonellosis can follow from inadequate cooking, mishandling, and recontamination. Raw meats can act as a source of cross-contamination of cooked meats, or other foods, in the kitchen or in meat processing plants.

Escherichia coli O157:H7. Generally a small percentage of cattle carry EHEC O157:H7 in the intestinal tract at slaughter (Table 1.4), but occasionally the percentage is high. Care taken during evisceration and hide removal can limit, but not entirely prevent, contamination of the carcass. Growth of the organism can occur if chilling, storage or transport conditions of the carcass are inadequate (temperature above about 7°C). Inadequately cooked ground beef, contaminated with *E. coli* O157:H7, has caused a number of outbreaks of bloody diarrhea (HC) and hemolytic uraemic syndrome (Doyle, 1991; Griffin and Tauxe, 1991). Cooking hamburgers to an internal temperature of 68°C has been recommended (Meng *et al.*, 1994). Undercooking of ground beef or hamburger patties is a common cause of outbreaks attributed to *E. coli* O157:H7. A large outbreak in the United States (1993) affected 732 people, undercooked hamburgers were the cause.

Ground beef has also been identified as carrying *E. coli* O157, published prevalences varying from 0 (Willshaw *et al.*, 1993; Lindqvist *et al.*, 1998; Tarr *et al.*, 1999) to 0.7 (Doyle and Schoeni, 1987), 1.3% (Kim and Doyle, 1992) and 2.4% (Sekla *et al.*, 1990). Quantitative analysis of some of the positive lots conducted by FSIS and CDC resulted in MPN values in the range of 1–4 cfu/g, with a single high value of 15 cfu/g (FSIS, 1993). In an Argentine study 11 *E. coli* O157:H7 isolates were detected in 6 (4%) out of 160 ground beef samples produced directly by retailers from different combinations of cuts and trimming of boneless beef (Chinen *et al.*, 2001). In the same study, 4 (8%) out of 83 fresh sausages were positive for *E. coli* O157:H7.

In a Swiss study, a total of 400 minced meat samples from 240 small butcheries were collected and analyzed for the presence of STEC and *L. monocytogenes* (Fantelli and Stephan, 2001). The samples comprised 211 samples of minced beef and 189 samples of minced pork. STEC was isolated from 7/400 (1.75%) samples. In particular, 5/211 (2.3%) minced beef samples and 2/189 (1%) minced pork samples were contaminated. Serotyping of the seven strains yielded five different serotypes, but none of the strains was O157:H7. Two STEC strains harbored *stx1* and *stx2* and five strains harbored *stx2c* genes. Furthermore, four strains harbored one or more additional virulence factors. However, none of the strains was positive for *eae*. *L. monocytogenes* was isolated from 43/400 (10.75%) samples. Nineteen of the 43 strains belonged to serotype 1/2a, two to serotype 1/2b, 12 to serotype 1/2c, and 10 to 4b. Forty-two strains harbored the *Lhly* and 43 strains the *plcA* genes. Macrorestriction analysis of the *L. monocytogenes* strains using *Sma*I yielded 12 different PFGE-patterns. The predominating pattern G was associated to the serotype 1/2c.

Ground beef patties inoculated with ca. 10^6 *E. coli* O157:H7 and frozen at −20°C for 24 h and then thawed at 4°C for 12 h, at 23°C for 3 h, or using microwave heating for 120 s at 700 W, showed variable destruction depending upon the strain, the recovery method and the thawing regimen (Sage and Ingham, 1998). Podolak *et al.* (1995) reported fumaric acid (1% and 1.5%) to be more effective than 1% lactic or acetic acid in reducing populations of *E. coli* O157:H7 in vacuum-packaged ground beef patties. According to guidelines of the UK Advisory Committee on the Microbiological Safety of Food (1995), however, target/temperature of 70°C for 2 min is recommended to ensure safety of minced meat products.

Campylobacter jejuni. After slaughter, thermophilic campylobacters have been found from 19% to 70% of sheep carcasses, from 2% to 32% of adult cattle carcasses and from 20% to 97% of calf carcasses. Thermophilic campylobacters (mostly *C. coli*) have been found on 20–60% of pork carcasses. In a Canadian survey (Lammerding *et al.*, 1988), campylobacters were found on 12% of pork, 15% of beef, and 35% of beef neck samples taken from carcasses before chilling.

Fecal samples from 94 dairy cows and 42 calves in three different herds were examined by a variety of techniques for campylobacters (Atabay and Corry, 1998). Seventy-nine percent of cattle in herd A carried campylobacters, 40% in herd B and 37.5% in herd C. Most animals carried only one species of *Campylobacter*: *C. hyointestinalis* was isolated most frequently (32% animals positive) with *C. fetus* subsp. *fetus* detected in 11% of animals and *C. jejuni* subsp. *jejuni* 7%. In addition, a novel biotype of *C. sputorum* was isolated from 60% of 47 cows tested in herd A.

During carcass chilling, there is a significant reduction in the prevalence and number of viable campylobacters, and, even on originally relatively heavily contaminated carcasses, numbers are usually <1 cfu/cm^2. Additional drying that occurs during storage and transport will further reduce campylobacters. By the time carcass meats reach retail markets the rate of contamination is considerably reduced

Table 1.17 *Campylobacter jejuni/coli* on meats at retail

Type	Country	Number	% Positive sampled	Reference
Beef	England	127	23.6	Fricker and Park, 1989
	USA	360	4.7	Stern *et al.*, 1985
	USA	230	0.4	Harris *et al.*, 1986
	USA	520	0.2	G. Schmid quoted in Harris *et al.*, 1986
	Japan	112	0.0	Ono and Yamamoto, 1999
	Kenya	50	2.0	Osano and Arimi, 1999
	Ireland	50	0.0	Madden *et al.*, 1998
Ground beef	USA	360	3.6	Stern *et al.*, 1985
Various ground meats	Italy	58	6.9	Baffone *et al.*, 1995
Ox liver	England	96	54.2	Kramer *et al.*, 2000
Beef offal	England	97	7.0	Bolton *et al.*, 1985
Pork	England	158	18.4	Fricker and Park, 1989
	USA	360	5.0	Stern *et al.*, 1985
	USA	149	0.7	Harris *et al.*, 1986
	USA	514	0.0	G. Schmid quoted in Harris *et al.*, 1986
	Japan	126	0.0	Ono and Yamamoto, 1999
	Ireland	50	0.0	Madden *et al.*, 1998
	Italy	27	3.7	Zanetti *et al.*, 1996
Pork (bacon)[b]	England	80	84.4	Phillips, 1998
	USA	384	1.3	Duffy *et al.*, 2001
Pork liver	England	99	71.7	Kramer *et al.*, 2000
Pork sausage	USA	360	4.2	Stern *et al.*, 1985
	Italy	200	0.5	Rindi *et al.*, 1986
Pork offal	England	26	11.5	Bolton *et al.*, 1985
Lamb	England	103	15.5	Fricker and Park, 1989
	USA	360	8.1	Stern *et al.*, 1985
	USA	37	0.0	Harris *et al.*, 1986
Lamb liver	England	96	72.9[a]	Kramer *et al.*, 2000
Ground meats	England	135	2.2	Bolton *et al.*, 1985
Sausage meat	England	143	0.7	Bolton *et al.*, 1985
	Italy	41	2.4	Zanetti *et al.*, 1996
	Italy	46	8.9	Baffone *et al.*, 1995
Offal—mixed	England	689	47.0	Fricker and Park, 1989

[a]Percent positive for *Campylobacter* spp.
[b]MAP samples positive for *Campylobacter* spp.

but can still, at times, be relatively high (Table 1.17). Contamination rates on offal meats at retail sale are higher than on carcass meats, probably owing to higher initial contamination at slaughter and better survival on offals that do not dry much during chilling.

Epidemiological evidence indicates that carcass meat of cattle, sheep, and pigs are relatively minor causes of human campylobacter infection.

Yersinia enterocolitica. The prevalence of pathogenic serotypes of *Y. enterocolitica* on pig carcasses can range from low (none detected on 210 carcasses, de Boer and Nouws, 1991; 2.5%, Mafu *et al.*, 1989) to very high (31%, Christensen, 1987; 63%, Nesbakken, 1988). Care taken during evisceration has a marked influence on the extent of contamination with *Y. enterocolitica* (Andersen *et al.*, 1991). Careful removal of the tongue and tonsils is also needed to prevent carcass contamination.

Epidemiological studies have associated human yersiniosis with the consumption of minced raw pork (Belgium, Tauxe *et al.*, 1987), of undercooked pork (Norway, Ostroff *et al.*, 1994) and, in the United States, of chitterlings (pork intestine, Lee *et al.*, 1991). Household preparation of chitterlings has also been implicated with an outbreak in children (Lee *et al.*, 1990). There is also a strong geographic correlation between the serotypes of strains isolated from pigs and from humans. Similarly, biochemical, phage-typing and DNA analysis have been unable to distinguish between porcine and human strains (Andersen *et al.*, 1991).

Fredriksson-Ahomaa *et al.* (2003) detected different *Y. enterocolitica* 4:O3 genotypes in pig tonsils from Southern Germany and Finland and concluded that 4:O3 genotypes are distributed geographically and can be used in epidemiological studies. Nesbakken *et al.* (2003) investigated the distribution of *Y. enterocolitica* in slaughter pigs in relation to dressing practices and inspection procedures.

Although pathogenic serotypes can grow at 1°C, growth is slow relative to that of spoilage microorganisms, and pathogenic strains are not good competitors. Consequently, the extent of growth in normal minced meat is restricted in the presence of the background flora at temperatures up to at least 15°C (Fukushima and Gomyoda, 1986; Kleinlein and Untermann, 1990).

However, several studies report that *Y. enterocolitica* is able to multiply in foods kept under chill storage and might even compete successfully (Hanna *et al.*, 1977; Stern *et al.*, 1980; Grau, 1981; Lee *et al.*, 1981; Gill and Reichel, 1989; Lindberg and Borch, 1994; Borch and Arvidsson, 1996; Bredholt *et al.*, 1999). The study of Bredholt *et al.* (1999) indicates that 10^4 cfu/g of *Y. enterocolitica* is able to multiply at 8°C in vacuum-packaged cooked ham and servelat sausage in the presence of 10^4–10^5 cfu/g of LAB. In the same experiments, these LAB cultures inhibited growth of *L. monocytogenes* and *E. coli* O157:H7.

A major problem may be the use of head-meat, particularly when remnants of tonsils and oral mucosa are present, in the production of minced raw pork.

Listeria monocytogenes. *L. monocytogenes*, and other *Listeria* spp., can contaminate carcasses from the hide or fleece or feces of cattle and sheep and from surfaces in the slaughter and dressing area (Lowry and Tiong, 1988; Gobat and Jemmi, 1991). Listeriae on pig carcasses arise particularly from contamination after singeing and during passage into the chiller (Gobat and Jemmi, 1991). There are opportunities for additional contamination as carcasses are moved through chillers, cold storage rooms and transport to retail sale. Some growth can also occur even under adequate chilling conditions (<5°C). Thorough cooking will destroy the organism. Raw meat can be one of the sources for contamination of ready-to-eat processed meats.

Staphylococcus aureus. During slaughter and dressing, carcasses of ruminants become contaminated from the skin of the animal, the equipment used (e.g. mesh protective gloves and aprons) and the hands of the workers (Peel *et al.*, 1975). Similarly, low numbers of *Staph. aureus* on pork carcasses (Takacs and Szita, 1984) are a mixture of survivors originally on the skin of the pig and contaminants acquired during slaughter (Rasch *et al.*, 1978). Chilling, storage or transport at temperatures <7°C will prevent growth. Even at high temperatures, on raw unprocessed meat, the organism is a poor competitor and is outgrown by other flora. For sufficient enterotoxin to be produced to cause food-poisoning, *Staph. aureus* must reach a population of at least 10^6 cfu/g. In addition, strains of *Staph. aureus* from animal sources are less likely than those of human origin to produce the enterotoxins (e.g. enterotoxin A) that are commonly implicated in food-poisoning (Devriese, 1990). Staphylococcal food-poisoning is not caused by raw unprocessed meat.

Clostridium botulinum. Most of the clostridia that occur in raw meats are harmless putrefactive mesophiles. However, from time-to-time *Cl. botulinum* occurs. Estimates of contamination range from <0.1 spore to 7 spores per kg (Lücke and Roberts, 1993). The incidence appears to be lower in beef and lamb than in pork. There is no means available to guarantee the absence of *Cl. botulinum* from raw meat. Most cases of meat-borne botulism have been from improperly preserved, home-produced processed meats that have been eaten without prior cooking (Tompkin, 1980). However, consumption of raw meat from marine mammals handled under conditions in which psychrotrophic non-proteolytic *Cl. botulinum* strains have multiplied, has caused outbreaks of botulism in the Inuit population of Northern Canada and Alaska. Apart from this, no cases of botulism from consumption of fresh meat have been reported (Lücke and Roberts, 1993).

Clostridium perfringens. *Cl. perfringens* is a common surface contaminant of beef, sheep and pig carcasses at slaughter (Smart *et al.*, 1979). It occurs in low numbers ($<$200 cfu/100 cm^2) and mainly as vegetative cells. Contamination is from fecal material and from soil and dust on the skin of the animal. The internal tissue of offal (e.g. liver) may contain small numbers of *Cl. perfringens* (Canada and Strong, 1964; Bauer *et al.*, 1981). Fresh meat is stored at temperatures ($<$15°C) that are too low to allow growth. Viable vegetative cells will tend to decrease in numbers during chill storage and will be destroyed in thorough cooking. Food poisoning results from the survival of spores in cooked meats and considerable growth (to $>10^5$ cfu/g) during inadequate cooling of the cooked product under anaerobic conditions (holding for some hours between 15°C and 50°C) and consumption without re-heating, which kills vegetative cells.

Toxoplasma gondii. Most human infections with *Toxoplasma gondii* are inapparent. However, serious consequences result when host cells are destroyed by multiplication of the protozoan (e.g. myocarditis, hepatitis and encephalitis). Infants borne to infected mothers may die or suffer blindness or brain damage. Reproductive disorders are caused by toxoplasmosis in sheep, goats, and pigs. Humans are infected by oocysts shed in cat feces, or ingestion of viable cysts in pig, sheep, goat or cattle tissue. A large European study identified inadequately cooked or cured meat as the main risk factor for infection with *Toxoplasma* (Cook *et al.*, 2000). Meat should be cooked to at least 60°C or frozen (−15°C for 3 days, −20°C for 2 days) to destroy cysts.

Sarcocystis hominis and Sarcocystis suihominis. Humans can be infected with *Sarcocystis hominis* by eating cysts present in raw beef muscle and *Sar. suihominis* by eating cysts in raw pork. Symptoms include nausea, stomach pain ,and diarrhea. Cooking or freezing meat destroys the cysts. Cattle and pigs become infected by ingesting sporocysts previously excreted by infected humans.

Trichinella spiralis. Trichinosis is acquired from ingestion of encysted larvae in the muscle of raw pork or wild animals (wild boar, fox, bear and walrus). In pork, larvae are found in all striated muscle, but the greatest density is in the diaphragm, tongue, masseter, and neck muscles. Larvae can be destroyed by cooking meat to at least 58°C, or by freezing (e.g. −15°C for 30 days).

The prevalence of outbreaks of trichinellosis appears to have increased in the EU Member States, the MEDLINE database recording 36 outbreaks of trichinellosis in the EU from 1966 to 1999. Those outbreaks involved several thousands of patients and were the result of consumption of imported horsemeat (urban outbreaks involving sometimes hundreds of patients), of hunted wild-boar meat (small outbreaks in families of hunters), of pork from pigs originating from small farms or grazing in wild areas (Table 1.18). Over 3 300 human cases of trichinellosis have been attributed to the consumption of imported horsemeat in France and Italy in the past 25 years. Most Spanish outbreaks were due to consumption of pork or wild-boar meat, and outbreaks from wild-boar meat are increasingly frequent in the Southeastern parts of France and in certain regions of Spain. Trichinellosis is not a notifiable disease in all Member States.

In the past 20 years, human trichinellosis due to the consumption of local domestic or wild animals has not been reported in Austria, Belgium, Denmark, Finland, Great Britain, Ireland, Luxembourg, Portugal, Sweden or the Netherlands (Pozio, 1998; 2001).

A survey in 1998 identified 10 outbreaks involving at least 791 patients in France, Germany, Spain, and Italy (Dupouy-Camet, 1999). Elsewhere, 623 cases were due to the consumption of horsemeat

Table 1.18 Sources of *Trichinella* infections in humans in countries of the European Union in the past 20 years (adapted from Pozio, 1998, and including 1998–2000 data)

Source	No. cases	Country
Pigs bred in small family farms	>1000	Spain
Pigs grazing in wild areas	800	France, Germany, Italy and Spain
Wild boars	1300[a]	France, Germany, Italy and Spain
Horses[b]	>3300	France and Italy
Total	>6400	

[a]Including some cases due to other game.
[b]Imported from non-EU countries.

imported from Yugoslavia and Poland in France (two outbreaks) and Italy (one outbreak) (Dupouy-Camet, 1999). In 1999, The Netherlands, Germany, Spain, France, UK, and Austria reported together 49 cases of human trichinellosis, of which ~50% had been acquired through the consumption of imported products. Non-imported cases were due either to wild-boar meat or to pork from organic farms. In 2000, the Netherlands, Germany, Spain, the UK and Italy reported 88 cases of human trichinellosis. These data are the best estimates available.

The apparent prevalence for *Trichinella* infections in pigs in the EU Member States is low, generally <1 in 100 000 animals, although higher in some regions.

In contrast, in the eastern parts of Croatia, *Trichinella* larvae were found in 0.5–1.5% of the swine examined in 1999 (Marinculic *et al.*, 2001).

Since 1975, several outbreaks of human trichinellosis have resulted from the consumption of raw or inadequately cooked horsemeat in France and Italy, identifying horsemeat as the most important source of human trichinellosis in Europe (Boireau *et al.*, 2000)

Besides rodents and wild animals, livestock raised under extensive management systems, e.g. free roaming pigs in forests, might play an important role in the transmission of the sylvatic genotypes.

Consumers expose themselves to the risk of trichinellosis when they consume raw or insufficiently cooked meat from improperly examined carriers of *Trichinella*, commonly raw or inadequately cooked horsemeat. German consumers are likely to be particularly exposed to *Trichinella*-contaminated pork, as consumption of raw pork or pork products is common (Schotte *et al.*, 1992).

There are many reports from many parts of the world on the close relationship between trichinellosis outbreaks and consumer habits involving the consumption of meat from domestic pigs, wild boars, horses, walruses, whales, bears, and dogs. A public health problem may arise from the introduction of new eating habits and methods of food preparation, or following the immigration of people from other cultures. International travel, and the subsequent adoption of traditional and unusual culinary habits, may explain imported trichinellosis cases. Some high-class restaurants serve barely cooked dishes as a mark of good quality, drawing attention to the retained freshness of their ingredients, which thus include undercooked meat. Additionally, globalization of trade is a risk factor, as animals or fresh meat products are exported from countries where trichinellosis is endemic among wild-life or domestic animals, e.g. a small outbreak of trichinellosis in Normandy, France, in 1998 was due to the consumption of vacuum-packed wild-boar meat imported from the United States (Dupouy-Camet, 2000).

The increasing trend to extensive pig farming, where pigs have access to natural pasture, increases the risk of infection with *Trichinella* spp.

Taenia saginata. Following the ingestion of the viable larval or cysticercus stage encysted in beef muscle, the beef tapeworm develops in the human jejunum. Cattle become infected from eggs shed in human feces. Freezing (−10°C for 10 days, −15°C for 6 days) or heating (>56°C) destroys cysticerci in beef.

Taenia solium. Humans are infected by ingestion of viable encysted cysticerci present in the liver, brain, skeletal muscle or myocardium of pigs. The adult tapeworm develops in the human ileum, and eggs shed in the feces of infected pigs. Humans can also become infected by accidental ingestion of eggs shed in human feces. In this case, cysticerci develop in the human and these can localize in a variety of tissues including the brain. Cerebral cysticercosis is one of the most common parasite diseases of the central nervous system. Freezing (e.g. −10°C for at least 10 days) or cooking inactivates cysticerci in meat.

E Control

There is no process in the conversion of livestock to carcasses that can ensure the absence of human pathogens. The main sources of microbial contamination of carcasses are the microorganisms in or on the live animal. Controls are needed during animal production, transport, and slaughter of animals and during chilling, storage, and transport of carcasses and offals (Simonsen *et al.*, 1987; Snijders, 1988; Kasprowiak and Hechelmann, 1992; Mackey and Roberts, 1993; NACMCF, 1993). Most of the control measures are aimed at reducing the initial level of the pathogens (H_0), however, steps aimed at reducing (ΣR) or preventing increase (ΣI) are also suggested.

E CONTROL (farm)

Summary

Significant hazards[a]	
Beef, lamb, and other ruminants:	• salmonellae. • VTEC. • *C. jejuni/coli.* • Parasites.
Pork:	• salmonellae. • *C. coli.* • *Y. enterocolitica.* • Parasites.
Control measures	
Initial level (H_0)	• Environment, feed, water (source and how supplied to animals), rodents, wild birds, insects, and invertebrates (salmonellae, thermophilic campylobacters, *L. monocytogenes*). • Movement of livestock from farm-to-farm, mixing of animals, and stress associated with stock sales/markets (salmonellae), silage (*L. monocytogenes*), manure control (VTEC).
Reduction (ΣR)	• Not relevant.
Increase (ΣI)	• Failing practices in H_0, above, would result in an increase.

(*continued*)

E CONTROL (Cont.)

Summary

Testing	• Mainly by assessing the disease/health status of the animals. • It is recommended to test water from boreholes for *E. coli*, *C. jejuni/coli*.
Spoilage	• Not relevant

[a]In particular circumstances, other hazards may need to be considered.

Hazards to be considered. See above.

As there are many sources of contamination of livestock during rearing, control is difficult to achieve, though there have been successes with some animal diseases (e.g. bovine tuberculosis, brucellosis, *S.* Cholerae-suis). Water, feed, environment, wild-life, farm animals (cats and dogs), replacement livestock and manure are sources of contamination. Control is especially important under intensive rearing conditions as pathogens spread rapidly.

Animals can acquire a number of pathogens from water (e.g. *Campylobacter* spp., salmonellae, VTEC, *Y. enterocolitica, Sarcocystis,* and *Taenia saginata*). Good quality water should be used in intensive units and water troughs regularly cleaned. Efforts should be made to limit contamination of water-courses by grazing animals, farm effluent and human sewage.

Control measures

Initial level of hazard (H_0). Good animal welfare will prevent overcrowding, poor nutrition, poor care and stress which increase the occurrence and spread of pathogens. Animals should be properly fed and receive appropriate veterinary surveillance and treatment. Information on diseases detected at slaughter as well as on pathological anatomical changes found by meat inspection (e.g. liver abscesses and *Fasciola hepatica*) should be used to adjust farming practice. Stress should be minimized by careful and quiet movement and transport of animals, by avoiding mixing of animals from different sources and by ventilation to prevent extremes of temperature and high concentrations of ammonia or carbon dioxide.

Replacement stock brought onto the farm should be kept isolated from other stock to prevent introduced pathogens from spreading. Calf-rearing units present a special problem where young susceptible animals are obtained from a variety of farms, often after being traded through markets. The time spent in transport and at markets needs to be kept short, and both need regular cleaning.

Increase of hazard (ΣI) and decrease of hazard (ΣR). Measures to eliminate salmonellae and other human pathogens from feeds of both animal and vegetable origin need to be improved and implemented. Properly pelleted feed reduces the prevalence of salmonellae. Special care needs to be taken with wet mashes to prevent growth of salmonellae in residues in troughs and piping. Feeding animals with "garbage" or "swill" increases the risk of acquiring salmonellae unless it is properly heat processed. The use of short-chain fatty acids has shown promise in reducing salmonellae in feeds. Proper ensiling of feed, where the pH is <4, reduces contamination with *Listeria* and *Salmonella* spp. Salmonellae in the saliva and tonsils of calves and other young animals contaminate drinking bowls and buckets, and regular cleaning of these is needed to limit spread of infection.

The design and construction of buildings in which animals are housed should allow effective cleaning and disinfection, and removal of manure. Removal and disposal of manure prevent recycling of excreted pathogens. Heavy soiling of animals is caused by poor housing conditions where there is irregular removal of manure, inadequate bedding and holding animals on muddy ground. Sometimes flanks and bellies of cattle are clipped before housing to reduce the accumulation of feces and other material on the hide. In some intensive units, automatic systems wash animals to remove manure and dirt from the skin. Cleaning and disinfection of intensive rearing premises should be carried out between batches of animals using an "all-in all-out" system of restocking. Insects, birds, rodents, cats, and dogs should be excluded from intensive units, tools, and equipment should be cleaned and disinfected, dedicated clothing and footwear should be used by animal attendants.

As there are many sources of contamination an integrated approach is needed, which involves many aspects in the breeding, raising, and transportation of livestock and feed production.

F CONTROL (transport and holding at abattoirs)

Summary

Significant hazards[a]	
Beef, lamb, and other ruminants:	• salmonellae. • VTEC. • *C. jejuni/coli.* • Parasites.
Pork:	• salmonellae. • *C. coli.* • *Y. enterocolitica.* • Parasites.
Control measures	
Initial level (H_0)	• Feed withdrawal (3–6 h before transportation) to reduce fecal excretion. • Avoiding intermittent feeding and starvation which increase susceptibility of ruminants to salmonellae infection. • Reducing intestinal fill, particularly of the rumen, to lessen the chance of gut breakage during slaughter. • Ante-mortem inspection, traceability back to farm. • Designing vehicles to make cleaning easier and to limit contamination, especially for multi-level trucks where feces can fall onto animals on lower decks. • Cleaning and sanitizing transport vehicles between loads. • Cleaning of holding pens and races to decrease external contamination and spread of enteric bacteria.
Reduction (ΣR)	• Not relevant.
Increase (ΣI)	• Limiting time and minimize stress in transit and holding to reduce spread of enteropathogens (e.g. salmonellae).
Testing	• Testing is not appropriate.
Spoilage	• Reducing long term stress and starvation prior to slaughter, and cold-wet conditions, which contribute to glycogen depletion and high pH (DFD) meat.

[a]In particular circumstances, other hazards may need to be considered.

G CONTROL (slaughter and dressing of cattle and sheep)

Summary

Significant hazards[a]	• salmonellae. • VTEC. • *C. jejuni/coli.* • Parasites.
Control measures	
Initial level (H_0)	• Cleaning and decontamination of equipment. • Hygiene of workers.
Reduction (ΣR)	• Decontamination with hot water and/or food grade acids (where permitted).
Increase (ΣI)	• Avoid contact between skin/fleece and carcass surface. • Avoid transfer of gut contents to carcass. • Prompt refrigeration minimizes increase.
Testing	• Slaughter hygiene monitored by sampling carcasses. • Verify cleaning and disinfection by microbiological, or other, tests. • It is recommended to test water from boreholes for *E. coli* and *C. jejuni/coli*.
Spoilage	• Not relevant.

[a]In particular circumstances, other hazards may need to be considered.

Hazards to be considered. See above.

Control measures

Initial level (H_0). Carcasses and offals can become contaminated by microorganisms from the hide, the hair or fleece and the intestinal tract. The care and skill of workers and the time available to them for hide-removal and evisceration are critical in limiting microbial contamination. Workers need training and an understanding of the hygienic consequences of operations during processing.

- Cleaning and decontaminating equipment before work begins (e.g. knives, steels, scabbards, aprons, brisket, and backbone saws and collection containers for offals). Provision of equipment and time for workers to wash their hands and fore-arms, and to wash and decontaminate instruments during slaughter and dressing.
- Clipping or tying the esophagus to prevent outflow of ruminal fluid.
- Removing udders of lactating animals before hide removal.
- Minimize contact of exposed tissue with external surfaces of the skin (e.g. by flapping and rolling of the skin).
- Minimize workers' handling external surfaces of the skin and then handling exposed tissue (mechanizing hide removal reduces alternate handling of skin and carcass).
- Taking special care when freeing the rectum and anal-sphincter to prevent transfer of fecal-contamination on the hide near the anus and on the tail (e.g. wide cut around the anal opening; preventing tail contacting tissue), and to limit transfer of contamination from the freed rectum of the rectal channel and abdominal cavity (tying rectal end and enclosing in a plastic bag).
- Removing rumen and intestinal tract without breakage (proper use of brisket saw and knife, cutting abdominal wall with blade turned outwards, use knife with bulb point).

- Cutting away visible contamination or any other treatment that can effectively remove this contamination, washing may spread contamination.
- Minimize contact of carcasses with fixed equipment of the slaughterhouse.

Reduction (ΣR)

- Decontaminating carcasses with water at high contact temperature or with solutions of food grade acids.

Increase (ΣI)

- Promptly removing carcasses, recovered meat (diaphragm, head-meat, and tails) and offals to refrigeration.
- Avoid holding offals and recovered meat in temporary containers in the slaughtering area because this facilitates growth of mesophilic pathogens (*Staph. aureus*, salmonellae).

Testing

- APC (30°C incubation) is a useful measure of slaughter hygiene and the bacteriological status of carcasses. Trend analysis is a useful measure of hygiene over time.
- This could be supplemented with testing for *E. coli* or Enterobacteriaceae (35–37°C incubation) as an indication of fecal contamination.
- Testing for salmonellae and other bacterial pathogens is less productive for improving hygiene at slaughter.

Spoilage. Not relevant.

H CONTROL (slaughter and dressing of pigs)

Summary

Significant hazards[a]	• salmonellae. • *C. jejuni/coli.* • *Y. enterocolitica.* • Parasites.
Control measures	
Initial level (H_0)	• Cleaning and decontamination of equipment. • Hygienic slaughter and dressing.
Reduction (ΣR)	• Use spray scalding or steam scalding in preference to vats.
Increase (ΣI)	• Avoid transfer of gut contents to carcass. • Prompt refrigeration minimizes increase.
Testing	• Slaughter hygiene monitored by sampling carcasses. • Verify cleaning and disinfection by microbiological tests.
Spoilage	• Not relevant.

[a]In particular circumstances, other hazards may need to be considered.

Hazards to be considered. See above.

Control measures

Initial level (H_0)

- Cleaning and decontaminating scalding, dehairing, scraping and polishing equipment before work begins (also knives, steels, scabbards, aprons, and collection containers for offals).
- Designing equipment to be cleaned effectively.
- Washing animals before scalding to reduce soiling scald water.
- Replacing water for scalding vats to prevent excessive accumulation of dirt.
- Using spray or steam scalding instead of vats.
- Scalding at temperatures between 60°C and 62°C without skin damage.
- Plugging the anus to prevent escape of feces into scalding water or during dehairing, scrapping or polishing.
- Promptly removing carcasses, recovered meat (diaphragm, head-meat, and tails) and offals to refrigeration.
- Avoid holding offals and recovered meat in temporary containers in the slaughtering area to minimize growth of mesophilic pathogens (*E. coli* and salmonellae).
- Minimize contact of carcasses with fixed equipment of the slaughterhouse.

Reduction (ΣR)

- Using water at 60–62°C in dehairing machinery to reduce carcass contamination.
- Using chlorinated water (ca. 50 ppm) during mechanical scraping and polishing (where permitted).
- Decontaminating carcasses with water at high contact temperature or with solutions of food grade acids.

Testing

- APC (30°C incubation) is a useful measure of slaughter hygiene and the bacteriological status of carcasses.
- Trend analysis is a useful measure of hygiene over time. This could be supplemented with testing for *E. coli* or Enterobacteriaceae (35–37°C incubation) as an indication of fecal contamination.
- Testing for salmonellae and other bacterial pathogens is less productive for improving hygiene at slaughter.

I CONTROL (chilling (cattle, sheep, and pigs))

Summary

Significant hazards[a]	• salmonellae. • VTEC. • *C. jejuni/coli*. • *Y. enterocolitica*. • Parasites.
Control measures	
Initial level (H_0)	• Hygienic slaughter and dressing.
Reduction (ΣR)	• Not appropriate.

(*continued*)

I CONTROL (Cont.)

Summary

Increase (ΣI)	• Optimize refrigeration and temperature distribution across chillers and carcasses.
Testing	• Slaughter hygiene monitored by sampling carcasses. • Verify cleaning and disinfection by microbiological tests.
Spoilage	• Not relevant.

[a]In particular circumstances, other hazards may need to be considered.

Hazards to be considered See above.

Control measures

Initial level of hazard (H_0)

- Design chillers to give uniform air distribution when loaded, and sufficient refrigeration capacity to reduce meat temperatures in a fully loaded chiller and limit condensation.
- Cleaning chiller floors, walls, and doors to reduce build-up of psychrotrophs.

Reduction of hazard (ΣR)

- Not relevant.

Increase of hazard (ΣI)

- Measuring and adjusting the flow and temperature of air during chilling (air flow should be at least 0.25 m/s over all parts of the carcass, and, except during loading, air temperatures should be <7°C and preferably <4°C).
- Spacing carcasses to allow unimpeded air flow over carcass surfaces (ideally spacing should allow 6 cm between nearest parts of carcasses).
- Placing lightest carcasses in regions of the chiller where slowest cooling occurs.
- Limiting time that chiller room doors are allowed to be open to prevent ingress of warm moist air.
- Loading warm carcasses into an area or room away from chilled carcasses to reduce the increase in surface temperature and moisture of the chilled carcasses. One method may be to start the cutting and packaging department early enough in the day to make space for the new carcasses.

Testing

- Microbiological testing is of limited use.
- Monitoring surface temperature (e.g. 1 mm deep at a specific carcass site) and/or deep tissue temperature (e.g. near the hip-joint or at a specified anatomical location) can be used to adjust air temperature and air speed and control the progress of chilling.
- Requirements that deep temperatures be reduced to 7°C within 24 h cannot be met with beef sides.
- Chilling beef sides to a deep-butt temperature of 12–15°C in 24 h is adequate to prevent growth of mesophiles and permit only small increases (≤10-fold) in psychrotrophs, but chilling must be continued for a further 24 h.

- Cooling offals and head or cheek meat on racks or in shallow pans to prevent growth of mesophilic pathogens (the insulating effects of thick layers permits growth in masses of meat).
- Cooling rates sufficient to limit or prevent growth of mesophilic pathogens (e.g. salmonellae) can be established by knowing the temperature history of the product from shortly after removal from the carcass till the temperature falls to about 7°C and calculating the expected growth (Gill and Jones, 1992).
- Monitoring air temperature and velocity at several locations in loaded chillers is recommended.
- APC (30°C incubation) is a useful measure of the bacteriological status of carcasses after chilling and accumulating. APCs as a trend analysis is a useful measure of hygiene and the effect of chilling over time.

Spoilage

- Not relevant.

J CONTROL (storage and transport)

Summary

Significant hazards[a]	• salmonellae. • VTEC. • *C. jejuni/coli.* • *Y. enterocolitica.* • Parasites.
Control measures	
Initial level (H_0)	• Cleaning chillers and transport vehicles when they are empty, before carcasses are loaded, to limit contamination with mesophiles and psychrotrophs. • Ensuring that the refrigeration unit in the transport is functioning correctly and set at the temperature required.
Reduction (ΣR)	• Not relevant.
Increase (ΣI)	• Loading carcasses when their mean temperature is close to the intended carriage temperature since some of the refrigeration systems in transport vehicles are not designed to reduce product temperature. • Maintaining storage and transport temperatures <7°C will prevent growth of mesophilic pathogens, but storage at -1 ± 0.5°C will give maximum shelf-life of chilled product. • Avoid passing over the dew-point.
Testing	• Monitoring air temperature at several locations during storage and transport is recommended. • APC (30°C incubation) gives an indication of the cleanliness of the chiller/truck.
Spoilage	• Not relevant.

[a]In particular circumstances, other hazards may need to be considered.

IV Carcass cutting and packaging

Carcasses are broken down to produce primal joints, retail cuts or boneless meat. The meat may then be stored chilled in air, in modified atmospheres or in vacuum packs.

A *Effects of processing on microorganisms*

The microbial flora on cuts of meat is determined by the previous history of the carcass and the hygienic conditions under which the cuts are prepared. Operations where carcasses have been cooled properly and butchered 1–2 days after slaughter typically have bacterial populations of $<10^4$ cfu/cm^2. Carcasses that have been stored for a week may have populations of 10^6 cfu/cm^2. Cutting room surfaces and equipment (knives, gloves, aprons, band saws, cutting table surfaces, and conveyor belts) spread spoilage organisms (e.g. *Pseudomonas* and *Broch. thermosphacta*) and pathogens present on carcasses to freshly exposed tissue and to meat from other carcasses. Equipment and surfaces will be a source of direct contamination when they have not been effectively cleaned or have been allowed to remain moist between cleaning and use. Significant increases in the prevalence and numbers of *Staph. aureus* during meat cutting have been shown to be caused by contamination from inadequately cleaned cutting surfaces, derinding equipment, and mesh gloves (Kasprowiak and Hechelmann, 1992). During cutting and boning, the microbial load on work surfaces and equipment may range from 10^4 to 10^6 cfu/cm^2 (Sheridan *et al.*, 1992) and will be a reflection of the care taken during cleaning and the microbial load on the carcasses being processed. Cuts that are handled most tend to show the highest level of contamination.

After cleaning, the equipment is kept dry to reduce subsequent growth of surviving organisms. The main flora on clean and dry surfaces tends to be micrococci. Damaged, cut or scored conveyor belts and cutting surfaces can be difficult to clean and should be replaced.

Often carcasses are butchered in rooms kept at 10°C. At this temperature, the doubling times of pathogens (e.g. *Salmonella* and *L. monocytogenes*) are at least 8 h (Mackey and Roberts, 1993), and of psychrotrophic spoilage bacteria are at least 2.8 h (Table 1.19). If the temperature of the meat is not >7–10°C, then growth of pathogens will not be significant during processing. At times, however, carcasses may be butchered when hot or warm (hot- or warm-boning), meat temperatures can then range up to 30–35°C. Growth of *Salmonella* and *E. coli* O157:H7 may be significant unless there is rapid transfer of the meat to cooling environments and the surface temperature of the meat is rapidly reduced to <7–8°C (Grau, 1987). The consequences of the time, meat surfaces spend above this temperature can be estimated from predictive equations for the growth of salmonellae and *E. coli* (Gill and Harrison, 1985; Smith, 1985; Mackey and Kerridge, 1988; Buchanan and Bagi, 1994).

Storage in air. Retail cuts are placed on trays and overwrapped with film of low moisture and high oxygen permeability, and stored chilled. Formation of metmyoglobin or bacterial spoilage limits storage-life to only a few days. Under these moist aerobic conditions, pseudomonads are the fastest growing

Table 1.19 Generation time (h) of psychrotrophs on meat in air

	Temperature (°C)			
	2	5	10	15
Pseudomonas (non-fluorescent)	7.6	5.1	2.8	2.0
Pseudomonas (fluorescent)	8.2	5.4	3.0	2.0
Acinetobacter spp.	15.6	8.9	5.2	3.1
Enterobacter spp.	11.1	7.8	3.5	2.4
Brochothrix thermosphacta	12.0	7.3	3.4	2.8

Gill, 1986.

organisms (Table 1.19) and dominate the final flora that can also include *Acinetobacter* and *Psychrobacter*. The rate of spoilage is determined by the temperature of storage, and the shelf-life by the number of pseudomonads present initially. Occasionally, high prevalence of *Broch. thermosphacta* on overwrapped meat may be caused by a slight build-up of carbon dioxide, which inhibits pseudomonads (Dainty and Mackey, 1992).

Liver has a high content of glucose (up to 6 mg/g) and, initially, a high pH (>6.2). During chilled storage, the pH slowly falls owing to tissue glycolysis. The surface flora becomes dominated by pseudomonads, but major components of the final flora can include *Acinetobacter, Enterobacter, Broch. thermosphacta,* and lactic acid bacteria (Gill, 1988). Tissue metabolism produces elevated carbon dioxide concentrations in the pack. Lactic acid bacteria are the main flora that develop within the liver. Kidneys have a high pH, and *Shewanella putrefaciens* and *Aeromonas* spp. are frequently found in the final flora.

Vacuum-packaged meat. Primal cuts, or lamb carcasses, are placed in bags made of film of low gas permeability (typically about 10–25 mL O_2/m^2/day/101 kPa, measured at 25°C), and the bags evacuated, sealed, and stored at −1°C to 0°C. Within the pack carbon dioxide is released from the tissue. Tissue respiration consumes residual oxygen and produces carbon dioxide. The gas atmosphere that develops contains 20–40% carbon dioxide and considerably <1% oxygen.

Growth of aerobic Gram-negative flora (*Pseudomonas, Acinetobacter* and *Psychrobacter*) is inhibited. Species of *Lactobacillus, Carnobacterium* and *Leuconostoc* dominate the flora of vacuum-packed meat reaching populations of 10^7–10^8 cfu/cm^2. The extent of growth of a number of other bacteria is determined by pH, film permeability and temperature. On lean meat of pH 5.4–5.8, the amount of growth of *Broch. thermosphacta* increases with increased film permeability to oxygen. On meat with pH ≥6.0, *Broch. thermosphacta* grows anaerobically to about 10^7 cfu/cm^2. *Shew. putrefaciens* is able to grow when the pH is ≥6.0, but fails to grow when the pH is ≤5.8. The proportion of Enterobacteriaceae in the final population is greater when storage is at 5°C, rather than at 0°C, or when the pH is high (>6.0). On vacuum-packed pork and lamb, *Broch. thermosphacta* and Enterobacteriaceae form a larger percentage of the developing flora than on vacuum-packed beef. With packs of lamb and pork there are more muscles than in packs of beef, and so a greater incidence of high-pH meat. The weep that accumulates in packed lamb carcasses has a pH >6.0. Psychrophilic clostridia (e.g. *Cl. laramie*) can grow on vacuum-packed normal pH beef stored at 2°C (Dainty *et al.*, 1989; Kalchayanand *et al.*, 1989, 1993; Broda *et al.*, 1996a,b, 1998a,b, 1999).

Growth of *L. monocytogenes* is poor on vacuum-packed meat of normal pH (5.5–5.7) stored at 0°C, but more extensive when the pH of the meat is high or when storage is at 5°C. *Yersinia* spp. have been isolated from vacuum-packed meats, and meat isolates of *Y. enterocolitica* grow on high pH, vacuum-packed meat at 0°C (Gill and Reichel, 1989). However, it is likely that these were environmental isolates. Growth of yersiniae, as of other Gram-negative bacteria, is poor on vacuum-packed, normal-pH meat stored near 0°C. Pathogenic serotypes of yersiniae appear to be inhibited by carbon dioxide at 4°C (Kleinlein and Untermann, 1990).

Modified atmospheres. To maintain the bright red color of meat, retail cuts are packed with an atmosphere high in oxygen (70–80%) and carbon dioxide (20–25%) in rigid plastic containers of low gas permeability. The high carbon dioxide concentration inhibits the Gram-negative aerobic flora and extends shelf-life. At temperatures near 0°C, lactic acid bacteria tend to dominate the final flora, though *Broc. thermosphacta* can be a significant proportion of this flora. Growth of Enterobacteriaceae and *Broch. thermosphacta* is favored by storage at temperatures near 5°C.

Primals and lamb carcasses are packed with an atmosphere of 100% carbon dioxide in plastic or foil laminated bags of very low gas permeability. To saturate the meat, 1–1.5 L of carbon dioxide is added per kg of meat. During storage at ~0°C, the growth rate of lactic acid bacteria is slowed and growth of *Broch. thermosphacta*, Enterobacteriaceae and *Shew. putrefaciens* is strongly inhibited, even with high-pH meat. With high-pH meat and insufficiently impermeable films, there can be some growth of *Broch. thermosphacta* and Enterobacteriaceae late in storage. No growth of *L. monocytogenes* or *Aeromonas hydrophila* occurs during 5 weeks storage at 5°C (Gill and Reichel, 1989).

Biogenic amines are formed in foods as a result of amino acid decarboxylation, catalyzed by bacterial enzymes. When consumed in sufficient quantities, these compounds may cause headache, hypertension, fever and heart failure. Technologies such as vacuum-packaging and carbon dioxide-modified atmosphere packaging (CO_2-MAP), when combined with low-temperature storage (−1.5 to 0.5°C), extend the shelf-life of meat. During low-temperature storage of pork in these packaging systems, LAB possessing the enzymes for biogenic amine formation, dominated the microflora. Monitoring levels of phenylethylamine, putrescine, cadaverine, histamine, tyramine, spermidine and spermine revealed that towards the end of shelf-life, tyramine, and phenylethylamine in pork of both packaging treatments approached levels considered to be potentially toxic (Nadon *et al.*, 2001).

B Spoilage

Growth of microorganisms on meat is at the expense of low molecular weight compounds (e.g. glucose, glucose-6-phosphate, ribose, glycerol, amino acids, and lactate), and spoilage is the result of changes in odor, flavor or appearance.

Storage in air. Initially, microorganisms growing on the meat surface use glucose preferentially. When the rate of glucose utilization exceeds the rate of diffusion from within the tissue, amino acids are degraded (Gill and Newton, 1977). The glucose content of meat is therefore a critical factor determining the onset of off-odors. On high-pH meat devoid of glucose, spoilage occurs when bacteria reach 10^6 cfu/cm^2, but occurs when numbers exceed 10^7 cfu/cm^2 on normal meat (Newton and Gill, 1978).

A complex mixture of esters, branched-chain alcohols, sulfur-containing compounds, amines, unsaturated hydrocarbons, and ketones have been found in meat stored in air (Dainty and Mackey, 1992; Table 1.20). *Ps. fragi* cluster 2 strains are the major producers of the sweet-fruity odors (ethyl esters), which are detected at incipient stages of spoilage. These and other pseudomonads also produce a range of sulfur-containing compounds, but not hydrogen sulfide, which are responsible for putrid, sulfury odors

Table 1.20 Bacterial metabolites—naturally contaminated meat in air

Ethyl acetate	Methanthiol
n-Propanoate	Dimethylsulphide
Isobutanoate	Dimethyldisulphide
2-Methylbutanoate	Dimethyltrisulphide
3-Methylbutanoate	Methylthioacetate
n-Hexanoate	Ammonia
n-Heptanoate	Putrescine
n-Octanoate	Cadaverine
Crotonate	Tyramine
3-Methyl-2-butenoate	Spermidine
Tiglate	Diaminopropane
Isopropyl acetate	Agmatine
Isobutyl acetate	1,4-Heptadiene
n-Propanoate	1-Undecene
n-Hexanoate	1,4-Undecadiene
Tiglate	Acetoin
Isopentyl acetate	Diacetyl
3-Methyl butanol	3-Methyl butanal
2-Methyl butanol	

Dainty and Mackey, 1992.

Table 1.21 Bacterial metabolites—naturally contaminated vacuum-packed meat

L(+)-Lactic acid	Ethyl acetate[a]
D(−)-Lactic acid	Ethyl n-propanoate[a]
Acetic acid	Ethyl 3-methylbutanoate[a]
Isobutanoic acid	Propyl acetate[a]
Isopentanoic acid	Hydrogen sulfide
Ethanol	Methanethiol
3-Methylbutanol	Dimethyldisulfide
Tyramine	Dimethyltrisulfide[a]
Putrescine	*Bis*(methylthio)methane[a]
Cadaverine	Methylthioacetate
	Methylthiopropanoate[a]

Dainty and Mackey, 1992.
[a]Found in high-pH meat only.

at advanced stages of spoilage. Unsaturated hydrocarbons are formed by pseudomonads, especially *Ps. lundensis*. Pseudomonads produce putrescine from arginine via ornithine. Branched-chain acids and alcohols (from branched-chain amino acids), acetoin, and diacetyl, are formed by *Broc. thermosphacta*. Enterobacteriaceae also produce branched-chain alcohols, but acetoin and diacetyl are only produced on meat of normal pH. Both *Broch. thermosphacta* and Enterobacteriaceae only produce 2,3-butanediol on meat of normal pH. Enterobacteriaceae (*Enterobacter agglomerans, Hafnia alvei, Serratia liquefaciens*) also produce sulfur-containing end products but differ from pseudomonads in not forming dimethyl sulfide and in producing hydrogen sulfide (Dainty and Mackey, 1992). Enterobacteriaceae form cadaverine from lysine.

Vacuum-packaged meat. Table 1.21 lists microbial end products detected in naturally contaminated, chilled meat stored in vacuum packs.

On meat of pH ≥6, a wider range of bacteria grow and *Broch. thermosphacta* and Gram-negative bacteria reach higher populations. This results in more rapid spoilage of high-pH meat and in the production of some bacterial end products not found in normal-pH meat (Table 1.20). Normal-pH meat tends to develop a relatively inoffensive sour acid odor (Dainty and Mackey, 1992). high-pH meat tends to develop sulphydryl, putrid and fecal odors.

Greening results from the formation of sulphmyoglobin from hydrogen sulfide reacting with oxymyoglobin (Carrico *et al.*, 1978). *Shew. putrefaciens* is a strong producer of hydrogen sulfide and causes rapid spoilage of high-pH meat. Enterobacteriaceae produces hydrogen sulfide and other sulfur-containing compounds (Dainty and Mackey, 1992). *Lactobacillus sake* forms hydrogen sulfide even on normal-pH (5.6–5.8) meat but produces it more rapidly on high-pH meat. The higher glucose content of normal-pH meat delays the utilization of cysteine by this organism (Egan *et al.*, 1989).

Growth of lactic acid bacteria, *Broch. thermosphacta*, Enterobacteriaceae, *Aeromonas* spp. and *Shew. putrefaciens* results in increased concentrations of acetic acid. *Broch. thermosphacta* produces isobutyric and isovaleric acids on high-pH meat. Higher concentrations of acetic, isobutyric, and isovaleric acids are found in stored high-pH meat than in normal-pH meat (Dainty *et al.*, 1979).

Tyramine appears to be formed by *Carnobacterium* spp. arginine is utilized by some strains of lactic acid bacteria, the resulting ornithine being converted to putrescine by Enterobacteriaceae (Dainty and Mackey, 1992). Enterobacteriaceae are responsible for the appearance of cadaverine.

Flavor changes (sour, acid and cheesy) can be detected in vacuum-packed normal pH beef after 10–12 weeks storage at 0°C (Egan, 1983). This is 4–6 weeks after the bacterial population has reached its maximum. Bitter and liver-like flavors develop after 14–16 weeks in the absence of contaminating flora.

With increased demands for long shelf-life, spoilage by psychrophilic clostridia has become a common problem. One type of spoilage is characterized by the production of hydrogen and carbon dioxide causing pack distension, and of butanol, butanoic acid, ethanol, acetic acid, and a range of sulfur-containing compounds (Dainty *et al.*, 1989; Kalchayanand *et al.*, 1993). The isolate/strain of Dainty *et al.* (1989) was shown by 16S rRNA sequencing to be a new species of the genus *Clostridium*, *Cl. estertheticum* (Collins *et al.*, 1992). *Cl. frigidicarnis*, was isolated from vacuum-packed, temperature abused, beef (Broda *et al.*, 1999). For other examples, see Broda *et al.* (1996a,b, 1998a,b). Rapid molecular methods to detect these clostridia are becoming available (Helps *et al.*, 1999).

Modified atmosphere. Consumer acceptability of raw meats stored in an elevated oxygen and carbon dioxide atmosphere is usually limited by color deterioration (metmyoglobin formation). Microbial spoilage is the result of growth of lactic acid bacteria and *Broch. thermosphacta*. Increased levels of a number of organic acids, acetoin, diacetyl, putrescine, and cadaverine have been found in such stored meats (Dainty and Mackey, 1992).

Storage in 100% carbon dioxide normally results in significant growth only of lactic acid bacteria and a long shelf-life (15 weeks; Gill and Penney, 1986). On high-pH meat, if there is insufficient carbon dioxide to saturate the meat, if the packing film is not impermeable to carbon dioxide, or if there is high initial contamination, Enterobacteriaceae and *Broch. thermosphacta* can bring about putrid or acid–aromatic spoilage odors and flavors (Gill and Penney, 1988; Gill and Harrison, 1989).

C Pathogens

If not properly cleaned, equipment and surfaces where the carcass is broken down can be a source of contamination. Growth of mesophilic pathogens on the meat is prevented by storage at temperatures <7°C. On chilled meat stored in air, the growth rates of psychrotrophic pathogens (*L. monocytogenes* and *Y. enterocolitica*) are much slower than that of spoilage organisms and increases are likely to be small. In general, vacuum-packaging is less inhibitory to the psychrotrophic pathogens than modified atmosphere packaging (Garcia de Fernando *et al.*, 1995). On vacuum-packed meat, growth of *Aeromonas* spp. and *L. monocytogenes* is strongly inhibited on normal-pH meat stored at 0°C, but significant growth can occur on high-pH meat and at higher temperatures (Dainty *et al.*, 1979; Palumbo, 1988; Grau and Vanderlinde, 1990). Similarly, in modified atmosphere-packaged meat, inhibition of psychrotrophic pathogens is due to the interaction of the low normal pH of meat, low temperature and elevated levels of carbon dioxide (Garcia de Fernando *et al.*, 1995). In the presence of 100% carbon dioxide, growth of these organisms even on high-pH meat is prevented at storage temperatures of 0–5°C (Gill and Reichel, 1989). Although strains of *Yersinia* grow on vacuum-packed meat, *Y. enterocolitica* serotypes O:3 and O:9 appear to be strongly inhibited by carbon dioxide at storage temperatures of 0–4°C (Kleinlein and Untermann, 1990). The available data suggest that vacuum-packaging and modified atmosphere packaging, if anything, enhances the microbiological safety of normal-pH meat by retarding the growth of psychrotrophic pathogens, but is only effective if the meat is adequately refrigerated.

Growth of the pathogens *Y. enterocolitica, L. monocytogenes, E. coli* O157:H7, and salmonellae was compared in ground beef packed in modified atmospheres of 60% CO_2/40% N_2/0.4% CO (high CO_2/low CO mixture), 70% O_2/30% CO_2 (high O_2 mixture) and in chub packs (Nissen *et al.*, 2000). Growth of *Y. enterocolitica* was almost totally inhibited both at 4°C and 10°C in the high CO_2/low CO mixture, but numbers in samples packed in high O_2 mixture increased from ca. 10^2-fold at 4°C, and ca. 10^3-fold at 10°C. *L. monocytogenes* showed very little growth at 4°C under all storage conditions.

Growth of *E. coli* O157:H7 in ground beef was almost totally inhibited at 10°C in both the high CO_2/low CO mixture and the high O_2 mixture. Growth in the chub packs reached 10^5 cfu/g on day 5. Strains of salmonellae in ground meat stored at 10°C for 5 and 7 days, grew to a higher number in the high CO_2/low CO mixture than in the high O_2 mixture emphasizing the importance of temperature control during storage.

D CONTROL (meat stored in air, vacuum-packed or packed in modified atmospheres.)

Summary

Significant hazards[a]	
Beef, lamb, and other ruminants:	• salmonellae. • VTEC. • *C. jejuni/coli.* • Parasites.
Pork:	• salmonellae. • *C. coli.* • *Y. enterocolitica.* • Parasites.
Control measures	
Initial level (H_0)	• Microbiological content of the meat to be broken down and packaged • Temperature control during butchery.
Reduction (ΣR)	• Not applicable.
Increase (ΣI)	• Place in chiller to allow efficient cooling to holding temperature. • Temperature control during storage. • Cleanliness of packing equipment; correct settings on packaging machine for evacuation, gas addition, and sealing; avoid fat smears and creases in sealing area; correct size and film type for vacuum or gas packing. • Tight packs with no head-space for vacuum packs; correct gas atmosphere in gas-flushed packages.
Testing	• pH of meat for vacuum-packaging (pH <6.0). • Check for leaking packs (faulty seals or pin-holes). • APC at 30°C at packaging gives useful indication of hygiene and expected refrigerated shelf-life; high APCs often indicate improperly cleaned equipment or unacceptable time–temperature history of raw materials. • Testing for pathogens at packaging does not effectively control safety of product.
Spoilage	• Spoilage by cold-tolerant clostridia can occur, even when the product is held under good temperature control.

[a]In particular circumstances, other hazards may need to be considered.

Comments. Control begins with the microbiological quality of the meat to be packaged. Meat with low numbers of spoilage organisms will have a longer shelf-life than meat with higher numbers.

V Frozen meat

Carcasses, sides or quarters, boneless meat in cartons and offal in cartons or tubs are frozen by a continuous blast of cold air (−10°C to −35°C; 1–12 m/s). Small pieces of meat may be frozen with liquid nitrogen, solid carbon dioxide or by plate freezing. Frozen meat is stored at ≤−10°C. Physical and biochemical changes gradually take place in frozen meat and reduce quality. Considerably lower temperatures (−18°C) may therefore be used to maintain high quality for long storage periods (e.g. for liver and pork).

A *Effects of processing on microorganisms*

Meat begins to freeze at about −1.5°C and about half the liquid in meat is frozen at −2.5°C. During cooling and initial freezing, the extent of bacterial growth is determined by the starting temperature of the meat and the rate of cooling. The rate of cooling varies with the insulating properties of the packaging material, the thickness of the meat to be frozen and the temperature and velocity of the cooling air flowing over the meat or pack. On carcasses, sides, and quarters, the microbial contamination is at the surface where freezing occurs first. In cartons and tubs, contaminated surfaces may be at the thermal center where the insulating properties of the meat delay freezing. When meat is packed into cartons or tubs at 20–35°C, there can be significant growth of mesophiles before the temperature is reduced to below about 7°C (Gill and Harrison, 1985). This can be aggravated if cartons are removed from freezers, palletized, and moved to storage before being completely frozen.

Freezing causes some reduction in microbial numbers, particularly of Gram-negative bacteria; it also results in sub-lethal damage to some cells. *Campylobacter* and vegetative cells of *Cl. perfringens* are especially sensitive to freezing. Although most of the reduction in viable counts occurs during freezing, slow death also occurs during frozen storage. In spite of some reduction in microbial numbers, both spoilage and potential pathogens (e.g. salmonellae and *E. coli* O157:H7) survive in frozen meats.

During thawing there is often a considerable increase in bacterial numbers because the meat surface temperature rises rapidly towards that of the thawing medium and remains near this temperature till thawing is complete. Thawing temperatures >10°C may allow growth of mesophiles (Bailey *et al.*, 1974; James *et al.*, 1977). Sometimes two-stage thawing procedures are used in which air temperatures are initially high (15–20°C), but are then reduced to about 5°C as the superficial temperature rises. For use in further processed meats (e.g. comminuted meats), meat is usually tempered to −5°C before chopping or grinding.

B *Spoilage*

Properly frozen meat (at ≤−10°C) does not spoil as the result of microbial growth. Although a number of mold species have been identified as the cause of black spot on meat, the evidence is that these could develop on frozen meat only when temperatures were relatively high for long periods. *Cladosporium cladosporioides* has a minimum growth temperature of about −5°C, and *Clad. herbarum* and *Penicillium hirsutum* have minimum growth temperatures between −5°C and −6°C (Gill and Lowry, 1982). At −3.5°C to −3.8°C, spore germination requires about 6 weeks. It was estimated that it would take at least 1 month at −2°C to form a mold colony of 1 mm and over 4 months at −5°C. On naturally contaminated meat stored at −5°C, yeasts (*Cryptococcus, Trichosporon,* and *Candida*) grew with a generation time of about 8 days reaching 10^6 cfu/cm^2 after 20 weeks (Lowry and Gill, 1984). Visible mold colonies (1–4 mm) of black or white spot (*Chrysosporium pannorum*) appeared after 40 weeks storage.

The rate of spoilage of thawed meat is similar to that of fresh meat.

C Pathogens

The bacterial pathogens are the same as on the meat before freezing. Salmonellae, *E. coli* O157:H7, *L. monocytogenes*, spores of *Cl. perfringens*, and some cells of *C. jejuni* survive. For instance, spores of *Cl. perfringens* were detected by enrichment in between a quarter and half of samples of frozen beef, mutton, horse and rabbit meat (Uemura *et al.*, 1985). Only 2% of isolates produced enterotoxin. Thawing >10°C may permit growth of salmonellae and EHEC. Drip exuded from thawed meat may contain pathogens and be a source of contamination.

Freezing for a sufficient duration inactivates *Toxoplasma*, *Sarcocystis*, *Trichinella* and *Taenia*.

D CONTROL (frozen meats)

Summary

Significant hazards[a]

Beef, veal, lamb, and other ruminants:	• salmonellae. • *E. coli.* • VTEC. • *C. jejuni.* • *Staph. aureus.* • *Cl. perfringens.* • Parasites.
Pork:	• salmonellae. • *C. coli.* • *Y. enterocolitica.* • *Staph. aureus.* • *Cl. perfringens.* • Parasites.
Control measures	
Initial level (H_0)	• The microbiological quality of the meat to be frozen. • Initial temperature of the meat to be frozen. • Other factors affecting the rate of freezing.
Reduction (ΣR)	• Freezing to kill parasites. • Elimination of bacterial pathogens cannot be guaranteed.
Increase (ΣI)	• Time and temperature of thawing.
Testing	• Microbiological testing of meat to be frozen is of limited value.
Spoilage	• Occurs only if the temperature exceeds −10°C for long periods. • Rancidity can occur depending on storage time and temperature.

[a]In particular circumstances, other hazards may need to be considered.

Hazards to be considered. Frozen meat presents no health hazards different from those associated with fresh raw meat.

Control measures

Initial level (H_0)

- The rate of cooling meat to freezing temperatures, the temperature in frozen storage and the time and temperature of thawing need to be controlled.
- The microbiological quality of the meat to be frozen: select meat of good microbiological quality from premises with a good record of hygienic production.
- Initial temperature of the meat to be frozen; thickness of the meat; spacing between carcasses, sides, quarters, boxes or tubs to allow free flow of cold air over the meat.
- Internal temperature of the meat when removed to frozen storage.

Reduction (ΣR)

- Although numbers of viable microorganisms may fall slightly upon freezing, elimination of pathogens cannot be guaranteed.

Increase (ΣI)

- Time and temperature of thawing (the meat surface will be at the temperature of the thawing medium and growth of pathogens may occur).

Testing. APC at 30°C is an appropriate test for frozen and/or thawed meat;

Spoilage. Only occurs if the temperature exceeds −10°C for long periods; Rancidity can occur depending on the storage time and temperature.

VI Raw comminuted meats

Comminuted meats may be made at retail stores from trimmings and meat left on bones after butchery, or at central processing plants from whole carcasses or from cartons of frozen manufacturing meat. Beef, veal, lamb, and pork are used alone or as mixtures. Fat is often added. Extenders, such as soy protein, non-fat dried milk or cereal products (bread crumbs and rusk), as well as salt and spices may also be mixed with meat. Products include ground meats, sausage meat, patties or burgers, or raw sausages. Although most of these products are stored and marketed in the chilled state, some are frozen (e.g. patties and burgers). Products are usually cooked before consumption, although, in some countries, raw ground meats may be eaten.

A *Effects of processing on microorganisms*

The bacteriology of comminuted meats is largely determined by the quality of the meat used. Comminuted meats prepared at retail stores often have a large bacterial load because they are made from scrap meat and trimmings from carcasses that have been stored several days. Centrally produced comminuted meats are produced from fresh or frozen meat which usually have lower counts than retail scrap meat. The mean aerobic viable count (incubated at 35°C) of 1 370 samples of ground beef collected from US slaughter establishments was $\log_{10}$ 4.6 cfu/g (Hogue *et al.*, 1993). During comminution, microorganisms originally present on the surface are distributed through the ground meat. Grinding will result in an increase in the temperature of the meat. The extent of this increase depends on the grinding process and not only can increase the growth rate of psychrotrophs but also may be sufficient to permit growth of mesophiles such as *E. coli* and salmonellae. Thus, it is important to control the temperature of the meat before comminution and after grinding. The grinder can be a significant source of contamination of the product with microorganisms including potential pathogens, if the equipment is not effectively cleaned before use.

After carcass boning, meat adhering to bones is removed manually or mechanically. Frequently, the recovered meat has a higher bacterial load than that on the carcass from which it is derived. The extra handling in manual recovery results in an increase in the total aerobic flora and in Enterobacteriaceae (Field, 1981). The time and temperature at which bones are held before recovery influence the microbial load on recovered meat. Mechanical methods crush or grind the bones and separate the meat as a paste. Some machines give product with a temperature as high as 35°C from bones at 1.1°C, whereas others produce recovered meat of 4.4°C (Swingler, 1982). Because bone-marrow forms part of the product, the pH of mechanically recovered meat is pH 6–7 (Field, 1981), and is an excellent medium for bacterial growth. Rapid cooling of recovered meat is required to prevent excessive microbial growth. Bacteria may grow on material retained within equipment operating at warm temperatures and serve as a continuous inoculum of product with pathogenic microorganisms (Gill, 1988). Effective cleaning of equipment before and after operation, and at frequent intervals during long periods of warm operation, is important.

For fresh sausage, meat is comminuted at ≤ -2°C to obtain a clean cut and optimum appearance, and the mixture is stuffed into hog, sheep or synthetic casings or formed into patties. Depending on the amount of salt and fat incorporated, the lowered a_w can inhibit the growth of *Pseudomonas, Psychrobacter,* and *Moraxella* spp. The shelf-life of such sausage will tend to be longer than that of simple ground meat.

In some countries, sulfur dioxide (added as sodium metabisulfite; limit 450 mg SO_2/kg) is added as an approved antimicrobial agent in raw ground products containing cereal. Sulfite represses growth of Gram-negative bacteria (*Pseudomonas* spp., Enterobacteriaceae, salmonellae), and the predominant developing flora consists of Gram-positive bacteria and yeasts (Dowdell and Board, 1968; Dillon and Board, 1991).

B Spoilage

In general, chilled comminuted raw products have a short shelf-life. Surveys in a number of countries almost invariably show high mean numbers of bacteria in retail ground meat (e.g. Table 1.22). Putrefactive spoilage is caused by pseudomonads and is usually restricted to regions near the surface of ground meat and low-salt sausages. Metabolic activity of the raw meat results in a reduced oxygen tension within the product, and the conditions are anaerobic. Inside comminuted products, spoilage is mainly by Gram-positive bacteria (*Broch. thermosphacta* and lactic acid bacteria). Psychrotrophic Enterobacteriaceae often form a significant part of the flora. In general, surface spoilage will be similar to that for aerobically stored retail meat, whereas within the product, spoilage will be like that for meats under modified atmospheres or in vacuum-packs. Some ground meat products (e.g. fresh pork sausage made with hot boned pork) are packaged in oxygen impermeable films as a means to achieve greater refrigerated shelf-life (e.g. 50 days at <5°C). The slightly lowered a_w of some sausages will retard growth of the aerobic Gram-negative flora and spoilage will be "souring" from growth of lactic acid bacteria and *Broch. thermosphacta*. Yeasts are a significant component of the flora of sulfited raw meat (sausage and ground meat or patties) and can reach populations of over 10^5 cfu/g. A film of yeasts can develop on the skins of stale sausages (Dillon and Board, 1991).

C Pathogens

Organisms of public health significance are frequently found in retail raw comminuted meats (e.g. Table 1.22).

Table 1.22 Bacterial numbers in ground beef at retail

Organism	Country	% Positive	Count	Reference
L. monocytogenes	Scotland	91.3	<20 cfu/g	Fenlon *et al.*, 1996
	Japan	12.2	10–<100 MPN/g	Inoue *et al.*, 2000
E. coli O157:H7	Spain	5.0		Blanco *et al.*, 1996a,b
Verocytotoxin *E. coli*	England	13[a]		Willshaw *et al.*, 1993
		22[b]		Willshaw *et al.*, 1993
	Switzerland	3.3		Gilgen *et al.*, 1998
		2.4		Baumgartner and Grand, 1995
	Argentina	28[b]		Parma *et al.*, 2000
		28[c]		Parma *et al.*, 2000
APC (cfu/g)	USA	21 000[d]		Roberts and Weese, 1998
		600[e]		Roberts and Weese, 1998
	USA		<3.0 $\log_{10}$ cfu/g[f]	Gamage *et al.*, 1997.
Coliforms	USA	60[d]		Roberts and Weese, 1998
		<10[e]		Roberts and Weese, 1998
E. coli	USA	<10[d]		Roberts and Weese, 1998
		<10[e]		Roberts and Weese, 1998
Staph. aureus	USA	<10[d]		Roberts and Weese, 1998
		<10[e]		Roberts and Weese, 1998

[a] Minced (ground) beef.
[b] Beefburgers.
[c] Ground (minced) beef.
[d] Trial 1 (meatpacking plant).
[e] Trial 2 (meatpacking plant)
[f] Chub-packed ground beef (meat-packing plant).

Prevalences of *Cl. perfringens* from 47 to 81% have been reported in samples of ground meats and fresh sausage (Ali and Fung, 1991). Numbers are usually <50 cfu/g but can reach 1000 cfu/g. The only significance of *Cl. perfringens* is that spores may survive cooking and subsequently grow to large numbers in some foods (e.g. stews, taco meat and chili) through temperature abuse. *Staph. aureus* is also often found in comminuted meats but, even with temperature abuse, does not grow sufficiently in these raw high a_w-value products to produce enough toxin to cause food poisoning. The reported prevalence of *L. monocytogenes* in samples of ground meats, patties, and sausage meat ranges from 10 to 92% (Johnson *et al.*, 1990; Farber and Peterkin, 1991). Numbers of listeriae are mostly <100 cfu/g but counts of about 1 000 cfu/g have been obtained. Even though the infective dose is not clear, cooking, adequate to destroy salmonellae and *E. coli* O157:H7, can provide consumer protection.

Fantelli and Stephan (2001) found STEC in 7 of 400 samples of minced meat (5/211 (2.3%) minced beef and 2/189 (1%) minced pork). The seven strains yielded five different serotypes, but none was O157:H7. Two STEC strains harbored *stx1* and *stx2* and five strains harbored *stx2c* genes. Furthermore, four strains harbored one or more additional virulence factors. However, none of the strains was positive for *eae*. *L. monocytogenes* was isolated from 43/400 (10.75%) samples, 19 belonging to serotype 1/2a, two to serotype 1/2b, 12 to serotype 1/2c, and 10 to 4b.

Pathogenic serotypes of *Y. enterocolitica* have been found in up to 30% of samples of retail ground pork (Andersen *et al.*, 1991) and in some samples of ground beef. However, other surveys have either failed to detect pathogenic serotypes (Bulte *et al.*, 1993) or have detected only a low incidence (1 in 400 samples; de Boer and Nouws, 1991). Contamination of ground beef appeared to have been caused by cross-contamination during grinding at retail stores. The high prevalence in ground pork may be due to the use of head-meat containing tonsillar or oral mucosal tissue. In some countries where raw ground pork is consumed, the use of head-meat has been prohibited. The extent of growth of pathogenic serotypes in ground meat is restricted in the presence of the normal flora (Fukushima and Gomyoda,

1986; Kleinlein and Untermann, 1990). At storage temperatures between 4°C and 15°C, increases in the number of pathogenic serotypes were not >100-fold by the time spoilage occurred.

Salmonellae, in low numbers, may be found in ground meat, sausage meat and raw sausages. The prevalence varies widely from about 1% (Table 1.22) to 30% of samples (Roberts *et al.*, 1975). A US survey of ground beef produced at slaughter establishments detected salmonellae in 3.4% of 1 370 frozen 25 g samples and 5.4% of 100 g samples (Hogue *et al.*, 1993). Outbreaks of salmonellosis have resulted from consumption of undercooked hamburger and eating raw ground meats.

The occurrence of EHEC has been reported in at least 20 countries (Griffin and Tauxe, 1991). *E. coli* O157:H7 has been detected in 2–4% samples of ground beef in some surveys (Doyle, 1991; Liore, 1994), but has not been found in other surveys (e.g. Willshaw *et al.*, 1993). The reported prevalence of VTEC also varies widely from survey to survey with incidences of up to 63% being reported for red meats (Samadpour *et al.*, 1994). At least part of this variation is due to the use of different isolation methods (ACMSF, 1995). A number of outbreaks of human illness have been linked to eating ground beef or hamburger patties which appeared to have been insufficiently cooked; one of these outbreaks involved >600 cases (Tarr, 1994a,b). US recommendations are that hamburger patties be cooked to an internal temperature of 68°C (Meng *et al.*, 1994). One report (Byrne *et al.*, 2002) suggests that product formulation and commercial production procedures can affect the heat resistance of *E. coli* O157:H7 in beef burgers, heat resistance in burgers using processed trimmings being higher than in burgers using fresh trimmings.

A large outbreak of *E. coli* O157:H7 infections occurred in the western USA and was associated with eating ground beef patties at restaurants of one fast-food chain (Tuttle *et al.*, 1999). Restaurants epidemiologically linked with cases had served patties produced on two consecutive dates; and cultures of recalled ground beef patties produced on those dates yielded *E. coli* O157:H7 strains indistinguishable from those isolated from patients. The median most probable number of organisms was 1.5 per g (range, <0.3–15) or 67.5 organisms per patty (range, <13.5–675) and testing meat from lots consumed by persons who became ill suggested an infectious dose of fewer than 700 organisms.

In Scotland, cases traced to environmental contamination confirmed that the infectious dose is low. The pathogen loading shed onto a field by sheep immediately prior to a scout camp where 18 scouts and two adults were infected with *E. coli* O157 led to an estimated dose ingested of 4–24 organisms (Strachan *et al.*, 2001).

Contamination of minced (ground) meat with salmonellae is still a major problem in some countries. The prevalence of salmonellae in minced meat produced in a European Union-approved slaughtering and cutting plant was 15.8% in pooled samples, corresponding to 6.3% of individual samples. Serotyping identified 69.6% as *S.* Typhimurium (Stock and Stolle, 2001).

Substantial research effort has been put into designing processes to reduce the numbers of *E. coli* O157:H7 on beef trim intended for production of ground beef. For example, Kang *et al.* (2001a) developed a "multiple-hurdle" antimicrobial process applied to the meat on a conveyer belt and monitored microbial profiles of inoculated lean beef trim tissue (BTL) and fat-covered lean beef trim (BTF) during prolonged refrigerated storage. Treatments included a water wash at 65 psi for five passes; water plus lactic acid [2% (vol/vol) room temperature lactic acid wash at 30 psi for three passes]; combination treatment 1 (water plus 65°C hot water at 30 psi for one pass plus hot air at 510°C for four passes plus lactic acid), combination treatment 2 (water plus hot water at 82°C for one pass plus hot air at 510°C for five passes plus lactic acid) and combination treatment 3 (water plus hot water at 82°C for three passes plus hot air at 510°C for six passes plus lactic acid). On BTL, microbial numbers were reduced about 100-fold. On BTF, microbial reductions were much greater.

The same authors (Kang *et al.*, 2001b) treated commercially produced, irregularly sized uninoculated beef trim by an optimized multiple antimicrobial process under spray system or hot air gun. After treatment, the trim was finely ground, vacuum-packaged and stored at 4°C for up to 20 days, when bacterial numbers were about 1.2–1.6 $\log_{10}$ lower than untreated ground beef. Whether such processes would be effective under commercial conditions is debatable. Decontamination with organic acids is not permitted in the European Union.

The addition of sodium lactate (up to 4.5%) had no effect on the rate of inactivation of a mixture of four strains of *E. coli* O157:H7 in ground beef (Huang and Juneja, 2003)

The effects of other treatments on survival of *E. coli* 0157:H7 in ground beef include the effect of gamma irradiation (Thayer and Boyd, 1993), temperature (Doyle and Schoeni, 1984) and acids (Abdul-Raouf *et al.*, 1993).

D CONTROL (raw comminuted meats)

Summary

Significant hazards[a]	
Beef, veal, lamb, and other ruminants:	• Salmonellae. • VTEC. • Parasites.
Pork:	• Salmonellae. • *C. coli.* • *Y. enterocolitica.* • Parasites.
Control measures	
Initial level (H_0)	• Slaughter and processing hygiene. • Selection of raw materials. • Initial temperature of the meat to be comminuted.
Reduction (ΣR)	• No reduction of pathogens occurs.
Increase (ΣI)	• Time and temperature control during comminution and during storage.
Testing	• Microbiological testing of comminuted meat is of limited value. • Some authorities recommend testing for salmonellae if the comminuted meat is to be eaten uncooked.
Spoilage	• Occurs rapidly if the temperature exceeds ca. 3–4°C.

[a]In particular circumstances, other hazards may need to be considered.

Hazards to be considered. See above.

Control measures

Initial level (H_0)

- Because the microbial content of freshly prepared comminuted meats is largely determined by the microbiological quality of the meat used, the time and temperature at which this meat is stored before grinding should be controlled. If frozen meat is used, it should be tempered to −2°C to −5°C under conditions that limit bacterial proliferation on the surfaces of the meat.

- Equipment coming in contact with the meat should be clean before use—especially equipment used for grinding, filling (sausage) or separation (i.e. mechanically separated meat).

Reduction (ΣR)

- Although numbers of viable microorganisms may fall slightly upon freezing, elimination of pathogens cannot be guaranteed.

Increase (ΣI)

- The temperature history between grinding and cooling to the storage temperature should be known and controlled.
- Periods spent >7–10°C may allow growth of salmonellae and *E. coli* O157:H7.
- The extent of this growth may be estimated by predictive equations (e.g. Mackey *et al.*, 1980; Mackey and Kerridge, 1988).
- Low meat temperatures during storage and distribution are essential to maintain shelf-life and to minimize growth of psychrotrophic pathogens.

Testing. When experience indicates that the prevalence of a pathogen is <1% in samples tested, routine microbiological testing for these pathogens is not recommended, but more intensive sampling plans can assure low levels of pathogens e.g. VTEC (ICMSF, 2002, Chapter 17, pp. 313–332): for example, failure to detect VTEC in 25 g samples from each of 30 sample units from a lot provides 95% confidence that the concentration of VTEC in the lot is no >1 cfu 250 g^{-1}.

VII Raw cured shelf-stable meats

This group includes raw hams and some low-acid dry sausage and high-acid fermented sausage where low a_w or a combination of low pH and reduced a_w provides microbial stability.

The reader is referred to Lücke (2000) for a comprehensive review of the technology and microbiology of fermented meats.

A Effects of processing on microorganisms

The processes used for raw ham production vary but include selection and careful cutting and trimming of the ham so that deep cuts are not made into the tissue, curing, ripening, drying, maturing, and ageing. For some products, bones are removed before curing and some hams are smoked. High-pH meat absorbs cure more slowly, dries more slowly, and there is a greater risk of extensive microbial growth than occurs with meat of normal pH. High-pH meat is therefore not usually used for shelf-stable raw ham production (Hechelmann and Kasprowiak, 1992).

The cure is applied as a dry salt, by pumping with pickle, by covering with pickle or by a combination of these treatments. When bones are removed, special care is needed in applying cure to folds and crevices. Bacteria in these crevices, and in any deep cuts, may grow and cause spoilage before the salt concentration is high enough to prevent spoilage. Initial curing and ripening is done at <5°C. Low temperatures are needed to prevent growth of undesirable microorganisms (e.g. non-proteolytic strains of *Cl. botulinum* type B), although the concentration of salt within the tissue is low (i.e. <4.5%; a_w >0.96; Lücke *et al.*, 1981). Toxin formation by non-proteolytic strains of *Cl. botulinum* has been demonstrated to occur in bone-in hams when the initial curing was done at 8°C.

Dry-cure hams, after salting and before final ripening, are washed to remove excess salt. If the washed hams are dried at elevated temperatures (e.g. 30°C), high numbers of *Staph. aureus* can develop on the surface (Untermann and Muller, 1992). When the a_w of all parts of the ham is <0.96, the drying temperature is raised initially to <15°C, and, as the a_w continues to fall, to 18–20°C. Smoking is done at temperatures of not >22°C. The final a_w of raw ham is <0.90.

The predominant flora developing on hams during curing and drying are Micrococcaceae with *Staph. xylosus, Staph. sciuri* and *Staph. equorum* as the major species (Molina *et al.*, 1990; Cornejo and Carrascosa, 1991; Rodriguez *et al.*, 1994). Lactic acid bacteria and yeasts are often relatively minor portions of the flora. These three groups of microorganisms are also found in the internal tissue of normal hams (Lücke, 1986; Rheinbaben and Seipp, 1986; Silla *et al.*, 1989). Enterotoxigenic strains of *Staph. aureus* (mostly enterotoxin A producers) can be found in low numbers, usually <100 cfu/g, on ham during its production. On properly prepared, and dried raw ham, *Staph. aureus* are often not found (Marin *et al.*, 1992). The major molds on ham during the early stages of production are *Penicillium* spp. Later, these are replaced by *Aspergillus* spp. (Rojas *et al.*, 1991; Casado *et al.*, 1992).

"Speck" is a specialty product traditionally produced in South Tyrol (Italy) and North Tyrol (Austria) by farmers, butcheries and meat industries. From 121 samples of Speck from North and South Tyrol, from different production types and geographic regions, 63 fungal species were isolated, of which only a few are typical colonizers (Peintner *et al.*, 2000). *Eurotium rubrum* and *Pen. solitum* were the dominating species in all types and parts of Speck (crust, meat and fat). Eight other *Penicillium* spp. were relatively frequent. The species diversity increased from industrially produced Speck to products from butcheries and farmers, and was greater in all types of South Tyrolean products. Among the typical mycobiota, *Pen. verrucosum, Pen. canescens* and *Pen. commune* are known as potentially mycotoxigenic.

Chinese sausage (La Chang), by contrast, is dried at a high temperature. Coarsely ground pork and fat mixed with sugar (1–10%), salt (2.8–3.5%), soy sauce, and nitrate is stuffed into small-intestine pig casings (16–28 mm). The filled casings are punctured to allow ready release of moisture during drying. The sausages are dried for 1–2 days at 45–50°C and then held at ambient temperature (2–3 days) for moisture equilibration. The microbial stability is due to the rapid reduction in a_w (<0.92 in 12 h) aided by the thin diameter of the sausage, the initial concentration of sugar and salt, the high drying temperature and the low relative humidity (Leistner, 1988). The final a_w is <0.80. Micrococci are the dominant flora followed by coagulase-negative staphylococci and lactic acid bacteria (Leistner, 1988; Guo and Chen, 1992). Vacuum-packing inhibits mold growth.

Some varieties of salami (e.g. Hungarian) depend on low a_w for their microbial stability since their pH is usually 5.8–6.2 (Incze, 1987, 1992). Ground pork and fat, mixed with 2.5–3% salt, nitrite or nitrate, and spices, are packed in moisture-permeable casings. Smoking, drying, and ripening are initially carried out at <10°C until the a_w falls <0.93–0.92. Drying and maturing are completed at 14–16°C to give a final a_w of <0.88. The initial a_w (0.96–0.97) and low temperature inhibits the growth of Gram-negative bacteria (pseudomonads, Enterobacteriaceae, and salmonellae). Furthermore, they slowly die during the long period of ageing and maturing. Lactic acid bacteria grow throughout the sausage but, because of the absence of added sugar, there is little change in pH. Micrococcaceae develop in the aerobic outer zone where they reduce nitrate to nitrite, and play a role in color stability by destroying peroxides, and contribute to flavor formation. When sausages are made with nitrate and are not smoked (e.g. traditional Italian salami), Micrococcaceae can dominate the flora (Lücke, 1986). During the latter stages of maturing, these organisms decline in number. Growth and enterotoxin formation by *Staph. aureus* are retarded by the low temperature and by the competitive flora. About mid-way through the ripening phase, mold develops on the sausage surface. The initial smoking retards mold growth until the a_w is low. The catalase activity of the white mycelial mat acts as a natural antioxidant preventing rancidity. Proteases and lipases play important roles in flavor and aroma formation (Bacus, 1986; Geisen *et al.*, 1993). Mycotoxins have not been found in normally ripened Hungarian salami (Incze, 1987). *Pen. chrysogenum* and *Pen. nalgiovense* may be inoculated onto mold-ripened sausages.

Fermented sausages rely on both a reduced pH (4.6–5.3) and a reduced a_w (<0.95) for microbial stability. Ground meat and fat together with salt (2.5–3%), sodium nitrite or nitrate, sugar (0.4–0.7%), and spices are fermented in moisture-permeable casings. The sausages are then dried and may be smoked. Semi-dry sausage loses up to 15% moisture, generally is smoked and receives a mild heat treatment (Bacus, 1986). Dry sausage is dried to remove 25–50% of the moisture. The temperature of fermentation can vary from <25°C to 43°C. European-style sausages are usually fermented at <25°C with a slower acidification rate than is used in North America where fermentation temperatures are 20–40°C and more tangy and lower pH (4.6–5.1) sausages are considered acceptable (Lücke, 1986; Bacus, 1986). In Hungary, for instance, the final pH of starter-fermented sausages is 5.1–5.3. Conditions in the sausage (initial a_w 0.96–0.97, low oxygen tension) select for lactic acid bacteria, which grow to reach about 10^8 cfu/g. Growth of lactic acid bacteria is stimulated by manganese in spices or in the culture medium. Fermentation of the added sugar to lactic acid by homofermentative strains causes the pH to fall to an ultimate pH of 4.6–5.3. Heterofermentative lactobacilli and leuconostocs are undesirable because of the production of carbon dioxide and acetic acid (Egan and Shay, 1991; Geisen *et al.*, 1993). During the initial phase of fermentation, Micrococcaceae (coagulase-negative staphylococci) also increase in numbers. These organisms reduce nitrate, if added during formulation, and destroy peroxides. When nitrite production from nitrate is required, or when the contribution of Micrococcaceae to aroma development is desired, the rate of pH fall should not be so rapid that the acid-sensitive Micrococcaceae are inactivated before they can bring about these desired changes (Incze, 1992; Geisen *et al.*, 1993). In this case, fermentation is done at <25°C. Numbers of Gram-negative bacteria decline rapidly during fermentation and drying. Yeasts (*Debaryomyces* spp.) develop in the aerobic outer zone of unsmoked or lightly smoked sausage (Bacus, 1986). Mold growth on the outer surface of fermented sausage results in an increase in pH. This can then permit the growth of *Staph. aureus* and food-poisoning (Hechelmann and Kasprowiak, 1992). In these sausages microbial stability needs both a reduced pH and a reduced a_w. In contrast, the stability of long-ripened sausage (e.g. Hungarian) is based on low a_w, and pH changes associated with surface mold do not result in staphylococcal enterotoxin formation.

The identity of "lactic acid bacteria" in foods is a slight concern because isolates from clinical infections proved identical, by 16S rRNA sequencing, to those found in fermented meat (and dairy) products and some were resistant to vancomycin (Aguirre and Collins, 1993). However, concerns have led to extensive studies on the lactic acid bacteria in foods and in the gut to evaluate any risk (Franz *et al.*, 1999; Klein, 2003; Ben Omar *et al.*, 2004).

Failure to achieve correct acid production can result in the growth of undesirable microorganisms (e.g. *Staph. aureus*). Control of the time and temperature to reach pH 5.3 is critical for controlling the growth of undesirable microorganisms (American Meat Institute, 1982). Frequently, starter cultures are used to ensure a predictable rate of pH fall. Different cultures are used depending on the temperature of fermentation, for instance, *Pediococcus acidilactici* for fermentations at 32–40°C, and *Lb. sake* and *Lb. curvatus* for fermentations below 25°C (Egan and Shay, 1991; Geisen *et al.*, 1993). Glucono-delta-lactone may be added to the meat emulsion so that hydrolysis to gluconic acid rapidly reduces pH. The starter cultures *Lb. plantarum* and *Lb. sake* can metabolize gluconic to acetic acid which imparts an altered flavor and odor (Egan and Shay, 1991). Encapsulated citric acid is now more commonly used as an acidulant in some countries. Additional information on raw cured meats and fermented sausages is available (Bacus, 1984; Campbell-Platt and Cook, 1995).

Concern over the occasional presence of biogenic amines in fermented sausages prompted Bover-Cid *et al.* (2000) to investigate several combinations of an amine-negative *Lb. sakei* strain, along with proteolytic *Staph. carnosus* or *Staph xylosus* strains as mixed starter cultures on biogenic amine production during the manufacture of dry fermented sausages. A mixed starter culture of *L. sakei* and *Staphylococcus* spp. (all amine-negative strains) drastically reduced tyramine, cadaverine and putrescine accumulation. Histamine, phenylethylamine and tryptamine were not produced. The polyamines, spermine, and spermidine, were found in raw materials and their levels decreased slightly in the spontaneously fermented batch. No correlation between proteolysis and biogenic amine production was observed. Conditions favoring starter development and using raw materials of good hygienic quality resulted in fermented sausages almost free of biogenic amines.

Subsequently, Bover-Cid *et al.* (2001) explored the effectiveness of an amine-negative starter culture (*Lb. sakei* CTC494) in reducing biogenic amine production during the ripening process. In sausages manufactured from good quality meat, extremely low biogenic amine levels resulted in the product (tyramine levels <15 mg/kg of dry matter and putrescine and cadaverine levels <5 mg/kg of dry matter). Sausages using poorer-quality raw materials contained much higher amine contents (308, 223 and 36 mg/kg of dry matter of cadaverine, tyramine and putrescine, respectively).

B *Spoilage*

Bacterial spoilage of these shelf-stable products can occur during production, before the a_w or pH values are low enough to prevent growth of spoilage organisms. Growth of undesirable molds on finished products give an unattractive appearance or can produce foreign aromas. Such mold growth can be prevented by vacuum-packaging, motified atmosphere packaging (e.g. sliced products) or by the development of the typical desirable mold (e.g. on mold-ripened sausages) during production.

Spoilage of raw hams can be caused by a variety of bacteria (Enterobacteriaceae, *Clostridium* spp.) growing within the meat before the salt concentration is sufficiently high and/or if the initial curing temperature is not sufficiently low to prevent multiplication (Gardner, 1983). Spoilage organisms gain access to the tissue through incisions during cutting and trimming, during pumping or by the circulatory system of the live animal. Sour or putrid odors, sometimes with pockets of gas, may be found near the hip-joint ("bone-taint"). Prevention is by proper chilling of the initial carcass, by using clean pumping pickle and equipment, by good temperature control during curing and ripening and by rapid cure diffusion.

Spoilage of Chinese sausage results from insufficient drying permitting growth of lactic acid bacteria and souring.

Growth of heterofermentative lactic acid bacteria during the fermentation of fermented sausage causes off flavors and aromas (e.g. from acetic acid) or gas pockets or pin-holes from the formation of carbon dioxide in the sausage.

C *Pathogens*

The pathogens to consider are *Cl. botulinum*, EHEC, *Staph. aureus*, salmonellae, and *L. monocytogenes*.

Non-proteolytic *Cl. botulinum* type B strains have been major agents of botulism from home- or farm-produced hams (Lücke and Roberts, 1993). These hams have been dry-salted and not brine injected. Botulism has resulted when the initial salting was done at temperatures $>5°C$ and/or hams were held at $>5°C$ before salt equilibration established an internal a_w <0.96.

EHEC have caused illness associated with the consumption of fermented sausage. In one of these incidents the organism concerned was a non-motile strain of *E. coli* (serotype 0111) and the product was a semi-dry fermented sausage; more than 50 persons were ill with 23 (median age 4 years) with hemolytic uremic symptoms (Anonymous, 1995a). A further incident linked to *E. coli* O157:H7 involved a dry-cured salami (CDCP, 1994, 1995; Anonymous, 1995b). Such incidents are perhaps not surprising in view of the recognized acid tolerance of these strains of *E. coli* (Benjamin and Datta, 1995). In laboratory experiments, growth of *E. coli* O157:H7 at 37°C was inhibited by 200 μg nitrite/mL at pH 5.0 (Tsai and Chou, 1996).

Shelf-stable raw cured meat products present no health risk for the consumer if a_w-values are <0.91, or ≤0.95 in combination with a pH-value <5.2 (Leistner and Rödel, 1976). *E. coli* O157:H7 survives in fermented meats and manufacturers should use meat that contains no, or very few, *E. coli* O157:H7 (Glass *et al.*, 1992; ICMSF, 2002). *E. coli* O157:H7 was isolated in 1 of 30 dry sausages in an Argentine study (Chinen *et al.*, 2001).

Processing of industrial fermented or dry or semi-dry sausages is characterized by a preliminary microbiological fermentation (~3 days at ±25°C), mainly by development of lactic acid bacteria, causing a decrease of pH (5.2–4.8), and further drying and ripening for several days or weeks at ± 18°C until a_w-values down to 0.88 are reached. The initially ground pork/beef and pork fat mixture regularly contains 2.5–3.0% sodium chloride, 0.5–0.7% glucose, 150–180 mg/kg nitrite or nitrate and spices (ICMSF, 1998a,b). In some instances, starter cultures composed by lactic acid bacteria (e.g. *Lb. plantarum*, *Lb. sake* and *Lb. curvatus*) and *Micrococcaceae* are added to meat/fat mixture.

After fermentation and drying and extended storage (2 weeks) at ambient temperature of commercially processed Lebanon bologna, a significant reduction in numbers of *E. coli* O157:H7 (ca. 1–2 $\log_{10}$) was observed (Faith *et al.*, 1997; 1998a, b). Similar reductions occurred in shorter extended storage times (20 h) at higher temperatures (e.g. 43°C) (Ellajosyula *et al.*, 1998). The use of relatively low heating temperatures, e.g. 54°C, for relatively short periods of time (30 min) has also been suggested as an effective process step to ensure the safety of fermented meats (Calicioglu *et al.*, 1997).

In Hungarian-style salami, selection of the starter cultures may prove to be an important means of controlling pathogens such as *L. monocytogenes* and *E. coli* O111 (Pidcock *et al.*, 2002).

Food-poisoning has been caused by *Staph. aureus* on raw hams (Marin *et al.*, 1992; Untermann and Muller, 1992). *Staph. aureus* is able to grow and produce enterotoxin on slices of raw ham if the a_w-value is >0.90. Values of a_w >0.90 are usually associated with the production of ham from high-pH meat. Raw hams of a_w >0.90 should be stored refrigerated.

Fermented sausages have been involved in outbreaks of both *Staph. aureus* food poisoning (Bacus, 1986) and salmonellosis (Taplin, 1982; Cowden *et al.*, 1989). Such food-borne illness is a consequence of improper fermentation allowing growth of *Staph. aureus* or growth of salmonellae. Both organisms can be expected to be present in the raw meat used. As growth is suppressed at pH ≤5.3, the time and temperature at which the fermenting sausage mix remains above this pH determines the extent of growth of *Staph. aureus*. Good manufacturing practice guidelines (American Meat Institute, 1982) have been developed, which limit the time the sausage meat is exposed to temperatures >15°C before pH 5.3 is reached. During the drying and ripening phases, the number of viable cells of *Staph. aureus* and salmonellae decrease faster with decreasing pH and increasing storage temperature (Smith *et al.*, 1975; Cowden *et al.*, 1989).

Staphylococcal food-poisoning has also occurred on fermented sausage when surface mold growth has been encouraged by an a_w that allows growth of *Staph. aureus* (Hechelmann and Kasprowiak, 1992).

L. monocytogenes could grow in sausage mix if there is a fermentation failure (Glass and Doyle, 1989a). However, during normal fermentation, drying, and ripening there is a 10–100-fold decrease in numbers (Johnson *et al.*, 1988; Junttila *et al.*, 1989). The small number of cells normally found in finished product cannot multiply. The risk of listeriosis from fermented dry sausage and other raw cured shelf-stable meats is very low.

D CONTROL (raw, cured shelf-stable meats)

Summary

Significant hazards[a]	• *Cl. botulinum.* • salmonellae. • VTEC. • *Staph. aureus.*
Control measures	
Initial level (H_0)	• Effective chilling of pig carcasses (deep muscle temperature 0°C after 24 h). • Use pork with pH not >5.8. • Care in processing (see below).
Reduction (ΣR)	• Reduction in numbers of pathogens cannot be guaranteed.
Increase (ΣI)	• Carry out salting/curing and salt equilibration at 5°C or below. • Keep hams at 5°C until internal water activity is <0.96. • Controlling temperature, time, and relative humidity during different stages of curing, drying, and ripening (maturation) as the initial salting temperature is increased from 5°C to the final ambient temperature for ageing. • A final a_w throughout the product of <0.90.
Testing	• No microbiological testing is recommended.
Spoilage	• Surface mold growth occurs at high relative humidity.

[a]In particular circumstances, other hazards may need to be considered.

Hazards to be considered. See above.

Control measures

Initial level (H_0)

- Careful cutting and trimming of the carcass to avoid deep incisions.
- Care in applying cure to folds and crevices where bones are removed.
- Use clean pickle and pickle pumping equipment.

Reduction (ΣR)

- Heating to 54°C for 30 min has been suggested but is not widely applied.

Increase (ΣI)

- During production, the temperature, a_w, and, for some products, the rate of decrease of pH of the meat, are controlled to prevent microbial spoilage and growth of pathogens.

Testing. During ageing/maturation/ripening, tests for spoilage of deep muscle are by inserting a probe (or bone) and smelling.

Spoilage

- Surface mold growth occurs at high relative humidity.
- Rancidity can occur depending on storage time and temperature.

CONTROL (Chinese sausages)

Summary

Significant hazards[a]	• *Cl. botulinum.* • salmonellae. • VTEC. • *Staph. aureus.* • *L. monocytogenes.*
Control measures	
Initial level (H_0)	• Careful cutting and trimming of the carcass to avoid deep incisions. • Care in applying cure to folds and crevices where bones are removed. • Use clean pickle and pickle pumping equipment.
Reduction (ΣR)	• Reduction in numbers of pathogens cannot be guaranteed.
Increase (ΣI)	• Salt and sugar in the formulation to reduce a_w. • Small diameter, pierced and permeable casings. • High temperatures (45–50°C) and low humidity (ca. 65%) during drying.
Testing	• No microbiological testing is recommended.
Spoilage	• Rancidity can occur depending on storage time and temperature.

[a]In particular circumstances, other hazards may need to be considered.

Hazards to be considered. See above.

Control measures

Initial level (H_0)

- Careful cutting and trimming of the carcass to avoid deep incisions.
- Care in applying cure to folds and crevices where bones are removed.
- Use clean pickle and pickle pumping equipment.

Reduction (ΣR)

- Reduction in numbers of pathogens cannot be guaranteed.

Increase (ΣI)

- During production, the temperature, a_w, and, for some products, the rate of decrease of pH of the meat are controlled to prevent microbial spoilage and growth of pathogens.

D CONTROL (dry salami, e.g., Hungarian)

Summary

Significant hazards[a]	• *Cl. botulinum.* • salmonellae. • VTEC. • *Staph. aureus.*
Control measures	
Initial level (H_0)	• Initial pathogen level has not been identified as a risk factor for this type of product.
Reduction (ΣR)	• Reduction in numbers of pathogens cannot be guaranteed.
Increase (ΣI)	• If the initial salt concentration is too low, excessive growth of Gram-negative bacteria may occur. • Initial light smoking dries the surface, prevents surface slime formation and delays mold growth until the a_w is low. • Slow removal of moisture prevents case-hardening which would reduce moisture loss from internal regions of sausage.
Testing	• The air temperature (<10°C until the a_w is <0.92–0.93), velocity, and relative humidity are measured and controlled, and the weight loss of the sausages is monitored. • Final a_w<0.88. • No microbiological testing is recommended.
Spoilage	• Rancidity can occur depending on storage time and temperature. • Surface mold growth occurs at high relative humidity.

[a]In particular circumstances, other hazards may need to be considered.

Hazards to be considered. See above.

Control measures

Initial level (H_0)

• For low-acid, dry salami (e.g. Hungarian), control is based on a slow rate of moisture removal over a long period of time at low temperature (initially <10°C). Initial pathogen level is not a risk factor.

Reduction (ΣR)

• Reduction in numbers of pathogens cannot be guaranteed.

Increase (ΣI)

• If the initial salt concentration is too low, excessive growth of Gram-negative bacteria may occur.
• Initial light smoking dries the surface, prevents surface slime formation and delays mold growth until the a_w is low.

D CONTROL (fermented, high acid sausages)

Summary

Significant hazards[a]	• *Cl. botulinum.* • Salmonellae. • VTEC. • *Staph. aureus.*
Control measures	
Initial level (H_0)	• If the initial salt concentration if too low, growth of Gram-negative bacteria may occur; if too high, fermentation may be inhibited.
Reduction (ΣR)	• Reduction of enteric pathogens occurs in validated processes.
Increase (ΣI)	• Selection of raw material of good microbiological quality (high numbers of *Staph. aureus* increase the risk of enterotoxin formation in the outer regions of the sausage and survival of enteric pathogens. • Select an active starter culture compatible with the temperature of fermentation, the carbohydrate to be fermented, the a_w, and the nitrite content of the sausage mix.
Testing	• Monitor the time and temperature of fermentation. • Microbiological testing depends on the process and product. • The rate of fermentation is monitored by pH measurement. • The temperature, relative humidity, and air speed in the drying and ripening rooms are measured and controlled, and the weight loss or a_w of the sausages is monitored.
Spoilage	• If the initial salt concentration is too low, growth of Gram-negative bacteria may occur; if too high, fermentation may be inhibited. • High numbers of heterofermentative lactic acid bacteria can cause objectionable flavors and aromas and gas pockets).

[a]In particular circumstances, other hazards may need to be considered.

Hazards to be considered. See above.

Control measures

Initial level (H_0)

- A rapid initial lowering of the pH to 5.3 is followed by a further lowering to 4.6–4.8 and a reduced a_w.

Reduction (ΣR)

- Sufficient fermentable carbohydrate to give the desired decrease in pH, or the addition of glucono-delta-lactone or encapsulated citric acid. Many of these products are heated before drying.

Increase (ΣI)

- Acid tolerance of salmonellae and VTEC is an unique concern with these particular products and should be addressed when developing a HACCP plan for the manufacture of these products.

Testing

- No routine microbiological testing is recommended. Product that ferments slowly, or not at all, should be sampled (outer 3 mm below casing) for *Staph. aureus* before applying a lethal heat treatment or drying. Product with high levels *Staph. aureus* (e.g. $>10^5$ cfu/g) should be tested for enterotoxin.
- Numbers of *Staph. aureus* should not exceed 10^4 cfu/g during fermentation.

VIII Dried meats

Dried meats are microbiologically stable at ambient temperature because of their low a_w-value. Traditional dried meats include Rou Gan (China), charqui (South America) and biltong (South Africa) (Osterhoff and Leistner, 1984; Prior, 1984).

Dried meats are prepared for use in formulated foods (e.g. soup mixes), for campers and the military, and as a snack food.

A Effects of processing on microorganisms

Biltong is made from strips (ca. 2.5 cm thick) of raw beef or raw meat from game animals. The strips are dry salted (2.5–4% salt), stored overnight at 4–5°C for salt equilibration, and then dried with heated air (e.g. 35°C, 30% R.H., 3 m/s) to a moisture content of $\leq$24%. During the drying process, bacterial numbers, particularly of Gram-positive micrococci, increase (van der Riet, 1982; Prior, 1984). On commercially produced biltong, aerobic mesophilic counts of 10^5–10^7 cfu/g and yeast and fungal counts of 10^5–10^6 cfu/g have been reported (van der Riet, 1982). Yeasts (e.g. *Candida, Rhodotorula*) outnumber molds. The most common molds are of the *Aspergillus glaucus* group. Growth of these molds has been shown to occur on biltong of a_w-value 0.72 in 67 days, but not at a_w 0.68 (van der Riet, 1976). Salmonellae have been detected on commercial biltong. *S. dublin* has been shown to maintain its viability for months in biltong made from the meat of infected animals.

In the production of Rou Gan, strips or slices of beef, pork or mutton are cooked with spices, sugar, salt, and soya sauce until the mixture is almost dry, and the pieces are dried on racks at 50–60°C to a final a_w of 0.60–0.69. Bacterial numbers on the final product are usually $<10^4$ cfu/g (Wang and Leistner, 1993).

Beef jerky is prepared as a snack food. Commercially produced beef jerky often includes a cooking step. Three procedures are used. In the first procedure, beef is trimmed of visible fat, connective tissue and tendons, pumped with a curing solution, massaged, filled into casings, cooked to an internal temperature of about 65°C, cooled, the logs sliced into discs 1.5–2 mm thick and the discs spread, dried, and smoked on mesh trays. Cooking destroys the normal vegetative flora of the raw meat. However, there are opportunities for contamination during handling, slicing, and loading into the drier. In the second procedure, raw beef is sliced into strips 2–3 mm thick and soaked in a curing solution before being hot-air dried and smoked on mesh trays. In the third method, raw meat is flaked with cure ingredients, formed into blocks or logs, tempered to −2 to −5°C, sliced, then dried, and smoked. In the second and third processes, flora from the original raw meat will still be present in the slices. Sometimes, temperatures high enough to cook the meat are used in smoking and drying. For the three processes, separation of the thin slices and rapid drying are needed to prevent microbial growth. The a_w is reduced to about 0.86 in 1–3 h and to a final water content of 15–20%. The USDA has established a moisture: protein ratio requirement of 0.75 for beef jerky.

Ground dried meat in the form of granules or powder is used in soup mixes. Meat, trimmed of fat, is cooked, minced and the mince is spread in thin layers and dried under controlled ventilation in hot-air tunnels (>50°C). Conditions in the drying tunnel are warm and moist until drying is well advanced. Pieces of meat must be small and sufficiently separated to ensure drying in 1–2 h. Microbial growth may occur in wet clumps of meat. Microorganisms on the dried product are those that have survived the initial cooking and those that were acquired during mincing and placing in the drying tunnel. Microbial numbers will increase if drying is not properly controlled.

Cooked or raw meat can be freeze-dried. The low temperatures employed during drying prevent microbial growth. There is little change in bacterial numbers during freeze-drying. The microbial quality of the dried meat is dependent on the meat used and contamination during packing.

B Spoilage

Microbial stability depends on the a_w of the products. Mold and yeasts may grow during storage, particularly when moisture is absorbed from the environment. This can be prevented by drying to an a_w <0.8 and vacuum-packing, or by drying and maintaining the a_w at ≤0.7.

C Pathogens

Clostridia, bacilli, *Staph. aureus*, salmonellae and other pathogenic bacteria, originally present on raw meat or as contaminants during preparation, may survive the drying process. During storage the number of vegetative cells will decline. Biltong (van de Riet, 1982) and commercially prepared beef jerky have caused salmonellosis (CDC, 1967). Home-made venison jerky has caused type F botulism (Midura *et al.*, 1972). Surviving pathogens will grow in re-hydrated product if held at temperatures permitting their development.

The safety of home-made jerky continues to be questioned. Producing a safe product that retains acceptable quality attributes is important. Lethality of salmonellae, *E. coli* O157:H7 and *L. monocytogenes*, as well as consumer acceptability and sensory attributes, of jerky prepared by four methods were examined (Harrison *et al.*, 2001). Drying marinated strips at 60°C (a traditional method), was compared with boiling strips in marinade or heating in an oven to 71°C prior to drying, and heating strips in an oven after drying to 71°C. Samples heated after drying and samples boiled in marinade prior to drying had slightly higher acceptability scores but were not statistically different from traditional samples. Numbers of the pathogens subjected to the four treatments showed a 5.8-, 3.9- and 4.6 $\log_{10}$ reduction of *E. coli* O157:H7, *L. monocytogenes* and salmonellae, respectively, even with traditional drying. Oven treatment of strips after drying had the potential to reduce pathogen numbers further by ca. 2 $\log_{10}$, suggesting that a safer, acceptable, home-dried beef jerky product can be produced by oven-heating jerky strips after drying.

D CONTROL (dried meats)

Summary

Significant hazards[a]	• *Cl. botulinum.* • salmonellae. • VTEC. • *Staph. aureus.* • *L. monocytogenes.* • *B. cereus.*

(*continued*)

D CONTROL (Cont.)

Summary

Control measures	
Initial level (H_0)	• Selection of raw materials. • If to be eaten without cooking, the microbiological quality of raw meat. • Avoiding contamination during preparation, loading and unloading of the drier, and packing.
Reduction (ΣR)	• Reduction in numbers of pathogens cannot be guaranteed unless the dried meat is cooked before consumption.
Increase (ΣI)	• GHP to avoid contamination of finished product. • Low a_w to control the growth of pathogens (moisture should be 20–24%). • Time–temperature history after re-hydration.
Testing	• No routine microbiological testing is recommended.
Spoilage	• Control of humidity and packaging to prevent the growth of yeasts and molds. • Protection, by suitable packaging, of the dried product from moisture reabsorption.

[a]In particular circumstances, other hazards may need to be considered.

Hazards to be considered. See above.

Control measures

Initial level (H_0)

- Selection of raw materials.
- If to be eaten without cooking, the microbiological quality of raw meat.
- Avoiding contamination during preparation, loading and unloading of the drier, and packing.

Reduction (ΣR)

- Reduction in numbers of pathogens cannot be guaranteed unless the dried meat is cooked before consumption.

Increase (ΣI)

- GHP to avoid contamination of finished product.
- Low a_w to control growth of pathogens (moisture should be 20–24%).
- Time–temperature history after re-hydration.

Testing. No routine microbiological testing is recommended.

Spoilage

- Control of humidity and packaging to prevent growth of yeasts and molds.

Protection, by suitable packaging, of the dried product from moisture reabsorption.

Comments. The most important control steps are strict regulation of the time–temperature profile during drying and drying to an appropriate low moisture content or a_w-value.

IX Cooked perishable uncured meats

The following paragraphs are applicable to both large-scale commercial and small-scale production of whole (intact) cuts.

Some meat may be only lightly cooked for consumption, giving a cooked appearance on the outside but leaving the meat essentially raw in the center. These will include "rare" cooked joints and comminuted meat products such as burgers, and flash-fried products. The color change of meat from red to grey associated with cooking, occurs at temperatures approaching 60°C, the precise temperature depending on the duration of the heating. The center of a piece of steak 15 mm thick, grilled "rare" barely attains 40°C even momentarily. Temperature in the range 40–60°C, especially of short duration, will not eliminate even relatively heat sensitive vegetative bacteria. Thus, many microbes will survive; organisms of concern including, salmonellae, *C. jejuni/coli*, pathogenic strains of *E. coli, Y. enterocolitica* and parasites. A further concern with this group of products is that partially cooked products may appear fully cooked to the person preparing the product for consumption and despite proper labeling may be mistaken for being fully cooked.

Most cooked perishable uncured meats are fully cooked prior to consumption. Some manufactured and pre-prepared meat products may be subjected to a defined heat treatment specifically designed to eliminate vegetative infectious pathogenic bacteria. These can be considered pasteurized meat products. Such products are usually cooked so that their slowest heating parts achieve temperatures between 60°C and 75°C. The minimum time–temperatures for cooking may be established by control authorities and specified in regulations. Examples can be found at the USDA website:

1. Appendix A Compliance Guidelines For Meeting Lethality Performance... Certain Meat And Poultry Products: http://www.fsis.usda.gov/OPPDE/rdad/FRPubs/95-033F/95-033F_Appendix%20A.htm
2. Processed Meat and Poultry Performance Standards—Lethality and Stabilization: http://www.fsis.usda.gov/OPPDE/rdad/FRPubs/97-013N/Engeljohn_Lethality/index.htm.
3. Compliance Guidelines to Control *L. monocytogenes* in Post-Lethality Exposed Ready-to-Eat Meat—Studies on Post-lethality Treatments (Mention... constitute endorsement by USDA) I. Steam Pasteurization and Hot Water Pasteurization...: http://www.fsis.usda.gov/oppde/rdad/FRPubs/97-013F/CompGuidelines.doc; http://www.fsis.usda.gov/oppde/rdad/FRPubs/97-013F/CompGuidelines.pdf.

Typical industrially produced cooked meats and meat products include, pre-cooked roast beef joints, hot and cold meat pies (including pasties and sausage rolls), and prepared meat-based ready-to-eat-meals and meal components (including sous-vide). These products are stored and distributed frozen or chilled, and may be used by food service establishments or at home. Some may be eaten cold, for instance roast beef in sandwiches, and some may be reheated for consumption (e.g. sliced meat in gravy).

A *Effects of processing on microorganisms*

The cooking process given to cooked meats produced industrially will effectively destroy all vegetative bacterial pathogens, viruses, parasites, and most vegetative spoilage microorganisms. Fecal streptococci and certain lactobacilli are relatively heat resistant and may survive commercially used processes.

To protect the consumer, specific heat treatments may be required. In the United States, minimum heat processes (any one of 16 time and temperature combinations equivalent to instant heating to 63°C) must be used for the manufacture of pre-cooked roast beef joints for chilled or frozen distribution (USDA, 1983). A similar requirement (equivalent to instant heating to 70°C) is mandatory for fully cooked meat patties (USDA, 1993). Whereas the former process is targeted at killing salmonellae, the latter is aimed at destruction of *E. coli* O157:H7.

Recognizing that such processes will not destroy bacterial spores, specific cooling requirements are also specified. For meat joints this requires them to be rapidly and continuously cooled such that the time between temperatures 48.9°C and 12.8°C does not exceed a total of 6 h, with cooling continuing until a temperature of 4.4°C is reached; this is intended to prevent the growth of spore-forming bacteria. In the UK, there are Department of Health guidelines for the manufacture of pre-cooked chilled and frozen foods for catering (Department of Health, 1989), and trade guidelines for pre-cooked chilled meals and meal components sold through retail outlets (CFA, 1993). These guidelines specify both cooking and cooling requirements and also hygienic practices to prevent contamination of the cooked product. Similar guidelines have been developed by other countries and can be useful sources of information (USDA 1988; FDA 1993). The cooking process required for products with a shelf-life of 10 days or less is a heat process equivalent to heating to a minimum of 70°C for 2 min. This will give at least a six decimal kill of *L. monocytogenes* (Gaze *et al.*, 1989).

Although *L. monocytogenes* is the target organism for such "short shelf-life" chilled foods, psychrotrophic strains of *Cl. botulinum* are the main organisms of concern in perishable, "extended shelf-life" chilled foods (shelf-life at chill longer than 10 days); these include the so-called "cuisine sous-vide" products. Thus, in the UK, the Advisory Committee on the Microbiological Safety of Foods in its report on vacuum-packaging and associated processes (ACMSF, 1992) made a series of recommendations on the safety of "extended shelf-life" vacuum-packed foods. These were the use of a heat treatment such that all components receive a minimum heat process of 10 min at 90°C (6 D reduction of group II strains (i.e., psychrotrophic) of *Cl. botulinum*); or to reduce the a_w of all components to 0.97 (equivalent to 3.5% w/w sodium chloride) or less; or to reduce the pH to 5.0; or to use a combination of extrinsic and intrinsic preservation treatments capable of giving equivalent security against group II strains of *Cl. botulinum*. Similar requirements to these are incorporated in the guidelines published by the European Chilled Foods Federation (ECFF, 1994).

Other national guidelines or regulations for vacuum-packed and other extended shelf-life meat products include those in the United States (NFPA, 1989); in Canada (AFSDAC, 1990); in Australia (AQIS, 1992); and in Europe (France, 1988; Benelux, 1991).

Hot and cold eating pies (a peculiarly British food) are cooked to temperatures between 71°C and 82°C (ICMSF, 1980b). This effectively destroys all vegetative pathogens and chilled storage prevents growth of surviving spore formers and post-cooking contaminants.

B Spoilage

The spoilage flora of this group of meat products depends on their composition, method of processing, packaging, and storage. Thus, the spoilage flora of a chilled stored, sliced cooked meat will depend on whether it is air-packed, in which case, spoilage will mainly be by Gram-negative psychrotrophic bacteria, including pseudomonads and members of the Enterobacteriaceae, or vacuum-packed, with spoilage mainly by lactic acid bacteria and *Broch. thermosphacta*. During chill storage of vacuum-packed pork, offensive, sickly odors developed and were shown to be due to a new species of the genus *Clostridium*, *Cl. algidicarnis* (Lawson *et al.*, 1994). Spoilage of cooked uncured meats by psychrotolerant clostridia resembling *Cl. laramie* has also been reported by Kalinowski and Tompkin (1999).

Spoilage of refrigerated meals will also depend on whether the product is handled after cooking and the time–temperature history of storage. Spoilage of cook-in-bag "sous-vide"' products will usually be by psychrotrophic spore-forming bacilli, whereas spoilage of products which are assembled and packed after cooking will depend on the microbial contaminants introduced during cooling, or subsequent handling, and will include lactic acid and psychrotrophic Gram-negative bacteria. The source of such organisms is often the non-meat components.

Meat pies usually spoil as a result of the growth of molds. Their spores usually contaminate the product during cooling and hence precautions are taken to minimize contamination during cooling and storage prior to packaging. Mold spoilage is usually first evident on the inside of the pastry where the moisture content is high, and on the outside when water has migrated through the pastry. *Mucor* species are the fastest growing molds on pastry and hence are the most common cause of spoilage followed by *Penicillium, Rhizopus,* and *Aspergillus* spp. (ICMSF, 1980b). Spoilage by mesophilic bacilli and clostridia may occur if pies are stored under warm conditions. Traditional British pork and meat pies are cooked whole, whereas the meat filling of other pies, such as shepherds pie, may be cooked in bulk. Bulk cooled meat is also used for frozen pies, which is then filled into a raw pastry case and lidded. Such pie meat will be subject to spoilage by clostridia or bacilli if not cooled rapidly after cooking (Sutherland and Varnum, 1982).

C Pathogens

Salmonellae that survive cooking, or when introduced by cross-contamination after cooking, are a recognized hazard of cooked meats (Bryan, 1980). Growth of *Staph. aureus* introduced by handling after cooking is also a hazard. Slow cooling of bulk-cooked meats and meat-containing products has been frequently associated with *Cl. perfringens* food poisoning and occasionally with *B. cereus* and *B. subtilis* food poisoning (Kramer and Gilbert, 1989). Kalinowski *et al.* (2003) considered the need for the cooling rate specified in US policies to control growth of *Cl. perfringens*, and concluded that their laboratory data and historical food safety data suggest a very low public health risk from commercially processed ready-to-eat meat (and poultry) products. Spore-forming pathogens (e.g. *Cl. botulinum* and *Cl. perfringens*) survive cooking and will multiply if the cooked food is held at $>10°C$ (*Cl. botulinum*) or $>15°C$ (*Cl. perfringens*). Large numbers of vegetative cells of *Cl. perfringens* ($\geq 10^7$) must be consumed for illness to occur (Brynestad and Granum, 2002). The number of *Cl. perfringens* spores present after cooking meat (and poultry) products in commercial operations is typically below detectable levels (Kalinowski *et al.*, 2003; Taormina *et al.*, 2003). There is no evidence that illness due to *Cl. perfringens* has occurred from commercially produced products in international commerce, although similar products have caused illness after mishandling, usually at foodservice or institutional catering. Incorrect storage times and temperatures were common in outbreaks attributed to *Cl. perfringens* (Brett and Gilbert, 1997; Brett, 1998; Bates and Bodnaruk, 2003; Brynestad and Granum, 2002). Outbreaks typically involve large pieces of meat or poultry. Cured meat products have rarely been implicated in *Cl. perfringens* owing to the relative sensitivity of *Cl. perfringens* to the combined effect of salt and sodium nitrite present (Roberts and Derrick, 1978; Gibson and Roberts, 1986). The general recommendation is to cool the cooked food as rapidly as possible through the range of temperatures that support multiplication of *Cl. perfringens* ($55°C$ to $<15°C$), although this presents a problem with large joints of meat (Kalinowski *et al.*, 2003; Taormina *et al.*, 2003; see also Chapter 2, Section V: Perishable cooked poultry products, at C Pathogens).

There are no reported cases of botulism from such products and although cooked meat products may be contaminated with low numbers of *L. monocytogenes* there are no reports of illness.

The importance of contamination of equipment/machinery was demonstrated by Lunden *et al.* (2002) when a dicing machine for cooked meat products was transferred from plant A to plant B, and then to plant C, resulting in *L. monocytogenes* of the PFGE type I, originally found only in plant A, being found in plants B and C.

There have been a small number of cases of botulism from frozen pies, after their inadvertent storage at warm temperatures for several days prior to consumption (CDC, 1983).

Closely related strains of *Listeria* spp. can persist in the environment of a meat processing plant (Senczek *et al.*, 2001). As part of a hygiene-monitoring programme, 131 *Listeria* isolates were detected after sampling different processing areas and meat products over a 2-year period. The isolates were differentiated by phenotypic characteristics, and the genomic *Apa*I and *Sma*I fragment patterns of all isolates were examined by pulsed-field gel electrophoresis (PFGE) which yielded 15 (*L. monocytogenes*), 20 (*L. innocua*), and 6 (*L. welshimeri*) pulsotypes. Of the environmental *L. monocytogenes* isolates PFGE-type B was clearly associated with processing area A, whereas PFGE-type E predominated in meat products.

D CONTROL (cooked perishable uncured meats)

Summary

Significant hazards[a]	• Salmonellae. • VTEC (beef). • *Staph. aureus.* • *L. monocytogenes.* • *Cl. perfringens.*
Control measures	
Initial level (H_0)	• Selection of raw materials. • Use high quality raw materials from suppliers using HACCP principles.
Reduction (ΣR)	• Cooking to eliminate vegetative cells of bacterial pathogens.
Increase (ΣI)	• GHP to prevent (re-)contamination of cooked products (cleaning- and disinfection programs in place, separation of raw materials and cooked product, training of personnel manually handling cooked products). • Chilling to prevent growth of clostridia. • Implementation of measures to prevent growth of *L. monocytogenes* during storage (frozen storage, addition of acetate/lactate, irradiation in-pack.
Testing	• Routine microbiological testing of products is not recommended. • APCs can indicate failures in hygiene during processing. • Environmental program surveilling *L. monocytogenes.* • If no prior knowledge is available, sampling and testing of products for *L. monocytogenes* may be appropriate.
Spoilage	• Selection of raw materials. • Time and temperature during transport, storage, and display.

[a]In particular circumstances, other hazards may need to be considered.

Hazards to be considered. See above.

Control measures

Initial level (H_0)

- Use of high quality raw materials from suppliers using HACCP.
- Complete physical separation of raw ingredients from cooked products.

Reduction (ΣR)

- Cooking to destroy appropriate vegetative and/or spore-forming microorganisms.

Increase (ΣI)

- Rapid cooling under conditions that prevent recontamination and growth of organisms surviving cooking, e.g. *Cl. perfringens*.
- Handling of cooked foods (e.g. slicing, portioning, and combining components) by trained food-handlers operating to high hygiene standards in areas physically separated from where raw meat and other uncooked foodstuff are handled.
- Hygienic design of equipment and work areas to minimize product contamination.
- Protection of product from environmental contamination.
- Packaging in clean packing materials.
- Storage and distribution under controlled temperatures.
- Label the product to advise on storage conditions and correct handling prior to consumption.
- Plant modifications (construction, moving or introducing equipment) increase the risk of pathogen contamination, if not properly controlled.

Testing

- Routine microbiological testing of end-products for pathogens is not an appropriate control measure. If there are shelf-life problems APC is appropriate.
- Monitoring environmental and line samples for listeriae or *L. monocytogenes* is recommended.
- APC or coliforms may be appropriate for in-line process monitoring for continuous systems like IQF cooked meats.

X Fully retorted shelf-stable uncured meats

Uncured meats and meat products (e.g. soups, meats with various cereal and vegetable additions) are packed in hermetically sealed containers (steel or aluminum cans, glass jars, semi-rigid metal foils or strong plastic). The product is heat treated in the container or aseptically packed after heat processing. As these products are "low-acid" foods, they must receive a minimum heat process designed to destroy *Cl. botulinum* (i.e. "a botulinum cook"; $F_0 = 2.5$) and they must be protected from post-processing contamination. In practice, heat treatments of $F_0 = 4$–6 are needed to prevent spoilage from the outgrowth of spores of higher heat resistance than those of *Cl. botulinum*.

A Effects of processing on microorganisms

Most of the microorganisms on the meat ingredients are vegetative and are readily inactivated by the heat treatment. Hygienically handled meat normally has less that one clostridial spore and <10 bacilli spores/g. Head-meat, diaphragm, and pig-skin frequently carry higher numbers of spores. The main source of spores is usually the non-meat ingredients (e.g. spices, soya, and milk protein, dried blood plasma) (Hechelmann and Kasprowiak, 1992). Raw spices may contain 10^6 spores/g. Decontaminated spices or spice extracts are recommended for the production of canned meat.

The establishment of the thermal process is a specialized field. The effectiveness of the heat treatment is influenced by many factors (e.g. heat transfer coefficient and initial temperature of the food, container size, head space and retort characteristics). The correct thermal process will result in commercial sterility. All *Cl. botulinum* spores and other pathogenic bacteria will be destroyed as well as spores of those microorganisms which could grow and produce spoilage under normal conditions of storage and distribution in temperate climates. Low numbers of thermophilic spores may survive but will not cause spoilage if the product is stored <40°C. F_0 treatments of 12–15 are needed to ensure destruction of spores of thermophilic bacilli and clostridia in products that are stored and distributed in tropical climates.

Post-processing contamination may occur through pin-holes from can defects or external corrosion, through faulty seams from improper manufacture or closure, or through the sealant where the lid is joined to the body. Some containers in the retort may have micro-leaks until cooled. The vacuum that develops in the container may suck cooling water into the product. Similarly, after containers leave the retort, water on the container or on runways may enter the container through openings in the container.

B Spoilage

If the product is spoiled before thermal processing, the heat treatment will kill the causative microorganisms but metabolites will remain to give an unacceptable product.

Spores of putrefactive anaerobes have a relatively high heat resistance. If heat processing is inadequate, surviving spores (e.g. *Cl. sporogenes*) can grow and cause spoilage after storage at ambient temperatures. Thermophilic spoilage may occur if the product is stored above about 43°C.

Microorganisms that gain access to the container after heat processing through pin-holes, seams, and from rough handling can cause a variety of spoilage defects. Generally, such spoilage is characterized by the presence of mixed cultures. However, the widespread use of chlorination of cannery cooling-waters has resulted in spore formers becoming more common as post-processing contaminants.

C Pathogens

Cl. botulinum is the main pathogen of concern. Home, canned or, home bottled meats have been responsible for a significant number of outbreaks of botulism (Tompkin, 1980). Commercially processed meat products have rarely been implicated. Commercially produced beef stew has caused botulism (Tompkin, 1980). Many commercial heat treatments used to prevent spoilage are more severe than needed to ensure a 12-D process for *Cl. botulinum*. Spores of *Cl. perfringens* and *B. cereus* are considerably less heat resistant than those of *Cl. botulinum*, and so an adequate "botulinum cook" will ensure that there are no surviving spores of these potential pathogens. Canned meat products that had been cooled in non-potable river water resulted in outbreaks of typhoid fever when they were imported into the UK (ICMSF, 1980a,b). Salmonellae, as post-processing contaminants, are thus a recognized hazard.

D CONTROL (fully retorted shelf-stable uncured meats)

Summary

Significant hazards[a]	• *Cl. botulinum.* • salmonellae. • *Staph. aureus.*
Control measures	
Initial level (H_0)	• Selection of raw materials.
Reduction (ΣR)	• Control temperature and time during cooking to inactivate spore-formers.

(*continued*)

D CONTROL (Cont.)

Summary

Increase (ΣI)	• Use of appropriate cans (e.g. resistant to corrosion). • Use potable water for processes. • Avoid manual handling of wet cans.
Testing	• No microbiological testing is recommended.
Spoilage	• Selection of raw materials (low spore count). • Control of temperature and time (spore destruction). • Control of can seams (to avoid recontamination).

[a]In particular circumstances, other hazards may need to be considered.

Hazards to be considered. See above.

Control measures. Process control is essential. Codes of practice for the production of low-acid canned foods have been established (e.g. FDA, 1973; CFPRA, 1977; FAO/WHO, 1979; AOAC/FDA, 1984; Thorpe and Barker, 1984), and the CCPs outlined (ICMSF, 1988a,b). Control is needed at pre-processing, thermal processing and post-processing.

Initial level (H_0)

- Material is produced under hygienic conditions to prevent spoilage before retorting and possible increase in spore level.

Reduction (ΣR)

- Containers used are resistant to corrosion from the contents, are of the required size, strength, and seam construction and are undamaged and free from defects.
- Containers are properly filled (e.g. weight, head-space, packing density, vacuum, and temperature of filled product, time between filling and retorting) and container closure equipment is functioning correctly.
- The retort is properly calibrated and operated (e.g. venting, loading pattern and operational parameters like steam pressure, temperature, time, water circulation and chain speed).

Increase (ΣI)

- Cooling is carried out carefully to avoid damage to the container integrity and contamination of the contents: such contamination by *Cl. botulinum*, salmonellae or *Staph. aureus* has occurred.
- Handling equipment and runways are clean and prevent physical shock to the container.
- Avoidance of manual handling of containers until they are dry.

Testing. Microbiological testing is not recommended.

Spoilage

- Selection of raw materials (low spore count).
- Control of temperature and time (spore destruction).
- Control of can seams (to avoid recontamination).

XI Cooked perishable cured meats

Cooked perishable cured meats contain about 125 mg/kg of nitrite before heating, are cooked to about 65–75°C and require refrigerated storage. Pasteurized cured meats may be cooked in the container in which they are marketed, or they may be repacked after cooking. They include such products as pate, bacon, pressed ham and emulsion-style sausages (e.g. frankfurters). They are distributed chilled and may, or may not, be heated before consumption. In some countries (e.g. North America), bacon is mildly heated to about 55°C and is intended to be cooked before serving.

A Effects of processing on microorganisms

Cured meats are made from intact muscles and cuts of meat (e.g. corned beef, ham and bacon); from pieces of meat that have been massaged and tumbled and then formed in casings or molds (e.g. pressed shoulder) or from fully comminuted meats that are extruded into casings or molds (e.g. emulsion-style sausages and pate). The cure is added to the meat by injection of brine, soaking in brine or blending during emulsion preparation. The cure contains sodium nitrite (the amount permitted is usually prescribed by regulation) and salt. Frequently, other components may be added such as sugar, ascorbate, phosphate, spices, and flavor enhancers. Antioxidants, chelates (e.g. citrate), sodium lactate and non-meat extenders also may be incorporated into cooked cured meats. Natural or liquid smoke may be added or applied to some products.

Salt reduces the a_w to about 0.98–0.96 and reduces the growth rate of a number of bacteria. Nitrite, as undissociated nitrous acid, also inhibits microbial growth, with inhibition more marked under anaerobic conditions (e.g. *Broch. thermosphacta*; Egan and Grau, 1981). Lactic acid bacteria are relatively resistant to nitrite. Ascorbate reduces the concentration of nitrite and acts as an oxygen scavenger while, at the same time, increases the anti-botulinal activity of nitrite when used in moderate amounts (ICMSF, 1980a). Alkaline phosphates, like sodium polyphosphate, raise the pH of cured meat. The higher pH results in more residual nitrite in the meat after cooking and in a slower rate of nitrite loss during storage, but in a reduced concentration of the undissociated form. While the addition of sodium lactate has some effect on lowering the a_w, it also has a more specific effect in inhibiting bacterial growth (Houtsma *et al.*, 1993). Both Gram-positive (*Broc. thermosphacta*, lactic acid bacteria, *Listeria* spp. and *Staph. aureus*) and Gram-negative bacteria are sensitive to lactate. Fermentation of sugars and some extenders (e.g. starch) by lactic acid bacteria can cause a significant fall in the pH of stored products.

Microorganisms will be distributed throughout the meat during cure injection, reforming or preparation of the emulsion.

As applies for cooked perishable uncured meats, the commercial processes used will normally destroy parasites, vegetative pathogens, and most vegetative spoilage bacteria. Some of the more heat-resistant streptococci may survive. Spores of bacilli and clostridia are little affected by the mild heat treatment. Products that are mildly heated (e.g. bacon heated to 55°C) may contain surviving pathogens (e.g. *Trich. spiralis* in North America and *Toxoplasma* in Europe). Different processing temperatures are established to meet the needs of the particular process or for microbial stability (Tompkin, 1986). Survival of some heat resistant strains of lactic acid bacteria (e.g. *Lb. viridescens*) can reduce shelf-life and require increased processing temperatures (Borch *et al.*, 1988). Excessive contamination during preparation, or growth before cooking, may result in increased numbers of such heat resistant vegetative bacteria and some may survive heating. Some of the added nitrite is destroyed during cooking and the remainder declines during storage. The rate of nitrite loss is faster at lower pH and at higher temperatures.

After cooking, the products are cooled in a shower of cold water and/or in air. Products sold in the container (e.g. can and plastic film) in which the meat was cooked, will contain microorganisms (e.g. sporeformers and thermotolerant lactic acid bacteria) which have survived the cooking process will be present. Products removed from the container (e.g. casing and mold) and repackaged, or sliced and packed, are subject to contamination during handling, slicing, and packing. The chillers can be a source of contamination with psychrotrophic microorganisms, such as lactic acid bacteria (Borch *et al.*, 1988), *Broc. thermosphacta* and *Listeria* spp. Contaminants can be transferred to the outer casing and from there to product surfaces. During storage of products in chillers, psychrotrophs can grow on the outer casing and exposed tissue of the cooked meat. Brine chill systems can be a significant source of spoilage organisms. Other major sources of contamination are the peeling and slicing machines, conveyor belts and personnel handling product during packing. Compressed or vacuumized air systems for packaging machines can be a major source of contamination. Whereas contamination with lactic acid bacteria may occur early in the post-cooking process (e.g. peeling of frankfurter-style sausages), contamination with Enterobacteriaceae can be associated with hand contact (Dykes *et al.*, 1991). Other products can be major sources of contamination. For example, lactic acid bacteria from fermented sausages that are stored, sliced, and packed in the same area as cooked product can significantly reduce the shelf-life of other cooked products (Makela, 1992). Some products (e.g. pates) may also be decorated after cooking with herbs, spices or gelatin. If these additives are not treated, they can be a major source of spoilage and potentially pathogenic microorganisms.

Products in this category must be stored and distributed under refrigeration (<5°C) to limit spoilage and to prevent the growth of mesophilic pathogens.

B Spoilage

These products are perishable and must be refrigerated. a statement normally appears on the package indicating the need for refrigeration. Products that are sold in the containers in which they are cooked (e.g. canned hams) are semi-perishable. They do not spoil microbiologically unless they are held above about 10°C. Products that are repackaged, sliced or otherwise exposed to contamination after cooking will spoil. The rate of spoilage at refrigeration temperatures (e.g. below about 10°C) is influenced by a variety of factors (Tompkin 1986, 1995).

The type of microbial spoilage and the rate at which it develops are determined by the composition of the product (e.g. a_w-value nitrite, pH, and lactate) and by the cooking and packaging conditions and the temperature of storage.

Product cooked in the final container and not re-contaminated after cooking can have a storage-life of 1–3 years at 0–5°C. However, if the heat treatment is inadequate to destroy relatively heat resistant psychrotrophs like *Lb. viridescens*, then survivors can cause spoilage (souring, gas formation or greening). Similarly, enterococci surviving heat processing can cause spoilage of products held at 5–7°C (Chyr *et al.*, 1981). Under commercial refrigeration conditions spoilage also has been caused by *Cl. putrefaciens*, particularly in products with low salt content (Tompkin, 1986). If stored at ambient temperatures, spoilage can involve a variety of thermotolerant mesophilic bacteria. In particular, mesophilic sporeformers can develop at temperatures >10°C. Surface softening and off-odors from surface growth of *B. cereus* or *B. licheniformis* (Bell and De Lacy, 1983) is dependent on oxygen, and, when films of low permeability are used, is restricted to regions beneath the film near the clip or seam (Bell and De Lacy, 1982).

In products repacked in films of low gas permeability, spoilage at <10°C is normally characterized by souring, discoloration, milky exudate, slime, and/or gas production. Spoilage is mostly by lactic acid bacteria (homo- or heterofermentative lactobacilli or leuconostocs). Slime may be a dextran produced by leuconostocs from sucrose, or be composed of glucose and galactose produced by leuconostocs and *Lb. sake* (Makela *et al.*, 1992). Greening is caused when hydrogen peroxide (produced by lactic acid bacteria in the presence of oxygen) oxidizes the porphyrin ring of nitrosohaemochrome to choleomyoglobin. Surface greening results from contamination and growth at the products surface. Center or core greening results from survival of *Lb. viridescens* in the center of the product. Spoilage by lactic acid bacteria is usually not evident until some time after maximum populations have been attained (Egan *et al.*, 1980), the time for spoilage to become evident being influenced by the temperature of storage.

Broch. thermosphacta can form a significant part of the spoilage flora of vacuum-packed products and spoil them more readily than lactobacilli (Egan *et al.*, 1980; Qvist and Mukherji, 1981). The amount of growth of *Broc. thermosphacta*, relative to that of lactic acid bacteria, is reduced by higher nitrite concentration, lower pH and lower film permeability (Egan and Grau, 1981; Nielsen, 1983a).

Enterobacteriaceae and halophilic vibrios have caused putrefactive spoilage (hydrogen sulfide and methanethiol) of vacuum-packed bacon and ham (Gardner, 1983). This spoilage is associated with high pH muscle (>6.0), low salt content (<4%) and high storage temperatures (>15°C). Growth of psychrotrophic Enterobacteriaceae on vacuum-packed sliced bologna-style sausage is inhibited by nitrite but increased by elevated storage temperatures and by the use of films with higher gas permeability (Nielsen, 1983b). Water activity is an important factor limiting their growth.

C Pathogens

Though cooking destroys vegetative pathogens, they can gain access to products during re-packaging, slicing, and handling. However, provided storage temperatures are kept below about 7°C, only psychrotrophic pathogens (e.g. *L. monocytogenes*) are likely to grow. The growth rate of psychrotrophic pathogens is affected by the a_w-value pH, nitrite, and lactate content, atmosphere, temperature, and the spoilage flora.

Control of growth of *Cl. botulinum* depends among the other factors that affect growth are salt, pH, lactate, type of meat, severity of the heat treatment, initial number of spores (Tompkin, 1986; Lücke and Roberts, 1993). Spores of non-proteolytic strains are likely to be reduced in number or at least sub-lethally damaged by the heat treatment.

There have been many outbreaks of staphylococcal food poisoning from cooked cured meats—particularly from ham. These have usually resulted from contamination of the cooked ham during slicing and then holding the ham at ambient temperature in air for some hours before consumption. This normally occurs at a foodservice establishment, group gatherings (e.g. picnic, church dinner or at home). Vacuum-packed cooked cured meat products are rarely implicated. Growth of *Staph. aureus* in these products is prevented by the combination of salt, nitrite, pH, anaerobic conditions and low storage temperatures. In addition, the faster growing lactic acid bacterial population in repacked products will limit the extent of staphylococcal growth even if the products are held at ambient temperature.

Similarly, the ingredients in cooked cured meats can retard the growth of salmonellae, *Y. enterocolitica,* and *B. cereus*, but they may not always prevent it under abuse conditions. Thus, though pathogenic serotypes of *Y. enterocolitica* did not grow at 27°C on some cooked cured meats (Raccach and Henningsen, 1984), on other vacuum-packed cooked cured meat the organisms grew slowly at 2°C and, at 5°C, increased by 100 000-fold (in the absence of other flora) in 14 days (Nielsen and Zeuthen, 1984). Cooked cured meats have not been implicated in human yersiniosis. Salmonellae are able to grow from about 12°C and *B. cereus* at from 8–15°C in the products they studied (Nielsen and Zeuthen, 1984; Asplund *et al.*, 1988). *Cl. perfringens* seems incapable of growing on vacuum-packed cooked cured meats (Nielsen and Zeuthen, 1984; Tompkin, 1986).

Pate (McLauchlin *et al.*, 1991) and jellied pork tongues (Goulet *et al.*, 1993; Savat *et al.*, 1995) have been implicated in widespread cases of listeriosis. *L. monocytogenes* can be expected to be destroyed during cooking (Mackey *et al.*, 1990; Zaika *et al.*, 1990) but can contaminate products after cooking. Listeriae can multiply in the moist environment of processing plants (e.g. floors, drains, chiller walls, and doors, frankfurter-peelers). Care is needed when cleaning to reduce the number of listeriae in the environment and when handling cooked product to prevent transfer of *L. monocytogenes* to the product. *L. monocytogenes* can grow on many chilled vacuum-packed cooked perishable cured meats (Glass and Doyle, 1989b; Schmidt and Kaya, 1990). The rate of growth depends on pH, a_w, nitrite content, and temperature. Growth of listeriae will be highest when the pH is near neutral, there is little nitrite and the a_w is ~0.99.

Recognition of *E. coli* O157:H7 as a serious hazard in meat products, and a lack of knowledge about its survival in many traditional processes has resulted in many publications. Chikthimmah and Knabel (2001) spread *E. coli* O157:H7, *S.* Typhimurium or *L. monocytogenes* on the surface of Lebanon bologna luncheon slices that were stacked, vacuum-packaged and stored at 3.6 or 13°C. Numbers of each pathogen declined during storage at 3.6 or 13°C, the higher temperature resulting in significantly faster destruction of *E. coli* O157:H7 and *L. monocytogenes*, with *E. coli* O157:H7 the most resistant to destruction.

Lebanon bologna is often processed at temperatures that do not exceed 48.8°C (120°F) to maximize eating quality. Chikthimmah *et al.* (2001a) investigated the influence of sodium chloride on growth of lactic acid bacteria and the subsequent destruction of *E. coli* O157:H7. Fermentation to pH 4.7 at 37.7°C reduced populations of *E. coli* O157:H7 by ca. 0.3 $\log_{10}$, in the presence or absence of curing salts. Subsequent destruction of *E. coli* O157:H7 during heating of the fermented product to 46.1°C was significantly reduced by the presence of 3.5% NaCl and 156 ppm $NaNO_2$, compared to product without curing salts. NaCl (5%) inhibited the growth of LAB, giving product with pH ca. 5.0, and further reduced the destruction of *E. coli* O157:H7. Lower concentrations of NaCl (0, 2.5%) resulted in bologna with higher counts of lactic acid bacteria and lower pH values. Low pH directly influenced destruction of *E. coli* O157:H7.

After numerous product recalls and one devastating outbreak of listeriosis in ready-to-eat product that claimed 21 lives, the FSIS published a federal register notice requiring manufacturers of ready-to-eat meat and poultry products to reassess their hazard analysis and critical control point plans for these products as specified in 9 CFR 417.4(a). Lebanon bologna is a moist, fermented ready-to-eat sausage. Because of undesirable quality changes, Lebanon bologna is often not processed >48.9°C (120°F). Research to validate the destruction of *L. monocytogenes* in Lebanon bologna batter in a model system (Chikthimmah *et al.*, 2001b) indicated that fermentation to pH 4.7 significantly reduced *L. monocytogenes* by 2.3 $\log_{10}$ cfu/g of the sausage mix ($P<0.01$). Heating the fermented mix to 48.9°C in 10.5 h destroyed at least 7.0 $\log_{10}$ cfu of *L. monocytogenes* per g of sausage mix. A combination of low pH (≤5.0) and heating to 43.3°C (115°F) or above destroyed >5 $\log_{10}$ cfu of *L. monocytogenes* per g of sausage mix during the processing of Lebanon bologna, validating the effectiveness of the existing commercial process, previously validated for destruction of *E. coli* O157:H7, for the destruction of >5 $\log_{10}$ cfu of *L. monocytogenes*.

Tompkin (2002) provides guidance to food processors in controlling *L. monocytogenes* in food-processing environments. The advice is based on experience that outbreaks of a few to several hundred scattered cases can result from an unusually virulent strain becoming established in a food-processing environment and contaminating multiple lots of food over days, or even months, of production. The information provides the basis for establishing an environmental sampling program, the organization and interpretation of the data generated by that program, and the action to be taken in the event of *Listeria*-positive results.

D CONTROL (cooked, perishable cured meats)

Summary

Significant hazards[a]	• Salmonellae. • VTEC (beef). • *Staph. aureus.* • *L. monocytogenes.*
Control measures	
Initial level (H_0)	• Selection and microbiological quality of raw materials.
Reduction (ΣR)	• Cooking to eliminate vegetative cells of bacterial pathogens.
Increase (ΣI)	• GHP to minimize post-process contamination (avoid mixing raw materials and cooked product, cleaning, and disinfection programs, training of personnel). • Formulation (nitrite and salt content). • Time–temperature at all stages.
Testing	• Testing for APC can be used for equipment (surveillance of hygiene) and products (just after packaging) as a measure of hygiene. • *L. monocytogenes* should be surveilled in the processing environment. • If no prior knowledge, testing product for *L. monocytogenes* may be used.
Spoilage	• Temperature and time of storage.

[a]In particular circumstances, other hazards may need to be considered.

Hazards to be considered. See above.

Control measures

Initial level (H_0)

• Time and temperature from preparation to cooking.

Reduction (ΣR)

• Cooking and cooling rates and target internal temperatures.

Increase (ΣI)

• Physical separation of raw materials from cooked products.
• Cleanliness of chillers and equipment used with cooked meat (e.g. slicers and conveyor belts).
• Cooked meat handled by trained staff in such a way as to minimize product contamination.
• Type of packaging and package integrity.
• Storage temperature.

Testing

• APC (30°C) should be used for shelf-life control.
• *Listeria*/*L. monocytogenes* monitoring of environmental and on line samples is also recommended.

XII Shelf-stable cooked cured meats

Shelf-stable cooked cured meats are sold in hermetically sealed containers after receiving a thermal process that is less than a "botulinum" cook of $F_0 = 2.5$. Stability and safety depend on the combined effect of heating and other factors that inhibit the growth of surviving spores, and a sealed container that prevents post-processing contamination.

A Effects of processing on microorganisms

For shelf-stable canned cured meats, the heat process destroys vegetative microorganisms, destroys some spores and sub-lethally damages other spores. Inhibition of the surviving spores is achieved primarily by the combined effect of salt and nitrite (Tompkin, 1986, 1993; Hauschild and Simonsen, 1985, 1986; Farkas and Andrássy, 1992; Lücke and Roberts, 1993). Safety and stability depends upon the combined effect of thermal destruction or injury of a low indigenous number of spores and inhibition of the survivors by an adequate amount of added salt and sodium nitrite. Spore counts in the raw meat mixture should be low (<3 clostridial spores/g or <100 mesophilic bacillary spores/g) (Hauschild and Simonsen, 1985, 1986). Guidelines for the heat treatments (F_0 values) and salt and nitrite concentrations for the safe production of shelf-stable canned cured ham, luncheon meat, and sausage have been developed (Hauschild and Simonsen, 1985, 1986).

Guidelines for liver, blood, and bologna-style sausages that do not require refrigeration have been developed (Hechelmann and Kasprowiak, 1992; Lücke and Roberts, 1993). Important factors to control are initial spore load, heat treatment, pH, a_w, and nitrite. For products, like Italian mortadella and German bruhdauerwurst, stability is achieved by heating to >75°C to inactivate vegetative cells, reducing the a_w to <0.95, heating in a sealed container or film to prevent re-contamination and maintaining a low redox potential to prevent the growth of the more a_w tolerant bacilli (Hechelmann and Kasprowiak, 1992; Lücke and Roberts, 1993).

Brawns are made shelf-stable by adjusting the pH to 5.0 with acetic acid and protecting the product from recontamination after heating. Gelder smoked sausage (a traditional Dutch product) is made shelf-stable by adjusting the pH to 5.4–5.6 with glucono-delta-lactone, reducing the a_w to 0.97, vacuum-packing and heating for 1 h to a center temperature of 80°C (Hechelmann and Kasprowiak, 1992; Lücke and Roberts, 1993).

B Spoilage

Spoilage is normally due to post-processing contamination through leaks in the container (e.g. in the seams of cans or through the clip-seals of plastic casings), or from the growth of *Bacillus* spp. at the surface. The extent of growth is determined mainly by the oxygen permeability of the casing or container.

C Pathogens

The major microorganism of concern is *Cl. botulinum*. The heat treatments and compositional formulae of the products in this class are designed to prevent growth of this organism. Other possible pathogens are vegetative that contaminate the product after heat treatment. Cans of corned beef contaminated by being cooled in river water have caused typhoid fever (Meers and Goode, 1965; Milne, 1967).

D CONTROL (shelf-stable cooked cured meats)

Summary

Significant hazards[a]	• *Cl. botulinum.* • salmonellae. • VTEC. • *Staph. aureus.*
Control measures	
Initial level (H_0)	• Selection of raw materials. • Microbial quality (spore load) of the raw ingredients. • Time and temperature from preparation to cooking
Reduction (ΣR)	• Control of temperature and time during cooking and cooling. • Cooking and cooling rates and target internal temperatures.
Increase (ΣI)	• Formulation (salt, nitrite and pH) to ensure multiplication is prevented. • Hermetically sealed packaging to prevent re-contamination. • Use water of appropriate quality. • Avoid manual handling of wet cans (risk of contamination if seals are faulty). • Prevention of damage to container integrity and of post-cooking contamination of contents.
Testing	• No routine microbiological testing of product is recommended. • Microbiological testing of some ingredients is a useful monitor of spore load. • Routine examination of can seams is recommended.
Spoilage	• Selection of raw materials to avoid lots with many spore-formers. • Temperature and time of cooking and cooling.

[a]In particular circumstances, other hazards may need to be considered.

Hazards to be considered. Food-borne illness has been traced to recontamination of the cooked product during cooling, e.g. with *S. typhi* in contaminated, unchlorinated and cooling water.

In general, the steps needed to ensure the microbial stability of shelf-stable cooked cured meats are similar to those for fully retorted "botulinum cook" meats.

XIII Snails

Snails are an important part of several food chains, surviving mainly on decaying vegetation and being consumed as food by fish and a variety of animals. Snails are considered a delicacy in many countries and are a staple part of the diet in parts of Asia where red meat and poultry are scarce sources of protein.

A Definition

Many varieties of edible snails are used. However, the Moroccan snail (*Helix aspersa*) is especially favored in Western European and Mediterranean countries and *Helix pomatia* in Central and Alpine

Europe (Andrews *et al.*, 1975). The tropical species, *Achtina fulica*, is processed largely for African and Southwest Asian markets.

B Production and processing

Snails are usually obtained from swamps and marshes, may contain parasites and pathogenic bacteria that they cannot purge. Several countries such as the UK and France are now raising snails by intensive farming in isolated trays, cages or pens (Daguzan, 1985; Runham, 1989). Snails are normally active only when feeding, and surround themselves with a membranous sheath while dormant (Andrews and Wilson, 1975). In the presence of moisture, they penetrate the sheath and emerge from their shells to forage for food. Snails that are imported are typically closely packed under conditions where there is ample moisture, hence the potential for cross-contamination is great. Snails inherently have high populations of indigenous bacteria and coliforms.

Fresh snails are usually eaten after cooking in hot water and prepared according to many regional recipes. Fresh or frozen snails are used as raw materials by the food industry for manufacturing traditional canned products. Ready-to-cook or cooked specialties such as snails in butter dressing with parsley and fricassee are being marketed. In addition, snail eggs prepared through an aromatic brining process are considered by some to be a gourmet delicacy (Pos, 1990).

C Pathogens

Snails are vectors of many parasites, including *Clonorchis sinensis, Eschinastoma ilocanum, Fasciola hepatica, Fasciola gigantica, Fasciolopsis buski, Opisthorchis felineus* and *Paragonimus mestermani* (Bryan, 1977), and pathogenic bacteria. Surveys of both food and aquarium snails have revealed *Salmonella* is a frequent contaminant, being isolated from 62% of *Achatina achatina* (edible land snail) from Nigeria (Obi and Nzeako, 1980), 43% of *H. aspersa* from Morocco (Andrews and Wilson, 1975) and 27% of *Ampullaria* spp. (aquarium snails) from Canada and Florida (Bartlett and Trust, 1976). *Salmonella* contamination of snails occurs predominantly on the surface of the shell and snail meat, but the pathogen may also penetrate the snail flesh (Andrews and Wilson, 1975). In addition, *A. hydrophila, Shigella* and *Arizona* spp. were isolated from 72, 38 and 46%, respectively, of *Achatina achatina* from Nigeria (Obi and Nzeako, 1980). High populations of coliforms (1.4×10^6/g) and fecal coliforms (1.2×10^3/g) have been found on aquarium snails (Bartlett and Trust, 1976).

D CONTROL (snails)

Summary

Significant hazards[a]	• Salmonellae. • *Shigella.* • VTEC. • Parasites.
Control measures	
Initial level (H_0)	• Contamination is not under control.
Reduction (ΣR)	• Snails should always be cooked before consumption. • If the concern is parasites, freezing is recommended.

(continued)

D CONTROL (Cont.)

Summary

Increase (ΣI)	• If cooked snails are replaced in the shells before serving, the shells should be heated to inactivate salmonellae, etc.
Testing	• Testing for salmonellae before or after harvest gives an indication of levels of contamination.
Spoilage	• Not relevant.

[a]In particular circumstances, other hazards may need to be considered.

Control measures. Snails are a common vehicle of parasitic or bacterial infections (e.g. *Salmonella, Shigella*) and should not be eaten raw or undercooked. Control measures include proper cooking and avoidance of recontamination after cooking. Testing for salmonellae may be appropriate using the method described by Andrews and Wilson (1975).

Parasites present in snails are readily killed by cooking or freezing, hence heating sufficiently to kill salmonellae also will kill parasites. Cooking snails according to typical gourmet recipes kills salmonellae (Andrews and Wilson, 1975). Handling raw snail meat and shells can potentially cause cross-contamination of surfaces in the kitchen and of other prepared foods. Many recipes suggest returning cooked snails to their original shells for serving (Andrews *et al.*, 1975). Normally this does not pose a hazard, because before insertion of the snail meat the shells are heated. This kills any salmonellae not removed during cleaning and washing of the shells.

XIV Froglegs

Froglegs have become an important food in international trade because they constitute a valuable export item from developing countries, especially in the Far East. In 1994, half a million kilos (value 3.8 million US Dollars) of froglegs were imported into the United States and between 1988 and 1992 froglegs with a commercial value of 30 million US dollars were imported into the European Union. Most of such imports originate in Indonesia. *Salmonella* in the product has been a matter of concern to public health authorities in both exporting and importing countries and many contaminated lots have been rejected (Pantaleon and Rosset, 1964; Andrews *et al.*, 1977; Shrivastava, 1977).

A Definition

Frogs of commercial significance include species such as *Rana tigrina* (the large Bullfrog).

B Production and processing

Frogs are harvested, often in the wild, and taken to be slaughtered. The processing procedure usually includes several steps to reduce the number of bacteria and to avoid cross-contamination (Garm, 1976). To remove soil, feces, slime, and other dirt before slaughter, live frogs are washed in holding tanks containing fresh water. The live frogs are then immersed into a 10% solution of salt to paralyze and anesthetize them; the solution frequently also contains 250 ppm of chlorine to reduce the superficial microbial level. After sacrificing, the hind-legs are cut at the abdomen above the waist such that the intestines are left intact. The legs are immediately washed in a 5% solution of salt containing up to

500 ppm of chlorine to remove blood and all extraneous material. To be presented skinned, the skin and the remaining portion of the cloaca are removed, the legs washed three times with a 200 ppm chlorine solution, trimmed by removing nails, loose flesh, and blood vessels and washed again several times with up to 250 ppm chlorine in order to reduce the number of salmonellae (Necker *et al.*, 1978; Rao *et al.*, 1978). The legs are finally wrapped in polyethylene bags and frozen.

C Pathogens

It is well known that reptiles and amphibia frequently harbor salmonellae (Ang *et al.*, 1973; Sharma *et al.*, 1974; Bartlett *et al.*, 1977; Shrivastava, 1978) and during transport from the point of capture to the local slaughterhouse, the level of contamination will increase.

As *Salmonella* will remain as a common organism in the regions of catch and as contamination during processing can be minimized but not eliminated, *Salmonella* will continue to be found in froglegs. Moreover, even when the production of froglegs is carried out under the most hygienic conditions, elimination of all salmonellae is difficult to achieve (Nickelson *et al.*, 1975). No other pathogens are a recognized concern.

D CONTROL (froglegs)

Summary

Significant hazards[a]	• salmonellae.
Control measures	
Initial level (H_0)	• Use healthy frogs of good quality. • Minimize time between capture and slaughter. • Use chlorinated water to reduce contamination. • Separate intact intestinal tract. • Clean and sanitize equipment and contact surfaces.
Increase (ΣI)	• Keep froglegs cooled during preparation and refrigerate product to <4°C quickly.
Testing	• Testing for salmonellae is a useful monitor of initial contamination.
Spoilage	• Temperature and time of storage.

[a]In particular circumstances, other hazards may need to be considered.

Hazards to be considered. See above.

Control measures

Initial level (H_0)

- Initial contamination reflects the growing environment, which is not under control, and hygiene after harvesting.

Reduction (ΣR)

- Irradiation, where permitted, effectively eliminates salmonellae (Rao *et al.*, 1978; Necker *et al.*, 1978), and *Vibrio cholerae* 01 (Sang *et al.*, 1987).

- *Salmonella* contamination of froglegs is derived from superficial contamination of live animals and from contamination during preparation. Unless froglegs are kept cooled during preparation and storage, *Salmonella* growth may occur. Control is therefore based on preventing contamination during preparation, reducing superficial contamination and temperature control.
- A rinsing method to determine the surface contamination by *Salmonella* has been advocated as there is danger of cross-contamination to other foodstuffs although blending will yield higher numbers of salmonella contaminated legs (Andrews *et al.*, 1977).

References

Abdul-Raouf, V.M., Beuchat, L.R. and Ammar, M.S. (1993) Survival and growth of *Escherichia coli* 0157:H7 in ground roasted beef as affected by pH, acidulants and temperature. *Appl. Environ. Microbiol.*, **59**, 2364–8.

Abouzeed, Y.M., Hariharan, H., Poppe, C. and Kibenge, F.S.B. (2000) Characterization of *Salmonella* isolates from beef cattle, broiler chickens and human sources on Prince Edward Island. *Comp. Immunol. Microbiol. Infect. Dis.*, **23**, 253–66.

ACMSF (Advisory Committee on the Microbiological Safety of Food, UK). (1992) Report on vacuum packaging and associated processes, advisory committee on the microbiological safety of foods. HMSO, London.

ACMSF (Advisory Committee on the Microbiological Safety of Food, UK). (1995) Report on verocytotoxin-producing *Escherichia coli*. HMSO, London. ISBN 0 11 321909 1.

Adesiyun, A.A. and Krishnan, C. (1995) Occurrence of *Yersinia enterocolitica* O:3, *Listeria monocytogenes* O:4 and thermophilic *Campylobacter* spp. in slaughter pigs and carcasses in Trinidad. *Food Microbiol.*, **12**, 99–107.

AFSDAC. (1990) Agriculture Canada—Canadian code of recommended practices for pasteurised/modified atmosphere packaged/refrigerated food, Agri-Food Safety Division, Agriculture Canada, March, 1990.

Aguirre, M. and Collins, M.D. (1993) Lactic acid bacteria and clinical infection. *J. Appl. Bacteriol.*, **75**(2), 95–107.

Ali, M.S. and Fung, D.Y.C. (1991) Occurrence of *Clostridium perfringens* in ground beef and ground turkey evaluated by three methods. *J. Food Safety*, **11**, 197–203.

American Meat Institute. (1982) Good manufacturing practices—fermented dry and semi-dry sausage, American Meat Institute, Washington, DC.

Andersen, J.K. (1988) Contamination of freshly slaughtered pig carcasses with human pathogenic *Yersinia enterocolitica*. *Int. J. Food Microbiol.*, **7**, 193–202.

Anderson, G.D. and Lee, D.R. (1976) *Salmonella* in horses: a source of contamination of horsemeat in a packing plant under federal inspection. *Appl. Environ. Microbiol.*, **31**, 661–3.

Anderson, E.S., Galbraith, N.S. and Taylor, C.E.D. (1961) An outbreak of human infection due to *Salmonella typhimurium* phage-type 20a associated with infection in calves. *Lancet*, **i**, 854–8.

Anderson, M.E., Marshall, R.T., Stringer, W.C. and Nauman, H.D. (1977) Combined and individual effects of washing and sanitizing on bacterial counts on meat—a model system. *J. Food Prot.*, **40**, 670–88.

Andersen, J.K., Sorensen, R. and Glensbjerg, M. (1991) Aspects of the epidemiology of *Yersinia enterocolitica*: a review. *Int. J. Food Microbiol.*, **13**, 231–8.

Anderson, R.C., Buckley, S.A., Callaway, T.R., Genovese, K.J., Kubena, L.F., Harvey, R.B. and Nisbet, D.J. (2001) Effect of sodium chlorate on *Salmonella typhimurium* concentrations in the weaned pig gut. *J. Food Prot.*, **64**(2), 255–8.

Andrews, W.H. and Wilson, C.R. (1975) Thermal susceptibility of *Salmonella* in the Moroccan food snail, *Helix aspersa*. *J. Assoc. Off. Anal. Chem.*, **58**, 1159–61.

Andrews, W.H., Wilson, C.R., Romero, A. and Poelma, P.L. (1975) The Moroccan food snail, *Helix aspersa*, as a source of *Salmonella*. *App. Microbiol.*, **29**, 328–30.

Andrews, W.H., Wilson, C.R., Poelma, P.L. and Romero, A. (1977) Comparison of methods for the isolation of *Salmonella* from imported frog legs. *Appl. Environ. Microbiol.*, **33**, 65–8.

Ang, O., Ozek, O., Cetin, E.T. and Toreci, K. (1973) *Salmonella* serotypes isolated from tortoises and frogs in Istanbul. *J. Hyg.*, **71**, 85–8.

Anonymous. (1995a) Community outbreak of haemolytic uremic syndrome attributable to *Escherichia coli* 0111:NM–South Australia, 1995. *MMWR*, **44**, 550, 551, 557, 558.

Anonymous. (1995b) *Escherichia coli* 0157:H7 outbreak linked to commercially distributed dry-cured salami—Washington and California 1994. *MMWR*, **44**, 157–60.

AOAC/FDA (Association of Official Analytical Chemists/United States Food and Drug Administration). (1984) Classification of visible can defects (exterior). Association of Official Analytical Chemists, Arlington.

AQIS. (1992) Australian Quarantine and Inspection Service, Code of Hygienic Practice for the Manufacture of Sous-Vide Products, AQIS, Department of Primary Industries and Energy, Australia.

Asplund, K., Nurmi, E., Hill, P. and Hirn, J. (1988) The inhibition of growth of *Bacillus cereus* in liver sausage. *Int. J. Food Microbiol.*, **7**, 349–52.

Atabay, H.I. and Corry, J.E. (1998) The isolation and prevalence of campylobacters from dairy cattle using a variety of methods. *J. Appl. Microbiol.*, **84**(5), 733–40.

Atalla, H.N., Johnson, R., McEwen, S., Usborne, R.W. and Gyles, C.L. (2000) Use of a Shiga toxin (Stx)-enzyme-linked immunosorbent assay and immunoblot for detection and isolation of Stx-producing *Escherichia coli* from naturally contaminated beef. *J. Food Prot.*, **63**(9), 1167–72.

Autio, T., Sateri, T., Fredriksson-Ahomaa, M., Rahkio, M., Lunden, J. and Korkeala, H. (2000) *Listeria monocytogenes* contamination pattern in pig slaughterhouses. *J. Food Prot.*, **63**(10), 1438–42.

Bacon, R.T., Belk, K.E., Sofos, J.N., Clayton, R.P., Reagan, J.O. and Smith, G.C. (2000) Microbial populations on animal hides and beef carcasses at different stages of slaughter in plants employing multiple-sequential interventions for decontamination. *J. Food Protect.*, **63**, 1080–6.

Bacus, J. (1984) *Utilization of Microorganisms in Meat Processing. A Handbook for Meat Plant Operators*, John Wiley & Sons, Inc., New York.

Bacus, J.N. (1986) Fermented meat and poultry products, in *Advances in meat research* (eds A.M. Pearson and T.R. Dutson), *volume 2, Meat and Poultry Microbiology*, AVI Publishing Co., Inc., Westport, CT, pp.123–64.

Baffone, W., Bruscolini, F., Pianetti, A., Biffi, M.R., Brandi, G., Salvaggio, L. and Albano, V. (1995) Diffusion of thermophilic *Campylobacter* in the Pesaro-Urbino area (Italy) from 1985 to 1992. *Eur. J. Epidemiol.*, **11**, 83–6.

Baggesen, D.L., Wegener, H.C., Bager, F., Stege, H. and Christensen, J. (1996) Herd prevalence of *Salmonella enterica* infections in Danish slaughter pigs determined by microbiological testing. Prevent. *Vet. Med.*, **26**, 201–13.

Bailey, C. (1971) Spray washing of lamb carcasses. in *Proceedings, 17th European Meeting of Meat Research Workers*, Bristol, pp. 175–81.

Bailey, C., James, S.J., Kitchell, A.G. and Hudson, W.R. (1974) Air-, water- and vacuum-thawing of frozen pork legs. *J. Sci. Food Agric.*, **25**, 81–97.

Barkocy-Gallagher, G.A., Arthur, T.M., Siragusa, G.R., Keen, J.E., Elder, R.O., Laegreid, W.W. and Koohmaraie, M. (2001) Genotypic analyses of *Escherichia coli* O157:H7 and O157 nonmotile isolates recovered from beef cattle and carcasses at processing plants in the Midwestern States of the United States. *Appl. Environ. Microbiol.*, **67**(9), 3810–8.

Bartlett, K.H. and Trust, T.J. (1976) Isolation of salmonellae and other potential pathogens from the freshwater aquarium snail *Ampullaria. Appl. Environ. Microbiol.*, **31**, 635–9.

Bartlett, K.H., Trust, T.J. and Lior, H. (1977) Small pet aquarium frogs as a source of *Salmonella. Appl. Environ. Microbiol.*, **33**, 1026–9.

Bates, J.R. and Bodnaruk, P.W. (2003) *Clostridium perfringens*, in *Foodborne Microorganisms of Public Health Significance*, 6th edn (ed A.D. Hocking), Australian Institute of Food Science and Technology Ltd (NSW Branch), Food Microbiology Group, Waterloo DC, New South Wales, Australia, pp. 479–504.

Bauer, F.T., Carpenter, J.A. and Reagan, J.O. (1981) Prevalence of *Clostridium perfringens* in pork during processing. *J. Food Prot.*, **44**, 279–83.

Baumgartner, A. and Grand, M. (1995) Detection of verotoxin-producing *Escherichia coli* in minced beef and raw hamburgers: comparison of polymerase chain reaction (PCR) and immunomagnetic beads. *Arch. Lebensmittelhyg.*, **46**, 127–30.

Bell, R.G. and De Lacy, K.M. (1982) The role of oxygen in the microbial spoilage of luncheon meat cooked in a plastic casing. *J. Appl. Bacteriol.*, **53**, 407–11.

Bell, R.G. and De Lacy, K.M. (1983) A note on the microbial spoilage of undercooked chub-packed luncheon meat. *J. Appl. Bacteriol.*, **54**, 131–4.

Ben Omar, N., Castro, A., Lucas, R., Abriouel, H., Yousif, N.M., Franz, C.M., Holzapfel, W.H., Perez-Pulido, R., Martinez-Canemero, M. and Galvez, A. (2004) Functional and safety aspects of Enterococci isolated from different Spanish foods. *Syst. Appl. Microbiol.*, **27**(1), 118–30.

Benelux. (1991) Code for the Production, Distribution and Sale of Chilled Longlife Pasteurised Meals.

Benjamin, M.M. and Datta, A.R. (1995) Acid tolerance of enterohemorrhagic *Escherichia coli. Appl. Environ. Microbiol.*, **61**, 1669–72.

Berends, B.R., Van Knapen, F., Snijders, J.M.A. and Mossel, D.A.A. (1997) Identification and quantification of risk factors regarding *Salmonella* spp. on pork carcasses. *Int. J. Food Microbiol.*, **36**, 199–206.

Bergdoll, M.S. (1989) *Staphylococcus aureus*, in *Food Borne Bacterial Pathogens* (ed M.P. Doyle), Marcel Dekker Inc., New York, pp. 463–523.

Besser, T.E., Hancock, D.D., Pritchett, L.C., McRae, E.M., Rice, D.H. and Tarr, P.I. (1997) Duration of detection of fecal excretion of *Escherichia coli* O157:H7 in cattle. *J. Infect. Dis.*, **175**, 726–9.

Beutin, L., Geier, D., Steinruck, H., Zimmermann, S. and Scheutz, F. (1993) Prevalence and some properties of verotoxin (Shiga-like toxin)-producing *Escherichia coli* in seven different species of healthy domestic animals. *J. Clin. Microbiol.*, **31**, 2483–8.

Biemuller, G.W., Carpenter, J.A. and Reynolds, A.E. (1973) Reduction of bacteria on pork carcasses. *J. Food Sci.*, **38**, 261–3.

Blanco, J.E., Blanco, M., Mora, A., Prado, C., Rio, M., Fernandez, L., Fernandez, M.J., Sainz, V. and Blanco, J. (1996a). Detection of enterohaemorrhagic *Escherichia coli* O157:H7 in minced beef using immunomagnetic separation. *Microbiologia*, **12**, 385–94.

Blanco, M., Blanco, J.E., Blanco, J., Gonzalez, E.A., Mora, A., Prado, C., Fernandez, L., Rio, M., J. and Alonso, M.P. (1996b) Prevalence and characteristics of *Escherichia coli* serotype O157:H7 and other verotoxin-producing *E. coli* in healthy cattle. *Epidemiol. Infect.*, **117**, 251–7.

Blanco, J., Gonzalez, E.A., Garcia, S., Blanco, M., Regueriro, B. and Bernardez, I. (1988) Production of toxins by *Escherichia coli* strains isolated from calves with diarrhoea in Galicia (North-western Spain). *Vet. Microbiol.*, **18**, 297–311.

Bliss, M.E. and Hathaway, S.C. (1996a) Effect of pre-slaughter washing of lambs on the microbiological and visible contamination of the carcases. *Vet. Rec.*, **138**, 82–6.

Bliss, M.E. and Hathaway, S.C. (1996b) Microbiological contamination of ovine carcasses associated with the presence of wool and faecal material. *J. Appl. Bacteriol.*, **81**, 594–600.

de Boer, E. and Nouws, J.F.M. (1991) Slaughter pigs and pork as a source of human pathogenic *Yersinia enterocolitica. Int. J. Food Microbiol.*, **12**, 375–8.

Boireau, P., Vallée, I., Roman, T., Perret, C., Mingyuan L., Gamble, H.R. and Gajadhar, A. (2000) *Trichinella* in horses: a low frequency infection with high human risk. *Parasitology*, **93**, 309–20.

Bolton, F.J., Dawkins, H.C. and Hutchinson, D.N. (1985) Biotypes and serotypes of thermophilic campylobacters isolated from cattle, sheep and pig offal and other red meats. *J. Hyg. (Camb.)*, **95**, 1–6.

Bolton, D., Doherty, A and Sheridan, J. (2001) Beef HACCP: intervention and non-intervention systems. *Int. J. Food Microbiol.*, **66**, 119–129.

Bolton, F.J., Dawkins, H.C. and Hutchinson, D.N. (1985) Biotypes and serotypes of thermophilic campylobacters isolated from cattle, sheep and pig offal and other red meats. *J. Hyg. (Camb.)*, **95**, 1–6.

Bonardi, S., Maggi, E., Bottarelli, A., Pacciarini, M.L., Ansuini, A., Vellini, G., Morabito, S. and Caprioli, A. (1999) Isolation of verocytotoxin-producing *Escherichia coli* O157:H7 from cattle at slaughter in Italy. *Vet. Microbiol.*, **67**, 203–11.

Borch, E. and Arvidsson, B. (1996) Growth of *Yersinia enterocolitica* O:3 in pork, in *Proceedings, Food Associated Pathogens, The International Union of Food Science and Technology*, Uppsala, Sweden, pp. 202–3.

Borch, E., Nerbrink, E. and Svensson, P. (1988) Identification of major contamination sources during processing of emulsion sausage. *Int. J. Food Microbiol.*, **7**, 317–30.

Bover-Cid, S., Izquierdo-Pulido, M. and Vidal-Carou, M.C. (2000) Mixed starter cultures to control biogenic amine production in dry fermented sausages. *J. Food Prot.*, **63**(11), 1556–62.

Bover-Cid, S., Izquierdo-Pulido, M. and Vidal-Carou, M.C. (2001) Effectiveness of a *Lactobacillus sakei* starter culture in the reduction of biogenic amine accumulation as a function of the raw material quality. *J. Food Prot.*, **64**(3), 367–73.

Brackett, R.E., Hao, Y.-Y., Doyle, M.P. (1994) Ineffectiveness of hot acid sprays to decontaminate *Escherichia coli* O157:H7 on beef. *J. Food Prot.*, **57**, 198–203.

Bredholt, S., Nesbakken, T. and Holck, A. (1999) Protective cultures inhibit growth of *Listeria monocytogenes* and *Escherichia coli* O157:H7 in cooked, sliced vacuum- and gas-packaged meat. *Int. J. Food Microbiol.*, **53**, 43–52.

Brett, M.M. (1998) 1566 Outbreaks of *Clostridium perfringens* food poisoning, 1970–1996, in *Proceedings, 4th World Congress on Foodborne Infections and Intoxications, volume 1*, Berlin, pp. 243–244.

Brett, M.M. and Gilbert, R.J. (1997) 1525 Outbreaks of *Clostridium perfringens* food poisoning, 1970–1996. *Review of Medical Microbiology*, **8** (suppl. 1), S64–5.

Broda, D.M., De Lacy, K.M., Bell, R.G. and Penney N. (1996a) Association of psychrotrophic *Clostridium* spp. with deep tissue spoilage of chilled vacuum-packed lamb. *Int. J. Food Microbiol.*, **29**(2–3), 371–8.

Broda, D.M., De Lacy, K.M., Bell, R.G., Braggins, T.J. and Cook, R.L. (1996b) Psychrotrophic *Clostridium* spp. associated with 'blown pack' spoilage of chilled vacuum-packed red meats and dog rolls in gas-impermeable plastic casings. *Int. J. Food Microbiol.*, **29**(2–3), 335–52.

Broda, D.M., De Lacy, K.M. and Bell, R.G. (1998a) Influence of culture media on the recovery of psychrotrophic *Clostridium* spp. associated with the spoilage of vacuum-packed chilled meats. *Int. J. Food Microbiol.*, **39**(1–2), 69–78.

Broda, D.M., De Lacy, K.M. and Bell, R.G. (1998b) Efficacy of heat and ethanol spore treatments for the isolation of psychrotrophic *Clostridium* spp. associated with the spoilage of chilled vacuum-packed meats. *Int. J. Food Microbiol.*, **39**(1–2), 61–8.

Broda, D.M., Lawson, P.A., Bell, R.G. and Musgrave, D.R. (1999) *Clostridium frigidicarnis* sp. nov., a psychrotolerant bacterium associated with "blown pack" spoilage of vacuum-packed meats. *Int. J. Sys. Bacteriol.*, **49**, 1539–50.

Brown, M. (ed). (2000) HACCP in the meat industry. Woodhead Publishing Ltd., Cambridge, UK. (ISBN 1 85573 448 6) pp 344.

Bryan, F.L. (1977) Diseases transmitted by foods contaminated by waste water. *J. Food Prot.*, **40**, 45–56.

Bryan, F. (1980) Foodborne diseases in United States associated with meat and poultry. *J. Food Prot.*, **43**, 140–50.

Bryner, J.H., O'Berry, P.A., Estes, P.C. and Foley, J.W. (1972) Studies of vibrios from gallbladder of market sheep and cattle. *Am. J. Vet. Res.*, **33**, 1439–44.

Brynestad, S. and Granum, P.E. (2002) *Clostridium perfringens* and foodborne infections. *Int. J. Food Microbiol.*, **74**(3), 195–202.

Buchanan, R.L. and Bagi, L.K. (1994) Expansion of response surface models for the growth of *Escherichia coli* 0157:H7 to include sodium nitrite as a variable. *Int. J. Food Microbiol.*, **23**, 317–32.

Bulte, M., Klein, G. and Reuter, G. (1993) Pig slaughter. Is the meat contaminated by *Yersinia enterocolitica* strains pathogenic to man ? *Fleischwirtsch. Int.*, **1**, 6–15.

Buncic, S. (1991) The incidence of *Listeria monocytogenes* in slaughtered animals, in meat, and in meat products in Yugoslavia. *Int. J. Food Microbiol.*, **12**, 173–80.

Byrne, C.M., Bolton, D.J., Sheridan, J.J., Blair, I.S. and McDowell, D.A. (2002) The effect of commercial production and product formulation on the heat resistance of *Escherichia coli* O157:H7 (NCTC 12900) in beef burgers. *Int. J. Food Microbiol.*, **79**, 183–92.

Caley, J.E. (1972) Salmonella in pigs in Papua New Guinea. *Aust. Vet. J.*, **48**, 601–4.

Calicioglu, M., Faith, N.G., Buege, D.R., Luchansky, J.B. (1997) Viability of *Escherichia coli* O157:H7 in fermented semidry low-temperature-cooked summer sausage. *J. Food Prot.*, **60**, 1158–62.

Campbell-Platt, G. and Cook, P.E. (eds). (1995) *Fermented Meats*. Blackie Academic and Professional, London.

Canada, J.C. and Strong, D.H. (1964) *Clostridium perfringens* in bovine livers. *J. Food Sci.*, **29**, 862–4.

Cantoni, C., D'Aubert, S. and Cattaneo, P.L. (1977) The bacterial flora of horse meat: its characteristics, its relation to temperature, and its role in spoilage. *Arch. Vet. Italiano*, **28**, 83–9.

Carneiro, D., Cassar, C., Miles, R. (1998) Trisodium phosphate increases sensitivity of Gram-negative bacteria to lysozyme and nisin. *J. Food Prot.*, **61**, 839–43.
Carrico, R.J., Peisach, J. and Alben, J.O. (1978) The preparation and some physical properties of sulfhemoglobin. *Journal of Biological Chemistry*, **253**, 2386–91.
Carse, W.A. and Locker, R.H. (1974) A survey of pH values at the surface of beef and sheep carcasses, stored in a chiller. *J. Sci. Food Agric.*, **25**, 1529–5.
Casado, M.-J.M., Borras, M.-A.D. and Aguilar, R.V. (1992) Fungal flora present on the surface of cured Spanish ham. Methodological study for its isolation and identification. *Fleischwirtsch. Int.*, **2**, 29–31.
Castillo, A., Lucia, L., Goodson, K., Savell, J. and Acuff, G. (1998a) Comparison of water wash, trimming, and combined hot water and lactic acid treatments for reducing bacteria of fecal origin on beef carcasses. *J. Food Prot.*, **61**, 823–8.
Castillo, A., Lucia, L., Goodson, K., Savell, J. and Acuff, G. (1998b) Use of hot water for beef carcass decontamination. *J. Food Prot.*, **61**, 19–25.
Castillo, A., Dickson, J.S., Clayton, R.P., Lucia, L.M. and Acuff, G.R. (1998c) Chemical dehairing of bovine skin to reduce pathogenic bacteria and bacteria of fecal origin. *J. Food Prot.*, **61**(5), 623–5.
Castillo, A., Lucia, L.M., Roberson, D.B., Stevenson, T.H., Mercado, I. and Acuff, G.R. (2001) Lactic acid sprays reduce bacterial pathogens on cold beef carcass surfaces and in subsequently produced ground beef. *J. Food Prot.*, **64**(1), 58–62.
Castillo, A., Hardin, M., Acuff, G. and Dickson, J. (2002) Reduction of microbial contamination on carcasses, in *Control of Foodborne Microorganisms* (eds V. Juneja and J. Sofos), *volume 114, Food Science and Technology*, Marcel Dekker, Inc., New York, Ch. 13, pp. 351–81.
CDC. (1967) Center for Disease Control. Salmonella surveillance report No. 67. National Communicable Disease Center, Athens, Ga.
CDC. (1983) Botulism and commercial pot pies-California. *Morb. Mortal. Wkly Rep.*, **32**, 39–40, 45.
CDCP (Centers for Disease Control and Prevention). (1994) *Escherichia coli* O157:H7 outbreak linked to commercially distributed dry-cured salami—Washington and California 1994. *Morb. Mortal. Wkly Rep.*, **44**, 157–60.
CDCP (Centers for Disease Control and Prevention). (1995) Community outbreak of haemolytic uremic syndrome attributable to *Escherichia coli* O11:NM—South Australia. *Morb. Mortal. Wkly Rep.*, **44**, 550–1, 557–8.
Cerqueira, A.M.F., Guth, B.E.C., Joaquim, R.M. and Andrade, J.R.C. (1999) High occurrence of Shiga toxin-producing *Escherichia coli* (STEC) in healthy cattle in Rio de Janeiro State, *Brazil. Vet. Microbiol.*, **70**, 111–21.
CFA. (1993) *Guidelines for the Manufacture, Distribution and Retail sale of Chilled Foods*, 2nd edn, Chilled Foods Association, London.
CFPRA. (1977) *Technical Manual No. 3—Guidelines for the Establishment of Scheduled Heat Process for Low Acid Canned Foods*. Campden Food Preservation Research Association, Chipping Campden, UK.
Chambers, P.G. and Lysons, R.J. (1979) The inhibitory effect of bovine rumen fluid on *Salmonella typhimurium*. *Research in Veterinary Science*, **26**, 273–6.
Chapman, P.A. (2000) Sources of *Escherichia coli* O157 and experiences over the past 15 years in Sheffield, UK. *J. Appl. Microbiol.*, (Symposium Suppl.), **88**, 51S-60S.
Chapman, P.A., Siddons, C.A., Wright, D.J., Norman, P., Fox, J. and Crick, E. (1993a) Cattle as a possible source of verotoxin-producing *Escherichia coli* O157 infections in man. *Epidemiol. Infect.*, **111**, 439–47.
Chapman, P.A., Wright, D.J. and Higgins, R. (1993b) Untreated milk as a source of verotoxigenic *E. coli* O157. *Vet. Rec.*, **133**, 171–2.
Chapman, P.A., Siddons, C.A. and Harkin, M.A. (1996) Sheep as a potential source of verocytotoxin-producing *Escherichia coli* O157. *Vet. Rec.*, **138**, 23–4.
Chapman, P.A., Siddons, C.A., Cerdan Malo, A.T. and Harkin, M.A. (1997) A 1-year study of *Escherichia coli* O157 in cattle, sheep, pigs and poultry. *Epidemiol. Infect.*, **119**, 245–50.
Chau, P.Y., Shortridge, K.F. and Huang, C.T. (1977) *Salmonella* in pig carcasses for human consumption in Hong Kong: a study on the mode of contamination. *J. Hyg.*, **78**, 253–60.
Chikthimmah, N. and Knabel, S.J. (2001) Survival of *Escherichia coli* O157:H7, *Salmonella typhimurium* and *Listeria monocytogenes* in and on vacuum packaged Lebanon bologna stored at 3.6 and 13.0 degrees C. *J. Food Prot.*, **64**(7), 958–63.
Chikthimmah, N., Anantheswaran, R.C., Roberts, R.F., Mills, E.W. and Knabel, S.J. (2001a) Influence of sodium chloride on growth of lactic acid bacteria and subsequent destruction of *Escherichia coli* O157:H7 during processing of Lebanon bologna. *J. Food Prot.*, **64**(8), 1145–50.
Chikthimmah, N., Guyer, R.B. and Knabel, S.J. (2001b) Validation of a 5-$\log_{10}$ reduction of *Listeria monocytogenes* following simulated commercial processing of Lebanon bologna in a model system. *J. Food Prot.*, **64**(6), 873–6.
Childers, A.B., Keahey, E.E. and Vincent, P.G. (1973) Sources of salmonellae contamination of meat following approved livestock slaughtering procedures. II. *J. Milk Food Technol.*, **36**, 635–8.
Childers, A.B., Keahey, E.E. and Kotula, A.W. (1977) Reduction of *Salmonella* and fecal contamination of pork during swine slaughter. *J. Am. Vet. Med. Assoc.*, **171**, 1161–4.
Chinen, I., Tanaro, J., Miliwebsky, E., Lound, L., Cillemi, G., Ledri, S., Bakschir, A., Scarpin, M., Manfredi, E. and Rivas, M. (2001) Isolation and characterization of *Escherichia coli* O157:H7 from retail meats in Argentina. *J. Food Prot.*, **64**, 1346–51.
Christensen, S.G. (1987) The *Yersinia enterocolitica* situation in Denmark. *Contr. Microbiol. Immunol.*, **9**, 93–7.
Chung, G.T. and Frost, A.J. (1969) The occurrence of salmonellae in slaughtered pigs. *Aust. Vet. J.*, **45**, 350–3.
Chyr, C.Y., Walker, H.W. and Sebranek, J.G. (1981) Bacteria associated with spoilage of braunschweiger. *J. Food Sci.*, **46**, 468–70.
Cizek, A., Alexa, P., Literák, I., Hamrík, J., Novák, P. and Smola, J. (1999) Shiga toxin-producing *Escherichia coli* O157 in feedlot cattle and Norwegian rats from a large-scale farm. *Lett. Appl. Microbiol.*, **28**(6), 435–9.

Clarke, R.C., Wilson, J.B., Read, S.C., Renwick, S., Rahn, K., Johnson, R.P., Alves, D., Karmali, M.A., Lior, H., McEwen, S.A., Spika, J. and Gyles, C.L. (1994) Verocytotoxin-producing *Escherichia coli* (VTEC) in the food chain: preharvest and processing perspectives, in *Recent Advances in Verocytotoxin-producing Escherichia coli Infections* (eds M.A. Karmali and A.G. Goglio), Elsevier Science B.V., Amsterdam, pp. 17–24.

Cobbold, R. and Desmarchelier, P. (2000) A longitudinal study of Shiga-toxigenic *Escherichia coli* (STEC) prevalence in three Australian diary herds. *Vet. Microbiol.*, **71**(1–2), 125–37.

Cobbold, R. and Desmarchelier, P. (2001) Characterisation and clonal relationships of Shiga-toxigenic *Escherichia coli* (STEC) isolated from Australian dairy cattle. *Vet. Microbiol. (Aust.)*, **79**(4), 323–35.

Collins, M.D., Rodriguez, U.M., Dainty, R.H., Edwards, R.A. and Roberts, T.A. (1992) Taxonomic studies on a psychrotrophic *Clostridium* from vacuum-packed beef: description of *Clostridium estertheticum* sp. nov. *FEMS Microbiol. Lett.*, **96**, 235–40.

Conedera, G., Chapman, P.A., Marangon, S., Tisato, E., Dalvit, P. and Zuin, A. (2001) A field survey of *Escherichia coli* O157 ecology on a cattle farm in Italy. *Int. J. Food Microbiol.*, **66**(1–2), 85–93.

Cook, A.J.C., Gilbert, R.E., Buffolano, W. *et al.* (on behalf of the European Research Network on Congenital Toxoplasmosis). (2000) Sources of Toxoplasma infection in pregnant women: European multicentre case–control study. *Br. Med. J.*, **321**, 142–7.

Cornejo, I. and Carrascosa, A.V. (1991) Characterization of *Micrococcaceae* strains selected as potential starter cultures in Spanish dry-cured ham processes. 1. Fast process. *Fleischwirtsch. Int.*, **2**, 58–60.

Cowden, J.M., O'Mahony, M., Bartlett, C.L.R., Rana, B., Smyth, B., Lynch, D., Tillett, H., Ward, L., Roberts, D., Gilbert, R.J., Baird-Parker, A.C. and Kilsby, D.C. (1989) A national outbreak of *Salmonella typhimurium* DT 124 caused by contaminated salami sticks. *Epidemiol. Infect.*, **103**, 219–25.

Cutter, C.N. and Siragusa, G.R. (1994) Efficacy of organic acids against *Escherichia coli* O157:H7 attached to beef carcass tissue using a pilot scale model carcass washer. *J. Food Prot.*, **57**, 97–103.

Cutter, C.N. and Siragusa, G.R. (1995) Application of chlorine to reduce populations of *Escherichia coli* on beef. *J. Food Safety*, **15**(1), 67–75.

Cutter, C.N. and Siragusa, G.R. (1996) Reduction of *Brochothrix thermosphacta* on beef surfaces following immobilization of nisin in calcium alginate gels. *Lett. Appl. Microbiol.*, **23**(1), 9–12.

Daguzan, J. (1985) Production of "Petit-gris" snails. *Ann. Zootechn.*, **34**, 127–48.

Cutter, C.N. and Rivera-Betancourt, M. (2000) Interventions for the reduction of *Salmonella typhimurium* DT 104 and non-O157:H7 enterohemorrhagic *Escherichia coli* on beef surfaces. *J. Food Prot.*, **63**(10), 1326–32.

Dainty, R.H. and Mackey, B.M. (1992) The relationship between the phenotypic properties of bacteria from chill-stored meat and spoilage processes. *J. Appl. Bacteriol.* (Symposium suppl.), **73**, 103S–14S.

Dainty, R.H., Shaw, B.G., de Boer, K.A. and Scheps, E.S.J. (1975) Protein changes caused by bacterial growth on beef. *J. Appl. Bacteriol.*, **39**, 72–81.

Dainty, R.H., Shaw, B.G., Harding, C.D. and Michanie, S. (1979) The spoilage of vacuum-packed beef by cold tolerant bacteria, in *Cold Tolerant Microbes in Spoilage and the Environment*, (eds A.D. Russell and R. Fuller) Academic Press, London, pp. 83–100.

Dainty, R.H., Edwards, R.A. and Hibbard, C.M. (1989) Spoilage of vacuum-packed beef by a *Clostridium* sp. *Journal of the Science of Food and Agriculture*, **49**, 473–86.

Dargatz, D.A., Fedorka-Cray, P.J., Ladely, S.R. and Ferris, K.E. (2000) Survey of *Salmonella* serotypes shed in feces of beef cows and their antimicrobial susceptibility patterns. *J. Food Prot.*, **63**(12), 1648–53.

Daube, G. (2001) Les plans de surveillance officiels de l'hygiène et des agents zoonotiques des filières de production carnée en Belgique: application à la définition de critères microbiologiques pour les carcasses de porcs et de bovins. Rapport, Université de Liège.

Davey, K.D. and Smith, M.G. (1989) A laboratory evaluation of a novel hot water cabinet for the decontamination of sides of beef. *Int. J. Food Sci. Technol.*, **24**, 305–16.

Davies, P.R., Bovee, F.G., Funk, J.A., Morrow, W.E., Jones, F.T. and Deen, J. (1998) Isolation of *Salmonella* serotypes from feces of pigs raised in a multiple-site production system. *J. Am. Vet. Med. Assoc.*, **15**, 1925–9.

Dell'Omo, G., Morabito, S., Quondam, R., Agrimi, U., Ciuchini, F., Macri, A. and Caprioli, A. (1998) Feral pigeons as a source of verocytotoxin-producing *Escherichia coli. Vet. Rec.*, **142**, 309–10.

Delmore, R.J., Jr., Sofos, J.N., Schmidt, G.R., Belk, K.E., Lloyd, W.R. and Smith, G.C. (2000) Interventions to reduce microbiological contamination of beef variety meats. *J. Food Prot.*, **63**, 44–50.

Department of Health. (1989) *Guidelines on Cook–Chill and Cook–Freeze catering systems.* HMSO, London.

Desmarchelier, P.M., Higgs, G.M., Mills, L., Sullivan, A.M., Vanderlinde, P.B. (1999) Incidence of coagulase positive *Staphylococcus* on beef carcasses in three Australian abattoirs. *Int. J. Food Microbiol.*, **47**(3), 221–9.

Devriese, L.A. (1990) Staphylococci in healthy and diseased animals, in *The Society for Applied Bacteriology Symposium Series No. 19* (eds R.G. Board and M. Sussman), Blackwell Scientific Publications, Oxford, England, pp. 71s-80s.

De Zutter, L., Uradzinski, J. and Pierard, D. (1999) Prevalence of enterohemorrhagic *E. coli* O157 in Belgian cattle. Abstracts of the Second International Symposium of the European Study Group on Enterohemorrhagic *Escherichia coli. Acta Clin. Belgica*, **54**, 48.

Dickson, J., Cutter, C., Siragusa, G., 1994. Antimicrobial effect of trisodium phosphate against bacteria attached to beef tissue. *J. Food Prot.*, **57**, 952–955.

Dillon, V.M. and Board, R.G. (1991) Yeasts associated with red meats. *J. Appl. Bacteriol.*, **71**, 93–108.

Dockerty, T.R., Ockerman, H.W., Cahill, V.R., Kunkle, L.E. and Weiser, H.H. (1970) Microbial level of pork skin as affected by the dressing process. *J. Ani. Sci.*, **30**, 884–90.
Dorn, C.R. and Angrick, E.J. (1991) Serotype O157:H7 *Escherichia coli* from bovine and meat sources. *J. Clin. Microbiol.*, **29**, 1225–31.
Dorn, C.R., Scotland, S.M., Smith, H.R., Willshaw, G.A. and Rowe, B. (1989) Properties of Vero cytotoxin-producing *Escherichia coli* of human and animal origin belonging to serotypes other than O157:H7. *Epidemiol. Infect.*, **103**, 83–95.
Dowdell, M.J. and Board, R.G. (1968) A microbiological survey of British fresh sausage. *J. Appl. Bacteriol.*, **31**, 378–96.
Doyle, M. and Schoeni, J. (1987) Isolation of *Escherichia coli* O157:H7 from fresh retail meats and poultry. *Appl. Environ. Microbiol.*, **53**, 2394–6.
Doyle, M.P. (1991) *Escherichia coli* O157:H7 and its significance in foods. *Int. J. Food Microbiol.*, **12**, 289–302.
Doyle, M.P. and Hugdahl, M.B. (1983) Improved procedure for recovery of *Yersinia enterocolitica* from meats. *Appl. Environ. Microbiol.*, **45**, 127–35.
Doyle, M.P. and Schoeni, J.L. (1984) Survival and growth characteristics of *Escherichia coli* associated with hemorrhagic colitis. *Appl. Environ. Microbiol.*, **48**, 855–6.
Doyle, M.P., Hugdahl, M.B. and Taylor, S.L. (1981) Isolation of *Yersinia enterocolitica* from porcine tongues. *Appl. Environ. Microbiol.*, **42**, 661–6.
Duffy, G., Garvey, P., Wasteson, Y., Coia, J.E., McDowell, D.A., (2001a) *Epidemiology of Verocytotoxigenic E. coli*. A technical booklet produced for an EU Concerted Action (CT98–3935). ISBN 1 84170 206 4; http://www.research.teagasc.ie/vteceurope/epitechbook.htm.
Duffy, E.A., Belk, K.E., Sofos, J.N., LeValley, S.B., Kain, M.L., Tatum, J.D., Smith, G.C. and Kimberling, C.V. (2001b) Microbial contamination occurring on lamb carcasses processed in the United States. *J. Food Prot.*, **64**(4), 503–8.
Duffy, E.A., Belk, K.E., Sofos, J.N. Bellinger, G.R., Pape, A. and Smith, G.C. (2001c). Extent of microbial contamination in United States pork retail products. *J. Food Prot.*, **64**, 172–8.
Dupouy-Camet, J. (1999) Is human trichinellosis an emerging zoonoses in the European community ? *Helminthologia*, **36**, 201–4.
Dupouy-Camet, J. (2000) Trichinellosis—a worldwide zoonosis. *Vet. Parasitol.*, **93**, 191–200.
Dykes, G.A., Cloete, T.E. and von Holy, A. (1991) Quantification of microbial populations associated with the manufacture of vacuum-packaged, smoked Vienna sausages. *Int. J. Food Microbiol.*, **13**, 239–48.
EC (European Commission). (2002) Trends and sources of zoonotic agents in animals, feedingstuffs, food and man in the European Union and Norway in 2000: Part 1. SANCO/927/2002. Prepared by the Community Reference Laboratory on the Epidemiology of Zoonoses, BgVV, Berlin, Germany.
ECFF. (1994) *European Chilled Foods Federation Guidelines*, ECFF, Paris.
Edel, W. and Kampelmacher, E.H. (1971) Salmonella infection in fattening calves at the farm. *Zbl. Vet. Med. B*, **18**, 617–21.
Edel, W., Guinee, P.A.M., van Schothorst, M. and Kampelmacher, E.H. (1967) *Salmonella* infections in pigs fattened with pellets and unpelleted meal. *Zen. fur. Vet. Med. B*, **14**, 393–401.
Edel, W., Guinee, P.A.M. and Kampelmacher, E.H. (1970) Effect of feeding pellets on prevention and sanitation of salmonella-infections in fattening pigs. *Zbl. Vet. Med. B*, **17**, 730–8.
Egan, A.F. (1983) Lactic acid bacteria of meat and meat products. *Antonie van Leeuwenhoek*, **49**, 327–36.
Egan, A.F. and Grau, F.H. (1981) Environmental conditions and the role of *Brochothrix thermosphacta* in the spoilage of fresh and processed meat, in *Psychrotrophic Microorganisms in Spoilage and Pathogenicity* (eds T.A. Roberts, G. Hobbs, J.H.B. Christian and N. Skovgaard), Academic Press, London, pp. 211–21.
Egan, A.F. and Shay, B.J. (1991) Meat starter cultures and the manufacture of meat products, in *Encyclopedia of Food Science and Technology*, J. Wiley & Sons, Inc., pp. 1735–45.
Egan, A.F., Ford, A.L. and Shay, B.J. (1980) A comparison of *Microbacterium thermosphactum* and lactobacilli as spoilage organisms of vacuum-packaged sliced luncheon meats. *J. Food Sci.*, **45**, 1745–8.
Egan, A.F., Shay, B.J. and Rogers, P.J. (1989) Factors affecting the production of hydrogen sulphide by *Lactobacillus sake* L13 growing on vacuum-packed beef. *J. Appl. Bacteriol.*, **67**, 255–62.
Elder, R.O., Keen, J.E., Siragusa, G.R., Barkocy-Gallagher, G.A., Koohmaraie, M. and Laegreid, W.W. (2000) Correlation of enterohemorrhagic *Escherichia coli* O157 prevalence in feces, hides, and carcasses of beef cattle during processing. *Proc. Natl Acad. Sci.*, **97**, 2999–3003.
Ellajosyula, K.R., Doores, S., Mills, E.W., Wilson, R.A., Anantheswaran, R.C. and Knabel, S.J. (1998) Destruction of *Escherichia coli* O157:H7 and *Salmonella typhimurium* in Lebanon bologna by interaction of fermentation pH, heating temperature and time. *J. Food Prot.*, **61**, 152–7.
El Moula, A.A. (1978) Incidence of zoonotic disease (salmonellosis) encountered in animals slaughtered in Egypt. M.V. Sc. Thesis, Fac. Vet. Med, Lyita, Egypt.
Empey, W.A. and Scott, W.J. (1939) Investigations on chilled beef. Part 1. Microbial Contamination Acquired in the Meatworks. Bulletin No. 126, Council for Scientific and Industrial Research, Melbourne, Australia.
Faith, N.G., Shere, J.A., Brosch, R., Arnold, K.W., Ansay, S.E., Lee, M.-S., Luchansky, J.B. and Kaspar, C.W. (1996) Prevalence and clonal nature of *Escherichia coli* O157:H7 on dairy farms in Wisconsin. *Appl. Environ. Microbiol.*, **62**, 1519–25.
Faith, N.G., Paniere, N., Larson, T., Lorang, T.D., Luchansky, J.B. (1997) Viability of *Escherichia coli* O157:H7 in pepperoni during manufacture of sticks and subsequent storage of slices at 21, 4 and −20°C under air, vacuum and CO_2. *Int. J. Food Microbiol.*, **37**, 47–54.
Faith, N.G., Paniere, N., Larson, T., Lorang, T.D., Kaspar, C.W., Luchansky, J.B., (1998a) Viability of *Escherichia coli* O157:H7 in salami following conditions of batter fermentation and drying of sticks and storage of slices. *Int. J. Food Prot.*, **61**, 377–382.

Faith, N.G., Wierzba, R.K., Ihnot, A.H., Roering, A.M., Lorang, T.D., Kaspar, C.W., Luchansky, J.B. (1998b) Survival of *Escherichia coli* O157:H7 in full and reduced fat pepperoni after manufacture of sticks, storage of slices at 4°C and 21°C under air and vacuum, and baking of slices on frozen pizza at 135°C, 191°C and 246°C. *Int. J. Food Prot.*, **61**, 383–9.

Fantelli, K. and Stephan, R. (2001) Prevalence and characteristics of Shigatoxin-producing *E. coli* (STEC) and *Listeria monocytogenes* strains isolated from minced meat in Switzerland. *Int. J. Food Microbiol.*, **70**, 63–9.

FAO/WHO. (1979) Recommended International Code of Practice for Low-Acid and Acidified Low-Acid Canned Foods (CAC/RCP 23–1979), Food and Agricultural Organization/World Health Organization.

Farber, J.M. and Peterkin, P.I. (1991) *Listeria monocytogenes*, a food-borne pathogen. *Microbiol. Rev.*, **55**, 476–511.

Farkas, J. and Andrássy, É. (1992) Combined effects of physical treatments and sporostatic factors on *Clostridium sporogenes* spores. I combined effects of heat treatment, nitrite reduced a_w and reduced pH in an anaerobic nutrient medium. *Acta Alimen.*, **21**, 39–48.

FDA. (1973) Thermally Processed Low-Acid Foods packed in Hermetically Sealed Containers GMP (Section 113:40). Federal Register 38 No. 16, 24 January 1973, 2398–2410, Food and Drug Administration USA.

FDA. (1993) Food Code. 1993 Recommendations of the United States Public Health Service and Food and Drug Administration. National Technical Information Service, Springfield, Virginia.

Fenlon, D.R., Wilson, J. and Donachie, W. (1996) The incidence and level of *Listeria monocytogenes* contamination of food sources at primary production and initial processing. *J. Appl. Bacteriol.*, **81**, 641–50.

Field, R.A. (1981) Mechanically deboned red meat. *Advan. Food Res.*, **27**, 23–107.

Food Safety Authority of Ireland. (1999) VTEC in pigeons, geese.

France. (1988) Prolongation of Life Span of Pre-cooked Food Modification of Procedures enabling Authorisation to be obtained. Veterinary Service of Food Hygiene, Service Note DGAL/SVHA/N88/No 8106, 31.5.88.

Franz, C.M., Holzapfel, W.H. and Stiles, M.E. (1999) Enterococci at the crossroads of food safety? *Int. J. Food Microbiol.*, **47**(1–2), 1–24.

Fredriksson-Ahomaa, M., Hielm, S. and Korkeala, H. (1999) High prevalence of yadA-positive *Yersinia enterocolitica* in pig tongues and minced meat at the retail level in Finland. *J. Food Prot.*, **62**, 123–7.

Fredriksson-Ahomaa, M., Korte, T. and Korkeala, H. (2000) Contamination of carcasses, offals, and the environment with yadA-positive *Yersinia enterocolitica* in a pig slaughterhouse. *J. Food Prot.*, **63**, 31–5.

Fredriksson-Ahomaa, M., Niskanen, T., Bucher, M., Korte, T., Stolle, A. and Korkeala, H. (2003) Different *Yersinia enterocolitica* 4:O3 genotypes found in pig tonsils in Southern Germany and Finland. *Syst. Appl. Microbiol.*, **26**(1), 132–7.

Fricker, C.R. and Park, R.W.A. (1989) A two-year study of the distribution of 'thermophilic' campylobacters in human, environmental and food samples from the Reading area with particular reference to toxin production and heat-stable serotype. *J. Appl. Bacteriol.*, **66**, 477–90.

Frost, A.J., O'Boyle, D. and Samuel, J.L. (1988) The isolation of *Salmonella* spp from feed lot cattle managed under different conditions before slaughter. *Australian Veterinary Journal*, **65**, 224–5.

FSIS (Food Safety and Inspection Service). (1993) Report on the *E. coli* O 157:H7 outbreak in the Western State. May 21, 1993, Food Safety and Inspection Service, United States Department of Agriculture.

Fukushima, H. and Gomyoda, M. (1986) Inhibition of *Yersinia enterocolitica* serotype O3 by natural microflora of pork. *Appl. Environ. Microbiol.*, **51**, 990–4.

Fukushima, H., Nakamura, R., Ito, Y., Saito, K., Tsubokura, M. and Otsuki, K. (1983) Ecological studies of *Yersinia enterocolitica*. I. Dissemination of *Y. enterocolitica* in pigs. *Vet. Microbiol.*, **8**, 469–83.

Fukushima, H., Maruyama, K., Omori, I., Ito, K. and Iorihara, M. (1991) Contamination of pigs with Yersinia at the slaughterhouse. *Fleischwirtsch. Int.*, **1**, 50–2.

Galland, J.C., House, J.K., Hyatt, D.R., Hawkins, L.L., Anderson, N.V., Irwin, C.K. and Smith, B.P. (2000) Prevalence of *Salmonella* in beef feeder steers as determined by bacterial culture and ELISA serology. *Vet. Microbiol.*, **76**, 143–51.

Galton, M.M., Smith, W.V., McElrath, H.B. and Hardy, A.V. (1954) *Salmonella* in swine, cattle and the environment of abattoirs. *J. Infect. Dis.*, **95**, 236–45.

Gamage, S.D., Faith, N.G., Luchansky, J.B., Buege, D.R. and Ingham, S.C. (1997) Inhibition of microbial growth in chub-packed ground beef by refrigeration (2oC) and medium-dose (2.2 to 2.4 kGy) irradiation. *Int. J. Food Microbiol..*, **37**, 175–82.

Garber, L.P., Wells, S.J., Hancock, D.D., Doyle, M.P., Tuttle, J., Shere, J.A. and Zhao, T. (1995) Risk factors for fecal shedding of *Escherichia coli* O157:H7 in dairy calves. *J. Am. Vet. Med. Assoc.*, **207**, 46–9.

Garcia de Fernando, G.D., Mano, S.B., Lopez, D. and Ordóñez, J.A. (1995) Effects of modified atmospheres on the growth of psychrotrophic pathogenic microorganisms on proteinacesus foods. *Microbiologia*, **11**, 7–22.

Gardner, G.A. (1971) A note on the aerobic microflora of fresh and frozen porcine liver stored at 5°C. *J. Food Technol.*, **6**, 225–31.

Gardner, G.A. (1983) Microbial spoilage of cured meats, in *Food Microbiology: Advances and Prospects, The Society of Applied Bacteriology Symposium Series No. 11.* (eds T.A. Roberts and F.A. Skinner). Academic Press, London.

Garm, R. (1976) Processing of froglegs for human consumption, FAO—Fisheries information note.

Gay, J.M., Rice, D.H. and Steiger, J.H. (1994) Prevalence of fecal *Salmonella* shedding by cull dairy cattle marketed in Washington State. *J. Food Prot.*, **57**, 195–7.

Gaze, J.E., Brown, G.D., Gaskell, D. and Banks, J.G. (1989) Heat resistance of *Listeria monocytogenes* in homogenates of chicken, beefsteak and carrot. *Food Microbiol.*, **6**, 251–9.

Gebreyes, W.A., Davis, P.R., Morrow, W.E.M., Funk, J.A. and Altier, C. (2000) Antimicrobial resistance of *Salmonella* isolates from swine. *J. Clin. Microbiol.*, **38**, 4633–36.

Geisen, R., Lücke, F.-K. and Krockel, L. (1993) Starter and protective cultures for meat and meat products. *Fleischwirtsch. Int.*, **1**, 34–44.

Gerats, G.G. (1987) What hygiene can achieve-how to achieve hygiene, in *Elimination of Pathogenic Organisms From Meat and Poultry* (ed F.J.M. Smulders), Elsevier Science Publishers B.V., Amsterdam, pp. 269–80

Gerats, G.E., Snijders, J.M.A. and van Logtestijn, J.G. (1981) Slaughter techniques and bacterial contamination of pig carcasses, in *Proceedings, 27th European Meeting of Meat Research Workers. Vienna, volume 1*, pp. 198–200.

Ghosh, A.C. (1972) An epidemiological study of the incidence of salmonellas in pigs. *J. Hyg. (Camb.)*, **70**, 151–60.

Gibson, A.M. and Roberts, T.A. (1986) The effect of pH, sodium chloride, sodium nitrite and storage temperature on the growth of *Clostridium perfringens* and faecal streptococci in laboratory media. *Int. J. Food Microbiol.*, **3**(4), 195-210.

Gilgen, M., Hübner, P., Höfelein, Lüthy, J. and Candrian, U. (1998) PCR-based detection of verotoxin-producing *Escherichia coli* (VTEC) in ground beef. *Res. Microbiol.*, **149**, 145–54.

Gill, C.O. (1976) Substrate limitation of bacterial growth at meat surfaces. *J. Appl. Bacteriol.*, **41**, 401–10.

Gill, C.O. (1986) The control of microbial spoilage in fresh meats, in *Advances in Meat Research* (eds A.M. Pearson and T.R. Dutson), *volume 2, Meat and Poultry Microbiology*, AVI Publishing Co. Inc., Westport, Conn, pp. 49–88.

Gill, C.O. (1987) Prevention of microbial contamination in the lamb processing plant, in *Elimination of Pathogenic Organisms From Meat and Poultry* (ed F.J.M. Smulders), Elsevier Science Publishers B.V., Amsterdam, pp. 203–19.

Gill, C.O. (1988) Microbiology of edible meat by-products, in *Edible Meat By-products. Advances in Meat Research* (eds A.M. Pearson and T.R. Dutson), *volume 5*, Elsevier Applied Science, London, pp. 47–82.

Gill, C.O. (1991) Microbial principles in meat processing, in *Microbiology of Animals and Animal Products* (ed J.B. Woolcock), World Animal Science, A6, Elsevier, Amsterdam, pp. 249–70.

Gill, C.O. and Baker, L.P. (1998) Assessment of the hygienic performance of a sheep carcass dressing process. *J. Food Prot.*, **61**, 329–33.

Gill, C.O. and Bryant, J. (1993) The presence of *Escherichia coli*, *Salmonella* and *Campylobacter* in pig carcass dehairing equipment. *Food Microbiol.*, **10**, 337–44.

Gill, C.O. and De Lacy, K.M. (1982) Microbial spoilage of whole sheep livers. *Appl. Environ. Microbiol.*, **43**, 1262–6.

Gill, C.O. and Harrison, J.C.L. (1985) Evaluation of the hygienic efficiency of offal cooling procedures. *Food Microbiol.*, **2**, 63–9.

Gill, C.O. and Harrison, J.C.L. (1989) The storage life of chilled pork packaged under carbon dioxide. *Meat Sci.*, **26**, 313–24.

Gill, C.O. and Jones, S.D.M. (1992) Evaluation of a commercial process for collection and cooling of beef offals by a temperature function integration technique. *Int. J. Food Microbiol.*, **15**, 131–43.

Gill, C.O. and Jones, T. (1997) Assessment of the hygienic performances of an air-cooling process for lamb carcasses and a spray-cooling process for pig carcasses. *Int. J. Food Microbiol.*, **38**, 85–93.

Gill, C.O. and Lowry, P.D. (1982) Growth at sub-zero temperatures of black spot fungi from meat. *J. Appl. Bacteriol.*, **52**, 245–50.

Gill, C.O. and Newton, K.G. (1977) The development of aerobic spoilage flora on meat stored at chill temperatures. *J. Appl. Bacteriol.*, **43**, 189–95.

Gill, C.O. and Newton, K.G. (1980) Growth of bacteria on meat at room temperatures. *J. Appl. Bacteriol.*, **49**, 315–23.

Gill, C.O. and Penney, N. (1986) Packaging conditions for extended storage of chilled dark, firm, dry beef. *Meat Sci.*, **18**, 41–53.

Gill, C.O. and Penney, N. (1988) The effect of the initial gas volume to meat ratio on the storage life of chilled beef packed under carbon dioxide. *Meat Sci.*, **22**, 53–663.

Gill, C.O. and Reichel, M.P. (1989) Growth of cold-tolerant pathogens *Yersinia enterocolitica*, *Aeromonas hydrophila* and *Listeria monocytogenes* on high-pH beef packaged under vacuum or carbon dioxide. *Food Microbiol.*, **6**, 223–30.

Gill, C.O., McGinnis, J.C. and Jones, T. (1999) Assessment of the microbiological conditions of tails, tongues, and head meats at two beef-packing plants. *J. Food Prot.*, **62**, 674–7.

Gill, C.O., Bryant, J. and Brereton, D.A. (2000a). Microbiological conditions of sheep carcasses from conventional or inverted dressing processes. *J. Food Prot.*, **63**, 1291–4.

Gill, C.O., Dussault, F., Holley, R.A., Houde, A., Jones, T., Rheault, N., Rosales, A. and Quessy, S. (2000b) Evaluation of the hygienic performances of the processes for cleaning, dressing and cooling pig carcasses at eight packing plants. *Int. J. Food Microbiol..*, **58**, 65–72.

Glass, K.A. and Doyle, M.P. (1989a) Fate and thermal inactivation of *Listeria monocytogenes* in beaker sausage and pepperoni. *J. Food Prot.*, **52**, 226–31.

Glass, K.A. and Doyle, M.P. (1989b) *Listeria monocytogenes* in processed meat products during refrigerated storage. *Appl. Environ. Microbiol.*, **55**, 1565–9.

Glass, K., Loeffelholz, J., Ford, P. and Doyle, M. (1992) Fate of *Escherichia coli* O157:H7 as affected by pH or sodium chloride in fermented dry sausage. *Appl. Environ. Microbiol.*, **58**, 2513–6.

Gobat, P.-F. and Jemmi, T. (1991) Epidemiological studies on *Listeria* spp. in slaughterhouses. *Fleischwirtsch. Int.*, **1**, 44–9.

Goulet, V., Lepoutre, A., Rocourt, J., Courtieu, A.L., Dehaumont, P. and Veit, P. (1993) Epidémie de listériosie en France: Bilan final et résultats de l'enquête épidémiologique. *Bull. Epidémiol. Hebdom.*, **4**, 13–4.

Grau, F.H. (1979) Fresh meats: bacterial association. *Archiv fur Lebensmittelhygiene*, **30**, 81–116.

Grau, F.H. (1981) Role of pH, lactate, and anaerobiosis in controlling the growth of some fermentative Gram-negative bacteria on beef. *Appl. Environ. Microbiol.*, **42**, 1043–50.

Grau, F.H. (1986) Microbial ecology of meat and poultry, in *Advances in Meat Research*, (eds A.M. Pearson and T.R. Dutson), *volume 2*, AVI publishing Co. Inc., Westport Conn. pp. 1–47.

Grau, F.H. (1987) Prevention of microbial contamination in the export beef abattoir, in *Elimination of Pathogenic Organisms from Meat and Poultry* (ed F.J.M. Smulders), Elsevier Science Publishers, Amsterdam, pp. 221–33.

Grau, F.H. (1988) *Campylobacter jejuni* and *Campylobacter hyointestinalis* in the intestinal tract and on the carcasses of calves and cattle. *J. Food Prot.*, **51**, 857–61.

Grau, F.H. and Smith, M.G. (1974) Salmonella contamination of sheep and mutton carcasses related to pre-slaughter holding conditions. *J. Appl. Bacteriol.*, **37**, 111–6.

Griffin, P.M. and Tauxe, R.V. (1991) The epidemiology of infections caused by *Escherichia coli* O157:H7, other enteropathogenic *E. coli*, and the associated hemolytic uremic syndrome. *Epidemiol. Rev.*, **13**, 60–98.

Grau, F.H. and Vanderlinde, P.B. (1990) Growth of *Listeria monocytogenes* on vacuum-packaged beef. *J. Food Prot.*, **53**, 739–43.

Guinee, P.A.M, Kampelmacher, E.H., van Keulen, A. and Hofstra, K. (1964) Salmonellae in healthy cows and calves in the Netherlands. *Zen. Vet. Reihe B*, **III**, 728–40.

Guo, S.-L. and Chen, M.-T. (1992) Studies on the microbial flora of Chinese-style sausage. 1. The microbial flora and its biochemical characteristics. *Fleischwirtsch. Int.*, **1**, 42–6.

Gutierrez, L.M., Memes, I., Garcia, M.L. Morena, B. and Bergdoll, M.S. (1982) Characterization and enterotoxigenicity of staphylococci isolated from mastitic ovine milk in Spain. *J. Food Prot.*, **45**, 1282–6.

Hamdy, M. (1991) Surface contamination of slaughtered camels. *Fleischwirtscaft*, **71**, 1311–2.

Hancock, D.D., Besser, T.E., Kinsel, M.L., Tarr, P.I., Rice, D.H. and Paros, M.G. (1994) The prevalence of *Escherichia coli* O157:H7 in dairy and beef cattle in Washington State. *Epidemiol. Infect.*, **113**, 199–207.

Hancock, D.D., Besser, T.E., Rice, D.H., Herriott, D.E. and Tarr, P.I. (1997) A longitudinal study of *Escherichia coli* O157 in fourteen cattle herds. *Epidemiol. Infect.*, **118**, 193–5.

Hancock, D.D., Besser, T.E., Rice, D.H., Ebel, E.D., Herriott, D.E., Carpenter, L.V. (1998) Multiple sources of *Escherichia coli* O157 in feedlots and dairy farms in the Northwestern United States. *Prev. Vet. Med.*, **35**, 245–50.

Hanna, M.O., Stewart, J.C., Zink, D.L., Carpenter, Z.L. and Vanderzant, C. (1977) Development of *Yersinia enterocolitica* on raw and cooked beef and pork at different temperatures. *J. Food Sci.*, **42**, 1180–4.

Hanna, M.O., Smith, G.C., Savell, J.W., McKeith, F.K. and Vanderzant, C. (1982) Microbial flora of livers, kidneys and hearts from beef, pork and lamb: effects of refrigeration, freezing and thawing. *J. Food Prot.*, **45**, 63–73.

Hansen, R., Rogers, R., Emge, S. and Jacobs, N.J. (1964) Incidence of *Salmonella* in the hog colon as affected by handling practices prior to slaughter. *J. Am. Vet. Med. Assoc.*, **145**, 139–40.

Hardin, M., Acuff, G., Lucia, L., Osman, J. and Savell, J. (1995) Comparison of methods for contamination removal from beef carcass surfaces. *J. Food Prot.*, **58**, 368–74.

Hariharan, H., Giles, J.S., Heaney, S.B., Leclerc, S.M. and Schurman, R.D. (1995) Isolation, serotypes, and virulence-associated properties of *Yersinia enterocolitica* from the tonsils of slaughter hogs. *Can. J. Vet. Res.*, **59**, 161–6.

Harris, N.V., Thompson, D., Martin, D.C. and Nolan, C.M. (1986) A survey of *Campylobacter* and other bacterial contaminants of pre-market chicken and retail poultry and meats, King County, Washington. *Am. J. Public Health*, **76**, 401–6.

Harrison, J.A., Harrison, M.A., Rose-Morrow, R.A. and Shewfelt, R.L. (2001) Home-style beef jerky: effect of four preparation methods on consumer acceptability and pathogen inactivation. *J. Food Prot.*, **64**(8), 1194–8.

Harvey, R.B., Young, C.R., Anderson, R.C., Droleskey, R.E., Genovese, K.J., Egan, L.F. and Nisbet, D.J. (2000) Diminution of *Campylobacter* colonization in neonatal pigs reared off-sow. *J. Food Prot.*, **63**(10), 1430–2.

Hauschild, A.H.W. and Simonsen, B. (1985) Safety of shelf-stable canned cured meats. *J. Food Prot.*, **48**, 997–1009.

Hauschild, A.H.W. and Simonsen, B. (1986) Safety assessment of shelf-stable canned cured meats-an unconventional approach. *Food Technol.*, **40**, 155–8.

Hechelmann, H. and Kasprowiak, R. (1992) Microbiological criteria for stable products. *Fleischwirtsch. Int.*, **1**, 4–18.

Helps, C.R., Harbour, D.A. and Corry, J.E. (1999) PCR-based 16S ribosomal DNA detection technique for *Clostridium estertheticum* causing spoilage in vacuum-packed chill-stored beef. *Int. J. Food Microbiol.*, **52**(1–2), 57–65.

Heuvelink, A.E., Van Den Biggelaar, F.L.A.M., De Boer, E., Herbes, R.G., Melchers, W.J.G., Huis In'T Veld, J.H.J. and Monnens, L.A.H. (1998) Isolation and characterization of verocytotoxin-producing *Escherichia coli* O157 strains from Dutch cattle and sheep. *J. Clin. Microbiol.*, **36**, 878–82.

Heuvelink, A., Roessink, G., Bosboom, K., De Boer, E. (2001) Zero-tolerance for fecal contamination of carcasses as a tool in the control of O157 VTEC infections. *Int. J. Food Microbiol.*, **66**, 13–20.

Hinton, M., Ali, E.A., Allen, V. and Linton, A.H. (1983) The excretion of *Salmonella typhimurium* in the faeces of calves fed milk substitute. *J. Hyg. (Camb.)*, **91**, 33–45.

Hogue, A.T., Dreesen, D.W., Green, S.S., Ragland, R.D., James, W.O., Bergeron, E.A., Cook, L.V., Pratt, M.D. and Martin, D.R. (1993) Bacteria on beef briskets and ground beef: correlation with slaughter volume and antemortem condemnation. *J. Food Prot.*, **56**, 110–3, 119.

Houtsma, P.C., de Wit, J.C. and Rombouts, F.M. (1993) Minimum inhibitory concentration (MIC) of sodium lactate for pathogens and spoilage organisms occurring in meat products. *Int. J. Food Microbiol.*, **20**, 247–57.

Howe, K., Linton, A.H. and Osborne, A.D. (1976) A longitudinal study of *Escherichia coli* in cows and calves with special reference to the distribution of O-antigen types and antibiotic resistance. *J. Appl. Bacteriol.*, **40**, 331–40.

Huang, L. and Juneja, V.K. (2003) Thermal inactivation of *Escherichia coli* O157:H7 in ground beef supplemented with sodium lactate. *J. Food Prot.*, **66**(4), 664–7.

Hudson, W.R., Roberts, T.A. and Whelehan, O.P. (1987) Bacteriological status of beef carcasses at a commercial abattoir before and after slaughterline improvements. *Epidemiol. Infect.*, **98**, 81–6.

Hume, M.E., Nisbet, D.J., Buckley, S.A., Ziprin, R.L., Anderson, R.C. and Stanker, L.H. (2001a) Inhibition of in vitro *Salmonella typhimurium* colonization in porcine cecal bacteria continuous-flow competitive exclusion cultures. *J. Food Prot.*, **64**(1), 17–22.

Hume, M.E., Harvey, R.B., Stanker, L.H., Droleskey, R.E., Poole, T.L. and Zhang, H.B. (2001b) Genotypic variation among arcobacter isolates from a farrow-to-finish swine facility. *J. Food Prot.*, **64**(5), 645–51.

Husu, J.R. (1990) Epidemiological studies on the occurrence of *Listeria monocytogenes* in the feces of dairy cattle. *J. Vet. Med., B*, **37**, 267–82.

Hunter, P.R. and Izsak, J. (1990) Diversity studies of salmonella incidents in some domestic livestock and their potential relevance as indicators of niche width. *Epidemiol. Infect.*, **105**, 501–10.

Hunter, D., Hughes, S. and Fox, E. (1983) Isolation of *Yersinia enterocolitica* from pigs in the United Kingdom. *Veterinary Record*, **112**, 322–3.

ICMSF (International Commission on Microbiological Specifications for Foods). (1980a) *Microbial Ecology of Foods, 1, Factors Affecting Life and Death of Microorganisms*, Academic Press, New York.

ICMSF (International Commission on Microbiological Specifications for Foods). (1980b) *Microbial Ecology of Foods, 2, Food Commodities*, Academic Press, New York.

ICMSF (International Commission on Microbiological Specifications for Foods). (1988a) *Microorganisms in Foods, 4, Application of the Hazard Analysis Critical Control Point (HACCP) System to Ensure Microbiological Safety And Quality*, Blackwell Scientific Publications, Oxford.

ICMSF (International Commission on Microbiological Specifications for Foods). (1998b) *Microorganisms in Foods, 6, Microbial Ecology of Food Commodities*, Academic Press, New York, USA.

ICMSF (International Commission on Microbiological Specifications for Foods). (2002) *Microorganisms in Foods, 7, Microbiological Testing in Food Safety Management*, (Chapter 17) *E. coli* O157:H7 in *Frozen Raw Ground Beef Patties*, Kluwer Academic/Plenum Publishers, New York, USA.

Iida, T., Kanzaki, M., Nakama, A., Kokubo, Y., Maruyama, T. and Kaneuchi, C. (1998) Detection of *Listeria monocytogenes* in humans, animals and foods. *J. Vet. Med. Sci.*, **60**, 1341–3.

Incze, K. (1987) The technology and microbiology of Hungarian salami. Tradition and current status. *Fleischwirtschaft*, **67**, 445–7.

Incze, K. (1992) Raw fermented and dried meat products. *Fleischwirtsch. Int.*, **2**, 3–12.

Ingham, S.C. and Schmidt, D.J. (2000) Alternative indicator bacteria analysis for evaluating the sanitary condition of beef carcasses. *J. Food Prot.*, **63**, 51–5.

Inoue, S., Nakama, A., Arai, Y., Kokubo, Y., Maruyama, T., Saito, A., Yoshida, T., Terao, M., Yamamoto, S. and Kumagai, S. (2000) Prevalence and contamination levels of *Listeria monocytogenes* in retail foods in Japan. *Int. J. Food Microbiol.*, **59**, 73–7.

James, S.J., Creed, P.G. and Roberts, T.A. (1977) Air thawing of beef quarters. *J. Sci. Food Agric.*, **28**, 1109–19.

Johanson, L., Underdahl, B., Grosland, K., Whelehan, O.P. and Roberts, T.A. (1983) A survey of the hygienic quality of beef and pork carcasses in Norway. *Acta Vet. Scand.*, **24**, 1–13.

Johnson, J.L., Doyle, M.P., Cassens, R.G. and Schoeni, J.L. (1988) Fate of *Listeria monocytogenes* in tissues of experimentally infected cattle and in hard salami. *Appl. Environ. Microbiol.*, **54**, 497–501.

Johnson, J.L., Doyle, M.P. and Cassens, R.G. (1990) *Listeria monocytogenes* and other *Listeria* spp. in meat and meat products. A review. *J. Food Prot.*, **53**, 81–91.

Jones, P.W., Collins, P., Brown, G.T.H. and Aitken, M. (1982) Transmission of *Salmonella mbandaka* to cattle from contaminated feed. *J. Hyg. (Camb.)*, **88**, 255–63.

Jones, B., Nilsson, T. and Sorqvist, S. (1984) Contamination of pig carcasses with scalding water. Continued studies with radiolabelled solutes and particles. *Fleischwirtschaft*, **64**, 1226–8.

Jordan, D., McEwen, S., Lammerding, A., McNab, W. and Wilson, B. (1999) Pre-slaughter control of *Escherichia coli* O157 in beef cattle: a simulation study. *Prev. Vet. Med.*, **41**, 55–74.

Junttila, J., Hirn, J., Hill, P. and Nurmi, E. (1989) Effect of different levels of nitrite and nitrate on the survival of *Listeria monocytogenes* during the manufacture of fermented sausage. *J. Food Prot.*, **52**, 158–61.

Kaddu-Mulindwa, D., Aisu, T., Gleier, K., Zimmermann, S. and Beutin, L. (2001) Occurrence of shiga toxin-producing *Escherichia coli* in fecal samples from children with diarrhea and from healthy zebu cattle in Uganda. *Int. J. Food Microbiol.*, **66**, 95–101.

Kalchayanand, N., Ray, B., Field, R.A. and Johnson, M.C. (1989) Spoilage of vacuum-packed beef by a *Clostridium*. *J. Food Prot.*, **52**, 424–6.

Kalchayanand, N., Ray, B. and Field, R.A. (1993) Characteristics of psychrotrophic *Clostridium laramie* causing spoilage of vacuum-packaged refrigerated fresh and roasted beef. *J. Food Prot.*, **56**, 13–7.

Kalinowski, R. and Tompkin, R.B. (1999) Psychrotrophic clostridia causing spoilage in cooked meat and poultry products. *J. Food Prot.*, **62**(7), 766–72.

Kalinowski, R.M., Tompkin, R.B., Bodnaruk, P.W. and Pruett, W.P., Jr. (2003) Impact of cooking, cooling, and subsequent refrigeration on the growth or survival of *Clostridium perfringens* in cooked meat and poultry products. *J. Food Prot.*, **66**(7), 1227–32.

Kampelmacher, E.H., Guinee, P.A.M., Hofstra, K. and van Keulen, A. (1963) Further studies on salmonella in slaughterhouses and in normal slaughter pigs. *Zen. Vet. Reihe B*, **10**, 1–27.

Kane, D.W. (1979) The prevalence of salmonella infection in sheep at slaughter. *NZ Vet. J.*, **27**, 110–3.

Kang, D.H., Koohmaraie, M., Dorsa, W.J. and Siragusa, G.R. (2001a) Development of a multiple-step process for the microbial decontamination of beef trim. *J. Food Prot.*, **64**(1), 63–71.

Kang, D.H., Koohmaraie, M. and Siragusa, G.R. (2001b) Application of multiple antimicrobial interventions for microbial decontamination of commercial beef trim. *J. Food Prot.*, **64**(2), 168–71.

Käsbohrer, A., Protz, D., Helmuth, R., Nöckler, K., Blaha, T., Conraths, F.J. and Geue, L. (2000) *Salmonella* in slaughter pigs of German origin: an epidemiological study. *Eur. J. Epidemiol.*, **16**, 141–6.
Kasprowiak, R. and Hechelmann, H. (1992) Weak points in the hygiene of slaughtering, cutting and processing firms. *Fleischwirtsch. Int.*, **2**, 32–40.
Kelly, C.A., Lynch, B. and McLoughlin, A.J. (1980) The microbiological quality of Irish lamb carcasses. *Irish J. Food Sci. Technol.*, **4**, 125–31.
Kelly, C.A., Dempster, J.F. and McLoughlin, A.J. (1981) The effect of temperature, pressure and chlorine concentration of spray washing water on numbers of bacteria on lamb carcases. *J. Appl. Bacteriol.*, **51**, 415–24.
Keteran, K., Brown, J. and Shotts, E.B. (1982) Salmonella in the mesenteric lymph nodes of healthy sows and hogs. *Am. J. Vet. Res.*, **43**, 706–7.
Kim, M. and Doyle, M. (1992) Dipstick immunoassay to detect enterohemorrhagic *Escherichia coli* O157:H7 in retail ground beef. *Appl. Environ. Microbiol.*, **58**, 2693–8.
Klein, G. (2003) Taxonomy, ecology and antibiotic resistance of enterococci from food and the gastro-intestinal tract. *Int. J. Food Microbiol.*, **88**(2–3), 123–31.
Kleinlein, N. and Untermann, F. (1990) Growth of pathogenic *Yersinia enterocolitica* strains in minced meat with and without protective gas with consideration of the competitive background flora. *Int. J. Food Microbiol.*, **10**, 65–72.
Kloos, W.E. (1980) Natural populations of the genus *Staphylococcus*. *Ann. Rev. Microbiol.*, **34**, 559–92.
Kramer, J.M. and Gilbert, R.J. (1989) *Bacillus cereus* and other *Bacillus* species, in *Foodborne Bacterial Pathogens* (ed M.P. Doyle) Marcel Dekker Inc., New York, pp. 22–70.
Kramer, J.M., Frost, J.A., Bolton, F.J. and Wareing, D.R.A. (2000) *Campylobacter* contamination of raw meat and poultry at retail sale: identification of multiple types and comparison with isolates from human infection. *J. Food Prot.*, **63**, 1654–9.
Kudva, I.T., Hatfield, P.G. and Hovde, C.J. (1997) Characterization of *Escherichia coli* O157:H7 and other Shiga toxin-producing *E. coli* serotypes isolated from sheep. *J. Clin. Microbiol.*, **35**, 892–9.
Kumur, S., Saxena, S.P. and Gupta, B.K. (1973) Carrier rate of salmonellas in sheep and goats and its public health significance. *J. Hyg. (Camb.)*, **71**, 43–7.
Laegreid, W.W., Elder, R.O. and Keen, J.E. (1999) Prevalence of *Escherichia coli* O157:H7 in range beef calves at weaning. *Epidemiol. Infect.*, **123**(2), 291–8.
Lammerding, A.M., Garcia, M.M., Mann, E.D., Robinson, Y., Dorward, W.J., Truscott, R.B. and Tittiger, F. (1988) Prevalence of *Salmonella* and thermophilic *Campylobacter* in fresh pork, beef, veal and poultry in Canada. *J. Food Prot.*, **51**, 47–52.
Laubach, C., Rathgeber, J., Oser, A. and Palumbo, S. (1998) Microbiology of the swine head meat deboning process. *J. Food Prot.*, **61**, 249–52.
Lawrie, R.A. (1985) *Meat Science*, 4th edn, Pergamon Press, Oxford.
Lawson, P., Dainty, R.H., Kristiansen, N., Berg, J. and Collins, M.D. (1994) Characterization of a psychrotrophic *Clostridium* causing spoilage in vacuum-packed cooked pork: description of *Clostridium algidicarnis* sp. nov. *Lett. Appl. Microbiol.*, **19**, 153–7.
Lee, J.A., Ghosh, A.C., Mann, P.G. and Tee, G.H. (1972) Salmonellas on pig farms and in abattoirs. *J. Hyg. (Camb.)*, **70**, 141–50.
Lee, W.H., Vanderzant, C. and Stern, N. (1981) The occurrence of *Yersinia enterocolitica* in foods, in *Yersinia enterocolitica* (ed E.J. Bottone), CRC Press, Inc., Boca Raton, Fla., pp. 161–71.
Lee, A.A., Gerber, A.R., Lonsway, D.R., Smith, J.D., Carter, G.P., Puhr, N.D., Parrish, C.M., Sikes, R.K., Fintarn, R.J. and Tauxe, R.V. (1990) *Yersinia enterocolitica* O:3 infections in infants and children, associated with the household preparation of chitterlings. *New England J. Med.*, **322**, 984–7.
Lee, L.A., Taylor, J., Carter, G.P., Quinn, B., Farmer, J.J., III and Tauxe, R.V. (1991) *Yersinia enterocolitica* O:3: an emerging cause of paediatric gastroenteritis in the United States. *J. Infect. Dis.*, **163**, 660–3.
Leistner, L. (1988) Shelf-stable oriental meat products, in *34th International Congress of Meat Science and Technology*, Brisbane, pp. 470–5.
Leistner, L. and Rödel, W. (1976) The stability of intermediate moisture foods with respect to microorganisms, in *Intermediate Moisture Foods* (eds R. Davies, G. Birch and K. Parker), Applied Science Publishers Ltd, London, pp. 120–37.
Letellier, A., Messier, S. and Quessy, S. (1999) Prevalence of *Salmonella* spp. and *Yersinia enterocolitica* in finishing swine at Canadian abattoirs. *J. Food Prot.*, **62**, 22–5.
Lindberg, C.W. and Borch, E. (1994) Predicting the aerobic growth of *Yersinia enterocolitica* O:3 at different pH values, temperatures and L-lactate concentrations using conductance measurements. *Int. J. Food Microbiol.*, **22**, 141–53.
Lindqvist, R., Antonsson, A., Norling, B., Persson, L., Ekstrom, A., Fager, U., Eriksson, E., Lofdahl, S. and Norberg, P. (1998) The prevalence of verocytotoxin-producing *Escherichia coli* (VTEC) and *E. coli* O157:H7 in beef in Sweden determined by PCR assays and an immunomagnetic separation (IMS) method. *Food Microbiol.*, **15**, 591–601.
Linton, A.H. (1979) Salmonellosis in pigs. *Br. Vet. J.*, **135**, 109–12.
Linton, A.H., Jennett, N.E. and Heard, T.W. (1970) Multiplication of *Salmonella* in liquid feed and its influence on the duration of excretion in pigs. *Res. Vet. Sci.*, **11**, 452–7.
Lior, H. (1994) *Escherichia coli* 0157:H7 and verotoxigenic *Escherichia coli* (VETC). *Dairy, Food Environ. Sanit.*, **14**, 378–82.
Loncarevic, S., Milanovic, A., Caklovica, F., Tham, W. and Danielsson-Tham, M.-L. (1994) Occurrence of *Listeria* species in an abattoir for cattle and pigs in Bosnia and Hercegovina. *Acta Vet. Scand.*, **35**, 11–5.
Lowry, P.D. and Gill, C.O. (1984) Development of a yeast microflora on frozen lamb stored at −5°C. *J. Food Prot.*, **47**, 309–11.
Lowry, P.D. and Tiong, I. (1988) The incidence of *Listeria monocytogenes* in meat and meat products: factors affecting distribution, in *Proceedings, 34th International Congress of Meat Science and Technology*, Part B, Brisbane, Australia, pp. 528–30.

Lücke, F.-K. (1986) Microbiological processes in the manufacture of dry sausage and raw ham. *Fleischwirtschaft*, **66**, 1505–9.
Lücke, F.-K. (2000) Fermented meats, Chapter 19, in *The Microbiological Safety and Quality of Food* (eds B.M. Lund, T.C. Baird-Parker and G.W. Gould), Aspen Publishers Inc., Gaithersburg MD, pp. 420–44.
Lücke, F.-K. and Roberts, T.A. (1993) Control in meat and meat products, in *Clostridium botulinum: Ecology and Control in Foods* (eds A.H.W. Hauschild and K.L. Dodds), Marcel Dekker Inc., New York, pp. 177–207.
Lücke, F.-K., Hechelmann, H. and Leistner, L. (1981) The relevance to meat products of psychrotrophic strains of *Clostridium botulinum*, in *Psychrotrophic Microorganisms in Spoilage and Pathogenicity* (eds T.A. Roberts, G. Hobbs, J.H.B. Christian and N. Skovgaard), Academic Press, London, pp. 491–7.
Lunden, J.M., Autio, T.J. and Korkeala, H.J. (2002) Transfer of persistent *Listeria monocytogenes* contamination between food-processing plants associated with a dicing machine. *J. Food Prot.*, **65**(7), 1129–33.
Mackey, B.M. and Derrick, C.M. (1979) Contamination of the deep tissue of carcasses by bacteria present on the slaughter instruments or in the gut. *J. Appl. Bacteriol.*, **46**, 355–66.
Mackey, B.M. and Kerridge, A.L. (1988) The effect of incubation temperature and inoculum size on growth of salmonellae in minced beef. *Int. J. Food Microbiol.*, **6**, 57–65.
Mackey, B.M. and Roberts, T.A. (1993) Improving slaughter hygiene using HACCP and monitoring. *Fleischwirtsch. Int.*, **2**, 40–5.
Mackey, B.M., Roberts, T.A., Mansfield, J. and Farkas, G. (1980) Growth of *Salmonella* on chilled meat. *J. Hyg. (Camb.)*, **85**, 115–24.
Mackey, B.M., Pritchet, C., Norris, A. and Mead, G.C. (1990) Heat resistance of *Listeria*: strain differences and effects of meat type and curing salts. *Lett. Appl. Microbiol.*, **10**, 251–5.
Madden, R.H., Moran, L. and Scates, P. (1998) Frequency of occurrence of *Campylobacter* spp. in red meats and poultry in Northern Ireland and their subsequent subtyping using polymerase chain reaction-restriction fragment length polymorphism and the random amplified polymorphic DNA method. *J. Appl. Microbiol.*, **84**, 703–8.
Mafu, A.A., Higgins, R., Nadeau, M. and Coustineau, G. (1989) The incidence of *Salmonella*, *Campylobacter*, and *Yersinia enterocolitica* in swine carcasses and the slaughterhouse environment. *J. Food Prot.*, **52**, 642–5.
Makela, P.M. (1992) Fermented sausage as a contamination source of ropy slime-producing lactic acid bacteria. *J. Food Prot.*, **55**, 48–51.
Makela, P., Schillinger, U., Korkeala, H. and Holzapfel, W.H. (1992) Classification of ropy slime-producing lactic acid bacteria on DNA–DNA homology, and identification of *Lactobacillus sake* and *Leuconostoc amelibiosum* as dominant spoilage organisms in meat products. *Int. J. Food Microbiol.*, **16**, 167–72.
Marin, M.E., de la Rosa, M.d.C. and Cornejo, I. (1992) Enterotoxigenicity of *Staphylococcus* strains isolated from Spanish dry-cured hams. *Appl. Environ. Microbiol.*, **58**, 1067–9.
Marinculic, A., Gaspar, A., Durakovic, E., Pozio, E. and La Rosa, G. (2001) Epidemiology of swine trichinellosis in Croatia. *Parasite*, **8**, 92–4.
Martin, D.R., Uhler, P.M., Okrend, A.J.G. and Chiu, J.Y. (1994) Testing bob calf fecal swabs for the presence of *Escherichia coli* O157:H7. *J. Food Prot.*, **57**, 70–2.
Mattila, T., Frost, A.J. and O'Boyle, D. (1988) The growth of salmonella in rumen fluid from cattle at slaughter. *Epidemiol. Infect.*, **101**, 337–45.
McCaughey, W.J., McClelland, T.G. and Hanna, J. (1971) Some observations on *Salmonella dublin* in clinically healthy beef cattle. *Br. Vet. J.*, **127**, 549–56.
McDonough, P.L., Rossiter, C.A., Rebhun, R.B., Stehman, S.M., Lein, D.H. and Shin, S.J. (2000) Prevalence of *Escherichia coli* O157:H7 from cull dairy cows in New York state and comparison of culture methods used during preharvest food safety investigations. *J. Clin. Microbiol.*, **38**, 318–22.
McEvoy, J.M., Doherty, A.M., Sheridan, J.J., Blair, I.S. and McDowell, D.A. (2001) Use of steam condensing at subatmospheric pressures to reduce *Escherichia coli* O157:H7 numbers on bovine hide. *J. Food Prot.*, **64**(11), 1655–60.
McLauchlin, J., Hall, S.M., Velani, S.K. and Gilbert, R.J. (1991) Human listeriosis and pate: a possible association. *Br. Med. J.*, **303**, 773–5.
McNamara, A.M. (1995) Establishment of baseline data on the microbiota of meats. *J. Food Saf.*, **15**, 113–9.
Meers, P.D. and Goode, D. (1965) *Salmonella typhi* in corned beef. *Lancet*, **i**, 426.
Meng, J., Doyle, M.P., Zhao, T. and Zhao, S. (1994) Detection and control of *Escherichia coli* O157:H7 in foods. *Trends Sci. Technol.*, **5**, 179–85.
Midura, T.F., Nygaard, G.S., Wood, R.M. and Bodily, H.J. (1972) *Clostridium botulinum* type F; Isolation from venison jerky. *Appl. Microbiol.*, **24**, 165–7.
Milne, D. (1967) The Aberdeen typhoid outbreak, 1964, *Report of the Departmental Committee of Enquiry*, Scottish Home and Health Department, Edinburgh.
Miyao, Y., Kataoka, T., Nomoto, T., Kai, A., Itoh, T. and Itoh, K. (1998) Prevalence of verotoxin-producing *Escherichia coli* harbored in the intestine of cattle in Japan. *Vet. Microbiol.*, **61**, 137–43.
Mohammad, A., Peiris, J.S.M., Wijewanta, E.A., Mahalingam, S. and Gunasekara, G. (1985) Role of verotoxigenic *Escherichia coli* in cattle and buffalo calf diarrhoea. *FEMS Microbiol. Lett.*, **26**, 281–3.
Molina, I., Silla, H. and Monzo, J.L. (1990) Study of the microbial flora in dry-cured ham. 2. *Micrococcaceae. Fleschwirtsch. Int.*, **2**, 47–8.
Montenegro, M.M., Bulte, M., Trumpf, T., Aleksic, S., Reuter, G., Bulling, E. and Helmuth, R. (1990) Detection and characterization of fecal verotoxin-producing *Escherichia coli* from healthy cattle. *J. Clin. Microbiol.*, **28**, 1417–21.

Moore, J.E. and Madden, R.H. (1998) Occurrence of thermophilic *Campylobacter* spp. in porcine liver in Northern Ireland. *J. Food Prot.*, **61**, 409–13.
Morgan, I.R., Krautil, F.L. and Craven, J.A. (1987a) Bacterial populations on dressed pig carcasses. *Epidemiol. Infect.*, **98**, 15–24.
Morgan, I.R., Krautil, F.L. and Craven, J.A. (1987b) Effect of time in lairage on caecal and carcass salmonella contamination of slaughter pigs. *Epidemiol. Infect.*, **98**, 323–30.
Moxley, R.A. and Francis, D.H. (1986) Natural and experimental infection with an attaching and effacing strain of *Escherichia coli* in calves. *Infect. Immun.*, **53**, 339–46.
Mukartini, S., Jehne, C., Shay, B. and Harper, C.M.L. (1995) Microbiological status of beef carcass meat in Indonesia. *J. Food Saf.*, **15**, 291–303.
Murray, K.A., Gilmour, A. and Madden, R.H. (2001) Microbiological quality of chilled beef carcasses in Northern Ireland: a baseline survey. *J. Food Prot.*, **64**(4), 498–502.
Nabbut, N.H. and Al-Nakhli, H.M. (1982) Incidence of salmonellae in lymph nodes, spleens and feces of sheep and goats slaughtered in the Riyadh public abattoir. *J. Food Prot.*, **45**, 1314–7.
NACMCF. (1993) Generic HACCP for raw beef: National Advisory Committee on Microbiological Criteria for Foods US Department of Agriculture. *Food Microbiol.*, **10**, 449–88.
NACMCF. (1995) *Campylobacter jejuni/coli. Dairy, Food Environ. Sanit.*, **15**, 133–53.
Nadon, C.A., Ismond, M.A. and Holley, R. (2001) Biogenic amines in vacuum-packaged and carbon dioxide-controlled atmosphere-packaged fresh pork stored at −1.5 degrees C. *J. Food Prot.*, **64**(2), 220–7.
Naylor, S., Low, J., Besser, T., Mahajan, A., Gunn, G., Pearce, M., McKendrick, I., Smith, D. and Gally, D. (2003) Lymphoid follicle-dense mucosa at the terminal rectum is the principal site of colonization of Enterohemorrhagic *Escherichia coli* O157:H7 in the bovine host. *Infect. Immun.*, **71**, 1505–12.
Nazer, A.H.K. and Osborne, A.D. (1976) Salmonella infection and contamination of veal calves: a slaughterhouse survey. *Br. Vet. J.*, **132**, 192–210.
Necker, D.P., Kumta, U.S. and Sreenivasan, A. (1978) Radiation processing for the control of *Salmonella* in frog legs. *Use Radiat. Radioisot. Stud. Anim., Proc. Radiat. Symp.*, 1975, Izatnagar, India.
Nesbakken, T. (1988) Enumeration of *Yersinia enterocolitica* O:3 from the porcine oral cavity, and its occurrence on cut surfaces of pig carcasses and the environment in a slaughterhouse. *Int. J. Food Microbiol.*, **6**, 287–93.
Nesbakken, T. and Kapperud, G. (1985) *Yersinia enterocolitica* and *Yersinia enterocolitica*-like bacteria in Norwegian slaughter pigs. *Int. J. Food Microbiol.*, **1**, 301–9.
Nesbakken, T., Nerbrink, E., Røtterud, O.-J. and Borch, E. (1994) Reduction of *Yersinia enterocolitica* and *Listeria* spp. on pig carcasses by enclosure of the rectum during slaughter. *Int. J. Food Microbiol.*, **23**, 197–208.
Nesbakken, T., Eckner, K., Høidal, H.K. and Røtterud, O.-J. (2003) Occurrence of *Yersinia enterocolitica* and *Campylobacter* spp. in slaughter pigs and consequences for meat inspection, slaughtering, and dressing procedures. *Int. J. Food Microbiol.*, **80**, 231–40.
Newton, K.G. and Gill, C.O. (1978) Storage quality of dark, firm, dry meat. *Appl. Environ. Microbiol.*, **36**, 375–6.
Newton, K.G., Harrison, J.C.L. and Wauters, A.M. (1978) Sources of psychrotrophic bacteria on meat at the abattoir. *J. Appl. Bacteriol.*, **45**, 75–82.
NFPA. (1989) *Guidelines for the Development, Production and Handling of Refrigerated Foods*, National Food Processors Association, US, Microbiology and Food Safety Committee.
Nickelson, R., Wyatt, L.E. and Vanderzant, C. (1975) Reduction of *Salmonella* contamination in commercially processed frog legs. *J. Food Sci.*, **40**, 1239–41.
Nielsen, H.-J.S. (1983a) Influence of nitrite addition and gas permeability of packaging film on the microflora in a sliced vacuum-packed whole meat product under refrigeration. *J. Food Technol.*, **18**, 573–85.
Nielsen, H.-J.S. (1983b) Influence of temperature and gas permeability of packaging film on development and composition of microbial flora in vacuum-packed bologna-type sausage. *J. Food Prot.*, **46**, 693–8.
Nielsen, H.-J.S. and Zeuthen, P. (1984) Growth of pathogenic bacteria in sliced vacuum-packed Bologna-type sausage as influenced by temperature and gas permeability of packaging film. *Food Microbiol.*, **1**, 229–43.
Nissen, H., Alvseike, O., Bredholt, S. and Nesbakken, T. (2000) Comparison between growth of *Yersinia enterocolitica*, *Listeria monocytogenes*, *Escherichia coli* O157:H7 and *Salmonella* spp. in ground beef packed by three commercially used packaging techniques. *Int. J. Food Microbiol.*, **59**, 211–20.
Nottingham, P.M. and Urselmann, A.J. (1961) Salmonella infection in calves and other animals. *NZ J. Agric. Res.*, **4**, 449–60.
Nottingham, P.M., Penney, N. and Wyborn, R. (1972) Salmonella infection in calves and other animals. III. Further studies with calves and pigs. *NZ J. Agric. Res.*, **15**, 279–83.
Obi, S.K.C. and Nzeako, B.C. (1980) *Salmonella, Arizona, Shigella* and *Aeromonas* isolated from the snail *Achatina achatina* in Nigeria. *Antonie van Leeuwenhoek*, **46**, 475–81.
Oblinger, J.L., Kennedy, J.E., Jr., Rothenberg, C.A., Berry, B.W. and Stern, N.J. (1982) Identification of bacteria isolated from fresh and temperature abused variety meats. *J. Food Prot.*, **45**, 650–4.
Offermann, U., Bodmer, T., Audigé, L. and Jemmi, T. (1999) Verbreitung von Salmonellen, Yersinien und Mykobakterien bei Schlachtschweinen in der Schweiz [The prevalence of salmonella, yersinia and mycobacteria in slaughtered pigs in Switzerland]. *Schweizer Archiv fur Tierheilkunde*, **141**, 509–515.
Ono, K. and Yamamoto, K. (1999) Contamination of meat with *Campylobacter jejuni* in Saitama, Japan. *Int. J. Food Microbiol.*, **47**, 211–9.
Oosterom, J. and Notermans, S. (1983) Further research into the possibility of salmonella-free fattening and slaughter of pigs. *J. Hyg. (Camb.)*, **91**, 59–69.

Oosterom, J., de Wilde, G.J.A., de Boer, E., de Blaauw, L.H. and Karman, H. (1983) Survival of *Campylobacter jejuni* during poultry processing and pig slaughtering. *J. Food Prot.*, **46**, 702–6 and 709.
Osano, O. and Arimi, S.M. (1999) Retail poultry and beef as sources of *Campylobacter jejuni*. *E. African Med. J.*, **76**, 141–3.
Osterhoff, D.R. and Leistner, L. (1984) [South African biltong–another close look] [Article in Afrikaans], J S Afr Vet Assoc. 1984 Dec;55(4), 201–2 [abstract in English].
Ostroff, S.M., Kapperud, G., Hutwagner, L.C., Nesbakken, T., Bean, N.H., Lassen, J. and Tauxe, R.V. (1994) Sources of sporadic *Yersinia enterocolitica* infections in Norway: a prospective case–control study. *Epidemiol. Infect.*, **112**, 133–41.
Palumbo, S.A. (1988) The growth of *Aeromonas hydrophila* K144 in ground pork at 5°C. *Int. J. Food Microbiol.*, **7**, 41–8.
Pantaleon, J. and Rosset, R. (1964) Sur la présence de *Salmonella* dans les grenouilles destinée a la consommation humaine. *Ann. Institut. Pasteur, Lille*, **15**, 225–7.
Parma, A.E., Sanz, M.E., Blanco, J.E., Blanco, J., Viñas, M.R., Blanco, M., Padola, N.L. and Etcheverría, A.I. (2000) Virulence genotypes and serotypes of verotoxigenic *Escherichia coli* isolated from cattle and foods in Argentina. *Eur. J. Epidemiol.*, **16**, 757–62.
Patterson, J.T. and Gibbs, P.A. (1978) Sources and properties of some organisms isolated in two abattoirs. *Meat Sci.*, **2**, 263–73.
Patterson, J.T. and Gibbs, P.A. (1979) Vacuum-packaging of bovine edible offal. *Meat Sci.*, **3**, 209–22.
Peel, B., Bothwell, J., Simmons, G.C. and Frost, A. (1975) A study of the number and phage patterns of *Staphylococcus aureus* in an abattoir. *Aust. Vet. J.*, **51**(3), 126–30.
Peintner, U., Geiger, J. and Poder, R. (2000) The mycobiota of speck, a traditional Tyrolean smoked and cured ham. *J. Food Prot.*, **63**(10), 1399–403.
Phebus, R., Nutsch, A., Schafer, D., Wilson, R., Riemann, M., Leising, J., Kastner, C., Wolf, J. and Prasai, R. (1997) Comparison of steam pasteurization and other methods for reduction of pathogens on surfaces of freshly slaughtered beef. *J. Food Prot.*, **60**, 476–84.
Phillips, C.A. (1998) The isolation of *Campylobacter* spp. from modified atmosphere packaged foods. *Int. J. Envir. Hlth Res.*, **8**, 215–21.
Phillips, D., Sumner, J., Alexander, J.F. and Dutton, K.M. (2001a) Microbiological quality of Australian beef. *J. Food Prot.*, **64**(5), 692–6.
Phillips, D., Sumner, J., Alexander, J.F. and Dutton, K.M. (2001b) Microbiological quality of Australian sheep meat. *J. Food Prot.*, **64**(5), 697–700.
Pidcock, K., Heard, G.M. and Henriksson, A. (2002) Application of non-traditional starter cultures in production of Hungarian salami. *Int. J. Food Microbiol.*, **76**, 75–81.
Pilon, J., Higgins, R. and Quessy, S. (2000) Epidemiological study of *Yersinia enterocolitica* in swine herds in Québec. *Can. Vet. J.*, **41**, 383–7.
Podolak, R.K., Zayas, J.F., Kastner, C.L. and Fung, D.Y.C. (1995) Reduction of *Listeria monocytogenes*, *Escherichia coli* O157:H7 and *Salmonella typhimurium* during storage of beef sanitized with fumaric, acetic and lactic acids. *J. Food Saf.*, **15**(3), 283–90.
Podolak *et al.* (1996) found fumaric acid at concentrations at concentrations of 1 and 1.5% to be more effective than 1% lactic or acetic acids in reducing populations of *E. coli* O157:H7 in vacuum packaged ground beef patties.
Pos, H.G. (1990) Production of clean snail eggs of *Helix aspersa* (Muller). *Snail Farming Res.*, **3**, 1–5.
Power, C.A., Johnson, R.P., McEwen, S.A., McNab, W.B., Griffiths, M.W., Usborone, W.R. and De Grandis, S.A. (2000) Evaluation of the Reveal and SafePath Rapid *Escherichia coli* O157 detection tests for use on bovine feces and carcasses. *J. Food Prot.*, **63**, 860–6.
Pozio, E. (1998) Trichinellosis in the European Union: epidemiology, ecology and economic impact. *Parisitol. Today*, **14**, 35–8.
Pozio, E. (2001) New patterns of *Trichinella* infection. *Vet. Parasitol.*, **98**, 133–48.
Pradel, N., Livrelli, V., De Champs, C., Palcoux, J.-B., Reynaud, A., Scheutz, F., Sirot, J., Joly, B. and Forestier, C. (2000) Prevalence and characterization of Shiga toxin-producing *Escherichia coli* isolated from cattle, food, and children during a one-year prospective study in France. *J. Clin. Microbiol.*, **38**, 1023–31.
Prieto, M., Garcia, M.L., Garcia, M.R., Otero, A. and Moreno, B. (1991) Distribution and evolution of bacteria on lamb carcasses during aerobic storage. *J. Food Prot.*, **54**, 945–9.
Prieto, M., Garcia-Armesto, M.R., Garcia-Lopez, M.L., Otero, A. and Moreno, B. (1992) Numerical taxonomy of Gram-negative, nonmotile, nonfermentative bacteria isolated during chilled storage of lamb carcasses. *Appl. Environ. Microbiol.*, **58**, 2245–9.
Prior, B.A. (1984) Role of micro-organisms in biltong flavour development. *J. Appl. Bacteriol.*, **56**, 41–5.
Skovgaard, N. and Nielsen, B.B., Public Health Laboratory Service Working Group. (1972) Salmonellas in pigs and animal feeding stuffs in England and Wales and in Denmark. *J. Hyg. (Camb.)*, **70**, 127–40.
Qvist, S. and Mukherji, S. (1981) *Brochothrix thermosphacta*—an important spoilage agent in vacuum-packaged sliced meat products, in *Psychrotrophic Microorganisms in Spoilage and Pathogenicity* (eds T.A. Roberts, G. Hobbs, J.H.B. Christian and N. Skovgaard), Academic Press, London, pp.223–30.
Raccach, M. and Henningsen, E.C. (1984) Role of lactic acid bacteria, curing salts, spices and temperature in controlling the growth of *Yersinia enterocolitica*. *J. Food Prot.*, **47**, 354–8.
Rahkio, M., Korkeala, H., Sippola, I. and Peltonen, M. (1992) Effect of pre-scalding brushing on contamination level of pork carcasses during the slaughter process. *Meat Sci.*, **32**, 173–83.
Ralovich, B. (1984) Listeria research—present situation and perspective. *Academiai Kiado*, Budapest.
Rao, D.N., Sreenivasamurthy, V. (1985) A note on the microbial spoilage of sheep meat at ambient temperature. *J. Appl. Bacteriol.*, **58**, 457–60.
Rao, N.M., Nandy, S.C., Joseph, K.T. and Santappa, M. (1978) Control of *Salmonella* in frog legs by chemical and physical methods. *Indian J. Exp. Biol.*, **16**, 593–6.

Rasch, B., Lie, O. and Yndestad, M. (1978) The bacterial flora in pork skin and the influence of various singeing methods on this flora. *Nord. Vet. Med.*, **30**, 274–81.

Rasmussen, M.A., Cray, W.C., Casey, T.A. and Whipp, S.C. (1993) Rumen contents as a reservoir of enterohemorrhagic *Escherichia coli*. *FEMS Microbiol. Lett.*, **114**, 79–84.

Reynolds, A.E. and Carpenter, J.A. (1974) Bactericidal properties of acetic and propionic acids on pork carcasses. *J. Ani. Sci.*, **38**, 515–9.

Rheinbaben, K.E.V. and Seipp, H. (1986) Studies on the microflora of uncooked hams with special reference to *Micrococcaceae*. *Chem. Mikrobiol. Technol. Lebensmittel*, **9**, 152–61.

Rice, D.H., Ebel, E.D., Hancock, D.D., Besser, T.E., Herriott, D.E. and Carpenter, L.V. (1997) *Escherichia coli* O157 in cull dairy cows on farm and at slaughter. *J. Food Prot.*, **60**, 1386–7.

Rindi, S., Cerri, D. and Gerado, B. (1986) Thermophilic campylobacter in fresh pork sausage. *Indust. Alimen.*, **25**, 648–50.

Rivas, T., Vizcaíno, J.A. and Herrera, F.J. (2000) Microbial contamination of carcasses and equipment from an Iberian pig slaughterhouse. *J. Food Prot.*, **63**, 1670–5.

Roberts, T.A. (1980) Contamination of meat. The effects of slaughter practices on the bacteriology of the red meat carcass. *R. Soc. Health J.*, **100**, 3–9.

Roberts, T.A. and Derrick, C.M. (1978) Sporulation of *Clostridium putrefaciens* NCTC 9836, and the resistance of the spores to heat, gamma radiation and curing salts. *J. Appl. Bacteriol.*, **38**, 33–7.

Roberts, W.T. and Weese, J.O. (1998) Shelf life of ground beef patties treated by gamma radiation. *J. Food Prot.*, **61**, 1387–9.

Roberts, D., Boag, K., Hall, M.L.M. and Shipp, C.R. (1975) The isolation of salmonellas from British pork sausage and sausage meat. *J. Hyg.*, **75**, 173–84.

Roberts, T.A., Britton, C.R. and Hudson, W.R. (1980a) The bacteriological quality of minced beef in the UK. *J. Hyg. (Camb.)*, **85**, 211–7.

Roberts, T.A., MacFie, H.J.H. and Hudson, W.R. (1980b) The effect of incubation temperature and site of sampling on assessment of the numbers of bacteria on red meat carcasses at commercial abattoirs. *J. Hyg. (Camb.)*, **85**, 371–80.

Roberts, T.A., Hudson, W.R., Whelehan, O.P., Simonsen, B., Olgaard, K., Labots, H., Snijders, J.M.A., van Hoof, J., Debevere, J., Dempster, J.F., Devereux, J., Leistner, L., Gehra, H., Gledel, H. and Fournaud, J. (1984) Number and distribution of bacteria on some beef carcasses at selected abattoirs in some member states of the European Communities. *Meat Sci.*, **11**, 191–205.

Robinson, R.A. and Loken, K.I. (1968) Age susceptibility and excretion of *Salmonella typhimurium* in calves. *J. Hyg. (Camb.)*, **66**, 207–16.

Rodriguez, M., Núñez, F. Córdoba, J.J., Sanabria, C., Bermúdez, E. and Asensio, M.A. (1994) Characterization of *Staphylococcus* spp. and *Micrococcus* spp. isolated from Iberian ham throughout the ripening process. *Int. J. Food Microbiol.*, **24**, 329–35.

Rojas, F.J., Jodral, M., Gosalvez, F. and Pozo, R. (1991) Mycoflora and toxigenic *Aspergillus flavus* in Spanish dry-cured ham. *Int. J. Food Microbiol.*, **13**, 249–56.

Rose, B.E., Hill, W.E., Umholtz, R., Ransom, G.M. and James, W.O. (2002) Testing for *Salmonella* in raw meat and poultry products collected at federally inspected establishments in the United States, 1998 through 2000. *J. Food Prot.*, **65**(6), 937–47.

Rosef, O. (1981) Isolation of *Campylobacter fetus* subsp. *jejuni* from the gallbladder of normal slaughter pigs, using an enrichment procedure. *Acta Vet. Scand.*, **22**, 149–51.

Runham, N.W. (1989) Snail farming in the United Kingdom. In Slugs and snails in world agriculture. *Br. Crop Prot. Council*, **41**, 49–55.

Sage, J.R. and Ingham, S.C. (1998) Survival of *Escherichia coli* O157:H7 after freezing and thawing in ground beef patties. *J. Food Prot.*, **61**(9), 1181–3.

Samadpour, M., Ongerth, J.E., Liston, J., Tran, N., Nguyen, D. Whittam, T.S., Wilson, R.A. and Tarr, P.I. (1994) Occurrence of shiga-like toxin-producing *Escherichia coli* in retail fresh seafood, beef, lamb, pork, and poultry from grocery stores in Seattle, Washington. *Appl. Environ. Microbiol.*, **60**, 1038–40.

Samelis, J., Sofos, J.N., Kendall, P.A. and Smith, G.C. (2001) Fate of *Escherichia coli* O157:H7, *Salmonella typhimurium* DT 104, and *Listeria monocytogenes* in fresh meat decontamination fluids at 4 and 10 degrees C. *J. Food Prot.*, **64**(7), 950–7.

Samuel, J.L., O'Boyle, D.A., Mathers, W.J. and Frost, A.J. (1980) Distribution of *Salmonella* in the carcasses of normal cattle at slaughter. *Res. Vet. Sci.*, **28**, 368–78.

Samuel, J.L., Eccles, J.A. and Francis, J. (1981) Salmonella in the intestinal tract and associated lymph nodes of sheep and cattle. *J. Hyg. (Camb.)*, **87**, 225–32.

Sang, F.C., Hugh-Jones, M.E. and Hagstad, H.V. (1987) Viability of *Vibrio cholerae* 01 on frog legs under frozen and refrigerated conditions and low dose radiation treatment. *J. Food Prot.*, **50**, 662–4.

Sarwari, A.R., Magder, L.S., Levinem, P., McNamara, A.M., Knower, S., Armstrong, G.L., Etzel, R., Hollingsworth, J. and Morris, J.G., Jr. (2001) Serotype distribution of *Salmonella* isolates from food animals after slaughter differs from that of isolates found in humans. *J. Infect. Dis.*, **183**, 1295–9.

Sauter, E.A., Jacobs, J.A., Parkinson, J.F. (1980) Growth of psychrophilic bacteria on lamb carcasses, in *Proceedings, 26th Meeting of Meat Research Workers, volume 2*, Colorado Springs, pp. 279–81.

Savat, G., Toquin, M.T., Michel, Y. and Colin, P. (1995) Control of *Listeria monocytogenes* in delicatessen industries: the lessons of a listeriosis outbreak. *Int. J. Food Microbiol.*, **25**, 75–81.

Schiemann, D.A. (1989) *Yersinia enterocolitica* and *Yersinia pseudotuberculosis*, in *Foodborne Bacterial Pathogens* (ed M.P. Doyle), Marcel Dekker, Inc. New York, pp. 601–72.

Schiemann, D.A. and Fleming, C.A. (1981) *Yersinia enterocolitica* isolated from throats of swine in eastern and western Canada. *Can. J. Microbiol.*, **27**, 1326–33.

Schmidt, U. and Kaya, M. (1990) Behaviour of *L. monocytogenes* in vacuum-packed sliced frankfurter-type sausage. *Fleischwirtschaft*, **70**, 1294–5.

Scholefield, J., Menon, T.G. and Lam, C.W. (1981) Psychrotrophic contamination of pig carcasses, in *Proceedings, 27th European Meeting of Meat Research Workers*, Vienna, Austria, pp. 621–4.

Schoonderwoerd, M., Clarke, R.C., van Dreumel, A.A. and Rawluk, S.A. (1988) Colitis in calves: natural and experimental infection with a verotoxin-producing strain of *Escherichia coli* O111:NM. *Can. J. Vet. Res.*, **52**, 484–7.

Schotte, M., Höfelschweiger, H. and Reuter, G. (1992) Investigations on human *Trichinella spiralis* infection in the German Federal Republic in 1987. *Arch. Lebensmittelhyg.*, **43**, 136–9.

Schraft, H., Kleinlein, N. and Untermann, F. (1992) Contamination of pig hindquarters with *Staphylococcus aureus*. *Int. J. Food Microbiol.*, **15**, 191–4.

Schurman, R.D., Hariharan, H. and Heaney, S. (2000) Prevalence and characteristics of Shiga toxin-producing *Escherichia coli* in beef cattle slaughtered on Prince Edward Island. *J. Food Prot.*, **63**, 1583–6.

Schweigert, B.S. (1987) The nutritional content and value of meat and meat products, in *The Science of Meat and Meat Products*, 3rd edn (eds J.F. Price and B.S. Schweigert), Food and Nutrition Press, Westport, Conn., pp. 275–305.

Scott, W.J. and Vickery, J.R. (1939) Investigations on chilled beef. Part II. *Cooling and Storage in the Meatworks*. Council for Scientific and Industrial Research, Australia, Bulletin No. 129.

SCVPH (Scientific Committee on Veterinary Measures related to Public Health, European Union) Opinion of the SCVPH on verotoxigenic *E. coli* (VTEC) in foodstuffs (adopted on 22–23 January 2003). www.europe.eu.int/.

Sekla, L., Milley, D., Stackiw, W., Sisler, J., Drew, J. and Sargent, D. (1990) Verotoxin-producing *Escherichia coli* in ground beef: Manitoba. *Can. Dis. Wkly Rep.*, **16**, 103–6.

Senczek, D., Stephan, R. and Untermann, F. (2001) Pulsed-field gel electrophoresis (PFGE) typing of Listeria strains isolated from a meat processing plant over a two-year period. *Int. J. Food Microbiol.*, **62**, 155–9.

Sharma, V.K., Kaura, Y.K. and Singh, I.P. (1974) Frogs as carriers of *Salmonella* and *Edwardsiella*. *Antonie van Leeuwenhoek, J. Microbiol. Serol.*, **40**, 171–5.

Sheridan, J.J. (1982) Problems associated with commercial lamb washing in Ireland. *Meat Sci.*, **6**, 211–9.

Sheridan, J.J. and Lynch, B. (1988) The influence of processing and refrigeration on the bacterial numbers on beef and sheep offals. *Meat Sci.*, **24**, 143–50.

Sheridan, J.J., Lynch, B. and Harrington, D. (1992) The effect of boning and plant cleaning on the contamination of beef cuts in a commercial boning hall. *Meat Sci.*, **32**, 185–94.

Shinagawa, K., Kanehira, M., Omoe, K., Matsuda, I., Hu, D.-L., Widiasih, D.A. and Sugii, S. (2000) frequency of Shiga toxin-producing *Escherichia coli* in cattle at a breeding farm and at a slaughterhouse in Japan. *Vet. Microbiol.*, **76**, 305–9.

Shiozawa, K., Nishina, T., Miwa, Y., Mori, T., Akahane, S. and Ito, K. (1991) Colonization in the tonsils of swine by *Yersinia enterocolitica*. *Contrib. Microbiol. Immunol.*, **12**, 63–7.

Shrivastava, K.P. (1977) Isolation and identification of *Salmonella* present in frozen frog legs. *Indian J. Microbiol.*, **17**, 54–7.

Sierra, M.-L., Sheridan, J.J. and McGuire, L. (1997) Microbial quality of lamb carcasses during processing and the acridine orange direct count technique (a modified DEFT) for rapid enumeration of total viable counts. *Int. J. Food Microbiol.*, **36**, 61–7.

Silla, H., Molina, I., Flores, J. and Silvestre, D. (1989) A study of the microbial flora of dry-cured ham. 1. Isolation and growth. *Fleischwirtschaft*, **69**, 1123–31.

Simonsen, B., Bryan, F.L., Christian, J.H.B., Roberts, T.A., Tompkin, R.B. and Silliker, J.H. (1987) Prevention and control of salmonellosis through application of HACCP. *Int. J. Food Microbiol.*, **4**, 227–47.

Sinha, B.K. and Mandar, L.N. (1977) Studies on the bacterial quality of market goat meat and its public health importance. *Indian J. Ani. Sci.*, 47, 478–81.

Siragusa, G.R. (1995) The effectiveness of carcass decontamination systems for controlling the presence of pathogens on the surfaces of meat animal carcasses. *J. Food Saf.*, **15**(3), 229–38.

Siragusa, G.R., Dickson, J.S. and Daniels, E.K. (1993) Isolation of *Listeria* spp. from feces of feedlot cattle. *J. Food Prot.*, **56**, 102–5.

Skelley, G.C., Fandino, G.E., Haigler, J.H. and Sherard, R.C., Jr. (1985) Bacteriology and weight loss of pork carcasses treated with a sodium hypochlorite solution. *J. Food Prot.*, **48**, 578–81.

Skjerve, E., Lium, B., Nielsen, B. and Nesbakken, T. (1998) Control of *Yersinia enterocolitica* in pigs at herd level. *Int. J. Food Microbiol.*, **45**, 195–203.

Skovgaard, N. and Morgen, C.-A. (1988) Detection of *Listeria* spp. in faeces from animals, in feeds, and in raw foods of animal origin. *Int. J. Food Microbiol.*, **6**, 229–42.

Skovgaard, N. and Nielson, B.B. (1972) Salmonellas in pigs and animal feeding stuffs in England and Wales and in Denmark. *J. Hyg. (Camb.)*, **70**, 127–40.

Skovgaard, N. and Norrung, B. (1989) The incidence of *Listeria* spp. in faeces of Danish pigs and in minced pork meat. *Int. J. Food Microbiol.*, **8**, 59–63.

Small, A., Reid, C.A., Avery, S.M., Karabasil, N., Crowley, C. and Buncic, S. (2002) Potential for the spread of *Escherichia coli* O157, *Salmonella*, and *Campylobacter* in the lairage environment at abattoirs. *J. Food Prot.*, **65**(6), 931–6.

Smart, J.L., Roberts, T.A., Stringer, M.F. and Shah, N. (1979) The incidence and serotypes of *Clostridium perfringens* on beef, pork and lamb carcasses. *J. Appl. Bacteriol.*, **46**, 377–83.

Smeltzer, T., Thomas, R. and Collins, G. (1980a) The role of equipment having accidental or indirect contact with the carcass in the spread of *Salmonella* in an abattoir. *Aust. Vet. J.*, **56**, 14–7.

Smeltzer, T., Thomas, R. and Collins, G. (1980b) Salmonellae on posts, handrails, and hands in a beef abattoir. *Aust. Vet. J.*, **56**, 184–6.
Smith, H.W. (1961) The development of the bacterial flora of the faeces of animals and man: the changes that occur during ageing. *J. Appl. Bacteriol.*, **24**, 235–41.
Smith, M.G. (1985) The generation time, lag time, and minimum temperature of growth of coliform organisms on meat, and the implications for codes of practice in abattoirs. *J. Hyg. (Camb.)*, **94**, 289–300.
Smith, M.G. and Graham, A. (1978) Destruction of *Escherichia coli* and salmonellae on mutton carcases by treatment with hot water. *Meat Sci.*, **2**, 119–28.
Smith, M.G. and Davey, K.R. (1990) Destruction of *Escherichia coli* on sides of beef by a hot water decontamination process. *Food Aust.*, **42**, 195–8.
Smith, J.L., Palumbo, S.A., Kissinger, J.C. and Huhtanen, C.N. (1975) Survival of *Salmonella dublin* and *Salmonella typhimurium* in Lebanon bologna. *J. Milk Food Technol.*, **38**, 150–4.
Smulders, F.J.M. (1987) Prospectives for microbial decontamination of meat and poultry by organic acids with special reference to lactic acid, in *Elimination of Pathogenic Organisms from Meat and Poultry* (ed F.J.M. Smulders), Elsevier, Amsterdam, pp. 319–44.
Snijders, J.M.A. (1975) Hygiene of pig slaughtering. I. Scalding. *Fleischwirtschaft*, **55**, 836–40.
Snijders, J. (1988) Good manufacturing practices in slaughter lines. *Fleischwirtschaft*, **68**, 753–6.
Snijders, J.M.A. and Gerats, G.E. (1976a) Hygiene of pig slaughtering. III. The effect of different factors on dehairing. *Fleischwirtschaft*, **56**, 238–41.
Snijders, J.M.A. and Gerats, G.E. (1976b) Hygiene of pig slaughtering. IV. Bacteriological status of carcasses at various stages of the slaughter line. *Fleischwirtschaft*, **56**, 717–21.
Sofos, J.N., Kochevar, S.L., Reagan, J.O. and Smith, G.C. (1999) Extent of beef carcass contamination with *Escherichia coli* and probabilities of passing US regulatory criteria. *J. Food Prot.*, **62**, 234–8.
Sorqvist, S. and Danielsson-Tham, M.-L. (1986) Bacterial contamination of the scalding water during vat scalding of pigs. *Fleischwirtschaft*, **66**, 1745–8.
Sorqvist, S. and Danielsson-Tham, M.-L. (1991) Survival of *Campylobacter*, *Salmonella* and *Yersinia* spp. in scalding water used in pig slaughter. *Fleischwirtsch. Int.*, **2**, 54–8.
Stephan, R., Stierli, F. and Untermann, F. (1997) Chemical attributes characterizing sticky post-mortem ageing in beef. *Meat Sci.*, **47**, 331–5.
Stephan, R., Annemüller, C., Hassan, A.A. and Lämmler, C. (2001) Characterization of enterotoxigenic *S. aureus* strains isolated from bovine mastitis in North-East Switzerland. *Vet. Microbiol.*, 373–82.
Stern, N.J. and Line, J.E. (2000) *Campylobacter* (Chapter 40), in *The Microbiological Safety and Quality of Food* (eds Lund, B.M., Baird-Parker, T.C. and Gould, G.W.), *volume II*, Aspen Publishers, Inc., Gaithersburg, MD, pp. 1040–56.
Stern, N.J., Pierson, M.D. and Kotula, A.W. (1980) Effects of pH and sodium chloride on *Yersinia enterocolitica* growth at room and refrigeration temperatures. *J. Fd. Sci.*, 45, 64–67.
Stern, N.J., Hernandez, M.P., Blankenship, L., Deibel, K.E., Doores, S., Doyle, M.P., Ng, H., Pierson, M.D., Sofos, J.N., Sveum, W.H. and Westhoff, D.C. (1985) Prevalence and distribution of *Campylobacter jejuni* and *Campylobacter coli* in retail meats. *J. Food Prot.*, **48**, 595–9.
Stock, K. and Stolle, A. (2001) Incidence of *Salmonella* in minced meat produced in a European Union-approved cutting plant. *J. Food Prot.*, **64**(9), 1435–8.
Stolle, A. (1981) Spreading of salmonellas during cattle slaughtering. *J. Appl. Bacteriol.*, **50**, 239–45.
Strachan, N.J., Fenlon, D.R. and Ogden, I.D. (2001) Modelling the vector pathway and infection of humans in an environmental outbreak of *Escherichia coli* O157. *FEMS Microbiol. Lett.*, **203**(1), 69–73.
Sutherland, J.P. and Varnum, A. (1982) Fresh meat processing, in *Meat Microbiology* (ed M.H. Brown), Applied Science Publisher, London, pp. 103–28.
Suthienkul, O., Brown, J.E., Seriwatana, J., Tienthongdee, S., Sastravaha, S. and Echeverria, P. (1990) Shiga-like-toxin-producing *Escherichia coli* in retail meats and cattle in Thailand. *Appl. Environ. Microbiol.*, **56**, 1135–9.
Swanenburg, M., Urlings, H.A., Keuzenkamp, D.A. and Snijders, J.M. (2001) *Salmonella* in the lairage of pig slaughterhouses. *J. Food Prot.*, **64**(1), 12–6.
Swingler, G.R. (1982) Microbiology of meat industry by-products, in *Meat Microbiology* (ed M.H. Brown), Applied Science Publishers Ltd, London, pp. 179–224.
Synge, B.A. (1999) Animal studies in Scotland, in *E. coli O157 in Farm Animals* (eds C.S. Stewart and H.J. Flint), CAB International, pp. 91–8; http://www.cabi-publishing.org/Bookshop/ReadingRoom/085199332X/085199332Xch7.pd.
Synge, B.A. and Hopkins, G.F. (1992) Verotoxigenic *Escherichia coli* O.157 in Scottish calves. *Vet. Rec.*, **130**, 583.
Takacs, I. and Szita, G. (1984) Studies on the technology-hygiene of pig slaughtering. *Acta Alimen.*, **13**, 272.
Tannock, G.W. and Smith, J.M.B. (1971) A *Salmonella* carrier state of sheep following intranasal inoculation. *Res. Vet. Sci.*, **12**, 371–3.
Taormina, P.J., Bartholomew, G.W. and Dorsa, W.J. (2003) Incidence of *Clostridium perfringens* in commercially produced cured raw meat product mixtures and behavior in cooked products during chilling and refrigerated storage. *J. Food Prot.*, **66**(1), 72–81.
Taplin, J. (1982) *Salmonella newport* outbreak—Victoria. *Commun. Dis. Intell.* (Aust.), **1**, 3–6.
Tarr, P.I. (1994) Review of *Escherichia coli* O157:H7 outbreak: Western United States. *Dairy Food Environ. Sanit.*, **14**, 372–3.

Tarr, P., Tran, N. and Wilson, R. (1999) *Escherichia coli* O157:H7 in retail ground beef in Seattle: results of a one-year prospective study. *J. Food Prot.*, **62**, 133–9.
Tauxe, R.V. (2002) Emerging foodborne pathogens. *Int. J. Food Microbiol.*, **78**, 31–42.
Tauxe, R.V., Vandepitte, J., Wauters, G., Martin, S.M., Goosens, V., DeMol, P., van Noyen, R. and Thiers, S. (1987) *Yersinia enterocolitica* and pork: the missing link. *Lancet*, **i**, 1129–32.
Teufel, P. (1982) *Campylobacter fetus* subsp. *jejuni*—excretion rates in the pig and survival in tap water and minced meat. *Fleischwirtschaft*, **62**, 1344–5.
Thayer, D.W. and Boyd, G. (1993) Elimination of *Escherichia coli* O157:H7 in meats by gamma-irradiation. *Appl. Environ. Microbiol.*, **59**, 1030–4.
Thorberg, B.M. and Engvall, A. (2001) Incidence of *Salmonella* in five Swedish slaughterhouses. *J. Food Prot.*, **64**(4), 542–5.
Thorpe, R.H. and Barker, P.M. (1984) *Visual Can Defects*. The Campden Food Preservation Research Association, Chipping Campden, UK.
Thran, B.H., Hussein, H.S., Hall, M.R. and Khaiboullina, S.F. (2001) Shiga toxin-producing *Escherichia coli* in beef heifers grazing an irrigated pasture. *J. Food Prot.*, **64**(10), 1613–6.
Timoney, J. (1970) Salmonella in Irish pigs at slaughter. *Irish Vet. J.*, **24**, 141–5.
Tkalcic, S., Brown, C.A., Harmon, B.G., Jain, A.V., Muellerm, E.P., Parks, A., Jacobsen, K.L., Martin, S.A., Zhao, T. and Doyle, M.P. (2000) Effects of diet on rumen proliferation and fecal shedding of *Escherichia coli* O157:H7 in calves. *J. Food Prot.*, **63**(12), 1630–6.
Tompkin, R.B. (1980) Botulism from meat and poultry products—a historical perspective. *Food Technol.*, **34**, 228–36, 257.
Tompkin, R.B. (1986) Microbiology of ready-to-eat meat and poultry products, in *Advances in Meat Research* (eds A.M. Pearson and T.R. Dutson), *volume 2*, Meat and Poultry Microbiology, AVI Publishing Westport, Connecticut, pp. 89–121.
Tompkin, R.B. (1993) Nitrite, in *Antimicrobials in Foods*, 2nd edn (eds A.L. Branen and P.M. Davidson), Marcell Dekker, Inc. New York, pp. 191–262.
Tompkin, R.B. (1995) The use of HACCP for producing and distributing processed meat and poultry products, in *Advances in Meat Research* (eds A.M. Pearson and T.R. Dutson), *volume* 10, *HACCP in Meat, Poultry, and Fish Processing*, Blackie Academic and Professional, London, pp. 72–108.
Tompkin, R.B. (2002) Control of *Listeria monocytogenes* in the food-processing environment. *J. Food Prot.*, **65**(4), 709–25.
Troeger, K. and Woltersdorf, W. (1990) Microbial contamination by scalding water of pig carcasses via the vascular system. *Fleischwirtsch. Int.*, **1**, 18–25.
Tsai, S. and Chou, C. (1996) Injury, inhibition and inactivation of *Escherichia coli* O157:H7 by potassium sorbate and sodium nitrite as affected by pH and temperature. *J. Sci. Food Agric.*, **71**, 10–2.
Tutenel, A., Pierard, D., Uradzinski, J., Jozwik, E., Pastuszczak, M., Van Hende, J., Uyttendaeale, M., Debevere, J., Cheasty, T., Van Hoof, J., De Zutter, L. (2002) Isolation and characterization of enterohemorrhagic *Escherichia coli* O157:H7 from cattle in Belgium and Poland. *Epidemiol. Infect.*, **129**, 41–7.
Tuttle, J., Gomez, T., Doyle, M.P., Wells, J.G., Zhao, T., Tauxe, R.V. and Griffin, P.M. (1999) Lessons from a large outbreak of *Escherichia coli* O157:H7 infections: insights into the infectious dose and method of widespread contamination of hamburger patties. *Epidemiol. Infect.*, **122**(2), 185–92.
Uemura, T., Kusunoki, H., Hosoda, K. and Sakaguchi, G. (1985) A simple procedure for the detection of small numbers of enterotoxigenic *Clostridium perfringens* in frozen meat and cod paste. *Int. J. Food Microbiol.*, **1**, 335–41.
Unnerstad, H., Romell, A., Ericsson, H., Danielsson-Tham, M.-L. and Tham, W. (2000) *Listeria monocytogenes* in faeces from clinically healthy dairy cows in Sweden. *Acta Vet. Scand.*, **41**, 167–171.
Untermann, F. and Muller, C. (1992) Influence of a_w value and storage temperature on the multiplication and enterotoxin formation of staphylococci in dry-cured raw hams. *Int. J. Food Microbiol.*, **16**, 109–15.
USDA (United States Department of Agriculture). (1983) Production requirements for cooked beef, roast beef and cooked corned beef. *Fed. Reg.*, **48**(106), 24314-8.
USDA (United States Department of Agriculture). (1988) *Time/Temperature Guidelines for Cooling Heated Products*. FSIS Directive 7110.3, Food Safety and Inspection Service, US Department of Agriculture, Washington, DC.
USDA (United States Department of Agriculture). (1993) *Heat-Processing Procedures; Cooling Instructions, and Cooking, Handling and Storage Requirements for Uncured Meat Patties*. Federal Register 58:41138 Docket Number 86-0141F.
USDA. (1994) *USDA Nationwide Beef Microbiological Baseline Data Collection Program: Steers and Heifers* (October 1992–September 1993). US Department of Agriculture, Food Safety and Inspection Service, Washington, DC.
Uyttendaele, M., Jozwik, E., Tutenel, A., De Zutter, L., Uradzinski, J., Pierard, D. and Debevere, J. (2001) Effect of acid resistance of *Escherichia coli* O157:H7 on efficacy of buffered lactic acid to decontaminate chilled beef tissue and effect of modified atmosphere packaging on survival of *Escherichia coli* O157:H7 on red meat. *J. Food Prot.*, **64**, 1661–6.
Vågsholm, I. (1999) EHEC än en gång, nytt GD dokument. Föredrag Veterinärmötet 1999, Uppsala 1999. Sveriges Veterinärförbund/Sveriges Veterinärmedisinska selskap, pp 87–91. ISSN 1402–9324.
van der Riet, W.B. (1976) Water sorption isotherms of beef biltong and their use in predicting critical moisture contents for biltong storage. *S. Afr. Food Rev.*, **3**, 93–6.
van der Riet, W.B. (1982) Biltong: a South African dried meat product. *Fleischwirtschaft*, **62**, 1000–1.
Vanderlinde, P.B., Shay, B. and Murray, J. (1998) Microbiological quality of Australian beef carcass meat and frozen bulk packed beef. *J. Food Prot.*, **61**, 437–43.

Vanderlinde, P.B., Fegan, N., Mills, L. and Desmarchelier, P.M. (1999) Use of pulse field gel electrophoresis for the epidemiological characterisation of coagulase positive *Staphylococcus* isolated from meat workers and beef carcasses. *Int. J. Food Microbiol.*, **48**(2), 81–5.
Vanderzant, C., Hanna, M.O., Ehlers, J.G., Savell, J.W., Griffin, D.B., Johnson, D.D., Smith, G.C. and Stiffler, D.M. (1985) Methods of chilling and packaging of beef, pork and lamb variety meats for transoceanic shipment: microbiological considerations. *J. Food Prot.*, **48**, 765–9.
Van Donkersgoed, J., Jericho, K., Grogan, H., Thorlakson, B. 1997) Preslaughter hide status of cattle and the microbiology of carcasses. *J. Food Prot.*, **60**, 1502–8.
Van Donkersgoed, J., Graham, T. and Gannon, V. (1999) The prevalence of verotoxins, *Escherichia coli* O157:H7, and *Salmonella* in the feces and rumen of cattle at processing. *Can. Vet. J.*, **40**, 332–8.
Van Donkersgoed, J., Berg, J., Potter, A., Hancock, D., Besser, T., Rice, D., LeJeune, J. and Klashinsky, S. (2001) Environmental sources and transmission of *Escherichia coli* O157 in feedlot cattle. *Can. Vet. J.*, **42**(9), 714–20.
van Renterghem, B., Huysman, F., Rygole, R. and Verstraete, W. (1991) Detection and prevalence of *Listeria monocytogenes* in the agricultural system. *J. Appl. Bacteriol.*, **71**, 211–7.
Vuddhakul, V., Patararungrong, N., Pungrasamee, P., Jitsurong, S., Morigaki, T., Asai, N. and Nishibuchi, M. (2000) Isolation and characterization of *Escherichia coli* O157 from retail beef and bovine feces in Thailand. *FEMS Microbiol. Lett.*, **182**, 343–7.
Walker, H.L., Chowdhury, K.A., Thaler, A.M., Petersen, K.E., Ragland, R.D. and James, W.O. (2000) Relevance of carcass palpation in lambs to protecting public health. *J. Food Prot.*, **63**(9), 1287–90.
Wang, W. and Leistner, L. (1993) Shafu: a novel dried meat product of China based on hurdle-technology. *Fleischwirtschaft*, **73**, 854–6.
Wauters, G., Goosens, V., Janssens, M. and Vandepitte, J. (1988) New enrichment method for isolation of pathogenic *Yersinia enterocolitica* serogroup O:3 from pork. *Appl. Environ. Microbiol.*, **54**, 851–4.
Weber, A., Potel, J., Schafer-Schmidt, R., Prell, A. and Datzmann, C. (1995) Studies on the occurrence of *Listeria monocytogenes* in fecal samples of domestic and companion animals. *Zen. Hyg. Umweltmedizin*, **198**, 117–23.
Wells, J.G., Shipman, L.D., Greene, K.D., Sowers, E.G., Green, J.H., Cameron, D.N., Downes, F.P., Martin, M.L., Griffin, P.M., Ostroff, S.M., Potter, M.E., Tauxe, R.V. and Wachsmuth, I.K. (1991) Isolation of *Escherichia coli* serotype O157:H7 and other shiga-like-toxin-producing *E. coli* from dairy cattle. *J. Clin. Microbiol.*, **29**, 985–9.
Wells, S.J., Fedorka-Cray, P.J., Dargatz, D.A., Ferris, K. and Green, A. (2001) Fecal shedding of *Salmonella* spp. by dairy cows on farm and at cull cow markets. *J. Food Prot.*, **64**, 3–11.
Whelehan, O.P., Hudson, W.R. and Roberts, T.A. (1986) Microbiology of beef carcasses before and after slaughterline automation. *J. Hyg. (Camb.)*, **96**, 205–16.
Williams, B.M. (1975) Environmental considerations in salmonellosis. *Vet. Rec.*, **96**, 318–21.
Williams, L.P. and Newell, K.W. (1967) Patterns of salmonella excretion in market swine. *Am. J. Public Health*, **57**, 466–71.
Williams, L.P. and Newell, K.W. (1970) Salmonella excretion in joy-riding pigs. *Am. J. Public Health*, **60**, 926–9.
Willshaw, G.A., Smith, H.R., Roberts, D., Thirlwell, J., Cheasty, T. and Rowe, B. (1993) Examination of raw beef products for the presence of vero cytotoxin producing *Escherichia coli*, particularly those of serogroup O157. *J. Appl. Bacteriol.*, **75**, 420–6.
Wilson, J.B., Clarke, R.C., Renwick, S.A., Rahn, K., Johnson, R.P., Karmali, M.A., Lior, H., Alves, D., Gyles, C.L., Sandu, K.S., McEwen, S.A. and Spika, J.S. (1996) Vero cytotoxigenic *Escherichia coli* infection in dairy farm families. *J. Inf. Dis.*, **174**, 1021–7.
Wood, R.L., Popischil, A. and Rose, R. (1989) Distribution of persistent *Salmonella typhimurium* infection in internal organs of swine. *Am. J. Vet. Res.*, **50**, 1015–21.
Wray, C., McLaren, I. and Pearson, G.R. (1989) Occurrence of "attaching and effacing" lesions in the small intestine of calves experimentally infected with bovine isolates of verocytotoxic *E coli*. *Vet. Rec.*, **125**, 365–8.
Wray, C., Todd, N., McLaren, I., Beedell, Y. and Rowe, B. (1990) The epidemiology of salmonella infection of calves: the role of dealers. *Epidemiol. Infect.*, **105**, 295–305.
Wray, C., Todd, N., McLaren, I.M. and Beedell, Y.E. (1991) The epidemiology of salmonella in calves: the role of markets and vehicles. *Epidemiol. Infect.*, **107**, 521–5.
Zaika, L.L., Palumbo, S.A., Smith, J.L., Del Carral, F., Bhaduri, S., Jones, C.O. and Kim, A.H. (1990) Destruction of *Listeria monocytogenes* during frankfurter processing. *J. Food Prot.*, **53**, 18–21.
Zanetti, F., Varoli, O., Stampi, S. and De Luca, G. (1996) Prevalence of thermophilic *Campylobacter* and *Arcobacter butzleri* in food of animal origin. *Int. J. Food Microbiol.*, **33**, 315–21.
Zhao, T., Doyle, M.P., Shere, J. and Garber, L. (1995) Prevalence of enterohemorrhagic *Escherichia coli* O157:H7 in a survey of dairy herds. *Appl. Environ. Microbiol.*, **61**, 1290–3.
Zweifel, C. and Stephan, R. (2003) Microbiological monitoring of sheep carcass contamination in three Swiss abattoirs. *J. Food Prot.*, **66**, 946–52.

2 Poultry products

I Introduction

In 2002, approximately 74 million metric tons of poultry meat were produced globally consisting of chicken (86%), turkey (7%), duck (4%), goose (3%), and other (<1%) (FAOSTAT, 2002). In 2001, approximately 9.6 million metric tons were exported (FAO, 2003). The 10 leading exporters were the USA (33%), Brazil (14%), The Netherlands (8%), France (8%), China–Hong Kong (7%), China (6%), Thailand (5%), Belgium (3.5%), United Kingdom (2%), and Germany (2%) (FAO, 2003). Products shipped in international trade include carcasses and parts of chickens, turkeys, ducks, geese, and a wide variety of cooked products.

Processors and control authorities of importing countries are interested in the microflora of these products as indices of good hygienic practices, storage conditions, remaining shelf-life, and potential risk to public health. This chapter describes: (a) sources of microorganisms that contaminate carcasses and products; (b) conditions that influence the spread of microorganisms during processing; (c) factors that affect survival and growth of microorganisms on poultry products; and (d) the significance of laboratory findings.

Most of the data in this chapter have been developed in intensive commercial operations where poultry are raised in large numbers, most frequently in poultry houses. Very few data are from small family operations where poultry roams freely or are raised in small numbers. Commercial operations tend to be highly regulated. Since regulations differ throughout the world and continually change, some information in this chapter will not be applicable to all regions or countries.

Campylobacter jejuni and *Salmonella* spp. are both pathogens of latent zoonoses with no obvious pathological changes in affected animals but presenting the main food-borne hazards in the poultry meat chain. The prevalence of thermophilic campylobacters is significantly greater in free-range poultry than in conventionally raised poultry because the former are exposed to the strong infection pressure on flocks due to wild birds, which cannot be controlled (Giessen *et al.*, 1996).

Thermophilic *Campylobacter* spp. are a major cause of bacterial gastrointestinal human infections in the USA (Altekruse *et al.*, 1999), in England and Wales (Frost *et al.*, 1998), and probably world-wide. Additionally, in recent years a rapidly increasing proportion of *Campylobacter* isolates all over the world have been found to be resistant to antibiotics used clinically (Velazquez *et al.*, 1995; Smith *et al.*, 1999; Van Looveren *et al.*, 2001).

Inadequately cooked poultry meat is one of the most common sources for epidemic and sporadic *Campylobacter* food-borne cases (Butzler and Oosterom, 1991; Altekruse *et al.*, 1999). Furthermore, cross-contamination of other foods from raw poultry meat during food preparation is also considered as important. To produce hygienic food under such conditions, appropriate measures must be adopted along the entire production chain according to the concept "from the farm to the fork" to reduce or exclude possible health hazards. Traditional meat inspection (inspection, palpation, and incision) at abattoirs is not sufficient in such circumstances. To prevent microbial contamination of the carcass, strict maintenance of slaughter hygiene is of crucial importance. It can be measured in daily practice by "slaughter-process-controls" and regular microbiological monitoring of carcasses. Moreover, to enable the risks involved to be estimated and appropriate measures to be taken, data for the feeding period should be complemented by collecting data relating to the shedding by the animals of such latent zoonotic pathogens.

A *Definitions*

Poultry meat is the muscle tissue, attached skin, connective tissue, and edible organs of avian species that are commonly used for food.

B *Important properties*

The water content of the edible portions of carcasses is about 71% for broiler chickens, 66% for roaster chickens, 56% for hen chickens, and 58% for medium fat turkeys (Mountney, 1976). The protein and fat content of chicken fryers are approximately 20.5% and 2.7%, respectively; chicken roasters, 20.2% and 12.6%; turkeys 20.1% and 20.2%; and duck 16.1% and 28.6% (Burton, 1976). A 1996 national survey in the United States of four skin-on chicken cuts found the mean yield of lean had not changed from 1979 values; whereas, the fat content had increased from 3% to 3.9% (Buege *et al.*, 1998). Although slight variation may have occurred in the proximate values for poultry meat, there have been significant changes in carcass composition. Increased demand for leaner poultry products based on breast meat has resulted in breeding higher body weight with a larger proportion of breast meat into broilers (Berri *et al.*, 2001) and turkeys.

Unlike red meat, in which fat is distributed throughout the tissues, most of the fat in poultry is found just under the skin and in the abdominal cavity. The relative ease of removing fat from poultry meat offers an advantage over pork or beef, when producing low-fat products. The amount of fat varies with age, sex, anatomy, and species.

In addition to the nutrients, other properties influence the growth of microorganisms in raw poultry meat. The water activity (a_w) ranges from 0.98 to 0.99 depending on if, and how long, the meat has been stored in dry air. The pH range of chicken breast muscle is from 5.7 to 5.9, while that of leg muscle is from 6.4 to 6.7. As chickens mature, the pH of the skin increases to an average of 6.6 for 9-week-old chickens and 7.2 for 25-week-old chickens (Adamcic and Clark, 1970). The redox potential of poultry meat is similar to that of beef, pork, and lamb. The skin, which harbors many microorganisms, nevertheless serves as a physical barrier to microorganisms that might contaminate the underlying muscle. Both poultry muscle and skin are excellent substrates for a wide variety of microorganisms.

C *Method of processing*

Freshly laid fertile eggs are collected and then incubated at a constant temperature for 21 and 28 days for chickens and turkeys, respectively. After hatching, the chicks or poults are delivered to farms, reared until ready for slaughter (6–15 weeks), and then transported to a processing plant. At the plant, they are killed and bled by severing the carotid arteries. Feathers, heads, feet, and viscera are removed and the carcasses are then washed, chilled, and distributed to markets packed in ice, refrigerated, or frozen. These general processing steps are illustrated in Figure 2.1. A significant proportion of poultry carcasses are used to produce a variety of raw and processed products (Figure 2.2).

D *Types of poultry products*

Poultry products include raw whole birds or parts and a wide variety of foods that contain poultry meat (Table 2.1).

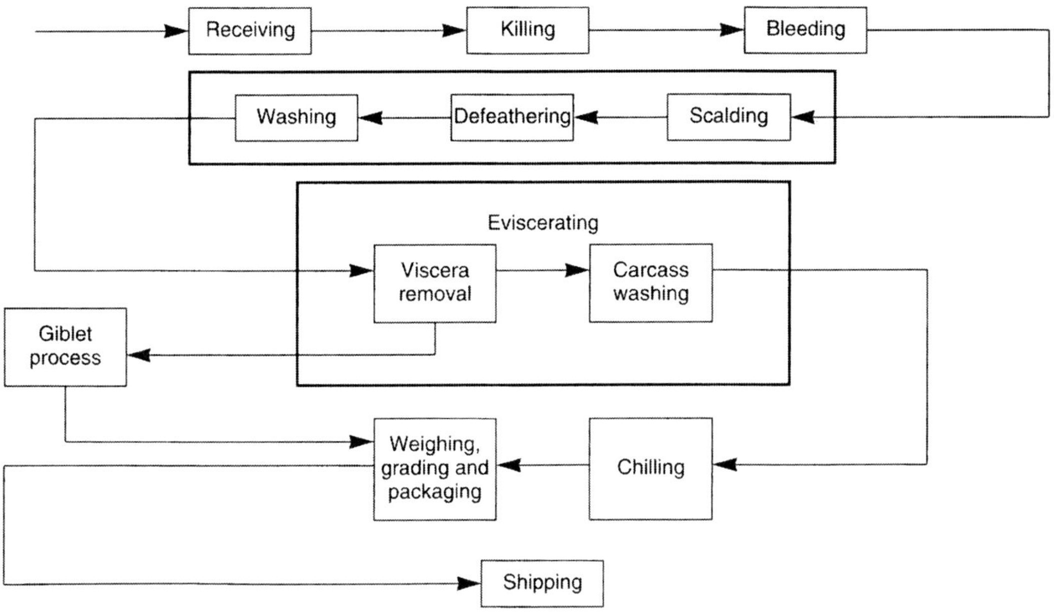

Figure 2.1 General process flow diagram for poultry processing.

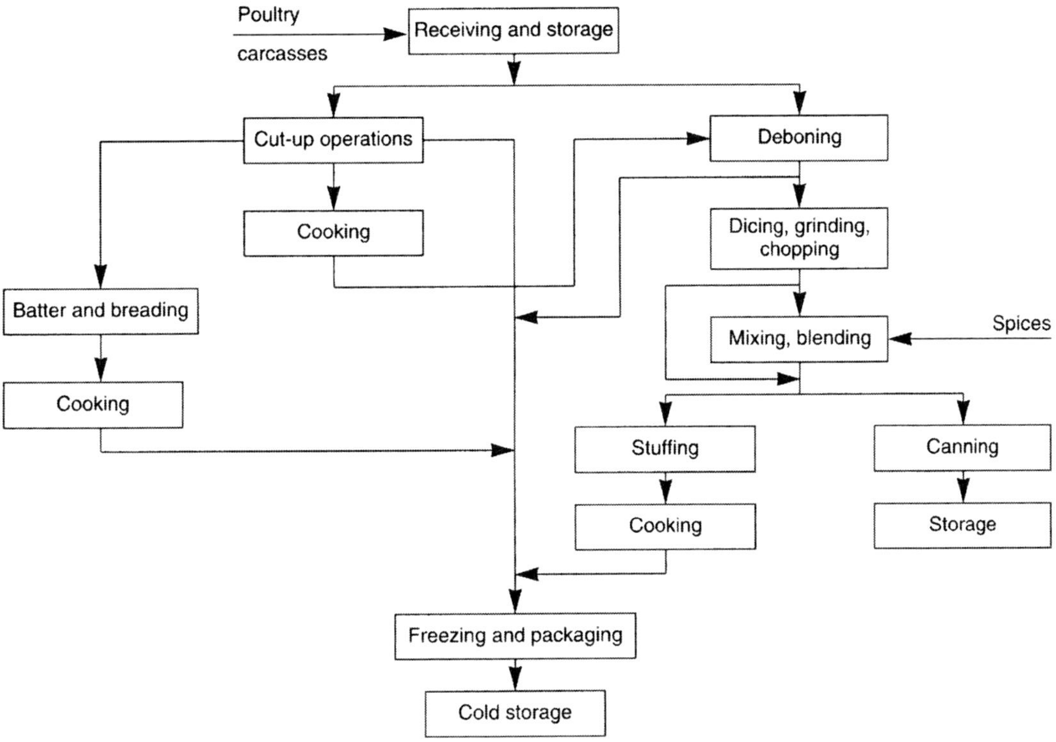

Figure 2.2 General process flow diagrams for further processing of chilled poultry carcasses.

Table 2.1 Examples of commercial poultry products

Raw fresh	Raw Frozen	Cooked, uncured	Cooked, cured	Dried
Whole birds	Whole birds	Products made from breast meat	Cured breast products	Freeze dried poultry meat
Parts-legs, thighs, breast, wings, backs	Parts-legs, thighs, breast, wings, backs	Barbecued poultry	Luncheon meat	Dry and semi-dry sausages
Uneviscerated poultry	Edible viscera-livers, gizzards, hearts	Baked whole birds	Frankfurters	Bouillon
Edible viscera-livers, gizzards, hearts	Mechanically deboned meat	Fried chicken portions	Ham, pastrami	Fat from cooking processes
Ground poultry meat	Ready-to-cook boneless poultry products	Canned products	Pate'	
Marinated poultry				

II Initial microflora (effect of farm practices)

Microbial contamination of the interior of eggs for hatching can occur either in the ovaries or oviducts during development of the egg or later as a result of eggshell penetration (see Chapter 15). After eggs hatch, the chicks acquire microorganisms from the hatchery cabinet and various environmental sources from the farm.

Some microorganisms (e.g. certain *Salmonella, Escherichia coli, Mycoplasma, Campylobacter*, vibrios, and enterococci) can infect the ovaries and the oviducts of hens. These organisms can pass from the ovary to the ovum or from the oviduct to the yolk, albumen, or membranes during egg formation. Gram-negative bacteria present in the egg, as a consequence of either ovarian transmission or penetration of the shell, can reach the yolk and subsequently infect the developing embryo.

Eggshell disinfection is normal practice at breeder hatcheries and may also be practiced at other hatcheries. Contaminated eggshells, fecal matter, and fluff from infected, newly hatched birds will contaminate incubators. Newly hatched chicks are readily colonized by microorganisms from their environment (Williams, 1978; Bailey *et al.*, 1992, 1994; Mitchell *et al.*, 2002).

Development of a normal intestinal flora can influence resistance to disease and establishment of human pathogens (e.g. salmonellae, campylobacters). The general health status of a flock also is a factor that can influence susceptibility to colonization by these pathogens. For example, there is evidence that coccidiosis in a flock can lead to a higher incidence of salmonellae infection (Arakawa *et al.*, 1992; Qin *et al.*, 1995). Also, feeds containing aflatoxin or ochratoxin A can cause higher mortality in young birds infected with salmonellae (Boonchuvit and Hamilton, 1975; Ellisalde *et al.*, 1994). The health status of poultry flocks also influences the level of pathogens on carcasses after processing. For example, flocks with airsacculitis will be lower in body weight and size. High-speed mechanical evisceration systems depend on uniformity of carcasses to minimize cuts and tears in the intestine and fecal contamination on the product (Russell, 2003).

Poultry feed and water can be a significant source of microorganisms on the farm. Certain animal-derived ingredients (e.g. rendered animal by-products, fish meal) have been the major sources of salmonellae. The kind and number of microorganisms found on the feathers and skin of live birds, and subsequently on dressed carcasses, can be influenced by the type and condition of the soil or litter on which the birds are raised. The types of soil and litter to which birds are exposed differ throughout the world. Litter becomes contaminated from droppings, feathers, and soil. As litter is used, increasing amounts of ammonia and pH can create unfavorable conditions for some microorganisms (e.g. salmonellae) (Turnbull and Snoeyenbos, 1973). However, old wet litter, droppings, and wet feed are good media for yeast and mold growth.

Insects, flies, and beetles can be reservoirs and vectors of microorganisms. Rodents and other small mammals spread microorganisms within poultry pens from their feet, hair, and feces (Lofton *et al.*, 1962). Wild birds that gain entrance to houses or the environment inhabited by domestic fowl can transmit salmonellae and other microorganisms from house-to-house and farm-to-farm (Snoeyenbos *et al.*, 1967). Reptiles, amphibians, pets, and livestock on the farm are additional reservoirs of certain pathogens. Farm workers can easily spread infectious agents with their boots or equipment and mismanage flocks in ways that spread microorganisms causing diseases.

Pathogens of concern to humans and poultry production can be readily transmitted among the birds in a house or flock. Young colonized chicks rapidly spread salmonellae and campylobacters to other young chicks. The infective dose of salmonellae for day-old chicks appears to be very low (e.g. fewer than 100 cells). Older birds become more resistant (Snoeyenbos *et al.*, 1969). Cannibalism in an infected flock spreads many pathogenic and saprophytic organisms that may be present in the intestinal tract or on the skin. Cannibalism has been responsible for outbreaks of botulism (type C) in broilers when dead birds are not removed from the houses on time (Blandford and Roberts, 1970). Excessive heat, cold, and other stresses lower the birds' resistance to infection.

Broilers and, to a large extent, turkeys are commonly raised in confinement with thousands of birds in a house. Pathogen transmission can be reduced by adequate spacing of houses; exhausting sufficient amounts of air; filtering incoming air; removing dead birds promptly; providing devices that keep birds out of feeders and water troughs or more, preferably, using nipple style drinkers that have become commonplace in most countries; cleaning watering devices daily; using pelleted feed; controlling access of wild birds, rodents, reptiles, and people to flocks; cleaning and disinfecting houses between successive flocks; wearing clean clothing; and disinfecting boots. Countries in which these practices prevail report low prevalence of salmonellae in live birds and on processed raw poultry (Lundbeck, 1974; Zecca *et al.*, 1977). These measures, no doubt, have an effect on the presence and number of other pathogens as well.

Preventing disease within a flock is important to human health because pathogens can be transmitted to workers during the slaughtering process (e.g. group A β-hemolytic streptococci, campylobacters, psittacosis) and to consumers via raw poultry products (e.g. salmonellae, *C. jejuni*). Diseased flocks also are less profitable for the farmer.

Commercially grown poultry flocks are normally collected on the farm, placed into crates, transported to the processing plant and slaughtered on the same day. In most countries each flock is slaughtered separately and not mixed with other flocks but in some cases a portion of a flock may be removed for slaughter and the remainder removed at a later date. Procedures for catching and transporting are designed to minimize stress to the birds.

III Primary processing (whole birds and parts)

A *Effects of processing on microorganisms*

Incoming flocks are the principal source of microorganisms found on poultry carcasses after processing. The feathers, feet, and bodies of the birds (Barnes, 1960, 1975) and the intestinal tract are all important sources of bacterial contamination.

Processing operations differ somewhat with the type of poultry being processed, the number of birds slaughtered, the type of processing equipment, and the intended use of the product. Regulations established by control authorities at the national level usually influence how birds are slaughtered and processed. In general, primary processing of poultry is as follows (Figure 2.1): birds are removed from crates, hung by the feet on shackles of a conveyor, stunned by electric shock, killed by cutting the carotid arteries, and allowed to bleed. Next, they are scalded, defeathered (i.e. plucked), and washed.

Heads, hocks, shanks, and oil glands (in some countries) are cut off; viscera are drawn, inspected, and removed; lungs are vacuumed away; and necks are cut off. The carcasses are then usually spray-washed and chilled. After chilling, they are graded and either packaged as whole birds or cut up and sold as parts.

Three general types of microorganisms constitute the microbial population of poultry carcasses: the natural flora of skin, the transient flora that happens to be on skin and feathers at the time of slaughtering, and contaminants to skin that are acquired during processing. The last group is acquired from processing equipment and water, other birds being processed, and, to a lesser extent, workers.

Numerous studies have measured the relative effect of each processing step on carcass contamination. Generally, the results show that an increase in aerobic count or Enterobacteriaceae is unusual. They either remain the same or decrease during processing. The data for salmonellae, however, are highly variable and appear to be influenced by the conditions of the incoming birds, processing, sampling, and the analytical method. The methods most commonly used for detecting the presence of salmonellae on poultry carcasses are quite sensitive. Although the prevalence may be high, the number of salmonellae/carcass is normally quite low such as fewer than 100 × cfu/100 g of skin (Mulder *et al.*, 1977b). The average percentage of carcasses testing positive for salmonellae in a national baseline survey in Canada was 21.1% for chicken broilers and 19.6 % for young turkeys. The number of salmonellae per carcass was estimated to be less than 100 cfu/carcass for 96.9% and 96.0% of the chickens and turkeys sampled, respectively (CFIA, 2000).

Retention, entrapment, and adhesion. Several mechanisms have been proposed to account for the presence of bacteria on poultry carcasses (Thomas *et al.*, 1987). A brief discussion would help visualize the problem of poultry contamination and the effects of processing.

"Retention" occurs when carcasses come into contact with water containing bacteria. A film of water is retained on the carcass surface. Thus, the level of contamination is related directly to the microbial concentration of the processing water (McMeekin and Thomas, 1978). Rinsing carcasses with water having a lower microbial population will reduce the microbial population. Estimates indicate that bacterial numbers on carcasses can be reduced by 90% through use of water sprays at selected points in processing (Thomas *et al.*, 1987). Soft scalded skin retains fewer microbial cells than hard scalded skin (Lillard, 1985).

"Entrapment" occurs when exposed tissues (e.g. skin, collagenous connective tissue layers of muscle) absorb water and begin to swell. Swelling exposes deep channels and crevices into which bacteria penetrate and become entrapped (Thomas and McMeekin, 1980, 1984; Lillard, 1988, 1989b; Benedict *et al.*, 1991). These bacteria cannot be removed by carcass sprays and are protected to some degree from chemicals used for decontamination. The method of scalding and defeathering determines the extent of physical damage to superficial layers of poultry skin. The greater the physical damage to epidermis and exposure of the dermal layer, the greater the risk of entrapment in and adhesion to the skin (Kim and Doores, 1993a,b; Kim *et al.*, 1993). With the passage of time, bacteria that are initially retained in the surface film of water may eventually become entrapped (Lillard, 1989a).

"Adhesion" occurs when microorganisms adhere to surface tissues. Only certain bacteria are capable of adhesion. In one study, all 13 strains of salmonellae tested adhered to chicken muscle fascia. In addition, one strain of *Campylobacter coli* and the fimbriated strains of *E. coli* adhered to the fascia. The fimbriae, however, of the *E. coli* strains probably played little or no role in the mechanism of adhesion (Campbell *et al.*, 1987). Cells grown on media to enhance capsular glycocalyx production were less able to adhere. It was suggested that glycocalyx can block the sites on bacteria cells that are responsible for adhesion.

The mechanism of adhesion into poultry carcasses is becoming clearer (Sanderson *et al.*, 1991). Adhesion occurs on the fascia or loose connective tissue that is under the skin and covers muscle.

Bacteria appear to adhere preferentially to connective tissue rather than to muscle fibers (Benedict *et al.*, 1991). In fact, adhesion to muscle fibers has not been observed. Fascia is a loose meshwork of collagen and elastin fibers embedded in a matrix of glycosaminoglycans (GAG). Bacteria apparently adhere to the GAG matrix rather than to the collagen fibers as had been previously reported (Thomas and McMeekin, 1991). Hyaluranon, a mucopolysaccharide, has been identified as the component of GAG to which bacterial cells adhere (Sanderson *et al.*, 1991). The optimal conditions for adhesion are neutral pH, very low ionic strength, prior immersion in water for a period of time (e.g. 20 min), and extended immersion in contact with the bacterial cells (Campbell *et al.*, 1987). Adding salts (e.g. sodium, magnesium, or calcium chloride) to the water can decrease adhesion and, to a limited degree, facilitate removal of bacterial cells (Thomas and McMeekin, 1981; Campbell *et al.*, 1987; Lillard, 1988; Benedict *et al.*, 1991). Microbial adhesion to carcasses during processing is not dependent upon flagella, fimbriae, or electrostatic attraction (Thomas and McMeekin, 1981, 1984; Lillard, 1989b; Sanderson *et al.*, 1991). Removal of bacteria by physiological saline is probably due to elution of hyaluranon from the fascia rather than separation of cells from the hyaluranon (Sanderson *et al.*, 1991). Zinc chloride reduces attachment, increases detachment and possibly is bactericidal to *Salmonella* Typhimurium (Nayak *et al.*, 2001).

While all three mechanisms (retention, entrapment, and adhesion) likely occur, the relative significance of each is uncertain (Lillard, 1986; Thomas *et al.*, 1987). Bacteria may become more closely associated with the outer skin surface as time elapses. This may be partially explained by entrapment and, perhaps, "non-specific" adhesion. This more general form of adherence differs from the specific adhesion described above for salmonellae on fascia and loose connective tissue.

The efficacy of different decontamination methods is influenced by the proportion of the microbial population that is retained, entrapped, or adhered. Furthermore, these factors must be considered when selecting sampling procedures to quantify or detect microorganisms on raw poultry. Firm adhesion of salmonellae to the skin of the live birds before they are processed may preclude the possibility of producing salmonellae-negative poultry (Lillard, 1989a). Research on entrapment and adhesion has been directed toward poultry carcasses that are chilled by water immersion. The relative importance of the three mechanisms has not been investigated in air chill systems.

Slaughter. Slaughtering consists of stunning birds with an electrical shock and cutting the carotid arteries as they are conveyed from the receiving area. These procedures have not been reported to have a significant impact upon microbial quality of the finished products.

Scalding. Carcasses are scalded (i.e. subjected to moist heat) to facilitate removal of feathers. Methods include hot-water immersion, hot-water spray, steam, and simultaneous hot-water spray and defeathering. Immersion scalding is most generally practiced. Time and temperature of scalding depend on whether the yellow epidermal layer (cuticle) is to be removed from the skin. Soft scalds (e.g. 52°C for about 3 min) will not remove epidermis from the skin. Hard scalds (e.g. 58°C for about 2.5 min) will remove the epidermis (Thomas *et al.*, 1987). Cuticle-free skin is a more suitable substrate for attachment of salmonellae (Kim *et al.*, 1993) and growth of spoilage organisms (Ziegler and Stadelman, 1955; Essary *et al.*, 1958; Clark, 1968; Berner *et al.*, 1969). Water in scald tanks is usually replaced at a rate of 0.2–1 L per carcass per minute to maintain an overflow of the scald tank. Turkeys are usually hard scalded to provide a white skin appearance. Poultry (e.g. spent layers) to be used for processing into cooked poultry products is usually hard scalded. Broilers are given a variety of scalding treatments. The treatment selected depends upon the appearance desired and method of chilling. In Europe, hard scalds are used for poultry that will be water chilled and sold frozen. Soft scalding is used for poultry that will be air chilled and sold fresh because hard scalds give the carcass an unattractive appearance (Richmond, 1990).

During scalding, soil, dust, and fecal matter from the feet, feathers, skin, and intestinal tract are released into the scald water. Also, water from the scald tank is picked up by feathers and skin during scalding. It is not surprising that a variety of bacteria (e.g. *Clostridium, Micrococcus, Proteus, Pseudomonas, Salmonella, Staphylococcus*, and *Streptococcus*) have been isolated from scald tank water or from carcasses or air sacs immediately after scalding (Fahey, 1955; Walker and Ayres, 1956; Surkiewicz *et al.*, 1969; Mead and Impey, 1970; Lillard, 1971; Lillard *et al.*, 1973; Mulder and Dorresteijn, 1977). Aerobic plate counts of scald water, however, usually remain at less than 50 000/mL (Walker and Ayres, 1956, 1959; Schmidhofer, 1969; Mead and Impey, 1970; Mulder and Veerkamp, 1974; Lundbeck, 1974). This is due to the combined effect of a continuous overflow, the introduction of clean replacement water, and the high temperature of scalding. The viable bacterial content depends upon the scalding conditions and the organic matter on the birds. For example, scald water at 52°C can have a total count of about 10^6/mL and Enterobacteriaceae of about 10^4/mL (Mulder, 1985).

After an initial increase during startup, the number per mL of scald water remains relatively constant throughout the day (Schmidhofer, 1969; Veerkamp, 1974; Mulder, 1976). Aerobic plate counts of the skin of broilers immediately after scalding are usually less than 16 000/cm^2 (Walker and Ayres, 1956; Clark and Lentz, 1969; Surkiewicz *et al.*, 1969, Notermans *et al.*, 1975a).

Time and temperature of scalding can influence the extent of microbial destruction and types of microbes that survive on carcasses (Salvat *et al.*, 1992b; Mead *et al.*, 1993). Scalding conditions are compared in Table 2.2. During soft scalding at 53°C for 128 s, Enterobacteriaceae and psychrotrophic counts decreased significantly, but slight variation occurred in aerobic plate count. During hard scalding at 60°C for 115 s, the counts in all three microbial groups decreased significantly.

In another study, soft scalding (about 55.5°C) caused a slight decrease in aerobic plate counts, Enterobacteriaceae and *E. coli*. The high numbers of survivors prevented an assessment of enteric contamination during subsequent stages of processing. When scalding at 60°C, however, nearly all the Enterobacteriaceae and *E. coli* were killed and their presence could then be used to indicate contamination during subsequent processing steps (Notermans *et al.*, 1977). Cross-contamination between carcasses could occur, if Enterobacteriaceae survives scalding (Mulder *et al.*, 1978).

A combination spray scalder and defeatherer (Veerkamp and Hofmans, 1973; Mulder, 1985), steam hot-water scalder (Patrick *et al.*, 1972, 1973), and batch-type scalder using steam at sub-atmospheric pressure (Klose *et al.*, 1971; Kaufman *et al.*, 1972; Lillard *et al.*, 1973; Patrick *et al.*, 1973) apparently reduce bacterial counts more effectively than immersion scalders, due to their higher temperatures and higher ratio of water per carcass. However, these scalders are not in general use because they either cause skin discoloration (Bailey, Personal communication) or offer no economic advantage over conventional methods (Mulder, 1985).

Defeathering (Plucking). Mechanical defeathering involves a series of machines with rotating rubber fingers that remove loosened feathers. Aerobic plate counts, but not psychrotrophic counts, are

Table 2.2 Comparison of scalding treatments on the microbial numbers ($\log_{10}$/g) of chicken pericloacal skin[a,b]

	60°C for 115 s		53°C for 128 s	
Organism or group	Before	After	Before	After
E. coli	3.5	0.8	>4.0	3.4
Enterobacteriaceae	4.9	<2.0	6.1	4.2
Psychrotrophs	4.3	<2.0	4.2	<2.0
Aerobic plate count	7.9	4.9	7.9	6.5

[a]Notermans *et al.* (1975a); Mulder and Dorresteijn (1977).
[b]Means of 8 samples.

Table 2.3 Microbial levels from the surface and subsurface of rubber fingers from three defeathering machines after cleaning

			Surface (log_{10} cfu/cm^2)		Subsurface (log_{10} cfu/g)	
Machine	Finger No.	Degree of wear	TVC	*Staph. aureus*	TVC	*Staph. aureus*
A	1	None	5.1	3.4	4.1	2.9
	2	None	6.1	3.5	3.6	<2.0
	3	Slight	5.6	3.1	5.5	2.6
B	1	Slight	5.0	3.4	5.5	3.3
	2	Moderate	>6.1	>6.0	>6.0	>5.0
	3	Substantial	5.6	3.8	5.3	3.0
C	1	Substantial	5.9	5.3	4.4	3.1
	2	Slight	5.8	5.1	5.5	4.8
	3	Substantial	5.0	4.2	5.5	4.6

Data of Thompson and patterson (1983). *Source*: Mead and Dodd (1990).
TVC = total viable count, 22°C.

significantly higher after defeathering (Walker and Ayres, 1956, 1959; Clark and Lentz, 1969). Higher aerobic plate counts and staphylococcal counts are related to the spread of microorganisms during defeathering and inadequate cleaning of the rubber fingers (Simonsen, 1975).

Heat from freshly scalded birds creates warm operating temperatures within defeathering machines. This warm, moist environment with an abundance of nutrients from carcasses provides favorable conditions for growth of *Staphylococcus aureus*. The rubber fingers are difficult to clean and are subject to wearing and cracking. Even before the fingers deteriorate, however, microorganisms can readily penetrate below the surface of the rubber (Table 2.3). The strains of *Staph. aureus* that colonize the equipment produce an extracellular slime and a tendency toward clumping (Dodd *et al.*, 1988). These properties facilitate attachment to the equipment and also are probably involved in increased resistance of these strains to low levels of chlorine (i.e. 1–5 μg/mL). Once established, these strains survive quite well, multiply as conditions permit and become indigenous to the equipment (Mead and Dodd, 1990). Consequently, *Staph. aureus* on raw poultry meat may consist of a mixture of strains from the live bird that survive scalding and processing plus indigenous strains acquired from defeathering equipment. There has been no effective means to eliminate the indigenous *Staph. aureus*. This contamination can be partially controlled by emphasizing adequate cleaning before disinfecting, replacing worn fingers, and avoiding excessive feather accumulation (Mead and Dodd, 1990). Providing a spray of water with sufficient chlorine (e.g. $\geq$50 μg/mL) and not shrouding the machines so heat can escape, also can be helpful (Purdy *et al.*, 1988).

When processing 14 naturally infected turkey flocks, samples from the rubber fingers of the first, second, and third feather pickers tested positive for salmonellae in 12, 11, and 7 flocks, respectively; and from the final spiral picker during 12 flocks. A chute and table over which the carcasses slid after defeathering tested positive during 10 and 12 of the flocks, respectively. Sixty-three percent of the carcasses sampled were positive for salmonellae after defeathering (Bryan *et al.*, 1968a). Isolates usually matched the serotypes associated with the incoming flocks (Table 2.4).

Clearly, defeathering can spread microorganisms from a few contaminated carcasses to many carcasses. It is during this early stage of processing that bacteria contaminate the skin or enter crevices in the skin and, perhaps, feather follicles and become difficult to remove at later stages of processing (Bryan *et al.*, 1968a; Notermans and Kampelmacher, 1974; Notermans *et al.*, 1975c). The microbial population on poultry carcasses closely reflects the microbial quality of the carcasses immediately after defeathering (Commission of the European Communities, 1976; McMeekin and Thomas, 1978; Salvat *et al.*, 1992b; Abu-Ruwaida *et al.*, 1994). (See Tables 2.5 and 2.6.).

Table 2.4 *Salmonella* serotypes isolated from turkey carcasses and from equipment during progressive stages of processing 15 consecutive flocks[a,b]

	Day I			Day II	Day III			Day IV[c]	Day V	Day VI				Day VII				VIII N	IX
Site	A	B	C	D	E	F	(EF)		G	H	I	(HI)	J	K	L	(KL)	M	N	
Fecal droppings, farm	–	E	g	g	i,j	b			k	–	b			b	–		–	–	
Water troughs, farm	–	–	g	g	i	–				k	–	–			–	–	–	–	–
Fecal droppings, truck	–	–	–	g	–	–			b	l,n	–		–	–	–		–	–	
Picker 1	a	–	a	a,g			c,i	g	m					i	m	m,p	b,p	p	–
Picker 2	–	c	a,e				b,c	–		m			–	m		b,g	b	p	–
Picker 3	–	–	e	e			g	–		m		–		–	–		–	–	–
Spiral Picker	b,c	a,c	b,d	d,f			a,b	b	–					l	–	g	g	a	–
Chute		–	a,e	b,d	g			b	–	–		l		b		g	g	c	q
Table (picking)	–	b,e	f	g			b	–	–					l	b	g	g	c	–
Carcasses after picking		b,e,f		g	c,h		b,g		l,m	n,o	g		d,g	b	g		c	a	
Carcasses after washing				–	–				–	g,l	–		–	e	g		–		
Gutter	–	–	f				b	–	l,m			b	b			–	g	c	–
Knives		–	f				–		l			–	–				–	–	
Head remover			f	b,g			–	–	–	n		–	c			–	–	–	–
Trussing table	b,d	e	f	g			e,i	–	–	n		–	b			–	g	–	–
Spin chill 1	–	–	b	g			–	–	–			b	–			–	–	–	–
Slide	a	–	f	f			–	–	–			–	–			–	g	–	–
Spin chill 2	–	–	–	–				–	–			–	–			–	g	–	–
Spin chill 3	–	–	–	–			–	–	–			–	–			–	p	–	–
Chute and grade table	–	–	–	–			–	–	–			–	–			–	–	–	–
Carcasses before icing	–	–	f	g	a	–			–		–	–			–	–	–	–	–
Chill tank[d]		–		–	–	–		–	–	–			–		–			–	–
Carcasses before packing[d]	c	e	f	g	i	f,i			–	–	–		–	–	–			–	–
Grading, packing table[d]		–	–	–			i		–			b	–			–	–	–	
Scales[d]			–	–			–		–			–	–			–	–	–	
Baggers[d]					–		–		–				b	–			–	–	–

[a]Bryan *et al.* (1968a); A–O indicate flock.
[b]Key: a. *S. infantis*; b. *S. anatum*; c. *S. chester*; d. *S. bredeney*; e. *S. typhimurium*; f. *S. cerro*; g. *S. sandiego*; h. *S.* derby; i. *S. newington*; j. *S. senftenberg*; k. *S. halmstad*; l. *S. muenchen*; m. *S. stanley*; n. *S. saint paul*; o. *S. blockley*; p. *S. schwarzengrund*; q. *S. montevideo*; –, negative for salmonellae; all blank spaces indicate that no samples were taken.
[c]After cleanup.
[d]Isolation made on the following day.

Table 2.5 Geometric means ($\log_{10}$) of bacterial counts/g of neck skin at different stages of processing chickens by chilled by immersion chilling[a]

Plant[b]	After defeathering	After evisceration	After spray washing	After immersion chilling	After packaging
Aerobic plate counts[c]					
1	5.24	5.42	5.15	4.51	4.72
2	4.32	5.14	5.16	5.06	5.02
3	4.43	5.44	5.00	5.14	5.15
4	5.36	5.37	5.25	5.05	5.04
5	5.15	5.34	5.25	4.99	4.98
Coliform counts[c]					
1	4.74	4.56	4.40	3.44	3.57
2	3.41	4.02	3.70	3.29	3.10
3	3.44	4.82	4.33	4.12	4.08
4	4.50	4.62	4.24	4.03	4.23
5	4.21	4.59	4.59	4.08	4.14

[a]Commission of the European Communities (1976).
[b]Plants 1, 2, and 3 used counter-current immersion chillers, and plants 4 and 5 used through-flow immersion chillers.
[c]During 8 sampling periods, 10 samples of neck skin were collected with the passage of 100 carcasses between samples from each sample station at each plant. These 10 samples were pooled for examination.

Table 2.6 Geometric means ($\log_{10}$) of bacterial counts/g of neck skin at different stages of processing chickens chilled by air-chilling[a]

Plant[b]	After defeathering	After evisceration	After spray washing	After air chilling	After packaging
Aerobic plate counts[c]					
1a	4.60	4.92	4.64	4.54	4.52
2a	4.84	5.21	4.66	4.73	4.74
3a	6.01	5.98	5.84	5.65	–
4a	6.22	6.24	6.39	6.12	–
5a	5.13	5.76	5.26	5.15	5.42
6a	5.36	5.37	5.25	5.37	5.15
Coliform counts[c]					
1a	3.31	3.70	3.80	3.71	3.53
2a	3.29	3.56	3.39	3.63	3.28
3a	4.96	5.07	4.70	4.68	–
4a	5.51	5.57	5.66	5.23	–
5a	4.26	5.06	4.49	4.41	4.88
6a	4.50	4.62	4.24	4.44	4.19

[a]Commission of the European Communities (1976).
[b]Plants 1a, 3a, 4a, 5a, and 6a used tunnel chilling, and plant 2a used a chilling room.
[c]During 8 sampling periods, 10 samples of neck skin were collected with the passage of 100 carcasses between samples from each sample station at each plant. These 10 samples were pooled for examination.

Evisceration. During evisceration, microorganisms are transferred from carcass-to-carcass by workers, inspectors, and by contact with equipment (Galton *et al.*, 1955; Wilder and MacCready, 1966; Bryan *et al.*, 1968a). Manual opening of the abdominal cavity and evisceration can give rise to considerable contamination, especially when intestines are cut. Manual evisceration involves pulling the intestines from the body cavity and draping them over the side of the carcass. Vacuuming the vent and cavity removes some fecal and other contamination. Mechanical evisceration requires proper maintenance and continuous cleaning of machinery to prevent an increase in the microbial population on poultry carcasses (Simonsen, 1975). Mechanical evisceration systems are set for specific bird sizes or weights. Variation in the size of the birds being processed increases the likelihood of torn, damaged intestines, and carcass contamination.

Table 2.7 Preventing increased Enterobacteriaceae on broiler skin by adding extra carcass sprays during evisceration

Plant Code	No. of extra sprays	$\log_{10}$ Enterobacteriaceae/g[a] After defeathering	After evisceration
A	0	3.93	4.70
A	3	3.93	3.87
B	0	2.98	3.98
B	3	2.98	3.00
C	0	3.66	4.97
C	1	3.57	4.38
C	2	3.57	3.88
C	3	3.66	3.70

From: Notermans *et al.* (1980).
[a]Mean values for 15–27 samples of cloacal skin.

Table 2.8 Reducing the incidence of salmonellae by adding carcass sprays during evisceration

	Percent of samples positive for salmonellae		
		After evisceration	
	After defeathering	No sprays	With extra sprays
Test 1	NT[a]	53[b] (71)[c]	18 (33)
Test 2	68 (84)	93 (100)	29 (50)

From: Notermans *et al.* (1980).
[a]not tested; [b]rinse of cloaca skin (8 g); [c]analysis of macerated cloaca skin (8 g).

Spray washing. In most countries, carcasses must be spray washed with water after evisceration. For example, in EU countries, at least 1.5 L of water must be applied per carcass weighing up to 2.5 kg (Mulder, 1985).

Spraying carcasses several times during evisceration is considered more effective for preventing an increase in Enterobacteriaceae and salmonellae than a single wash after evisceration (Mulder, 1985; Tables 2.7 and 2.8). In a study of three plants using a combination of manual and mechanical evisceration procedures, adding extra sprays was beneficial (Notermans *et al.*, 1980).

The sprays can decrease the aerobic plate count, Enterobacteriaceae, and coliforms by 50–90% (Stewart and Patterson, 1962; Sanders and Blackshear, 1971; May, 1974; Mulder and Veerkamp, 1974; Mulder, 1976). In addition, the prevalence of salmonellae can be decreased by spray washing (Bryan *et al.*, 1968a; Morris and Wells, 1970). Certain *Pseudomonas* spp., however, can contaminate carcasses during spraying (Lahellec *et al.*, 1973). Adding organic acids or chlorine to these sprays does not extend the shelf-life of fresh poultry. Lower bacterial numbers, however, can be achieved when chlorine is added to both the spray water and chill tank water (Jul, 1986).

In most facilities, mechanical inside–outside carcass washers are used to wash both the cavity and the outer surface of carcasses before chilling. In a limited number of plants, two inside–outside carcass washers are used in series to remove surface contaminants. An evaluation of washing turkey cavities led to the conclusion that a hand-held water sprayer is not beneficial (Wesley and Bovard, 1983).

Chilling. Chilling is necessary to delay growth of psychrotrophic spoilage bacteria and prevent growth of food-borne pathogens. The method of chilling has been highly controversial and influenced by the relative impact of economics, hygiene, and local regulations. The ideal method will provide chilling at the least cost without causing microbial contamination or permitting multiplication. Preferably, chilling will involve decontamination to further improve microbial safety and quality. Many countries have established limits for water uptake. Chilling systems continue to evolve as industry strives to achieve greater efficiency, reduce cost, improve microbial control, and adjust to changing regulatory requirements.

Carcasses can be chilled by circulation of cold air, sprays of cold water, or by immersion in tanks of cold water with or without addition of ice. In Europe, wet-chilled poultry must be frozen; whereas, air-chilled and spray-chilled poultry can be distributed refrigerated. In other regions of the world, the majority of poultry is wet chilled and may be distributed refrigerated or frozen. There is no evidence that air, spray, or water immersion chilling results in a lower prevalence or concentration of salmonellae on carcasses when these systems are properly controlled. Other factors (e.g. levels of salmonellae on

Table 2.9 Temperatures (°C) of chicken carcasses during processing[a,b]

Plant[b]	After evisceration	After spray washing	After air chilling	After packaging
Air chilled				
1a	27.5	26.3	12.2	13.4
2a	30.5	32.5	2.0	3.4
3a	39.8	38.8	8.2	–
4a	28.0	27.0	5.0	–
5a	34.4	32.1	0.3	2.8
6a	34.0	32.0	7.6	8.0
Immersion chilled				
1	30.5	29.5	7.3	9.3
2	31.6	26.6	10.1	–
3	28.6	27.5	7.7	9.3
4	34.0	32.1	8.9	8.7
5	30.0	28.4	7.8	9.1

[a]Commission of the European Communities (1976).
[b]Average of 8 experimental periods. Ten carcasses placed into an insulated box, thermocouples inserted among the carcasses in the center of the box, and temperatures determined after an equilibration period of 30 min. In plant 4a, temperatures of carcasses were measured by thermocouples inserted into the deep breast musculature.
[c]Plants 1a, 3a, 4a, 5a, and 6a used tunnel chilling and plant 2a used a chilling room. Plants 1, 2, and 3 used counter-current immersion chillers. Plants 4 and 5 used through-flow immersion chillers.

incoming birds; scalding, defeathering, and evisceration conditions) prior to chilling have greater influence on the levels of contamination.

Typical temperatures of chicken carcasses after evisceration, chilling, and packaging are shown in Table 2.9. Many countries have established chilling requirements for poultry. For example, regulations in the United States require carcasses to be chilled to 4.4°C or lower in 4, 6, or 8 h for carcasses weighing less than 4, 4–8, or over 8 pounds, respectively (CFR, 1992). This requirement has been cited as a critical point in the slaughtering process for broilers (Codex, 2002), but the origin of this requirement is uncertain. Most likely, the chill rates were based on what was considered a good commercial practice in the USA at the time they were established. There is no evidence the time–temperature requirements were based on a scientific assessment of the conditions necessary to control pathogens.

Air chilling. Air chilling consists of blowing dry or moist air into and over carcasses. Various combinations of time, temperature, and humidity can be used. One method has used dry air operated at 0 to −2°C (Grey and Mead, 1986; Jordan, 1991). Another commercial system involved 1 h at −17°C and then overnight holding at 4°C (Schmitt *et al.*, 1988). Dry air chilling dehydrates skin and although this might be expected to retard microbial growth, this benefit may not be realized due to rehydration of carcass surfaces after packaging (Grey and Mead, 1986). Moist air chilling involves spraying carcasses with water while blowing air over them to cause evaporative chilling. Moist air chilling can be followed by dry air (Jul, 1986). By the mid-1990s, pre-chilling for 1 h at 5°C followed by 1.5 h at 0°C had become more commonly applied for air chilling (Ristic, 1997).

When air chilling is applied, birds are soft scalded at about 50°C to minimize skin discoloration due to dehydration. This scald temperature does not appreciably reduce enteric microorganisms. Microbial quality of the air must be controlled to avoid carcass contamination. An EU-sponsored study showed no commercial benefit in shelf-life between air chilling and water chilling. Mean shelf-life values of 8.8 and 8.6 days, respectively, were obtained at 2°C with the two methods (Grey and Mead, 1986). Others also have reported no difference in shelf-life (Mulder and Bolder, 1987; Schmitt *et al.*, 1988). No difference in the shelf-life of broilers at 0°C was found as a consequence of the combined effects of scalding (50°C and 57°C) and chilling method (air, water, and evaporative chilling) (Mulder and Bolder, 1987). In another study, comparable aerobic counts of $\log_{10}$ 3.91 and 4.57 and Enterobacteriaceae counts of $\log_{10}$ 2.5 and 2.58 were obtained for moist air and water-chilled carcasses, respectively (Ristic, 1997).

Spray chilling. This procedure consists of spraying chilled water onto carcasses as they are suspended in air (Leistner *et al.*, 1972; Leistner, 1973; Mulder and Veerkamp, 1974; Ristic, 1997; Mielnik *et al.*, 1999). In various investigations, 0.3–15 L of water (at 0–11°C) have been sprayed for each carcass, for 15–30 min to reduce carcass temperatures, usually to 7°C or below. The large amount of water necessary to spray-chill poultry has reduced commercial acceptance of this method. Frozen chickens from a plant using this system were found to have salmonellae contamination rate of over 85%. Thus, the system does not necessarily improve the hygienic state of poultry over other systems (Eisgruber and Stolle, 1992). Spray chilling has been reported to reduce weight loss (1.8%) compared to dry-air chilling while improving color. Microbiological quality and rate of spoilage for poultry chilled by the two methods were the same (Mielnik *et al.*, 1999).

Continuous immersion chilling. This method of chilling involves moving carcasses through a tank or series of tanks containing chilled water. Water agitation can be achieved by mechanical means or with compressed air. The flow of fresh cold water should be against the direction of the carcasses. The bacteriological quality of the water at the exit end of the chiller determines the bacterial population that is retained on the carcasses. Continuous immersion chilling is very efficient and relatively inexpensive. Agitation can remove some microorganisms from the carcasses; however, they may be transferred to other carcasses if the conditions of chilling are not controlled.

A relationship between the number of salmonellae in chill tank water and the prevalence of salmonellae on carcasses has been reported (Table 2.10). Almost 55% of the water samples were negative for salmonellae even though 19% of the corresponding carcasses were positive. About 8% of the water samples had more than 100 salmonellae/mL (Green, 1987). The data indicate that water samples can be a convenient method for estimating the prevalence of salmonellae on carcasses after chilling. In another study, immersion chilling caused about a 10-fold reduction in total count and Enterobacteriaceae on broiler carcasses (Table 2.11). The prevalence of salmonellae, however, increased from 10–13% to 28–38% (Lillard, 1990).

Twelve investigations have shown that aerobic plate counts and indicator organisms, particularly those of fecal origin, were reduced during continuous immersion chilling (Thomson *et al.*, 1974). In

Table 2.10 Relationship between the incidence of salmonellae on carcasses and overflow water from chill tanks[a]

No. (%) of water samples	Estimate of salmonellae in 100 mL water	% of carcasses positive
870 (54.5)	<1	19[b]
190 (11.9)	1–4	45
192 (12.0)	5–10	53
221 (13.9)	11–100	55
122 (7.6)	>100	69

Source: Green (1987).
[a]Survey data from U.S. broiler plants involving 1 719 carcasses and 1 595 water samples.
[b]Carcasses were sampled by a whole carcass rinse.

Table 2.11 Effect of immersion chilling on the microbial levels of broiler carcasses from two commercial plants

	Mean $\log_{10}$/ml of carcass rinse				%Positive Salmonellae	
	Aerobic count		Enterobacteriaceae			
	A	B	A	B	A	B
Pre-Chill	4.69	4.67	4.01	4.09	13	10
Pre-Chill	3.78	3.94	2.97	2.97	28	38*

Adapted from Lillard (1990).
*Significantly different from pre-chill ($P = <0.05$). All data are based upon 40 whole carcass rinse samples from each site at each plant (plants A and B). Chill water was not chlorinated.

three investigations, one or more of these counts increased. Factors accounting for the differences include: (a) extent of bacterial contamination on the carcasses before chilling; (b) amount of fresh water replacement per carcass; and (c) number of carcasses in relation to the volume of chiller water. If insufficient water is used per carcass, microorganisms can accumulate in the water and increase on carcasses rather than decrease. Research during the 1970s and 1980s provided information that led to improved modifications of immersion chilling in Europe (Jul, 1986).

Properly controlled modern counter-flow immersion chilling procedures can reduce the level of salmonellae, improve carcass appearance by washing the outer surfaces, minimize cross-contamination, and control organic matter in chill water. Regulations, such as those in the EU and North America, specify minimum requirements for hygienic operation of continuous immersion chilling.

Slush-ice (static-tank) chilling In most countries, this method of chilling has been replaced by more efficient and hygienic methods. Chilling in slush-ice with approximately equal amounts of ice, water, and carcasses, takes advantage of the cooling capacity of melting ice. Carcasses can be held in tanks for 4–24 h, during which time psychrotrophic bacteria can multiply on the carcasses and in water (Barnes and Shrimpton, 1968). Chlorination at 5–20 μg/mL has been used in some plants to prevent multiplication of psychrotrophic bacteria.

Packaging whole carcasses or parts. Chilled poultry can be packaged as whole carcasses, cut and sold as parts, or processed into a wide variety of products. Microbial contamination can occur during handling and processing in the plant. The shelf-life of raw poultry is significantly affected by the extent of contamination with psychrotrophic bacteria. The hygienic condition of conveyors and other equipment (particularly knives, tables, tubs, and scales) that contact the chilled poultry is often a major factor influencing the rate of spoilage.

Equipment can be a means for cross-contamination with enteric pathogens during further processing.

Short-term holding. After chilling, poultry (e.g. whole carcasses, parts, boned meat, and giblets) may be held for additional processing within the plant or shipped to another plant for further processing. This method of holding is influenced by local regulations. One common method is to place the poultry between layers of ice in tanks. Boned meat can be placed into plastic bags and sealed to prevent uncontrolled moisture absorption from the melting ice. Another method is to add dry ice (carbon dioxide snow or pellets) to the poultry as it is being put into tanks. The holding step also can be used to finish the chilling process and reduce the poultry meat to less than 5°C. The cleanliness of tanks and time and temperature of holding are major factors influencing microbial quality.

Edible viscera. During evisceration, edible viscera (hearts, livers, and gizzards) are removed from carcasses, chilled, and conveyed to a viscera-preparation and wrapping area, where they are cleaned, sorted, and packaged in bulk or put into bags. Bulk packaged viscera are normally frozen in boxes for distribution. Small bags containing edible viscera are inserted into the cavity of dressed carcasses. Equipment used to convey, chill, process, or wrap edible viscera can be a significant source of contamination if not properly controlled.

Uneviscerated (Effilée or New York dressed) poultry. Uneviscerated poultry can be processed by several methods. Carcasses can be uneviscerated or only the intestines are removed. Feathers may be dry-picked, wet-picked, or left on the bird (e.g. game birds). The head or feet may or may not be removed. Chilling may be in air blast refrigerators or rooms. When uneviscerated carcasses are dry processed and held in a cool and well-ventilated area, bacterial growth will occur in the intestines.

Skin surfaces are too dry to permit rapid bacterial growth. Traditionally, these carcasses have been held close to 10°C. The sensory quality of pheasants held at 10°C is preferred over 5 or 15°C (Barnes *et al.*, 1973). When held at 10°C, pheasants develop the desired flavor and tenderness after 13 days (Griffiths, 1975). Uneviscerated carcasses, however, keep much longer without spoilage if they are held at lower temperatures, preferably below 5°C (Barnes and Impey, 1975). The inspection procedures for eviscerated poultry can be satisfactorily modified for uneviscerated processing (Johnson *et al.*, 1992).

Manually deboned meat. The majority of poultry removed from the carcass is removed manually with knives, a wide variety of other hand tools, and machines. This typically involves a large number of people stationed along a conveying system that moves carcasses past the people with each person assigned to make specific cuts or remove certain portions. The process of manual deboning is normally done in a chilled room. The various portions of deboned meat are accumulated and placed into boxes, bins, or large shipping containers for holding and/or distribution. There is a low risk of pathogen contamination while deboning occurs provided the room is chilled; however, cross-contamination may occur leading to an increase in prevalence across the material being produced. Contamination with psychrotrophs that persist in the cold working environment ultimately can influence the shelf-life of the raw product.

Mechanically deboned meat (MDM). Mechanically deboned poultry meat is produced from chilled necks, backs, and skeletal frames from cut-up or boning operations. A certain amount of meat remains on the bones of poultry carcasses after manual deboning. This residual meat can be removed mechanically by grinding the bones and passing the mixture through a fine screen to separate the meat from the bone fragments. Bone material is collected for rendering and eventual use in animal feed (see Chapter 4). The MDM can be passed through a heat exchanger or some other means, to reduce the temperature to 4°C or below (Mulder and Dorresteijn, 1975). Low aerobic plate counts (e.g. 10^4/g) are possible in plants that slaughter, chill, debone carcasses, and produce MDM on the same day.

The MDM can be used directly as a raw material, frozen in boxes or blended with certain ingredients for shipment at refrigerated temperatures to other plants for further processing. It is the primary ingredient of some poultry products (e.g. frankfurters, bologna). The MDM can be incorporated into almost any cooked product where a fine comminuted emulsion is an acceptable ingredient. The use of MDM is controlled by local regulations. Frozen material will more likely appear in international trade.

The microbiological and chemical quality (e.g. rancidity, color stability) of MDM reflects the hygienic conditions of slaughtering, manual deboning, mechanical separation, and the times and temperatures to which the material is exposed during processing. When processing steps occur in more than one plant, microbial growth will occur on unfrozen raw materials during storage and transit. When it is known that MDM will be used to make a cured product, some or all of the intended salt and sodium nitrite can be added to retard spoilage and extend the time the MDM can be held before use.

A WHO consultation noted that MDM has several important characteristics (WHO, 1989).

(1) The bacterial content depends largely on the raw material used. For example, product harvested from chicken backs differs greatly from that harvested from chicken legs.
(2) MDM has the highest surface-to-volume ratio of all meat products and, as bacteria tend to multiply on surfaces, MDM is a highly perishable product.
(3) As MDM is a homogenate, contaminating bacteria are spread throughout the product.
(4) MDM is frequently contaminated with pathogens from animal carcasses (e.g. *Salmonella*, *Campylobacter*, *Listeria* spp., and *Clostridium perfringens*), and may have a relatively high count of coliform and other enteric bacteria.
(5) In most countries, MDM is used for further processed and heated products and is not offered directly to the public.

Table 2.12 Effect of temperature on generation times of psychrotrophic spoilage bacteria for growth of pure cultures on meat

	Generation time (h)							
	Aerobic (°C)				Anaerobic (°C)			
	2	5	10	15	2	5	10	15
Pseudomonas spp.								
Non-fluorescent	7.6	5.1	2.8	2.0	–	–	–	–
Fluorescent	8.2	5.4	3.0	2.0	–	–	–	–
Acinetobacter spp.	15.6	8.9	5.2	3.1		–	–	–
Enterobacter spp.	11.1	7.8	3.5	2.4	55.7	23.2	8.5	5.4
Broc. thermosphacta	12.0	7.3	3.4	2.8	32.8	20.1	9.7	6.8
Lactobacillus spp.	–	–	–	–	8.4	6.5	4.6	3.8

Source: Adapted from Gill (1986) by Lambert *et al.* (1991).

The consultation noted that there have been no reports that MDM has been responsible for foodborne disease (WHO, 1989). MDM is widely used by commercial manufacturers to manufacture cooked products and products to be cooked by the end user.

B Spoilage

Carcasses and portions. Post-mortem biochemical changes that occur in poultry meat are similar to those in beef, pork, and lamb. Thus, it is not surprising that spoilage of poultry meat is influenced by the same factors influencing spoilage of red meats (Newton and Gill, 1981). Poultry breast meat has a final pH of about 5.8 but a higher pH can result from pre-slaughter stress. The spoilage flora of breast meat is essentially the same that develops on normal beef. On the other hand, poultry leg meat has lower glycogen content, resulting in a much higher final pH (6.4–6.7). Poultry leg meat is similar to dark, firm, dry beef, which, due to deficiencies in glucose and glycolytic intermediates, has a high pH (>6.0) (Newton and Gill, 1981). This relatively high pH permits the growth of bacteria (e.g. *Shewanella putrefaciens*) that are sensitive to lower pH values and produce very objectionable sulfide compounds. Poultry skin has a pH of about 6.6 or higher.

Raw poultry meat held at refrigeration temperatures (−2 to 5°C) will eventually spoil. Four major factors determine the rate of spoilage; (a) storage temperature; (b) numbers and types of psychrotrophic bacteria; (c) pH; and (d) type of packaging.

Storage temperature is the most important factor influencing the rate of spoilage. For example, the time for spoilage of raw turkey can be extended from 14 to 23–38 days by reducing the temperature of storage from 2 to 0 to −2°C, respectively (Barnes *et al.*, 1978). The longest storage life for unfrozen poultry is obtained when held as close as possible to its freezing point. *Pseudomonas* species are the major cause of spoilage at −2 to 5°C. As the holding temperature exceeds 5°C, a wider variety of bacteria are involved in spoilage. This is evident from the generation times for the common spoilage flora of poultry (Table 2.12).

The number and type of bacteria on freshly packaged raw poultry is the second most important factor affecting shelf-life. The optimal temperatures for growth of psychrotrophic bacteria are 20–30°C, but they can multiply below 5°C. Their ability to multiply at low temperatures is essential to their involvement in spoilage (Figure 2.3). Psychrotrophic spoilage bacteria enter poultry processing plants on birds' feathers and feet and in water and ice supplies. More importantly, they multiply on soiled surfaces of equipment (e.g. chill-water tanks, conveyors, knives, wire mesh gloves, tubs, tables). The air temperature in processing rooms influences the number and type of bacteria. Rooms with temperatures of 15°C or below may select for a higher percentage of psychrotrophic bacteria in the indigenous

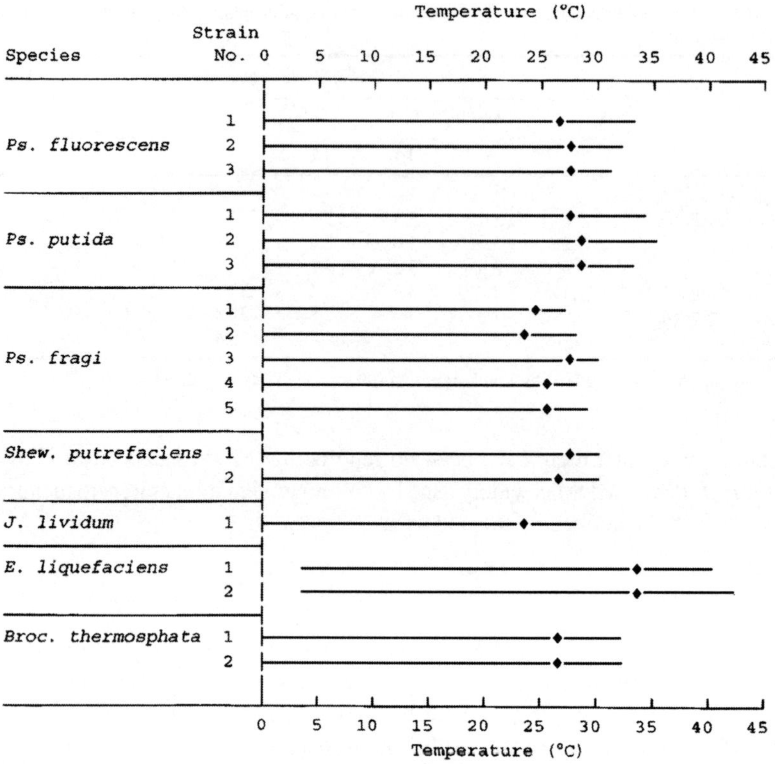

Figure 2.3 Temperature limits and optional growth temperatures (♦) of 19 bacteria isolated from broiler skin (Gall *et al.*, 1988).

population of the environment. In rooms of 25°C and above, the rapid rate of bacterial multiplication must be considered. The microbial population on equipment can be controlled by effective cleaning and disinfecting procedures. Some evidence exists that yeasts (e.g. *Yarrowia lipolytica*, *Candida zeylanoides*) may play a role in the spoilage of chicken when stored at ≤5°C (Ismail *et al.*, 2000; Hinton *et al.*, 2002).

The initial population of psychrotrophs on poultry significantly influences product shelf-life (Russell, 1997). Spoilage (i.e. off-odors) can be detected when the microbial population reaches about 10^7–10^8/cm^2 (Ayres *et al.*, 1950; Barnes, 1976; Studer *et al.*, 1988). Spoilage occurs when bacteria metabolize certain nutrients in meat and produce end products that are organoleptically unacceptable. The major nutrients and metabolites are listed in Table 2.13. Glucose is the preferred nutrient (Nychas *et al.*, 1988) and as glucose is depleted, other small molecular-weight compounds (e.g. amino acids) are metabolized. Metabolism of amino acids leads to production of sulfide compounds that are readily detected in low quantities and the perception of spoilage (McMeekin, 1975). Spoilage of raw poultry occurs on the surface. The few bacteria, if any, in the interior of poultry muscle multiply slowly or not at all during low-temperature storage.

The inherent pH of poultry meat is the third most important factor influencing the type of bacteria that multiply and their rate of multiplication. The pH of poultry leg muscle is 6.4–6.7 while breast muscle is much lower, pH 5.7–5.9. Pseudomonads, the major cause of spoilage, grow equally well on both types of meat (Barnes and Impey, 1968). Although chicken leg and breast meat have been reported to spoil at comparable rates when packaged under aerobic conditions (Clark, 1970), in commercial practice, poultry leg meat spoils more quickly than breast meat. Certain spoilage bacteria (e.g. *Acinetobacter* and

Table 2.13 Substrates used for growth and metabolic by-products of major meat spoilage microorganisms

	Substrates used for growth[a]		Major end products of metabolism	
Microorganism	Aerobic	Anaerobic	Aerobic	Anaerobic
Pseudomonas	Glucose[1] Amino acids[2] Lactic acid[3]		Slime, sulfides, esters, acids, amines	
Acinetobacter/ Moraxella	Amino acids[1] Lactic acid[2]		Esters,nitriles, oximes, sulfides	
Shewanella putrefaciens	Glucose[1] Amino acids[2] Lactic acid[3]	Glucose[1] Amino acids[1]	Volatile sulfides	H_2S
Brochothrix thermosphacta	Glucose[1] Ribose[2]	Glucose[1]	Acetic acid Acetoin Isovaleric acid Isobutyric acid	Lactic acid, volatile fatty acids
Enterobacter	Glucose[1] Glucose 6 phosphate[2] Amino acids[3] Lactic acid[4]	Glucose[1] Glucose 6-phosphate[2] Amino acids[3]	Sulfides Amines	Lactic acid, CO_2, H_2 H_2S Amines
Lactobacillus	–	Glucose[1] Amino acids[1]	–	Lactic acid, Volatile fatty acids

[a]The superscript number indicates the order of substrate utilization.
Adapted from: Lambert *et al.* (1991).

Shew. putrefaciens) can multiply on leg muscle but only slowly or not at all on breast muscle (Barnes, 1976; McMeekin, 1975, 1977). The high pH of skin must be considered. The increasing availability of marinated chicken parts can alter the rate of spoilage, particularly when the marinade is low in pH (Perko-Mäkelä *et al.*, 2000a; Buses and Thompson, 2003) but not affect the survival of pathogens such as campylobacters (Perko-Mäkelä *et al.*, 2000a).

Packaging material and method (e.g. aerobic, vacuum, modified gas atmosphere) is the fourth most important factor influencing the rate of spoilage. In oxygen-permeable films, *Pseudomonas* spp. (*Pseudomonas fluorescens, Ps. putida, Ps. fragi*, and related species), and to a lesser extent *Shew. putrefaciens*, *Acinetobacter*, and *Moraxella*, are the principle causes of spoilage (Barnes, 1976; McMeekin, 1975, 1977; Sawaya *et al.*, 1993). In oxygen-impermeable films, accumulation of CO_2 in the package suppresses pseudomonads and shifts the cause of spoilage toward less oxygen-dependent bacteria (e.g. lactobacilli, leuconostocs, *Shew. putrefaciens, Broc. thermosphacta*, Enterobacteriaceae, *Aeromonas*). In the absence of air, slower growth rates and fewer offensive spoilage odors result in an extended shelf-life (Barnes and Melton, 1971; Mead *et al.*, 1986; Studer *et al.*, 1988; Sawaya *et al.*, 1993). The flora that develops is dependent on the combined effect of all the major factors affecting spoilage. For example, when packaging conditions (e.g. anaerobic) limit growth of pseudomonads, other species become more important. *Shewanella putrefaciens*, a strong sulfide producer, becomes a major component of the spoilage flora under anaerobic conditions (Newton and Gill, 1981; Gill *et al.*, 1990). The growth of *Shew. putrefaciens* may be limited to tissues of higher pH (e.g. leg meat, skin). Marinades containing acids can alter the pH of packaged poultry. A survey of marinated modified atmosphere packaged chicken strips on the sell-by date (7–9 days) at retail level in Finland found the dominant flora was *Leuconostoc gasicotatum* (Susiluoto *et al.*, 2003). Further extension of shelf-life may be attained by eliminating residual oxygen during packaging such as with vacuum packaging (Jones *et al.*, 1982; Gill *et al.*, 1990; Sawaya *et al.*, 1993). Data in Table 2.14 demonstrates the effect of two factors on

Table 2.14 Effect of vacuum packaging on the shelf life of raw turkey parts held at 1°C

	Days until spoilage	
	Air	Vacuum
Breast meat	16	25
Drumsticks	14	20

From: Jones *et al.* (1982).

poultry shelf-life: restricting oxygen (e.g. vacuum packaging) and the greater susceptibility of dark meat (drumsticks) compared to light meat (breast).

Combining carbon dioxide with packaging in an oxygen impermeable film is even more effective than vacuum packaging (Gill *et al.*, 1990). Broilers vacuum packaged with film of low-oxygen transmission spoiled by 2 weeks at 3°C and 3 weeks at −1.5°C due to a flora dominated by *Enterobacter*. Broilers packaged in an oxygen impermeable film with added carbon dioxide were initially dominated by lactobacilli and eventually *Enterobacter* spp. with the development of a putrid spoilage odor. This method of packaging extended shelf-life to 7 weeks at 3°C and 14 weeks at −1.5°C. In studies with ground chicken and turkey meat packaged with two different gas mixtures (nitrogen and carbon with and without oxygen) and held for 20 days at 1°C, the composition of the meat (e.g. fat and myoglobin content) influenced the degree of redness and acceptable appearance (Saucier *et al.*, 2000). Through 15 days, a dominant spoilage flora was yet to emerge in either of the gas mixtures.

Several commercial methods for packaging are used which retain a desirable fresh appearance and retard spoilage. These methods involve the use of various gas mixtures and incorporate carbon dioxide at levels of about 20% or higher to inhibit pseudomonad growth. Low oxygen levels reduce product acceptance due to a poor color. In the absence of oxygen, poultry meat becomes darker (e.g. purple) due to deoxygenation of myoglobin, the major pigment of muscle tissue. With 20% or more carbon dioxide, oxygen can be added and the normal fresh color of the meat is retained. Nitrogen is typically added as inert filler. Gas mixtures with high levels of carbon dioxide can cause some packages to collapse or develop a vacuum packaged appearance as carbon dioxide is absorbed into the meat.

Another method of packaging raw poultry is to place trays of packaged poultry into a large oxygen impermeable bag in a box. The bag is then vacuumized, backfilled with a gas mixture of 20% or more carbon dioxide, and then sealed. When received at the retail store, the trays are removed and placed onto display. Due to the oxygen-permeable film over the retail packages, oxygen can penetrate into the product during display and the normal fresh meat appearance returns. For maximum shelf-life, the retailer should not break the seal on the bag until the product is to be displayed (Bohnsack *et al.*, 1988).

Another form of retail packaging involves placing poultry meat into a tray, overwrapping with an oxygen-impermeable film, vacuumizing, backfilling with a gas mixture of 20% or more carbon dioxide and a high level of oxygen, and then sealing the film. These trays, containing a modified atmosphere, are shipped to stores for direct display.

Dark meat usually has a shorter shelf-life than light meat when packaged in modified atmosphere. The differences noted in Table 2.14 for vacuum packaging are typical of what might be observed with modified atmosphere packaging. Carbon dioxide also is used in the shipment of bulk quantities of raw poultry to retailers (Hart *et al.*, 1991) or to other plants for further processing. The use of carbon dioxide in solid form as dry ice to chill poultry meat may have some residual effect for delaying spoilage.

Edible viscera. Muscle tissues generally have a longer shelf-life than giblets. Giblets placed into the cavities of poultry carcasses can spoil more rapidly and cause product rejection. This can be avoided by controlling the rate of chilling, the level of contamination and not holding giblets over to a later date for inserting into carcasses.

Uneviscerated (Effilee or New York dressed) poultry. The microorganisms that multiply on uneviscerated poultry depend on the holding temperature. In domestic fowl (e.g. chicken and turkey), bacteria in the intestines produce hydrogen sulfide that diffuses into the muscle tissue and combines with hemoglobin to produce a green discoloration. Greening first develops near the vent and across the abdomen immediately above where the tips of the caeca lie and then spreads to the back, ribs, and neck (Barnes *et al.*, 1973). The rate of green discoloration is decreased considerably by holding carcasses at low temperatures (e.g. 4°C). Greening may not develop in certain fowl. This is reportedly due to differences in diet and low numbers of certain bacteria in the intestine during post-slaughter holding (Barnes, 1979). Except in instances where the muscle may have been damaged (e.g. from shooting wild fowl), microbial growth has not been detected in the muscle tissue of uneviscerated birds (Barnes and Shrimpton, 1957; Barnes *et al.*, 1973; Barnes and Impey, 1975). Psychrotrophs (e.g. pseudomonads) on the skin multiply slowly or not at all as long as the surface is dry and intact. When cuts are made to remove the head, neck, and viscera, spoilage organisms contaminate the cut muscle surfaces and subsequently multiply and produce off-odors (Barnes, 1976).

C Pathogens

The consumption or handling of raw or undercooked poultry meat has a long history as a source of foodborne illness (Bryan, 1980; Bean and Griffin, 1990; Beckers, 1988; Todd, 1992; Studahl and Andersson, 2000; WHO/FAO, 2002; Stern *et al.*, 2003). Outbreaks involving large numbers of people are usually due to *Salmonella*, *Cl. perfringens*, and *Staph. aureus*. Campylobacteriosis typically occurs as sporadic cases rather than outbreaks. Worldwide, campylobacters and salmonellae are by far the most important human pathogens associated with poultry products.

Salmonella. The primary source of salmonellae in the processing plant is the flock that is being processed. Thus, the type of salmonellae on raw poultry carcasses primarily reflects the serovars of salmonellae in the flock being slaughtered (Bryan *et al.*, 1968a; McBride *et al.*, 1978, 1980; Rigby, 1982; Olsen *et al.*, 2003, Table 2.4). During processing, carcasses become contaminated and salmonellae can be spread from carcass to carcass by equipment, tools, and workers (Bryan *et al.*, 1968a; Morris *et al.*, 1969; Notermans *et al.*, 1975a, Finlayson, 1978; McBride *et al.*, 1980; Lillard, 1990). Control of salmonellae contamination must begin at the farm because there are no steps in poultry processing that eliminate salmonellae (Simonsen *et al.*, 1987). The extent of cross-contamination can be, however, minimized by implementing and controlling certain hygienic practices during slaughtering and chilling. When present on broiler carcasses, salmonellae are usually in small numbers, i.e. about 10–20 but occasionally more than 1400 cfu/100 g of skin (Notermans *et al.*, 1975b; Mulder *et al.*, 1977b). In a study of five plants over 12–13 weeks, mean values of less than 10 cells/carcass were found even though the prevalence ranged from 9% to 77% at each plant (Table 2.15) (Waldroup *et al.*, 1992a). Because of the low numbers of salmonellae, large surface areas are normally sampled to detect their presence. An internationally accepted method for sampling and analysis has not been agreed upon and it is difficult to compare prevalence data for salmonellae on raw poultry from different countries. In addition, the location in the food chain where samples are collected can yield very different values (Simmons *et al.*, 2003).

Single rinses of whole carcasses do not remove all the salmonellae (Rigby, 1982) and repeated samplings of whole carcasses may fail to detect salmonellae even though they may be present (Lillard, 1989a).

Outbreaks of salmonellosis are usually due to inadequate cooking, recontamination of cooked poultry and/or cross-contamination of ready-to-eat foods. It has been estimated that a 50% reduction in the prevalence of contaminated chicken would result in a 50% reduction in the expected risk per serving. Furthermore, a 40% reduction in the concentration of salmonellae cells on chicken carcasses exiting the chill tank would result in a 65% reduction in risk per serving (WHO/FAO, 2002).

It is well established that antibiotic resistant strains of salmonellae, *Campylobacter* and enterococci have been linked to livestock and/or poultry (Anderson *et al.*, 2003). This has led to a growing international consensus that antibiotics should not be used at low doses as additives in animal feeds to enhance the rate of growth. Antibiotics, however, do have value for treating diseased animals and flocks as per accepted veterinary practice. Veterinarians should avoid, where possible, using antibiotics that are important for treating human disease. Information on the status of antibiotic resistant pathogens in poultry and humans is available through a variety of sources, such as DANMAP(2002, 2003), European Antimicrobial Resistance Surveillance System and the World Health Organization Antimicrobial Resistance Information Bank (CDC, 2003; DVI, 2003; EARSS, 2003; EC, 2002; WHO, 2003).

Campylobacter. Raw poultry is considered an important risk factor for human campylobacteriosis (Richmond, 1990; Kapperud *et al.*, 1992; ACMSF, 1993; Hernandez, 1993; Pearson *et al.*, 1993; Jacobs-Reitsma *et al.*, 1994; NACMCF, 1994; Stern *et al.*, 2003). In one study, 82% of the isolates from chicken and 98% of the isolates from humans with gastroenteritis were of the same *C. jejuni* biotype, suggesting an epidemiological link (Shanker *et al.*, 1982). More recent research has identified other factors (e.g. consumption of unpasteurized milk and untreated or contaminated water, having contact with farm animals and pets) to be of comparable or greater importance than poultry meat (Friedman *et al.*, 2000; Studahl and Andersson, 2000; Tenkate and Stafford, 2001; WHO/FAO, 2002). Two case-control studies concluded that eating and handling chicken could account for less than 10% of human cases (Neal and Slack, 1997; Neimann *et al.*, 1998), but this proportion may differ in other countries or regions of the world. Cross-contamination from raw poultry to ready-to-eat foods in the home or in restaurants is another risk factor to consider (Effler *et al.*, 2001; Rodrigues *et al.*, 2001; WHO/FAO, 2002). A reduction in prevalence of *Campylobacter* spp. in infected chickens to a level of 10–15% in Sweden did not result in a reduction of domestically acquired human infections. The increase in human infections occurring during that time may have reflected an increase in the consumption of chicken, particularly refrigerated chicken (Studahl and Andersson, 2000). A three-year study of *Campylobacter* isolates in Finland showed fluctuation in pulsed field gel electrophoresis (PFGE) genotypes among the human population and chicken meat sampled at retail. Most of the predominant PFGE genotypes among the human isolates were the same as in chicken. While this suggests that chicken can be a source of human campylobacteriosis, the data also could mean that the genotypes are common in many sources and that the predominant genotypes circulate and colonize various host animals (Hänninen *et al.*, 2000). Subsequent analysis of isolates by serotyping and several genotyping methods confirmed that the common *C. jejuni* genotypes form genetic lineages that colonize both humans and chickens (Hänninen *et al.*, 2001).

The minimum infective dose is not certain, but a few hundred cells are thought to be sufficient to cause human infection (Robinson, 1981; Hernandez, 1993; NACMCF, 1994). The location (restaurant, home) where chicken is prepared and consumed can influence the degree of risk (Friedman *et al.*, 2000; Rodrigues *et al.*, 2001). In temperate climates campylobacteriosis among humans is highly seasonal with peak periods of infection occurring in the warmer months (Hänninen *et al.*, 2000; Nylen *et al.*, 2002). This may be related to the higher prevalence of *Campylobacter* spp. in poultry during the warmer months (Kapperud *et al.*, 1993; Jacobs-Reitsma *et al.*, 1994; Willis and Murray, 1997; Hänninen *et al.*, 2000; Wedderkopp *et al.*, 2000, 2001; Refrégier-Petton *et al.*, 2001). Examination of the FoodNet results from the USA (CDC, 1999) does not reveal a significant increase in campylobacteriosis following a national holiday in November (i.e. thanksgiving) when a majority of the population traditionally consumes cooked turkey. Case-control studies have not identified consuming turkey meat as a major risk factor and very little research has explored the role of turkey and other non-chicken meat as sources of human campylobacteriosis. A relatively small number of turkey isolates collected from one farm over a 3-year period were found to match those from humans in one UK study (Fitzgerald *et al.*, 2001). When completed, the WHO/FAO risk assessment should provide a better understanding of the relative role of chicken in human campylobacteriosis, identify research needs, and provide direction for potential risk management strategies.

A preliminary risk assessment from Denmark examined three approaches to reduce the probability of exposure and illness: (i) by reducing the prevalence of *Campylobacter* positive flocks; (ii) by reducing the concentration of *Campylobacter* on the contaminated chickens; or (iii) by improving the relative level of hygiene during food handling in private kitchens. The simulations showed that all three approaches could reduce the probability of illness caused by *Campylobacter* in chickens. For example, to obtain a reduction in human cases by a factor 25, flock prevalence should be reduced by a factor 25. A similar reduction in the number of human cases could be obtained by reducing the concentration of *Campylobacter* on the contaminated chickens by a factor 100 (2 log cfu/g) or by improving the level of food hygiene in private kitchens by a factor 25 (Christensen *et al.*, 2001; Rosenquist *et al.*, 2003).

Campylobacter jejuni cannot multiply at the refrigeration temperatures used for storing raw poultry. Freezing can cause a 10–100-fold reduction in *Campylobacter* cells compared to carcasses maintained under refrigeration (Stern *et al.*, 1985). Changing market conditions that led to a decrease in frozen chicken and an increase in refrigerated chicken was considered to be an important factor for an increase in campylobacteriosis among consumers in Iceland (Stern *et al.*, 2003). Many strains of *Campylobacter* that colonize poultry are not pathogenic to humans and some human isolates do not readily colonize poultry (Korolik *et al.*, 1995; Koenraad *et al.*, 1995; Corry and Atabay, 2001). With the development of genotyping methods for *Campylobacter* spp. (Wassenaar and Newell, 2000) the relationship between certain strains of *C. jejuni* in poultry and campylobacteriosis in humans is being revealed (Nadeau *et al.*, 2002; Perko-Mäkelä *et al.*, 2002). In Quebec, Canada, for example, approximately 20% of the human *Campylobacter* isolates were genetically related to genotypes found in poultry (Nadeau *et al.*, 2002). In The Netherlands analysis by AFLP found only 16 of 64 chicken isolates to match the 67 human clinical isolates (Duim *et al.*, 2000). Analysis of DNA fragment patterns from *C. jejuni* and *C. coli* from chicken flocks and humans in Australia indicated only a small proportion of *C. jejuni* from chickens can cause disease in humans (Korolik *et al.*, 1995).

In some countries, negative flocks are more common at the time of slaughter and flocks that are positive are colonized with a limited number of subtypes of *Campylobacter* (Ayling *et al.*, 1996; Berndstrom *et al.*, 1996; Jacobs-Reitsma, 1997; Shreeve *et al.*, 2000, 2002; Newell *et al.*, 2001; Petersen and Wedderkopp, 2001; Nadeau *et al.*, 2002; Perko-Mäkelä *et al.*, 2002a,b). The prevalence of *Campylobacter* within positive flocks is highly variable at the time they are presented for slaughter. The diversity of strains decreases during slaughtering, evisceration, and chilling (Newell *et al.*, 2001; Hiett *et al.*, 2002).

The vast majority of *Campylobacter* isolates obtained from live birds and carcasses during slaughter are *C. jejuni*. In most countries, the majority (50–80%) of the chicken carcasses sold at retail are contaminated with *C. jejuni* (Hernandez, 1993). In a study of five plants over 12–13 weeks, mean values of 10^2–10^3 cfu/carcass were detected. The percent of carcasses positive for *C. jejuni/coli* averaged from 67% to 100% (Table 2.15) (Waldroup *et al.*, 1992a). In other studies, *C. jejuni* has been detected at levels of 10^3–10^6 cfu/carcass (Shanker *et al.*, 1982) and 10^3 cfu/g of pericloacal skin from air-chilled carcasses (Oosterom *et al.*, 1983a). *Campylobacter* concentrations of at least 10^3 and 10^6 cfu/g have been found in crop and cecal samples, respectively (Stern *et al.*, 1995; Achen *et al.*, 1998; Berrang *et al.*, 2000a,b; Musgrove *et al.*, 2001). The crop can be a significant source of carcass contamination during the evisceration process (Sarlin *et al.*, 1998). There is evidence that measures implemented to control

Table 2.15 Combined effect of six processing changes[a] on the number and incidence of certain bacteria on broiler carcasses after chilling (From. Waldroup *et al.*, 1992a.)

	No. of Samples	$\log_{10}$/ml of carcass rinse				
		A	B	C	D	E
Aerobic plate count						
Control[b]	112	3.76	3.83	4.45	3.94	3.55
Test[c]	96	3.27*	3.23*	4.04*	3.64*	3.34*
Coliform						
Control[b]	112	2.50	2.17	2.87	2.75	2.35
Test[c]	96	2.26	1.91	2.69*	2.68	2.42
E. coli						
Control[b]	112	2.02	1.80	2.36	2.23	1.97
Test[c]	96	1.67	1.48	2.06*	2.17	11.98
		$\log_{10}$/carcass[d] (% positive carcasses)				
Salmonellae						
Control[b]	112	0.11 (26)	0.17 (32)	0.74 (77)	0.19 (38)	0.85 (30)
Test[c]	96	0.11 (17)	0.15 (28)	0.22* (48)*	0.15 (24)	0.23* (9)*
L. monocytogenes						
Control[b]	112	0.48(1)	0.48 (1)	1.01 (76)	0.90 (59)	0.48 (0)
Test[c]	96	0.48 (0)	0.48 (0)	0.54* (29)*	0.54* (19)*	0.48 (0)
C. jejuni/coli						
Control[b]	96	3.65 (96)	2.37 (67)	3.86 (96)	4.19 (98)	3.45 (98)
Test[c]	96	2.83* (78)*	1.84* (78)*	3.46* (91)	4.26 (100)	2.70* (82)*

*$P = \leq 0.05$ or lower.
[a]Changes tested: counter current scalder; bird wash after scalder; 20 ppm chlorine added to water sprays for carcasses after defeathering, transfer belt between defeathering and evisceration, and carcasses before chilling; and chlorination of chill tank water to provide free chlorine (1–5 ppm) in overflow water. Data were collected at five commercial plants (Plants A–E).
[b]Control carcasses collected over 6–7 weeks.
[c]Test carcasses collected over 6 weeks after implementing the changes.
[d]MPN procedures used. Lower limits of detection were used for statistical analysis (i.e. salmonellae = 1/carcass or $\log_{10} = 0$; *C. jejuni/coli* and *L. monocytogenes* = 3/carcass or $\log_{10} = 0.477$).

salmonellae can lead to reductions in *Campylobacter* on chicken carcasses (Stern and Robach, 2003). One study found no relationship between aerobic plate counts and the prevalence of *Campylobacter* or salmonellae on broiler carcasses and thus an aerobic plate count is not suitable for use as an index for the presence of these pathogens (Cason *et al.*, 1997).

Staphylococcus aureus. Live poultry can carry staphylococci in bruised tissues, infected lesions, nasal sites, arthritic joints, and on skin surfaces. The majority of the resident staphylococci on the surface of the live bird are destroyed during scalding, however, the carcasses can become recontaminated with new, plant specific strains of *Staph. aureus* from the defeathering machines (Gibbs *et al.*, 1978b). Thus, the conditions of scalding and defeathering can influence both the number and type of *Staph. aureus* on poultry carcasses. Many of the strains isolated from raw poultry meat are non-typable with the international set of phages that have been selected for their activity against human strains. There are conflicting reports concerning the enterotoxigenicity of *Staph. aureus* isolated from raw poultry (Gibbs *et al.*, 1978a; Isigidi *et al.*, 1992). Enterotoxin production by strains of *Staph. aureus* associated with live poultry and poultry carcasses appears to be uncommon and thus, these strains are generally not of public health concern (Hajek and Marsalek, 1973; Shiozawa *et al.*, 1980; Isigidi *et al.*, 1992). The poultry isolates have not been reported to be an important cause of food-borne illness. Low storage temperatures and competitive spoilage flora prevent staphylococcal growth and enterotoxin production in raw poultry products.

Listeria monocytogenes. *L. monocytogenes* is often present on raw poultry (Table 2.15) (Waldroup *et al.*, 1992a). The potential for multiplication of *L. monocytogenes* on poultry meat is likely to be affected by the same factors as in red meats. In one study, multiplication did not occur on skinless chicken breast meat (pH 5.8) at 1°C under any packaging condition. At 6°C, a 10-fold increase occurred before spoilage in aerobically packaged breast meat but not in any of the products packaged with carbon dioxide (30% CO_2 + air, 30% CO_2 + N_2, or 100% CO_2) (Hart *et al.*, 1991). The higher pH of leg meat may provide more favorable conditions for multiplication but this has not been evaluated. There is no evidence that multiplication of *L. monocytogenes* in raw poultry during storage is a factor in human listeriosis. An early case control study suggested that undercooked raw poultry is involved in human listeriosis among individuals susceptible to listeriosis (Schuchat *et al.*, 1992).

Clostridium perfringens. *Cl. perfringens* occurs on the surfaces of raw poultry but usually in small numbers (Strong *et al.*, 1963; Hall and Angelotti, 1965; Mead and Impey, 1970; Bryan and Kilpatrick, 1971), the primary source being fecal matter and soil. Raw poultry is ordinarily stored at temperatures too low for *Cl. perfringens* growth. Qualitative or quantitative specifications for *Cl. perfringens* in raw poultry are not recommended.

Cl. botulinum. *Cl. botulinum* type C can cause high mortality in poultry flocks but human botulism due to *Cl. botulinum* type C has never been confirmed. This concern was discussed critically by Roberts and Gibson (1979). Even if humans were sensitive to type C toxin, the likelihood of toxin occurring in the meat of a freshly slaughtered bird would be very remote. Furthermore, the toxin is relatively heat sensitive and would likely be inactivated during cooking (Smart and Rush, 1987). The precautions normally used to prevent other forms of botulism will prevent outgrowth and toxin production from surviving spores (Roberts and Gibson, 1979). *Clostridium botulinum* types A and B have caused human botulism and death when poultry or foods containing poultry are cooked and improperly stored at high temperatures (Tompkin, 1980).

At the farm and hatchery. There is no indication that husbandry practices on the farm will influence the shelf-life (i.e. rate of spoilage) of raw poultry. Husbandry practices on the farm, however, can have considerable influence on the prevalence of *C. jejuni* and salmonellae on raw poultry. More effort has gone into studying and discussing how to eliminate salmonellae from poultry than any other poultry-associated human pathogen. More recent improvements in methodology for detecting, quantifying, and typing have greatly increased the knowledge base for *C. jejuni.* A major hindrance in controlling *C. jejuni* and salmonellae at the farm level is that *C. jejuni* and the primary salmonellae serovars of concern to human health do not cause economic loss to poultry farmers. Thus, there is little economic reason for farmers to eradicate these human pathogens from their flocks, except for selected salmonellae serovars. Despite this, concern for human health has led several countries to establish official programs for salmonellae control on farms (Wierup, 1991; Wray and Corkish, 1991; Bisgaard, 1992; Dawson, 1992; Hirn *et al.*, 1992; Balzer, 1993; Edel, 1994; Wierup *et al.*, 1995; Wegener *et al.*, 2003). Some countries have instituted official monitoring programs to measure the prevalence of *Campylobacter* spp. in their flocks, detect trends, and identify risk factors.

Vertical transmission of *C. jejuni* from breeder flocks through hatcheries to farms does not occur or is not significant in relation to horizontal transmission on the farm (Acuff *et al.*, 1982; Lindblom *et al.*, 1986; Pokamunski *et al.*, 1986; Hoop and Ehrsam, 1987; Annan-Prah and Janc, 1987; Jones *et al.*, 1991b; Pearson *et al.*, 1993; Humphrey *et al.*, 1993; van de Giessen *et al.*, 1998; Fitzgerald *et al.*, 2001; Petersen *et al.*, 2001). This assessment has been questioned (Cox *et al.*, 2002). Although the oviduct can be colonized with *C. jejuni* (Camarda *et al.*, 2000), the evidence does not support this as being a risk factor for flock infection. Horizontal transmission on farms is widely accepted as being the primary route of flock infection and is of major importance (Genigeorgis *et al.*, 1986; Lindblom *et al.*, 1986; Pokamunski *et al.*, 1986; Hoop and Ehrsam, 1987; Shanker *et al.*, 1990; Jones *et al.*, 1991; van de Giessen *et al.*, 1992, 1998; Pearson *et al.*, 1993; Humphrey *et al.*, 1993; Shreeve *et al.*, 2000, 2002; Hiett *et al.*, 2002). When young chicks are exposed to *C. jejuni*, infection spreads rapidly with certain strains showing greater virulence and pathological impact to the chicks (Clark and Bueschkens, 1988). Multiple genotypes of *C. jejuni* can exist within a broiler flock and is most likely due to exposure to multiple environmental sources and genetic drift within the *Campylobacter* population (Thomas *et al.*, 1997). Attempts to determine the sources of flock infection with *C. jejuni* have met with mixed success. Potential sources are generally similar to those identified for the transmission of salmonellae on farms (e.g. flies, insects, wild birds, rodents, farmer's boots, contaminated litter or houses from previous infected flocks, other livestock on the farm, surrounding environment). Although contaminated water may be an important source, feed has not been implicated. The evidence indicates that control measures should be directed toward husbandry and hygiene practices that can prevent introduction of *C. jejuni* to flocks on the farm. Most research on farms has focused on transmission of *C. jejuni* among birds within an enclosed building. There is potential value in the use of competitive exclusion to prevent flock infection.

Broilers from a farm implicated in an outbreak of campylobacteriosis continued to be the source of sporadic cases among the population served by the farm for another 18 months. A contaminated water supply was identified as the source of the *C. jejuni* in the broilers. The proportion of birds colonized with campylobacters was reduced from 81% to 7% by chlorinating the water, cleaning, and disinfecting the drinking system, and withdrawing furazolidone from the feed. These measures also led to a 10^3–10^4 reduction in campylobacters from the carcasses after processing. When these measures were discontinued, both bird colonization and the number of *Campylobacter* on the carcasses returned to their former levels. *Campylobacter jejuni* in the water were viable but non-culturable and only could be detected by direct microscopy and fluorescent antibody methods (Pearson *et al.*, 1993). In another study involving 176 farms, it was concluded that disinfecting the drinking water was the factor most likely to have the greatest impact toward reducing the prevalence of *Campylobacter* spp. in broilers (Kapperud *et al.*, 1993). Other factors included not tending to other poultry or pigs before entering the broiler house. Dipping boots into disinfectant before entering the broiler house can reduce or delay the risk of flock colonization (Humphrey *et al.*, 1993; Shreeve *et al.*, 2000). Standardization of biosecurity measures (e.g. cleaning and disinfecting houses before stocking, hygienic practices of personnel) resulted in more than a 50% reduction in flock infection among 13 test flocks when compared with 25 control flocks, at 42 days of age. Among the factors identified as being significant control measures were replenishing the boot disinfectant twice weekly and daily sanitization of the water supply. Houses with fans mounted on the side resulted in lower risk, possibly due to their ease of cleaning and disinfecting compared to roof mounted fans (Gibbens *et al.*, 2001).

Molecular typing (*fla*A and PFGE) has been used to detect carry-over from one flock to the next. Although the flocks in 60 out of 100 houses had positive consecutive flocks, the evidence from genotyping indicated carry-over was a minor source of colonization for the subsequent flocks. None of the environmental samples from 20 houses before flock replacement yielded *Campylobacter*. It was concluded that the practice in the UK of removing spent litter between flocks and cleaning and disinfecting the houses is adequate to limit carry-over. It appears that sources of *Campylobacter* that are external to the poultry house are of greater importance (Shreeve *et al.*, 2002).

A study of Danish broiler farms having a high proportion of successive positive flocks (i.e. at least 5 positive flock rotations in one year) and one other farm used a combination of genotyping and serotyping to measure the persistence of certain strains. It was found that certain *C. jejuni* (e.g. clones related to *fla* type 1/1 serotype O:2) have the ability to cause persistent infections on farms. This is strong evidence that local on-farm reservoirs of such clones are of greater importance for colonization of flocks than sporadic isolates that may be transitory in nature (Petersen and Wedderkopp, 2001). In flocks that are simultaneously colonized by more than one clone, the different clones coexist rather than exclude each other (Petersen *et al.*, 2001).

Transmission of salmonellae among flocks can occur both horizontally and vertically; thus, a multifaceted plan that addresses all phases of production from breeder farms through transport of the live birds to slaughter must be implemented (Bailey *et al.*, 2001, 2002; Russell, 2002; Fluckey *et al.*, 2003). Principal sources of infection are contaminated feed, parent stock, hatchery environment, and residual salmonellae contamination in the buildings in which previous flocks were housed (Lahellec and Colin, 1985; Hinton *et al.*, 1987; McCapes *et al.*, 1991; Bailey *et al.*, 2001; Heyndrickx *et al.*, 2002). A multivariate analysis of 111 broiler breeder flocks in 32 farms over a 5-year period led to the conclusion that flocks housed in farms with a foot disinfection tub for workers entering the poultry house, houses with good hygiene barriers (i.e. facilities for changing boots and clothing, orderliness, adequate cleaning, and disinfection procedures), and feed received from a large feed mill were 46 times less likely to have salmonellae-positive flocks (Henken *et al.*, 1992). The population density of birds in poultry houses has not been a factor influencing the prevalence of salmonellae on carcasses after evisceration (Waldroup *et al.*, 1992b; Angen *et al.*, 1996). In France, the risk of positive flocks was found to increase with warm weather, static air (i.e. no fans), when two or more people cared for the flocks, farms with three or more houses, when the drinking water was acidified and the presence of a large number of litter beetles (Refrégier-Petton *et al.*, 2001). Failure to adhere to biosecurity measures was considered an important factor as the environment was considered the main source of *Campylobacter*.

Some control measures that have been proposed are of questionable value due to conflicting reports of efficacy. In some cases, this is due to the complexity of the problem on farms and the interdependency of causative factors. Procedures that repeatedly have been shown to be effective can be rendered ineffective if other important factors are not controlled. Following is a discussion of practices that have been reported to reduce the risk of salmonellae and *Campylobacter* and should be considered when developing a control program for the farm and hatchery.

Eggs for hatching. The greatest concern to human health is ovarian transmission of *Salmonella* Enteritidis (Humphrey, 1994; Mason, 1994; Edel, 1994; WHO/FAO, 2002). Maintaining salmonellae negative breeder flocks that produce salmonellae-free eggs for hatching is critical if raw poultry meat is not to contain salmonellae (Edel, 1994; Skov *et al.*, 1999a,b, 2002; Cox *et al.*, 2000, 2001; Gruber and Köfer, 2002). In many European countries, all parent flocks must be vaccinated for *S.* Enteritidis (Gruber and Köfer, 2002). In Denmark, breeder flocks are tested using a combination of bacteriological analysis of fecal samples, hatchery fluff or organ samples, and serological analysis of serum or yolk samples (Skov *et al.*, 2002). In some countries, breeder flocks are tested and diseased birds (reactors) are removed or treated (Edel, 1994). New breeder stock may be quarantined to prevent the introduction of infected birds into a flock.

Eggshells also can become contaminated with microorganisms from the intestines, when the eggs pass through the cloaca and when the eggs contact nest materials, litter, or incubator surfaces (Board, 1969; Williams *et al.*, 1968; Humphrey, 1994; Camarda *et al.*, 2000; WHO/FAO, 2002). Contamination of the shell can be reduced by providing an ample quantity of clean nest materials, gathering eggs frequently, and cleaning and/or disinfecting eggs promptly after gathering (Williams, 1978). See Chapter 15 for current recommendations.

Collecting, cleaning, and disinfecting eggs for hatching must be controlled to minimize transmission of salmonellae from breeder flocks to the hatchery and then to the farm (Davies and Wray, 1994). Eggs also may be fumigated after reaching the hatchery. Fumigation with formaldehyde has been commonly used for disinfection of eggs and hatchery equipment. Due to concerns of possible carcinogenicity to workers, an alternative such as ozone or liquid disinfectant may be necessary (Whistler and Sheldon,

1989a,b,c; Brake and Sheldon, 1990; Kuhl, 1990; Patterson *et al.*, 1990; Sheldon and Brake, 1991; Davies and Wray, 1994). Disinfecting eggs for hatching is of questionable value for control of salmonellae (Cox *et al.*, 2000). Collectively, the foregoing procedures and others that follow can help eliminate breeder flocks as a source of human and/or poultry pathogens.

Air currents can distribute microorganisms throughout a hatchery. Negative air pressure should be provided in dirty areas and incoming air should be filtered. Hatchery layout and control of personnel, equipment, air, and waste materials are important in preventing the spread of pathogens among the eggs and infection in the newly hatched poults.

Hatchery hygiene. Control of salmonellae in the hatchery is a critical factor that must be controlled through an effective hygiene program (Davies and Wray, 1994; Bailey *et al.*, 2001, 2002). Certain strains of salmonellae can persist in a hatchery for an extended period of time and infect newly hatched chicks (Davies *et al.*, 2001; Bailey *et al.*, 2002; Wilkens *et al.*, 2002; Liebana *et al.*, 2002). The significance of hatchery hygiene to human health was confirmed when a hatchery was identified as the source of human salmonellosis. A strain of *S.* Infantis with the same PFGE pattern was isolated from hatchery samples, birds, and 19 patients in 1999. Five additional PFGE-linked cases were identified in the spring of 2000, indicating persistence of the strain in the hatchery environment (Wilkins *et al.*, 2002). A strong hatchery hygiene program, however, cannot prevent spread of salmonellae during hatching, if the incoming eggs are contaminated (Bailey *et al.*, 1994). The prevalence of salmonellae in eggs for hatching has been reported to be as low as 1 in 10,000 (Wilding and Baxter-Jones, 1985). One contaminated egg, however, can lead to substantial spread of salmonellae throughout a hatching cabinet (Bailey *et al.*, 1992, 1994, 1998). The warm and moist conditions required for hatching eggs is very favorable for multiplication of salmonellae. One study found that the samples of egg fragments, belting material, and paper pads that have been used to line chick delivery trays, tested positive (71–84%) for salmonellae. In 38 of 40 samples tested, the number of salmonellae exceeded 1 000 cfu/g, thus, indicating multiplication in the hatching cabinet (Cox *et al.*, 1990). This information has led to research to control salmonellae during the hatching process. For example, atomizing hydrogen peroxide (2.5%) during hatching reduced salmonellae levels in hatching cabinets and colonization of the ceca of 7-day-old chicks (Bailey *et al.*, 1996). An electrostatic space charge system installed in a hatching cabinet reduced the amount of airborne dust and bacteria. The number of salmonellae per gram of caecal contents of 7-day-old chicks also was reduced (Mitchell *et al.*, 2002).

Poultry house hygiene. Broiler houses positive for salmonellae before chick placement result in more positive flocks at the end of rearing (Rose *et al.*, 1999). Cleaning and disinfecting poultry houses between flocks can influence the prevalence of salmonellae in the environment, when new chicks are placed and, eventually, the serotypes present in the flock when slaughtered (Lahellec *et al.*, 1986). Procedures have been described for cleaning and disinfecting turkey breeder houses between flocks. Results from monitoring several producers have demonstrated that the procedures can be effective for controlling salmonellae over a period of several years (Poss, 1985). Litter that is high in moisture ($\geq 0.85\ a_w$, $\geq 35\%$ moisture) increases the potential for salmonellae multiplication. Controlling leaks from watering devices and from outside the poultry house, along with providing adequate ventilation can help maintain drier litter conditions (Mallison *et al.*, 1998; Hayes *et al.*, 2000; De Rezende *et al.*, 2001). Disinfecting poultry houses by a combination of moist heat and formalin has been common practice in Denmark (Gradel, 2002). Preliminary tests indicate that poultry houses can be disinfected by heating to 60°C with steam (100% RH) for 24 h. Applying a phenolic disinfectant spray followed by fogging with a formaldehyde solution also has been found to be effective (Davies *et al.*, 2001). Detailed guidelines have been developed for the management of poultry house hygiene in Scotland, including recommendations for disinfection using formalin-based disinfectants (DEFRA, 1997). A study of 86 broiler houses in

France found that houses were less likely to test positive before placement of new chicks when the houses were given a final disinfection with a formaldehyde- or gluteraldehyde-based disinfectant and having the disinfectant applied by a professional contractor. Furthermore, houses in which the previous flock required treatment for disease were more likely to test positive for salmonellae after cleaning and disinfecting (Rose *et al.*, 2000).

Litter management differs in various regions of the world. In many countries, each house is cleaned and disinfected and the litter is changed before placement of new chicks. In the UK, removing spent litter between flocks and cleaning and disinfecting the houses has been shown to be adequate to limit carry-over of *Campylobacter* spp. from one flock to the next (Shreeve *et al.*, 2002). The data from Europe and the UK indicate lower flock prevalence rates, following improved implementation of control measures to control the *S.* Enteritidis epidemic and other strains of salmonellae that were attributable to poultry are the evidence that *Campylobacter* is controllable on the farm. In one study, improved preventive measures instituted by a farmer successfully prevented flock infection in a series of three sequential flocks and on another farm six sequential flocks were negative for *C. jejuni* (van de Giessen *et al.*, 1992). In regions where the same emphasis has not prevailed flock prevalence rates remain higher, the diversity of strains within and between flocks is greater and certain strains can be found in different houses on the same farm (Hiett *et al.*, 2002).

The time between flock placement in poultry houses influences whether subsequent flocks will be positive for *Campylobacter* spp. In Denmark, 50% of the broiler houses were restocked within 12 days. Restocking in less than six days was found to be associated with a higher risk of the flock being positive (Wedderkopp *et al.*, 2000). In Finland, the time between flock replacement is about 2 weeks (Perko-Mäkelä *et al.*, 2002b). In France, 64% of the farmers reported that time between the first application of disinfectant to cleaned houses and when new poults are placed was $\leq$15 days (Refrégier-Petton *et al.*, 2001). Maintaining strict biosecurity measures can limit introduction of *Campylobacter* into poultry houses, but once introduced *Campylobacter* quickly spreads throughout the flock (Pattison, 2001). A common experience among researchers is the difficulty in identifying the source(s) for *Campylobacter* and clearly discerning whether the source(s) are from outside or inside the houses (Nesbit *et al.*, 2001).

Investigation of farm management practices on 12 commercial squab (young pigeon) farms found the probability of *Campylobacter* increased with exclusion of manure from the nesting material, more frequent disinfection of the water, cleaning the boards on which the birds land, having a culture positive parent, and not cleaning the shipping crates. Since some of the findings contradict practices that may be appropriate for other species of birds, this suggests the need to verify best management practices for unique species (Jeffrey *et al.*, 2001).

Feed. Feed has not been identified as a source of *Campylobacter* spp., probably because campylobacters are sensitive to dehydration (Humphrey *et al.*, 1993; Whyte *et al.*, 2003). Providing salmonellae-negative feed, however, is essential (MacKenzie and Bains, 1976; Zecca *et al.*, 1977; Hinton *et al.*, 1987; Jones *et al.*, 1991a; McCapes *et al.*, 1991; Angen *et al.*, 1996; Davies *et al.*, 2001; Corry *et al.*, 2002; Chadfield *et al.*, 2001). The prevalence of salmonellae in grain constituents is normally very low, although they can be a source of salmonellae (Bains and MacKenzie, 1974). Ingredients of animal origin (e.g. rendered meat and bone meal) have a history of being more heavily contaminated with salmonellae. Omitting feed ingredients of animal origin can reduce the risk of salmonellae in poultry at the time of slaughter (Jacobs-Reitsma *et al.*, 1994). Certain strains of salmonellae can become established in feed mills and persist for extended periods of time (Davies *et al.*, 2001; Liebana *et al.*, 2002). Pelleting can eliminate salmonellae from feed but the systems must be adequately controlled (Davies *et al.*, 2001). Pelleted feed must be protected from recontamination. Providing non-pelleted feed to chicks placed on the farm increased the probability that the resulting flocks would test positive for salmonellae at the end of the rearing period (Rose *et al.*, 1999).

Reports that adhesion of salmonellae to the intestine of day-old chicks and to the caeca of 7–10 day-old broilers is inhibited by adding D-mannose, lactose, or complex carbohydrates to the feed has not been demonstrated to consistently influence the prevalence or numbers of salmonellae in live birds or in poultry carcasses when sampled prior to chilling (Izat *et al.*, 1990b; Waldroup *et al.*, 1992c; Chambers *et al.*, 1997). Adding the yeast, *Saccharomyces boulardii*, to feed reduced the prevalence of salmonellae-positive caeca in young chickens (Line *et al.*, 1998).

Drinking water. Water in open troughs becomes contaminated by dust, litter, feed, feathers, and from feet, beaks, and feces (Patterson and Gibbs, 1977). Open drinking water systems (e.g. troughs, bell shaped drinkers) are prone to fecal contamination and should be replaced by nipple-style drinkers (Renwick *et al.*, 1992). Drinking water should be treated with a disinfectant (Kapperud *et al.*, 1993; Pearson *et al.*, 1993). Water spilled onto litter or feed permits microbial multiplication in these materials. Chlorination of drinking water was not found to be an effective control measure for decreasing colonization by campylobacters in broilers (Stern *et al.*, 2002).

Dead bird disposal. Timely disposal of dead birds by incineration or complete removal from the premises is important (Renwick *et al.*, 1992) to avoid flock mortality from *Cl. botulinum* type C.

Wild bird, rodent control. Maintaining an effective wild bird and rodent control program is necessary to restrict entrance of birds and rodents in poultry houses. Wild birds and rodents are reservoirs of salmonellae and campylobacters.

Vaccination. One of the most effective control measures for salmonellae in live poultry has been vaccination. Reductions in *S.* Enteritidis infections from eggs and poultry meat in the UK and Europe have been achieved through vaccination of breeder and layer flocks (Soo *et al.*, 2002). German regulations now require vaccination of all pullets in houses of more than 250 birds (Schroder, 2002). Vaccines can be live or inactivated cells and are being further developed to improve performance and broaden the spectrum of activity to other *Salmonella* spp. (Schroder, 2002; Springer *et al.*, 2002; Clifton-Hadley *et al.*, 2002a,b; Woodward *et al.*, 2002; Oostenbach, 2002; Witvlieft, Mols-Vorstermans and Wijnhoven, 2002; Gruber and Köfer, 2002). Flagella have been shown to be necessary for colonization by *Campylobacter* in a variety of animals, including chickens. A vaccine directed toward the flagella and preventing colonization may lead to an effective strategy for controlling campylobacters in poultry Nachamkin *et al.*, 1993).

Competitive exclusion (CE). Using CE in young poults to assist in early establishment of a favorable intestinal flora can preclude, under certain conditions, establishment of salmonellae. Several commercial preparations have become available and shown in several countries to be effective for broilers (Aho *et al.*, 1992; Hirn *et al.*, 1992; Wierup *et al.*, 1992; Salvat *et al.*, 1992a; Schneitz *et al.*, 1998; Ferreira *et al.*, 2003). This concept should be adaptable to other poultry such as turkeys and ducks (Schneitz *et al.*, 1992; Schneitz and Nuoto, 1992). Preparations derived from the epithelial wall of caeca have led to reductions of salmonellae and *Campylobacter* in broilers (Stern *et al.*, 2001; Chen and Stern, 2001) and on broiler carcasses from treated flocks (Blankenship *et al.*, 1993a). An avian-specific probiotic containing *Lactobacillus acidophilus* and *Streptococcus faecium* given to poults from days 11–33 resulted in reduced colonization and fecal shedding of *C. jejuni* at time of slaughter (Morishita *et al.*, 1997). When used in combination with *S.* Typhimurium specific antibodies the extent of colonization of broilers with *S.* Typhimurium was reduced (Promsopone *et al.*, 1998). Although CE is not a panacea for pathogen control, it can be effective in combination with other husbandry practices (Nurmi *et al.*, 1992). The efficacy of CE when administered at the farm level is substantially less when poults have been pre-exposed to salmonellae in the hatchery. Continued research has led to improvements and broader application of the concept by industry (Stavric and D'Aoust, 1993; Hollister *et al.*, 1994; Behling and Wong, 1994; Nisbet, 2002; Ferreira *et al.*, 2003).

Stress. Minimizing stress in young chicks during transfer from the hatchery and during the first week or so on the farm when they are most susceptible to infection is very important (Bailey, 1988). Similarly, stress should be minimized during capture for transport to the plant. Birds subjected to stress during catching and transportation are more likely to shed salmonellae when they arrive at the plant for processing (Bhatia and McNabb, 1980).

Feed withdrawal. Withdrawing feed at the appropriate time (e.g. 8–12 h) before transporting poultry to the slaughtering plant is important (Wabeck, 1972; Rigby and Pettit, 1981; Papa and Dickens, 1989; Papa, 1991; Duke *et al.*, 1997). Intestinal tracts that are full of feed are more prone to rupturing during evisceration. Timing for feed withdrawal varies with the type of bird, whether the birds will be eviscerated by hand or mechanically and the type of equipment employed. Salmonellae and *Campylobacter* levels may increase in the crop during feed withdrawal (Ramirez *et al.*, 1997; Byrd *et al.*, 1998; Willis *et al.*, 2000). Feeding a fermentable carbohydrate (e.g. glucose) increased the level of lactics in the crop, decreased crop pH, and resulted in lower levels of salmonellae (Hinton *et al.*, 2000). Adding lactic acid to the drinking water about 8 h before transportation to the slaughtering facility reduced the prevalence of salmonellae and *Campylobacter* in crops and on carcasses after slaughter (Byrd *et al.*, 2001). Feeding the yeast, *Sacch. boulardii*, during the final 60 h before capture and transport was found to reduce the prevalence of salmonellae-positive caeca in chickens after transport (Line *et al.*, 1997).

Monitoring flocks. Many countries have instituted monitoring programs to identify flocks that are positive for one or more species of salmonellae (e.g. *S.* Enteritidis) and impose special requirements for management and/or disposition of the flocks. Such programs have proven to be an effective means to monitor the status of a country's control program and work towards continuous improvement (Wegener *et al.*, 2003; Rose *et al.*, 2003). The methods for sampling and analysis should be standardized within the country and be sufficiently sensitive to detect positive flocks (Skov *et al.*, 1999a,b). Through multivariate analysis of such databases, it is possible to identify risk factors where improved control is needed. Equally important, wasting limited resources on factors that are not significant can be avoided (Angen *et al.*, 1996; Skov *et al.*, 1999; Rose *et al.*, 2003). Flocks that test positive for salmonellae and/or *C. jejuni* should be processed after negative flocks (Sarlin *et al.*, 1998; Codex, 2003).

Transportation from farm to plant. The stress associated with feed withdrawal, capturing at the farm, loading into crates, transportation, and holding at the plant can increase the prevalence of salmonellae within a flock when processed (Line *et al.*, 1997). Contamination of feathers with microorganisms of fecal origin increases as birds are confined in crates for transport to the plant (Seligmann and Lapinsky, 1970; Patterson and Gibbs, 1977; Rigby and Pettit, 1980; Rigby *et al.*, 1980a,b; 1982). Microorganisms in feces and on feathers can be spread from bird-to-bird within the crates. During hanging and bleeding, wing flapping creates aerosols and dust. Salmonellae and *Campylobacter* have been isolated from air in poultry plant unloading zones (Zottola *et al.*, 1970; Abu-Ruwaida *et al.*, 1994).

Contaminated transport crates can be a significant source of salmonellae on carcasses after processing (Rigby *et al.*, 1980a,b, 1982; Carr *et al.*, 1999; Bailey *et al.*, 2001; Slader *et al.*, 2002). Flocks on the farm that test negative can yield salmonellae-positive carcasses due to infections acquired during transport, probably influenced by stress. This indicates that where possible, crates should be cleaned and disinfected after delivering birds to the plant but raises practical issues (e.g. attempting to clean crates during the winter). Research has been conducted to develop a cost-effective method to clean and disinfect crates (El-Assaad *et al.*, 1990, 1993; Ramesh *et al.*, 2003). Further demonstration of the benefit of this practice in reducing pathogens would be useful. The effect of inadequately cleaned transport crates on the prevalence of salmonellae on carcasses, during processing, will vary depending on the effectiveness of the controls (Davies and Bedford, 2002; Corry *et al.*, 2002). Crates also are a source of *Campylobacter* spp. (van de Giessen *et al.*, 1998; Stern *et al.*, 2001) with evidence that certain subtypes present on crates can be detected on carcasses after evisceration and chilling (Newell *et al.*, 2001; Slader *et al.*, 2002). The process of catching and putting birds in crates significantly ($P < 0.001$) increased the chance of contamination by *Campylobacter* (Slader *et al.*, 2002). *Campylobacter* levels increase during transport and holding prior to slaughter (Stern *et al.*, 1995).

At the processing plant. The conditions of slaughtering, defeathering, evisceration, chilling, and packaging will influence the extent to which processed poultry will be contaminated with pathogens and spoilage bacteria. In plants with effective hygiene programs, the origin of the food-borne pathogens on raw poultry is the live bird; not the processing facility. Thus, elimination of *C. jejuni* and salmonellae on farms is essential to prevent contamination during processing and produce raw poultry meat that is free of these pathogens.

Live birds often harbor human pathogens; thus, the processing conditions will strongly influence the extent of contamination on raw poultry. The potential for contamination begins with the knife used for severing the carotid arteries and ends when raw poultry is sealed in a package for shipment from the plant. The overall objective of processing is to convert live birds into ready-to-cook poultry meat with minimal microbial contamination. Processing conditions should be designed and controlled to minimize contamination from the external surfaces of the live bird, the intestinal tract and from carcass-to-carcass. This can be achieved by proper equipment design, maintenance, and hygiene; plant layout; operator training and skill; and controlling conditions used in certain processing steps (e.g. scalding, defeathering, chilling). The importance of equipment design, plant layout and hygiene has been discussed (ICMSF, 1988). The following practices, if controlled and properly performed, can reduce the level of enteric pathogens (salmonellae, *C. jejuni*) and other forms of microbial contamination.

Scalding. Hard scalding systems (e.g. 58°C for about 2.5 min) are more bactericidal to non-sporeforming bacteria than soft scalding systems. Scalding conditions should employ the most bactericidal conditions that yield the desired product characteristics for the type of bird and the intended market. Overscalding, however, must be avoided because the skin can become cooked and tear during defeathering. Carcasses should be sprayed with water before scalding to reduce the microbial load entering the scald tank (Mulder, 1985) and the flow of water should be counter to the direction of the carcasses (Mulder, 1985; Waldroup *et al.*, 1992a). Salmonellae reductions occur during the washing/scalding process but some salmonellae may survive. For example, samples collected from a three tank counterflow scalder over an eight day-period, contained viable salmonellae for seven and two days in the first two scald tanks and in the third scald tank, respectively (Cason *et al.*, 2000).

Bacteria are more heat resistant when attached to skin (Notermans and Kampelmacher, 1974, 1975; Humphrey *et al.*, 1984; Yang *et al.*, 2001). Increasing scald water pH increases the death rate of bacteria on chicken skin. The $D_{52^\circ C}$ for *S.* Typhimurium attached to chicken skin decreased from 61.7 ± 0.4 at pH 6.0 to 26.6 ± 3 min at pH 9.0 (Humphrey *et al.*, 1984). In other research, salmonellae have been detected on chicken carcasses after scalding at 55°C for 105 s, but not at 60°C for 200 s (Notermans *et al.*, 1975a; Mulder and Dorresteijn, 1977). Salmonellae, however, have been isolated from turkey carcasses after a 60°C scald treatment (Bryan *et al.*, 1968a).

Scalding is neither a site of significant contamination nor a step that can be relied upon to reduce the prevalence of *C. jejuni* on poultry carcasses (Yusufu *et al.*, 1983; Oosterom *et al.*, 1983b; Wempe *et al.*, 1983; Genigeorgis *et al.*, 1986).

Decimal reduction values of $D_{52^\circ C} = 150\,s$ and $D_{62^\circ C} = 8\,s$ have been reported for salmonellae isolated from poultry plants (Mulder *et al.*, 1978). These values, however, may underestimate their heat resistance in scald water during operation because turbulence of the water removes blood, fecal matter, and debris from the carcasses. This results in a high organic solid content, possibly due to the fecal material, and a pH of about 6.0 (Humphrey, 1981). The increase in organic solids does not affect the heat resistance of salmonellae, but the decrease in pH to about 6.0 is very important. For example, the $D_{52^\circ C}$ values for seven salmonellae isolated from chicken ranged from 10.2 to 34.5 min in scald water having a pH of about 6.0 (Humphrey, 1981). Increasing the pH to 9.0 reduced the heat resistance of *S.* Typhimurium from 34.5 to 1.25 min at 52°C (Humphrey *et al.*, 1984). Thus, maintaining an alkaline pH in the scald tank water can be an effective means to reduce the level of salmonellae on poultry during scalding.

Defeathering. Defeathering is a major source of contamination with *Campylobacter* spp. (Yusufu *et al.*, 1983; Oosterom *et al.*, 1983a; Wempe *et al.*, 1983; Dromigny *et al.*, 1985; Genigeorgis *et al.*, 1986), salmonellae (Bryan *et al.*, 1968a; Morris and Wells, 1970; Notermans *et al.*, 1975a), and *E. coli* (van Schothorst *et al.*, 1972; Mulder *et al.*, 1977a; Whyte *et al.*, 2001; Berrang *et al.*, 2001). The high prevalence and concentration of *Campylobacter* is due to leakage from the cloaca during the process of defeathering (Berrang *et al.*, 2001). Carcass contamination during defeathering can be minimized through proper equipment maintenance and hygiene, avoiding excessive feather accumulation, providing a spray of chlorinated water (e.g. $\geq$50 μg/mL) and allowing heat to escape during operation. Carcasses should be sprayed with water after scalding and again after defeathering (Mulder, 1985). Plugging and suturing the cloacal cavity reduced both the prevalence and concentration of *Campylobacter* on carcasses that occurred during defeathering (Berrang *et al.*, 2001). Patents have been issued for plugging the cloacal opening, but none have been commercialized (Singh, 1998; Anderberg, 2000).

Evisceration. Training operators in proper technique can reduce the risk of extensive contamination due to cutting or breaking the viscera during manual evisceration. Proper maintenance and adjustment of automated eviscerating equipment and speed of the line are important for the same reason. In the event intestines are cut or broken during evisceration, the contaminated carcasses can be washed and/or trimmed to remove the visible contamination. This practice can reduce bacteria of enteric origin (e.g. coliforms, *E. coli*, salmonellae) to numbers that are comparable or below those on normal carcasses (Blankenship *et al.*, 1975, 1990, 1993b; Walsh and Thayer, 1990; Bilgili *et al.*, 1992; Waldroup *et al.*, 1993). Examination of carcasses with and without visible fecal contamination found that *E. coli* was the predominant species of Enterobacteriaceae present (Jiménez *et al.*, 2003). The extent of contamination during evisceration can influence microbial levels during water chilling and subsequent handling before packaging. Inside–outside carcass washers can prevent an increase in the number of Enterobacteriaceae and the prevalence of salmonellae. In general, the type of salmonellae on poultry carcasses and equipment reflects the types in the live birds being processed. Transfer of salmonellae from one flock to subsequent flocks can occur, however, during the slaughtering process and existing methods for cleaning and disinfecting may not remove salmonellae from equipment and prevent contamination of flocks processed on subsequent days (Olsen *et al.*, 2003).

Chilling. Controlling the chilling process can minimize contamination with psychrotrophic spoilage bacteria and pathogens. During immersion chilling, several factors can be controlled to provide a washing effect and decrease microbial numbers on carcasses: (i) number of carcasses in relation to the volume of water; (ii) amount of replacement water; (iii) multiple chillers and counter flow chilling so carcasses move progressively through cleaner water as they are chilled; (iv) chlorination of water in the final chill tank; (v) hygiene of the chilling equipment; and (vi) microbial quality of the ice, if ice is used for chilling. Chlorination is more effective, when maintained at an acidic pH (e.g. pH 6.5). The pH of the water can be adjusted when necessary by injecting carbon dioxide or adding an acidic solution.

During air and spray evaporative chilling, the hygienic status of the environmental factors (e.g. air, equipment, water, condensate) to which the carcasses are exposed must be controlled.

Equipment hygiene. Effective cleaning and disinfecting procedures must be applied to prevent microbial build up on equipment. Microorganisms can be transferred from feathers, feet, and skin to other carcasses via utensils and equipment. Such transfers occur when birds are hung on metal shackles and killed with knives, as feathers are removed by rubber fingers in picking machines, as feet and heads are cut off by machines or knives, as evisceration is done by knives or machines, as carcasses slide across tables and down chutes and contact the sides or bottoms of chill tanks and coolers, and as they are re-hung onto metal shackles. The transfer continues as chilled carcasses are dropped onto and slide across tables, weighed before packaging, and dropped into chutes that direct them into bins or shipping boxes. Edible viscera become contaminated by surfaces of chutes, tables, gizzard skinning machines, knives, and scissors.

Equipment that is not cleaned and disinfected at the end of a day's operation will have residual fat, blood, and meat that provide nutrients for microbial growth that can then be transferred to products that are processed the following day. Equipment, particularly from chilling to packaging, is a major source of spoilage bacteria. The principles of cleaning and disinfecting are discussed in ICMSF Book 4 (ICMSF, 1988). Properly cleaned and sanitized equipment should not be a source for contamination of carcasses with pathogens associated with the previous day's flock(s). This is evident in Table 2.4 and in tests with *E. coli*. During a 6-week period, *E. coli* isolates from the equipment at four different sites were biotyped. Over 90% of the biotypes at each of the four sites were recovered on one occasion only and none were found on more than three of the six weekly samplings. It was concluded that the *E. coli* were transients associated with the birds being processed and the equipment had not become colonized with *E. coli*. Thorough cleaning and disinfection of processing equipment can prevent colonization of *E. coli* (Cherrington *et al.*, 1988) and, very likely, other Enterobacteriaceae. An extensive study of broilers being processed in a plant that slaughters about 100,00 birds per day on automated equipment found occasions when equipment was not adequately cleaned and resulted in contamination of carcasses from flocks processed later the same day or on subsequent days (Olsen *et al.*, 2003).

Worker performance and hygiene. Training workers to perform their jobs correctly is very important to minimize product contamination. Bacteria are transferred from poultry carcasses to the hands or gloves of workers and then to other carcasses. Workers risk getting pathogens on their hands, in cuts or other lesions, and inhaling them from contaminated carcasses. Workers can shed bacteria from their skin, hair, nose, and throat as they work, particularly if they have skin or respiratory tract infections, or if they practice poor personal hygiene. Carcass contamination from workers, however, is minor compared to the microflora that is indigenous to the live bird and to contamination from certain processing equipment (e.g. defeathering machines). During a 15-year survey of over 41 000 fecal specimens from poultry plant personnel, only 0.085% yielded salmonellae (Lundbeck, 1974).

Product temperature. Holding, distributing, and displaying fresh poultry at low temperatures is the most important factor affecting the shelf-life of raw poultry. The lower the temperature, the longer the shelf-life. Most pathogens cannot multiply below 6°C, but psychrotrophic spoilage bacteria can multiply to below 0°C. At −2°C, psychrotrophic bacteria have extended lag periods and longer generation times. Therefore, to increase shelf-life, poultry should be stored below 3°C or, if feasible, at −2°C. A lower temperature will freeze the product; however, some processors do freeze poultry and then ship at refrigerated temperatures. Shelf-life is generally extended by the length of time the product was frozen. Predictive modeling can be useful for estimating the impact of various storage temperatures and times upon product quality and acceptability.

Decontamination. Traditionally, agencies responsible for poultry inspection have emphasized preventing contamination with visible soil during processing. Reliance upon hygienic procedures, however, has not always been a successful means of producing microbiologically safe poultry. In the absence of a kill step (e.g. heating, irradiating), the prevalence of certain enteric pathogens (e.g. salmonellae, *C. jejuni*) on raw poultry continues to be unacceptable. The need for a decontamination procedure has been recognized and investigated for decades. An acceptable method would be rapid, economical, effective, employ additives that are safe for consumers and plant employees, not cause a loss in product quality, and consumers would consider the method acceptable. The conditions of testing can influence the interpretation of test data and the assessment of efficacy. Preferably, the procedure should be evaluated in commercial facilities with natural levels of contamination. Six general methods will be discussed: (1) chlorination of processing water; (2) organic acid sprays or dips; (3) acidified solutions; (4) alkaline solutions; (5) heat; and (6) irradiation.

Chlorination of processing water. Some countries prohibit the use of chlorinated water on poultry. Numerous investigations have been conducted on the use of chlorine for reducing and preventing the spread of bacteria during poultry processing (Thomson *et al.*, 1974; Tompkin, 1977). The results vary, probably due to the conditions of application. Short-term exposure of carcasses to chlorinated water as in a spray cabinet is of virtually no benefit over untreated water. Longer term exposure, however, to chlorinated water by adding chlorine to multiple sprays and to chill tank water can provide the exposure time required for microbial reductions to occur.

A limitation of chlorine is its rapid inactivation by organic material. Chlorine, however, can prevent a buildup of bacterial slime on equipment surfaces and eliminate microorganisms, particularly psychrotrophic bacteria, present in the incoming water supply or in the water of chill tanks (Barnes, 1965; Mead and Thomas, 1973; Mead *et al.*, 1975). One recommendation has been to add chlorine ($\geq 50\,\mu$g/mL) to the water used in defeathering machines (Purdy *et al.*, 1988). Adding chlorine to the water sprays of automatic eviscerating machines also is beneficial. In one study, the aerobic plate count of the machinery was 40 000/cm^2 after 4 h of operation. Addition of chlorine at 20, 40, and 70 μg/mL resulted in counts of 400/cm^2, 20/cm^2, and 3/cm^2, respectively (Bailey *et al.*, 1986). Use of chlorinated inside–outside carcass washers also can be effective for reducing microbial numbers.

Concentrations of 45–50 μg/mL residual chlorine in chill tanks using 5 L of water per carcass or 25–30 μg/mL when using 8 L of water per carcass have been recommended (Mead and Thomas, 1973). Providing 20 or 34 μg/mL free chlorine to water entering the chill tank significantly reduced aerobic counts, fecal coliforms and salmonellae and extended the shelf-life of raw poultry (Table 2.16) (Lillard, 1980).

Chlorinating processing water at several locations can help control the microbial content of poultry carcasses (Mead *et al.*, 1995). For example, it was considered beneficial to add 20 μg/mL chlorine to the carcass wash after defeathering, water used on the transfer belt between defeathering and evisceration, and the final carcass wash after evisceration; and to add chlorine to the chill water to provide 1–5 μg/mL free chlorine in the overflow (Table 2.15) (Waldroup *et al.*, 1992a).

Although chlorination of chill tank water should not be relied upon to eliminate salmonellae, it can aid in reducing salmonellae and other microorganisms of concern (Lillard, 1980; Green, 1987; James

Table 2.16 Effect of chlorination on microbial levels in chill tank water and on breast skin after exiting the chill tank

Chill water	Mean $\log_{10}$/ml or /g Aerobic count	Fecal coliform	Salmonellae no. positive/ no. tested	Days of shelf life
Untreated	3.41[a] (4.60)[b]	2.38 (3.07)	25/60[c] (8/56)	20.6[d]
20 ppm Cl_2[e]	2.89 (3.55)	1.11 (1.39)	9/52 (1/52)	28.2
34 ppm Cl_2	2.61 (3.83)	<1.00 (1.68)	0/44 (2/44)	32.3

Aerobic count and fecal coliform data are mean values for 44-60 samples per variable collected over 11–15 operating days.
Source: Lillard (1980).
[a]Count/ml of chill tank water.
[b]Count/g of breast skin.
[c]25 positive out of 60 water samples.
[d]Mean no. of days at 2°C until off odor for carcasses packaged immediately after chill tank.
[e]Concentration of free chlorine in water entering chill tank (source of chlorine: gas).

et al., 1992). With the establishment of a performance standard for salmonellae on poultry after chilling the use of pH-controlled chlorinated chill water has been an important in-plant control measure to ensure compliance and has led to reductions in the overall prevalence of salmonellae in poultry in the United States (Russell, 2002). In a comparison of campylobacters on turkey carcasses at two plants, both of which had prevalence rates of 40–42% before chill, the plant that did not chlorinate its chill water had 37.6% post-chill while the plant that used 20 μg/mL was 19.8% (Logue *et al.*, 2003).

Organic acid sprays and dips. During the past 30 years there has been extensive research on the use of organic acids (Dickson and Anderson, 1992). Adding organic acids to scald tank water does not improve the microbial quality of poultry carcasses (Lillard *et al.*, 1987; Izat *et al.*, 1990a), but applying organic acids before, during, or after chilling may be beneficial.

Lactic acid dips have been tested at other stages in broiler processing. A lactic acid dip (1 or 2% lactic acid at 19°C for 10 s) before air chilling was beneficial. While successive dips at different steps during processing did not reduce bacterial numbers, spoilage was delayed when carcasses were dipped after chilling. Dipping in 2% was no better than 1% lactic acid. The time to reach 10^7–10^8/g at 0°C was 11–15 and 18–22 days for the untreated and treated carcasses, respectively (van der Marel *et al.*, 1988). In another study, legs treated with 2% lactic acid had a shelf-life of only 8 days at 6°C (Zeitoun and Debevere, 1990).

In general, bactericidal activity of organic acids reportedly increases linearly with increasing concentration (Tamblyn and Conner, 1997). Using a skin-attachment model (Conner and Bilgili, 1994), salmonellae that are either loosely or firmly attached to poultry skin have increased resistance to organic acids. Concentrations of ≥4% of the organic acids studied were necessary to achieve ≥2 $\log_{10}$ reduction of *S.* Typhimurium (Tamblyn and Conner, 1997).

A major problem with organic acids is skin discoloration. The time and temperature of exposure influences the degree of discoloration and destruction of salmonellae (Izat *et al.*, 1990a). Skin discoloration was reduced by mixtures of lactic acid and sodium lactate prepared to provide 5%, 7.5%, and 10% lactic acid and adjusted to pH 3.0. As lactic acid concentration increased, the shelf-life improved for chicken legs held at 6°C in polyethylene bags. Legs treated with 10% buffered lactic acid were judged better after 12 days than untreated legs after 6 days.

Combining buffered lactic acid and modified atmosphere packaging (90% CO_2:10% O_2) further extended shelf-life. Chicken legs treated with 0, 2, 5, 7.5, and 10% buffered lactic acid had shelf-lives of 13, 14, 15, 16, and 17 days at 6°C, respectively (Zeitoun and Debevere, 1992).

A surface application of sodium diacetate powder to the surface of chicken after chilling (Moye and Chambers, 1991) has been proposed for decontamination.

Acidified solutions. Acidified sodium chlorite has been demonstrated to be an effective antimicrobial for broiler carcasses (Kemp *et al.*, 2000, 2001; Schneider *et al.*, 2002; Kemp and Schneider, 2002). In a five plant study, acidified sodium chlorate solution (pH 2.5, 14–18°C) sprayed onto the outer and inner surfaces of eviscerated broilers for 15 s was more effective for reducing the levels of *E. coli*, salmonellae, and *Campylobacter* on fecally contaminated carcasses than when they were removed from the line for reprocessing (Kemp *et al.*, 2001).

Alkaline solutions. In some countries, a solution of trisodium phosphate is used in commercial operations as a spray for poultry in continuous on-line processing systems in lieu of removing carcasses from the line to remove visible fecal and crop contamination (Russell, 2002). Trisodium phosphate reduces the level of salmonellae and *Campylobacter* on poultry carcasses (Bender and Brotsky, 1992; Greene, 1992; Whyte *et al.*, 2001). Another means of application could involve dipping carcasses for 15 s in a 12% (w/w) aqueous solution of trisodium phosphate (pH 11.5) at 10°C after chilling.

Heat. Moist heat such as hot water dipping is being used on a very limited basis commercially to reduce the prevalence of salmonellae on raw poultry. This necessitates time and temperature combinations that cause denaturation at the surface. Thus, the use of such treatments must be limited to specific applications where denaturation does not cause rejection of the product by the consumer. In absence of denaturation, reductions are not likely to be achieved (Göksoy *et al.*, 2001; Berrang *et al.*, 2000 a,b).

Irradiation. Irradiation up to 10 kGy does not introduce any specific toxicological, nutritional, or microbiological hazard to food (WHO, 1981, 1994), but lower levels are necessary to avoid deleterious impact on the quality and acceptance of raw poultry meat. Doses within this range can be used to inactivate pathogens of importance to raw poultry products (e.g. salmonellae, *Campylobacter*) as well as spoilage bacteria. Microorganisms vary greatly in their susceptibility to irradiation. Doses of 2–7 kGy should be more than adequate to destroy the number of pathogens in hygienically produced raw poultry (WHO, 1989). Packaged poultry can be irradiated either when chilled or frozen. Products irradiated at a dose of 7 kGy or less must be kept refrigerated or frozen because it gives only a pasteurization (radurization) treatment. In 1992, 11 countries had approved irradiation of raw poultry for the destruction of certain food-borne pathogens (Israel Official Gazette, 1982; Kampelmacher, 1984; Lynch *et al.*, 1991; Cross, 1992; ICGFI, 1992). Although approved, irradiation of raw poultry meat has not gained commercial acceptance for three major reasons. The first is the adverse perception that many consumers have with regard to the technology (Frenzen *et al.*, 2001). The second involves color and flavor defects that can occur when raw poultry is irradiated (Nam and Ahn, 2002; McKee, 2002; Liu *et al.*, 2003; Yoon, 2003). The third is that while campylobacteriosis and salmonellosis associated with raw poultry are major concerns these diseases do not have the same degree of severity, public health impact, and consumer reaction that has driven some manufacturers in the US to apply electron beam irradiation to a portion of their ground beef production to destroy *E. coli* O157:H7.

Irradiation can be an effective means to increase shelf-life. As with other methods of preservation, irradiation is not perfect, but it does offer certain advantages that are not available with other methods (Kampelmacher, 1984). After irradiation at 2.5 kGy or less, changes in color and odor of the raw chicken disappear on cooking or roasting and no change in taste is detectable but under commercial conditions this may yield a variable result. Higher doses (e.g. up to 7.5 kGy) should be applied at freezing temperatures (e.g. −18°C) to avoid changes in product quality (Mulder, 1984). Concern has been expressed that irradiation might be used to correct or cover deficiencies in hygiene during the slaughtering process.

This concern has led to the recommendation of a guideline to determine if poultry has been processed according to internationally recognized good manufacturing practices and is, therefore, acceptable for irradiation. The guideline consisted of an aerobic plate count at 20°C of $n = 5, c = 3, m = 5 \times 10^5$ and $M = 10^7$ (WHO, 1989).

Gamma rays from Co^{60} or Cs^{137} or fast electrons of up to 10 MeV can be used to irradiate packaged poultry products. A dose of 2–7 kGy prolongs shelf-life, while 5–7 kGy significantly reduces the number of pathogenic microorganisms. A dose of about 1 kGy is required to affect a 10-fold reduction in number of Enterobacteriaceae, including salmonellae, on broiler carcasses (Mossel *et al.*, 1968; Mulder, 1975). A dose of 2.5 kGy results in a 25-fold reduction of these organisms on frozen carcasses (Mulder, 1975). One group concluded that a dose of 2 kGy would be adequate for the number of salmonellae that may normally be found on raw chicken (Kamat *et al.*, 1991). Others have recommended a maximum dose of 3 kGy for non-frozen poultry to provide microbial control and acceptable organoleptic quality (Kahan and Howker, 1978).

Irradiation is less effective for killing salmonellae on frozen carcasses than on chilled carcasses (Matsuyama *et al.*, 1964; Mulder, 1984). If frozen products are to be irradiated, 5 kGy or larger doses should be given (FAO/IAEA/WHO Expert Committee, 1999).

When raw poultry meat is subjected to irradiation and storage under refrigeration, *Moraxella* spp. become the primary cause of spoilage. Enterococci possess considerable resistance to irradiation and also can be found in large numbers in spoiled products. In packages in which the atmosphere is anaerobic, lactic acid bacteria can be an important cause of spoilage.

An irradiation dose of 2.5 kGy should provide an extension of shelf-life of at least seven days (Mulder, 1984). Raw whole chicken carcasses γ-irradiated with 2.5 kGy and stored at 1°C had a shelf-life of 11–6 days. Packaged chicken legs irradiated with about 3.7 kGy and stored at 1–2°C were satisfactory for at least one week but less so after about three weeks. Breast meat was satisfactory for about 3 weeks but less so after about four weeks (Basker *et al.*, 1986). Turkey breast meat irradiated with 2.5 kGy and stored at 1°C for 21 days in a barrier film showed negligible microbial growth. This treatment, however, caused an intense pink color and unacceptable odors, which were not of microbial origin. Under the same conditions, except for packaging in an oxygen-permeable film, microbial growth occurred but not to the extent of spoilage. In addition, the products' acceptance was little different from the frozen control samples. Thus, the use of an oxygen barrier film can be detrimental for irradiated poultry (Lynch *et al.*, 1991). A similar pink discoloration was noted by others (Coleby *et al.*, 1960; Rhodes, 1965; Kahan and Howker, 1978). Electron beam irradiation combined with packaging with elevated levels of carbon dioxide can lead to greater shelf-life than with either treatment, alone (Grandison and Jenning, 1993). Ground chicken patties irradiated at 3.1 kGy had a distinct irradiation odor.

D_{10} values for salmonellae on poultry meat or skin, ranges from 0.18–1.29 kGy (Mulder, 1984). Irradiation doses of 5–7 kGy are likely to reduce salmonellae on poultry by a factor of approximately 10 000 (International Atomic Energy Agency, 1968). This should be sufficient because there are usually fewer than 100 salmonellae per gram of skin or per 500 mL of thaw water (Mulder *et al.*, 1977b). The D_{10} values for four strains of *Listeria monocytogenes* on poultry meat were 0.417–0.553 kGy (Patterson, 1989). Thus, irradiation offers an effective method for eliminating salmonellae, *L. monocytogenes* and other non-sporeforming pathogens from raw poultry carcasses. A review of four studies on the effects of irradiation on *Cl. botulinum* led to the conclusion that enough normal flora survived on poultry irradiated at 3 kGy so that spoilage occurred before toxin was detected. Thus, irradiation of chicken at a dose of 3 kGy or less will not result in any additional health hazard for *Cl. botulinum* types A, B, or E (Hoeting, 1990).

Packaging before irradiation will avoid subsequent contamination. Storage of poultry at 4°C or lower will retard the multiplication of surviving microorganisms and will inhibit the outgrowth of spores.

D CONTROL (primary processing, whole birds and parts)

Summary

Significant hazards[a]	• Salmonellae, thermophilic campylobacters.
Control measures	
Initial level (H_0)	• Controls applied from the breeder through slaughtering.
Reduction (ΣR)	• Very limited opportunities exist to reduce the level of salmonellae and *Campylobacter* on raw poultry meat.
Increase (ΣI)	• The low storage temperatures used for raw poultry should prevent multiplication of salmonellae and campylobacters. • Cross-contamination can occur during handling and storage; thereby, leading to an increase in prevalence by the time the raw poultry reaches the retail market.
Testing	• Monitoring for salmonellae and/or *Campylobacter* spp. at selected stages in the chain from the breeder farms through processing can provide information about the strengths and weaknesses of control measures as applied by a company, industry, or within a country.
Spoilage	• Four factors influence the rate and type of spoilage of raw poultry at refrigeration temperatures: storage temperature, numbers and types of psychrotrophic bacteria, pH, and type of packaging.

[a]In particular circumstances, other hazards may need to be considered.

Hazards to be considered. The hazards of primary concern to human health are salmonellae and *Campylobacter*.

Control measures. Certain controls have been shown to be effective at the breeder farm, hatchery, on the grow-out farm, during transportation and during slaughter that can help prevent vertical and horizontal transfer of salmonellae at the various stages in the chain from the breeder farm through the slaughtering process. Horizontal transfer of *Campylobacter* begins mainly at the grow-out farm level.

Initial level of hazard (H_0). The initial level of salmonellae and *Campylobacter* on raw poultry meat is influenced strongly by the prevalence and concentrations of these pathogens in and on the live birds that are slaughtered and the extent to which the slaughtering system can minimize contamination of the carcasses.

Reduction of hazard (ΣR). There are limited opportunities to reduce the levels of salmonellae and *Campylobacter* on raw poultry.

Increase of hazard (ΣI). The low temperatures used for raw poultry should prevent multiplication of salmonellae and thermophilic campylobacters on raw poultry. Cross-contamination can occur during handling and storage; thereby, leading to an increase in prevalence at the retail market.

Testing. Monitoring for salmonellae and/or *Campylobacter* spp. at selected stages in the chain from the breeder farms through processing can provide information about the strengths and weaknesses of control measures as applied by a company, industry, or within a country.

Spoilage. Raw poultry meat held at refrigeration temperatures (−2°C to 5°C) will spoil at a rate influenced by four major factors: (a) storage temperature, (b) numbers and types of psychrotrophic bacteria, (c) pH, (d) type of packaging.

IV Frozen poultry products

Chilled poultry has a shelf-life measured in days when held near the point of freezing, whereas frozen poultry at or near −18°C will not spoil from microbial activity. Poultry products are frozen by exposure to a continuous blast of cold air in tunnels or rooms, immersion of packaged poultry in cold brine or propylene glycol solution, exposure to still cold air, exposure to liquefied or solidified gasses (such as nitrogen or carbon dioxide), plate freezing (usually prepared dinners), or various combinations of these methods. Typical products are listed in Table 2.1. The following material relates to the freezing of raw poultry. Similar effects of freezing on survival and death observed in raw poultry also apply generally to microorganisms found on cooked frozen poultry products. The information provided for cooked poultry products in Section V should be applicable after the frozen cooked products are thawed.

A Effects of processing on microorganisms

Freezing and frozen storage reduces the number of certain microorganisms on poultry; some are killed, while others are only sublethally damaged (Kraft, 1992; Bailey *et al.*, 2000). At temperatures below −10°C, sublethally damaged microorganisms die over time, but above this temperature, recovery can occur (Mulder, 1973).

Aerobic plate counts on poultry skin may decrease by 10–99% during the process of freezing; further death then occurs during frozen storage but at a slower rate. An 84–98% decrease in aerobic plate count on turkey skin occurred during brine immersion and 95–99% during air blast freezing. Fluorescing bacteria were reduced 99.9% by both methods of freezing. Coliforms, enterococci, and staphylococci were reduced 99%, 97%, and 96%, respectively, during air-blast freezing (Kraft *et al.*, 1963). In another study, no decrease in the aerobic plate count and only a slight decrease in Enterobacteriaceae occurred when poultry was stored at −20°C. The bacteria killed by freezing were largely those in the surface water on the carcass and not those attached to or in the skin (Notermans *et al.*, 1975d). Enough psychrotrophic bacteria survive on frozen poultry to multiply during thawing and storage under refrigeration. Spoilage may occur if the time and temperature of the thawing process are not controlled. Micrococci and enterococci survive in frozen poultry for considerable periods of time (Straka and Combes, 1951; Wilkerson *et al.*, 1961). In a study of naturally contaminated chicken salmonellae prevalence and concentration did not change when the chicken was held for 14 days within the range of 4–8°C (Bailey *et al.*, 2000).

B Spoilage

Frozen poultry products do not usually undergo microbial spoilage. Perhaps for this reason, there is very little published information on microbial spoilage. Some yeasts and molds multiply on frozen meat at temperatures as low as −7°C. Organisms occasionally implicated are *Cladosporium herbarum*, causing black spots; *Thamnidium elegans*, causing whisker-like growth; and *Sporotrichum carnis*, causing white spots. Mold growth can occur on the surface of frozen products (e.g. frozen pies containing poultry meat) while freezer units are in the defrost cycle (Gunderson, 1962). Spoilage will occur if products are thawed and stored at refrigerator temperatures for sufficient time. The rate of spoilage is similar to that of chilled poultry (Elliott and Straka, 1964) and is affected by the same factors. From a bacteriological standpoint, the practice of freezing raw whole birds or parts and then thawing them for retail sale has a distinct advantage over distributing chilled poultry because freezing shifts all the chilled shelf-life to the retail and consumer levels (Elliott and Straka, 1964). Enzymatic activity can cause off-flavor in frozen poultry products that are stored for prolonged periods. The extent of product deterioration varies with the type of product, method of processing and packaging, and storage conditions.

C Pathogens

Raw frozen poultry harbors many of the same pathogens as chilled products and present the same problems of cross-contamination after thawing. Water released during thawing is potentially hazardous because it may contain pathogens. Many surfaces can become contaminated from this water and the raw poultry (van Schothorst *et al.*, 1976). Freezing does not kill all the salmonellae, and, therefore, they can be frequently found on frozen raw poultry (Gunderson and Rose, 1948, Kraft *et al.*, 1963; Bryan *et al.*, 1968c). *Campylobacter jejuni* and the vegetative cells of *Cl. perfringens*, however, are quite sensitive to freezing. The rate of death is influenced by the conditions of freezing and time of storage (Oosterom *et al.*, 1983b; Rayes *et al.*, 1983; Gill and Harris, 1984). Research and experience in Iceland suggest that freezing raw poultry can be an effective control measure for reducing the risk of campylobacteriosis (Chan *et al.*, 2001; Stern *et al.*, 1985, 2003). The time and temperature at which poultry is frozen influences the reduction of *Campylobacter*. After 52 weeks at −20°C a 4 $\log_{10}$ reduction in *Campylobacter* occurred; whereas, at −86°C the reduction was only 0.5 $\log_{10}$ (Zhao *et al.*, 2003). This is significant because frozen poultry is commonly held for weeks or months rather than days at temperatures near −20°C. The spores of *Cl. perfringens* and *Cl. botulinum* are very resistant to freezing.

D CONTROL (frozen poultry product)

Summary

Significant hazards[a]	• Salmonellae.
Control measures	
Initial level (H_0)	• It should be assumed that salmonellae and thermophilic campylobacters will be present on frozen raw poultry.
Reduction (ΣR)	• Cooking will kill salmonellae and thermophilic campylobacters.
Increase (ΣI)	• Thawing at temperatures that prevent growth of salmonellae.
Testing	• Not normally recommended.
Spoilage	• Microbial spoilage does not occur at freezer temperatures

[a]In particular circumstances, other hazards may need to be considered.

Hazards to be considered. Salmonellae and *Campylobacter* are frequently present on raw poultry.

Control measures

Initial level of hazard (H_0). The concentration and prevalence of salmonellae and thermophilic campylobacters are variable on raw poultry depending on the source and controls at the farm and during processing. These values will influence the concentration and prevalence in frozen raw poultry.

Reduction of hazard (ΣR). The time and temperature of storage influence the rate of reduction in campylobacters but salmonellae levels are not likely to change. Raw poultry that has been frozen should be cooked before serving.

Increase of hazard (ΣI). Raw poultry must be thawed under conditions that prevent an increase in salmonellae levels and contamination of ready-to-eat foods.

Testing. Some countries have adopted testing procedures for frozen raw poultry. The microbiological criteria and sampling plans should be equivalent (i.e. not more stringent or lenient) as those applied to domestic production.

Spoilage. The microbiological quality of frozen poultry depends on the factors previously described for chilled poultry. Chilled carcasses should be packaged, promptly frozen, and held at or near −18°C. Carcasses must be thawed at temperatures and times that prevent an unacceptable increase in microorganisms that can cause decomposition.

V Perishable, cooked poultry products

A wide variety of perishable cooked ready-to-eat poultry products are available (Table 2.1), prepared by many different processes (Tompkin, 1986, 1995b; Acuff *et al.*, 2001), but most involve three general processes. Whole carcass or part of it may be injected with a solution containing desired ingredients and then placed onto racks or into stockinettes for cooking. If desired, smoke is applied during cooking. These products may be frozen or refrigerated for distribution and sale.

Large, boneless products can be made by injecting the meat; tumbling or massaging to extract soluble protein; stuffing into a desired mold, casing, stockinet, or plastic barrier film package; and then cooking. These products may be given additional treatments during or after cooking (e.g. smoking, caramel coating, oven browning, browning in hot oil, or applying spices or condiments). Some products are sliced and packaged. These products may be frozen or refrigerated for distribution and sale.

Cooked products consisting of ground poultry are usually made on continuous systems that involve grinding meat, adding other ingredients, forming to the desired size and shape, applying a batter and breading, cooking, freezing, and packaging. Large quantities of chicken parts (breasts, wings, drumsticks) are cooked in oil on continuous systems, seasoned, frozen, and then packaged. Frozen cooked ground products and parts are normally distributed, frozen, and then thawed at the retail level and sold refrigerated. Some frozen cooked poultry products may be placed at refrigerated temperature for display and sale at the retail level.

A *Effects of processing on microorganisms*

The microbial content of cooked products is influenced by the method of processing, packaging, and storage (Denton and Gardner, 1982; Tompkin, 1986; Tompkin, 1995b; Acuff *et al.*, 2001). All the products in this category should be cooked at a time and temperature to obtain a cooked appearance, appropriate tenderness, and other desirable organoleptic qualities. In addition, the processes should be adequate to ensure destruction of enteric pathogens. For example, in contrast to some beef products (e.g. ground beef patties, rare roast beef), thermal processes for cooked poultry products can be designed and controlled to ensure the destruction of 10^7 salmonellae/g in the coldest area of the product and still have acceptable organoleptic quality. Thermal processes of this nature should be more than adequate to destroy the levels of salmonellae, *C. jejuni*, *L. monocytogenes*, and *Staph. aureus* that normally occur in raw poultry meat. To achieve the organoleptic characteristics expected, poultry products are usually cooked more extensively than comparable beef products.

Vegetative bacteria and, perhaps, some spores on the surface of poultry products are killed during cooking, but some in the center of the product may survive depending on the thermal process (e.g. enterococci, *Lactobacillus viridescens*). The thermal processes are not adequate to assure destruction of sporeforming pathogens (*Cl. perfringens*, *Cl. botulinum*). Cooked products are subjected to post-process contamination during slicing and packaging.

Barbecued poultry is cooked on a grill or in a rotisserie often at temperatures below those used for baking poultry, but for longer periods of time. Microbial growth can occur if the cooked barbecued products become contaminated and stored for several hours at abnormal temperatures (Pivnick *et al.*, 1968; Seligmann and Frank-Blum, 1974).

During the frying of chicken in oil, temperatures at the geometric center usually reach 93°C or higher, which is lethal to vegetative bacteria but not spores. These products are also subject to post-process contamination during subsequent handling and packaging.

B Spoilage

Cooked poultry products that are stored, distributed, and displayed at refrigeration temperatures will eventually spoil. The rate and type of microbial spoilage depends on how the product has been processed, packaged, and stored.

Products that are cured (i.e. contain sodium nitrite) often have higher salt and lower moisture and will generally spoil in the same manner as similar cured meats made of beef, pork, lamb, or veal (see Chapter 1). When exposed to the processing environment before final packaging, the spoilage flora is mixed. If oxygen is excluded, e.g. by vacuum or modified atmosphere packaging, the spoilage flora often becomes dominated by lactic acid bacteria, such as *Lactococcus*, *Carnobacterium*, and *Lactobacillus* spp. (Barakat *et al.*, 2000). In the absence of oxygen, these packaged products may develop a milky exudate; gassiness; and, if a fermentable carbohydrate has been added, sourness with decreased pH.

Many cooked breast meat products do not contain sodium nitrite. If these products become contaminated after cooking (e.g. during slicing) and packaged in the absence of oxygen, a substantial population of Gram-negative bacteria develops that includes a variety of species within the family Enterobacteriaceae. Spoilage is characterized by very strong offensive odors that may not be detected until the packages are opened. There is a potential for biogenic amine formation. This defect can be prevented by applying a shorter code date (e.g. days rather than weeks), reducing contamination through more effective GHP and/or by adding adequate levels of sodium lactate and/or sodium diacetate to the product. There have been reports that sodium lactate can delay botulinal toxin production in products of non-cured poultry products. At the level (e.g. 2% w/w) that is normally used to extend shelf-life, sodium lactate is of marginal value as an antibotulinal agent. There have been no reports of botulism from non-cured breast meat products, since they were moved from the frozen to refrigerated storage and distribution in the 1980s.

Cooked non-cured breast meat products are also cooked in an oxygen impermeable film and then chilled. These cook-in-bag products have a relatively long shelf-life (e.g. >60 days) when stored at less than 4°C. Occasionally, spoilage occurs due to the growth of psychrotrophic sporeformers that survive the cooking process (Kalinowski and Tompkin, 1999; Meyer *et al.*, 2003). Products of this type also may be frozen for storage and distribution.

Two other defects of microbial origin have been observed in cooked non-cured poultry products. Both can be due to microbial growth in the raw meat before cooking. One defect involves the formation of small holes similar to those found in certain cheeses. This defect has been observed in products that have been injected and held for some time at refrigeration temperatures before cooking and is due to multiplication of certain bacteria (e.g. *Aeromonas* spp.) that can multiply at $\leq 10°C$ in breast muscle injected with a solution containing salt and phosphate. The holes are thought to be due to the production of carbon dioxide that is released from the meat and expands during cooking. Apart from the small holes in the cooked product, there is no other loss of quality. The products are otherwise normal in appearance and odor and do not contain recoverable microorganisms. This defect can be confused with small holes that develop when an excessive quantity of carbon dioxide in the form of dry ice (i.e. carbon dioxide in frozen, solid form) has been used to chill the meat at a step prior to cooking. Controlling the time and temperature of holding the raw meat prior to cooking prevents the defect. It is possible that the time-temperature factor after injecting the salt–phosphate solution may be important for two reasons. The injection process contaminates the interior of the meat and, perhaps, certain ingredients play a role.

The second defect is a pink discoloration throughout the interior of the product. There are several potential causes for the development of a pink color in cooked non-cured poultry breast meat (Cornforth, 1991; Cornforth *et al*., 1998; Schwarz *et al*., 1999). One involves the reduction of nitrate in processing water to nitrite, by the microbial flora of raw poultry meat. The best means to prevent this cause of pink discoloration is to remove nitrate from the water. Pink discoloration of microbial origin in cook-in-bag breast products can be due to the growth of psychrotrophic clostridia. In addition to the pink discoloration, a strong hydrogen sulfide odor is produced. Since no gas is produced, the product in the package appears normal until opening. The combined offensive odor and internal pinkness results in product rejection (Kalinowski and Tompkin, 1999; Meyer *et al*., 2003).

C *Pathogens*

Certain microbial pathogens should be considered in a hazard analysis for the production and use of perishable cooked poultry products. Salmonellae present on raw poultry meat can survive as a result of undercooking or contaminate products that are exposed to the processing environment and handled before final packaging. Proper controls for cooking can prevent survival of salmonellae. Likewise, proper layout of the production facility to separate raw meat processing areas from cooking, chilling, storing, and packaging areas can virtually eliminate the risk of recontamination by salmonellae. This is evident from two extensive surveys conducted at manufacturing plants. During an extensive survey involving 6 606 analyses of cooked poultry products collected over four years from processing plants in the United States, only four analyses were positive for salmonellae (Green, 1993). Ready-to-eat meat and poultry products sampled at manufacturing plants for regulatory compliance during the years 2001 and 2002 yielded only 23 samples positive for salmonellae in over 14 000 tested (USDA-FSIS, 2003). The few positive products could have been prevented through effective application of GHP and HACCP principles (Simonsen *et al*., 1987; ICMSF, 1988; Tompkin, 1990, 1994, 1995a,b).

The potential exists for food-borne illness from sporeforming pathogens (e.g. *Cl. botulinum*, *Cl. perfringens*) that survive the cooking of perishable poultry products. Food-borne illness from these anaerobic pathogens could occur only after raw poultry is cooked and then held at $>10°C$ (*Cl. botulinum*) or $>15°C$ (*Cl. perfringens*) for sufficient time for extensive multiplication to occur. Large numbers of vegetative cells of *Cl. perfringens* ($\geq 10^7$) must be consumed for illness to occur (Brynestad and Granum, 2002). The number of *Cl. perfringens* spores present after cooking meat and poultry products in commercial operations are typically below detectable levels (Kalinowski *et al*., 2003; Taormina *et al*., 2003). There does not appear to be any documented evidence that *Cl. perfringens* food-borne illness has occurred with the commercially produced products that are likely to appear in international commerce. These products, however, have been implicated when they are mishandled at the user level (e.g. food-service, retailer, homes, institutional cooking). Historically, it has been at this level that food-handling errors in storage times and temperatures have occurred and perishable ready-to-eat foods have been implicated in food-borne illness. For example, from 1970 to 1996, 94% of 1 525 of the reported outbreaks of *Cl. perfringens* illness in the UK were due to mishandling at the food-service level (Brett and Gilbert, 1997; Brett, 1998). A similar experience has been noted in Australia and Norway (Bates and Bodnaruk, 2003; Brynestad and Granum, 2002) and very likely represents the more common scenario for many other countries. Outbreaks attributed to *Cl. perfringens* have involved a wide variety of prepared foods, including large pieces of meat or poultry and others containing meat or poultry as one of many ingredients. The significance of poultry meat as the source of *Cl. perfringens* in foods containing gravies, spices and other ingredients is not known. Cured meat or poultry products have rarely been implicated in *Cl. perfringens* illness. This should not be unexpected considering the relative sensitivity of *Cl. perfringens* to the combined effect of salt and sodium nitrite present in these products (Roberts and Derrick, 1978; Gibson and Roberts, 1986). There are conflicting views on the rate of chilling and what might be considered a tolerable level of *Cl. perfringens* (e.g. performance objective) in cooked poultry products manufactured in commercial establishments. Inoculated studies (Juneja and Marmer, 1996) suggest a need for greater caution and rapid chilling than manufacturing facilities generally have used historically but epidemiologic data and commercial experience do not support this conclusion (Kalinowski *et al.*, 2003; Taormina *et al.*, 2003). A risk assessment might help clarify this issue.

Contamination with *Staph. aureus*, usually at low levels, can occur during handling and packaging; however, proper refrigeration will prevent multiplication and the risk of enterotoxin production in these products. Staphylococcal food-borne illness is primarily associated with poultry products that are cooked and served in homes or food-service establishments where contamination occurs and the foods are then improperly held at higher temperatures for sufficient time for enterotoxin production to occur. Almost all staphylococcal food-borne illness from poultry has been due to recontamination of cooked meat by a food handler rather than from the raw poultry (Bryan, 1968, 1980; Cox and Bailey, 1987; Mead, 1992).

The final group involves psychrotrophic pathogens that can establish themselves and multiply in the environment of manufacturing plants where operating temperatures for chilling and handling cooked products is too low for the growth of the above pathogens. In this group, only *L. monocytogenes* has been identified as a significant contaminant of commercially produced cooked poultry products. Surveys of cooked poultry products indicate that post-processing contamination can occur. For example, 12% of 527 cooked poultry products from retail stores were positive for *L. monocytogenes* (Gilbert *et al.*, 1989). Cooked poultry products collected during the years 1990 through 1999 from manufacturing plants in the United States showed prevalence rates in the range of 1.0 –3.2% (Levine *et al.*, 2001). A survey of ready-to-eat sliced luncheon meat and poultry from retail stores found the prevalence rate among 9 199 samples was 0.89% (Gombas *et al.*, 2003). Commercially manufactured ready-to-eat poultry products have been implicated in isolated cases and several large outbreaks of listeriosis (McLauchlin, 1991; CDC, 1999a,b, 2000, 2002; USDA-FSIS, 2002; Tompkin, 2002). The outbreaks involved persistent strains of *L. monocytogenes* that had become established in the manufacturing plant and contaminated the products between cooking and packaging. The existence of persistent strains in equipment and the environment in which ready-to-eat foods are exposed to contamination is a critical feature of this issue (Tompkin, 2002; Lundén *et al.*, 2002; Autio *et al.*, 2002; Berrang *et al.*, 2002). A case control study in the Unite States identified undercooked chicken, presumably prepared in the home or food-service establishments, as a risk factor for human listeriosis (Schuchat *et al.*, 1992). This and other reports of listeriosis attributed to a wide variety of ready-to-eat foods have led to modifications in industry practices (Tompkin *et al.*, 1992; Tompkin, 1995a) and the establishment of regulatory requirements and standards for *L. monocytogenes* in ready-to-eat poultry products.

The FDA-FSIS risk assessment identified "deli meats" as the food group of highest risk for listeriosis in the USA (FDA-FSIS, 2003). This estimate was influenced by two major outbreaks involving non-cured cooked turkey breast products that were sliced and sold through the deli counter of grocery stores. The combined data from the FDA-FSIS risk ranking of food groups and the FAO-WHO risk assessment for *L. monocytogenes* in ready-to-eat foods (FAO-WHO, 2004a,b) will influence national and international policies. Exporters should be aware of the regulatory requirements for each importing country.

The microbiological safety and quality of cooked products depend upon the use of GHP and HACCP to: (a) assure the use of a thermal process that destroys non-sporeforming pathogens; (b) control the chilling step to prevent the multiplication of mesophilic sporeforming pathogens; (c) prevent cross-contamination from raw meats to cooked product; (d) control the environment and handling of cooked products to minimize contamination with *L. monocytogenes*; (e) control storage and distribution times and temperatures to ensure microbiological safety and, where appropriate; and (f) provide food handling and preparation procedures to the end user (Tompkin, 1995b).

Three general approaches have evolved to help manage the risk of *L. monocytogenes* in those ready-to-eat poultry products that are stored and handled under conditions that permit growth. The first approach involves managing the manufacturing environment where cooked ready-to-eat products are exposed and subject to contamination. A considerable amount of information is being developed through research and experience that can be used to provide guidance on controlling *L. monocytogenes* in the environment (Tompkin *et al.*, 1999). This is an evolving field with improvements being made in plant layout and construction, equipment design and procedures for cleaning and disinfection. In the second approach, research has identified additives (e.g. sodium lactate, sodium diacetate) that can be used to reduce, if not prevent, the growth of *L. monocytogenes* when stored at refrigeration temperatures (Seman *et al.*, 2002). A wider variety of additives than is currently available should be become available through research. The third approach is to treat products after final packaging with a listericidal process (e.g. heat, ultra high pressure) before releasing the product for distribution (Muriana *et al.*, 2002; Murphy and Berrang, 2002; Murphy *et al.*, 2003a,b). A number of in-pack pasteurization systems are commercially available and used by producers of ready-to-eat poultry products in a variety of countries. As experience is gained with these systems improvements in functionality and reductions in production cost per package should lead to wider application of this approach.

D CONTROL (perishable, cooked poultry products)

Summary

Significant hazards[a]	• *Listeria monocytogenes*, salmonellae, *Cl. perfringens*.
Control measures	
Initial level (H_0)	• *L. monocytogenes*, salmonellae and *Cl. perfringens* spore concentrations are low in raw poultry. • The low temperatures and short times used to prevent spoilage of raw poultry also prevent an increase in their initial numbers.
Reduction (ΣR)	• Cooking destroys vegetative pathogens but not spores. • Some packaged poultry products are in-pack pasteurized.
Increase (ΣI)	• Preventing recontamination of cooked ready-to-eat poultry products with *L. monocytogenes* is critical if the food supports growth under the normal conditions of storing and distribution. • Freezing and/or additives can be used to control growth of *L. monocytogenes* in packaged products.

(*continued*)

D CONTROL (Cont.)

Summary

	• Sufficiently rapid chilling after cooking should control germination and outgrowth of *Cl. perfringens*. • Contamination with salmonellae is a concern when the system for GHP is inadequate or ineffectively applied.
Testing	• Tests for *L. monocytogenes* may be appropriate when the conditions of manufacture are unknown and the food can support growth. • Tests for the concentration of *Cl. perfringens* can provide guidance on product disposition when the rate of chilling is in question.

[a]In particular circumstances, other hazards may need to be considered.

Hazards to be considered. Significant hazards for cooked ready-to-eat products include *L. monocytogenes*, salmonellae, and *Cl. perfringens*. Other pathogens may be of concern in regions where certain diseases may be endemic in the population (e.g. shigellosis) and testing procedures are available. It is prudent to consider the risk of *Cl. perfringens* when a deviation has occurred and the product has been stored at higher temperature that could permit growth.

Control measures

Initial level of hazard (H_0). *L. monocytogenes* and salmonellae concentrations are generally low (e.g. <10 cfu/g) in raw poultry. Spore concentrations are typically <1 cfu/g in raw poultry meat. The low temperatures and short time used to prevent spoilage of raw poultry used in making cooked poultry products also prevent an increase in these initial numbers. *Campylobacter* concentrations may be higher in freshly produced raw poultry, but few data exist for the raw poultry meat used for manufacturing cooked products. *Campylobacter* has not been linked to the commercially cooked poultry that appears in commerce.

Reduction of hazard (ΣR). The low concentrations of *L. monocytogenes* and salmonellae normally present are readily destroyed by the time–temperature combinations used to manufacture cooked poultry products. Increasing amounts of cooked poultry products are in-pack pasteurized as a precaution against *L. monocytogenes*.

Increase of hazard (ΣI). Recontamination with *L. monocytogenes* of cooked poultry products that support growth is a significant risk factor. The risk further increases in those products with extended expiration dates; thereby, providing the time necessary for the concentration to increase. In the event a product is contaminated with *L. monocytogenes*, the potential increase in numbers during refrigerated storage can be controlled by the addition of acceptable inhibitors. An increase in salmonellae should not occur at the refrigerated temperatures used for perishable cooked poultry products. An increase in the concentration of *Cl. perfringens* may occur during the chilling process following cooking but there is little, or no, evidence that this has led to sufficiently high numbers to cause food-borne illness in commercial products that are likely to appear in international trade.

Testing. Tests for *L. monocytogenes* may be appropriate for cooked poultry products when the conditions of manufacture are unknown and the food can support growth. Tests for the concentration of *Cl. perfringens* can provide guidance on product disposition when the rate of chilling is in question.

Spoilage. The time to spoilage is dependent on the degree of contamination with psychrotrophs while the cooked product is exposed and handled before final packaging and is longer at lower storage temperatures.

VI Fully retorted ("botulinum-cooked") poultry products

The processes used for canned shelf-stable poultry products are similar to those of other for low-acid foods. The poultry meat is packed into cans with other ingredients, hermetically sealed under vacuum, retorted at temperatures near 115°C for sufficient time to achieve "commercial sterility", and then cooled. The time–temperature combinations for retorting must destroy all the spores that can germinate and multiply during subsequent conditions to which the product will be subjected. Although readily available in the marketplace, there is only limited information available on the production of canned shelf-stable poultry products (Tompkin, 1986; Acuff *et al.*, 2001). This reflects the favorable safety record and microbiological quality of these products.

Additional information on shelf-stable canned foods can be found elsewhere throughout this text (e.g. Chapter 1 (Section X) and Chapter 5 (Section 6)).

A CONTROL (fully retorted shelf-stable poultry products)

Summary

Significant hazards[a]	• *Cl. botulinum*, enteric pathogens, *Staph. aureus*.
Control measures	
Initial level (H_0)	• Selection of raw materials.
Reduction (ΣR)	• Control temperature and time during retorting to inactivate spore-formers.
Increase (ΣI)	• Use of appropriate containers and lids (e.g. resistant to corrosion). • Ensure integrity of the cans with satisfactory seals; avoid damage to containers from poor handling. • Use potable water for cooling. • Avoid manual handling of wet cans.
Testing	• No microbiological testing is recommended; routine checking of seals.
Spoilage	• Selection of raw materials. • Control of temperature and time during retorting (spore destruction). • Control of can seams (to avoid recontamination). • Hygienic handling of wet cans while cooling and drying.

[a]In particular circumstances, other hazards may need to be considered.

Hazards to be considered. The primary hazard of concern is *Cl. botulinum* that may occur in the raw materials (e.g. ingredients, raw poultry meat). With good process control and canning practices, the risk of *Cl. botulinum* is controllable. There have been instances where other microbiological hazards have occurred in low-acid canned foods, such as recontamination through the container seals during cooling due to contaminated cooling water (salmonellae) or manual handling of wet cans (*Staph. aureus*). Very rarely, recontamination with *Cl. botulinum* has occurred and resulted in illness and deaths because *Cl. botulinum* was present in the post-retorting environment.

Control measures. Process control is essential. Codes of practice for the production of low-acid canned foods have been established (e.g. FDA, 1973; FAO/WHO, 1979; CFPRA, 1977; AOAC/FDA, 1984; Thorpe and Barker, 1984), and the CCPs outlined (ICMSF, 1988). Control is needed at pre-processing, thermal processing, and post-processing.

Initial level (H_0). Raw materials are stored and prepared under hygienic conditions to prevent spoilage before retorting and possible increase in spore level. Certain ingredients (e.g. dehydrated onions and other root crops) naturally may contain botulinal spores.

Reduction (ΣR). Containers must be resistant to corrosion from the contents, of the required size, strength, and seam construction, undamaged, and free from defects.

Containers are properly filled (e.g. weight, head-space, packing density, vacuum and temperature of filled product, time between filling and retorting) and container closure equipment is functioning correctly.

The retort is properly calibrated and operated (e.g. venting, loading pattern and operational parameters like steam pressure, temperature, time, water circulation, and chain speed).

Increase (ΣI). Cooling is carried out carefully to avoid damage to the integrity of the container and contamination of the contents.

Handling equipment and runways are clean and prevent physical shock to the container.

Avoid manual handling of containers until they are dry.

Testing. Microbiological testing of product is not recommended.

Cooling water should be checked for disinfectant concentration and, if uncertain, potable quality.

Monitoring the microbiological quality of cooling water is useful.

Integrity of can seals must be routinely checked.

Spoilage. Spoilage of low-acid canned foods, including canned poultry products is controllable and should rarely occur.

VII Dried poultry products

Several methods have been used to prepare dried poultry meat products (Mountney, 1976). Finely comminuted chicken meat has been sprayed-dried for use in soups. Thin layers of ground cooked meat on trays have been dried in ovens or vacuum chambers. Chunk-size poultry meat has been dried in conventional air dryers. Poultry meat products have also been dried on heated rollers and by heating in edible oil.

Freeze dehydration offers another means to produce dried poultry meat. During preparation for freeze-drying, carcasses are cooked, skinned, and then boned. The meat is then drained, cooled, diced, frozen, and put into a chamber under vacuum where the temperature is raised so that sublimation occurs.

The pressure is then equalized with nitrogen. The product is then packaged in an oxygen- and moisture-impermeable package or can. Freeze-dried meat can be blended with other dry ingredients to produce dry soup mixes and similar products.

A *Effects of processing on microorganisms*

Most vegetative forms of microorganisms are killed during heating in ovens and oil, but the more sophisticated drying methods are designed to preserve cellular structure and, thus, microbes may survive. Post-processing contamination during packaging should not be a serious problem.

B *Spoilage*

Dried poultry meat containing <10% moisture and packaged in a water-impermeable container will not support microbial growth and can be held for prolonged periods at ambient temperature. If the moisture content is slightly above 10%, molds and possibly yeasts may grow and spoil the product.

C *Pathogens*

Sporeforming pathogens (e.g. *B. cereus, Cl. perfringens*) can survive cooking, freezing, drying, and subsequently germinate and multiply if the products are reconstituted and held at temperatures permitting their multiplication. Cooked dried products can become contaminated with salmonellae, staphylococci, listeriae, and other microorganisms if they are processed in an environment where cross-contamination from raw poultry can occur or if the processing conditions are not controlled.

D CONTROL (dried poultry products)

Summary

Significant hazards[a]	• Salmonellae.
Control measures	
Initial level (H_0)	• It must be assumed that salmonellae will be present on raw poultry that is used to manufacture dried poultry products.
Reduction (ΣR)	• Cooking kills salmonellae.
Increase (ΣI)	• Preventing recontamination, maintaining low water activity.
Testing	• Not normally necessary but if source is unknown, test for salmonellae.
Spoilage	• Microbial spoilage does not occur at low-water activity.

[a]In particular circumstances, other hazards may need to be considered.

Hazards to be considered. Salmonellae are frequently present on raw poultry.

Control measures

Initial level of hazard (H_0). Raw poultry is perishable and should be stored at sufficiently low temperatures to avoid deterioration in quality. The low temperatures necessary (e.g. <5°C) to maintain quality also prevent an increase in salmonellae.

Reduction of hazard (ΣR). Dehydrated poultry that is manufactured for dried soup mixes and similar foods involve a cooking step before drying. The cooking process must be validated to kill salmonellae.

Increase of hazard (ΣI). The layout of the plant and other aspects of GHP should be adequate to prevent contamination of the cooked poultry throughout drying and packaging.

Testing. Testing for salmonellae may be appropriate when the conditions of manufacture are not known or dehydrated cooked poultry is to be used as an ingredient in a food that may be consumed without a subsequent kill step (e.g. as an ingredient in a dip for snacks).

Spoilage. The quality of dehydrated poultry is to ensure rapid chilling of the poultry after slaughter and control the length of time before processing.

References

Abu-Ruwaida, A.S., Sawaya, W.N., Dashti, B.H., Murad, M. and Al-Othman, H.A. (1994) Microbiological quality of broilers during processing in a modern commercial slaughterhouse in Kuwait. *J. Food Protect.*, **57**, 887–92.

Achen, M.T., Morishita, T.Y. amd Ley, E.C. (1998) Shedding and colonization of *Campylobacter jejuni* in broilers from day-of-hatch to slaughter age. *Avian Dis.*, **42**, 732–7.

ACMSF (Advisory Committee on the Microbiological Safety of Food) (1993) Interim Report on *Campylobacter*. HMSO, London. ISBN 0 11 321662 9 (£10.65).

Acuff, G.R., Vanderzant, C., Gardner, F.A. and Golan, F.A. (1982) Examination of turkey eggs, poults and brooder facilties for *Campylobacter jejuni*. *J. Food Protect.*, **45**, 1279–81.

Acuff, G.R., McNamara, A.M. and Tompkin, R.B. (2001) Meat and poultry products, in *Compendium of Methods for the Microbiological Examination of Foods* (eds F.P. Downes and K. Ito), 4th edn, American Public Health Association, Washington, DC, pp. 463–71.

Adamcic, M. and Clark, D.S. (1970) Bacteria-induced biochemical changes in chicken skin stored at 5°C. *J. Food Sci.*, **35**, 103–6.

Aho, M., Nuotio, L., Nurmi, E. and Kiiskinen, T. (1992) The competitive exclusion of campylobacters from poultry with K-bacteria and Broilact. *Int. J. Food Microbiol.*, **15**, 265–75.

Altekruse, S.F., Stern, N.J., Fields, P.I. and Swerdlow, D.L. (1999) *Campylobacter jejuni*—an emerging foodborne pathogen. *Emerg. Infect. Dis.*, **5**, 28–35.

Anderson, A.D., McClellan, J., Rossiter, S. and Angulo, F.J. (2003) *Public Health Consequences of Use of Antimicrobial Agents in Food Animals in the United States.* Foodborne and Diarrheal Diseases Branch, Division of Bacterial and Mycotic Diseases, Centers for Disease Control and Prevention, Atlanta. http://www.cdc.gov/narms/pub/publications/a_anderson.pdf.

Angen, Ø., Skov, M.N., Chriél, M., Agger, J.F. and Bisgaard, M. (1996) A retrospective study on salmonella infection in Danish broiler flocks. *Prev. Vet. Med.*, **26**, 223–37.

Annan-Prah, A. and Janc, M. (1987) The mode of spread of *Campylobacter jejuni/coli* to broiler flocks. *J. Vet. Med. Ser. B*, **35**, 11–8.

Arakawa, A., Fukata, T., Baba, E., McDougald, L.R., Bailey, J.S. and Blankenship, L.C. (1992) Influence of coccidiosis on *Salmonella* colonization in broiler chickens under floor-pen conditions. *Poult. Sci.*, **71**, 59–63.

Autio, T., Lundén, J., Fredriksson-Ahomaa, M., Björkroth, J., Sjöberg, A-M. and Korkeala, H. (2002) Similar *Listeria monocytogenes* pulsotypes detected in several foods originating from different sources. *Int. J. Food Microbiol.*, **77**, 83–9.

Ayling, R.D., Woodward, M.J., Evans, S. and Newell, D.G. (1996) Restriction fragment length polymorphism of polymerase chain reaction products applied to the differentiation of poultry campylobacters for epidemiological investigations. *Res. Vet. Sci.*, **60**, 168–72.

Ayres, J.C., Ogilvy, W.S. and Stewart, G.F. (1950) Postmortem changes in stored meats. I. Microorganisms associated with the development of slime on eviscerated cut-up poultry. *Food Technol.*, **4**, 199–205.

Bailey, J.S. (1988) Integrated colonization control of *Salmonella* in poultry. *Poult. Sci.*, **67**, 928–32.

Bailey, J.S., Thomson, J.E., Cox, N.A. and Shackelford, A.D. (1986) Chlorine spray washing to reduce bacterial contamination of poultry processing equipment. *Poult. Sci.*, **65,** 1120–3.

Bailey, J.S., Cason, J.A. and Cox, N.A. (1992) Ecology and implications of Salmonella-contaminated hatching eggs, in *Proceedings XIX World's Poultry Congress*, Amsterdam, pp. 72–5.

Bailey, J.S., Cason, J.A. and Cox, N.A. (1998). Effect of *Salmonella* in young chicks on competitive exclusion treatment. *Poult. Sci.*, **77**, 394–9.

Bailey, J.S., Cox, N.A. and Berrang, M.E. (1994) Hatchery-acquired salmonellae in broiler chicks. *Poult. Sci.*, **73**, 1153–7.

Bailey, J.S., Buhr, R.J., Cox, N.A. and Berrang, M.E. (1996) Effect of hatching cabinet sanitation treatments on *Salmonella* cross-contamination and hatchability of broiler eggs. *Poult. Sci.*, **75**, 191–6.

Bailey, J.S., Lyon, B.G., Lyon, C.E. and Windham, W.R. (2000) The microbiological profile of chilled and frozen chicken. *J. Food Protect.*, **63**, 1228–30.

Bailey, J.S., Stern, N.J., Fedorka-Cray, P., Craven, S.E., Cox, N.A., Cosby, D.E., Ladely, S. and Musgrove, M.T. (2001). Sources and movement of *Salmonella* through integrated poultry operations: a multistate epidemiological investigation. *J. Food Protect.*, **64**, 1690–7.

Bailey, J.S., Cox, N.A., Craven, S.E., Cosby, D.E. (2002) Serotype tracking of Salmonella through integrated broiler chicken operations. *J. Food Protect.*, **65**, 742–5.

Balzer, J. (1993) Preharvest pathogen reduction efforts in Denmark, in *Proceedings of U.S. Department of Agriculture, Food Safety and Inspection Service*, World Congress on Meat and Poultry Inspection—1993, pp. A2.01–A2.04.

Barakat, R.K., Griffiths, M.W. and Harris, L.J. (2000) Isolation and characterization of *Carnobacterium*, *Lactococcus* and *Enterococcus* spp. from cooked, modified atmosphere packaged refrigerated poultry meat. *Int. J. Food Microbiol.*, **62**, 83–94.

Barnes, E.M. (1960) The sources of the different psychrophilic organisms on chilled eviscerated poultry, in *Proceedings 10th International Congress of Refrigeration*, Copenhagen **3**, 97–100.

Barnes, E.M. (1965) The effect of chlorinating chill tanks on the bacteriological condition of processed chicken. *Suppl. Bull. Inst. Int. Froid, Commission 4, Karlsruhe*, Annexe **1965-1**, 219–25.

Barnes, E.M. (1975) The microbiological problems of sampling a poultry carcass. Quality Poultry Meat, in *Proceedings of the 2nd European Symposium on Poultry Meat Quality*, Oosterbeek, Netherlands. pp. (23) 1–8.

Barnes, E.M. (1976) Microbiological problems of poultry at refrigerator temperatures—a review. *J. Sci. Food Agric.*, **27**, 777–82.

Barnes, E.M. (1979) The intestinal microflora of poultry and game birds during life and after storage. *J. Appl. Bacteriol.*, **46**, 407–19.

Barnes, E.M. and Impey, C.S. (1968) Psychrophilic spoilage bacteria of poultry. *J. Appl. Bacteriol.*, **3**, 97–107.

Barnes, E.M. and Impey, C.S. (1975) The shelf-life of uneviscerated and eviscerated chicken carcasses stored at 10°C and 4°C. *Br. Poul. Sci.*, **16**, 319–26.

Barnes, E.M. and Melton, W. (1971) Extracellular enzymic activity of poultry spoilage bacteria. *J. Appl. Bacteriol.*, **34**, 599–609.

Barnes, E.M. and Shrimpton, D.H. (1957) Causes of greening of uneviscerated poultry carcasses during storage. *J. Appl. Bacteriol.*, **20**, 273–85.

Barnes, E.M. and Shrimpton, D.H. (1968) The effect of processing and marketing procedures on the bacteriological condition and shelf life of eviscerated turkeys. *Br. Poult. Sci.*, **9**, 243–51.

Barnes, E.M., Mead, G.C. and Griffiths, N.M. (1973) The microbiology and sensory evaluation of pheasants hung at 5, 10 and 15°C. *Br. Poult. Sci.*, **14**, 229–40.

Barnes, E.M., Impey, C.S., Geeson, J.D. and Buhagiar, R.W.M. (1978) The effect of storage temperature on the shelf life of eviscerated air-chilled turkeys. *Br. Poult. Sci.*, **19**, 77–84.

Basker, D., Klinger, I., Lapidot, M. and Eisenberg, E. (1986) Effect of chilled storage of radiation-pasteurized chicken carcasses on the eating quality of the resultant cooked meat. *J. Food Technol.*, **21**, 437–41.

Bates, J.R. and Bodnaruk, P.W. (2003) *Clostridium perfringens*, in *Foodborne Microorganisms of Public Health Significance* (ed A.D. Hocking),6th edn, Australian Institute of Food Science and Technology Ltd (NSW Branch), Food Microbiol. Group, Waterloo DC, New South Wales, Australia, pp. 479–504.

Bean, N.H. and Griffin, P.M. (1990) Foodborne disease outbreaks in the United States, 1973–1987: pathogens, vehicles, and trends. *J. Food Protect.*, **53**, 804–17.

Beckers, H.J. (1988) Incidence of foodborne diseases in the Netherlands: annual summary 1982 and an overview from 1979 to 1982. *J. Food Protect.*, **51**, 327–34.

Behling, R.G. and Wong, A.C.L. (1994) Competitive exclusion of *Salmonella enteritidis* in chicks by treatment by a single culture plus dietary lactose. *Int. J. Food Microbiol.*, **22**, 1–9.

Bender, F.G. and Brotsky, E. (1992) Process for treating poultry carcasses to control salmonellae growth. U.S. patent No. 5,143,739.

Benedict, R.C., Schultz, F.J. and Jones, S.B. (1991) Attachment and removal of *Salmonella* spp. on meat and poultry tissues. *J. Food Safety*, **11**, 135–48.

Berndtson, E., Emanuelsson, U., Danielsson-Tham, M.-L. and Engvall, A. (1996) One year epidemiological study of campylobacters in eighteen Swedish chicken farms. *Prev. Vet. Med.*, **26**, 167–85.

Berner, H., Kluberger, A. and Bresse, M. (1969) Investigations into a new method of chilling poultry. III. Testing aspects of hygiene for the new method. *Fleischwirtschaft*, **49**, 1617–20, 23.

Berrang, M.E., Buhr, R.J. and Cason, J.A. (2000) *Campylobacter* recovery from external and internal organs of commercial broiler carcasses prior to scalding. *Poult. Sci.*, **79**, 286–90.

Berrang, M.E., Dickens, J.A. and Musgrove, M.T. (2000) Effects of hot water application after defeathering on the levels of *Campylobacter*, coliform bacteria and *Escherichia coli* on broiler carcasses. *Poult. Sci.*, **79**, 1689–93.

Berrang, M.E., Buhr, R.J., Cason, J.A. and Dickens, J.A. (2001) Broiler carcass contamination with *Campylobacter* from feces during defeathering. *J. Food Protect.*, **64**, 2063–66.

Berrang, M.E., Meinersmann, R.J., Northcutt, J.K. and Smith, D.P. (2002) Molecular characterization of *Listeria monocytogenes* isolated from a poultry further processing facility and from fully cooked product. *J. Food Protect.*, **65**, 1574–9.

Berri, C., Wacrenier, N., Millet, N. and Le Bihan-Duval, E. (2001) Effect of selection for improved body composition on muscle and meat characteristics of broilers from experimental and commercial lines. *Poult. Sci.*, **80**, 833–8.

Bhatia, T.R.S. and McNabb, G.D. (1980) Dissemination of *Salmonella* in broiler chicken operations. *Avian Dis.*, **24**, 616–24.

Bilgili, S.G., Jetton, J.F., Conner, D.E., Kotrola, J.S. and Moran, E.T. (1992) Microbiological quality of commercially processed broiler carcasses: The influence of fecal contamination during evisceration, in *Proceedings of Salmonella and Salmonellosis*, CNEVA Ploufragan/Saint-Brieuc, France, pp. 118.
Bisgaard, M. (1992) A voluntary *Salmonella* control programme for the broiler industry, implemented by the Danish Poultry Council. *Int. J. Food Microbiol.*, **15**, 219–24.
Blandford, T.B. and Roberts, T.A. (1970) An outbreak of botulism in broiler chicken. *Vet. Rec.*, **87**, 258–61.
Blankenship, L.C., Cox, N.A., Craven, S.E., Mercuri, A.J. and Wilson, R.L. (1975) Comparison of the microbiological quality of inspection-passed and fecal contamination-condemned broiler carcasses. *J. Food Sci.*, **40**, 1236–8.
Blankenship, L.C., Bailey, J., Cox, N.A., Rose, M., Dua, M., Berrang, M., Musgrove, M. and Wilson, R. (1990) Broiler carcass reprocessing reevaluated. *Poult. Sci.*, **69** (Suppl. 1), 158.
Blankenship, L.C., Bailey, J.S., Cox, N.A., Stern, N.J., Brewer, R. and Williams, O. (1993a) Two-step mucosal competitive exclusion flora treatment to diminish salmonellae in commercial broiler chickens. *Poult. Sci.*, **72**, 1667–72.
Blankenship, L.C., Bailey, J.S., Cox, N.A., Musgrove, M.T., Berrang, M.E., Wilson, R.L., Rose, M.J. and Dua, S.K. (1993b) Broiler carcass reprocessing, a further evaluation. *J. Food Protect.*, **56**, 983–5.
Board, R.G. (1969) Microbiology of the hen's egg. *Adv. Appl. Microbiol.*, **11**, 245–81.
Bohnsack, U., Knippel, G. and Höpke, H.-U. (1988) The influence of a CO_2 atmosphere on the shelf-life of fresh poultry. *Fleischwirtschaft*, **68**, 1553–7.
Boonchuvit, B. and Hamilton, P.B. (1975) Interaction of aflatoxin and paratyphoid infections in broiler chickens. *Poult. Sci.*, **54**, 1567–73.
Brake, J. and Sheldon, B.W. (1990) Effect of a quaternary ammonium sanitizer for hatching eggs on their contamination, permeability, water loss, and hatchability. *Poult. Sci.*, **69**, 517–25.
Brett, M.M. (1998) 1566 outbreaks of *Clostridium perfringens* food poisoning, 1970–1996, in *Proceedings of 4th World Congress on Foodborne Infections and Intoxications*, Berlin, *volume 1*, pp. 243–244.
Brett, M.M. and Gilbert, R.J. (1997) 1525 outbreaks of *Clostridium perfringens* food poisoning, 1970–1996. *Rev. Med. Microbiol.*, **8** (Supplement 1) S64–S65.
Brown, D. (1993) Evolution of world trade. *Broiler Ind.* **56**, 72, 74, 76.
Bryan, F.L. (1968) What the sanitarian should know about staphylococci and salmonellae in non-dairy products. I. Staphylococci, *J. Milk Food Technol.*, **31**, 110–6.
Bryan, F.L. (1980) Foodborne diseases in the United States associated with meat and poultry. *J. Food Protect.*, **43**, 140–50.
Bryan, F.L. and Kilpatrick, E.G. (1971) *Clostridium perfringens* related to roast beef cooking, storage, and contamination in a fast food service restaurant. *Am. J. Public Health*, **61**, 1869–85.
Bryan, F.L., Ayres, J.C. and Kraft, A.A. (1968a) Contributory sources of salmonellae on turkey products. *Am. J. Epidemiol.*, **87**, 578–91.
Bryan, F.L., Ayres, J.C. and Kraft, A.A. (1968c) Destruction of salmonellae and indicator organisms during thermal processing of turkey rolls. *Poult. Sci.*, **47**, 1966–78.
Brynestad, S. and Granum, P.E. (2002) *Clostridium perfringens* and foodborne infections. *Int. J. Food Microbiol.*, **74**, 195–202.
Burton, B.T. (1976) *Human Nutrition*, 3rd ed, McGraw-Hill, New York.
Buses, H. and Thompson, L. (2003) Dip application of phosphates and marinade mix on shelf life of vacuum-packaged chicken breast fillets. *J. Food Protect.*, **66**, 1701–03.
Butzler, J.P. and Oosterom, J. (1991) *Campylobacter*: pathogenicity and significance in foods. *Int. J. Food Microbiol.*, **12**, 1–8.
Byrd, J.A., Corrier, D.E., Hume, M.E., Bailey, R.H., Stanker, L.H. and Hargis, B.M. (1998) Incidence of *Campylobacter* in crops of preharvest market-age broiler chickens. *Poult. Sci.*, **77**, 1303–05.
Byrd, J.A., Hargis, B.M., Caldwell, D.J., Bailey, R.H., Herron, K.L., McReynolds, J.L., Brewer, R.L., Anderson, R.C., Bischoff, K.M., Callaway, T.R. and Kubena, L.F. (2001) Effect of lactic acid administration in the drinking water during preslaughter feed withdrawal on *Salmonella* and *Campylobacter* contamination on broilers. *Poult. Sci.*, **80**, 278–83.
Camarda, A., Newell, D.G., Nasti, R. and Di Modugno, G. (2000) Genotyping *Campylobacter jejuni* strains isolated from the gut and oviduct of laying hens. *Avian Dis.*, **44**, 907–12.
Campbell, S., Duckworth, S., Thomas, C.J. and McMeekin, T.A. (1987) A note on adhesion of bacteria to chicken muscle connective tissue. *J. Appl. Bacteriol.*, **63**, 67–71.
Carr, L., Rigakos, C., Carpenter, G., Berney, G. and Joseph, S. (1999) An assessment of livehaul poultry transport container decontamination. *Dairy, Food Environ. Sanitation*, **19**, 753–9.
Cason, J.A., Bailey, J S., Stern, N.J., Whittemore, A.D. and Cox, N.A. (1997) Relationship between aerobic bacteria, salmonellae and *Campylobacter* on broiler carcasses. *Poult. Sci.*, **76**, 1037–41.
Cason, J.A., Hinton, A. and Ingram, K.D. (2000) Coliform, *Escherichia coli* and salmonellae concentrations in a multiple-tank, counterflow poultry scalder. *J. Food Protect.*, **63**, 1184–8.
CDC (Centers for Disease Control and Prevention). (1999a) Incidence of foodborne illnesses: preliminary data from the foodborne diseases active surveillance network (FoodNet)—United States, 1998. *Morb. Mortal. Wkly Rep.*, **48**, 189–95.
CDC (Centers for Disease Control and Prevention). (1999b) Update: multistate outbreak of listeriosis—United States, 1998–1999. *Morb. Mortal. Wkly Rep.*, **47**, 1117–8.
CDC (Centers for Disease Control and Prevention). (2000) Multistate outbreak of listeriosis—United States, 2000. *Morb. Mortal. Wkly Rep.*, **49**, 1129–1130.
CDC (Centers for Disease Control and Prevention). (2002) Outbreak of listeriosis—Northeastern United States, 2002. *Morb. Mortal. Wkly Rep.*, **51**, 950–951.

CFIA (Canadian Food Inspection Agency). (2000) Canadian microbiological baseline survey of chicken broiler and young turkey carcasses, June 1997–May 1998. Canadian Food Inspection Agency, Food of Animal Origin Division. http://www.cfia-acia.agr http://www.cfia-acia.agr

CFR (1992) Temperatures and chilling and freezing requirements. General chilling requirements. *Code of Federal Regulations. Animals and Animal Products*. 9CFR381.66b2. U.S. Government Printing Office, Washington, pp. 435.

Chadfield, M., Skov, M., Christensen, J., Madsen, M. and Bisgaard, M. (2001) An epidemiological study of *Salmonella enterica* serovar 4, 12:b:—in broiler chickens in Denmark. *Vet. Microbiol.*, **82**, 233–47.

Chan, K.F., Tran, H.L., Kanenaka, R.Y. and Kathariou, S. (2001) Survival of clinical and poultry-derived isolates of *Campylobacter jejuni* at a low temperature (4°C). *Appl. Environ. Microbiol.*, **67**, 4186–91.

Chambers, J.R., Spencer, J.L. and Modler, H.W. (1997) The influence of complex carbohydrates on *Salmonella typhimurium* colonization, pH and density of broiler ceca. *Poult. Sci.*, **76**, 445–51.

Chen, H-C. and Stern, N.J. (2001) Competitive exclusion of heterologous *Campylobacter* spp. in chicks. *Appl. Environ. Microbiol.*, **67**, 848–51.

Cherrington, C.A., Board, R.G. and Hinton, M. (1988) Persistence of *Escherichia coli* in a poultry processsing plant. *Lett. Appl. Microbiol.*, **7**, 141–3.

Christensen, B., Sommer, H., Rosenquist, H. and Nielsen, N. (2001) Risk assessment on *Campylobacter jejuni* in chicken products, First edn, January 2001. The Danish Veterinary and Food Administration. http://www.foedevaredirektoratet.dk/Foedevare/Mikrobiologiske_forureninger/Campylobacter/Forside.htm

Clark, A.G. and Bueschkens, D.H. (1998) Horizontal spread of human and poultry-derived strains of *Campylobacter jejuni* among broiler chicks held in incubators and shipping boxes. *J. Food Protect.*, **51**, 438–41.

Clark, D.S. (1968) Growth of psychrotolerant pseudomonads and *Achromobacter* on chicken skin. *Poult. Sci.*, **47**, 1575–78.

Clark, D.S. (1970) Growth of pschrotolerant pseudomonads and achromobacter on various chicken tissues. *Poult. Sci.*, **49**, 1315–18.

Clark, D.S. and Lentz, C.P. (1969) Microbiological studies in poultry processing plants in Canada. *Can. Inst. Food Technol. J.*, **2**, 33–6.

Clifton-Hadley, F.A., Breslin, M., Venables, L.M., Sprigings, K.A., Cooles, S.W., Houghton, S. and Woodward, M.J. (2002) A laboratory study of Salenvac®T, an inactivated and bivalent iron restricted *Salmonella enterica* serovar Enteritidis and Typhimurium dual vaccine, against Typhimurium challenge in poultry, in *Salmonella & Salmonellosis 2002, Proceedings*, May 29–31, Saint-Brieuc (eds P. Colin and G. Clement), ISPAIA-ZOOPOLE development, Ploughfragan, France, pp. 617–618.

Clifton-Hadley, F.A., Dibb-Fuller, M.P., Venables, L.M., Sprigings, K.A., Venables, L.M., Cooles, S.W., Osborn, M.K., Houghton, S. and Woodward, M.J. (2002) Efficacy of Salenvac®T, an inactivated and bivalent iron restricted *Salmonella enterica* serovar Enteritidis and Typhimurium dual vaccine, in reducing other Salmonella group B infections in poultry, in *Salmonella & Salmonellosis 2002, Proceedings*, May 29–31, Saint-Brieuc (eds P. Colin and G. Clement), ISPAIA-ZOOPOLE development, Ploughfragan, France, pp. 619–620.

Codex (Codex Alimentarious Commission). (2002) Discussion paper on risk management strategies for Salmonella spp. in poultry. CX/FH 03/5-Add.1, November 2002, Joint FAO/WHO Food Standards Programme, Codex Committee on Food Hygiene, Food and Agriculture Organization of the United Nations, Rome.

Codex (Codex Alimentarious Commission). (2003) Discussion paper on risk management strategies for Campylobacter spp. in poultry. CX/FH 02/X—Joint FAO/WHO Food Standards Programme, Codex Committee on Food Hygiene, Food and Agriculture Organization of the United Nations, Rome.

Coleby, B., Ingram, M. and Shepherd, H.J. (1960) Treatment of meats with ionizing radiations. III. Radiation pasteurization of whole, eviscerated chicken carcasses. *J. Sci. Food Agric.*, **11**, 61–71.

Commission of the European Communities. (1976) *Evaluation of the Hygienic Problems Related to the Chilling of Poultry Carcasses*, Information on Agric. No. 22. EEC, Brussels.

Cornforth, D.P. (1991) Methods for identification and prevention of pink color in cooked meat, in *44th Am. Recip. Meat Conf.* (no editor), National Livestock and Meat Board, Chicago, IL, pp. 53–58.

Cornforth, D.P., Rabovitser, J.K., Ahuja, S., Wagner, J.C., Hanson, R., Cummings, B. and Chudnovsky, Y. (1998) Carbon monoxide, nitric oxide and nitrogen dioxide levels in gas ovens related to surface pinking of cooked beef and turkey. *J. Agric. Food Chem.*, **46**, 255–61.

Corry, J.E.L. and Atabay, H.I. (2001) Poultry as a source of *Campylobacter* and related organisms. *J. Appl. Microbiol.*, **90**, 96S–114S.

Corry, J.E.L., Allen, V.M., Hudson, W.R., Breslin, M.F. and Davies, R.H. (2002) Sources of salmonella on broiler carcasses during transportation and processing: modes of contamination and methods of control. *J. Appl. Microbiol.*, **92**, 424–32.

Cox, N.A. and Bailey, J.S. (1987) Pathogens associated with processed poultry, in *The Microbiology of Poultry Meat Products* (eds F.E. Cunningham and N.A.Cox), Academic Press, Inc., New York, pp. 293–316.

Cox, N.A., Bailey, J.S., Mauldin, J.M. and Blankenship, L.C. (1990) Presence and impact of *Salmonella* contamination in commercial broiler hatcheries. *Poult. Sci.*, **69**, 1606–09.

Cox, N.A., Berrang, M.E. and Cason, J.A. (2000) *Salmonella* penetration of egg shells and proliferation in broiler hatching eggs—a review. *Poult. Sci.*, **79**, 1571–74.

Cox, N.A., Berrang, M.E. and Mauldin, J.M. (2001) Extent of salmonellae contamination in primary breeder hatcheries in 1998 as compared to 1991. *J. Appl. Poult. Res.*, **10**, 202–5.

Cox, N.A., Stern, N.J., Hiett, K.L and Berrang, M.E. (2002) Identification of a new source of *Campylobacter* contamination in poultry: transmission from breeder hens to broiler chickens. *Avian Dis.*, **46**, 535–41.

Cross, H.R. (1992) Irradiation of poultry products. 9CFR Part 381. *Fed. Reg.*, **57**, 43588–600.
DANMAP 2001 (2002). Use of antimicrobial agents and occurrence of antimicrobial resistance in bacteria from food animals, foods and humans in Denmark. ISSN 1600–2032. (http://www.vetinst.dk www.vetinst.dk) (G-1606).
DANMAP (2003) Danish Integrated Antimicrobial resistance Monitoring and Research Programme. http://www.vetinst.dk/high_uk.asp?page_id=180.
Davies, R.H. and Wray, C. (1994) An approach to reduction of salmonella infection in broiler chicken flocks through intensive sampling and identification of cross-contamination hazards in commercial hatcheries. *Int. J. Food Microbiol.*, **24**, 147–60.
Davies, R., Breslin, M., Corry, J.E.L., Hudson, W. and Allen, V.M. (2001) Observations on the distribution and control of salmonellae species in two integrated broiler companies. *Vet. Rec.*, **149**, 227–32.
Davies, R. and Bedford, S. (2002) Intensive investigation of *Salmonella* contamination in a poultry processing plant, in *Salmonella & Salmonellosis 2002, Proceedings*, May 29–31, Saint-Brieuc (eds P. Colin and G. Clement), ISPAIA-ZOOPLE development, Ploughfragan, France, pp. 275–282.
Dawson, P.S. (1992) Control of *Salmonella* in poultry in Great Britain. *Int. J. Food Microbiol.*, **15**, 21–7.
DEFRA (Department for Environment, Food and Rural Affairs). (1997) *Code of Practice for the Prevention and Control of Salmonella in Chickens reared for meat on farm.* Scottish Executive Environment and Rural Affairs Department, DEFRA Publications, London.
Denton, J.H. and Gardner, F.A. (1982) Effect of further processing systems on selected microbiological attributes of turkey meat products. *J. Food Sci.*, **47**, 21–7.
Dickson, J.S. and Anderson, M.E. (1992) Microbiological decontamination of food animal carcasses by washing and sanitizing systems:a review. *J. Food Protect.*, **55**, 133–40.
Dodd, C.E.R., Chaffey, B.J. and Waites, W.M. (1988) Plasmid profiles as indicators of the source of contamination of *Staphylococcus aureus* endemic within poultry processing plants. *Appl. Environ. Microbiol.*, **54**, 1541–9.
Dromigny, E., Vachine, I. and Jouve, J.L. (1985) *Campylobacter* in turkey hens at the slaughterhouse:contamination during various steps in processing. *Rev. Med.*, **136**, 713–20.
Duim, B., Ang, C.W., van Belkum, A., Rigter, A., van Leeuwen, N.W.J., Endtz, H.P. and Wagenaar, J.A. (2000) Amplified fragment length polymorphism analysis of *Campylobacter jejuni* strains isolated from chickens and from patients with gastroenteritis or Guillain-Barré or Miller Fisher syndrome. *Appl. Environ. Microbiol.*, **66**, 3917–23.
Duke, G.E., Basha, M. and Noll, S. (1997) Optimum duration of feed and water removal prior to processing in order to reduce the potential for fecal contamination in turkeys. *Poult. Sci.*, **76**, 516–22.
EARSS (European Antimicrobial Resistance Surveillance System). (2003) http://www.earss.rivm.nl/.
EC (European Commission) (2002). Trends and Sources of Zoonotic Agents in Animals, Feedstuffs, Food and Man in the European Union and Norway in 2000 to the European Commission in accordance with Article 5 of the Directive 92/117/EEC, prepared by the Community Reference Laboratory on the Epidemiology of Zoonoses, BgVV, Berlin, Germany. Working document SANCO/927/2002, Part 1.
Edel, W. (1994) *Salmonella enteritidis* eradication programme in poultry breeder flocks in The Netherlands. *Int. J. Food Microbiol.*, **21**, 171–8.
Effler, P., Ieong, M.-C., Kimura, A., Nakata, M., Burr, R., Cremer, E. and Slutsker, L. (2001) Sporadic *Campylobacter jejuni* infections in Hawaii: associations with prior antibiotic use and commercially prepared chicken. *J. Infect. Dis.*, **183**, 1152–5.
Eisgruber, H. and Stolle, A. (1992) *Salmonella* in commercial chickens cooled by the air/water spray system, in *Proceedings 3rd World Congress*. Foodborne Infections and Intoxications, June 16–19, *Volume 1*, Berlin, Robert von Ostertag-Institut (BGA), Berlin, pp. 327–330.
El-Assaad, F.G., Stewart, L.E. and Carr, L.E. (1990) Disinfection of poultry transport cages. Presented at the Am. Soc. Agric. Eng., Columbus, Ohio, June 24–27, Paper no. 90–6015 (24 pages).
El-Assaad, F.G., Stewart, L.E., Mallinson, E.T., Carr, L.E., Joseph, S.W. and Berney, G. (1993) Decontamination of poultry transport cages. Presented at the Am. Soc. Agric. Eng., Spokane, Washington, June 20–23, Paper no. 933010 (26 pages).
Elissalde, M.H., Ziprin, R.L., Huff, W.E., Kubena, L.F. and Harvey, R.B. (1994) Effect of ochratoxin A on *Salmonella*-challenged broiler chicks. *Poult. Sci.*, **73**, 1241–8.
Elliott, R.P. and Straka, R.P. (1964) Rate of microbial deterioration of chicken meat at 2°C after freezing and thawing. *Poult. Sci.*, **43**, 81–6.
Essary, E.O., Moore, W.E.C. and Kramer, C.Y. (1958) Influence of scald temperatures, chill times, and holding temperatures on the bacterial flora and shelf-life of freshly chilled, tray-packed poultry. *Food Technol.*, **12**, 684–7.
Fahey, J.E. (1955) Some observations on "air sac" infection in chickens. *Poult. Sci.*, **34**, 982–4.
FAOSTAT. (2002) FAOSTAT Agriculture Data. Food and Agriculture Organization of the United Nations, Rome. http://apps.fao.org/page/collections?subset=agriculture
FAO/IAEA/WHO (1999) Report of a Joint Study Group on "High-Dose Irradiation: Wholesomeness of Food Irradiated with Doses above 10 kGy", WHO Technical Report Series, no. 890, WHO, Geneva.
FAO/WHO (Food and Agriculture Organization of the United Nations/ World Health Organization) (2004a) Risk Assessment of Listeria monocytogenes in ready-to-eat foods. *Interpretative Summary. Microbiological Risk Assessment Series No. 4.* Food and Agriculture Organisation, Rome and World Health Organisation, Geneva.
FAO/WHO (2004b) Risk Assessment of *Listeria monocytogenes* in Ready-to-Eat Foods, Technical Report. *Microbiological Risk Assessment Series No. 5*, Food and Agriculture Organisation, Rome and World Health Organisation, Geneva.
FDA (2001) Risk assessment on the human health impact of fluoroquinolone resistant *Campylobacter* Associated with the consumption of chicken, October 18, 2000, revised January 5, 2001.

FDA–FSIS (Food and Drug Administration–Food Safety and Inspection Service) (2003) Quantitative assessment of the relative risk to public health from foodborne *Listeria monocytogenes* among selected categories of ready-to-eat foods. Food and Drug Administration, Center for Science and Applied Nutrition, College Park, Maryland.

Ferreira, A.J.P., Ferreira, C.S.A., Knobl, T., Moreno, A.M., Bacarro, M.R., Chen, M., Robach, M. and Mead, G.C. (2003) Comparison of three commercial competitive-exclusion products for controlling *Salmonella* colonization of broilers in Brazil. *J. Food Protect.*, **66**, 490–2.

Finlayson, M. (1978) Salmonellae in Alberta poultry products and their significance in human infection. *Proc. Int. Symp. Salmonella Prospects Control*, Univ. Guelph, pp. 156–180.

Fitzgerald, C., Stanley, K., Andrew, S. and Jones, K. (2001) Use of pulsed-field gel electrophoresis and flagellin gene typing in identifying clonal groups of *Campylobacter jejuni* and *Campylobacter coli* in farm and clinical environments. *Appl. Environ. Microbiol.*, **67**, 1429–36.

Fluckey, W.M., Sanchez, M.X., McKee, S.R., Smith, D., Pendelton, E. and Brashears, M.M. (2003) Establishment of a microbiological profile for an air-chilling poultry operation in the United States. *J. Food Protect.*, **66**, 272–9.

Food and Agriculture Organization/International Atomic Energy Agency/World Health Organization (FAO/IAEA/WHO) Expert Committee. (1977) Wholesomeness of Irradiated Food, *Tech. Rep. Ser.*, No. 604. WHO, Geneva.

Frenzen, P.D., DeBess, E.E., Hechemy, K.E., Kassenborg, H., Kennedy, M., McCombs, K. and McNees, A. (2001) Consumer acceptance of irradiated meat and poultry in the United States. *J. Food Protect.*, **64**, 2020–6.

Friedman, C., Reddy, S., Samual, M., Marcus, R., Bender, J., Desai, S., Shiferaw, B., Helfrick, D., Carter, M., Anderson, B., Hoekstra, M. and the EIP Working Group. (2000) Risk factors for sporadic *Campylobacter* infections in the United States: a case-control study on FoodNet sites. Presented at the Int. Conference on Emerging Infectious Diseases, Atlanta, GA, July 16–19, 2000. http://www.cdc.gov/foodnet/pub/iceid/2000/friedman_c.htm

Frost, J.A., Oza, A.N., Thwaites, R.T. and Rowe, B. (1998) Serotyping scheme for *Campylobacter jejuni* and *Campylobacter coli* based on direct agglutination of heat-stable antigens. *J. Clin. Microbiol.*, **36**, 335–9.

Galton, M.M., Mackel, D.C., Lewis, A.L., Haire, W.C. and Hardy, A.V. (1955) Salmonellosis in poultry and poultry processing plants in Florida. *Am. J. Vet. Res.*, **16**, 132–7.

Genigeorgis, C., Hassuneh, M. and Collins, P. (1986) *Campylobacter jejuni* infection on poultry farms and its effect on poultry meat contamination during slaughtering. *J. Food Protect.*, **49**, 895–903.

Gibbs, P.A., Patterson, J.T. and Harvey, J. (1978a) Biochemical characteristics and enterotoxigenicity of *Staphylococcus aureus* strains isolated from poultry. *J. Appl. Bacteriol.*, **44**, 57–74.

Gibbs, P.A., Patterson, J.T. and Thompson, J.K. (1978b) The distributin of *Staphylococcus aureus* in a poultry processing plant. *J. Appl. Bacteriol.*, **44**, 401–10.

Giessen A.W., Bloemberg, B.P. va de, Ritmeester, W.S. and Tilburg, J.J. (1996) Epidemiological study on risk factors and risk reducing measures for Campylobacter infections in Dutch broiler flocks. *Epidemiol. Infect.*, **117**, 245–50.

Gilbert, R.J., Miller, K.L. and Roberts, D. (1989) *Listeria monocytogenes* and chilled foods. *Lancet*, **i**, 383–4.

Gibbons, J.C., Pascoe, S.J.S., Evans, S.J., Davies, R.H. and Sayers, A.R. (2001) A trial of biosecurity as a means to control *Campylobacter* infection of broiler chickens. *Prev. Vet. Med.*, **48**, 85–99.

Gibson, A.M. and Roberts, T.A. (1986) The effect of pH, sodium chloride, sodium nitrite and storage temperature on the growth of *Clostridium perfringens* and fecal streptococci in laboratory media. *Int. J. Food Microbiol.*, **3**, 195–210.

Gill, C.O. (1986) The control of microbial spoilage in fresh meats, in *Meat and Poultry Microbiology* (eds A.M. Pearson and T.R. Dutson), Advances in Meat Research, *Volume 2*, AVI Publishing Co., Inc., Westport, CT, pp. 49–88.

Gill, C.O. and Harris, L.M. (1984) Hamburgers and broiler chickens as potential sources of human *Campylobacter* enteritis. *J. Food Protect.*, **47**, 96–9.

Gill, C.O., Harrison, J.C.L. and Penney, N. (1990) The storge life of chicken carcasses packaged under carbon dioxide. *Int. J. Food Microbiol.*, **11**, 151–8.

Göksoy, E.O., James, C., Corry, J.E.L. and James, S.J. (2001) The effect of hot water immersions on the appearance and microbiological quality of skin-on chicken breast pieces. *Int. J. Food Sci. Technol.*, **36**, 61–9.

Gombas, D.E., Chen, Y., Clavero, R.S. and Scott, V.N. (2003) Survey of *Listeria monocytogenes* in ready-to-eat foods. *J. Food Protect.*, **66**, 559–69.

Gradel, K.O. (2002) Heat treatment of persistently Salmonella infected poultry houses, in *Salmonella & Salmonellosis 2002, Proceedings*, May 29–31, Saint-Brieuc (eds P. Colin and G. Clement), ISPAIA-ZOOPOLE development, Ploughfragan, France, pp. 595–595.

Grandison, A.S. and Jennings, A. (1993) Extension of the shelf life of fresh minced chicken meat by electron beam irradiation combined with modified atmosphere packaging. *Food Control*, **4**, 83–8.

Green, S.S. (1987) Results of a national survey:*Salmonella* in broilers and overflow chill tank water, 1982–1984. Science Division, Food Safety and Inspection Service, U.S. Department of Agriculture, Washington, DC.

Green, S.S. (1993) Personal communication. Food Safety and Inspection Service, U.S. Department of Agriculture, Washington, DC.

Greene, J. (1992) FSIS Backgrounder. FSIS permits trisodium phosphate in poultry plants. U.S. Department of Agriculture, Food Safety and Inspection Service, Washington, DC.

Grey, T.C. and Mead, G.C. (1986) The effects of air and water chilling on the quality of poultry carcasses, in *Meat Chilling*, 1986, Bristol, September 10–12, Int. Inst. Refrigeration, Paris, pp. 95–99.

Griffiths, N. (1975) Sensory evaluation of pheasants hung at 10°C for up to 15 days. *Br. Poult. Sci.*, **16**, 8–7.

Gruber, H. and Köfer, J. (2002) Results of five years *Salmonella* prevention in parent flocks in Styria (Austria), in *Salmonella &*

Salmonellosis 2002, Proceedings, May 29–31, Saint-Brieuc (eds P. Colin and G. Clement), ISPAIA-ZOOPOLE development, Ploughfragan, France, pp. 589–590.
Gunderson, M.F. (1962) Mold problem in frozen foods, in Proc. Low Temperature Microbiology Symposium—(1961) Camden, New Jersey. Campbell Soup Company, Camden, pp. 299–312.
Gunderson, M.F. and Rose, K.D. (1948) Survival of bacteria in a precooked fresh-frozen food. *Food Res.*, **13**, 254–63.
Hajek, V. and Marsalek, E. (1973) The occurrence of enterotoxigenic *Staphylococcus aureus* strains in hosts of different animal species. *Zbl. Bakt. Hyg., I. Abt. Orig.*, **A223**, 63–8.
Hall, H.E. and Angelotti, R. (1965) *Clostridium perfringens* in meat and meat products. *Appl. Microbiol.*, **13**, 352–7.
Hänninen, M-L., Perko-Mäkelä, P., Pitkälä, A. and Rautelin, H. (2000) A three-year study of *Campylobacter jejuni* genotypes in humans with domestically acquired infections and in chicken samples from the Helsinki area. *J. Clin. Microbiol.*, **38**, 1998–2000.
Hart, C.D., Mead, G.C. and Norris, A.P. (1991) Effects of gaseous environment and temperature on the storage behaviour of *Listeria monocytogenes* on chicken breast meat. *J. Appl. Bacteriol.*, **70**, 40–6.
Hayes, J.R., Carr, L.E., Mallinson, E.T., Douglass, L.W. and Joseph, S.W. (2000) Characterization of the contribution of water activity and moisture content to the population distribution of *Salmonella* spp. in commercial poultry houses. *Poult. Sci.*, **79**, 1557–61.
Henken, A.M., Frankena, K., Goelema, J.O., Graat, E.A.M. and Noordhuizen, J.P.T.M. (1992) Multivariate epidemiological approach to salmonellosis in broiler breeder flocks. *Poult. Sci.*, **71**, 838–843.
Hernandez, J. (1993) Incidence and control of *Campylobacter* in foods. *Microbiologia SEM*, **9**, 57–65.
Heyndrickx, M., Vandekerchove, D., Herman, L., Rollier, I., Grijspeerdt, K. and de Zutter, L. (2002) Routes for salmonella contamination of poultry meat: epidemiological study from hatchery to slaughterhouse. *Epidemiol. Infect.*, **129**, 253–65.
Hiett, K.L., Stern, N.J., Fedorka-Cray, P., Cox, N.A., Musgrove, M.T. and Ladely, S. (2002) Molecular subtype analyses of *Campylobacter* spp. from Arkansas and California poultry operations. *Appl. Environ. Microbiol.*, **68**, 6220–36.
Hinton, A., Buhr, R.J. and Ingram, K.D. (2000) Reduction of *Salmonella* in the crop of broiler chickens subjected to feed withdrawal. *Poult. Sci.*, **79**, 1566–70.
Hinton, A., Cason, J.A. and Ingram, K.D. (2002) Enumeration and identification of yeasts associated with commercial poultry processing and spoilage of refrigerated broiler carcasses. *J. Food Protect.*, **65**, 993–8.
Hinton, M.H., Al-Chalaby, Z.A.M. and Hinton, A.H. (1987) Field and experimental investigations into the epidemiology of *Salmonella* infections in broiler chickens, in *Elimination of pathogens from meat and poultry* (ed F.J.M. Smulders), Elsevier Science Publishers, Amsterdam, The Netherlands, pp. 27–37.
Hirn, J., Nurmi, E., Johansson, T. and Nuotio, L. (1992) Long-term experience with competitive exclusion and salmonellas in Finland. *Int. J. Food Microbiol.*, **15**, 281–5.
Hoeting, A.L. (1990) Irradiation in the production, processing and handling of food. 21CFR Part 179. *Fed. Reg.*, **55**, 18538–18544.
Hollister, A.G., Corrier, D.E., Nisbet, D.J., Beier, R.C. and DeLoach, J.R. (1994) Effect of cecal cultures lyophilized in skim milk or reagent 20 on *Salmonella* colonization in broiler chicks. *Poult. Sci.*, **73**, 1409–16.
Hoop, R. and Ehrsam, H. (1987) A study of the epidemiology of *Campylobacter jejuni* and *Campylobacter coli* in broiler production. *Schweiz. Arch. Tierheilk.*, **129**, 193–203.
Humphrey, T.J. (1981) The effects of pH and levels of organic matter on the death rates of Salmonellas in chicken scald-tank water. *J. Appl. Bacteriol.*, **51**, 27–39.
Humphrey, T.J. (1994) Contamination of egg shell and contents with *Salmonella enteritidis*: a review. *Int. J. Food Microbiol.*, **21**, 31–40.
Humphrey, T.J., Lanning, D.G. and Leeper, D. (1984) The influence of scald water pH on the death rates of *Salmonella typhimurium* and other bacteria attached to chicken skin. *J. Appl. Bacteriol.*, **57**, 355–9.
Humphrey, T.J., Henley, A. and Lanning, D.G. (1993) The colonization of broiler chickens with *Campylobacter jejuni*: some epidemiological investigations. *Epidemiol. Infect.*, **110**, 601–7.
IAEA (1968) *Elimination of Harmful Organisms from Food and Feed by Irradiation*, STI/PUB/200. Int. Atomic Energy, Vienna.
ICGFI (1992) Ninth meeting of the Int. consultative group on food irradiation. Inventory of product clearances. *Int.* Consultative Group on Food irradiation. Joint FAO/IAEA Division, Atomic Energy Agency, Vienna.
ICMSF (1988) *Microorganisms in Foods. 4. Application of the Hazard Analysis Critical Control Point System to Ensure Microbiological Safety and Quality*, Blackwell Scientific Publications, London.
ICMSF (1994) Choice of sampling plan and criteria for *Listeria monocytogenes*. Int. Commission on Microbiological Criteria for Foods. *Int. J. Food Microbiol.*, **22**, 89–96.
Isigidi, B.K., Mathieu, A.- M., Devriese, L.A., Godard, C. and van Hoof, J. (1992) Enterotoxin production in different *Staphylococcus aureus* biotypes isolated from food and meat plants. *J. Appl. Bacteriol.*, **72**, 16–20.
Ismail, S.A.S., Deak, T., Abd El-Rahman, H.A., Yassein, M.A.M. and Beuchat, L.R. (2000) Presence and changes in populations of yeasts on raw and processed poultry products stored at refrigeration temperature. *Int. J. Food Microbiol.*, **62**, 113–21.
Israel Official Gazette. (1982) Public health ordinance. Preservation of foodstuffs (amended). *Collection of Regulations*, **4354**, 1073.
Izat, A.L., Colberg, M., Thomas, R.A., Adams, M.H. and Driggers, C.D. (1990a) Effects of lactic acid in processing waters on the incidence of salmonellae on broilers. *J. Food Quality*, **13**, 295–306.
Izat, A.L., Hierholzer, R.E., Kopek, J.M., Adams, M.H., Reiber, M.A. and McGinnis, J.P. (1990b) Effects of D-mannose on incidence and levels of salmonellae in ceca and carcass samples of market age broilers. *Poult. Sci.*, **69**, 2244–7.

Jacobs-Reitsma, W. (1997) Aspects of epidemiology of *Campylobacter* in poultry. *Vet. Q.*, **19**, 113–7.

Jacobs-Reitsma, W.F., Bolder, N.M. and Mulder, R.W.A.W. (1994) Cecal carriage of *Campylobacter* and *Salmonella* in Dutch broiler flocks at slaughter: a one-year study. *Poult. Sci.*, **73**, 1260–6.

James, W.O., Brewer, R.L., Prucha, J.C., Williams, W.O. and Parham, D.R. (1992) Effects of chlorination of chill water on the bacteriologic profile of raw chicken carcasses and giblets. *J. Am. Vet. Med. Assoc.*, **200**, 60–3.

Jeffrey, J.S., Atwill, E.R. and Hunter, A. (2001) Farm and management variables linked to fecal shedding of *Campylobacter* and *Salmonella* in commercial squab production. *Poult. Sci.*, **80**, 66–70.

Jimenéz, S.M., Tiburzi, M.C., Salsi, M.S., Pirovani, M.E. and Moguilevsky, M.A. (2003) The role of visible faecal material as a vehicle for generic *Escherichia coli*, coliform, and other enterobacteria contaminating poultry carcasses during slaughtering. *J. Appl. Microbiol.*, **95**, 451–6.

Johnson, P.L., Baker, T., Getz, M., Lynch, J. and Brodsky, M. (1992) Health risk assessment of undrawn (New York dressed) poultry in Ontario, in *Proc. 3rd World Congress Foodborne Infections and Intoxications, volume 1*, pp. 252–255.

Jones, J.M., Mead, G.C., Griffiths, N.M. and Adams, B.W. (1982) Influence of packaging on microbiological, chemical and sensory changes in chill-stored turkey portions. *Br. Poult. Sci.*, **23**, 25–40.

Jones, F.T., Axtell, R.C., Rives, D.V., Scheideler, S.E., Tarver, F.R., Walker, R.L. and Wineland, M.J. (1991a) A survey of *Salmonella* contamination in modern broiler production. *J. Food Protect.*, **54**, 502–7.

Jones, F.T., Axtell, R.C. Rivers, D.V., Scheideler, S.E., Tarver, F.R., Walker, R.L. and Wineland, M.J. (1991b) A survey of *Campylobacter jejuni* contamination in modern broiler production and processing systems. *J. Food Protect.*, **54**, 259–62.

Jordan, P. (1991) On line chilling for poultry. *Meat Int.*, **1**, 66–7.

Jul, M. (1986) Chilling broiler chicken: an overview, in *Meat Chilling 1986*, Bristol, September 10–12, Int. Institute of Refrigeration, Paris, pp. 83–93.

Juneja, V.K. and Marmer, B.S. (1996) Growth *Clostridium perfringens* from spore inocula in *sous-vide* turkey products. *Int. J. Food Protect.*, **32**, 115–23.

Kahan, R.S. and Howker, J.J. (1978) Low-dose irradiation of fresh non-frozen chicken and other preservation methods for shelf-life extension and for improving its public-health quality, in *Food Preservation by Irradiation, Volume 2*, Int. Atomic Energy Agency, Vienna, pp. 221–242.

Kalinowski, R.M. and Tompkin, R.B. (1999). Psychrotrophic clostridia causing spoilage in cooked meat and poultry products. *J. Food Protect.*, **62**, 766–72.

Kalinowski, R.M., Tompkin, R.B., Bodnaruk, P.W. and Pruett, W.P. (2003) Impact of cooking, cooling, and subsequent refrigeration on the growth or survival of *Clostridium perfringens* in cooked meat and poultry products. *J. Food Protect.*, **66**, 1227–32.

Kamat, A.S., Alur, M.D., Nerkar, D.P. and Nair, P.M. (1991) Hygienization of Indian chicken meat by ionizing radiation. *J. Food Safety*, **12**, 59–71.

Kampelmacher, E.H. (1984) Irradiation of food. A new technology for perserving and ensuring the hygiene of foods. *Fleischwirtschaft*, **64**, 322–7.

Kapperud, G., Skjerve, E., Bean, N.H., Ostroff, S.M. and Lassen, J. (1992) Risk factors for sporadic *Campylobacter* infections: results of a case-control study in Southeastern Norway. *J. Clin. Microbiol.*, **30**, 3117–21.

Kapperud, G., Skjerve, E., Vik, L., Hauge, K., Lysaker, A., Aalmen, I., Osteroff, S.M. and Potter, M. (1993) Epidemiogical investigation of risk factors for Campylobacter colonization in Norwegian broiler flocks. *Epidemiol. Inf.*, **111**, 245–55.

Kaufman, V.F., Klose, A.A., Bayne, H.G., Pool, M.F. and Lineweaver, H. (1972) Plant processing of sub-atmospheric steam scalded poultry. *Poult. Sci.*, **51**, 1188–94.

Kemp, G.K., Aldrich, M.L. and Waldroup, A.L. (2000) Acidified sodium chlorite antimicrobial treatment of broiler carcasses. *J. Food Protect.*, **63**, 1087–92.

Kemp, G.K., Aldrich, M.L., Guerra, M.L. and Schneider, K.R. (2001) Continuous online processing of fecal- and ingesta-contaminated poultry carcasses using an acidified sodium chlorite antimicrobial intervention. *J. Food Protect.*, **64**, 807–12.

Kemp, G.K. and Schneider, K.R. (2002) Reduction of *Campylobacter* contamination on broiler carcasses using acidified sodium chlorite. *Dairy, Food Environ. Sanit.*, **22**, 599–602.

Kim, J.-W. and Doores, S. (1993a) Influence of three defeathering systems on microtopography of turkey skin and adhesion of *Salmonella typhimurium*. *J. Food Protect.*, **56**, 286–91.

Kim, J.-W. and Doores, S. (1993b) Attachment of *Salmonella typhimurium* to skins of turkey that had been defeathered through three different systems: scanning electron microscopic examination. *J. Food Protect.*, **56**, 395–400.

Kim, J.-W., Knabel, S.J. and Doores, S. (1993) Penetration of *Salmonella typhimurium* into turkey skin. *J. Food Protect.*, **56**, 292–6.

Kim, J-L., Slavik, M.F., Griffis, C.L. and Walker, J.L. (1993) Attachment of *Salmonella typhimurium* to skins of chicken scalded at various temperatures. *J. Food Protect.*, **56**, 661–5.

Klose, A.A., Kaufman, V.F. and Pool, M.F. (1971) Scalding poultry by steam at subatmospheric pressures. *Poult. Sci.*, **50**, 302–04.

Koenraad, P.M., Ayling, R., Hazeleger, W.C., Rombouts, F.M. and Newell, D.G. (1995) The speciation and subtyping of *Campylobacter* isolates from sewage plants and waste water from a connected poultry abattoir using molecular techniques. *Epidemiol. Inf.*, **115**, 485–94.

Kraft, A.A. (1992) Refrigeration and freezing, in *Psychrotrophic Bacteria in Foods* (ed A.A. Kraft), CRC Press, Inc., London, pp. 241–64.

Kraft, A.A., Ayres, J.C., Weiss, K.F., Marion, W.W., Balloun, S.L. and Forsythe, R.H. (1963) Effect of method of freezing on survival of microorganisms on turkey. *Poult. Sci.*, **42**, 128–37.

Kuhl, H.Y. (1990) Washing and sanitizing hatching eggs. *Int. Hatch. Pract.*, **3**, 29–33.
Lahellec, C. and Colin, P. (1985) Relationship between serotypes of *Salmonella* from hatcheries and rearing farms and those from processed poultry carcasses. *Br. Poult. Sci.*, **26**, 179–86.
Lahellec, G., Meurier, C. and Bennejean, G. (1973) *J. Res. Avic.* Cunic. December (Cited in Barnes, 1974).
Lahellec, C., Colin, P., Bennejean, G., Paquin, J., Guillerm, A. and Debois, J.C. (1986) Influence of resident *Salmonella* on contamination of broiler flocks. *Poult. Sci.*, **65**, 2034–39.
Lambert, A.D., Smith, J.P. and Dodds, K.L. (1991) Shelf life extension and microbiological safety of fresh meat—a review. *Food Microbiol.*, **8**, 267–97.
Leistner, L. (1973) Sprüh-Luft Kühlung von Schlachthahnchen ein Alternativ-Verfahren zum Spinchiller. *Proceedings Poultry Meat Symp.*, Roskilde, Denmark, A13, pp. 1–8.
Leistner, L., Rossmanith, E. and Woltersdorf, W. (1972) Rationalisierung des Sprüh-Kühlverfahrens für Schlachthahnchen. *Fleischwirtschaft*, **52**, 362–4.
Liebana, E., Crowley, C.J., Garcia-Migura, L., Breslin, M.F., Corry, J.E., Allen, V.M. and Davies, R.H. (2002) Use of molecular fingerprinting to assist the understanding of the epidemiology of *Salmonella* contamination within broiler production. *Br. Poult. Sci.*, **43**, 38–46.
Lillard, H.S. (1971) Occurrence of *Clostridium perfringens* in broiler processing and further processing operations. *J. Food Sci.*, **36**, 1008–10.
Lillard, H.S. (1980) Effect on broiler carcasses and water of treating chiller water with chlorine or chlorine dioxide. *Poult. Sci.*, **59**, 1761–66.
Lillard, H.S. (1985) Bacterial cell characteristics and conditions influencing their adhesion to poultry skin. *J. Food Protect.*, **48**, 803–07.
Lillard, H.S. (1986) Distribution of "attached" *Salmonella typhimurium* cells between poultry skin and a surface film following water immersion. *J. Food Protect.*, **49**, 449–54.
Lillard, H.S. (1988) Effect of surfactant or changes in ionic strength on the attachment of *Salmonella typhimurium* to poultry skin and muscle. *J. Food Sci.*, **53**, 727–30.
Lillard, H.S. (1989a) Incidence and recovery of salmonellae and other bacteria from commercially processed poultry carcasses at selected pre- and post-evisceration steps. *J. Food Protect.*, **52**, 88–91.
Lillard, H.S. (1989b) Factors affecting the persistence of *Salmonella* during the processing of poultry. *J. Food Protect.*, **52**, 829–32.
Lillard, H.S. (1990) The impact of commercial processing procedures on the bacterial contamination and cross-contamination of broiler carcasses. *J. Food Protect.*, **53**, 202–4.
Lillard, H.S., Klose, A.A., Hegge, R.I. and Chew, V. (1973) Microbiological comparison of steam (at sub-atmospheric pressure) and immersion-scalded broilers. *J. Food Sci.*, **38**, 903–4.
Lillard, H.S., Blankenship, L.C., Dickens, J.A., Craven, S.E. and Shackelford, A.D. (1987) Effect of acetic acid on the microbiological quality of scalded picked and unpicked broiler carcasses. *J. Food Protect.*, **50**, 112–4.
Lindblom, G.-B., Sjögren, E. and Kaijser, B. (1986) Natural campylobacter colonization in chickens raised under different environmental conditions. *J. Hyg.*, **96**, 385–91.
Line, J.E., Bailey, J.S., Cox, N.A. and Stern, N.J. (1997) Yeast treatment to reduce *Salmonella* and *Campylobacter* populations associated with broiler chickens subjected to transport stress. *Poult. Sci.*, **76**, 1227–31.
Line, J.E., Bailey, J.S., Cox, N.A., Stern, N.J. and Tompkins, T. (1998) Effect of yeast-supplemented feed on *Salmonella* and *Campylobacter* populations in broilers. *Poult. Sci.*, **77**, 405–10.
Liu, Y., Fan, Z., Chen, Y-R. and Thayer, D.W. (2003) Changes in structure and color characteristics of irradiated chicken breasts as a function of dosage and storage time. *Meat Sci.*, **63**, 301–07.
Lofton, C.B., Morrison, S.M. and Leiby, R.D. (1962) The Enterobacteriaceae of some Colorado small mammals and birds and their possible role in gastroenteritis in man and domestic animals. *Zoonoses Res.*, **1**, 277–93.
Logue, C.M., Sherwood, J.S., Elijah, L.M., Olah, P.A. and Dockter, M.R. (2003) The incidence of *Campylobacter* spp. on processed turkey from processing plants in the midwestern United States. *J. Appl. Microbiol.*, **95**, 234–41.
Lundbeck, H. (1974) *Prevention of* Salmonella *infections in the chicken industry (Mimeo.) VI Latin American Congress of Microbiology*, Caracas. Unpublished.
Lundén, J.M., Autio, T.J. and Korkeala, H.J. (2002) Transfer of persistent *Listeria monocytogenes* contamination between food-processing plants associated with a dicing machine. *J. Food Protect.*, **65**, 1129–33.
Lynch, J.A., Macfie, H.J.H. and Mead, G.C. (1991) Effect of irradiation and packaging type on sensory quality of chill-stored turkey breast fillets. *Int. J. Food Sci. Technol.*, **26**, 653–68.
MacKenzie, M.A. and Bains, B.S. (1976) Dissemination of *Salmonella* serotypes from raw feed ingredients to chicken carcasses. *Poult. Sci.*, **55**, 957–60.
Mallinson, E.T., Joseph, S.W. and Carr, L.E. (1998) *Salmonella*'s Achilles' heel. Broiler Industry (**December**), 22, 24, 26, 30, 32.
Mason, J. (1994) *Salmonella enteritidis* control programs in the United States. *Int. J. Food Microbiol.*, **21**, 155–69.
Matsuyama, A., Thornley, M.J. and Ingram, M. (1964) The effect of freezing on the radiation sensitivity of vegetative bacteria. *J. Appl. Bacteriol.*, **27**, 110–24.
May, K.N. (1974) Chilling of poultry meat. 3. Changes in microbiological numbers during final washing and chilling of commercially slaughtered broilers. *Poult. Sci.*, **53**, 1282–85.
McBride, G.B., Brown, B. and Skura, B.J. (1978) Effect of bird type, growers, and season on the incidence of salmonellae in turkeys. *J. Food Sci.*, **43**, 323–26.

McBride, G.B., Skura, B.J., Yada, R.Y. and Bowmer, E.J. (1980) Relationship between incidence of *Salmonella* contamination among pre-scalded, eviscerated and post-chilled chickens in a poultry processing plant. *J. Food Protect.*, **43**, 538–42.

McCapes, R.H., Osburn, B.I. and Riemann, H. (1991) Safety of foods of animal origin: Model for elimination of *Salmonella* contamination of turkey meat. *J. Am. Vet. Med. Assoc.*, **199**, 875–80.

McKee, S.R. (2002) Effect of electron beam irradiation on poultry meat safety and quality. *Poult. Sci.*, **81**, 896–03.

McLauchlin, *J*. (1991) The epidemiology of listeriosis in Britain, in *Proc. Int. Conf. on Listeria and Food Safety.* (ed A. Amgar), Aseptic Processing Association, Laval, France, pp. 38–47.

McMeekin, T.A. (1975) Spoilage association of chicken breast muscle. *Appl. Microbiol.*, **29**, 44–47.

McMeekin, T.A. (1977) Spoilage association of chicken leg muscle. *Appl. Environ. Microbiol.*, **33**, 1244–46.

McMeekin, T.A. and Thomas, C.J. (1978) Retention of bacteria on chicken skin after immersion in bacterial suspensions. *J. Appl. Bacteriol.*, **45**, 383–87.

Mead, G.C. (1975) Hygiene aspects of the chilling process. Qual. Poult. Meat, Proc. Eur. Symp. Poult. Meat, 2nd, Oosterbeck, Neth. pp. (35) 1–8.

Mead, G.C. (1992) Colonization of poultry processing equipment with staphylococci: an overview, in *Prevention and Control of Potentially Pathogenic Microorganisms in Poultry and Poultry Meat Processing*. Proceedings 10. The Attachment of Bacteria to the Gut. (A).

Mead, G.C. and Dodd, C.E.R. (1990) Incidence, origin and significance of staphylococci in processed poultry. *J. Appl. Bacteriol.* (Symposium Supplement) 81S–91S.

Mead, G.C. and Impey, C.S. (1970) The distribution of clostridia in poultry processing plants. *Br. Poult. Sci.*, **11**, 407–14.

Mead, G.C. and Thomas, N.L. (1973) Factors affecting the use of chlorine in the spin-chilling of eviscerated poultry. *Br. Poult. Sci.*, **14**, 99–117.

Mead, G.C., Adams, B.W. and Parry, R.T. (1975) The effectiveness of in-plant chlorination in poultry processing. *Br. Poult. Sci.*, **16**, 517–26.

Mead, G.C., Griffiths, N.M., Grey, T.C. and Adams, B.W. (1986) The keeping quality of chilled duck portions in modified atmosphere packs. *Lebens.-Wiss. u. Technol.*, **19**, 117–21.

Mead, G.C., Hudson, W.R. and Hinton, M.H. (1993) Microbiological survey of five poultry processing plants in the UK. *Br. Poult. Sci.*, **34**, 497–503.

Mead, G.C., Hudson, W.R. and Hinton, M.H. (1995). Effect of changes in processing to improve hygiene control on contamination of poultry carcasses with *Campylobacter*. *Epidemiol. Inf.*, **115**, 495–500.

Meyer, J.D., Cerveny, J.G. and Luchansky, J.B. (2003) Inhibition of nonproteolytic clostridia and anaerobic sporeformers by sodium diacetate and sodium lactate in cook-in-bag turkey breast. *J. Food Protect.*, **66**, 1474–78.

Mielnik, M.B., Dainty, R.H., Lundby, F. and Mielnik, J. (1999) The effect of evaporative air chilling and storage temperature on quality and shelf life of fresh chicken carcasses. *Poult. Sci.*, **78**, 1065–73.

Mitchell, B.W., Buhr, R.J., Berrang, M.E., Bailey, J.S. and Cox, N.A. (2002) Reducing airborne pathogens, dust and *Salmonella* transmission in experimental hatching cabinets using an electrostatic space charge system. *Poult. Sci.*, **81**, 49–55.

Morishita, T.Y., Aye, P.P., Harr, B.S., Cobb, C.W. and Clifford, J.R. (1997) Evaluation of an avian-specific probiotic to reduce the colonization and shedding of *Campylobacter jejuni* in broilers. *Avian Dis.*, **41**, 850–55.

Morris, G.K. and Wells, J.G. (1970) *Salmonella* contamination in a poultry-processing plant. *Appl. Microbiol.*, **19**, 795–99.

Morris, G.K., McMurray, B.L., Galton, M.M. and Wells, J.G. (1969) A study of the dissemination of salmonellosis in a commercial broiler chicken operation. *Am. J. Vet. Res.*, **30**, 1413–21.

Mossel, D.A.A., van Schothorst, M. and Kampelmacher, E.H. (1968) *Prospects for Salmonella Radicidation of Some Foods and Feeds with Particular Reference to the Estimation of the Dose Required. Elimination of Harmful Organisms from Food and Feed by Irradiation*, IAEA, Vienna, pp. 43–57.

Mountney, G.J. (1976) *Poultry Products Technology*, 2nd ed, AVI, Westport, CT.

Moye, C.J. and Chambers, A. (1991) Poultry processing. An innovative technology for salmonella control and shelf life extension. *Food Aust.*, **43**, 246–249.

Mulder, R.W.A.W. (1973) *Shelf Life of Thawed Poultry Meat*, Rep. No. 9873. Spelderholt Inst. Poult. Res., Beekbergen, Netherlands.

Mulder, R.W.A.W. (1975) Radiation-inactivation of *Salmonella panama* and *Escherichia coli* K 12 present on deepfrozen poultry carcasses. Qual. Poult. Meat, Proc. Symp. Poult. Meat Qual., 2nd, Oosterbeek, The Netherlands, pp. (14) 1–7.

Mulder, R.W.A.W. (1976) Microbiological aspects of poultry processing. Vleesdistrib. *Vleestechnol.*, **2**, 20–22 (in Dutch).

Mulder, R.W.A.W. (1984) Ionising energy treatment of poultry. *Food Technol., Aust.*, **36**, 418–20.

Mulder, R.W.A.W. (1985) Decrease microbial contamination during poultry processing. *Poultry*, **March**, 52–55.

Mulder, R.W.A.W. and Bolder, N.M. (1987) Shelf life of chilled poultry after various scalding and chilling treatments. *Fleischwirtschaft*, **67**, 114–6.

Mulder, R.W.A.W. and Dorresteijn, L.W.J. (1975) Microbiological quality of mechanically deboned poultry meat. Qual. Poult. Meat, Proc. Eur. Symp. Poult. Meat Qual., 2nd, Oosterbeek, Neth. pp. (50) 1–7.

Mulder, R.W.A.W. and Dorresteijn, L.W.J. (1977) Hygiene beim brühen von Schlachtgeflügel. (Hygiene during the scalding of broilers.) *Fleischwirtschaft*, **57**, 2220–2.

Mulder, R.W.A.W. and Veerkamp, C.H. (1974) Improvements in poultry slaughterhouse hygiene as a result of cleaning before cooling. *Poult. Sci.*, **53**, 1690–4.

Mulder, R.W.A.W., Dorresteijn, L.W.J. and van der Brock, J. (1977a) Cross-contamination during the scalding and plucking of broilers. *Br. Poult. Sci.*, **19**, 61–70.

Mulder, R.W.A.W., Notermans, S. and Kampelmacher, E.H. (1977b) Inactivation of salmonellae on chilled and deep frozen broiler carcasses by irradiation. *J. Appl. Bacteriol.*, **42**, 179–85.
Mulder, R.W.A.W., Dorresteijn, L.W.J. and Van Der Broek. (1978) Cross-contamination during the scalding and plucking of broilers. *Br. Poult. Sci.*, **19**, 61–70.
Muriana, P.M., Quimby, W., Davidson, C.A. and Grooms, J. (2002) Postpackage pasteurization of ready-to-eat meats by submersion heating for reduction of *Listeria monocytogenes*. *J. Food Protect.*, **65**, 963–9.
Murphy, R.Y. and Berrang, M.E. (2002) Effect of steam and hot water post process pasteurization on microbial and physical property measures of fully cooked vacuum packaged chicken breast strips. *J. Food Sci.*, **67**, 2325–9.
Murphy, R.Y., Duncan, L.K., Driscoll, K.H. and Marcy, J.A. (2003a) Lethality of *Salmonella* and *Listeria innocua* in fully cooked chicken breast meat products during postcook in-package pasteurization. *J. Food Protect.*, **66**, 242–8.
Murphy, R.Y., Duncan, L.K., Driscoll, K.H, Marcy, J.A. and Beard, B.L. (2003b) Thermal inactivation of *Listeria monocytogenes* on ready-to-eat turkey breast meat products during postcook in-package pasteurization with hot water. *J. Food Protect.*, **66**, 1618–22.
Musgrove, M.T., Berrang, M.E., Byrd, J.A., Stern, N.J. and Cox, N.A. (2001) Detection of *Campylobacter* spp. in ceca and crops with and without enrichment. *Poult. Sci.*, **80**, 825–8.
Nachamkin, I., Yang, X.-H. and Stern, N.J. (1993) Role of *Campylobacter jejuni* flagella as colonization factors for three-day-old chicks: analysis with flagellar mutants. *Appl. Environ. Microbiol.*, **59**, 1269–73.
NACMCF (National Advisory Committee on Microbiological Criteria for Foods). (1994) *Campylobacter jejuni/coli*. The National Advisory Committee on Microbiological Criteria for Foods. *J. Food Protect.*, **57**, 1101–21.
Nadeau, É., Messier, S. and Quessy, S. (2002) Prevalence and comparison of genetic profiles of *Campylobacter* strains isolated from poultry and sporadic cases of campylobacteriosis in humans. *J. Food Protect.*, **65**, 73–8.
Nam, K.C. and Ahn, D.U. (2002). Carbon monoxide–heme pigment is responsible for the pink color in irradiated raw turkey beast meat. *Meat Sci.*, **60**, 25–33.
Nayak, R., Kenney, P.B. and Bissonnette, G.K. (2001) Inhibition and reversal of *Salmonella typhimurium* attachment to poultry skin using zinc chloride. *J. Food Protect.*, **64**, 456–61.
Neal, K.R. and Slack, R.C.B. (1997) Diabetes mellitus, anti-secretory drugs and other risk factors for *Campylobacter* gastroenteritis in adults: a case-control study. *Epidemiol. Inf.*, **119**, 307–11.
Neimann, J., Engberg, J., Molbak, K. and Wegener, H.C. (1998) Foodborne risk factors associated with sporadic campylobacteriosis in Denmark.. *Dan. Veterinaertidsskr.*, **81**, 702–5.
Newell, D.G., Shreeve, J.E., Toszeghy, M., Domingue, G., Bull, S., Humphrey, T. and Mead, G. (2001) Changes in the carriage of *Campylobacter* strains by poultry carcasses during processing in abattoirs. *Appl. Environ. Microbiol.*, **67**, 2636–40.
Newton, K.G. and Gill, C.O. (1981) The microbiology of DFD fresh meats: a review. *Meat Sci.*, **5**, 223–32.
Nesbit, E.G., Gibbs, P., Dreesen, D.W. and Lee, M.D. (2001) Epidemiologic features of *Campylobacter jejuni* isolated from poultry broiler houses and surrounding environments as determined by use of molecular strain typing. *Am. J. Vet. Res.*, **62**, 190–4.
Nisbet, D. (2002) Defined competitive exclusion cultures in the prevention of enteropathogen colonisation in poultry and swine. *Antonie Van Leeuwenhoek*, **81**, 481–6.
Notermans, S. and Kampelmacher, E.H. (1974) Attachment of some bacterial strains to the skin of broiler chickens. *Br. Poult. Sci.*, **15**, 573–85.
Notermans, S. and Kampelmacher, E.H. (1975) Further studies on the attachment of bacteria to skin. *Br. Poult. Sci.*, **16**, 487–96.
Notermans, S., van Leusden, F.M., van Schothorst, M. and Kampelmacher, E.H. (1975a) Salmonella-contaminatie van Slachtkuikens Tijdens Het Slachtproces in Enkele Pluimveeslachterijen. (Contamination of broiler chicken by *Salmonella* during processing in a number of poultry-processing plants.) *Tijdschr. Diergeneesk.*, **100**, 259–64.
Notermans, S., van Schothorst, M., van Leusden, F.M. and Kampelmacher, E.H. (1975b) Onderzoekingen over het Kwantitatief Voorkomen van Salmonellae bij Diepvrieskuikens. (Quantitative studies for the presence of salmonellae in deep frozen broiler chickens.) *Tijdschr. Diergeneesk.*, **100**, 648–53.
Notermans, S., van Schothorst, M. and Kampelmacher, E.H. (1975c) Der Einfluss des Keimgehaltes des Spinchiller-Wassers auf den Keimgehalt des Tauwassers von Gefrierhähnchen. (The influence of the bacterial content of spin-chiller water on the bacterial content of thaw water from frozen chickens.) *Fleischwirtschaft*, **55**, 1087–90.
Notermans, S., van Schothorst, M. and Kampelmacher, E.H. (1975d) Der Einfluss des Keimgehaltes des Spinchiller-Wassers auf den Keimgehalt des Tauwassers von Gefrierhähnchen (The influence of the bacterial content of spin-chiller water on the bacterial count of thaw water from frozen chickens). *Fleischwirtschaft*, **55**, 1087–90.
Notermans, S., van Leusden, F.M. and van Schothorst, M. (1977) Suitability of different bacterial groups for determining faecal contamination during post scalding stages in the processing of broiler chickens. *J. Appl. Bacteriol.*, **43**, 383–9.
Notermans, S., Terbijhe, R.J. and van Schothorst, M. (1980) Removing faecal contamination of broiler chickens by spray cleaning during evisceration. *Br. Poult. Sci.*, **21**, 115–21.
Nurmi, E., Nuotio, L. and Schneitz, C. (1992) The competitive exclusion concept: development and future. *Int. J. Food Microbiol.*, **15**, 237–40.
Nychas, G.J., Dillon, V.M. and Board, R.G. (1988) Glucose, the key substrate in the microbiological changes occurring in meat and certain meat products. *Biotechnol. Appl. Biochem.*, **10**, 203–31.
Nylen, G., Dunstan, F., Palmer, S.R., Andersson, Y., Bager, F., Cowden, J., Feierl, G., Galloway, Y., Kapperud, G., Megraud, F., Molbak, K., Peterson, L.R. and Ruutu, P. (2002) The seasonal distribution of *Campylobacter* infection in nine European countries and New Zealand. *Epidemiol. Inf.*, **128**, 383–390.

Olsen, J.E., Brown, D.J., Madsen, M. and Bisgaard, M. (2003) Cross-contamination with salmonella on a broiler slaughterhouse line demonstrated by use of epidemiological markers. *J. Appl. Microbiol.*, **94**, 826–35.

Oostenbach, P. (2002) The use of Nobilis Salenvac, Nobilis SG9R and flavomycin to control *Salmonella* in poultry, in *Salmonella & Salmonellosis 2002, Proceedings*, May 29–31, Saint-Brieuc (eds P. Colin and G. Clement), ISPAIA-ZOOPOLE development, Ploughfragan, France, pp. 613–614.

Oosterom, J., Notermans, S., Karman, H. and Engles, G.B. (1983a) Origin and prevalence of *Campylobacter jejuni* in poultry processing. *J. Food Protect.*, **46**, 339–44.

Oosterom, J., de Wilde, G.J.A., de Boer, E., de Blaauw, L.H. and Karman, H. (1983b) Survival of *Campylobacter jejuni* during poultry processing and pig slaughtering. *J. Food Protect.*, **46**, 702–6.

Papa, C.M. and Dickens, J.A. (1989) Lower gut contents and defacatory responses of broiler chickens as affected by feed withdrawal and electrical treatment at slaughter. *Poult. Sci.*, **68**, 1478–84.

Papa, C.M. (1991) Lower gut contents of broiler chickens withdrawn from feed and held in cages. *Poult. Sci.*, **70**, 375–80.

Patrick, T.E., Goodwin, T.L., Collins, J.A., Wyche, R.C. and Love, B.E. (1972) Steam versus hot-water scalding in reducing bacterial loads on the skin of commercially processed poultry. *Appl. Microbiol.*, **23**, 796–8.

Patrick, T.E., Collins, J.A. and Goodwin, T.L. (1973) Isolation of *Salmonella* from carcasses of steam- and water-scalded poultry. *J. Milk Food Technol.*, **36**, 34–6.

Patterson, M. (1989) Sensitivity of *Listeria monocytogenes* to irradiation on poultry meat and in phosphate-buffered saline. *Lett. Appl. Microbiol.*, **8**, 181–4.

Patterson, J.T. and Gibbs, P.A. (1977) Incidence and sources of Enterobacteriaceae found on frozen broilers. Proceedings of the 3rd European Symposium on Poultry Meat Quality, pp. 69–75.

Patterson, P.H., Ricke, S.C., Sunde, M.L. and Schrefer, D.M. (1990) Hatching eggs sanitized with chlorine dioxide foam: egg hatchability and bactericidal properties. *Avian Dis.*, **34**, 1–6.

Pattison, M. (2001) Practical intervention strategies for *Campylobacter*. *J. Appl. Microbiol.*, **90**, 121S–125S.

Pearson, A.D., Greenwood, M., Healing, T.D., Rollins, D., Shahamat, M., Donaldson, J. and Colwell, R.R. (1993) Colonization of broiler chickens by waterborne *Campylobacater jejuni*. *Appl. Environ. Microbiol.*, **59**, 987–96.

Perko-Mäkelä, P., Koljonen, M., Miettinen, M. and Hänninen, M.-L. (2002a) Survival of *Campylobacter jejuni* in marinated and nonmarinated chicken products. *J. Food Saf.*, **20**, 209–16.

Perko-Mäkelä, P., Hakkinen, M., Honkanen-Buzalski, T. and Hänninen, M.-L. (2002b) Prevalence of campylobacters in chicken flocks during the summer of 1999 in Finland. *Epidemiol. Inf.*, **129**, 187–92.

Petersen, L. and Wedderkopp, A. (2001) Evidence that certain clones of *Campylobacter jejuni* persist during successive broiler flock rotations. *Appl. Environ. Microbiol.*, **67**, 2739–45.

Petersen, L., Nielsen, E.M. and On, S.L.W. (2001) Serotype and genotype diversity and hatchery transmission of *Campylobacter jejuni* in commercial poultry flocks. *Vet. Microbiol.*, **82**, 141–54.

Pivnick, H., Erdman, I.E., Manzatiuk, S. and Pommier, E. (1968) Growth of food poisoning bacteria on barbecued chicken. *J. Milk Food Technol.*, **31**, 198–201.

Pokamunski, S., Kass, N., Borochovich, E., Marantz, B. and Rogol, M. (1986) Incidence of *Campylobacter* spp. in broiler flocks monitored from hatching to slaughter. *Avian Pathol.*, **15**, 83–92.

Poss, P.E. (1985) Cleaning and disinfection programs in the turkey breeder industry, in *Proc. Int. Symp. Salmonella* (ed G.N. Snoeyenbos), American Assoc. Avian Path., Kennett Square, PA., pp. 134–141.

Promsopone, B., Morishita, T.Y., Aye, P.P., Cobb, C.W., Veldkamp, A. and Clifford, J.R. (1998) Evaluation of an avian-specific probiotic and *Salmonella typhimurium*-specific antibodies on the colonization of *Salmonella typhimurium* in broilers. *J. Food Protect.*, **61**, 176–80.

Purdy, J., Dodd, C.E.R., Fowler, D.R. and Waites, W.M. (1988) Increase in microbial contamination of defeathering machinery in a poultry processing plant after changes in the method of processing. *Lett. Appl. Microbiol.*,**6**, 35–8.

Qin, Z.R., Fukata, T., Baba, E. and Arakawa, A. (1995) Effect of *Eimeria tenella* infection on *Salmonella enteritidis* infection in chickens. *Poult. Sci.*, **74**, 1–7.

Ramesh, N., Joseph, S.W., Carr, L.E., Douglass, L.W. and Wheaton, F.W. (2003) Serial disinfection with heat and chlorine to reduce microorganism populations on poultry transport containers. *J. Food Protect.*, **66**, 793–7.

Ramirez, G.A., Sarlin, L.L., Caldwell, D.J., Yezak, C.R., Hume, M.E., Corrier, D.E., Deloach, J.R. and Hargis, B.M. (1997) Effect of feed withdrawal on the incidence of *Salmonella* in the crops and ceca of market age broiler chickens. *Poult. Sci.*, **76**, 654–6.

Rayes, H.M., Genigeorgis, C.A. and Farver, T.B. (1983) Prevalence of *Campylobacter jejuni* on turkey wings at the supermarket level. *J. Food Protect.*, **46**, 292–4.

Refrégier-Petton, J., Rose, N., Denis, M. and Salvat, G. (2001) Risk factors for *Campylobacter* spp. contamination in French broiler-chicken flocks at the end of the rearing period. *Prev. Vet. Med.*, **50**, 89–100.

Renwick, S.A., Irwin, R.J., Clarke, R.C., McNab, W.B., Poppe, C. and McEwen, S.A. (1992) Epidemiological associations between characteristics of registered broiler chicken flocks in Canada and the *Salmonella* culture status of floor litter and drinking water. *Can. Vet. J.*, **33**, 449–58.

Rhodes, D.N. (1965) The radiation pasteurization of broiler chicken carcasses. *Br. Poult. Sci.*, **6**, 265–71.

Richmond, M. (1990) The Microbiological Safety of Food. Part 1. (Campylobacter). *Report of the Committee on the Microbiological Safety of Food*. HMSO, London, pp. 45–58 and 130–131.

Rigby, C.E. (1982) Most probable number cultures for assessing *Salmonella* contamination of eviscerated broiler carcasses. *Can. J. Compar. Med.*, **46**, 279–82.

Rigby, C.E. and Pettit, J.R. (1980) Changes in the salmonella status of broiler chickens subjected to simulated shipping conditions. *Can. J. Compar. Med.*, **44**, 374–81.
Rigby, C.E. and Pettit, J.R. (1981) Effects of feed withdrawal on the weight, fecal excretion, and *Salmonella* status of market age broiler chickens. *Can. J. Compar. Med.*, **45**, 363–5.
Rigby, C.E. and Pettit, J.R., Baker, M.F., Bentley, A.H., Salomons,M.D. and Lior, H. (1980a) Sources of salmonellae in an uninfected commercially-processed broiler flock. *Can. J. Compar. Med.*, **44**, 267–74.
Rigby, C.E., Pettit, J.R., Baker, M.F., Bentley, A.H., Salomons, M.O. and Lior, H. (1980b) Flock infection and transport as sources of salmonellae in broiler chickens and carcasses. *Can. J. Compar. Med.*, **44**, 328–37.
Rigby, C.E., Pettit, J.R., Bentley, A.H., Spencer, J.L., Salomons, M.O. and Lior, H. (1982) The relationships of salmonellae from infected broiler flocks, transport crates or processing plants to contamination of eviscerated carcasses. *Can. J. Compar. Med.*, **46**, 272–8.
Ristic, M. (1997) Application of chilling methods on slaughtered poultry. *Die Fleischwirtschaft*, **77**, 810–1.
Roberts, T.A. and Derrick, C.M. (1978) The effect of curing salts on the growth of *Clostridium perfringens* (*welchii*) in a laboratory medium. *J. Food Technol.*, **13**, 349–53.
Roberts, T.A. and Gibson, A.M. (1979) The relevance of *Clostridium botulinum* type C in public health and food processing. *J. Food Technol.*, **14**, 211–26.
Robinson, D.A. (1981) Infective dose of *Campylobacter jejuni* in milk. *Br. Med. J.*, **282**, 1584.
Rodrigues, L.C., Cowden, J.M., Wheeler, J.G., Sethi, D., Wall, P.G., Cumberland, P., Tompkins, D.S., Hudson, M.J., Roberts, J.A. and Roderick, P.J. (2001) The study of infectious intestinal disease in England: risk factors for cases of infectious intestinal disease with *Campylobacter jejuni* infection. *Epidemiol. Inf.*, **127**, 185–93.
Rose, N., Beaudeau, F., Drouin, P., Toux, J.Y., Rose, V. and Colin, P. (1999) Risk factors for *Salmonella enterica* subsp. *enterica* contamination in French broiler-chicken flocks at the end of the rearing period. *Prev. Vet. Med.*, **39**, 265–77.
Rose, N., Beaudeau, F., Drouin, P., Toux, J.Y., Rose, V. and Colin, P. (2000) Risk factors for *Salmonella* persistence after cleansing and disinfection in French broiler-chicken houses. *Prev. Vet. Med.*, **44**, 9–20.
Rose, N., Mariani, J.P., Drouin, P., Toux, J.Y., Rose, V. and Colin, P. (2003) A decision-support system for *Salmonella* in broiler-chicken flocks. *Prev. Vet. Med.*, **59**, 27–42.
Rosenquist, H., Nielsen, N.L., Sommer, H.M., Norrung, B. and Christensen, B.B. (2003) Quantitative risk assessment of human campylobacteriosis associated with thermophilic *Campylobacter* species in chickens. *Int. J. Food Microbiol.*, **83**, 87–103.
Russell, S.M. (1997). Rapid prediction of the potential shelf-life of fresh broiler chicken carcasses under commercial conditions. *J. Appl. Poult. Res.*, **6**, 163–8.
Russell, S.M. (2002) Intervention strategies for reducing *Salmonella* prevalence. *WATT Poultry*, 28–45.
Russell, S.M. (2003) The effect of airsacculitis on bird weights, uniformity, fecal contamination, processing errors and populations of *Campylobacter* spp. and *Escherichia coli*. *Poult. Sci.*, **82**, 1326–31.
Salvat, G., Lalande, F., Humbert, F. and Lahellec, C. (1992a) Use of a competitive exclusion product (Broilact) to prevent *Salmonella* colonization of newly hatched chicks. *Int. J. Food Microbiol.*, **15**, 307–11.
Salvat, G., Colin, P. and Allo, J.C. (1992b) Evolution of microbiological contamination of poultry carcasses during slaughtering: A survey on 12 French abattoirs, in *Proc. 8. Other pathogens of concern (no salmonella and campylobacter)* (eds E. Nurmi, P. Colin and R.W.A.W. Mulder), DLO Centre for Poultry Research and Information Services, Beekbergen, The Netherlands, pp. 25–35.
Sanders, D.H. and Blackshear, C.D. (1971) Effect of chlorination in the final washer on bacterial counts of broiler chicken carcasses. *Poult. Sci.*, **50**, 215–9.
Sanderson, K., Thomas, C.J. and McMeekin, T.A. (1991) Molecular basis of the adhesion of *Salmonella* serotypes to chicken muscle fascia. *Biofouling*, **5**, 89–101.
Sarlin, L.L., Barnhart, E.T., Caldwell, D.J., Moore, R.W., Byrd, J.A., Caldwell, D.Y., Corrier, D.E., Deloach, J.R. and Hargis, B.M. (1998) Evaluation of alternative sampling methods for *Salmonella* critical control point determination at broiler processing. *Poult. Sci.*, **77**, 1253–7.
Saucier, L., Gendron, C. and Gariépy, C. (2000) Shelf life of ground poultry meat stored under modified atmosphere. *Poult. Sci.*, **79**, 1851–6.
Sawaya, W.N., Abu-Ruwaida, A.S., Hussain, A.J., Khalafawi, M.S. and Dashti, B.H. (1993) Shelf life of vacuum-packaged eviscerated broiler carcasses under simulated market storage conditions. *J. Food Saf.*, **13**, 305–21.
Schmidhofer, T. (1969) Hygiene bei der Geflügelschlachtung. *Wien. Tieraerztl. Montatsschr.*, **56**, 402–10.
Schmitt, R.E., Gallo, L. and Schmidt-Lorenz, W. (1988) Microbial spoilage of refrigerated fresh broilers. IV. Effect of slaughtering procedures on the microbial association of poultry carcasses. *Lebensm. Wiss. u.-Technol.*, **21**, 234–8.
Schneider, K.R., Kemp, G.K. and Aldrich, M.L. (2002). Antimicrobial treatment of air chilled broiler carcasses. Acidified sodium chlorite antimicrobial treatment of air chilled broiler carcasses. *Dairy, Food Environ. Sanit.*, **22**, 102–8.
Schneitz, C. and Nuotio, L. (1992) Efficacy of different microbial preparations for controlling salmonella colonization in chicks and turkey poults by competitive exclusion. *Br. Poult. Sci.*, **33**, 207–11.
Schneitz, C., Nuotio, L., Mead, G. and Nurmi, E. (1992) Competitive exclusion in the young bird: challenge models, administration and reciprocal protection. *Int. J. Food Microbiol.*, **15**, 241–4.
Schneitz, C., Kiiskinen, T., Toivonen, V. and Nash, M. (1998) Effect of Broilac®on the physicochemical conditions and nutrient digestibility in the gastrointestinal tract of broilers. *Poult. Sci.*, **77**, 426–32.
Schröder, I. (2002) A contribution to consumer protection: TAD *Salmonella* vac®E—a new oral vaccine for chickens against

Salmonella Enteritidis, in *Salmonella & Salmonellosis 2002, Proceedings*, May 29–31, Saint-Brieuc (eds. P. Colin and G. Clement), ISPAIA-ZOOPOLE development, Ploughfragan, France, pp. 571–575.

Schuchat, A., Deaver, K.A., Wenger, J.D., Plikaytis, B.D., Mascola, L., Pinner, R.W., Reingold, A.L. and Broome, C.V. (1992) Role of foods in sporadic listeriosis. I. Case-control study of dietary risk factors. *J. Am. Med. Assoc.*, **267**, 2041–5.

Schwarz, S.J., Claus, J.R., Wang, H., Marriott, N.G., Graham, P.P. and Fernandes, C.F. (1999) Inhibition of pink color development in cooked, uncured turkey breast through ingredient incorporation. *Poult. Sci.*, **78**, 255–66.

Seligmann, R. and Frank-Blum, H. (1974) Microbial quality of barbecued chickens from commercial rotisseries. *J. Milk Food Technol.*, **37**, 473–6.

Seligmann, R. and Lapinsky, Z. (1970) *Salmonella* findings in poultry as related to conditions prevailing during transportation from the farm to the processing plant. *Res. Vet.*, **27**, 7–14.

Seman, D.L., Borger, A.C., Meyer, J.D., Hall, P.A. and Milkowski, A.L. (2002) Modeling the growth of *Listeria monocytogenes* in cured ready-to-eat processed meat products by manipulation of sodium chloride, sodium diacetate, potassium lactate, and product moisture content. *J. Food Protect.*, **65**, 651–8.

Shanker, S., Rosenfield, J.A., Davey, G.R. and Sorrell, T.C. (1982) *Campylobacter jejuni*: incidence in processed broilers and biotype distribution in human and broiler isolates. *Appl. Environ. Microbiol.*, **43**, 1219–20.

Shanker, S., Lee, A. and Sorrell, T.C. (1990) Horizontal transmission of *Campylobacter jejuni* amongst broiler chicks: experimental studies. *Epidemiol. Inf.*, **104**, 101–10.

Sheldon, B.W. and Brake, J. (1991) Hydrogen peroxide as an alternative hatching egg disinfectant. *Poult. Sci.*, **70**, 1092–8.

Shiozawa, K., Kato, E. and Shimizu, A. (1980) Enterotoxigenicity of *Staphylococcus aureus* strains isolated from chickens. *J. Food Protect.*, **43**, 683–5.

Shreeve, J.E., Toszeghy, M., Pattison, M. and Newell, D.G. (2000) Sequential spread of *Campylobacter* infection in a multipen broiler house. *Avian Dis.*, **44**, 983–8.

Shreeve, J.E., Toszeghy, M., Ridley, A. and Newell, D.G. (2002) The carry-over of *Campylobacter* isolates between sequential poultry flocks. *Avian Dis.*, **46**, 378–85.

Simmons, M., Fletcher, D.L., Cason, J.A. and Berrang, M.E. (2003) Recovery of *Salmonella* from retail broilers by a whole-carcass enrichment procedure. *J. Food Protect.*, **66**, 446–50.

Simonsen, B. (1975) Microbiological aspects of poultry meat quality. *Qual. Poult. Meat, Proc. Eur. Symp. Poultry Meat Qual., 2nd, Oosterbeek, Neth.*, **2**, 1–10.

Simonsen, B., Bryan, F.L., Christian, J.H.B., Roberts, T.A., Tompkin, R.B. and Silliker, J.H. (1987) Prevention and control of food-borne salmonellosis through application of Hazard Analysis Critical Control Point (HACCP). *Int. J. Food Microbiol.*, **4**, 227–47.

Singh, P.S. (1998) Method for reducing fecal leakage and contamination during meat processing. U. S. Patent Number 5,733,185.

Skov, M.N., Angen, Ø., Chriél, M., Olsen, J.E. and Bisgaard, M. (1999a) Risk factors associated with *Salmonella enterica* Serovar *typhimurium* infection in Danish broiler flocks. *Poult. Sci.*, **78**, 848–54.

Skov, M.N., Carstensen, B., Tornøe, N. and Madsen, M. (1999b) Evaluation of sampling methods for the detection of *Salmonella* in broiler flocks. *J. Appl. Microbiol.*, **86**, 695–700.

Skov, M.N., Feld, N.C., Carstensen, B. and Madsen, M. (2002) The serologic response to *Salmonella enteritidis* and *Salmonella typhimurium* in experimentally infected chickens, followed by an indirect lipopolysaccharide enzyme-linked immunosorbent assay and bacteriologic examinations through a one-year period. *Avian Dis.*, **46**, 265–73.

Slader, J., Domingue, G., Jørgensen, F., McAlpine, K., Owen, R.J., Bolton, F.J. and Humphrey, T.J. (2002) Impact of transport crate reuse and of catching and processing on *Campylobacter* and *Salmonella* contamination of broiler chickens. *Appl. Environ. Microbiol.*, **68**, 713–9.

Smart, J.L. and Rush, P.A.J. (1987) *In-vitro* heat denaturation of *Clostridium botulinum* toxins types A, B and C. *Int. J. Food Sci. Technol.*, **22**, 293–8.

Smith, K.E., Besser, J.M., Hedberg, C.W., Leano, F.T., Bender, J.B., Wicklund, J.H., Johnson, B.P., Moore, K.A. and Osterholm, M.T. (1999) Quinolone-resistant *Campylobacter jejuni* infections in Minnesota, 1992–1998. *Invest. Team. N. Engl. J. Med.*, **340**, 1525–32.

Snoeyenbos, G.H., Morin, E.W. and Wetherbee, D.K. (1967) Naturally occurring *Salmonella* in "blackbirds" and gulls. *Avian Dis.*, **11**, 642–6.

Snoeyenbos, G.H., Carlson, V.L., Smyster, C.F. and Olesiuk, O.M. (1969) Dynamics of *Salmonella* infection in chicks reared on litter. *Avian Dis.*, **13**, 72–83.

Soo, S.S., Evans, S.J., O'Brien, S.J., Velander, N.Q. and Ward, L.R. (2002) The United Kingdom *Salmonella* Enteritidis epidemic one decade on: controlling infection in poultry has reduced human disease, in *Salmonella & Salmonellosis 2002, Proceedings*, May 29–31, Saint-Brieuc (eds P. Colin and G. Clement), ISPAIA-ZOOPOLE development, Ploughfragan, France, pp. 555–9.

Springer, S. Lehman, J., Lindner, T. Alber, G. and Selbitz, H-J. (2002) A new live *Salmonella* Enteritidis vaccine for chicken—experimental evidence of its safety and efficacy, in *Salmonella & Salmonellosis 2002, Proceedings*, May 29–31, Saint-Brieuc (eds P. Colin and G. Clement), ISPAIA-ZOOPOLE development, Ploughfragan, France, pp. 609–10.

Stavric, S. and D'Aoust, J.-Y. (1993) Undefined and defined bacterial preparations for the competitive exclusion of *Salmonella* in poultry—a review. *J. Food Protect.*, **56**, 173–80.

Stern, N.J. and Robach, M.C. (2003) Enumeration of *Campylobacter* spp. in broiler feces and in corresponding processed carcasses. *J. Food Protect.*, **66**, 1557–63.

Stern, N.J., Clavero, M.R.S., Bailey, J.S., Cox, N.A. and Robach, M.C. (1995) *Campylobacter* spp. in broilers on the farm and after transport. *Poult. Sci.*, **74**, 937–41.

Stern, N.J., Cox, N.A., Bailey, J.S., Berrang, M.E. and Musgrove, M.T. (2001) Comparison of mucosal competitive exclusion and competitive exclusion treatment to reduce *Salmonella* and *Campylobacter* spp. colonization in broiler chickens. *Poult. Sci.*, **80**, 156–60.

Stern, N.J., Fedorka-Cray, P., Bailey, J.S., Cox, N.A., Craven, S.E., Hiett, K.L., Musgrove, M.T., Ladely, S., Cosby, D. and Mead, G.C. (2001) Distribution of *Campylobacter* spp. in selected U.S. poultry production and processing operations. *J. Food Protect.*, **64**, 1705–10.

Stern, N.J., Robach, M.C., Cox, N.A. and Musgrove, M.T. (2002) Effect of drinking water chlorination on *Campylobacter* spp. colonization of broilers. *Avian Dis.*, **46**, 401–4.

Stern, N.J., Hiett, K.L., Alfredsson, G.A., Kristinsson, K.G., Reiersen, J., Hardardottir, H., Briem, H., Gunnarsson, E., Georgsson, F., Lowman, R., Berndtson, E., Lammerding, A.M., Paoli, G.M. and Musgrove, M.T. (2003) *Campylobacter* spp. in Icelandic poultry operations and human disease. *Epidemiol. Inf.*, **130**, 23–32.

Stewart, D.J. and Patterson, J.T. (1962) Bacteriology of processed broilers. 2. Experiments in broiler processing. *Northern Ireland Ministry of Agriculture Rec. Exp. Res.*, **11**, Part 1, 65–71.

Straka, R.P. and Combes, F.M. (1951) The predominance of micrococci in the flora of experimental frozen turkey meat steaks. *Food Res.*, **16**, 492–3.

Strong, D.H., Canada, J.C. and Griffiths, B.B. (1963) Incidence of *Clostridium perfringens* in American foods. *Appl. Microbiol.*, **11**, 42–4.

Studahl, A. and Andersson, Y. (2000) Risk factors for indigenous campylobacter infection: a Swedish case-control study. *Epidemiol. Inf.*, **125**, 269–75.

Studer, P., Schmitt, R.E., Gallo, L. and Schmidt-Lorenz, W. (1988) Microbial spoilage of refrigerated fresh broilers. II. Effect of packaging on microbial association of poultry carcasses. *Lebensm.-Wiss. u.-Technol.*, **21**, 224–8.

Surkiewicz, B.F., Johnston, R.W., Moran, A.B. and Krumm, G.W. (1969) A bacteriological survey of chicken eviscerating plants. *Food Technol.*, **23**, 1066–9.

Susiluoto, T., Korkeala, H. and Bjorkroth, K.J. (2003) *Leuconostoc gasicomitatum* is the dominating lactic acid bacterium in retail modified atmosphere packaged marinated broiler meat strips on sell-by-day. *Int. J. Food Microbiol.*, **80**, 89–97.

Tamblyn, K.C. and Conner, D.E. (1997) Bactericidal activity of organic acids against *Salmonella typhimurium* attached to broiler chicken skin. *J. Food Protect.*, **60**, 629–33.

Taormina, P.J., Bartholomew, G.W and Dorsa, W.J. (2003) Incidence of *Clostridium perfringens* in commercially produced cured raw meat product mixtures and behavior in cooked products during chilling and refrigerated storage. *J. Food Protect.*, **66**, 72–81.

Tenkate, T.D. and Stafford, R.J. (2001) Risk factors for campylobacter infection in infants and young children: a matched case-control study. *Epidemiol. Inf.*, **127**, 399–404.

Thomas, C.J. and McMeekin, T.A. (1980) Contamination of broiler carcass skin during commercial processing procedures: an electron microscopic study. *Appl. Environ. Microbiol.*, **40**, 133–44.

Thomas, C.J. and McMeekin, T.A. (1981) Attachment of *Salmonella* spp. to chicken muscle surfaces. *Appl. Environ. Microbiol.*, **42**, 130–4.

Thomas, C.J. and McMeekin, T.A. (1984) Effect of water uptake by poultry tissues on contamination by bacteria during immersion in bacterial suspensions. *J. Food Protect.*, **47**, 398–402.

Thomas, C.J. and McMeekin, T.A. (1991) Factors which affect retention of *Salmonella* by chicken muscle fascia. *Biofouling*, **5**, 75–87.

Thomas, L.M., Long, K.A., Good, R.T., Panaccio, M. and Widders, P.R. (1997) Genotypic diversity among *Campylobacter jejuni* isolates in a commercial broiler flock. *Appl. Environ. Microbiol.*, **63**, 1874–7.

Thomas, C.J., McMeekin, T.A. and Patterson, J.T. (1987) Prevention of microbial contamination in the poultry processing plant, in *Elimination of Pathogenic Organisms from Meat and Poultry* (ed F.J.M. Smulders), Elsevier Science Publishers, Amsterdam, The Netherlands, pp. 163–179.

Thomson, J.E., Whitehead, W.K. and Mercuri, A.J. (1974) Chilling poultry meat—A literature review. *Poult. Sci.*,**53**, 1268–81.

Thompson, J.K. and Patterson, J.T. (1983) *Staphylococcus aureus* from a site of contamination in a broiler processing plant. *Rec. Agric. Res.*, **31**, 45–53.

Todd, E.C.D. (1992) Foodborne disease in Canada—a 10-year summary from 1975 to 1984. *J. Food Protect.*, **55**, 123–32.

Tompkin, R.B. (1977) Control by chlorination, in Proceedings of International Symposium on Salmonella and Prospects for Control (ed D.A Barnum), University of Guelph, Guelph, Ontario, June 8–11, pp. 122–130.

Tompkin, R.B. (1980) Botulism from meat and poultry products—a historical perspective. *Food Technol.*, **34**, 229–36, 257.

Tompkin, R.B. (1986) Microbiology of ready-to-eat meat and poultry products, in *Advances in Meat Research* (ed A.M. Pearson and T.R. Dutson), *Volume 2*, pp. 89–121, AVI Publishing Co., Westport, CT.

Tompkin, R.B. (1990) Use of HACCP in the production of meat and poultry products. *J. Food Protect.*, **53**, 795–803.

Tompkin, R.B. (1994) HACCP in the meat and poultry industry. *Food Control*, **5**, 153–61.

Tompkin, R.B. (1995a) The hazard analysis critical control point (HACCP) system, in *Proceedings of International Meat Poult HACCP Alliance Symposium* (eds S.C. Ricke and G.R. Acuff), Texas A&M University, College Station, Texas (In press).

Tompkin, R.B. (1995b) The use of HACCP for producing and distributing processed meat and poultry products, in *Advances in Meat Research* (eds A.M. Pearson and T.R. Dutson), *Volume 10*, Blackie Academic & Professional, London, pp. 72–108.

Tompkin, R.B. (2002) Control of *Listeria monocytogenes* in the food-processing environment. *J. Food Protect.*, **65**, 709–25.

Tompkin, R.B., Christiansen, L.N., Shaparis, A.B., Baker, R.L. and Schroeder, J.M. (1992) Control of *Listeria monocytogenes* in processed meats. *Food Aust.*, **44**, 370–6.

Tompkin, R.B., Scott, V.N., Bernard, D.T., Sveum, W.H. and Gombas, K.S. (1999) Guidelines to prevent post-processing contamination from *Listeria monocytogenes*. *Dairy, Food Environ. Sanit.*, **19**, 551–62.

Turnbull, P.C.B. and Snoeyenbos, G.H. (1973) The role of ammonia, water activity, and pH in the salmonellacidal effect of long-used poultry litter. *Avian Dis.*, **17**, 72–86.

USDA-FSIS (United States Department of Agriculture, Food Safety and Inspection Service). (2002) New Jersey firm recalls poultry products for possible *Listeria* contamination. Recall Release, November 20:FSIS-RC-098-2002. http://www.fsis.usda.gov/oa/recalls/prelease/pr098-2002.htm.

USDA-FSIS (United States Department of Agriculture, Food Safety and Inspection Service). (2003) Electronic reading room: microbiological testing program. Table 9. Prevalence (%) of Salmonella in RTE meat and poultry products, CY 2001–2002 (combined results). http://www.listeria/FSIS salmonella data 2001 and 2002.htm.

van de Giessen, A., Mazurier, S-I., Jacobs-Reitsma, W., Jansen, W., Berkers, P., Ritmeester, W. and Wernars, K. (1992) Study on the epidemiology and control of *Campylobacter jejuni* in poultry broiler flocks. *Appl. Environ. Microbiol.*, **58**, 1913–7.

van de Giessen, A.W., Tilburg, J.J.H.C., Ritmeester, W.S. and van der Plas, J. (1998). Reduction of *Campylobacter* infections in broiler flocks by application of hygiene measures. *Epidemiol. Inf.*, **121**, 57–66.

van der Marel, G.M., van Logtestijn, J.G. and Mossel, D.A.A. (1988) Bacteriological quality of broiler carcasses as affected by in-plant lactic acid decontamination. *Int. J. Food Microbiol.*, **6**, 31–42.

Van Looveren, M., Daube, G., De Zutter, L., Dumont, J.M., Lammens, C., Wijdooghe, M., Vandamme, P., Jouret, M., Cornelis, M. and Goossens, H. (2001) Antimicrobial susceptibilities of *Campylobacter* strains isolated from food animals in Belgium. *J. Antimicrob. Chemother.*, **48**, 235–40.

van Schothorst, M., Notermans, S. and Kampelmacher, E.H. (1972) Hygiene in poultry slaughter. *Fleischwirtschaft*, **6**, 749–52.

van Schothorst, M., Northholt, M.D., Kampelmacher, E.H. and Notermans, S. (1976) Studies on the estimation of the hygienic condition of frozen broiler chickens. *J. Hyg.*, **76**, 57–63.

Veerkamp, C.H. (1974) The simultaneous scalding and plucking of broiler carcasses compared with an industrial method of processing. *Proceedings of the 15th World Poultry Congress*, New Orleans, pp. 450–451.

Veerkamp, C.H. and Hofmans, G.J.P. (1973) New development in poultry processing, simultaneous scalding and plucking. *Poult. Int.*, **12**, 16–18.

Velazquez, J.B., Jimenez, A., Chomon, B. and Villa, T.G. (1995) Incidence and transmission of antibiotic resistance in *Campylobacter jejuni* and *Campylobacter coli*. *J. Antimicrob. Chemother.*, **35**, 173–178.

Wabeck, C.J. (1972) Feed and withdrawal time relationship to processing yield and potential fecal contamination of broilers. *Poult. Sci.*, **51**, 1119–21.

Waldroup, A.L., Rathgeber, B.M., Forsythe, R.H. and Smoot, L. (1992a) Effects of six modifications on the incidence and levels of spoilage and pathogenic organisms on commercially processed postchill broilers. *Appl. Poult. Sci.*, **1**, 226–234.

Waldroup, A.L., Skinner, J.T., Hierholzer, R.E., Kopek, J.M. and Waldroup, P.W. (1992b) Effects of bird density on *Salmonella* contamination of prechill carcasses. *Poult. Sci.*, **71**, 844–9.

Waldroup, A.L., Yamaguchi, W., Skinner, J.T. and Waldroup, P.W. (1992c) Effects of dietary lactose on incidence and levels of salmonellae on carcasses of broiler chickens grown to market age. *Poult. Sci.*, **71**, 288–95.

Waldroup, A.L., Rathgeber, B.M., Hierholzer, R.E., Smoot, L., Martin, L.M., Bigili, S.F., Fletcher, D.L., Chen, T.C. and Wabeck, C.J. (1993) Effects of reprocessing on microbiological quality of commercial prechill broiler carcasses. *J. Appl. Poult. Res.*, **2**, 111–6.

Walker, H.W. and Ayres, J.C. (1956) Incidence and kinds of microorganisms associated with commercially dressed poultry. *Appl. Microbiol.*, **4**, 345–9.

Walker, H.W. and Ayres, J.C. (1959) Microorganisms associated with commercially processed turkeys. *Poult. Sci.*, **38**, 1351–5.

Walsh, J.L. and Thayer, S.G., (1990) Acid treatments for on-line reprocessing of contaminated poultry. *Technical Report, Volume 1*, Southeastern Poultry & Egg Association. Decatur, Georgia.

Wasseneaar, T.M. and Newell, D.G. (2000) Genotyping of *Campylobacter* spp. *Appl. Environ. Microbiol.*, **66**, 1–9.

Wedderkopp, A., Rattenborg, E. and Madsen, M. (2000) National surveillance of *Campylobacter* in broilers at slaughter in Denmark in 1998. *Avian Dis.*, **44**, 993–9.

Wedderkopp, A., Gradel, K.O., Jorgenson, J.C. and Madsen, M. (2001) Pre-harvest surveillance of *Campylobacter* and *Salmonella* in Danish broiler flocks: a 2-year study. *Int. J. Food Microbiol.*, **68**, 53–9.

Wegener, H.C., Hald, T., Wong, D.L.F., Madsen, M., Korsgaard, H., Bager, F., Gerner-Smidt, P. and Mølbak, K. (2003) *Salmonella* control programs in Denmark. *Emerg. Infect. Dis.*, **9**, 774–80.

Wempe, J.M., Genigeorgis, C.A., Farver, T.B. and Yusufu, H.I. (1983) Prevalence of *Campylobacter jejuni* in two California chicken processing plants. *Appl. Environ. Microbiol.*, **45**, 355–9.

Wesley, R.L. and Bovard, K.P. (1983) The effect of hand-held inside bird washers on turkey carcass hygienic quality. *Poultry Sci.*, **62**, 338–40.

Whistler, P.E. and Sheldon, B.W. (1989a) Biocidal activity of ozone *versus* formaldehyde against poultry pathogens in a prototype setter. *Poult. Sci.*, **68**, 1068–73.

Whistler, P.E. and Sheldon, B.W. (1989b) Bactericidal activity, eggshell conductance, and hatchability effects of ozone versus formaldehyde disinfection. *Poult. Sci.*, **68**, 1074–7.

Whistler, P.E. and Sheldon, B.W. (1989c) Comparison of ozone and formaldehyde as poultry hatchery disinfectants. *Poult. Sci.*, **68**, 1345–50.

Wilkins, M.J., Bidol, S.A., Boulton, M.L., Stobierski, M.G., Massey, J.P. and Robinson-Dunn, B. (2002) Human salmonellosis associated with young poultry from a contaminated hatchery in Michigan and the resulting public health interventions. *Epidemiol. Inf.*, **129**, 19–27.

Willis, W.L. and Murray, C. (1997) *Campylobacter jejuni* seasonal recovery observations of retail market broilers. *Poult. Sci.* **76**, 314–7.
WHO (World Health Organization). (1981) Wholesomeness of irradiated food. Report of a Joint FAO/IAEA/WHO Expert Committee. *Technical Report Series No. 659*. World Health Organization, Geneva.
WHO (World Health Organization). (1989) Consultation on microbiological criteria for foods to be further processed including by irradiation. Int. Consultative Group on Food Irradiation. World Health Organization, Geneva.
WHO (World Health Organization). (1994) Safety and nutritional adequacy of irradiated food. World Health Organization, Geneva.
WHO (World Health Organization). (2003) WHO Antimicrobial Resistance Information Bank, Geneva. http://oms2.b3e.jussieu.fr/arinfobank/
WHO/FAO (World Health Organization /Food and Agriculture Organization of the United Nations). (2002) Risk assessments of Salmonella in eggs and broiler chickens. Interpretive summary, in *Microbiological Risk Assessment Series No. 1*. Food Safety Department, World Health Organization, Geneva.
Whyte, P., Collins, J.D., McGill, K., Monahan, C. and O'Mahony, H. (2001) Quantitative investigation of the effects of chemical decontamination procedures on the microbiological status of broiler carcasses during processing. *J. Food Protect.*, **64**, 179–183.
Whyte, P., McGill, K. and Collins, J.D. (2003) A survey of the prevalence of *Salmonella* and other enteric pathogens in a commercial poultry feed mill. *J. Food Saf.*, **23**, 13–24.
Wierup, M. (1991) The control of salmonella in food producing animals in Sweden, in *Proc. Symp. Diagnosis and Control of Salmonella* (ed G.H. Snoeyenbos), Carter Printing Co., Richmond, VA, pp. 65–77.
Wierup, M., Wahlström, H. and Engström, B. (1992) Experience of a 10-year use of competitive exclusion treatment as part of the *Salmonella* control programme in Sweden. *Int. J. Food Microbiol.*, **15**, 287–291.
Wierup, M., Engström, B., Engvall, A. and Wahlström, H. (1995) Control of *Salmonella enteritidis* in Sweden. *Int. J. Food Microbiol.*, **25**, 219–226.
Wilder, A.N. and MacCready, R.A. (1966) Isolation of *Salmonella* from poultry. Poultry products and poultry processing plants in Massachusetts. *New England J. Medicine*, **274**, 1453–1460.
Wilding, G.P. and Baxter-Jones, C. (1985) Egg borne salmonellosis: is prevention feasible? in *Proceedings of an Int. Symposium on Salmonella* (ed G.N. Snoeyenbos), American Assoc. Avian Path., Kennett Square, PA, pp. 126–133.
Wilkerson, W.B., Ayres, J.C. and Kraft, A.A. (1961) Occurrence of enterococci and coliform organisms on fresh and stored poultry. *Food Technol.*, **15**, 286–92.
Williams, J.E. (1978) Paratyphoid infection, in *Diseases of Poultry* (eds M.S. Hofstad, B.W. Calnek, C.F. Helmboldt, W.M. Reid and H.W. Yoder, Jr.), 7th ed, Iowa State Univ. Press, Ames.
Williams, J.E., Dillard, L.H. and Hall, G.O. (1968) The penetration patterns of *Salmonella typhimurium* through the outer structures of chicken eggs. *Avian Dis.*, **12**, 445–66.
Willis, W.L., Murray, C. and Talbott, C. (2000) Effect of delayed placement on the incidence of *Campylobacter jejuni* in broiler chickens. *Poult. Sci.*, **79**, 1392–5.
Wray, C. and Corkish, J.D. (1991) Salmonella control programmes in the United Kingdom, in *Proceedings of a Symposium on Diagnosis and Control of Salmonella* (ed G.H. Snoeyenbos), Carter Printing Co., Richmond, VA, pp. 59–64.
Yang, H., Li, Y. and Johnson, M.G. (2001) Survival and death of *Salmonella typhimurium* and *Campylobacter jejuni* in processing water and on chicken skin during poultry scalding and chilling. *J. Food Protect.*, **64**, 770–6.
Yoon, K.S. (2003) Effect of gamma irradiation on the texture and microstructure of chicken beast meat. *Meat Sci.*, **63**, 273–7.
Yusufu, H.I., Genigeorgis, C., Farver, T.B. and Wempe, J.M. (1983) Prevalence of *Campylobacter jejuni* at different sampling sites in two California turkey processing plants. *J. Food Protect.*, **46**, 868–72.
Zecca, B.C., McCapes, R.H., Dungan, W.W., Holte, R.J., Worcester, W.W. and Williams, J.E. (1977) The Dillon Beach project: a five-year epidemiological study of naturally occurring *Salmonella* infection in turkeys and their environment. *Avian Dis.*, **21**, 141–59.
Zeitoun, A.A.M. and Debevere, J.M. (1990) The effect of treatment with buffered lactic acid on microbial decontamination and on shelf life of poultry. *Int. J. Food Microbiol.*, **11**, 305–12.
Zeitoun, A.A.M. and Debevere, J.M. (1992) Decontamination with lactic acid/sodium lactate buffer in combination with modified atmosphere packaging effects on the shelf life of fresh poultry. *Int. J. Food Microbiol.*, **16**, 89–98.
Ziegler, F. and Stadelman, W.J. (1955) Increasing shelf-life of fresh chicken meat by using chlorination. *Poult. Sci.*, **34**, 1389–91.
Zhao, T., Ezeike, G.O.I., Doyle, M.P., Hung, Y.-C. and Howell, R.S. (2003) Reduction of *Campylobacter jejuni* on poultry by low temperature treatment. *J. Food Protect.*, **66**, 652–5.
Zottola, E.A., Schmeltz, D.L. and Jezeski, J.J. (1970) Isolation of salmonellae and other airborne microorganisms in turkey processing plants. *J. Milk Food Technol.*, **33**, 395–9.

3 Fish and fish products

I Introduction

Finfish and shellfish are second only to meat and poultry as staple animal protein foods for most of the world. The range of fish products is very large and includes foods prepared by a broad spectrum of both traditional and modern food technology methods. In some countries, fish are a major source of protein. In the last two decades, there has been an extensive expansion in finfish- and shellfish-production primarily due to developments in aquaculture. The catches of wild fish have stagnated at ~90 million metric tones since 1990, whereas aquaculture production has increased to ~30 million metric tones in year 2000 (Figure 3.1; FAO, 1998). Today 20–30% of fish and shellfish used for human consumption are reared in aquaculture.

An increasing proportion of fish in international trade comes from non-industrialized (developing) countries and are fished from or produced in warm tropical waters. Most studies on quality changes and safety aspects have been conducted on fish from temperate waters, and more research in fish from tropical waters is called for. International agreements related to the national ownership of fish stocks has changed the pattern of supply so that foreign vessels, which formerly dominated the catching sector, now also play an expanded role as receiving and processing facilities for fish caught by local fishermen. In both industrialized and non-industrialized nations, considerable amounts of finfish and shellfish are caught by small, short-voyage vessels where little or no ice may be used and on-board sanitary conditions are often not adequate. These fish enter both domestic and international markets.

Finfish and shellfish are harvested from both deep waters remote from land and shallow waters adjacent to coastlines. Estuaries, where ocean and fresh waters intermingle (often a large river), are frequently rich fishing areas for finfish and shellfish, which can be subjected to bacterial contamination from human and animal sources. Commercial fishing and shellfish gathering also takes place in rivers and lakes whose waters can range from pristine to contaminated. Consequently, even though the dominant population of bacteria on finfish and shellfish is remarkably consistent and made-up mainly of saprophytic species, the level of contamination of living fish with bacteria of significance for both public health and sensory quality can vary greatly among localities. Finfish, crustaceans, and molluscan shellfish are normally considered separately because of their different physiologies, modes of life, feeding, and handling/processing requirements after harvest. Subdivision by source into freshwater vs. seawater species, nearshore vs. open ocean species, and warm water (tropical) vs. cold-water species is also helpful for categorizing microbiological characteristics and public health risks. As mentioned, aquaculture is a major source of fish and shellfish in international trade. Some of the circumstances associated with aquaculture rearing require that its products be considered separately from the same products derived from wild stocks.

A Definitions

Foods discussed in this chapter include all products from aquatic and marine animals commonly marketed for human consumption excluding marine mammals, amphibians, reptiles, some convenience foods and animal feeds. The term "fish" is used commonly both as: (1) a specific term for "finfish" (free swimming members of *Pisces* and *Elasmobranchii*), and (2) a generic term to encompass all edible aquatic and marine finfish, molluscan shellfish, and crustaceans. However, to avoid confusion in this

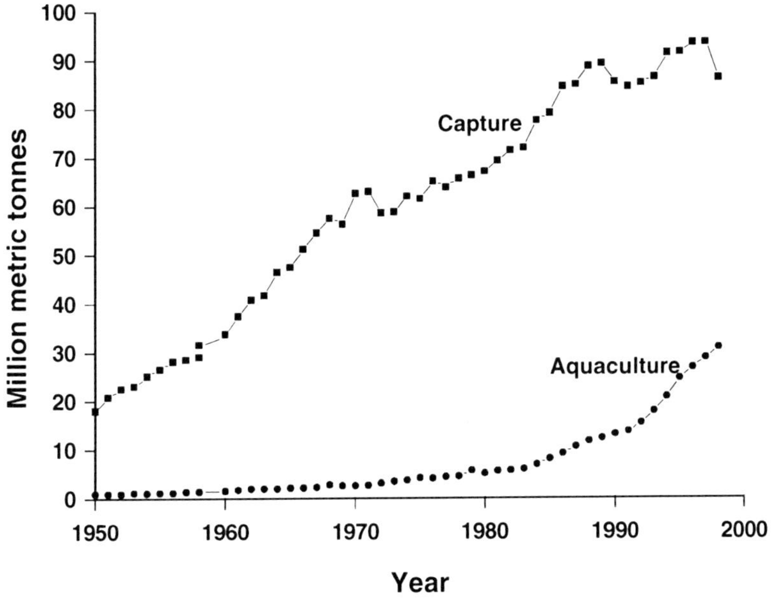

Figure 3.1 Global fish production—wild catches and aquaculture.

chapter, fish will be restricted to the generic definition and finfish will be applied to the more specific definition.

The "crustacea" include shrimp, prawns, crabs, lobsters, crayfish, and related animals having a chitinous exoskeleton.

The terms "mollusc" and "molluscan shellfish" include mussels, cockles, scallops, clams, oysters, abalone, and other aquatic animals with a calcareous shell. These animals are mainly sessile or display a limited range of movement. Their filter feeding behavior has significant impact on the potential health hazards for the consumer. Terrestrial snails are discussed in Chapter 1.

B Important properties

The edible muscle of aquatic animals, like meat and poultry, is high in protein and water. In contrast to warm-blooded animals, fish do not accumulate glycogen and are thus very low in carbohydrate. Finfish may be divided into lean and fatty species. Lean species, such as cod, only accumulate lipid in the liver, whereas fatty species, such as mackerel, can accumulate lipid in the muscles. The gross composition of fish can vary greatly with season and is related to spawning cycles. Lipid and water content rise and fall inversely. This affects both texture and flavor, and for instance the appropriate ripening of herring used in semi-preserved fish products is only possible during times when the lipid content is high. In contrast, the lipid content has little influence on the microbial properties of the product. Typical ranges of the proximate composition of various raw marine finfish are listed in Table 3.1. Gross composition of the several hundred species of aquatic and marine animals used as human food varies widely, although, in most cases, the edible striated muscle portions are roughly comparable.

Crustaceans mainly consists of water and protein and the cooking tends to dehydrate crustaceans and other fish, increasing the relative portions of other macronutrients. Mollusca differ from other fish in having a significant carbohydrate content (Table 3.1).

Table 3.1 Typical proximate analysis (g/100 g) of various raw fish, crustaceans, and mollusc

	% Composition (w/wet weight)				
Fish	Water	Protein	Lipid	Carbohydrate	Ash
Marine finfish					
Cod	80	18	0.7	<0.5	1.2
Tuna	70	23	1	<0.5	1.3
Herring	72	18	9	<0.5	1.5
Salmon	68–78	20	3.5–11	<0.5	1.2–2.5
Freshwater fish					
Nile perch	75–79	15–20	1–10	<0.5	Nd
Trout	72	20	3–6	<0.5	1.3
Crustacean					
Crab	80	18	0.6–1.1	<0.5	1.8
Shrimp	76	20	1	<0.5	1.5
Mollusc					
Oyster	82–85	7–10	2.5	4–5	1.3
Clam	82	13	1	2.6	1.9

Table 3.2 Typical composition of the non-protein nitrogen (NPN) fraction of fish and crustacean

	mg per 100 g wet weight			
Compound	Cod	Herring	Shark	Lobster
Amino acids (total)	75	300	100	3000
Arginine	<10	<10	<10	750
Glycine	20	20	20	100–1000
Glutamic acid	<10	<10	<10	270
Histidine	<1	86	<1	<1
Proline	<1	<1	<1	750
Creatine	400	400	300	0
Betaine	0	0	150	100
Trimethylamine oxide	350	250	500–1000	100
Anserine	150	0	0	0
Carnosine	0	0	0	0
Urea	0	0	2000	0
NPN (total)	1200	1200	3000	5500

Gram and Huss (2000) modified from Shewan (1974).

Fish flesh provides an excellent substrate for the growth of most heterotrophic bacteria with compositional attributes that affect bacterial growth and the related biochemical activities. The carbohydrate content of finfish and crustaceans is negligible, limiting the pH depression associated with the production of lactic acid during *rigor mortis*. Tuna and halibut can reach a pH as low as 5.4, whereas *post-rigor* cod has a pH between 6.0 and 7.0 (Kelly *et al.*, 1966). Molluscs contain 2–5% glycogen (Bremner and Statham, 1983), which supports a substantial decline in pH as the carbohydrate is metabolized by the muscle tissue.

Fish tissues contain high levels of free non-protein nitrogen (NPN) compounds, which are readily available to support the *post-mortem* bacterial growth (Table 3.2). Trimethylamine oxide (TMAO), an odor-less compound, is present in some finfish, and is typically reduced to trimethylamine by spoilage bacteria, producing the characteristic "fishy" smell of spoiled fish. Urea occurs at high levels in elasmobranchs and is metabolized to ammonia. The composition of free amino acids in fish flesh significantly influences the pattern of spoilage and can impact human health, e.g. through formation of biogenic amines (see later).

Levels of nucleotides in fish tissues are important because of their relationship to quality and spoilage. The proportional occurrence of different nucleotides is used as an index of quality, particularly in Japan. The proportion present as ATP, ADP, or AMP initially is largely determined by the physical condition of the fish at the time of capture. Exhausted fish have low levels of ATP and pass quickly through *rigor mortis*. Bacteria are also involved in the degradation of nucleotides, and levels of their breakdown products have been correlated to acceptability, spoilage, and organoleptic deterioration (Manthey *et al.*, 1988; Fletcher *et al.*, 1990; Boyd *et al.*, 1992).

II Initial microflora

The population of microorganisms associated with living fish reflects the microflora of the environment at the time of capture or harvest, but is modified by the ability of different microorganisms (mainly bacteria) to multiply in the sub-environments provided by the skin/shell surfaces, gill areas, and the alimentary canal. Shellfish taken from waters near human habitations will tend to have higher bacterial loads and a more diverse microflora compared with those taken from isolated areas (Faghri *et al.*, 1984). The muscle tissue and internal organs of freshly caught, healthy finfish and molluscan shellfish are normally sterile, but bacteria may be found on the skin, chitinous shell, gills of fish, as well as in their intestinal tract (Baross and Liston, 1970; Shewan, 1977). The circulatory system of some crustaceans is not "closed" and the hemolymph of crabs can harbor substantial levels of bacteria, particularly members of the genus *Vibrio*. Examples of the relative populations of various aerobic bacteria on finfish and crustaceans from different sources are shown in Table 3.3.

Microbial levels vary depending on water conditions and temperature. Finfish and crustaceans from colder (<10–15°C) waters generally yield counts of 10^2–10^4 cfu/cm^2 of skin and gill surface, whereas animals from warm waters have levels of 10^3–10^6 cfu/cm^2. Tropical shrimps carry higher numbers of bacteria, 10^5–10^6 cfu/g, than cold-water species, 10^2–10^4 cfu/g. Counts for intestinal contents vary widely from as low as 10^2 cfu/g in non-feeding fish to 10^8 cfu/g in actively feeding species. Counts in molluscs show marked variation with water temperature from $\leq 10^3$ cfu/g in cold unpolluted water to $\geq 10^6$ cfu/g in warm waters or when bacterial pollution levels are high.

A Saprophytic microorganisms

After capture or slaughter and death, finfish are normally stored in crushed ice or chilled brines, giving rise to changes in the microflora. The most important environmental factor influencing composition of fish microflora is temperature. Typically, bacterial populations on finfish and shellfish from temperate waters are predominantly psychrotrophic, reflecting water temperatures of $\leq 10^\circ$C in the main water mass. However, surface water temperatures may rise during prolonged hot weather and under those conditions pelagic species of finfish (e.g. mackerel and herring) may show increased skin counts and elevated levels of mesophilic bacteria. Pelagic finfish and other surface dwelling creatures in the tropical oceans typically have higher levels of mesophilic bacteria. Both psychrotrophs and mesophiles grow well at ambient temperature (20–35°C), and spoilage occurs in 1–2 days at temperatures above 15°C. Although psychrotrophs are present in significant proportions on fish from warm tropical waters (Gram *et al.*, 1989), iced storage of tropical fish typically leads to a long shelf life (Lima dos Santos, 1981; Deveraju and Setty, 1985; Gram *et al.*, 1989).

The microflora of marine finfish and shellfish is often incorrectly referred to as being predominantly halophilic. In most cases, the microflora are not true obligate halophiles. Instead, the microorganisms are predominantly halotolerant; able to grow over a wide range of salt concentrations, but displaying optimal growth at sodium chloride concentrations of 1–3%. This is enhanced by the common use of ice

Table 3.3 Bacterial genera associated with various raw finfish and crustaceans, and their percentage of the total microflora

Fish type	% Composition									Reference
	Pseudomonas	*Vibrionaceae*	*Acinetobacter–Moraxella*	*Flavobacterium–cytophaga*	Other Gram-negative	Coryneforms	Gram-positive cocci	*Bacillus*	Other or not identified	
Marine fish (temperate)										
North Sea Fish 1932	5	1	56	11	5	–	23	–	–	Shewan (1971)
North Sea Fish 1960	16	2	23	27	10	18	4	–	–	Shewan (1971)
North Sea Fish 1970	22	1	41	10	21	–	1	–	–	Shewan (1971)
Haddock (North Atlantic)	26	2	45	15	–	4	4	2	–	Laycock and Regier (1970)
Flatfish (Japan)	21	29	22	13	–	13	2	–	–	Simidu *et al.* (1969)
"Pescada" (Brazil)	32	–	35	5	18	4	4	–	3	Watanabe (1965)
Shrimp (North Pacific)	10	–	47	22	–	3	7	4	8	Harrison and Lee (1969)
Scampi (UK)	3	–	11	2	–	81	–	–	3	Walker *et al.* (1970)
Marine fish, (tropical)										
Mullet (Australia)	18	–	9	8	–	12	51	2	–	Gillespie and Macrae (1975)
Prawn (India)	11	10	23	6	–	13	6	–	–	Surendran *et al.* (1985)
Sardine (India)	20	28	30	4	–	–	7	–	11	Surendran *et al.* (1989)
Shrimp (Texas Gulf)	22	2	14	9	1	40	11	–	2	Vanderzant *et al.* (1970)
Shrimp (Texas, pond)	2	–	15	25	12	43	3	0.5	1	Vanderzant *et al.* (1971)
Freshwater fish, (temperate)										
Pike (Spain)	10		15		55	10		5	5	González *et al.* (1999)
Brown trout (Spain)		6			40		15		34	
Trout (Spain, reared)	11		7		26	5	45		6	González *et al.* (1999)
Trout (DK, reared)	26	7	47	–	–	15	4	–	–	Spanggaard *et al.* (2001)
Freshwater fish, (tropical)										
Nile perch (Kenya)	6	2	43	–	9	5	30	5	–	Gram *et al.* (1990)
Catfish (India)	–	–	10	–	–	–	50	40	–	Venkataranan and Sreenivasan (1953)
Carp (India)	–	–	20	11	30	–	39	–	–	Venkataranan and Sreenivasan (1953)

Table 3.4 Culturable microorganisms on gills, skin, and in the intestinal tract sampled from a total of 49 rainbow trout

	Number of isolates			
Group/genera	26 Skin samples	38 Gill samples	33 Gut samples	Total number of isolates
Pseudomonas spp.	27	96	23	146
Acinetobacter/Moraxella	73	133	32	238
Enterobacteriaceae	0	46	168	214
Vibrionaceae	4	59	100	163
Other Gram-negative	26	53	77	156
Gram-positive	15	23	38	76
Yeasts	0	1	23	24
Total	145	412	461	1 018

Modified from Spanggaard *et al.* (2001).

to chill fish and shellfish, which exposes the bacterial population to decreasing salinity during storage, favoring the survival and growth of halotolerant species. One example of the effect of salinity is the bacteria found in the intestines of finfish, halotolerant *Vibrio* spp. often being reported as dominant in marine species, with *Aeromonas* spp. dominant in fresh water fish. These genera alternate in anadromous fish, which spawn in fresh water, but spend their adult life in the ocean.

The number and variety of microorganisms in the intestine are determined by the quantity and origin of food consumed by fish. Non-feeding fish have very low levels of bacteria in the intestine. During feeding, fermentative Gram-negative bacteria often become dominant. Bacteria living on the surfaces of marine animals are phenotypically capable of living on carbon sources such as amino acids, peptides, and other non-carbohydrate sources (Table 3.4). Utilization of these substrates typically leads to the production of slightly alkaline conditions in stored fish products. Members of the Enterobacteriaceae commonly found in the intestine of warm-blooded and reptilian species are not normally isolated from finfish captured away from coastlines.

Typically, bacteria from skin and gills are predominantly aerobic, although facultative bacteria, particularly *Vibrio* spp., may occur in high numbers on pelagic fish (Simidu *et al.*, 1969). Obligately anaerobic bacteria are uncommon on the surface of fish but can occur in significant numbers in the intestine (Matches and Liston, 1973; Matches *et al.*, 1974; Huber *et al.*, 2004). Lactic acid bacteria, in particular carnobacteria, are also commonly isolated from fish gut (Ringø and Gatesoupe, 1998).

The bacteria on finfish and shellfish are predominantly Gram-negative for fish from temperate waters. A higher proportion of Gram-positive cocci and *Bacillus* spp. can be found on some fish from warm, tropical, waters and some studies report as much as 50–60% of the microflora being of these types (Shewan, 1977). However, the microflora of fish from warm, tropical, waters may also be dominated by Gram-negative bacteria (Table 3.3, Gram *et al.*, 1990). The microflora of living fish from temperate waters is remarkably consistent, and commonly includes members of the genera *Psychrobacter, Moraxella, Pseudomonas, Acinetobacter, Shewanella* (previously *Alteromonas*), *Flavobacterium, Cytophaga*, and *Vibrio/Aeromonas. Corynebacterium* and *Micrococcus*. The Gram-negative bacteria on warm water finfish are similar to those on cold-water fish. Fresh-water fish show similar patterns except that *Aeromonas* replaces *Vibrio. Psychrobacter–Acinetobacter–Corynebacterium*, and *Micrococcus* dominate on crustaceans with lesser proportions of *Pseudomonas*. The microflora of molluscs is similar to that of fish, but *Vibrio* spp. are more prominent and often dominate the microflora of oysters. As molluscs are commonly associated with inshore environments, their microflora may reflect terrestrial influences, and Enterobacteriaceae and Streptococcacae occur. Run-off from land can have consequences for the microflora of fish cultured close to land. Thus, whilst *Listeria monocytogenes* cannot be detected on newly caught fish from open waters, the incidence of the bacterium may be high on fish caught in waters with agricultural run-off (Ben Embarek, 1994a; Jemmi and Keusch, 1994; Huss *et al.*, 1995).

In addition to bacteria, yeasts, such as *Rhodotorula*, *Torulopsis*, *Candida* spp., and occasionally fungi, are reported from finfish and shellfish (Table 3.4). However, literature on their occurrence on living fish is scant (Morris, 1975; Sikorski *et al.*, 1990). Yeasts are fairly widespread in fresh and salt waters but are present at much lower levels than bacteria. Fungi, except for some specialized planktonic forms, appear restricted mainly to estuarine and fresh waters. Their occurrence on fish is probably adventitious except in a few instances where they are parasitic, e.g. for salmonid fish (usually when debilitated by high water temperatures or spawning) and crustaceans by chitinolytic fungi, especially *Fusarium solani*, have been reported.

B Pathogens and toxicants

Fish and fish products may cause a variety of food-borne diseases in humans and accounted for 19% of 967 food-borne outbreaks (with known cause) in the United States from 1993 to 1997 (Table 3.5; Olsen *et al.*, 2000). No food commodity was identified in almost two-third (1 784 of 2 751) of the outbreaks. Shellfish (molluscs and crustaceans) caused as many cases (~1 900) as did poultry, which is consumed in much larger quantities. The major etiological agents are bacteria, virus, parasites, aquatic toxins, and biogenic amines (Table 3.6). Viral diseases and shellfish toxins are typically carried by shellfish, whereas intoxication by the marine toxin, ciguatera, and biogenic amines are the major causes of disease from fin fish.

In Japan, seafoods were involved in several reported outbreaks (Table 3.7). In particular, molluscan shellfish were causing disease being responsible for two-third of the diseases caused by seafood. Also, several cases of pufferfish toxin poisoning caused deaths (Japanese Ministry of Health, Labour and Welfare, 2002). These cases were typically associated with preparation of pufferfish in the home and not in restaurants. The etiological agents identified included especially Norwalk virus (and other small round structured viruses) and *Vibrio parahaemolyticus*. Also, in 1998, soy sauce marinated salmon roe caused a major outbreak of *Escherichia coli* O157:H7 (IARS, 1998).

Table 3.5 Food implicated in food-borne disease in the United States 1993–1997

	Outbreaks		Cases		Deaths	
Food	Number	%	Number	%	Number	%
Meat	66	2.4	3 205	3.7	4	13.8
Pork	28	1.0	988	1.1	1	3.4
Poultry	52	1.9	1 871	2.2	0	0.0
Other meat	22	0.8	645	0.7	2	6.9
Shellfish	47	1.7	1 868	2.2	0	0.0
Fish	140	5.1	696	0.8	0	0.0
Egg	19	0.7	367	0.4	3	10.3
Dairy products	18	0.7	313	0.4	1	3.4
Ice cream	15	0.5	1 194	1.4	0	0.0
Bakery goods	35	1.3	853	1.0	0	0.0
Fruits and vegetables	70	2.5	12 369	14.4	2	6.9
Salads	127	4.6	6 483	7.5	2	6.9
Other	66	2.4	2 428	2.8	0	0.0
Several foods	262	9.5	25 628	29.8	1	3.4
Total (known foods)	967	35.2	58 908	68.5	16	55.2
Known food	967	35.2	58 908	68.5	16	55.2
Unknown food	1 784	64.8	27 150	31.5	13	44.8
TOTAL	2 751	100.0	86 058	100.0	29	100.0

Modified from Olsen *et al.* (2000).

Table 3.6 Etiological agents implicated in disease from fish and shellfish (crustaceans and molluscs) from 1993 to 1997

	Number of outbreaks (% of total) caused by	
Agent	Shellfish	Fin fish
Bacteria	5 (11%)	2 (1%)
Virus	11 (23%)	1 (<1%)
Parasites	0 (0%)	0 (0%)
Aquatic biotoxins	3 (6%)	55 (39%)
Histamine	2 (4%)	66 (47%)
Known	21 (45%)	124 (89%)
Unknown	26 (55%)	16 (11%)
Total	47 (100%)	140 (100%)

Modified from Olsen *et al.* (2000).

Table 3.7 Food implicated in reported food-borne diseases in Japan 1999–2001. Number of reported outbreaks, cases and deaths modified from Japanese Ministry of Health, Labour and Welfare (2002)

	Outbreaks		Outbreaks[a]		Cases		Cases[a]		Deaths[a]	
Food	NO	%	NO	%	NO	%	NO	%	NO	%
Meat, poultry, pork	137	2.0	137	2.0	2 826	2.7	2 826	3.0	0	0.0
Fish, total	594	8.6	594	8.6	10 212	9.8	10 212	10.8	6	40.0
Molluscan	305	4.4	305	4.4	5 749	5.5	5 749	6.1	1	6.7
Pufferfish	80	1.2	80	1.2	126	0.1	126	0.1	5	33.3
Fish products	46	0.7	46	0.7	3 316	3.2	3 316	3.5	0	0.0
Egg	115	1.7	115	1.7	2 479	2.4	2 479	2.6	0	0.0
Dairy products	11	0.2	10	0.1	14 246	13.6	4 246	4.5	0	0.0
Cereal, grain	67	1.0	67	1.0	1 143	1.1	1 143	1.2	0	0.0
Vegetable	245	3.6	245	3.6	3 048	2.9	3 048	3.2	4	26.7
Snacks	47	0.7	47	0.7	1 792	1.7	1 792	1.9	0	0.0
Several foods	274	4.0	274	4.0	10 367	9.9	10 367	11.0	2	13.3
Others	1 305	19.0	1 305	19.0	39 720	38.1	39 720	42.1	2	13.3
Total known	2 841	41.3	2 840	41.3	89 149	85.4	79 149	83.9	14	93.3
Unknown	4 031	58.7	4 031	58.7	15 234	14.6	15 234	16.1	1	6.7
Total	6 872	100.0	6 871	100.0	104 383	100.0	94 383	100.0	15	100.0

[a]Excluding a major outbreak (10 000 cases) from year 2000 of staphylococcal enterotoxin caused by milk powder.

Bacteria. Pathogenic microorganisms reported to be associated with seafood are listed in Table 3.8. These microorganisms are categorized according to whether they originate in the aquatic environment (i.e. potentially on the raw material), the general environment, or whether they are the result of contamination/pollution from the human/animal resevoir.

Some bacteria are both frank pathogens for humans and indigenous members of the normal microflora of the marine environment or marine animals. These include psychrotrophic types of *Clostridium botulinum* and various *Vibrio* species (Hackney and Dicharry, 1988). *Plesiomonas shigelloides* and *Aeromonas hydrophila* are both aquatic bacteria that have been associated with gastroenteritis in humans. Although they can be isolated in high numbers from some cases of diarrhea, feeding studies with volunteers have failed to reproduce symptoms and their precise role in human illness remains unclear.

Cl. botulinum is derived most commonly from sediments, and can be assumed to be present on whole fish (Dodds, 1993; Dodds and Austin, 1997; Pullela *et al.*, 1998). Disease is caused by a neurotoxin and the different types of toxin are used to distinguish *Cl. botulinum* serovars; from type A to type G. Serotype E and non-proteolytic strains of types B and F can be isolated from the intestine and occasionally from the skin of marine fish (Hobbs, 1976). Psychrotrophic *Cl. botulinum* are typically present only in low numbers on marine fish, but has been reported to occur at relatively high levels in pond-raised trout (Huss *et al.*, 1974a), and type E is considered a truly aquatic bacterium (Huss, 1981). Finfish may

Table 3.8 Pathogenic agents potentially transmitted to man from fish and fish products

	Agents of disease from fish and fish products				
Natural habitat	Bacteria	Virus	Parasites	Aquatic toxins	Biogenic amines
Aquatic sources	*Cl. botulinum* E (B and F), *V. parahaemolyticus*, *V. cholerae*, *V. vulnificus*, *Aeromonas* spp., *Plesiomonas*		Nematodes, (*Anisakis*, *Pseudoterranova*), Cestodes, Trematodes	Ciguatera, Tetrodotoxin, PSP, ASP, DSP	Enterobacteriaceae, *Photobacterium*
General environment	*L. monocytogenes*, *Cl. botulinum* A and B				
Animal–man-resevoir	*Staph. aureus*, *Salmonella*, *Shigella*, *E. coli*	Noro, Hep A, B, SRV, Rotavirus			Enterobacteriaceae

Modified from Huss *et al.* (2001).

themselves die of botulism as a result of consuming dead finfish (Eklund *et al.*, 1982b). In animals raised by aquaculture, poor management practices aggravate the occurrence of pathogenic microorganisms, e.g. increased incidence of *Cl. botulinum* in trout raised in ponds with earth bottoms (Cann *et al.*, 1975), and botulism epidemics in young salmon populations associated with cannibalism (Eklund *et al.*, 1982b). Increased incidence of botulism in finfish often appears to be related to excessive feed levels.

Mesophilic *Vibrio* species have been isolated from both pelagic and bottom dwelling fish (Simidu *et al.*, 1969; Baross and Liston, 1970; Sera and Ishida, 1972; Joseph *et al.*, 1982). Among the potentially pathogenic *Vibrio* occurring naturally on finfish and shellfish, *Vibrio parahaemolyticus* is most widespread. During recent years, serotypes causing seafood-borne disease have changed and for instance serotype O3:K6 is now commonly associated with disease (CAOF, 2000; Chowdhury *et al.*, 2000). During cholera outbreaks, *Vibrio cholerae* can be isolated from fish, but it is also an indogenous marine species and can be isolated from waters with no outbreaks (Rogers *et al.*, 1980; Feachem, 1981, 1982; Feachem *et al.*, 1981). The mesophilic *Vibrio* species are most commonly found in in-shore waters with reduced salinity. For example, *V. vulnificus* is commonly found in estuarine fish, particularly bottom feeders, but is less common in off-shore fish (DePaola *et al.*, 1994). *Vibrio* is the genus most often implicated in diseases of bacterial origin resulting from eating contaminated shellfish (Janda *et al.*, 1988; Levine *et al.*, 1993).

The incidence and levels of mesophilic vibrios present on marine animals is greatly affected by water temperature (Kaneko and Colwell, 1973; Kelly, 1982; Williams and LaRock, 1985; West, 1989; O'Neil *et al.*, 1990). Typically, they multiply rapidly at temperatures between 20°C and 40°C. This is reflected in the large numbers of the organisms isolated from molluscan shellfish when water temperatures rise to 30°C, and their virtual absence from molluscs taken from cold waters (IOM, 1991), although they have also been isolated from crabs taken from cold waters (Faghri *et al.*, 1984). *Vibrio parahaemolyticus* is widespread when water temperatures exceed 15°C (Kaneko and Colwell, 1973; Liston and Baross, 1973) and incidence of disease is highly seasonal (Figure 3.2). *Vibrio* levels in the estuarine environment are dependent on the time of day, depth, and tidal levels (Koh *et al.*, 1994). *Vibrio parahaemolyticus*, *V. cholerae*, and *V. vulnificus* are routinely part of the microflora of crustaceans captured from estuarine waters (Davis and Sizemore, 1982; Faghri *et al.*, 1984; Molitoris *et al.*, 1985; Varma *et al.*, 1989). In Western countries, seafood-related illness caused by pathogenic *Vibrio* species is commonly associated with crustaceans or molluscan shellfish, whereas finfish are a common vehicle for outbreaks in Japan and other Asian countries. Disease caused by *V. vulnificus* is particularly associated with the consumption of live oysters but a Japanese outbreak caused by squilla (a crustacean) has also been reported (Ono *et al.*, 2001). The major form of the disease is primary septicaemia, i.e. septicaemia with no apparent infectious focus (Levine and Griffin, 1993; Oliver and Kaper, 1997). Other presentations are wound infections or gastrointestinal infection. Of these disease forms gastrointestinal infection is very rare,

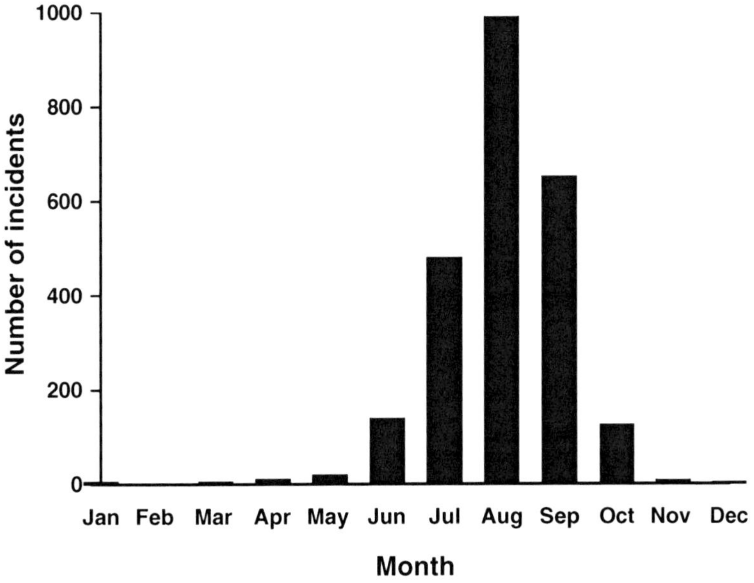

Figure 3.2 Number of reported outbreaks of *Vibrio parahaemolyticus* gastroenteritis from 1989 to 1999 as a function of time of year (CAOF, 2000).

only accounting for 3% of the reported cases involving *V. vulnificus* in the United States (Evans *et al.*, 1999). There are no reports of any gastrointestinal infections with *V. vulnificus* as the etiological agent from Europe. Detailed information on *V. parahaemolyticus* can be found in a recent US FDA risk assessment (FDA, 2001a).

Listeria monocytogenes occurs in the general environment (Farber, 1991; Dillon and Patel, 1992; Farber and Peterkin, 2000) and can be isolated from finfish and shellfish; particularly from fish caught or cultured close to land with agricultural run-off (Huss *et al.*, 1995). It is often isolated from ready-to-eat foods in low numbers. Although its occurrence in some fish products is explained by the lack of a listericidal process during processing, the organism does have an almost unique ability to colonize food processing environments, and the same DNA-type had been isolated from a fish smoke house over a 4-year period (Fonnesbech Vogel *et al.*, 2001). Smoked mussels and gravad trout were recently implicated in cases of human listeriosis (Mitchell, 1991; Misrachi *et al.*, 1991; Ericsson *et al.*, 1997; Brett *et al.*, 1998) and the bacterium has also been a cause of febrile gastroenteritis in otherwise healthy individuals consuming cold-smoked trout (Miettinen *et al.*, 1999). It is of particular importance as a food-borne pathogen in ready-to-eat products (WHO/FAO, 2001; FDA/FSIS, 2003).

Other environmental microorganisms like mesophilic, halotolerant *Cl. botulinum* (types A and B) can also cause disease from fish products and are particularly important in salted products kept at high temperatures.

Clostridium perfringens is frequently isolated from fish but is probably derived from human and animal fecal pollution (Matches *et al.*, 1974). *Campylobacter jejuni*, *Yersinia enterocolitica*, *E. coli*, *Shigella* spp., and *Salmonella* spp. have all been isolated from finfish or shellfish taken from waters subject to human sewage pollution or terrestrial run-off. These organisms typically originate from the human–animal reservoir; however, some studies have shown that once introduced into warm waters, *Salmonella* and *E. coli* may persist for long periods of time (Jiminez *et al.*, 1989). Consequently, in some instances, it is unclear if the microorganisms are inherent in the fish or acquired during capture or harvesting. When evaluating the presence and/or level of enteric organisms, it should also be remembered that these bacteria may be sub-lethally damaged when exposed to stressful conditions and escape enumeration on selective media. Further, some studies indicate that enteric bacteria and some *Vibrio* spp. can enter a so-called "viable-but-non-culturable" stage in which they survive the low-nutrient concentrations or the low temperatures in the environment and escape detection also on standard non-selective media. Subsequently, when conditions become more favorable, they may revert to a culturable stage (Turner *et al.*, 2000; Ohtomo and Saito, 2001).

Viruses. Viruses of public health concern have been isolated from seafood, particularly molluscan shellfish, which concentrate and retain the virus particles (Jaykus, 2000). Viruses are a natural part of the marine environment and represent the most abundant life form in the sea. However, none of these are pathogenic to man (Lees, 2000). The viruses causing seafood-borne disease all have their natural niche in the human gastrointestinal tract and are derived from human feces (Cliver, 1988). Their presence in seafood is a consequence of poor hygiene because they are transmitted either by contaminated water or by food handlers.

Viral diseases transmitted by human enteric viruses either cause viral gastroenteritis or viral hepatitis (Caul, 2000). Most seafood-borne outbreaks are caused by so-called Norovirus or by Hepatitis A virus (Morse *et al.*, 1986; Gerba, 1988; IOM, 1991; ACMSF, 1995; EEC, 2002). Viruses are divided into groups depending on organization and transcription of DNA or RNA (Table 3.9). The largest reported outbreak of shellfish-borne hepatitis A involved almost 300 000 cases in Shanghai, China, in 1988 and was attributed to clams (Halliday *et al.*, 1991; Cheng *et al.*, 1992).

Parasites. Parasites can occur extensively in finfish and crustaceans where their natural life cycle typically involve passing through several hosts with marine or terrestrial mammals as final hosts. Few of the many helminthic parasites are capable of infecting humans. Several reviews can be consulted for more detailed information (Higashi, 1985; Olson, 1987; WHO, 1995; Cross, 2001; Orlandi *et al.*, 2002). The dominant cause of illness is the consumption of raw or mildly processed seafoods (IOM, 1991; WHO, 1995) in which the parasites are not killed. Approximately 50 helminthic parasites from fish may cause disease in man. Three groups of parasites are important as human pathogens from fish and fish products; namely nematodes (round worms), trematodes (flukes), and cestodes (tape worms). Most of the species have lifecycles involving several stages developing in several hosts. For instance, the nematode *Anisakis simplex* ("herring worm") is transmitted from shrimp to fish, to its normal final host in marine mammals. Several trematodes have terrestrial mammals as final hosts and their life cycles invole snails and fish as intermediate hosts. Whilst parasitic infections are mostly limited to particular

Table 3.9 Groups of virus causing gastrointestinal disease from seafood

Virus	Type	Family	Associated with seafood-borne disease	Comment
Noro	SS RNA	Caliciviridae	Frequently	
Hepatitis A	SS RNA	Picornaviridae	Frequently	
Hepatitis E	SS RNA	Caliciviridae?	Not documented	Cause of enteric non-A and non-B hepatitis. Outbreaks associated with drinking water
Astrovirus	SS RNA	Astroviridae	Astrovirus from oysters were suspected in *one* outbreak	Few food-borne cases
Rotavirus	DS RNA	Reoviridae	Not documented	Isolated from sewage
Adenovirus	DS DNA	Adenoviridae	Not documented	Isolated from sewage

From Huss *et al.* (2004) based on Lees (2000) and Caul (2000).
SS, single stranded and DS, double stranded.

Table 3.10 Prevalence of *A. simplex* in reared and wild caught marine fish species

Fish	Origin	Number of samples	% Positive	Rereference
Farmed salmon	Washington	50	0	Deardorff and Kent (1989)
Farmed salmon	Norway	2 832	0	Angot and Brasseur (1993)
Farmed salmon	Scotland	867	0	Angot and Brasseur (1993)
Farmed coho salmon	Japan	249	0	Inoue *et al.* (2000)
Farmed rainbow trout	Japan	40	0	Inoue *et al.* (2000)
Wild salmon	Washington	237	100	Deardorff and Kent (1989)
Wild salmon	North Atlantic	62	65	Bristow and Berland (1991)
Wild salmon	West Atlantic	334	80–100	Beverley-Burton and Pippy (1978)
Wild salmon	East Atlantic	34	82	Beverley-Burton and Pippy (1978)
Wild coho salmon	Japan	40	100	Inoue *et al.* (2000)
Sardines	Mediterranean	7	14	Pacini *et al.* (1993)
Herring	Mediterranean	4 948	86	Declerck (1988)
Herring	Pacific Ocean	127	88	Myers (1979)
Kulmule	Mediterranean	40	25	Pacini *et al.* (1993)
Cod	Pacific Ocean	509	84	Myers (1979)

geographical areas where the parasites have adapted to specific hosts and niches, these patterns may slowly be changing. The increase in international travel and international food trade contributes to transfer of food-borne parasitic diseases (Orlandi, 2002).

In cold-water regions, tapeworms of the genus *Diphyllobothrium* and round worms of the genera *Anisakis* and *Pseudoterranova* are reported most frequently. For example, Adams *et al.* (1994) isolated *Anisakis* from salmon and mackerel sushi at a rate of 10% and 5%, respectively, but did not isolate the nematode from tuna or rockfish. In another investigation, *Anisakis* was isolated from 40% of the sides of raw chum salmon examined at levels of 1–3 parasites per 200 g (Gardiner, 1990). *Anisakis simplex* larvae occur in a variety of marine fish (Table 3.10), however, they are not found in farmed fish as the parasites originate from the live feed used by wild fish. Salmon have caused cases of diphyllobothriasis (Ruttenber *et al.*, 1984), and several fish species have been the cause of anisakiasis (Muraoka *et al.*, 1996; Machi *et al.*, 1997).

It is estimated that ~40 million people are infected with trematodes by food-borne infection and the vast majority (~38 million) of these originates from fish and fish products (WHO, 1995). The major causes of fish-borne trematode infections are the liver flukes *Opisthorchis felineus*, *Opisthorchis viverrini*, and *Clonorchis sinensis*, and the lung flukes belonging to *Paragonimus* spp. Intestinal flukes from the families Heterophyidae and Echinostomatidae also cause infections (WHO, 1995). Trematodes are typical of some warm water areas and are endemic in several areas in particular in China and South-East Asia. Carps are important intermediate hosts for liver flukes, whereas lung flukes are transmitted by crustaceans (freshwater crabs) and intestinal flukes are carried by a range of aquatic organisms (fish, molluscs, and crustaceans).

The larvae of *Ani. simplex* are, to some extent, resistant to curing and marinating, but can be inactivated by °deep-freezing at −17° to −20°C for 24 h (Ganowiak, 1990). At sufficient levels of NaCl and acid, as used in a range of European fish products, the larvae are inactivated. Karl *et al.* (1995) reported that *Ani. simplex* larvae were not killed in 17 weeks by 4–5% NaCl in 2.5% acetic acid, whereas 8–9% NaCl in the same amount of acid killed the larvae in 5–6 weeks (ICMSF, 1996).

Trematode infections can be limited by a range of control measures, including general improvement of sanitation, and elimination of the use of animal waste as fertilizers in fish-ponds or the use of "night soils" as fertilizers on land (WHO, 1995). As all the trematodes require several hosts, removal of the intermediate hosts (snail) is a control procedure to be investigated further. Very few studies have been conducted on eliminating trematodes from fish products. However, several food-processing steps are effective in eliminating trematodes. For example, *O. viverrini* is eliminated by heating (5 min at 80°C), by exposure to organic acids (1–2 h in 4%) or by salting (10% in 4 h). Also, freezing (−20°C for 3–4 days or 28°C for 1 day) kills the parasites (WHO, 1995).

Protozoans such as *Cryptosporidium parvum*, *Cyclospora*, and *Giardia (lamblia) intestinalis* are mainly transmitted by contaminated water and have caused several food-borne outbreaks due to consumption vegetables or fruits being grown or washed with the water. There have been no cases so far linking protozoan diseases to fish consumption.

Aquatic toxins. In recent years, seafood toxins have been responsible for ~30% of all seafood-borne outbreaks of illness in the United States, with the majority of cases being associated with consumption of finfish (Olson *et al.*, 2000; Table 3.5). Available epidemiological data suggest that a similar situation exists globally. The principal seafood-associated intoxications having a microbiological origin include paralytic shellfish poisoning (PSP), diarrhetic shellfish poisoning (DSP), neurotoxic shellfish poisoning (NSP), amnesic shellfish poisoning (ASP) (also known as domoic acid poisoning), and ciguatera. Of these, ciguatera is the most common, and together with scombroid toxicity it accounts for almost two-third of seafood borne disease in the United States (Ragelis, 1984; Taylor, 1986, 1988; Table 3.5). PSP, DSP, and NSP are all caused by toxins produced by dinoflagellates, and ASP by a toxin produced by a diatom. These intoxications typically result from consumption of bivalve molluscs that have been feeding on toxigenic algae, accumulating the toxin in their organs. Planktonivorous fish may consume large numbers of toxigenic algae, particularly those associated with PSP and DSP. If such fish are eaten without the removal of the internal organs, they can be quite poisonous and human PSP fatalities have been reported in the Philippines and Guam (IOM, 1991). The concentration of algal toxins like PSP in the animals decreases when the algal blooms diminish. This clearance may, at least partially, be caused by bacterial degradation of PSP (Smith *et al.*, 2001).

The toxin(s) causing ciguatera originate from toxigenic microalgae growing in and around tropical coral reefs. The toxins are passed up the marine food chain through herbivorous reef finfish to more far-ranging carnivorous species (see below). The responsible toxigenic microorganisms, their toxins,

Table 3.11 Toxigenic algae and related conditions associated with seafood

Disease	Microorganisms	Toxin	Incriminated seafood
Paralytic shellfish poisoning	*Alexandrium catenella* *Alexandrium tamarensis* Other *Alexandrium* spp. *Pyrodinium bahamense* *Gymnodinium catenatum*	Saxitoxin Neosaxitoxin Gonyautoxins Other saxitoxin derivatives	Mussels, oysters, clams, scallops, planktonivorous fish
Diarrhetic shellfish poisoning	*Dinophysis fortii*, *Dinophysis acuminate*, *Dinophysis acuta*, *Dinophysis mitra*, *Dinophysis norvegica*, *Dinophysis sacculus*, *Prorocentrum lima*, Other *Prorocentrum* spp.	Okadaic Acid Dinophysistoxin Pectenotoxin Yessotoxin	Mussels, scallops, clams
Neurotoxic shellfish poisoning	*Gymnodinium breve*	Brevetoxins	Oysters, mussels, clams, scallops
Amnesic shellfish poisoning	*Pseudonitzschia pungens*	Domoic Acid	Mussels, clams
Ciguatera	*Gambierdiscus toxicus* *Ostreopsis lenticularis*	Ciguatoxin Maitotoxin Scaritoxin	Reef-associated fish
Pufferfish poisoning	??	Tetrodotoxin	Pufferfish (Tetraodontia)

and the seafood commonly involved are shown in Table 3.11. Humans typically become intoxicated from eating the toxic fish. Algal toxins include a diverse group of chemical structures.

The toxins produced by algae and dinoflagellates are typically associated with specific localities and often associated with algal blooms (Taylor, 1990). There are indications that the incidence of harmful algal blooms may be increasing globally (Hallegraeff, 1993). All the seafood toxins are heat resistant and are not destroyed by cooking. Nor can they be detected organoleptically or by routine microbiological analysis. Although there are continuing improvements in methods for the direct detection of algal toxins (Croci *et al.*, 1994), most surveillance is based on the identification of toxigenic phytoplankton organisms. The identification is based on morphological characteristics determined by microscopic examination. Precise identification of toxigenic species usually requires expertise based on familiarity with the algal groups involved. Descriptions of toxigenic species are in various specialist publications (Anderson *et al.*, 1985; Graneli *et al.*, 1990). The toxigenic organisms are not normally detectable in the seafood, and identification depends on toxin analyses. Surveillance and investigation of outbreaks are primarily based on analysis of indicator animals (e.g. mussels) or suspected toxic commercial species. In certain locations, water samples (plankton tow samples) are periodically examined for probable toxigenic algae. The last decade has seen an improvement of analytical techniques for detection of some of the aquatic toxins, including cell bio-assays for detection of PSPs (Manger *et al.*, 1993, 1995), and HPLC analyses for PSPs, DSPs, and other toxins. However, in most countries the official method is the mouse bioassay.

Ciguatera has been described as a syndrome rather than a specific disease because patients experience a wide variety of symptoms, mainly neurological and often bizarre symptoms' because a number of different toxins have been implicated. However, the toxins most consistently present, ciguatoxins and maitotoxin, have been isolated from both *Gambierdiscus toxicus* and incriminated finfish (IOM, 1991). The finfish most commonly involved are inhabitants of coral reefs or predators of such fish (Lehane and Lewis, 2000). Toxin is accumulated by herbivorous species grazing on algae due to incidental ingestion of *G. toxicus*. The toxin is passed up the food chain when these fish are eaten by predators, effectively concentrating toxin levels. Some of the tropical species that may be toxic are shown in Table 3.12. Large predatory species such as amberjack, moray eel, snappers, barracuda, and groupers tend to be most toxic. Toxicity fluctuates among species and fishing areas, with many species such as groupers

Table 3.12 Fish associated with ciguatera

Common fish name	Genus or family name
Amberjack	*Seriola* spp.
Snapper	*Lutjanidae*
Grouper	*Serranidae*
Goatfish	*Mullidae*
Jacks	*Carangidae*
Barracuda	*Sphyrenidae*
Ulua	*Caranx* spp.
Wrasse	*Labridae*
Surgeonfish	*Acanthuridae*
Moray Eel	*Muraeinidae*
Pampano	*Trachinotus* spp.
Rabbit Fish	*Signidae*
Parrot Fish	*Scaridae*
Spanish Mackerel	*Scomberomorus*
Triggerfish	*Balistidae*
Angelfish	*Pomancanthidae*
Hogfish	*Lachnolaimus*
Sardines	*Sardina, Sardinops*
Miscellaneous Reef Fish	

and some snappers being non-toxic most of the time. "Hot spots" occur on reef areas where toxicity levels are persistently high, although toxicity in these areas may persist for long periods or disappear relatively quickly. It is not currently possible to screen reef areas for the toxic microalgae because analytical techniques for detecting the toxin in suspected toxic fish are poorly developed and detection of the microalgae in water samples is complicated by their growth in close association with sessile brown macroalgae. The present approach to control is embargoing potentially toxic species caught near known "hot spot" areas and discouraging commercial and sport fishing in these areas. The ciguatera problem is greatest in fisheries of the Caribbean region and around islands in the tropical Pacific. However, mainland regions abutting tropical oceans such as Australia, India, Bangladesh, South-East Asian countries, Central and (some of) South America and South-Eastern United States encounter cases. Sporadic cases of ciguatera have been observed increasingly among tourists returning home from tropical island vacations and from fish imported to northern countries from endemic toxic areas (IOM, 1991).

Puffer fish poisoning is caused by the presence of tetrodotoxin in species of *Tetraodontiae*. This has traditionally been a problem in Japan (Table 3.7) where these species are considered a delicacy. The toxin accumulates in specific organs like the gonads. Tetrodotoxin has been assumed to be an endogenous ichthyosarcotoxin. However, it has also been suggested that the compound, or a derivative of it, may be produced by common marine bacteria, particularly *Vibrio* spp. (Sugita *et al.*, 1989; Lee *et al.*, 2000). It has been suggested that the bacteria may be the ultimate source of the toxin, but evidence available at present is insufficient to confirm that the compound in puffer fish is of microbiological origin, and some researchers believe that the detection of tetrodotoxin in bacterial cultures is an artifact (Matsumura, 1995, 2001). There are also reports of production of some marine algal toxins by bacteria, but again the evidence is insufficient to incriminate them as agents of human intoxication associated with finfish or shellfish. Any member of the order *Tetraodontiformes* is potentially poisonous. The EU has prohibited import of fish from this family (EEC, 1991b).

The severity of illness due to intoxications depends on the dose ingested and the general state of health of the patient. Death can result from PSP or puffer fish poisoning, but there are no *sequelae* in surviving patients. Death is less likely from ciguatera, although extreme cases involving respiratory failure have been reported (Gillespie *et al.*, 1986). However, disabling effects are long lasting and may recur after apparent recovery, even after several years. Also, okadaic acid may act as a carcinogen and promote tumor formation (Fujiki and Suganuma, 1999). Deaths have not been reported for NSP or DSP, and symptoms seem to be self-resolving. Fatalities have been associated with domoic acid poisonings, as has serious mental impairment of elderly patients.

For a more detailed discussion of seafood toxins of microbiological origin, the reader is referred to Hallegraeff (1993), Lehane and Lewis (2000), and Whittle and Gallacher (2000). A well-illustrated guide for the identification of potentially toxic marine algae was compiled by Larsen and Moestrup (1989).

Scombroid toxin. Scombroid poisoning is caused by consumption of fish containing high levels of histamine (and other biogenic amines) resulting from histidine decarboxylase activity of bacteria multiplying on the fish after death.

Scombroid poisoning differs from other seafood-associated intoxications because the toxins are the result of the bacterial production of biogenic amines resulting from decarboxylation of their parent compounds (i.e. amino acids). Studies on the mechanism of scombroid poisoning have focused on the role of histamine produced via histidine decarboxylase from histidine in muscle tissue. However, there is uncertainty concerning the levels of histamine in fish that are necessary to cause symptoms in humans because of protoxic effects of other biogenic amines produced simultaneously. James and Olley (1985) emphasized the probable role of combinations of biogenic amines, discussing the potential toxic synergy between histamine and cadaverine. Fortunately, scombroid poisoning is usually a mild

Table 3.13 Fish associated with scombroid poisoning

Common name	Genus
Mahi mahi	*Coryphaena*
Tuna	*Thunnus*
Bluefish	*Pomatomus*
Marlin	*Makaira*
Mackerel	*Scomber*
Blue Ulua	*Caranx*
Opelu	*Dicapterus*
Redfish	*Sebastes*
Salmon[a]	*Onchorhynchus /Salmo*

[a]High levels of histamine has been reported on cold-smoked salmon, however, no reports of scombroide toxicity have been related to salmon.

disease with symptoms of vomiting, diarrhea, and allergy-like reactions (including puffiness around the eyes and mouth, tingling, and itching). Recovery is complete and the disease is not a significant cause of death. Some studies with volunteers failed to link high levels of histamine (and amines) to scombrotoxicoses (Clifford *et al.*, 1991; Ijomah *et al.*, 1991) and it has been suggested that endogenous histamine released from mast cells can be involved in the condition (Ijomah *et al.*, 1991).

Like algal toxins, histamine is not apparent to the consumer and cannot be destroyed by cooking. Cases of scombroid poisoning have been reported worldwide (Taylor, 1986). The finfish most commonly associated with scombroid poisoning include tuna, bonito, mackerel, dolphin fish (mahi mahi), and bluefish. A list of finfish species implicated in outbreaks of scombroid poisoning is given in Table 3.13. All of these fish have naturally high levels of free histidine in the muscle. The proportions of other biogenic amines produced vary among the fish species. Many of the fish implicated in scombroid poisoning are caught in tropical and subtropical waters and may not be cooled quickly after capture.

It is presumed that scombroid toxins arise from bacterial metabolism of amino acids to biogenic amines. The bacteria most frequently incriminated include mesophilic Enterobacteriaceae such as *Morganella morganii, Proteus* spp., *Klebsiella pneumoniae*, and *Hafnia alvei*. The origin of these bacteria is still an open question. Psychrotrophic Enterobacteriaceae are normal members of the fish microflora (Table 3.4), however, the mesophilic species would not be expected to be part of the natural microflora of marine fish, and it is likely that they arise from handling during and after catching. It has been reported that a toxic state can arise within 4 h, toxicity being associated with bacterial populations in excess of 10^6 cfu/g (Okuzumi *et al*, 1984). Once the decarboxylases are formed, their activity will continue, albeit at a lower level, even if temperatures are reduced to levels that do not support bacterial growth (Klausen and Huss, 1987).

The U.S. Food and Drug Administration considers a histamine level in tuna of $\geq$5 mg/100 g of fish muscle indicative of mishandling and a level of 50 mg/100 g as a health hazard. This latter level is close to the threshold toxic dose of 60 mg/100 g estimated by Simidu and Hibiki (1955). Temperature abuse times sufficient for significant toxin accumulation may not result in the production of odors and flavors typical of spoilage. Thus, the usual organoleptic indicators of spoilage may not be present to act as a warning to the consumer. A number of psychrotrophic bacteria, e.g. *Photobacterium phosphoreum*, which is normally found on fish, also produce biogenic amines at temperatures $\leq$5°C, but they do so slowly. In this instance, the amines are considered a product of normal decomposition and have been suggested as a spoilage index. However, occasionally levels can be high, e.g. Ritchie and Mackie (1980) found 35 mg/100 g histamine and 23 mg/100 g in mackerel stored at 1°C for 15 days. Generally, high levels of histamine are preventable by good process control, particularly temperature control from harvest to consumption.

III Primary processing

A Finfish of marine and freshwater origin

Capture and initial processing. Finfish are caught by trawl, net, hook and line, or traps in bodies of water more or less remote from the processing plants. Because of the catching methods, which may extend over several hours, and the unstable and difficult working conditions at sea, there is little control over either the condition of the animals at death or the time of death. This contrasts with red meat and poultry industries in which animals in good physiological conditions are brought live to the processing abattoir, where they are killed quickly and with minimal stress. An example of a typical process flow diagram for fresh finfish is provided in Figure 3.3.

After capture, finfish must be protected from spoilage to the extent possible during transport to the processing plant to ensure both microbiological quality and safety. The periods involved vary from a few hours to 3 weeks or more. Storage is normally in melting ice at 0°C or chilled brine (or sea water) at −2°C. Increasingly, fishing vessels are equipped with facilities to freeze all or part of the catch at sea, thus effectively halting microbial spoilage.

Finfish may be eviscerated on-board the vessel prior to packing in ice. This is a common practice in European fishing vessels, at least for larger finfish. In other parts of the world, particularly where sailing time between fishing grounds and port is short (e.g. Pacific Coast of the United States), fish are stored uneviscerated until reaching port. Small fish such as herring are not normally eviscerated at the time of capture.

There continues to be controversy concerning the advantages and disadvantages of gutting at sea. The process certainly removes a large reservoir of potential spoilage bacteria, but the cutting opens flesh surfaces to direct bacterial attack. Products of bacterial growth in the gut and the action of intestinal enzymes and fecal material can cause discoloration in flesh adjacent to the belly cavity and off-flavors.

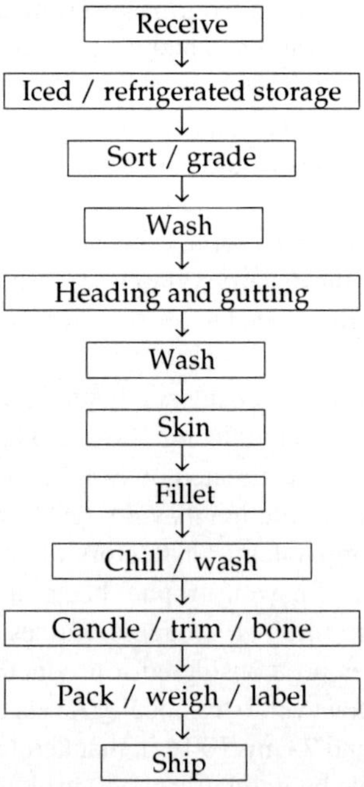

Figure 3.3 Example of a process flow diagram for raw fresh finfish.

In times of heavy feeding, the belly wall may digest in uneviscerated stored fish, causing the so-called belly-burst. Because of this, it appears best to gut immediately on capture and wash the carcass thoroughly.

The changes particularly in temperature when chilling alter the microbial environment. In general, microbial changes occurring in fish stored in ice or chilled sea-water are similar on shipboard and on land. However, ideal storage is frequently more difficult to maintain on board ships and may give rise to undesirable conditions. Fish pressed against wooden pen boards may become "bilgy" as a result of anaerobic bacterial growth. A bilgy odor may also develop around the gills of fish due to localized anaerobic growth caused by inadequately circulated chilled brine.

While considerable quantities of fish are sold to consumers whole or eviscerated, the bulk of finfish undergo further processing in the raw condition. These operations may be entirely by hand or by a combination of hand and mechanical processing. Methods for mechanically separating flesh from bones and skin (i.e. deboning machines) yield a minced flesh product quite different from traditional fillets, steaks, and chunks of meat. Descriptions of grading, washing, scaling, deheading, gutting, filleting, skinning, and meat separation are given by Bykowski (1990).

Spoilage. There is no doubt that the accumulated metabolic products of bacteria are the primary causes for the organoleptic spoilage of raw finfish. Endogenous biochemical changes, particularly in the nucleotide fraction, occur after death (Pedrosa-Menabrito Hultin, 1992; and Regenstein, 1988) and these changes reduce the flavors and odors associated with fresh caught fish. However, these changes are not the cause of the fishy, ammonia, and sulfide odors and slimy and pulpy texture of spoiled fish. These defects are the result of microbial action such as reduction of TMAO to trimethylamine, oxidative deamination of amino acids and peptides to ammonia, release of fatty acids, and breakdown of sulfur-containing amino acids to methyl mercaptan, dimethylsulfhide, and hydrogen sulfide (Gram, 1992; Kraft, 1992) (Table 3.14).

From an initial heterogenous population of bacteria, iced storage results in a selection of a few species. At the point of spoilage, the microflora is dominated by *Pseudomonas* spp., *Shewanella* spp. (marine fish), and *Acinetobacter/Moraxella*. This change in microflora occurs in all fish species, independent of the temperature and salinity of the water in which the fish is caught. Only some of the bacteria present are responsible for the spoilage off-odors and flavors. Specific spoilage bacteria have the ability to produce spoilage compounds in amounts and under conditions similar to the naturally spoiling product. The bacteria most commonly identified with spoilage of fresh, iced fish are species of *Shewanella* and *Pseudomonas* (Liston, 1980; Jørgensen and Huss, 1989), with *Shewanella putrefaciens*-like bacteria predominating in marine fish stored at 0–2°C (Gram, 1992; Kraft, 1992). *Shewanella putrefaciens*-like bacteria has been isolated consistently from spoiled chilled fish throughout the world and probably deserves the epithet "fish spoilage bacterium". The number of *Shew. putrefaciens* (H_2S-producing bacteria) on iced marine fish is linearly correlated with the remaining shelf life (Figure 3.4). *Shewanella putrefaciens* reduces TMAO to TMA in an anaerobic respiration and is capable of producing H_2S

Table 3.14 Off-odors and substrate in spoiling seafood products

Sensory impression	Spoilage product	Spoilage substrate	Product example	Specific spoilage organism
Slime	EPS (dextran)	Sucrose	Brined shrimp	*Leuconostoc*
Fishy off-odor	Trimethylamine	Trimethylamine oxide	Cod	*Shewanella, Photobacterium*
Ammonia (putrid)	NH_3	Amino acids	Several products	Many microorganisms
Sulfidy off-odor	H_2S	Cysteine	Cod, smoked salmon	*Shewanella, L. saké, L. curvatus*
Sulfydryl off-odor	$(CH_3)_2S_2$	Methionine	Cod	*Pseudomonas*
Fruity off-odor	Esters	Phospholipid	Nile perch	*Ps. fragi*

Modified from Gram *et al.* (2002).

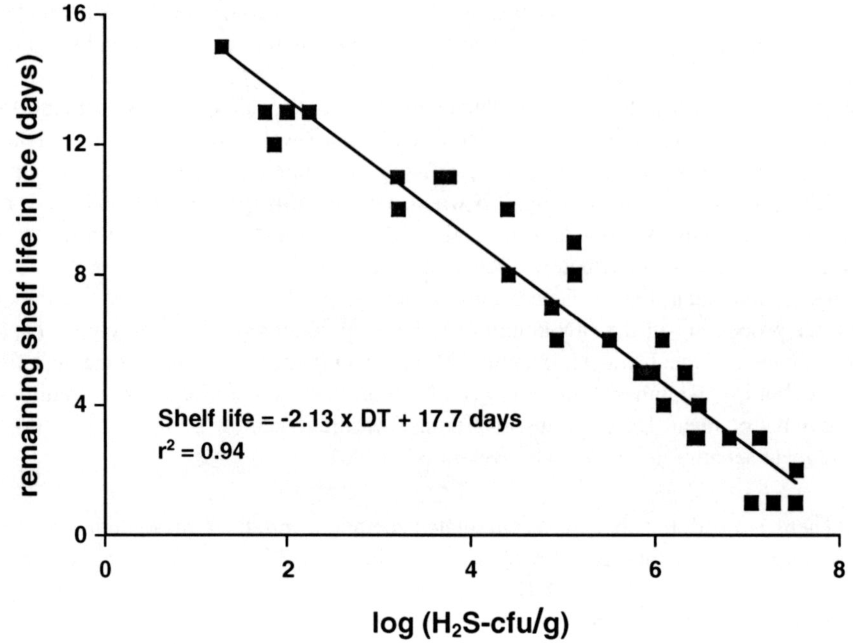

Figure 3.4 Correlation between H_2S-counts and remaining shelf life of iced cod.

and other sulfides. Spoilage of chilled (iced) fresh-water fish again involves Gram-negative bacteria as dominant forms with *Pseudomonas* spp. being the dominant specific spoilage bacteria (Gram *et al.*, 1989). *Aeromonas* spp. or other members of the Vibrionaceae may dominate the spoilage microflora of fish held above 5°C (Barile *et al.*, 1985; Gram *et al.*, 1987; Liston, 1992; Wong *et al.*, 1992). The spoilage at low temperatures is not characterized by hydrogen sulfide and trimethylamine but rather by sweet sulfydryl odors (Gram *et al.*, 1989). Trimethylamine oxide is primarily associated with some marine fish species and has been believed not to occur in freshwater fish. However, Nile perch and Tilapia from Lake Victoria both contain rather high levels of TMAO (Anthoni *et al.*, 1990). Despite the occurrence of TMAO, the spoilage is dominated by pseudomonads that do not reduce TMAO.

Packing of fish in modified atmospheres (including CO_2) will inhibit growth of the respiratory bacteria like *Pseudomonas* and *Shewanella*. Despite this inhibition, marine packed fish spoils almost as rapidly as aerobically packed fish. The spoilage is caused by growth of the CO_2-tolerant *P. phosphoreum,* which per cell produces about 30 times as much trimethylamine as *Shew. putrefaciens* (Dalgaard, 1995). This marine bacterium occurs naturally in the intestine of several marine fish species, often in high numbers (Dalgaard *et al.*, 1993). Its role in spoilage has only recently been recognized because it is heat sensitive and is killed by molten agar during pour-plating or incubation of agar plates at temperatures >25°C. If *P. phosphoreum* is eliminated, e.g. by a freezing step, before chill storage in CO_2 atmosphere, lactic acid bacteria dominate the microflora (Emborg *et al.*, 2002). Little is known about spoilage of CO_2-packed tropical fish, although some studies indicate that lactic acid bacteria become dominant. Such a change in microflora is likely to extend iced shelf life significantly. It is not known if this bacterium, which is a marine organism, plays any role in the spoilage of CO_2-packed freshwater fish.

The primary factor controlling the spoilage of raw finfish is temperature and substantial extensions of shelf life can be achieved with storage at −2°C to 0°C (Boyd *et al.*, 1992). Chilling exerts a selective pressure on the bacterial populations and the psychrotrophic strains increase at the expense of mesophilic bacteria in accordance with their capability to compete. The specific temperature of storage influences the composition of the microflora (Kraft, 1992). In general, when finfish from temperate regions are held at ca. 0°C, there is no or only a short lag phase from 1 to 5 days, exponential growth from 6 to 14 days, and stationary growth thereafter. Under most commercial conditions, shelf life of fish from temperate waters as measured organoleptically shows rapidly deteriorating quality from about 12 days onward. This corresponds to the accumulation of bacterial metabolic products, with some variation in different fish species. The counts on tropical fish chilled rapidly and held near 0°C show some extension of the lag phase, and a slower growth than observed in iced fish from temperate waters. This results in a shelf life $\geq$30 days for a number of species (Poulter and Nicolaides, 1985; Gram *et al.*, 1989).

Finfish caught by nets or hook and line are commonly killed or die relatively rapidly when brought into the air. There is no substantial evidence that agonal invasion of adjacent tissues or blood vessels by intestinal bacteria occurs in finfish. However, bacteria may gain access through puncture wounds and even through areas bruised during the death struggle, and may multiply rapidly in these localized areas (Tretsven, 1964). Contamination of fish with spoilage bacteria from handling and stowage procedures on board the vessel can affect subsequent spoilage (Thrower, 1987; Ward and Baj, 1988). Under tropical conditions, finfish must be chilled quickly after capture. This is particularly important for fish with high levels of endogenous free histidine (e.g. tuna, mackerel, and mahimahi) that could result in dangerous levels of histamine (see below). Delays in chilling when ambient temperatures are high can significantly shorten shelf life during subsequent refrigerated storage (Gram, 1989; Liston, 1992). In temperate regions, where ambient temperatures on boat decks at sea are usually cool, short delays in chilling of fish, though accelerating spoilage, rarely have much effect on the ultimate composition of the spoilage flora. When ambient temperatures are high, even short delays in chilling will affect storage life and the dominance of bacterial species during spoilage. Thus, Barile *et al.* (1985) reported mackerel spoiling after 15 days in ice due to *Shew. putrefaciens* and *Pseudomonas* spp. when iced immediately. However, when held for 9 h at 26°C before icing, the mackerel spoiled after 5 days due to *Pseudomonas* and mesophilic *Bacillus* spp.

Most steps following the catching of fish pose some risk of further contamination. Consequently, ice, boxes, operations during off-loading, and handling during auctioning may contribute to the bacterial load. However, Huss *et al.* (1974b) demonstrated that a 1000-fold difference in initial bacterial load only caused a reduction in iced shelflife of plaice and cod from 12 to 9 days. Thus, despite numerous handling operations, the most important factor in ensuring an appropriate quality is to ensure chill storage, preferably in ice, and rapid handling/transfer.

In many countries, finfish are sold by public auction, requiring that they must be displayed in wooden, metal, or plastic containers on an open or covered quayside or in an auction shed. There are several potential bacteriological dangers in this process. Exposure to abuse temperatures permits spoilage bacteria to multiply (Shamshad *et al.*, 1992).

At the fish-processing plant, there will be frequently a sorting and washing process prior to temporary iced or refrigerated storage or actual processing. As in all food processing operations, potable water would be used.

Net increases in bacterial counts on products undergoing simple processing are indicative of poor manufacturing practices. However, as initial counts on fish may vary from 10^2 to 10^7 cfu/g, a total aerobic count can only be used as an indication of time–temperature conditions during storage if sufficient data are known about a particular fish species. Processing is a wet operation, and contamination is minimized in most plants by copious use of water to wash away product residues on equipment, tables, floors, and machines during processing. Any water coming into contact with foods or food processing surfaces should be potable (e.g. by chlorination) but in some areas of the world, well or seawater is used during primary processing, potentially being a source of pathogens. Workers should wear protective clothing and washable gloves and should be provided with disinfectant dips for hands and equipments (e.g. filleting knives). Cuts and abrasions on workers' hands should be covered with waterproof dressings (Ganowiak, 1990).

Spoilage bacteria grow almost entirely on the surfaces of finfish. Therefore, spoilage proceeds slowly when the surface/volume ratio is low, and rapidly when the ratio is high. Thus, the spoilage rate increases as one passes from whole finfish to gutted, to fillets or steaks, and finally to minced (ground) muscle tissue. An exception may be finfish caught while feeding heavily. On occasion, the intensive enzyme activity in the intestinal tract of these animals digests the gut wall soon after death, resulting in belly-burst and permitting localized microbial entry into the flesh surrounding the belly (Pigoff and Tucher, 1990a).

Some products of spoilage have been used as measures of the quality of stored finfish. Nucleotides, particularly ATP and its decomposition products, are often measured as an index of quality since nucleotide breakdown is largely an endogenous biochemical process independent of bacterial activity in fish post-mortem. However, spoilage strains of *Pseudomonas* and *Shewanella* decompose inosine monophosphate, inosine, and hypoxanthine, casting some doubt on the explicitly endogenous basis of nucleotide degradation (van Spreekens, 1987; Fletcher and Statham, 1988a,b; Gram, 1989).

Pathogens. Very few bacterial pathogens are of concern in fresh fish. The pathogens associated with chilled raw finfish at the end of initial processing are essentially the same as those associated with the initial microflora or acquired during the early phases of capture. As indicated earlier, it must be assumed that *Cl. botulinum*, including non-proteolytic serotypes B, E, and F, are present on whole fish. The number of spores is typically low (Dodds, 1993), in the range of 10–500 spores per kg (Dodds and Austin, 1997), although levels as high as 5 300 spores per kg have been reported in Danish farmed trout (Huss *et al.*, 1974a). *Clostridium botulinum* has been shown to be present on trout and salmon fingerlings raised in freshwater ponds (Huss *et al.*, 1974a). The psychrotrophic types are of particular interest because of their ability to grow and produce toxin at temperatures as low as 3.3°C. Particular care must be taken with finfish that are thermally processed at temperatures below those needed to inactivate *Cl. botulinum* spores, and then held (anaerobically) for long periods at low temperatures above 3°C. This concern also holds for lightly preserved fish and fresh fish stored under reduced oxygen conditions such as vacuum or modified atmosphere packaging unless strict temperature ($\leq$3°C) control is maintained (Genigeorgis, 1985; Statham and Bremner, 1989; Reddy *et al.*, 1992, 1996). Three cases of botulism due to consumption of grilled surgeon-fish in Hawaii indicate that the hazard exists for raw fish when temperature control is poor (MMWR, 1991c). The fish had been held uneviscerated on ice for 6–16 days in a non-functional freezer case at a temperature of ~11°C. Apparently, *Cl. botulinum* type B grew in the gut or visceral cavity. When the fish was grilled whole, toxin persisted and was concentrated in the intestine and surrounding area of flesh. This supports the desirability of gutting fish before storage and certainly before cooking, avoiding eating intestines, cooling adequately, and holding fish at $\leq$2°C. Much concern has been raised on the risk of botulism from packed fresh fish, however, despite the slight extension of shelf life, most fish will spoil prior to toxin formation, even when packed. In a study of ~1100 samples of commercially prepared vacuum-packed fish, none became toxic before sensory spoilage (Lilly and Kautter, 1990).

Adequate temperature control is also critical for preventing or controlling the growth of the wide range of marine and terrigenous pathogens that potentially may be present on finfish after initial processing, including *Vibrio* spp., *Salmonella* spp., *Staphylococcus aureus*, *Bacillus cereus*, *E. coli*, *Cl. perfringens*, *L. monocytogenes*, and *Shigella* spp. None of these organisms present a hazard on chill stored and cooked fish. However, in areas where fish is consumed raw, strict hygienic conditions must be maintained and contamination of bacteria with low infectious dose must be avoided. Thus, the high rate of *V. parahaemolyticus* gastroenteritis in Japan and other Asian countries is related to the widespread consumption of raw or very lightly processed finfish, with high ambient temperatures being an apparent contributing factor (Sakazaki, 1969; Chan *et al.*, 1989). The microorganism grows very rapidly at temperatures between 25°C–42°C, and can reach infectious levels of $>10^5$ cfu/g in 2–3 h at these elevated temperatures (Katoh, 1965; Liston, 1974). *Vibrio* species are readily inactivated by cooking (Delmore and Crisley, 1979; Ama *et al.*, 1994). *Listeria, Shigella,* and *Salmonella* have been isolated from fresh waters (Ward, 1989; Singh and Kulshrestha, 1993). *Campylobacter jejuni* has been also isolated from the surface of fresh water finfish (Khalafalla, 1992).

Inadequate cold storage is also a major factor contributing to scombroid fish poisoning as temperature abuse allows the mesophilic Enterobacteriaceae (e.g. *Morganella morganii*) to decarboxylate amino acids. The biogenic amines are heat stable and will not be inactivated by cooking. As is evident from Table 3.6, histamine (or biogenic amine) poisoning is a common cause of sea-food borne disease.

Non-cultured fish must be assumed to carry parasitic microorganisms and, as some may be pathogenic to humans, such parasites constitute a risk if the fish is consumed raw and/or if the parasites are not removed or destroyed before consumption. Fish to be consumed raw or semi-raw must, according to the EU legislation, be exposed to freezing at −20°C for 24 h (EEC, 1991b). The potential pathogens on fresh-water fish again vary with species and water condition.

Interrelations. The psychrotrophic microorganisms that spoil raw finfish generally grow well from temperatures near 0–30°C. Mesophiles, which include most of the pathogens, generally grow well between 8°C and 42°C. Thus, seafood refrigerated below 5°C usually will not support the growth of pathogens. Further, in the temperature range where both groups grow well (8–30°C), typically psychrotrophs have shorter lag periods (i.e. more rapid onset of growth) and faster growth rates, and generally spoil finfish before the pathogens grow to dangerous levels (Michener and Elliot, 1964; Elliot and Michener, 1965). While this is generally true, it should be noted that mesophilic bacteria allowed to multiply to high numbers due to delays in chilling of tropical fish after capture can survive during refrigerated storage (Barile *et al.*, 1985). Further, enzymes, such as histidine decarboxylase, produced during mesophilic growth can continue to be active under subsequent refrigeration.

There are exceptions to the temperature ranges described above: *Cl. botulinum* types E and F can grow at 3.3°C; *Salmonella* spp. can grow competitively on fish held at or above 8°C (Matches and Liston, 1968); *V. parahaemolyticus* can grow on finfish competing well with spoilage bacteria (Katoh, 1965); psychrotrophic pathogens such as *L. monocytogenes*, *Y. enterocolitica*, and *A. hydrophila* can grow at temperatures below 5°C. Conversely, staphylococci are generally poor competitors, particularly at temperatures ≤30°C.

CONTROL (finfish of marine and freshwater origin)

Summary

Significant hazards[a]	• Aquatic biotoxins (ciguatera). • Histamine. • *Cl. botulinum* type E (some types of packed fish). • Parasites (if consumed raw). • *Vibrio* spp. (if consumed raw). • Enteric pathogens (if consumed raw).
Control measures	
Initial level (H_0)	• Avoid fish from certain (tropical) areas or areas with algae blooms. • Avoid fish (for raw consumption) from areas with contaminated water.

(*continued*)

CONTROL (Cont.)

Summary

Reduction (ΣR)	• Freeze to inactivate parasites. • Cooking will kill vegetative pathogens and may destroy type E toxin.

Increase (ΣI)	• Time–temperature—control growth of:
	  • Histamine forming bacteria;
	  • *Cl. botulinum* type E;
	  • *Vibrio* spp.

Testing	• Surveying catching waters for algal blooms. • Sensory evaluation to determine temperature abuse.
Spoilage	• Growth of spoilage bacteria. • Time × −temperature control;. • Sensory evaluation.

[a]In particular circumstances, other hazards may need to be considered.

Hazards to be considered. Based on eidemiological data and the microbial ecology of the product, the major hazards in fresh fish are aquatic toxins (ciguatera) and histamine (Olsen *et al.*, 2000). When fish is packaged under vacuum or in CO_2 atmosphere, psychrotrophic *Cl. botulinum* must be considered a hazard. Only if the fish is to be consumed raw or semi-raw, parasites (e.g. *A. simplex*), indigenous bacterial pathogens (e.g. *V. parahaemolyticus*) and enteric organisms (salmonellae, virus) become potential risks.

Control measures

Intial level of hazard (H_0). Ciguatera and algal intoxications are best handled through the control of finfish sources, and harvesting from "hot spots" or during an algal bloom should be avoided. This is currently achieved through periodic assessment of fishing grounds for toxic algae and the avoidance of certain large fish from high-risk areas. The potential risk (in fish consumed raw) from enteric organisms with low infectious dose can only be controlled at H_0, by avoiding waters contaminated with human or animal waste. Using GHP, e.g. potable water, is important in controlling initial levels of several pathogens.

Reduction of hazard (ΣR). Parasitic organisms must be assumed to be present in wild fish and control of this hazard is obtained through a reduction step, e.g. by cooking. Finfish to be consumed raw should be frozen for a sufficient amount of time to assure inactivation of parasites (−20°C for at least 24 h; EEC, 1991b). This is particularly important for fishing grounds where there is a high incidence of finfish containing parasites that can infect humans. The presence of *Vibrio* spp. on raw fish cannot be avoided due to their natural presence in aquatic (marine environments). In cooked products, these organisms (and the potential risk) are controlled by elimination.

Increase of hazard (ΣI). Although some bacteria may produce biogenic amines at low temperatures, rapid chilling, preferably in melting ice (~0°C), of finfish immediately after capture and the subsequent maintenance of the low temperature are the primary means for preventing biogenic amine (scombroid) toxicity. Maintaining refrigerated storage temperatures as low as possible (≤2°C) also prevents the growth of psychrotrophic pathogens (non-proteolytic *Cl. botulinum)* in packed, fresh products. Thus for both hazards, it is not the presence of the toxin/amine producing organisms that constitutes a risk but the increase (*I*) during storage that must be controlled. In products intended for raw consumption, maintaining low temperature controls, e.g. *V. parahaemolyticus* by preventing increase. Also, the spread of *Vibrio* spp. should be minimized through proper handing ensuring that cross-contamination does not take place.

Testing. Testing for algal toxins is difficult and time consuming and is not a feasable way of controlling this hazard. In fresh fish where histamine formation can occur, a sensory evaluation will indicate often if temperature abuse has occurred. Testing for bacterial pathogens is not recommended for fresh fish. Visual inspection can be used to determine presence/absence of parasites and is used as a quality check also for fish intended for freezing.

Spoilage. The best way to control spoilage is to ensure rapid chilling of the fish immediately after catch/slaughter. Although chemical measurements (e.g. determintation of trimethylamine in gadoid species) can indicate the degree spoilage, sensory evaluation is by far the best to determine degree of spoilage. Bacterial counts, e.g. aerobic plate counts, cannot indicate quality or degree of spoilage. For some fish, the number of spoilage bacteria (e.g. H_2S-producing bacteria on iced, fresh cod) can predict remaining shelf life, however, the number does not *per se* give any indication of quality. Ice and water can be a source of spoilage (and pathogenic) bacteria and should be of good (potable) quality. Ice should not be re-used.

B Crustacea

Crustaceans like shrimp, crab, and lobster are related to the invertebrates and do not have a skeleton-like bony fish. Thus, their outer shell is their skeleton. Crustaceans are typically high-value products and culturing of shrimps and prawns has become a major export item, in particular for East and South-East Asian countries where 1.5 million tones of crustaceans were produced in 1998. Also India, producing 415 000 tones in 1998, is becoming a major producer of crustaceans. Bangladesh has seen a dramatic rise in the amount of crustaceans cultured almost tripling from 1988 to 1998 (from ~32 000 tonnes to–97 000 tones) (FAOSTAT, 2001).

Catching and initial processing. Crabs and lobsters are trapped in baited pots or cages and transported live, sometimes in refrigerated sea-water, to holding tanks or directly to processing plants. Only live animals are processed and dead animals are discarded. In deeper water fisheries, crab processing may be done at sea. The initial step in processing is cooking, and as such the processing of crab and lobster is described in the section on Cooked crustaceans.

Shrimps are commonly captured by trawlers and iced or held in refrigerated sea-water for transport to processing plants. In large shrimp fisheries, the animals may be "headed" on board the vessel. This procedure removes the edible tail section from the head, gills, and thorax, which contain organs and some of the viscera. This removes a large external source of bacterial contamination (Novak, 1973), but exposes flesh at the broken surfaces. The microbiological benefits derived from this process appear to be minimal, but it does eliminate a major source of enzymes responsible for the development of black spot (melanosis).

Unlike crabs and lobsters, shrimps die quickly on capture and cannot be held live. In a few fisheries, shrimp are caught in traps and held live, usually for local consumption. An increasing proportion of the shrimps supply is produced by aquaculture. These shrimps are either processed directly at the rearing ponds or shipped to distant processing plants.

At the processing plants, shrimps are washed and sorted for size. After heading, shrimps may be peeled and deveined. This can be done on either raw or cooked shrimps. "Picking", i.e. the separation of the edible tail meat from the carapace and removal of the "sand vein" (lower intestine), is commonly done by peeling machines that are quite sanitary, provided they are adequately maintained. However, a large amount of shrimps is are still hand peeled, often under less than ideal sanitary conditions. This has been a source of contamination by *Staph. aureus* and *Salmonella* and other pathogens, particularly for shrimps from specific tropical regions in which peeling is a cottage industry. Sometimes freshly caught shrimps are held for a day in ice to facilitate shell separation. Most shrimps are then frozen with or without preliminary cooking and/or breading.

Saprophytes and Spoilage. The microflora of raw crustacea reflects the quality of the water from which the animals were harvested, and is also affected by the on-ship and in-plant environment, as well as the duration and type of refrigerated storage (Vanderzant *et al.*, 1970, 1971; Faghri *et al.*, 1984; Heinsz *et al.*, 1988). The microflora of live crabs and lobster is largely associated with the chitinous shell and the intestinal tract. Crabs taken from waters near human habitations will tend to have higher bacterial loads and a more diverse microflora than those taken from isolated areas (Faghri *et al.*, 1984).

Spoilage microorganisms are of little importance at this point since the live muscle tissue remains sterile. However, as noted earlier, the hemolymph system of crabs is open and may contain bacteria, particularly *Vibrio* spp. Bacterial invasion of tissues following death is of little importance since dead carcasses are discarded before processing. Crabs and lobsters may undergo endogenous biochemical changes during live storage that may affect their quality, but there do not appear to be significant changes due to bacteria.

Shrimp spoilage is different since the animals die immediately upon capture. The initial bacterial level at the processing plant will be a function of the quality and extent of shipboard storage. Thus, most shrimps have high counts (10^5–10^7 cfu/g) at the time of receipt at the processing plant. Shrimps from tropical waters tend to have higher counts (Cann, 1977). In a well-controlled operation, mesophilic counts generally decrease by 7–10-fold during the initial processing, largely due to the removal of bacteria by washing (Surkiewcziz *et al.*, 1967; Duran *et al.*, 1983). However, in plants with poor sanitation, bacterial levels increase. Bacterial levels tend to be lower in operations employing mechanical peelers (Ridley and Slabyj, 1978). Refrigerated storage selects for a psychrotrophic microflora; the dominant spoilage bacteria appear to be members of the *Acinetobacter—Moraxella* group. However, *Pseudomonas* and coryneform bacteria are commonly associated with the spoilage microflora. Frozen storage tends to reduce the number of Gram-negative bacteria, and subsequent storage at chill temperatures (0–5°C) can cause growth of Gram-positive micrococci and lactic acid bacteria.

Figure 3.5 shows how the numbers of bacteria on raw shrimp are influenced by the various steps associated with processing. More frequently than not, the processed products from sub-standard plants will have higher bacterial levels than the unprocessed product.

During spoilage, shrimp flesh undergoes many of the same biochemical changes seen in finfish. Volatile basic substances increase, and the pH rises. When the temperature of stored shrimps is allowed to rise, or initial cooling after capture is delayed, indole is produced and its presence is used by some regulatory agencies as an index of decomposition. Indole is a product of bacterial degradation of tryptophan. Not all shrimps contain tryptophan and thus indole can only be used as spoilage indicator for some types of shrimps (Table 3.15). Formaldehyde level has been reported to be a good indicator of spoilage in banana shrimp (*Penaeus merguiensis*) (Yamagata and Low, 1995).

Pathogens. Crustaceans are a source for *V. parahaemolyticus* outbreaks in the United States and other countries (Barker *et al.*, 1974), and *V. parahaemolyticus*, *V. cholerae*, and *V. vulnificus* are routinely a part of the microflora of raw crustaceans harvested from estuarine waters (DePaola *et al.*, 1994).

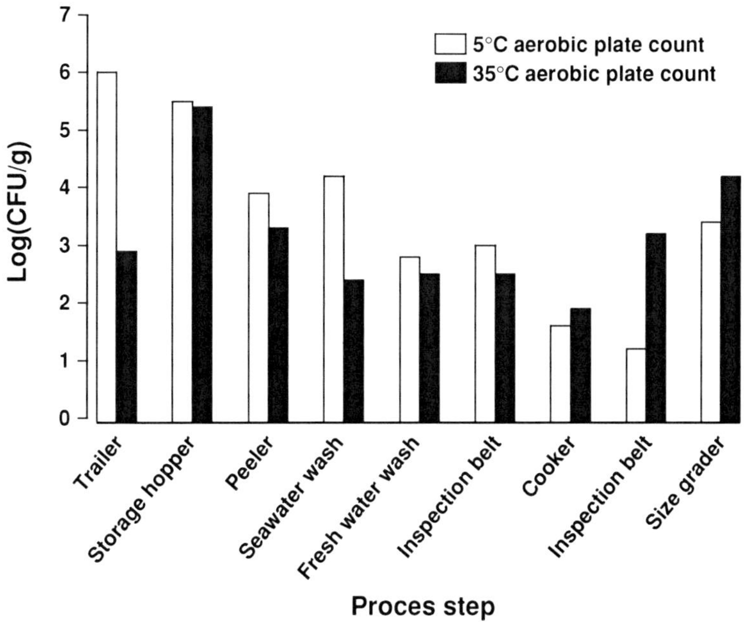

Figure 3.5 Psychrotrophic and mesophilic aerobic plate counts during different steps in the processing of raw shrimps (modified from Ridley and Slabyi, 1978).

Table 3.15 Indole formation in shrimp/prawns

Species	Temperature (°C)	Time (days)	Indole (μg/100 g)	Reference
Penaeus merguiensis	0–4	9–13	4	Shamshad *et al.* (1990)
Penaeus setiferus or	0–4	8–13	10–15	Chang *et al.* (1983)
Penaeus duorarum	12–22	1–2	>100	
Pandalus platycens	0	14–21	30–40	Layrisse and Matches (1984)
Pandalus jordani	0	10	65	Matches (1982)
	11	3	130	
	22	2	623	
Pandalus borealis	0	10	4	Solberg and Nesbakken (1981)
	22	1	1	

In Japan, three *V. vulnificus* deaths were caused by consumption of squilla (crustacean). A survey of raw shrimps in commerce found the incidence of *V. parahaemolyticus* and *V. vulnificus* was 37% and 17%, respectively, with 54% of the isolates being resistant to at least one antibiotic (Berry *et al.*, 1994). *Vibrio* and *Aeromonas* spp. are commonly associated with the shells of crustacea because of the microorganisms' ability to utilize chitin as a carbon/energy source (Baross and Liston, 1970; Colwell, 1970; Nalin *et al.*, 1979; Cann *et al.*, 1981; Yu *et al.*, 1991; Platt *et al.*, 1995).

Crustaceans trapped in estuarine or in-shore marine waters may be contaminated with potentially pathogenic bacteria from sewage, and thus may become contaminated with a variety of fecal bacteria and viruses. The incidence of *Salmonella* and *Shigella* spp. in raw, wild marine crustaceans is typically low, although this varies with the geographical location (D'Aoust *et al.*, 1980; Fraiser and Koburger, 1984; Gerigk, 1985; Sedik *et al.*, 1991). The incidence of salmonellae is often higher in aquaculture-reared shrimps (see sSection IV). The ubiquitous association of *Staph. aureus* with food handlers means that excessive handling of raw crustaceans is likely to lead to increased incidence of this pathogen. Inspection and product packaging are two operations associated with the introduction of *Staph. aureus* (Ridley and Slabyj, 1978). This also appears to be one of the sources of salmonellae contamination, particularly for shrimps from specific tropical regions where peeling is a cottage industry. Similarly, hand peeling of shrimp under insanitary conditions could introduce a variety of human bacterial pathogens such as salmonellae. *L. monocytogenes* is present in a high percentage of fresh and low-salinity water samples (Colburn *et al.*, 1990) and can be isolated routinely from raw shrimp, crab and crawfish (Weagant *et al.*, 1988; Hartemink and Georgsson, 1991; Motes, 1991; Adesiyun, 1993). A survey of commercial raw shrimp imported from a number of countries indicated that the incidence of salmonellae was 8% (ranging from 3% to 100% on a country basis) and *L. monocytogenes* was 7% (ranging from 4% to 60%) (Gecan *et al.*, 1994). An irradiation dose of 5 kGy was insufficient to completely eliminate an initial level of 8.5×10^4 cfu/g of *L. monocytogenes* in frozen shrimp (Brandao-Areal *et al.*, 1995).

Crab and shrimp have been reported as a vehicle for classical cholera in outbreaks in the United States and Latin America (Blake *et al.*, 1980b; Morris and Black, 1985; MMWR, 1991a, b, 1992a; Popovic *et al.*, 1993). It appears that *V. cholerae* O1 can become established in in-shore and estuarine waters and seafood must be considered primary vehicles for transmission of cholera, particularly in endemic areas (Blake *et al.*, 1980b; Shandera *et al.*, 1983; Lin *et al.*, 1986; Kaysner *et al.*, 1987).

Psychrotrophic strains of *Cl. botulinum* probably occur on crustaceans. *Cl. botulinum* types A and C have been isolated from shrimp, and serotypes C and F from crab and shrimp in the tropics. The risk from these microorganisms appears negligible under normal handling and processing conditions. White and brown shrimp (*Penaeus* spp.) inoculated with *Cl. botulinum* serotype E supported toxin production at 10°C, but not 4°C (Garren *et al.*, 1994). Crustacean shellfish may also act as vehicles of viral diseases.

A potential hazard related to domoic acid contamination of shellfish was recently identified in crabs sampled on the West Coast of the United States (Horner and Postel, 1993). Crabs feeding on contaminated shellfish, or perhaps directly on plankton, had contaminated viscera posing a risk to consumers of whole crab, or by cross-contamination to eaters of leg and body meat. Similar contamination was shown previously with PSP toxins but so far no human illness has been reported (Horner and Postel, 1993).

CONTROL (crustacea)

Summary

Significant hazards[a]	• *Vibrio* spp. (if consumed raw). • Enteric pathogenes (if consumed raw).
Control measures	
Initial level (H_0)	• Avoid catching in contaminated waters.
Reduction (ΣR)	• Cooking destroy vegetative pathogens.
Increase (ΣI)	• Time × −temperature.
Testing	• Sensory evaluation.
Spoilage	• Growth of spoilage bacteria. • Time × −temperature;. • Sensory evaluation.

[a]In particular circumstances, other hazards may need to be considered.

Hazards to be considered. The health hazards of raw crustaceans are in some ways similar to other raw fish products. However, aquatic toxins are generally not accumulated, no formation of biogenic amines occurs and parasites have not been reported. Both viral and bacterial pathogens are potential hazards if the product is eaten, raw or if re-contamination occurs after a cooking step. If the product is hand peeled, pathogens can be introduced. Most hand peeling is probably done on cooked shrimp, and the specific hazards related to this are discussed in a later section. If the product is peeled in the raw state, subsequent cooking will eliminate pathogenic organisms.

Control measures. Most of the controls for maintaining the microbiological integrity of raw shrimp are the same as those outlined for finfish (see above), including harvesting from quality fishing grounds, rapid chilling after capture, maintenance of sanitation, and avoidance of cross-contamination.

Initial level of hazard (H_0). Because shrimp are harvested by trawling, the animals should be washed immediately with fresh sea-water to remove mud and sediment, reducing initial levels of contamination (H_0).

Reduction of hazard (ΣR). Excessive handling of shrimp by humans can be a source of salmonellae, *Staph. aureus*, and other pathogens. Fortunately, such bacteria are usually destroyed by the normal cooking procedures to which crab, lobsters, and most shrimp are subjected, relying on a reduction step (ΣR) for safety. However, cooking may not destroy all bacteria if the initial levels are high. There are reports of bacterial counts as high as 3.7 10^5 cfu/g in crab boiled for 30 min and 4.5×10^4 cfu/g oin tropical prawns after boiling for 3 min, however, re-contamination (ΣI) of cooked products is the major issue for control.

Increase of hazard (ΣI). If the shrimp are to be headed at capture, it should be done at once. Immediately after capture, the shrimp should be iced on board ship and maintained on ice or under refrigeration to avoid increase of pathogenic bacteria (ΣI). The temperature should be reduced to $<5°C$ within 2 h after catch. Picking, when done by adequately maintained peeling machine, is generally a sanitary processing step preventing re-contamination, i.e. increase (ΣI). Sanitary design and proper cleaning of machines is needed to avoid accumulation of shrimp parts on the equipment. An equivalent level of control is difficult to achieve when shrimp are hand peeled, particularly if carried out under less than ideal sanitary conditions. A primary control for crabs and lobsters is maintaining them alive and healthy until cooking, thus preventing increase (ΣI) of pathogenic or spoilage bacteria. This means careful handling; storage longer than a few hours requires immersion in cool aerated sea-water. During storage, temperatures should be checked and there should be monitoring for dead animals, which should be removed promptly.

Testing. Microbiological indicators do not appear to be effective for predicting the presence of salmonellae in shrimp (D'Aoust *et al.*, 1980). *Vibrio* spp. are indigenous, aquatic, bacteria and must be assumed to be present. Hence, processing and handling should ensure that there is no cross-contamination to cooked products. As for fresh finfish, testing of raw crustaceans for *Vibrio* spp. is not useful for assuring safety.

Spoilage. Raw, iced crustaceans spoil in the same manner as fresh finfish, and time-temperature control is the most important control measure. Preferably, the crustaceans should be kept in melting ice at $\sim 0°C$.

C Mollusca

The animals included in this section are all bivalve molluscs, which feed by selectively filtering out small planktonic organisms, including bacteria, from sea-water. They include principally clams, mussels, cockles, oysters, and scallops.

Harvesting and processing procedures. Clams grow buried a few centimeters to a meter or more in marine and fresh water sediments. They are harvested by mechanical dredges, by scuba divers with air or water jets, and by digging by hand at low tide. Most are wild populations that depend on natural processes for seeding and renewal, but in recent years there has been increased use of artificial enhancement of natural populations and closed cycle aquaculture of clams. Mussels grow naturally attached to rocks, cliffs, and other surfaces that are tidally submerged. They are cultivated throughout the world on submerged strings suspended from rafts or other floating structures from which they are readily harvested. They are harvested by hand. Oysters occur naturally in estuarine areas in and below tidal areas, growing on the surface of sediments. They have been cultivated by selective relaying of captured spat for centuries, and in recent years controlled production of seed in hatcheries has led to more controlled aquaculture. However, growing oysters depend on natural food sources in sea-water. They are harvested by hand picking, tonging, raking, or mechanical dredges. Often, oysters are kept in clean (disinfected) water for some days to allow clearance of some pathogenic agents (see below).

All these animals are naturally sessile in the adult form but have the capacity to close their shells tightly enabling them to survive for long periods out of water. They are normally held in a live state until processed or through distribution and retailing in the case of unprocessed animals. They are frequently eaten live after removal from the shell (shucking). Transport on ice is recommended. An example of a typical process flow diagram for molluscs is provided in Figure 3.6.

Scallops are somewhat more mobile animals but generally congregate in limited areas, moving short distances in fairly shallow water by forcing a jet of water through partly closed shells. They are caught by dredges and may be shucked on-board the vessel or brought ashore in the shell. Typically, only the adductor muscle is eaten in North America and much of Europe. This is the major product in international trade, but whole scallops are commonly eaten by a number of cultures and by gourmets.

Oysters, mussels, and clams may be marketed as whole live animals in the shell or as shucked meat. Whole clams may be frozen in the shell. Shucked meats are sold from bulk stocks with individual wrapping for the customer or they may be packaged in plastic containers or glass jars at the processing plant and shipped under refrigeration. Shucking is mainly a hand operation. Oyster processing also involves a mechanical "blowing" operation, which involves tumbling the shucked meats in vigorously aerated fresh water. This removes shell fragments and "plumps" the oysters.

Saprophytes and spoilage. Molluscs carry a resident bacterial population that in the case of oysters fluctuates between 10^4 and 10^6 cfu/g of tissue, the higher counts occurring when water temperatures are high. The microflora is dominated by Gram-negative bacteria of the genera *Vibrio–Pseudomonas–Acinetobacter—Moraxella*, *Flavobacterium*, and *Cytophaga* (Colwell and Liston, 1960; Lovelace *et al.*, 1968; Vanderzant *et al.*, 1973). This seems to be the resident population for molluscs. Smaller numbers of Gram-positive bacteria may also be present.

When molluscs feed on polluted water, they concentrate contaminating bacteria, including enteric pathogens and viruses, if present. As oysters and other molluscs harvested for human consumption are normally acquired from estuarine areas that receive some waste from point (sewage) and non-point (run-off) land sources (Stelma and McCabe, 1992; Ekanem and Adegoke, 1995), it is not

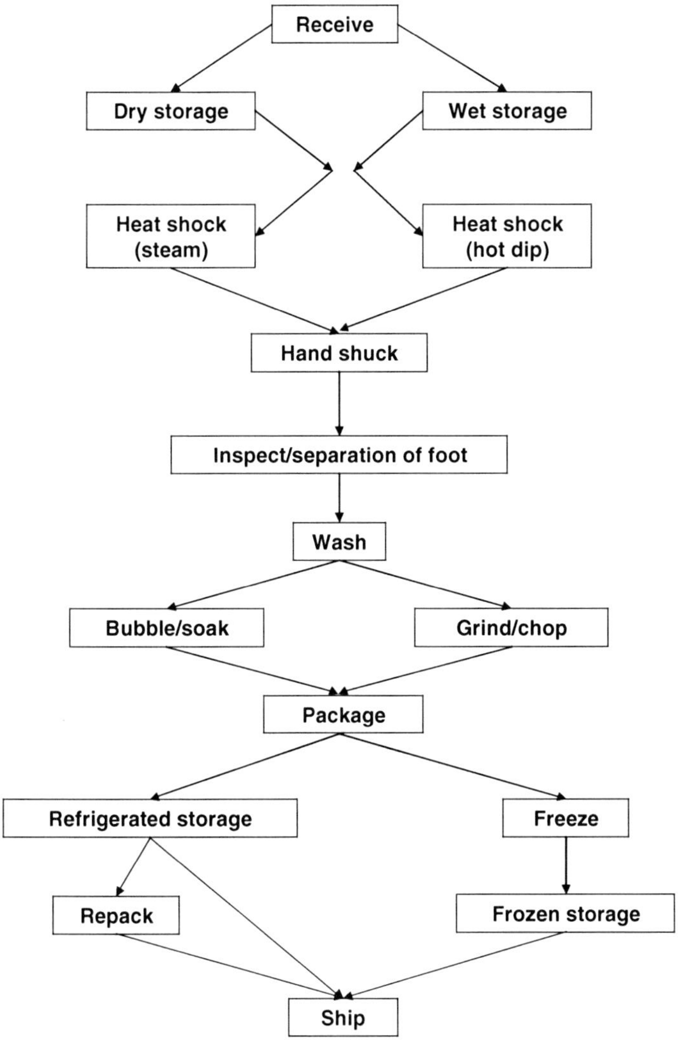

Figure 3.6 Example of flow diagram for shucked mollusca (NMFS, 1990b).

uncommon for small numbers of mesophilic coliforms to be found. However, they are not part of the normal resident bacterial populations. A possible exception is the presence of large numbers of marine mammals, particularly pinnipeds (seals) and sea-lions, which can contribute to the presence of fecal contamination (Smith *et al.*, 1986).

During spoilage of shucked molluscs, bacterial populations normally increase to 10^7 cfu/g or more. Gram-negative proteolytic bacteria, usually *Pseudomonas* and *Vibrio* spp., are prominent in the spoilage flora. In addition, saccharolytic bacteria are active, fermenting the glycogen in the tissues to various organic acids. *Lactobacillus* spp. were reported as a major component of the spoilage microflora and identified as the fermenting organism (Shiflett *et al.*, 1966). However, lactobacilli may not occur consistently in marine products, and when lactobacilli were added to vacuum packed scallops, they did not inhibit the growth of *Vibrio* spp. (Bremner and Statham, 1983). However, lactic acid (1.25%) rinses have been used experimentally to retard spoilage (Kator and Fisher, 1995).

Biochemically, spoilage includes both proteolytic and saccharolytic activity. Ammonia and other amines accumulate, but so do acids. The pH of molluscs typically falls during spoilage (in contrast to finfish and crustacea in which pH increases). Fresh oysters have pH values from about 6.2 to 6.5, but this decreases to 5.8 or below during spoilage.

Pathogens. Raw molluscan shellfish are probably the most common vehicle of seafood-related foodborne disease (Table 3.2). This is easily explained by the fact that the animals are filter feeders and accumulate pathogenic agents and that they are eaten with no heat treatment.

Molluscan shellfish consistently carry mesophilic *Vibrio* species. The most commonly found is *Vibrio alginolyticus* (Baross and Liston, 1970; Matté *et al.*, 1994; Suñén *et al.*, 1995), but *V. cholerae* non-O1, *V. parahaemolyticus,* and *V. vulnificus* are frequent, particularly when water temperatures are high. Indeed, in warm water regions these microorganisms may be present at a level of hundreds of thousands or even millions per gram of animal meat (Kaspar and Tamplin, 1993), though typically the levels are substantially lower (Matté *et al.*, 1994). In warm water regions where *V. cholerae* O1 is endemic, the microorganism is found sporadically in molluscs. However, in epidemic situations they can be present in comparably high numbers. Information pertaining to the incidence of *V. cholerae* O139 (a new epidemic serovar) in molluscan shellfish is not yet available.

Vibrio parahaemolyticus may increase in the live animal if held at high temperature (26°), thus numbers in live oysters increased with 1.7 $\log_{10}$ units in 10 h and with 2.9 $\log_{10}$ units in 24 h at this temperature (Gooch *et al.*, 2002). In contrast, the number of *V. parahaemolyticus* decreased (0.8 $\log_{10}$ unit in 14 days) if kept at refrigeration temperature.

Outbreaks of disease have been reported among individuals eating raw or lightly processed shellfish involving each of these vibrios (Blake, 1983; Klontz *et al.*, 1993; Popovic *et al.*, 1993). In addition, *V. mimicus*, *V. hollisae,* and *V. furnissii* have been reported associated with occasional shellfish associated outbreaks. The severity of most vibrio infections is mild and patients normally recover quickly and spontaneously after a few days of diarrhea. However, *V. cholera* O1 can be life threatening if not properly treated, and severe septicemic conditions have been recorded from *V. cholerae* non-O1, *V. parahaemolyticus*, and *Vibrio hollisae*, though uncommon. A particularly dangerous microorganism for individuals with underlying liver disorders and certain other diseases (e.g. cirrhosis, haemochromatosis, and diabetes) is *V. vulnificus*. In such patients, the species can cause a fulminating primary septicemia that can result in death in 24–48 h. Fortunately, the microorganism does not usually cause disease, or produces a mild and self-resolving condition, in healthy individuals. The microorganism is particularly associated with oysters, with nearly all cases resulting from consumption of raw oysters. There are indications that encapsulation of *V. vulnificus* may increase its survival in oysters (Harris-Young *et al.*, 1995).

Molluscan shellfish are notoriously liable to become contaminated with bacteria derived from human sewage because of their sessile estuarine habitat. In years past, typhoid fever was a common consequence of raw oyster consumption in Europe and North America. This disease and other infections from sewage-derived bacteria including non-specific salmonellosis, shigellosis, etc. have become less common in industrialized countries. This is a consequence of both improved methods of sewage treatment and effective surveillance of shellfish growing areas for human fecal contamination. Nevertheless, even in countries with good regulatory programes some cases still occur, often from contaminated shellfish illegally harvested from closed areas. In Japan, in 2001–2002, raw oysters were the cause of a shigellosis outbreak (Miyahara and Konuma, 2002; Konuma, 2002). It is often not possible to demonstrate a direct relationship between the presence of indicator organisms in the growing water and the presence of specific pathogens in the shellfish (Hood *et al.*, 1983; Martinez-Manzanares *et al.*, 1992). *Campylobacter jejuni* has been implicated as a cause of shellfish-associated gastroenteritis outbreaks (Griffin *et al.*, 1980; Abeyta *et al.*, 1993). The organism is believed to originate from sea gull feces (Teunis *et al.*, 1997). Some studies failed to detect *Campylobacter* spp. in molluscs (Ripabelli *et al.*, 1999), whilst others have found as many as 42% of samples positive (Wilson and Moore, 1996). In general, there is a paucity of data and further research in needed to determine the magnitude of the problem. *Staph. aureus* has been isolated from a substantial percentage of shellfish meat samples (Ayulo *et al.*, 1994), presumably due to handling during shucking; however, due to the large numbers of other bacteria, it is not likely that *Staph. aureus* will cause disease.

The most common microbiological problem for consumers of molluscs is now viral infections (Liston, 1990; Lees, 2000). Viruses are more resistant to water treatment procedures and therefore more likely to survive in effluent water. Moreover, viruses may survive in marine environments for months, and perhaps years, when water temperatures are low (Boardman and Evans, 1986; Gerba, 1988). This has resulted, at least in the United States, in a pattern of viral disease mainly from clams harvested in cooler northern waters and vibrio gastroenteritis from oysters grown in warmer southern waters. However, shellfish-derived viral and vibrio diseases can occur in both temperature zones. Hepatitis virus (A, non-A/non-B, and unidentified) are transferred by bivalves (IOM, 1991) but Norwalk and Norwalk-like, and other small round viruses are the major cause of gastroenteritis transmitted by foods, including seafoods (ACMSF, 1994; Lees, 2000).

The majority of marine toxins important from a food standpoint are produced by microalgae in the phytoplankton, which is the source of food organisms for molluscs. Toxic algae are not consistently present, at least at levels that could cause dangerous toxicity. Instead, they occur as blooms in response to changes in physical and chemical conditions, which are not entirely understood (Hallegraeff, 1993). The four toxin groups significant to consumers of molluscs are those causing PSP, NSP, DSP, and ASP. The principal toxigenic microorganisms involved are the dinoflagellates *Alexandrium* (*Gonyaulax*), *Gymnodium* and *Dinophysis* and the pennate diatom *Pseudonitzschia*. PSP-causing *Alexandrium* appear to be globally distributed and show a seasonal occurrence along the coasts of North America where blooms occur (unpredictably) from May to September. However, occasional episodes have been reported in winter allegedly due to resuspension of cyst forms in sediment by storms. The pattern appears much more sporadic in other parts of the world. NSP seems only to have been reported in the Gulf of Mexico and Southeast Atlantic coastlines of the United States, but *Gymnodinium* is quite widespread, as are characteristic red tides, so the hazard may be more extensive than appears. DSP on the other hand seems to be widespread along Japanese and European coastlines but has not been reported in North America. However, the implicated *Dinophysis* species do occur in both Atlantic and Pacific waters. Domoic acid poisonings (ASP) have been reported from northeastern Canada and the coastline of California, Oregon, and Washington, but the *Pseudonitzschia* diatoms are probably global in incidence. This disease seems to be associated with cold (5°C) water temperatures, but information on this is limited. Toxigenic species have been spread from one geographical area to another in the ballast of ocean going ships (Hallegraeff, 1993).

Interrelations. The presence of a large natural population of bacteria does not seem to affect the uptake and short-term survival of potentially pathogenic bacteria and viruses. However, there is not much information available on the growth rates of such organisms during the spoilage of oysters, when the pH drops below 6.0. Palumbo *et al.* (1985) noted a decline in the numbers of *Aero. hydrophila* during the refrigerated storage of shucked oysters that corresponded to the pH decline. However, it would be unwise to rely on this pH decline to destroy pathogens because some pathogens can survive for long periods in oysters stored in the shell.

CONTROL (mollusca)

Summary

Significant hazards[a]	• Shellfish toxins. • Virus. • Enteric bacterial pathogens. • *Vibrio* spp.
Control measures	
Initial level (H_0)	• Avoid harvesting from contaminated areas.
Reduction (ΣR)	• Depuration (partially effective). • Cooking.
Increase (ΣI)	• Keep animal alive.
Testing	• Surveying water quality for algal blooms and preferably for enteric virus. • Animals tested for *Salmonella.*
Spoilage	• Spoilage does not occur if animal is kept alive.

[a]In particular circumstances, other hazards may need to be considered.

Hazards to be considered. Live molluscs to be consumed raw (e.g. oysters) pose several severe risks to the consumer. The filter feeding nature of the animal allows it to concentrate pathogenic organisms that pose a very low risk when present in low numbers. Virus, shellfish toxins, *Vibrio*s and enteric bacteria are all major hazards in this product.

Control measures. Assuring the safety of molluscs for the consumer lies in actions taken at two levels: (1) control of the initial level of the pathogenic agents (e.g. controlling H_0), and (2) reducing levels (ΣR), e.g. by a heating step.

Initial level of hazard (H_0). Although H_0 is controlled by growth conditions, and control of harvest and distribution, the reducing step will typically be part of the final preparation. None of the control measures that can be installed to "control" H_0 will be able to guarantee a completely safe, raw product. Thus, the product will always pose some level of risk to the consumer. Health warning labels have been used in retail establishments in some states in the United States to caution at-risk individuals concerning the hazards associated with the consumption of raw shellfish.

Reduction of hazard (ΣR). The consumer can greatly influence the health risk from eating molluscs by introducing a reduction step (ΣR). A decision to eat shellfish only if adequately cooked reduces dramatically the risks associated with the consumption of these foods (IOM, 1991). This can also have an impact on non-bacterial concerns. For example, the kinetics of heat inactivation of paralytic shellfish poison in soft-shell clams was found to be similar to those of most bacteria (Gill *et al.*, 1985).

The shellfish should be harvested from areas free from contaminating sewage or other undesirable land run-off (e.g. animal waste), and the water should not contain sufficient toxic microalgae that would render the shellfish hazardous. The approach used by the U.S. National Shellfish Sanitation Program Manual of Operations Parts I & II (FDA, 1989a,b) has been found to be very effective when diligently applied. Most molluscs are harvested from sediments and should be washed immediately after capture with clean seawater to remove mud and debris. However, even with thorough washing, molluscs, as filter feeders, can be assumed to contain a high level of bacteria in their digestive tract. This can be reduced somewhat by providing the live animal with a source of clean seawater with low bacterial counts, and allowing it to purge itself. The most widely used methods are depuration and relaying. Each relies on the mollusc's inherent ability to clean itself of contamination when supplied with a source of pathogen-free water. Depuration involves shellfish being placed in a controlled flow of disinfected water for a relatively short period of time (usually 24–48 h). With relaying, the shellfish are transferred to new beds that have naturally clean seawater. They are kept there for periods up to several weeks and then re-harvested. Even following such a depuration, pathogenic agents may persist in the animals. Although these methods are reasonably effective in eliminating or reducing greatly Enterobacteriaceae within the mollusc, they are substantially less effective for viruses and toxins, and have little effect on vibrios, including *V. cholerae* O1 (Richards, 1988; Croci *et al.*, 1994; Murphree and Tamplin, 1995) and *V. vulnificus*. Reducing the storage temperature of the live animals after harvest, while awaiting processing, or during distribution if marketed live, helps reduce the potential for growth of mesophilic spoilage organisms and pathogens.

The presence of naturally occurring, potentially pathogenic, vibrios in shellfish at harvest is difficult to deal with because in most cases it is a question of both numbers of particular vibrios and consumer susceptibility. Where cholera is endemic or epidemic, the safest procedure is to warn consumers against eating raw shellfish and to close harvesting on shellfish beds where *V. cholerae* O1 is regularly isolated from water. This requires water samples to be tested specifically for the microorganism. The evidence concerning *V. vulnificus,* suggests that risk is partly related to numbers of the species present in the shellfish which, in turn, usually relates to water temperature (Ruple and Cook, 1992). It has been suggested that harvesting be restricted when water temperatures exceed a particular value associated with the development of high *V. vulnificus* populations in oysters (NACMCF, 1992). Presently, some producers in high temperature water areas are labeling their packaged oysters to warn susceptible consumers of the potential hazard of eating oysters raw. Chilling oysters immediately after harvest may significantly reduce levels of *V. vulnificus*. (Cook and Ruple, 1989; Ruple and Cook, 1992). Depuration also appears to help reduce *V. vulnificus* levels in oysters (Groubert and Oliver, 1994). A range of technologies, such as mild heating, freezing, or high-pressure treatment may also assist in reducing levels of pathogenic bacteria.

If oysters are schucked, cross-contamination during shucking must be avoided, and adequate cooling applied to prevent growth of small numbers of pathogens to more infective numbers prior to eating (Cook, 1994; Bouchriti *et al.*, 1995). These two factors can give rise to a number of controls that are similar to those discussed earlier for finfish and crustaceans.

Increase of hazard (ΣI). Animals must be kept alive and at low temperature to avoid increase in numbers of pathogenic bacteria.

Testing. The control of shellfish sources requires ongoing assessment of the quality of the growing waters, although it must be emphasized that testing and evaluating water quality cannot guarantee that the animals are free of pathogens. Procedures involved include establishing the suitability of a water area by examining water drainage access, bacteriological, and viral analysis of the waters and shellfish to provide baseline data, and regular surveillance of the areas thereafter. Surveillance includes water analysis and analysis of indicator molluscs for algal toxins when deemed necessary. Shellfish beds are opened or closed to harvest on the basis of analytical results and there are provisions for the identification of shellfish according to area of harvest throughout shipment. Also, the EU legislation (EEC, 1991a) has special requirements for the growing waters of molluscs and classifies waters as category A, B, or C depending on levels of e.g. fecal coliforms (EEC, 1991a; Lees, 2001).

Typically, water surveillance involves periodic examination for the presence of indicators of faecal contamination, such as members of the Enterobacteriaceae. While these have generally been useful, they have limited efficacy as index organisms for the presence of specific enteric pathogens (Hood *et al*, 1983; Martinez-Manzanares *et al.*, 1992). Further, there is little relationship between enteric indicators in growing waters and the presence of *Vibrio* spp., viruses, or algal toxins in molluscs. As previously mentioned, algal toxins require identification of the toxic species. Similarly, viruses require specialized analyses.

The periodic examination of shucked shellfish for microbiological indicators, such as APC or *E. coli* counts, can be used to verify the overall effectiveness of process control programmes over time. However, their use in the routine examination of production lots is of limited value for assuring either the microbiological safety or quality of raw molluscan shellfish.

Realizing that surveillance of water quality cannot guarantee a safe product, several agencies have set microbiological limits e.g. for *E. coli* or salmonellae, for mollusc (EEC, 1991a, 1993). However, such end product controls can never eliminate a certain level of risk (ICMSF, 2001) and this should be made clear to the consumer.

Spoilage. Spoilage is not an issue for molluscs destined for live consumption, as the immune system of the animal will prevent degradation of the meat. For shucked, chilled oysters, sensory evaluation is the best way of assessing spoilage, and low-temperature storage (0–2°C) should be maintained to retard spoilage.

IV Aquaculture

Though wild finfish and shellfish still make up the bulk of the seafood in commercial distribution, the most rapidly growing segment of the fisheries industry is aquaculture products. Aquaculture is estimated currently to account for 25% of the total world seafood supply, including a wide range of aquatic, estuarine, and marine species of finfish, molluscs, and crustaceans (FAO, 1998). This includes popular items such as catfish, salmon, trout, shrimp, crayfish, and oysters. Aquaculture is international in scope, with catfish being produced in Southeastern United States; salmon in Northern Europe, Tasmania, and the Maine and Pacific Northwest coasts of the United States; and shrimp and prawns from Asia and South America (IOM, 1991). Future gains in seafood availability are likely to be the result of increased use of aquaculture; harvesting of wild seafood appears to be reaching its maximum sustainable level (Reilly *et al.*, 1992).

A Initial microflora

The initial microbial populations on freshly harvested, properly handled pond-reared finfish will be similar in composition to wild caught fish and consist of a diverse mixture of Gram-negative and Gram-positive genera including *Acinetobacter*, *Aeromonas*, *Citrobacter*, *Enterobacter*, *Escherichia*, *Flavobacterium*, *Micrococcus*, *Moraxella*, *Pseudomonas*, *Staphylococcus*, *Streptococcus*, and *Vibrio*. Pond-reared animals, particularly those from tropical areas, often have significantly higher incidence of Enterobacteriaceae than equivalent marine products (Reilly *et al.*, 1986). However, this appears to be a function of locality and use of manure for fertilizing ponds (Christopher *et al.*, 1978; Reilly *et al.*, 1992; Dalsgaard *et al.*, 1995).

B Spoilage

The icing of fish raised in warm-water ponds tends to produce a decline in initial populations, favoring the survival of Gram-positive species. However, upon extended refrigerated storage, the Gram-negative species, particularly *Pseudomonas* and *Acinetobacter* are selected (Acuff *et al.*, 1984; Wempe and Davidson, 1992; Nedoluha and Westhoff, 1993) and the products spoil similarly to wild caught fish.

C Pathogens

The presence of pathogenic species in aquaculture products is dependent on a number of factors including the source of nutrients used to enrich ponds, the extent of feeding, the population density in the ponds, and the methods used to harvest, process, and distribute the product. Pond-reared trout have been reported to have a high incidence of *Cl. botulinum* (Huss *et al.*, 1974a; Cann *et al.*, 1975), which may be aggravated by excessive feed levels. It has been suggested that the use of "wet fish" as trout feed, led to the very high levels of spores found in Danish trout ponds (Huss, 1980). This is no longer allowed as feed and extruded dry feed is not likely to carry high levels of bacteria.

The use of animal and human excreta to enrich ponds, as well as run-off from adjacent agricultural lands, can significantly increase the incidence of enteric pathogens in both the growing waters and the harvested animals (Ward, 1989; Reilly *et al.*, 1992; Twiddy, 1995). This practice has been associated with certain food-borne parasitic infections, particularly trematodes (flukes). A high incidence of classic enteric pathogens such as *Salmonella* and *Shigella* is common in water receiving manure or sewage run-off (Wyatt *et al.*, 1979; Saheki *et al.*, 1989; Reilly *et al.*, 1992; Twiddy, 1995), and is likely to be reflected in an increased incidence of enteric pathogens on raw aquaculture products and in the processing environment (Iter and Varma, 1990; Reilly *et al.*, 1992; Reilly and Twiddy, 1992; Ward, 1989). However, others did not find that chicken manure increased the isolation of salmonellae (Dalsgaard *et al.*, 1995). Even in the absence of overt faecal contamination, a low incidence of salmonellae in ponds and pond-reared aquaculture products is not uncommon. Some of the factors thought to affect their presence in pond water include water temperature, organic content, salinity, pH, stocking level, and fish size. Fish feeds, as well as amphibia and aquatic birds, have been implicated as potential sources. Also important in this context may be findings that *Salmonella* and *E. coli* survive well in warm, tropical waters (Jiménez *et al.*, 1989). A high percentage of salmonellae isolated from pond-reared finfish and crustaceans were resistant to multiple antibiotics (Hatha and Lakshmanaperumalsamy, 1995; Twiddy, 1995).

A high incidence of pathogenic *Vibrio* spp., including *V. cholerae* non-01, *V. parahaemolyticus*, and *V. vulnificus* can also occur in both the growing waters and raw products from rearing ponds in tropical regions (Christopher *et al.*, 1978; Leangphibul *et al.*, 1986; Varma *et al.*, 1989; Nair *et al.*, 1991; Reilly and Twiddy, 1992; Wong *et al.*, 1992; Dalsgaard *et al.*, 1995). Elevated incidence of *V. cholerae* non-O1, *V. parahaemolyticus* and *V. vulnificus* has been reported in aquaculture products from sewage-enriched waters (Varma *et al.*, 1989; Nair *et al.*, 1991). The presence of *V. cholerae* non-O1 in tropical shrimp culture areas could, however, not be correlated to water salinity, temperature, dissolved oxygen, or pH (Dalsgaard *et al.*, 1995).

Listeria spp. appear to be common in the freshwater aquaculture environment (Jemmi and Keusch, 1994) but is not typically isolated from marine units in open waters (Huss *et al.*, 1995). If net pens are close to land where heavy rainfall results in run-off, the incidence of *Listeria* spp. may be very high.

In addition to the aquaculture-specific concerns listed above, care must be exercised to control the pathogens normally associated with the wild finfish and shellfish. It has been suggested that special care be exercised when finfish and shellfish derived from wastewater aquaculture are destined for human consumption (Hejkal *et al.*, 1983), particularly those that will be consumed raw. However, there appears to be little indication that aquaculture products that are subsequently cooked have any inherent increased risk in relation to food-borne disease compared with other raw meat or poultry (Reilly *et al.*, 1992). Considering the increased potential for control, aquaculture products reared using sound management practices would be expected to have increased microbiological safety and quality.

One of the major constraints of aquaculture is disease. Whilst good management practices and vaccines are important measures to control disease, treatment with antibiotics remains the method of choice in several places. Antibiotic residues are seldom found because fish must be held for specific periods of time without antibiotic treatment before slaughter. Use of antibiotics will cause development of resistance in the bacterial population (Spanggaard *et al.*, 1993; DePaola *et al.*, 1995). Thus, unlimited use of antibiotics (as prophylactics) has led to collapse of fish farms because antibiotic-resistant fish pathogens could no longer be eliminated (Karunasagar *et al.*, 1994). Of more concern to the consumer, is the potential spread of such antibiotic resistance to human pathogenic bacteria (Twiddy, 1995). Of 187 *Salmonella* isolated from imorted foods in the US in 2000, 15 were resistant to one or several antibiotics and 10 of these 15 were derived from seafoods (Zhao *et al.*, 2001). Whilst this could be caused by excessive use of antibiotics in agriculture, or clinics filtering into the growing/catching waters, it could also be caused by use of antibiotics in the ponds or netpins. To our knowledge, the use of antibiotics in aquaculture has not been directly linked to treatment failures in humans. However, a *Salmonella enterica* DT104 out-break in Denmark in 1998 involved 25 people and two deaths and the quinolone-resistant type was traced to swine herds (Mølbak *et al.*, 1999).

D CONTROL (aquaculture)

Summary

Significant hazards[a]	• Algal toxins. • Bacterial and viral pathogens as for "fresh fish" (when consumed raw). • Antibiotic residues. • Parasites (in some areas; not in reared salmon and other fish fed with dry feeds).
Control measures	
Initial level (H_0)	• Avoid harvesting during/after algal blooms. • Avoid animal/human fertilizers. • Holding periods after antibiotic treatments.
Reduction (ΣR)	• Freezing. • Cooking.
Increase (ΣI)	• Time × −temperature.
Testing	• surveying pond water for fecal contamination.
Spoilage	• Time × −temperature. • Sensory evaluation.

[a]Under particular circumstances, other hazards may need to be considered.

Hazards to be considered. Aquaculture products differ in some ways from wild caught fish, but overall the same rationale in terms of identification of hazards as used for marine and freshwater fish applies. However, aquatic toxins like ciguatera is not a safety issue being associated with tropical reefs, whereas shellfish toxins can accumulate if algal blooms occur in the ponds or ocean areas where the net pens are placed. Parasites are not tranferred by dried fish feed and is only a potential risk if live fish (or fish offal) is used as feed. The same types of indigenous, potentially pathogenic, bacteria, as described for wild fish, will be present on reared fish. Enteric organisms may occur in higher numbers, particularly in areas where ponds are situated in fields with agricultural run off or if manure is used to fertilise the ponds. *L. monocytogenes* is more common in fresh water ponds with agricultural run-off. Antibiotic residues can occur and inappropriate use of antibiotics can lead to the selection of antibiotic-resistant bacteria.

Control measures

Intial level of hazard (H_0). Algal shellfish toxins are controlled by limiting initial concentrations (H_0) through surveillance of culture waters for blooms. Antibiotic residues can be avoided by allowing the animals a clearing period without treatment before slaughter. The exact time will depend on the size of the animal and the temperature of the water. Most legislation will provide guidance on time $\times$ temperature requirements for antibiotic clearance.

Reduction of hazard (ΣR). In cooked products, potential enteric pathogens and vibrios are controlled by the reduction step (ΣR) during cooking. Maintaining the microbiological quality of the rearing sites, limits the presence of pathogens. Use of animal or human excreta for enriching rearing-ponds should be discouraged unless subsequent controls insure that there is no increased risk of the product. This includes increased care to prevent cross-contamination of cooked (or otherwise processed) products produced from aquaculture sources known to have a high incidence of pathogenic enteric microorganisms. Presence of low levels of *Vibrio* spp. – and from some areas also *Salmonella* spp. – is to be expected and, as for indigenous bacteria, such as *Cl. botulinum*, processing procedures must ensure that these organisms do not become a hazard. Raw seafood does not differ from other raw commodities in that a low level of potential human pathogenic organisms are present and that the product should be handled accordingly.

Increase of hazard (ΣI). Controls for psychrotrophic spore formers in packed fresh fish are similar to wild fish, e.g. controlling increase (ΣI) during storage by temperature control.

Testing. Periodic sampling of pond waters or sediment for indicators of faecal pollution (e.g. Enterobacteriaceae, *E. coli*) or *Vibrio* spp. may be useful for assessing the classes (e.g. vibrios versus enterics) and relative extent of pathogens that are likely to be encountered with aquaculture products harvested from those waters (Leung *et al.*, 1992). In freshwater ponds it may be useful to know if *L. monocytogenes* are present in large numbers so that further treatment and processing can be designed to reduce numbers and/or ensure that growth does not occur in the product.

Spoilage. Products from aquaculture do not differ from products caught from a wild population and time–temperature control is the most important controlling factor limiting increase (ΣI) of spoilage bacteria.

V Frozen raw seafood

A *Freezing process*

Raw seafood may be frozen as whole fish or clams in the shell, eviscerated and butchered fish, tuna loins, fillets, steaks or portions, shucked shellfish, *etc.* Some of these products are breaded prior to freezing. In most cases, seafood are frozen, unwrapped to facilitate rapid freezing, but for some purposes products may be packaged before freezing. Fish are frozen on-board fishing vessels, processing ships, and in land-based processing plants. Finfish destined for shaped products (e.g. fish sticks) are frozen in blocks and later cut with saws. All types of freezing systems are used for seafood including contact plate or shelves, brine, and other direct contact refrigerant systems, continuous moving belt air freezing systems and passive air blast freezers as well as traditional sharp freezers (Pigoff and Tucker, 1990b). The rate of freezing is as rapid as possible, but with large whole fish such as tuna and shark freezing this may take several days. Frozen seafood is taken down to a temperature below −18°C and more commonly with modern practices to even lower temperatures. The faster the initial freezing, the lesser the damage to the protein fraction and the better is the product. Whole tuna destined for secondary processing are frozen on shipboard in brine tanks where the system cannot reach such low temperatures. Storage of frozen seafood is at −20°C or lower to maintain product quality. Fish frozen before *rigor mortis* are often held at −7°C for a few days to enhance quality.

B Saprophytes and spoilage

Freezing halts bacterial growth and metabolism and the major cause of quality changes during frozen storage is non-microbial changes in the protein and lipid fractions of the fish flesh.

Bacterial counts on frozen products to some extend reflect the bacteriological quantity of the raw material and contamination or its removal during processing. Additional bacteria may be introduced as the result of batter and breading (Surkiewicz *et al.*, 1967; Duran *et al.*, 1983). The reduction in count resulting from freezing and storage in the frozen state is highly variable, and this makes the assessment of prefreezing quality difficult in some cases (DiGirolamo *et al.*, 1970). The psychrotrophic bacteria in fish are not particularly resistant to freezing stress, but response is so strain specific that no general rule can be given. Freeze injury is generally more pronounced with Gram-negative bacteria than with Gram-positive species.

Spoilage microorganisms may grow in raw fish products if held too long before freezing, frozen at a grossly slow rate, thawed too slowly, or held thawed too long. No microorganisms grow below −10 to −12°C. Because thawing without cooking is an inherently slow process, this step selects for a psychrotrophic population. Once thawed, the biochemical and microbiological changes are similar to those described earlier for raw refrigerated seafood (see Section III).

Seafood held at improperly elevated frozen storage temperatures (−10 to −5°C, for example) are likely to support very slow mould growth. A few moulds, and possibly yeasts can grow in that range, whereas bacteria only grow at somewhat higher temperatures. While bacteria do not grow in frozen foods, they are able to varying degrees survive extended frozen storage. Frozen thawed fish spoil about as fast as fish that has never been frozen. One exception is CO_2 storage of frozen, thawed marine fish. CO_2 storage selects, as described above, for the CO_2 resistant spoilage bacterium *Photobacterium phosphoreum* (Dalgaard *et al.*, 1993). However, this organism is freeze-sensitive and subsequent CO_2 storage of frozen-thawed fish may extend shelf-life dramatically (Bøknæs *et al.*, 2000). Bacteria populations tend to decline over time during frozen storage, though the rate is highly dependent on bacterial species. For example, enterococci are highly resistant to extended frozen storage. In general Gram-negative bacteria die off more rapidly during frozen storage than Gram-positive. Spores are even more resistant.

C Pathogens

As noted above, freezing will bring about a general reduction of the bacterial populations on seafood. This is true for pathogens as well as psychrotrophic spoilage organisms. Generally Gram-negative pathogens such as *Salmonella* and other Enterobacteriaceae are sensitive to freezing injury and there is also some mortality of mesophilic vibrios. However, because products are generally maintained at refrigeration temperatures prior to freezing, it is unlikely that the extent of freeze injury will be great. In all cases there is measurable survival and the extent of actual mortality is highly variable. Spores are unaffected by freezing and vegetative cells of Gram-positive bacteria including *Staphylococcus* and

Listeria usually survive well. During storage of frozen seafood, there is a continued die off of vegetative bacteria at rates corresponding to the specific species' sensitivity and the temperature regime in the storage chamber.

Freezing followed by frozen storage will normally destroy all fish parasites dangerous to man. This procedure is recommended for raw products to be eaten as sushi. Freezing does not affect marine toxins accumulated in the living animal nor bacterial toxins produced during improper storage before freezing.

Survival of bacteria in seafood during frozen storage can have importance for infective organisms such as *Salmonella*, *Shigella*, *Listeria*, *V. cholerae*, and other *Vibrio* species since these may be transmitted without further growth and infectivity is dose related. Most reports suggest that *V. cholerae* tends to be reduced to very low levels after about 3–6 weeks of storage, but *V. parahaemolyticus* can persist for several months (Johnson and Liston, 1973).

It is important to note again that although freezing halts the production of histidine decarboxylase by bacteria, any pre-formed enzyme continues to be active. This can result in significant elevation of histamine levels during long term frozen storage, particularly when freezer temperatures are too high. The enzyme activity observed would be sufficient to raise the histamine levels to above the spoilage limit of 5 ppm or even the danger level of 50 ppm.

D CONTROL (frozen raw seafood)

Summary

Significant hazards[a]	• Aquatic toxins. • Histamine (histidine containing species).
Control measures	
Initial level (H_0)	• Avoid fishing during algal blooms. • Avoid fishing in particular areas.
Reduction (ΣR)	
Increase (ΣI)	• Time–temperature control.
Testing	• Sensory evaluation.
Spoilage	• Quality of fish before freezing (see above). • Time–temperature. • Sensory evaluation.

[a]Under particular circumstances, other hazards may need to be considered.

Hazards to be considered. Frozen raw seafood presents no health hazards different from the fresh, raw seafood.

Control measures. In addition to the controls associated with raw fish that have already been discussed (e.g. aquatic toxins and histamine), frozen seafoods have two other factors that must be considered: rate of freezing and temperature control during frozen storage. These factors primarily affect sensory quality of the product, although the low temperature is the principal means to stop microbial activity in frozen seafood. Freezing should be as rapid as possible, and once frozen the product should be held at or below -18°C. The thawing of the product can also have a strong influence on the microbiological quality and safety of the product. Thawing should be as rapid as possible and should avoid having the exterior surface of the product exposed to abuse temperatures while waiting for the centre to thaw.

Testing. Microbial analysis of product that has continuously maintained at frozen temperatures can only to a limited extend provide information on the bacteriological quality of the fish before it was frozen. Loss of temperature control during frozen storage can be investigated by the use of psychrotrophic counts, which would be expected to increase if the product had been exposed for any substantial duration to thawing conditions. However, the use of this tool is dependent on having historical data on the normal levels of psychrotrophs in the product and the immediate analysis of the sample so that the results are not confounded by exposure to thawing condition during sample acquisition or processing.

If *Vibrio* spp. are tested in raw, frozen seafood produced from fish from warm waters it should be taken into account that the organisms are indigenous to the aquatic environment and must be expected to be present.

VI Minced fish and surimi products

A significant world trade exists in finfish blocks, minced fish, surimi, and the products made from them. In addition to making finfish blocks from fillets, they can also be prepared from fillet pieces or minced finfish, which are compressed and frozen. Like the blocks from fillets, these are used to prepare fish fingers (sticks) or fish portions, frequently after being breaded and battered. These products are commonly precooked or partially cooked. The prepared products are commonly refrozen, packaged and stored and shipped as frozen foods. Minced finfish blocks are prepared by separating the flesh from skin and skeleton of eviscerated animal using deboning machines and are often prepared from residues of filleting processes. Surimi is also prepared by mechanical deboning, but is reduced essentially to muscle protein fibers by repeated washing. The evisceration of fish for surimi is ususally done manually. The fibers are mixed with cryoprotectants before compression into blocks and freezing. Minced finfish blocks generally have higher bacterial populations than fillet blocks because of both the raw material source and the mixing process during formation of the mince. Surimi varies in bacterial content depending on the pretreatment handling and length of storage of the fish and processing control. In either case, counts of the order of 10^6 cfu/g are not uncommon.

Surimi is the basis of a variety of simulated (imitation) products, which are made by a mechanized procedure. This involves partial thawing to just below freezing, mixing with salt and other additives to yield a smooth paste, which is extruded onto a moving belt as a flat ribbon. The material then passes through various heating steps, which effectively eliminate all but bacterial spores. The material is then texturized to form a rope. This is further treated to form desired shapes. Most commonly, the product simulates crab, shrimp, or scallops. In most processes, but not all, there is a terminal hot-water pasteurization process before the product is frozen.

Surimi is also served as kamaboko which is surimi cooked on a wooden tray and sliced into thin pieces.

Surimi products, though made from finfish, are quite different in composition from most seafood. They contain none of the soluble nitrogen-based compounds, which are the main substrates of spoilage bacteria, but have significant amounts of added sugars and often egg products. Though minced fish that is used to make surimi has a microflora very similar to that of raw fish, surimi products carry only minimal populations (10^1–10^2 cfu/g) of mostly spore-forming bacteria (Elliot, 1987; Matches *et al.*, 1987; Yoon *et al.*, 1988). When thawed and held at refrigeration temperatures the products have a very long shelf life as measured organoleptically. However, it has been shown experimentally that surimi products provide an excellent substrate for the growth of a number of pathogenic bacteria. Thus, there is a potential hazard if such products are subject to temperature abuse and this is compounded by the absence of the usual highly odoros spoilage indicators normally associated with mishandled fish products. *Aeromonas hydrophila* effectively competed with *Pseudomonas fragi* on low-salt, but not high-salt surimi (Ingham and Potter, 1988).

There is no record of food-borne illness from consumption of products derived from minced fish or surimi. The products must, as all foods, be produced observing GHP to avoid cross-contamination for the human–animal resevoir.

VII Cooked crustaceae (frozen or chilled)

A substantial proportion of crustacea in international trade are cooked prior to marketing. This includes products such as shrimp, lobsters, and more recently crayfish. Because the first step in crab processing is a cooking step, all crab products fall into this category. Once cooked, these products are either distributed in a refrigerated state, or more likely they are frozen. Frozen products are either marketed as such or they are thawed prior to retail display. Breaded frozen shrimp products may be only partially cooked prior to freezing. A substantial portion of crabmeat is pasteurized for extended refrigerated storage (see section on pasteurized products). Examples of process flow diagrams for cooked shrimp and blue crabs are provided in Figures 3.7 and 3.8, respectively.

A Cooking, picking, and packaging

Crabs routinely receive a thermal process prior to picking or other processing, and are appropriately classified as a cooked product. A number of other crustacea (e.g. shrimp, lobster, langostinos, crayfish, etc.) are often marketed as refrigerated or frozen, cooked, ready-to-eat products. The cooking process can be blanching (shrimp, 95–100°C), boiling, or steaming under pressure (lobsters, crabs, >100°C). The duration of the cook is generally short to minimize quality loss; typically the equivalent of 20 min in boiling sea-water for crabs. Smaller species are cooked whole but larger crabs (e.g. King crab and snow crab) are split before cooking.

Meat, particularly claw meat, from cooked lobsters and some species of crabs (e.g. blue crab) is picked by hand or with minimal mechanization. Crabmeat is frequently separated from small pieces of shell by flotation in salt brine, then washed in fresh water before packaging and chilling or freezing. Cooked shrimp are marketed both peeled and in-shell. Peeled shrimp are typically but not exclusively cooked prior to removal of the shell. Peeling of shrimp is increasingly done by machine, but hand peeling still accounts for much shrimp in world trade, particularly from non-industrialized countries, and most processing plants will have a manual "fine peeling" step following a mechanical peeler.

The shelf life of crabmeat can be extended by packing the "picked" meat in containers and pasteurizing so that the internal temperature reaches 85°C (185°F) for at least 1 min (FDA, 2001b). The product is then held under refrigeration (≤2°C). This method allows the crabmeat to be held in bulk storage for up to 6 months. Cooked, peeled shrimp is packaged and frozen without further treatment, or after a batter and breading dip.

B Saprophytes and spoilage

The cooking step reduces significantly the bacterial counts in crustaceans, destroying vegetative cells of both spoilage and pathogenic species (Ingham and Moody, 1990). However, care must be exercised to ensure that there are no "cold spots" that would receive less than sufficient thermal processing.

Crab picking and shell separation, which may involve brine flotation, often recontaminates the picked meat with a variety of microorganisms. Refuse containers for picking waste, insects, and cross-contamination from live crab carts have been identified as sources of recontamination after cooking. Counts of 10^5 cfu/g in picked meat are common, and microorganisms present will include Gram-positive rods and cocci, Gram-negative rods, and yeast. Storage of picked meat under refrigeration normally

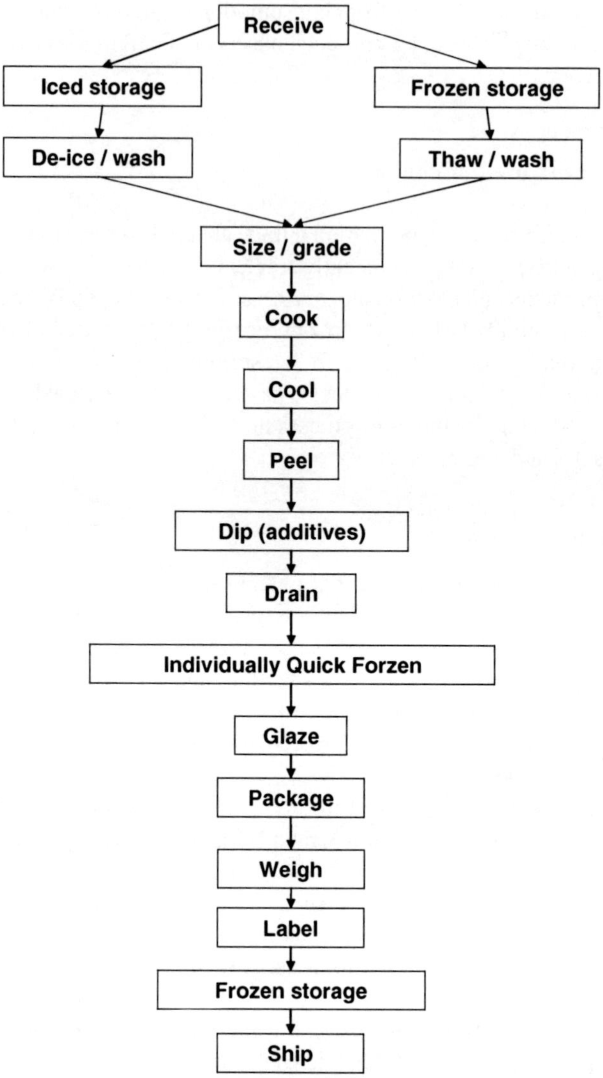

Figure 3.7 Example of a process flow diagram for cooked shrimp (NMFS, 1989).

results in a progressive increase in counts to $>10^7$ cfu/g in 7 days. The microflora is usually dominated by Gram-negative rods, typically *Pseudomonas* and *Acinetobacter–Moraxella* spp. Spoilage typically involves the production of volatile amines.

The arrival of shrimp at the processing plant in a non-living state makes their processing somewhat different from crabs and lobster. Their initial bacterial level at the processing plant will be a function of the quality and extent of shipboard storage, with counts ranging from 10^3 to 10^7 cfu/g. Shrimp taken from tropical waters will tend to have higher counts (Cann, 1977). Shrimp cooked by boiling or steaming for a few minutes either prior to or after peeling show ~100- and 10 000-fold reductions in mesophilic and psychrotrophic plate counts, respectively (Ridley and Slabyj, 1978). However, there tends to be a rapid return to bacterial levels equivalent to those prior to cooking as the product is held and handled during inspection and grading, though there may be a shift to a more mesophilic microflora.

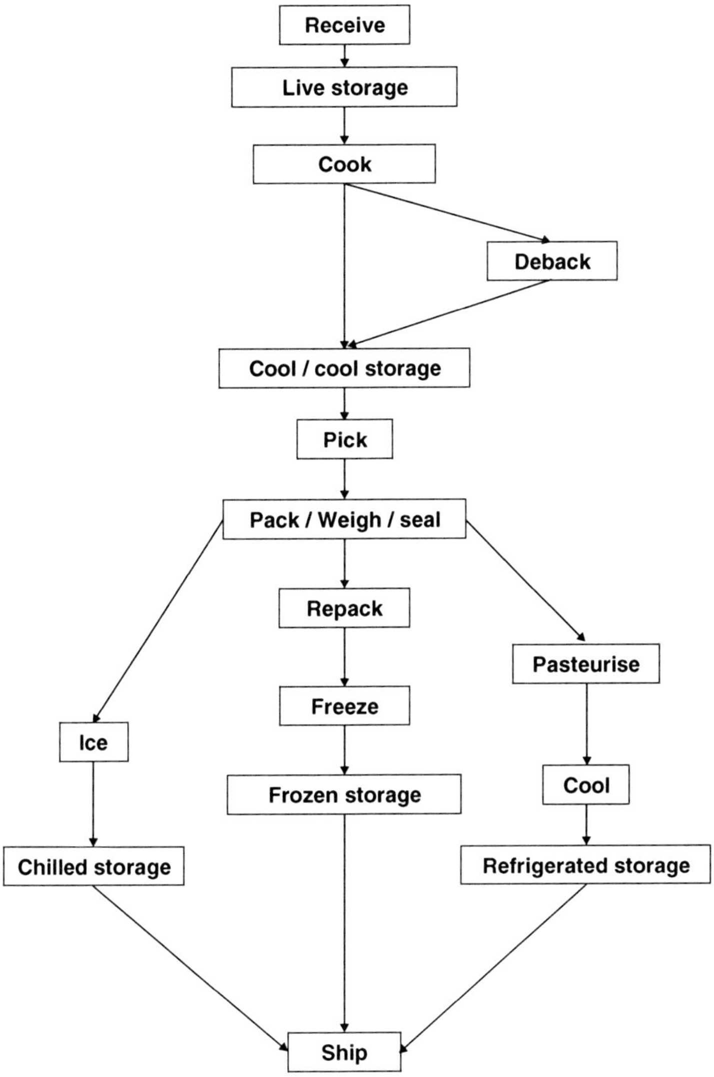

Figure 3.8 Example of a process flow diagram for hard blue crabs (NMFS, 1990c).

In general, the level of microorganisms post-processing is a reflection of the level on the in-coming raw material (Høegh, 1986). Somewhat smaller recontamination rates are observed with brine-cooked shell-on shrimp (Ridley and Slabyj, 1978). Because of handling, low levels of coliforms, *E. coli*, and staphylococci may be present in finished products. When cooked shrimp are stored chilled rather than frozen, the spoilage flora is dominated by *Pseudomonas* or *Acinetobacter–Moraxella* and sometimes coryneform bacteria (Cann, 1977) or *Aeromonas* (Palumbo *et al.*, 1985; Palumbo and Buchanan, 1988).

C Pathogens

The potential for the presence and growth of pathogenic bacteria in cooked crustacea is a function of the adequacy of the thermal process, the extent of post-processing contamination, and the maintenance of proper refrigerated or frozen storage. Temperature abused crabmeat, shrimp, etc., can support the growth of a wide range of species, particularly if competitive microorganisms have been eliminated as a result of the thermal processing. The extent of manual manipulation of product often provides ample opportunity for the introduction of a variety of human pathogens. This is reflected in the reported incidence of food-borne outbreaks associated with cooked crustacea (Bryan, 1980). *Vibrio parahaemolyticus*, a common member of the microflora associated with raw shrimp and crab, is a cause of food poisoning associated with these products. As an example, boiled crab caused 691 cases of *V. parahaemolyticus* gastroenteritis in Japan in 1996 (Japanese Ministry of Health, Labour and Welfare, 1999). Cholera cases have been attributed to the consumption of crab from epidemic areas (Finelli *et al.*, 1992). However, a significant percentage of the outbreaks linked to cooked shrimp, lobster, and crabmeat products have been attributed to long-established food-borne pathogens including *Staph. aureus*, *Salmonella*, *Shigella*, and virus. Maintenance of proper storage temperatures is a prime concern, particularly under conditions where growth of pathogens could occur before overt spoilage of the product (Ingham *et al.*, 1990).

As the picking process is by hand, the possibilities for transfer of pathogenic bacteria from humans to crabmeat is high. The extent of human handling with shrimp varies from extensive in operations where product is hand peeled to minimal where peeling, inspection and packaging are mechanized. The ubiquitous association of *Staph. aureus* with food handlers often results in the introduction of low numbers into raw and cooked products, particularly crabmeat (Ridley and Slabyj, 1978; D'Aoust *et al*, 1980; Hackney *et al.*, 1980; Swartzentruber *et al.*, 1980; Wentz *et al.*, 1985), though the former should be eliminated during the cooking step. Reintroduction of the pathogen may be exacerbated by the use of brine flotation as a means of removing shell fragments from the product; the high salt levels select for halotolerant staphylococci over Gram-negative spoilage organisms. Fortunately, staphylococci tend not to compete well with the normal flora of crabmeat, particularly at marginal abuse temperatures, and generally die slowly during refrigerated storage (Slabyj *et al.*, 1965). However, significant growth of staphylococci can occur in temperature-abused product, particularly if the spoilage microflora has been suppressed as a result of the thermal processing (Gerigk, 1985; Buchanan *et al.*, 1992). This has resulted in occasional staphylococcal food poisoning outbreaks, particularly in crabmeat (Bryan, 1980). Hand peeling of cooked shrimp can introduce *Staph. aureus* on the product and if temperature-abused, it will grow well and can produce enterotoxin. Potentially, the use of processes, such as modified atmosphere packaging, that alter significantly the microflora of the product could impact the potential for significant outgrowth of *Staph. aureus* before overt spoilage of temperature abused product. A variety of other *Staphylococcus* spp. has also been isolated from frozen crabmeat (Ellender *et al.*, 1995).

The potential for temperature-abused cooked crustaceans serving as a source for a variety of food-borne pathogens focuses attention on the need to maintain adequate sanitation and temperature control after initial thermal processing. This is reinforced by the linking of cooked crustaceans to occasional outbreaks of enteric pathogens such as *Salmonella* and *Shigella* (Bryan, 1980; Mazurkiewicz *et al.*, 1985). The incidence of *Salmonella* and *Shigella* in raw and cooked products is typically low; however, this can vary among geographical locations (D'Aoust *et al.*, 1980; Gilbert, 1982; Fraiser and Koburger, 1984; Gerigk, 1985; Sedik *et al.*, 1991). There are indications that modified atmosphere storage slows the growth of salmonellae in marginally temperature abused (11°C) cooked crab; however, this effect is lost if the product was subjected to higher abuse temperatures (Ingham *et al.*, 1990).

The levels of enteric pathogens tend to decline during refrigerated storage; however, this is neither rapid nor sufficient to ensure elimination (Taylor and Nakamura, 1964). *Yersinia enterocolitica*, a psychrotrophic enteric pathogen, is an exception in that the species will grow in both cooked shrimp and crabmeat at refrigeration temperatures (Peixotto *et al.*, 1979). It can be isolated in low numbers from raw shrimp and crab (Peixotto *et al.*, 1979; Faghri *et al.*, 1984). However, it is readily inactivated by cooking, and declines during frozen storage (Peixotto *et al.*, 1979).

Vibrio parahaemolyticus, *V. cholerae*, and *V. vulnificus* are routinely part of the microflora of raw crustaceans harvested from estuarine waters (Davis and Sizemore, 1982; Faghri *et al.*, 1984; Molitoris *et al.*, 1985; Varma *et al.*, 1989) and are important bacterial pathogens in cooked crustaceans. The presence of these species in the environment has been correlated with elevated seawater temperatures (Kelly, 1982; Williams and LaRock, 1985; O'Neil *et al.*, 1990); however, they have also been isolated from crabs taken from cold waters (Faghri *et al.*, 1984).

The high incidence of *Vibrio* in raw crustacea makes adequate thermal inactivation and prevention of cross-contamination important measures for preventing pathogenic *Vibrio* in cooked product. Prevention of cross-contamination is particularly important (Hackney *et al.*, 1980; Karunasagar *et al.*, 1984), since *Vibrio* are considered sensitive to thermal inactivation (Delmore and Crisley, 1979; Shultz *et al.*, 1984). However, there have been reports that standard heating procedures for primary treatment of crabs may not be sufficient to destroy *V. cholerae* in the internal tissues, and more rigorous treatments have been proposed (Blake *et al.*, 1980a). Numbers of *Vibrio* spp. decline slowly during refrigerated and higher frozen storage temperatures (i.e. 4°C and −20°C), but are stable at lower storage temperatures (i.e. −80°C) (Bradshaw *et al.*, 1974; Oliver, 1981; Boutin *et al.*, 1985; Oliver and Wanucha, 1989). A cryoprotective agent for *V. cholerae* has been identified in prawn shells (Shimodori *et al.*, 1989), and chitin has been reported to enhance the pathogen's acid tolerance (Nalin *et al.*, 1979), but not its thermotolerance (Platt *et al.*, 1995). Non-O1 *V. cholerae* strains with increased tolerance to refrigerated and frozen storage have been observed (Wong *et al.*, 1995).

Commonly employed microbiological indicator tests, including APC, coliform, fecal coliform, and enterococci assays, are of limited use for indicating the presence of *V. parahaemolyticus* (Hackney *et al.*, 1980).

Identification of the role that food-borne transmission plays in the etiology of human listeriosis, along with the ability of *L. monocytogenes* to grow at refrigeration temperatures, has prompted considerable study of its characteristics in cooked shrimp, crabmeat, and crayfish. For example, the microorganism grew readily on cooked crayfish tail meat at 6°C, or after short term abuse at 12°C, but did not grow if the product was maintained at 0°C (Dorsa *et al.*, 1993a). Pasteurized crabmeat supports the growth of *L. monocytogenes* at 1°C, and increasing storage temperatures to 5°C increases growth rates substantially (Rawles *et al.*, 1995). *Listeria monocytogenes* is present in a high percentage of fresh and low salinity water samples (Colburn *et al.*, 1990) and can be isolated routinely from raw shrimp and crab (Motes, 1991). The microorganism is isolated at approximately the same rate (10%) from cooked peeled shrimp and cooked crabmeat (Weagant *et al.*, 1988; Hartemink and Georgsson, 1991; Rawles *et al.*, 1995). It is unclear whether this is due to insufficient thermal inactivation, cross-contamination, or post-processing contamination from environmental or food handler sources. The species is often associated with processing environments and Destro *et al.* (1996) found that several different RAPD/PFGE types were associated with a processing plant producing frozen, un-cooked shrimp. *Listeria monocytogenes* is considerably more heat resistant than *Vibrio* or enteric pathogens, particularly at lower thermal processing temperatures (Harrison and Huang, 1990; McCarthy *et al.*, 1990; Dorsa *et al.*, 1993a). For example, its $D_{60^\circ C}$ value in crayfish tail meat homogenate was 4.7 min (Dorsa and Marshall, 1995). This was reduced to $D_{60^\circ C} = 2.4$ min by treating the meat with 1% lactic acid. *Listeria monocytogenes* is also substantially more tolerant of elevated sodium chloride levels than enterics (Buchanan *et al.*, 1989), which could favor its presence during brine flotation of crabmeat. Spraying crayfish tails with citric acid or potassium sorbate was not effective as a means of reducing *L. monocytogenes* on cooked crayfish (Dorsa *et al.*, 1993b). Modified atmosphere packaging can partially retard the growth of *L. monocytogenes* at refrigeration temperatures, but this effect is lost at abuse temperatures (Oh and Marshall, 1995). Treatment of crayfish tails with ≥1.5% lactic acid increased the effectiveness of modified atmosphere packaging (Pothuri *et al.*, 1996).

Listeria monocytogenes achieved substantial growth at 5°C, reaching levels in crabmeat in excess of 10^6 cfu/g within 14–21 days in inoculated pack studies (Brackett and Beuchat, 1990; Buchanan and Klawitter, 1992). The microorganism can survive in frozen foods essentially unchanged for extended periods (Harrison *et al.*, 1991; Palumbo and Williams, 1991). While *L. monocytogenes* has been isolated from both cooked shrimp and crabmeat, and has survival and growth characteristics of concern, it is important to note that there has been no direct link between cooked crustacean products and the etiology of food-borne listeriosis.

Strains of *Cl. botulinum* (usually serotype E) have been isolated from crabs, but so far there is no evidence of a specific botulism problem. Pasteurized crabmeat is stored for several months in cans that can be expected to develop anaerobic conditions. The pasteurization process standard adopted by the U.S. National Blue Crab Industry Association is sufficient to inactivate non-proteolytic, but not proteolytic *Cl. botulinum* spores. The product should be refrigerated to prevent germination and outgrowth of spores that could impact both microbiological quality and safety. It would be reasonable to assume that temperature-abuse of this product occurs occasionally; however, in the 50 years that the pasteurization process has been used there have been no food-poisoning cases attributable to the practice.

D CONTROL (cooked crustaceae, frozen or chilled)

Significant hazards[a]	• *Staph. aureus* (post-heating contamination). • Bacterial enteric pathogens (post-heating contamination). • Viral enteric pathogens (post-heating contamination). • *Cl. botulinum* (mesophilic in canned crab meat).
Control measures	
Initial level (H_0)	• Not applicable.
Reduction (ΣR)	• Heating/cooking destroys pathogens.
Increase (ΣI)	• Observe GHP to avoid cross-contamination. • Cold storage of canned crab meat.
Testing	• *Staph. aureus* in cooked crustaceans. • Salmonellae.
Spoilage	• Time × −temperature.

[a]Under particular circumstances, other hazards may need to be considered.

Hazards considered. As outlined in the section on fresh crustaceans, aquatic toxins, histamine, and parasites are not major safety issues. Pathogenic bacteria, both indigenous and contaminating, are killed by the cooking step leaving a product, which in principle is sterile. As cooked crustaceans typically are handled both by peeling machines and manually following the cooking step, contamination with human borne pathogens, e.g. *Staph. aureus*, salmonellae, or virus may occur. Also, cross-contamination by *Vibrio* species present in the raw material has been observed.

Control measures

Intial level of hazard (H_0). See section on raw shellfish.

Reduction of hazard (ΣR). The presence of enteric pathogens such as salmonellae, *Shigella* spp. or virus is associated with insufficient thermal processing (lack of control of ΣR).

Cooked product should be cooled promptly and held at $\leq 2°C$. Flotation brines should be refrigerated, and should be changed or treated on a scheduled basis to adequately control microbial loads. At each change, the tank should be cleaned and disinfected. Processing temperatures, cooling rates, and refrigerated/frozen storage temperatures should be monitored on a continuous basis.

Products intended to have an extended refrigerated shelf life must include controls that inhibit growth of psychrotrophic pathogens such as *L. monocytogenes* and *Cl. botulinum* (see section on lightly preserved fish products).

Increase of hazard (ΣI). High numbers of contaminating *Staph. aureus* are associated with excessive handling by humans and subsequent growth of the organisms. This recontamination (ΣI) is controlled through training of personnel and minimizing direct contact with cooked product (e.g. use of gloves or utensils). This pathogen's halotolerance may also allow its build up in brines; this can be controlled by refrigeration of the brine and its periodic replacement. Further storage and use of cooked, peeled crustaceans must prevent growth (ΣI) of the bacterium to levels where enterotoxin is formed. Enteric organisms can occur as result of recontamination (ΣI) after thermal processing, often due to improper handling by personnel. *Vibrio* spp. are most often associated with cross-contamination (ΣI) of cooked product. This can be controlled through proper design of processing facilities, control of the movement of product, personnel and equipment, and adherence to sanitation programs.

Testing. Growth and toxin production is required for *Staph. aureus* to cause illness and if no prior knowledge of processing (e.g. GHP and HACCP programmes) is available, some assurance of safety can be acquired by microbiological testing. Thus the EU has a microbiological standard for cooked crustaceans for *Staph. aureus* (EEC, 1991a, 1993) with $n = 5$, $c = 2$, $m = 100$ cfu/g and $M = 1000$ cfu/g. The EU legislation (EEC, 1991a, 1993) also contains a standard for *Salmonella* which must be absent in 25 g with $n = 5$ samples. The legislation also contains standards for either *E. coli* or thermotolerant coliforms. Cooked crustaceans may be tested for presence of *Vibrio* spp., which will only appear if recontamination has occurred. If detected, subsequent use of the product should ensure that the organisms are either inactivated (e.g. by re-cooking) or that growth in prevented, e.g. by brining in NaCl–acid-type brines with preservatives.

Spoilage. Cooked crustaceans must be chilled rapidly to control growth of spoilage bacteria. The product is normally frozen and quality deterioration related to protein degradation during frozen storage.

VIII Lightly preserved fish products

A *Introduction*

Approximately 15% of the world catch of fish is still processed by curing (drying, salting, or smoking), often by traditional methods such as air- or sun-drying, hand salting, and smoking over fires. The so-called lightly preserved fish products are products preserved by a combination of preservation parameters such as light salting, cold-smoking, lowering of pH, and cooling. These are products in which the normal Gram-negative spoilage flora is somewhat inhibited and shelf life thus extended as compared to fresh iced fish. The products are not sterile and do allow microbial growth and typically do spoil because of microbial activity. Examples of such products are cold-smoked fish, pickled ("gravad") fish, roe marinated in soy sauce and brined, and cooked crustaceans in brine.

Smoking of seafood is done principally to produce a food with attractive appearance and flavour. Most smoked seafood are only lightly brined (<6%) and smoked, so they are capable of supporting bacterial growth. Cold-smoked products (<30°C) typically retain a mixed population of bacteria and depending on storage temperature and atmosphere of packaging different spoilage scenarios prevail. Hot-smoked seafoods (>60°C) have substantially reduced bacterial populations, but provide an excellent substrate for potential pathogen growth because of elimination of competitive bacteria. Hot-smoked products are categorized as pasteurized products and are discussed below.

Several fish species are used for cold-smoking, but salmon (typically reared salmon) is the most common. The fish are filleted and salted either by dry salting, brining or injection of brine. After a short "freshening" the fillets are cold-smoked. The cold-smoking typically lasts 6–24 h and takes place at 22–30°C. These lower temperatures are used to avoid coagulation of the fish proteins; thus the fillets retain a raw appearance. After cooling, the fillets are packed or sliced and packed. The shelf life is dependent on the NaCl content but a vacuum-packed product can keep for 3–8 weeks at refrigeration temperatures (Figure 3.9).

Pickled (gravad) fish are typical of the Scandinavian countries and are made from salmon, trout, or halibut. Fish fillets are sprinkled with salt, sugar, and spices (dill) and placed under pressure for 1–3 days at refrigerated temperature (5°C). The fillets are sliced and the product stored chilled. Shelf life is short; typically 1–2 weeks. This product is different from another type of pickled fish (so-called marinated herring) in which salted or acid brined herring is placed in an acetic acid–NaCl–spice marinade. These products have a much longer shelf life and are described below.

Japanese examples of this category include the marinating of for instance salmon roe in soy sauce followed by packaging and frozen storage.

Brined cooked crustaceans are typically made from shrimp or crab meat. The cooked meat is placed in a brine with salt (4–6%), citric acid (pH 4.5–5.5), sugar (or sweetener), and benzoic and sorbic acid. The products are stored chilled and can be stable for 4–6 months.

B Saprophytes and spoilage

Curing typically results in a shift from a population dominated by Gram-negative bacteria in the raw product to one in which Gram-positive organisms dominate. The extent of this shift is related to the severity of the process. Generally species of spore-forming *Clostridium* and *Bacillus*, Gram-positive cocci, and lactic acid bacteria are the dominant survivors, but some Gram-negative bacteria (e.g. psychrotrophic Enterobacteriaceae) will also normally survive. During subsequent refrigerated storage, these bacteria will multiply and cause spoilage. In these products, packaging that provides a barrier to oxygen transfer is likely to select for microaerophilic Gram-positive and fermentative Gram-negative species. For example, packages of cold-smoked Canadian and Norwegian salmon stored at 0–2°C for 80 days had lactobacilli as the predominant spoilage organism (Parisi *et al.*, 1992). The microbial ecology of these products is somewhat similar to packed meat products and minimally processed vegetables stored in CO_2 atmosphere. Tyramine-producing strains of *Carnobacterium piscicola* and *Lactobacillus viridescens* have been isolated from sugar-salted fish during refrigerated storage, and this biogenic amine has been proposed as an index of microbial quality (Leisner *et al.*, 1994). Rapid growth of lactic acid bacteria is seen during chill storage of vacuum-packed cold-smoked fish. The spoilage flora is variable but typically fall into one of the following three groups: (i) dominated by lactic acid bacteria at $\sim 10^8$ cfu/g, (ii) a combination of lactic acid bacteria (10^7–10^8 cfu/g), and Enterobacteriaceae (10^6 cfu/g), or (iii) marine vibrios (e.g. *P. phosphoreum*) often with lactic acid bacteria (Truelstrup Hansen *et al.* 1998). Spoilage is characterized by development of multiple volatile compounds (Jørgensen *et al.*, 2000) and is caused by *Lactobacillus* species (*curvatus* or *saké*) and some members of the Gram-negative flora. The *Lb. curvatus/saké* group is capable of degrading sulfur-containing amino acids to hydrogen sulfide.

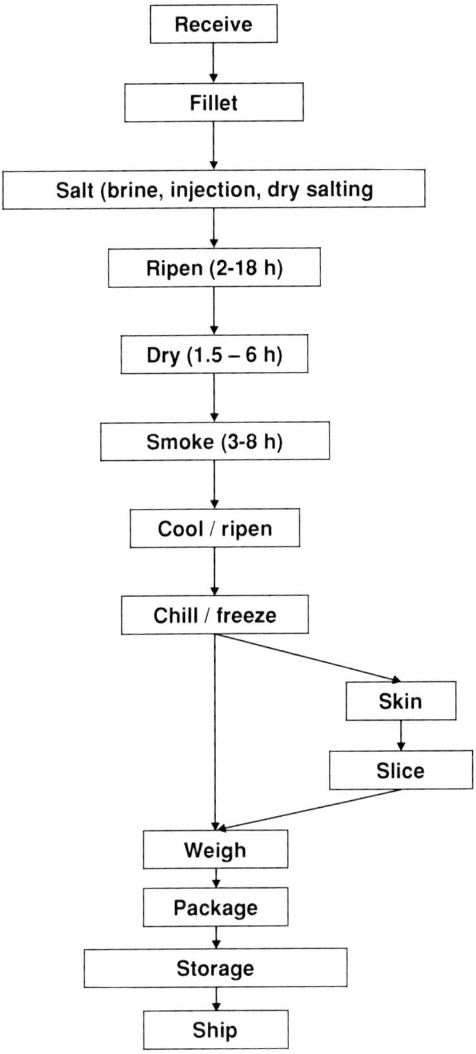

Figure 3.9 Example of process diagram for production of cold-smoked salmon (Huss *et al.*, 1995).

In contrast, *C. piscicola*, which often dominates the microflora appears to have no adverse effects on sensory quality (Paludan-Müller *et al.*, 1998).

Lactic acid bacteria will also become dominant during chill storage of brined crustaceans, although growth is typically very slow. *Leuconostoc* may cause a ropy slime spoilage if carbohydrates like sucrose are used for sweetening. This may be avoided by using artificial sweeteners.

C *Pathogens*

Cold-smoked finfish and shellfish as well as other lightly preserved fish products have been implicated in outbreaks of food poisoning. A hazard analysis of the process/product reveals four major agents: *Cl. botulinum* type E, *L. monocytogenes*, histamine, and parasites, assuming that raw materials from areas with aquatic toxins are not used (IFT, 2001; Huss *et al.*, 1995). However, pathogens from the human-animal reservoir may also be transferred to the products, particularly if manually handled. Hence, in Japan, 69 cases of *E. coli* O157:H7 were caused by soy sauce marinated salmon roe (Japanese Ministry for Health, Labour and Welfare, 1998).

Clostridium botulinum type E (its spores) may be present on the raw material and will survive the processing. Despite the widespread occurrence of *Cl. botulinum* in the aquatic environment (Dodds, 1993; Dodds and Austin, 1997), it is not commonly found on smoked fish. Heinitz and Johnson (1998) failed to detect any spores on 201 commercial vacuum-packed smoked fish samples.

Neither of the preservation parameters on its own, NaCl or low temperature, can prevent growth and toxin formation in vacuum-packed products. Several studies have evaluated the combinations of NaCl and low temperature required to inhibit growth and toxin formation (Graham *et al.*, 1997). Recently, Dufresne *et al.* (2000) inoculated cold-smoked trout and found that products stored at 4–12°C with 1.7% NaCl (water phase salt) spoiled before becoming toxic. At 12°C, the vacuum-packed trout spoiled in 11–12 days and toxin was detected after 14 days. No toxin was detected at 4°C for 28 days at which point, the product was considered spoiled. A combination of 3.5% NaCl (WPS) and 5°C will prevent toxin formation for at least 4 weeks (ACMSF, 1992). Using trout naturally contaminated with *Cl. botulinum* spores for production of hot-smoked fish, Cann and Taylor (1979) found no toxin-production in fish with 3% NaCl stored for 30 days at 10°C. Finnish studies reported surprisingly rapid toxin production by *Cl. botulinum* type E inoculated into cold-smoked trout (Hyytiia *et al.*, 1997). Toxin was detected in vacuum-packed cold-smoked trout containing 3.4% NaCl after 4 weeks at 4°C and after 3 weeks at 8°C. Oddly, levels of *Cl. botulinum* appeared to decrease during the same period from 140 to 70–80 cfu/g (Hyytiia *et al.*, 1997). The authors do not discuss the unusual results, but it is possible that the toxin detected did not originate from multiplying cells, but may have been released from cells that lysed.

Recently, there has been a great deal of interest in the potential for growth of *L. monocytogenes* in cold-smoked seafood and other lightly preserved seafood products. Cold-smoked mussels containing $>10^7$ *L. monocytogenes*/g were implicated as the cause of a cluster of listeriosis cases in Tasmania (Mitchel, 1991; Misrachi *et al.*, 1991). Also, a small out-break of listeriosis in Sweden was traced to cold-smoked and gravad trout (Ericsson *et al.*, 1997). Quantitative risk assessments (FDA/FSIS, 2003) have identified cold-smoked fish as a high-risk product with respect to listeriosis.

The pathogen has been isolated from both hot- and cold-smoked finfish (Rørvik *et al.*, 1992; Dillon *et al.*, 1994) and the processing environment (Jemmi and Keusch, 1994; Fonnesbech Vogel *et al.*, 2001). The incidence in newly processed cold-smoked fish varies from 0% in some smoke houses to 100% in others (Heinitz and Johnson, 1998; Jørgensen and Huss, 1998). A recent US survey (Kraemer, 2001) found 3% of cold-smoked fish (at retail level) to be positive for *L. monocytogenes*. Hot-smoked fish also was positive at a 3% prevalence and values were identical for imported and domestically produced smoked fish. In an even more recent US survey (Gombas *et al.* 2003), 4–5% of samples of smoked fish were positive for *L. monocytogenes*.

The microorganism survives cold-smoking and can subsequently grow to high levels during refrigerated storage, particularly in salmon (Guyer and Jemmi, 1991; Dillon and Patel, 1993; Embarek and Huss, 1993; Peterson *et al.*, 1993; Eklund *et al.*, 1995). It should be emphasized that whilst the bacterium easily grows to levels of 10^8 cfu/g in inoculated packs (Nilsson *et al.*, 1999), the growth in naturally contaminated products appears to be much slower (lower growth rate) and reach lower maximum numbers (Jørgensen and Huss, 1998). Whilst the vast majority of cold-smoked fish at retail level only contains low levels, a few samples were found to contain 10^5 *L. monocytogenes* per gram (Kraemer, 2001). Testing more than 30 000 ready-to-eat food samples, only two contained levels between 10^4 and 10^6 *L. monocytogenes* per gram—both being smoked fish (Gombas *et al.*, 2003). Sanitation and cleanup procedures can eliminate the microorganism; however, recontamination occurs soon after restarting processing (Eklund *et al.*, 1995). Several studies have found that although the raw fish is probably the original

Table 3.16 Number of samples, number of *L. monocytogenes* positive and randomly amplified polymorphic DNA (RAPD) type of *L. monocytogenes* of a salmon smoke house

				No of samples with RAPD type (2, 3, . . . , X)									
year	Sample	Number of samples	Number of positives	2	3	5	6	7	12	13	15	110	X
1995	Product	20	17			2		4	25				
1996	Product	20	12						12				
1998	Raw fish	36	0										
	Raw fish environment	239	55	4	1		36	4	1	1	6		2
	Smoking environment	8	0										
	Slicing environment (1)	150	80	3			1		63	1	10		2
	Slicing environment (2)	147	39						37		1	1	
	Product	40	15	7					7		1		
1999	Raw fish	12	0										
	Raw fish environment	105	17		3			5			6		3
	Smoking environment	2	0										
	Slicing environment (1)	75	9						6		3		
	Slicing environment (2)	100	3						2		1		
	Product	148	15					6	7		7		

X = unique types, isolated only once each (Fonnesbech Vogel *et al.*, 2001).

source of the bacterium, the processing equipment (e.g. slicers) are the most important immediate source of product contamination. In one smoke-house, a particular DNA-type of *L. monocytogenes* was isolated from the product and from slicing environments over a 4-year period (Fonnesbech Vogel *et al.*, 2001; Table 3.16). Freezing, sodium nitrite and modified atmosphere packaging, as well as sodium lactate in combination with sodium nitrite and sodium chloride, has been reported to help retard the growth of *L. monocytogenes* in vacuum packaged salmon (Pelroy *et al.*, 1994a,b). The lack of growth in naturally contaminated products as compared to inoculated trials has been explained by the anti-listerial action of the lactic acid bacteria growing on the product. Thus experiments have shown that growth of *L. monocytogenes* in cold-smoked salmon may be inhibited by the addition of high levels of carnobacteria (Nilsson *et al.*, 1999; Duffes *et al.*, 1999).

Histamine and other biogenic amines can be found in gravad and cold-smoked fish (Leisner *et al.*, 1994; Jørgensen *et al.*, 2000) and concern has been raised particularly while cold-smoking of tuna, which contains high levels of histidine—the precursor for histamine. Although histamine may be formed during chill storage of cold-smoked salmon e.g. by *P. phosphoreum*, nothing is known about potential histamine formation during the cold-smoking process.

Cold-smoking does not inactivate *Anisakis* in salmon (Gardiner, 1990) and freezing pre- or post-processing is needed to assure elimination of the parasite. Thus EU legislation prescribes that wild salmon be frozen for at least 24 h at −18°C at some stage during processing (EEC, 1991b). As discussed (Table 3.6), farmed fish do not contain parasites.

D CONTROL (lightly preserved fish products)

Summary

Significant hazards[a]	• Aquatic toxins. • Parasites. • *Cl. botulinum* type E. • *L. monocytogenes.* • Enteric human pathogens.

(continued)

D CONTROL (Cont.)

Summary

Control measures	
Initial level (H_0)	• Avoid fish from areas of algal blooms. • Avoid fish from specific (tropical) areas.
Reduction (ΣR)	• Freeze to kill parasites.
Increase (ΣI)	• NaCl and low temperature (type E). • NaCl and low temperature (delay *L. monocytogenes*). • GHP to prevent cross-contamination. • Frozen storage, lactate or other additions may prevent growth of *L. monocytogenes*.
Testing	• NaCl content. • *L. monocytogenes* if no knowledge of prior history. • Sensory evaluation.
Spoilage	• Time × −temperature. • Sensory evaluation.

[a]Under particular circumstances, other hazards may need to be considered.

Hazards considered. Several of the hazards identified for finfish are also relevant for lightly preserved products. Aquatic toxins can be present in the raw material and thus also in further processed products. Parasites will typically not be eliminated by the processing steps used for lightly preserved products and must be considered as a hazard. Histamine may be formed in the raw material or during processing/storage of some products, e.g. cold-smoked tuna. Psychrotolerant pathogenic bacteria that are not eliminated by the processing steps are potential hazards in these products. This includes *Cl. botulinum* and *L. monocytogenes*. If appropriate sanitary conditions are not in place, enteric human pathogens can be transferred to the products during handling. Manual handling can introduce *Staph. aureus*, however, due to the associate microlflora, this organism will not be able to grow to hazardous (e.g. enterotoxin producing) levels.

Control measures

Initial level of hazard (H_0). Aquatic toxins can be controlled by not harvesting fish from areas of algal blooms.

Reduction of hazard (ΣR). The risk from parasites must be controlled by elimination of the organisms (ΣR) if fish from a wild population are used. This can be done introducing a freezing step (−18°C for at least 24 h) at any time during processing. Freezing is not necessary (nor required) if reared salmon is used (EEC, 1991b).

All cold-smoked fish should be promptly refrigerated and held below 5°C during subsequent storage and transportation. Keeping the storage temperatures as low as possible is particularly important for cold-smoked fish since the smoking process cannot be relied upon to inactivate psychrotrophic pathogens.

Increase of hazard (ΣI). Psychrotrophic *Cl. botulinum* are not eliminated by the processes and growth (ΣI) and toxin production must be prevented during storage. Achieving a minimum salt concentration of 3.5% water phase salt will prevent growth and toxin formation at 5°C for at least 4 weeks (Graham *et al.*, 1997). *L. monocytogenes* is primarly introduced as a process-contaminant (ΣI) and its growth (ΣI) should be prevented or limited during storage. Process contamination is minimized by efficient cleaning and sanitation programmes. In particular, the processing environment should be monitored for the presence of niches (slicers, drains, brine) where *L. monocytogenes* may reside. Current salt/temperature levels cannot guarantee against growth of this organisms. Thus assurance against growth should be based on other measures, e.g. freezing.

Testing. Checking of NaCl-levels in brine and in finished product is important to ensure that *Cl. botulinum* growth can be inhibited if proper temperature control is installed. A food safety objective of 100 *L. monocytogenes* per gram at time of consumption has been suggested (ICMSF, 1994) and sampling plans designed around this. Surveillance of the processing environment for this organism is an important part of a control program.

Spoilage. As described above, spoilage of these types of products is complex and often caused by (or influenced by) specific groups of lactic acid bacteria whereas other groups of lactic acid bacteria has no influence on sensory quality. Low temperature ($\leq$5°C) delays spoilage. Aerobic counts are often high already after 2 weeks of storage and cannot be used to evaluate quality or spoilage. The best way to assess spoilage is through sensory evaluation.

IX Semi-preserved fish products

A *Introduction*

This group of fish products is heterogenous but typically the products contain NaCl, acid, and preservatives like benzoic acid, sorbic acid or nitrate. Often the products are not exposed to heat either during processing or by the consumer, and thus inactivation of pathogenic agents such as parasites must rely on combinations of NaCl and acid. The Scandinavian marinated herrings, the German roll-mopps or the anchovies of Southern Europe are typical examples of this group of products. Several of the fermented fish products are, strictly speaking, semi-preserved products, but are discussed separately below.

Marinated herrings are made from either barrel-salted, partially gutted herrings that have been stored for 6–12 months or from acid-brined herring fillets. A softening (ripening) of the herring fillets takes place during the barrel-salting; this ripening is caused by enzymes from the fish gut. The herrings are placed in marinades containing acetic acid, NaCl and, often, some sugar and spices. The products are stored at refrigeration temperatures.

Anchovies are similar to the barrel-salted herring in that partially gutted (or non-gutted) whole fish are placed in salt and brine and left to ripen for months. These are typically sold in cans but have not been heat-sterilized and must be kept chilled.

Also, different kinds of fish roe are barrel salted (at 15–25% NaCl) for months before further processing to caviar products. As the anchovies, caviar products are typically packed in cans but not heat-treated.

B *Saprophytes and spoilage*

Not many microorganisms are capable of growth in the combinations of acid and NaCl used for the semi-preserved products. The salting that typically takes place at 15–20% NaCl allows for little microbial activity. During the marinating, the NaCl concentration is reduced to 5–10%. If acidification is not sufficient, yeasts and lactic acid bacteria may grow and cause gas and off odor development but keeping pH at 4.5 and including preservatives ususally controls this development.

During the barrel salting, yeasts and anaerobic spore formers may grow and cause either a fruity or a putrid type of spoilage (Knøchel and Huss, 1984). This problem is rare and is typically seen if poor raw materials are used.

C Pathogens

Few pathogens are of real concern in this type of seafood products. As the products are not heat treated, parasites must be inactivated by the combination of NaCl and acid used. NaCl and acetic acid will ensure a die-off of *Anisakis* larvae (Karl *et al.*, 1984). In particular, the NaCl level is crucial, as larvae may survive for 14 weeks at 4.3% NaCl with 3.2% acetic acid, whereas 6.3% at the same level of acetic acid caused a kill of all larvae in 6 weeks.

Due to the anaerobic conditions during processing and storage, growth, and toxin production by clostridia must be considered. The psychrotrophic *Cl. botulinum* will not grow at the NaCl concentrations used, but the mesophilic, proteolytic types tolerate much higher concentrations. However, sufficient levels of NaCl and chilling can prevent this problem.

Histamine may be formed in the raw material prior to salting and processing.

D CONTROL (semi-preserved fish products)

Summary

Significant hazards[a]	• Parasites. • *Cl. botulinum.* • *L. monocytogenes.*
Control measures	
Initial level (H_0)	• Not applicable.
Reduction (ΣR)	• Appropriate levels of NaCl and acid kill parasites.
Increase (ΣI)	• Appropriate levels of NaCl and acid. • Storage at low temperature.
Testing	• Not applicable.
Spoilage	• Appropriate levels of acid and NaCl. • Use fresh raw materials.

[a]Under particular circumstances, other hazards may need to be considered.

Hazards to be considered. These products are typically based on raw fish and as the process does not involve a heating or freezing step, parasites are a hazard. Histamine can be formed in some fish species used and is also a potential hazard. Typically, the levels of acetic acid, NaCl and/or preservatives (sorbic acid, benzoic acid, etc.) are so high that bacterial pathogens are inactivated and not of concern.

Control measures. Appropriate combinations of NaCl and acid will control potential microbial hazards and spoilage of these products. Control of parasites is achieved by combinations of acetic acid and NaCl resulting in a kill (ΣR) of the organisms.

The safety of semi-preserved fish products depends on how effectively the secondary barriers (e.g. pH, a_w, antimicrobials) inhibit the germination and outgrowth (ΣI) of *Cl. botulinum* spores or the vegetative cells of more resistant vegetative pathogens such as *L. monocytogenes*. At the production level, control is a function of careful development and verification of formulations and processes that insure that each of the multiple barriers are at their effective levels. The specific steps to be controlled, and the extent to which a step must be controlled, can vary substantially depending on the process and formulation involved. In most cases assurance of safety includes refrigeration of the products during distribution, marketing, and home use. Post-processing control should focus on adequate temperature control. Instructional labels can help the consumer handle the product in a safe manner.

Testing. NaCl and acidity must be measured and controlled during production, but no microbiological tests are appropriate for assurance of safety.

Spoilage. Just as assurance against bacterial pathogens depends on acidity, a_w and, for some products, low temperature, also spoilage organisms (yeasts and lactic acid bacteria) are controlled by these parameters.

X Fermented fish products

Various types of fermented fish products are found in the world. A range of fish–rice and fish–vegetable products are produced in South East Asia and the Pacific Islands and depend on acid production by endogenous lactic acid bacteria (and possibly yeast) for safety and stability of the product. The fermentation, which is run at ambient tropical temperature (30°C), rapidly becomes anaerobic. These products typically contain low levels of NaCl (less than 10%). A range of products with much higher NaCl-content is also found in South East Asia and Japan. These products, fish sauces and fish pastes, are not fermented by microorganisms but rather are the result of slow hydrolysis by endogenous enzymes. The microbiology of these products is discussed in the chapter on spices. In Alaska and the Canadian Arctic territories, Inuits prepare a variety of fermented fish and marine mammal products by simple means including burying them in the ground or storing them behind a warm stove. Several of these products have been the cause of botulism (Wainwright *et al.*, 1988).

Little is know about the microbiology of these products. A mixture of lactic acid bacteria and yeasts are isolated from South East Asian products (Adams *et al.*, 1985; Paludan-Müller *et al.*, 2002) and *L. plantarum* are probably responsible for the acidification of several of these products. Fish contains very little carbohydrate and the fermentation therefore requires the addition of carbohydrates. Addition of rice and cassava to the products has been thought to serve this purpose, however, it has recently been found that garlic (or rather the fructan of garlic) is the important carbohydrate source for the fermentation of several of the South East Asian products (Paludan-Müller *et al.*, 2002; Figure 3.10).

These products rely on NaCl and acidification and/or putrification and particularly the products from arctic territories have caused botulism. The problem of botulism in fermented seafood products is related to failures to control the process. The main categories of fish products that cause trouble are produced by traditional, poorly controlled techniques, which are intrinsically dangerous. If acid is produced too slowly *Cl. botulinum* may grow and produce toxin. Thus, a fish-rice fermented product was the cause of type E botulism in Japan in 1951 (Dolman and Iida, 1963; Iida, 1970).

The changes in protein/lipid fractions are of questionable value in stabilizing the product or protecting against growth of *Cl. botulinum* type E. Consequently, there are a significant number of cases of botulism each year among these Eskimo and Inuit people (Wainwright *et al.*, 1988; MMWR, 2001).

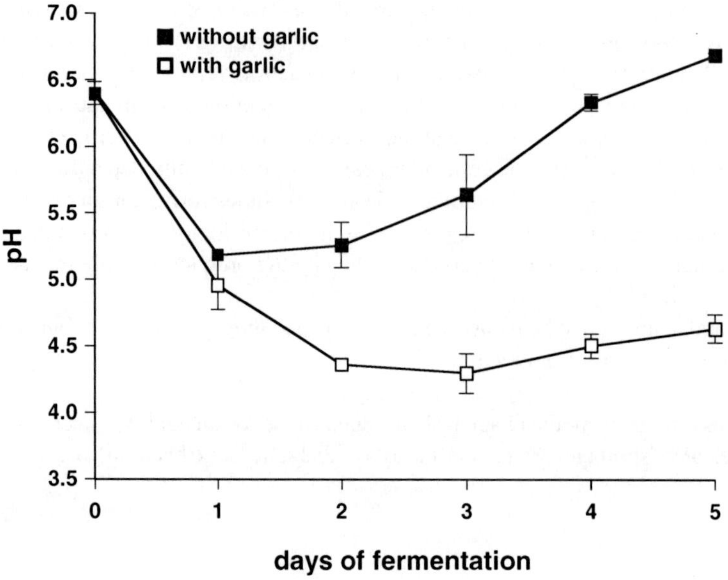

Figure 3.10 Fermentation of Thai *som-fak* with and without the addition of 4% garlic (Paludan-Müller *et al.*, 2002).

A third ethnic fish food type causing botulism are whole fish, which are salted and air dried with viscera intact. Where this is done without refrigeration, there is good possibility that *Cl. botulinum* will grow. Examples of such outbreaks have been reported in the United States involving kapchunka and moloha (MMWR, 1985, 1992b; Badhey *et al.*, 1986; Slater *et al.*, 1989; Telzak *et al.*, 1990), and a massive outbreak of botulism occured in Egypt where 91 people were hospitalized following consumtion of faseikh—an uneviscerated, salted mullet fish (Weber *et al.*, 1993). A related source of botulism is home prepared ceviche, which was responsible for three cases in Puerto Rico in 1978 (CDC, 1981).

Little is know about disease caused by parasites transferred by fermented products, but one must assume that the processes (and time) involved are not sufficient to inactivate parasites. Thus unless the products are consumed cooked, there is a high risk from parasites.

A CONTROL (fermented fish products)

Summary

Significant hazards[a]	• Parasites. • *Cl. botulinum.* • *Vibrio* spp. • Enteric human pathogens.
Control measures	
Initial level (H_0)	• Do not use fish from contaminated waters. • Do not use animal/human fertilizers for fish ponds.

(*continued*)

A CONTROL (Cont.)

Summary

Reduction (ΣR)	• Cook product. • Use frozen raw material.
Increase (ΣI)	• Ensure rapid drop in pH and sufficient NaCl.
Testing	• Measure pH and NaCl-content.
Spoilage	• Ensure rapid decrease of pH. • Starter cultures can control spoilage organisms.

[a]Under particular circumstances, other hazards may need to be considered.

Safety of fermented products relies on an initial inhibition (ΣI) of growth of pathogenic bacteria, e.g. by NaCl and spices and a rapid increase in acidity to stabilize the product. The pH should be reduced to 4.5 in two days (Adams *et al.*, 1985). *Clostridium botulinum* is a major hazard in these products. Typically, the process does not include an inactivation of parasites, which are considered a risk. They can be controlled by using frozen raw materials or by cooking by the consumer (ΣR). The presence of enteric pathogens on fish from contaminated waters is a risk and must be controlled by elimination at the source, controlling H_0 and by ensuring a rapid decrease in pH to prevent increase (ΣI).

XI Fully dried or salted products

Fully cured, shelf-stable products have a_w values low enough to prevent growth of all microorganisms except some xerophilic fungi and halophilic bacteria. Hard smoked products (e.g. kippered salmon) are dried to a very low a_w, making them stable to unrefrigerated storage. However, contaminating pathogenic bacteria may persist on such products depending on their resistance to desiccation. On a practical basis, there seems to be an insignificant microbiological hazard from such products, but there are a limited number of reports of mycotoxins, particularly aflatoxins, being detected in traditionally dried fish products.

The production and spoilage of dried fish products has been reviewed by Doe and Olley (1990). Spoilage fungi can grow on salted, dried fish, but only at a_w above 0.7 due to the inhibitory effect of salt. The main cause of spoilage of salted, dried fish in temperate zones has been reported to be *Wallemai sebi* (= *Sporendonema epizoum*) (Frank and Hess, 1941). This fungus grows poorly under tropical conditions (Wheeler *et al.*, 1988), where the main cause of spoilage is *Polypaecilum pisce* (Wheeler *et al.*, 1986).

Reports of aflatoxin production on dried fish have been limited. In the production of salted dried fish in Southeast Asia, *Aspergillus flavus* is capable of rapid growth during the early stages of drying, but ceases when the fish dry below 0.87–0.85 a_w. *Aspergillus flavus* was found quite frequently in dried fish samples examined from Indonesia, but all of these were below 0.8 a_w. Visible growth was not observed, and aflatoxin was not found (Pitt, 1995). However, in Africa, where fresh-water fish are smoked and dried in the absence of salt, aflatoxin has been reported in quite large amounts (Jonsyn and Lahai, 1992; Mugula and Lyimo, 1992; Diyaolu and Adebajo, 1994). This appears to be aggravated by warm climates and product being exposed to insects.

Air-drying must be carried out rapidly enough so that excessive bacterial growth does not occur in the early phases of the process. In traditional drying of fish, there is a substantial risk of contamination of fish with microorganisms from humans, particularly where the fish are laid on beaches or low wooden drying racks. This can only be dealt with by a change in the drying procedure to separate the fish from sources of contamination and by the individuals involved recognising the importance of good personal hygiene. Fortunately, most dried fish of this type is eaten after cooking which greatly reduces risk for the consumer.

If dried fish are eaten without processing (ready-to-eat), contamination with pathogens from the human-animal reservoir and subsequent food-borne disease can be a risk. Thus, dried squid has been the cause of salmonellosis (*Salmonella oranienburgh* and Chester) in Japan emphasizing the ability of this organism to survive in dry environments (Japanese Ministry for Health, Labour and Welfare, 1999).

A CONTROL (fully dried or salted fish products)

Summary

Significant hazards[a]	• Mycotoxin formation.
Control measures	
Initial level (H_0)	• Limit spore contamination where possible.
Reduction (ΣR)	• None applicable.
Increase (ΣI)	• Dry as rapidly as possible; careful control of a_w during storage.
Testing	• Not required in salted fish, dried fish. Advisable for fresh water smoked fish.
Spoilage	• Ensure rapid drying. • Avoid storage in humid areas that would allow fungal growth.

[a]Under particular circumstances, other hazards may need to be considered.

In these products, a_w is so low that bacterial pathogens or parasites are not a risk. Health hazards primarily include mycotoxin formation.

XII Pasteurized products

A Introduction

Other seafood products than the cooked, canned, or cold-smoked products, receive a heat treatment. Several products are pasteurized and typical of such products are the hot-smoked fish and sous-vide products.

In Europe, Canada, and North America, hot-smoked fish are typically prepared from fatty species like mackerel, herring, or salmon. The fish are salted by dry-salting or brining and smoked at 60–70°C for half to one hour. Hot smoking is a common way of preserving fish in Africa and typically takes place over longer periods of time. The meat of hot-smoked fish, as opposed to cold-smoked products, appear cooked.

Recently the sous-vide (French for "under vacuum") processing has become popular. These products are pre-cooked at temperatures between 65°C and 90°C and stored under vacuum and distributed chilled. These refrigerated and processed foods with extended durability (REPFEDs) have become popular in the catering sector.

B Saprophytes and spoilage

The hot smoking of fish results in a marked reduction of bacterial numbers. If stored aerobically, mould growth on stored smoked fish is quite common and may be a major cause for rejection of the product (Efiuvwevwere and Ajiboye, 1996). If stored vacuum-packed at refrigerated temperatures, very little microbial growth is seen and sensory rejection is often caused by rancidity or textural changes.

Little is know about spoilage of sous-vide fish products. Ben Embarek (1994b) found that after 3 weeks of storage at 5°C, some sous vide cooked cod developed putrid offensive off odors due to growth of spore forming Gram-positive bacteria. Similarly, clostridia have been implicated in spoilage of pasteurized crab-meat (Cockey and Chai, 1991).

C Pathogens

Two bacterial pathogens are of great concern in the pasteurized fish products: *Cl. botulinum* and *L. monocytogenes*. Botulism outbreaks from hot-smoked fish were a major problem in the sixties and seventies. The outbreaks were due to post-processing growth and toxin production by *Cl. botulinum* that were probably on the fish at the time of capture. The botulism cases involved hot-smoked fish, which were insufficiently heated to destroy spores of *Cl. botulinum* serotype E. NaCl concentrations were low and after processing the products were held at storage temperatures high enough to allow growth (Pace and Krumbiegel, 1973). Hot-smoking processes reaching temperatures of 60–92°C can inactivate spores of non-proteolytic *Cl. botulinum*, but were insufficient to inactivate spores of proteolytic strains (Eklund *et al.*, 1988). The redox potential of smoked fish flesh is sufficiently low enough to permit growth of *Cl. botulinum* even in the presence of an oxygen containing environment. Hot-smoking may encourage *Cl. botulinum* growth by elimination of competing bacteria. An associate microflora may sometimes enhance toxin production. Thus, Huss *et al.* (1980) showed that co-inoculation of hot-smoked herring with spoilage bacteria led to faster toxin formation, persumably because the spoilage bacteria used oxygen and created a favorable anaerobic atmosphere. An important risk factor is inadequate cleaning and sanitation of smoking facilities. This can allow build up of populations of *Cl. botulinum*, thereby ensuring recontamination of smoked fish. A post-smoking heat-pasteurization process has been described (Eklund *et al.*, 1988). Hot-smoked fish is usually eaten without further cooking making effective control of the potential botulism hazard especially important. It has been recommended that such products be labelled to indicate the need to keep refrigerated (Eklund *et al.*, 1988).

The D-values and z-values for *L. monocytogenes* are within the range reported for other foods, but are dependent on the fish species (Ben Embarek and Huss, 1993). For example, the D-values at 56°C, 58°C, 59°C, 60°C, and 62°C in brined green mussels were 48.1, 16.3, 9.5, 5.5, and 1.9 min, respectively, with a z-value of 4.3°C (Bremer and Osborne, 1995). *Listeria monocytogenes* does not appear to survive hot smoking in trout, but grew to $> 10^7$ cfu/g when inoculated at low levels after thermal processing (Jemmi and Keusch, 1992). High NaCl levels ($>5\%$) in combination with vacuum packaging inhibited growth in smoked salmon at 5°C, but not 10°C (Peterson *et al.*, 1993). The incidence of *L. monocytogenes* does not appear to reflect the smoking temperature. Heinitz and Johnson (1998) found 18% of cold-smoked fish and 8% of hot-smoked fish positive whereas a recent US study found that 3.1% of retail hot-smoked fish was positive for *L. monocytogenes*, which was similar to the 3.3% reported for cold-smoked fish (Kraemer, Personal communication). This indicates that post-process contamination takes place. It is noteworthy that *Aero. hydrophila*, a much more heat sensitive organisms, has been found in a substantial portion of hot- and cold-smoked fish products (Gobat and Jemmi, 1993), indicating significant potential for post-heating recontamination.

D CONTROL (pasteurized fish products)

Summary

Significant hazards[a]	• Aquatic toxins. • Histamine. • *Cl. botulinum.* • *L. monocytogenes* (if handled after heating). • Human enteric pathogens (if handled after heating).
Control measures	
Initial level (H_0)	• Avoid fish from algal blooms. • Avoid fish from specific (tropical) areas.
Reduction (ΣR)	• Time × –temperature during heating.
Increase (ΣI)	• Time × –temperature during storage. • GHP to prevent cross-contamination.
Testing	• *L. monocytogenes* (if no knowledge of prior history).
Spoilage	• Time × –temperature. • Sensory evaluation

[a]Under particular circumstances, other hazards may need to be considered.

Hazards to be considered. Both aquatic toxins and biogenic amines are hazards in these types of products and if formed in the raw material are not removed by the heating step. Vegetative bacterial cells are typically killed by the pasteurization process but spores may survive and *Clostridium botulinum* is a major hazard in these products if vacuum-packaged. Post-process contamintion with *L. monocytogenes* can occur and constitutes a health hazard in vacuum-packed, chill-stored products such as hot-smoked fish.

Control measures

Intial level of hazard (H_0). As described under fresh finfish, aquatic toxins should be controlled by surveillance of catching waters thus controlling H_0. Avoiding problems with *Cl. botulinum* in hot-smoked products requires action at all levels of the process. Frequent and effective clean up will keep ambient contamination by *Cl. botulinum* at low levels (H_0).

Reduction of hazard (ΣR). Heating to an internal temperature in excess of 82°C during the process greatly reduces the chances for survival and subsequent growth of *Cl. botulinum* serotype E (Eklund *et al.*, 1982a; Pelroy *et al.*, 1982). The temperatures used in hot smoking and sous-vide treatment are sufficient to kill *L. monocytogenes* (Ben Embarek and Huss, 1993).

Increase of hazard (ΣI). Biogenic amines are controlled by keeping raw materials chilled ($\leq$2°) avoiding growth of decarboxylating bacteria (controlling ΣI). *Clostridium botulinum* control also requires barriers against growth. Control of formulations, particularly salt and nitrite levels, and of temperature is essential to prevent growth (ΣI). As for lightly preserved products, a water phase salt level of 3.5% combined with storage at temperatures at or below 5°C will prevent growth and toxin production for at least 4 weeks. Since the temperatures used are sufficient to kill *L. monocytogenes*, proper cleaning and sanitation of the proces environment is crucial to avoid post-process contamination (ΣI).

Testing. Measurements of NaCl percentage and temperature are important to document control of psychrotrophic *Cl. botulinum*. Plant contamination with *L. monocytogenes* must be evaluated and if the organism is present, testing of finished product can determine if a food safety objective (e.g. 100 *L. monocytogenes* per gram at consumption) can be met.

Spoilage. Ensuring appropriate heating steps and, if required, subsequent chill storage are control measures to limit microbial spoilage. No specific testings are recommended for this purpose.

XIII Canned seafood

A *Processing*

Canned seafood products such as salmon, tuna, and sardines, which are given a full retort process, should be "commercially sterile" and free from living bacteria that are potentially pathogenic. For such foods, the bacteriological hazards are the same as those for other low-acid canned foods and relate, with one exception, to problems of improper or inadequate processing or leakage. Adequate process control is the key to ensure a safe final product.

The one exception relates to scombroid poisoning. The toxin responsible (see above) is resistant to heat, and cases of scombroid poisoning have resulted from consumption of canned tuna (MMWR, 1975). Although scombroid poisoning from canned tuna and bonito is mainly due to the development of histamine in the fish prior to processing, the condition may be exacerbated by increased histidine decarboxylase activity during processing. Significant increases in histamine levels were observed experimentally after bonito was processed at 116°C to an $F_0 = 6.0$ (Pan and James, 1985). However, excess histamine in fish destined for canning does not appear to be a significant problem (Lopez-Sabater *et al.*, 1994).

B CONTROL (canned seafood)

Summary

Significant hazards[a]	• Aquatic toxins. • Histamine. • *Cl. botulinum*.
Control measures	
Initial level (H_0)	• Avoid fish from areas with algal blooms. • Avoid fish from specific (tropical) areas.
Reduction (ΣR)	• Time × −temperature control during sterilization.
Increase (ΣI)	• Time × −temperature during storage of raw material.
Testing	• Histamine testing (sensory analysis; HPLC to confirm).
Spoilage	• Sensory evaluation.

[a]Under particular circumstances, other hazards may need to be considered.

The general HACCP control program for low-acid canned foods, which is described elsewhere in the book and is widely used on a worldwide basis, applies equally well to fully processed canned fish products. The only specific factor for fish is the safety of the raw material particularly in relation to marine toxins and histamine. Knowledge of the origin of the fish, their species, and the conditions under which the raw material was held is important information for prevention of algal, ciguatera, and scombroid intoxications. Wherever justified suspicions concerning the source of fish is required, analyses for specific toxins may be needed.

References

Abeyta, C., Jr., Deeter, F.G., Kaysner, C.A., Stott, R.R. and Wekell, M.M. (1993) *Campylobacter jejuni* in a Washington state shellfish growing bed associated with illness. *J. Food Prot.*, **56**, 323–25.

ACMSF (Advisory Committee on the Microbiological Safety of Food). (1992) Report on Vacuum Packaging and Associated Processes, HMSO, London, UK.

ACMSF (Advisory Committee on the Microbiological Safety of Food). (1995) Workshop on Foodborne Viral Infections, HMSO, London, ISBN 0 11 321961 X.

Acuff, G.A., Izat, L. and Finne, G. (1984) Microbial flora of pondreared tilapia. *J. Food Prot.*, **47**, 778–80.

Adams, A.N., Leja, L.L. Jinneman, K., Beer, J., Yuen, G.A. and Wekell, M.M. (1994) Anisakid parasites, *Staphylococcus aureus* and *Bacillus cereus* in sushi and sashimi from Seattle area restaurants. *J. Food Prot.*, **57**, 311–17.

Adams, M.R., Cooke, R.D. and Rattagool, P. (1985) Fermented fish products of South East Asia. *Trop. Sci.*, **25**, 61–73.

Adesiyun, A.A. (1993) Prevalence of *Listeria* spp., *Campylobacter* spp., *Salmonella* spp., *Yersinia* spp., and toxigenic *Escherichia coli* on meat and seafoods in Trinidad. *Food Microbiol.*, **10**, 395–403.

Ama, A.A., Hamdy, M.K. and Toledo, R.T. (1994) Effects of heating, pH, and thermoradiation on the inactivation of *Vibrio vulnificus*. *Food Microbiol.*, **11**, 215–27.

Anderson, D.M., White, A.W. and Baden, D.G. (1985) *Toxic Dinoflagellates*, Elsevier Science Publishers, NYC.

Angot, V. and Brassuer, P. (1993) European farmed Atlantic salmon (*Salmo salar* L.) are safe from anisakid larvae. *Aquaculture*, **118**, 339–44.

Anthoni, U., Børresen, T., Christophsersen, C., Gram, L. and Nielsen, P.H. (1990) Is trimethylamine oxide a reliable indicator for the marine origin of fishes? *Comp. Biochem. Physiol.*, **97B**, 569–71.

Ayulo, A.M.R., Machado, R.A. and Scussel, V.M. (1994) Enterotoxigenic *Escherichia coli* and *Staphylococcus aureus* in fish and seafood from the southern region of Brazil. *Int. J. Food Microbiol.*, **24**, 171–8.

Badhey, H., Cleri, D.J., D'Amato, R.F., Veinni, V., Tessler, J., Wallman, A.A., Mastellone, A.J., Giuliani, M. and Hochstein, L. (1986) Two fatal cases of type E adult food-borne botulism with early symptoms and terminal neurologic signs. *J. Clin. Microbiol.*, **23**, 616–8.

Barile, L.E., Milla, A.D., Reilly, A. and Villadsen, A. (1985) Spoilage patterns of mackerel (*Rastrelliger faughni*) 1. Delays in icing, in *Spoilage of Tropical Fish and Product Development*, A. Reilly, FAO Fish. Rep., 317 (Suppl.), pp. 29–40.

Barker, W.H., Weaver, R.E., Morris, G.K. and Martin, W.T. (1974) Epidemiology of *Vibrio parahaemolyticus* infections in humans, in *Microbiology—1974* (ed D. Schlessinger), Amer. Soc. Microbiol., Washington, DC, pp. 257–62.

Baross, J. and Liston, J. (1970) Occurrence of *Vibrio parahaemolyticus* and related hemolytic vibrios in marine environments of Washington state. *Appl. Microbiol.*, **20**, 179–86.

Ben Embarek, P.K. (1994a) Presence, detection and growth of *Listeria monocytogenes* in seafoods: a review. *Int. J. Food Microbiol.*, **23**, 17–34.

Ben Embarek, P.K. (1994b) Microbial safety and spoilage of sous vide fish products. *PhD Thesis*, Technological Laboratory of the Danish Ministry of Agriculture and Fisheries & the Royal Veterinary and Agricultural University.

Ben Embarek, P.K. and Huss, H.H. (1993) Heat resistance of *Listeria monocytogenes* in vacuum packaged pasteurized fish fillets. *Int. J. of Food Microbiol.*, **20**, 85–95.

Berry, T.M., Park, D.L. and Lightner, D.V. (1994) Comparison of the microbial quality of raw shrimp from China, Ecuador, or Mexico at both wholesale and retail levels. *J. Food Prot.*, **57**, 150–3.

Beverley-Burton, M. and Pippy, J.H.C. (1978) Distribution, prevalence and mean number of larval *Anisakis simplex* (Nematoda: Ascaridoidea) in Atlantic salmon, *Salmo salar* L. and their use as biological indicators of host stocks. *Enviorn. Biol. Fishes*, **3**, 211–22.

Blake, P.A. (1983) Vibrios on the half shell: What the walrus and the carpenter didn't know. *Ann. Intern. Medi.*, **99**, 558–9.

Blake, P.A., Weaver, R.E. and Hollis, D.G. (1980a) Diseases of humans (other than cholera) caused by vibrios. *Ann. Rev.Microbiol.*, **34**, 341–67.

Blake, P.A., Allegra, D.T., Synder, J.D., Barrett, T.J., McFarland, L., Caraway, C.T., Feeley, J.C., Craig, J.P., Lee, J.V., Puhr, N.D. and Feldman, R.A. (1980b) Cholera—a possible endemic focus in the United States. *N. Engl. J. Medi.*, **302**, 305–9.

Boardman, G.D. and Evans, S.M. (1986) Detection and occurrence of waterborne viruses. *J. Water Pollut. Control Fed.*, **58**, 717–21.
Bøknæs, N., Østerberg, C., Nielsen, J. and Dalgaard, P. (2000) Influence of freshness and frozen storage temperature on quality of thawed cod fillets stored in modified atmosphere packaging. *Food Sci. Technol.*, **33**, 244–8.
Bouchriti, N., El Marrakchi, A., Goyal, S.M. and Boutaib, R. (1995) Bacterial loads in Moroccan mussels from harvest to sale. *J. Food Prot.*, **58**, 509–12.
Boutin, B.K., Reyes, A.L., Peeler, J.T. and Twedt, R.M. (1985) Effect of temperature and suspending vehicle on the survival of *Vibrio parahaemolyticus* and *Vibrio vulnificus*. *J. Food Prot.*, **48**, 875–78.
Boyd, L.C., Green, D.P. and LePors, L.A. (1992) Quality changes of pond-raised hybrid striped bass during chillpack and refrigerated storage. *J. Food Sci.*, **57**, 59–62.
Brackett, R.E. and Beuchat, L.R. (1990) Pathogenicity of *Listeria monocytogenes* grown on crabmeat. *Appl. Environ. Microbiol.*, **56**, 1216–20.
Bradshaw, J.G., Francis, D.W. and Twedt, R.M. (1974) Survival of *Vibrio parahaemolyticus* in cooked seafood at refrigeration temperatures. *Appl. Microbiol*, **27**, 657–61.
Brandao-Areal, H., Charbonneau, R. and Thibault, C. (1995) Effect of ionization on *Listeria monocytogenes* in contaminated shrimps. *Sci.Aliments*, **15**, 261–72.
Bremer, P.J. and Osborne, C.M. (1995) Thermal-death times of *Listeria monocytogenes* in green shell mussels (*Perna canaliculus*) prepared for hot smoking. *J. Food Prot.*, **58**, 604–8.
Bremner, H.A. and Statham, J.A. (1983) Spoilage of vacuum packed chill-stored scallops with added lactobacilli. *Food Technol., Aust.*, **35**, 284–7.
Brett, M.S.Y., Short, P. and McLauchlin, J. (1998) A small outbreak of listeriosis associated with smoked mussels. *Int. J. Food Microbiol.*, **43**, 223–9.
Bristow, G.A. and Berland, B. (1991) A report on some metazoan parasites of wild marine salmon (*Salmo salar* L.) from the west coast of Norway with comments on their interactions with farmed salmon. *Aquaculture*, **98**, 311–8.
Bryan, F.L. (1980) Epidemiology of foodborne diseases transmitted by fish, shellfish, and marine crustaceans in the United States, 1970–1978. *J. Food Prot.*, **43**, 859–76.
Buchanan, R.L. and Klawitter, L.A. (1992) Effectiveness of *Carnobacterium piscicola* LK5 for controlling the growth of *Listeria monocytogenes* Scott A in refrigerated foods. *J. Food Safety*, **12**, 219–36.
Buchanan, R.L., Stahl, H.G. and Whiting, R.C. (1989) Effects and interactions of temperature, pH, atmosphere, sodium chloride, and sodium nitrite on the growth of *Listeria monocytogenes*. *J. Food Protect.*, **52**, 844–51.
Buchanan, R.L., Schultz, F.J., Golden, M.H., Bagi, L.A. and Marmer, B. (1992) Feasibility of using microbiological indicator assays to detect temperature abuse in refrigerated meat, poultry, and seafood products. *Food Microbiol.*, **9**, 279–301.
Bykowski, P.J. (1990) The preparation of the catch for preservation and marketing, in *Seafood: Resources, Nutritional Composition, and Preservation* (ed Z.E. Sikorski), CRC Press, Inc. Boca Raton, FL, pp. 77–92.
Cann, D.C. (1977) Bacteriology of shellfish with reference to international trade, *Handling, Processing and Marketing of Tropical Fish*, Trop. Prod. Inst., London, pp. 377–94.
Cann, D.C. and Taylor, L.Y. (1979) The control of the botulism hazard in hot-smoked trout and mackerel. *J. Food Technol.*, **14**, 123–9.
Cann, D.C., Taylor, L.V. and Hobbs, G. (1975) The incidence of *Clostridium botulinum* in farm raised trout raised in Great Britain. *J. Appl. Bacteriol.*, **39**, 331–6.
Cann, D.C., Taylor, L.V. and Merican, Z. (1981) A study of the incidence of *Vibrio parahaemolyticus* in Malaysian shrimp undergoing processing for exports. *J. Hyg.*, **87**, 485–91.
CAOF (Committee on Animal Origin of Foods). (2000) Report on Preventive Measures for *Vibrio Parahaemolyticus* Foodborne Infections. Committee on Animal Origin of Foods under the Food Sanitation Investigation Council, Japan.
Caul, E.O. (2000) Foodborne viruses, in *The Microbiological Safety and Quality of Foods* (eds B.M Lund, T.C. Baird-Parker and G.W. Gould), Gaithersburg, Aspen, pp. 1457–89.
CDC (Centers for Disease Control). (1981) *Annual Summary of Foodborne Disease, 1978*, U.S. Department of Health and Human Services, Atlanta, GA, 53 pp.
Chan, K., Woo, M.L., Lam, L.Y. and French, G.L. (1989) *Vibrio parahaemolyticus* and other vibrios assoicated wiht seafood in Hong Kong. *J. Appl. Bacteriol.*, **66**, 57–64.
Chang, O., Cheuk, W.L., Nichelson, R., Martin, R. and Finne, G. (1983) Indole in shrimp: effect of fresh storage temperature, freezing and boiling. *J. Food Sci.*, **48**, 813–6.
Cheng, X.K., Lai-Yi, k. and Moy, G.G. (1992) An epidemic of foodborne hepatitis A in Shanghai, in *Proc. 3rd Worlg Cong. Foodborne Infec. Intoxications*, Robert von Ostertag Instit., Berlin, p. 119.
Chowdhury, N.R., Chakraborty, S., Ramamurthy, T., Nishibuchi, M., Yamasaki, S., Takeda, Y. and Nair, G.B. (2000) Molecular evidence of clonal *Vibrio parahaemolyticus* pandemic strains. *Emerging Infect. Dis.*, **6**, 631–6.
Christopher, F.M., Vanderzant, C., Parker, J.D. and Conte, F.S. (1978) Microbial flora of pond-reared shrimp (*Penaeus stylirostris*, *Penaeus vannamei*, and *Penaeus setiferus*). *J. Food Prot.*, **41**, 20–3.
Clifford, M.N., Walker, R., Wright, J., Ijomah, P., Hardy, R., Murray, C.K. and Rainsford, K.D. (1991) Evidence of histamine being the causative toxin in scombroid poisoning. *N. Engl. J Med.*, **325**, 515–6.
Cliver, D. (1988) Virus transmission via foods. *Food Technol.*, **42**, 241–8.

Cockey, R.R. and Chai, T. (1991) Microbiology of crustaceae processing: crabs, in *Microbiology of Marine Food Products* (eds D.R. Ward and C.R. Hackney), Van Nostrand Reinhold, New York, London, pp. 41–63.

Colburn, K.G., Kaysner, C.A., Abeyta, C., Jr. and Wekell, M.M. (1990) *Listeria* species in a California coast estuarine environment. *Appl. Environ. Microbiol.*, **56**, 2007–11.

Colwell, R.R. (1970) Polyphasic taxonomy of the genus Vibrio: Numerical taxonomy of *Vibrio cholerae*, *Vibrio parahaemolyticus*, and related *Vibrio* species. *J. Bacteriol*, **104**, 410–33.

Colwell, R.R. and Liston, J. (1960) Microbiology of shellfish: Bacteriological study of the natural flora of Pacific oysters (*Crassostrea gigas*). *Appl. Microbiol.*, **8**, 104–9.

Cook, D.W. (1994) Effect of time and temperature on multiplication of *Vibrio vulnificus* in postharvest gulf coast shellstock oysters. *Appl. Environ. Microbiol.*, **60**, 3483–4.

Cook, D.W. and Ruple, A.D. (1989) Indicator bacteria and Vibrionaceae multiplication in post-harvet shellstock oysters. *J. Food Prot.*, **52**, 343–49.

Croci, L., Toti, L., de Medici D. and Cozzi, L. (1994) Diarrhetric shellfish poison in mussels: Comparison of methods of detection and determination of the effectiveness of depuration. *Int. J. Food Microbiol.*, **24**, 337–42.

Cross, J.H. (2001) Fish and invertebrate-borne helminths, in *Foodborne Disease Handbook* (eds Y.H. Hui, S.A. Sattar, K.D. Murell, W.K. Nip, and P.S. Stanfield), 2nd edn, *Volume 2*, Marcel Dekker Inc., NY Basel, pp. 249–88.

Dalgaard, P. (1995) Qualitative and quantitative characterization of spoilage bacteria from packed fish. *Int. J. Food Microbiol.*, **26**, 319–33.

Dalgaard, P., Gram L. and Huss, H.H. (1993) Spoilage and shelf life of cod fillets packed in vacuum or modified atmospheres. *Int. J. Food Microbiol.*, **19**, 283–94.

Dalsgaard, A., Huss, H.H., H-Kittikun, A. and Larsen, J.L. (1995) Prevalence of *Vibrio cholerae* and *Salmonella* in a major shrimp production area in Thailand. *Int. J. Food Microbial.*, **28**, 101–13.

D'Aoust, J.Y., Gelinas, R. and Maishment, C. (1980) Presence of indicator organisms and the recovery of *Salmonella* in fish and shellfish. *J. Food Prot.*, **43**, 679–82.

Davis, J.W. and Sizemore, R.K. (1982) Incidence of *Vibrio* species associated with blue crabs (*Callinectes sapidus*) collected from Galveston Bay, Texas. *Appl. Environ. Microbiol.*, **43**, 1092–97.

Deardorff, T.L. and Kent, M.L. (1989) prevalence of larval *Anisakis simplex* in pen-reared and wild-caught salmon (salnidae) from Pugets Sound, Washington. *J. Wildlife Dis.*, **25**, 416–9.

Declerck, D. (1988) Présence de larves de *Anisakis simplex* dans le hareng (*Clupea harengus* L.) *Rev l Agric.*, **41**, 971–80.

Delmore, R.P., Jr. and Crisley, F.D. (1979) Thermal resistance of *Vibrio parahaemolyticus* in clam homogenate. *J. Food Prot.*, **42**, 131–4.

DePaola, A., Capers, G.M. and Alexander, D. (1994) Densities of *Vibrio vulnificus* in the intestines of fish from the U.S. Gulf coast. *Appl. Environ. Microbiol.*, **60**, 984–8.

DePaola, A., Peeler, J.T. and Rodrick, G.E. (1995) Effect of oxytetracycline-medicated feed on antibiotic resistance of Gram-negative bacteria in catfish ponds. *Appl. Environ. Microbiol.*, **61**, 2335–40.

Destro, M.T., Leitão, M.F.F. and Farber, J.M. (1996) Use of molecular typing methods to trace the dissimination of *Listeria monocytogenes* in a shrimp processing plant. *Appl. Environ. Microbiol.*, **62**, 705–11.

Devaraju, A.N. and Setty, T.M.R. (1985) Comparative study of fish bacteria from tropical and cold/temperate marine waters, in *Spoilage of Tropical Fish and Product Development* (ed A. Reilly), FAO Fish. Rep., 317 (Suppl.), pp. 97–107.

DiGirolamo, R., Liston, J. and Matches, J. (1970) The effects of freezing on the survival of *Salmonella* and *E. coli* in Pacific oysters. *J. Food Sci.*, **35**, 13–6.

Dillon, R. and Patel, T.R. (1992) *Listeria* in seafood: a review. *J. Food Prot.*, **55**, 1009–15.

Dillon, R. and Patel, T. (1993) Effect of cold smoking and storage temperatures on *Listeria monocytogenes* inoculated cod fillets (*Gadus morhus*). *Food Res. Int.*, **26**, 97–101.

Dillon, R., Patel, T. and Ratnam, S. (1994) Occurrence of *Listeria* in hot and cold smoked seafood products. *Int. J. Food Microbiol.*, **22**, 73–77.

Diyaolu, S.A. and Adebajo, L.O. (1994) Effects of sodium chloride and relative humidity on growth and sporulation of moulds isolated from cured fish. *Nahrung*, **38**, 311–7.

Dodds, K.L. (1993) *Clostridium botulinum* in the environment, in *Clostridium botulinum: Ecology and Control in Foods* (eds A.H.W. Hauschild and K.L. Dodds), Marcel Dekker, New York, pp. 21–52.

Dodds. K.L. and Austin, J.W. (1997) *Clostridium botulinum*, in *Food Microbiology—Fundamentals and Frontiers* (eds M.P. Doyle, L.R. Beuchat and T.J. Montville), American Society for Microbiology, Washington DC, pp. 288–304.

Doe, P. and Olley, J. (1990) Drying and dried fish products, in *Seafood: Resources, Nutritional Composition, and Preservation* (ed Z.E. Sikorski), CRC Press, Inc. Boca Raton, FL, pp. 125–46.

Dolman, C.E. and Iida, H. (1963) Type E botulism: Its epidemiology, prevention and specific treatment. *Can. J. Public Health.*, **54**, 293–308.

Dorsa, W.J. and Marshall, D.L. (1995) Influence of lactic acid and modified atmosphere on thermal destruction of *Listeria monocytogenes* in crawfish tail meat homogenate. *J. Food Safety*, **15**, 1–9.

Dorsa, W.J., Marshall, D.L., Moody, M.W. and Hackney, C.R. (1993a) Low temperature growth and thermal inactivation of *Listeria monocytogenes* in precooked crawfish tail meat. *J. Food Prot.*, **56**, 106–9.

Dorsa, W.J., Marshall, D.L. and Semien, M. (1993b) Effect of potassium sorbate and citric acid sprays on growth of *Listeria monocytogenes* on cooked crawfish (*Procambarus clarkii*) tail meat at 4°C. *J. Appl. Bacteriol.*, **26**, 480–2.

Duffes, F., Corre, C., Leroi, F., Dousset, X. and Boyaval, P. (1999) Inhibition of *Listeria monocytogenes* by in situ produced and semipurified bacteriocins on *Carnobacterium* spp. on vacuum-packed, refrigerated cold-smoked salmon. *J. Food Prot.*, **62**, 1395–403.
Dufresne, I., Smith, J.P., Liu, J.N., Tarte, I., Blanchfield, B. and Austin, J.W. (2000) Effect of films of different oxygen transmission rate on toxin production by *Clostridium botulinum* type E in vacuum packaged cold and hot smoked trout fillets. *J. Food Safety*, **20**, 251–68.
Duran, A.P., Wentz, B.A., Lanier, J.M., McClure, F.D., Schwab, A.H., Swartzentruber, A., Barnard, R.J. and Read, R.B., Jr. (1983) Microbiological quality of breaded shrimp during processing. *J. Food Prot.*, **46**, 974–7.
EEC. (1991a) Council directive 91/492/EEC of 15th July laying down the health conditions for the production and the placing on the market of live bivalve molluscs. *Off. J. Eur. Commun.*, No. **L268**, 1.
EEC. (1991b) Council directive 91/493/EEC of 22nd July 1991 laying down the health conditions for the production and the placing on the market of fishery products. *Off. J. Eur. Commun.*, No. **L268**, 15.
EEC. (1993) Commission decision 93/51/EEC of 15 December 1992 on the microbiological criteria applicable to the production of cooked crustaceans and molluscan shellfish. *Off. J. Eur. Commun.*, No. **L013**, 11–3.
EEC. (2002) Opinion of the Scientific Committee on Veterinary Measures relating to public health on Norwalk-like viruses. European Commission. Health and Consumer Protection Directorate-General. Adopted on 30–31 January 2002.
Efiuvwevwere, B.J.O. and Ajiboye, M.O. (1996) Control of microbiological quality and shelf-life of catfish (*Clarias gariepinus*) by chemical preservatives and smoking. *J. Appl. Bacteriol.*, **80**, 465–70.
Ekanem, E.O. and Adegoke, G.O. (1995) Bacteriological study of West African clam (*Egeria radiata* Lamarch) and their overlying waters. *Food Microbiol.*, **12**, 381–5.
Eklund, M.W., Pelroy, G.A., Paranjpye, R., Peterson, M.E. and Teeny, F.M. (1982a) Inhibition of *Clostrodium botulinum* types A and E toxin production by liquid smoke and NaCl in hot-process smoke-flavored fish. *J. Food Prot.*, **45**, 935–41.
Eklund, M.W., Peterson, M.E., Poysky, F.T., Peck L.W. and Conrad, J.F. (1982b) Botulism in juvenile Coho salmon (*Oncorynchus kisutch*) in the United States. *Aquaculture*, **27**, 1–11.
Eklund, M.W., Peterson, M.E., Paranjpuke R. and Pelroy, G. (1988) Feasibility of a heat-pasteurization process for the inactivation of non-proteolytic *Clostridium botulinum* types B and E in vacuum-packaged hot-process (smoked) fish. *J. Food Prot.*, **51**, 720–6.
Eklund, M.W., Poysky, F.T., Paranjpye, R.N., Lashbrook, L.C., Peterson, M.E. and Pelroy, G.A. (1995) Incidence and sources of *Listeria monocytogenes* in cold-smoked fishery products and processing plants. *J. Food Prot.*, **58**, 502–8.
Ellender, R.D., Huang, L., Sharp, S.L. and Tettleton, R.P. (1995) Isolation, enumeration, and identification of Gram-positive cocci from frozen crabmeat. *J. Food Prot.*, **58**, 853–7.
Elliot, E.L. (1987) Microbiological quality of Alaska pollock surimi, in *Seafood Quality Determination* (eds D.E. Kramer and J. Liston), Elsevier Science Publishers, Amsterdam, pp. 269–82.
Elliott, R.P. and Michener, H.D. (1965) *Factors affecting the growth of psychrophilic microorganisms in foods—a review*, Tech. Bull. No. 1320. U.S. Dept. Agric., Albany, CA.
Emborg, J., Laursen, G., Rathjen, T. and Dalgaard, P. (2002) Microbiology and spoilage of CO_2 packed fresh and frozen/thawed salmon. *J. Appl. Microbiol.*, **92**, 790–9.
Ericsson, H., Eklöw, A., Danielsson-Tham, M.L., Loncarevic, S., Mentzing, L.O., Persson, I., Unnerstad, H. and Tahm, W. (1997) An outbreak of listeriosis sustpected to have been caused by rainbow trout. *J. Clin. Microbiol.*, **35**, 2904–7.
Evans, M.C., Griffin, P.M. and Tauxe, R.V. (1999) *Vibrio* surveillance system. Summary data 1997–1999. Letter of information dated October 4th from Public Health Service, CDC, Atlanta, USA.
Faghri, M., Pennington, C.L., Cronholm, L.S. and Atlas, R.M. (1984) Bacteria associated with crabs from cold waters with emphasis on the occurrence of potential human pathogens. *Appl. Environ. Microbiol.*, **47**, 1054–61.
FAO (Food andAgricultural Organization) (1998) *The State of World Fisheries and Aquaculture*. FAO, Rome, Italy.
FAOSTAT. (2001) FAOSTAT. http://www.fao.org/fi/statist/FISOFT/FISHPLUS.asp.
Farber, J.M. (1991) *Listeria monocytogenes* in fish products. *J. Food Prot.*, **54**, 922–924.
Farber, J.M and Peterkin, P.I. (2000) *Listeria monocytogene*, in *The Microbiological Safety and Quality of Foods* (eds B.M. Lund, A.C. Baird-Parker and G.W. Gould), Chapman & Hall, London, pp. 1178–232.
FDA (Food and Drug Administration). (1989a) Revision of Sanitation of shellfish growing areas. *National Shellfish Sanitation Program Manual of Operations Part I.*, Center for Food Safety and Applied Nutrition, Division of Cooperative Programs, Shellfish Sanitation Branch, Washington, DC.
FDA (Food and Drug Administration). (1989b) Revision of Sanitation of the harvesting, processing and distribution of shellfish, in *National Shellfish Sanitation Program Manual of Operations Part II.*, Center for Food Safety and Applied Nutrition, Division of Cooperative Programs, Shellfish Sanitation Branch, Washington, DC.
FDA (Food and Drug Administration) (2001a) *Draft Risk Assessment on the Public Health Impact of Vibrio parahaemolyticus in Molluscan Shellfish*, FDA, Center for Food Safety and Applied Risk Assessment.
FDA (Food and Drug Administration) (2001b) *Fish and Fishery Products, Hazards and Controls Guidance*, 3rd edn, US Food and Drug Administration, Center for Food Safety and Applied Nutrition, Washington DC, USA.
FDA/FSIS (Food and Drug Administration). (2003) Quantitative Assessment of the Relative Risk to Public Health from Foodborne *Listeria monocytogenes* Among Selected Categories of Ready-To-Eat Foods. Food and Drug Administration, Center for Science and Applied Nutrition, College Park, Maryland.

Feachem, R.G. (1981) Environmental aspects of cholera epidemiology I. A review of selected reports of endemic and epidemic situations during 1961–1980. *Trop. Dis. Bull.*, **78**, 675–98.

Feachem, R.G. (1982) Environmental aspects of cholera epidemiology. III. Transmission and Control. *Trop. Dis. Bull.*, **79**, 1–47.

Feachem, R.G., Miller, C. and Drasar, B. (1981) Environmental aspects of cholera epidemiology. II. Occurrence and survival of *Vibrio cholerae* in the environment. *Trop. Dis. Bull.*, **78**, 865–80.

Frank, M. and Hess, E. (1941) Studies of salt fish. V. Studies of *Sporendonema epizoum* from "dun" salt fish. *J. Fish. Res. Bd. Can.*, **5**, 276–86.

Finelli, L., Swerdlow, D., Mertz, K., Regazzoni, H. and Spitalny, K. (1992) Outbreak of cholera associated with crab brought from an area with epidemic disease. *J. Infec. Dis.*, **166**, 1433–35.

Fletcher, G.C. and Statham, J.A. (1988a) Shelf-life of sterile yellow-eyed mullet (*Aldrichetta forsteri*). *J. Food Sci.*, **53**, 1030–35.

Fletcher, G.C. and Statham, J.A. (1988b) Deterioration of sterile chill-stored and frozen trumpeter fish (*Latridopsis forsteri*). *J. Food Sci.*, **53**, 1336–9.

Fletcher, G.C., Bremner, H.A., Olley, J. and Statham, J.A. (1990) The relationship between inosine monophosphate, hypoxanthine and Smiley scales for fish flavor. *Food Rev. Int.*, **6**, 489–503.

Fonnesbech Vogel, B., Ojeniyi, B., Ahrens, P., Huss, H.H. and Gram, L. (2001) Elucidation of *Listeria monocytogenes* contamination routes in cold-smoked salmon processing plants detected by DNA-based typing methods. *Appl. Environ. Microbiol.*, **68**, 2586–95.

Fraiser, M.B. and Koburger, J.A. (1984) Incidence of salmonellae in clams, oysters, crabs, and mullet. *J. Food Prot.*, **47**, 343–5.

Fujiki, H. and Suganuma, M. (1999) Unique features of the okadaic acid activity class of tumor promoters. *J. Cancer Res. Clin. Oncol.*, **125**, 150–5.

Ganowiak, Z.M. (1990) Sanitation in marine food industry, in *Seafood: Resources, Nutritional Composition and Preservation* (ed Z.E. Sidorski), CRC Press, Inc. Boca Raton, FL, pp. 211–30.

Gardiner, M.A. (1990) Survival of *Anisakis* in cold smoked salmon. *Can. Inst. Food Sci. Technol. J.*, **23** (2/3), 143–4.

Garren, D.M., Harrison, M.A. and Huang, Y.-W. (1994) *Clostridium botulinum* type E outgrowth and toxin production in vacuum-skin packaged shrimp. *Food Microbiol.*, **11**, 467–72.

Gecan, J.S., Bandler, R. and Staruszkiewicz, W.F. (1994) Fresh and frozen shrimp: A profile of filth, microbiological contamination, and decomposition. *J. Food Prot.*, **57**, 154–8.

Genigeorgis, C. (1985) Microbial safety of the use of modified atmospheres to extend the storage life of fresh meat and fish. A review. *Int. J. Food Microbiol.*, **1**, 237–41.

Gerba, C.P. (1988) Viral transmission by seafood. *Food Technol.*, **42**(3), 99–103.

Gerigk, V.K. (1985) Microbiologische Untersuchungen von gekochten, geschalten, tiefgefrorenen Garnelenschwanzen (shrimps). *Arch. Lebensmittelhyg.*, **36**, 40–3.

Gilbert, R.J. (1982) The microbiology of some foods imported into England through the port of London and Heathrow (London) Airport, in *Control of the Microbial Contamination of Foods and Feeds in International Trade: Microbial Standards and Specifications* (eds H. Kurata and C.W. Hesseltine), Saikon Pub. Co. Tokyo, pp. 105–19.

Gill, T.A., Thompson, J.W. and Gould, S. (1985) Thermal resistance of paralytic shellfish poison in soft-shell clams. *J. Food Prot.*, **48**, 659–62.

Gillespie, N.C. and Macrae, I.C. (1975) The bacterial flora of some Queensland fish and its ability to cause spoilage. *J. Appl Bacteriol.*, **39**, 91–100.

Gillespie, N.C., Lewis, R.J., Pearn, J.H., Bourke, A.T.C., Holmes, M.J., Bourke, J.B. and Shields, W.J. (1986) Ciguatera in Australia: occurrence, clinical features, pathophysiology and management. *Med. J. Aust.*, **145**, 584–90.

Gobat, P.-F. and Jemmi, T. (1993) Distribution of mesophilic *Aeromonas* species in raw and ready-to-eat fish and meat products in Switzerland. *Int. J. Food Microbiol.*, **20**, 117–20.

Gombas, D.E., Chen, Y., Clavero, R.S. and Scott, V.N. (2003). Survey of *Listeria monocytogenes* in ready-to-eat foods. *J. Food Prot.*, **66**, 559–69.

Gooch, J.A., DePaola, A., Bowers, J. and Marshall, D.L. (2002) Growth and survival of *Vibrio parahaemolyticus* in postharvest American oysters. *J. Food Prot.*, **65**, 970–4.

Graham, A.F., Mason, D.R., Maxwell, F.J. and Peck, M.W. (1997) Effect of pH and NaCl on growth from spores of non-proteolytic *Clostridium botulinum* at chill temperature. *Lett. Appl. Microbiol.*, **24**, 95–100.

Gram, L. (1989) Identification, characterization and inhibition of bacteria isolated from tropical fish. *PhD Thesis*, Technological Laboratory, Ministry of Fisheries, Lyngby, Denmark.

Gram, L. (1992) Evaluation of the bacteriological quality of seafood. *Int. J. Food Microbiol.*, **16**, 25–39.

Gram, L. and Huss, H.H. (2000) Fresh and processed fish and shellfish, in *The Microbiological Safety and Quality of Foods* (eds B.M. Lund, A.C. Baird-Parker and G.W. Gould), Chapman & Hall, London, pp. 472–506.

Gram, L., Ravn, L., Bruhn, J.B., Rasch, M., Christensen, A.B. and Givskov, M. (2002) Food spoilage—interactions between food spoilage bacteria. *Int. J. Food Microbiol.*, **78**, 79–97.

Gram, L., Wedell-Neergaard, C. and Huss, H.H. (1990) The bacteriology of spoiling Lake Victorian Nile perch (*Lates niloticus*). *Int. J. Food Microbiol.*, **10**, 303–16.

Gram, L., Oundo, J.O. and Bon, J. (1989) Storage life of Nile perch (*Lates niloticus*) in relation to temperature and initial bacterial load. *Trop. Sci.*, **29**, 221–36.

Gram, L., Trolle, G. and Huss, H.H. (1987) Detection of specific spoilage bacteria on fish stored at high (20°C) and low (0°C) temperatures. *Int. J. Food Microbiol.*, **4**, 65–72.

Graneli, E., Sundstrom, B., Edler, L. and Anderson, D.M. (1990) *Toxic Marine Phytoplankton*, Elsevier Science Publishers, Amsterdam, p. 553.
Griffin, M.R., Dalley, E., Fitzpatrick, M. and Austin, S.H. (1980) *Campylobacter gastroenteritis* associated with raw clams. *J. Med. Soc. New Jersy.*, **80**, 607–9.
Groubert, T.N. and Oliver, J.D. (1994) Interaction of *Vibrio vulnificus* and the eastern oyster, *Crassostrea virginica*. *J. Food Prot.*, **57**, 224–8.
Guyer, S. and Jemmi, T. (1991) Behavior of *Listeria monocytogenes* during fabrication and storage of experimentally contaminated smoked salmon. *Appl. Environ. Microbiol.*, **57**, 1523–7.
Hackney, C.R. and Dicharry, A. (1988) Seafood borne bacterial pathogens of marine origin. *Food Technol.*, **42**(3), 104–9.
Hackney, C.R., Ray, B. and Speck, M.L. (1980) Incidence of *Vibrio parahaemolyticus* in and the microbiological quality of seafood in North Carolina. *J. Food Protect.*, **43**, 769–3.
Hallegraeff, G.M. (1993) A review of harmful algal blooms and their apparent global increase. *Phycotogia*, **32**, 79–99.
Halliday, M.L., Kang, L.Y., Zhou, T.K., Hu, M.D., Pan, Q.C., Fu, T.Y., Huang, Y.S. and Hu, S.L. (1991) An epidemic of hepatitis a attributable to the ingestion of raw clams in Shanghai, China. *J. Infect. Dis.*, **164**, 852–9.
Harrison, J.M. and Lee, J.S. (1969) Microbial evaluation of Pacific shrimp processing. *Appl. Microbiol.*, **18**, 188–92.
Harrison, M.A. and Huang, Y.-W. (1990) Thermal death times for *Listeria monocytogenes* (Scott A) in crabmeat. *J. Food Prot.*, **53**, 878–80.
Harrison, M.A., Huang, Y.W., Chao, C.-H. and Shineman, T. (1991) Fate of *Listeria monocytogenes* on packaged, refrigerated, and frozen seafood. *J. Food Prot.*, **54**, 524–7.
Harris-Young, L., Tamplin, M.L., Mason, J.W., Aldrich, H.C. and Jackson, J.K. (1995) Viability of *Vibrio vulnificus* in association with hemocytes of the American oyster (*Crassostrea virginica*). *Appl. Environ. Microbiol.*, **61**, 52–7.
Hartemink, R. and Georgsson, F. (1991) Incidence of *Listeria* species in seafood and seafood salads. *Int. J. Food Microbiol.*, **12**,189–96.
Hatha, A.A.M. and Lakshmanaperumalsamy, P. (1995) Antibiotic resistance of *Salmonella* strains isolated from fish and crustaceans. *Lett. Appl. Microbiol.*, **21**, 47–9.
Heinitz, M.L. and Johnson, J.M. (1998) The incidence of *Listeria* spp., *Salmonella* spp., and *Clostridium botulinum* in smoked fish and shellfish. *J. Food Prot.*, **61**, 318–23.
Heinsz, L.J., Harrison, M.A. and Leiting, V.A. (1988) Microflora of brown shrimp (*Penaeus aztecus*) from Georgia coastal waters. *Food Microbiol.*, **5**, 141–5.
Hejkal, T.W., Gerba, C.P., Henderson, S. and Freeze, M. (1983) Bacteriological, virological and chemical evaluation of a wastewater aquaculture system. *Water Res.*, **17**, 1749–56.
Higashi, G.H. (1985) Foodborne parasites transmitted to man from fish and other aquatic foods. *Food Technol.*, **39**, 69.
Hobbs, G. (1976) *Clostridium botulinum* and its importance in fishery products. *Adv. Food Res.*, **22**, 135–85.
Hood, M.A., Ness, G.E. and Blake, N.J. (1983) Relationship among fecal coliforms, *Escherichia coli* and *Salmonella* spp. in shellfish. *App. Environ. Microbiol.*, **45**, 122–26.
Horner, R.A. and Postel, J.R. (1993) Toxic diatoms in western Washington waters (U.S. west coast). *Hydrobiologia*, **269/270**, 197–205.
Huber, I., Spangaard, B., Nielsen, J., Appel, K.F., Nielsen, T.F. and Gram, L. (2004) Phylogenetic analysis and *in situ* identification of the intestinal microflora of rainbow trout (*Onchorhynchus mykiss*, Walbaum). *J. Appl. Microbiol.*, **96**, 117–32.
Hultin, H.O. (1992) Biochemical deterioration of fish muscle, in *Quality Assurance in the Fish Industry* (eds H.H. Huss, M. Jakobsen and J. Liston), Elsevier Science Publishers, Amsterdam, pp. 125–38.
Huss, H.H. (1980) Distribution of *Clostridium botulinum*. *Appl. Environ. Microbiol.*, **39**, 764–9.
Huss, H.H. (1981) *Clostridium botulinum* type E and botulism. *DSc Thesis*, Lyngby (DK) Technical University,Technological Laboratory of the Danish Ministry of Fisheries. 58 pages.
Huss, H.H., Pedersen, A. and Cann, D.C. (1974a) The incidence of *Cl. botulinum* in Danish trout farms. II: measures to reduce the contamination of the fish. *J. Food Technol.*, **9**, 451–8.
Huss, H.H., Dalsgård, D., Hansen, L., Ladefoged, H., Pedersen, A. and Zittan, L. (1974b) The influence of hygiene in catch handling on the storage of cod and plaice. *J. Food Technol.*, **9**, 213–21.
Huss, H.H., Schaeffer, I., Pedersen, A. and Jepsen, A. (1980) Toxin production by *Clostridium botulinum* Type E in smoked fish in relation to the measured oxidation reduction (Eh) potential, packaging method and the associated microflora, in *Advances in Fish Science and Technology* (ed J.J. Connell), Fishing News Books Ltd, England, pp. 476–9.
Huss, H.H., Ben Embarek, P.K. and From Jeppesen, V. (1995) Control of biological hazards in cold-smoked salmon production. *Food Control*, **6**, 335–40.
Huss, H.H., Reilly, A. and Ben Embarek, P.K. (2001) Prevention and control of hazards in seafood. *Food Control*, **11**, 149–56.
Huss, H.H., Ababouch, L. and Gram, L. (2004) Assessment and Management of Seafood Safety and Other Quality Aspects. FAO Fish. Techn. Pap. 444.
Hyytia, E., Eerola, S., Hielm, S. and Korkeala, H. (1997) Sodium nitrite and potassium nitrate in control of non-proteolytic *Clostridium botulinum* outgrowth and toxigenesis in vacuum-packed cold-smoked rainbow trout. *Int. J. Food Microbiol.*, **37**, 63–72.
Høegh, L. (1986) Bacteriological quality control of cooked, peeled Greenland shrimp [In Danish]. *MSc Thesis*, Danish Institute for Fisheries Research.
IARS (Infectious Agents Surveillance Report). (1998) [In Japanese]. October, Volume 19.

ICMSF (International Commission for the Microbiological Specifications for Foods). (1994) Choice of sampling plan and criteria for *Listeria monocytogenes*. *Int. J. Food Microbiol.*, **22**, 89–96.
ICMSF (International Commission for the Microbiological Specifications for Foods). (1996) Microorganisms in Foods 5, *Characteristics of Microbial Pathogens*, Blackie Academic & Professional, London, UK.
ICMSF (International Commission for the Microbiological Specifications for Foods) (1998) Microorganisms in Foods 6, *Microbial Ecology of Food Commodities*, Blackie Academic & Professional, London, UK.
ICMSF (International Commission for the Microbiological Specifications for Foods) (2001) Microorganisms in Foods 7, *Managing the Microbiological Safety of Foods*, Aspen.
IFT (Institute of Food Technologists). (2001) Processing Parameters Needed to Control Hazards in Cold-smoked Fish. Task Order #2.
Iida, H. (1970) Epidemiological and clinical observations of botulism outbreaks in Japan, in *Proceedings of the First US-Japan Conference on Toxic Microorganisms, Mycotoxin, Botulism* (ed Herzberg), UJNR Jount Panels on Toxic Microorganisms and the U.S. Department of the Itnerior.
Ijomah, P., Clifford, M.N., Walker, R., Wright, J., Hardy, R. and Murray, C.K. (1991) The importance of endogenous histamine relative to dietary histamine in the aetiology of scombrotoxicosis. *Food Addit. Contam.*, **8**, 531–42.
Ingham, S.C. and Moody, M.W. (1990) Enumeration of aerobic plate count and *E. coli* during blue crab processing by standard methods, petrifilm, and redigel. *J. Food Prot.*, **53**, 423–4.
Ingham, S.C. and Potter, N.N. (1988) Growth of *Aeromonas hydrophila* and *Pseudomonas fragi* on mince and surimis from Atlantic pollock and stored under air or modified atmosphere. *J. Food Prot.*, **51**, 966–70.
Ingham, S.C., Alford, R.A. and McCown, A.P. (1990) Comparative growth rates of *Salmonella typhimurium* and *Pseudomonas fragi* on cooked crab meat stored under air and modified atmosphere. *J. Food Prot.*, **53**, 566–7.
Inoue, K., Oshima, S.-I., hirata, T. and Kimura, I. (2000). Possibility of anisakid larvae infection in farmed salmon. *Fish. Sci.*, **66**, 1049–52.
IOM (Institute of Medicine). (1991) *Seafood Safety* (ed F.E. Ahmed), National Academy of Sciences, Washington DC, USA.
Iter, T.S.G. and Varma, P.R.G. (1990) Sources of contamination with *Salmonella* during processing of frozen shrimps. *Fish Technol.*, **27**, 60–3.
James, D. and Olley, J. (1985) Summary and future research needs, in *Histamine in Marine Products: Production by Bacteria, Measurement and Prediction of Formation* (eds B.S. Par and D. James), *FAO* Fisheries Technical Paper 252, pp. 47–50.
Janda, J.M., Powers, C., Bryant, R.G. and Abbott, S.L. (1988) Current perspectives on the epidemiology and pathogenesis of clinically significant *Vibrio* spp. *Clin. Microbiol. Rev.*, **1**, 245–67.
Japanese Ministry for Health, Labour and Welfare. (1999) National Institute of Infectious Diseases and Infectious Diseases Control Division. *Infectious Agents Surveillance Report vol 20*, July 1999.
Japanese Ministry for Health, Labour and Welfare. (1998) National Institute of Infectious Diseases and Infectious Diseases Control Division. *Infectious Agents Surveillance Report vol. 19*, October 1998.
Japanese Ministry for Health, Labour and Welfare. (2002) Japanese statistics for foodborne outbreaks.
Jaykus, L. (2000) Enteric viruses as 'emerging agents' of foodborne disease. *Irish J. Agric. Food Res.*, **39**, 245–55.
Jemmi, T. and Keusch, A. (1992) Behavior of *Listeria monocytogenes* during processing and storage of experimentally contaminated hot-smoked trout. *Int. J. Food Microbiol.*, **15**, 339–46.
Jemmi, T. and Keusch, A. (1994) Occurrence of *Listeria monocytogenes* in freshwater fish farms and fish-smoking plants. *Food Microbiol.*, **11**, 309–16.
Jiménez, L., Muniz, I, Toranzos, G.A. and Hazen, T.C. (1989) Survival and activity of *Salmonella typhimurium* and *Escherichia coli* in tropical waters. *J. Appl. Bacteriol.*, **67**, 61–9.
Johnson, H.C. and Liston, J. (1973) Sensitivity of *Vibrio parahaemolyticus* to cold in oysters, fish fillets and crabmeat. *J. Food Sci*, **38**, 437–41.
Joseph, S.W., Colwell, R.R. and Kaper, J.B. (1982) *Vibrio parahaemolyticus* and related halophilic vibrios. *CRC Crit. Rev. Microbiol.*, **10**, 77–124.
Joslyn, F.E. and Lahai, G.P. (1992) Mycotoxic flora and mycotoxins in smoke-dried fish from Sierra Leone. *Nahrung* **36**, 485–9.
Jørgensen, B.R. and Huss, H.H. (1989) Growth and activity of *Shewanella putrefaciens* isolated from spoiling fish. *Int. J. Food Microbiol.*, **9**, 51–62.
Jørgensen, L.V. and Huss, H.H. (1998) Prevalence and growth of *Listeria monocytogenes* in Danish Seafood. *Int. J. Food Microbiol.*, **42**, 127–31.
Jørgensen, L.V., Huss, H.H. and Dalgaard, P. (2000) The effect of biogenic amine production by single bacterial cultures and metabiosis on cold-smoked salmon. *J. Appl. Microbiol.*, **89**, 920–34.
Kaneko, T. and Colwell, R.R. (1973) Ecology of *Vibrio parahaemolyticus* in Chesapeake Bay. *J. Bacteriol.*, **113**, 24–32.
Karl, H., Roepstorff, P., Huss, H.H. and Bloemsma, B. (1995) Survival of *Anisakis* larvae in marinated herring fillets. *Int. J. Food Sci. Technol*, **29**, 661–70.
Karunasagar, I., Venugopal, M.N. and Karunasagar, I. (1984) Levels of *Vibrio parahaemolyticus* in Indian shrimp undergoing processing for export. *Can. J. Microbiol.*, **30**, 713–5.
Karunasagar, I., Pai, R., Malathi, G.R. and Karunasagar, I. (1994) Mass mortality of *Penaeus monodo* larvae due to antibiotic-resistant *Vibrio harveyi* infection. *Aquaculture*, **128**, 203–9.
Kaspar, C.W. and Tamplin, M.L. (1993) Effects of temperature and salinity on the survival of *Vibrio vulnificus* in seawater and shellfish. *Appl. Environ. Microbiol.*, **59**, 2425–9.

Katoh, H. (1965) Studies on the growth rate of various food bacteria. III. The growth of *Vibrio parahaemolyticus* in raw fish meat. *Nippon Saikingaku Zasshi* **20**, 541–4.
Kator, H. and Fisher, R.A. (1995) Bacterial spoilage of processed sea scallop (*Placopecten magellanicus*) meats. *J. Food Prot.*, **58**, 1351–6.
Kaysner, C.A., Abeyta, C., Wekell, M.M., DePaola, A., Stott R.F. and Leitch, J.M. (1987) Incidence of *Vibrio cholerae* from estuaries of the United States West coast. *Appl. Environ. Microbiol.*, **53**, 1344–8.
Kelly, M.T. (1982) Effect of temperature and salinity on *Vibrio* (Beneckea) *vulnificus* occurrence in a Gulf Coast environment. *Appl. Environ. Microbiol.*, **44**, 820–4.
Kelly, K., Jones, N.R., Love, R.H. and Olley, J. (1966) Texture and pH in fish muscle related to 'cell fragility' measurements. *J. Food Technol.*, **1**, 9–15.
Khalafalla, F.A. (1992) *Campylobacter jejuni* as surface contaminant of fresh water fish, *in Proc. 3rd World Congr. Foodborne Infections and Intoxications, Volume 1*, Robert von Ostertag-Institute, Berlin, pp. 458–60.
Klausen, N.K. and Huss, H.H. (1987) Growth and histamine production by *Morganella morganii* under various temperature conditions. *Int. J. Food Microbiol.*, **5**, 147–56.
Klontz, K.L., Williams, L., Baldy, L.M. and Campos, M. (1993) Raw oyster-associated *Vibrio* infections: Linking epidemiologic data with laboratory testing of oysters obtained from a retail outlet. *J. Food Prot.*, **56**, 977–9.
Koh, E.G.L., Huyn, J.-H. and LaRock, P.A. (1994) Pertinence of indicator organisms and sampling variables to *Vibrio* concentrations. *Appl. Environ. Microbiol.*, **60**, 3897–900.
Kraft, A.A. (1992) *Psychrotrophic Bacteria in Foods: Disease and Spoilage*, CRC Press, Inc. Boca Raton, FL.
Larsen, J. and Moestrup, O. (1989) *Guide to Toxic and Potentially Toxic Marine Algae*, The Fish Inspection Service, Ministry of Fisheries, Dronningens Tvaergade 21, P.O. Box 9050, DK-1022 Copenhagen K, Denamark (ISBN 87 983238 0 6).
Laycock, R.A. and Regier, L.W. (1970) Pseudomonads and achromobacters in the spoilage of irradiated haddock of different preirradiation quality. *Appl. Microbiol.*, **20**, 333–41.
Layrisse, M.E. and Matches, J.R. (1984) Microbiological and chemical changes of spotted shrimp (*Pandalus platyceros*) stored under modified atmosphere. *J. Food Prot.*, **47**, 453–7.
Leangphibul, P., Nilakul, C., Sornchai, C., Tantimavancih, S. and Kasemsuksakul, K. (1986) Investigation of pathogenic bacteria from shrimp farms. *Kasetsart J.*, **20**, 333–7.
Lee, M.J., Jeong, D.Y., Kim, W.S., Kim, H.D., Kim, C.H., Park, W.W., Kim, K.S., Kim H.M. and Kim, D.S. (2000) A tetrodotoxin-producing *Vibrio* strain, LM-1 from the puffer fish *Fugu vermicularis radiatus*. *Appl. Environ. Microbiol.*, **66**, 1698–701.
Lees, D. (2000) Viruses and bivalve shellfish. *Int. J. Food Microbiol.*, **59**, 81–116.
Lehane, L. and Lewis, R.J. (2000) Ciguatera: recent advances but the risk remains. *Int. J. Food Microbiol*, **61**, 91–125.
Leisner, J.J., Millan, J.C., Huss, H.H. and Larsen, L.M. (1994) Production of histamine and tyramine by lactic acid bacteria isolated from vacuum-packed sugar-salted fish. *J. Appl. Bacteriol.*, **76**, 417–23.
Leung, C.-K., Huang, Y.-W. and Pancorbo, O.C. (1992) Bacterial pathogens and indicators in catfish and pond environments. *J. Food Prot.*, **55**, 424–7.
Levine, W.C., Griffin, P.M. and the Gulf Coast *Vibrio* Working Group. (1993) *Vibrio* infections on the Gulf Coast: Results of first year of regional surveillance. *J. Infect. Dis.*, **67**, 479–83.
Lilly, T., Jr. and Kautter, D.A. (1990) Outgrowth of naturally occurring *Clostridium botulinum* in vacuum-packaged fresh fish. *J. Assoc. Offical Anal. Chem.*, **73**, 211–2.
Lima dos Santos, C.A.M. (1981) The storage of tropical fish on ice–a review. *Tropi Sci.* **23**, 97–127.
Lin, F.-Y.C., Morris, J.G., Jr., Kaper, J.B., Gross, T., Michalski, J., Morrison, C., Libonati, J.P. and Israel, E. (1986) Persistance of cholera in the United States: isolation of *Vibrio cholerae* O1 from a patient with diarrhea in Maryland. *J. Clin. Microbiol.*, **23**, 624–26.
Liston, J. (1974) Influence of seafood handling procedures on *Vibrio parahaemolyticus*, in *International Symposium on Vibrio parahaemolyticus* (eds T. Fujino, G. Sakaguchi, R. Sakazaki and Y. Takeda), Saikon, Tokyo.
Liston, J. (1980) Microbiology in fishery science, in *Advances in Fishery Science and Technology*, Fishing News Books Ltd., Farnham, Surrey, England, pp. 138–57.
Liston, J. (1990) Microbial hazards of seafood consumption. *Food Technol.*, **44**, 56, 58–62.
Liston, J. (1992) Bacterial spoilage of seafood, in *Quality Assurance in the Fish Industry* (eds H.H. Huss, M. Jacobsen and J. Liston), Elsevier Science Publishers, Amsterdam, pp. 93–105.
Liston, J. and Baross, J. (1973) Distribution of *V. parahaemolyticus* in the marine environment. *J. Milk Food Technol.*, **36**, 113–7.
Lopez-Sabater, E.I., Rodriguez-Jerez, J.J., Roig-Sagues, A.X. and Mora-Ventura, M.A.T. (1994) Bacteriological quality of tuna fish (*Thunnus thynnus*) destined for canning: Effect of tuna handling on presence of histidine decarboxylase bacteria and histamine level. *J. Food Prot.*, **57**, 318–23.
Lovelace, T.E., Tubiash, H. and Colwell, R.R. (1968) Quantitative and qualitative commensal bacterial flora of *Crassostrea virginica* in Chesapeake Bay. *Proc. Natl Shellfish Assoc.*, **58**, 82–87.
Machi, T., Okino, S., Saito, Y., Horita, Y., Taguchi, T., Nakazawa, T., Nakamura, Y., Hirai, H., Miyamori, H. and Kitagawa, S. (1997) Severe chest pain due to gastric anisakiasis. *Intern. Med.*, **36**, 28–30.
Manger, R.L., Leja, L.S., Lee, S.Y., Hungerford, J.M. and Wekell, M.M. (1993) Tetrazolium-based cell bioassay for neurotoxin active on voltage-sensitive sodium channels: semiautomated assay for saxitoxins, brevetoxins and ciguatoxins. *Anal. Biochem.*, **214**, 190–4.

Manger, R.L., Leja, L.S., Lee, S.Y., Hungerford, J.M., Hokama, Y., Dickey, R.W., Granade, H.R., Lewis, R., Yasumoto, T. and Wekell, M.M. (1995) Detection of sodium channel toxins: directed cytotoxicity assays of purified ciguatoxins, brevetoxins, saxitoxins and seafood extracts. *J. Am. Organization Anal. Chem.*, **78**, 521–7.

Manthey, M., Karnop, G. and Rehbein, H. (1988) Quality changes of European catfish (*Silurus glanis*) from warm-water aquaculture during storage on ice. *Int. J. Food Sci Technol.*, **23**, 1–9.

Martinez-Manzanares, E., Morinigo, M.A., Castro, D., Baledona, M.C., Munoz, M.A. and Borrego, J.J. (1992) Relationship between indicators of fecal pollution in shellfish-growing water and the occurrence of human pathogenic microorganisms in shellfish. *J. Food Prot.*, **55**, 609–14.

Matches, J.R. (1982) Effects of temperature on the decomposition of Pacific coast shrimp (*Pandalus jordani*). *J. Food Sci.*, **47**, 1044–7, 1069.

Matches, J.R. and Liston, J. (1968) Low temperature growth of *Salmonella*. *J. Food Sci.*, **33**, 641–5.

Matches, J.R. and Liston, J. (1973) Methods and techniques for isolation and testing of clostridia from the estuarine environment, in *Estuarine Microbial Ecology* (eds L.H. Stevenson and R.R. Colwell), University of South Carolina Press, SC, pp. 345–62.

Matches, J.R., Liston, J. and Curran, D. (1974) *Clostridium perfringens* in the environment. *Appl. Microbiol.*, **28**, 655–66.

Matches, J.R., Raghubeer, E., Yoon, I.H. and Martin, R.E. (1987) Microbiology of surimi based products, in *Seafood Quality Determination* (eds D.E. Kramer and J. Liston), Elsevier, Amsterdam, pp. 373–87.

Matsumura, K. (1995) Reexamination of tetrodotoxin production by bacteria. *Appl. Environ. Microbiol.*, **66**, 3468–70.

Matsumura, K. (2001) Letter to the editor: No ability to produce tetrodotoxin in bacteria. *Appl. Environ. Microbiol.*, **67**, 2393–4.

Matté, G.R., Matté, M.H., Rivera, I.G. and Martins, M.T. (1994) Distribution of potentially pathogenic vibrios in oysters from a tropical region. *J. Food Prot.*, **57**, 870–3.

Mazurkiewicz, E., Tordoir, B.T., Oomen, J.M.V., de Lange, L. and Mol, H. (1985) *Bacillaire dysenterie* in Utrecht; twee epedemieen. *Ned Tijdschr. Genneskd.*, **129**, 895–9.

McCarthy, S.A., Motes, M.L. and McPhearson, R.M. (1990) Recovery of heat-stressed *Listeria monocytogenes* from experimentally and naturally occurring contaminated shrimp. *J. Food Prot.*, **53**, 22–5.

Michener, H.D. and Elliott, R.P. (1964) Minimum growth temperatures for food-poisoning, fecal-indicator and psychrophilic microorganisms. *Adv. Food Res.*, **13**, 349–96.

Miettinen, M.K., Siitonen, A., Heiskanen, P., Haajanen, H., Björkroth, K.J. and Korkeala, H.J. (1999) Molecular epidemiology of an outbreak of febrile gastroenteritis caused by *Listeria monocytogenes* in cold-smoked rainbow troug. *J. Clin. Microbiol.*, **37**, 2358–60.

Misrachi, W.A., Watson, A.J. and Coleman, D. (1991) *Listeria* in smoked mussels in Tasmania. *Commun. Dis. Intell.*, **15**, 427.

Mitchell, D.L. (1991) A case cluster of listeriosis in Tasmania. *Commun. Dis. Intell.*, **15**, 427.

Miyahara, M. and Konuma, H. (2002) Detection of *Shigella sonnei* from imported frozen oysters [in Japanese]. *Bokin Bobai* (*The Society for Antibacterial and Antifungal Agents, Japan*), **30**, 299–302.

MMWR (Morbidity and Mortality Weekly Report). (1975) Scombroid Poisoning—New York City. *MMWR*, **24**, 342, 347.

MMWR (Morbidity and Mortality Weekly Report). (1985) Botulism associated with commercially distributed kapchunka—New York City. *MMWR*, **34**, 546–7.

MMWR (Morbidity and Mortality Weekly Report). (1991a) Update: cholera outbreak—Peru, Ecuador, and Columbia. *MMWR*, **40**, 225–7.

MMWR (Morbidity and Mortality Weekly Report). (1991b) Update: cholera—Western Hemisphere, and recommendations for treatment of cholera. *MMWR*, **40**, 562–5.

MMWR (Morbidity and Mortality Weekly Report). (1991c) Fish botulism—Hawaii, 1990. *MMWR*, **40**, 412–4.

MMWR (Morbidity and Mortality Weekly Report). (1992a) Cholera associated with international travel, 1992. *MMWR*, **41**, 664–8.

MMWR (Morbidity and Mortality Weekly Report). (1992b) Outbreak of type E botulism associated with an uneviscerated, salt-cured fish product—New Jersy, 1992. *MMWR*, **41**, 1–2.

MMWR (Morbidity and Mortality Weekly Report) (2001) Botulism outbreak associated with eating fermented food—Alaska, 2001. *MMWR*, **50**, 680–2.

Molitoris, E., Joseph, S.W., Krichevsky, M.I., Sindhuhardja, W. and Colwell, R.R. (1985) Characterization and distribution of *Vibrio alginolyticus* and *Vibrio parahaemolyticus* isolated from Indonesia. *Appl. Environ. Microbiol.*, **50**, 1388–94.

Morris, E.O. (1975) Yeasts from the marine environment. *J. Appl. Bacteriol.*, **38**, 211–23.

Morris, J.G. and Black, R.E. (1985) Cholera and other vibrioses in the United States. *N. Engl. J. Med.*, **312**, 343–50.

Morse, D.L., Guzewich, J.J., Hanrahan, J.P., Strigof, R., Shayegani, M., Deibel, R., Herrmann, J.E., Cukor G. and Blacklow, N.R. (1986) Widespread outbreaks of clam- and oyster-associated gastroenteritis. Role of Norwalk virus. *N. Engl. J. Med.*, **314**, 678–81.

Motes, M.L., Jr. (1991) Incidence of *Listeria* spp. in shrimp, oysters, and estuarine waters. *J. Food Prot.*, **54**, 170–5.

Mugula, J.K. and Lyimo, M.H. (1992) Microbiological quality of traditional market cured fish in Tanzania. *J. Food Safety*, **13**, 33–41.

Muraoka, A., Suehiro, I., Fujii, M., Nagata, K., Kusunoki, H., Kumon, Y., Shirasaka, D., Hosooka, T. and Murakami, K. (1996) Acute gastric anisakiasis: 28 cases during the last 10 years. *Dig. Sci.*, **41**, 2362–5.

Murphree, R.L. and Tamplin, M.L. (1995) Uptake and retention of *Vibrio cholerae* O1 in the eastern oyster, *Crassostrea virginica*. *Appl. Environ. Microbiol.*, **61**, 3656–60.

Myers, B.J. (1979) Anisakine nematodes in fresh commercial fish from waters along the Washington, Oregon and California coasts. *J. Food Prot.*, **42**, 380–4.

Mølbak, K., Baggesen, D.L., Aarestrup, F.M., Ebbesen, J.M., Engberg, J., Frydendahl, K., Gerner-Smidt, P., Munk Petersen, A. and Wegener, H.C. (1999) An outbreak of multidrug-resistant, quinolone-resistant *Salmonella enterica* serotype Typhimurium DT104. *N. Engl. J. Med.*, **341**, 1420–5.
NACMCF (National Advisory on Microbiological Criteria for Foods). (1992) Microbiological criteria for raw molluscan shellfish. *J. Food Prot.*, **55**, 463–80.
Nair, G.B., Bhadra, R.K., Ramamurthy, T., Ramesh, A. and Pal, S.C. (1991) *Vibrio cholerae* and other vibrios associated with paddy field cultured prawns. *Food Microbiol.*, **8**, 203–8.
Nalin, D.R., Daya, V., Levine, M.M and Cisneros, L. (1979) Adsorption and growth of *Vibrio cholerae* on chitin. *Infect. Immun.*, **25**, 768–70.
Nedoluha, P.C. and Westhoff, D. (1993) Microbiological flora of aquacultured hybrid striped bass. *J. Food Prot.*, **56**, 1054–60.
Nilsson, L., Gram, L. and Huss, H.H. (1999) Growth control of *Listeria monocytogenes* on cold-smoked salmon using a competitive lactic acid bacteria flora. *J. Food Prot.*, **62**, 336–42.
NMFS (National Marine Fisheries Service). (1989) *Model Seafood Surveillance Project: HACCP Regulatory Model: Cooked Shrimp*, National Seafood Inspection Laboratory, Pascagoula, MS.
NMFS (National Marine Fisheries Service). (1990a) *Model Seafood Surveillance Project: HACCP Regulatory Model: Raw Fish*, National Seafood Inspection Laboratory, Pascagoula, MS.
NMFS (National Marine Fisheries Service). (1990b) *Model Seafood Surveillance Project: HACCP Regulatory Model: Molluscan Shellfish*, National Seafood Inspection Laboratory, Pascagoula, MS.
NMFS (National Marine Fisheries Service). (1990c) *Model Seafood Surveillance Project: HACCP Regulatory Model: Blue Crab*, National Seafood Inspection Laboratory, Pascagoula, MS.
Novak, A.F. (1973) Microbiological considerations in the handling and processing of crustacean shellfish, in *Microbial Safety of Fishery Products* (eds C.O. Chichester and H.D. Graham), Academic Press. New York, pp. 59–73.
Oh, D.-H. and Marshall, D.L. (1995) Influence of packaging method, lactic acid and monolaurin on *Listeria monocytogenes* in crayfish tail meat homogenate. *Food Microbiol.*, **12**, 159–63.
Ohtomo, R. and Saito, M. (2001) Increase in the culturable cell number of *Escherichia coli* during recovery from saline stress: possible implication for resuscitation for the VBNC state. *Microbial Ecol.*, **42**, 208–14.
Okuzumi, M., Yamanaka, H., Kubozuka, T., Ozaki H. and Matsubara, K. (1984) Changes in numbers of histamine forming bacteria on/in common mackerel stored at various temperatures. *Bull. Jpn Soc. Sci. Fish.* **50**, 653–65.
Oliver, J.D. (1981) Lethal cold stress of *Vibrio vulnificus* in oysters. *Appl. Environ. Microbiol.*, **41**, 710–7.
Oliver, J.D. and Kaper, J.B. (1997) *Vibrio* species. in *Food Microbiology—Fundamentals and Frontiers* (eds M.P. Doyle, L.R. Beauchat and T.J. Montville), ASM Press, Washington, DC, pp. 228–64.
Oliver, J.D. and Wanucha, D. (1989) Survival of *Vibrio vulnificus* at reduced temperatures and elevated nutrient. *J. Food Safety*, **10**, 79–86.
Olsen, S.J., MacKinnon, L.C., Goulding, J.S., Bean, N.H. and Slutsker, L. (2000) Surveillance for foodborne-disease outbreaks—United States, 1993–1997. *Morbidity Mortality Weekly Report CDC Surveill. Summ.*, **49**, 1–62.
Olson, R.E. (1987) Marine fish parasites of public health importance, in *Seafood Quality Determination* (eds D.E. Kramer and J. Liston), Elsevier Science Publishers, Amsterdam, pp. 339–55.
O'Neil, K.R., Jones, S.H. and Grimes, D.J. (1990) Incidence of *Vibrio vulnificus* in northern New England water and shellfish. *FEMS Microbiol. Lett.*, **72**, 163–8.
Ono, T., Inoue, Y., Yokoyama, M., Sakae, J., Goto, K., Kawazu, T., Hirano, Y., Minami, R., Matsui, T., Komatsuzaki, M., Oyama, T. and Okabe, T. (2001) *Vibrio vulnificus* infections reported in Kumamoto Prefecture, Japan in Japanese. *Infect. Agents Surveill. Rep.*, **22**, 249–9.
Orlandi, P.A., Chu, T.D.-M., Bier, J.W. and Jackson, G.J. (2002) Parasites and the food supply. *Food Technol.*, **56**, 72–81.
Pace, P.J. and Krumbiegel, E.R. (1973) *Clostridium botulinum* and smoked fish production: 1963–1972. *J. Milk Food Technol.*, **36**, 42–9.
Pacini, R., Panizzi, L., Galleschi, G., Quagli, E., Galassi, R., Fatighenti, P. and Morganti, R. (1993) Presenza di larve di anisakidi in prodotti ittici freschi e congelati del commercio. *Ind. Alimentari*, **32**, 942–4.
Paludan-Müller, C., Dalgaard, P., Huss, H.H. and Gram, L. (1998) Evaluation of the role of *Carnobacterium piscicola* in spoilage of vacuum- and modified-atmosphere-packed cold-smoked salmon stored at 5°C. *Int. J. Food Microbiol.*, **39**, 155–66.
Paludan-Müller, C., Valyasevi, R., Huss, H.H. and Gram, L. (2002) Genotypic and phenotypic characterisation of garlic fermenting lactic acid bacteria isolated from *som fak*, a Thai low-salt fermented fish product. *J. Appl. Microbiol.*, **92**, 307–14.
Palumbo, S.A. and Buchanan, R.L. (1988) Factors affecting the growth or survival of *Aeromonas hydrophila* in foods. *J. Food Safety*, **9**, 37–51.
Palumbo, S.A. and Williams, A.C. (1991) Resistance of *Listeria monocytogenes* to freezing in foods. *Food Microbiol.*, **8**, 63–8.
Palumbo, S.A., Maxino, F., Williams, A.C., Buchanan, R.L. and Thayer, D.W. (1985) Starch ampicillin agar for the quantitative detection of *Aeromonas hydrophila* in foods. *Appl. Environ. Microbiol.*, **50**, 1027–30.
Pan, B.S. and James, D. (1985) Histamine in marine products: production by bacteria, measurement and prediction of formation. FAO Fisheries Technical Paper 252, Rome, Italy.
Parisi, E., Civera, T. and Giacone, V. (1992) Changes in the microflora and in chemical composition of vacuum-packed smoked salmon during storage, in *Proc. 3rd World Congr. Foodborne Infections and Intoxications, Volume 1*, Robert von Ostertag-Institute, Berlin, pp. 257–60.
Pedrosa-Menabrito, A. and Regenstein, J.M. (1988) Shelf-life extension of fresh fish: A review. I. Spoilage of fish. *J. Food Qual.*, **11**, 117–27.

Peixotto, S.S., Finne, G., Hanna, M.O. and Vanderzant, C. (1979) Presence, growth and survival of *Yersinia enterocolitica* in oysters, shrimp, and crab. *J. Food Prot.*, **42**, 974–81.

Pelroy, G., Eklund, M.W., Paranjpye, R.N., Suzuki, E.M. and Peterson, M.E. (1982) Inhibition of *Clostridium botulinum* types A and E toxin formation by sodium nitrite and sodium chloride in hot-process (smoked) salmon. *J. Food Prot.*, **45**, 833–41.

Pelroy, G., Peterson, M., Paranjpye, R., Almond, J. and Eklund, M.W. (1994a) Inhibition of *Listeria monocytogenes* in cold-process (smoked) salmon by sodium lactate. *J. Food Prot.*, **57**, 108–13.

Pelroy, G., Peterson, M., Paranjpye, R., Almond, J. and Eklund, M.W. (1994b) Inhibition of *Listeria monocytogenes* in cold-process (smoked) salmon by sodium nitrite and packaging method. *J. Food Prot.*, **57**, 114–9.

Peterson, M.E., Pelroy, G.A., Paranjpye, R.N., Poysky, F.T., Almond, J.S. and Eklund, M.W. (1993) Parameters for control of *Listeria monocytogenes* in smoked fishery products: Sodium chloride and packaging methods. *J. Food Prot.*, **56**, 938–43.

Pigoff, G.M. and Tucker, B.W. (1990a) Methods that cause fish to die while struggling, in *Seafood: Effects of Technology on Nutrition*, Marcel Dekker, Inc., New York, pp. 26–7.

Pigoff, G.M. and Tucker, B.W. (1990b) Adding and removing heat, in *Seafood: Effects of Technology on Nutrition*, Marcel Dekker, Inc. New York, pp. 104–35.

Pitt, J.L. (1995) Fungi from Indonesian dried fish, in *Fish Drying in Indonesia. Proceedings of an international workshop held at Jakarta*, Indonesia on 9–10 February 1994 (eds B.R. Chapm and E. Highley), Canberra, A.T.C., Australian Centre for International Agricultural Research, ACIAR Proceedings, Vol. 59, pp. 89–96.

Platt, M.W., Rich, M.D. and McLaughlin, J.C. (1995) The role of chitin in the thermoprotection of *Vibrio cholerae*. *J. Food Prot.*, **58**, 513–4.

Popovic, T., Olsvik, O., Blake, P.A. and Wachsmuth, K. (1993) Cholera in the Americas: foodborne aspects. *J. Food Prot.*, **56**, 811–21.

Pothuri, P., Marshall, D.L. and McMillin, K.W. (1996) Combined effects of packaging atmosphere and lactic acid on growth an survival of *Listeria monocytogenes* in crayfish tail meat at 4°C. *J. Food Prot.*, **59**, 253–6.

Poulter, N.H. and Nicolaides, L. (1985) Quality changes in Bolivian Fresh-water fish species during storage in ice, in *Spoilage of Tropical Fish and Product Development* (ed A. Reilly), FAO Fish. Rep. 317 Suppl, pp. 11–28.

Pullela, S., Fernandes, C.F., Flick, G.J., Libey, G.S., Smith, S.A. and Coale, C.W. (1998) Indicative and pathogenic microbiological quality of aquacultured finfish grown in different production systems. *J. Food Prot.*, **61**, 205–10.

Ragelis, E.P. (1984) Ciguatera seafood poisoning: An overview, in *Seafood Toxins* (ed E.P. Ragelis), American Chemical Society, Washington, DC, pp. 25–36.

Rawles, D., Flick, G., Pierson, M., Diallo, A., Wittman, R. and Croonenberghs, R. (1995) *Listeria monocytogenes* occurrence and growth at refrigeration temperatures in fresh blue crab (*Callinectes sapides*) meat. *J. Food Prot.*, **58**, 1219–21.

Reddy, N.R., Armstrong, D.J., Rhodehamel, E.J. and Kautter, D.A. (1992) Shelf-life extension and safety concerns about fresh fishery products packaged under modified atmospheres: a review. *J. Food Safety*, **12**, 87–118.

Reddy, N.R., Paradis, A., Roman, M.G., Solomon, H.M. and Rhodehamel, E.J. (1996) Toxin development by *Clostridium botulinum* in modified atmosphere-packaged fresh tilapia fillets during storage. *J. Food Sci.*, **61**, 632–5.

Reilly, P.J.A. and Twiddy, D.R. (1992) *Salmonella* and *Vibrio cholerae* in brackishwater cultured tropical prawns. *Int. J. Food Microbiol.*, **16**, 293–301.

Reilly, A., Dangla, E. and Dela Cruz, A. (1986) Postharvest spoilage of shrimp (*Penaeus monodon*), in *The First Asian Fisheries Forum* (eds J.L. Maclean, L.B. Dizon and L.V. Hosillos), Asian Fisheries Society, Manila, Philippines.

Reilly, P.J.A., Twiddy, D.R. and Fuchs, R.S. (1992) Review of the occurrence of *Salmonella* in cultured tropical shrimp. *FAO Fisheries Circular No. 851*, United Nations FAO, Rome.

Richards, G.P. (1988) Microbial purification of shellfish: a review of depuration and relaying. *J. Food Prot.*, **51**, 218–51.

Ridley, S.C. and Slabyj, B.M. (1978) Microbiological evaluation of shrimp (*Pandalus borealis*) processing. *J. Food Prot.*, **41**, 40–3.

Ringø, E. and Gatesoupe, J. (1998) Lactic acid bacteria in fish: a review. *Aquaculture*, **160**, 177–203.

Ripabelli, G., Sammarco, M.L., Grasso, G.M., Fanelli, I., Crapioli, A. and Luzzi, I. (1999) Occurrence of *Vibrio* and other pathogenic bacteria in *Mytilus galloprivincialis* (mussels) harvested from Adriatic Sea, Italy. *Int. J. Food Microbiol.*, **49**, 43–48.

Ritchie, A.H. and Mackie, I.M. (1980) The formation of diamines and polyamines during storage of mackerel (*Scomber iscombris*), in *Advances in Fish Science and Technology* (ed. J.J. Cornell), Fishery News Books, Ltd. Farnham, UK.

Rogers, R.C., Cuffe, R.G.C.J., Cossins, Y.M., Murray, D.M. and Bourke, A.T.C. (1980) The Queensland cholera incident of 1977. 2. The epidemiological investigation. *Bull. WHO*, **58**, 665.

Rørvik, L.M., Yndestad, M., Caugant, D.A. and Heidenreich, B. (1992) *Listeria monocytogenes* and *Listeria* spp.: Contamination in a salmon slaughtery and smoked salmon processing plant, in *Proc. 3rd World Congr. Foodborne Infections and Intoxications, Volume 2*, Robert von Ostertag-Institute, Berlin, p. 1090.

Ruple, A.D. and Cook, D.W. (1992) *Vibrio vulnificus* and indicator bacteria in shellstock and commercially processed oysters from the Gulf Coast. *J. Food Prot.*, **55**, 667–71.

Ruttenber, A.J., Weniger, B.G., Sorvillo, F., Murray, R.A. and Ford, S.L. (1984) Diphyllobothriasis associated with salmon consumption in Pacific coast states. *Am. J. Trop. Med. Hyg.*, **33**, 455–59.

Saheki, K., Kobayashi, S. and Kawanishi, T. (1989) *Salmonella* contamination of eel culture ponds. *Nippon Suisan Gakkaishi*, **55**, 675–679.

Sakazaki, R. (1969) Halophilic vibrio infections, in *Food-Borne Infections and Intoxications* (ed H. Riemann), Academic Press, New York, pp. 115–129.

Sedik, M.F., Roushdy, S.A., Khalafalla, F.A. and Awad, H.A.E. (1991) Microbiological status of Egyptian prawn. *Die Nahrung*, **35**, 33–38.

Sera, H. and Ishida, Y. (1972) Bacterial flora in the digestive tract of marine fish. III. Classification of isolated bacteria. *Nippon Suisan Gakkaishi*, **38**(8), 853–858.

Shamshad, S.I., Zuberi R. and Quadri, R.B. (1992) Bacteriological status of seafood marketed in Karachi, Pakistan, in *Quality Assurance in the Fish Industry* (eds H.H. Huss, M. Jakobsen and J. Liston), Elsevier Science Publishers, Amsterdam, pp. 315–319.

Shandera, W.X., Hafkin, B., Martin, D.L., Taylor, J.P., Maserang, D.L., Wells, J.G., Kelly, M., Ghandi, K., Kaper, J.B., Lee, J.V. and Blake, P.A. (1983) Persistance of cholera in the United States. *Am. J. Trop. Med. Hyg.*, **32**, 812–17.

Shewan, J.M. (1971) The microbiology of fish and fishery products—a progress report. *J. Appl. Bacteriol.*, **34**, 299–315.

Shewan, J.M. (1977) The bacteriology of fresh and spoiling fish and some related chemical changes induced by bacterial action, in *Handling Processing and Marketing of Tropical Fish*, Tropical Products Institute, London, pp. 51–66.

Shiflett, M.A., Lee, J.S. and Sinnhuber, R.O. (1966) Microbial flora of irradiated Dungeness crabmeat and Pacific oysters. *Appl. Microbiol.*, **14**, 411–15.

Shimodori, S., Moriya, T., Kohashi, O., Faming, D. and Amako, K. (1989) Extraction from prawn shells of substances cryoprotective for *Vibrio cholerae*. *Appl. Environ. Microbiol.*, **55**, 2726–28.

Shultz L.M., Rutledge, J.E., Grodner, R.M. and Biede, S.L. (1984) Determination of the thermal death time of *Vibrio cholerae* in blue crabs (*Callinectes sapidus*). *J. Food Prot.*, **47**, 4–6.

Sikorski, Z.E., Kolakowska, A. and Burt, J.R. (1990) Postharvest biochemical and microbial changes, in *Seafood: Resources, Nutritional Composition, and Preservation* (ed Z.E. Sikorski), CRC Press, Inc. Boca Raton, FL, pp. 55–76.

Simidu, W. and Hibiki, S. (1955) Studies on the putrefaction of aquatic products. XXIII On the critical concentration of poisoning for histamine. *Bull. Japan. Soc. Sci. Fish.*, **21**, 365–367.

Simidu, W., Kaneko, E. and Aiso, K. (1969) Microflora of fresh and stored flatfish *Kareius bicoloratus*. *Nippon Suisan Gakkaishi*, **35**(1), 77–82.

Singh, B.R. and Kulshrestha, S.B. (1993) Prevalence of *Shigella dysentriae* group A type in fresh water fishes and seafood. *J. Food Sci. Technol.*, **30**, 52–53.

Slabyj, B.M., Dollar, A.M. and Liston, J. (1965) Post irradiational survial of *Staphylococcus aureus* in seafoods. *J. Food Sci.*, **30**, 344–350.

Slater, P.E., Addiss, D.C., Cohen, A., Leventhal, A., Chassis, G., Zehavi, H., Bashari, A. and Costin, C. (1989) Foodborne botulism: an international outbreak. *Int. J. Epidemiol.*, **18**, 693–96.

Smith, A.W., Skilling, D.E., Barlough, J.E. and Berry, E.S. (1986) Distribution in the North Pacific Ocean, Bering Sea and Artic Ocean of animal populations known to carry pathogenic calici viruses. *Dis. Aquatic Org.*, **2**, 73–80.

Smith, E.A., Grant, F., Ferguson, C.M.J. and Gallacher, S. (2001) Biotransformation of paralytic shellfish toxin by bacteria isolated from bivalve molluscs. *Appl. Environ. Microbiol.*, **67**, 2345–53.

Spanggaard, B., Jørgensen, F., Gram, L. and Huss, H.H. (1993) Antibiotic resistance in bacteria isolated from three freshwater fish farms and an unpolluted stream in Denmark. *Aquaculture*, **115**, 195–207.

Spanggaard, B., Huber, I., Nielsen, J., Sick, E.B., Bressen Pipper, C., Martinussen, Slierendrecht, W.J. and Gram, L. (2001) The probiotic potential against virbiosis of the indigenous microflora of rainbow trout. *Environ. Microbiol.*, **3**, 755–65.

Statham, J.A. and Bremner, H.A. (1989) Shelf-life extension of packaged seafoods: a summary of a research approach. *Food Technol. Aust.*, **41**, 614–20.

Stelma, G.N., Jr. and McCabe, L.J. (1992) Nonpoint pollution from animal sources and shellfish sanitation. *J. Food Prot.*, **55**, 649–56.

Sugita, H., Iwata, J., Miyajima, C., Kubo, T., Noguchi, T., Hashimoto, K. and Deguchi, Y. (1989) Changes in microflora of a puffer fish *Fugu niphobles* with different water temperatures. *Marine Biol.*, **101**, 299–304.

Suñén, E., Acebes, M. and Fernádez-Astorga, A. (1995) Occurrence of potentially pathogenic vibrios in molluses (mussels and clams) from retail outlets in the north of Spain. *J. Food Safety*, **15**, 275–81.

Surendran, P.K., Joseph, J., Shenoy, A.V., Perigreen, P.A., Mahadeva, I. and Gopakumar, K. (1989) Studies on spoilage of commercially important tropical fishes under iced storage. *Fish. Res.*, **7**, 1–9.

Surendran, P.K., Mahadeva, I.K. and Gopakumar, K. (1985) Succession of bacterial genera during iced storage of three species of tropical prawns, *Penaeus indicus, Matapenaeus dobsoni* and *M. affinis*. *Fish. Technol.*, **22**, 117–20.

Surkiewicz, B.F., Hyndman, J.B. and Yancey, M.V. (1967) Bacteriological survey of the frozen prepared foods industry. II. Frozen breaded raw shrimp. *Appl. Microbiol.*, **15**, 1–9.

Swartzentruber, A., Schwab, A.H., Duran, A.P., Wentz, B.A. and Read, R.B. Jr. (1980) Microbiological quality of frozen shrimp and lobster tail in the retail market. *Appl. Environ. Microbiol.*, **40**, 765–69.

Taylor, S.L. (1986) Histamine food poisoning:toxicology and clinical aspects. *CRC Crit. Rev. Toxicol.*, **17**, 91–128.

Taylor, S.L. (1988) Marine toxins of microbial origin. *Food Technol.*, **42**(3), 94–8.

Taylor, F.J.R. (1990) Red tides, brown tides and other harmful algal blooms: the view into the 1990's, in *Toxic Marine Phytoplankton* (eds E. Graneli, B. Sundstrom, L. Edler and D.M. Anderson), Elsevier, Amsterdam, pp. 527–33.

Taylor, B.C. and Nakamura, M. (1964) Survival of shigellae in food. *J. Hyg., Camb.*, **62**, 303–11.

Taylor, S.L. (1986) Histamine food poisoning:toxicology and clinical aspects. Telzak, E.E., Bell, E.P., Kautter, D.A., Crowell, L., Budnick, L.D., Morse, D.L. and Schultz, S. (1990) An international outbreak of type E botulism due to uneviscerated fish. *J. Infect. Dis.*, **161**, 340–42.
Teunis, P., Havelaar, A., Vliegenthart, J., Roessink, C. (1997) Risk assessment of *Campylobacter* species in shellfish: identifying to unknown. *Water Sci. and Technol.*, **35**, 29–34.
Thrower, S.J. (1987) Handling practices on inshore fishing vessels: effect on the quality of finfish products. *CSIRO Food Res.*, **47**, 50–55.
Tretsven, W.I. (1964) Bacteriological survey of filleting processes in the Greater Northwest. III. Bacterial and physical effects of pughing fish incorrectly. *J. Milk and Food Technol.*, **27**, 13–17.
Truelstrup Hansen, L., Drewes Røntved, S. and Huss, H.H. (1998) Microbiological quality and shelf life of cold-smoked salmon from three different processing plants. *Food Microbiol.*, **15**, 137–150.
Turner, K., Porter, J., Pickup, R. and Edwards, C. (2000) Changes in viability and macromolecular content of long-term batch cultures of *Salmonella typhimurium* measured by flow cytometry. *J. Appl. Microbiol.*, **89**, 90–99.
Twiddy, D.R. (1995) Antibiotic-resistant human pathogens in integrated fish farms. *ASEAN Food J.*, **10**, 22–29.
Vanderzant, C., Mroz, E. and Nickelson, R. (1970) Microbial flora of Gulf of Mexico and pond shrimp. *J. Milk and Food Technol.*, **33**, 346–50.
Vanderzant, C., Nickelson, R. and Judkins, P.W. (1971) Microbial flora of pond reared brown shrimp (*Peneaus aztecus*). *Appl. Microbiol.*, **21**, 916–21.
Vanderzant, C., Thompson, C.A., Jr. and Ray, S.M. (1973) Microbial flora and level of *Vibrio parahaemolyticus* of oysters (*Crassostrea virginica*), water and sediment from Galveston Bay. *J. Milk Food Technol.*, **36**, 447–52.
van Spreekens, K. (1987) Histamine production by the psychrophilic flora, in *Seafood Quality Determination* (eds D.E. Kramer and J. Liston), Elsevier Science Publisher, Amsterdam, pp. 309–18.
Varma, P.R.G., Iter, T.S.G., Joseph, M.A. and Zacharia, S. (1989) Studies on the incidence of *Vibrio cholerae* in fishery products. *J. Food Sci. and Technol.*, **26**, 341–42.
Venkataraman, R. and Sreenivasan, A. (1953) The bacteriology of freshwater fish. *Ind. J. Med. Res.*, **41**, 385–92.
Wainwright, R.B., Heyward, W.L., Middaugh, J.P., Hatehway, C.L., Harpster, A.P. and Bender, T.R. (1988) Food-borne botulism in Alaska, 1947–1985: epidemiology and clinical findings. *J. Infect. Dis.*, **157**, 1158–62.
Walker, P., Cann, D. and Shewan, J.M. (1970) The bacteriology of "scampi" (*Nephrops norvegicus*). I. Preliminary bacteriological, chemical, and sensory studies. *J. Food Technol.*, **5**, 375–85.
Ward, D.R. (1989) Microbiology of aquaculture products. *Food Technol.*, **43**(11), 82–86.
Ward, D.R. and Baj, N.J. (1988) Factors affecting the microbiological quality of seafoods. *Food Technol.*, **42**(3), 85–89.
Watanabe, K. (1965) Technological problems of handling and distribution of fresh fish in southern Brazil, in *The Technology of Fish Utilization* (ed R. Kreuzer), Fishing News, London, pp. 44–46.
Weagant, S.D., Sado, P.N., Colburn, K.G., Torkelson, J.D., Stanley, F.A., Krane, M.H., Shields, S.C. and Thayer, C.F. (1988) The incidence of *Listeria* species in frozen seafood products. *J. Food Prot.*, **51**, 655–57.
Weber, J.T., Hibbs, R.G., Jr., Darwish, A., Mishu, B., Corwin, A.L., Rakha, M., Hatheway, C.L., el Sharkawy, S., el-Rahim, S.A., al-Hamd, M.F. *et al.* (1993) A massive outbreak of type E botulism associated with traditional salted fish in Cairo. *J. Infect. Dis.*, **167**, 451–54.
Wempe, J.W. and Davidson, P.M. (1992) Bacteriological profile and shelflife of white amur (*Ctenopharyngodon idella*). *J. Food Sci.*, **57**, 66–68, 102.
Wentz, B.A., Duran, A.P., Swartzentruber, A., Schwab, A.H., McClure, F.D., Archer, D., Read, R.B., Jr. (1985) Microbiological quality of crabmeat during processing. *J. Food Prot.*, **48**, 44–49.
West, P.A. (1989) The human pathogenic vibrios—A public health update with environmental perspectives. *Epidemiol. Infect.*, **103**, 1–34.
Wheeler, K.A., Hocking, A.D., Pitt, J.L. and Anggawati, A. (1986) Fungi associated with Indonesian dried fish. *Food Microbiol.*, **3**, 351–57.
Wheeler, K.A., Hocking, A.D. and Pitt, J.L. (1988) Effects of temperature and water activity on germination and growth of *Wallemia sebi*. *Trans. Br. Mycol. Soc.*, **90**, 365–68.
Whittle, K. and Gallacher, S. (2000) Marine toxins. *Br. Med. Bull.*, **56**, 236–53.
WHO (World Health Organization). (1995) Control of foodborne trematode infections. WHO Technical Report Series No. 849, Geneva, Switzerland.
WHO/FAO (World Health Organization/Food and Agricultural Organization). (2001) Joint FAO/WHO Expert Consultation on risk assessment of microbiological hazards in foods. Risk characterization of *Salmonella* spp. in eggs and broiler chickens and *Listeria monocytogenes* in ready-to-eat product. 30th April–4th May 2001, Rome, Italy (as of April 2002: http://www.fao.org/WAICENT/FAOINFO/ECONOMICS/ ESN/pagerisk/reportSL.pdf).
Williams, L.A. and LaRock, P.A. (1985) Temporal occurrence of *Vibrio* species and *Aeromonas hydrophila* in estuarine sediments. *Appl. Environ. Microbiol.*, **50**, 1490–95.
Wilson, I.G. and Morre, J.E. (1996) Presence of *Salmonella* spp. and *Campylobacter* spp. in shellfish. *Epidemiol.Infect.*, **116**, 147–53.
Wong, H.-C., Ting, S.-H. and Shieh, W.-R. (1992) Incidence of toxigenic vibrios in foods available in Taiwan. *J. Appl. Bacteriol.*, **73**, 197–202.
Wong, H.-C., Chen, L.-L. and Yu, C.-M. (1995) Occurence of vibrios in frozen seafoods and survival of psychrotrophic *Vibrio cholerae* in broth and shrimp homogenate at low temperature. *J. Food Prot.*, **58**, 263–67.

Wyatt, L.E., Nickelson, R. and Vanderzant, C. (1979) Occurrence and control of *Salmonella* in freshwater catfish. *J. Food Sci.*, **44**, 1067–1069, 1073.

Yamagata, M. and Low, L.K. (1995) Banana shrimp, *Penaeus merguiensis*, quality changes during iced and frozen storage. *J. Food Sci.*, **60**, 721–26.

Yoon, I.H., Matches, J.R. and Rasco, B. (1988) Microbiological and chemical changes of surimi-based imitation crab during storage. *J. Food Sci.*, **53**, 1343–1346, 1426.

Yu, C., Lee, A.M., Bassler, B.L. and Roseman, S. (1991) Chitin utilization by marine bacteria. A physiological function for bacterial adhesion to immobilized carbohydrates. *J. Biol. Chem.*, **266**, 24260–267.

Zhao, S. (2001) Antimicrobial resistance of *Salmonella* isolated from imported food. *Paper presented at the 101st Annual Meeting of the American Society for Microbiology*, Orlando, Florida, 19th–23rd May 2001.

4 Feeds and pet foods

I Introduction

A feeding-stuff may be defined as any component of a ration that serves some useful function (Church, 1979). Most ingredients of feeding-stuffs provide a source of one or more nutrients although some may be included to improve acceptability or as preservatives. Feeding-stuffs are usually classified into six categories (National Research Council, 1972):

- roughages include pasture and feed from green plants;
- dry forages and roughages (hay) and silage;
- concentrates used primarily for energy such as cereal grains, milling by-products, seed and mill screenings, animal, vegetable and marine fats, molasses, and others;
- protein concentrates containing more than 20% protein, including plant-protein sources (oilseed meals, corn, gluten, distillery and brewery by-products, and dried legumes); proteins of animal origin (meat meal, meat and bone meal, miscellaneous animal and poultry by-products, fish meal, milk and milk by-products); and single-cell sources;
- mineral and vitamin supplements; and
- non-nutritive additives (e.g. antibiotics, antioxidants, buffers, colors and flavors, emulsifying agents, enzymes, and bacterial preparations).

Feed is the major cost of intensive animal production systems and, therefore, the conditions of feed preparation and processing are controlled to supply an adequate nutritional diet with high efficiency of feed conversion while encouraging consumption without waste. Feed may be processed simply to improve the capability of machinery to handle it or to add other ingredients. Feeds also may be processed to alter the physical form or particle size, isolate specific components, improve palatability, detoxify poisons, or prevent spoilage (Church, 1979). Feedstuff and feed processing may involve a large variety of mechanical methods (e.g. grinding, crumbling or milling, blocking, cake making, pelleting, flaking, extruding), thermal processing (e.g. conditioning, dehydrating, heat processing, steaming, toasting), chemical processes (e.g. solvent extraction, acid preservation), and microbiological fermentation (e.g. silage making).

The quality of feeding-stuffs depends on available energy, digestibility, nutrient content, and absence of toxins. In addition, quality depends on microbiological condition because spoilage and decay of feedstuffs can impair the efficiency of feed utilization.

Feeds containing pathogens or toxins, especially mycotoxins, can be direct sources of animal disease. Epidemiological data have occasionally linked contaminated feed to human health problems as a result of handling animals or consuming animal-derived products. Human salmonellosis is the principal pathogen of concern. Reported outbreaks of listeriosis, botulism, or bovine spongiform encephalopathy (BSE) among livestock further emphasize the potential of contaminated animal feeds for disseminating pathogens (Hinton and Bale, 1990; Anonymous, 1996; Brown *et al.*, 2001).

The probability of human disease arising from toxicity of animal products as a result of mycotoxin ingestion by animals in feeds is dependent on both toxin and animal type. Ruminants, in particular, are able to detoxify mycotoxins quite effectively by microorganisms in their rumen. Levels of aflatoxin that must be ingested by various domestic animals to produce a residue level in edible portions of 1 μg/kg total aflatoxins are given in Table 4.1. This level, 1 μg/kg, is considered to be acceptable and is rarely

Table 4.1 Levels of aflatoxin in feeds resulting in detectable levels (1 μg/kg) of aflatoxin residue in edible portions of domestic animals

Animal	Tissue	Level in feed (μg/kg)	Residue (μg/kg)
Beef steer	Muscle	700	1
	Liver	14 000	1
	Kidney	5 500	1
Dairy cow	Milk	300	1
Pig	Muscle	400	1
	Liver	800	1
Laying hen	Egg	2 200	1
Chicken	Muscle	2 000	1
	Liver	1 200	1

Modified from Stoloff (1977).

reached. In particular, poultry are quite sensitive to aflatoxins, so many poultry-producing countries have monitoring programs for aflatoxins in poultry feeds, particularly when contaminated maize is a problem, and consequently, levels in poultry meat and eggs are usually low.

The major cause for concern is the possibility of aflatoxin M1 excretion in milk. This toxin is produced by hydroxylation of the dominant naturally occurring aflatoxin, B1, when it is ingested in feeds. Aflatoxin M1 has been detected in many countries where animals are fed compounded feeds (Van Egmond, 1989), particularly in cows with a high milk production (Veldman *et al.*, 1992a). Consequently, levels of aflatoxin permitted in feeds for dairy cows are very low in developed countries (Van Egmond, 1989). The potential also exists for mycotoxins to be present in other dairy products, especially cheese, as a result of mycotoxins in feeds (Scott, 1989).

Monogastric animals are less able to detoxify mycotoxins than ruminants. Pigs, in particular, are sensitive to mycotoxins in feeds, as little as 1–5 mg/kg of the trichothecene toxin deoxynivalenol (vomitoxin) can cause feed refusal and vomiting (Vesonder and Hesseltine, 1981; D'Mello *et al.*, 1999). Rapid metabolization and low transmission rates into tissues ensure that carry-over of trichothecene toxins into edible portions is usually not a problem (Prelusky, 1994; Bauer, 1995). However, ochratoxin A, which is commonly produced in grains in north temperate climates by the growth of *Penicillium verrucosum* during storage, accumulates in kidneys and depot fats in pigs (Hald, 1991a). In areas of Europe where the consumption of pig meat is high, the potential for human disease has to be controlled (Höhler, 1998); many Europeans carry detectable levels of ochratoxin A in their blood (Hald, 1991b; Zimmerli and Dick, 1995).

It is not feasible to cover all types of feeding-stuff that may be used for animal feeding due to their tremendous variety. Rather, the text will focus on a few major ingredients, which may be indirect sources of human pathogens or mycotoxins. Categories of feeding-stuffs considered will include roughage (particularly silage), ingredients of animal origin (meat meal, meat and bone meal, fish meal, and other specific animal by-products), compounded feeds, and pet foods.

II Roughages

Roughage consists of plant material, excluding the seeds and roots, which is used for feeding grazing and browsing animals such as domestic ruminants and horses. Roughage is a bulky feed that has low weight per unit of volume. Roughages are highly variable in physical and chemical composition and nutritional quality. They range from very good (lush young grass, legumes, and high-quality silage) to very poor nutrient sources (straws, hulls, and some browse) (Church, 1979). Plant materials may be used with no preparation (e.g. pasture, grazed forage) or collected and fed loosely on the ground or in feeding troughs.

Roughages (e.g. hay) can be stored dry for feeding animals when grazing is not possible, i.e. in winter or during times of drought. In traditional hay making, herbage is cut at an optimal stage of development, field dried, and then moved to storage either in loose form or, more often, in bales. Chopping, grinding, pelleting, or cubing can facilitate transportation, handling, and feeding. In some instances, drying may be completed in the barn by circulating hot air through the hay. Other crops (e.g. alfalfa, maize, and beet pulp) are also dried for storage and subsequent feeding. Some chopped herbage is dried in rotating drums with hot air (800–1000°C in modern high-temperature processes). After drying, the material is ground and stored in bulk, under inert gas, pressed or pelleted.

A large quantity of herbage is converted to silage by anaerobic fermentation. Whole maize plants, grass, and legumes are widely used to make silage. Ensiling can be also used for a large variety of herbaceous material such as waste from canneries, processing food crops (sweet corn, green beans, green peas), or vegetative residues (e.g. beet tops).

A *Effects of processing on microorganisms*

When making hay, the moisture content is reduced to permit storage without marked nutritional change or microbial activity. When cut, herbage usually contains 70–80% moisture that must be reduced to 15–20% for storage. Rapid drying to 14–15% moisture results in the least change in chemical composition and microbial activity. Drying, however, does not inactivate bacterial or fungal spores and many non-spore-forming microorganisms may survive.

Ensiling involves a natural anaerobic fermentation by lactic acid producing bacteria to convert soluble carbohydrates to lactic acid and, thereby, decrease the pH and inhibit enzyme and microbial activity (McDonald *et al.*, 1991; Driehuis and Oude Elferink, 2000). If the pH is low enough (about 3.8–4.5, depending on crop properties) and the silage is stored under oxygen-limiting conditions, its longevity increases. While making silage, the herbage is harvested, chopped, and placed into silos where it gets packed and sealed to restrict oxygen and encourage a desirable fermentation.

The initial dominant microbial population of standing or freshly harvested forage crops are various aerobic microorganisms, but they do not contribute to the silage fermentation and their growth is inhibited soon after the silo is sealed. During the early phase of ensiling, different obligate or facultative anaerobes (lactic acid bacteria, enterobacteria, clostridia, and yeasts) compete for available nutrients. In well-preserved silage, lactic acid bacteria rapidly dominate the fermentation. Lactic acid bacteria are present in relatively low numbers (usually 10^3–10^5 cfu/g) on green forage. Under normal processing conditions, they multiply rapidly and in less than 8 days reach levels up to 10^{10} cfu/g. These bacteria are mainly lactobacilli (facultatively heterofermentative *Lactobacillus plantarum*, *Lb. casei*, *Lb. curvatus* or obligate heterofermentative *Lb. brevis*, *Lb. buchneri*) and to a lesser extent species of *Enterococcus*, *Leuconostoc*, and *Pediococcus*. They convert the soluble carbohydrates to lactic acid (up to 10% of total dry matter in well-preserved silage), and to a lesser extent, to acetic acid and propionic acid. The pH decreases to about 3.8–4.5, depending on crop properties, which eventually stops the fermentation. The presence of volatile fatty acids (e.g. acetic, propionic acid, and butyric acid) in silage assists in inhibiting fungi when silage is exposed to air (Moon, 1983; Driehuis *et al.*, 1999).

B *Spoilage*

When making hay, appreciable changes may occur if drying is slow in the fields, stacks, or bales. Excessive moisture in the hay results in a microbial fermentation causing an increase in temperature, browning, and development of thermophilic fungi. Severe toxicity from the consumption of moldy hay sometimes occurs, more commonly in monogastric animals such as horses (Lacey, 1991).

When making silage, an abnormal fermentation may occur resulting in a slow or insufficient decrease in pH, thus permitting abundant growth of enterobacteria, yeasts, or clostridia. This may occur, for example, when ensiling a very wet material that is low in water-soluble carbohydrates or a highly buffered plant material (e.g. legumes). The two main sources of spores of clostridia in silage are soil and animal manure. During cutting and harvesting of a silage crop, contamination of the crop with soil particles is unavoidable. Spores of clostridia can survive the passage through the alimentary tract of livestock consuming the silage, leaving the animal in feces. In many situations, animal manure is used as organic fertilizer for silage crops. This practice adds to the pool of clostridia spores in the soil and on the crop. However, the initial level of spores is of only minor influence on the final level of spores under conditions that permit growth of clostridia in silage. *Clostridium* species typically associated with silage include both saccharolytic species (e.g. *Clostridium tyrobutyricum* and *Cl. butyricum*) and proteolytic species (e.g. *Cl. sporogenes* and *Cl. bifermentans*) (McDonald *et al.*, 1991). Growth of clostridia in silage is associated with a high pH, which may lead to instability of the silage, and relatively large amounts of butyric acid, ammonia, and biogenic amines, such as tryptamine and histamine, causing losses in nutritional value and low palatability (Van Os, 1997).

Another type of spoilage relates to the proliferation of (facultative) aerobic microorganisms as a result of infiltration of air. Aerobic spoilage of silage is manifested in two ways. Firstly, the deterioration of surface layers, often visible by the development of molds. Secondly, deterioration of material deeper inside the silo, usually observed as heating of the silage. Acid-tolerant yeasts, oxidizing residual sugars, and lactic acid as substrates usually initiate this deterioration process. As this process proceeds, the pH rises, which in turn allows the growth of many other spoilage microorganisms, including molds, bacilli, enterobacteria, and *Listeria* (Woolford, 1990; McDonald *et al.*, 1991).

C *Pathogens*

Hay and silage may contain relatively high levels of *Cl. botulinum*, particularly when containing soil or when the herbage has been collected from fields treated with sewage sludge and particularly poultry waste. There is evidence that *Cl. botulinum* is commonly found in cultivated or manured soil or sludge (Mitscherlich and Marth, 1984). Another possible origin of *Cl. botulinum* in hay or silage is the presence of a cadaver (e.g. of a mouse or bird), which entered the hay or silage during the harvesting operation. As *Cl. botulinum* does not grow below pH 5.3, growth of this microorganism in well-fermented silage is unlikely. However, growth in deteriorating parts cannot be excluded. Spores and toxins of *Cl. botulinum* have been detected in the outer layers of wrapped big bale silage (Ricketts *et al.*, 1984; Wilson *et al.*, 1995). Botulism has been diagnosed in cattle and horses fed "big bale silage" (Hinton and Bale, 1990). Botulism traced to silage, particularly in cattle, has been reviewed in depth by Roberts (1988) and by Kehler and Scholz (1996).

Escherichia coli O157 and other strains of *E. coli* that have been involved in cases or outbreaks of hemolytic enteritis and hemolytic uraemic syndrome (HUS) in humans may, in theory, be transmitted via grass and other unprocessed or re-contaminated feeding-stuffs. Survival for months and even multiplication of these strains in wet feed and feed-troughs has been demonstrated (Hancock *et al.*, 2001). However, the role of feed, and particularly different types of feed, has not been well established (Herriott *et al.*, 1998; Buchko *et al.*, 2000). Numbers of Enterobacteriaceae in silage decline sharply when the pH falls below 4.5. However, infiltration of oxygen during ensilage prolongs their survival (Donald *et al.*, 1995), and extensive growth of Enterobacteriaceae during aerobic deterioration of silage was described by Lindgren *et al.* (1985).

The role of silage in the transmission of *Listeria monocytogenes* to ruminants was documented in 1960s (Gray, 1960). Since then, others have found silage to be a major source of listeriae on the farm (Grønstøl, 1979). During normal silage production, the pH falls below 4.2, which is bacteriostatic for listeriae. If air is excluded, the listeriae rapidly die. If the fermentation is not optimal *L. monocytogenes* can multiply. *L. monocytogenes* has often been reported to occur in poor quality silage at levels of more than 12 000 cells/g (Fenlon, 1986). Furthermore, *L. monocytogenes* can survive in silage for years (Dijkstra, 1975).

Insufficient acidification and aerobic deterioration may be the most important factors permitting multiplication of *L. monocytogenes* in silage (Hird and Genigeorgis, 1990; Perry and Donnelly, 1990). The frequency of listeriae isolations increases with increasing silage pH. In one study, *L. monocytogenes* was isolated from 22% of the samples with pH <4, 37% of the samples with pH 4–5, and 56% with pH >5 (Graostal, 1979). In another study, listeriae were isolated from 7.9% of the 114 silage samples with pH below 4, 52.9% of the 70 samples with pH 5.0–5.9 and all five samples with pH above 6.0. *L. innocua* and *L. monocytogenes* (84.3% and 15.7%, respectively) were the only listeriae isolated (Perry and Donnelly, 1990). Factors affecting silage pH (i.e. dry matter content, available carbohydrates, and buffering capacity), the rate of the decrease in pH, and the stability of the silage (i.e. thorough packing of the ensiled material and an air-tight seal) influence the growth of *L. monocytogenes*. If oxygen can enter the ensiled material, listeriae can withstand acidity as low as pH 3.8 for long periods (Fenlon, 1989). An experiment demonstrated that for a silage at 26% dry matter, the crucial level of oxygen that permits an increase of *L. monocytogenes* to levels significant for pathogenesis lies between 0.5% and 0.1%. At lower levels of 0.1–0%, *L. monocytogenes* perishes (Donald *et al.*, 1993). Even in good-quality silage, oxygen may enter at the surface or along the sides, permitting growth of fungi, which may cause the pH to increase and, in turn, permit the growth of listeriae.

There are many sources for *L. monocytogenes* in silage. It can survive, and even multiply, in soil (Botzler *et al.*, 1974), and is present in rivers, lakes, canals, and surface waters, flooding meadows, rivers, lakes, and canals (Brackett, 1988). Consequently, *L. monocytogenes* is often found on pasture grasses (Weishimer, 1968; Weiss and Seeliger, 1975) and is likely to be present when vegetation is harvested for making silage. *L. monocytogenes* also has been recovered from raw and treated sewage and vegetation treated with sewage sludge (Al-Ghazaii and Al-Azawi, 1986). Wild birds and scavengers may be vectors for contaminating silage with *L. monocytogenes* (Fenlon, 1985).

A variety of toxigenic fungi have been isolated from moldy silage, but outbreaks of mycotoxicosis attributable to silage are surprisingly rare (Lacey, 1991; Scudamore and Livesey, 1998). Most well-authenticated toxicoses are attributable to *Aspergillus* species, especially *Aspergillus fumigatus* (Yamazaki *et al.*, 1971; Cole *et al.*, 1977; Lacey, 1991). Aflatoxins may occur in silage, but are not considered to be a major problem (Lacey, 1991).

D CONTROL (roughages)

Summary

Significant hazards	• *L. monocytogenes* in silage making.
Control measures	
Initial level of contamination (H_0)	• Do not use grass or other raw material on which animals with listeriosis were kept.
Increase of contamination level (ΣI)	• Assure proper fermentation, limit air, add fermentable carbohydrates, acids and/or starters.
Reduction of contamination (ΣR)	• Proper silage making with limited air infiltration and low pH (4.2 for silage with 25% dry matter) reduces the number of *L. monocytogenes* sufficiently.

(*continued*)

CONTROL (Cont.)

Summary

Testing	• Observation, smelling, and pH (4.2 for silage with 25% dry matter).
Spoilage	• Silage and other roughages may become moldy when it remains or gets too wet.

In general, microbial pathogens can be controlled in roughages by appropriate cultivation and rapid drying of the harvested vegetation. Special consideration should be given to fields treated with manure or sewage sludge. Machinery designed to crush the stems of plants having high water content (e.g. alfalfa) can be used to speed-up the drying process. Eliminating decayed or moldy material reduces the risk of animal disease and/or transmission to humans.

Although in theory, many animal and human pathogens may be present in roughage, the hazardous concern is *L. monocytogenes*, particularly while making silage. Certain steps can be taken during the ensiling process to avoid spoilage and proliferation of *L. monocytogenes*. Thus, when harvesting vegetation, inclusion of soil with the vegetation as much as possible should be avoided, and the vegetation should be chopped to provide a more uniform material and facilitate the release of plant juices containing water-soluble carbohydrates. Best results are obtained when the material is chopped in range 0.5–2 cm and then filled directly into a silo. Limiting the entry of air during ensiling encourages a good fermentation. To that aim, the material should be thoroughly packed as soon as possible after chopping and hermetically sealed, such as in a sealed tower silo or in a bunker and covered with plastic sheeting.

Under these conditions, the quality of the silage will depend mainly on the water content of the ensiled crops and pH control by appropriate additives. For most crops, a dry matter (solids) content of 25–45% is near optimal for silage making. If the dry matter is low, especially with lower than normal soluble carbohydrates (i.e. $\sim$15% of the dry matter), insufficient acid will be produced. This is particularly a problem for legumes, which have lower water-soluble carbohydrate levels and are more highly buffered than grasses. In areas where the collected material is very wet or where legume or grass–legume mixtures are ensiled, it is highly advisable to wilt the herbage before ensiling.

Wilting can be done mechanically in the field, much similar to hay making, and dried in barns with the aid of heat. Wilting increases the dry matter content and can significantly reduce the level of listeriae (Fenlon, 1989).

Silage fermentation can be controlled by additives, which can be categorized in four categories. The first category consists of additives that provide extra substrates for fermentation and comprises primarily molasses and polysaccharide-degrading enzymes. The second category consists of chemical additives that inhibit the growth of spoilage microorganisms during the acidification phase and comprises primarily products based on (salts of) formic acid. In addition, products containing nitrites or sulfites are available. In a few countries, products containing formaldehyde are allowed. The third category consists of additives that stimulate lactic acid fermentation and comprises cultures of lactic acid bacteria. Some of these products additionally contain polysaccharide-degrading enzymes. The fourth category consists of additives that inhibit aerobic deterioration and comprises both chemical products and bacterial cultures. The chemical products of this category usually contain compounds with antimycotic activity, such as propionic acid and benzoic acid, whereas the bacterial products include cultures of lactic acid bacteria, propionic acid bacteria, and *Bacillus* species.

Extensive research has been conducted on the use of biological additives such as lactic acid bacteria inoculants (Weinberg *et al.*, 1993). These inoculants ensure a rapid, efficient fermentation of the water-soluble carbohydrates into lactic acid by homolactic fermentation. This ensures a rapid decrease in pH and improved silage stability. Homofermentative lactic acid bacteria inoculants that have been recommended include *Lb. plantarum*, *Pediococcus acidilactici* and *Lactococcus lactis*, often with the addition of *Enterococcus faecium*. Many studies have shown the advantage of such inoculants for suppressing undesirable microorganisms by the production of lactic acid rather than bacteriocins (Perry and Donnelly, 1990). Various studies, however, have shown that silage inoculants that enhance homofermentative lactic acid fermentation can impair aerobic stability (Weinberg *et al.*, 1993). The background of this effect is that the use of such inoculants often results in a reduction in the concentration of undissociated acetic acid. Recently, it was discovered that the heterofermentative lactic acid bacterium, *Lb. buchneri*, is a very effective inhibitor of aerobic spoilage. This effect of *Lb. buchneri* appears mainly due to its capability to anaerobically degrade lactic acid to acetic acid and 1,2-propanediol, which in turn causes a significant reduction in numbers of fungi in silage (Driehuis *et al.*, 1999).

The growth and survival of *Listeria* in silage are dependent on the degree of anaerobiosis and on the pH. *L. monocytogenes* added to grass at ensiling rapidly disappeared under strict anaerobic conditions and at a pH lower than 4.4. However, at an oxygen tension of 0.5% (v/v) survival was prolonged, and growth was observed even at a pH as low as 4.2. Higher oxygen tensions strongly encouraged *L. monocytogenes* growth (Donald *et al.*, 1995). So, proper silage making is necessary to prevent survival or multiplication of *L. monocytogenes*.

Control of fungal development in silage usually relies on maintenance of anaerobic conditions. Careful attention to sealing of silos and other containers used for storage of silage is important. Most chemicals promoted as mold inhibitors are only marginally effective against common fungi at commercially feasible levels.

No single test is available for accurate assessment of silage quality. Smell and visual appearance are good methods for those familiar with silage. Silage should be green, not brown or black, have a firm texture without sliminess, be free of objectionable odors such as ammonia or butyric acid, and not moldy. The pH is commonly used, but by itself is unreliable, since the optimal pH for silage depends on the dry matter content. Good silage with 25% dry matter should have a pH of 4.2 or less. With low dry-matter content the pH should be lower (Church, 1979).

III Animal by-products

Feed ingredients derived from warm-blooded animals have been used for many years as supplemental protein sources. Raw materials include inedible offal, viscera, bones, blood, feathers and condemned carcasses or parts collected from slaughtering plants, trimmings and bones from meat processing plants and retail stores, and dead animals or combinations of these.

All such animal-protein materials are made into meals and other by-products (e.g. fats) by rendering. Rendering is a process that involves heat and other procedures to separate water, fat, and protein. The basic process involves collecting and preparing the raw material; heating generally between 110°C and 150°C removing the water; removing the fat by centrifuging, pressing, or solvent extraction; cooling; drying; milling; screening; sorting; blending; and storing. Many variations of these operations have been developed and are used according to the type of raw material, machinery, and the facility.

Meals derived from warm-blooded animals consist mainly of meat meal or meat and bone meal. These products have been defined by professional organizations and feed control officials. In practice, meat meal is the finely ground dry residue remaining after rendering carcasses or animal trimmings exclusive of hair, hooves, horn, hide trimmings, viscera, manure, feathers, hydrolyzed feathers, egg shells, or non-animal materials. If the product contains more than 4.4% of phosphorus, then it is designated as meat and bone meal. Nearly all meat meal and meat and bone meal are blended to obtain standardized blends of protein, fat and ash content. The blends usually have about 50%, 55%, or 60% protein, which determines the grade, and an ash content of about 24–36%. The fat content depends on the process used. Solvent extraction leaves a fat content of 2–4.5% whereas pressing leaves 7–12%. Moisture content is usually low, from 5% to a maximum of 10%. Meat meal and meat and bone meal are used as sources of protein in mixed feeds for monogastric farm livestock and pets. Blood meal and feather meal also are used as protein supplements. Blood meal is produced from dried, ground blood and has a protein content of 85%. Feather meal (hydrolyzed feather meal) produced by heating feathers under pressurized steam, has an average protein content of 85–87%. Since BSE has been linked to the use of processed animal proteins in animal feeding, the above-mentioned meals cannot be fed to all animals, which are kept, fattened or bred for the production of food according to the EU Regulation No. 1234/2003. In other countries, similar or less restrictive bans are implemented.

A *Effects of processing on microorganisms*

The initial microflora in the raw meat and poultry is described in Chapters 1 and 2. Inedible raw materials contain large numbers of microorganisms, and possibly, parasites, pathogenic bacteria, fungi, or viruses.

Before rendering, the raw materials are pre-crushed and reduced in size for cooking. The dust and aerosols generated during crushing can spread contaminating microbes throughout the factory, including areas where the finished product is handled (Swingler, 1982).

Most raw materials from warm-blooded animals are processed by dry rendering. After pre-crushing, the material is transferred into a cooker (or dry-melter), a steam-jacketed tank equipped with an internal agitator. During rendering, the material is batch-heated under pressure (generally at 115–150°C) until most of the moisture has been driven off as steam (20 min to 4 h according to temperature). The dried material is then emptied onto perforated pans to allow the fat to drain away. After further processing to remove additional fat, the meat meal is ready for sale (Wilder, 1971). If the fat is to be removed by solvent extraction, the raw material is ground and then mixed with solvent (70°C for 8 h). The solvent and fat then undergo an evaporating–condensing process to separate the fat and recycle the solvent. The extracted solids are heat-treated in a jacketed tank (110–120°C for 45–60 min) and finally washed with steam. When fat is extracted by high-speed centrifugation or pressing, the rendered material is not subjected to any further high temperatures. Whatever the process, defatting leaves the organic matter as a hard cake, which is milled to a meal, screened, sorted, blended with other meals or ingredients, stored and bagged.

The conditions of the heat treatment used for hot rendering are such that most bacteria and parasites are destroyed (Hess *et al.*, 1970). However, the time–temperature combinations used commercially for heating may vary greatly from 87.8°C for a few minutes to 140.6°C for 20 min (Swingler, 1982). In addition, it has been shown that the high-fat and low-water environment of batch dry rendered material will protect bacterial spores against thermal inactivation (Lowry *et al.*, 1979, quoted by Swingler, 1982). Therefore, some marginal processing conditions may result in non-sterile products containing spores or other heat-resistant biological material. After the heat treatment, other operations are performed at high temperature (e.g. fat extraction stage by solvent), which can destroy heat-sensitive contaminants. After the heating process, however, there are many opportunities for recontamination by cross-contamination with raw materials, from common machinery, dust, aerosols from grinders, flies, or rodents. As a result, microbial levels in the finished products can be frequently high.

In one study aimed at assessing the microbial load of several commercial batches of meat meal and meat and bone meal, it was found that the aerobic plate count was (per gram) from 3×10^4 to 3×10^6 sulfite-reducing clostridia 10^4; Enterobacteriaceae 10^6, enterococci 10^2 to 3×10^3; and molds 10–10^4 (Milanovic and Beganovic, 1974). Contamination with Enterobacteriaceae may vary widely depending upon the country of origin (Reusse *et al.*, 1976; Van Schothorst and Oosterom, 1984). The most serious problem related to post-heat treatment contamination is contamination with salmonellae.

B Spoilage

The various meals of animal origin are microbiologically stable products. They rely on low-water content and a_w for preservation. Commercial meat and bone meal has an a_w between 0.3 and 0.45 with a moisture content of about 5%. Such conditions do not support microbial growth provided the meals are stored in dry conditions. Thus, microbial spoilage is not normally an important factor. However, if the product gets wet (e.g. during transportation or storage), rapid microbial multiplication and deterioration may occur. Normally, spoilage results from molds capable of growth at relatively low a_w.

There is no evidence of significant microbiological problems with fats and oils because they are low in moisture, unsuitable as growth media, and have undergone extensive physical and chemical refining.

C Pathogens

The production of animal by-products and their use as feed ingredients may cause problems unless safe processes are employed and recontamination after heat processing is avoided. Animal disease (e.g. foot and mouth disease, swine fever, and swine vesicular disease) traced to meat meal (Hinton and Bale, 1990) has been recognized worldwide. For this reason, the treatment of animal products is controlled by legislation or regulations. As mentioned earlier, however, large variations in the time–temperature of processing and/or poor sanitation practices continue to result in the presence of pathogens. Epidemiological evidence suggests that anthrax and BSE have occurred in livestock due to inappropriate heat treatment. Salmonellae are of concern due to recontamination after rendering.

Bacillus anthracis, the cause of anthrax, is a spore-forming bacterium, which may survive insufficient time–temperature conditions. In some regions of the world, carcasses of diseased animals may be inappropriately heat-treated and *B. anthracis* has been isolated from bone meal (Morehouse and Wedman, 1961; Davies and Harvey, 1972).

Bovine spongiform encephalopathy was first recognized in the United Kingdom in 1986. Initial epidemiological studies indicated an extended common-source epidemic. Later studies, taking into account the disease's similarity to scrapie, identified the more likely source to be cattle exposed to a transmissible spongiform encephalopathies (TSE) agent from feed containing meat and bone meal derived from ruminants (Wilesmith *et al.*, 1988; WHO, 1992). A key factor explaining the emergence of the disease was changes in rendering practices in the United Kingdom from 1981–1982. Some rendering plants had replaced the batch process with continuous processing, which used a temperature of about 100°C. At the same time, fat extraction with solvents was replaced with high-speed centrifugation. After 1981, only 15% of the meat and bone meal had been produced by solvent extraction (Wilesmith *et al.*, 1991).

The former production system included several factors important for inactivation of TSE agents: higher time–temperature conditions for cooking the raw material, activity of the organic solvent, additional heat applied for solvent extraction, and steam injection under pressure to produce meals of low-fat content. The absence of such protective factors in the new process probably contributed to survival of the agents and spread of disease. It was observed that Scotland and Northern England counties using meals produced with batch-heating and solvent extraction had a lower incidence of BSE than Southern England where meals were prepared at lower temperatures and fat extraction by centrifugation (Garcia, 1992). The background and general guidance can be found in ACDP (1996) and Brown *et al.* (2001).

Information prepared by specialist groups on BSE, in particular the development of the epidemic, geographical incidence, nature of the disease and its transmission, precautions, and control measures has been published by the OIE (Office International des Epizooties) and the World Health Organization (Anonymous, 1996; WHO, 2000; WHO, 2002) and via websites listed in Chapter 1.

Table 4.2 Incidence of *Salmonella* in feeds of animal origin and compound feed

Material	No. of samples analyzed	Positive samples (%)	References
Meat and bone meal	982	21.3	Skovgaard and Nielsen (1972)
Fish meal	30	3.3	
Meat and bone meal	242	7	Patterson (1972)
Blood meal	36	5.6	
Poultry offal meal	101	9	
Feather meal	414	7.2	
Meat and bone meal	15	60	Jones *et al.* (1991)
Poultry meal	10	40	
Fish meal	18	33.3	
Meat meal	5276	2.2	Eld *et al.* (1991)
Bone meal	3205	3.1	
Feather meal	2624	0.6	
Fish meal 1988	893	16	Vielitz (1991)
Fish meal 1989	1564	24	
Rendered animal protein			
1990 winter	4935	30-9	Smittle *et al.* (1992)
1990 fall	5186	37	
1991 spring	6089	21	
1992 winter	6610	26	
Meat and bone meal	146	8.9	Jardy and Michard (1992)
Fish meal	92	4.3	
Compound feeds			
1999	2416	0.4	Anonymous (2002)
2000	2516	0.3	
2001	2616	0.2	

Clostridium perfringens is still an important meat- and poultry-borne pathogen (Craven *et al.*, 1999). The microorganism can be found in feeding-stuffs (Xylouri *et al.*, 1997) and can multiply in the intestinal tract and become a source of contamination during slaughter.

Salmonellae are the main pathogens of concern and their presence in meals from animal origin is abundantly documented in the scientific literature (Table 4.2). Rendering yields products free of salmonellae. However, after heating, contamination may occur. Raw animal-waste (meat, bones, offals, and feathers) frequently harbors salmonellae. Although the press cakes are free of *Salmonella* after heating, they will become recontaminated if contact with raw materials occurs. Raw materials will also contaminate the areas where they are unloaded and prepared for rendering. The extent of the contamination of these areas may be so high that normal hygienic practice is inadequate to prevent the entry of salmonellae into the plant and spread to finished product areas. In addition, raw materials are crushed or ground before cooking. Dust and aerosols generated by these operations may spread contamination throughout the factory, including where they can directly contaminate the flow of finished product (Quevedo and Carranza, 1966; Loken *et al.*, 1968). Dust and aerosols can also inoculate pockets of condensate or other moist spots in or near the processing line (Clise and Swecker, 1965). In a survey of a plant using hot solvent extraction, salmonellae were not isolated from the final product, although many isolations were made in the percolation areas (Timoney, 1968). There is an abundance of nutrients in rendering establishments. Thus, high numbers of microorganisms, including salmonellae, may be found in moist niches, establishing new loci of infection. As rendered products pass through cooling, draining, drying, milling, screening, mixing, bagging, and conveying to final product storage areas, varying amounts of highly contaminated material may slough-off from the growth niches and fall into the product (Gabis, 1990, quoted by Jones *et al.*, 1991). Insects such as flies or cockroaches can also transport salmonellae from such niches to the final product. Excrements of birds and rodents often test positive for salmonellae and can serve as vectors of contamination (Jones *et al.*, 1991).

Due to this large variety of environmental sources, and despite improvements in hygienic conditions in the processing plants, salmonellae are still found occasionally (Anonymous, 2002) and some plants have indigenous serovars that are very difficult to eradicate (Eld *et al.*, 1991). Similar problems are encountered in the production of feedstuffs of vegetable origin, such as rapeseed cakes and soy cakes.

Table 4.2 is a limited listing of surveys on the incidence of salmonellae in meals and a few other animal by-products.

Reported incidences vary widely partly due to real variation and partly due to differences in analytical methods. In dry products of this kind, salmonellae are injured and require resuscitation (pre-enrichment) (Van Schothorst *et al.,* 1979). The size of sample is another important factor contributing to variation. In one study, a significant decrease in the incidence of positive samples from a 1985-summer survey to a 1990-winter survey (55.1% and 30.9%, respectively) was attributed to the decrease in the size of the analytical sample from 75 to 25 g (Smittle *et al.,* 1992).

Reported estimates of the number of salmonellae in meals or other animal feed ingredients often range from 1 to 40/g, although higher numbers are found occasionally. No significant relationship between number of salmonellae and percent of positive samples has been demonstrated (Patterson, 1972; Jardy and Michard, 1992; Smittle *et al.*, 1992).

A wide variety of serotypes have been recovered from meals. Studies have traced certain serotypes isolated from animals back to the feeds consumed (Rowe, 1973; Mackenzie and Bains, 1976; Jones *et al.*, 1982). However, the serotypes of salmonellae most frequently isolated from feeds are often not those associated with salmonellosis in animals (HMSO, 1995; Veldman, 1995). In particular, the incidence of *Salmonella* Dublin and *S*. Typhimurium in feeds is very low. For example, the incidence of *S*. Typhimurium in both imported and domestic feed samples analyzed in the United States from 1974 to 1985 was less than 2% (Wagner and McLaughlin, 1986). This is low compared to their prevalence in cattle (Eld *et al.*, , 1991). The same occurs with *S*. Enteritidis, *S*. Thompson and *S*. Typhimurium, which are among the main invasive serotypes found in poultry. They are recovered at low frequency from poultry feeds (Hinton and Bale, 1990). Non-invasive serovars, however, are more prevalent (Williams, 1981). Consequently, animal feeds contaminated with salmonellae may be less important than one might think. Animal infection may originate from other sources with only a few serotypes originating from feed. Eliminating salmonellae from feeds will not guarantee the absence of salmonellae in animals (Oosterom *et al.*, 1982). Nevertheless, to assist the control of salmonellae efforts should be made to reduce contamination of animal by-products as this will avoid introduction of further salmonellae into the farm environment (Edel *et al.*, 1970; Davies *et al.,* 2001).

The practice of recycling animal waste as animal feed can also produce health risks. In particular, poultry manure (i.e. wet cage-layer excreta, dried poultry waste, ensiled poultry litter) is used for cattle feed. Concern over feeding such waste has been raised due to the risk of transmitting pathogens such as *Cl. botulinum* or salmonellae. *Clostridium botulinum* type C has been isolated from deep litter on which broiler chickens have been reared (Smart *et al.*, 1987; Neill *et al.*, 1989). Cattle have suffered botulism (usually type C) after being fed a feed compounded with poultry waste and/or poultry litter (Neill *et al.*, 1989).

Ample documentation indicates that salmonellae are found in poultry litter. Salmonellae isolation from litter has been found to be a reliable indicator of both flock and carcass contamination (Bhatia *et al.*, 1979). In Denmark, "sock-samples" are used to detect contaminated rearing and layer flocks. A "sock-sample" consists of 15-cm pieces of tube gauze mounted onto the footwear of poultry house inspectors. In one experiment, *S*. Typhimurium was isolated from wet cage-layer excreta used for feeding purpose, thereby recycling infection. However, it was not recovered from cage-layer excreta treated with propionic acid or acetic acid or in the mesenteric lymph nodes and gallbladders of cattle fed with feed compounded with poultry waste and/ or litter (Smith *et al.*, 1978).

D CONTROL (animal by-products)

Summary

Significant hazards	• Salmonella and, in certain regions, BSE.
Control measures	
Initial level of contamination (H_0)	• Do not use animals with anthrax or BSE, processed animal proteins or high-risk material (for BSE).
Reduction of contamination (ΣR)	• Heat the raw material at 130°C for 20 min at 3 bars to achieve a 3 decimal reduction of the BSE agent. This heat treatment will also reduce the number of *B. anthracis* and salmonellae to safe levels.
Increase of contamination level (ΣI)	• Assure adequate separation between raw and processed material, keep line-environment dry and free of salmonellae.
Testing	• Test environmental samples for the presence of salmonellae or determine the level of Enterobacteriaceae to detect niches where multiplication may have occurred.
Spoilage	• Spoilage is not a problem as long as the meal remains dry.

While producing meals derived from warm-blooded animals, the primary concern is to avoid passing heat-resistant pathogens such as *B. anthracis* and BSE from feed to cattle to man and to preventing recontamination with salmonellae. It is nearly impossible to control the initial levels of the first two hazards, because their multiplication in the host is the cause of their slaughtering or death and thus for rendering the animals. However, infected carcasses or parts thereof (see further on) can be sent to incineration plants instead of rendering plants in order to reduce the initial level of these hazards. In the production of meat and bone meal, it is very important to inactivate these pathogens in animal tissues through appropriate processing procedures.

To destroy *B. anthracis* spores, a heat treatment of 140°C for 20 min has been established in some countries (Genigeorgis and Riemann, 1979, quoted by Swingler, 1982).

As regards the BSE agent, a treatment of 133°C, for 20 min at 3 bars (of particles <50-mm in size), or equivalent conditions are necessary to reduce the infectivity by 1000-folds (SSC, 1999). However, in BSE affected countries, where the risk of heavily contaminated animal by-products with BSE agent is significant, treatment of the raw material is not considered as a reliable method for the eradication of the BSE epidemic and ensuring feed safety. Therefore, the approach that has been taken is based on the minimization of the risk of BSE agent entering the feed chain. For this purpose, the United Kingdom and subsequently the EU countries have introduced a number of legislative measures aiming at preventing exposure of cattle to potentially contaminated feedstuff. First in 1988, United Kingdom prohibited the feeding of ruminant-derived protein to any ruminant animals. Subsequently, the EU Commission took, in 1994, the decision of banning the use of mammalian derived proteins to any ruminant animals.

However, despite these measures, and the fact that the incidence of BSE began to decline after 1993, the ban did not prove to be fully effective in containing the BSE epidemic. A number of cows born after the ban (BAB) were still affected. Subsequent investigation showed that at a number of points along the feed chain (feed mill, delivery lorries, and farms), cross-contamination occurred between the ruminant-derived protein for use in pig and poultry rations and the cattle feedstuffs. Consequently, in

March 1996, the United Kingdom banned the use of mammalian meat and bone meal (MBM) from the feed of all farm animals. The EU also introduced similar measures and in December 2000, the Council of the European Union prohibited, on a temporary basis, the feeding of processed animal proteins to all farmed animals. The EU Regulation No. 1234/2003 banned most commercially used processed animal proteins for feeding all animals which are kept, fattened or bred for the food production. Under certain conditions, exceptions are made for feeding fish meal, gelatine and hydrolyzed proteins to non-ruminants, and milk and milk products to all farm animals. Considering that cats were also shown to be susceptible to BSE agent, this provision was also made for pet food. Preventive measures have also been introduced in some other countries. For instance, in August 1997, the US Food and Drug Administration prohibited the use of most mammalian protein in the manufacture of animal feeds given to ruminants.

In addition to the feed ban, other public health measures have been introduced in BSE affected countries. Depending on the country, these measures include (EC, 2001a,b) the following:

- Statutory slaughter of all suspected cases of BSE, and depending on the country, all other ruminants of the holding or the cohort of the animal (same class of age and exposed to the same risk) in which the disease was confirmed.
- Ban of use of specified risk material (SRM), which refers to tissues of cattle, sheep, and goats that are known to, or might, harbor detectable BSE infectivity in infected animals. This may vary depending on the country. The European Commission defines SRM as: (i) the skull including the brain and eyes, the tonsils, the spinal cord of bovine animals aged over 12 months, the intestines from the duodenum to the rectum of bovine animals of all ages; (ii) the skull including the brains and eyes, the spinal cord of ovine and caprine animals aged over 12 months or the ones that have a permanent incisor erupted through the gum, and the spleen of ovine and caprine animals of all ages (additional provision has been made for United Kingdom and Portugal).
- Interdiction of use of bones of the head, vertebral column of bovine, ovine and caprine animals in the manufacture of mechanically recovered meat.
- Surveillance of BSE and screening all bovines aged over 30 months for BSE and 24 months for cases of special emergency slaughter.

Import restriction for countries where BSE is confirmed or likely. A number of measures are also recommended for ensuring safety of gelatine, tallow, cosmetic, and pharmaceuticals. The fundamental principle in ensuring safety of these products is safe sourcing of raw material (WHO, 2001).

Dry rendering or wet rendering, where the products are subjected to conditions of high temperature and pressure appropriate to inactivate spores, will result in the destruction of salmonellae. Equipment used may be quite sophisticated and include, for example, an automatic feed mechanism coupled with an exit gate control that allows only cooked material to exit the cooker or end point controls to signal the point in the cooking cycle at which the products have been sufficiently cooked and dried (Wilder, 1971). With all commonly used rendering processes, a sufficient reduction in numbers of salmonella can, and should, be obtained. In case of doubt, process validation and trials should be carried out. Moreover, additional treatments may be given to reduce the numbers of salmonellae that may have recontaminated the product during further drying and processing. Heating and pelleting can reduce the level of salmonellae 100–1000-fold (Stott *et al.*, 1975); expansion and extrusion are even more efficient. Ionizing irradiation is effective for destroying salmonellae in these products (Mossel *et al.*, 1967). An alternative technique is to add chemical additives to the feed (Hinton and Linton, 1988). Such processes, which require specific control, are discussed in Section V.

Control procedures for preventing recontamination involve complete separation of raw and processed material, good hygiene, and sanitation procedures. Complete separation of raw material and finished product is critical to avoid reintroducing salmonellae into the heated or extruded material.

The areas for raw and processed materials should be physically separated by a leak-proof wall. In addition, the two areas should have completely separate equipment and personnel for processing and maintenance.

Good hygiene and sanitation procedures are important. The processed material area should be kept scrupulously clean. Special care must be taken to prevent spreading dust from raw materials area to the finished products. Steam and water should be kept away from the processed material. Condensation of water should be avoided, especially in places where a fall in temperature is likely to occur during processing or transportation. If this is a major problem, control may require increasing air movement or insulating certain critical areas. Dry cleaning is preferred to wet cleaning in order to prevent possibility for growth of salmonellae in the environment.

Conveying systems should be as short as possible. Fresh air for cooling work rooms should not be from within the plant but from outside and as far as possible from the raw material area. Sweepings should not be reprocessed. Persons who enter the "clean" finished product area should put on clean clothing, wash hands, and clean the soles of their shoes (or change their shoes). Birds, flies, rodents, and other vermin should be eliminated from this area.

Finished product parameters for animal by-products depend upon the expected storage time. For storage of 2–3 years, an a_w of 0.65 and moisture content of 8.5% or less is necessary. For storage of up to 5 months, an a_w of 0.7 and moisture content of 9% is preferred. If storage will not exceed 5 weeks, an a_w of 0.75 and moisture content of 10.5% is acceptable (Thalmann and Wolf, quoted by Boloh, 1992). Increasing numbers of the hazards of concern due to multiplication is prevented by these levels of water activity.

Verification of the plant's performance is necessary. Testing for the presence of salmonellae in environmental samples should be done at frequent intervals (MAFF, 1989). Positive findings should be followed by a thorough investigation to eliminate the source of contamination. Samples of end products may be directly examined for salmonellae. Since salmonellae are not uniformly dispersed in the material, occur in low numbers, and may be injured as a consequence of heat stress or osmotic shock, detection depends largely on the method of sampling, size of samples, and use of pre-enrichment procedures. Instead of testing for salmonellae, samples may be cultured for indicators. In The Netherlands, regulations required analysis for *Cl. perfringens* and a review of the thermal processing to verify effectiveness of the cook. In certain countries, a test for Enterobacteriaceae has been advocated to locate potential contamination points along the processing line (Quevedo, 1965; Van Schothorst and Oosterom, 1984). It may also be applied to the final meal or feed to assess cross-contamination and growth (Cox *et al.*, 1988). In one survey, the correlation coefficient between Enterobacteriaceae and salmonellae was 0.81 with confidence limits (95%) of 0.35–0.95. It was concluded that Enterobacteriaceae are not good indicators for the presence of salmonellae, but they can be used to assess the hygienic quality of animal by-products (Michanie *et al.*, 1985).

IV Fish meal

Between 20% and 30% of the total world catch of fish is used to manufacture animal feeds. The greater tonnage comes from processing whole fish that are not suitable for human consumption because they are too bony, too oily, or otherwise unsatisfactory; these fish are sometimes called "industrial fish". Examples of fish used for fish meal include capelin, menhaden (*Brevoortia* spp.), sand eel, sprat, Norway pout, blue whiting, horse mackerel, Atlantic herring (*Clupea* spp.), anchovy (*Engraulis* spp.), pilchard, and related species. In the United States, for example, the entire menhaden catch goes to rendering. A secondary source is the waste (offal) from fish and shellfish operations. South America, especially Peru and Chile, is a big producer with a yearly catch between 5 and 15 million tonnes. Amounts have fluctuated partly due to the El Nino. Several European countries (Denmark, Norway, Iceland amongst others) process ~6 million tonnes per year and the United States process 1 million tonnes.

Fish meal is basically prepared through three processes: boiling, separation, and drying. Industrial fish measuring 100 kg produces ~20 kg of fish meal and 2–10 kg of fish oil. Whole or chopped fish are boiled in its own juice using an indirect supply of steam. The cooked material then passes into a screw press, which separates the liquid fraction (the press water) from the solid fraction—the press cake. The former contains about 50% of the water and most of the oil. The liquid passes a decanter in which the solids are separated and returned to the press cake and subsequently the oil is removed by centrifugation. The remaining liquid (the "stick-water") is concentrated through evaporation, e.g. by falling film evaporators, and the product, "the solubles" added to the press cake. The press cake is then dried to a water content of 5–10%. A drying temperature of 90–100°C is used in the conventional process but some plants also produce LT meal, which had been dried at low temperature of ~70°C. The material cools during further processing. The meal is sometimes cured before grinding and bagging by stacking it in a shed to allow oxidation to proceed. In most cases, antioxidants such as butylated hydroxytoluene (BHT) or ethoxyquin are mixed into the meal as it leaves the dryer. Stabilized meal passes directly from the dryer through a hammer mill, which reduces the particle size, and then into bags or bulk storage. In some parts of the world (e.g. Angola), the cooked, pressed fish are simply allowed to dry in the sun.

An increasing amount of ensiled fish is being produced as animal feed and the extent of current interest in the product suggests that production will increase even more. Ensiling involves liquefying the fish under acid conditions so that the final pH is below 4.5, producing a bacteriologically stable product. In one system, the chopped or comminuted fish is mixed directly with mineral (e.g. sulfuric) or organic (e.g. formic) acid and allowed to liquefy at temperatures above 20°C. Another process involves mixing a fermentable carbohydrate such as molasses or cereal meal with the minced or chopped fish, inoculating with lactic acid bacteria (e.g. *Lb. plantarum* or *Streptococcus lactis*) and allowing the fermentation to proceed, optimally at temperatures near 30°C. In both procedures, oil is removed from the final product by skimming or centrifuging.

Fish solubles or fish concentrates are either a by-product of fish meal or a primary product of enzymatic digestion of entire fish.

Fish meals are used widely in animal rations for their high protein content (60–70%). The fat level varies but is usually about 10%. Fish meal contains all the necessary nutrients for microbial growth except moisture, which is generally between 7% and 10%. On a global scale, some 7 million tones of fish meal and 1.5 million tones of fish oil are produced. During the last decade, aquaculture production has increased significantly (see Chapter 3) and ~50% of fish meal and 90% of the fish oil is used for fish feed. Fish meal is also used for poultry and pork feed. A significant part of the fish oil is used for human consumption (IFFO, 2003).

A Effects of processing on microorganisms

Fish may harbor different kinds of microorganisms, mostly reflecting the microbiology of the aquatic environment (see Chapter 3). The heat treatment reduces the number of microorganisms to a low level (the actual number depending on the initial flora and the time–temperature combination used). However, Enterobacteriaceae and particularly *Salmonella* may be present in varying degrees due to recontamination after the heating. A study of fish meal from five factories found a mean aerobic plate count of 10^4, sulfite-reducing clostridia at approximately 3×10^3, Enterobacteriaceae and enterococci ranging from 10^2 to 10^4, and molds from 10^2 to 10^5/g (Milanovic and Beganovic. 1974). The extent of recontamination with Enterobacteriaceae, including *Salmonella*, may differ widely among production sites and countries of origin (Van Schothorst *et al*., 1966; Reusse *et al*., 1976).

B Spoilage

Fish meal is a microbiologically stable product because its a_w (0.33–0.65) is below the value that will support growth. Thus, in most cases, microbial spoilage is not important. If the product gets wet (e.g. during transport or storage), deterioration will occur due to rapid multiplication of fungi and/or bacteria.

There is no evidence of significant microbial problems with oils because they are unsuitable as growth media and undergo extensive physical and chemical refining.

C Pathogens

Fish meal has been recognized as a source of salmonellae in animal feeds since the early 1950s, when several serotypes such as *S.* Agona were introduced into many countries due to import of Peruvian fish meal. Salmonellae in feed are the principal problem not because they cause disease in the animals, but because they may ultimately cause food-borne illness in people who either handle or consume the products derived from the animals (Wiseman and Cole, 1990). An investigation showed that Enterobacteriaceae, including salmonellae, could be isolated from the product at all stages of processing (Quevedo, 1965). The poor sanitation of many fish meal-rendering plants contributes to the spread of contamination. Table 4.2 lists the percentages of fish meal samples found positive in several investigations. *Erysipelothrix insidiosa* has been recovered from Irish fish meal (Buxton and Fraser, 1971). If fish meal were to get wet, it would probably become moldy with the consequent potential for mycotoxin productions (Mossel, 1972; Gedek, 1973).

D CONTROL (fish meal)

Summary

Significant hazards	• Salmonellae.
Control measures	
Initial level of contamination (H_0)	• Levels of salmonellae in fish will usually be very low.
Reduction of contamination (ΣR)	• The raw material is heated at 100°C or above, which reduces the number of salmonellae sufficiently.
Increase of contamination level (ΣI)	• Assure adequate separation between raw and processed material, keep line-environment dry and free of salmonellae.
Testing	• Test environmental samples for the presence of salmonellae or determine the level of Enterobacteriaceae to detect niches where multiplication may have occurred.
Spoilage	• This is not a problem as long as the meal remains dry.

Salmonellae are the main hazard of concern. The initial levels are not very important, because it is not a fish pathogen, and the time–temperature used for processing will achieve a sufficient number of decimal reductions. Formerly, it was common to find the floors, walls, and equipment of fish meal plants covered with a layer of fine meal and fish caught in equipment or lying in corners. It was thought that the heat of cooking and rendering would produce a "sterile" product, and that, therefore, good sanitation was unnecessary. Now, however, under pressure from regulatory agencies, and faced with buyer specifications for *Salmonella*-negative product, fish meal manufacturers in advanced countries, at least, have come to recognize that they are engaged in a type of food processing and that recontamination has to be controlled. Many processors have taken steps to protect their product from salmonellae contamination with varying degrees of success (EC, 2002).

Control procedures for fish meal manufacture are essentially the same as those for animal by-products (Section II) and keeping the meal dry during transport and storage to prevent microbial growth. The control of pathogens in feed requires carefully planned and properly supervised intervention at several points in the chain of production and distribution, i.e. the application and implementation of the HACCP system (Wiseman and Cole, 1990).

Testing of environmental samples for the presence of salmonellae and for Enterobacteriaceae as indicators of the effectiveness of control measures is recommended. Testing of end products for the presence of salmonellae may be required to meet buyer and legal requirements, but its effectiveness to detect contamination is very limited (see Book 7).

V Compounded feeds

During the last decades, compounded feeds have been developed extensively to improve animal performance. They include meals or cakes of appropriate formulation for use as complete feeds for poultry or swine or as complements for specific animal situations (e.g. dairy cows). Compounded feeds are prepared from a large variety of ingredients including cereal grains (e.g. wheat, barley, maize), milling by-products, oilseed meals or cakes (e.g. soybean, rapeseed, sunflower, safflower, cotton), dehydrated alfalfa, root crops (e.g. manioc), animal by-products, fats, and miscellaneous by-products from the food industry such as citrus pulp. These materials are shipped to feed mills, stored in silos, blended according to appropriate formulations, and ground to provide feeds, which can be stored in bulk or in bags. More often, meals are further treated and pelleted, transported and stored in bulk on the farm, in silos, or less frequently in bags.

Basically, the quality of compounded feeds depends on appropriate formulation for specific animal production and nutritional requirements, and the quality of raw materials, in terms of fungal degradation and the presence or absence of mycotoxins. Cereals used in the manufacture of compounded feeds can be of relatively poor quality, particularly when they are downgraded from food grade due to fungal growth or damage by weather. Mycotoxins can often be found, also due to the availability of very sensitive analytical methods (Veldman *et al.*, 1992b) (see also Chapter 8).

The bacteriological status of compounded (mixed) feeds is also important. Pathogens may be present in the ingredients as a result of cross-contamination during processing and/or contamination during storage. Salmonellae are the most serious bacteriological hazard.

The UK codes of practice for the storage, handling and transportation of raw materials intended for direct use or incorporation into animal feeds (MAFF, 1995a) and conditions used for the design of processes and premises for the production of feed for livestock (MAFF, 1995b) contain the basic hygiene requirements for the production of *Salmonella*-free feeds. Application of the HACCP principles is strongly recommended (Butcher and Miles, 1995) and applied in certain countries such as The Netherlands.

A Effects of processing on microorganisms

Some major ingredients such as cereal grains, milling by-products, seeds, or other plant products are used as ingredients without prior heat treatment, thereby introducing microorganisms into the feed mixture. Other plant ingredients, such as oilseeds products, are processed either mechanically (e.g. expeller process) with the generation of a large amount of heat, by solvent extraction, or a combination of these two. These processing conditions eliminate vegetative microorganisms but not bacterial or fungal spores. Animal by-products are processed under high temperature conditions (see Section III). Both oilseed products and animal by-products may be subjected to contamination after processing or during storage, and become a source of various bacteria, including salmonellae, to finished mixed feeds.

Before processing, ingredients are stored, blended, and ground. During these operations, the lot can become contaminated from ingredients, unclean silos and machinery, dust and the environment. Grinding and adding liquid ingredients results in hot and moist conditions, which may favor bacterial or fungal development.

Pelleting is a basic process in mixed feed manufacturing. It is accomplished by grinding the feed, and forcing it through a die. Usually, but not always, feedstuffs are conditioned by steaming or moistening before pelleting, as this makes it easier to make a firm pellet, which holds its form (Church, 1979). The heat involved in conditioning and pelleting may reduce vegetative bacteria by up to 1000-fold and, thereby, result in pasteurizing the feed (Stott *et al.*, 1975). Enterobacteriaceae, however, have been recovered from pelleted feeds probably as a result of recontamination (Cox *et al.*, 1988). Whether the pelleting process eliminates pathogens such as salmonellae depends upon feed moisture, time and temperature of conditioning, efficiency of heat and moisture transfer in the feed, and the number and thermal resistance of microorganisms (Blankenship *et al.*, 1985; Himathongkham *et al.*, 1996).

Pelleting is followed by rapid cooling and removal of excess moisture in a vertical or horizontal cooling system. Such systems may represent a weak point in the operation since insufficient cooling could allow warm, wet material to be transferred into storage bins or silos, with subsequent microbial growth. Also, the air used for drying may be a source of contamination. Dust and deposits in the cooling system, on the conveyor line and in storage bins or silos may constitute growth niches leading to contamination of the finished mixed feed. Thus, the microflora of mixed feeds reflects the initial flora of the ingredients and the conditions of manufacturing.

Pigmented bacteria of the genera *Erwinia* and *Enterobacter* (*Enterobacter agglomerans*) are commonly present in cereals and cereal by-products of good quality and are typically found in mixed feeds containing high levels of these ingredients. Other Enterobacteriaceae (*Citrobacter, Klebsiella, Serratia, Escherichia, Proteus* spp.) and *Pseudomonas* are present at various levels. Gram-positive species of *Staphylococcus*, *Sarcina*, *Bacillus*, *Clostridium*, and *Enterococcus* may be found in lower numbers.

The fungi found will reflect those in the raw material unless a substantial heat process is applied. When heat treatments have been applied, or when preservative ingredients have been added, sporulating *Bacillus* and *Clostridium* species dominate the microflora. Fungi are found in low numbers (e.g. $<10^2$ cfu/g) in mixed feeds (IAG, 1993).

B Spoilage

The reduced moisture content and low a_w help preserve the mixed feed and prevent microbial growth. Moistened or rehydrated feeds, however, may spoil rapidly due to bacterial and/or fungal growth. Such conditions may result from insufficient cooling and drying in the final stages of processing. More often moistening of feedstuffs is a consequence of diurnal variation in temperature during storage in farm silos resulting in water migration and condensation, creating wet spots, which permit bacterial and fungal growth, spoilage, and toxin production.

The spoilage microflora of mixed feeds is characterized by the development of bacteria and molds present in low numbers in good-quality feeds. Important bacteria include *Micrococcus, Staphylococcus, Bacillus, Streptomyces*, and *Thermoactinomyces*. Mold spoilage during storage is usually initiated by xerophiles including *Eurotium* species and *Aspergillus penicilloides*. As a_w increases, other *Aspergillus* spp. may develop, together with *Penicillium* spp. If heating occurs, *Asp. candidus*, *Asp. flavus*, and later *Thermoascus* spp. may develop (Sauer *et al.*, 1992; Ominski *et al.*, 1994, see Chapter 8). At higher a_w values *Paecilomyces*, *Scopulariopsis*, *Trichoderma*, *Chaetomium*, *Mucor*, and *Rhizopus* may develop. Such genera are indicators of advanced decay of feedstuffs (IAG, 1993).

C Pathogens

Documented evidence demonstrates that mixed feeds may be a source of salmonellae in broiler or swine production (Edel *et al.*, 1967; Smeltzer *et al.*, 1980; Hinton *et al.*, 1987; Eld *et al.*, 1991; Jones *et al.*, 1991). In some instances, the presence of salmonellae in mixed feeds is traceable to the ingredients used. The importance of contaminated animal by-products has been discussed in Section III.C. Vegetable ingredients have normally been considered to be contaminated to a lesser extent although several investigations have implicated these materials as an important source of salmonellae. Thus, salmonellae were isolated from sunflower and soybean cakes at rates of 22% and 16%, respectively (Mackenzie and Bains, 1976). Imported soybean, sunflower, and rapeseed cakes have been found positive for salmonellae at rates of 3.1, 10.1, and 1.9%, respectively (Jardy and Michard, 1992). Such a frequency of contamination is of particular concern because plant ingredients are used in larger proportions than animal by-products, i.e. up to 50% for soybean by-products; up to 20% for rapeseed or sunflower by-products; 5–6% for animal by-products (if they may be used); 1% for fish meal. The UK information also indicates the importance of oilseed by-products as a source of *Salmonella*.

The pelleting process has been found to be effective for reducing salmonellae in feeds (Eld *et al.*, 1991). However, in one survey, salmonellae were isolated from mash feeds at a rate of 35% and from pelleted feeds at a rate of 6.3%, indicating that the effectiveness of the pelleting process has to be validated or that recontamination needs to be prevented (Jones *et al.*, 1991).

If salmonellae survive pelleting, they can contaminate the rest of the feed mill. If they become established, and even multiply, a focus of contamination will result. The cooling system is an important source of salmonellae when wet surfaces exist because these surfaces offer an excellent substrate for growth. As in the case of animal by-products, dust and feed deposit along the conveyor line and in silos can become niches of contamination for the feed. In Sweden, during 1983–1987, 94 (40.3%) of 233 strains of salmonellae were isolated from dust and scrapings in processing plants (Eld *et al.*, 1991).

Although mycotoxin formation is possible in compounded feeds after manufacture (as discussed earlier), this requires gross abuse and is usually limited to inadequate storage on farms. More serious and widespread problems arise from the use of poor quality cereal grains. In many parts of the world, cereal grains that fail to meet acceptable standards for human consumption are used in animal feeds. Such grain may have deteriorated due to mold growth before harvest relating to wet harvest conditions or geographical factors (See Chapter 8 for further details).

D CONTROL (compounded feeds)

Summary

Significant hazards	• Salmonellae.
Control measures	
Initial level of contamination (H_0)	• Various ingredients of compounded feeds may be contaminated with *Salmonella*, usually in low numbers. • Keeping the ingredients dry to prevent multiplication is thus important. • Selection of *Salmonella*-free ingredients for a dry blended meal is recommended but may be difficult to achieve. • Moldy ingredients should not be used.
Increase of contamination level (ΣI)	• Assure adequate separation between raw and processed material, keep line-environment dry and free of salmonellae.

(continued)

CONTROL (Cont.)

Summary

Reduction of contamination (ΣR)	• Pelletizing the meal can achieve a 3–4D reduction of the level of salmonellae. • When meal is fed wet, addition of acids may achieve some reduction in contamination.
Testing	• Test environmental samples for the presence of salmonellae or determine the level of Enterobacteriaceae to detect niches where multiplication may have occurred. Testing of ingredients for the presence of salmonellae may assist in detecting highly contaminated lots.
Spoilage	• This is not a problem as long as the meal remains dry. Molds may grow when the feed becomes (accidentally) wet, in that case mycotoxins may become a problem.

Salmonellae are the major pathogens of concern. It is accepted that elimination of salmonellae from feeds will not guarantee the absence of *Salmonella* in animals since there are many sources for this pathogen on farms. Nevertheless, control is necessary to minimize the risk of salmonellae contamination in feed. Important control points for feedstuff production have been reviewed in comprehensive surveys (Williams, 1981; Skovgaard, 1989). Successful control requires basic interventions such as:

- Avoiding the use of *Salmonella*-contaminated ingredients. Testing for the presence of salmonellae at port of entry (imported ingredients) and/or on receipt at the feed mill (domestic materials) may identify sources of highly contaminated ingredients.
- Appropriate plant layout to avoid cross-contamination between raw materials and finished feedstuffs. Separate silos for ingredients and separate lines for cattle, pig, and poultry feed contribute to that aim and should be established where possible.
- Correct use of bactericidal treatments like conditioning and pelleting or alternative preservation techniques such as chemical disinfection (see below for discussion).
- Correct operation of equipment, good hygienic housekeeping, and cleaning routine. Control condensation to prevent moistening feeds in coolers, especially at the top of cooling towers. Special care must be taken to prevent wet surfaces and feed deposits in silos, mixers, cooling systems, elevators, and transport systems; prevent the spread of contamination via feed dust; air intakes near "unclean" areas should be avoided or closed; air used to cool pelleted products should be filtered; insects, rodents, and other vermin should be eliminated; staff must be specifically trained in proper cleaning and disinfection procedures.
- Environmental and line sampling to identify sources of contamination and where improvements are needed has proven to be more effective for salmonellae control than end-product testing.

Appropriate storage conditions in farm silos with specific consideration to control of temperature and humidity (material of construction, appropriate ventilation, insulation where necessary), appropriate use and housekeeping (regularity of flow to avoid coatings and feed deposits, complete evacuation of feeds, thorough cleaning after emptying, disinfection at regular intervals), monitoring temperature and humidity, periodic examination for salmonellae and molds. As decontamination plays an important role in control, the following discussion will focus on this topic.

For conditioning and pelleting to reduce or eliminate salmonellae effectively, appropriate temperature control is necessary. Experimental data (Van der Wal, 1979; Skovgaard, 1989) suggest that the conditioner temperature must not fall below 74°C and the pelleting press must increase the product temperature in range 81–85°C, so that the temperature in the pellets exiting the pelleter will not be less than 81°C. Monitoring can be accomplished by continuous recording or by measuring the temperature in the conditioner and the finished product at least every 2 h. Variations of these conditions have been developed to improve the heat treatment, without altering the nutritional value of the feed. Conditioning with direct, live steam (up to 85°C for meals with 14.5% moisture content), extrusion (93–183°C under a pressure of 50–100 bars) or expansion (140°C for a few seconds under a pressure of 40 bars) before pressing can enhance the reduction of salmonellae.

Irradiation after pelleting has been proposed to produce *Salmonella*-free feed (Mossel *et al.*, 1967), but due to technical, regulatory, and cost considerations its use is restricted to producing specialized diets such as for laboratory animals.

Adding chemical preservatives is an alternative technique to heating feedstuffs (reviewed by Hinton and Bale, 1990). Short-chain fatty acids (e.g. lactic acid, formic acid, formic acid, or propionic acid) have been shown to reduce the incidence of salmonellae infection in chicks consuming artificially contaminated acid-treated feed (Hinton and Linton, 1988). Commercial products consist, mainly of propionic acid, its salts, or a mixture of these in liquid or solid form. Such preservative is also effective for inhibiting mycotoxin formation during storage. They appear to be inactive or only slightly active in dry feed and then exert their antimicrobial effect when a_w is high enough to liberate free fatty acids from the salts. Since feeds are usually stored under dry conditions, they exert their antibacterial activity either when the feed becomes hydrated or even after it has been consumed by the animals (Hinton *et al.*, 1991). The effectiveness of acid treatment is dependent upon the moisture content of the feed, temperature of storage, ambient humidity, level of microbial contamination, and type of animal. Acid protection is particularly beneficial in feedstuffs during the rearing period. Acids are ineffective for preventing infection if the animals are exposed to other sources of infection or if the feed contains large numbers of microorganisms. They are most useful in feeds with low-moisture content (<12–13%) that are stored in hot, wet climatic conditions, and in feeds with high moisture content (13.5–14% or above). Feeding fermented feeds may be another possibility to reduce *Salmonella* contamination (Van Winsen, 2001).

Testing for salmonellae in final feeds may be necessary for legal or commercial reasons.

Control of mycotoxins in compounded feeds after manufacture relies on good storage, minimizing temperature fluctuations, and access to moisture.

VI Pet foods

Commercially processed pet foods include high-moisture canned products, intermediate-moisture, semi-moist (a_w 0.80–0.90) products, and dried products (a_w <0.60).

Dry and semi-moist pet foods are mixtures of cereals and animal by-products. Occasionally, they contain "feed-grade" non-fat dry milk, dried yeast, and soybean meal. The major source of protein in dry pet foods is (or was) rendered animal by-products, whereas in semi-moist foods it is raw animal offal. Since cats have been shown to be susceptible to prion diseases (Feline Spongiform Encephalopathy), use of animal-by products in the production of pet food has been subject to some restrictions in some of the BSE affected countries or by certain companies (WHO, 2001). For the same reason, "high-risk" animal tissues are sometimes banned for use in pet foods.

In some countries, raw animal offal is used directly as pet foods. These products will not be discussed, and it is unnecessary to discuss the microbiology of canned pet foods, which should comply with low-acid canned food specifications. Indeed, this is essential because humans are sometimes consumers of canned pet foods.

A *Effects of processing on microorganisms*

In dry pet foods, the two major ingredients, cooked cereals and meat and bone meal (or other animal by-products), are sometimes mixed without heat treatment, resulting in a microflora that is more or less a reflection of the initial flora of the ingredients. In most cases, however, the material is heated by an expansion–extrusion process that raises the temperature above 100°C, killing all vegetative microorganisms, including salmonellae. Intermediate-moisture foods may or may not be expanded and the temperatures achieved vary. One of the following three processes is used:

1. A single-stage process in which the material is mixed and then passed through a cooker extruder.
2. A two-stage process in which preliminary pasteurization of the meat and liquid ingredients is followed by cooking and extruding.
3. A cold extrusion process that is preceded by pasteurization. In most processes, the highest product temperature reached is around 95°C.

The required stability is achieved by a combination of low a_w, low pH, and preservatives. The reduction of a_w is achieved by humectants, some of which have antimicrobial activity (e.g. propane, 1,2-diol; butane, 1,3-diol; sorbitol, or diol esters). The microbial flora of the end product depends on the initial flora and the processing techniques that have been used. Because these can vary so greatly, it is difficult to describe the "normal" microbiological condition of these intermediate-moisture pet foods.

B *Spoilage*

As long as the water activity of dry or semi-moist pet foods remains low, spoilage is no problem, but moistened and rehydrated foods will spoil rapidly due to fungal and/or bacterial growth.

C *Pathogens*

Many of the ingredients (e.g. animal by-products, non-fat dry milk, dried yeast, and soybean meal) have been reported to contain salmonellae. Although the heat applied in the expansion and extrusion process is sufficient to destroy salmonellae, contamination of the finished product can occur from cross-contamination or recontamination from environmental niches.

Human beings can be infected when *Salmonella*-contaminated pet foods are brought into the kitchen. Although some people eat such foods, the greater risk is cross-contamination to human foods in which growth can occur and, thereby, become more hazardous. Pets with clinical or sub-clinical disease, acquired from eating contaminated pet food, can spread pathogens to humans. Pet foods are commonly rehydrated at the time of feeding and then held for extended periods at ambient temperatures. Thus, the risk of infection increases to pets and, subsequently, to humans, particularly children, in the same household (Morse *et al.*, 1976).

A case of *S.* Havana infection in a 2 ½-month-old child led to an investigation of 25 samples of dried dog food, representing four different manufacturers and two retail store brands for salmonellae (Pace *et al.*, 1977). *Salmonella* serotypes were isolated from seven of eight bags of commercially processed dried dog food. Each of 11 samples from one manufacturer contained one or more salmonellae serotypes. Eight of the samples contained *S.* Havana that had antibiotic sensitivity patterns similar to or identical to those of 9 of 10 *S.* Havana isolates recovered from the dog food, child, and mother.

Semi-moist pet foods have not been a serious health problem, because the product is sufficiently cooked during manufacturing to destroy enteric pathogens such as salmonellae, a_w and pH are reduced, and humectants and preservatives are added to prevent microbial growth. These foods are fed to pets without rehydration, so they will not support microbial growth unless they get wet accidentally.

Improperly processed canned pet foods could theoretically be a danger to the elderly and impoverished people who eat them. However, when processed under good manufacturing practices canned pet foods are commercially sterile and as safe as other low-acid canned foods.

D CONTROL (pet foods)

Summary

Significant hazards	• Salmonellae in dry pet food.
Control measures	
Initial level of contamination (H_0)	• Usually the numbers of salmonellae in the ingredients is low.
Reduction of contamination (ΣR)	• Pelletization and particularly extrusion can achieve 4–6 decimal reduction of salmonellae.
Increase of contamination level (ΣI)	• Assure adequate separation between raw and processed material, keep line-environment dry and free of salmonellae. • Add a sufficient quantity of a humectant to prevent multiplication in a semi-moist food.
Testing	• Test environmental samples for the presence of salmonellae or determine the level of Enterobacteriaceae to detect niches where multiplication may have occurred.
Spoilage	• This is not a problem as long as the pet food remains dry, or at the most a semi-moist food.

While producing dry unheated pet foods, controlling the ingredients is essential to avoid health hazards. When producing all pet foods, applying Good Manufacturing Practices and HACCP can ensure safe products. *Salmonella* testing may be appropriate for dry pet foods, and particularly those intended to be rehydrated before use.

References

ACDP (Advisory Committee on Dangerous Pathogens) (1996) *BSE (Bovine Spongiform Encephalopathy). Background and General Occupational Guidance*, HSE Books, London (ISBN 07176 12120).

Al-Ghazali, M.R. and Al-Azawi, S.K. (1986) Detection and enumeration of *Listeria monocytogenes* in a sewage treatment plant in Iraq. *J. Appl. Bacteriol.*, **60**, 251–4.

Anonymous (1996) Bovine Spongiform Encephalopathy: an update. *Revue Scientifique et Technique de l'Office International des Epizooties*, **15**(3), 1087–118.

Anonymous (2002) Annual report on Zoonosis in Denmark, Danish Zoonosis Center, Copenhagen, Denmark.

Bauer, J. (1995) Zum Metabolismus von Trichothecenen beim Schwein. *Deutsche Tierärztliche Wochenschrift*, **102**, 50–52.

Bhatia, T.R.S., McNabb, G.D., Wyman, H. and Nayar. G.P. (1979) *Salmonella*: isolation from litter as an indication of flock infection and carcass contamination. *Avian Dis.*, **23**, 838–47.

Blankenship, U.C., Shackelford, D.A., Cox. N.A., Burdick, D., Bailey, J.S. and Thomson, J.E. (1985) Survival of Salmonellae as a function of poultry feed processing conditions, in *Proceedings of International Symposium on Salmonella* (ed. G. Snoeyenbos), American Association of Avian Pathologists, University of Pennysylvania Publication, Kennett Square. PA. pp. 211–20.

Boloh, Y. (1992) Farines de viande. *Rev. Ind. Agric. Alim.*, **459**, 60–2.

Botzler, R.G., Cowan, A.B. and Wetzler. T.F. (1974) Survival of *Listeria monocytogenes* in soil and water. *J. Wild Dis.*, **10**, 204–12.

Brackett, R.E. (1988) Presence and persistence of *Listeria monocytogenes* in food and water. *Food Technol.*, **42**, 162–4, 178.

Brown, P., Will, R.G., Bradley, R., Asher, D.M. and Detwiler, L. (2001) Bovine Spongiform Encephalopathy and variant Creutzfeld–Jacob Disease: background, evolution, and current concerns. *Emerg. Infect. Dis.*, **7**, 6–16

Buchko, S.J., Holley, R.A., Olson, W.O., Gannon, V.P. and Veira, D.M. (2000) The effect of different grain diets on fecal shedding of *Escherichia coli* O157:H7 by steers. *J. Food Prot.*, 63(11), 1467–74.

Butcher, G.D. and Miles, R.D. (1995) Minimizing microbial contamination in feed mills producing poultry feed. Florida Cooperative Extension Service, Institute of Food and Agricultural Sciences, University of Florida, VM-93.

Buxton, A. and Fraser. G. (1977) *Animal Microbiology, Volume 1*, Blackwell Scientific Publications, Oxford.

Church, D.C. (1979) *Livestock Feeds and Feedings*, 4th edn, O & B Books, Corvalis, Oregon.

Clise, J.D. and Swecker, E.E. (1965) Salmonellae from animal by-products. *Public Health Rep.*, **80**, 899–905.

Cole, R.J., Kirksey, J.W., Dorner, J.W., Wilson, D.M., Johnson, J.C., Johnson, A.N., Bedell, D.M., Springer, J.P., Chexal, K.K., Clardy, J.C. and Cox, R.H. (1977) Mycotoxins produced by *Aspergillus fumigatus* species isolated from moulded silage. *J. Agric. Food Chem.*, **25**, 826–30.

Cox, L.J., Keller, N. and van Schothorst, M. (1988) The use and misuse of quantitative determination of Enterobacteriaceae in food microbiology. *J. Appl. Bacteriol.*, **65** (Symposium Supplement), S237–49.

Craven, S.E., Stern, N.J., Fedorka-Cray, P., Bailey, J.S. and Cox. N.A. (1999) Epidemiology of *Clostridium perfringens* in the production and processing of broiler chickens, in *Proceedings of the 17th ICFMH*, Veldhoven, The Netherlands, September 13–17, 357–360.

Davies, D.G. and Harvey. R.W.S. (1972) Anthrax infection in bone meal from various countries of origin. *J. Hyg.*, **70**(3), 455–7.

Davies, P.R., Morrow, W.E.M., Jones, F.T., Deen, J., Fedorka-Cray, P.J. and Gray, J.T. (2001) Elevated risk of *Salmonella* carriage by market hogs in a barn with open flush gutters. *JAVMA*.

Dijkstra, R.G. (1975) Recent experiences of the survival times of *Listeria* bacteria in suspension of brain, tissue, silage, faeces and in milk, in *Problems of Listeriosis* (ed M. Woodbine), Leicester University Press Publication, Surrey, p. 71.

D'Mello, J.P.F., Placinta, C.M. and Macdonald, A.M.C. (1999) *Fusarium* mycotoxins: a review of global implications for animal health, welfare and productivity. *Animal Feed Sci. Technol.*, **80**, 183–205.

Donald. S., Fenlon, D.R. and Seddon, B. (1993) A novel system for monitoring the influence of oxygen tension on the microflora of grass silage. *Lett. Appl. Microbiol.*, **17**, 253–5.

Donald, A.S., Fenlon, D.R. and Seddon, B. (1995) The relationships between ecophysiology, indigenous microflora and growth of *Listeria monocytogenes* in grass silage. *J. Appl. Bacteriol.*, **79**, 141–8.

Driehuis, F. and Oude Elferink, S.J.W.H. (2000) The impact of the quality of silage on animal health and food safety: a review. *Vet. Q.*, **22**, 212–7.

Driehuis, F., Oude Elferink, S.J.W.H. and Spoelstra, S.F. (1999) Anaerobic lactic acid degradation in maize silage inoculated with *Lactobacillus buchneri* inhibits yeast growth and improves aerobic stability. *J. Appl. Microbiol.*, **87**, 583–94.

EC (2002) Trends and sources of zoonotic agents in animals, feedstuffs, food and man in the European Union and Norway in 2000. Part 1. SANCO/927/2002.

EC (2001a) Regulation (EC) No. 999/2001 of the European Parliament and of the Council of 22 May 2001, laying down the rules for the prevention, control and eradication of certain transmissible spongiform encephalopathies. *J. Eur. Commun.*, L 147/1–40.

EC (2001b) Regulation (EC) No. 1234/2003 of the European Parliament and of the Council and Regulations (EC) No 1236/2001 as regards transmissible spongiform encephalopathies and animal feeding. *J. Eur. Commun.*, L 173/6–7.

Edel, W., Guinée, P.A.M., van Schothorst, M. and Kampelmacher, E.H. (1967) *Salmonella* infections in pigs fattened with pellets and unpelleted meal. *Zbl. Vet. Med. B*, **14**, 393–401.

Edel, W., van Schothorst, M., Guinee, P.A.M. and Kampelmacher, E.M. (1970) Effect of feeding pellets on the prevention and sanitation of salmonella infections in fattening pigs. *Zbl. Vet. Med. B*, **17**, 730–8.

Eld, K., Gunnarsson, A., Holmberg, T., Hurvell, B. and Wierup, M. (1991) *Salmonella* isolated from animals and feedstuffs in Sweden during 1983–1987. *Acta Vet. Scand.*, **32**, 261–77.

Fenlon, D.R. (1985) Wild birds and silage as reservoirs of *Listeria* in the agricultural environment. *J. Appl Bacteriol.*, **59**, 537–44.

Fenlon, D.R. (1986) Rapid quantitative assessment of the distribution of *Listeria* in silage implicated in a suspected outbreak of listeriosis in calves. *Vet. Rec.*, **118**, 240–2.

Fenlon, D.R. (1989) The influence of gaseous environment and water availability on the growth of *Listeria*. *Microbiol. Alim. Nut.*, **7**, 165–9.

Garcia, V. (1992) L'encephalopathie spongiforme bovine. Rapport de la Commission de l'Agriculture, de la Peche et du Developpement Rural au Parlement Européen. no. A3-0368/92.

Gedek, B. (1973) Futtermittelverderb durch Bakterien und Pilze und seine nachteiligen Folgen. Uebers. *Tierernährung*, **1**, 45–46.

Gray, M.L. (1960) Isolation of *Listeria monocytogenes* from natural silage. *Science*, **132**, 1767–8.

Grønstøl, H. (1979) Listeriosis in sheep. Isolation of *Listeria monocytogenes* from grass silage. *Acta Vet. Scand.*, **20**, 492–7.

Hald, B. (1991a) Porcine nephropathy in Europe, in *Mycotoxins, Endemic Nephropathy and Urinary Tract Tumors* (eds M. Castegnaro, R. Plestina, G. Dirheimer, I.N. Chermozensky and H. Bartsch), International Agency for Research on Cancer, Lyon, France, pp. 49–56.

Hald, B. (1991b) Ochratoxin A in human blood in European countries, in *Mycotoxins, Endemic Nephropathy and Urinary Tract Tumors* (eds M. Castegnaro, R. Plestina, G. Dirheimer, I.N. Chermozensky and H. Bartsch), International Agency for Research on Cancer, Lyon. France, pp. 159–64.

Hancock, D., Besser, T., Lejeune, Davis, M. and Rice, D. (2001) The control of VTEC in the animal reservoir. *Int. J. Food Microbiol.*, **66**, 71–78.

Hess, G.W., Moulthrop, J.I. and Norton, H.R., II (1970) New decontamination efforts and techniques for elimination of *Salmonella* from animal protein rendering plants. *J. Am. Vet. Med. Assoc.*, **157**, 1975–80.

Herriott, D.E., Hancock, D.D., Ebel, E.D., Carpenter, L.V., Rice, D.H. and Besser, T.E. (1998) Association of herd management factors with colonization of dairy cattle by Shiga toxin-positive *Escherichia coli* O157. *J. Food Prot.*, **61**(7), 802–7.

Himathongkham, S.M., Des Gracas Pereira, M. and Riemann, H. (1996) Heat destruction of *Salmonella* in poultry feed: effect of time, temperature and moisture. *Avian Dis.*, **40**, 72–77.

Hinton, M. and Bale. M.J. (1990) Animal pathogens in feed, in *Feedstuff Evaluation* (eds J. Wiseman and D.J.A. Cole), Butterworths Publications, London, pp. 429–44.

Hinton, M. and Linton, A.H. (1988) Control of *Salmonella* infections in broiler chickens by the acid treatment of their feed. *Vet. Rec.*, **123**, 416–21.

Hinton. M., Al Chalaby, Z.A.M. and Linton, A.H. (1987) Field and experimental investigations into the epidemiology of *Salmonella* infections in broiler chickens, in *Elimination of Pathogenic Organisms from Meat and Poultry* (ed F.J. Smulders), Elsevier, Amsterdam, pp. 27–36.

Hinton, M., Cherrington, C.A. and Chopra, I. (1991) Acid treatment of feed for the control of *Salmonella* infections in poultry. *Vet. Annu.*, **31**, 90–5.

Hird, D.W. and Genigeorgis, C. (1990) Listeriosis in food animals: clinical signs and livestock as a potential source of direct (non foodborne) infection for humans, in *Foodborne Listeriosis* (eds A.S. Miller, J.L. Smith and G.A. Somkuti), Elsevier Science, Amsterdam, pp. 31–9.

HMSO (1995) *Steering Group on Microbiological Safety of Foods: Annual Report 1994*, HMSO, London.

Höhler, D. (1998) Ochratoxine A in food and feed: occurrence, legislation and mode of action. *Z. Ernährungswissenschaft*, **37**, 2–12.

IAG (International Analytical Group) (1993) Microbiological analysis and quality assessment of mixed feeds: the IAG concept. Working document of the Section on Feed Microbiology, Posieux, Switzerland.

IFFO (International Fishmeal & Fish Oil Organisation) (2003) http://www. iffo.org.uk.

Jardy, N. and Michard. J. (1992) *Salmonella* contamination in raw feed components. *Microbial. Alim. Nut.*, **10**, 233–40.

Jones. P.W., Collins, P., Brown, G.T.H. and Aitkin, M. (1982) Transmission of *Salmonella* mbandaka to cattle from contaminated feed. *J. Hyg.*, **88**, 255–63.

Jones, F.T., Axtell, R.C., Rives, D.V., Scheideler, S.E., Tarver, F.R., Walker. R.L. and Wineland, M.J. (1991) A survey of *Salmonella* contamination in modern broiler production. *J. Food Prot.*, **54**, 502–7.

Kehler, W. and Scholz, H. (1996) Botulismus des Rindes. *Übersichten zur Tierernährung* **24**, 83–91.

Lacey. J. (1991) Natural occurrence of mycotoxins in growing and conserved forage crops, in *Mycotoxins and Animal Foods* (eds J.E. Smith and R.S. Henderson), CRC Press, Boca Raton, Florida, pp. 363–97.

Lindgren S., Petterson, K., Kaspersson, A., Jonsson, A. and Lingvall, P. (1985) Microbial dynamics during aerobic deterioration of silages. *J. Sci. Food Agric.*, **36**, 765–74.

Loken. K.I., Culbert, K.H., Solee, R.E. and Pemeroy, B.S. (1968) Microbiological quality of protein feed supplements produced by rendering plants. *Appl. Microbiol.*, **16**, 1002–5.

Mackenzie. M.A. and Bains, B.S. (1976) Dissemination of *Salmonella* serotypes from raw feed ingredients to chicken carcasses. *Poulty Sci.*, **55**, 957–60.

MAFF (1989) UK Processed and Animal Protein Order (SI 1989/661).

MAFF (1995a) Code of practice for the control of *Salmonella* during the storage handling, transportation of raw materials intended for incorporation into, or direct use as, animal feeding stuff. UK Ministry of Agriculture, Fisheries and Food Publication.

MAFF (1995b) Code of practice for the control of *Salmonella* in the production of feed for livestock. UK Ministry of Agriculture, Fisheries and Food Publication.

McDonald, P., Henderson, A.R. and Heron, S.J.E. (1991) *The Biochemistry of Silage*, Chalcombe Publications, Marlow.

Michanie, S., Isequilla, P., Lasta. J. and Quevedo, F. (1985) Enterobacteriaceae as indicators of the presence of *Salmonella* and the hygienic quality in meat and bone meal. *Rev. Argent. Microbiol.*, **21**, 43–6.

Milanovic. A. and Beganovic, A. (1974) Microflora of fodder of animal origin. *Veterinaria (Sarajevo)*, **23**, 467–75.

Mitscherlich, E. and Marth, E.H. (1984) *Microbial Survival in the Environment*, Springer Verlag, Berlin.

Moon, N.J. (1983) Inhibition of the growth of acid tolerant yeasts by acetate, lactate and propionate and their synergistic mixture.*J. Appl. Bacteriol.*, **55**, 453–60.

Morehouse, L.G. and Wedman. E.E. (1961) *Salmonella* and other disease-producing organisms in animal by-products. A survey. *J. Am. Vet. Med. Assoc.*, **139**, 989–95.

Morse, E.V., Duncan, M.A., Estep, D.A., Riggs, W.A. and Blackburn, B.O. (1976) Canine Salmonellosis: a review and report of dog to child transmission of *Salmonella enteritidis*. *Am. J. Public Health*, **66**, 82–3.
Mossel, D.A.A. (1972) Hygiene of food and fodder in South America. *Scienza dell Alimentazione*, **18**(5), 172–81.
Mossel, D.A.A., van Schothorst, M. and Kampelmacher, E.H. (1967) Comparative study on decontamination of mixed feeds by radicidation and by pelletisation. *J. Sci. Food Agric.*, **18**, 362–7.
National Research Council (NRC) (1972) *United States–Canada Tables of Feed Composition*, N.R.C. Publishers, Washington, DC.
Neill, S.D., McLaughlin, M.F. and McIlroy, S.G. (1989) Type C botulism in cattle being fed ensiled poultry litter. *Vet. Rec.*, **124**, 558–60.
Ominski, K.H., Marquardt, R.R., Sinha. R.N. and Abramson, D. (1994) Ecological aspects of growth and mycotoxin production by storage fungi, in *Mycotoxins in Grain: Compounds other than Aflatoxin* (eds J.D. Miller and H.L. Trenholm), Eagan Press, St Paul, Minnesota, pp. 287–312.
Oosterom, J., van Erne, E.H.W. and van Schothorst, M. (1982) Epidemiological studies on *Salmonella* in a certain area ('Walcheren Project'). *Zbl. Bakt. Hyg. I Abt. Orig. A*. **252**, 490–506.
Pace, P.J., Silver, K.H. and Wisniewski, H.J. (1977) *Salmonella* in commercially produced dried dog food: possible relationship to a human infection caused by *Salmonella enteritidis* serotype Havana. *J. Food Prot.*, **40**, 317–21.
Patterson, J.T. (1972) Salmonellae in animal feedingstuffs. *North. Ireland Minist. Agric. Rec. Agric. Res.*, **20**, 27–33.
Perry, C.M. and Donnelly, C.W. (1990) Incidence of *Listeria monocytogenes* in silage and its subsequent control by specific and nonspecific antagonism. *J. Food Prot.*, **53**, 642–7.
Prelusky, D.B. (1994) Residues in food products of animal origin, in *Mycotoxins in Grain: Compounds other than Aflatoxin* (eds J.D. Miller and H.L. Trenholm), Eagan Press, St Paul, Minnesota, pp. 405–19.
Quevedo, F. (1965) Les enterobacteriaceae dans la farine de poisson. *Ann. Inst. Pasteur, Lille*, **16**, 157–62.
Quevedo, F. and Carranza, N. (1966) Le role des mouches dans la contamination des aliments au Perou. *Ann. Inst. Pasteur, Lille*, **17**, 199–202.
Reusse, U., Meyer. A. and Tillack, J. (1976) Zur Methodik des Salmonellen Nachweises aus gefrorenem Geflügel. *Arch. Lebensmittelhyg.*, **27**, 98–100.
Ricketts, S.W., Greet, T.R.C. and Glyn, P.J. (1984) Thirteen cases of botulism in horses fed big bale silage. *Equine Vet. J.*, **16**, 515–8.
Roberts, T.A. (1988) Botulism, in *Silage and Health* (eds B.A. Stark and J.M. Wilkinson), Chalcombe Publications, Marlow, Bucks, UK, pp. 35–43.
Rowe, B. (1973) Salmonellosis in England and Wales, in *The Microbiological Safety of Food* (eds B.C. Hobbs and J.M.B. Christian), Academic Press, London, pp. 165–80.
Sauer, D.B., Meronuck. R.A. and Christensen, C.M. (1992) Microflora, in *Storage of Cereal Grains and their Products* (ed D.B. Sauer), American Association of Cereal Chemists, St Paul, Minnesota, pp. 313–40.
Scott, P.M. (1989) Mycotoxigenic fungal contaminants of cheese and other dairy products, in *Mycotoxins in Dairy Product* (ed H.P. van Egmond), Elsevier Applied Science, Amsterdam, pp. 193–259.
Scudamore, K.A. and Livesey, C.T. (1998) Occurrence and significance of mycotoxins in forage crops and silage: a review. *J. Sci. Food Agric.*, **77**, 1–17.
Skovgaard, N. (1989) *Salmonella*: critical control points for feedstuff production, in *Report of WHO Consultation on Epidemiological Emergency in Poultry and Egg Salmonellosis*, WHO/CDS/VPH/89-82, WHO Pub., Geneva, Switzerland.
Skovgaard, N. and Nielsen, B.B. (1972) Salmonellas in pigs and animal feedingstuffs in England and Wales and in Denmark. *J. Hyg.*, **70**, 127–40.
Smart, J.L., Jones, T.O., Clegg, F.G. and McMurtry, M.J. (1987) Poultry waste associated type C botulism in cattle. *Epidemiol. Infect.*, **98**, 73–9.
Smeltzer, T., Thomas, R., Tranter, G. and Klemm, J. (1980) Microbiological quality of Queensland stockfeeds with special reference to *Salmonella*. *Aust. Vet. J.*, **56**, 335–8.
Smith, O.B., Macleod, G.K. and Usborne, W.R. (1978) Organoleptic, chemical and bacterial characteristics of meat and offals from beef cattle fed wet poultry excreta. *J. Food Prot.*, **41**(9), 712–16.
Smittle, R.B., Kornacki, J.L. and Flowers, R.S. (1992) *Salmonella* survey in rendered animal proteins in the USA and Canada, in *Proc. 3rd World Cong. Foodborne Infect. Intox*, Berlin, Inst. Vet. Med., R. von Ostertag Inst. Pub., Berlin, pp. 997–1001.
Stoloff, L. (1977) Aflatoxins—an overview, in *Mycotoxins in Human and Animal Health* (eds J.V. Rodricks, C.W. Hesseltine and M.A. Mehlman), Pathotox Publishers, Park Forest South, Illinois, pp. 7–28.
Stott, J.A., Hodgson, J.E. and Chaney, J.C. (1975) Incidence of salmonellae in animal feed and the effect of pelleting on content of Enterobacteriaceae. *J Appl. Bacteriol.*, **39**, 41–6.
Swingler, G.B. (1982) Microbiology of meat industry by-products, in *Meat Microbiology* (ed M.H. Brown), Applied Science, London.
Timoney, J. (1968) The sources and extent of *Salmonella* contamination in rendering plants. *Vet. Rec.*, **83**, 541–3.
Van Egmond, H.P. (1989) Aflatoxin M1: occurrence, toxicity, regulation, in *Mycotoxins in Dairy Products* (ed H.P. van Egmond), Elsevier Applied Science, Amsterdam, pp. 11–55.
Van der Wal, P. (1979) *Salmonella* control of feedstuffs by pelleting or acid treatment. *World's Poult. Sci. J.*, **35**, 70–8.
Van Os, M. (1997) Role of ammonia and biogenic amines in intake of grass silage by ruminants. Thesis, Wageningen University, The Netherlands.

Van Schothorst, M. and Oosterom, H. (1984) Enterobacteriaceae as indicators of good manufacturing practice in rendering plants. *Antonie van Leeuwenhoek*. **50**, 1–6.
Van Schothorst, M., Mossel, D.A.A., Kampelmacher, E.H. and Drion, E.F. (1966) The estimation of the hygienic quality of feed components using an Enterobacteriaceae enrichment test. *Zentralbl. Veterinaermed., Reihe B*, **13**, 273–85.
Van Schothorst, M., Van Leusden, F.M., De Gier, F., Rijnierse, V.F.M. and Veen, A.J.D. (1979) Influence of reconstitution on isolation of *Salmonella* from dried milk. *J .Food Prot.*, **42**, 936–7.
Van Winsen, R.L. (2001) Contribution of fermented feed to porcine gastrointestinal microbial ecology: influence on the survival of *Salmonella*, Universiteit Utrecht, VVDO, Fac Diergeneeskunde, ISBN 90-393-2810-2.
Veldman, A., Meijs, J.A.C., Borggreve, G.J. and Heeres-van der Tol, J.J. (1992a) Carry-over of aflatoxin from cow's food to milk, *Animal Prod.* **55**, 163–8.
Veldman, A., Borggreve, G.J., Mulders, E.J. and van de Lagemaat, D. (1992b) Occurence of the mycotoxins ochratoxin A, zearalenone and deoxynivalenol in feed components. *Food Addit. Contam.*, **9**, 647–5.
Veldman, A., Vahl, H.A., Borggreve, G.J. and Fuller, D.C. (1995) A survey of the incidence of *Salmonella* species and Enterobacteriaceae in poultry feeds and feed components. *Vet. Rec.*, **136**, 169–72.
Vesonder, R.F. and Hesseltine, C.W. (1981) Vomitoxin: natural occurrence on cereal grains and significance as a refusal and emetic factor to swine. *Process Biochem.*, **16**, 12, 14–15, 44.
Vielitz, F. (1991) EG Zoonosen Verordung—neue Vorschriften für die Bekämpfung von Geflügelkrankheiten, Lohmann Information, March–April, 9–13.
Wagner, D.E. and McLaughlin, S. (1986) *Salmonella* surveillance by the Food and Drug Administration: a review 1974–1985. *J. Food Prot.*, **49**, 734–8.
Weinberg, Z.O., Ashbell, G., Hen, Y. and Azrieli, A. (1993) The effect of applying lactic acid bacteria at ensiling on the aerobic stability of silage. *J. Appl. Bacteriol.*, **75**, 512–18.
Weiss, J. and Seeliger, H.P.R. (1975) Incidence of *Listeria monocytogenes* in nature. *Appl. Microbiol.*, **30**, 29–32.
Welishimer, H.J. (1968) Isolation of *Listeria monocytogenes* from vegetation. *J. Bacteriol.*, **95**, 300–3.
WHO (World Health Organization) (1992) Public health issues related to animal and human spongiform encephalopathies: memorandum from a WHO meeting. *Bull. World Health Org.*, 183–90.
WHO (2000) Consultation on Public Health and Animal Transmissible Spongiform Encephalopathies: Epidemiology, Risk and Research Requirements, Document WHO/CDS/CSR/APH 2000.2. World Health Organisation, Geneva.
WHO (2001) Joint WHO/FAO/OIE Technical Consultation on BSE: public health, animal health and trade, OIE Headquarters, Paris, 11–14 June 2001. WHO, Geneva, Switzerland.
WHO (2002) Bovine Spongiform Encephalopathy, Fact sheet No 113, Revised November 2002, http://www.who.int/mediacentre/factsheets/fs 113/en.
Wilder, D.M.H. (1971) Feeds, in *The Science of Meat and Meat Products* (eds J.F. Price and B.S. Schweigert), 2nd edn, W.H. Freemann, San Francisco.
Wilesmith, J.W., Wells, G.A., Cranwell, M.P. and Ryan, J.B. (1988) Bovine spongiform encephalopathy: epidemiological studies. *Vet. Rec.*, **123**, 364–8.
Wilesmith. J.W., Ryan, J.B. and Atkinson. M.J. (1991) Bovine spongiform encephalopathy: epidemiological studies on the origin. *Vet. Rec.*, **128**, 199–203.
Williams, J.E. (1981) *Salmonella* in poultry feeds. A world wide review. *World Poult. Sci. J.*, **37**, 6–25, 97–105.
Wilson, R.B., Boley, M.T. and Corvin, B. (1995) Presumptive botulism in cattle associated with plastic packaged hay. *J. Vet. Diag. Invest.*, **7**, 167–9.
Wiseman, J. and Cole, D.J. (eds) (1990) Feedstuff evaluation, in *Animal Pathogens in Feeds*.
Woolford, M.K. (1990) The detrimental effects of air in silage. *J. Appl. Bacteriol.*, **68**, 101–16.
Xylouri, E., Papadopoulou, C., Antoniadis, G. and Stoforos, E. (1997) Rapid identification of *Clostridium perfringens* in animal feedstuffs. *Anaerobe*, **3**, 191–3.
Yamazaki, M., Suzuki, S. and Miyaki, K. (1971) Tremorgenic toxins from *Aspergillus fumigatus*. *Chem. Pharmacol. Bull.*, **19**, 1739–40.
Zimmerli, B. and Dick, R. (1995) Determination of ochratoxin A at the ppt level in human blood, serum, milk and some foodstuffs by high-performance liquid chromatography with enhanced fluorescence detection and immunoaffinity column cleanup methodology and Swiss data. *J. Chromatogr. B*, **666**, 85–99.

5 Vegetables and vegetable products

I Introduction

A Definitions and important properties

A vegetable is the edible component of a plant including leaves, stalks, roots, tubers, bulbs, flowers, fruits, and seeds. In mushrooms, the fruiting body is usually the organ of interest. Although considered by some to be a vegetable, tomatoes are fruits and are included in that chapter. With the exception of certain seeds, plant tissues are low in protein. Water, fiber, starch, certain vitamins, minerals, and some lipids are the principal components. In general, the pH of vegetable tissue is in the range 5–7. Since the overall composition and pH are very favorable, growth of numerous microbial species can be expected if adequate moisture is present.

Virtually, all vegetables in their natural state are susceptible to spoilage by microorganisms, at a rate depending on various intrinsic and extrinsic factors. Drying, salting, freezing, refrigeration, canning, fermentation, irradiation, and packing under vacuum or modified atmospheres are used to preserve plant materials. In some instances, two or more processes are combined. Treatments alter the flora, the environment, or both, in a fashion that extends the stability of the food.

In general, raw vegetables in prime condition support less growth of food-borne pathogens than animal products because the intact cell structure provides a protective barrier. Raw vegetables may support growth of pathogens once cell integrity has been lost through wilting, aging, or injury such as chopping, shredding, bruising, or juicing. Unless protective conditions such as modified atmosphere packaging (MAP) are employed, vegetables are rapidly subject to spoilage when cell integrity is lost, and this reduces the likelihood of consumption. Following the trend in consumer preference for minimally processed foods, and the greater potential for microbiological hazards and spoilage that may occur when preservative steps are minimized, the scientific literature on vegetables in the past decade has almost exclusively concerned fresh or minimally processed commodities.

Produce plays a major role in international trade. Consumption of fresh produce in the United States increased 24% between 1970 and 1997, and outbreaks also increased (Tauxe *et al.*, 1997; Burnett and Beuchat, 2000). A similar trend has been observed in Canada (Sewell and Farber, 2001) and Europe (ECSCF, 2002). In the United States, 1% of food-borne illness outbreaks in the 1970s were associated with produce, compared with 6% in the 1990s (Sivapalasisngam and Friedman, 2001). Outbreaks previously involved fewer individuals, i.e. median of 4 cases/outbreaks in the 1970s compared with 43 cases/outbreak in the 1990s. Outbreaks of human disease due to transmission of pathogenic microorganisms on vegetables usually result from fecal contamination of produce through the use of raw sewage or manure fertilizer (Wachtel *et al.*, 2002); contaminated water for irrigation, cooling, or washing; contaminated ice for transport; or from unhygienic handling. Pathogenic microorganisms, viruses, and parasites can survive for months or years in sewage sludge and soil, or on vegetables (ICMSF, 1988). Spray irrigation with wastewater disperses enteric bacteria through aerosolization (Katzenelson *et al.*, 1997) but contamination is reduced when surface irrigation is used (Solomon *et al.*, 2002). In addition, certain pathogens such as salmonellae can multiply on or in some vegetables.

"Organically grown" foods are increasingly popular in the United States and Europe and are produced in a manner that avoids the use of man-made fertilizers, pesticides, and growth promoters (IFST, 2001). The use of raw animal manure as a fertilizer on organically grown and other vegetable crops increases the potential for the presence of pathogens (RCP, 1996; Nicholson *et al*., 2000). Because of the broad use of manure for organically grown crops, the microbiological safety of these crops is questioned (Tauxe, 1997). Manure has to be properly composted to eliminate pathogenic bacteria (Lung *et al*., 2001). In addition, enteric viruses (which may come from manure compost or sewage sludge) can survive on vegetables for up to 30 days (Badawy *et al*., 1985). The EU regulation 2092/91 strictly governs the cultivation, production, processing, packaging, and labeling of organic crops in Europe, and defines inspection and certification requirements for producers and processors. National standards in the United States went into effect in 2002, and various state and private certification bodies exist.

Vegetarian or vegan diets are also gaining in popularity. Although vegetable products are a main component of a vegetarian diet, fruits, nuts, grains or cereals, and soy-based products are also important. This chapter covers only vegetable products. Special consideration is given to sprouted seeds, which have been the source of food-borne outbreaks around the world (NACMCF, 1999; IFT, 2001). Refer to appropriate chapters for information on other vegetarian food products.

Mushrooms and cassava are crops of great economic importance in many parts of the world. Since their methods of production or processing are rather different to the main classes of vegetables described in this chapter, these crops are considered separately. Conversely, fresh herbs, such as basil and parsley, share microbial ecology similar to that for other leafy vegetables and are addressed in appropriate sections.

II Initial microflora (including field practices and harvest)

Soil, water, air, insects, and animals all contribute to the microflora of vegetables, but their relative importance as sources differs with the structural entity of the plant; e.g. leaves will have greater exposure to air, whereas root crops will have greater exposure to soil. Human activities and agricultural practices are particularly important. For example, the use of pesticides to control insects may limit the spread of microorganisms. Cultivation, either by hand or mechanically, introduces and/or distributes microorganisms into ecological niches from which previously they were absent, and introduction of human or other animal waste material into the water or soil has an obvious impact on the microbial flora of vegetables.

The means by which microbes penetrate plant tissues is not clearly established, but their presence usually is not deleterious to the growing plant. An equilibrium or coexistence occurs, although it can be broken, and spoilage can develop under particular circumstances. External surfaces of vegetables in the field are likely to bear the heaviest microbial load but interior tissues may occasionally contain bacteria that, by various mechanisms, have gained entry. For example, cucumbers may be invaded by *Pseudomonas lachrymans* via the fruit stomata (Wiles and Walker, 1951) or from wounds at the junctions of stems and peduncles (Pohronezny *et al*., 1978). Agents such as insects, fungi, nematodes, animals, birds, rain or hail can facilitate entry of spoilage bacteria (Lund, 1983). Irrigation waters contaminated with parasites and microsporidia can also be sources of contamination for fruits and vegetables. In a survey of irrigation water used for crops in the United States and several Central American countries, 28% of water samples tested positive for microsporidia, 60% tested positive for *Giardia* cysts, and 35% tested positive for *Cryptosporidium* oocysts (Thurston-Enriquez *et al*., 2002). The hands of personnel involved in picking, trimming, sorting, tying, and packaging and the equipment used in these operations also contribute to the number of microbes and their distribution on the product. Harvesting often injures the produce resulting in release of nutrients that enhance microbial growth and points of entry for microorganisms, which would result in the spoilage of the produce. Containers and vehicles used to transport vegetables are an additional source of microorganisms, especially when they are used repeatedly without cleaning.

Table 5.1 Common fungi causing spoilage of market vegetables

Vegetable	Genus	Type of spoilage
Carrots	*Alternaria*	Black rot
Celery	*Sclerotinia*	Watery soft rot
Lettuce	*Bremia, Phytophthora*	Downy mildew
Onions	*Aspergillus*	Black mold rot
	Colletotrichum	Smudge (anthracnose)
Asparagus	*Fusarium*	Fusarium rot
Green beans	*Rhizopus*	Rhizopus soft rot
	Pythium	Wilt
Potatoes	*Fusarium*	Tuber rot
Cabbages	*Botrytis*	Gray mold rot
Cauliflower	*Alternaria*	Black rot
Spinach	*Phytophthora*	Downy mildew

A *Saprophytic microorganisms*

The microorganisms present on fresh vegetables are derived from the soil, air, and water. Most spoilage is caused by fungi, chiefly members of the genera *Penicillium*, *Sclerotinia*, *Botrytis*, and *Rhizopus* (Wu *et al.*, 1972; Pendergrass and Isenberg, 1974; Goodliffe and Heale, 1975). Table 5.1 lists some of the diseases (often manifested by spoilage) of market vegetables and the microorganisms responsible. Most leafy decay in the field is caused by bacteria, for example, *Ps. cichorii*, *Ps. marginalis*, and *Erwinia carotovora* (Grogan *et al.*, 1977; Ohata *et al.*, 1979; Tsuchiya *et al.*, 1979; Miller, 1980), or by fungi such as *Sclerotinia* spp., or *Botrytis cinerea* (Nguyen-the and Carlin, 2000). It has also been reported that *Sclerotinia sclerotiorum*, the cause of pink rot of celery, produces phytotoxins, which cause a blistering cutaneous reaction in field workers who handle the produce (Wu *et al.*, 1972).

Other saprophytes include coryneforms, lactic acid bacteria, spore-formers, coliforms, and micrococci. Although lactic acid bacteria usually are present only in low numbers, their importance increases both in relation to spoilage after processing, and in the fermentation of vegetables by natural processes. Of the yeasts, *Rhodotorula* spp., *Candida* spp., and *Kloeckera apiculata* tend to predominate (Nguyen-the and Carlin, 2000). Fungi, including *Sclerotinia* spp., *Bot. cinerea*, *Aureobasidium pullulans*, *Fusarium* spp., and *Alternaria* spp. are often present, but in lower numbers than bacteria. The strictly anaerobic microorganisms that also may be present have not been well characterized, except for certain heat-resistant spore-formers important in spoilage of canned vegetables.

B *Pathogens*

Generally, fresh vegetables that have not been exposed to raw human or other raw animal waste material do not contain animal and human pathogens, with the exception of those which occur naturally in soil, decaying vegetable matter, etc. *Bacillus cereus*, *Listeria monocytogenes*, and *Clostridium botulinum* are found in soil and are present on fresh produce. Irrigation and fertilization of vegetable crops with raw human and raw animal wastes or contaminated surface water can contribute the etiological agents of infectious hepatitis, typhoid fever, shigellosis, salmonellosis, listeriosis, viral gastroenteritis, cholera, amoebiasis, giardiasis, and other enteric as well as parasitic diseases (Beuchat and Ryu, 1997; Robertson and Gjerde, 2001). Root crops and low-growing leaf and stalk crops are heavily contaminated by using sewage effluent or contaminated irrigation water (Geldreich and Bordner, 1971; Nichols *et al.*, 1971; Solomon *et al.*, 2002). A leak of partially treated sewage water into a creek used for irrigation of commercial produce resulted in contamination of a crop of cabbages with at least six different serotypes of *Escherichia coli* (Wachtel et al., 2001). Viruses from sewage do not bind readily with soil particles and can enter groundwaters leading to contamination of water sources (Seymour and Appleton, 2001). Travelers are frequently warned to avoid eating raw vegetables in certain regions of the world to prevent traveler's diarrhea. Manure management and the potential impact on the presence of pathogens on vegetables have also been questioned in developed countries in relation to disposal of large quantities of animal waste from feed lots. Use of Good Agricultural Practices greatly minimizes the potential for contamination of vegetable crops with pathogens (FDA, 1998; ECSCF, 2002). Specific pathogens are discussed in Section III.

C Good agricultural practices

Pre-harvest

The initial flora of raw vegetables is greatly influenced by pre-harvest agricultural practices. As the incidence of food-borne disease associated with contamination of produce increases even in developed countries, there is need to focus on the use of good agricultural practices for growing vegetables that are intended to be consumed in the raw state. The EU Council Directive 86/278 regulates the use of sewage sludge in the EU (MAFF, 1998). The EPA regulations 40 CFR Part 503 (EPA, 2002) outlines the standards for the use or disposal of sewage sludge in the US. This also covers the use of treated sewage sludge as a soil amendment. Among practices that may be restricted are grazing of animals or harvesting foraging crops within 3 weeks of application and the harvesting of fruit or vegetable crops within 10 months of application. Treatments required for sewage sludge to reduce the public health impact from human pathogens include pasteurization, anaerobic digestion, aerobic digestion, composting, lime stabilization, liquid storage, or dewatering and storage. Pre-harvest agricultural practice that can minimize the potential for crop contamination with human pathogens and include the following (IFPA and WGA, 1997; FDA, 1998; Brackett, 1999):

Soil. Evaluate previous use of fields to minimize the potential for initial contamination. Avoid use of grazing pastures for crops that are consumed raw. Minimize animal access to fields. Avoid contamination of crops from run-off water by eliminating grazing pastures that are up hill from crop fields or by controlling run-off during rain.

Fertilizer. Monitor manure and compost during the break down process to assure that pathogens are eliminated prior to use on crops intended to be consumed in the raw state. Pathogens can survive for months in manure and manure slurry (Wang *et al.*, 1996; Nicholson *et al.*, 2000). Jiang *et al.* (2001) reported that *E. coli* O157:H7 survived for 6 months at 15°C or 21°C when contaminated manure was mixed with soil. *E. coli* O157:H7 and *Salmonella* Enteritidis inoculated at 10^7 cfu/g of manure did not survive for 72 and 48 hr, respectively, when composted at 45°C; however, no change in population was observed when held at room temperature (Lung *et al.*, 2001). The UK regulations require that manure be composted for a minimum of 3 months and reach a temperature of 60°C prior to use; however, the effectiveness of this practice for the entire compost pile remains to be demonstrated (IFST, 2001). A recent review of the effect of manure management practices on the microbial profile of produce is available in the IFT task force report (IFT, 2001).

Irrigation. Evaluate the water source for the potential presence of pathogens. Surface water is more readily contaminated than well water; however, even well water must be protected against agricultural run-off to prevent contamination. Testing for *E. coli* may be a useful indicator of potential fecal contamination for surface- and well water. Choice of the optimum irrigation method may vary with the water source. Spray irrigation spreads potential contamination through aerosols, whereas drip irrigation reduces the potential for contamination.

Field workers. Education of field workers on the potential for contamination of crops is essential to minimize the potential for crop contamination. Ill workers should not work in the fields. Adequate toilet and hand washing facilities are necessary to prevent the contamination of crops that are consumed raw. Facilities should be well maintained in an area that will not cause water run-off into the crop field. Portable toilets should be removed from the field area to be emptied if possible, and creation of aerosols of the human waste must be prevented. The US regulation 29 CFR 1928.110 details the minimum requirements for field sanitation, including handwashing facilities, toilet facilities, potable drinking water, maintenance of handwashing and toilet facilities, etc. The Occupational Safety and Health Administration (OSHA) of the labor department enforces these regulations. Some US states have requirements that are more stringent than the Federal OSHA standards.

Post-harvest

Removal of damaged or dirty areas of vegetables can decrease the numbers of microorganisms and should be a primary step in harvesting and preparation (Adams *et al.*, 1989; IFPA and WGA, 1997). Field containers should be washed to remove soil prior to cooling to minimize the potential for pathogens in the final washed vegetables (IFPA and WGA, 1997).

Studies of vegetables at the time of harvest showed mean mold population ranging from $<10^3$–6.7×10^4 cfu/g (or cm^2) of tissue. The number of molds on the plant tissue increased when rainfall immediately preceded harvest and when the temperature was below 24°C (Webb and Mundt, 1978). The quantity of soil adhering to the plant at the time of sampling strongly affects the microbial load. Total aerobic bacterial population range from 10^3–10^7/g (Table 5.2). Counts vary broadly among and between vegetable types. Nguyen-the and Carlin (2000) reviewed initial aerobic mesophilic bacterial population on fresh and processed vegetables.

III Raw and minimally processed vegetables

Raw vegetables may be harvested and shipped directly to the retail market, or retained in storage for varying periods of time before sale. Several general commodity types based on raw or relatively unprocessed vegetables are definable, although the distinctions between them are not always clear. Raw vegetables may be placed into bags, boxes, or crates with little or no washing. Minimally processed vegetables are chopped, shredded, peeled, etc., before being presented for sale. Blending and extracting are also used to prepare vegetable juices. Prepared salads contain chopped, shredded, or whole vegetables in a dressing (Brocklehurst, 1994). The common mayonnaise or vinaigrette dressings give products with pH ranges of 3.2–5.1 or 3.7–4.0, respectively, whereas with other dressings the pH may be 4.0–5.7 (Rose, 1984) or as high as 6.6 (Pace, 1975; Terry and Overcast, 1976). Some prepared salads contain dressing in a separate package. Generally, microbes do not survive well in mayonnaise, salad dressings, and sauces although *E. coli* O157:H7 appears to be inactivated at a slower rate than other *E. coli* strains (Smittle, 2000).

MAP may be used to control the rate of senescence of vegetables and increase the shelf life. Generally, this is accomplished by reducing the oxygen content and increasing the CO_2 content. Because raw vegetable tissue continues to respire, precise atmosphere control is difficult to achieve. Temperature control is essential for maintaining quality, especially for minimally processed vegetables.

A Effects of transportation, processing, and storage on microorganisms

The microbial population on raw vegetables is influenced by many factors. The importance of agricultural and harvesting practices was discussed in the previous section. Initial flora in the field remains associated with raw vegetables during transportation, further processing, and storage.

Table 5.2 Total bacterial population on vegetables

Vegetable	$\log_{10}$ count/g	Reference
Asparagus	4.0–5.0	Berrang *et al.* (1990)
Beans (green, pieces)	6.0–7.6	Swanson (1990)
Beets	6.5[a]	Splittstoesser (1970)
Broccoli	4.0	Berrang *et al.* (1990)
	6.7	Brackett (1989)
Broccoli (florets)	3.9–6.7	Swanson (1990)
Cabbage (nappa, leaves and sliced)	5.5–7.4	Swanson (1990)
Cabbage (red, sliced)	3.6–5.8	Swanson (1990)
Cabbage (white)	3.6–6.3[a]	Splittstoesser (1970)
	4.3	Garg *et al.* (1990)
Carrots	5.6[a]	Splittstoesser (1970)
Carrots (cut)	3.7–7.3	Swanson (1990)
Cauliflower	4.0–5.0	Berrang *et al.* (1990)
Cauliflower (florets)	3.9–6.7	Swanson (1990)
Corn	5.0–7.0[a]	Splittstoesser (1970)
Kale	6.1–7.0[a]	Splittstoesser (1970)
Lettuce	4.3	Garg *et al.* (1990)
Lima beans	3.0–5.2[a]	Splittstoesser (1970)
Mushrooms (sliced)	5.3–8.9	Swanson (1990)
Onions (green, sliced)	6.3–7.7	Swanson (1990)
Onions (red, sliced)	3.0–6.9	Swanson (1990)
Peas (green)	5.3–7.5[a]	Splittstoesser (1970)
Peas (trimmed)	4.9–5.9	Swanson (1990)
Peas (snow, trimmed)	3.8–7.7	Swanson (1990)
Peppers (green, red, yellow, cut)	6.0–7.8	Swanson (1990)
Peppers (whole)	3.3–4.1	Swanson (1990)
Potatoes	4.9–7.5[a]	Splittstoesser (1970)
Snap beans	5.8–6.5[a]	Splittstoesser (1970)
Spinach	6.3–7.4[a]	Splittstoesser (1970)
Squash (yellow cut)	4.6–7.1	Swanson (1990)
Zucchini (cut)	6.9–8.4	Swanson (1990)
Zucchini (whole)	4.1–7.4	Swanson (1990)

[a]Data from seven investigations.

There are relatively small differences between the microbial population on protected vegetables like peas in a pod and unprotected vegetables like potatoes (Table 5.2). Similarly, there are few differences in the population between root and leaf crops, though this may reflect the greater surface area per gram of leaf vegetables. Washing can remove up to 1–2 logs of the surface flora, but those microorganisms trapped in mucilaginous exudate on the vegetable will remain. Washing can be deleterious in that any residual water on processed or unprocessed vegetables will allow rapid multiplication of resident microbes.

Additional handling occurs when the produce reaches wholesale and retail markets. This should be done under adequate refrigeration and with proper attention to hygiene to maintain optimal quality and food safety. Produce may be unpacked, trimmed, remoistened, repackaged, and displayed for sale. Some vegetables are cut or chopped for use in ready to eat products. Microorganisms multiply faster on cut produce, owing to the greater availability of nutrients and water. Substantial differences in the bacterial population of cut and whole peppers and zucchini have been observed (Swanson, 1990; Table 5.2).

Further handling of produce affords the opportunity for contamination (including pathogens) originating from the handler, work surfaces, or utensils that were previously exposed to other food materials. Sufficient moisture, appropriate temperature, and adequate time will ensure a continuing increase in the bacterial population, although it is difficult to estimate the magnitude of the change in microbial load during these manipulations. The use of gloves by food handlers can reduce the transmission of food-borne pathogens if properly used. However, a majority of gloves were shown to be permeable to bacteria in a simulation of actual use, a permeability that may increase over time of usage (Montville *et al.*, 2001).

B Saprophytes and spoilage

The predominant microorganisms on healthy raw vegetables are usually bacteria, although significant numbers of molds and yeasts may be present (Koburger and Farhat, 1975; Splittstoesser *et al.*, 1977; Swanson, 1990). Hydrocooling may also introduce spoilage flora if water quality is not maintained (NFPA, 2001). Certain vegetables can internalize organisms if washed in water that is colder than the produce (FDA, 1998). This has led to the recommendation that the rinse water for certain commodities should be hyperchlorinated and at least 10°C warmer than the incoming product.

Bacteria contribute significantly to post-harvest microbial spoilage. Such spoilage may be due to bacteria that cause soft rots and other rots, spots, blights, and wilts. Soft rots, occurring during transport and storage, are usually caused by coliforms, *Er. Carotovora*, and certain pseudomonads, e.g. *Ps. fluorescens* (*marginalis*). Microorganisms causing spoilage other than soft rot include corynebacteria, xanthomonads, and pseudomonads (Brocklehurst and Lund, 1981). Contamination by soft-rot microorganisms often occurs in the field, but invasion of the plant tissue occurs following trauma induced by subsequent transport and storage. Soft rot of potatoes by *Clostridium* spp. has also been reported (Lund and Nicholls, 1970).

The soft rot by which most fresh-cut vegetables deteriorate was investigated in "ready-to-use" fresh salads of endive and chicory that had been washed, peeled, sliced or shredded, packed, stored below 10°C, and sold within 8–10 days. Deterioration was mainly due to fluorescent pectinolytic pseudomonads (specifically *Ps. marginalis*). A clear relationship existed between levels of *Ps. marginalis* (which is widespread on plant leaves) inoculated onto salads and shelf life. *Erwinia herbicola* was the next most commonly isolated organism (Nguyen-the and Prunier, 1989).

Total microbial population on carrots in Quebec were found to average 9×10^6 cfu/g with mold population of 4×10^3/g. These population were reduced from10- to 100-fold by washing, brushing, and rinsing, or by washing, brushing, complete peeling, and rinsing (Munsch *et al.*, 1982).

A study of raw, minimally processed vegetables showed a wide range of aerobic mesophilic population varying from 70 microorganisms/g (peeled potato) though carrot, celery, endive, and minestrone mix, to as high as 3×10^8 microorganisms/g (wheat sprouts). *Staphylococcus aureus* was absent, and contamination levels by Enterobacteriaceae, yeasts, and molds were very low. After storage for 6 days at 4°C and 20°C, aerobic colony counts of mixed salad vegetables reached 3×10^6/g and 3×10^8/g, whereas coliform counts were <100/g and 1×10^6/g, respectively (Masson, 1988).

In coleslaw and chopped lettuce, Gram-negative rods (principally *Pseudomonas* spp.) were the predominant microflora, with population of mesophiles ranging from 3×10^4/g to 1×10^6/g and psychrotrophs from 3×10^3 to 3×10^5/g in packaged produce. Peeling carrots and onions, or removing the outer leaves of lettuce, cabbage, and cauliflower, reduced the level of the intrinsic flora, whereas immersion in ice-cold chlorinated water (target level of 300 mg/L free Cl_2) had a marginal effect. Production equipment hygiene can also have a significant effect. The most serious sites for microbial build-up were the shredders and slicers, where counts tended to increase 100-fold. Only low numbers of lactic acid bacteria and (except for carrot sticks) fungi were found (Garg *et al.*, 1990). Many microbial species multiply rapidly in cut vegetables. For example, at the end of their shelf life after storage at 7–10°C, prepared vegetables had counts of *Pseudomonas* spp. and lactic bacteria in excess of 10^8/g (Brocklehurst *et al.*, 1987; Nguyen-the and Prunier, 1989).

Extending the shelf life of minimally processed, refrigerated prepackaged vegetables through use of modified atmosphere packaging may be potentially hazardous, since growth of many pathogens can occur in the absence of other controls. MAP slowed the growth of spoilage microorganisms during cold storage of chicory endive, but *L. monocytogenes* were not inhibited (Bennick *et al.*, 1996). An extension of shelf life of fresh-cut produce to $\geq 50\%$ was achieved by MAP, whereas growth of *L. monocytogenes* and *Aeromonas* spp. proceeded (Jacxsens *et al.*, 1999).

Refrigeration is an important microbial control factor throughout processing and storage. Microbial growth on cabbage or unacidified coleslaw is essentially prevented at 1°C, but at higher temperatures deterioration is primarily due to tissue break down rather than to microorganisms (King *et al.*, 1976). Because certain pathogens can survive or even grow at 5°C (Berrang *et al.*, 1989a,b), the pros and cons of extending the storage life of minimally processed vegetables must be considered. The ability to maintain effective refrigeration varies considerably, and in some regions it is very common for refrigeration conditions to exceed 4°C during distribution, retail display, or in the home. Processing or packaging may delay growth of normal spoilage microorganisms at elevated temperatures, but not the growth of some psychrotrophic pathogens such as *L. monocytogenes* (Brackett, 1987). Other pathogens capable of growth at higher temperatures may become a problem, for example, if refrigerated vegetables are more grossly temperature abused (King and Bolin, 1989).

A second or third control factor in addition to temperature (e.g. acidity, preservative, atmosphere, bacteriocin) may be desirable to prevent the growth of pathogens. The atmosphere may be controlled through such means as careful choice of packaging to control diffusion of gases, or by MAP. Preservatives such as sulfite, ascorbic acid, and other additives are not uncommon. Optimum MAP conditions are very specific to the food commodity, and numerous papers describe modifications; most of which deal with texture, flavor, and nutritional factors, completely ignoring microbial growth. Nguyen-the and Carlin (2000) reviewed microbial considerations for modified atmosphere storage of fresh vegetables.

Optimum temperature for preventing spoilage depends on the commodity. Injury from low-temperature storage may enhance microbial rotting in tropical or subtropical produce such as cucumbers, tomatoes, bell peppers, and potatoes (Nguyen-the and Carlin, 2000).

C Pathogens

Pathogens can be introduced onto vegetables as a result of agricultural practices previously discussed in the initial flora section. Contamination of produce by pathogens should be expected in countries where animal wastes are used as fertilizer. The efficacy of subsequent washing with chlorinated water to remove pathogens is questionable (Nichols *et al.*, 1971; Zhang and Farber, 1996). Washing lettuce in warm chlorinated water (100 ppm), can even favor the growth of food-borne pathogens such as *L. monocytogenes* and *E. coli* O157:H7 (Delaquis *et al.*, 2001).

Survival times for coliforms, bacterial pathogens, and enteric viruses on most raw vegetables are moisture- and temperature-dependent and extend significantly beyond the useful life of the products (Geldreich and Bordner, 1971; Nichols *et al.*, 1971; Konowalchuk and Speirs, 1975). The presence of pathogens on market produce is amply documented (Papavassiliou *et al.*, 1967; Garcia-Villanova Ruiz *et al.*, 1987; Kaneko *et al.*, 1999; IFT 2001). Even when night soil is not used, contamination of vegetables with human pathogens can occur. Reviews list more than 30 kinds of vegetables from which pathogens have been isolated (Beuchat, 1996; Nguyen-the and Carlin, 2000; IFT, 2001). A risk profile for microbial contamination of raw vegetables was recently published (ECSCF, 2002).

Shigella and Salmonella. Outbreaks of shigellosis have involved lettuce in several countries (Martin *et al.*, 1986; Davis *et al.*, 1988; Frost *et al.*, 1995; Kapperud *et al.*, 1995), green onions (Cook *et al.*, 1995), and parsley (Crowe *et al.*, 1999). Parsley was contaminated with *Shigella sonnei* (Naimi *et al.*, 2003). In each of the outbreak-associated restaurants, parsley was chopped, held at room temperature, and used as an ingredient or garnish for multiple dishes, whereas infected food workers at several restaurants could have contributed to the propagation of the outbreak, most of the implicated parsley was sourced from a particular farm in Mexico. *Shigella* spp. also have been isolated from Egyptian raw vegetables and salads (Saddick *et al.*, 1985; Satchell *et al.*, 1990). *Shigella sonnei* survives or proliferates in shredded cabbage when packaged and stored under vacuum or modified atmosphere, as well as under aerobic conditions (Satchell *et al.*, 1990). *S. sonnei* population on parsley decreased 1 log per week at 4°C, but increased 1–3 logs when held at 21°C, illustrating the effectiveness of temperature control (Crowe *et al.*, 1999). *Salmonella* spp. were detected in 5 of 3852 samples of retail bagged prepared ready-to-eat salad vegetables. Of these five samples, two were "wild rocket" salads, one four-leaf salad, and one organic salad (little gem, lollo rosso, rocket, and mizuna), all from the same grower in Italy. The other salad was a four-leaf salad from a grower in Spain, and it was this salad that contained *Salmonella* Newport PT33. A further investigation uncovered a national outbreak of salmonellosis, with the same strain of *S.* Newport PT33 being isolated from 19 individuals in England and Wales (Sagoo *et al.*, 2003).

A range of *Salmonella* serovars have been isolated from vegetables (Tamminga *et al.*, 1978a,b; Garcia-Villanova Ruiz *et al.*, 1987; Jerngklinchan and Saitanu, 1993; D'Aoust, 1994; NACMCF, 1998; IFT, 2001). Extensive outbreaks of salmonellosis attributable to contaminated bean sprouts (O'Mahony *et al.*, 1990), alfalfa sprouts (Mahon *et al.*, 1997; Van Beneden, 1996; NACMCF, 1998; Winthrop *et al.*, 2003), celery, lettuce, cabbage, endive, and watercress have occurred (Beuchat, 1996; Nguyen-the and Carlin, 2000). Salmonellae also have been found in a wide range of "organically grown" products including beans, peas, sunflower seeds, and alfalfa (Andrews *et al.*, 1979). In produce grown in the United States, 1 of 62 cilantro samples contained *Salmonella* and *Shigella*, 2 of 111 lettuce samples contained *Salmonella*, and 3 of 66 green onion samples contained *Shigella* (FDA, 2001). Out of 767 commodities analyzed, 6 samples each were positive for *Salmonella* or *Shigella*. A total of 1003 imported fruits and vegetables from 21 countries were sampled, with 4.4% testing positive for *Salmonella* or *Shigella* (FDA, 2001).

Salmonella inoculated onto fresh broccoli and cauliflower had a similar growth rate at 25°C, with a generation time of 2–4 h and a lag of 3 h. At 15.5°C, growth was more rapid on cauliflower than on broccoli with generation times of ca. 7–8 and 13–25 h, respectively, and lag times of ca. 24 h. *Salmonella* population declined ca. 1.5 logs in 13 days at 7.5°C (Swanson, 1987).

Disinfection dips and treatments can reduce *Salmonella* population on minimally processed raw vegetables; however, complete inactivation cannot be assured (Beuchat and Ryu, 1997; Weissinger *et al.*, 2000; IFT, 2001). The use of bacteriophage as a biocontrol method for *Salmonella* on fresh-cut vegetables has been examined. Although effective, phages have to be strain-specific and phage delivery must be optimized to be fully effective in reducing or eliminating *Salmonella* (Leverentz *et al.*, 2001).

A study by Caldwell *et al.* (2003) showed that free-living nematodes can ingest pathogens and protect them against produce sanitizers. Using *S.* Poona ingestion by the nematode *Caenorhabditis elegans* as a model, it was found that *Caen. elegans* containing *S.* Poona protected the bacteria against sanitizers when inoculated onto the surface of lettuce, but not when inoculated onto the surface of cantaloupes.

E. coli. Raw vegetables, especially lettuce, have been identified as a common cause of travelers' diarrhea (Merson *et al.*, 1976; Beuchat, 1996). Although outbreaks of *E. coli* O157:H7 are normally associated with ground beef products, salad bar vegetables have occasionally been implicated, presumably due to cross-contamination from beef (Barnett *et al.*, 1995; Beuchat, 1996). Lettuce was the source of an outbreak of *E. coli* O157:H7 illness involving more than 100 people in Montana during 1995 (CDC, 1995) and in Ontario, involving 21 people that same year (Sewell and Farber, 2001). A very large *E. coli* O157:H7 outbreak in Japan, involving over 6 000 cases, was associated with raw radish sprouts. Multiple PFGE patterns were isolated and the source of the contamination is unknown (Michino *et al.*, 1996). Carrots were also involved in an enterotoxigenic *E. coli* outbreak (CDC, 1994). A survey of 63 salads served at 31 foodservice outlets found *E. coli* (but no O157:H7) in 8 of the samples and *L. monocytogenes* in 1 sample (Lin *et al.*, 1996).

Survival or growth of *E. coli* O157:H7 on shredded lettuce was not affected by packaging under modified atmospheres (Abdul-Raouf *et al.*, 1993). *E. coli* O157:H7 tolerates acidic environments more effectively than *Salmonella*. Because of this, it may persist in environments previously thought to be safe. The organism survived in sheep manure for more than 21 months under fluctuating temperatures (Kudva *et al.*, 1995). A study looking at the survival of *E. coli* in bovine manure added to soil showed that the organism can survive at least 19 weeks at 9–21°C (Lau and Ingham, 2001) and can be detected in soil 21 weeks after manure application (Natvig *et al.*, 2002). However, proper composting of manure should be able to eliminate all *E. coli* O157:H7 present (Lung *et al.*, 2001).

Goodburn (1999) reviewed potential implications and appropriate controls for verotoxigenic *E. coli* in minimally processed foods. Good agricultural practices are essential because of the low infectious dose for the organism and the lack of effective processing treatments to eliminate the organism in minimally processed vegetables. Transmission of *E. coli* to lettuce has been demonstrated to occur through irrigation regardless of the method of irrigation (Solomon *et al.*, 2002). However, Little *et al.* (1999) found little or no contamination with *E. coli* on imported lettuce, and when the organism was detected, the lettuce had been bought from a location other than a supermarket. A recent study by Solomon *et al.* (2002) suggests that *E. coli* O157:H7 in irrigation water and manure can be internalized into lettuce plant tissue. However, this study may not reflect typical growing conditions, and more work needs to be done in this area.

L. monocytogenes. Because it is able to grow under refrigerated storage conditions, *L. monocytogenes* has been studied extensively in vegetables. The organism is widely distributed in soil and on plant vegetation, where it can persist for long periods (Beuchat *et al.*, 1990; Beuchat, 1996). Pinner *et al.* (1992) reported that 11% of 683 vegetables sampled from listeriosis patients' refrigerators were contaminated with *L. monocytogenes*. Although *L. monocytogenes* is commonly present in the environment, outbreaks associated with vegetables are rare (FDA and FSIS, 2001). An outbreak of listeriosis was attributed to *L. monocytogenes*-contaminated cabbage (probably fertilized by sheep manure) used to make coleslaw (Schlech *et al.*, 1983). Celery, tomatoes, or lettuce were also implicated (although not conclusively) on epidemiological grounds (Ho *et al.*, 1986). Other vegetables have been associated epidemiologically or in sporadic cases (FDA and FSIS, 2001).

L. monocytogenes population on ready-to-eat vegetables is likely to be low, with the median percentage of contaminated servings with <1 cfu/g at 91% (FDA and FSIS, 2001). *L. monocytogenes* (mainly serotype 1) was isolated from market cabbage, cucumbers, potatoes, and radishes in the United States (Heisick *et al.*, 1989), from 11 of 25 samples of fresh-cut vegetables in The Netherlands (Beckers *et al.*, 1989), from 7 of 66 samples of salad vegetables in Northern Ireland (Harvey and Gilmour, 1993), and from high percentages of bean sprouts and leafy vegetables in Malaysia (Arumugaswamy *et al.*, 1994). Guerra *et al.* (2001) surveyed ready-to-eat and processed foods produced in Portugal. Vegetables were purchased wrapped from local markets or supermarkets. One of the 23 samples of ready-to-eat vegetables contained *L. innocua* and none of the 14 samples of raw vegetables contained any pathogens. *Listeria* spp. were not detected in 110 samples of lettuce, celery, tomatoes, and radishes analyzed in Canada (Farber *et al.*, 1989) or in freshly marketed vegetables in the United StatesUS (Petran *et al.*, 1988) or in 27 raw vegetables cut for salad in Japan (Kaneko *et al.*, 1999). *L. monocytogenes* and a number of *Listeria* spp. were found on 126 samples of selected fresh produce from retail markets in the Washington, D.C. area (Thunberg *et al.*, 2002). In another study, Porto and Eiroda (2001) examined 250 vegetable samples and found an incidence of *Listeria* of 3.2%. Sagoo *et al.* (2001) surveyed the microbiological quality of 3 200 samples of ready-to-eat organic vegetables. They found the incidence of *Listeria* spp. (not including *L. monocytogenes*) to be 0.2% (6/3 200) with four samples containing levels $>10^2$ cfu/g. The other samples were below the limit of detection (20 cfu/g in a 25 g sample). *L. monocytogenes* was not found in any of the 3 200 samples. The frequency of *L. monocytogenes* contamination is variable with harvest area, temperature, fertilizers, washing procedures, and contact with soil, all affecting contamination (NACMCF, 1991).

Growth of *L. monocytogenes* on vegetables is slow at refrigeration temperatures, with a median growth rate of 0.07 logs/day at 5°C (FDA and FSIS, 2001) and 0.5–1.5 logs in 7 days on lettuce and endive at 10°C (Carlin and Nguyen-the, 1994). *L. monocytogenes* increased about 10^4-fold within 6 days on asparagus, broccoli, and cauliflower stored at 15°C in air or under modified atmosphere, although at 4°C the organism decreased or increased only 10-fold (Beuchat *et al.*, 1986; Berrang *et al.*, 1989a; Beuchat and Brackett, 1990b). *L. monocytogenes* also multiplied in shredded lettuce stored at 5°C, 12°C, and 25°C, although the increase was only 10-fold after 14 days at 5°C and 12°C, and 1000-fold at 25°C (Steinbruegge *et al.*, 1988). Various salads left at 4°C for 4 days supported small (two-fold) increases in *L. monocytogenes* (Sizmur and Walker, 1988). Farber *et al.* (1998) found that *L. monocytogenes* inoculated onto fresh-cut vegetables and stored at 4°C for 9 days showed little to no growth. The least growth was on carrots, but considerable growth occurred on butternut squash. The same study found good growth of *L. monocytogenes* on all of the fresh-cut vegetables tested when stored at 10°C for 9 days, except for chopped carrots, where the population decreased ~2 log units by the end of the test time. The microorganism increased about 1 log in numbers at 5°C in shredded cabbage stored under normal atmosphere and modified with 70% carbon dioxide and 30% nitrogen (Kallender *et al.*, 1991). Certain vegetables may contain natural inhibitors of *L. monocytogenes* (Beuchat and Brackett, 1990a; NACMCF, 1991). Carrots had an inhibitory effect on *L. monocytogenes* during storage at 5°C or 15°C, though this inhibition disappeared if the carrots were cooked. Even 1% of raw carrot juice in broth culture media inhibited growth of this microorganism (Beuchat and Brackett, 1990a; Beuchat *et al.*, 1994; Beuchat and Doyle, 1995). Heat treatment (20°C or 50°C) of lettuce can lead to the enhancement of *L. monocytogenes* growth during subsequent storage at 5°C or 15°C (Li *et al.*, 2002). Controlled atmosphere storage did not appear to influence growth rates of *L. monocytogenes* (Berrang *et al.*, 1989a).

Yersinia. Isolation of *Yersinia enterocolitica* only gains significance when bioserogroups known to be human pathogens are detected. Illness is more commonly associated with animal products than with vegetables (ICMSF, 1996). *Y. enterocolitica* was isolated from ~50% of raw vegetables sampled in France; however, serotypes O:3 and O:9 that cause human illness were not found (Delmas and Vidon, 1985). Other studies reported the presence of *Y. enterocolitica* in salads and raw vegetables in The Netherlands (de Boer *et al.*, 1986), grated carrot from restaurants in France (Catteau *et al.*, 1985), and in vegetable salad mixes (Brocklehurst *et al.*, 1987), however, serotypes were not specified.

Spore-formers. Spore-forming pathogens cannot be eliminated by any of the available processing technologies currently in use for raw vegetables. It should be assumed that these pathogens may be present and appropriate storage conditions should be used. *Bacillus cereus* food-borne illness normally involves rice, but green salads, cress, mustard sprouts, and mung bean sprouts have been implicated (Roberts *et al.*, 1982; Portnoy *et al.*, 1976). Although fruits and vegetables have been implicated in 24% of cases of botulism (Hauschild, 1989), these have mostly been due to improperly canned products. Because vegetables are often in contact with the soil, they are easily contaminated with spores of *Cl. botulinum*. Hauschild *et al.* (1975) found 15 *Cl. botulinum* spores per 100 g of unwashed mushrooms and 41 *Cl. botulinum* spores per 100 g of washed mushrooms. Solomon and Kautter (1986, 1988) isolated *Cl. botulinum* from onion skins and fresh garlic clove skins, and Solomon *et al.* (1990) isolated *Cl. botulinum* type A from the outer leaves of fresh cabbage. Concern that modified atmosphere packing of vegetables might provide an anaerobic environment conducive to growth of *Cl. botulinum* led to examination of 1 118 samples of vegetables packed under modified atmospheres, in 454 g lots (Lilly *et al.*, 1996). The incidence of *Cl. botulinum* spores was only 0.36%.

Raw vegetables containing *Cl. botulinum* are unlikely to be hazardous unless they have been prepared or stored under conditions that will lead to multiplication and toxin formation. For example, chopped garlic in soybean oil, with pH >4.6 and stored unrefrigerated, was responsible for a large outbreak of type B botulism (Health and Welfare Canada, 1986; FDA, 1989). This outbreak followed temperature abuse of commercial products, e.g. chopped garlic in oil was unheated, contained no acidulants or preservatives, and relied solely upon refrigeration for safety (Solomon and Kautter, 1988; St. Louis *et al.*, 1988).

Vegetable tissues damaged by cutting, particularly those cut rather than torn, have a higher respiration rate than undamaged tissues (Brocklehurst, 1994). This higher rate could cause lower O_2 concentrations and increase the risk of *Cl. botulinum* if the products are improperly stored at elevated temperatures (Sugiyama and Yang, 1975; Tamminga *et al.*, 1978a). Laboratory studies have confirmed the ability of *Cl. botulinum* to produce toxin in fresh mushrooms wrapped in film with no holes (Kautter *et al.*, 1978), and bottled chopped garlic (Solomon and Kautter, 1988). The potential risk of botulism from cut vegetables such as cabbage and lettuce is questionable and dependent upon storage temperature and the organoleptic quality of the product when toxin is produced (Solomon *et al.*, 1990 and Petran *et al.*, 1995). The risk of botulinal toxin being produced prior to spoilage in modified atmosphere packaged produce has been estimated as being <1 in 10^5 (Larson *et al.*, 1997). Austin *et al.* (1998) conducted a study of vegetables inoculated with *C. botulinum* spores and incubated at 5°C, 10°C, 15°C, and 25°C in MAP for up to 21 days, or as soon as the samples became toxic. Results showed that *C. botulinum* could grow on a variety of MAP vegetables if the storage temperature is conducive to growth. The reduced O_2 and increased CO_2 in the packages are beneficial for the growth of *C. botulinum*. Incubation temperature was a critical factor in the growth of this pathogen and the production of toxic samples.

Viruses. A wide range of human pathogenic viruses can be transmitted by fresh produce (Bagdasargan, 1964; Herrmann and Cliver, 1968; Badawy *et al.*, 1985). A food handler was implicated in 8 of 14 reported viral gastroenteritis outbreaks; salads were the implicated source in 5 cases (Hedberg and Osterholm, 1993). Norwalk virus was transmitted from food handlers (who later became ill) via green salads in two outbreaks (Griffin *et al.*, 1982; Gross *et al.*, 1989). Salad items were also implicated in 6 of 12 outbreaks of viral gastroenteritis attributed to Norwalk or similar viruses in Minnesota (Karitsky *et al.*, 1984). Commercially distributed lettuce was linked to 202 cases of hepatitis A infection in Kentucky (Rosenblum *et al.*, 1990). Green onions were implicated in a hepatitis A outbreak involving 43 cases (Dentinger *et al.*, 2001). For the years 1990–1996, Norovirus was the most common agent associated with fresh produce outbreaks in Minnesota, causing 54% of the outbreaks. Most of these outbreaks were caused by an ill food handler who contaminated the produce at the point of preparation (NACMCF, 1998).

Poliovirus survives well under refrigerated storage conditions in certain produce. No decline in population was observed for green onions, whereas a 90% reduction was observed in 11.6 days for lettuce and 14.2 days for white cabbage under refrigerated conditions (Kurdziel *et al.*, 2001).

Herpes Simplex-Virus type 1 in saliva has been shown to remain infective over a 1 h period at 2°C when on the surface of lettuce or tomato, even though there was a 2-log drop in titer between 30 and 60 min after inoculation (Bardell, 1997). Bidawid *et al.* (2001) inoculated hepatitis A virus onto lettuce in normal and modified atmospheres for 12 days at 4°C and ambient temperature. The virus survival rate at 4°C was between 47.5% (normal atmosphere) and 83.6% (70% CO_2), whereas at ambient temperature, it was between 0.01% (normal atmosphere) and 42.8% (70% CO_2).

Parasites. The role of fresh vegetables in the transmission of parasitic disease is an emerging area of study. Epidemiological linkage is the tool to investigate the role of parasites in produce-associated outbreaks due to the lack of cost-effective methods to easily detect these organisms. Protozoan parasites *Cryptosporidium parvum*, *Cyclospora cayetanensis*, and *Giardia lamblia* have been involved with outbreaks associated with raw or minimally processed vegetable (IFT, 2001). *Cyclospora cayetanensis* has caused outbreaks associated with mesclun lettuce (CDC, 1996) and basil containing products (CDC, 1997). *Cryptosporidium parvum* was suspected in an outbreak involving salad with celery (Besser-Wiek *et al.*, 1996). *Giardia lamblia* was associated with a lettuce and onion outbreak, possibly due to contamination with water used to wash the vegetables (CDC, 1989a). *Fasciola hepatica* infection has been linked to consumption of contaminated watercress. Other water-borne parasites can be transmitted to water crops such as watercress, cater caltrop, water chestnut, and water bamboo when cultivated in ponds subject to run-off from grazing pastures (WHO, 1995).

Twenty-five samples of water used for irrigation were recently tested (22 in Central America, 3 in the United States), with all containing *Giardia* and *Cryptosporidium* cysts (Thurston-Enriquez *et al.*, 2002). An examination for parasites in fruits and vegetables in Norway (including pre-cut salad mix) showed that of 475 fruits and vegetables, 4% were contaminated with Cryptosporidium and 2% with *Giardia* cysts (Robertson and Gjerde, 2001.

Other pathogen concerns. *Campylobacter jejuni* is generally associated with animal products rather than vegetables. A *C. jejuni* outbreak associated with mushrooms is described in the mushroom section. A survey of fresh produce from retail supermarkets or farmers markets in the Washington, D.C. area by Thunberg *et al.* (2002) found no evidence of *Campylobacter* or *Salmonella* in 127 samples. The potential for vegetables to carry opportunistic pathogens into situations where they can become dangerous should not be ignored. For example, raw vegetables have been suspected as vehicles for introducing microorganisms such as *Ps. aeruginosa* and *Klebsiella* spp. into hospital environments, where their exposure to individuals suffering from burn wounds or recovering from surgery is potentially dangerous.

D CONTROL (raw and minimally processed vegetables)

Summary

Significant hazards	• Enteric bacterial pathogens. • Viruses. • Parasites. • *L. monocytogenes.* • *Cl. botulinum.*
Control measures	
Initial level (H_0)	• Good agricultural practices.
Increase (ΣI)	• Store and transport under refrigeration (<4°C). • Packaging to avoid condensation. • For *Cl. botulinum*, package to avoid anaerobic conditions. • Clean and disinfect daily, avoiding aerosol formation for *L. monocytogenes.* • Use sharp cutting equipment to minimize tissue damage. • Minimize surface moisture on produce. • Separate field produce from processed produce to avoid cross contamination. • Education of food handlers on the importance of personal hygiene for viruses, parasites, and enteric pathogen control. • Acidify vegetables packed in oil to pH <4.6 to control *Cl. botulinum.*

(*continued*)

CONTROL (Cont.)

Summary

Reduction (ΣR)	• Wash using clean, cold water, ozonated water, chlorine, chlorine dioxide, or other disinfectants where appropriate.
Testing	• Routine microbiological testing of vegetables is not recommended. • Aerobic colony counts can monitor effectiveness of processing. • *E. coli* can be used as an indicator of enteric pathogens; coliforms and fecal coliforms are not useful. • Environmental monitoring for *L. monocytogenes* or *Listeria* spp. is useful. • Monitor equipment hygiene using tests such as ATP.
Spoilage	• Control measures for pathogens will control most spoilage organisms, except lactic acid bacteria. • With most perishable produce, spoilage will occur before pathogens grow to hazardous levels.

Control measures. The following control measures are needed for production of raw and minimally processed vegetables.

Initial level of hazards (H_0)

Good agricultural practices. The microbiological quality and safety of raw and minimally processed vegetables is highly dependent on agricultural and harvesting practices discussed in the previous section because further processing cannot eliminate microorganisms. Control of pathogens and spoilage microorganisms on raw vegetables post-harvest requires effective procedures for cleaning and sanitizing equipment and control of the environment [i.e. temperature, relative humidity (RH), atmospheric composition] in which raw vegetables are stored. Only in this way can a substantial microbial build-up be prevented. The general principles for controlling bacterial spoilage in minimally processed vegetables are as follows (IFPA and WGA,1997; FDA, 1998; Nguyen-the and Carlin, 2000).

Increase of hazards (ΣI)

Temperature control. Temperature control is the primary measure used to maintain the quality and safety of raw and minimally processed vegetables. Cooling of vegetables can be done by several methods. Vacuum cooling reduces pressure to the point where water evaporates to cool the product and is applicable to hardy vegetables. Pressure cooling pulls air over the product. Hydrocooling removes heat from the field by cascading water over the product. Vacuum hydrocooling combines vacuum and hydrocooling by spraying water on the product just prior to the "flash point" of the vacuum cycle. The speed and uniformity of cooling produce by vacuum hydrocooling has outstanding advantages in achieving quality and longer shelf life. Produce that benefits from such cooling procedures includes lettuce, asparagus, broccoli, Brussels sprouts, cabbage, celery, sweet corn, and peas (Ryall and Lipton, 1979).

Manipulation of both temperature and gaseous atmosphere during vegetable storage can be very effective in retarding spoilage. In theory, storage near 0°C prevents or greatly retards most of the spoilage flora. The Institute of Food Science and Technology, for example, recommends a storage temperature of 0–5°C for dressed salads of pH $\leq$5.0, and prepared salads without a dressing (IFST,

1991). Not all produce can be stored at such a low temperature, as low temperatures can inflict tissue damage in peppers, tomatoes, potatoes, and cucumbers. Low-temperature sweetening of potatoes occurs with storage below 7–0°C.

Packaging and modified atmosphere. Simply reducing the temperature is rarely sufficient when vegetables are put into storage. Equally important is maintenance of a high RH to prevent dehydration and maintain structural integrity. Many vegetables are maintained at 90–95% RH. Lower RH causes moisture loss and quality deterioration. Proper control of humidity and air movement is necessary, however, to prevent condensation of moisture in the environment or on the vegetable surface, which can support localized microbial growth.

Preventing or retarding mold spoilage is accomplished primarily by maintaining a healthy physiological condition of the vegetables, so they are more resistant to attack rather than by direct action on the fungus itself. Vegetable tissue is living, respiring matter. During storage, O_2 is consumed and CO_2 is generated. A combination of reduced O_2 and elevated CO_2 concentrations is an effective means to prevent mold spoilage. To maintain optimum storage conditions, continuous monitoring and control of the level of each gas is essential. The use of nitrogen gas packaging did not significantly alter the growth of microbes (total aerobic bacteria, coliforms, *B. cereus*, and psychrotrophic bacteria) on fresh-cut lettuce and cabbage when stored at 1°C, 5°C, or 10°C for 5 days (Koseki and Itoh, 2002).

The facultatively anaerobic coliforms that cause soft rot grow under modified atmospheres. Low temperature will not control these bacteria since the lower growth limits for *Er. carotovora* and *Er. carotovora* var. *atroseptica* are 4.0°C and 1–2.8°C, respectively. Consequently, minimizing moisture deposition on product surfaces, achieved by movement and interchanges of air, is needed to retard spoilage.

An interesting concept that has been explored is the incorporation of nisin and lysozyme, or EDTA, into biodegradable packaging films. These compounds retain their bactericidal properties through the heat-press and cast-forming processes (Padgett *et al.*, 1998).

Although concern has been expressed for the potential growth of *Cl. botulinum* in minimally processed, cut vegetables packaged for retail and food service, very few products have been associated with botulism, despite their increasing volume and distribution. Petran *et al.* (1995) observed that lettuce was organoleptically unacceptable before toxin production. However, outbreaks of botulism have occurred with other raw vegetable products such as garlic in oil. The risk of botulism from products such as garlic in oil can be prevented by adequate acidification. Introduction of new products in anaerobic packaging with extended shelf life should be studied carefully to assure that *Cl. botulinum* is adequately controlled.

Facility hygiene and equipment maintenance. In addition to washing and chlorinating vegetables to reduce their microbial content, facility layout and process flow should be designed to minimize opportunities for cross-contamination from unwashed/untreated vegetables to the finished, cleaned products. Effective cleaning and hygiene practices should be applied with adequate frequency to prevent the processing or packaging equipment and environment from having a negative impact on the microbiological quality and safety of the product. Care must be taken during processing and cleaning to avoid creating aerosols with high-pressure hoses. These aerosols can contaminate product or clean surfaces if not controlled.

Particular attention should be given to control *L. monocytogenes*, a common natural contaminant of raw vegetables. Its ability to survive and multiply at low temperatures makes it a difficult microorganism to control in the refrigerated, moist environments of food processing establishments. It also can exist in biofilms, which impede the effectiveness of sanitation processes. To control *L. monocytogenes*, the various potential avenues of entry and cross-contamination must be considered (NACMCF, 1991).

Hygiene of cooling systems is particularly important because the large volume of water involved supports the growth of bacteria that can spread these to product being cooled. For all types of cooling, the water reservoir should be inspected and fresh water should be used daily. Vacuum hoses should be inspected for residue and cleaned to prevent growth of potential spoilage or pathogenic organisms (IFPA and WGA, 1997). Condensation drip pans for each system should be inspected and cleaned regularly to prevent growth of bacteria, especially *L. monocytogenes*.

For fresh-cut vegetables, damage to plant cells along cut surfaces leads to reduced shelf life, whereas thorough washing to remove the free cellular contents released by cutting can prolong it (Bolin *et al.*, 1977). Therefore, equipment design and maintenance to minimize damage of vegetable tissues are important for microbiological control. For example, the stability of shredded lettuce is affected by the method of cutting. A slicing action with a sharp knife can yield twice the storage life than a chopping action. Storage life is further maximized by the combined effects of minimizing cellular damage, minimizing microbial load, and storing dry at a temperature slightly above freezing (Bolin *et al.*, 1977). Also, shreds 1 mm thick had a shorter shelf life than shreds 3 mm thick (Bolin and Huxsoll, 1991).

As previously discussed, food handlers can be a source of enteric pathogens, viruses, and parasites in minimally processed vegetables. Education on the importance of personal hygiene in control of transmission of these hazards is important to minimize the potential for transmission of disease. The use of glove barriers was useful to prevent the transfer of bacteria from food to foodservice workers hands and transfer to other foods (Montville *et al.*, 2001).

Formulation. Minimally processed vegetables containing dressings provide an opportunity to use formulation as a control measure. Many dressings are acidic, which minimizes the potential for microbial growth. For example, three different strains of *E. coli* O157:H7 did not survive when inoculated into retail mustards. The only exception to this was a Dijon mustard with mayonnaise where two of the three strains were able to survive (Mayerhauser, 2001). *E. coli* O157:H7 population decreased from 0.1 to 0.5 log in numbers in commercially prepared coleslaw (pH 4.3 and 4.5) stored at 4–21°C, with greater declines occurring at 21°C (Wu *et al.*, 2002). Acidification of minimally processed vegetables packed in oil to pH <4.6 is important for control of *Cl. botulinum* as described in the previous section.

Reduction of hazards (ΣR)

Produce washing and disinfection. Beuchat (1998) reviewed surface decontamination of raw and minimally processed vegetables. Chlorination of water for washing or cooling the produce can reduce bacterial levels if carefully controlled. Chlorine (added as chlorine gas or hypochlorite salt) remains only partly as free, available chlorine; the remainder is immobilized by combining with organic and other impurities in the water. The most active bactericidal form of chlorine is undissociated hypochlorous acid with the proportion of chlorine in this form being dependent on the pH of the solution. Concentrations of free, available chlorine >10 mg/L at neutral pH are sufficient to kill vegetative bacteria within minutes, but this level may be difficult to maintain; thus, continuous monitoring of chlorine levels is necessary. Owing to the instability of chlorine in the presence of organic matter, it is doubtful that low levels of chlorine do more than killing the spoilage bacteria that might be in the water. The effectiveness of chlorination depends on pH, temperature, time, and the amount of organic matter such as may occur when the water is recycled during use. Adjusting the pH of a 100 mg/L chlorine solution from 9 to 4.5–5.0 increased effectiveness 1.5–4.0-fold (Adams *et al.*, 1989). Dipping fresh produce in 200 mg/L chlorine reduced *L. monocytogenes* levels 2 logs for Brussels sprouts, 1.3–1.7 logs for cut lettuce, and 0.9–1.2 logs for cut cabbage (Alzamora *et al.*, 2000). It has also been suggested that product surface structures and damage may play a role in protecting pathogens during disinfection (Burnett and Beuchat, 2000; Takeuchi and Frank, 2001).

The addition of water for washing or cooling can contribute to the proliferation of spoilage microorganisms. At levels of 50–100 mg/L total available chlorine, chlorination reduces the microbial load of wash water, but the benefits to the packaged vegetable product may be variable (Lund, 1983). For example, although the incidence of bacterial soft rot due to *Er. carotovora* on potatoes or carrots immersed in processing waters was reduced by maintaining 2–25 mg/L chlorine (Segall and Dow, 1973), 100 mg/L chlorine failed to reduce decay of packaged spinach, even though as little as 5 mg/L reduced the bacteria population in the water and on the leaves (Friedman, 1951). In addition, washing lettuce with warm chlorinated water contributed to the growth of *E. coli* O157:H7 and *L. monocytogenes* (Delaquis *et al.*, 2002). Residual water left on produce surfaces after washing can provide the moisture necessary for microbial growth and provides a vehicle to spread potential spoilage throughout the product. Dewatering must be done gently to prevent tissue damage that can also promote spoilage.

Although chlorine is the primary disinfection agent used for raw and minimally processed vegetables, its use is prohibited in certain countries, and other agents may be used. Xu (1999) reviewed the use of ozone for fresh vegetables. Peroxyacetic acid mixed with hydrogen peroxide and 1-hydroxyethylidene-1,1-di-phosphoric acid is approved for use on vegetables in the United States. Chlorine dioxide gas can achieve a 5 log reduction of *E*.coli O157:H7 in green peppers (Han *et al.*, 2001). Many washing and disinfection procedures are based on proprietary agents, their actual method of use evolving through experience of the manufacturer. Some agents may adversely affect the flavor, texture, or appearance of certain vegetable products (Huxsoll and Bolin, 1989). Other methods of decontamination that have been explored have been the use of ozone (Kim *et al.*, 1999) and acidic electrolyzed water (Koseki *et al.*, 2001; Koseki *et al.*, 2002) on lettuce. Electrolyzed oxidizing water and acidified chlorinated water reduced *E. coli* O157:H7 and *L. monocytogenes* on lettuce 2.4–2.6 logs over that for water treatment alone without altering quality (Park *et al.*, 2001). Continued investigation of new techniques of disinfection is needed.Various means of sanitizing have been studied for their effectiveness on reducing bacterial numbers. Hydrogen peroxide used in combinations with lactic acid or mild heat has been shown to reduce *E. coli* O157:H7, *Salmonella*, and *L. monocytogenes* levels on lettuce. Treatment with lactic acid reduced the sensory quality of lettuce, but mild heat with hydrogen peroxide was better at maintaining high product quality (Lin *et al.*, 2002). The use of apple cider vinegar, white vinegar, bleach and reconstituted lemon juice were tested for reducing the numbers of *E. coli* O157:H7 on iceberg lettuce. Although white vinegar was the most effective, it also imparted a poor sensory quality to the lettuce, making it sour and slightly wilted (Vijaykumar and Wolf-Hall, 2002). Free chlorine at a concentration of 50 mg/L without pH control and 100 mg/L free chlorine at pH 7.0 were found to be optimum for eliminating microflora on artichoke and borage, respectively (Sanz *et al.*, 2002).

A comparison of commonly used disinfectants and their ability to inactivate calicivirus on strawberry, lettuce, and a food-contact surface revealed that none of the disinfectants used was effective at the manufacturer's recommended strength. However, phenolic compounds at two to four times their recommended concentrations appeared to be effective at decontaminating environmental surfaces, and a combination of quaternary ammonium compound and sodium carbonate was effective at twice the recommended concentration. For artificially contaminated strawberries and lettuce, the only effective disinfection formulation was peroxyacetic acid and hydrogen peroxide at four times the manufacturer's recommended concentration for 10 min (Gulati *et al.*, 2001).

Testing. The rate of spoilage of raw and minimally processed vegetables is influenced by the initial number and type of microorganisms on the freshly cut or packaged product. Tests for aerobic plate count can be useful to determine the general impact of processing and handling. Rapid methods, such as ATP measurement, are a useful tool to measure equipment hygiene.

The presence of enteric pathogens is the major food safety concern. Testing for all the possible pathogens mentioned above is not recommended. It may be appropriate, however, to use *E. coli* as an indicator of the hygienic conditions of growing, harvesting, transporting, and processing. Enterobacteriaceae, coliforms, or "fecal coliforms" are not effective indicators because they occur naturally in the field and plant environment (Nguyen-the and Carlin, 2000; Bracket and Splittstoesser, 2001).

For raw and minimally processed vegetables that support the growth of *L. monocytogenes*, microbiological testing of the processing environment may be appropriate. Environmental monitoring of non-food contact surfaces is useful to identify potential harborage sites for the organism. Drains, standing water, cracks, and crevices are all potential sampling sites. Samples could also be collected from product contact surfaces, particularly from equipment used after the vegetables have passed through bacterial reduction steps (e.g. washing and disinfecting). The data can be used to assess control of this pathogen in the processing environments and the risk of product contamination.

IV Cooked vegetables

A *Effects of processing on microorganisms*

Many vegetables are cooked prior to consumption. Because this process disrupts the plant cell structure, microorganisms grow readily at the favorable pH and moisture content of most vegetables. Conventional cooking processes such as boiling, steaming, baking, and frying destroy vegetative cells, with more extensive heat treatments achieving greater lethality. However, spore-formers can survive. Once thawed, frozen vegetables have properties similar to cooked vegetables. Canned vegetables after opening also have similar properties; however, mesophilic spore formers will not be present.

B *Saprophytes and spoilage*

Few microorganisms are present on freshly cooked vegetables. However, cross-contamination between raw and cooked vegetables can readily occur in restaurant, home, and food processing environments. The type, variety, and numbers of spoilage organisms present depend on the extent of cross-contamination, temperature of storage, and packaging conditions. Sous-vide products that are cooked in the final package have low initial population, whereas vegetables handled extensively after cooking have increased loads and a shorter shelf life.

C *Pathogens*

A range of home-processed foods has been responsible for outbreaks of botulism, as have foods mishandled in foodservice establishments. These include a hot sauce containing jalapeno peppers prepared without adequate heating (Terranova *et al.*, 1978), potato salad prepared with foil-wrapped baked potatoes stored at room temperature, and sautéed onions that had been stored unrefrigerated and then served on a patty-melt sandwich (MacDonald *et al.*, 1985). Solomon and Kautter (1986) demonstrated that *Cl. botulinum* produces toxin in sautéed onions.

Baked potatoes have been involved in several separate episodes of botulism in the United States since 1978, some of which have been published (Seals *et al.*, 1981; MacDonald *et al.*, 1986; Hauschild 1989; Brent *et al.*, 1995). In each instance, the potatoes were baked in foil and allowed to stand at room temperature for an extended period. The potatoes were then used to make potato salad or another dish. A two-step pasteurization (Tyndallization) process for raw potatoes did not prevent toxin formation by *Cl. botulinum* inoculated into vacuum-packed peeled potato within 5–9 days at 25°C. This indicates that if peeled raw potatoes are not given a heat treatment at least equivalent to an $F_0 = 3$ process, they should be stored below 4°C (Lund *et al.*, 1988). Botulism from vacuum-packed lotus root further illustrates the importance of vegetables as a source of botulism (Otofuji *et al.*, 1987). In virtually all cases of botulism ascribed to vegetables, type A or B toxins have been involved (Notermans, 1993).

Vegetative pathogens can grow in cooked vegetables that are held at abusive temperatures if they are recontaminated after cooking. For example, a salad made with canned corn was involved in a non-invasive listeriosis outbreak (Aureli *et al.*, 2000). The corn was initially commercially sterile; therefore, the organism was introduced during preparation. The *L. monocytogenes* population were around 10^6 cfu/g in retained samples. Laboratory studies demonstrated that this level could be reached in inoculated corn after 10 h at room temperature.

D CONTROL (cooked vegetables)

Summary

Significant hazards	• *L. monocytogenes.* • *Cl. botulinum (Anaerobically Packaged).*
Control measures	
Initial level (H_0)	• Separation of cooked and raw ingredients to prevent cross-contamination.
Increase (ΣI)	• Store and transport at <5°C or >60°C. • pH <4.6 for *Cl. botulinum.* • Education regarding *Cl. botulinum* in potatoes due to the repeated outbreaks associated with temperature abuse. • Establish shelf life to prevent 1–2 log increase of *L. monocytogenes* prior to consumption. • Clean and disinfect equipment and production environment daily, avoiding aerosol formation.
Reduction (ΣR)	• Population reduction depends on cooking time and temperature.
Testing	• Routine microbiological testing of vegetables is not recommended. • Aerobic colony count and/or coliforms may be used to monitor process control. • Manufacturing environmental monitoring for *L. monocytogenes* or indicators minimizes the potential for contamination. • Monitor equipment hygiene using tests such as ATP.
Spoilage	• Control measures for pathogens will control most spoilage organisms, except lactic acid bacteria.

V Frozen vegetables

A Effects of processing on microorganisms

Most vegetables are blanched before freezing to inactivate plant enzymes and thereby to stabilize the product during subsequent frozen storage. However, the organoleptic quality of some vegetables (e.g. peppers, leeks, and parsley) is better if blanching is avoided. During the blanching process, a belt or screw conveyor moves vegetables through a bath of water at 95–99°C for 1–5 min. Blanching can reduce microbial numbers 10–10^4-fold (Splittstoesser, 1970). While temperatures of blanching are capable of inactivating vegetative cells, the microflora of the frozen product is a reflection of the handling the product received after blanching. Air-borne microorganisms from the handling of raw vegetables settle on post-blanch surfaces (Mundt *et al.*, 1966; Mundt and Hammer, 1968). Microorganisms from slicers, cutters, choppers, conveyor belts, flumes, lifts, hoppers, and fillers contaminate the product. The contribution from each source depends on the frequency and effectiveness of cleaning these pieces of equipment.

Bacterial growth in or on product before freezing is rarely a problem because of the relatively short time elapsed between blanching and freezing, but there are exceptions. For example, it may be necessary to add product manually to packages that have been mechanically filled to meet weight requirements. Any 'weighing reserves' that are held at room temperature for a substantial period of time can have high microbial levels that will be reflected in the final product, therefore time and/or temperature should be managed to reduce the potential for microbial growth.

The freezing step is generally not a lethal process; however, a portion of the microflora present may be injured, which may influence the distribution of particular species in frozen foods. Gram-negative bacteria are more easily killed or injured by freezing than Gram-positive bacteria. Protracted storage in the frozen state may further reduce the numbers, the extent dependent on time, the nature of the microorganisms, the nature of the food, and the temperature of storage. However, a proportion of the microorganisms always is likely to survive.

B *Saprophytes and spoilage*

In most frozen vegetables, the predominant microorganisms are the lactic acid bacteria (Mundt *et al.*, 1967; Splittstoesser, 1970). Significant numbers of *Leuconostoc mesenteroides* and enterococci are often found. Micrococci and Gram-positive and Gram-negative rods (including coliforms) also constitute a considerable portion of the total microflora of certain products. A study to assess what the frozen vegetable industry in the United States could accomplish at that time found that aerobic plate population of frozen blanched vegetables ranged from 10^3–10^6 cfu/g (Splittstoesser and Corlett, 1980); peas yielded some of the lowest counts and chopped broccoli the highest. After blanching, coliforms, streptococci, and leuconostocs may recontaminate vegetables; >40% of frozen blanched vegetables were positive for fecal coliforms, but only 12% were positive for *E. coli* (Splittstoesser *et al.*, 1983).

Microbial spoilage of frozen vegetables is rare. In the frozen state, spoilage is essentially precluded by the low temperature and reduced a_w. The few reports of spoilage of thawed products show that the spoilage rate is temperature-dependent (White and White, 1962; Michener *et al.*, 1968).

Aerobic bacterial population ranging from 10^1 to over 10^5 cfu/g in frozen vegetables can be observed. Maintenance of good hygienic practices can achieve aerobic colony population below 10^5 cfu/g on a routine basis.

C *Pathogens*

Frozen vegetables are rarely involved in food poisoning incidents because: (i) non-spore-forming pathogens generally do not survive blanching; (ii) any pathogens contaminating the product post-blanch cannot grow at the temperature of the frozen food; and (iii) most frozen vegetables are cooked before being consumed. This treatment will also inactivate viruses. Because of their excellent public health record, there have been relatively few surveys of the incidence and numbers of pathogens in frozen vegetables. Insalata *et al.* (1970) found *Cl. botulinum* spores in 6 of 60 pouches of frozen, vacuum-packaged spinach; however, freezing inhibits the growth of the organism, which provides effective control. Salmonellae were not found during a survey of a limited number of samples of frozen peas, green beans, and corn (Splittstoesser and Segen, 1970). Coagulase-positive staphylococci did not exceed 10 cfu/g in any of 112 samples of frozen peas, beans, and corn (Splittstoesser *et al.*, 1965). For these reasons, food-borne illnesses resulting from microorganisms carried on frozen vegetables are unlikely. Control of contamination and time–temperature abuse after preparation is needed to prevent illness.

D CONTROL (frozen vegetables)

Summary

Significant hazards	• None.
Control measures	
Initial level (H_0)	• Controls used for raw vegetables.
Increase (ΣI)	• Freezing prevents growth of microorganisms. • Time–temperature control ($<5°C$ or $>60°C$) needed after preparation.
Reduction (ΣR)	• Blanching reduces microbial load. • Freezing inactivates parasites.
Testing	• Aerobic colony count and/or coliforms may be used to monitor process control. • Environmental monitoring for *L. monocytogenes* or indicators minimizes the potential for contamination. • Monitor equipment hygiene using tests such as ATP.
Spoilage	• Freezing prevents the growth of spoilage organisms. • Insipient spoilage may occur if adequate sanitation and temperature control measures are not followed.

Comments. For most products, controlling microbial levels in frozen vegetables is influenced by the blanching step, which is done to destroy enzymes. Temperatures achieved also eliminate the vegetative microbial cells. For example, a blanching process sufficient to inactivate catalase and peroxidase in peas will inactive 5 logs of *L. monocytogenes* (Mazzotta, 2001). Cleaning and sanitizing the equipment between the blancher and freezer frequently and thoroughly are necessary to maintain relatively low microbial levels. Conveyor belts used for transporting blanched vegetables can be difficult to clean and a source of recontamination (Splittstoesser *et al.*, 1961, Surkiewicz *et al.*, 1967; Splittstoesser, 1983).

For long-term storage, microbial spoilage is controlled at $-16°C$ or lower. At temperatures just below 0°C, growth can occur (Geiges and Schuler, 1988). Since many microorganisms contaminating vegetables can survive freezing, psychrotrophic microorganisms may cause spoilage if the temperature is permitted to rise to near the freezing point during storage or transport. Storage of peas at temperatures below $-18°C$ is claimed to be necessary to prevent deterioration by the pink yeast *Rhodotorula glutinis* (Collins and Buick, 1989).

VI Canned vegetables

A Effects of processing on microorganisms

The canning process is designed to make a variety of products, including vegetables, "shelf stable." To achieve this, the thermal process must destroy those microbial forms that are capable of growth under ambient conditions of storage (e.g. up to 35°C). The most heat-resistant forms are the spores of mesophiles and thermophiles. Among the mesophiles, there are numerous spoilage types, and one pathogen of concern (i.e. *Cl. botulinum*). The thermophiles are not pathogenic, but are generally more heat-resistant than mesophiles. However, thermophiles cause spoilage only under very specific conditions, namely, abnormally high storage temperatures such as may occur in tropical climates or in storage facilities that can become very hot. Under-processing of low-acid vegetables by home canners is the major cause of botulism. For example, in the United States between 1950 and 1996, botulism outbreaks were attributed to vegetables more frequently than any other commodity (CDC, 1998). Most were attributed to string beans; other produce included peppers, potatoes, beans, beets, mushrooms, corn, carrots, olives, and celery. Since the implementation of US Low Acid Canned Food regulations in the 1970s, cases of botulism from commercial shelf stable canned vegetables have been rare.

Most canned vegetables are "low acid", so defined because their pH is >4.6. Low-acid canned vegetables are given a heat treatment sufficient to destroy the more heat-resistant mesophiles, such as *Cl. sporogenes,* to render the product "commercially sterile". Thermal processes for canned vegetables are based on the rate of heat penetration into containers for each commodity type, container size, and equipment used for heating. Generally, the heat treatment required for commercial sterility exceeds that required for safety, i.e. a "botulinum cook" ($F_0 > 3.0$ min) based on the "12D concept", which destroys *Cl. botulinum.* In some countries, the antibiotic nisin can be added to certain canned foods that are subject to temperature extremes to prevent the growth of thermophilic spore-forming bacteria (Jarvis and Morisetti, 1969).

Vegetables may be acidified before heat-processing (e.g. salsa or pickled vegetables). If the pH of such products is ≤4.6, *Cl. botulinum* will not grow and the concerns are spoilage by acid-tolerant clostridia and bacilli such as *Cl. pasteurianum* and *B. coagulans* and non-spore-forming acid-tolerant microorganisms. The heat process applied to achieve commercial sterility depends mainly on the pH of the acidified vegetables. If the pH value is <3.8, acid-tolerant spore-formers are unable to grow and less heat is required for commercial sterility. It is important that the conditions used for acidification are established such that the pH in all components/ingredients is uniform before or shortly after completion of the thermal process (<24 h). The equilibration pH of an acidified food is an important control point and must be monitored (Codex Alimentarius, 1983; U.S. Department of Health and Human Services, 2002).

B Saprophytes and spoilage

Commercially sterile canned vegetables may contain viable thermophilic spores that are unable to grow under normal storage conditions. The source of thermophiles may be water used in flumes conveying the product, starch or sugar used as ingredients, or the raw vegetables themselves.

Canned vegetables may spoil from an insufficient thermal process, container leakage, or high-temperature storage. Insufficient processing could permit the survival of mesophilic spores that often cause off-odors and swelling of the can. This form of spoilage is important because it indicates that *Cl. botulinum* may have survived and produced toxin in the product. Under-processing can be attributed to faulty equipment (e.g. inaccurate thermometers, inadequate steam supply, or inaccurate timing devices) or poor operating procedures (e.g. poor fill control, inadequate venting, failure to place product in the retort). In addition, the consistency of a product may vary from batch to batch. An especially viscous batch may not permit rapid heat penetration as expected, so that the centers of the cans are under processed. Rehydration of dried vegetables such as peas must be carefully controlled to avoid such problems.

After the cans are processed, spoilage microorganisms in cooling water or wet can conveyor tracks can enter through faulty can seams or punctures, or through the hot, still soft seam mastic. It is common to find a mixed population of microorganisms such as micrococci, lactobacilli, and streptococci, in cans spoiled due to leakage. If the cooling water is chlorinated, leaker cans may contain primarily spore-forming microorganisms (Denny and Parkinson, 2001).

Prolonged high-temperature storage can cause three types of spoilage in vegetables:

1. Flat sour spoilage is caused by facultative anaerobic microorganisms that produce acid without gas, such as *B. stearothermophilus* and *B. coagulans.* The spores of these microorganisms germinate

only >40°C, but the vegetative cells can grow over a wider range. The minimum temperature for growth of *B. stearothermophilus* is 30–45°C and the maximum 65–75°C. Maximum and minimum temperatures are 15–25°C and 55–60°C for *B. coagulans* (Buchanan and Gibbons, 1974).

2. Thermophilic anaerobic spoilage is caused by obligately thermophilic spore-forming anaerobes, such as *Cl. thermosaccharolyticum*, which produces large quantities of hydrogen and carbon dioxide.
3. Sulfide "Sulfide stinker" spoilage is caused by the obligately thermophilic spore-forming anaerobe *Desulfotomaculum nigrificans*, which produces hydrogen sulfide. The cans remain flat but an odor of hydrogen sulfide is detectable, and the food may become blackened if iron is present.

C *Pathogens*

Despite the widespread occurrence of spores of *Cl. botulinum* in the raw material, commercially canned vegetables have, in general, had an excellent safety record. In certain rare instances, however, botulism resulting from under-processing (Lynt *et al.*, 1975) and staphylococcal food poisoning due to post-processing leaker contamination (Bashford *et al.*, 1960) or growth prior to processing of mushrooms (Hardt-English *et al.*, 1990) have been directly attributed to canned vegetables. Home-canned foods have been responsible for most of the botulism outbreaks. Commercially produced vegetables have caused botulism outbreaks when insufficient control over the heating step resulted in under-processing, including canned asparagus (Notermans, 1993) and bottled vegetable soup (Bruno, 1998). While canning manuals establish times and temperatures of processes for given recipes and size of container, home canners frequently alter recipes, use larger food containers without adjusting the process sufficiently, use pressure cookers improperly or not at all, and they may taste questionable foods. Sometimes vegetables with *Cl. botulinum* toxin have normal, or near-normal, odor, and appearance. Various reported cases of botulism from consumption of improperly home-bottled tomato juice, home-canned jalapeno peppers, and home-canned olives are cited by Notermans (1993). Vegetables support the growth and toxin production of Group I (proteolytic) strains (Notermans, 1993) and Group II (saccharolytic) strains (Carlin and Peck, 1995) of *Cl. botulinum*.

D CONTROL (canned vegetables)

Summary

Significant hazards	• *Cl. botulinum*.
Control measures	
Initial level (H_0)	• None.
Increase (ΣI)	• pH <4.6 for acidified products. • Proper container closure.
Reduction (ΣR)	• Validated thermal process.
Testing	• Microbiological testing of canned vegetables is not recommended. • Thermophile testing of starch and sugar ingredients may be appropriate.
Spoilage	• Heat treatments for spoilage are generally higher than those required for safety.

Hazards to be considered. Development of canning processes requires extensive technical expertise to assure destruction of *Cl. botulinum*. Other food-borne pathogens will not survive thermal process treatments required for commercial sterility. Microbial toxins, however, may survive certain treatments. Because of this, use of wholesome ingredients is important for product safety.

Control measures. Detailed texts should be consulted for specific information on canned product controls. For additional control information, see Codex Alimentarius (1983), Chapter II in ICMSF (1988), Denny and Parkinson (2001) and U.S. Department of Health and Human Services (2002). A summary of controls follows.

Initial level of hazard (H_0). *Cl. botulinum* occurs naturally in the soil and low levels may be present on incoming produce. There are no effective treatments to reduce initial levels.

Increase of hazards (ΣI). The equilibrium pH of acidified vegetables should be <4.6. This will prevent the outgrowth of *Cl. botulinum* spores with prolonged storage of the product. Effective continer closures prevent the entry of organisms during cooling, when the vaccum created due to the temperature differential may pull potential contaminants into containers that are not properly sealed.

Reduction of hazards (ΣR). Retorting time–temperature schedules appropriate to the product, type and size of the container are essential for production of a commercially sterile product. These must be developed by an individual with extensive training in the area thermal processing.

Testing. Routine culturing of canned vegetables is not recommended (Deibel and Jantschke, 2001), however, the following monitoring is essential for effective thermal processing.

- Initial temperature of the product entering the retort meets minimum requirements of scheduled process.
- Automatic retort controls and recordings are reviewed to assure correct time and temperature.
- Validation of the retort temperature distribution in the retort.
- Monitoring pH for acidified products.
- Routine evaluation of container closer effectiveness.

Monitoring thermophile levels in starch or sugar ingredients may be useful if products are likely to be held at elevated temperatures.

Spoilage. Heat treatments required to destroy heat-resistant spoilage organisms generally exceed those needed to destroy *Cl. botulinum.* In addition to the controls identified above, the following are needed to control spoilage organisms.

- Ensuring that the raw material is of good quality and minimally contaminated with microorganisms.
- Good sanitation in preparing the food for canning, particularly in blanchers where thermophilic microorganisms grow.
- Pre- and post-process equipment is adequately cleaned and sanitized.
- Choice of ingredients (e.g. spices) containing few thermophilic spores.
- The time between filling and processing is as short as possible.
- Controlled cooling with chlorinated water.
- Careful handling of cans after processing including no manual handling of wet containers.
- The integrity of the container is maintained by proper seam/seal control and handling.
- Storage at moderate temperatures, unless the product is given a thermal treatment sufficient to destroy thermophiles (e.g. $F_0 > 18.0\,\text{min}$).

VII Dried vegetables

This section covers only artificially dried vegetables such as peas, onions, garlic, potatoes, carrots, etc. Dried vegetables are inherently stable due to reduced water activities and are rarely involved in food-borne illness. They are important items of commerce where they are commonly used as ingredients in processed foods. There is relatively little recent published information on the microbiology of dried vegetables.

A *Effects of processing on microorganisms*

Some vegetables are blanched before drying whereas others (e.g. onions and garlic) are not. Therefore, the microbial flora of dried vegetables depends to a considerable degree on whether the flora of the raw, cleaned product has been largely destroyed by blanching. Vegetables for drying may support the growth of microorganisms if they are held too long at ambient temperatures. They may also become contaminated by unclean equipment (Vaughn, 1951).

Most vegetables are dried by blowing heated air over trays or through perforated belts in drying tunnels. Loading of the belts or trays is critical because uneven loading can lead to improper air circulation and temporary "wet spots" that will permit microbial growth. Because water evaporates during drying, the product temperature rarely exceeds 35–45°C, whereas the air temperature may be 80–100°C. Drying rarely reduces the number of microorganisms and may apparently increase the number because they are concentrated in a smaller amount of product (Murphy, 1973). Population between 10^3–10^4 cfu/g have been reported on product dried by blowing hot air through vegetables on perforated trays or belts (Dennis, 1987). For peas freeze-dried within 15–20 min after blanching, population of 10^2 cfu/g were found. Organisms commonly isolated from dried vegetables include lactic acid bacteria, *Enterococcus faecalis*, staphylococci, spores of *Bacillus* spp., yeasts, and molds (*Penicillium* and *Aspergillus* spp.).

Most microorganisms on onions reside on the skin, root area, and tops. Many of these can be removed by trimming, as is done with onions that will be sliced. On the other hand, diced, chipped, or flaked onions often are not trimmed extensively, therefore, the bacterial content in the dried product is higher than when sliced. Pre-treating raw onions in a 24% salt brine before dehydration reduced microbial numbers in the final product (Firstenberg *et al.*, 1974).

B *Saprophytes and spoilage*

The number and types of microorganisms depend upon the type of vegetable and the conditions of growing and harvesting. For example, dried celery has a high microbial content due to the growing conditions. Onions and garlic also tend to have high population due to their being grown in soil. The saprophytic flora of vegetables that have been blanched before drying consists of those microorganisms that grow well on product-contact surfaces of the equipment. As with frozen vegetables, the lactic acid bacteria predominate. The saprophytes associated with the dried vegetables that have not been blanched will more closely approximate the flora of the raw product. The method of drying also affects the flora. For example, belt-dried onions have lower microbial population than tray-dried onions. The flora of the former consists mainly of bacterial spores, whereas the latter has a high "lactic" population (Sheneman, 1973). Coliforms, enterococci, and clostridia are commonly found in dried onions (Clark *et al.*, 1966; Vaughn, 1970) because these microorganisms are normally associated with raw vegetables. Apparently, onion juice is toxic to *E. coli*, which is, therefore, usually absent from this food (Vaughn, 1951; Sheneman, 1973).

Spoilage of most dried vegetables is rare due to their low a_w. If moisture is inadvertently added to the finished material, spoilage could develop, but this rarely occurs.

C Pathogens

Vegetative cells of bacterial pathogens are rarely present in dried vegetables, however, spores of *B. cereus*, *Cl. botulinum*, or *Cl. perfringens*, if present in the soil, are likely to carry through onto the final dried product. They remain harmless unless permitted to grow when the dried vegetable is rehydrated.

Non-sporulating microorganisms like *E. coli* or salmonellae are destroyed by blanching, but dried vegetables not previously blanched could contain these microorganisms if they had been grown in soils contaminated with animal or human wastes.

Many dried vegetables are reconstituted with boiling water, as occurs when dried soups are prepared for serving. They also are used as ingredients in the production of processed foods that are cooked. For example, dried bell peppers and onions are added during the formulation and processing of cooked luncheon meats. Thus, in products such as these, vegetative pathogens that might be present would be killed during cooking.

D CONTROL (dried vegetables)

Summary

Significant hazards	• None when using fresh vegetable controls.
Control measures	
Initial level (H_0)	• Controls for raw vegetables.
Increase (ΣI)	• Prompt drying to a_w <0.6. • Cleaning of equipment. • Even dryer loading. • Moisture control in the processing environment. • Time–temperature control (<5°C or >60°C) after rehydration.
Reduction (ΣR)	• Blanching, when applicable, reduces microbial load.
Testing	• Routine culturing is not recommended. • Aerobic colony count for hygiene and process control. • Coliforms are not considered a useful indicator. • *E. coli* may indicate cause for concern.
Spoilage	• a_w <0.6 prevents growth.

Hazards to be considered. Microbial food safety hazards are effectively controlled using methods discussed above for fresh and frozen vegetables.

Control measures

Initial level of hazard (H_0). It is important to use controls described for raw vegetables when procuring ingredients for dried vegetable production. The drying process frequently is not effective in reducing numbers and therefore pathogens present on incoming ingredients will frequently survive.

Increase of hazards (ΣI). By removal of water, the drying process will concentrate microbes present on the raw materials, contributing to an apparent increase in microbial numbers. The basis for dried vegetable preservation is the reduction of a_w to levels that inhibit microbial growth, i.e. <0.6. Minimizing the time of storage of the cleaned cut vegetables before drying removes the potential for population increases prior to drying. As with other vegetable processes, frequent and thorough cleaning of equipment prevents contamination of vegetables with microbes that grow in equipment. Hygienic handling of the dried product is also needed to prevent introduction of microorganims to dried product.

Proper loading of the product into the dryer is necessary to attain even drying. Pockets of moist product can lead to subsequent spoilage moisture is retained. Similarly, moisture control in the processing environment is an important factor to minimize the risk of recontamination of dried vegetables. Once moisture is restored to the vegetables, microbial growth can occur. It is therefore important to practice time–temperature control (<5°C or >60°C) after rehydration.

Reduction (ΣR). Blanching, when applicable, reduces the microbial load. The extent of the reduction depends upon the product and conditions used for the produces. The primary purpose of blanching is to inactivate enzymes, and not all dried vegetables can be blanched because significant quality issues may result.

Testing. Routine culturing of dried vegetables to determine their bacterial content is not recommended, although aerobic colony counts may be a useful measure of hygiene and process control. The microbial population will vary according to the type of vegetable and the conditions of growing and processing. For these reasons, typical numbers of bacteria differ widely among the various dried vegetables that are commercially available. Coliforms are not considered a useful indicator of fecal contamination as they occur naturally on many dried vegetables, sometimes in high numbers (e.g. onions). However, the presence of *E. coli* may indicate cause for concern.

Dried vegetables intended for sensitive population (e.g. infants) could be tested in a similar manner as infant formulae.

Spoilage. The general controls for food safety concerns also control potential spoilage. Once dried to below an a_w value of <0.60, microbial growth is effectively controlled. Product should be stored to preclude the entry of moisture, which may lead to subsequent fungal growth.

VIII Fermented and acidified vegetables

Vegetables can be preserved or processed by salting, acidifying, or fermentation (Fleming *et al.*, 2001). In the Orient, blends of vegetables known as kimchi, safur, asin, nukamiso, dua chua, and paw tsay are fermented. Often such blends include fish, nuts, and the liquid remaining after washing rice before cooking (Orillo *et al.*, 1969; Pederson, 1979; Steinkraus, 1983). The most commonly fermented vegetable in North America and Western Europe is cabbage, but cucumbers, cauliflower, carrots, radishes, beets, beans, green tomatoes, peppers, chard, and turnips are also preserved in this way. The processes vary depending on the type of product desired and its intended use, i.e. direct consumption or garnish.

Fully cured, salt stock pickles may be made into a variety of products by leaching out much of the salt. Addition of vinegar makes them "sour pickles"; addition of vinegar and sugar makes them "sweet pickles". Pickles made by direct acidification are termed "fresh pack" and are not fermented. They are combined with vinegar and spices and then pasteurized.

A Effects of processing on microorganisms

The flora involved in the fermentation of vegetables is derived from the raw vegetables and from the processing plant equipment used to prepare them for the fermentation tank. The vegetables are not blanched and thus retain the epiphytic lactic acid bacteria that were associated with them in the field.

The vegetable is first salted with brine or with dry salt crystals. Salting has a pronounced effect in selecting the predominant microflora. A complex interaction of factors governs the sequence of microorganisms that develop during fermentation, and during spoilage, if it occurs, i.e. pH, salt, organic acids, a_w, temperature, redox potential, oxygen, and carbon dioxide. With minor variations, the fermentation of all plant foods follows a similar path, consisting of a sequential growth of lactic acid bacteria, including *Leuc. mesenteroides*, *Lactobacillus brevis*, *Pediococcus acidilactici*, *Ped. pentosaceus*, and *Lb. plantarum*. Other "lactics", e.g. *Enterococcus faecalis*, have been observed, but are not important in the fermentation.

The fermentation of cabbage is typical. It can be fermented intact, i.e. whole heads submerged in a cover brine (Pederson *et al.*, 1962), but more commonly, it is chopped and dry salted. The salt draws out the plant juices containing fermentable carbohydrates and other nutrients, forming brine. Large numbers of soil bacteria and microorganisms from the harvesting and transport equipment remain. Additional microorganisms enter from spices or condiments. Maintaining the temperature between 20°C and 24°C is important, though not critical. Higher temperatures often permit the ascendancy of undesirable microorganisms that give an inferior or unmarketable product. Lower temperatures slow acid development and pre-dispose the product to spoilage. The fermentation vats are covered to minimize the development of oxidative microorganisms that utilize the acid generated during the fermentation process. Sauerkraut can be successfully produced in winter at temperatures as low as 7.5°C if anaerobic conditions are maintained. *Leuc. mesenteroides*, which grow at lower temperatures than the other lactic acid bacteria, provide the conditions necessary for preservation. When the mass of sauerkraut warms in spring, fermentation is completed by other lactic bacteria; the process requires ≥6 months to complete but yields a superior quality product (Pederson and Albury, 1969).

In the initial stages of fermentation, i.e. before significant quantities of acid have been generated by the lactic acid bacteria, Gram-negative facultative anaerobes such as coliforms multiply rapidly. This is a normal occurrence and is inconsequential unless the lactic fermentation fails. Under normal circumstances, *Leuc. mesenteroides* is the predominant lactic microorganism in the early stages of fermentation. The pH drops to 4.6–4.9 and *Leuc. mesenteroides* is followed by the more acid-tolerant lactic acid bacteria, i.e. the pediococci and lactobacilli. In a matter of a few weeks, the fermentation is complete and the final product has a pH of 3.5–3.8 and a titratable acidity (as lactic acid) of 1.8%. The interaction of salt and acid and the absence of dissolved oxygen preclude the growth of aerobic and many Gram-negative bacteria. The finished sauerkraut may be marketed raw, pasteurized, or canned.

Cucumbers are processed into pickles in a variety of ways, including fermentation and direct acidification. The mildest fermentation process begins with raw cucumbers in a low salt (2.6–4.0%) brine. Lactic fermentation takes place for a few days at room temperature and continues during refrigerated storage. Alternatively, "salt stock" pickles are fermented and cured at high-salt concentrations. Normally, the fermentation process is begun at an intermediate salt level (8–10%), and the salt is increased gradually to 15%. However, if the weather is cool when tanks are being filled, the initial salt concentration may be as low as 6% to enhance the rate of fermentation. The initial flora may be largely coliforms, but pediococci and *Lb. plantarum* soon predominate in the fermentation.

Pure culture fermentation of cucumbers yields the best product (Etchells *et al.*, 1964, 1966). If the fermentation is carried out in containers up to 5 gallons (22 L) in capacity, the cucumbers are heat-pasteurized to destroy the natural flora. For large-scale production, pasteurization is impractical, and the natural flora is suppressed by acidifying the cucumber brine to pH 3.3. The use of chlorination to control the flora (Etchells *et al.*, 1973) has largely been discontinued due to production of off-flavors.

A pure culture of *Lb. plantarum*, or a mixture of *Lb. plantarum* and *Ped. cerevisiae*, is then added, and fermentation takes place.

B Saprophytes and spoilage

Spoilage of fermenting vegetables can occur in a number of ways. One of the principal causes is uneven distribution of salt. If the salt concentration is very high in a localized area, certain yeasts (Pederson and Kelly, 1938) or lactobacilli (Stamer *et al.*, 1973) may grow and turn the product pink; if the salt is low, sauerkraut may soften from coliform bacteria. Fermented vegetables can be rendered microbiologically stable provided that essentially all fermentable carbohydrates are removed during primary fermentation, sufficient acid is present to prevent growth of spore-forming spoilage bacteria, and oxygen is excluded from products to prevent surface growth of yeasts, molds and spoilage bacteria (Fleming *et al.*, 1983). If oxygen is present, oxidative yeasts can grow rapidly and utilize the developing lactic acid. This will increase the pH and allow the growth of the less acid-tolerant spoilage forms.

The shelf life of raw sauerkraut is governed by the temperature of storage. Pasteurized or canned sauerkraut will remain microbiologically stable until the container is opened, whereupon oxidative yeasts may enter and spoil the product at a rate that is a function of temperature and the number and types of yeasts entering.

The saprophytic and spoilage flora of pickled cucumbers, depends on the method of manufacturing the product. In salt stock pickles with high-salt concentrations (15%), yeasts, obligate halophiles, and coliforms may develop if the acidity generated by the lactic acid bacteria is not sufficient. Dill pickles manufactured by the low-salt brine ($<5\%$) process can show a bloating effect due to yeasts, heterofermentative lactics, and coliforms if the proper fermentation flora do not develop (Etchells *et al.*, 1968). Softening of the fruit flesh also can occur due to enzymatic action of yeasts or *Bacillus* spp. growing in the brine, or to enzymes carried into the fermentation on the cucumber. These latter enzymes originate from filamentous fungi that grow on the flower-end of the fruit in the field (Etchells *et al.*, 1958). The yeast population in the brine can be altered qualitatively and quantitatively by the addition of sorbic acid. The species of *Brettanomyces*, *Pichia*, and *Saccharomyces* that normally predominate are then repressed, and if the salt content of the brine is low, species of *Candida* will predominate (Etchells *et al.*, 1961). Thus, the benefit derived from use of sorbic acid is questionable.

Sweet and sour pickles (non-pasteurized) made from salt stock pickles are preserved by vinegar and/or sugar. Should either the acid or sugar level be insufficient, spoilage due to lactic acid bacteria or yeasts will develop (Fleming *et al.*, 2001).

If pasteurization of fresh pack pickles is inadequate, yeasts and lactic acid bacteria will spoil the product. If the acidity is insufficient in the cover brine, spores of butyric anaerobes will germinate, grow, and spoil the product. High spore population in products of this kind indicate inadequate washing of the cucumbers before brining, but are of no public health significance if the acetic acid content is adequately high and the pH adequately low.

Certain vegetables intended for use in soups may be dry-salted or brined (20% NaCl) and stored at 1.7–4.4°C (Fleming *et al.*, 2001). Such products often contain halophiles, cocci, and spores, but usually are used before problems develop.

C Pathogens

Proper acidification effectively controls microbial pathogens, thus there are no documented cases of food poisoning from the ingestion of commercially pickled foods (Fleming *et al.*, 2001). However, the ability of the combination of salt and acidity in raw fermented products to preclude the growth of the vegetative cells of any of the food-borne pathogens should not be taken for granted. Vegetative pathogens tend to die at a rate influenced by pH, salt concentration, and temperature.

D CONTROL (fermented and acidified vegetables)

Summary

Significant hazards	• None.
Control measures	
Initial level (H_0)	• Controls for raw vegetables.
Increase (ΣI)	• Reduced pH controls growth of pathogens.
Reduction (ΣR)	• Certain acid levels may reduce pathogens.
Testing	• Routine culturing is not recommended.
Spoilage	Depends on type of product and is controlled by: • Proper distribution of salt or brine make-up. • Proper fermentation sequence (in fermented pickles). • Maintenance of the appropriate temperature during fermentation. • Destruction or inhibition of the activity of oxidative yeasts. • The use of starter cultures.

Hazards to be considered. Pathogenic microorganisms are not a significant issue for fermented vegetables.

Spoilage. Yeast inhibition can be achieved by securely covering the vat surface to exclude oxygen. Alternatively, short-wave ultraviolet light directed to the brine surface during fermentation is likewise effective.

Thorough washing to remove dirt and reduce the level of undesirable microorganisms and softening enzymes allows the starter to hold the natural microflora in check (Fleming, 1982; Vaughn, 1985).

Use of a starter culture of *Lb. plantarum* or a mixture with *Ped. cerevisiae* rather than relying on the natural flora is recommended, and gas formers can be controlled by purging with N_2 gas to sweep CO_2 from the brine until fermentation is complete (Etchells *et al.*, 1973; Fleming *et al.*, 1975).

Disposal of brines is strongly controlled in many areas. Recycled brine should be heated to inactivate enzymes and microorganisms, followed by precipitation, flocculation, and filtration. Excess acid can be removed by purging with air to encourage oxidative yeasts.

Pasteurization of non-fermented pickles is accomplished by heating to an internal temperature of 73°C for 15 min followed by a rapid cooling to prevent overcooking and concomitant softening. In some instances, sodium benzoate is used as a preservative if a milder brine or pasteurization treatment is desired. However, even benzoate will not prevent spoilage of pickles prepared under insanitary conditions that introduce large numbers of sugar- or vinegar-tolerant yeasts.

Shelf stability of the finished product depends on proper refrigeration of raw fermented vegetables, or appropriate pasteurization or canning.

IX Sprouts

Seed sprouts, including alfalfa, chick peas, cress, fenugreek, soya, lentils, sunflower, radish, and other vegetables, are traditionally consumed raw for their nutritional value and flavor. Mung beans may also be consumed raw, but are frequently cooked. The NACMCF (1999) reviewed current US industry practices and microbial ecology of food-borne pathogens during sprout production. Numerous food-borne outbreaks have been associated with sprouted seeds (Taormina *et al.*, 1999; Sewell and Farber, 2001). For example, sprouts were the vehicle for >50% of California's multicounty food-borne outbreaks with confirmed vehicles from 1996 to 1998 (Mohle-Boetani *et al.*, 2001). The seeds used for sprouting are an item of international commerce and have been the source of food-borne illness in the importing countries (NACMCF, 1999).

A Effects of harvesting, transportation, processing, and storage

The moist, warm growth conditions needed for seed germination and the availability of nutrients from seed coats or damaged seeds encourage microbial growth, thus high counts of non-pathogenic bacteria on sprouts do not necessarily indicate a public health problem or a lack of quality. However, there also exists a real potential for even low levels of pathogenic microorganisms to multiply and cause problems (Brown and Oscroft, 1989; Hara-Kudo *et al.*, 1997; Itoh *et al.*, 1998, NACMCF, 1999).

Seeds are soaked in water or preferably sanitizing solution prior to germination and growth. Sanitizing solutions discussed in the control section may reduce microbial loads, including pathogens, but may not be effective in eliminating microorganisms. Mung beans are sprouted by soaking (8–18 h) in containers of various dimensions, and are then kept moist in the dark until hypocotyls reach the desired size. Sprouts are usually grown at 20–30°C (25°C optimum) for 3–8 days. Lower temperatures favor root growth and higher temperatures reduce hypocotyl thickness. The heat produced by the respiratory activity of sprouts makes temperature control essential. Frequent watering by submerging or sprinkling sprouts every 4–8 h is commonly used, preferably with automatic timing. Sprinklers with irregular water distribution may cause decay problems. Smaller spouts such as alfalfa, broccoli, clover, and radish, are frequently grown in rotating drums with automatic spray systems. They may also be grown in trays. Rotating drums automatically provide air circulation and water sprays. Typically, seeds are soaked, rinsed, germinated, and grown for 3–7 days at ~25°C. Cleaning and hygiene of the drums and processing areas are essential to prevent spoilage of the crop during germination and sprouting.

Though seed coats are not toxic, they may impart off-flavors. After sprouts are harvested, they are generally washed in agitated water. The sprouts float and undesirable material sinks. Washing may be done by hand, in commercial reel-type washers, or mechanical husk-removers. Dewatering must prevent damage to the sprouts while removing as much moisture as possible to prolong shelf life. Sprouts respire rapidly, and this may cause rapid wilt and decay problems. Refrigeration is important to delay spoilage. If a refrigerated vehicle is not available, they may be film-wrapped and moved in insulated cases with sufficient ice to maintain a cool environment.

B Saprophytes and spoilage

Seeds have aerobic plate counts of 10^3–10^7 cfu/g (NACMCF, 1999) and coliform counts up to 1×10^4 cfu/g prior to germination (Prokopowich and Blank, 1991). Aerobic plate counts of mung bean and other sprouts commonly are in the range of 10^8–10^9 cfu/g, psychrotrophs 10^7 cfu/g, and coliforms 10^6–10^7 cfu/g (Patterson and Woodburn, 1980: Andrews *et al.*, 1982; Sly and Ross, 1982; Splittstoesser *et al.*, 1983). Coliform counts may be due to the growth of the bean microflora rather than to insanitary conditions. In one study, the coliforms were mainly *Klebsiella pneumoniae* and *Enterobacter aerogenes*, while only one of 32 "fecal" coliform isolates was *E. coli* (Splittstoesser *et al.*, 1983). Most microbial growth occurs during the first 2 days of germination. Although high microbial population are associated with sprouts, this does not necessarily indicate potential public health concern.

Bean sprouts do not keep well. The hypocotyl wilts rapidly, turns brown, and flavor changes may occur, particularly if sprouts are exposed to light. In the refrigerator, sealed in polyethylene bags, the shelf life is about 7–12 days (Buescher and Chang, 1982). Sprouts also suffer from slimy decay, but there is little published material on the organisms responsible.

C Pathogens

Numerous international outbreaks involving salmonellae or *E. coli* O157:H7 associated with alfalfa, clover, mung bean, cress, and radish sprouts have been reported (NACMCF, 1999; Taormina *et al.*, 1999; IFT, 2001; Sewell and Farber, 2001). The largest sprout associated outbreak involved *E. coli* O157:H7 contaminated radish sprouts, with over 6 000 people in Japan infected (Watanabe and Ozasa, 1997; Michino *et al.*, 1999). *S.* Saint-paul identical to that isolated in an outbreak in the United Kingdom was found in unopened sacks of mung beans, and cases of *S.* Virchow PT34 infections also have been associated with mung bean sprouts (O'Mahony *et al.*, 1990). An outbreak involving 492 cases in Finland and Sweden was caused by *S.* Bovismorbificans in alfalfa sprouts using seeds imported from Australia (Pönkä *et al.*, 1995). In 1995, 242 persons in 17 US States and Finland became ill from infections of *S.* Stanley associated with the consumption of alfalfa sprouts made using seed from the same seed shipper (Mahon *et al.*, 1997). It is estimated that sprouts caused 22 800 cases of gastrointestinal illness or urinary tract infection, and two deaths, in the years 1996–1998 in California alone (Mohle-Boetani *et al.*, 2001).

In most of the above outbreaks, contaminated seed was a contributing factor. If present, salmonellae, *E. coli* O157:H7, or *Bacillus cereus* multiply during sprout germination (Andrews *et al.*, 1982; Brown and Oscroft, 1989; Hara-Kudo *et al.*, 1997; Itoh *et al.*, 1998; NACMCF, 1999). *Salmonella* Typhi and *Vibrio cholerae* also grew to $>10^5$ cfu/g within 24 h when inoculated onto alfalfa seeds at the beginning of germination; however, numbers decreased 1–2 logs when inoculated 24 h after germination (Castro-Rosas and Escartín, 2000). *Salmonella* present at <1 MPN/g on naturally contaminated alfalfa seeds grew to 10^2–10^4 MPN/g during the sprouting process (Stewart *et al.*, 2001). The source of pathogens could be the seeds, contaminated water, or workers. Pathogens may be protected by cracks and crevices in the seed coat, and seeds damaged by crushing, insect infestation, or molds may shield salmonellae so that they are not detectable during routine normal analysis, yet are capable of multiplying during the prolonged germinating period (Jaquette *et al.*, 1996). High levels of organic matter in sprout suspensions quickly reduce the available chlorine in washes, and massive initial levels of chlorine appear to be needed in an initial soak if salmonellae are to be destroyed. No chemical sanitizing treatment of seeds that prevents pathogen growth during sprouting has been described, although a highly effective treatment for reducing pathogen levels has been soaking seeds in 20 000 ppm calcium hypochlorite (Taormina and Beuchat, 1999; Weissinger and Beuchat, 2000; Lang *et al.*, 2000; Holliday *et al.*, 2001). A single lot of alfalfa seeds were involved in a sprout associated outbreak of *S. mbandaka* in 1998–1999. Three sprouters that used the seed with a 2000–20 000 ppm calcium hypochlorite pre-soak had no reported illnesses, while two sprouters with inconsistent soaking procedures were implicated (NACMCF, 1999). In an outbreak attributed to *S.* Muenchen in 1999, however, implicated seeds had been pre-treated with 20 000 ppm calcium hypochlorite (Proctor *et al.*, 2001). Because pathogen growth is rapid under the sprouting conditions, detectable levels are achieved in spent irrigation water prior to harvest (Fu, 2001; Stewart *et al.*, 2001; Howard and Hutcheson, 2003). Sprout growers have been advised to test spent irrigation water for the presence of pathogens (FDA, 1999).

Bacillus cereus has also been associated with one home-grown sprout outbreak involving soy, mustard, and cress seeds (Portnoy *et al.*, 1976). . Growth of *B. cereus* was supported during sprouting of alfalfa, mung bean, rice, and wheat seeds (Harmon *et al.*, 1987; Piernas and Giraud, 1997). Because of the lack of epidemiological evidence on other sprout outbreaks, *B. cereus* does not appear to present a significant hazard for commercial production of certain sprouts.

D CONTROL (sprouts)

Summary

Significant hazards	• *Salmonella*. • *E. coli* O157:H7. • *B. cereus* (home-production only).
Control measures	
Initial level (H_0)	• Good agricultural practices for production of seed used for sprouting. • Good hygienic practices in seed cleaning and storage.
Increase (ΣI)	• Train food handlers on personal hygiene. • Good hygienic practices for sprout production.
Reduction (ΣR)	• Seed treatments for 5D reduction. • Train food handlers on seed disinfection. • Cook sprouts prior to consumption, if possible.
Testing	• Aerobic colony counts or coliforms are not useful indicators. • Test spent irrigation water for pathogens and/or *E. coli* prior to harvest. • Environmental monitoring for *L. monocytogenes* or *Listeria* spp. • Monitor equipment hygiene using tests such as ATP.
Spoilage	• Control measures for pathogens control most spoilage organisms. • Strict control of sanitary conditions is also needed to prevent crop failure due to growth of plant pathogens. • Rinse sprouted seeds in chilled water to cool quickly. • Store at <5°C to prevent spoilage.

General considerations. Discussions of the control of mung bean sprout quality are found in Buescher and Chang (1982) and Brown and Oscroft (1989). Additionally, the NACMCF (1999) and IFT (2001) discuss controls appropriate for other types of sprouts.

Control of hazards begins with the grower, by following good agricultural practices, protecting seed crops from contamination by birds, animals, insects, etc., in maintaining clean and disinfected equipment and sampling microbiologically before shipping dried seeds (NACMCF, 1999). It is important to recognize that seeds are seldomly produced exclusively for sprouting purposes. Rather, most seed is produced for agricultural purposes. Therefore, programs that enhance seed grower and producer knowledge on potential hazards and appropriate controls are important to mitigate potential risks.

Once received, seeds should be stored in a clean, dry environment to prevent mold and bacterial growth. Foreign matter should be removed before soaking. A pre-soak in mild detergent and chlorinated (2–4 ppm free chlorine) water will remove residual foreign matter that could reduce the effectiveness of further chlorination. Since seeds are a raw agricultural commodity that may be contaminated with pathogens, disinfection of water used for soaking or growing is a critical aspect of control. No matter how carefully seeds are chosen, they could be contaminated with pathogenic organisms capable of multiplying during growing. Soaking presents the first opportunity to reduce pathogens; however, treatments may not reliably eliminate pathogens (NACMCF, 1999). Soaking alfalfa seeds for 45 min in a solution of 0.5% sodium hypochlorite in a commercial facility was insufficient for preventing an outbreak of salmonellosis in Finland (Pönkä *et al.*, 1995). An initial soak in water with a total chlorine level of 20 000 ppm has been proposed to reduce microbial loads (NACMCF, 1999). Treatment of mung bean seed with buffered 3.0% (wt/vol) $Ca(OCl)_2$ reduced *Salmonella* spp. and *E*.coli O157:H7 4–5 logs (Fett, 2002). Excess chlorine should be removed by washing in regular chlorinated tap water until the rinse is clear. Chlorine treatments are not allowed in some regions (e.g. Europe), therefore other decontamination methods should be used when appropriate. Irradiation of seeds has been approved in the United States (FDA, 2000). Fumigation of alfalfa seeds and mung bean seeds with ammonia gas has been shown to cause a 2–3 log destruction of *E. coli* O157:H7 and *Salmonella* Typhimurium (Himathongakham *et al.*, 2001). Salmonellae may also be inactivated by heating seeds in water at 57–60°C for 5 min. However, since only a narrow range of times and temperatures are able to kill *S.* Stanley without serious reduction in seed germination rates, decontamination of seeds by heat may have limited commercial practicality (Jaquette *et al.*, 1996). However, the combination of dry heat and irradiation has been found to be an effective means of reducing or eliminating *E. coli* O157:H7 from alfalfa, radish, and mung bean seeds (Bari *et al.*, 2003).

For seed germination, total chlorine levels of 100–200 ppm are recommended (Brown and Oscroft, 1989) in disinfected containers, for times dependent on the temperature (longer times are required when cool). Water for irrigation during germination (at 20–30°C for 4–7 days) should contain at least 2–4 ppm free chlorine, preferably maintained by an automatic system. Chlorine is preferred over ultraviolet treatment of irrigation water because it disinfects the whole system. All equipment and rooms should be easily cleaned and disinfected, and have adequate drainage for rinse waters which frequently contain high levels of microorganisms.

Water for initial washing of harvested sprouts should be chilled, and should contain 100–200 ppm total chlorine; final washing stages (5 ± 2°C) should contain 2–4 ppm free chlorine. If water is recirculated, seed coats, roots, etc. should be filtered out.

The rapid respiration rate of sprouts must be slowed to prevent rapid wilting and decay in storage. They are best washed in cold water to chill them quickly, and held at 0.6–4.4°C (Buescher and Chang, 1982). Freshness can be prolonged in polyethylene, cellophane, or PVDC film bags. At 0–5°C, the shelf life is 10–11 days and storage life decreases by 50% after even only 1 h at 20°C (Tajiri, 1979a,b). Exposure to light should be minimized. Visual checks should be made for wilting, browning, or rot, and samples should be tested for off-flavors. During transport, film-wrapped cartons should be maintained cool, and must be transferred gently. Re-usable trays or boxes should be cleaned and disinfected after every use.

Finally, consumers can reduce the risk of enteric pathogens, if necessary, by pasteurizing sprouts using a brief dip in boiling water. In 1988, concerns about the safety of seed sprouts led the UK Department of Health and Social Security to advise the public to boil bean sprouts for 15 s before consumption. Dipping green sprouts such as alfalfa for 10 s in boiling water just prior to use reduces pathogen levels with minimal impact on quality (R. Buchanan, personal communication).

Testing. Since extensive microbial growth occurs during germination and sprouting, high aerobic plate counts and *Enterobacteriaceae* are expected. Thus, tests of this nature are of little or no value for assessing hygiene or the microbial quality of the product. Furthermore, this information does not help in predicting the rate of spoilage.

Microbiological testing of the seeds has not reliably rejected lots contaminated with salmonellae. Testing the sprouted seeds (Pönkä *et al.*, 1995) or spent irrigation water (Fu, 2001; Howard and Hutcheson,

2003) has been more effective for detecting positive lots. Other means for controlling the presence of pathogens, particularly enteric pathogens, are necessary. Manufacturing environmental monitoring for *L. monocytogenes* or *Listeria* spp. is appropriate to identify potential harborage sites to minimize the potential for contamination.

Spoilage. Control measures for pathogens will generally control most spoilage organisms. Strict control of sanitary conditions is also needed to prevent crop failure due to growth of plant pathogens. Rinsing of sprouted seeds in chilled water to cool quickly will reduce product temperature, which reduces spoilage. It is important to remove as much water as possible without damaging the sprouts to prevent spoilage. Product should be stored at <5°C to prevent spoilage.

X Mushrooms

Mushrooms are produced and sold as fresh, dried, marinated, or canned. Mushroom spores germinate to form a mass of mycelium, which colonizes its substrate. The reproductive stage begins as vegetative growth ceases. The substrate is fully colonized, and the fruiting body (pileus and cap and gill tissues) forms. In many countries, production is dominated by the white mushroom *Agaricus bisporus*. Most of the knowledge of mushroom biology relates to *Ag. bisporus* (Hayes, 1985).

Aseptic growth conditions are not used for mushrooms and frequently other microorganisms play an essential role. *Ag. bisporus* and *Coprinus comatus* (Shaggy Cap) require complex soils and compost beds to fruit. The main nutrient source for *Agaricus* and *Coprinus* is compost derived from wheat or rice straw, and horse, water buffalo, or chicken manure, which is fermented for several days to provide a stable medium with low levels of nutrients to inhibit competing microorganisms. Modern systems use a primary interval of stacked straw, manure, and fertilizers, followed by controlled pasteurization in heated rooms designed to maintain aerobic conditions at a temperature high enough to kill pests and mushroom-pathogen contaminants and select for a microbial population which rapidly uses up free ammonia. The temperature during fermentation may reach 63°C (Harper and Miller, 1992) or as much as 80°C (Derikx *et al.*, 1990). The compost is then cooled to 23–28°C and inoculated with spawn.

Lentinus edodes (Shiitake or Japanese Forest Mushroom) is widely cultivated on wood logs; spawn is inoculated into holes in logs of various woods which are kept at correct moisture and a temperature of 24–28°C for 8 months, then transferred to shady (12–20°C) areas and watered regularly; fruits develop over a period of about 6 years. *Volvariella volvaceae* (Chinese straw mushroom) and *Pleurotus* (oyster mushroom) are grown on rice or other straw. *Vol. volvaceae* is traditionally inoculated into stacks of bundles of wet straw between successive rice crops. Temperatures inside stacks increase to 40–50°C and fruits form in 12–14 days. Productivity under the local climatic conditions is variable, and rodents, snails, and slugs are significant pests. In recent years, a more industrial approach to cultivation has seen introduction of indoor growing beds, fluorescent lamps, and short pasteurization procedures.

A Effects of harvesting, transportation, processing, and storage on microorganisms

Mushrooms are contaminated with microorganisms from their growth substrate. They may additionally be contaminated from rodent droppings and insects. Also, since harvesting is almost entirely done by hand, human pathogens may be transferred to the fruiting bodies. The soft flesh is easily bruised, accelerating the spread of spoilage microorganisms, for example, *Pseudomonas* spp., yeasts, and molds. Fresh mushrooms spoil quickly when stored aerobically at ambient temperatures; *Agaricus* is generally sold packaged under plastic film. Mushrooms may also contain spore-forming bacteria, of which *Cl. botulinum* is the greatest concern.

B Saprophytes and spoilage

The microbiological interactions during growth are complex and there are various fungal, bacterial, or viral pathogens of mushrooms. For example, *Verticillium fungicola* infections cause malformed stipes and fruiting body. *Ps. tolaasii* causes brown blotching of the pileus. Defects such as La France, watery stipe and dieback disease, resulting from viral infections, previously caused major crop losses, but are largely prevented by proper pasteurization and disinfection of the growth menstruum. Other problems arise from competitive molds, pests such as flies and mites in the attractive breeding area of mushroom substrates, nematodes from the soil and compost, and rodents which are attracted to the grain used as a base medium for the spawn.

Microbial contaminants on fresh mushrooms include mainly *Ps. fluorescens*, yeasts and molds, presumably from the soil. Washing with water at the collection station removes most of the contamination, and the remainder is nearly eliminated by the blanching step. However, population of all microorganisms increases rapidly during processing, probably through recontamination from dirty tables, conveyors, and screens (Su and Lee, 1987).

C Pathogens

Mushroom production virtually ensures the presence *of Cl. botulinum* type A (Kautter *et al.*, 1978) or type B (Hauschild *et al.*, 1975). Anaerobic storage conditions could lead to toxin formation in food commodities containing mushroom. Several outbreaks of botulism have been traced to consumption of home or restaurant canned (CDC, 1973a; Health and Welfare Canada, 1987), commercially marinated mushrooms (CDC, 1973b; Pivnick *et al.*, 1973; Todd *et al.*, 1974) and commercially canned mushrooms (Lynt *et al.*, 1975). In a botulinum outbreak from canned mushrooms from China, a change in processing methods from hand packing to machine packing with vibration caused the mushrooms to be filled more compactly. Although the filled weight was the same, the rate of heat penetration decreased, resulting in insufficient processing.

The atmosphere inside film-wrapped *Agaricus* has a decreased oxygen content which prolongs shelf life but favors the growth of *Cl. botulinum*. Evidence that toxin can be formed in air-tight packs inoculated with *Cl. botulinum* (Sugiyama and Yang, 1975; Kautter *et al.*, 1978; Sugiyama and Rutledge, 1978) led the US Food and Drug Administration to require mushroom packs to be pierced. Both the danger and need for this practice have been challenged because of the heavy spoilage flora on mushrooms and the fact that low temperature storage is common (Notermans *et al.*, 1989).

During 1989, canned mushrooms from the People's Republic of China were implicated in a series of outbreaks of staphylococcal food poisoning occurring in the US (CDC, 1989b). Theories were expressed that *Staph. aureus* grew either during prolonged holding of mushrooms in brine before they reached the cannery, to form enterotoxins that were not inactivated during retorting, or as a result of defective can seams (Hardt-English *et al.*, 1990). One study (Park *et al.*, 1992) suggested that an enzyme-linked immunosorbent assay kit procedure used to test samples for staphylococcal enterotoxin could have given false-positive results. Subsequent research has demonstrated that preformed enterotoxin can be rendered serologically inactive by the extensive thermal process required for low acid canned foods, such as for canned mushrooms. This interferes with the conventional serological tests for detection even though the enterotoxin is still biologically active and can cause illness. Methods have been developed to restore serological detection of the enterotoxin (Bennett, 1992, 1994; Bennett *et al.*, 1993).

Fresh mushrooms were epidemiologically linked to an outbreak of *C. jejuni* in the Seattle, Washington area. It was hypothesized that the mushrooms were contaminated from an infected mushroom picker or packer (Seattle-King County Department of Public Health, 1984). *C. jejuni* was isolated from 1.5% of 200 packages of retail mushrooms (Doyle and Shoeni, 1986).

Dried mushrooms have been the vehicle for salmonellosis in Germany (BgVV, 2002). Soaking the mushrooms in lukewarm water allows the salmonellae to grow, and this could cause cross-contamination with salads, spouts, and other foods that are not cooked before consumption.

D CONTROL (mushrooms)

Summary

Significant hazards	• *Cl. botulinum.* • *Staph. aureus* enterotoxin (canned mushrooms). • Enteric bacterial pathogens.
Control measures	
Initial level (H_0)	• Pasteurize compost for mushroom growing beds. • Monitor environment for pathogens.
Increase (ΣI)	• Maintain aerobic packaging for fresh mushrooms (*Cl. botulinum*). • Prohibit or chill (<10°C) brine for mushroom harvest (*Staph. aureus*). • Train food handlers on importance of personal hygiene. • Control hydration time for dried mushrooms.
Reduction (ΣR)	• Use appropriate canning procedures for canned mushrooms. • Use in cooked applications if source does not control for pathogens.
Testing	• *E. coli* may be an indicator of fecal contamination for fresh. • No other microbiological tests are recommended for routine surveillance.
Spoilage	• Controls used for pathogens effectively control spoilage organisms, especially pasteurization of compost growing beds.

General considerations. Correct preparation of the growing bed through proper fermentation and heat treatment to pasteurize the compost is essential to prevent crop loss from mushroom pathogens. During mushroom growth, steps should be taken to prevent access of flies, mites, and rodents to the compost bed. At the end of the growing cycle, the entire growing structure should be disinfected using live steam, to eliminate viruses and other mushroom pathogens.

When harvesting, personal hygiene must be practiced by pickers to prevent transfer of human pathogens. Mushrooms must be handled carefully at all stages to prevent bruising. Where brine is used to preserve fresh mushrooms until they reach canning collection stations, it should be chilled (<10°C) to prevent growth of *Staph. aureus*. Only clean washing water should be used at collection stations. Frequent cleaning and disinfection of tables, conveyors, and screens are essential to prevent the build-up of debris and spread of spoilage microorganisms.

An acid vacuum-chelation process is claimed to improve color and texture of canned mushrooms by permitting a slightly reduced process, without risk of botulinal survival and toxigenesis. It involves acidification of vacuum-hydrated mushrooms by blanching in 0.05 M citric acid solution (pH 3.5) and the addition of the equivalent of 200 ppm $CaNa_2$-EDTA to the canning brine (Okereke *et al.*, 1990).

Testing. If fresh mushrooms are to be used without cooking, tests for *E. coli* as an indicator of fecal contamination may be appropriate. No other microbiological tests are recommended for routine surveillance of fresh, dried, marinated, or canned mushrooms.

Spoilage. Controls used for pathogens effectively control spoilage organisms, especially pasteurization of compost growing beds.

XI Cassava

Cassava (tapioca) is a staple food in many countries and there are many recipes for its use, including fresh cassava, beiju (fermented), sour cassava starch, chips, and gari.

Cassava processing is generally done under conditions where the application of sophisticated control technologies is impractical. Gari goes through fermentation and other processing steps before drying (Steinkraus, 1989a). In Zaire, cassava tubers are dug, peeled, retted (rotted in water), drained, pounded and pulped, transported for defibering, and made into bread (chikwangue) by kneading, precooking, kneading again, wrapping in leaves, and cooking for up to 120 min (Regez *et al.*, 1987). Another fermented product made from cassava is "Tapai ubi" (Malaysia) or "Tape ketella" (Indonesia). This is produced using cassava as the substrate with a starter culture ("ragi"). Another substrate commonly used is glutinous rice (Steinkraus (1983, 1989b).

A *Effects of harvesting, transportation, processing, and storage on microorganisms*

Sampling of the cassava processing from three regions of Zaire revealed average aerobic colony population of 2×10^8 cfu/g at the retted tuber stage, diminishing to $<1 \times 10^2$ cfu/g for fresh bread, and rising to 1×10^9 cfu/g after 7 days of storage. During retting, the population of *Bacillus* spp., lactic bacteria (*Leuconostoc* spp. and *Lactobacillus* spp.), and *Corynebacterium* spp. reached levels of 1×10^9 cfu/g, almost disappeared on cooking, and then rose slowly to about 5×10^4 cfu/g after 3 days of storage (Okafor *et al.*, 1984). Microorganisms play an important role in the texture and flavor of traditional cassava products; therefore, high levels of aerobic and lactic bacteria during processing are not viewed as a sign of bad processing or a health risk (Okafor *et al.*, 1984). Lactic acid bacteria predominate, although yeasts and clostridia are observed (Brauman *et al.*, 1996). The lactic acid fermentation generated ethanol and lactic acid (Brauman *et al.*, 1996).

Leaves used for wrapping the dough are often dirty, and are an obvious source of contamination. Up to 4×10^5 cfu/mL coliforms (*Klebsiella* spp.) were detected in all samples of water used for retting, and up to 8×10^3 cfu/g in cassava pulp, but not in fresh or stored bread. This is not unusual because coliforms occur naturally in plant material and baking reaches temperatures that destroy vegetative microbes.

The high fungal population in retted tubers (3×10^8 cfu/g) fell below 1×10^2 cfu/g after cooking, but rose to 3×10^7 cfu/g after 7 days of storage. Molds are the main factor limiting the shelf life of cassava products, but there is also the risk of the production of mycotoxins.

In Tanzania, use of a *Lb. plantarum* starter culture for fermentation of cassava flour resulted in a higher quality product and reduced levels of *Enterobacteriaceae*, yeast, and molds during fermentation, when compared with the use of 5% inoculum from the previous spontaneous fermentation (Kimaryo *et al.*, 2000). Starter culture usage may provide more consistent control in the production of cassava flour.

Aflatoxins have been reported in various cassava products (Masimango and Kalengayi, 1982), but these and other reported detection of aflatoxins appear to be artifacts of the analytical method.

B CONTROL (cassava)

Summary

Significant hazards	• Information too limited to identify.
Control measures	
Initial level (H_0)	• None.
Increase (ΣI)	• Use uncontaminated water for retting. • Proper fermentation.
Reduction (ΣR)	• Cooking of retted product.
Spoilage	• Time–temperature control.

Comments. To reduce the number of potential pathogens introduced into cassava, the quality of retting should be improved by retting in barrels containing the best quality water obtainable. A good lactic fermentation, which lowers the pH, also slows down multiplication of *Aspergillus flavus* and *Asp. parasiticus* (Holmquist *et al.*, 1983).

If artificial cooling is not possible, cassava bread for sale at markets should at least be protected from direct sunshine, to minimize the rise in temperature. Several practices have been recommended (Regez *et al.*, 1987) to increase the shelf life and reduce growth of microorganisms during storage of cassava bread.

The bread should be cooked for at least 2 h; this practice is often neglected in the interest of selling the product rapidly.

Leaves used for wrapping should be soaked in boiling water to reduce contamination levels.

Cassava bread sold at local markets should be regularly checked in order to detect inadequately processed material.

References

Abdul-Raouf, U.M., Beuchat, L.R. and Ammar, M.S. (1993) Survival and growth of *Escherichia coli* O157:H7 on salad vegetables. *Appl. Environ. Microbiol.*, **59**, 1999–2006.

Adams, M.R., Hartley, A.D. and Cox, J.L. (1989) Factors affecting the efficacy of washing procedures used in the production of prepared salads. *Food Microbiol.*, **6**, 69–77.

Alzamora, S.M., Tapia, M.S. and López-Malo, A. (2000) *Minimally Processed Fruits and Vegetables*, Aspen Publishers, Inc., pp. 86.

Andrews, W.H., Wilson, C.R., Poelma, P.L., Rornero, A. and Mislivec, P.B. (1979) Bacteriological survey of sixty health foods. *Appl. Environ. Microbiol.*, **37**, 559–66.

Andrews, W.H., Mislivec, P.B., Wilson, C.R., Bruce, V.R., Poelma, P.L., Gibson, R., Trucksess, M.W. and Young, K. (1982) Microbial hazards associated with bean sprouting. *J. Assoc. Off. Anal. Chem.*, **65**, 241–8.

Arumugaswamy, R.K., Rusul Rahamat Ali, G. and Hamid, S.N.B.A. (1994) Prevalence of *Listeria monocytogenes* in foods in Malaysia. *Int. J. Food Microbiol.*, **23**, 117–21.

Aureli, P., Fiorucci, G.C., Caroli, D., Marchiaro, G., Novara, O., Leone, L. and Salmaso, S. (2000) An outbreak of febrile gastroenteritis associated with corn contaminated by *Listeria monocytogenes*. *N. Engl. J. Med.*, **342**, 1236–41.

Austin, J.W., Dodds, K.L., Blanchfield, B. and Farber, J.M. (1998) Growth and toxin production by *Clostridium botulinum* on inoculated fresh-cut packaged vegetables. *J. Food Prot.*, **61**, 324–8

Badawy, A.S., Gerba, C.P. and Kelley, L.M. (1985) Survival of rotavirus SA-11 on vegetables. *Food Microbiol.*, **2**, 199–205.

Bagdasargan, G.A. (1964) Survival of viruses of the enterovirus group (poliomyelitis, ECHO, Coxsackie) in soil and on vegetables. *J. Hyg. Epidemiol. Microbiol. Immunol.*, **8**, 497–505.

Bardell, D. (1997) Survival of herpes simplex virus type I on some common foods routinely touched before consumption. *J. Food Prot.*, **60**, 1259–61.

Bari, M.L., Nazuka, E., Todoriki, S. and Isshiki, K. (2003) Chemical and irradiation treatments for killing *Escherichia coli* O157:H7 on alfalfa, radish and mung bean seeds. *J. Food Prot.*, **66**, 767–74.

Barnett, B.J., Schwartze, M., Sweat, D., Lea, S., Taylor, J., Bibb, B., Pierce, G. and Hendricks, K. (1995) Outbreak of *Escherichia coli* O157:H7, Waco, Texas. *Epidemic Intelligence Service* 44th, March 27–31, 1995, Centers for Disease Control, Atlanta, GA, pp.17–18.

Bashford, T.E., Gillespy, T.J. and Tomlinson, A.J.H. (1960) Report of the Fruit and Vegetable Canning and Quick Freezing Association, Chipping Camden, England.

Beckers, H.J., in't Veld, P.H., Soentoro, P.S.S. and Delfgou-van Asch, E.H.M. (1989) The occurrence of *Listeria* in food, in *Proceedings, Symposium on Foodborne Listeriosis*, Wiesbaden, Germany, September 1988, Behr's Verlag, Hamburg, pp. 85–97.

Bennett, R.W. (1992) The biomolecular temperament of staphylococcal enterotoxin in thermally processed foods. *J. AOAC Int.*, **75**, 6–12.

Bennett, R.W. (1994) Urea renaturation and identification of *Staphylococcal enterotoxin*, in *Rapid Methods and Automation in Microbiology and Immunology*. (eds R.C. Spencer, E.P. Wright and S.W.B. Newsom), Intercept Ltd., Hampshire, UK, pp. 401–11.

Bennett, R.W., Sullivan, T., Catherwood, K., Lukey, L.J. and Abhayaratna, N. (1993) Behavior and serological identification of *Staphylococcal enterotoxin* in thermally processed mushrooms, in *Mushroom Biology and Mushroom Products* (eds S.-T. Chang, J.A. Buswell and S.-W. Chiu), The Chinese University Press, The Chinese University of Hong Kong, Hong Kong, pp. 193–207.

Bennick, M.H.J., Peppelenbos, H.W., Nguyen-the, C., Carlin, F., Smid, E.J. and Gorris, L.G.M. (1996) Microbiology of minimally processed, modified atmosphere packaged chicory endive. *Postharvest Biol. Technol.*, **9**, 209–21.

Berrang, M.E., Brackett, R.E. and Beuchat, L.R. (1989a) Growth of *Listeria monocytogenes* on fresh vegetables stored under controlled atmosphere. *J. Food Prot.*, **52**, 702–5.

Berrang, M.E., Brackett, R.E and Beuchat, L.R. (1989b) Growth of *Aeromonas hydrophila* on fresh vegetables stored under a controlled atmosphere. *Appl. Environ. Microbiol.*, **55**, 2167–71.

Berrang, R.E., Brackett, R.E and Beuchat, L.R. (1990) Microbial, colour and textural qualities of fresh asparagus, broccoli, and cauliflower stored under controlled atmosphere. *J. Food Prot.*, **53**, 391–5.

Besser-Wiek, J.W., Forfang, J., Hedberg, C.W., *et al.* (1996) Foodborne outbreak of diarrheal illness associated with *Cryptosporidium parvum*—Minnesota, 1995. *Morb. Mortal. Wkly Rep.*, **45**, 783–4.

Beuchat, L.R. (1996) Pathogenic microorganisms associated with fresh produce. *J. Food Prot.*, **59**, 204–16.

Beuchat, L.R. (1998) Surface decontamination of fruits and vegetables eaten raw: a review. World Health Organization (WHO/FSF/FOS/98.2).

Beuchat, L.R. and Brackett, R.E. (1990a) Inhibitory effects of raw carrots on *Listeria monocytogenes*. *Appl. Environ. Microbiol.*, **56**, 1734–42.

Beuchat, L.R. and Brackett, R.E. (1990b) Survival and growth of *Listeria monocytogenes* on lettuce as influenced by shredding, chlorine treatment, modified atmosphere packaging and temperature. *J. Food Sci.*, **55**, 755–8, 870.

Beuchat, L.R. and Doyle, M.P. (1995) Survival and growth of *Listeria monocytogenes* in foods treated or supplemented with carrot juice. *Food Microbiol.*, **12**, 73–80.

Beuchat, L.R. and Ryu, J.-H. (1997) Produce handling and processing practices. *Emerg. Infect. Dis.*, **3**, 459–65.

Beuchat, L.R., Brackett, R.E., Han, D.Y.Y. and Conner, D.E. (1986) Growth and thermal inactivation of *Listeria monocytogenes* in cabbage and cabbage juice. *Can. J. Microbiol.*, **32**, 791–5.

Beuchat, L.R., Berrang, R.E. and Brackett, R.E. (1990) Presence and public health implications of *Listeria monocytogenes* on vegetables, in *Foodborne Listeriosis* (eds A.L. Miller, J.L. Smith and G.A. Somkuit), Elsevier, Amsterdam, pp. 175–81.

Beuchat, L.R., Brackett, R.E. and Doyle, M.P. (1994) Lethality of carrot juice to *Listeria monocytogenes* as affected by pH, sodium chloride and temperature. *J. Food Prot.*, **57**, 470–4.

BgVV. (2000) German Institute for Health and Consumer Protection and Veterinary Medicine. Caution when using dried mushrooms! http://www.bfr.bund.de/cms5w/sixcms/detail.php/1411

Bidawid, S., Farber, J.M. and Sattar, S.A. (2001) Survival of heatitis A virus on modified atmosphere packaged (MAP) lettuce. *Food Microbiol.*, **18**, 95–102.

Bolin, H.R., Stafford, A.E., King, A.D., Jr. and Huxsoll, C.C. (1977) Factors affecting the storage stability of shredded lettuce. *J. Food Sci.*, **42**, 1319–21.

Bolin, H.R. and Huxsoll, C.C. (1991) Effect of preparation procedures and storage parameters on quality retention of salad-cut lettuce. *J. Food Sci.*, **56**, 60–7.

Brackett, R.E. (1987) Microbiological consequences of minimally processed fruits and vegetables. *J. Food Qual.*, **10**, 195–306.

Brackett, R.E. (1989) Changes in the microflora of packaged fresh broccoli. *J. Food Qual.*, **12**, 169–81.

Brackett, R.E. (1999) Incidence, contributing factors, and control of bacterial pathogens in produce. *Postharvest Biol. Technol.*, **15**, 305–11.

Brackett, R.E. and Splittstoesser, D.F. (2001) Fruits and vegetables, in *Compendium of Methods for the Microbiological Examination of Foods* (eds F.P. Downes and K. Ito), 4th edn, Am. Public Health Assoc., Washington, DC, pp. 515–20.

Brauman, A., Keleke, S., Malonga, M., Miambi, E. and Ampe, F. (1996) Microbiological and biochemical characterization of cassava retting, a traditional lactic acid fermentation for foo-foo (cassava flour) production. *Appl. and Environ. Microbiol.*, **62**, 2854–8.

Brent, J., Gomez, H., Judson, F., Miller, K., Rossi-Davis, A., Shillam, P., Hatheway, C., McCrosky, L., Mintz, E., Kallander, K., McKee, C., Romer, J., Singleton, E., Yager, J. and Sofos, J. (1995) Botulism from potato salad. *J. Food Prot.*, **15**, 420–2.

Brocklehurst, T.F. (1994) Delicatessen salads and chilled prepared fruit and vegetables, in *Shelf life Evaluation of Foods* (eds C.M.D. Man and A.A. Jones), Blackie Academic and Professional, London, pp. 87–126.
Brocklehurst, T.F and Lund, B.M. (1981) Properties of pseudomonas causing spoilage of vegetables stored at low temperature. *J. Appl. Bacteriol.*, **50**, 259–66.
Brocklehurst, T.F., Zaman-Wong, C.M. and Lund, B.M. (1987) A note on the microbiology of retail packs of prepared salad vegetables. *J. Appl. Bacteriol.*, **63**, 409–15.
Brown, K.L. and Oscroft, C.A. (1989) Guidelines for the hygienic manufacture and retail sale of sprouted seeds with particular reference to Mung beans. Technical Manual No. 25, Campden Food and Drink Research Association, Chipping Campden, UK.
Bruno, S. (1998) Botulism caused by Italian bottled vegetables. *Lancet*, **352**, 884.
Buchanan, R.E. and Gibbons, N.E. (eds) (1974) *Bergey's Manual of Determinative Bacteriology*, 8th edn, Williams & Wilkins, Baltimore, MA, USA.
Buescher, R.W. and J.-S. Chang. (1982) Production of mung bean sprouts. *Arkansas Farm. Res.*, **31**, 13.
Burnett, S.L. and Beuchat, L.R. (2000) Human pathogens associated with raw produce and unpasteurized juices, and difficulties with decontamination. *J. Ind. Microbiol. Biotechnol.*, **25**, 281–7.
Caldwell, K.K., Adler, B.B., Anderson, G.L., Williams, P.L. and Beuchat, L.R. (2003) Ingestion of *Salmonella enterica* serotype poona by a free-living nematode, *Caenorhabditis elegans*, and protection against inactivation by produce sanitizers. *Appl. Environ. Microbiol.*, **69**, 4103–10.
Carlin, F. and Nguyen-the, C. (1994) Fate of *Listeria monocytogenes* on four types of minimally processed green salads. *Lett. Appl. Microbiol.*, **18**, 222–6.
Carlin, F. and Peck, M.W. (1995) Growth and toxin production by non-proteolytic and proteolytic *Clostridium botulinum* in cooked vegetables. *Lett. Appl. Microbiol.*, **20**, 152–6.
Castro-Rosas, J. and Escartín, E.F. (2000) Survival and growth of *Vibrio cholerae* O1, *Salmonella typhi*, and *Escherichia coli* O157:H7 in alfalfa sprouts. *J. Food Sci.*, **65**, 162–5.
Catteau, M., Krembel, C. and Wauters, G. (1985) *Yersinia enterocolitica* in raw vegetables. *Sci. Aliment.*, **5**, 103–6.
CDC (Centers for Disease Control and Prevention) (1973a) Botulinal toxin in a commercial food product. *Morbid. Mortal. Wkly Rep.*, **22**, 57–58.
CDC (1973b) Botulism traced to commercially canned mushrooms. *Morbid. Mortal. Wkly Rep.*, **22**, 241–2.
CDC (1989a) Epidemiologic notes and reports common-source outbreak of giardiasis—New Mexico. *Morbid. Mortal. Wkly Rep.*, **38**, 405–7.
CDC (1989b) Multiple outbreaks of staphylococcal food poisoning caused by canned mushrooms. *Morbid. Mortal. Wkly Rep.*, **38**, 417–8.
CDC (1994) Foodborne outbreaks of enterotoxigenic *Escherichia coli*—Rhode Island and New Hampshire, 1993. *Morbid. Mortal. Wkly Rep.*, **43**, 81, 87–9.
CDC (1995) Outbreak of *E. coli* O157:H7, Northwestern Montana, EPI-AID 95–68.
CDC (1996) Update: outbreaks of *Cyclospora cayetanensis* infections—U.S. and Canada, 1996. *Morbid. Mortal. Wkly Rep.*, **45**, 611–2.
CDC (1997) Outbreak of cyclosporiasis—Northern Virginia—Washington, DC—Baltimore, Maryland, metropolitan area, 1997. *Morbid. Mortal. Wkly Rep.*, **46**, 689–91.
CDC (1998) Botulism in the United States, 1899–1996. *Handbook for Epidemiologists, Clinicians and Laboratory Workers*, Center for Disease Control and Prevention, Atlanta, GA, USA.
Clark, W.S., Jr., Reinbold, G.W. and Rambo, R.S. (1966) Enterococci and coliforms in dehydrated vegetables. *Food Technol.*, **20**, 1353–6.
Codex Alimentarius (1983) Recommended international code of practice for low-acid and acidified low-acid canned foods. CAC/RCP 23-1979, Joint FAO/WHO Food Standards Programme, FAO, Rome.
Collins, M.A. and Buick, R.K. (1989) Effect of temperature on the spoilage of stored peas by *Rhodotorula glutinis*. *Food Microbiol.*, **6**, 135–41.
Cook, K.A., Boyce, T., Langkop, C., Kuo, K., Schwartz, M., Ewert, D., Sowers, E., Wells, J. and Tauxe, R. (1995) Scallions and shigellosis: a multistate outbreak traced to imported green onions, *Epidemic Intelligence Service 44th Annu. Conf.*, March 27–31, CDC, Atlanta, GA, pp. 36.
Crowe, L., Lau, W., McLeod, L., *et al.* (1999) Outbreaks of *Shigella sonnei* infection associates with eating fresh parsley—United States and Canada, July–August 1998. *Morbid. Mortal. Wkly Rep.*, **48**, 285–9.
D'Aoust, J.-Y. 1994. *Salmonella* and the international food trade. *Int. J. Food Microbiol.*, **24**, 11–31.
Davis, H.J., Taylor, P., Perdue, J.N., Stelma, G.N., Humphreys, J.M., Rowntree, R. and Greene, K.D. (1988) A shigellosis outbreak traced to commercially distributed lettuce. *Am. J. Epidemiol.*, **128**, 1312–21.
de Boer, E., Seldam, W.M. and Oosterom, J. (1986) Characterization of *Yersinia enterocolitica* and related species isolated from foods and porcine tonsils in the Netherlands. *Int. J. Food Microbiol.*, **3**, 217–24.
Deibel, K.E. and Jantschke, M. (2001) Canned foods-tests for commercial sterility, in *Compendium of Methods for the Microbiological Examination of Foods* (eds F.P. Downes and K. Ito), 4th edn. Am. Public Health Assoc., Washington, DC., pp. 577–82.
Delaquis, P., Stewart, S., Cazaux, S. and Tiovonen, P. (2002) Survival and growth of *Listeria monocytogenes* and *Escherichia coli* O157:H7 in ready-to-eat iceberg lettuce washed in warm chlorinated water. *J. Food Prot.*, **65**, 459–64.
Delmas, C.L. and Vidon, D.J.-M. (1985) Isolation of *Yersinia enterocolitica* and related species from foods in France. *Appl. Environ. Microbiol.*, **50**, 767–71.

Dennis, C. (1987) Microbiology of fruits and vegetables, in *Essays in Agricultural and Food Microbiology* (eds J.R. Norris and G.L. Pettipher), J. Wiley & Sons, UK, pp. 227–60.
Denny, C.B. and Parkinson, N.G. (2001) Canned foods-tests for cause of spoilage, in *Compendium of Methods for the Microbiological Examination of Foods* (eds F.P. Downes and K. Ito), 4th Edition, Am. Pub Health Assoc., Washington, DC. pp. 583–600.
Dentinger, C.M., Bower, W.A., Nainan, O.V., Cotter, S.M., Myers, G.M., Dubusky, L.M., Fowler, S., Salehi, E.D. and Bell, B.P. (2001) An outbreak of hepatitis a associated with green onions. *J. Infect. Dis.*, **183**, 1273–6.
Derikx, P.J.L., op den Camp, H.J.M., van der Drift, C., van Griensven, L.J.L.D. and Vogels, G.D. (1990) Biomass and biological activity during the production of compost used as a substrate in mushroom cultivation. *Appl. Environ. Microbiol.*, **56**, 3029–34.
Doyle, M.P. and Shoeni, J.L. (1986). Isolation of *Camplylobacter jejuni* from retail mushrooms. *Appl. Environ. Microbiol.*, **51**, 449.
EPA (Environmental Protection Agency) (2002) Code of Federal Regulations. Title, 40 CFR Part 503 and 114, US Gov. Print. Off., Washington DC, USA.
Etchells, J.L., Bell, TA., Monroe, R.J., Masley, P.M. and Demain, A.L. (1958) Populations and softening enzyme activity of filamentous fungi on flowers, ovaries, and fruit of pickling cucumbers. *Appl. Microbiol.*, **6**, 427–40.
Etchells, J.L., Borg, A.F. and Bell, T.A. (1961) Influence of sorbic acid on populations and species of yeasts occurring in cucumber fermentations. *Appl. Microbiol.*, **9**, 139–44.
Etchells, J.L., Costilow, R.N., Anderson, T.E. and Bell, T.A. (1964) Pure culture fermentation of brined cucumbers. *Appl. Microbiol.*, **12**, 523–35.
Etchells, J.L., Borg, A.F., Kittel, ID., Bell, T.A. and Fleming, H.P. (1966) Pure culture fermentation of green olives. *Appl. Microbiol.*, **14**, 1027–41.
Etchells, J.L., Borg, A.F. and Bell, T.A. (1968) Bloater formation by gas-forming lactic acid bacteria in cucumber fermentations. *Appl. Microbiol.*, **16**, 1029–35.
Etchells, J.L., Bell, T.A., Fleming, H.P., Kelling, R.E. and Thompson, R.L. (1973) Suggested procedure for the controlled fermentation of commercially brined pickling cucumbers—the use of starter cultures and reduction of carbon dioxide accumulation. *Pickle Pack. Sci.*, **3**, 4–14.
ECSCF (European Commission Scientific Committee on Foods) (2002) *Risk Profile on the Microbiological Contamination of Fruits and Vegetables Eaten Raw.* Adopted April 2002 (http://europa.eu.int/comm/food/fs/sc/scf/out125_en.pdf).
Farber, J.M., Sanders, G.W. and Johnston, M.A. (1989) A survey of various foods for the presence of *Listeria* species. *J. Food Prot.*, **52**, 456–8.
Farber, J.M., Wang, S.L.M., Cai, Y. and Zhang, S. (1998) Changes in populations of *Listeria monocytogenes* inoculated on packaged fresh-cut vegetables. *J. Food. Prot.*, **61**, 192–5.
FDA (Food and Drug Administration) (1989) HHS News, P89-20, United States Food and Drug Administration, Washington DC, USA.
FDA (1998) *Guidance for Industry—Guide to Minimize Microbial Food Safety Hazards for Fresh Fruits and Vegetables*, United States Food and Drug Administration, Washington DC, USA (http://www.foodsafety.gov/~dms/prodguid.html.
FDA (1999) Guidance for Industry: Reducing Microbial Food Safety Hazards for Sprouted Seed and Guidance for Industry: Sampling and Microbial Testing of Spent Irrigation Water During Sprout Production. Fed. Reg. 64, 57893–57902.
FDA (2000) Irradiation in the Production, Processing and Handling of Food. Final Rule. Fed. Reg. 65, 64605–7.
FDA (2001) Survey of domestic produce (http://www.cfsan.fda.gov/~dms/prodsur8.html).
FDA and FSIS (Food Safety Inspection Service, United States Department of Agriculture). (2001) Draft Assessment of the Relative Risk to Public Health from Foodborne *Listeria monocytogenes* Among Selected Categories of Ready-to-Eat Foods, January 2001, Washington DC, USA.
Fett, W.F. (2002) Reduction of *Escherichia coli* O157:H7 and *Salmonella* spp. On laboratory-inoculated mung bean seed by chlorine treatment. *J. Food Prot.*, **65**, 848–52.
Firstenberg, R., Mannheim, C.H. and Cohen, A. (1974) Microbial quality of dehydrated onions. *J. Food Sci.*, **39**, 685–8.
Fleming, H.P. (1982) Fermented vegetables. in *Economic Microbiology* (ed A.H. Rose), *volume 7*, Fermented Foods, Academic Press, London, pp. 227–58.
Fleming, H.P., Etchells, J.L., Thompson, R.L. and Bell, T.A. (1975) Purging of $C0_2$ from cucumber brines to reduce bloater damage. *J. Food Sci.*, **40**, 1304–10.
Fleming, H.P., McFeeters, R.F., Thompson, R.L. and Sanders, D.C. (1983) Storage stability of vegetables fermented with pH control. *J. Food Sci.*, **48**, 975–81.
Fleming, H.P., McFeeters, R.F. and Breidt, F. (2001) Fermented and acidified vegetables, in *Compendium of Methods for the Microbiological Examination of Foods* (eds F.P. Downes and K. Ito) 4th edn, Am. Public Health Assoc., Washington, DC, pp. 521–32.
Friedman, B.A. (1951) Control of decay in prepackaged spinach. *Phytopathology*, **41**, 709–13.
Frost, J.A., McEvoy, M.B., Bentley, CA., Andersson, Y. and Rowe, A. (1995) An outbreak of *Shigella sonnei* infection associated with consumption of iceberg lettuce. *Emerg. Infect. Dis.*, **1**, 26–9.
Fu, T., Stewart, D., Reineke, K., Ulaszek, J., Schlesser, J. and Tortorello, M. (2001) Use of spent irrigation water for microbiological analysis of alfalfa sprouts. *J. Food Protect.*, **64**, 802–6.
Garcia-Villanova Ruiz, B., Galvez Vargas, R. and Garcia-Villanova, R. (1987) Contamination on fresh vegetables during cultivation and marketing. *Int. J. Food Microbiol.*, **4**, 285–91.

Garg, N., Churey, J.J. and Splittstoesser, D.F. (1990) Effect of processing conditions on the microflora of fresh-cut vegetables. *J. Food Prot.*, **53**, 701–3.

Geiges, O. and Schuler, U. (1988) Behaviour of microorganisms during long-term storage of food products at sub-zero temperatures. *Microbiol. Alim. Nutr.*, **6**, 249–57.

Geldreich, E.E. and Bordner, R.H. (1971) Fecal contamination of fruits and vegetables—a review. *J. Milk Food Technol.*, **34**, 184–95.

Goodburn, K. (1999) VTEC and agriculture, European Chilled Foods Federation.

Goodliffe, J.P. and Heale, J.B. (1975) Incipient infections caused by *Botrytis cinerea* in carrots entering storage. *Ann. Appl. Biol.*, **80**, 243–6.

Griffin, MR., Sarowiec, J.J., McCloskey, D.L., Capuano, B., Pierzynski, B., Quinn, M., Wajuarski, R., Parkin, W.E., Greenberg, H. and Gary, G.W. (1982) Foodborne Norwalk virus. *Am. J. Epidemiol.*, **115**, 178–84.

Grogan, R.G., Misaghi, I.J., Kimble, K.A., Greathead, A.S., Rivie, D. and Bardin, R. (1977) Varnish spot, destructive disease of lettuce in California caused by *Pseudmoonas cichorii*. *Phytopathology*, **67**, 957–60.

Gross, T.P., Ceade, J.G., Gary, G.W., Harting, D., Goeller, D. and Israel, E. (1989) An outbreak of acute infectious nonbacterial gastroenteritis in a high school in Maryland. *Public Health Rep.*, **104**, 164–9.

Guerra, M.M., McLauchlin, J. and Bernardo, F.A. (2001) *Listeria* in ready-to-eat and unprocessed foods produced in Portugal. *Food Microbiol.*, **18**, 423–9.

Gulati, B.R., Allwood, P.B., Hedberg, C.W. and Goyal, S.M. (2001) Efficacy of commonly used disinfectants for the inactivation of calicivirus on strawberry, lettuce and a food-contact surface. *J. Food Prot.*, **64**, 1430–4.

Han, Y., Floros, J.D., Linton, R.H., Nielsen, S.S. and Nelson, P.E. (2001) Response surface modeling for the inactivation of *Escherichia coli* O157:H7 on green peppers (*Capsicum annuum* L.) by chlorine dioxide gas treatments. *J. Food Prot.*, **64**, 1128–33.

Hara-Kudo, Konuma, H., Iwaki, M., Kasuga, F., Sugita-Konishi, Y., Ito, Y. and Kumagai, Y. (1997) Potential hazard of radish sprouts as a vehicle of *Escherichia coli*. *J. Food Prot.*, **60**, 1125–7.

Hardt-English, P., York, G., Stier, R. and Cocotas, P. (1990) Staphylococcal food poisoning outbreaks caused by canned mushrooms from China. *Food Technol.*, **44**, 74–8.

Harmon, S.M., Kautter, D.A., Solomon, H.M. 1987. *Bacillus cereus* contamination of seeds and vegetable sprouts grown in a home sprouting kit. *J. Food Prot.*, **50**, 62–5.

Harper, F. and Miller, F.C. (1992) Physical management and interpretation of an environmentally controlled composting ecosystem. *Aust. J. Exp. Agric.*, **32**, 657–67.

Harvey, J. and Gilmour, A. (1993) Occurrence and characteristics of *Listeria* in foods produced in Northern Ireland. *Int. J. Food Microbiol.*, **19**, 193–205.

Hauschild, A.H.W. (1989) *Clostridium botulinum*, in *Foodborne Bacterial Pathogens* (ed. M.P. Doyle), Marcel Dekker, New York, pp.111–89.

Hauschild, A.H.W., Aris, B.J. and Husheimer, R. (1975) *Clostridium botulinum* in marinated products. *Can. Inst. Food Sci. Technol. J.*, **8**, 84–7.

Hayes, W.A. (1985) Biology and technology of mushroom culture, in *Microbiology of Fermented Foods* (ed. B.J.B. Wood) *Volume I*, Elsevier Applied Science, London, pp. 295–321.

Health and Welfare Canada (1986) Botulism in Canada—summary for 1985. *Can. Dis. Wkly Rep.*, **12**, 53–4.

Health and Welfare Canada (1987) Restaurant-associated botulism from in-house bottled mushrooms—British Columbia. *Can. Dis. Wkly. Rep.*, **13**, 35–6.

Hedberg, C.W. and Osterholm, M.T. (1993) Outbreaks of foodborne and waterborne viral gastroenteritis. *Clin. Microbiol. Rev.*, **6**, 199–210.

Heisick, J.E., Wagner, DE., Nierman, M.L. and Peeler, J.T. (1989) *Listeria* spp. found on fresh market produce. *Appl. Environ. Microbiol.*, **55**, 1925–7.

Herrmann, J.E. and Cliver, D.O. (1968) Methods of detecting foodborne enteroviruses. *Appl. Microbiol.*, **16**, 1564–9.

Himathongkham, S., Nuanualsuwan, S., Riemann, H. and Cliver, D.O. (2001) Reduction of *Escherichia coli* O157:H7 and *Salmonella* Typhimurium in artificially contaminated alfalfa seeds and mung beans by fumigation with ammonia. *J. Food Prot.*, **64**, 1817–9.

Ho, J.L., Shands, K.N., Friedland, G., Eckind, P. and Fraser, D.W. (1986) An outbreak of type 4b *Listeria monocytogenes* infection involving patients from eight Boston hospitals. *Arch. Int. Med.*, **146**, 520–4.

Holliday, S.L., Scouten, A.J. and Beuchat, L.R. (2001) Efficacy of chemical treatments in eliminating *Salmonella* and *Escherichia coli* O157:H7 on scarified and polished seeds. *J. Food Protect.*, **64**, 1489–95.

Holmquist, G.U., Walker, H.W. and Stahr, H.M. (1983) Influence of temperature, pH, water activity and antifungal agents on growth of *Aspergillus flavus* and *Aspergillus parasiticus*. *J. Food Sci.*, **48**, 778–82.

Howard, M.B. and Hutcheson, S.W. (2003) Growth dynamics of *Salmonella* Enterica strains on alfalfa sprouts and in waste seed irrigation water. *Appl. Environ. Microbiol.*, **69**, 548–53.

Huxsoll, C.C. and Bolin, H.R. (1989) Processing and distribution alternatives for minimally processed fruits and vegetables. *Food Technol.*, **43**(2), 124–8.

ICMSF (International Commission on Microbiological Specifications for Foods). (1988) *Microorganisms in Foods 4: Application of the Hazard Analysis Critical Control Point (HACCP) System to Ensure Microbiologicol Safety and Quality*, Blackwell Scientific Publications, Oxford.

ICMSF. (1996) *Microorganisms in Foods 5: Characteristics of Microbial Pathogens*, Blackie Academic & Professional, London.

IFST (Institute of Food Science and Technology). (1991) *Guidelines for the Handling of Chilled Foods*, 2nd edn, Institute of Food Science and Technology, London, UK.

IFST. (2001) *ISFT Position Statement: Organic Foods*, March 16, 2001 (http://www.ifst.org/hottop24.htm).

Insalata, N.F., Witzeman, S.J. and Berman, J.H. (1970) The problems and results of an incidence study of the spores of *Clostridium botulinum* in frozen vacuum-pouch-pack vegetables. *Dev. Ind. Microbiol.*, **11**, 330–4.

IFPA (International Fresh-cut Produce Association) and WGA (Western Growers Association). (1997) *Voluntary Food Safety Guidelines for Fresh Produce*, Washington DC.

IFT (Institute of Food Technologists). (2001) *Analysis and Evaluation of Preventive Control Measures for the Control and Reduction/Elimination of Microbial Hazards on Fresh Produce and Fresh-Cut Produce*, Contract 223-98-2333, Task Order No. 3 for The US Food and Drug Administration.

Itoh, Y., Sugita-Konishi, Y., Kasuga, F., Iwaki, M., Hara-Kudo, Y. Saito, N., Noguchi, Y., Konuma, H. and Kumagai, S. (1998) Enterohemorrhagic *Escherichia coli* O157:H7 present in radish sprouts. *Appl. Environ. Microbiol.*, **64**, 1532–5.

Jacxsens, L., Devlieghere, F., Falcato, P. and Debevere, J. (1999) Behavior of *Listeria monocytogenes* and *Aeromonas* spp. on fresh-cut produce packaged under equilibrium-modified atmosphere. *J. Food Prot.*, **62**, 1128–35.

Jaquette, C.B., Beuchat, L.R. and Mahon, B.E. (1996) Efficacy of chlorine and heat treatment in killing *Salmonella stanley* inoculated onto alfalfa seeds and growth and survival of the pathogen during sprouting and storage. *Appl. Environ. Microbiol.*, **62**, 2212–15.

Jarvis, B. and Morisetti, M.D. (1969) The use of antibiotics in food preservation. *Int. Biodeterior. Bull.*, **5**, 39–61.

Jerngklinchan, J. and Saitanu, K. (1993) The occurrence of salmonellae in bean sprouts in Thailand. *Southeast Asian J. Trop. Med. Pub. Health*, **24**, 114–8.

Jiang, X., Morgan, J.M. and Doyle, M.P. (2001) Fate of *Escherichia coli* O157:H7 in cow manure-amended soil, in *8th Annual Meeting of the Univ. of Georgia Center for Food Safety*, March 6–7, 2001, Atlanta, GA.

Kallender, K.D., Hitchins, A.D., Lancette, G.A., Schmieg, J.A., Garcia, G.R., Solomon, H.M. and Sofos, J.N. (1991) Fate of *Listeria monocytogens* in shredded cabbage stored at 5 and 25°C under a modified atmosphere. *J. Food Prot.*, **54**, 302–4.

Kapperud, G., Rorvik, L.M., Hasseltvedt, V., *et al.* (1995) Outbreak of *Shigella sonnei* infection traced to imported iceberg lettuce. *J. Clin. Microbiol.*, **33**, 609–14.

Karitsky, J.N., Osterholm, M.T., Greenberg, H.B., *et al.* (1984) Norwalk gastroenteritis: a community outbreak associated with bakery product consumption. *Ann. Intern. Med.*, **100**, 519–21.

Katzenelson, E., Telch, B. and Shuval, H.I. (1997) Spray irrigation with wastewater: the problem of aerosolization and dispersion of enteric microorganisms. *Prog. Water Technol.*, **9**, 1–11.

Kautter, D. A., Lilly, T., Jr. and Lynt, R. (1978) Evaluation of the botulism hazard in fresh mushrooms wrapped in commercial polyvinylchloride film. *J. Food Prot.*, **41**, 120–1.

Kim, J-G., Yousef, A.E. and Chism, G.W. (1999) Use of ozone to inactivate microoganisms on lettuce. *J. Food Saf.*, **19**, 17–34.

Kimaryo, V.M., Massawe, G.A., Olasupo, N.A., Holzapfel, W.H. (2000) The use of a starter culture in the fermentation of cassava for the production of kivunde, a traditional Tanzanian food product. *Int. J. Food Microbiol.*, **56**, 179–190.

King, A.D., Jr. and Bolin, H.R. (1989) Physiological and microbiological storage stability of minimally processed fruits and vegetables. *Food Technol.*, **43**, 132–6, 139.

King, A.D., Jr., Michener, H.D., Bayne, H.G. and Mihara, K.L. (1976) Microbial studies on shelf life of cabbage and cole slaw. *Appl. Environ. Microbiol.*, **31**, 404–7.

Koburger, J.A. and Farhat, B.Y. (1975) Fungi in foods. VI. A comparison of media to enumerate yeasts and moulds. *J. Milk Food Technol.*, **38**, 466–8.

Konowalchuk, J. and Speirs, J.I. (1975) Survival of enteric viruses on fresh vegetables. *J. Milk Food Technol.*, **38**, 469–72.

Koseki, S. and Itoh, K. (2002) Effect of nitrogen gas packaging on the quality and microbial growth of fresh-cut vegetables under low temperatures. *J. Food Prot.*, **65**, 326–32.

Koseki, S. Yoshida, K., Isobe, S. and Itoh, K. (2001) Decontamination of lettuce using acidic electrolyzed water. *J. Food Prot.*, **64**, 652–8.

Kudva, I.T., Hatfield, P.G. and Hovde, C.J. (1995) Effect of diet on the shedding of *E. coli* O157:H7 in a sheep model. *Appl. Environ. Microbiol.*, **61**, 1363–70.

Kurdziel, A.S., Wilkinson, N., Langton, S. and Cook, N. (2001) Survival of poliovirus on soft fruit and salad vegetables. *J. Food Prot.*, **64**, 706–9.

Lang, M.M., Ingham, B.H. and Ingham, S.C. (2000) Efficacy of novel organic acid and hypochlorite treatments for eliminating *Escherichia coli* O157:H7 from alfalfa seeds prior to sprouting. *Int. J. Food Microbiol.*, **58**, 73–82.

Larson, A.E., Johnson, E.A., Barmore, C.R. and Hughes, M.D. (1997) Evaluation of the botulism hazard from vegetables in modified atmosphere packaging. *J. Food Prot.*, **60**, 1208–14.

Lau, M.M. and Ingham, S.C. (2001) Survival of faecal indicator bacteria in bovine manure incorporated into soil. *Lett. Appl. Microbiol.*, **33**, 131–6.

Li, Y., Brackett, R.E., Chen, J. and Beuchat, L.R. (2002) Mild heat treatment of lettuce enhances growth of *Listeria monocytogenes* during subsequent storage at 5°C or 15°C. *J. Appl. Microbiol.*, **92**, 269–75.

Lilly, T., Solomon, H.M. and Rhodehamel, E.J. (1996) Incidence of *Clostridium botulinum* in vegetables packaged under vacuum of modified atmosphere. *J. Food Prot.*, **59**, 59–61.

Lin, C.-M., Fernando, S.Y. and Wei, C-i. (1996) Occurrance of *Listeria monocytogenes*, *Salmonella* spp., *Escherichia coli* and *E. coli* O157:H7 in vegetable salads. *Food Contam.*, 7, 135–40.

Lin, C.-M., Moon, S.S., Doyle, M.P. and McWatters, K.H. (2002) Inactivation of *Escherichia coli* O157:H7, *Salmonella enterica*

serotype Enteritidis, and *Listeria monocytogenes* on lettuce by hydrogen peroxide and lactic acid and by hydrogen peroxide with mild heat. *J. Food Prot.*, **65**, 1215–20.
Little, C., Roberts, D., Youngs, E. and de Louvis, J. (1999) Microbiological quality of retail imported unprepared whole lettuces: a PHLS food working group study. *J. Food Prot.*, **62**, 325–8.
Lund, B.M. (1983) Bacterial spoilage, in *Post-Harvest Pathology of Fruits and Vegetables* (ed C. Dennis), Academic Press, London, pp. 219–57.
Lund, B.M. and Nicholls, J.C. (1970) Factors influencing the soft-rotting of potato tubers by bacteria. *Potato Res.*, **13**, 210–4.
Lund, B.M., Graham, A.F. and George, S.M. (1988) Growth and formation of toxin by *Clostridium botulinum* in peeled, inoculated, vacuum-packed potatoes after a double pasteurisation and storage at 25°C. *J. Appl. Bacteriol.*, **64**, 241–6.
Lung, A.J., Lin, C.-M., Kim, J.M., Marshall, M.R., Nordstedt, R., Thompson, N.P. and Wei, C.I. (2001) Destruction of *Escherichia coli* O157:H7 and *Salmonella enteritidis* in cow manure composting. *J. Food Prot.*, **64**, 1309–14.
Lynt, R.K., Kautter, D.A. and Read, R.B. (1975) Botulism in commercially canned foods. *J. Milk Food Technol.*, **38**, 546–50.
MacDonald, K.L., Spengler, R.F., Hatheway, C.L., Hargrett, N.T. and Cohen, M.L. (1985) Type A botulism from sautéed onions. *J. Am. Med. Assoc.*, **253**, 1275–8.
MacDonald, K.L., Cohen, M.L. and Blake, P.A. (1986) The changing epidemiology of adult botulism in the United States. *Am. J. Epidemiol.*, **124**, 794–9.
MAFF (Ministry of Agriculture, Fisheries and Food). (1998) *Code of Good Agricultural Practice for the Protection of Soil, October 1998*, UK.
Mahon, B.E., Pönkä, A., Hall, W., Komatsu, K., Beuchat, L., Shiflett, S., Siitonen, A., Cage, G., Lambert, M., Hayes, P., Bean, N., Griffin, P. and Slutsker, L. (1997). An international outbreak of *Salmonella* infections caused by alfalfa sprouts grown from contaminated seed, *J. Infect. Dis.*, **175**, 876–82.
Martin, D.L., Gustafson, T.L., Pelesi, J.W., Suarez, L. and Pierce, G.V. (1986) Contaminated produce—a common source for two outbreaks of *Shigella gastroenteritis*. *Am. J. Epidemiol.*, **124**, 299–305.
Masimango, N.T. and Kalengayi, M.M.R. (1982) Aflatoxins in foods and foodstuffs in Zaire, in *Proceedings of the International Symposium on Mycotoxins* (eds K. Naguib, D.L. Park, M.M. Naguib and A.E. Pohland) Cairo, Egypt, 1981, pp 431–5.
Masson, A. (1988) Microbiologie des legumes frais predecoupes (Microbiology of fresh ready cut vegetables). *Microbiol. Alim. Nutr.*, **6**,197–9.
Mayerhauser, C.M. (2001) Survival of enterohemorrhagic *Escherichia coli* O157:H7 in retail mustard. *J. Food Prot.*, **64**, 783–7.
Mazzotta, A. (2001) Heat resistance of *Listeria monocytogenes* in vegetables: evaluation of blanching processes. *J. Food Prot.*, **64**, 385–7.
Merson, M.H., Morris, G.K., Sack, D.A., Wells, J.E., Feeley, J.C., Sack, R.B., Creech, W.B., Zapikan, A.Z. and Gangarosa, E.J. (1976) Travelers' diarrhea in Mexico. *N. Engl. J. Med.*, **294**, 1299–305.
Michener, H.D., Boyle, F.P., Notter, G.K. and Guadagni, D.G. (1968) Microbiological deterioration of frozen parfried potatoes upon holding after thawing. *Appl. Microbiol.*, **16**, 759–61.
Michino, H., Araki, K., Minami, S., Takaya, S. and Sakai, N. (1996) Investigation of large-scale outbreak *Escherichia coli* O157:H7 infection among school children in Sakai City, 1996, in *Abstracts of the 32nd Joint Conference US-Japan Cooperative Medical Science Program, Cholera and Related Diarrheal Diseases Panel*, 1996, Nagasaki, Japan, pp. 84–9.
Michino, H., Araki, K., Minami, S., Takaya, S., Sakai, N., Miyazaki, M., Ono, A. and Yanagawa, H. (1999) Massive outbreak of *Escherichia coli* O157:H7 infection in school children in Sakai City, Japan, associated with consumption of white radish sprouts. *Am. J. Epidemiol.*, **150**, 787–96.
Miller, S.A. (1980) Susceptibility of lettuce cultivars to marginal leaf blight caused by *Pseudomonas marginalis*. *N.Z. J. Exp. Agric.*, 8, 169–71.
Mohle-Boetani, J.C., Farrar, J.A., Werner, S.B., Minassian, D., Bryant, R., Abbott, S., Slutsker, L. and Vugia, D.J. (2001) *Escherichia coli* O157 and *Salmonella* infections associated with sprouts in California. Ann. Intern. Med., **135**, 239–47.
Montville, R., Chen, Y. and Schaffner, D.W. (2001) Glove barriers to bacterial cross-contamination between hands to food. *J. Food Prot.*, **64**, 845–9.
Mundt, J.O. and Hammer, J.L. (1968) Lactobacilli on plants. *Appl. Microbiol.*, **16**, 1326–30.
Mundt, J.O., Anandam, E.J. and McCarty, I.E. (1966) Streptococceae in the atmosphere of plants processing vegetables for freezing. *Health Lab. Sci.*, **3**, 207–13.
Mundt, J.O., Graham, W.F. and McCarty, I.E. (1967) Spherical lactic acid-producing bacteria of southern-grown raw and processed vegetables. *Appl. Microbiol.*, **15**, 1303–8.
Munsch, M.H., Simard, R.E. and Girard, J.M. (1982) Effect of further treatments on the microflora of commercially washed stored carrots. *Can. Inst. Food Sci. Technol. J.*, **15**, 322–4.
Murphy, R.P. (1973) Microbiological contamination of dried vegetables. *Process. Biochem.*, **8**, 17–9.
NACMCF (National Advisory Committee on Microbiological Criteria for Foods). (1991) *Listeria monocytogenes*. Recommendations by the National Advisory Committee on Microbiological Criteria for Foods. *Int. J. Food Microbial.*, **14**, 185–246.
NACMCF (1998). Microbiological safety evaluations and recommendations on fresh produce. *Food Control.*, **10**, 321–47.
NACMCF (1999). Microbiological Safety Evaluations and Recommendations on Sprouted Seeds, Adopted May 28, 1999, US Food and Drug Administration.
Naimi, T.S., Wicklund, J.H., Olsen, S.J., Krause, G., Wells, J.G., Bartkus, J.M., Boxrud, D.J., Sullivan, M., Kassenborg, H., Besser, J.M., Mintz, E.D., Osterholm, M.T. and Hedberg, C.W. (2003) Concurrent outbreaks of *Shigella sonnei* and enterotoxigenic *Escherichia coli* infections associated with parsley: implications for surveillance and control of foodborne illness. *J. Food Prot.*, **66**, 535–41.

Natvig, E.E., Ingham, S.C., Ingham, B.H., Cooperbrand, L.R. and Roper, T.R. (2002) *Salmonella enterica* serovar Typhimurium and *Escherichia coli* contamination of root and leaf vegetables grown in soils with incorporated bovine manure. *Appl. Environ. Microbiol.*, **68**, 2737–44.

NFPA (National Food Processors Association), International Fresh-cut Produce Association, United Fresh Fruit and Vegetable Association. (2001) *Field Cored Lettuce Best Practices*, NFPA, Washington, DC.

Nguyen-the, C. and Carlin, F. (2000) Fresh and processed vegetables, in *The Microbiological Safety and Quality of Food* (eds B.M. Lund, T.C. Baird-Parker and G.W. Gould), Aspen Publishers, Gaithersburg, MD, pp. 620–84.

Nguyen-the, C. and Prunier, J.P. (1989) Involvement of pseudomonads in deterioration of 'ready-to-use' salads. *Int. J. Food Sci. Technol.*, **24**, 47–58.

Nichols, A.A., Davies, P.A., King, K.P., Winter, E.J. and Blackwall, F.L.C. (1971) Contamination of lettuce irrigated with sewage effluent. *J. Hortic. Sci.*, **46**, 425–33.

Nicholson, F.A., Hutchison, M.L., Smith, K.A., Keevil, C.W., Chambers, B.J. and Moore, A. (2000) *A Study on Farm Manure Applications to Agricultural Land and an Assessment of the Risks of Pathogen Transfer into the Food Chain*, MAFF.

Notermans, S.H.W. (1993) Control in fruits and vegetables, in *Costridium botulinum: Ecology and Control in Foods* (eds A.H.W. Hauschild and K.L. Dodds), Marcel Dekker, New York. NY, pp 233–60.

Notermans, S., Dufrenne, J. and Gerrits, J.P.G. (1989) Natural occurrence of *Clostridium botulinum* on fresh mushrooms (*Agaricus bisporus*). *J. Food Prot.*, **52**, 733–6.

Ohata, K.T., Tsuchiya, Y. and Shirata, A. (1979) Difference in kinds of pathogenic bacteria causing head rot of lettuce of different cropping trees. *Ann. Phytopathol. Soc. Jpn*, **45**, 333–8.

Okafor, N., Ijoma, B. and Oyulu, C. (1984) Studies on the microbiology of cassava retting for foo-foo production. *J. Appl. Bacteriol.*, **56**, 1–13.

Okereke, A., Beelman, R.B. and Doores, S. (1990) Control of spoilage of canned mushrooms inoculated with *Clostridium sporogenes* PA3679 spores by acid-blanching and EDTA. *J. Food Sci.*, **5**, 1331–3, 1337.

O'Mahony, M., Cowden, J., Smyth, B., *et al.* (1990) An outbreak of *Salmonella saint-paul* infection associated with beansprouts. *Epidemiol. Infect.*, **104**, 229–35.

Orillo, C.A., Sison, E.C., Luis, M. and Pederson, C.D. (1969) Fermentation of Philippine vegetable blends. *Appl. Microbiol.*, **17**, 10–3.

Otofuji, T., Tokiwa, H. and Takahashi, K. (1987) A food poisoning incident caused by *Clostridium botulinum* toxin A in Japan. *Epidemiol. Infect.*, **99**, 167–72.

Padgett, T., Han, I.Y. and Dawson, P.L. (1998) Incorporation of food-grade antimicrobial compounds into biodegradable packaging films. *J. Food Prot.*, **61**, 1330–5.

Pace, P.J. (1975) Bacteriological quality of delicatessen foods: are standards needed? *J. Milk Food Technol.*, **38**, 347–53.

Papavassiliou, J., Tzannetis, S., Leka, H. and Michopoulos, G. (1967) Coli-aerogenes bacteria on plants. *J. Appl. Bacteriol.*, **30**, 219–23.

Park, C.E., Akhtar, M. and Rayman, M.K. (1992) Nonspecific reactions of a commercial enzyme-linked immunosorbent assay kit (TECRA) for detection of staphylococcal enterotoxins in foods. *Appl. Environ. Microbiol.*, **58**, 2509–12.

Park, C.M., Hung, Y.C., Doyle, M.P., Ezeike, G.O.I. and Kim, C. (2001) Pathogen reduction and quality of lettuce treated with electrolyzed oxidizing and acidified chlorinated water. *J. Food Sci.*, **66**, 1368–72.

Patterson, J.E. and Woodburn, M.J. (1980) *Klebsiella* and other bacteria on alfalfa and bean sprouts at the retail level. *J. Food Sci.*, **45**, 492–5.

Pederson, C.S. (1979) *Microbiology of Food Fermentations*, 2nd edn, AVI, Westport, CT, USA.

Pederson, C.S. and Albury, M.N. (1969) The sauerkraut fermentation. New York Sta. Agric. Exp. Sta. Bull., 824.

Pederson, C.S. and Kelly, C.D. (1938) Development of pink color in sauerkraut. *Food Res.*, **3**, 583–8.

Pederson, CS., Niketic, G. and Albury, M.N. (1962) Fermentation of the Yugoslavian pickled cabbage. *Appl. Microbiol.*, **10**, 86–9.

Pendergrass, A. and Isenberg, F.M.R. (1974) The effect of relative humidity on the quality of stored cabbage. *Hortic. Sci.*, **9**, 226–7.

Petran, R.L., Zottola, E.A. and Gravini, R.B. (1988) Incidence of *Listeria monocytogenes* in market samples of fresh and frozen vegetables. *J. Food Sci.*, **53**, 1238–40.

Petran, R.L., Sperber, W.H. and Davis, A.B. (1995) *Clostridium botulinum* toxin formation in Romaine lettuce and shredded cabbage: effect of storage and packaging conditions. *J. Food Prot.*, **58**, 624–7.

Piernas, V. and Giraud, J.P. (1997) Disinfection of rice seeds prior to sprouting. *J. Food Sci.*, **62**, 611–5.

Pinner, R.W., Schuchat, A., Swaminathan, B., *et al.* (1992) Role of foods in sporadic listeriosis: II. Microbiologic and epidemiologic investigation. JAMA, **267**, 2046–50.

Pivnick, H., Chang, P.C. and Rioti, J.F. (1973) Botulism. *Quebec Epidemiol. Bull.* Health Welfare Can., **17**, 88.

Pohronezny, K., Larsen, P.O. and Leben, P.O. (1978) Observations on cucumber fruit invasion by *Pseudomonas lachrymans. Pl. Div. Reptr*, **62**, 306–9.

Pönkä, A., Andersson, Y., Siltonen, A., de Jong, B., Jahkola, M., Halkala, O., Kuhmonen, A. and Pakkala, P. (1995) *Salmonella* in alfalfa sprouts. *Lancet*, **345**, 462–3.

Portnoy, J.E., Goepfert, J.M. and Harmon, S.M. (1976) An outbreak of *Bacillus cereus* food poisoning resulting from contaminated vegetable sprouts. *Am. J. Epidemiol.*, **103**, 589–93.

Porto, E. and Eiroa, M.N.U. (2001) Occurrence of *Listeria monocytogenes* in vegetables. *Dairy Food Environ. Sanit.*, **21**, 282–6.

Proctor, M.E., Hamacher, M., Tortorello, M.L., Archer, J.R. and Davis, J.P. (2001) Multistate outbreak of *Salmonella* serovar

Muenchen infections associated with alfalfa sprouts grown from seeds pretreated with calcium hypochlorite. *J. Clin. Microbiol.*, **39**, 3461–5.

Prokopowich, D. and Blank, G. (1991) Microbiological evaluation of vegetable sprouts and seeds. *J. Food Prot.*, **54**, 560–2.

Regez, P.F., Ifebe, A. and Mutinsumu, M.N. (1987) Microflora of traditional cassava foods during processing and storage: the cassava bread (Chikwangue) of Zaire. *Microbiol. Alim. Nutr.*, **5**, 303–11.

Roberts, D., Watson, G.N. and Gilbert, R.J. (1982) Contamination of food plants and plant products with bacteria of public health concern, in *Bacteria and Plants* (eds M.E. Rhodes-Roberts and F.A. Skinner), Academic Press, London, pp.169–95.

Robertson, L.J. and Gjerde (2001) Occurrence of parasites on fruits and vegetables in Norway. *J. Food Prot.*, **64**, 1793–8.

Rose, S.A. (1984) Studies of the microbiological status of prepacked delicatessen salads collected from retail chill cabinets. Campden Food Preservation Research Association, Technical Memorandum, No. 371.

Rosenblum, L.S., Mirkin, I.R., Allen, D.T., Safford, S. and Badler, S.C. (1990) A multifocal outbreak of hepatitis A traced to commercially distributed lettuce. *Am. J. Public Health*, **80**, 1075–9.

RCP (Royal Commission on Environmental Pollution). (1996) *Sustainable Use of Soil*, HMSO, ISBN 0-10-131652–6.

Ryall, A.L. and Lipton, W.J. (1979) *Handling Transportation and Storage of Vegetables. I. Vegetables and Melons*, AVI Publishing Inc., Westport, CT, USA.

Saddick, M.F., El-Sherbeeny, M.R. and Bryan, F.L. (1985) Microbiological profiles of Egyptian raw vegetables and salads. *J. Food Prot.*, **48**, 883.

Sagoo, S.K., Little, C.L. and Mitchell, R.T. (2001) The microbiological examination of ready-to-eat organic vegetables from retail establishments in the United Kingdom. *Lett. Appl. Microbiol.*, **33**, 434–9.

Sagoo, S.K., Little, C.L., Ward, L., Gillespie, I.A. and Mitchell, R.T. (2003) Microbiological study of ready-to-eat salad vegetables from retail establishments uncovers a national outbreak of salmonellosis. *J. Food Prot.*, **66**, 403–9.

Sanz, S., Gimenez, M., Olarte, C., Lomas, C. and Portu, J. (2002) Effectiveness on chlorine washing disinfection and effects on the appearance of artichoke and borage. *J. Appl. Microbiol.*, **93**, 986–93.

Satchell, F.B., Stephenson, P., Andrews, W.H., Estela, L. and Allen, G. (1990) The survival of *Shigella sonnei* in shredded cabbage. *J. Food Prot.*, **53**, 558–62, 624.

Schlech, W.F., III, Lavigne, P.M., Bortulussi, R.A., *et al.* (1983) Epidemic listeriosis—evidence for transmission by food. *N. Engl. J. Med.*, **308**, 203–6.

Seals, J.E., Snyder, J.D., Edell, TA., Hatheway, CL., Johnson, C.J. and Hughes, J.M. (1981) Restaurant-associated type A botulism: transmission by potato salad. *Am. J. Epidemiol.*, **113**, 436–44.

Seattle-King County Department of Public Health. (1984) Surveillance of the flow of *Salmonella* and *Camplylobacter* in a community, FDA Bureau of Veterinary Medicine, Washington DC.

Segall, R.H. and Dow, A.T. (1973) Effects of bacterial contamination and refrigerated storage on bacterial soft rot of potatoes. *Pl. Dis. Reptr.*, **57**, 896–9.

Sewell, A.M. and Farber, J.M. (2001) A review: foodborne outbreaks in Canada linked to produce. *J. Food Prot.*, **64**, 1863–77.

Seymour, I.J. and Appleton, H. (2001) A review: foodborne viruses and fresh produce. *J. Appl. Microbiol.*, **91**, 759–73.

Sheneman, J. (1973) Survey of aerobic mesophilic bacteria in dehydrated onion products. *J. Food Sci.*, **38**, 206–9.

Sivapalasisngam, S. and Friedman, C.R. (2001) Sprouts, salads ciders: the growing challenge of fresh produce-associated foodborne infections. In *8th Annual Meeting of the Univ. of Georgia Center for Food Safety*, March 6–7, 2001, Atlanta, GA.

Sizmur, K. and Walker, C.W. (1988) *Listeria* in prepacked salads. Lancet, **i**(8595), 1167.

Sly, T. and Ross, E. (1982) Chinese foods: relationship between hygiene and bacterial flora. *J. Food Prot.*, **45**, 115–8.

Smittle, R.B. (2000) Microbiological safety of mayonnaise, salad dressings and sauces in the United States: a review. *J. Food Prot.*, **63**, 1144–53.

Solomon, H.M. and Kautter, D.A. (1986) Growth and toxin production by *Clostridium botulinum* in sautéed onions. *J. Food Prot.*, **49**, 618–20.

Solomon, H.M. and Kautter, D.A. (1988) Outgrowth and toxin production by *Clostridium botulinum* in bottled chopped garlic. *J. Food Prot.*, **51**, 862–5.

Solomon, H.M., Kautter, D.A., Lilly, T. and Rhodehamel, E.J. (1990) Outgrowth of *Clostridium botulinum* in shredded cabbage at room temperature under a modified atmosphere. *J. Food Prot.*, **53**, 831–3, 845.

Solomon, E.B., Potenski, C.J. and Matthews, K.R. (2002) Effect of irrigation method on transmission to and persistance of *Escherichia coli* O157:H7 on lettuce. *J. Food Prot.*, **65**, 673–6.

Splittstoesser, D.F. (1970) Predominant microorganisms on raw plant foods. *J. Milk Food Technol.*, **33**, 500–5.

Splittstoesser, D.F. (1983) Indicator organisms on frozen blanched vegetables. *Food Technol.*, **37**(6), 105–6.

Splittstoesser, D.F. and Corlett, D.A., Jr. (1980) Aerobic plate counts of frozen blanched vegetables processed in the United States. *J. Food Prot.*, **43**, 717–9.

Splittstoesser, D.F. and Segen, B. (1970) Examination of frozen vegetables for *Salmonella*. *J. Milk Food Technol.*, **33**, 111–3.

Splittstoesser, D.F., Wettergreen, W.P. and Pederson, C.S. (1961) Control of microorganisms during preparation of vegetables for freezing. *Food Technol.*, **15**, 332–4.

Splittstoesser, D.F., Hervey, G.F.R., II and Wettergreen, W.P. (1965) Contamination of frozen vegetables by coagulase-positive staphylococci. *J. Milk Food Technol.*, **28**, 149–51.

Splittstoesser, D.F., Groll, M., Downing, D.L. and Kaminski, J. (1977) Viable counts versus the incidence of machinery mold (*Geotrichum*) on processed fruits and vegetables. *J. Milk Food. Technol.*, **40**, 402–5.

Splittstoesser, D.F., Queale, D.T. and Andaloro, B.W. (1983) The microbiology of vegetable sprouts during commercial production, *J. Food Saf.*, **5**, 79–86.

Stamer, J.R., Hrazdina, G. and Stoyla, B.O. (1973) Induction of red color formation in cabbage juice by *Lactobacillus brevis* and its relationship to pink sauerkraut. *Appl. Microbiol.*, **26**, 161–6.
Steinbruegge, E.G. Maxcy, R.B. and Liewen, M.B. (1988) Fate of *Listeria monocytogenes* on ready-to-serve lettuce. *J. Food Prot.*, **51**, 596–9.
Steinkraus, K.H. (ed) (1983) *Handbook of Indigenous Fermented Foods, volume 9*, Microbiology Series, Marcel Dekker Inc. New York.
Steinkraus, K.H. (1989a) Industrialization of gari fermentation, in *Industrialization of Indigenous Fermented Foods* (ed. K.H. Steinkraus), Marcel Dekker Inc., pp. 208–20.
Steinkraus, K.H. (1989b) Tapai processing in Malaysia: a technology in transition, in *Industrialization of Indigenous Fermented Foods*, (ed. K.H. Steinkraus), Marcel Dekker Inc., pp. 169–90.
St. Louis, M.E., Peck, S.H.S., Bowering, D., *et al.* (1988) Botulism from chopped garlic: delayed recognition of a major outbreak. *Ann. Intern. Med.*, **108**, 363–8.
Stewart, D.S., Reineke, K.F., Ulaszek, J.M. and Tortorello, M.L. (2001) Growth of *Salmonella* during sprouting of alfalfa seeds associated with salmonellosis outbreaks. *J. Food Protect.*, **64**, 618–22.
Su, J.-M. and Lee, H.C. (1987) Studies on microbial contamination of fresh mushroom during postharvest handling and processing. *Food Sci. China*, **14**, 95–104.
Sugiyama, H. and Rutledge, K.S. (1978) Failure of *Clostridium botulinum* to grow in fresh mushrooms packaged in plastic film overwraps with holes. *J. Food Prot.*, **41**, 348–50.
Sugiyama, H. and Yang, K.H. (1975) Growth potential of *Clostridium botulinum* in fresh mushrooms packaged in semipermeable plastic film. *Appl. Microbiol.*, **30**, 964–69.
Surkiewicz, B.F., Groomes, R.J. and Padron, A.P. (1967) Bacteriological survey of the frozen prepared foods industry. III. Potato products. *Appl. Microbiol.*, **15**, 1324–31.
Swanson, K.M.J. (1987) Growth of *E. coli*, *Salmonella*, and *Staph. aureus* on fresh broccoli and cauliflower, personal communication.
Swanson, K.M.J. (1990) Microbial counts on fresh vegetables prior to processing, personal communication.
Tajiri, T. (1979a) Studies on production and keeping quality of bean sprouts. Effect of storage temperature and its fluctuation on keeping quality of mung bean sprouts. *J. Jpn. Soc. Food Sci. Technol.*, **26**, 18–24.
Tajiri, T. (1979b) Studies on production and keeping quality of bean sprouts, II. Effect of packaging materials and storage temperatures on the keeping quality of mung bean sprouts. *J. Jpn. Soc. Food Sci. Technol.*, **26**, 542–6.
Takeuchi, K. and Frank, J.F. (2001) Quantitative determination of the role of lettuce leaf structures in protecting *Escherichia coli* O157:H7 from chlorine disinfection. *J. Food Prot.*, **64**, 147–51.
Tamminga, S.K., Beumer, R.R., Kiejbets, M.J.H. and Kampelmacher, E.H. (1978a) Microbial spoilage and development of food poisoning bacteria in peeled completely or partly cooked vacuum packaged potatoes. *Arch. Litbensmittelhyg.*, **29**, 215–9.
Tamminga, S.K., Beumer, R.R. and Kampelmacher, E.H. (1978b) The hygienic quality of vegetables grown in or imported into the Netherlands: a tentative survey. *J. Hyg. (Cambridge)*, **80**, 143–54.
Taormina, P.J. and Beuchat, L.R. (1999) Comparisons of chemical treatments to eliminate enterohaemorrhagic *Escherichia coli* O157:H7 from alfalfa seeds. *J. Food Protect.*, 62, 318–24.
Taormina, P.J., Beuchat, L.R. and Slutsker, L. (1999) Infections associated with eating seed sprouts: an international concern. *Emerg. Infect. Dis.*, **5**, 626–34.
Tauxe, R.V. (1997) Does organic gardening foster foodborne pathogens? *J. Am. Med. Assoc.*, **277**(21), 1680.
Tauxe, R., Kruse, H., Hedberg, C., Potter, M., Madden, J. and Wachsmuth, K. (1997) Microbial hazards and emerging issues associated with produce. A preliminary report to the National Advisory Committee on Microbiological Criteria for Foods. *J. Food Prot.*, **60**, 1400.
Terranova, W., Bremen, R.R., Locey, R.P. and Speck, S. (1978) Botulism type B: epidemiological aspects of an extensive outbreak. *Am. J. Epidemiol.*, **108**, 150–6.
Terry, R.C. and Overcast, W.W. (1976) A microbiological profile of commercially prepared salads. *J. Food Sci.*, **41**, 211–3.
Thunberg, R.L., Tran, T.T., Bennett, R.W., Matthews, R.N. and Belay, N. (2002) Microbial evaluation of selected fresh produce obtained at retail markets. *J. Food Prot.*, **65**, 677–82.
Thurston-Enriquez, J.A., Watt, P., Dowd, S.E., Enriquez, R., Pepper, I.L. and Gerba, C.P. (2002) Detection of protozoan parasites and microsporidia in irrigation waters used for crop protection. *J. Food Prot.*, **65**, 378–82.
Todd, E.C.D., Pivnick, H., Chang, P.C., Sharpe, A.N., Park, C. and Riou, J. (1974) *Clostridium botulinum* in commercially marinated mushrooms. *Can. J. Public Health*, **65**, 63–4.
Tsuchiya, Y., Ohata, K., Iemura, H., Sanematsu,T., Shirata, A. and Fujii, H. (1979) Identification of causal bacteria of head rot of lettuce. *Bull. Natl Inst. Agric. Sci. C.*, **33**, 77–99.
U.S. Department of Health and Human Services (2002) Food and Drug Administration. Code of Federal Regulations. Thermally Processed Low-Acid Foods Packaged in Hermetically Sealed Containers and Acidified Foods, 21 CFR 113 and 114. US Gov. Print. Off., Washington DC, USA.
Van Beneden, C.A. (1996) A prolonged multistate outbreak of *Salmonella newport* infections—sprouts from a health food gone bad, in *Abst. Epidemic Intelligence Service 45th Annu. Conf.*, April 22–26, CDC, Atlanta, GA.
Vaughn, R.H. (1951) The microbiology of dehydrated vegetables. *Food Res.*, **16**, 429–38.
Vaughn, R.H. (1970) Incidence of various groups of bacteria in dehydrated onions and garlic. *Food Technol.*, **24**, 189–91.
Vaughn, R.H. (1985) The microbiology of vegetable fermentations, in *Microbiology of Fermented Foods, volume 1* (ed B.J.B. Wood), Elsevier Applied Science, London, pp. 49–109.

Vijaykumar, C. and Wolf-Hall, C.E. 2002. Evaluation of household sanitizers for reducing levels of *Escherichia coli* O157:H7 on iceberg lettuce. *J. Food. Prot.*, **65**,1646–50.
Wachtel, M.R., Whitehand, L.C. and Mandrell, R.E. (2002) Prevalence of *Escherichia coli* associated with a cabbage crop inadvertently irrigated with partially treated sewage wastewater. *J. Food Prot.*, **65**, 471–5.
Wang, G., Zhoa, R. and Doyle, M.P. (1996) Fate of enterohemorrhagic *Escherichia coli* O157:H7 in bovine feces. *Appl. Environ. Microbiol.*, **62**, 2567–70.
Watanabe, Y. and Ozasa, K. (1997) An epidemiological study on an outbreak of *Escherichia coli* O157:H7 infection. *Rinsho Byori.*, **45**, 869–74.
Webb, T.A. and Mundt, J.O. (1978) Moulds on vegetables at the time of harvest. *Appl. Environ. Microbiol.*, **35**, 655–8.
Weissinger, W.R. and Beuchat, L.R. (2000) Comparison of chemical treatments to eliminate *Salmonella* on alfalfa seeds. *J. Food Prot.*, **63**, 1475–82.
Weissinger, W.R., McWatters, K.H. and Beuchat, L.R. (2001) Evaluation of volatile chemical treatments for lethality to *Salmonella* on alfalfa seeds and sprouts. *J. Food Prot.*, **64**, 442–50.
White, A. and White, H.R. (1962) Some aspects of the microbiology of frozen peas. *J. Appl. Bacteriol.*, **25**, 62–71.
WHO (1995) Control of foodborne trematode infections, WHO Technical Report Series 849, Geneva, Switzerland.
Wiles, A.B. and Walker, J.C. (1951) The relation of *Pseudomonas lachrymans* to cucumber fruits and seeds. *Phytopathology*, **41**, 1059–64.
Winthrop, K.I., Palumbo, M.S., Farrar, J.A., Mohle-Boetani, J.C., Abbott, S., Beatty, M.E., Inami, G. and Werner, S.B. (2003) Alfalfa sprouts and *Salmonella* Kottbus infection: a multistate outbreak following inadequate seed disinfection with heat and chlorine. *J. Food. Prot.* **66**, 13–7.
Wu, C.M., Koehler, P.E. and Ayres, J.C. (1972) Isolation and identification of xanthotoxin (8-methoxypsoralen) and bergapten (5-methoxypsoralen) from celery infected with *Sclerotinia sclerotiorum*. *Appl. Microbiol.*, **23**, 852–6.
Wu, F.M., Beuchat, L.R., Doyle, M.P., Garrett, V., Wells, J.G. and Swaminathan, B. (2002) Fate of *Escherichia coli* O157:H7 in coleslaw during storage. *J. Food Prot.*, **65**, 845–7.
Xu, L. (1999) Use of ozone to improve the safety of fresh fruits and vegetables. *Food Technol.* **53**, 58–62.
Zhang, S. and Farber, J.M. (1996) The effects of various disinfectants against *Listeria monocytogenes* on fresh-cut vegetables. *Food Microbiol.*, **13**, 311–21.

6 Fruits and fruit products

I Introduction

A *Definitions*

Fruits are defined in general terms as "the portions of plants which bear seeds". Such a definition includes true fruits such as citrus, false fruits such as apples and pears, and compound fruits such as berries. The definition includes tomatoes, olives, chilies, capsicum, eggplant, okra, peas, beans, squash, and cucurbits such as cucumbers and melons although, for culinary purposes, a number of these fruits are classified as vegetables. For the purpose of the current chapter, tomatoes, olives, cucumbers, and melons will be considered fruits whereas egg plant, okra, peas, beans, squash, chilies, and capsicum will be considered as either vegetables or spices.

B *Important properties*

Most fruits are high in organic acids, and hence have a low pH (Table 6.1). However, melons and some tropical fruits such as durian (*Durio* spp.) have a pH near neutrality. The principal acid in citrus fruits and berries is citric acid; in pome and stone fruits, malic acid; and in grapes and carambola, tartaric and malic acids. Care must be exercised in interpreting the pH values cited for most fruits. The pH values for fruits are typically determined by homogenizing an intact fruit and determining the pH of the expressed juice or pulp. However, this is not the microenvironment that a microorganism experiences when invading an intact fruit. For example, in an intact orange, the acidic juice is maintained within juice sacs, leaving the surrounding tissue with pH values closer to neutrality. The traditional interpretation of the acidity of many fruits, is being modified as recent research with apples, tomatoes, and oranges, has demonstrated the growth of pathogenic enteric bacteria within intact or wounded fruit (Asplund and Nurmi, 1991; Wei *et al.*, 1995; Janisiewicz *et al.*, 1999a; Dingman, 2000; Liao and Sapers, 2000).

Fruits are divided into two categories, climacteric and non-climacteric, based on their respiration pattern during ripening. Climacteric fruits are those exhibiting moderate respiration that peaks during climacteric stage. They have a short-life after ripening commences (1 week) with moderate total post-harvest life (a few days to a few weeks). They have fairly critical harvest maturity, and are sensitive to damage and microbial infection after climacteric peak in respiration. Examples are strawberries, grapes, and litchi. Non-climacteric fruits exhibit low to moderate respiration rates with a short to medium storage life. If harvested mature, but unripe, they have the potential for a long shelf-life if ripening can be delayed. Examples are bananas, papayas, and avocado (ASEAN-COFAF, 1984).

Fruits are of substantial nutritional value as a major source of vitamin C. Some provide useful levels of potassium, calcium, magnesium, and contain significant levels of vitamin A, thiamine and niacin (Holland *et al.*, 1992).

C *Methods of processing*

Due to their low pH most fruits are more susceptible to damage from fungi (and yeast) rather than from bacteria. This low pH also means that most fruit-based products require only pasteurization to be

Table 6.1 Representative pH values for fresh fruit[a]

Fruit	pH range	Fruit	pH range
Apple	2.9–3.9	Mango	3.8–4.7
Apricot	3.3–4.4	Olive	3.6–3.8
Banana	4.5–4.7	Orange	3.0–4.0
Blackberry	3.0–4.2	Passionfruit	1.9–2.2
Blueberry	3.2–3.4	Peach	3.3–4.2
Cantaloupe	6.2–6.5	Pear	3.4–4.7
Cherry	3.2–4.0	Pineapple	3.4–3.7
Cranberry	2.5–2.7	Plum	3.2–4.0
Fig	4.6–5.0	Raspberry	2.9–3.5
Grape	3.0–4.5	Squash	5.0–5.4
Grapefruit	2.9–3.4	Strawberry	3.0–3.9
Honeydew	6.3–6.7	Tomato	4.0–4.5
Lemon	2.2–2.6	Watermelon	5.2–5.6
Lime	2.3–2.4		

[a]Adapted from Beuchat (1978), Splittstoesser (1987), CRC (1990) and Brackett (1997).

microbiologically stable. Examples of exceptions include olives, cucumbers, melons, and some varieties of tomatoes. Pasteurization alone may not be sufficient for fruit based products prepared under tropical conditions.

Fruits may be processed by canning, freezing, sun-drying, or dehydration by reducing their water activity through concentration–removal of water or the addition of salt and/or sugar. The pH of tomatoes is reduced to below 4.5 by adding acids during processing, while olives, chilies, cucumbers, and durian are often pickled or fermented with lactic acid bacteria to produce microbiologically stable products that no longer need a low–acid canning process to retard spoilage.

D Types of final products

Fresh fruits are commonly sold after minimal processing and packaging treatments and may be chilled or refrigerated. Common processing steps for fresh fruits may include washing, dipping, waxing, or wrapping in paper impregnated with preservatives against mold. In some countries, certain fruits such as apples may be held for several months under refrigeration or controlled atmosphere storage, and then sold as fresh produce. Fresh fruits that have been pre-peeled, pre-sliced or otherwise prepared and packaged for convenience are increasingly being sold at retail.

Fruits are also frequently sold as canned, frozen or dried products. Moistened dried-fruit packaged with an added preservative have become increasingly popular. Dried fruits are also used in a variety of other products, e.g., confectionery bars, biscuits, chocolates, breads, mueslis, and other cereal-based products. The microbiology of these products usually differs little from that of unprocessed dried fruit and will not be considered further here.

Chopped fresh, frozen, or canned fruit may be sold in fruit salads and related products, or incorporated in dairy products such as yoghurts.

Tomatoes are canned as juice, as whole, peeled or diced fruits with or without added juice, or as concentrated purees, pastes or soups; dried as whole or halved fruit, or as a powder; or formulated into products such as tomato sauce (catsup or ketchup) preserved with vinegar, or in the form of chilli sauce. Canned Salsa products consisting of chopped tomatoes and various vegetables are also common.

Stone fruits (e.g. peaches) are sometimes infused with glucose syrups and partially dried to produce glacé confectionery.

II Initial microflora (fresh fruits)

The initial microflora of fruits comes from the field, and from harvesting and transportation equipment. Field sources include soil, insects, air, birds, animals, and fruit exudates. Soil is the primary source of heat resistant fungal ascospores, especially *Byssochlamys* species. Insects carry a variety of microbes, and species that puncture or otherwise injure a fruit are important carriers of spoilage microorganisms. For example, piercing insects are responsible for inoculation of figs with yeasts and other fungi including *Aspergillus flavus*. Insects are a potential means by which human enteric pathogens could be transmitted to fruit (Janisiewicz *et al.*, 1999a,b). Plant pathogens such as fire blight can be disseminated among pear and apple trees by rainwater runoff (van der Zwet and van Buskirk, 1984). Irrigation water can be an important source of microorganisms including enteric pathogens such as *Salmonella*. Fruit exudates provide nutrients for yeasts, especially pigmented Basidiomycetous yeasts such as *Rhodotorula* species. Some fungal pathogens of fruit, including *Lasiodiplodia theobromae*, may be present systemically in trees and invade developing fruit through the stems (Johnson *et al.*, 1991, 1992). Plant pathogens that cause crop losses or "market diseases" have developed a variety of means for invading plants. For example, *Pseudomonas syringae* and *Ps. solanacearum*, two species that infect tomato, gain entry via the stromata and roots, respectively (Getz *et al.*, 1983; Vasse *et al.*, 1995).

While microorganisms are largely restricted to the surface of intact, healthy fruit, low levels of largely Gram-negative bacteria can be routinely isolated from the interior of fruit, particularly species of *Pseudomonas*, *Xanthomonas*, *Enterobacter*, and *Corynebacterium*. The frequency and location of the bacteria varies with the fruit and the stage of maturity, with frequency of internal bacteria being high in tomatoes and cucumbers, occasional for melons and bananas, and rare for citrus fruit, grapes, peaches, and olives (Samish *et al*, 1961, 1962; Samish and Etinger-Tulezynska, 1963).

Tropical fruits may contain high numbers of microorganisms. Mangoes, collected from the field, supermarkets and wet markets were reported to contain total plate counts of 10^4–10^6 cfu/g and yeasts and mold of 10^3–10^4 cfu/g (Anonymous, 1999a).

III Primary processing

A Effects of processing on microorganisms

The processes of harvesting, cleaning, sorting, packing, and initial storage of fruit usually have little effect on the initial microflora. Examples of long used techniques designed to reduce fungal load and infection include waxing, dipping in warm (40°C–50°C) water which may contain fungicides such as benomyl, thiabendazole, or sodium *o*-phenylphenate (SOPP), and washing with 200 ppm chlorine. After dipping, fruit may be individually wrapped in paper impregnated with fungicides such as biphenyl, or packed in trays that ensure separation of individual fruit. With the increase in fresh produce microbial food safety concerns, there has been increased interest in the identification of dry or wet cleaning processes for the reduction of microbial populations. This includes evaluation of a number of different sanitizers such as trisodium phosphate, hydrogen peroxide, peroxyacetic acid, chlorine dioxide, acidified sodium chlorite (Zhuang and Beuchat, 1996; Pao and Brown, 1998; Buchanan *et al.*, 1999; Park and Beuchat, 1999; Sapers *et al.*, 1999; Liao and Sapers, 2000; Pao *et al.*, 2000; Wisniewsky *et al.*, 2000; Du *et al.*, 2002) and the evaluation of new antimicrobial coatings (Zhuang *et al.*, 1996; McGuire and Hagenmaier, 2001). Such treatments have limited effectiveness, with microbial reduction generally limited to the range of 1–3 log cycles.

Such primary processes may effectively delay spoilage, but may also cause damage to some fruit and hence hasten infection and ultimate spoilage. The impact of immersion of fruit in water, during operations such as fluming, washing, and hydrocooling, on the internalization of microorganisms has been studied extensively both in regard to spoilage and transmission of human pathogens. The immersion of warm fruit in cold water, results in a pressure differential that results in the uptake of water. A similar pressure differential can occur, when fruits are immersed in water of sufficient depth. This has been observed experimentally with a number of fruits. This inward movement of water has been experimentally associated with the uptake of spoilage bacteria and fungi in apples, pears, and tomatoes (Segall *et al.*, 1977; Bartz and Showalter, 1981; Bartz, 1982; Sugar and Spotts, 1993; Bartz, 1999), and identified as a contributing factor for core rot in apples. Infiltration of enteric pathogenic bacteria into intact fruit has also been demonstrated experimentally with apples, tomatoes, oranges, grapefruit, mangoes, and cantaloupes (Buchanan *et al.*, 1999; Burnett *et al.*, 2000; Walderhaug *et al.*, 2000).

B Spoilage

Bacteria, yeasts and molds account for up to 15% of post harvest spoilage of fresh produce. Consequently, microbial spoilage represents significant economic loss throughout the fruit distribution chain (Brackett, 1994). Spoilage of fruit is often classified according to where or when it becomes evident. Pre-harvest or field spoilage refers to spoilage occurring before the fruits are harvested, while spoilage that expressed itself after harvest, is often termed post-harvest spoilage. However, some microorganisms can cause spoilage to occur both pre- and post-harvest, and some microbiological problems that become evident after harvest often begin before harvest (Wiley, 1994). Spoilage conditions within are often referred to as market diseases and are typically named after the overt characteristics of the conditions and not the microorganism responsible. Thus, *Alternaria citri* is responsible for black rot in oranges, *Asp. niger* in figs, and *Alt. alternata* in tomatoes and melons.

Although there are several important bacterial causes of market diseases, particularly bacterial soft rots that are caused by *Erwinia carotovora*, fruits' lower pH, due to naturally present acids, often inhibits the growth of bacteria. Consequently, fungi (both yeast and molds) are the dominant microorganisms in many fruits and include both spoilage and innocuous types. Common genera include members of *Aspergillus, Penicillium, Mucor, Alternaria, Cladosporium* and *Botrytis spp.* (Brackett, 1994). Although fungi are largely responsible for fruit spoilage, not all fungi that have been insolated from fruit are spoilage fungi.

Yeasts occurring on fruits are evenly divided between ascosporogeneous and imperfect species. *Saccharomyces, Hanseniaspora, Pichia, Kloeckera, Candida*, and *Rhodotorula* are among the most common genera (Splittstoesser, 1987). Populations of yeasts on fruits, can be high, e.g., averages of 38,000–680,000 cfu/g were isolated from grapes, and damaged or defective fruits can contain as many as 10 million cfu/g of fruits. In contrast, sound apple contains only about 1000 yeast cells/g (Brackett, 1994).

Defense mechanisms in fruits appear to be highly effective against nearly all fungi. Only a relatively few genera and species are able to invade a particular fruit type and cause serious losses. Some fungi are highly specialized pathogens, attacking only one or two kinds of fruits; others have a more general ability to invade fruit tissue. Common spoilage fungi found in fresh fruits are listed in Table 6.2. The most important fungal diseases of fruits are briefly described below by fruit type.

While spoilage of fruits is distinct and separate from safety considerations, it has been noted that that there is an apparent association of *Salmonella* contamination of fresh fruits and vegetables with bacterial soft rot (Wells and Butterfield, 1997). The survival and growth of *Escherichia coli* O157:H7 in apples, appears to be enhanced by wounds from plant pathogens, physical damage, or insects (Janisiewicz *et al.*, 1999a; Dingman, 2000; Riordan *et al.*, 2000), while preventing post-harvest decay of apples helps prevent the growth of the *E. coli* 0157:H7 (Janisiewicz *et al.*, 1999a).

Citrus fruits. Rotting of citrus fruit throughout the world is commonly caused by *Penicillium italicum* and *Pen. digitatum*, termed blue rot and green rot, respectively. Infection can occur at any stage after

Table 6.2 Common fungi spoiling fresh fruits[a]

Fruit	Fungus	Spoilage
Citrus	*Penicillium digitatum*	Green rot
Oranges, lemons	*Penicillium italicum*	Blue rot
	Alternaria citri	Stem end, black rot
	Geotrichum candidum	Sour rot
	Penicillium ulaiense	Whisker mould
Pome fruits		
Apples, pears	*Penicillium expansum*	Blue rot
	Penicillium solitum	Blue rot
	Phlyctema vagabanda	Bulls eye rot
	Rhizopus stolonifer	Transit rot
Stone fruits		
Peaches, apricots	*Alternaria* sp.	Black to brown spots
Cherries, plums	*Monilinia fructicola*	Brown rot
Nectarines	*Rhizopus stolonifer*	Transit rot
	Trichothecium sp.	Pink rot
Bananas	*Lasiodiplodia theobromae*	Cushion rot
	Colletotrichum sp.	Crown rot
	Nigrospora oryzae	Squirter rot
	Fusarium semitectum	Soft rot
Figs	*Aspergillus niger*	Black rot
	Fusarium spp.	Soft rot
	Hanseniaspora uvarum	Souring
Tropical fruits	*Lasiodiplodia theobromae*	Stem end rot
Avocadoes	*Colletotrichum* spp.	Anthracnose
Mangoes	*Phomopsis* sp.	Stem end rot
Papayas	*Diplodia* sp.	Stem end rot
Soft fruits	*Botrytis cinerea*	Grey rot
Strawberries	*Rhizopus stolonifer*	Leaking rot
Raspberries	*Mucor piriformis*	Leaking rot
	Phytophthora cacotorum	Leather rot
Grapes	*Botrytis cinerea*	Grey rot
Pineapples	*Fusarium* sp.	Brown rot
	Penicillium sp.	Brown rot
Tomatoes	*Alternaria alternata*	Black rot
	Rhizopus stolonifer	Watery rot
	Geotrichum candidum	Sour rot
Melons	*Colletotrichum lagenarium*	Anthracnose
	Alternaria alternata	Black rot

[a]From Hall and Scott (1977), Beuchat, 1978, Ryall and Pentzer (1982), Splittstoesser (1987), and Snowdon (1991).

harvest. Initiation requires damage to skin tissue, which readily occurs in modern bulk handling systems. Decay spreads from fruit-to-fruit by contact (Snowdon, 1990).

In lemons and limes, *Geotrichum candidum* causes sour rot, a pale, and soft area of decay that later develops into a creamy and slimy surface growth (Butler *et al.*, 1965; Morris, 1982; Snowdon, 1990). Infection usually occurs in over-mature fruit after long and high temperature storage. Black centre rot of oranges, caused by *Alt. citri*, appears as an internal blackening of the fruit.

Penicillium ulaiense is a recently identified citrus pathogen. Closely related to *Pen. italicum*, this species has caused losses in areas such as California, where fungicidal control of *Pen. italicum* has been effective (Holmes *et al.*, 1993, 1994). Less common and usually less serious spoilage of citrus can be produced by a variety of fungi (Table 6.3).

Pome fruits. The most destructive fungal spoilage agent of apples and pears is *Pen. expansum*, which causes a blue rot. *Penicillium expansum* grows at low temperatures, so cold storage only retards, rather than prevents, spoilage (Hall and Scott, 1977). Damaged and over mature fruits are the most susceptible.

Table 6.3 Other fungi spoiling fresh fruits[a]

Fruit	Fungus	Spoilage
Citrus	*Aspergillus niger*	Black rot
Oranges, lemons	*Botrytis cinerea*	Grey mould rot
	Diaporthe sp.	Stem end rot
	Sclerotinia spp.	Cottony rot
	Septoria sp.	Septoria rot
	Trichoderma sp.	Cocoa-brown rot
	Fusarium sp.	Brown rot
	Phytophthora sp.	Brown rot
	Diplodia sp.	Stem end rot
	Phomopsis sp.	Stem end rot
Pome fruits		
Apples, pears	*Botrytis cinerea*	Grey mould rot
	Phytophthora sp.	Brown rot
	Venturia sp.	Black spot
	Physalospora obtusa	Black rot
	Alternaria sp.	Black to brown spots
Stone fruits		
Peaches, apricots	*Cladosporium herbarum*	Grey black rot
Cherries, plums	*Diplodia* sp.	Watery, tan rot
Nectarines	*Geotrichum candidum*	Sour rot
	Aspergillus niger	Black rots
	Botrytis cinerea	Grey rot
	Penicillium expansum	Blue rot
	Monilinia fructicola	Brown rot
Figs	*Alternaria* spp.	Brown to black spot
	Botrytis cinerea	Grey mould rot
	Penicillium spp.	Blue mould
	Kloeckera apiculata	Souring
Tropical fruits	*Lasiodiplodia theobromae*	Stem end rot
Avocadoes	*Colletotrichum* spp.	Anthracnose
Mangoes	*Phomopsis* sp.	Stem end rot
Papayas	*Diplodia* sp.	Stem end rot
Soft fruits	*Cladosporium* spp.	Grey black rot
Strawberries	*Sclerotinia* spp.	Watery white rot
Raspberries		
Grapes	*Cladosporium* spp.	Black rot
	Penicillium spp.	Blue mould
	Rhizopus stolonifer	Watery soft rot
Blueberries	*Alternaria* spp.	Woolly mould
	Botrytis cinerea.	Grey mould rot
	Monilinia spp.	Mummification
Tomatoes	*Cladosporium herbarum*	Grey black rot
	Botrytis cinerea	Grey mould rot
	Rhizoctonia solani	Soft rot
Melons	*Cladosporium* spp.	Black rot
	Fusarium spp.	Pink rot
	Penicillium spp.	Blue mould rot

[a]From Hall and Scott (1977), Ryall and Pentzer (1982), Splittstoesser (1987), Snowdon (1991).

With the advent of effective fungicidal control of *Pen. expansum*, *Pen. solitum* has emerged as a serious problem in apples (Pitt *et al.*, 1991).

Core rot in apples, is the growth of any variety of molds initially in the seed cavity and then spreading to the mesocarp tissue. Molds that have been associated commonly with this type of spoilage are *Alt. alternata*, *Botrytis cinerea*, *Penicillium* spp., *Coniothyrium* spp., *Pleospora herbarum*, and *Pestalotia laurocerasi* (Combrink *et al.*, 1985).

Botrytis cinerea causes grey rot in cold-stored pears and, less commonly in apples (Hall and Scott, 1977). The rot is firmer than blue mould rot, and becomes covered in ash–grey spore masses. The mould invades through wounds or abrasions and can spread rapidly in packed fruit. Other fungi that commonly cause rots of pome fruits are listed in Table 6.3.

Fire blight is a bacterial disease of pear and apple trees that is caused by *Erwinia amylovora*. It is treated by the use of antibiotics such as streptomycin, tetracycline or oxytetracycline. However, it appears that transfer of resistance genes is transposon-mediated leading to an increased incidence of antibiotic resistance among commensal bacteria associated with the fruit (Schnabel and Jones, 1999).

Stone fruits. Stone fruits (peaches, plums, apricots, nectarines, and cherries) are all susceptible to brown rot caused by *Monilia fructicola* or the closely-related species *Mon. fructigena* and *Mon. laxa*. Brown rot colonises water-soaked spots on the fruit, which within 24 h, becomes brown enlarging and deepening, rapidly, then producing a dusting of pale brown conidia. The whole fruit may rot in 3–4 days (Hall and Scott, 1977; Snowdon, 1990).

Transit rot, caused by *Rhizopus stolonifer*, is so named because it usually develops in the high humidity conditions established in boxed fruit during transport. It produces a soft rot in the fruit, which then becomes surrounded by a coarse and loose "nest" of mycelium. Growth spreads rapidly, engulfing fruit adjacent to the originally infected one, and sometimes all the fruits in a box, in only 2–3 days.

Penicillium expansum causes blue mould rot in cherries and plums, but is uncommon in other types of stone fruits (Ryall and Pentzer, 1982). Other spoilage fungi of stone fruits are listed in Table 6.3.

An important bacterial pathogen of peaches and apricots is *Xanthomonas campestris* pv. *pruni*. It causes a market disease referred to as bacterial spot. Like fire blight, it is treated with streptomycin, tetracycline, or oxytetracycline, which has lead to concerns about the emergence of antibiotic resistance among commensal microorganisms associated with these fruits (Schnabel and Jones, 1999).

Grapes. *Botrytis cinerea* is regarded as the highly desirable "noble rot" in certain wine grapes (Coley-Smith *et al.*, 1980), but it is by far the most serious cause of spoilage in table grapes (Ryall and Pentzer, 1982). In the early stages of invasion, the fungus develops on stems and inside the berry; later growth then expands rapidly through tight bunches where humidity is high and large "nests" of rot may quickly develop.

Infection by black *Aspergillus* species, *Asp. niger*, *Asp. carbonarius*, and *Asp. aculeatus*, occurs on damaged grapes in warmer climates. This damage may result from insect or mechanical penetration, splitting due to rain before harvest, or infection by pathogenic fungi such as *Botrytis* and *Rhizopus* (Snowdon, 1990; Leong *et al.*, 2004).

Penicillium species do not usually attack grapes, before harvest, but are common in stored grapes (Barkai-Golan, 1974; Hall and Scott, 1977; Ryall and Pentzer, 1982).

Berries/soft fruit. The two principal fungal rots observed with most berry and related soft fruit crops are caused by *Botr. cinerea* and *Rhiz. stolonifer* (Dennis, 1983a). *Botrytis* causes soft rots in cane berries such as raspberries and loganberries, but a firm, dry rot in strawberries. In both cases, the fruit becomes covered with a growth of grey mould. Losses in strawberries can be high as the fungus spreads by contact and forms "nests" of rotting fruit. Initial contamination generally occurs in the field and can lead to substantial crop losses if not treated with fungicides, particularly during flowering. *Botrytis cinerea* infections at the stem end of the fruit appear to be related to an initial contamination of the flower, with the mold lying dormant until maturation of the fruit. This accounts for approx., 15% of strawberry spoilage by *Bot. cinerea*. The majority of spoilage is due to the mold infecting multiple points on the surface of the fruit post-harvest. The mold is able to grow on strawberries at temperatures as low as 0°C, so refrigeration retards, but does not prevent spoilage. Shelf-life can be extended through the use of low dose ionizing radiation and/or modified atmosphere packaging. *Botrytis cinerea* is also an important cause of spoilage in raspberries, blackberries, loganberries, and gooseberries.

Rhizopus stolonifer causes a large proportion of marketing losses of all berry fruits, and in some regions, is a major cause of market losses in strawberries harvested late in the season. The mold is associated with a market disease of strawberries referred to as "leak" disease, wherein the rotting fruit has a wet appearance and ultimately collapses completely, exuding juice. This infection is favored by holding the fruit at temperatures above 20°C, allowing the fungus to spread rapidly.

Mucor piriformis causes a similar condition in strawberries. Like *Bot. cinerea*, initial infection with this mold can occur during flowering or on the surface of intact fruit. Additional mold species that are linked to spoilage of soft fruits are *Colletotrichum gloeosporioides*, *Mucor hiemalis*, *Rhizopus sexualis*, and various species of the genera *Penicillium*, *Cladosporium*, *Alternaria*, *and Stemphylium*.

Yeasts are normal colonizers of strawberries, being present at up to 10^5 CFU/g in macerates of mature berries (Buhagiar and Barnett, 1971). Despite the presence of a wide variety of yeast species on strawberries, spoilage of this soft fruit by yeasts is rare (Dennis, 1983b).

Figs. The invasion of Smyrna figs by yeasts was documented by Miller and Phaff (1962). This type of fig is pollinated by the fig wasp, which at the same time introduces yeasts (e.g., *Candida guilliermondii*) and bacteria (*Serratia* species). These microorganisms do not cause spoilage themselves, but at maturity attract *Drosophila* flies, which carry the spoilage yeasts *Hanseniaspora uvarum*, *Kloeckera apiculata* and *Torulopsis stellata*. These spoilage yeasts produce "souring" of the figs due to acid production.

Growth of *Asp. flavus* and the associated production of aflatoxins in figs, has been recognized as a serious problem (Buchanan *et al.*, 1975). Black *Aspergillus* species can also infect figs, with the possibility of ochratoxin A formation (Özay and Alperden, 1991, Doster *et al.*, 1996).

Tomatoes. With an internal pH of 4.0–4.5, tomatoes can be affected by fungal and bacterial market diseases. The primary spoilage bacterium is *Erwinia carotovora* subsp. *carotovora*, which causes bacterial soft rot. A number of market diseases are also associated with bacterial plant pathogens that lead to crop losses or defects that reduce the economic value of a crop. For example, "bacterial speck" on mature tomatoes is associated with infection of the plant with *Xan. campestris* pv. *vesicatoria*, *Xan. campestris* pv. *tomato*, and *Cornyebacterium michiganense* pv. *michiganense* (Getz *et al.*, 1983).

Several of those produced by fungi are important. *Alternaria* rots of tomatoes appear as dark brown to black, smooth, and slightly sunken lesions, which are of firm texture and can become several centimetres in diameter. The cause is *Alt. alternata*, which attacks fruit damaged by mechanical injury, cracking due to excessive moisture during growth, or chilling (Snowdon, 1991).

Chilling injury also allows the entry of other fungi also. *Cladosporium* rot caused by *Cladosporium herbarum* and grey mould rot due to *Bot. cinerea* can both be potentiated by chilling injury. *Botrytis cinerea* can also affect mechanically-damaged green fruit, on which it forms "ghost spots", small whitish rings, often with darker centers. Rot can spread rapidly at higher temperatures, during packing and transport (Ryall and Lipton, 1979; Snowdon, 1991).

Rhizopus species appear to be able to attack almost any kind of fruit or vegetable, and the tomato is no exception. "In severe cases of *Rhizopus* rot, as there seems to be no mild ones, the fruit resembles a red, water filled balloon" (Ryall and Lipton, 1979). When the fruit collapses, grey mycelium, a fermented odor and white to black spore masses become visible. The disease starts in cracked or injured fruit, but may spread by contact thereafter.

Sour rot in tomatoes is caused by *Geo. candidum*. Lesions are a light greenish grey and may extend as a sector from end to end of the fruit. Tissues remain firm at first, but later weaken and emit a sour odor. This disease invades only damaged or cracked fruit, and is disseminated by *Drosophila* flies (Ryall and Lipton, 1979).

Tomatoes grown without stakes or trellises can develop soil rot caused by *Rhizoctonia solani*. Small brown spots of this disease develop concentric rings, when they grow to 5 mm or more in diameter. Injury is not necessary for the development of this rot, but soil contact is (Snowdon, 1991).

Melons. The relatively neutral pH values of most melons make them susceptible to spoilage by both bacteria and fungi. Soft rot is the primary bacterial spoilage condition and is most often associated with *Erw. carotovora.*

Watermelons sometimes develop anthracnose from *Col. lagenarium.* This disease forms circular or elongated welts that are initially dark green and later become brown, disfiguring the melon surface. Pink *Colletotrichum* conidia may become visible, if humidity remains high (Snowdon, 1991).

Cantaloupes and rock melons may be affected by several different diseases, the most important being *Alternaria* rot due to *Alt. alternata.* Mould invasion usually takes place at the stem scar, producing dark brown to black lesions and eventually invading the flesh, forming firm, and adherent areas.

Cladosporium species can also invade melons through the stem scar, forming a rot similar to that caused by *Alternaria.* In both cases, prompt shipping and correct cool storage will limit the losses from these diseases.

Several *Fusarium* species can invade melons, especially when storage temperatures are high or storage periods become excessive. *Penicillium* species may also occasionally cause problems under these conditions (Ryall and Lipton, 1979; Snowdon, 1991).

Tropical fruit. Fruits from tropical areas are susceptible to a different array of diseases than those grown in subtropical or temperate climates. Study of such diseases is still a developing science with many pressing problems, not the least being that most tropical and subtropical fruits are injured by low temperatures and so disease control cannot be assisted by refrigeration. Tropical fruits generally do not tolerate temperatures below 8–10°C and suffer from chill injury when stored at temperatures of 1–2°C to kill off fruit flies (Chan, 1997).

Bananas are the most important tropical fruit in international trade. Most post-harvest diseases of bananas are due to fungal rots in the stalks and crowns, rather than on the sides of the fruit (Eckert *et al.*, 1975). A comprehensive study of bananas shipped from the Windward Islands to England showed that nearly 20 fungal species can cause crown rots. The most important were *Col. musae* (synonym *Gloeosporium musarum*) and *Fusarium semitectum*, with several other *Fusarium* species also significant (Wallbridge, 1981) (Table 6.3).

The major rots of other tropical fruits are usually anthracnoses, brown, or black spots on the skin that reduce crop value and may eventually destroy the fruit. Anthracnoses are usually caused by *Colletotrichum* species (often referred to as *Gloeosporium* species in the literature). Stem end rots due to *Las. theobromae* (= *Botryodiplodia theobromae*) occur in most tropical tree fruits (Snowdon, 1990).

Among the viruses, the papaya ringspot virus is the most widespread and damaging virus infecting papaya and cucurbits worldwide. This economic burden provided the impetus for the recent development of viral resistant transgenic papaya in Hawaii (Swain and Powell, 2001).

C Pathogens

Bacterial pathogens. Pathogenic bacteria are not normally associated with fruit; however, it is possible for pathogens to be present due to faecal contamination. Historically, fruits were considered as low risk food, and had been implicated with illness only on isolated occasions. This in part reflects the fact that fruits are traditionally considered acidic foods that would not support the growth of most foodborne pathogens. However, as discussed above, on fruits with lower acid contents, such as melons, apples, and tomatoes, survival of enteric pathogens may be prolonged or growth may even occur (Escartin *et al.*, 1989; Asplund and Nurmi, 1991; Madden, 1992; Golden *et al.*, 1993). Growth in particular, is likely to occur on cut surfaces of fruit. The potential for fruits to serve as a potential route of transmission was reinforced during the 1990s, when a series of foodborne illness outbreaks was attributed to the consumption of unpasteurised apple juice/cider and orange juice. These outbreaks were caused by three different pathogenic microorganisms: *E. coli* O157:H7, *Salmonella* spp., and *Cryptosporidium parvum* (see Chapter 14). These outbreaks provided the impetus for a more detailed examination of the potential for fruit to serve as a vehicle for foodborne disease (Conner and Kotrola 1995; Semanchek and Golden, 1996; NACMCF, 1999; Sewell and Farber, 2001).

A number of outbreaks of salmonellosis have been associated with fresh tomatoes (Wood *et al.*, 1991; CDC, 1993; Beuchat, 1996), watermelon (Gayler *et al.*, 1955; Lawson *et al.*, 1979; Blostein, 1993), and cantaloupes (Anonymous, 1993, Del Rosario and Beuchat, 1995; Beuchat, 1996; Sewell and Farber, 2001; Anderson *et al.*, 2002). In the latter case, there have been a series of outbreaks associated with *S*. Poona (Sewell and Farber, 2001; Anderson *et al.*, 2002) and to a lesser extent S. Chester (Ries *et al.*, 1990) and *S*. Oranienburg (Sewell and Farber, 2001). Investigations of these outbreaks largely traced the cantaloupes to a farm in Mexico and identified contaminated water used for irrigation and post-harvest washing and cooling as a source of the *Salmonella*. Recently, an outbreak of *S*. Newport infections in the United States was traced to mangoes (Sivapalasingam *et al.*, 2003). In this instance, the source of the microorganisms appeared to be cooling water used after the fruit had been heated to eliminate larva of the Medfly before entry into the country.

Growth of *Clostridium botulinum* and toxin production was reported to be possible in stored tomatoes, if growth of mold (*Alternaria* or *Fusarium*, but not *Rhizoctonia*) also occurs (Draughon *et al.*, 1988). However, in a later experiment with tomatoes inoculated with spores of *C. botulinum* (type A and type B) and *Alternaria*, and stored under passively modified (sealed, 1.0–2.9% O_2) or controlled atmosphere (1% O_2, 20% CO_2, and balance N_2) storage, botulinum toxin was not detected until after tomatoes became inedible due to mould growth (Hotchkiss *et al.*, 1992). It was concluded that the risk of botulism from consumption of stored tomatoes was insignificant. The growth of *C. botulinum* on pre-cut pieces of cantaloupe and honeydew melons under different atmospheres, indicated that properly refrigerated (<7°C) melon pieces did not support neurotoxin formation (Larson and Johnson, 1999). Toxin production did not occur during abuse temperature (15°C), before gross spoilage. However, if the competing microflora was reduced by UV treatments, toxin production at 27°C did occur with only marginal spoilage.

Viruses, protozoan and parasites. Enteric viruses may be present on fruits, as a result of human faecal contamination either before or after harvest. Noroviruses (formerly Norwalk and Norwalk-like viruses) and Small Round Structured Viruses (SRSVs) and hepatitis A virus are the major concerns. Workers preparing fruits for consumption are an important source of contamination, but workers at other points in production and processing or polluted water may also be sources. There have been few studies of the survival of hepatitis A or Noroviruses, however, the major influences on survival of pathogenic bacteria mentioned above also determine the survival time of enteric viruses (Cliver, 1983). Some substances present in fruits cause reversible inactivation of viruses (Cliver and Kostenbader, 1979). However, a study of survival of poliovirus in fresh raspberries indicated no loss of viability over a 2-week refrigerated storage period (Kurdziel *et al.*, 2001). A variety of fruits and fruit juices have transmitted hepatitis A (Cliver, 1983) or viral gastroenteritis (Caul, 1993).

Potentially, fruit could serve as the vehicle for a variety of protozoan diseases transmitted by an oral–faecal route, such as apple cider, which has been associated with *Cryp. parvum* outbreaks. However, the most prominent outbreaks during the 1990s, involved the yearly association of *Cyclospora cayetanensis* infections with raspberries imported from Guatemalan to Canada and the United States (CDC, 1997b; Herwaldt and Ackers, 1997; Shellabear and Shah, 1997; Soave *et al.*, 1998; Herwaldt *et al.*, 1999; Sterling and Ortega, 1999; Sewell and Farber, 2001). The protozoan is most closely related to the genus *Eimeria*, a common cause of diarrheal disease in a wide variety of birds and animals (Soave *et al.*, 1998). However, the only known reservoir for *Cyc. cayetanensis* to date is humans. Thus, the most likely source of contamination of the raspberries associated with the outbreaks is either the farm or packinghouse workers or the water used for irrigation or treatment of the plants. However, to date, the source of the protozoa in the production and packing house environment, has not been identified.

Mycotoxins. The mycotoxin problems likely to occur in fresh fruits, are the formation of patulin in apples by *Pen. expansum*, aflatoxins in figs by *Asp. flavus*, and ochratoxin A in grapes from *Asp. carbonarius*. Apple rots and mouldy grapes are conspicuous, and consumers can be expected to remove them before fruit is eaten fresh. Hence, patulin and ochratoxin A are more of a problem in juice manufacture, and are discussed in Chapter 22.

Aflatoxins carry over from contaminated figs in the production of fig wine (Möller and Nilsson, 1991). Degradation of aflatoxins occur at a constant rate, with a half-life calculated to be 115 days. However, most figs are consumed after drying, so aflatoxins in figs are discussed below.

D CONTROL (fresh fruits)

Summary

Significant hazards[a]	• Bacteria: *E. coli* O157:H7, salmonellae • Protozoa, parasites, viruses: *Cryptosporidium parvum, Cyclospora cayetanensis*, hepatitis A virus, Noroviruses. • Mycotoxins: Patulin in apples (*Pen. expansum*), aflatoxin in figs (*Asp. flavus*), ochratoxin A in grapes (*Asp. carbonarius*).
Control measures	
Initial level (H_0)	• Avoid hazardous agricultural practices. • Avoid applying untreated or improperly treated manure in fruit orchards: treat manure to kill bacteria before it is used as fertilizer. • Keep contaminated irrigation water away from fruits. • Use potable or treated water for washing and processing fruits.; • Avoid using dropped or windfall fruits.; • Discard badly bruised or rotten fruits immediately.; • Assure microbiological safety and quality of fresh fruit by good agricultural practices and proper handling on the farm, and effective processing and handling at packinghouses and during shipment and subsequent marketing.
Reduction (ΣR)	• Few measures are available other than sorting and washing to reduce hazards.
Increase (ΣI)	• Minimise increase by appropriate storage. • Train workers to wash their hands properly and make sure they have access to toilets. • Keep packing facilities clean and free of pests.
Testing	• Routine microbiological testing of fruit is not recommended. • Aerobic colony counts may be useful to monitor effectiveness of processing on the microbial populations. • *E. coli* may be used as an indicator of faecal contamination. • Equipment hygiene may be monitored using rapid tests such as ATP.
Spoilage	• Spoilage of most fruits is caused by fungal growth. Most of the practices preventing food safety hazards will also prevent spoilage.

[a]In particular circumstances, other hazards may need to be considered.

Control measures. A correctly constructed HACCP plan for production of fruit can significantly reduce the risk that fruit would be a source of foodborne disease (Rushing *et al.*, 1996). The HACCP should take account of the possibility that fruit may carry pathogenic microorganisms. Fruits and fruit products are frequently consumed raw without having been exposed to a process that can reliably eliminate pathogens. Therefore, it is essential that measures to prevent contamination of fruit with pathogens of faecal origin, are in place at all points in the growing, processing, storage, distribution, and preparation chain (Brackett, 1992). These measures include avoiding hazardous agricultural practices, for example inappropriate use of organic fertilisers or the harvesting of fallen apples that may have been exposed to animal faeces. Good sanitation must be practised during picking, packing, and grading. Washing of fruit in water, brings about a small reduction in the microbial load if done correctly. The addition of chlorine to the wash water or dipping in chlorine solutions, can significantly reduce the concentration of pathogens on the surface of fruit, and therefore reduce hazards to the consumer, but cannot be relied upon to eliminate pathogens (Beuchat, 1996). Inadequate sanitation of processing environments and poor hygiene among workers are particularly hazardous when fruit is being peeled and sliced or otherwise prepared for consumption.

Many of the same practices that improve the microbiological safety of fresh fruits, are the ones that help prevent spoilage. Care must be exercised to minimize damage to the fruit, maintain appropriate storage temperatures and conditions, wash and cool fruit in a manner that minimizes infiltration of microorganisms, and maintain adequate sanitation. Segregation of contaminated fruit at the earliest, will help prevent the spread of contamination to adjacent pieces.

Fungi. Before harvest, fungicides and insecticides are effective in reducing the incidence of fungal invasion of fruit. Control depends on adequate conditions for handling, transport, and storage. However, implementing control measures is made difficult by (i) the variety of spoilage fungi which may occur; (ii) limitations due to the physiology of the fruit; and (iii) conditions favourable to senescence and ripening of fruit are also favourable for fungal growth (Smith, 1962).

Control of temperature, relative humidity, atmospheric composition, and the use of chemical inhibitors are all important in preservation of fresh fruit. Again, avoidance of physical damage to the fruit and the prevention of internalization of mold spores are important. Some details are given below.

Citrus fruit. Control of fungal spoilage in citrus fruit relies primarily on careful handling. Post-harvest treatments are based on washes heated to 40–50°C and containing detergents, weak alkali, and/or fungicides such as thiabendazole or sodium o-phenylphenate (SOPP). After dipping, fruits may be individually wrapped in waxed paper containing biphenyl or packed in trays that ensure separation of individual fruits (Ryall and Pentzer, 1982). Control of *Geo. candidum* relies on storage at temperatures below 5°C.

Pome fruits. Control measures include careful handling and the use of fungicides such as benomyl, dichloran or SOPP. Because rots spread by contact from fruit to fruit, it is also common practice to individually wrap fruits in waxed papers containing a fungicide such as biphenyl. Properly controlled atmospheric storage can effectively maintain unspoiled fruit, particularly apples, for extended periods of time.

Stone fruits. Infection by *Monilia fructicola* commences in the orchard. Rigorous pre-harvest spray programs with benomyl or similar benzimidazole fungicides are necessary. Storage temperatures below 5°C assist in control (Snowdon, 1990).

Dichloran is an effective fungicide against *Rhizopus*. A combined benomyl and dichloran pre-harvest spray program for the control of both *Monilia* and *Rhizopus*, has been used extensively for peaches, apricots, and nectarines (Hall and Scott, 1977; Ryall and Pentzer, 1982).

Grapes. Fungal control involves the use of pre-harvest sprays with benomyl and rapid transfer of fruits to cold storage, after picking. Post-harvest treatments with sulphur dioxide or benomyl, are also effective (Hall and Scott, 1977).

Berry fruits. Pre-harvest spray programs with benomyl, are important for control, as is refrigerated storage. For most soft fruits, avoidance of wetting is critical for extending shelf-life. Post- harvest antifungal treatments are of little benefit (Ryall and Pentzer, 1982), however, low dose irradiation has proven highly effective for delaying mold spoilage (Thayer and Rajkowski, 1999).

Tomatoes and melons. *Alternaria alternata* can be transmitted through seeds, so fungicide treatment before planting can be beneficial for both tomatoes and melons. *Alternaria alternata* grows at all acceptable handling temperatures, so post-harvest invasion can be avoided only by rapid marketing. Careful irrigation and handling to avoid cracking and post-harvest fungicidal dips will reduce losses (Snowdon, 1991).

Rhizopus rot in tomatoes can be effectively reduced by improved hygienic practice in the field: removal of fallen or culled fruit to reduce spore dispersal, disinfection of field boxes by heating, and careful harvesting (Snowdon, 1991). Post-harvest hot water or fungicide dips, irradiation, waxing, and wrapping are all of value.

Sour (*Geotrichum*) rot control requires control of insects, that puncture fruit, reduction in mechanical damage, and culling of cracked fruit (Snowdon, 1991).

Anthracnose of melons must be controlled by strict hygienic conditions, by the removal of all crop debris and cleaning of containers. Planting on plastic sheets, also decreases contact with soil. Inoculating young plants with the fungus will reduce disease severity and fungicide sprays can be of value (Snowdon, 1991). The use of resistant cultivars can reduce or overcome *Cladosporium* rot.

Fusarium rots are difficult to control, sometimes requiring soil disinfection, seed dressing, systemic fungicides, and careful inspection in the packing house (Snowdon, 1991).

Tropical fruits. Benomyl and thiabendazole, chlorine, and hot water have all been quite successfully used for the control of banana rot (Eckert *et al.*, 1975). Control of anthracnoses relies on benomyl or a variety of other fungicides. *Colletotrichum* conidia appears to be especially heat sensitive, and hot water dips for 5 min at about 55°C have been beneficial for preservation of mangoes (Smoot and Segall, 1963) and other fruits. However, after such heat treatments, care must be taken that only appropriately treated water is subsequently used to cool the fruit. Contaminated cooling water has been linked to mango-associated cases of salmonellosis. For further information on the control of diseases of tropical fruits, see Eckert *et al.* (1975) and Champ *et al.* (1994).

IV Pre-cut (minimally processed) fruit

A *Effects of processing on microorganisms*

Pre-cut (P-C) fruits include ready-to-use/eat, pre-cut, lightly processed, and fresh-cut fresh fruits. These are sold under refrigerated storage in supermarkets, retail food stores and restaurants or chilled on ice by roadside fruit stalls in many developing countries. P-C fruits are products that contain live tissues or those that have been only slightly modified from the fresh condition and are like fresh in character and quality (Wiley, 1994). Such minimally processed refrigerated fruits meet consumer demands, for e.g., "like-fresh" fruit products with extended shelf-life, that ensure food safety and maintain nutritional and sensory quality. Typical production of P-C fruits often includes preservation methods in which some, but not all species of microorganisms are reduced in count and specific enzyme systems may be partially or fully inactivated in the package or prior to packaging (Wiley, 1994).

In processing P-C fruits, operations such as peeling remove important natural barriers to contamination. Likewise, the cutting of the fruit exposes new surface to microbial contamination and generally releases nutrients that enhance the growth of microorganisms. The process of cutting/slicing can transfer microorganisms from the exterior surfaces onto the newly exposed cut surfaces (Lin and Wei, 1997). For a number of fruits, the newly exposed interior surfaces can support the growth of a variety of microorganisms including enteric bacteria that are pathogenic for humans (Escartin *et al.*, 1989; Liao and Sapers, 2000; Larson and Johnson, 1999; Riordan *et al.*, 2000; Weissinger *et al.*, 2000). Thus, an important step in the successful preparation of P-C fruits are effective cleaning and sanitization of the fruit surfaces, prior to cutting and the maintenance of a high level of sanitation throughout processing and packaging. Typically, the fruits undergo extensive washing and disinfection with 200 ppm chlorine or other sanitizing agents (see below).

B Spoilage

The type and importance of P-C fruit spoilage reflects the intended use of the product and the adequacy of the cold chain. For street vendors, where the shelf-life of the product is a few hours and the product is generally not refrigerated or packaged for extended storage, spoilage is generally not an issue. As the shelf-life of the product becomes increasingly extended, it becomes more important to adequately refrigerate. With a product that has a 7–14 day shelf-life, the microorganisms of concern are psychrotrophs that are capable of growing at 2–4°C, and typically have an optimal growth at temperatures between 20–30°C (Brackett, 1994).

Properly processed P-C fruits contain low microbial loads as shown for durian and dragon fruits (Anonymous, 1999a,b). For P-C jack fruit, the microbial counts (TPC, yeast, mould, and coliforms) (see a) need to be reduced by 1–3 log cycles (Faridah *et al.*, 1999). Microbial contamination of P-C durian sampled from wet markets and supermarkets was reported to be low and no pathogens were detected. Samples from the wet market were reported to have TPC and yeasts ranging from 10^2 to 10^4 cfu/g, mold too few to count, coliforms from <3 to 23 MPN/g and *E. coli* of <3 MPN/g. Those obtained from supermarkets yielded the following results: TPC from not detected to 10^3 cfu/g, yeasts from not detected to 10^2 cfu/g, mold too few to count, coliforms from <3 to 9 MPN/g and *E. coli* at <3 MPN/g (Chudhangkura *et al.*, 1999). An evaluation of 30 samples of street-vended sliced papaya observed TPC between 10^3–10^7 cfu/g, coliforms were detectable in 70% of the samples at a range of <3–160 cfu/g, and *E. coli* was verified in approximately half of the samples containing coliforms (Mukhopadhyay *et al.*, 2002).

C Pathogens

Pathogens can be introduced into fruits from soil, water, animal waste, insects, and fruit-handlers. Fruits can serve as vehicles for almost any foodborne pathogenic microorganisms that can result in disease under favorable conditions. However, only a relatively few pathogenic microorganisms would normally be considered a serious threat to refrigerated fruits (Brackett, 1994). A number of outbreaks have occured with precut fruits, such as salmonellosis associated with pre-cut watermelons (Gayler *et al.*, 1955; Lawson *et al.*, 1979; CDC, 1979) and pre-cut cantaloupes (CDC, 1991). There have been several outbreaks of salmonellosis associated with pre-cut tomato products (O'Mahoney *et al.*, 1990; Cummings *et al.*, 2001), and at least one report of *Campylobacter* infections associated with pre-sliced cucumbers and tomatoes (Kirk *et al.*, 1997).

More recently, concerns have been raised about growth of *Listeria monocytogenes* in fruits such as melons where pH is between 5.5 and 6.5 (Ukuku and Fett 2002; Leverentz *et al.* 2003), and some studies

(Leverentz *et al.*, 2003) have demonstrated that rapid growth of the organism can occur during storage at 10°C.

An outbreak caused by *E. coli* O157 was recently traced in cucumbers in a cucumber salad (Duffell *et al.* 2003) and fresh fruits may need to be considered as a vehicle of this organism.

D CONTROL (pre-cut [minimally processed] fruit)

Summary

Significant hazards[a]	• Pathogenic enteric bacteria, protozoa, *Listeria monocytogenes* (melons)
Control measures	
Initial level (H_0)	• Start with high quality fruit.
Reduction (ΣR)	• Fresh fruits are typically treated with disinfectants to reduce external contamination prior to be processed to P-C fruits. • Techniques used are similar to those for P-C vegetable products (see Chapter 5) and for sanitizing fruit before juice production (see Chapter 13). • Acidification of the finished product is common both as a means of controlling bacterial growth and preventing browning of the fruit. • A number of the organic acids may have fungicidal or fungistatic activity (Wiley, 1994).
Increase (ΣI)	• Maintenance of the cold chain throughout the distribution and marketing system is critical to achieve desired shelf lives and microbial safety. • Control of sanitation in processing facilities.
Testing	• Primary process control assures that key steps to minimize microbial contamination are applied. • Verify that numbers of defective fruit are as low as possible. • Verify that sorting and washing activities are effective via periodic review of employee training and performance. • Verify that concentrations of disinfectant solutions are correct and active, and that plant sanitation programs are operational. • Supervision of worker hygiene.
Spoilage	• Spoilage is mostly caused by fungal growth, therefore control of time and temperature can prevent spoilage.

[a]Under particular circumstances, other hazards may need to be considered.

Control measures. A number of different disinfectants have been evaluated for their effectiveness against various pathogenic enteric bacteria, including chlorine, acidified sodium chlorite, Tsunami™, hydrogen peroxide, chlorine dioxide, and hot water (Pao and Brown, 1998; Park and Beuchat, 1999; Sapers *et al.*, 1999; Liao and Sapers, 2000; Pao *et al.*, 2000; Wisniewsky *et al.*, 2000; Fleischman *et al.*, 2001; Du *et al.*, 2002; Ukudu and Fett, 2002).

Starting with high quality fruit is critical to the successful production of P-C fruit. An approved supplier program should be developed for the fresh fruit suppliers to ensure good agricultural practices and proper handling are being followed to meet food safety requirements. Upon receipt, the fruit should be thoroughly washed and then inspected to ensure that the level of defective fruit is acceptable for the production of P-C products. Windfalls or dropped fruit should not be used to produce P-C products.

Packaging and modified atmosphere storage and refrigeration. Packaging serves to protect and provide a suitable atmosphere for P-C fruits, but does not constitute the actual act of preserving. Packaging maintains the desired relative humidity to prevent dehydration. Refrigeration, on the other hand, is the active act of preservation to reduce adverse quality and nutritional changes and extend shelf-life of fruits. The actual preservation method may precede packaging or may occur by treating the fruit and packaging simultaneously (Wiley, 1994). Other options include MAP treatment immediately after packaging or continuously during the shelf-life period. In the latter case, it is the modified atmosphere developed during packaging that result in preservation, together with reduced temperature provided by refrigeration.

Modified atmosphere packaging (MAP) combines modified atmospheres and chilling temperatures to retard microbial spoilage and delay fruit senescence. Microbial growth is affected by the amounts of oxygen and carbon dioxide present in the package (Day *et al.*, 1990). Care must be taken in selecting the MAP to be employed, since P-C fruits is an actively respiring system and certain gas combinations will adversely affect the fruit metabolism and thus its shelf-life.

Facility hygiene and equipment maintenance

Monitoring. Tests for aerobic plate count can be useful to determine the general impact of processing and handling. Rapid methods, such as ATP measurement, are a useful tool to measure equipment hygiene.

The presence of enteric pathogens is the major food safety concern, but testing for all the possible pathogens mentioned above is not recommended. It may be appropriate, however, to use *E. coli* as an indicator of the hygienic conditions of growing, harvesting, transporting, and processing. Enterobacteriaceae, coliforms, or 'fecal coliforms' are not effective indicators, because they occur naturally in the field and plant environment and may not be directly linked to the attributes being controlled to assure microbial safety and quality (See Chapter on Vegetables).

For minimally processed fruits that support the growth of *L. monocytogenes*, microbiological testing of the processing environment may be appropriate.

V Frozen fruits

A Effects of processing on microorganisms

Fruits to be preserved by freezing are sometimes pre-treated by blanching to inactivate enzymes. This effectively destroys the surface vegetative microflora. Sometimes chemicals are added, such as ascorbic acid to control oxidation or citric acid to control browning. It is unlikely that such treatments have any substantial effect on the microflora of the product. Some fruits, such as peaches, which brown rapidly are thinly sliced and dipped in sulphite for several hours. Such treatments greatly reduce the microbial load.

Fungi, especially yeasts, proliferate on equipment used to prepare product for freezing. Some are killed or injured by the freezing process, and the numbers slowly decline further in storage. Provided product is handled correctly after thawing, such contamination is of no consequence.

B Spoilage

The normal microflora of frozen fruits consists mainly of fungi, especially yeasts. Growth and spoilage are influenced by storage temperature: partial or complete thawing will frequently lead to yeast spoilage, from gas production. However, if adequately maintained at frozen temperatures, spoilage is generally due to non-microbial causes.

C Pathogens

Pathogenic bacteria can survive for extended periods in frozen fruit, particularly if the fruit is non-acidic or if the microorganism is acid resistant or pre-adapted to acid tolerance. While generally not considered a major source of foodborne disease, there are reported cases associated with frozen fruit. For example, contamination of frozen mamey with *S.* Typhi, led to an outbreak of typhoid fever in the United States (Katz *et al.*, 2002). Enteric viruses may survive sufficiently well in frozen fruit to cause outbreaks. For example, frozen strawberries have been linked to outbreaks of hepatitis A in the United States (Ramsay and Upton, 1989; CDC, 1997a), and frozen raspberries have been linked to a calicivirus outbreak in Finland (Pönkä *et al.*, 1999).

D CONTROL (frozen fruits)

Summary

Significant hazards[a]	• No major hazards.
Control measures	
Initial level (H_0)	• Microbial populations on fruits to be frozen are best controlled by: adequate washing, removal of obviously diseased fruit, careful handling to prevent bruises, frequent cleaning and sanitation of handling and conveying equipment and prompt freezing of the prepared fruit.
Reduction (ΣR)	• Sulphuring is of limited effectiveness.
Increase (ΣI)	• Frozen storage below −10°C will prevent all microbial growth, but does not necessarily lead to inactivation of microorganisms. • Time/temperature control needed before, during and after preparation, and during transportation, storage and sale.
Testing	• Routine microbiological testing of fruit is not recommended. • Aerobic colony count and/or coliforms may be used to monitor process control or to monitor potential temperature abuse. • Environmental testing for *L. monocytogenes* or indicators monitors the potential for contamination. • Monitor equipment hygiene using tests such as ATP.
Spoilage	• No problems with microbial spoilage.

[a]Under particular circumstances, some hazards need to be considered.

VI Canned fruits

A Effects of processing on microorganisms

Pasteurising heat processes are used for nearly all fruit products. Care is needed with low acid fruits such as tomatoes, which in some countries are acidified to ensure a pH below 4.5 before processing (Lopez, 1971; Schoenemann and Lopez, 1975). Fruit products with pH values above 4.6 or a_w above 0.85 are subjected to the sterilization process for low acid canned food, to kill spores of *Cl. botulinum*.

B Spoilage

Bacteria. The saprophytic bacterial flora of canned fruits is made up of mesophilic and thermoduric spores, as vegetative cells have been destroyed by pasteurisation. Bacterial spoilage of canned fruits is rare, and due to butyric or thermophilic anaerobes. Butyric anaerobes such as *Cl. pasteurianum* can grow at pH 3.8 in syrups, and cause spoilage of pears by production of butyric acid, hydrogen, and carbon dioxide (Jakobsen and Jensen, 1975).

Underprocessing of tomato products can result in growth of thermoduric facultative anaerobes such as *Bacillus coagulans*, leading to flat sour spoilage. Spoilage of tomato juice by *B. coagulans* is accompanied by pH reduction, but not gas production and the taste of such spoiled products has been described as "medicinal", "phenolic", and "fruity" (Pederson and Becker, 1949). *B. coagulans* is common in soil and readily contaminates tomato processing lines. There is a direct relationship between the concentration of soil particles and number of *B. coagulans* spores in tank water used to wash tomatoes. This species can multiply in washing equipment, if introduction of cold water is insufficient causing water temperatures to rise to 27°C–32°C (Fields, 1970; Segmiller and Evancho, 1992). The vegetative cells of some strains of *B. coagulans* can grow in tomato juices of pH 4.2, although heat-treated spores can only germinate and grow if juice pH is above 4.3 (Pederson and Becker, 1949).

Occasionally, heat tolerant lactobacilli have caused spoilage of canned tomato products (Gould, 1974).

Heat resistant fungi. Traditionally, mild heat processes have been used for acid foods such as fruits and fruit products. Pasteurisation at temperatures of 70°C–75°C is effective, as it inactivates most enzymes, yeasts and the spores of common contaminant fungi. However, fungi that produce ascospores are capable of surviving such processes and causing spoilage.

In practice, only a few species of heat resistant fungi have been isolated from fruit products after a heat process, and still fewer have been recorded as causing spoilage. *Byssochlamys fulva* and *Bys. nivea* head the list of species that spoil strawberries in cans or bottles (Hull, 1939; Put and Kruiswijk, 1964; Richardson, 1965), blended juices that contain passionfruit, and fruit gel baby foods (Hocking and Pitt, 1984). *Neosartorya fischeri* has also been repeatedly isolated from strawberries (Kavanagh *et al.*, 1963; McEvoy and Stuart, 1970) and other products, but has rarely been reported to cause spoilage. *Talaromyces flavus*, *Tal. Bacillisporus*, and *Eupenicillium* species are other potential causes of spoilage in heat processed products (Hocking and Pitt, 1984).

The source of heat resistant fungal ascospores is soil. Juices particularly at risk from heat resistant fungi, are those made from pineapples, passionfruit and berries, where fruit frequently come in contact with soil before or during harvest (Cartwright and Hocking, 1984).

C Pathogens

Canned fruits are among the safest processed foods. However, outbreaks of botulism have been recorded from home canned fruits including pears, apricots, peaches, and tomatoes (Odlaug and Pflug, 1978). The usual underlying cause is an initial fruit pH above 4.5 coupled with under-processing. However, some instances of botulism from home canned fruits have resulted from growth of other microorganisms causing an increase in pH permitting *Cl. botulinum* to grow (Odlaug and Pflug, 1978). In commercial operations, pH is controlled to ensure a safe heat process and prevent such problems.

Although tomatoes are not a good substrate for growth of *L. monocytogenes*, cells inoculated into commercially processed tomato juice remained viable for periods exceeding the normal shelf-life (Beuchat and Brackett, 1991).

The presence of mycotoxins in canned fruit products is largely confined to the formation of patulin in apple juice by *Bys. nivea* (Roland and Beuchat, 1984). *Byssochlamys* species have been reported to produce patulin in spoiled canned fruit, but only at very low levels. Tomato products can potentially become toxic due to tenuazonic acid from growth of *Alternaria* species, but such occurrences have not been well documented.

D CONTROL (canned fruits)

Summary

Significant hazards[a]	• If product has a pH > 4.5, then *Cl. botulinum* must be considered. • Patulin in apple juice (*Byss. nivea*)
Control measures	
Initial level (H_0)	• Control of raw materials through approved supplier programme. • Apply GMP throughout the handling and processing chain. • Bacterial spoilage of correctly processed products only occurs when initial spore numbers are exceptionally high. • Adequate cleaning of process lines and equipment should prevent such occurrences.
Reduction (ΣR)	• Application of HACCP principles for canning of low acid and acidified fruits.
Increase (ΣI)	• Holding canned products at elevated temperatures and observing for swells before release.
Testing	• Testing suspected cans to identify cause of spoilage. • Monitoring levels of thermopiles in raw material and ingredients is useful if products are likely to be held at elevated temperatures.
Spoilage	• *Cl. pasteurianum*, *B. coagulans*, heat resistant fungi • Control of spoilage of canned pears by *Cl. pasteurianum* relies on a combination of adjusting the pH to 3.8–4.0, reducing a_w to 0.97–0.98 by use of sugar syrup, and an appropriate heat treatment (Jakobsen and Jensen, 1975).

[a]Under particular circumstances, other hazards may need to be considered.

Control measures

Fungi. Some types of fruit, e.g. berries, pineapple, mango, and passionfruit, may become contaminated with soil and hence with heat resistant fungal ascospores. It is usually not practical to increase pasteurisation processes to a level where heat resistant fungal spores can be destroyed, because the organoleptic properties of the fruit will be impaired. For the fruits mentioned, control requires the monitoring of juices used as raw materials for the presence of heat resistant ascospores. Specifications for raw materials are often set as low as one heat resistant ascospore per 100 ml. Raw materials not meeting such specifications, are either rejected or usedjin alternative ways such as in frozen products (Cartwright and Hocking, 1984; Hocking and Pitt, 1984).

Monitoring raw materials for heat resistant fungi involves heating samples at 75°C for 30 min, then incubating the samples, with or without the addition of an agar medium, for up to 4 weeks. Heat resistant spoilage fungi, such as *Byssochlamys, Talaromyces, Neosartorya*, and *Eupenicillium* species can be selectively detected in this manner. Culture methods were outlined by Beuchat and Rice (1979). Two methods are recommended: a plating method, first described by Murdock and Hatcher (1978) adapted for larger samples (Pitt and Hocking, 1997) and the second, a direct incubation method (Hocking and Pitt, 1984; Pitt and Hocking, 1997).

VII Dried fruits

A Effects of processing on microorganisms

Sulphured tree and vine fruits. Some types of fruits, e.g. peaches, apricots, pears, and bananas, are treated with high levels of sulphur dioxide before drying, which is essential to preserve fruit appearance by preventing browning from the Maillard reaction The SO_2 also completely eliminates the microflora, even during prolonged storage. Rehydration to produce moist packs usually does not change this situation. Such products have no microbiology and will not be considered further. However, some "natural" fruits of these types are dried without sulphur dioxide and the remarks in the following sections apply. Such fruits are identifiable by a general brown colouration due to products of the Maillard reaction.

Unsulphured fruits. Fruits that are not treated with SO_2 include prunes, dates, figs, and vine fruits. The microbiology of processing these fruits is described below.

Fruits are often washed before processing and some types are treated with alkaline detergent solutions to strip surface wax and speed water removal. The subsequent dehydration process used influences the microflora of the dried product. Sun drying, used extensively for certain fruits throughout the world, is subjected to the vagaries of the weather. Strong sunlight will greatly reduce initial microflora. Only black spored *Aspergillus* species are capable of survival, with consequent potential for ochratoxin A production. However poor drying conditions can cause proliferation of yeasts and filamentous fungi, especially *Penicillium* species.

Mechanical dehydration reduces the total microbial load, but the extent of the reduction depends both on the type of fruit and the severity of the process. For example, low temperature drying of figs, at 54°C–60°C, reduced but did not eliminate yeasts (Natarajan *et al.*, 1948). In contrast, prunes are dried at 70°C–80°C, a process that results in commercial sterility (Miller *et al.*, 1963). However prunes are readily recontaminated during subsequent handling (Pitt and Christian, 1968; Pitt and Hocking, 1997).

Figs are usually sun dried, often on the ground where they fall, allowing ample time for growth of fungi. *Aspergillus flavus* is important, as aflatoxins are readily produced in figs. Inoculation of firm, ripe figs with spores of *Asp. flavus* produced fungal growth and aflatoxin formation within two days (Boudra *et al.*, 1994).

Moist packs. To satisfy consumer demands for more palatable ready-to-eat foods, dried fruits are often rehydrated to 0.85–0.90 a_w, before packing. The dried fruits are reprocessed in hot or boiling water baths, which effectively destroys the microflora. Sulphured fruits will usually retain enough SO_2 to remain stable, if well packed. However, dates, figs, and prunes processed without SO_2 are prone to recontamination after cooking. Many countries now permit the addition of weak acid preservatives such as sorbic or benzoic acid to such packs to ensure microbial stability.

Glacé fruits. Glacé fruits are produced by infusing blanched fruit with increasingly concentrated glucose syrups that contain SO_2 as a preservative. After this treatment, fruits are usually given a low temperature dehydration process to reduce a_w to about 0.85. The initial microflora is largely destroyed by this process.

B Spoilage

Unsulphured fruits. Fruits that are not treated with SO_2, are susceptible to spoilage by xerophilic fungi. However, if fruits are properly dried and stored, the extent of damage should be slight.

Australian vine fruits that are sun dried and not preserved with SO_2 are usually contaminated with *Asp. niger* and closely related species, which may grow to some extent during drying (King *et al.*, 1981, Leong *et al.*, 2004). Other fungi are much less common. Californian vine fruits are also dried in the sun, but without the elevated drying racks used in Australia as a protection against rain and surface water. Losses in the occasional rainy drying season in inland California can be catastrophic.

Poor factory hygienic conditions may result in contamination of dried fruits, during packaging. In particular, the extreme xerophile *Xeromyces bisporus*, which is able to grow quite rapidly at 0.70–0.75 a_w, may build up on conveyers and other equipment, be transferred to the fruits and then cause spoilage of products, which is safe from all other fungi (Pitt and Hocking, 1982, 1997).

Mature figs are always contaminated in the seed cavity by yeasts (Miller and Phaff, 1962). Spoilage of dried figs sometimes occurs, if these contaminant yeasts include xerophilic species.

Ready to eat packs. Nearly every known xerophilic fungus was isolated from Australian dried and high moisture prunes by Pitt and Christian (1968). At that time, microbial stability of this product relied on hot filling into packages. The most common fungi isolated were *Eurotium* species, especially *Eur. herbariorum, Xer. bisporus*, and xerophilic *Chrysosporium* species. Moistened ready-to-eat prunes and dates are now commonly preserved with benzoic or sorbic acids, which prevent fungal growth.

Glacé fruits. Sulphur dioxide used as a preservative in the glacé fruit process, is only partially effective, as free SO_2 is rapidly bound by glucose. Partially prepared glacé pineapple may spoil due to the growth of the yeast *Schizosaccharomyces pombe*. This species apparently possesses a unique combination of resistance to SO_2 and ability to grow at reduced a_w, enabling it to grow at a particular point in the infusion process (Pitt and Hocking, 1997).

C Pathogens

Bacterial pathogens. Survival of pathogenic bacteria on dried fruits is usually poor, and limited to a few weeks. Relatively long storage periods before sale, normal for such products, further minimises risks. However, *E. coli* O157:non-H7 has been isolated from one sample of conventionally grown imported raisin and one sample of organically grown imported apricot (Johannessen *et al.* 1999).

Mycotoxins. The possibility of mycotoxin production in high moisture unsulphured dried fruit (above 0.85 a_w) exists, but has not been reported to be significant.

Growth of *Asp. carbonarius*, a black *Aspergillus* species, on grapes before or during drying, can lead to production of ochratoxin A. Formation of ochratoxin A by *Asp. carbonarius* (and a few isolates of the closely related species *Asp. niger*) was discovered only recently (Abarca *et al.*, 1994; Téren *et al.*, 1996; Heenan *et al.*, 1998). Infection of grapes by black *Aspergillus* species occurs in the vineyard as the result of insect or mechanical penetration, splitting due to rain before harvest, or infection by pathogenic fungi such as *Botrytis* or *Rhizopus* (Snowdon, 1990; Leong *et al.*, 2004).

Aflatoxins have been a cause for concern in dried figs for a number of years, because the fungus carried in by insects, is able to penetrate the fruits before harvest (Buchanan *et al.*, 1975). Unacceptable levels have been reported in dried figs from Turkey and Greece (Masson and Meier, 1988; Reichert *et al.*,

1988; Boyacioglu and Gonul, 1990; Özay and Alperden, 1991; Sharman *et al.*, 1991) and Syria (Haydar *et al.*, 1990), but not Pakistan (Shah and Hamid, 1989). As with peanuts, aflatoxins are not uniformly distributed, usually being present in a low proportion of the figs. Percentages containing significant levels (usually 10 ug/kg or more) were reported as 1% (Steiner *et al.*, 1988), 2–4% (Boyacioglu and Gonul, 1990), or sometimes higher, 7% (Masson and Meier, 1988), 24% (Sharman *et al.*, 1991), or 29% (Özay and Alperden, 1991). The highest levels of aflatoxin B_1 found in individual fruit ranged from 12 μg/kg (Haydar *et al.*, 1990) to 63 μg/kg (Özay and Alperden, 1991), 112 μg/kg (Boyacioglu and Gonul, 1990) and 165 μg/kg (Sharman *et al.*, 1991).

Figs can also be contaminated by ochratoxin A. Figs with "fig smut", which have been infected with black *Aspergillus* species, including *Asp. carbonarius*, sometimes contain unacceptable levels of this toxin (Özay and Alperden, 1991; Doster *et al.*, 1996).

D CONTROL (dried fruits)

Summary

Significant hazards[a]	• Ochratoxin A in dried vine fruits, and aflatoxins and ochratoxin A in figs.
Control measures	
Initial level (H_0)	• Minimize time of storage of the cleaned, cut fruits before drying.
Reduction (ΣR)	• Blanching, when applicable, reduces microbial load.
Increase (ΣI)	• Frequent and thorough cleaning of equipment. • Prompt drying to low a_w, either by sundrying or dehydration. • Appropriate loading of the product into the dryer to achieve even drying. • Hygienic handling of the dried product. • Storage of the dried product to preclude entry of moisture. • Moisture control is an important factor to minimize the risk of recontamination of dried fruits.
Testing	• Aerobic plate count is a useful measure of hygiene and process control. • The microbial population will vary according to the type of fruits and the conditions of growing and processing. • The presence of coliforms is not a useful indicator of faecal contamination. • The presence of *E. coli* may indicate cause for concern.
Spoilage	• Dried fruits may spoil due to growth of filamentous fungi.

[a]Under particular circumstances, other hazards (e.g. bacterial pathogens) may need to be considered.

Control measures

Ready-to-eat packs. Most countries now permit the addition of weak acid preservatives such as sorbate or benzoate to high moisture prunes, figs, and other similar products. Even so, frequent and careful cleaning of processing and filling lines and equipment is essential to prevent build up of fungi, especially *Xer. bisporus* and xerophilic *Chrysosporium* species (Pitt and Hocking, 1997).

Glacé fruits. Control of *Schiz. pombe* during production of glacé fruits is not easy. The only effective technique is careful cleaning of processing equipment.

Dried figs. Control of the formation of aflatoxins in figs, is very difficult. However, screening fruits for the presence of aflatoxin by short-wave length UV light, is a useful procedure. As with maize, aflatoxins in figs can be detected by bright greenish yellow fluorescence under short-wave length (365 nm) UV light (Steiner *et al.*, 1988). By this means, total aflatoxin in a 56 kg batch of figs was reduced from the original level of 23–0.3 μg/kg. This technique is now used in many packing plants, permitting reasonable process control.

Destruction of aflatoxin by sulphur dioxide with or without assistance from heat or other processes has been investigated (Altug *et al.*, 1990; Icibal and Altug, 1992). Sodium bisulphite (1% solution) combined with 0.2% hydrogen peroxide or exposure to 2 g/kg sulphur dioxide gas plus 65°C heat and 0.2% hydrogen peroxide reduced aflatoxin levels in figs considerably. However such techniques are unlikely to find commercial application.

As with other products containing aflatoxins, final control is exercised by aflatoxin assays. The large size of individual pieces makes sampling figs for aflatoxin detection and control, particularly difficult. Sampling plans based on 10–20 kg lots have been developed (Bruland *et al.*, 1992; Hussain and Vojir, 1993; Sharman *et al.*, 1994).

Little has been written about the control of ochratoxin A in figs.

Dried vine fruits. Because *Asp. carbonarius* finds entry into grapes through damage from a variety of causes, control of ochratoxin A formation is very difficult. Reduction in damage, by reducing insect infestation, disease control, and careful handling before drying are important. Rapid drying, where this is practicable, will also reduce the level of ochratoxin.

As with aflatoxins in figs, the final control measure is assaying and sorting. No techniques exist for sorting individual berries on the basis of ochratoxin content, however sorting out damaged or dark berries will reduce the toxin load, often quite effectively.

VIII Fermented and acidified fruits

A *Effects of processing on microorganisms*

The fruits most commonly fermented are cucumbers, olives, nutmeg *(Myristica fragrance)*, green papaya (*Carica papaya)*, green mangoes (*Mangifera* spp.), and a number of under utilized fruits of south east Asia, such as *kedondong (Spondias cytherea), belimbing asam (Averrhoa bilimbi), chermai (Eugeria michelii)* and lime *(Citrus aurantifolia)* and smaller quantities of green peppers, chilies, and tomatoes are also processed this way (Merican *et al.* 1984; Saono *et al.*, 1986). The processes used vary with the final product type.

Olives. There are three main types of table olives: Spanish-style green olives, Greek-style naturally black olives, and Californian-style ripe olives. The production of olives varies with olive type but in general the process is similar. Olives at harvest, contain a bitter principle, oleuropein, which breaks down into compounds inhibitory to lactic acid bacteria (Fleming *et al.*, 1973), especially *Leuconostoc* species, the dominant bacteria at the start of the fermentation process. Oleuropein is removed before fermentation, by treating the olives with dilute alkali (1.8–2.5% lye). When the lye has penetrated almost to the pit, the alkali is leached out with water. This process also leaches out most of the fermentable carbohydrate so a source must be supplied (Beuchat, 1978). Most of the normal flora of the fresh olive is removed by the alkali treatment, but recontamination occurs during leaching and subsequent processing.

The desirable fermentation organisms are similar to those responsible for sauerkraut fermentation and include *Leuconostic mesenteroides, Lactobacillus brevis*, and *Lb. plantarum*. *Enterococcus* spp. is also common in naturally occurring fermentations along with *Ent. casseliflavus* and *Lb. pentosus* as a starter culture for Spanish-style olives (de Castro *et al.*, 2002). The primary factors influencing the fermentation of olives, is availability of fermentable substrates, salt content, pH, aeration, and temperature (Garrido Fernandez *et al.*, 1997; Duran Quintana *et al.*, 1999; Tassou *et al.*, 2002). Olives are fermented using 6–10% brine (Beuchat, 1978; Tassou *et al.*, 2002). Decreasing the salt content to 3–4% results in softening of the olives and the formation of gas pockets, which appear to be associated with the growth of three yeast *Saccharomyces kluyveri, Sac. oleaginosus*, and *Hansenula anomala* (Beuchat, 1978). Altering the salt content and temperature of fermentation results in changes in the absolute amounts and ratios among the six organic acids produced during the fermentation, the amount of ethanol production, and the ratio of D- and L-lactic acid (Tassou *et al.*, 2002).

Fermented green olives are covered with 7%–8% brine containing 0.6%–0.9% lactic acid, vacuum packed, usually in glass jars, and pasteurised. Black (ripe) olives require several washing treatments to completely remove the alkali from the oleuropein treatment.

Cucumbers. Cucumbers are processed into pickles in several ways, including fermentation and direct acidification. Mild fermentation begins with raw cucumbers in a low salt (2.6%–4.0%) brine. Lactic acid fermentation takes place for a few days at room temperature and continues during refrigerated storage. An alternative fermentation commences with higher salt levels (8%–10%) and salt is then gradually increased to 15%. The initial flora may be very varied, but *Pediococcus* species and *Lb. plantarum* soon dominate.

Pure culture fermentation of cucumbers yields the best product (Etchells *et al.*, 1964, 1966). If fermentation is carried out in small containers, pasteurisation can be used to destroy the surface flora of the cucumbers. However, in large-scale plants, the natural flora is suppressed by acidifying the initial brine to pH 3.3. Pure cultures of *Lb. plantarum*, or a mixture of *Lb. plantarum* and *Pediococcus cerevisiae*, are then added and fermentation commences immediately (Etchells *et al.*, 1973).

Pickles may also be made by direct acidification. This process combines cucumbers with vinegar and spices, followed by a pasteurisation step.

Fully cured, salted pickles are often leached to remove salt before the addition of acetic acid with or without sugar, to produce sweet or sour pickles, respectively.

Fruits of south east Asia. The basic pickling process involves salting whole or cut pieces of fruits, at various salt concentration and lengths of time (Merican *et al.* 1984; Saono *et al*, 1986). A lactic acid fermentation takes place and fermentable carbohydrates may or may not be added. Microorganisms isolated from the pickled vats include *Lb. brevis, Lb. plantarum* and *Leuc. mesenteroides* (Saono *et al*, 1986). When the product is ready for harvest, the salt is leached out if high salt has been added, and the pickled fruit is either packed in acidified syrup or used as an ingredient in spicy "*acar*". For some fruits such as nutmeg, bisulphite is used in addition to salt, to prevent browning and preserve the colour.

B Spoilage

Olives. Inadequate leaching of alkali from treated olives inhibits the normal *Leuconostoc* fermentation, permitting spoilage by yeasts and coliforms (West *et al.*, 1941; Etchells *et al.*, 1966). Olive fermentation is slow, providing time for problems to occur. Gassiness and bloating due to heterofermentative lactics, yeasts, and coliforms are most frequent and a cheesy odour may develop (Pederson, 1971). Storage of olives in low salt acidified brines may cause gas defects due to growth of *Pichia* and *Saccharomyces* species. Softening of fermented green olives may be caused by growth of *Rhodotorula* species that are pectinolytic (Vaughn *et al.*, 1969a, 1972). Small white spots on the surfaces of olives may be due to *Lb. plantarum* (Thompson *et al.*, 1955; Vaughn *et al.*, 1953) or various yeasts (Pederson, 1971). A variety of molds, including *Aspergillus, Geotrichum, Paecilomyces, Verticillium*, and *Penicillium*, can grow extensively on the surface of brine in open vats (Beuchat, 1978), though their impact on the quality of the product is unclear.

Spoilage of green olives can occur in improperly sealed containers. The dominant initial flora is lactate oxidising, obligately aerobic yeasts. Growth and acid utilization may raise the pH and permit growth of other salt tolerant yeasts and bacteria.

Softening and sloughing of skin and flesh from black olives being prepared for canning, has been observed (Vaughn *et al.*, 1969b). Bacteria with pectinolytic activity may be the cause.

Cucumbers. The saprophytic and spoilage flora of pickled cucumbers depends on the process used. High salt (15%) brined pickles may support growth of yeasts, halophiles, and coliforms, if acid production is low. Dill pickles from low salt (5% or less) brines can bloat due to yeasts, heterofermentative lactic acid bacteria or coliforms, if the desirable fermentation flora fails to develop adequately (Samish *et al.*, 1957; Etchells *et al.*, 1968). Various bacteria, including enterics, occur naturally within intact fermenting tomatoes and cucumbers (Samish and Dimant, 1959; Samish and Etinger-Tulczynsky, 1962; Meneley and Stanghellini, 1974) and may contribute to bloating. Additional microorganisms can be internalised, if untreated water is used to hydrocool newly harvested cucumbers (Reina *et al.*, 1995).

Softening of flesh can also occur due to enzymes from yeasts or fungi growing in the cucumber flowers in the field (Etchells *et al.*, 1958). Mold growth within brined cucumber tissue during purging with high volumes of air can lead to softening of the product.

The addition of weak acid preservatives during brining will alter the flora. The normal yeast flora may be suppressed and replaced by undesirable species (Etchells *et al.*, 1961). Thus the use of preservatives in processing may be of little value.

Sweet and sour pickles, which are not pasteurised but rely on acetic acid for preservation, are susceptible to spoilage by preservative resistant yeasts, especially film formers such as *Can. krusei*, and *Pichia membranaefaciens* (Pitt and Hocking, 1997).

C Pathogens

Food poisoning from correctly fermented fruit products, is very unlikely. No mycotoxin problems are known.

D CONTROL (fermented and acidified fruits)

Summary

Significant hazards[a]	• Foodborne illness from correctly fermented or acidified fruits is extremely rare. • Formation of mycotoxins has not been reported.
Control measures	
Initial level (H_0)	• Does not apply.
Reduction (ΣR)	• Does not apply.
Increase (ΣI)	• Does not apply.
Testing	• Does not apply.
Spoilage	• Microbial spoilage can occur in some products and control depends on appropriate processing techniques.

[a]Under particular circumstances, other hazards may need to be considered.

Control measures. Many of the control measures are similar to the good manufacturing practices used for vegetables and the reader should consult the chapter on vegetables.

Olives. Control of spoilage in fermented green olive production, depends initially on effective alkali treatment followed by its complete removal. This is necessary to ensure adequate acid generation by lactic acid bacteria in low salt brines. Proper fermentation is also dependent on the controlled availability of a fermentable carbohydrate source. Proper brine concentrations and fermentation temperatures are needed to get the correct end-products of the fermentation. Maintenance of anaerobic conditions is essential to prevent growth of oxidative yeasts.

Cucumbers. Control of spoilage depends on correct brine concentrations, acidification, and pure starter cultures. Pasteurisation treatments need to be adequate and controlled. Clean handling is necessary to prevent excessive numbers of unwanted bacteria and yeasts that may overload the desirable fermentative bacteria and lead to spoilage. Preservative-resistant yeasts pose a special problem in unpasteurised products; scrupulous cleanliness in the packing plant is the only effective control. Bloating is controlled through the purging of fermentation vats to remove CO_2 and to introduce O_2 into the interior of the cucumber, as a means of accelerating gas exchange and thus increasing the rate of fermentation (Daeschel and Fleming, 1981; 1983).

IX Canned Tomato products

The main tomato products are canned foods such as canned tomatoes (whole, peeled or diced) with or without added juice, or packed in tomato purée; tomato concentrates including tomato juice and tomato pastes; tomato powder; and formulated products such as salsa, tomato sauce (catsup or ketchup), soup and chili sauce.

A Effects of processing on microorganisms

The tomato is considered to be an acidic raw material with a pH value generally 4.6 or less and so a relatively mild heat treatment should render tomato products commercially sterile. The acid content of the products to prevent growth of bacterial spores, including those of *Cl. botulinum*. However, the acidity of tomatoes has decreased in recent decades and mechanical harvesting results in a higher load of microorganisms than in handpicked fruit. If spoilage is to be prevented, tomatoes would have to be processed sufficiently such that adverse quality changes will result. Acidification of canned tomato products has proved a practical solution to this problem (Lopez, 1971; Schoenemann and Lopez, 1975; Powers, 1976).

For processing of tomato juice various methods are employed including: retorting of packaged product under pressure, atmospheric processing under agitation or in still conditions, and hot filling followed by atmospheric processing in a steam tunnel (Gould, 1974). Agitating product and processing under pressure results in shorter process times. Atmospheric processing destroys microorganisms with low heat resistance, but is inadequate to destroy heat resistant flat sour bacterial spores. Bulk sterilisation, i.e. use of a flash steriliser followed by hot filling, holding, and water-cooling, is used to destroy flat sour spores. More recently aseptic packing after sterilisation has also been used.

B Spoilage

Spoilage of canned tomatoes may occur because of under processing or leakage.

The bacteria usually found on tomatoes are lactobacilli and other relatively heat sensitive genera. Therefore, spoilage due to under processing is caused predominantly by aciduric spore formers such as the gas producing anaerobe *Cl. pasteurianum* and more commonly the flat sour species *B. coagulans*.

The taste of tomato juice spoiled by aciduric flat sour spoilage is described as "medicinal", "phenolic" or "fruity", and this taste is usually accompanied by a reduction in pH. Ends of cans of spoiled product remain flat.

Bacillus coagulans is a common soil bacterium. The vegetative cells of some strains can grow in tomato juices of pH 4.15–4.25, however, heated spores could not germinate and grow in tomato juice adjusted to a pH lower than 4.3 (Pederson and Becker, 1949).

A definite relationship exists between the concentration of soil particles and the number of flat sour spores in water from tomato soaking tanks. These bacteria have been found to multiply in equipment for tomato washing where the volume of cold water is insufficient and water temperatures rise to 27°C 32°C. Spores have been isolated from various points in tomato canneries (Fields, 1970; Segmiller and Evancho, 1992).

"Machinery mould", *Geo. candidum* (= *Oidium lactis*) may be a contaminant in tomato juice plants with poor sanitation and is the dominant mould on tomato processing equipment. *Bacillus coagulans* may be able to grow and produce spores in tomato juice in which this mould has grown (Fields, 1962).

Less frequently, other types of microbial spoilage occur, in particular swollen cans caused by some heat tolerant species of *Lactobacillaceae* (Gould, 1974).

C Pathogens

The heat treatment of tomato products should be designed to inactivate vegetative and spore forming bacteria capable of multiplication in the product. With product of pH 4.5 or less, the growth of *Cl. botulinum* spores will be inhibited. However, aciduric *Bacillus* species capable of elevating the pH of tomato juice, can sometimes be isolated from soil and natural vegetable materials (Al-Dujaili and Anderson, 1991). Instances of botulism from home canned tomatoes were apparently related to visible mould growth that had caused a pH rise sufficient to permit growth of *Cl. botulinum* spores.

Tomatoes are not a favourable substrate for growth of *L. monocytogenes*. Nevertheless, cells of *L. monocytogenes* inoculated into commercially processed tomato juice and sauce, remained viable for periods exceeding normal product shelf-life (Beuchat and Brackett, 1991).

D CONTROL (tomato products)

Summary

Significant hazards	• *Cl. botulinum* • *L. monocytogenes*
Control measures	
Initial level (H_0)	• Washing to keep spore count low.
Reduction (ΣR)	• Heat treatment to reduce vegetative cells.
Increase (ΣI)	• pH control (acidification) is crucial to prevent increase of surviving organisms.
Testing	• Holding canned products at elevated temperatures and observing for swelling (from "Canned" section)
Spoilage	• *Cl. pasteurianum* • *B. coagulans*

Control measures. Tomatoes must be thoroughly washed before processing, because bacterial spore counts are closely related to residual soil on the fruit (Mercer and Olson, 1969). All processing equipment should be flushed clean routinely to minimise the risk of spoilage from thermophilic spore forming bacteria accumulating in the equipment.

Critical to successful processing are (i) the destruction of microorganisms capable of reproduction in the product; and (ii) achievement of reduced pH or increased acidity which will prevent growth of those microorganisms that do survive.

This barrier must control not only the acid generating flat sour bacteria, but also acid utilising microorganisms whose growth might result in pH elevation to the point where *Cl. botulinum* can grow. The *B. coagulans* heat process for tomato juice is 0.7 min at 121.1°C (250°F) or equivalent (NRC, 1985).

Control of bacteria in tomatoes, is commonly accomplished by the addition of an edible organic acid, usually citric acid; solid sweeteners are often added as well to compensate for tartness due to the added acid (Gould, 1974; Wahem, 1990).

References

Abarca, M.L., Bragulat, M.R., Castellá, G. and Cabañes, F.J. (1994) Ochratoxin A production by strains of *Aspergillus niger* var. *niger*. *Appl. Environ. Microbiol.*, **60**, 2650–52.

Al-Dujaili, F. and Anderson, R.E. (1991) Aciduric, pH elevating *Bacillus* which cause non effervescent spoilage of underprocessed tomatoes. *J. Food Sci.*, **56**, 1611–13.

Altug, T., Yousef, A.E. and Marth, E.H. (1990) Degradation of aflatoxin B_1 in dried figs by sodium bisulfite with or without heat, ultraviolet energy or hydrogen peroxide. *J. Food Prot.*, **53**, 581–2.

Anderson, S. M., Verchick, L., Sowadsky, R., *et al.* (2002) Multistate outbreaks of *Salmonella* serotype Poona infections associated with eating cantaloupes from Mexico – United States and Canada. *Morb. Mortal. Wkly Rep.*, **51**, 1044–7.

Anonymous. (1993) Cantaloupe appears to be source of Oregon *E. coli* outbreaks. *Food Chemical News*, August 30th, 1993, 14–5.

Anonymous. (1999a) *Minimal Processing and Food Safety of Harumanis Mango*. Center for Food and Nutrition Studies, Bogor Agriculture University, Indonesia.

Anonymous. (1999b) Project Terminal Report. Vietnam. Annex J. *Development of additional MP technologies needed for QA systems for MP fruits*. Association of South East Asian Nation (ASEAN) Australian Economic Cooperation Programme (AAECP) III Quality Assurance Systems for ASEAN Fruits (QASAF) Project.

ASEAN-COFAF (1984). ASEAN Horticultural Produce Handling Workshop Report. Association of South East Asian Nations Committee on Food, Agriculture and Forestry Secretariat, Jakarta, Indonesia.

Asplund, K. and Nurmi, E. (1991) The growth of salmonellae in tomatoes. *Int. J. Food Microbiol.*, **13**, 177–82.

Barkai-Golan, R. (1974) Species of *Penicillium* causing decay of stored fruits and vegetables in Israel. *Mycopathol. Mycol. Appl.*, **54**, 141–5.

Bartz, J. A. (1982) Infiltration of tomatoes immersed at different temperatures to different depths in suspensions of *Erwinia carotovora* subsp. *carotovora*. *Plant Dis.*, **66**, 302–6.

Bartz, J. A. (1999) Washing fresh fruits and vegetables: lessons from treatment of tomatoes and potatoes with water. *Dairy Plant Environ. Sanit.*, **19**, 853–64.

Bartz, J. A. and Showalter, R.K. (1981) Infiltration of tomatoes by aqueous bacterial suspensions. *Phytopathology*, **71**, 515–8.

Beuchat, L.R. (1978) *Food and Beverage Mycology*, AVI Publishing Co., Inc. Westport, CT. pp. 83–109.

Beuchat, L.R. (1996) Pathogenic microorganisms associated with fresh produce. *J. Food Prot.*, **59**, 204–16.

Beuchat, L.R. and Brackett, R.F. (1991) Behaviour of *Listeria monocytogenes* inoculated into raw tomatoes and processed tomato products. *Appl. Environ. Microbiol.*, **57**, 1367–71.

Beuchat, L.R. and Rice, S.L. (1979) *Byssochlamys* spp. and their importance in processed fruits. *Adv. Food Res.*, **25**, 237–88.

Blostein, J. (1993) An outbreak of *Salmonella javiana* associated with consumption of water melon. *J. Environ. Health*, **56**, 29–31.

Boudra, H., LeBars, J., LeBars, P. and Dupuy, J. (1994) Time of *Aspergillus flavus* infection and aflatoxin formation in ripening of figs. *Mycopathologia*, **127**, 29–33.

Boyacioglu, D. and Gonul, M. (1990) Survey of aflatoxin contamination of dried figs grown in Turkey in 1986. *Food Addit. Contam.*, **7**, 235–7.

Brackett, R.E. (1992) Shelf stability and safety of fresh produce as influenced by sanitation and disinfection. *J. Food Prot.*, **55**, 808–14.

Brackett, R.E. (1994) Microbiological spoilage and pathogens in minimally processed refrigerated fruits and vegetables, in *Minimally Processed Refrigerated Fruits and Vegetables* (ed. R.C. Wiley) Chapman & Hall, New York, pp. 269–312.

Breidt, F. and Fleming, H.P. (1997) Using lactic acid bacteria to improve the safety of minimally processed fruits and vegetables. *Food Technol.*, **51**(9), 44–8, 51.

Bruland, H.G., Matthiaschk, G., Sanitz, W., Vierkotter, S., Weber, R. and Wenzel, H. (1992) Aflatoxins in dried figs – sampling techniques (in German). *Deutsche Lebensmittel-Rundschau*, **88**, 183–5.

Buchanan, J.R., Sommer, N.F. and Fortlage, R.J. (1975) *Aspergillus flavus* infection and aflatoxin production in fig fruits. *Appl. Microbiol.*, **30**, 238–41.

Buchanan, R.L., Edelson, S.G., Miller, R.L. and Sapers, G.M. (1999) Contamination of intact apples after immersion in an aqueous environment containing *Escherichia coli* O157:H7. *J. Food Prot.*, **62**, 444–50.

Buchanan, J.R., Sommer, N.F. and Fortlage, R.J. (1975) *Aspergillus flavus* infection and aflatoxin production in fig fruits. *Appl. Microbiol.*, **30**, 238–41.

Buhagiar, R.W.M. and Barnett, J.A. (1971) The yeasts of strawberries. *J. Appl. Bacteriol.*, **34**, 727–39.

Burnett, S.L., Chen, J. and Beuchat, L.R. (2000) Attachment of *Escherichia coli* O157:H7 to the surfaces and internal structures of apples as detected by confocal scanning laser microscopy. *Appl. Environ. Microbiol.*, **66**, 4679–87.

Butler, E.E., Webster, R.K. and Eckert, J.W. (1965) Taxonomy, pathogenicity, and physiological properties of the fungus causing sour rot of citrus. *Phytopathology*, **55**, 1262–8.

Cartwright, P. and Hocking, A.D. (1984) *Byssochlamys* in fruit juices. *Food Technol. Aust.*, **36**, 210–11.

Caul, E.O. (1993) Outbreaks of gastroenteritis associated with SRSV's. *Public Health Laboratory Service Microbiology Digest*, **10**(1), 2–8.

de Castro, A., Montano, A., Casado, F.-J. Sanchez, and Rejano, L. (2002) Utilization of *Enterococcus casseliflavus* and *Lactobacillus pentosus* as starter cultures for Spanish-style green olive fermenation. *Food Microbiol.*, **19**, 637–44.

Caul, E.O. (1993) Outbreaks of gastroenteritis associated with SRSV's. Public Health Laboratory Service, *Microbiol. Digest*, **10** (1), 2–8.

CDC (Centers for Disease Control and Prevention). (1979) *Salmonella oranienburg* gastroenteritis associated with consumption of precut water melon. Illinois *Morb. Mortal. Wkly Rep.* **28**, 522–3.

CDC (Centers for Disease Control and Prevention). (1991) Multistate outbreak of *Salmonella poona* infections. US and Canada *Morb. Mortal. Wkly Rep.* **40**, 549–52.

CDC Centers for Disease Control and Prevention). (1993) Multistate outbreak of *Salmonella* serotype Montevideo infections. EPI-AID 79–93.

CDC (Centers for Disease Control and Prevention). (1997a) Hepatitis A associated with consumption of frozen strawberries – Michigan 1997. *Morb. Mortal. Wkly Rep.* **46**, 288, 295.

CDC (Centers for Disease Control and Prevention). (1997b) Outbreak of cyclosporiasis. Northern Virginia-Washington DC-Baltimore Maryland Metropolitan Area *Morb. Mortal. Wkly Rep.* **46**, 689–91.

Champ, B.R., Highley, E. and Johnson, G.I. (eds). (1994). *Postharvest Handling of Tropical Fruits.* ACIAR Proceedings No. 50 Australian Centre for International Agricultural Research, Canberra, A.C.T.

Chan, H.T. (1997). Heat shocking fruits for heat and cold tolerance, in *Proceedings of the Sixth Association of South East Asian Nations food Conference.* Singapore Institute of Food Technology, Singapore, pp. 99–106.

Chudhangkura A, Maneepun S., Varanyanond W. Satonsaovapak S., Saiyudthong S., Japakaset J., Anantraksakul P. and Wattanasiritham L. (1999). Minimal processing and food safety in Durian. *ASEAN Australian Economic Cooperation Programme (AAECP) III Quality Assurance Systems for ASEAN Fruits (QASAF) Project.* Institute of Food Research and Product Development, Kasetsart University, Bangkok, Thailand.

Cliver, D.O. (1983) Manual on Food Virology. World Health Organisation, Geneva.

Cliver, D.O. and Kostenbader, K.D. (1979) Antiviral effectiveness of grape juice. *J. Food Prot.*, **42**, 100–4.

Coley-Smith, J.R., Verhoeff, K. and Jarvis, W.R. (eds). (1980) in *The Biology of Botrytis.* Academic Press, London.

Combrink, J.C., Kotzé, J.M., Wehner, F.C. and Grobbelaar, C.J. (1985). Fungi associated with core rot of Starking apples in South Africa. *Phytophylactica*, **17**, 81–3.

Conner, D.E. and Kotrola, J.S. (1995) Growth and survival of *Escherichia coli* O157:H7 under acidic conditions. *Appl. Environ. Microbiol.*, **61**, 382–5.

Cummings, K., Barrett, E., Mohle-Boetani, J.C., Brooks, J.T., Farrar, J., Hunt, T., Fiore, A., Komatsu, K., Werner, S.B, and Slutsker, L. (2001) A multistate outbreak of ***Salmonella enterica*** serotype baildon associated with domestic raw tomatoes. *Emerg. Infect. Dis.*, **7**, 1046–8.

Daeschell, M.A. and Fleming, H.P. (1983) Rapid and specific staining for routes of liquid entry into cucumber fruit. *J. Am Soc. Hortic. Sci.*, **108**, 481–3.

Day, N.B., Skura, B.J. and Powrie, W.D. (1990) Modified atmosphere packaging of blueberries: microbiological changes. *Can. Inst. Food Sci. Technol. J.*, **23**, 59–65.

de Castro, A., Montano, A., Casado, F. -J. Sanchez and Rejano, L. (2002) Utilization of *Enterococcus casseliflavus* and *Lactobacillus pentosus* as starter cultures for Spanish-style green olive fermentation. *Food Microbiol.*, **19**, 637–644.

Del Rosario, B.A. and Beuchat, L.R. (1995) Survival and growth of enterohemorrhagic *Escherichia coli* O157:H7 in cantaloupe and watermelon. *J. Food Prot.*, **58**, 105–7.

Dennis, C. (1983a) Soft fruit, in *Post-Harvest Pathology of Fruits and Vegetables* (ed. C. Dennis), Academic Press, London, pp. 23–42.

Dennis, C. (1983b) Yeast spoilage of fruit and vegetable products. *Indian Food Packer*, **37**, 38–53.

Dingman, D.W. (2000) Growth of *Escherichia coli* O157:H7 in bruised apple (*Malus domestica*) tissue as influenced by cultivar, date of harvest, and source. *Appl. Environ. Microbiol.*, **66**, 1077–83.

Doster, M.A., Michailides, T.J. and Morgan, D.P. (1996) *Aspergillus* species and mycotoxins in figs from California orchards. *Plant Dis.*, **80**, 484–9.
Draughon, F.A., Chen, S. and Mundt, J.O. (1988) Metabolic association of *Fusarium*, *Alternaria*, and *Rhizoctonia* with *Clostridium botulinum* in fresh tomatoes. *J. Food Sci.*, **53**, 120–3.
Du, J., Han, Y. and Linton, R.H. (2002) Inactivation by chlorine dioxide gas (ClO_2) of *Listeria monocytogenes* spotted onto different apple surfaces. *Food Microbiol.*, **19**, 481–90.
Duffell, E., Espie, E., Nichols, T., Adak, G.K., De Valk, H., Anderson, K. and Stuart, J.M. (2003). Investigation of an outbreak of *E.coli* O157 infections associated with a trip to France of schoolchildren from Somerset, England. *Eurosurveillance*, **8**, 81–6
Duran Quintana, M.C., Garcia Garcia, P. and Garrido Fernandez, A. (1999). Establishment of conditions for green table olive fermentation at low temperature. *Int. J. Food Microbiol.*, **51**, 133–43.
Eckert, J.W., Rubio, P.P., Mattoo, A.K. and Thompson, A.K. (1975) Diseases of tropical fruits and their control, in *Postharvest Physiology, Handling and Utilization of Tropical and Subtropical Fruits and Vegetables* (ed. E.B. Pantastico) AVI Publishing Co., Westport, CT, pp. 415–43.
Escartin, E.F., Ayala, A.C. and Lozano, J.S. (1989) Survival and growth of *Salmonella* and *Shigella* on sliced fresh fruit. *J. Food Prot.*, **52**, 471–2.
Etchells, J.L., Bell, T.A., Monroe, R.J., Masley, P.M. and Demain, A.L. (1958) Populations and softening enzyme activity of filamentous fungi on flowers, ovaries, and fruit of pickling cucumbers. *Appl. Microbiol.*, **6**, 427–40.
Etchells, J.L., Borg, A.F. and Bell, T.A. (1961) Influence of sorbic acid on populations and species of yeasts occurring in cucumber fermentations. *Appl. Microbiol.*, **9**, 139–44.
Etchells, J.L., Costilow, R.N., Anderson, T.E. and Bell, T.A. (1964) Pure culture fermentation of brined cucumbers. *Appl. Microbiol.*, **12**, 523–35.
Etchells, J.L., Borg, A.F., Kittel, I.D., Bell, T.A. and Fleming, H.P. (1966) Pure culture fermentation of green olives. *Appl. Microbiol.*, **14**, 1027–41.
Etchells, J.L., Borg, A.F. and Bell, T.A. (1968) Bloater formation by gas-forming lactic acid bacteria in cucumber fermentations. *Appl. Microbiol.*, **16**, 1029–35.
Etchells, J.L., Bell, T.A., Fleming, H.P., Kelling, R.E. and Thompson, R.L. (1973) Suggested procedure for the controlled fermentation of commercially brined pickling cucumbers – the use of starter cultures and reduction of carbon dioxide accumulation. *Pickle Packing Sci.*, **3**, 4–19.
Faridah, M.S., Latifah, M.N., Asiah A.S. and Mahmud M. (1999). Microbiological changes in minimally processed jackfruit packed in different packaging system. *ASEAN Australian Economic Cooperation Programme (AAECP) III Quality Assurance Systems for ASEAN Fruits Project, Regional Technical Workshop 2–4 Dec.* Malaysian Agricultural Research and Development Institute, Kuala Lumpur, Malaysia.
Fields, M.L. (1962) The effect of *Oidium lactis* on the sporulation of *Bacillus coagulans* in tomato juice. *Appl. Microbiol.*, **10**, 70–3.
Fields, M.L. (1970) The flat sour bacteria. *Adv. Food Res.*, **18**, 163–217.
Fleischman, G.J., Bator, C., Merker, R. and Keller, S.E. (2001) Hot water immersion to eliminate *Escherichia coli* O157:H7 on the surface of whole apples: thermal effects and efficacy. *J. Food Prot.*, **64**, 451–5.
Fleming, H.P., Walter, W.M. and Etchells, J.L. (1973) Antimicrobial properties of oleuropein and products of its hydrolysis from green olives. *Appl. Microbiol.*, **26**, 777–81.
Garrido Fernandez, A., Dams, M. and Fernandez Diez, M.J. (1997) *Table Olives: Production and Processing*, Kluwer Academic, New York.
Gayler, G.W., MacCready, R.A., Reardon, J.P. and McKernan, B.F. (1955) An outbreak of salmonellosis traced to water melon. *Public Health Rep.*, **70**, 311–3.
Getz, S., Fulbright, D.W. and Stephens, C.T. (1983) Scanning electron microscopy of infection sites and lesion development on tomato fruit infected with *Pseudomonas syringae* pv. *tomato*. *Phytopathology*, **73**, 39–43.
Golden, D.A., Rhodehamel, E.J. and Kautter, D.A. (1993) Growth of *Salmonella* spp. in cantaloupe, watermelon, and honeydew melons. *J. Food Prot.*, **56**, 194–6.
Gould, W.A. (1974) *Tomato Production, Processing and Quality*. AVI Publishing Co., Westport, CT.
Hall, E.G. and Scott, K.J. (1977) *Storage and Market Diseases of Fruit*, Commonwealth Scientific and Industrial Research Organisation, Melbourne, Australia.
Haydar, M., Bennelli, L. and Brera, C. (1990) Occurrence of aflatoxins in Syrian foods and foodstuffs: a preliminary study. *Food Chem.*, **37**, 261–8.
Hedberg, C.W., Angulo, F.J., White, K.E., Langkop, C.W., Schell, W.L., Stobierski, M.G., Schuchat, A., Besser, J.M., Dietrich, S., Helsel, L., Griffin, P.M., McFarland, J.W. and Osterholm, M.T. (1999) Outbreaks of salmonellosis associated with eating uncooked tomatoes: implications for public health. *Epidemiol. Infect.*, 122(**3**), 385–93.
Heenan, C.N., Shaw, K.J. and Pitt, J.I. (1998). Ochratoxin A production by *Aspergillus carbonarius* and *A. niger* isolates and detection using coconut cream agar. *J. Food Mycol.*, **1**, 63–72.
Herwaldt, B.L. and Ackers, M.L. (1997) An outbreak of cyclosporiasis associated with imported raspberries. *New Engl. J. Med.*, **336**, 1548–56
Herwaldt, B.L., Beach, M.J. and the Cyclospora Working Group (1999) The return of *Cyclospora* in 1997: another outbreak of cyclosporiasis in North America associated with imported raspberries. *Ann. Intern. Med.*, **130**, 210–20.
Hocking, A.D. and Pitt, J.I. (1984) Food spoilage fungi. II. Heat resistant fungi. *CSIRO Food Res. Q.*, **44**, 73–82.

Holland, B., Unwin, I.D. and Buss, D.H. (1992) Fruit and nuts, *First Supplement to the Fifth Edition of McCance and Widdowson's The Composition of Foods*, Royal Society of Chemistry, Cambridge.

Holmes, G.J, Eckert, J.W. and Pitt, J.I. (1993) A new postharvest disease of citrus in California caused by *Penicillium ulaiense*. *Plant Dis.*, **77**, 537.

Holmes, G.J, Eckert, J.W. and Pitt, J.I. (1994) Revised description of *Penicillium ulaiense* and its role as a pathogen of citrus fruits. *Phytopathology*, **84**, 719–27.

Hotchkiss, J.H., Banco, M.J., Busta, F.F., Genigeorgis, C.A., Kociba, R., Rheaume, L., Smoot, L.A., Schuman, J.D. and Sugiyama, H. (1992) The relationship between botulinal toxin production and spoilage of fresh tomatoes held at 13 and 23 degrees under passively modified and controlled atmospheres and air. *J. Food Prot.*, **55**, 522–7.

Hull, R. (1939) Study of *Byssochlamys fulva* and control measures in processed fruits. *Ann. Appl. Biol*, **26**, 800–22.

Hussain, M. and Vojir, F. (1993) A sampling plan for the control of aflatoxin B_1 in imported dried figs (in German). *Dtsch. Lebensmittel-Rundsch.*, **89**, 379–83.

Icibal, N. and Altug, T. (1992) Degradation of aflatoxins in dried figs by sulphur dioxide alone and in combination with heat, ultraviolet energy and hydrogen peroxide. *Lebensmittel-Wiss. Technol.*, **25**, 294–6.

Jakobsen, M. and Jensen, H.C. (1975) Combined effect of water activity and pH on the growth of butryic anaerobes in canned pears. *Lebensmittel-Wiss. Technol.*, **8**, 158–60.

Janisiewicz, W.J., Conway, W.S., Brown, M.W., Sapers, G.M., Fratamico, P. and Buchanan R.L. (1999a) Fate of *Escherichia coli* O157:H7 on fresh-cut apple tissue and its potential for transmission by fruit flies. *Appl. Environ. Microbiol.*, **65**, 1–5.

Janisiewicz, W.J., Conway, W.S. and Leverentz, B. (1999b) Biological control of postharvest decays of apple can prevent growth of *Esherichia coli* O157:H7 in apple wounds. *J. Food Prot.*, **62**, 1372–5.

Johannessen, G.S., Kruse, H. and Torp, M. (1999) Occurrence of bacteria of hygienic interest in organically grown fruits and vegetables. *Proceedings of the Food Micro and Food Safety into Next Millennium*, pp. 377–80.

Johnson, G.I., Mead, A.J., Cooke, A.W. and Dean, J.R. (1991) Mango stem-end rot pathogens - infection levels between flowering and harvest. *Ann. Appl. Biol.*, **119**, 465–73.

Johnson, G.I., Mead, A.J., Cooke, A.W. and Dean, J.R. (1992) Mango stem-end rot pathogens - fruit infection by endophytic colonisation of the inflorescence and pedicel. *Ann. Appl. Biol.*, **120**, 225–34.

Katz, D.J., Cruz, M.A., Trepka, M.J., Suarez, J.A., Fiorella, P.D. and Hammond, R.M. (2002) An outbreak of typhoid fever in Florida associated with an imported frozen fruit. *J. Infect. Dis.*, **186**, 234–9.

Kavanagh, J., Larchet, N. and Stuart, M. (1963) Occurrence of a heat-resistant species of *Aspergillus* in canned strawberries. *Nature*, **198**, 1322.

King, A.D., Hocking, A.D. and Pitt, J.I. (1981) Mycoflora of some Australian foods. *Food Technol. Aust.*, **33**, 55–60.

Kirk, M., Waddel, R., Dalton, C., Creaser, A. and Rose, N. (1997) A prolonged outbreak of *Campylobacter* infection in a training facility. *Commun. Dis. Intell.*, **21**, 57–61.

Kurdziel, A.S., Wilkinson, N., Langton, S. and Cook, N. (2001) Survival of poliovirus on soft fruit and salad vegetables. *J. Food Prot.*, **64**, 706–9.

Larson, A.E. and Johnson, E.A. (1999) Evaluation of botulinal toxin production in packaged fresh-cut cantaloupe and honeydew melons. *J. Food Prot.*, **62**, 948–52.

Lawson, A., Wallis, J., Lewandowski, C., Jenson, D., Potsic, S., Nickels, M.K., Lesko, M., Langkap, C., Martin, R.J., Endo, T., Ehrhard, H.B. and Francis, B.J. (1979) *Salmonella oranienburg* gastroenteritis associated with consumption of precut water melons – Illinois. *Morb. Mortal. Wkly Rep.*, **28**, 522–3.

Leong, S.-L., Hocking, A.D. and Pitt, J.I. (2004) Occurrence of fruit rot fungi (*Aspergillus* section *Nigri*) on some drying varieties of irrigated grapes. *Aust. Grape Wine Res.* **10**(1) 83–8.

Leverentz, B., Conway, W.S., Camp, M.J., Janisiewicz, W.J., Abuladze, T., Yang, M., Saftner, R. and Sulakvelidze, A. (2003) Biocontrol of *Listeria monocytogenes* on fresh-cut produce by treatment with lytic bacteriophages and a bacteriocin. *Appl. Environ. Microbiol.*, **69**, 4519–26.

Liao, C.-H. and Sapers, G.M. (2000) Attachment and growth of *Salmonella* Chester on apple fruits and in vivo response of attached bacteria to sanitzer treatments. *J. Food Prot.*, **63**, 876–83.

Lin, C.-M. and Wei, C.-I. (1997) Transfer of *Salmonella* Montevideo onto the interior surfaces of tomatoes by cutting. *J. Food Prot.*, **60**, 858–63.

Lopez, A. (1971) Updating developments in acidification of canned whole tomatoes. *Canning Trade*, April 12th, p. 8.

Madden, J.M. (1992) Microbial pathogens in food produce – the regulatory perspective. *J. Food Prot.*, **55**, 821–3.

Masson, A. and Meier, P. (1988) Contamination of spices, dried or frozen mushrooms and Turkish figs by mold (in French) *Microbiol. Alim. Nutr.*, **6**, 403–6.

McEvoy, I.J. and Stuart, M.R. (1970) Temperature tolerance of *Aspergillus fischeri* var. *glaber* in canned strawberries. *Irish J. Agric. Res.*, **9**, 59–67.

McGuire, R.G. and Hagenmaier, R.D. (2001) Shellac formulations to reduce epiphytic survival of coliform bacteria on citrus fruit postharvest. *J. Food Prot.*, **64**, 1756–60.

Meneley, J.C. and Stanghellini, M.E. (1974) Detection of enteric bacteria within locular tissue of healthy cucumbers. *J. Food Sci.*, 39, 1267–8.

Mercer, W.A. and Olson, W.A. (1969) Tomato Infield Washing Station Study. National Canners' Assn, Research Laboratory Report D-2167. National Canners' Association, Washington, DC.

Merican, Z., Yeoh, Q.L. and Idrus, A.Z. (1984) Malaysian Fermented Foods. ASEAN Protein Project Occasional Paper No. 10. Science Council of Singapore, Singapore.

Miller, M.W. and Phaff, H.J. (1962) Successive microbial populations of Calimyrna figs. *Appl. Microbiol.*, **10**, 394–400.
Miller, M.W., Fridley, R.B. and McKillop, A.A. (1963) The effects of mechanical harvesting on the quality of prunes. *Food Technol.*, **17**, 1451–3.
Möller, T. and Nilsson, K. (1991) Aflatoxins in fig wine (in Swedish). *Vaar Foda*, **43**, 111–3.
Morris, S.C. (1982). Synergism of *Geotrichum candidum* and *Penicillium digitatum* in infected citrus fruits. *Phytopathology*, **72**, 1336–9.
Mukhopadhyay, R., Mitra, A., Roy, R. and Guha, A.K. (2002) An evaluation of street-vended sliced papaya (*Carica papaya*) for bacteria and indicator micro-organisms of public health significance. *Food Microbiol.*, **19**, 663–7.
Murdock, D.I. and Hatcher, W.S. (1978) A simple method to screen fruit juices and concentrates for heat-resistant mold. *J. Food Prot.*, **41**, 254–6.
NACMCF (National Advisory Committee for Microbiological Criteria for Foods) (1999) Microbiological safety evaluation and recommendations on fresh produce. *Food Control*, **10**, 117–43.
Natarajan, C.P., Chari, C.N. and Mrak, E.M. (1948) Yeast populations in figs during drying. *Fruit Products J.*, **27**, 242–3, 267.
NRC (U.S. National Research Council) (1985) *An Evaluation of the Role of Microbiological Criteria for Foods and Food Ingredients*. National Academy Press, Washington, DC, p. 269.
Odlaug, T.E. and Pflug, I.J. (1978) *Clostridium botulinum* and acid foods. *J. Food Prot.*, **41**, 566–73.
O'Mahoney, M., Barnes, H., Stanwell-Smith, R. and Dickens, T. (1990) An outbreak of *Salmonella* Heidelberg associated with a long incubation period. *J. Public Health Med.*, **12**, 19–21.
ÖPzay, G. and Alperden, I. (1991) Aflatoxin and ochratoxin - contamination of dried figs (*Ficus carina* L.) from the 1988 crop. *Mycotoxin Res.*, **7**, 85–91.
Pao, S. and Brown, G.E. (1998) Reduction of microorganisms on citrus fruit surfaces during packinghouse processing. *J. Food Prot.*, **61**, 903–6.
Pao, S., Davis, C.L. and Kelsey, D.F. (2000) Efficacy of alkaline washing for decontamination of orange fruit surfaces inoculated with *Escherichia coli*. *J. Food Prot.*, **63**, 961–4.
Park, C.M. and Beuchat, L.R. (1999) Evaluation of sanitizers for killing *Escherichia coli* O157:H7, *Salmonella*, and naturally occurring microorganisms on cantaloupes, honeydew melons, and asparagus. *Diary Food Environ. Sanit.*, **19**, 842–7
Pederson, C.S. (1971) *Microbiology of Food Fermentations*. AVI, Westport, CT.
Pederson, C.S. and Becker, M.E. (1949) Flat Sour Spoilage of Tomato Juice. *N.Y. State Agric. Exp. Sta. Tech. Bull. 287*. New York State Agricultural Experiment Station, Geneva, N.Y.
Pitt, J.I. and Christian, J.H.B. (1968) Water relations of xerophilic fungi isolated from prunes. *Appl. Microbiol.*, **16**, 1853–8.
Pitt, J.I. and Hocking, A.D. (1982) Food spoilage fungi. I. *Xeromyces bisporus* Fraser. *CSIRO Food Res. Q.*, **42**, 1–6.
Pitt, J.I. and Hocking, A.D. (1997). *Fungi and Food Spoilage*, 2nd ed., Blackie Academic and Professional, London.
Pitt, J.I., Spotts, R.A., Holmes, R.J. and Cruickshank, R.H. (1991) *Penicillium solitum* revived, and its role as a pathogen of pomaceous fruit. *Phytopathology*, **81**, 1108–12.
Pönkä, A., Maunula, L., von Bonsdorff, C.H. and Lyytikäinen, O. (1999) Outbreak of calicivirus gastroenteritis associated with eating frozen raspberries. *Eurosurveillance Monthly*, **4**(6), 66–9.
Powers, J.J. (1976) Effect of acidification of canned tomatoes on quality and shelf life. *CRC Crit. Rev. Food Sci. Nutr.*, **7**, 371–95.
Put, H.M.C. and Kruiswijk, J.T. (1964) Disintegration and organoleptic deterioration of processed strawberries caused by the mould *Byssochlamys nivea*. *J. Appl. Bacteriol.*, **27**, 53–8.
Ramsay, C.N. and Upton, P.A. (1989) Hepatitis A and frozen raspberries. *Lancet*, **i**(8628), 43–4.
Reichert, N., Steinmeyer, S. and Weber, R. (1988) Determination of aflatoxin B_1 in dried figs by visual screening, thin-layer chromatography and ELISA(in German). *Z. Lebensmittel-Untersuchung Forschung*, **186**, 505–8.
Reina, L.D., Fleming, H.P. and Humphries, E.G. (1995) Microbiological control of cucumber hydrocooling water with chlorine dioxide. *J. Food Prot.*, **58**, 541–6.
Richardson, K.C. (1965) Incidence of *Byssochlamys fulva* in Queensland-grown canned strawberries. *Queensland J. Agric. Anim. Sci.*, **22**, 347–50.
Ries, A.A., Zaza, S., Langkop, C., Tauxe, R.V., and Blake, P.A. (1990) A multistate outbreak of*Salmonella* Chester linked to imported cantaloupe (Abstract). in *American Society for Microbiology. Program and abstracts of the 30th Interscience Conference on Antimicrobial Agents and Chemotherapy*. American Society for Microbiology, Washington, DC, p. 238.
Riordan, D.R., Sapers, G.M. and Annous, B.A. (2000) The survival of *Escherichia coli* O157:H7 in the presence of *Penicillium expansum* and *Glomerella cingulata* in wounds on apple surfaces. *J. Food Prot.*, **63**, 1637–42.
Roland, J.O. and Beuchat, L.R. (1984) Influence of temperature and water activity on growth and patulin production by *Byssochlamys nivea* in apple juice. *Appl. Environ. Microbiol.*, **47**, 205–7.
Rushing, J.W., Angulo, F.J. and Beuchat, L.R. (1996) Implementation of a HACCP program in a commercial fresh-market tomato packinghouse: A model for the industry. *Dairy Food Environ. Sanit.*, **16**, 549–53.
Ryall, A.L. and Lipton, W.J. (1979) *Handling, Transportation and Storage of Fruits and Vegetables*. AVI Publishing Co, Westport, CT.
Ryall, A.L. and Pentzer, W.T. (1982) Handling, Transportation and Storage of Fruits and Vegetables, *Fruits and Tree Nuts, volume 2*, 2nd ed. AVI Publishing Co, Westport, CT.
Samish, Z. and Dimant, D. (1959) Bacterial population in fresh, healthy cucumbers. *Food Manufact.*, **34**, 17–20.
Samish, Z., Dimant, D. and Marani, T. (1957) Hollowness in cucumber pickles. *Food Manufact.*, **32**, 501–6.
Samish, Z. and Etinger-Tulezynska, R. (1962) Bacteria within fermenting tomatoes and cucumbers. in *Proceedings of the First International Congress on Food Science and Technologies* (ed. J.M. Leitch), Gordon and Breach, New York. pp. 373–84.

Samish, Z. and Etinger-Tulezynska, R. (1963) Distribution of bacteria within the tissue of healthy tomatoes. *Appl. Microbiol.*, **11**, 7–10.

Samish, Z., Etinger-Tulezynska, R. and Bick, M. (1961) Microflora within healthy tomatoes. *Appl. Microbiol.*, **9**, 20–5.

Samish, Z., Etinger-Tulezynska, R. and Bick, M. (1962) The microflora within the tissue of fruits and vegetables. *J. Food Sci.*, **28**, 259–66.

Saono, S., Hull, R.R. and Dhamcharee, B. (1986). *A Concise Handbook of Indigenous Fermented Foods in the Asca Countries*, LIPI, Jakarta, Indonesia. pp. 107–12.

Sapers, G.M., Miller, R.L. and Mattrazzo, A.M. (1999) Effectiveness of sanitizing agents in inactivating *Escherichia coli* in golden delicious apples. *J. Food Sci.*, **64**, 734–6.

Schnabel, E.L. and Jones, A.L. (1999) Distribution of tetracycline resistance genes and transposons among phylooplane bacteria in Michigan apple orchards. *Appl. Environ. Microbiol.*, **65**, 4898–907.

Schoenemann, D.R. and Lopez, A. (1975) Heat processing effects on physical and chemical characteristics of acidified canned tomatoes. *J. Food Sci.*, **40**, 195.

Segall, R.H., Henry, F.E. and Dow, A.T. (1977) Effect of dump-tank water temperature on the incidence of bacterial soft rot of tomatoes. *Proc. Florida State Hortic. Soc.*, **90**, 204–5.

Segmiller, J.L. and Evancho, G.M. (1992). Aciduric flat sour spore formers, in *Compendium of Methods for the Microbiological Examination of Foods* (eds C. Vanderzant and D.F. Splittstoesser), American Public Health Association, Washington, DC, pp. 291–7.

Semanchek, J.J. and Golden, D.A. (1996) Survival of *Escherichia coli* O157:H7 during fermentation of apple cider. *J. Food Prot.*, **59**(12), 1256–9.

Sewell, A.M. and Farber, J.M. (2001) Foodborne outbreaks in Canada linked to produce. *J. Food Prot.*, **64**, 1864–77.

Shah, F.H. and Hamid, A. (1989) Aflatoxins in various foods and feed ingredients. *Pak. J. Sci. Ind. Res.*, **32**, 733–6.

Sharman, M., Macdonald, S., Sharkey, A.J. and Gilbert, J. (1994) Sampling bulk consignments of dried figs for aflatoxin analysis. *Food Addit. Contam.*, **11**, 17–23.

Sharman, M., Patey, A.L., Bloomfield, D.A. and Gilbert, J. (1991) Surveillance and control of aflatoxin contamination of dried figs and fig paste imported into the United Kingdom. *Food Addit. Contam*, **8**, 299–304.

Sharman, M., Macdonald, S., Sharkey, A.J. and Gilbert, J. (1994) Sampling bulk consignments of dried figs for aflatoxin analysis. *Food Addit. Contam.*, **11**, 17–23.

Shellabear, C.K. and Shah, A.J. (1997) *Cyclospora cayatenensis:* an emerging food pathogen. in *Proceedings of the Sixth ASEAN Food Conference*. SIFST, Singapore, pp 99–106.

Sivapalasingam, S., Barrett, E., Kimura, A., et al., (2003) A multistate outbreak of *Salmonella enterica* serotype Newport infection linked to mango consumption: impact of water-dip disinfection technology, *Clin. Infect. Dis.* **37**, 1585–90.

Smith, W.L. (1962) Chemical treatments to reduce postharvest spoilage of fruits and vegetables. *Bot. Rev.*, **28**, 411–45.

Smoot, J.J. and Segall, R.H. (1963) Hot water as a postharvest treatment of mango anthracnose. *Plant Dis. Rep.*, **47**, 739–42.

Snowdon, A.L. (1990) *A Colour Atlas of Post-harvest Diseases and Disorders of Fruits and Vegetables, volume 1*, General Introduction and Fruits. Wolfe Scientific, London.

Snowdon, A.L. (1991) *A Colour Atlas of Post-harvest Diseases and Disorders of Fruits and Vegetables, volume 2*, Vegetables. Wolfe Scientific, London.

Soave, R., Herwaldt, B.L. and Relman, D.A. (1998) Cyclospora. *Infect. Dis. Clin. N. Am.*, **12**, 1–13.

Splittstoesser, D.F. (1987) Fruits and fruit products, in *Food and Beverage Mycology*, 2nd ed. (ed. L.R. Beuchat), Van Nostrand Reinhold, New York, pp. 101–28.

Steiner, W.E., Rieker, R.H. and Battaglia, R. (1988) Aflatoxin contamination in dried figs: distribution and association with fluorescence. *J. Agric. Food Chem.*, **36**, 88–91.

Sterling, C.R. and Ortega, Y.R. (1999) *Cyclospora*: an enigma worth unravelling. *Emerg. Infect. Dis.*, **5**, 48–53.

Sugar, D. and Spotts, R.A. (1993) The importance of wounds in infection of pear fruit by *Phialophora malorum* and the role of hydrostatic pressure in spore penetration of wounds. *Phylopathology*, **83**, 1083–6.

Swain, S. and Powell, D.A. (2001) Papaya Ringspot Virus Resistant Papaya: A Case Study. Technical report. http://www.plant.uoguelph.ca/safefood/gmo/papayarep.htm.

Tassou, C.C., Panagou, E.Z. and Katsaboxakis, K.Z. (2002) Microbiological and physicochemical changes of naturally occurring black olives fermented at different temperatures and NaCl levels in the brines. *Food Microbiol.*, **19**, 605–15.

Tēren, J., Varga, J., Hamari, Z., Rinyu, E. and Kevei, F. (1996) Immunochemical detection of ochratoxin A in black *Aspergillus* strains. *Mycopathologia*, **134**, 171–6.

Thayer , D.W. and Rajkowski, K T. (1999) Developments in irradiation of fresh fruits and vegetables. *Food Technol.*, **53**(11), 62–5.

Thompson, T.L, Engelhead, W.E. and Pivnick, H. (1955) Pustule formation by lactobacilli on fermented vegetables. *Appl. Microbiol.*, **3**, 314–6.

Ukuku, D.O. and Fett, W. (2002). Behavior of Listeria monocytogenes inoculated on cantaloupe surfaces and efficacy of washing treatments to reduce transfer from rind to fresh-cut pieces. *J. Food Prot.*, **65**, 924–30.

Van der Zwet, T. and van Buskirk, P. D. (1984) Detection of endophytic and epiphytic *Erwinia amylovora* in various pear and apple tissues. *Acta Horticul.*, **151**, 69–77.

Vasse, J., Frey, P. and Trigalet, A. (1995) Microscopic studies of intercellular infection and protoxylem invasion of tomato roots by *Pseudomonas solanacearum*. *Mol. Plant Microbe Interact.*, **8**, 241–51.

Vaughn, R.H., Won, W.D., Spencer, F.B., Pappagianus, D., Foda, I.O. and Krumperman, P.H. (1953) *Lactobacillus plantarum*, the cause of "yeast spots" on olives. *Appl. Microbiol.*, **1**, 82–5.

Vaughn, R.H., Jakubczyk, T., MacMillan, J.D., Higgins, T.E., Davé, B.A. and Crampton, V.M. (1969a) Some pink yeasts associated with softening of olives. *Appl. Microbiol.*, **18**, 771–5.

Vaughn, R.H., King, A.D., Nagel, C.W., Ng, H., Levin, R.E., MacMillan, J.D. and York, G.K. (1969b) Gram negative bacteria associated with sloughing, a softening of California ripe olives. *J. Food Sci.*, **34**, 771–5.

Vaughn, R.H., Stevenson, K.E., Davé, B.A. and Park, H.C. (1972) Fermenting yeasts associated with softening of olives. *Appl. Microbiol.*, **23**, 316–20.

Wahem, I.A. (1990) The effects of acidification and sugar addition on quality attributes of canned tomatoes. *J. Food Process. Preserv.*, **14**, 1–15.

Walderhaug, M.O., Edelson-Mammel, S.G., DeJesus, A.J., Eblen, B.S., Miller,A.J. and Buchanan, R.L. (2000). Routes of infiltration, survival, and growth of *Salmonella enterica* serovar hartford and *Escherichia coli* O157:H7 in oranges, *Abstract, Presented at International Association for Food Protection Meeting*, August 6–9, 2000.

Wallbridge, A. (1981) Fungi associated with crown-rot disease of boxed bananas from the Windward Islands during a two-year survey. *Trans. Br. Mycol. Soc.*, **77**, 567–77.

Wei, C.I., Huang, T.S., Kim, J.M., Lin, W.F., Tamplin, M.L. and Bartz, J.A. (1995) Growth and survival of *Salmonella* Montevideo on tomatoes and disinfection with chlorinated water. *J. Food Prot.*, **58**, 829–36.

Weissinger, W.R., Chantarapanont, W. and Beuchat, L.R. (2000) Survival and growth of *Salmonella* Baildon in shredded lettuce and diced tomatoes and the effectiveness of chlorinated water as a sanitizer. *Int. J. Food Microbiol.*, **62**, 123–31.

Wells, J.M. and Butterfield, J.E. (1997). *Salmonella* contamination associated with bacterial soft rot of fresh fruits and vegetables in the market place. *Plant Dis.*, **81**, 867–72

West, N.S., Gililland, J.R. and Vaughn, R.H. (1941) Characteristics of coliform bacteria from olives. *J. Bacteriol.*, **41**, 341–53.

Wiley, R.C. (1994) Introduction to minimally processed refrigerated fruits and vegetables. in *Minimally Processed Refrigerated Fruits and Vegetables* (ed. R.C. Wiley) Chapman & Hall, New York, pp. 1–14.

Wisniewsky, M.A., Glatz, B.A., Gleason, M.L. and Reitmeier, C.A. (2000) Reduction of *Escherichia coli* O157:H7 counts on whole fresh apples by treatment with sanitizers. *J. Food Prot.* **63**, 703–8.

Zhuang, R.-Y. and Beuchat, L.R. (1996) Effectiveness of trisodium phosphate for killing *Salmonella* Montevideo on tomatoes. *Lett. Appl. Microbiol.*, **22**, 97–100.

Zhuang, R., Beuchat, L.R., Chinnan, M.S., Shewfelt, R.L. and Huang, Y.-W. (1996) Inactivation of *Salmonella* Montevideo on tomatoes by applying cellulose-based edible films. *J. Food Prot.*, **59**, 808–12.

van der Zwet, T. and van Buskirk, P.D. (1984) Detection of endophytic and epiphytic *Erwinia amylovora* in various pear and apple tissues. *Acta Hortic.*, **151**, 69–77.

7 Spices, dry soups, and oriental flavorings

This chapter deals with spices, herbs, and dry vegetable seasonings, and covers some oriental flavorings such as soy sauces, fish pastes, and shrimp sauces. It also outlines the microbiology of dry soups and gravy mixes.

I Spices, herbs, and dry vegetable seasonings

A Definitions

The International Standard Organization (ISO) defines spices as "vegetable products or mixtures thereof, free from extraneous matter, used for flavoring, seasoning, and imparting aroma in foods" (ISO, 1995).

In the broadest sense "spices" are any parts of various aromatic plant products, with the exception of the leaves, used primarily to season, flavor, or to impart an aroma to foods. The term applies equally to the spices in the whole, broken, or ground form. Most are fragrant, aromatic, and pungent, consisting of rhizome, root, bark, leaf, flowers, fruit, seed, and other parts of the plant.

The so-called "true spices" are products of tropical plants and may be fruits—peppers, allspice, coriander; arils—mace; flower buds—cloves; rhizomes—ginger; or barks—cassia, cinnamon. Spice seeds (e.g. nutmeg, fenugreek, mustard, caraway, celery, and aniseed) may be from either tropical or temperate areas.

Spice essential oils are the volatile aromatic substances prepared by steam distillation of ground spices. Spice oleoresins comprise both the volatile and non-volatile resins present in spices and prepared by solvent extraction of coarsely ground spices using suitable food grade solvents like hexane and ethylene-dichloride.

Condiments are spices alone, or blends of spices, which have been formulated with other flavor potentiators to enhance the flavor of foods. They can be either simple such as celery, garlic, or onion salt, or complex mixtures such as chili sauces, mustard, or chutney. Compounding flavors are mixtures of essential oils and synthetic aromatics used in flavor compositions for flavoring non-traditional candies, biscuits, and soft drinks.

The characteristics and nomenclature of all recognized spices and condiments were reviewed by Pruthi (1983).

Herbs are generally defined as leafy parts of soft-stemmed plants (e.g. oregano, marjoram, basil, curry leaves, mints, rosemary, and parsley) of various perennial and annual plants. Herbs are classified as culinary or medicinal herbs depending on their usage. Culinary herbs may, or may not, be strongly aromatic in character, but those used in flavoring of food have distinctive aromatic characteristic. Herbs in medicinal use refer to all plants with medicinal value.

B Important properties

Histologically and chemically, spices and herbs are too diverse to be described here; concise presentations of these qualities are available (Peter, 2001; Tainter and Grenis, 2001).

Spices are of interest to microbiologists for three main reasons. They may (i) exhibit antimicrobial activity and occasionally aid in preservation; (ii) support mold growth if improperly dried or allowed to become moist in storage, leading to spoilage and sometimes mycotoxin production; or (iii) contain excessive numbers of microorganisms that may cause spoilage or more rarely, disease, when introduced into food.

Antimicrobial activities of spices and their effect in foods. The role of spices and herbs as antimicrobial agents is discussed in numerous publications. Such studies can be subdivided into four categories: (i) screening studies, (ii) studies on combinations of specific food-borne bacteria (usually pathogens) and specific spices, (iii) studies on antifungal activities; and (iv) studies on specific active components (Tainter and Grenis, 2001).

Spices and herbs containing the most inhibitory essential oils are cloves, thyme, oregano, cinnamon, allspice, cumin, and caraway (Table 7.1). Few others such as onions and garlic are also known to contain such compounds. However, most herbs and many spices show only little or no antimicrobial activity. The specific antimicrobial agents vary depending on the spice or herb but are often identical with the most important flavor compounds (Peter, 2001). They are usually related to major components such as eugenol and eugenol derivatives, but others such as allicin, allyl isothiocyanate, and anethol cinnamic aldehyde have been described and are summarized by Farkas (2000). The composition and content of essential oils vary from spice to spice and even within the same spice depending on agricultural practices, geographic and climatic conditions during the growing season (Lawrence, 1978).

Table 7.1 Concentrations of essential oils in some spices and antimicrobial activity of active components

Spice	Essential oil in whole spice (%)	Antimicrobial compounds in distillate or extract: Compound	%	Antimicrobial concentration (ppm) lab media	Organisms
Allspice (*Piementa dioica*)	3.0–5.0	Eugenol Methyl eugenol	73–78 9.6	1000 (G) 150 (I)	Yeast, *Acetobacter*,[a] *Cl. botulinum* 67B[b]
Cassis (*Cinnamomum cassis*)	1.2	Cinnamic aldehyde Cinnamyl acetate	75–90	10–100 (G)	Yeast, *Acetobacter*[a]
Clove (*Syzgium aromaticum*)	16.0–19.0	Eugenol Eugenol acetate	72–92	1000 (G) 150 (I)	Yeast, *Cl. botulinum*,[b] *V. parahaemolyticus*[c]
Cinnamon bark (*Cinnamomum zeylanicum*)	0.5–1.0	Cinnamic aldehyde Eugenol	65–76 4–10	10–1000 (G) 100 (I)	Yeast, *Acetobacter*,[a] *Cl. botulinum* 67B,[b] *L. monocytogenes*[d,e]
Garlic (*Allium sativum*)	0.3–0.5	Allyl sulfonyl Allyl sulfide		10–100 (I)	*Cl. botulinum* 67B,[b] *L. monocytogenes*,[d-f] Yeast, bacteria[e]
Mustard (*Sinapis nigra*)	0.5–1.0	Allyl isothionate	90	22–100	Yeast, *Acetobacter*,[a] *L. monocytogenes*[d]
Oregano (*Origanum vulgare*)	0.2–0.8	Thymol Carvacrol	 60–85	100 (G) 100–200 (I)	*V. parahaemolyticus*,[g] *Cl. botulinum* A, B, E[f]
Paprika (*Capsicum annuum*)		Capsicidin		100 (I)	*Bacillus*
Thyme (*Thymus vulgaris*)	2.5	Thymol Carvacrol		100 (G) 100 (I)	*V. parahaemolyticus*,[g] *Cl. botulinum* 67B,[b] Gram + bacteria,[h] *Asp. parasiticus*, *Asp. flavus*,[c,d,i] aflatoxin B_1 and G_1

[a]Blum and Fabian (1943). [b]Ismaiel and Pierson (1990a,b). [c]Farag *et al.* (1989b). [d]Karapinar and Aktug (1987). [e]Tynecka and Gos (1973). [f]Bahk *et al.* (1990). [g]Beuchat (1976). [h]Gál (1968, 1969). [i]Farag *et al.* (1989a).
(G) = Germicidal; (*I*) = Inhibitory

The degree of inhibition depends on several factors such as the concentration of the active substance(s), the food matrix, the method used to determine the inhibitory activity, the solubility of the components in the different components of the food or the target microorganisms. In view of the very large number of specific studies (hundreds) published, it is certainly not possible to provide an exhaustive and balanced summary here. However, Shelef (1983) reviewed publications prior to the 1980s, and more recent reviews by Hirasa and Takemasa (1998), Smith-Palmer *et al.* (1998), Hammer *et al.* (1999), Dorman and Deans (2000), Tainter and Grenis (2001) and Kalemba and Kunicka (2003) provide good summaries of older and recent research and a starting point to access more detailed and specific papers.

There is little available literature on the practical utilization of spices as antimicrobials in foods. This is probably due to the fact that, in general, the concentrations needed to achieve efficient inhibition frequently impact negatively on the organoleptic characteristics of the food. Consequently, the concentrations of essential oils in spiced foods (Table 7.1) are generally too low to prevent microbial growth (Salzer *et al*, 1977; Zaika, 1988) and levels found inhibitory in laboratory media are often insufficient to cause inhibition in food matrices (Evert Ting and Deibel, 1992).

The mechanisms of inhibition of germination of spores, or of growth of vegetative forms, by active components of essential oils are varied, reflecting their chemical diversity. Most studies have been devoted to the effect of phenolic compounds such as thymol, carvacrol, or eugenol (Ultee *et al.*, 1998,1999; Lambert *et al.*, 2001; Walsh *et al.*, 2003).

C *Methods of processing and preservation*

Herbs and spices have traditionally been traded as dry products allowing for easy transportation and storage. This is still the case today. Numerous herbs and spices are, however, grown in humid tropical regions, which can make drying difficult, in particular if there is limited availability of equipment for mechanical drying.

The main steps of processing are cleaning, curing, drying, grinding, and pulverizing. Other steps, such as fermentation, are applied in a few instances, for example, for cassia bark to facilitate removal of outer layers, or for allspice berries to develop color and appearance.

Cleaning is required to remove insects, stones, twigs, and soil. Equipment used for this operation utilizes the physical differences between spices and foreign material. Magnets, sifters, de-stoners, air tables and separators, indent and spiral separators are the most frequently used pieces of equipment. Depending on the herbs and spices handled, several items of equipment are used in combination to eliminate different types of foreign material. A detailed description of the different possibilities, as well as of the subsequent grinding operations, is provided by Tainter and Grenis (2001).

In addition to dried products, fresh or frozen herbs are commercialized as well. In the case of fresh products, distribution is often restricted to local areas utilising refrigeration or freezing of herbs, wider distribution is possible. Spices such as garlic are also processed into shelf stable products, using usually mild heat processing, alone or in combination with acidification.

D *Types of final products*

Spices are frequently used whole or ground. For many industrially prepared foods, concentrated spice extracts, either volatile oils or oleoresins, are used. These extracts have numerous advantages: they can be standardized for flavor strength or color, they do not represent a problem in terms of foreign materials, and are almost sterile. Such extracts are generally obtained by grinding the spice, extracting with water or solvents which are then removed by drying or distillation. Such extracts are either used as solutions or dried, or encapsulated on carriers such as salt, dextrose, maltodextrin, or gum arabic. The manufacture of such extracts is described in detail by Tainter and Grenis (2001) and Peter (2001).

E Initial microflora

There is little information on the initial microflora of herbs and spices in the field before harvest. In the absence of such data, it is assumed that the initial microflora is similar to that of other agricultural products harvested under similar soil and climate conditions. The relatively few studies of the microbiology of spice plants are largely restricted to the etiological agents of diseases.

Spices and herbs are presumed to contain those microorganisms indigenous to the soil and plants in which they are grown and that are capable of surviving the drying process. Sources of contamination are dust and soil, fecal material from birds, rodents, and other animals, and possibly the water used in some processes such as the soaking of peppercorns to prepare white pepper.

Microbial counts vary according to the region of origin, the year of production, and the harvest and storage conditions prior to drying. Observed counts are thus a reflection of the original bioload, of growth, as well as of die-off. Drying as well as subsequent storage reduces the number of vegetative cells and this die-off is probably enhanced by oxidation and the presence of active compounds in herbs and spices (Farkas, 2000). The remaining flora consists mainly of spore-forming bacteria and molds because of their ability to survive over prolonged periods in dry materials.

Results of surveys of microbial contamination of untreated spices sampled in processing establishments or at import are shown in Table 7.2 or in an updated version published by Farkas (2000), which,

Table 7.2 Distribution (%) of Aerobic Plate Counts (APC) and mold counts in untreated spices[a,b]

	Aerobic Plate Count (cfu/g)								Mold count (cfu/g)						
Spice	*N*[c]	<2[d]	2–3	3–4	4–5	5–6	6–7	>7	*N*	<2	2–3	3–4	4–5	5–6	6–7
Allspice	33	–	–	3	7	46	42	3	27	37	22	15	18	7	–
Anise	22	–	–	23	36	36	5	–	16	56	25	6	13	–	–
Basil	21	–	–	–	14	48	38	–	17	65	24	6	6	–	–
Bay	41	5	5	46	34	7	3	–	35	34	29	26	11	–	–
Capsicum (chili)	57	–	2	9	28	18	31	12	59	44	15	22	7	7	5
Caraway	17	–	12	35	29	18	6	–	14	57	7	29	7	–	–
Cardamon	15	7	13	13	27	7	33	–	15	67	13	20	–	–	–
Cassia	36	6	66	11	14	3	–	–	20	55	5	25	5	10	–
Cinnamon	42	2	5	19	48	21	2	2	51	18	33	43	6	–	–
Cloves	28	32	21	25	18	4	–	–	26	88	–	8	4	–	–
Coriander	30	–	3	3	30	37	13	13	23	4	–	61	26	9	–
Cumin	12	–	–	–	33	42	25	–	8	25	–	62	–	13	–
Fennel	16	13	6	13	43	13	13	–	11	73	27	–	–	–	–
Fenugreek	10	–	–	20	30	30	20	–	8	13	37	25	–	25	–
Garlic	32	–	–	16	47	28	9	–	15	60	49	–	–	–	–
Ginger	33	3	9	21	15	45	7	–	28	57	14	18	11	–	–
Mace	28	–	–	43	50	7	–	–	22	59	14	23	4	–	–
Marjoram	21	–	5	–	19	43	28	5	14	–	7	64	29	–	–
Mustard	67	9	30	33	18	9	1	–	63	86	6	6	2	–	–
Nutmeg	45	11	20	44	16	4	4	–	33	52	27	12	9	–	–
Oregano	56	2	4	21	41	23	9	–	48	35	25	33	9	–	–
Paprika	80	–	–	–	11	9	62	18	61	44	44	7	5	–	–
Pepper (black)	108	–	–		3	5	50	42	82	32	10	28	5	2	23
Pepper (white)	42	–	–	12	26	57	5	–	44	2	2	34	36	23	2
Sage	17	–	–	11	41	41	6	–	14	7	21	21	50	–	–
Savory	10	–	–	40	50	10	–	–	6	–	67	33	–	–	–
Thyme	19	–	–	5	11	32	53	–	16	6	–	6	81	6	–
Turmeric	24	–	–	4	–	21	46	29	32	87	3	6	3	–	–

[a]Spices not treated with microbiocidal agents.

[b]Collated from published and unpublished data for dried spices analyzed in North America, Europe, the Middle East, and Japan

[c]*N* = number of samples: APC = 962, mold count = 808: most of the samples examined for mold count were also examined for APC.

[d]Numbers are $\log_{10}$: <2 = <100; 2–3 = 100–999; 3–4 = 1000–9999, etc.

Table 7.3 Total aerobic bacterial counts and spore counts in various spice samples[a]

	$\log_{10}$ (cfu/g) at 30°C	
Spice	Aerobic count	Spore count
Allspice	5.8	5.9
Caraway seed	5.2	3.4
Chili	6.0	5.8
Coriander I	6.4	5.9
Coriander II	6.0	4.5
Ginger	8.4	7.9
Marjoram	6.5	4.8
Mustard	5.8	5.7
Nutmeg	5.7	5.7
Paprika I	7.0	7.1
Paprika II	6.0	5.7
Paprika III	5.4	5.4
Paprika IV	5.0	4.5
Paprika V	4.8	4.3
Pepper, Black I	8.0	8.1
Pepper, Black II	7.5	7.4
Pepper, Black III	7.4	7.4
Pepper, White I	5.6	4.1
Pepper, White II	5.6	5.2
Pepper, White III	3.5	3.5
Mixed Spices	6.3	6.2

[a]From Neumayr *et al.* (1983).

however, does not differ greatly. Similar contamination patterns have been shown in other surveys performed in different regions (Hartgen and Kahlan, 1985; Shamshad *et al.*, 1985; Pafumi, 1986; Garcia *et al.*, 2001).

In many spices, most of the microbial flora consists of aerobic mesophilic spores, often representing >50% of the mesophilic aerobic counts (Table 7.3). Species most frequently found include *Bacillus subtilis, B. licheniformis, B. megaterium, B. pumilus, B. brevis, B. polymyxa*, and *B. cereus* (Goto *et al.*, 1971; Julseth and Deibel, 1974; Palumbo *et al.*, 1975; Seenappa *et al.*, 1979; Baxter and Holzapfel, 1982; Fábri *et al.*, 1985; Ito *et al.*, 1985; Shamshad *et al.*, 1985). For example, Sheneman (1973) reported that the bacterial flora of dried onions consists mainly of *Bacillus* species, such as *B. subtilis, B. licheniformis, B. cereus*, and *B. firmus*. The proportion of obligately anaerobic spore-formers is usually small (Inal *et al.*, 1975; Fábri *et al.*, 1985; Kovács-Domján, 1988).

Thermophilic anaerobes and aerobes are found occasionally, sometimes in moderate numbers (Kadis *et al.*, 1971, Pruthi, 1983, Kovács-Domján, 1988). Consequently, some spices are potentially prolific sources of high heat-resistant spores of bacteria, including thermoduric flat sours, putrefactive anaerobes, and "sulfide stinkers" (Krishnaswamy *et al.*, 1973), which reduce the stability of canned foods stored at tropical ambient temperatures.

Psychrotrophic or psychrophilic spore-formers are not common in spices or herbs, even in those having high mesophilic counts (Michels and Visser, 1976). Psychrotrophic non-spore-forming bacteria (i.e. capable of growth at <7°C) are generally less numerous in spices and herbs than mesophiles. de Boer *et al.* (1985) found that the psychrotrophic count was $<10^5$ cfu/g in 88% of 143 samples and $<10^3$ cfu/g in 53% of samples. Psychrotrophic counts of $\geq 10^6$ cfu/g were found in some samples of thyme, dill, coriander, basil, chervil, and liquorice.

A wide variety of mesophilic non-sporing bacteria may be present in spices (Julseth and Deibel, 1974). Coliforms are often found (Pafumi, 1986; see also Table 7.4) but *Escherichia coli* is less frequent (Baxter and Holzapfel, 1982; Schwab *et al.*, 1982). However, ~30% of black and white peppercorns

Table 7.4 Frequency of occurrence of coliforms and *Escherichia coli* in untreated spices[a–c]

	Coliforms		*Escherichia coli*	
cfu/g	Number of samples	%	Number of samples	%
$<10^{-1}$	110	48	180	79
10^{-1}–10^{1}	29	13	11	5
10^{1}–10^{2}	31	14	22	10
10^{2}–10^{3}	21	9	10	4
10^{3}–10^{4}	20	9	4	2
10^{4}–10^{5}	10	4	1	–
10^{5}–10^{6}	7	3	–	–
Total	228		228	

[a]Whole and ground spices.
[b]*E. coli* was found in the following untreated spices: basil, bay, capsicum, celery seed, coriander, cumin, dill, fennel, garlic, ginger, onion, oregano, parsley, pepper (black), rosemary, sage, and thyme.
[c]Collated from published and unpublished data for dried spices analyzed in North America, Europe, the Middle East, and Japan.

contained *E. coli* (Pafumi, 1986). Of 64 samples of parsley from retail and manufacturing outlets and market gardens in Germany, 30 contained *E. coli* (Käferstein, 1976). In a British survey, 42% of 100 samples from 10 different spices and herbs contained *E. coli* at levels of less than 10 cfu/g (Roberts *et al.*, 1982). In another study, 23% of 53 samples of spices and herbs contained more than cfu/g 10^4 Enterobacteriaceae (de Boer *et al.*, 1985). A survey on four selected spices (black peppercorns, white peppercorns, coriander, and fennel seed) imported into the United States did not show any relationship between the enteric microflora found in spices and that found in associated fecal pellets (Satchell *et al.*, 1989).

Fecal streptococci occur in about half of spice samples, usually in low numbers, and rarely exceeding 10^4 cfu/g (Masson, 1978; Baxter and Holzapfel, 1982). Staphylococci and lactic acid bacteria are rare in spices (Baxter and Holzapfel, 1982; Masson, 1978). Flannigan and Hui (1976) found up to 4×10^3 cfu/g thermophilic actinomycetes (mainly *Thermoactinomycetes vulgaris*) in 6 of 20 samples of ground spices.

Spices can play a major source of mold contamination in meat products (Eschmann, 1965; Christensen *et al.*, 1967; Hadlok, 1969). Mold counts of spices are not correlated with aerobic plate counts (Table 7.2). White pepper, black pepper, chili, and coriander seem to be most heavily contaminated with molds. Although the types of molds isolated from spices can vary greatly, *Eurotium* species, *Aspergillus niger*, and *Penicillium* spp. are usually most prevalent (Table 7.5).

Yeasts have been found in spices in low numbers only (Masson, 1978; Baxter and Holzapfel, 1982). *Candida huminicola*, *Can. parapsilosis*, and *Can. tropicalis* were reported to be the main yeast species in Indian spices (Krishnaswamy *et al.*, 1973).

F Primary processing

Harvesting and initial processing. The various spices require widely diverse harvest and post-harvest methods, which have an impact on the initial microbial content. The main processing factor influencing the microbiological quality of spices is the rapidity of drying to prevent spoilage while retaining their desirable characteristics.

Descriptions, characteristics, production, harvesting and initial processing, usage and functional properties of all major herbs and spices are discussed in great detail by Tainter and Grenis (2001) and Peter (2001).

Table 7.5 Main components of the mold flora of untreated spices as percentage of mold count[a,b]

	Mold[c] (cfu/g)	*Absidia* spp.	*Asp. can.*	*Asp. flav.*	*Asp. fum.*	*Asp. gl.*	*Asp. nid.*	*Asp. nig.*	*Asp. tam.*	*Asp. terr.*	*Asp. ver.*	*M.p* spp.	*Pen.* spp.	*Rhiz.* spp.
Allspice	7.0×10^4	3	+[d]	1	–	9	–	80	–	–	1	–	6	–
Anise	9.5×10^3	–	–	1	1	55	–	3	–	1	2	–	33	–
Cardamon	1.6×10^3	–	–	3	–	64	12	12	–	–	–	–	–	9
Capsicum (chili)	3.9×10^4	–	–	4	–	69	1	17	–	–	1	1	1	1
Cinnamon	8.7×10^4	33	–	+	–	–	–	62	+	–	+	–	2	–
Coriander	1.3×10^5	4	7	5	1	67	1	1	–	2	10	–	2	–
Cumin	1.5×10^3	–	–	–	–	62	7	7	–	7	–	–	17	–
Fennel	6.7×10^3	2	3	2	2	62	4	6	2	2	5	–	–	–
Fenugreek	2.5×10^3	–	2	–	–	16	6	8	–	2	–	2	60	2
Ginger	1.7×10^3	–	15	3	–	32	–	9	–	–	–	–	35	3
Mace	8.0×10^2	12	–	–	–	88	–	–	–	–	–	–	–	–
Nutmeg	6.2×10^4	–	–	4	–	70	–	12	8	2	–	–	3	–
Paprika	5.5×10^2	27	–	–	–	27	–	10	–	–	–	–	18	18
Pepper (black)	6.4×10^5	–	2	1	–	92	–	+	+	–	1	–	2	–
Pepper (white)	6.5×10^4	–	16	7	8	2	13	11	2	3	12	–	21	–
Turmeric	2.0×10^1	–	–	–	–	100	–	–	–	–	–	–	–	–

[a]From Flannigan and Hui (1976); see also Pal and Kundu (1972), Moreau and Moreau (1978), Dragoni (1978).
[b]Minor components were *Thermoascus crustaceus* in black pepper; *Talaromyces dupontii* in fenugreek; *Thermomyces lanuginosus* in white pepper, *Alternaria altenata* in red pepper, *Fusarium poae* in fennel, *Syncephalastrum racemosum* in ginger and nutmeg.
[c]*Asp., Aspergillus*; *can., candidus*; *flav., flavus*; *fum., fumigatus*; *gl., glaucus* (group); *nid., nidulans*; *nig., niger*; *tam., tamarii*; *terr., terreus*; *ver., versicolor*; *M.p, Mucor pusillus*; *Pen., Penicillium*; *Rhiz., Rhizopus*.
[d] +, present; –, not present.

A Codex Alimentarius Code of Hygienic Practice for Spices and Dried Aromatic Plants has been published. It describes requirements for environmental hygiene in the production/harvesting area, establishment design and facilities, personnel hygiene and health requirements, hygienic processing requirements, and end-product specifications. It is emphasized that, in addition to good agricultural practices, raw spices should be protected from contamination by human, animal, and other wastes which might constitute a hazard to health of the consumer through spices.

Spoilage

Spoilage of spices. Virtually no bacterial spoilage of spices occurs subsequent to harvesting and drying. However, fungal spoilage may occur prior to drying, or during storage and shipping if relative humidity and temperature are high, or localized wetting occurs.

Mold counts of untreated spices (turmeric, rosemary, and white pepper) held in polyethylene pouches with humidities >80% increased up to 10^8 cfu/g during 1–3 months of storage at 30–35°C (Ito *et al.*, 1985). Seenappa and Kempton (1980a,b) observed that during storage of dried whole red peppers at 70% relative humidity, *Eurotium* species grew and colonized stalks, pods, and seeds. At 85% relative humidity, *Eurotium* species were replaced by *Asp. niger*, *Asp. flavus*, and *Asp. ochraceus*. At 95% relative humidity, *Asp. flavus* and *Asp. ochraceus* or *Asp. flavus* alone were the predominant fungal contaminant.

Insect infestation may contribute to the biodeterioration of spices, by serving as a source of mold spores (Seenappa *et al.*, 1979).

Emulsions of mixed essential oils can sometimes support bacterial growth to 10^7–10^8 cfu/mL in the aqueous phase when inhibitory molecules (e.g. cinnamic aldehyde in oil of cinnamon) partition into a non-inhibitory oil phase (e.g. oil of nutmeg). Adjustment to pH 4 with lactic acid effectively controls this problem (Pirie and Clayson, 1964).

Spoilage of foods by microorganisms from spices. Spices containing excessive numbers of bacterial spores have been associated with the spoilage of canned foods (Bean and Salvi, 1970; Julseth and Deibel, 1974). Moldy spices introduce viable spores that may grow in some products, and may introduce moldy off-flavors and enzymes that impact the textural properties of foods (*e.g.* pectinase and protease). The quality of processed meats may be adversely affected by the bacteria and molds introduced with spices (Palumbo *et al.*, 1975). However, the potential for spoilage depends on whether the meat is canned and pasteurized or heated to obtain a shelf stable product; fermented; or cooked and refrigerated. Even when there is little likelihood of contaminated spice causing spoilage (*e.g.* in dry gravy bases or dehydrated soups), spices may introduce microorganisms considered undesirable to industrial or regulatory interests (Kadis *et al.*, 1971; Surkiewicz *et al.*, 1972, 1976).

Pathogens. Herbs and spices are not major contributors to food-borne disease. However, they occasionally contain bacteria that can cause infections. Spices are frequently contaminated with toxigenic molds and may sometimes even contain mycotoxins.

Bacteria. Spore-forming microorganisms that are capable of causing gastroenteritis when ingested in large numbers are found in spices, but usually in low numbers. A typical example is *B. cereus*, which is often reported (de Boer *et al.*, 1985; Powers *et al.*, 1976; Roberts *et al.*, 1982; Pafumi, 1986; Kovács-Domján, 1988), and which must multiply to at least 10^5–10^6 cfu/g in the food to which the spice is added. From 110 samples of various spices tested for prevalence and levels of *B. cereus*, the organism was found in 53% of the samples (50–8500 cfu/g) with 89% of the isolates being enterotoxigenic (Powers *et al.*, 1976). In extreme cases, counts up to 10^5 cfu/g have been found (Baxter and Holzapfel, 1982; Pafumi, 1986). *B. subtilis* and *B. licheniformis* commonly found in many spices have been linked to food-borne gastroenteritis in a few cases (Kramer *et al.*, 1982) but none of them directly to spices.

A relatively high incidence of *Clostridium perfringens* has also been found in several spices (Powers *et al.*, 1975; Leitao *et al.*, 1973–1974; de Boer *et al.*, 1985; Roberts *et al.*, 1982; Salmeron *et al.*, 1987) but usually with numbers <500 cfu/g and rarely >1000 cfu/g. Since spores of *Cl. perfringens* can survive cooking temperatures and will grow in foods held at room temperatures or up to 50°C, spices must be considered as a potential issue for such foods.

Clostridium botulinum has been at the origin of outbreaks related to spices in oil such as garlic (St. Louis *et al.*, 1988; Morse *et al.*, 1990; Lohse *et al.*, 2003) or mustard prepared with fried lotus rhizomes (Otofuji *et al.*, 1987).

Salmonellae have been found in several herbs and spices (Guarinao, 1972; Leitao *et al.*, 1973–1974; Bockemühl and Wohlers, 1984; Pafumi, 1986; Satchell *et al.*, 1989; Bruchmann, 1995) with prevalences ranging between 2% and7% being reported. The presence of salmonellae is of particular concern when herbs and spices are consumed raw or added to prepared foods without further cooking (D'Aoust, 1994, 2000). An oubreak was caused by *S.* Thompson fresh cilantro (Campbell *et al.*, 2001). The closely related *Citrobacter freundii* was at the origin of an outbreak with sandwiches prepared with green butter containing contaminated parsley (Tschappe *et al.*, 1995). Black and white pepper have been implicated as vehicles for the spread of *Salmonella* Weltevreden causing serious cases of salmonellosis (Laidley *et al.*, 1974; Severs, 1974; WHO, 1973, 1974). Black pepper contaminated with *S.*Oranienburg was also at the origin of an outbreak in Norway in 1981–1982 involving over 120 patients and causing one fatality (CDC, 1982; WHO, 1982; Gustavsen and Breen, 1984).

In 1993, a nationwide outbreak of salmonellosis occurred in Germany, which was traced to contaminated paprika originating from South America, and paprika-powdered potato chips with as few as ca. 0.04 salmonellae per gram (Lehmacher *et al.*, 1995). Of the estimated 1000 cases, children below 14 years of age were principally affected. *S.* Saint Paul, *S.* Rubislaw, S. Javiana, and monophasic and non-motile strains of rare *Salmonella* O-groups were isolated from both paprika products and patients.

Staphylococcus aureus is rarely found in dry spices (Julseth and Deibel, 1974; Powers *et al.*, 1976). This is true also for *Listeria monocytogenes*, which was found only at levels <0.04 cfu/g in different herbs and spices (Benezet *et al.*, 2001).

The levels of pathogenic microorganisms reported for a number of spices probably reflects an under reporting of bacteria of public health importance, due to methodological limitations. The presence of inhibitory compounds in a number of spices require special techniques (e.g. resuscitation of injured cells and overcoming inhibition at low dilutions), but still may underestimate the prevalence (Wilson and Andrew, 1976). It has been suggested that diluted spice pre-enrichment ratios of 1:1000 are necessary for cloves, pimento, cinnamon, oregano, and mustard seed to isolate *Salmonella* with confidence (Pafumi, 1986).

Molds. A relatively high incidence of toxigenic molds, including *Asp. flavus, Asp. parasiticus, Asp. fumigatus, Asp. ochraceus, Penicillium citrinum*, and *Pen. islandicum*, has been reported in some spices (Christensen, 1972; Mislivec *et al.*, 1972; Pal and Kundu, 1972; Shank *et al.*, 1972a,b; Bhat *et al.*, 1987; Anisa Ath-kar *et al.*, 1988).

Aflatoxins have been detected in a range of spices, e.g. black pepper, ginger, turmeric, celery seed, nutmeg, coriander, red pepper, cumin seed and mustard seed (Scott and Kennedy, 1973, 1975; Flannigan and Hui, 1976; Seenappa and Kempton, 1980a,b; Awe and Schranz, 1981; Emerole *et al.*, 1982; Misra, 1987; Misra *et al.*, 1989; Sahay and Prasad, 1990), although the levels recorded were generally low. While certain spices and herbs, especially cinnamon, cloves, and possibly oregano, inhibit mycelial growth and subsequent toxin production, others, particularly sesame seed, ginger, and rosemary, appear to be conducive to aflatoxin production (Llewellyn *et al.*, 1981; Buchanan and Shepherd, 1982).

Nutmeg and red pepper appear to be especially prone to aflatoxin production (Seenappa and Kempton, 1980b), but levels reported are usually $<25\ \mu g/kg$ (Beljaars *et al.*, 1975). In a survey of 21 different imported spices by the U.S. Food and Drug Administration (FDA), nutmeg and chili were found to contain detectable levels of aflatoxins most frequently (Wood, 1989). Total aflatoxin levels as high as $700\ \mu g/kg$ in Nigeria (Emerole *et al.*, 1982) and $966\ \mu g/kg$ in Thailand (Shank *et al.*, 1972b) have been reported. A French survey implicated pepper as a source of toxigenic *Asp. flavus*, which produced high levels of aflatoxin in sausages and pepper cheese (Jacquet and Teherani, 1974). Care must to be taken when extrapolating the results of laboratory studies in which spices were crushed and/or sterilized because this may make them more susceptible to toxigenic molds than when in the natural state.

Although the widespread use of spices makes it important to control contamination by aflatoxigenic and other mycotoxin-producing fungi, the actual ingestion of mycotoxins with spices is generally low in relation to that which occurs in staples such as cereals.

G Processing

Large quantities of herbs and spices are traded without further processing. However, depending on the final use, for example, in ready-to-eat products or in products where growth is possible, spices free of pathogens and very low counts may be required. A number of technologies are known and are, or have been, applied to destroy pathogens, in particular vegetative pathogens such as *Salmonella*, and to reduce microbial loads.

Summaries of available technologies are provided by Gerhardt (1994), Hirasa and Takemasa (1998), Peter (2001), and Tainter and Grenis (2001).

Gas treatment. In some countries, treatment of spices with low levels of methyl bromide or ethylene oxide has been applied to destroy insects. For many years, the most widely used method to destroy microorganisms in dry food ingredients was fumigation with ethylene oxide or, to a much lesser extent, propylene oxide (Mayr and Suhr, 1972; Gerhardt and Ladd Effio, 1982, Farkas, 1998). Because of the extreme flammability of ethylene oxide, various non-flammable mixtures of ethylene oxide in inert gases (carbon dioxide or a chlorinated hydrocarbon) were applied. These inert gases do not add to or detract from the biocidal activity of ethylene oxide. Such fumigation treatments were normally carried out in specially designed vacuum chambers. The concentration range of ethylene oxide used in treatments for microorganisms in dry food commodities is 400–1000 mg/L. Spices thus treated are frequently, but erroneously, termed "sterile".

Under industrial conditions, ethylene oxide treatment reduces the aerobic plate count (per gram) of spices by 10^1–10^4-fold, depending on the type of spice, the composition of the microflora, and conditions of the treatment. Bacterial spores are only marginally more resistant than vegetative cells (Blake and Stumbo, 1970; Werner *et al.*, 1970). Decreases in mold counts are usually in the range of 10^2–10^3-fold per gram, and occasionally greater. The rate of destruction of microbial cells depends on the concentration of the fumigant, temperature, relative humidity of the atmosphere in the fumigation chamber, the moisture content of the product treated (degree of dryness of the microbial cells), the porosity of the product, and the permeability of the packaging material (Hoffman, 1971; Russell, 1971).

The moisture content of the spice to be treated should be as high as possible but compatible with keeping quality (Guarino, 1972). The temperature should be elevated slightly to ~25–30°C, to increase the rate of destruction of microorganisms (Coretti and Inal, 1969). At 20–25°C, fumigation for 6–7 h is required, although this varies depending on the microflora present (Hadlok and Toure, 1973).

Chamber temperatures of 50–60°C may be used; but it is unlikely that the center of bags or barrels of spice in the chamber attain these temperatures. A mixture of ethylene oxide and methyl formate was recommended for those few spices (turmeric and mustard seed) whose color and flavor are adversely affected by ethylene oxide (Mayr and Suhr, 1972).

In some countries, propylene oxide is used in preference to ethylene oxide. On a weight basis, it is less bactericidal than ethylene oxide, but it is less likely to form toxic by-products.

Due to toxicological considerations, the use of these gases is more and more discouraged (Gerhardt and Ladd Effio, 1983; Neumayr *et al.*, 1983; OSHA, 1984; EEC, 1989). For a more extensive discussion of gas treatment, see also ICMSF, (1980a, Vol. 1, Chapter 10).

Irradiation. Ultraviolet irradiation has little penetrating power and has limited effectiveness in reducing bacteria on spices (Walkowiak *et al.*, 1971) even with continuous agitation to expose surfaces (Eschmann, 1965).

Research and development over some 40 years on a large variety of dry food ingredients and herbs has proved that ionizing radiation is an effective process for destroying contaminating organisms (Farkas, 1988; Steele, 2001). For practical reasons, ionizing radiation applied to food is limited to gamma rays from isotopic sources such as ^{60}Co and ^{137}Cs, machine produced X-rays (of energies up to 5 MeV), or accelerated electrons (with energies up to 10 MeV). Gamma rays and X-rays have a high penetrating capacity when compared with accelerated electrons. Penetration depends on the kinetic energy of photons or electrons and on the density of the product to be treated. Except for differences related to penetration and exposure time, electromagnetic ionizing radiations and electron beams are equivalent in food irradiation and can be used interchangeably (Josephson and Peterson, 1982–1983; Urbain, 1986).

Depending on the design of the irradiation facility and requirements, products can be treated in bulk. Spices are best packaged before irradiation to avoid re-contamination after radiation treatment. Because treatment of food with ionizing radiation causes almost no temperature rise in the product, irradiation can be applied through packaging materials including those that cannot withstand heat.

The primary aim of radiation processing of spices is the inactivation of bacterial and mold spores. Depending on the number and type of microorganisms and the chemical composition of the commodity, a radiation dose of up to 20 kGy may be required to achieve commercial "sterility" (*i.e.* a total viable cell count of <10 cfu/g) in natural spices and herbs. However, doses of 3–10 kGy can reduce viable cell counts to a satisfactory level (from 10^5–10^7 cfu/g to $<10^3$–10^4 cfu/g) without affecting quality attributes (Zehnder and Ettel, 1982; Sugimoto *et al.*, 1986; Munasiri *et al.*, 1987; Farkas, 1988; Singh *et al.*, 1988; Narvaiz *et al.*, 1989; Ito and Islam, 1994, Nieto-Sandoval *et al.*, 2000). The number of bacterial spores normally decreases by at least 10^2-fold after irradiation with 5 kGy. There is little difference in the radiation resistance of aerobic spores most frequently occurring in spices (with an apparent overall D-value varying between <1.7 and 2.7 kGy). This is not much affected, at least from the practical point of view, by the water activity of the spice. In fact, D-values derived from irradiation of spice samples are similar to those obtained for related pure strains of aerobic spore-formers in aqueous systems (Briggs, 1966; Härnulv and Snygg, 1973). Slightly higher D_{10} values were obtained for electron beams or converted X-rays irradiation than for gamma-ray irradiation which was explained by the much higher dose rate of machine sources imparting less oxidation damage to microorganisms (Ito and Islam, 1994).

Sulfite-reducing clostridia, usually present in low ($<10^3$ cfu/g) populations, can be eliminated by 4 kGy (Neumayr *et al.*, 1983). Thermophilic spore-forming bacteria, of great importance to the canning industry, can be essentially eliminated with the same radiation doses as those necessary to cause a sufficient reduction of the total aerobic viable counts. Bacteria of the family Enterobacteriaceae are relatively radiation sensitive, even in dry ingredients, and in most cases a dose of ~5 kGy is sufficient for their elimination. A dose of 4–5 kGy can eliminate molds at least as effectively as ethylene oxide treatment. The germicidal efficiency of irradiation is much less dependent on moisture content or humidity than is ethylene oxide (Farkas *et al.*, 1973; Farkas and Andrássy, 1984).

Table 7.6 illustrates the effect of irradiation on the microbial content of spices and the effectiveness of radiation for the treatment of black pepper.

No post-irradiation recovery of surviving microorganisms has been observed during storage of irradiated spice samples. On the contrary, a further decrease of survivors has been observed in some cases (Bachman and Gieszczynska, 1973). The surviving microflora of spices treated with "pasteurizing" doses of ionizing radiation has lower heat and salt resistance, and is more fastidious relative to pH, moisture, and growth temperature requirements than that of untreated spices, which reduce the microflora's ability to survive and grow in processed food products (Farkas *et al.*, 1973; Kiss and Farkas, 1981; Farkas and Andrássy, 1985). The heat-sensitizing effect of irradiation increases with increase in radiation dose, and the weakening of the surviving microflora in irradiated dry ingredients is permanent and does not diminish during normal storage of the ingredients (Farkas and Andrássy, 1984).

A code of good irradiation practice for the control of pathogens and other microflora in spices, herbs, and other vegetable seasonings has been prepared by the International Consultative Group on Food Irradiation (ICGFI, 1988). For a more extensive discussion of irradiation treatments, see ICMSF (1980a, Vol. 1, Chapters 2 and 3).

Table 7.6 Microbial decontamination of black pepper by gamma radiation[a]

	$\log_{10}$ cfu/g at a dose of (kGy)					
Group of microorganisms	0	2	4	6	8	10
Total aerobic mesophiles	8.0	6.2	5.2	3.9	2.1	<1.8
Aerobic mesophilic spores						
(a) Surviving 1 min at 80°C	7.7	6.6	4.7	3.0	1.8	<1.8
(b) Surviving 20 min at 100°C	6.0	2.9	0.2	–	–	–
Anaerobic mesophilic spores						
(a) Surviving 1 min at 80°C	7.5	6.1	3.1	<1.8	<1.8	<1.8
(b) Surviving 20 min at 100°C	5.9	<1.8	<1.8	<1.8	<1.8	<1.8
Enterobacteriaceae	4.7	2.8	1.7	1.1	<−0.5	
Lancefield Group D streptococci	4.9	1.7	0.4	<−0.5	<−0.5	
Molds	4.6	<1.8	–	–	–	

[a]From Soedarman *et al.* (1984).

Other decontamination methods. Because of heat sensitivity of the delicate flavor and other essential components of many spices and herbs or some other specific functional properties, the normal heat sterilization process cannot be applied (Thiessen and Hoffmann, 1970; Thiessen, 1971; Maarse and Nijssen, 1980). Hot ethanol vapor has been suggested as an alternative "gaseous sterilization" treatment of natural spices and other foods (Wistreich *et al.*, 1975). This treatment may result in the desired antimicrobial effect with whole seeds, i.e. whole pepper, but it is not feasible for ground or leafy spices (Neumayr and Leistner, 1981). Another chemical method is the reduction of viable cell counts by acidification with hydrochloric acid followed by neutralization, i.e. *in situ* salt formation (Scharf, 1967). Although such a procedure has been examined for decontaminating paprika, and the resultant paste is suitable for use in some meat products (Huszka *et al.*, 1973), this treatment is applicable only over a very limited range of conditions and commodities.

Microwave treatment has little practical utility because microwave heating is seriously hampered at the low moisture content of dry commodities (Vajdi and Pereira, 1973). Due to the heterogeneity of products and the dielectric field, the heating effect is very uneven in these materials. This results in significant adverse changes in sensory quality (Neumayer *et al.*, 1983; Dehne *et al.*, 1990).

A process utilizing superheated steam created interest in the treatment of spice seed, berries, and roots or rhizomes (Dehne *et al.*, 1992a,b). However, heat/steam is less useful for products such as leafy herbs, which lose flavor and color, and ground products such as onion and garlic, which cake and must be re-milled. Another thermal method proposed to decontaminate spices is to treat them in an extruder (USP, 1985). Different pressure–time–temperature combinations have been tested, and the method is in commercial operation. Cooking extrusion according to a UK patent (G.B. 2236 6588, published 1993) achieves $>10^4$-fold reduction of viable counts by virtue of shear force, temperature shock, pressure differential, and probably the antimicrobial effect of spice oils within a high stress environment. This process can successfully handle color-sensitive materials, *e.g.* herbs and paprika, due to the short time at high temperature in the extruder (Tuley, 1991). Studies performed by Almela *et al.* (2002) have made use of HTST treatments under overpressure to treat paprika with only minimal losses in color.

A steam treatment has been developed for whole spices (Sorensen, 1987, 1989), which utilizes a specific fraction of extract from food grade beef bones as coating material to prevent losses of volatiles during the thermal process. The treatment is claimed to be feasible for most spices, with some exceptions where excessive darkening (paprika and garlic), loss of green color (dill), or development of roasted/cooked taste (onion) may cause problems.

Because thermal treatments required for a large reduction of viable cell counts result in serious losses of sensory or functional properties of products (Modlich and Weber, 1993), the various alternative modern heat processing methods described in this chapter are suitable mostly for cell-count reduction of slightly to moderately contaminated and unground spices, or, for special cases such as a reduction of molds. As noted by Dehne and Bögl (1993), the problem of highly contaminated natural spices cannot be addressed simply through the decontamination processes, and must include the introduction of suitable hygiene measures in the producer countries.

H CONTROL (spices, herbs, and dry vegetable seasonings)

Summary

Significant hazards	• *Salmonella* spp., others such as *Cl. botulinum*, *Cl. perfringens*, and *B. cereus* may be significant depending on the final use of the herbs and spices.

(*continued*)

H CONTROL (Cont.)

Summary

Control measures	
Initial level (H_0)	• Low levels of pathogens are likely to be present in untreated herbs and spices.
Reduction (ΣR)	• Cleaning, curing, drying, grinding, and pulverizing of herbs and spices have no effect on the initial flora. Reductions can only be achieved applying processes described above.
Increase (ΣI)	• Inappropriate drying conditions may lead to increases of the initial flora, in particular of molds. For treated herbs and spices, post-treatment contamination needs to be considered and appropriate preventive measures implemented.
Testing	• The establishment of microbiological specifications should be related to the final use of herbs and spices. Testing untreated herbs and spices will not provide a guarantee of absence of pathogens. It is therefore important to select and collaborate with suppliers to achieve the desired quality. Testing for indicators such as total viable counts or Enterobacteriaceae provides some information on deviations from good hygiene practices and their extent.
Spoilage	• Spoilage with molds is possible, in particular before and during the drying steps if the humidity is poorly controlled.

II Dry soup and gravy mixes

A *Definitions*

Dry soup and gravy mixes, including bouillon cubes and consommés, have many ingredients in common. Some, such as beef-flavor mix can be used as either a soup or a gravy (Komarik *et al.*, 1974). The main ingredients are meats (Chapter 1), poultry (Chapter 2), seafoods (Chapter 3), vegetables (Chapter 5), flours, starches and thickeners (Chapter 8), fats (Chapter 11), sugars (Chapter 12), milk (Chapter 16), eggs (Chapter 15), and seasonings (this Chapter) (Binsted and Devey, 1970).

Dry soups and gravy mixes are made by mixing the different ingredients according to recipes. Some raw materials have to be made suitable for use in such dry mix processes, e.g. by agglomeration, drying, fat coating, or milling. In this type of processes, no heat-treatments (killing steps) are usually applied. The final products, either powders or pastes, are packed in different formats, either pressed as cubes, or filled in laminated sachets or other moisture-proof containers.

Depending on the manufacturer, products intended for the same purpose may be manufactured as shelf stable liquid concentrates, intermediate moisture or low moisture pastes.

B *Initial microflora*

Since most of these formulated foods are simple blends, the initial microflora depends on the flora of their dry ingredients (Karlson and Gunderson, 1965). Manufacturers can select ingredients with low microbial levels, especially for the preparation of the so-called instant soups, which need not be

Table 7.7 Occurrence of different microorganisms in dry soups

Microorganism	Samples positive (%)	Numbers (per gram)
Enterobacteriaceae	39[a]	Tested in 0.1 g
Coliforms	80[b]	10^1–6.4×10^3 (mean 9.2×10^1)
Escherichia coli	3–18[a,b,d]	1×10^2–5×10^2
Aerobic spores	50–71[b,d]	<10[c]–10^4[d]
Yeasts and molds	–	2.1×10^2–3.9×10^3[c]

[a]Krugers-Dagneux and Mossel (1968). [b]Catsaras *et al.* (1961). [c]Fanelli *et al.* (1965). [d]Coretti and Müggenburg (1967).

cooked before consumption. For instance, some firms use spice extracts instead of dry spices and others use specially processed ingredients (e.g. decontaminated spices and freeze-dried vegetables) (Anema and Michels, 1974). Nevertheless, dry mixes generally contain a wide variety of microorganisms (Table 7.7). Aerobic plate counts generally range between 10^3 and 10^5 cfu/g (Catsaras *et al.*, 1961; Fanelli *et al.*, 1965; Krugers-Dagneux and Mossel, 1968; Kadis *et al.*, 1971; Anema and Michels, 1974). Microbiological specifications for soups and bouillons have been proposed by the International Soup Commission on the basis of detailed studies of ingredients and whether the formulation is for so-called instant and normal soups (AIIBP, 1992).

C *Primary processing*

Effects of processing on microorganisms. Mixing and packing under dry hygienic conditions has no effect on the microflora. However, there is a need to design ingredient storage vessels to prevent the ingress of moisture, and to make sure that fillers are kept clean and dry. Whenever possible, wet cleaning should be avoided as the presence of water and dry mixes in these processing environments provides a rich growth medium for potential contaminants.

Spoilage. These foods are shelf stable. Their chemical and microbial stability depends on maintenance of the moisture level at ~7% or below (corresponding to an $a_w = 0.1$–0.35), during storage, mixing and filling, and packaging in moisture impervious packages.

Depending on their pH and water activity, liquid concentrates or pastes may be prone to spoilage with xerophilic yeasts or molds. As with other foodstuffs, products should have a water activiy <0.7 to be stable.

Pathogens. Staphylococcus aureus or, more frequently, spore-formers such as *Cl. perfringens* or *B. cereus*, are often present in low numbers in dry mixes (Nakamura and Kelly, 1968; Keoseyan, 1971; Fallesen, 1976). If the reconstituted product is held warm, particularly between 30°C and 50°C for 6 h or more, these pathogens may grow to population that can cause illness (Tuomi *et al.*, 1974; Fallesen, 1976; Gilbert and Taylor, 1976; Jephcott *et al.*, 1977; Craven, 1980). Outbreaks with this type of microorganisms have indeed mainly been associated with prepared dishes (*e.g.* meat and gravy) held under inappropriate conditions.

Although rare, contamination with salmonellae can occur in dry soups and gravy (Powers *et al.*, 1971; Anonymous, 1974, 1979; Sveum and Kraft, 1981), the origin being either contaminated ingredients or contamination from the processing environment during the different operations. Salmonellosis outbreaks have occurred from dry mixes containing cottonseed flour and dried yeasts (McCall *et al.*, 1966).

The microbiological quality of the so-called instant soups, which need not be cooked before consumption, is of particular importance (Anema and Michels, 1974; AIIBP, 1992).

D CONTROL (dry soup and gravy mixes)

Summary

Significant hazards	• Salmonellae; spp., others such as *Cl. perfringens* and *B. cereus* may be significant depending on the final use of the soups or preparations.
Control measures	
Initial level (H_0)	• Depending on the composition of the products, low levels of pathogens are likely to be present if untreated herbs and spices used as ingredients. Low levels are ensured by adherence to GHP and the application of HACCP.
Reduction (ΣR)	• For dry-mixed products, no reduction during processing. In the case of heat-processed products, killing is achieved, the extent being dependent on the initial levels.
Increase (ΣI)	• In dry products, pasty or liquid products with very low water activity, low pH, or high salt content increase is only caused by contamination from the processing environment.
Testing	• Testing is only done to verify compliance to pre-established specifications, e.g. according to AIIBP (1992).
Spoilage	• Spoilage with molds is possible, in particular for pasty and liquid products at water activities >0.7.

III Soy sauces

A Definition

Soy sauces are important seasonings in Eastern Asia. The prototype soy sauce originated in China ~2000 years ago, and has not been substantially changed since it was brought into Japan. Soy sauces are salty liquid seasonings ranging from amber to dull brown color, produced from soy beans/defatted soy grits with or without wheat, barley, and/or rice by a two-stage fermentation, an aerobic solid-state fungal fermentation is followed by an anaerobic mixed lactic-yeast, submerged fermentation in a strong salt brine.

B Important properties

There are two main types of soy sauces, the Chinese and the Japanese, which differ in organoleptic qualities and composition, depending on the proportion of raw materials and types of fermentations. Soy sauces produced in various South-East Asian countries most closely resemble the Chinese type and their production varies from one region to another. The salt content of Japanese style soy sauces is 16–18% and the water activity is as low as 0.80 or below. Chinese and various South-East Asian soy sauces vary in salt content between 10% and 26%, except Indonesian "kecaps", which has 6–7% NaCl. The pH of Japanese soy sauces is 4.7–5.0 (Yokotsuka, 1986b). The pH values of some commercial brands of soy sauce from the Philippines were reported to be in the range of 5.2–6.1 (Soriano and Pardo, 1978). The pH values of various types of soy sauces from Thailand and Malaysia range from 4.0–5.5 (Merican, 1978; Sundhagul *et al*., 1978).

The Chinese style soy sauces are generally prepared from soybeans with little or no wheat, and have very low alcohol content. Good quality Chinese "chiang-yiu" is dark in color and has a high specific gravity, viscosity, and high nitrogen content. The Japanese soy sauces are produced from soybeans and roasted wheat in different proportions, and their alcohol contents are higher (up to ~3%) than those of Chinese styles. Their viscosity and nitrogen content is lower than those of the Chinese type, but their amino acid content may be higher. There are five main types of Japanese soy sauce (called shoyu): koikuchi (produced in the largest proportion), usukuchi, tamari, shiro, and saishikomi. Koikuchi is reddish-brown and has a strong flavor. Usukuchi is light and has a maximum total nitrogen content of 1.2%. Both are made from approximately equal parts of soy beans and wheat kernels. Koikuchi and usukuchi type shoyus have higher alcohol (1–3%) and higher lactic acid (0.8–1.5%) content. The tamari-type shoyu is essentially a Chinese type soy sauce made primarily of cooked soybeans and small proportions of wheat or barley flour. It contains less alcohol (<0.5%) and somewhat less lactic acid (0.5–1.2%) than the previous types. Shiro shoyu is made mostly from wheat kernels with a very small amount of soybeans and is very light. Shaishikono shoyu is prepared from the product of the solid-state fungal fermentation (koji) and unpasteurized shoyu instead of salt water. It is very dark and high in solid content (Yokotsuka, 1986a,b).

C *Methods of processing and preservation*

Koikuchi-type Japanese soy sauce. The industrial production of the Japanese-type soy sauce (koikuchi shoyu) can be divided into the following stages (Yokotsuka, 1986a; Hose, 1992):

- Treatment of raw materials.
- Koji production.
- Moromi preparation and fermentation (aging).
- Pressing of aged mash.
- Pasteurization and refining.

These are illustrated in Figure 7.1.

Treatment of raw materials. The soybeans/defatted grits are soaked in water and cooked with steam under pressure. This process greatly influences the digestibility of soy protein and, therefore, affects both the koji and moromi stages (see below) of fermentation. The wheat kernels are roasted briefly at 160–180°C, and then coarsely crushed.

Koji fermentation. The cooked soybean-roasted wheat mixture is inoculated with a starter culture of *Asp. oryzae* and/or *Asp. soyae*, and spread in 30–40-cm thick layer on large perforated stainless steel plates in the koji room. This mass is aerated by moisture-controlled air for 2–3 days at ~30°C, allowing the molds to grow throughout the mass to develop their enzymes necessary to hydrolyze the proteins and starch of the raw materials (Tochikura and Nakadai, 1988). This mold-cultured material is called koji.

Moromi stage. Koji is mixed with cold salt water with 22–23% salt content to a final volume of 120–130% of that of the raw materials. The mash, called moromi, is kept in fermentation tanks for 4–8 months, depending on the temperature, and occasionally agitated by bubbling compressed air through the moromi, to promote microbial growth. In moromi undergoing normal fermentation, lactic acid bacteria grow first and produce lactic acid and acetic acid to lower the pH, enabling the main fermentation yeast to grow. This is followed by a "maturation" fermentation by a second group of yeasts. The order and extent of growth of these three groups of organisms and their balance are crucial as they affect the quality of the soy sauce (Noda *et al.*, 1980).

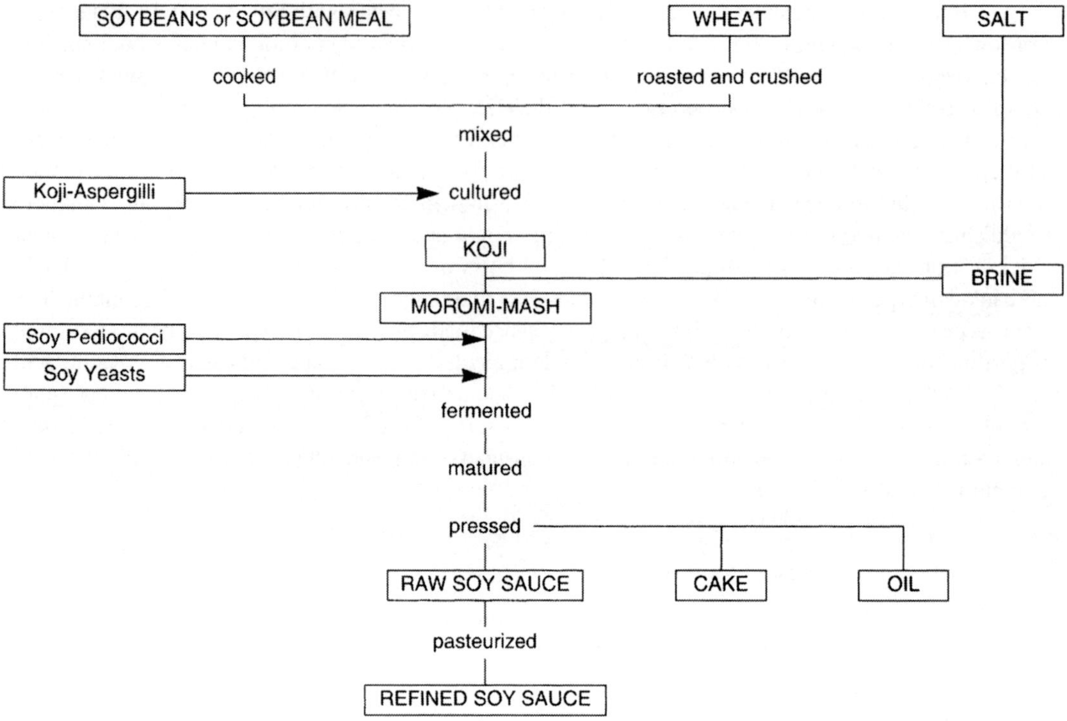

Figure 7.1 Flow sheet for manufacture of Koikuchi-type soy sauce.

During this fermentation period, koji mold enzymes hydrolyze most of the proteins to amino acids and oligopeptides, the starch remaining after koji production is also hydrolyzed. A considerable proportion of sugar is fermented to lactic acid, acetic acid, and alcohol by lactic acid bacteria and yeasts. Current practice in the Japanese industry uses pure cultures of *Pediococcus halophilus* and *Zygosaccharomyces rouxii*, which are added to the mash. Initially, the pediococci decrease the pH from 6.5–7.0 to 4.7–4.9. The lactic acid fermentation is gradually superceded by yeast fermentation.

Soy sauce yeasts grow vigorously at pH 4.0–5.0 in the presence of high brine concentrations (Ohnishi, 1957). However, in practice, to ensure sufficient yeast growth and good fermentation, the yeasts (at concentrations of $\sim 10^6$ cfu/g of moromi) are added when the pH of the moromi is between pH 5.0 and 5.2 (Matsumoto and Imai, 1981). This is followed by vigorous agitation of the moromi using compressed air blown through it at appropriate timing to enhance yeast growth and, hence, fermentation.

Soy sauce yeasts include two osmophilic groups, the main fermentation yeasts such as *Zygosaccharomyces rouxii* and maturation yeasts such as *Candida versatilis* and *Can. etchellsii*. *Z. rouxii* grows at water activity between 0.78 and 0.81 and in media containing 24–26% salt (Yoshii, 1979). The water activity of moromi with 17–18% salt is >0.80, making it ideal for *Z. rouxii* to grow and effect fermentation. The maturation yeasts, *Can. versatilis* and *Can. etchellsii*, also have high salt tolerance and are capable of growing in moromi with water activity of 0.78 or salt content of 26% (*w*/*v*) (Yoshii, 1979).

Pressing. The fermented aged moromi mash is filtered under high pressure through nylon cloth filters to separate into filtrate and cake. The filtrate is stored in tanks to allow sedimentation and separation of the oil layer. Both the sediments and oil layers are removed to obtain raw soy sauce.

Pasteurization and refining. The raw soy sauce is adjusted to obtain a standard concentration of brine and total nitrogen, then pasteurized at 70–80°C for 20–30 min. Alternatively, the raw sauce can be subjected to a high temperature short time treatment of 120°C for a few seconds to affect killing of bacterial spores that may be present in the raw sauce. After pasteurization, the sauce is stored (without stirring) at 50–60°C for 3–4 days to obtain better color and aroma. Further, sedimentation will occur and the clarified sauce is then packed in appropriate containers or further treated depending on the types.

Packaging and shelf life. The industrially produced clarified sauce is bottled or dehydrated by various drying processes, usually after pasteurization. Refined and pasteurized, preservative-free "shoyu" packaged in fiber drums has a shelf life of 1 year at room temperature when stored unopened. If stored opened at room temperature, it has a shelf life of about 2 weeks. The Japanese Soy Sauce Association's official method requires a shelf life of 3 years in glass bottles and 1.5 years in plastic containers. During storage the color of soy sauce will gradually darken due to the amino-carbonyl reaction between amino acids and sugars. The flavor will also deteriorate.

Traditional soy sauces produced in China and South-East Asian Countries. The fermentation of soy sauce in China and Asian countries other than Japan traditionally employs "natural" fermentations using fungi in the environment from previous batches. In South-East Asia, the initial process involves the breakdown of protein and starch as a solid-state fermentation of boiled soybeans mixed with wheat flour and spread on bamboo trays. The traditional Korean fermentation involves kneading steamed soybeans into brick-like blocks and leaving it at ambient temperature to allow a variety of fungi to grow. The fermentation for these types of soy sauce depends on chance contamination of the necessary molds. Usually *Aspergillus* spp. and other molds are present on the bamboo tray used for a prior fermentation. The microorganisms commonly present include *Asp. oryzae* and other aspergilli; *Rhizopus oligosporus, R. oryzae,* and other *Rhizopus* spp.; *Mucor* spp.; and occasionally *Penicillium* spp. The use of traditional fermentation techniques is increasingly being replaced by use of pure cultures of *Asp. oryzae* to inoculate the bean–flour mixture in modern sterile koji rooms, or in stainless steel trays.

The brine phase (moromi) stage of the Chinese-type soy sauce fermentation depends on halotolerant bacteria and yeasts that are naturally present. Species reported to be present in the submerged brine fermentation are *B. subtilis, B. pumilus, B. citreus, B. licheniformis, Sarcina maxima,* and *Z. rouxii* (Korea); *Pediococcus halophilus, Ped. soyae, Pichia* spp., *Candida* spp., and *B. licheniformis* (Malaysia); *Hansenula anomala, H. subpelliculosa,* and *Lb. delbrueckii* (Philippines); *Saccharomyces* spp. and *Lactobacillus* spp. (Singapore); and *Ped. halophilus, Staphylococcus* spp., and *Bacillus* spp., (Thailand). Unlike the Japanese method, these microorganisms are not generally added intentionally; the fermentation is dependent on the selection of desirable organisms from the environment by specific physical and chemical parameters present in the various phases of the fermentation process. The moromi stage for the South-East Asian and Korean sauces is left under the sun for $\geq$2 months to mature. South-East Asian soy sauces have a shelf life of at least 1 year. Generally, their shelf life depends mainly on their salt content and added preservatives. When a preservative is used, it is most often sodium benzoate (400–1000 mg/kg).

D *Types of final products*

Soy sauces or their dehydrated versions are used as natural seasoning agents to enhance the flavor of foods such as soups, meat and poultry, vegetables, and seafoods. They provide a meaty, hearty flavor when added to various sauces, gravies, salad dressings, and other condiments.

A range of products can be produced using soy sauce as the main ingredient, and in many South-East Asian countries these fall under the generic name of soy sauce. Thick sauces in this area are concentrated soy sauces prepared traditionally by evaporation in the sun, but nowadays the product is often thickened with caramel, molasses, and/or starch. Light or "white" sauces are prepared by diluting soy sauce with water to the required consistency, whereas sweet soy sauce such as Indonesian "kecap manis" is prepared from black soybeans (without wheat) in the same manner as the other South-East Asian soy sauces, with the addition of palm sugar and herb extract to the mature sauce, before filtering, concentrating by heating, and packed.

Blended soy sauce, i.e. soy sauce blended with hydrolyzed vegetable protein, is produced in many South-East Asian countries. These products, as well as products using hydrolyzed vegetable protein as an ingredient such as hydrolyzed vegetable protein sauce, soups, gravy mixes, and bouillon cubes, were reported to contain 3-monochloropropane-1,2-diol (3-MCPD) in levels higher than that allowed in many countries (Hamlet, 1999; MAFF, 1999). 3-MCPD should not be present in naturally fermented soy sauce.

E Initial microflora

The initial microflora of the raw ingredients of soy sauce, soybeans, wheat, barley, and rice, is described in detail in Chapter 8 (Cereals and Cereal Products). Mesophilic vegetative microorganisms including coliforms are present, commonly in low populations, along with low numbers of spores of *Bacillus* spp. and *Clostridium* spp. Psychrotrophic bacteria and *Actinomycetes* are almost always present (ICMSF, 1980b, Vol. 2, p. 672). Soy grits also can contain thermophilic bacteria, including spores and fecal streptococci, coliforms and *E. coli* (Hose, 1992, Table 7.8).

In traditional technologies, the main source of contamination during koji fermentation is the bamboo fermentation tray. The trays are never washed and are depended upon to inoculate new batches. Another source of contamination is the crude salt that is used in the second stage of fermentation. It can be a source of halophilic/halotolerant microorganisms that are required for fermentation and aroma development.

F Primary processing

Effects of processing on microorganisms. The heat treatments of raw materials of koji fermentation eliminate most of their initial microflora which is then replaced by the natural microflora in the manufacturing environment or by the purposeful addition of starter cultures. In the industrial production of soy sauce, "tane koji", the fungal starter for the soybean/wheat mixture, is produced by culturing selected strains of *Asp. oryzae* or *Asp. sojae* on either steamed, polished rice or a mixture of wheat bran and soybean meal (Yong and Wood, 1976; Beuchat, 1984). Microbiological changes during production of koji are illustrated in Figure 7.2 (Hose, 1992).

Microflora changes in moromi fermentation of shoyu mash are shown by Figure 7.3 (Tamagawa *et al.*, 1975). The microorganisms which grow in the brine mash are limited to halotolerant lactic acid bacteria and yeasts such as *Ped. halophilus* (previously *Ped. soyae*) and yeasts including *Z. rouxii*, *Torulopsis halophilus*, *Tor. nodaensis* and *Tor. halonitratophila* (Ho *et al.*, 1984), and *Candida* spp. such as *Can.*

Table 7.8 Ranges of bacterial counts in nine batches of soy grits[a]

Type of microbial count	$\log_{10}$ cfu/g
Total mesophilic bacteria	2.9–5.5
Total thermophilic bacteria	2.6–4.1
Mesophilic spores	<1.0–4.1
Thermophilic spores	<1.0–3.8
Fecal streptococci	<1.0–3.7

[a]Hose (1992).

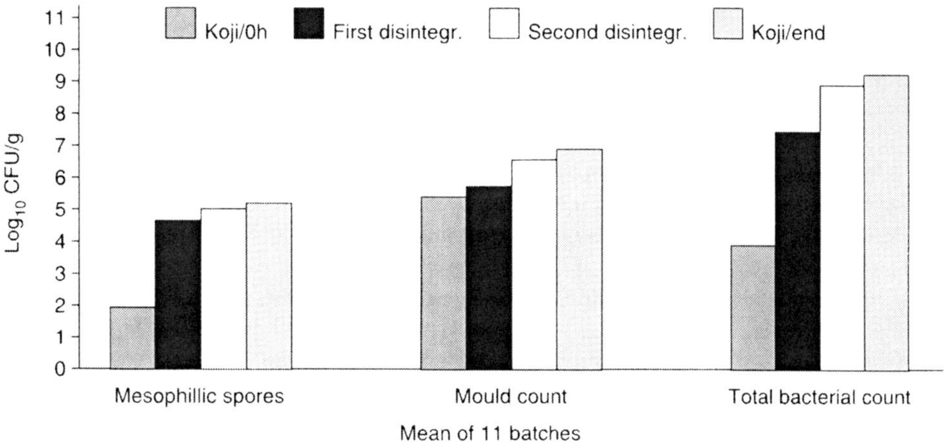

Figure 7.2 Change in mold, total bacterial and mesophilic spore counts during production of tane koji. (From Hose, 1992.)

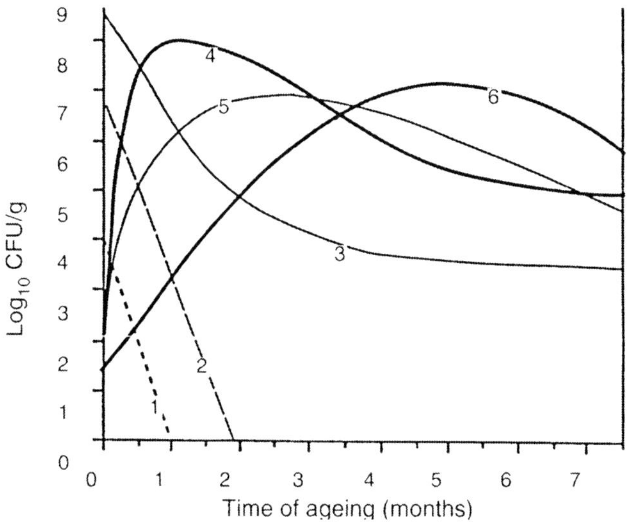

Figure 7.3 Change in microorganisms in Moromi-mash. Curve 1: Wild yeasts; 2: *Staphylococcus* spp.; 3: *Bacillus* spp.; 4: *Pediococcus halophilus*; 5: *Zygosaccharomyces rouxii*; 6: *Candida versatilis*.

versatilis and *Can. etchellsii* (Fukushima, 1989), which are tolerant to the high salt concentrations. *Bacillus licheniformis* and *B. subtilis* were also found growing in Malaysian soy sauce fermentation (Ho *et al.*, 1984). Molds may appear on the surface of the mash, but they are believed to have no relation to proper fermentation at this stage or during aging (Yokotsuka, 1960).

The use of pure culture inocula at all stages of modern production of soy sauces has reduced the risk of carrying unwanted contaminants from one batch to the next, and shortened the fermentation time. Selection parameters of starter microorganisms for use in the fermentation of soy sauce include, among others, non-mycotoxigenicity of the *Asp. oryzae*/*Asoyae* strains selected. *Pediococcus halophilus* must produce lactic acid and other organic acids, the yeast strains should produce alcohol and/or desired flavor substances in a high salt concentration brine (Sugiyama, 1984).

Contaminating microorganisms reported to be isolated from koji include *Micrococcus*, *Streptococcus*, *Lactobacillus*, and *Bacillus* (Fukushima, 1989). Most of these contaminants are not salt tolerant and are thus not able to survive the brine fermentation. The initial acetic acid concentrations present during the shoyu-koji-making process aid to suppress the growth of non-acid tolerant contaminating bacteria of the koji substrate (Hayashi *et al.*, 1979). Probably only the spores of contaminating bacilli can survive the brine.

Spoilage. In traditional products, non-pasteurized products or products with a low salt content, aspergilli, film and pellicle-forming yeasts may cause spoilage (Roling *et al.*, 1994). In low-salt (<15%) sauces, some types of spoilage bacteria may also grow unless the pH is low or preservatives (e.g. sodium benzoate) are present.

The principal contaminants of koji are coagulase-negative *Staphylococcus* spp. and *B. subtilis* (Chiba, 1977). The contaminating *Staphylococcus* spp. grow symbiotically with the koji mold, and the combination becomes a problem when koji fermentation takes place at low temperatures (at or below 25°C), whereas *B. subtilis* grows in competition with the koji molds, especially at higher temperatures.

Wild salt-tolerant lactic acid bacteria may grow in soy sauce and produce biogenic amines such as tyramine and histamine (Uchida, 1982; Stratton *et al.*, 1991). Some wild salt-tolerant lactic acid bacteria may produce ornithine by decomposing arginine in an abnormal fermentation, resulting in the accumulation of citrulline as an intermediate product. When such raw soy sauce is pasteurized, ethyl carbamate can be produced by the reaction between citrullin and alcohol (Matsudo *et al.*, 1993).

A salt-tolerant wild yeast *Z. rouxii* var. *halomembranis* can grow in moromi that has progressed to the maturation process. This wild yeast is harmless to health, but has a high salt tolerance, and grows to form a membranous film on the surface of moromi or soy sauce resulting in deterioration of the soy sauce aroma and flavor.

Pathogens. There have been no reports of illnesses due to enteropathogenic microorganisms associated with soy sauces. Carry over of mycotoxins from contaminated raw materials does not appear to be a problem. Badly molded soybeans and grains are not used because of their impact on product quality. The strains of *Asp. oryzae* and *Asp. soyae* used by the Japanese industry do not produce aflatoxin, and testing them for production of other kinds of mycotoxins indicate no serious problems (Steinkraus *et al.*, 1983). It is important to select strains that do not produce mycotoxins such as cyclopiazonic acid, kojic acid, β-nitropropionic acid or aspergillic acid. However, traditional fermentation does not use pure mold cultures, and a variety of fungi, including undesirable types, can be present. Hence, the possibility of low levels of mycotoxin contamination of traditional soy sauce cannot be ruled out.

Fermented shoyu was determined to have bactericidal activity for some enteropathogenic bacteria, including, *S.* Typhi-Shikata, *Shigella flexnerii*, and *Vibrio cholerae*-inaba. *Clostridium botulinum* type A and B spores inoculated into shoyu survived, but did not grow within 3 months at 30°C (Steinkraus *et al.*, 1983).

G Control

Under normal circumstances, pH, water activity (salt content), and the presence of competing harmless or desirable microorganisms may prevent growth of undesirable microbes during and after the fermentation may of soy sauces and thereby ensure preservation during the primary storage in closed containers. The main control factor is high salt content. Pasteurization is becoming more widely accepted as a means of prolonging microbiological stability (Husin and Yeoh, 1989).

Important control points in the industrial production of soy sauce are as follows.

Raw materials and their heat treatment. Soy beans/grits contain substantial populations of mesophilic aerobic bacteria and mesophilic and thermophilic aerobic spores. Undesirable growth of these bacteria can occur during soaking, so the time of soaking needs to be considered. The soaking water must be changed at least every 2–3 h, otherwise undesirable spore-forming *Bacillus* spp. can grow to high levels (Beuchat, 1984). Autoclaving soybeans after soaking should be monitored by time–temperature measurement, *e.g.* 1 h at 1 kg/cm^2 steam pressure (Steinkaraus *et al.*, 1983). The water content of steamed soybeans should not be allowed to exceed 62% to prevent bacterial spoilage.

The function of the wheat flour is to coat the surface of the soybeans, thereby suppressing bacterial growth. Therefore, if crushed wheat is used, its particle size should be small enough to perform this function. The time–temperature of roasting of wheat (170°C ± 5°C for 40–50 s.) before crushing and particle size (30% less than 30 mesh size) can be considered as control points.

Cooling. Recontamination of raw materials by equipment following heat treatment can occur. If cooling is inadequate, growth of contaminating microorganisms is possible before controlled fermentation can be initiated. Insufficient cooling could also have an adverse effect on germination of the mold starter culture (tane koji). Time of cooling to 30°C and flow-rate of cooling air should be monitored.

Koji fermentation. Although substrates for the koji fermentation undergo a heat treatment before fermentation, contamination from equipment plus contamination by bacteria present in the mold starter culture can occur. Conditions in the koji rooms (machines) must therefore favor the koji molds. This requires control of temperature, humidity, and air flow in the koji room, and testing the germination rate of "tane koji". Good growth of koji molds is achieved by incubating koji at 30–32°C at the beginning of the fermentation when enzyme production starts. After the initial 24 h, the incubation temperature of the koji room is lowered to between 28°C and 30°C. If a conveyor belt is used to transport raw materials and equipment to the koji room, it should be washed thoroughly and disinfected.

Brine fermentation (moromi phase). In the event of incorrect addition of salt (e.g. <22% salt content in brine), undesirable growth of microorganisms other than the salt-tolerant bacteria and yeasts of the moromi starter is possible. Correct addition of salt and water, and adjusting the dry matter content is a control point. The sodium chloride content of the mash should be >17% to prevent growth of undesirable putrefactive bacteria during fermentation. A concentration of sodium chloride in excess of 23%, however, can retard the growth of desirable halophilic bacteria and yeasts (Beuchat, 1984). Controlling the moromi fermentation requires monitoring temperature (30°C ± 2°C), aeration timing, and flow rate. It is also important to ensure that the transportation route of moromi, the fermentation tanks, and all contact surfaces are kept clean and sanitized as a routine maintenance procedure.

Pasteurization of raw sauce. After moromi fermentation/ageing and pressing, the sauce should undergo a pasteurization process (e.g. heating to 70–80°C for 20–30 min). At this low temperature treatment *B. subtilis* spores will not be killed but be reduced by filtration using celite as filter aid (Haga, 1971). Pasteurization of raw sauce should be monitored by time–temperature measurement. Pasteurization or sterilization of soy sauce and canned food flavored with it should take into account the properties of thermophilic and facultative anaerobic bacilli, which are reported to be present in the brine fermentation of soy sauce in Malaysia (Ho *et al.*, 1984).

Post-pasteurization steps. Recontamination of soy sauce following pasteurization during sedimentation, filtration, and bottling or spray-drying is also possible by contact with contaminated equipment or with air-borne flora. Preservatives such as sodium benzoate (0.4–1.0 g/L) can be added to prevent the growth of yeasts during storage, depending on the pH of the product. During bottling, it is desirable to sterilize the nozzle of the bottling equipment using steam or alcohol, which is practised in Japanese plants.

IV Fish and shrimp sauces and pastes

A *Definitions*

Fish sauces and pastes are traditional products obtained by maceration and/or autolysis of muscles, with or without microbial action, in the presence of high salt concentrations. These products usually have a strong characteristic aroma and are used as flavorings in many oriental foods. Under this category are shrimp and fish pastes, fish sauces, and pickled shrimp or fish sauces.

B *Important properties*

Fish sauces and pastes are produced by hydrolysis in the presence of high salt concentrations. Under such conditions, the proteolytic activities of the fish enzymes may be low, and the fermentation lengthy. For example, fermentation of fish sauce can take 6–12 months to complete.

Hydrolysis mainly occurs by proteolytic enzymes in the flesh and entrails of raw fish (Amano, 1962).

The substrates have to be properly handled and thoroughly mixed with salt in the initial stages of processing. Usually factors such as pH, acidity, and temperature are not monitored during processing. Salt concentration is a critical factor. Traditional processes rely heavily on the high salt concentration, *e.g.* 13–15% in Malaysian shrimp paste (Merican *et al.*, 1984) and 20–25% in Philippines's "bagoong" (Soriano *et al.*, 1986), and low a_w, *e.g.* 0.67 for Malaysian shrimp paste (Adnan and Owen, 1984), to prevent the growth of pathogenic and spoilage microorganisms. Methods of preparation have been perfected through generations, and include safety features in manufacturing that take into consideration climate, environment, and handling problems. The products generally have good keeping quality. In addition, they may be pasteurized prior to bottling. Blending with other foods may be practised.

C *Methods of processing and preservation*

Fish/shrimp sauces and pastes are principally fermented by the action of enzymes in the flesh and entrails, in the presence of high salt concentrations. Microorganisms are assumed to contribute to the characteristic flavors because these flavors are lacking in products prepared under sterile conditions (Amano, 1962). In some products such as shrimp sauce, fermentation is influenced by the combined effects of shrimp enzymes and microbial enzymes.

Fish/shrimp paste. Fish/shrimp paste is made from fish or shrimp with 6–10% salt by a process of salt-controlled anaerobic fermentation. The manufacture of shrimp paste is seasonal, depending on the availability of *Acetes* shrimp.

The fish or shrimp is first washed in sea-water at coastal landing sites. This step can introduce a high level of contamination if the coastal waters are heavily polluted. Sorting to remove foreign materials is then carried out. This is typically done by hand which can also introduce additional microorganisms. The fish/shrimp are drained, mixed thoroughly with salt and spread out on mats to dry in the sun for 5–8 h, until the moisture content is about 50%. The fish/shrimp are then minced, packed tightly in tubs, and allowed to ferment under anaerobic conditions for about 7 days. The paste is pressed tightly to ensure the total exclusion of air and to prevent oxidation, which would result in a putrid product. The paste is taken out in lumps and dried further in the sun for 5–8 h. This is followed by a second mincing, pressing into tubs, and fermenting for another 7 days. The process of drying, mincing, pressing, and fermenting is repeated 6–7 times, depending on the aroma and texture required. The longer the fermentation, the stronger is the aroma. If stronger aroma is desired, the shrimp are mixed with salt and held overnight at ambient temperature before the initial drying and mincing steps (Adnan, 1984).

Fish sauce. Fish sauce is traditionally prepared by autolysis and fermentation of small fish that are salted and stored under anaerobic conditions for 6–12 months. In Malaysia, fish sauce is prepared from small anchovies, i.e. *Anchoviella commersonii* and *Anch. indicus*. When larger fish are used such as in Thailand, the fish are comminuted before being mixed with salt, and may be supplemented with hepatopancreatic tissues (Raksakulthai and Haard, 1992).

Shrimp sauce. Acetes shrimp is prepared for fermentation in much the same way as shrimp paste. After salting (10–20%), the shrimp are allowed to drain overnight before adding cooked rice (6–15%). The mixture is fermented in an enclosed container for 7 days. The ferment is then stirred to ensure homogeneity and the fermentation is continued for another 2 weeks. The mature product (known as cinkaluk in Malaysia) may be blended with tomato ketchup and cooked before bottling.

D Types of final products

Under this category of fish and shrimp, sauces and pastes are products of South-East Asian countries known by different names in different localities, with some variations in the methods of processing and utilization of the final products. Examples of these products are as follows.

- "Bagoong" of the Philippines, which is the partially or completely fermented product of small fish or small shrimp and salt, with or without added condiments, flavoring materials or coloring matter (Soriano *et al.*, 1986). Bagoong is either eaten raw or cooked, and is generally used as a flavoring or condiment in many traditional recipes. It is also used as an appetizer, served in a mixture with onions, garlic, tomatoes, or green mangoes.
- Fish or shrimp paste of Indonesia (trassi), Malaysia (belacan), and Thailand (kapi) is the macerated mince of small fish or shrimp and salt that has undergone an anaerobic fermentation. This product is used as an ingredient in a spicy "dipping sauce", or in traditional dishes.
- The fish sauces, "budu" (Malaysia), "nampla" (Thailand, Laos), "nuoc-mam" (Cambodia, Vietnam), and "patis" (Philippines), are the amber to brown liquid supernatant of a salted fish fermentation. The products vary in consistency from country to country and are used in much the same way as soy sauce, i.e. as a table condiment, as a dipping sauce or as ingredient in cooking traditional dishes.
- The fermented (pickled) fish or shrimp sauces, "cincaluk" (shrimp)(Malaysia) and "burong isda" or "burong dalag" (fish) (Philippines) are a fermented shrimp/fish-salt mixture with cooked rice. The product undergoes a lactic acid fermentation (Soriano *et al.*, 1986). These sauces are also used as condiments, as dipping sauces or as ingredients in cooking. The Philippine product (burong isda) is eaten as is or with rice. The fish may still be whole and intact. Some sauce comes from partial hydrolysis of the fish and rice but the product is largely solid.

E Initial microflora

The microflora, including both identity, numbers, and distribution, of the final products are influenced by the initial microflora, human contact such as the hands of fishermen and handlers, and the equipment used in these operations.

The initial microflora is that present on the raw materials, which reflects that of the marine environment and salt. When fish and shrimp are caught in coastal waters, polluting microorganisms such as *Streptococcus faecalis* have been reported to be present during the early stages of fermentation (Orillo and Pederson, 1968; Ohhira *et al.*, 1990). For an account of the microorganisms of fish and shrimp, please refer to Chapter 3. See earlier comments in this chapter regarding the microflora of salt preparations.

Poor handling of raw materials and an unsanitary environment during processing and storage are the most critical problems associated with these products, *e.g.* drying of salted shrimp or fish in the sun exposes the products to flies and other pests.

F Primary processing

Effects of processing on microorganisms. The high salt content and microaerophilic or anaerobic conditions during fermentation encourage the proliferation of lactic acid bacteria, e.g. *Leuconostoc mesenteroides* subsp. *mesenteroides* and *Lactobacillus plantarum* in Malaysian shrimp pastes, *Leuc. mesenteroides* subsp. *mesenteroides* in Indonesian shrimp paste; *Lb. plantarum* in Malaysian fish sauce; *Leuc. mesenteroides* subsp. *mesenteroides* in pickled shrimp; and *Strep. faecium* and *Leuc. mesenteroides* subsp. *mesenteroides* in pickled fish (Ohhira *et al.*, 1990). Burong dalag undergoes a lactic fermentation in which *Leuc. mesenteroides, Ped. cerevisiae*, and *Lb. plantarum* play major roles (Soriano *et al.*, 1986).

Studies on Philippines' bagoong revealed that bacteria actually decreased during the course of fermentation at ambient temperature from an initial count of 3.5×10^6–5.8×10^4 cfu/g after 1 week, to 6.2×10^3 cfu/g at 2 weeks, to 100 cfu/g after 8 weeks (Amano, 1962). In contrast, "shiokara", a Japanese product increased in bacterial populations from an initial count of 9.0×10^4 to 3.2×10^7 cfu/g in 41 days (Amano, 1962). During the course of shiokara fermentation, a succession of organisms appeared, viz. *Bacillus, Micrococcus*, and *Lactobacillus* at the beginning, giving rise to salt tolerant *Micrococcus* at later stages of fermentation. *Vibrio, Achromobacter/Moraxella*, and *Flavobacterium* were also isolated from commercial shiokara, with total bacterial counts ranging from 1.5×10^3 to 2.4×10^7 cfu/g. *Vibrio, Flavobacterium, Achromobacter/Moraxella*, and *Micrococcus* were all able to grow in 10–20% NaCl concentrations, whereas *Bacillus* spp. only grew well at 5% NaCl (Amano, 1962). All these isolates were capable of producing ammonia.

Growth of anaerobes in the presence of high salt appears to contribute to the typical flavor of fermented fish products. Products such as pickled shrimp and fish to which a carbohydrate source has been added undergo a lactic acid fermentation. This favors the growth of *Lb. plantarum, Lb. pentoaceticus, Strep. faecium*, and yeasts (Amano, 1962).

Fermented fish products such as pickled shrimp and shrimp paste may contain histamine >500 mg/kg (Azudin and Saari, 1990), with a 10-year-old fish sauce having as much as 700 mg/kg (Amano, 1962). Although it has been reported that salted bonito had been implicated in bacterially formed histamine poisoning, properly prepared fish pastes and sauces have not been linked to a serious problem with histamine toxicity. However, the practice of using semi-spoiled fish as raw materials is a cause of concern.

Spoilage. Salt plays a significant role in these products and acts as preservative to inhibit the growth of spoilage and pathogenic microorganisms. Although the high salt content prevents the growth of most spoilage microorganisms, moderately halophilic bacteria such as *Bacillus* and *Staphylococcus* spp. have been isolated in spoiled fish sauce (Mabesa *et al.*, 1986), and extremely halophilic strains of *Halobacterium salinarum* have been reported as spoilage agents in salted fish (Sanderson *et al.*, 1988). These bacteria are associated with poor hygienic practices. Halophilic microorganisms surviving in fermented fish tend to produce ammonia and thus contribute to off-flavors (Amano, 1962).

Pathogens. The high salt content inhibits the growth of pathogens. Although reports of food poisoning from fermented fish products are rare, the trend of reducing the salt content of these products may pose a serious problem. A home-made product, made with reduced salt, was reported to cause *Cl. botulinum* type E poisoning in Japan (Amano, 1962).

Shrimp paste does not support the growth of molds and attempts to detect *Asp. flavus* and aflatoxin have yielded negative results (Sim *et al.*, 1985).

G Control

Under normal circumstances, the high salt concentration of fish sauces and pastes is sufficient to inhibit the growth of spoilage and pathogenic microorganisms. However, quality control and standardization of the process are necessary to ensure a hygienically consistent product. A code of practice should be developed to guide processors on good manufacturing practices. Pasteurization or cooking is usually practised as a means of prolonging shelf life. In products such as fish/shrimp pastes and blended fish sauces, benzoic acid/sodium benzoate is added to extend shelf life.

Control points that should be considered or used include the following.

1. Avoiding harvesting fish and shrimp from polluted waters. Use only clean water, salt, and utensils. Monitoring by visual inspection the catch, equipment, and raw ingredients. Use only fish of good quality for fermentation.
2. Enforcing good hygienic practices and proper sanitation and hygiene of food handlers involved in the preparation and processing of products. Monitor by inspection the cleanliness of food handlers, operations, and premises.
3. Control a_w through sodium chloride addition and/or drying, ($a_w < 0.85$ for fish/shrimp sauces and $a_w < 0.75$ for fish/shrimp pastes). Assure that sufficient levels of salt are added to decrease aw. Monitor a_w to ensure maximum values are not exceeded.
4. Proper control the anaerobic fermentation, including the use of correctly designed equipment. Traditional methods and equipment currently in use are often not properly designed for maintaining anaerobic conditions. Monitor by visual inspection areas of discoloration indicative of aerobic spoilage. Ideally, plants should invest in proper equipment.

References

Adnan, N.M.A. (1984) Studies on belacan from *Acetes*. Biotechnology Programme Meeting, Food Technology Div., MARDI, Kuala Lumpur, unpublished.

Adnan, N.M.A. and Owens, J.D. (1984) Technical note: microbiology of oriental shrimp paste. *J. Food Technol.*, **19**, 499–502.

AIIBP (Association Internationale de l'Industrie des Bouillons et Potages, Commission Technique) (1992) New microbiological specifications for dry soups and bouillons. Alimenta, **31**, 62–5.

Almela, L., Nieto-Sandoval, J.M., Fernandez Lopez , J.A. (2002) Microbial inactivation of paprika by a high- temperature short-X time treatment. Influence on color properties. *J. Agric. Food. Chem.* **50**, 1435–40.

Amano, K. (1962) The influence of fermentation on the nutritive value of fish with special reference to fermented fish products of South-East Asia, in *Fish in Nutrition* (eds H. Heen and R. Kreuzer), Fishing News (Book) Ltd., London, pp. 180–200.

Anema, P.J. and Michels, M.J.M. (1974) Microbiology of instant dry soup mixes, in *Proceedings of an International Symposium on Food Microbiology*, 2, Fed. Assoc. Sci. Tec. (FAST), Milan, (in Italian, English abstracts), 165–82.

Anisa Ath-kar, M., Prakash, H.S. and Shetty, H.S. (1988) Mycoflora of Indian spices with special reference to aflatoxin producing isolates of *Aspergillus flavus*. *Indian J. Microbiol.*, **28**(1 and 2), 125–7.

Anonymous (1974) Soup mix recalled because of *Salmonella* contamination. *Food Chem. News*, October 14th, p. 52.

Anonymous (1979) Soup mix recalled because of *Salmonella*. *Food Chem. News*, June 27th, p. 8.

Awe, M.J. and Schranz, I.L. (1981) High pressure liquid chromatographic determination of aflatoxin in spices. *J. Assoc. Off. Anal. Chem.*, **64**, 1377–82.

Azudin, M.N. and Saari, N. (1990) Histamine content in fermented and cured fish products in Malaysia, in *FAO—Fisheries Report, No.401, Supplement*, Seventh Indo-Pacific Fishery Commission Working Party on Fish Technology and Marketing, pp. 105–11.

Bachman, S. and Gieszczynska, J. (1973) Studies on some microbiological and chemical aspects of irradiated spices, in *Aspects of the Introduction of Food Irradiation in Developing Countries*, IAEA, Vienna, p. 33.

Bahk, J., Yousef, A.E. and Marth, E.H. (1990) Behaviour of *Listeria monocytogenes* in the presence of selected spices. *Lebensm. Wiss. u. Technol.*, **23**, 66–9.

Baxter, R. and Holzapfel, W.H. (1982) A microbial investigation of selected spices, herbs and additives in South Africa. *J. Food Sci.*, **47**, (570–8).

Bean, P.G. and Salvi, A. (1970) The bacteriological quality of some raw materials used in the Italian canning industry. *Ind. Alimenta* (Pinerolo, Italy), **9** (4), 547–63 (in Italian).

Beljaars, P.R., Schumans, J.C.H.M.K. and Koken, P.J. (1975) Quantitative fluorodensitometric determination and survey of aflatoxins in nutmeg. *J. Assoc. Off. Anal. Chem.*, **58**, 263–71.

Benezet, A,. de la Osa, J.M, Pedregal, E., Botas, M., Olmo, N. and Pérez Flórez, F. (2001) Presencia de *Listeria monocytogenes* en especias. *Alimentaria*, **321**, 41–3.

Beuchat, L.R. (1976) Sensitivity of *Vibrio parahaemolyticus* to spices and organic acids. *J. Food Sci.*, **41**, 899–902.

Beuchat, L.R. (1984) Fermented soybean foods. *Food Technol.*, **64**(6), 66–70.

Bhat, R., Geeta, H. and Kulharni, P.R. (1987) Microbial profile of armin seeds and chili powder sold in retail shops in the city of Bombay. *J. Food Protect.*, **50**, 418–9.

Binsted, R. and Devey, J.D. (1970) *Soup Manufacture. Canning, Dehydration and Quick Freezing*, 3rd edn, Food Trade Press, London.

Blake, D.F. and Stumbo, C.R. (1970) Ethylene oxide resistance of microorganisms important in spoilage of acid and high-acid foods. *J. Food Sci.*, **35**, 26–9.

Blum, H.B. and F.W. Fabian (1943) Spice oils and their components for controlling microbial surface growth. *Fruit Prod. J.*, **22**, 326–9, 347.

Bockemühl, J. and Wohlers, B. (1984) Zur Problematik der Kontamination unbehandelter Trockenprodukte der Lebensmittelindustrie mit Salmonellen. *Zbl. Bakt. Hyg. Abt. Orig. B*, **178**, 535–41.

Briggs, A. (1966) The resistance of spores of the genus *Bacillus* to phenol, heat and radiation. *J. Appl. Bacteriol.*, **29**, 490–504.

Bruchmann, M. (1995), Salmonella-Kontamination von Gewurzen. Untersuchungsergebnisse 1993 im Bundesland Brandenburg, *Arch. Lebensm. Hyg.*, **46**, 1–24.

Buchanan, R.L. and Shepherd, A.J. (1982) Inhibition of *Aspergillus parasiticus* by thymol. *J. Food Sci.*, **46**, 976–7.

Campbell, J.V., Mohle-Boetani, J., Reporter, R., Abbott, S., Farrar, J., Brandl, M., Mandrell, R. and Werner, S.H. (2001) An outbreak of *Salmonella* serotype Thompson associated with fresh cilantro. *J. Infect. Dis.*, **183**, 984–7.

Catsaras, M., Sampaio Ramos, M.H. and Buttiaux, R. (1961) Étude microbiologique des potages déshydrates ou concentrés du marché francais. *Ann. Inst. Pasteur, Lille*, **12**, 163–74.

CDC (1982) Outbreak of *Salmonella oranienburg* infection—Norway. *Morb. Mortal. Wkly Rep. Centers Dis. Control*, **31**, 655–6.

Chiba, H. (1977) Bacterial contamination in soy sauce koji and its counterplan. *J. Soc. Brewing* (Japan Nihon Jozo Kyokai Zasshi), **72**, 410.

Christensen, C.M. (1972) Pure spices—how pure? *Am. Soc. Microbiol. News* (ASM News), **38**, 165.

Christensen, C.M., Fanse, H.A., Nelson, G.H., Bates, F. and Mirocha, C.J. (1967) Microflora of black and red pepper. *Appl. Microbiol.*, **15**, 622–6.

CAC (Codex Alimentarius Commission) (1995) Code of Hygienic Practice for Spices and Dried Aromatic Plants, CAC/RCP 42-1995.

Coretti, K. and Inal, T. (1969) Rückstandsprobleme bei der Kaltentkeimung von Gewürzen mit T-Gas (Äthylenoxid). *Fleischwirtschaft*, **49**, 599–604.

Coretti, K. and Müggenburg, H. (1967). Keimgehalt van Trockensuppen und seine Beurteilung. *Feinkostwirtschaft* **4**, 76–8.

Craven, S. (1980) Growth and sporulation of *Clostridium perfringens* in foods. *Food Technol.*, **34**(4), 80–7, 95.

D'Aoust, J.Y. (1994) *Salmonella* and the international food trade. *Int. J. Food Microbiol.*, **24**, 11–31.

D'Aoust, J.-Y. (2000) *Salmonella*, in *The Microbiological Safety and Quality of Food* (eds B.M Lund, T.C. Baird-Parker and G.W. Gould), *volume II*, Chapter 45, pp. 1233–99.

de Boer, E., Spiegelenberg, W.M. and Janssen, F.W. (1985) Microbiology of spices and herbs. *Antonie Van Leeuwenhoek*, **51**(4), 435–8.

Dehne, L.I. and Bögl, K.W. (1993) Pasteurization of spices by microwave and high frequency. *Food Market. Technol*, **7**(5), 35–8.

Dehne, L.I., Reich, E. and Bögl, K.W. (1990) Zum Stand der Entkeimung von Gewürzen mittels Mikrowellen und Hochfrequenz. *Soz.Ep. Hefte* 4/1990. Inst. f. Sozialmedizin und Epidemiologie des Bundesgesundheitsamtes, Berlin.

Dehne, L.I., Wirz, J. and Bögl, K.W. (1992a) Unterschungen zur Entkeimung von Gewürzen mittels Dampf. *Soz. Ep. Hefte* 1/1992. Inst. f. Sozialmedizin und Epidemiologie des Bundesgesundheitsamtes, Berlin.

Dehne, L.I., Wirz, J. and Bögl, K.W. (1992b) Ergänzende Untersuchungen zur Dampfentkeimung von Gewürzen als Alternative zur Bestrahlung. *Soz. Ep. Hefte* 2/1992. Inst. f. Sozialmedizin und Epidemiologie des Bundesgesundheitsamtes, Berlin.

Dorman, H.J. and Deans, S.G. (2000) Antimicrobial agents from plants: antibacterial activity of plant volatile oils. *J. Appl. Microbiol.*, **88**, 308–16.

EEC (1989) Directive 89/365. Ref. L159 of 10 June 1989. Official Journal of the European Communities.

Emerole, G.O., Uwaifo, A.O., Thabrew, M I. and Bababunni, E.A. (1982) The presence of aflatoxin and some polycyclic aromatic hydrocarbons in human foods. *Cancer Lett.*, **15**, 123–9.

Eschmann, K.H. (1965) Gewürze—eine Quelle bakteriologischer Infektionen. *Alimenta*, **4**(3), 83–7.

Evert Ting, W.T. and Deibel, K.E. (1992) Sensitivity of *Listeria monocytogenes* to spices at two temperatures. *J. Food Saf.*, **12**, 129–37.

Fábri, I., Nagel, V., Tabajdi-Pintér, V., Zalavári, Zs., Szabad, J. and Deák, T. (1985) Qualitative and quantitative analysis of aerobic spore-forming bacteria in Hungarian paprika, in *Fundamental and Applied Aspects of Bacterial Spores* (eds G.J. Dring, D.J. Ellar and G.W. Gould), Academic Press, London, pp. 455–62.

Fallesen, K.B. (1976) The bacteriological quality of reconstituted soups, in relation to that of the dried soups from which they are prepared. *Dan. Veterinaertidskr.*, **59** (17), 714–7.

Fanelli, M.J, Peterson, A.C. and Gunderson, M.F. (1965) Microbiology of dehydrated soups. I. A survey. *Food Technol.*, **19**, 83–4.
Farag, D.S., Daw, Z.Y., Hewedi, F.M. and El-Baroty, G.S. (1989a) Antimicrobial activity of some Egyptian spice essential oils. *J. Food Prot.*, **52**, 665–7.
Farag, D.S., Daw, Z.Y. and Abo-Raya, S.H. (1989b) Influence of some spice essential oils on *Aspergillus parasiticus* growth and production of aflatoxins in synthetic medium. *J. Food Sci.*, **54**, 74–6.
Farkas, J. (1988) *Irradiation of Dry Food Ingredients*, CRC Press, Inc., Boca Raton, Florida.
Farkas, J., (1998) Irradiation as a method for decontaminating food. A review. *Int. J. Food Microbiol.*, **44**, 189–204.
Farkas, J. (2000) Spices and Herbs, in *The Microbiological Safety and Quality of Foods* (eds B.M. Lund, T.C. Baird-Parker and G.W. Gould), *volume 1*, Aspen Publication.
Farkas, J. and Andrássy, É. (1984) Comparative investigations of some effects of gamma radiation and ethylene oxide on aerobic bacterial spores in black pepper, in *Proceedings of the of IUMS—ICFMH 12th International Symposium: Microbial Associations and Interactions in Food*, (eds I. Kiss, T. Deák, and K. Incze), Akadémiai Kiadó, Budapest, p. 393.
Farkas, J. and Andrássy, É. (1985) Increased sensitivity of surviving bacterial spores in irradiated spices, in *Fundamental and Applied Aspects of Bacterial Spores*. (eds G.J. Dring, D.J. Ellar and G.W. Gould), Academic Press, London, pp. 397–407.
Farkas, J., Beczner, J., Incze, K. (1973) Feasibility of irradiation of spices with special reference to paprika, in *Radiation Preservation of Food*, IAEA, Vienna, p. 389.
Flannigan, B. and Hui, S.C. (1976) The occurrence of aflatoxin-producing strains of *Aspergillus flavus* in the mould floras of ground spices. *J. Appl. Bacteriol.*, **41**, 411–8.
Fukushima, D. (1989) Industrialization of fermented soy sauce production centering around Japanese shoyu, in *Industrialization of Indigenous Fermented Foods*, (ed K.H. Steinkraus), Marcel Dekker, Inc., New York and Basel, pp. 1–88.
Gál, I.E. (1968). Über die antibakterielle Wirksamkeit von Gewürzpaprika. Aktivitätsprüfung von Capsicidin und Capsaicin. *Z. Lebensm.- Unters. -Forsch.*, **138**, 86–92.
Gál, I.E. (1969). The bacteriostatic effect of capsaicine (in Hungarian). *Élelmiszervizsgálati Közl.*, **15**(2), 80–5.
Garcia, S., Iracheta, F., Galvan, F. and Heredia, N. (2001) Microbiological survey of retail herbs and spices from Mexican markets. *J. Food Protect.*, **64**, 99–103.
Gerhardt, U. (1994) *Gewürze in der Lebensmittelindustrie*, Behr Verlag, Hamburg.
Gerhardt, U. and Ladd Effio, J.C. (1982) Äthyleneoxidanwendung in der Lebensmittelindustrie. Ein Situationsbericht über Für und Wider. *Fleischwirtschaft*, **62**, 1129.
Gerhardt, U. and Ladd Effio, J.C. (1983) Rückstandverhalten von Äthylenoxid in Gewürzen. *Fleischwirtschaft*, **63**, 606–8.
Gilbert, R.J. and Taylor, A.J. (1976) *Bacillus cereus* food poisoning, in *Microbiology in Agriculture, Fisheries and Food* (eds F.A. Skinner and J.G. Carr), Academic Press,London, pp. 197–213.
Goto, A., Yamazaki, K. and Oka, M. (1971) Bacteriology of radiation sterilization of spices. *Food Irrad.*, **6**(1), 35–42.
Guarino, P.A. (1972) Microbiology of spices, herbs and related materials, in *Proceedings of the Annual Symposium on Fungi in Foods*, Sect., Inst. Food Technol., Rochester, NY, pp. 16–18.
Gustavsen, S. and Breen, O. (1984) Investigation of an outbreak of *Salmonella oranienburg* infections in Norway, caused by contaminated black pepper. *Am. J. Epidemiol.*, **119**, 806–12.
Hadlok, R. (1969) Schimmelpilzkontamination von Fleischerzeugnissen durch naturbelassene Gewürze. *Fleischwirtschaft*, **49**, 1601–9.
Hadlok, R. and Toure, B. (1973) Mykologische und bakteriologische Untersuchungen entkeimter Gewürze. *Arch. Lebensmittel Hyg.*, **24**(1), 20.
Haga, H. (1971) Pasteurization and preservation of shoyu, soy sauce. *J. Soc. Brewing* (Japan Nihon Jozo Kyokai Zasshi), **66**(11), 1034–7.
Hamlet, C. (1999) Analysis of hydrolysed vegetable protein for chloropropandiols using selected ion storage. Varian-Chromatography Systems—GCMS Applicat.
Hammer, K.A., Carson, C.F. and Riley, T.V. (1999) Antimicrobial activity of essential oils and other plant extracts. *J. Appl. Microbiol.*, **86**, 985–90.
Härnulv, B.G. and Snygg, B.G. (1973) Radiation resistance of spores of *Bacillus subtilis* and *B. stearothermophilus* at various water activities. *J. Appl. Bacteriol.*, **36**, 677–82.
Hartgen, H. and Kahlan, D.I. (1985) Bedeutung der Koloniezahl bei Haushaltsgewürzen. *Fleischwirtschaft*, **65**(1), 99.
Hayashi, K., Tereda, M., Mizunuma, T. and Yokotsuka, T. (1979) Retarding effect of acetic acid on growth of contaminating bacteria during shoyu-koji making process. *J. Food Sci.*, **44**, 359–62.
Hirasa, K. and Takemasa, M. (1998) *Spice Science and Technology*, Dekker, Basel.
Ho, C.C., Toh, S.E., Ajan, N. and Cheah, K.P. (1984) Isolation and characterization of halophilic yeasts and bacteria involved in soy sauce fermentation in Malaysia. *Food Technol. Aust.*, **36**, 227–30.
Hoffman, R.K. (1971) Toxic gases, in *Inhibition and Destruction of the Microbial Cell* (ed W.B. Hugo), Academic Press, London, pp. 225–58.
Hose, H. (1992) *Soya sauce*, Nestlé Research Centre, Lausanne, Switzerland, unpublished.
Husin, A. and Yeoh, Q.L. (1989) Progress in oriental food science and technology, in *Trends in Food Science and Technol., Singapore, Oct. 1987* (ed H.G. Ang), Singapore Inst. of Food Science and Technology, Singapore, pp. 291–9.
Huszka, T., Cséfalvy, I. and Incze, K. (1973) Sterilization of powdered paprika by means of hydrochloric acid (in Hungarian). *Konzerv és Paprikaipar*, **6**, 213.
ICGFI (1988) Code of Good irradiation practice for the control of pathogens and other microflora in spices, herbs and other vegetable seasonings, International Consultative Group on Food Irradiation, Document No. 5.

ICMSF (International Commission on Microbiological Specifications for Foods). (1980a) Microbial Ecology of Foods, *Factors Affecting Growth and Death of Microorganisms, volume 1*, Academic Press, New York.

ICMSF (International Commission on Microbiological Specifications for Foods). (1980b) Microbial Ecology of Foods, *Food Commodities, volume 2*, Academic Press, New York.

Inal, T., Keskin, S., Tolgay, Z. and Tezcan, I. (1975) Gewürzsterilization durch Anwendung von Gamma-Strahlen. *Fleischwirtschaft*, **55**, 675–7.

ISO, 1995. Spices and condiments - Botanical nomenclature. ISO 676.

Ismaiel, A.A. and Pierson, M.D. (1990a) Inhibition of germination, outgrowth and vegetative growth of *Clostridium botulinum* 67B by spice oils. *J. Food Prot.*, **53**, 755–8.

Ismaiel, A.A. and Pierson, M.D. (1990b) Effect of sodium nitrite and origanum oil on growth and toxin production of *Clostridium botulinum* in TYG broth and ground pork. *J. Food Prot.*, **53**, 958–60.

Ito, H. and Islam, M. (1994) Effect of dose rate on inactivation of microorganisms in spices by electron-beams and gamma-rays irradiation. *Radiat. Phys. Chem.*, **43**, 545–50.

Ito, H., Watanake, H., Bagiawati, S., Muhamad, L.J. and Tamura, N. (1985) Distribution of microorganisms in spices and their decontamination by gamma-irradiation, in *Food Irradiation Processing*, IAEA, Vienna, pp. 171.

Jacquet, J. and Teherani, M. (1974) An unusual presence of aflatoxin in certain animal products: possible role of pepper. *Bull. Acad. Vet. Fr.*, **47**, 313.

Jephcott, A.E., Barton, B.W., Gilbert, R.J. and Shearer, C.W. (1977) An unusual outbreak of food-poisoning associated with meals-on-wheels. *Lancet*, **ii**(8029), 129–30.

Josephson, S. and Peterson, M.S. (eds) (1982–1983) *Preservation of Food by Ionizing Radiation, volume 1–3*, CRC Press, Inc., Boca Raton, Florida.

Julseth, R.M. and Deibel, R.H. (1974) Microbial profile of selected spices and herbs at import. *J. Milk Food Technol.*, **37**, 414–9.

Kadis, V.W., Hill, D.A. and Pennifold, K.S. (1971) Bacterial content of gravy bases and gravies obtained in restaurants. *Can. Inst. Food Technol. J.*, **4**, 130–2.

Käferstein, F.K. (1976) The microflora of parsley. *J. Milk Food Technol.*, **39**, 837–40.

Kalemba D. and Kunicka A. (2003) Antibacterial and antifungal properties of essential oils. *Curr. Med. Chem.*, **10**, 813–29

Karapinar, M. and Aktug, S.E. (1987) Inhibition of foodborne pathogens by thymol, eugenol, menthol and anethole. *Int. J Food Microbiol.*, **4**, 130–2.

Karlson, K.E. and Gunderson, M.F. (1965) Microbiology of dehydrated soups. II. Adding machine approach. *Food Technol.*, **19**(1), 86–90.

Keoseyan, S.A. (1971) Incidence of *Clostridium perfringens* in dehydrated soup, gravy and spaghetti mixes. *J. Assoc. Off. Anal. Chem.*, **54**, 106–8.

Kiss, I. and Farkas, J. (1981) Combined effect of gamma irradiation and heat treatment on microflora of spices, in *Combination Processes in Food Irradiation*, IAEA, Vienna, p. 107.

Komarik, S.L., Tressler, D.K. and Long, L. (1974) *Meats, Poultry, Fish and Shellfish, volume 1*, Food Products Formulary Series, AVI, Westport, Connecticut.

Kovács-Domján, H. (1988) Microbiological investigations of paprika and pepper with special regard to spore formers including *B. cereus*. *Acta Alimentaria*, **17**, 257–64.

Kramer, J.M., Turnbull, P.C.B., Munshi, G. and Gilbert, R.J. (1982) Identification and characterisation of *Bacillus cereus* and other *Bacillus* species associated with foods and food poisoning, in *Isolation and Identification Methods for Food Poisoning Organisms* (eds J.E.L. Corry, D. Roberts and F.A. Skinner), Technical Series No. 17, Society of Applied Bacteriology, Academic Press, London.

Krishnaswamy, M.A., Patel, J.D., Pathasarathy, N. and Nair, K.K.S. (1973) Some of the types of coliforms, aerobic mesophilic spore formers, yeasts and moulds present in spices. *J. Plant Crops*, **1** (Supplement), 200–3.

Krugers-Dagneux, E.L. and Mossel, D.A.A. (1968) The microbiological condition of dried soups, in *The Microbiology of Dried Foods* (eds E.H. Kampelmacher, M. Ingram and D.A.A. Mossel), Grafische Ind., Haarlem, The Netherlands, pp. 411–25.

Laidley, R., Handzel, S., Severs, D. and Butler, R. (1974) *Salmonella weltevreden* outbreak associated with contaminated pepper. *Epidemiol. Bull.*, **18**, 62.

Lambert, R.J., Skandamis, P.N., Coote, P.J. and Nychas, G.J. (2001) A study of the minimum inhibitory concentration and mode of action of oregano essential oil, thymol and carvacrol. *J. Appl. Microbiol.*, **91**, 453–62.

Lawrence, B.M. (1978) Recent progress in essential oils. *Perfum. Flavorist*, **2**(12), 44–9.

Lehmacher, A., Bockemuhl, J. and Aleksic, S. (1995) Nationwide outbreak of human salmonellosis in Germany due to contaminated paprika and paprika-powdered potato chips. *Epidemiol. Infect.*, **115**, 501–11.

Leitao, M.F., Delazari, I. and Mazzoni, H. (1973–1974) Microbiology of dehydrated foods (in Port.). *Coletanea Inst. Technol. Aliment.*, **5**, 223–241 [Food Sci. Technol. Abstr., **7** (9B72), 1975].

Llewellyn, G.C., Burkett, M.L. and Eadie, T. (1981) Potential mold growth, aflatoxin production and antimycotic activity of selected natural spices and herbs. *J. Assoc. Off. Anal. Chem.*, **64**, 955–60.

Lohse, N., Kraghede, P.G. and Molbak, K. (2003) Botulism in a 38-year-old man after ingestion of garlic in chilli oil. *Ugeskr Laeger*, **165**, 2962–3.

Maarse, H. and Nijssen, L.M. (1980) Influence of heat sterilization on the organoleptic quality of spices. *Nahrung*, **24**, 29–38.

Mabesa, R.C., Lagtapon, S.C. and Villaral, M.J.A. (1986) Characterization and identification of some halophilic bacteria in spoiled fish sauce. *Philippine J. Sci.*, **115**(4), 329–34.

MAFF. (1999) Press release. *Industry Alerted to Contaminant Levels in Soy Sauce*, Min. of Agric., Fisheries and Food, U.K.

Masson, A. (1978) Hygienic quality of spices. *Mitteilungen aus dem Gebiete der Lebensmitteluntersuchung und Hygiene*, **69**(4), 544–9.

Matsudo, T., Aoki, T., Abe, K., Fukada, N., Higuchi, T., Sasaki, M. and Uchida, I. (1993) Determination of ethyl carbamate in soy sauce and its possible precursor. *J. Agric. Food Chem.*, **41**, 352–6.
Matsumoto, I. and Imai, S. (1981) Influence of yeasts on lactic acid fermentation in miso and soy sauce (shoyu). *J. Soc. Brewing* (Japan Nihon Jozo Kyokai Zasshi), **76**, 696–700.
Mayr, G.E. and Suhr, H. (1972) Preservation and sterilization of pure and mixed spices, in *Proceedings of a Conference on Spices*, Tropical Products Institute, London, pp. 201–7.
McCall, C.E., Collins, R.N., Jones, D.B., Kaufmann, A.F. and Brachman, P.S. (1966) An interstate outbreak of salmonellosis traced to a contaminated food supplement. *Am. J. Epidemiol.*, **84**, 32–9.
Merican, Z. (1978) Status of soya sauce research in Malaysia, in *Proceedings of ASEAN Workshop on Soya Sauce Manufacturing Techniques*, Sub-Committee on Protein, Singapore, pp. 101–5.
Merican, Z., Yeoh, Q.L. and Idrus, A.Z. (1984) Malaysian fermented foods. Occasional Papers No. 10, Science Council of Singapore.
Michels, M.J. and Visser, F.M.W. (1976) Occurrence and thermoresistance of spores of psychrophylic and psychrotophic aerobic sporeformers in soil and foods. *J. Appl. Bacteriol.*, **41**, 1–11.
Mislivec, P.B., Douglas, R.G. and Kautter, D.A. (1972) Toxic moulds in black and white peppercorns. *Abstr. Annu. Meet. Am. Soc. Microbiol.*, **72**, 27.
Misra, N. (1987) Mycotoxins in spices. III. Investigation on the natural occurrence of aflatoxins in *Coriandrum sativum* L. *J. Food Sci. Technol.*, **24**, 324–5.
Misra, N., Rathore, A. and Miskra, D. (1989) Mycotoxins in spices. II. Investigation on natural occurrence of aflatoxins in *Cuminum cyminum* L. *Int. J. Tropical Plant Dis.*, **7**, 81–3.
Modlich, G. and Weber, H. (1993) Vergleich verschiedener Verfahren zur Gewürzentkeimung. Mikrobiologische und Sensorische Aspekte. *Fleischwirtsch.*, **73**, 337–43.
Morse, D.L., Pickard, L.K., Guzewich, J.J., Devine, B.D. and Shayegani, M. (1990) Garlic-in-oil associated botulism: episode leads to product modification. *Am. J. Public Health*, **80**(11), 1372–3.
Munasiri, M.A., Parte, M.N., Ghanehar, A.S., Sharma, A., Padwal-Desai, S.R. and Nadkarni, G.B. (1987) Sterilization of ground prepackaged Indian spices by gamma irradiation. *J. Food Sci.*, **52**, 823–4, 826.
Nakamura, M. and Kelly, K.D. (1968) *Clostridium perfringens* in dehydrated soups and sauces. *J. Food Sci.*, **33**, 424–6.
Narvaiz, P., Lescano, G., Kairiyama, E. and Kaupert, N. (1989) Decontamination of spices by irradiation. *J. Food Saf.*, **10**, 49–61.
Neumayr, L. and Leistner, L. (1981) Mitteilungsbl. Bundesanstalt f. Fleischforsch., Kulmbach, 72, 4600.
Neumayr, L., Promeuschel, L., Arnold, I. and Leistner, L. (1983) Gewürzentkeimung. Verfahren und Notwendigkeit. Abschlussbericht für die Adalbert-Raps-Stiftung zum Forschungsvorhaben, Institut für Fleischforschung, Kulmbach.
Nieto-Sandoval, J.M., Almela, L., Fernandez-Lopez, J.A. and Munoz, J.A. (2000) Effect of electron beam irradiation on color and microbial bioburden of red paprika. *J. Food Prot.*, **63**, 633–7.
Noda, F., Hayashi, K. and Mizunuma, T. (1980) Antagonism between osmophilic lactic acid bacteria and yeasts in brine fermentation of soy sauce. *Appl. Environ. Microbiol.*, **40**, 452–7.
Ohhira, I., Jeong, C.M., Miyamoto, T. and Kataoka, K. (1990) Isolation and identification of lactic acid bacteria from traditional fermented sauce in Southeast Asia. *J. Dairy Food Sci.*, **39**(5), 175–82.
Ohnishi, H. (1957) Studies on osmophilic yeasts. Part II. Factors affecting growth of soy yeasts and others in the environment of a high concentration of sodium chloride. *Bull. Agric. Chem. Soc. Jpn*, **21**, 143.
Orillo, C.A. and Pederson, C.S. (1968) Lactic acid bacterial fermentation of burong. *Appl. Microbiol.*, **16**, 1669–71.
OSHA (1984) Occupational exposure to ethylene oxide. *Occup. Saf. Health Admin. Fed. Reg.*, 49 (122), 25734.
Otofuji, T., Tokiwa, H. and Takahashi K. (1987) A food-poisoning incident caused by *Clostridium botulinum* toxin A in Japan. *Epidemiol. Infect.*, **99**, 167–72.
Pafumi, J. (1986) Assessment of the microbiological quality of spices and herbs. *J. Food Protect.*, **49**, 958–63.
Pal, N. and Kundu, A.K. (1972) Studies on *Aspergillus* spp. from Indian spices in relation to aflatoxin production. *Sci. Cult.*, **38**, 252–4.
Palumbo, S.A., Rivenburgh, A.I., Smith, J. L. and Kissinger, J.C. (1975) Identification of *Bacillus subtilis* from sausage products and spices. *J. Appl. Bacteriol.*, **38**, 99–105.
Peter, K.V. (2001) *Handbook of Herbs and Spices*, CRC Press, Woodhead Publishing Limited.
Pirie, D.G. and Clayson, D.H.F. (1964) Some causes of unreliability of essential oils as microbial inhibitors in foods, in *Microbial Inhibitors in Foods* (ed N. Molin), Almqvist & Wiksell, Stockholm, pp. 145–50.
Powers, E.M., Ay, C., El-Bisi, H.M. and Rowley, D. B. (1971) Bacteriology of dehydrated space foods. *Appl. Microbiol.*, **22**(3), 441–5.
Powers, E.M., Lawyer, R. and Masuoka, Y. (1975) Microbiology of processed spices. *J. Milk Food Technol.*, **38**, 683–7.
Powers, E.M., Latt, T.G. and Brown, T. (1976) Incidence and levels of *Bacillus cereus* in processed spices. *J. Milk Food Technol.*, **39**, 668–70.
Pruthi, J.S. (1983) *Spices and Condiments Chemistry, Microbiology, Technology*, Academic Pess, New York.
Raksakulthai, N. and Haard, C.S. (1992) Correlation between the concentration of peptides and amino acids and the flavour of fish sauce. *ASEAN Food J.*, **7**(2), 86–90.
Roberts, D., Watson, G.N. and Gilbert, R.J. (1982) Contamination of food plants and plant products with bacteria of public health significance, in *Bacteria and Plants* (eds M.E. Rhodes-Roberts and F.A. Skinner), The Society for Applied Bacteriology Symposium Series No. 10, Academic Press, London, pp. 169–95.
Roling, W.F.M., Timotius, K.H., Stouthamer, A.A. and van Verseveld, H.W. (1994) Physical factors influencing microbial interactions and biochemical changes during the Baceman stage of Indonesian kecap (soy sauce) production. *J. Ferm. Bioeng.*, **77**(3), 293–300.

Russell, A.D. (1971) The destruction of bacterial spores, in *Inhibition and Destruction of the Microbial Cell* (ed W.B. Hugo), Academic Press, London, pp. 451–612.

Sahay, S.S. and Prasad, T. (1990) The occurrence of aflatoxins in mustard and mustard products. *Food Addit. Contam.*, **7**, 509–13.

Salmeron, J., Jordano, R., Ros, G. and Pozo-Lora, R. (1987) Microbiological quality of pepper (*Piper nigrum*) II. Food poisoning bacteria. *Microbiol. Aliments Nutr.*, **5**, 83–6.

Salzer, U., Bröker, U., Klie, H. and Liepe, H. (1977) Wirkung von Pfeffer und Pfefferinhaltsstoffen auf die Mikroflora von Wurstware. *Fleischwirtschaft*, **57**, 2011–4, 2017–21.

Sanderson, K., McMeekin, T.A., Indriati, N., Anggawati, A.M. and Sudrajat, Y. (1988) Taxonomy of halophilic and halotolerant bacteria from Indonesian fish and brine samples. *ASEAN Food J.*, **4**(1), 31–7.

Satchell, F.B., Bruce, V.R., Allen, G. Andrews, W.H. and Gerber, H.R. (1989) Microbiological survey of selected imported spices and associated fecal pellet specimens. *J. Assoc. Off. Anal. Chem.*, **72**, 632–7.

Scharf, M.M. (1967) Sterilization of spices by in situ salt formation. US Patent 3,316,100, Ser. no. 455, 327, April 25, 1967.

Schwab, A.H., Harpestad, A.D., Schwarzentruber, A., Lanier, J.M., Wentz, B.A., Duran, A.P., Barnard, R.J. and Read, R.B., Jr. (1982) Microbiological quality of some spices and herbs in retail markets. *Appl. Environ. Microbiol.*, **44**, 627–30.

Scott, P.M. and Kennedy, B.P.C. (1973) Analysis and survey of ground black, white and capsicum peppers for aflatoxins. *J. Assoc. Off. Anal. Chem.*, **56**, 1452–7.

Scott, P.M. and Kennedy, B.P.C. (1975) The analysis of spices and herbs for aflatoxins. *Can. Inst. Food Sci. Technol. J.*, **8**, 124–5.

Seenappa, M. and Kempton, A.G. (1980a) *Aspergillus* growth and aflatoxin production on black pepper. *Mycopathologia*, **70**, 135–7.

Seenappa, M. and Kempton, A.G. (1980b) Application of minicolumn detection method for screening spices for aflatoxin. *J. Environ. Sci. Health*, **15**, 219–31.

Seenappa, M., Stobbs, L.W. and Kempton, A.G. (1979) The role of insects in the biodeterioration of Indian red peppers by fungi. *Int. Biodeterioration Bull.*, **15**, 96–102.

Severs, D. (1974) Salmonella food poisoning from contaminated white pepper. *Epidemiol. Bull.*, **18**, 80.

Shamshad, S.I., Zuberi, R. and Qadri, K. (1985) Microbiological studies on some commonly used spices in Pakistan. *Pak. J. Sci. Ind. Res.*, **28**(6), 395–9.

Shank, R.C., Wogan, G.N. and Gibson, J.F. (1972a) Dietary aflatoxins and human liver cancer. I. Toxigenic moulds in foods and foodstuffs of tropical South East Asia. *Food Cosmet. Toxicol.*, **10**(1), 51–60.

Shank, R.C., Wogan, G.N., Gibson, J.B. and Nondasuta, A. (1972b) Dietary aflatoxins and human liver cancer. II. Aflatoxins in market foods and foodstuffs of Thailand and Hong Kong. *Food Cosmet. Toxicol.*, **10**, 61–79.

Shelef, L.A. (1983) Antimicrobial effect of spices. *J. Food Saf.*, **6**, 29–44.

Sheneman, J. (1973) Survey of aerobic mesophilic bacteria in dehydrated onion products. *J. Food Sci.*, **38**, 206–9.

Sim, T.S., Teo, T. and Sim, T.F. (1985) A note on the screening of dried shrimps, shrimp paste and raw groundnut kernels for aflatoxin-producing *Aspergillus flavus*. *J. Appl. Bacteriol.*, **59**(1), 29–34.

Singh, L., Mohan, M.S., Padwal-Desai, S.R., Sankaran, R. and Sharma, T.R. (1988) The use of gamma irradiation for improving microbiological qualities of spices. *J. Food Sci. Technol.*, **25**, 357–60.

Smith-Palmer, A., Stewart, J. and Fyfe, L. (1998) Antimicrobial properties of plant essential oils and essences against five important food-borne pathogens. *Lett. Appl. Microbiol.*, **26**, 118–22.

Soedarman, H., Stegeman, H., Farkas, J. and Mossel, D.A.A. (1984) Decontamination of black pepper by gamma radiation, in *Microbial Associations and Interactions in Foods* (eds I. Kiss, T. Deak, and K. Incze). D. Reidel Publishing Co., Dordrecht, The Netherlands, pp. 401–8.

Sorensen, S. (1987) Spice encapsulation, safe and sound. *Food*, **9**(1), 41–3.

Sorensen, S. (1989) Heat sterilization of spices. Paper presented at the First Meeting of Pepper Exporters/ Importers/ Traders/ Grinders, organized by the International Pepper Community, Bali, Indonesia, June 1–2, 1989.

Soriano, M.R. and Pardo, L.V. (1978) Studies on the improvement of soy sauce manufacture in the Philippines, in *Proceedings of the ASEAN Workshop on Soya Sauce Manufacturing Techniques*, Sub-Commitee on Protein, Singapore, pp. 121–34,

Soriano, M.R., Navarro, N.S. and Parel, S.O. (1986) Solid Substrate Food Fermentation Technology in the Philippines, in *Proceedings of the 2nd ASEAN Workshop on Solid Substrate Fermentation*, ASEAN S/C on Proteins, pp. 82–94.

Steele, J.H. (2001) Food irradiation: a public health challenge for the 21st century. *Clin. Infect. Dis.*, **33**, 376–7.

Steinkraus, K.H., Cullen, E.C., Pederson, C.S., Nellis, L.F. and Cavitt, B.K. (eds) (1983) *Handbook of Indigenous Fermented Foods*. Marcel Dekker, Inc., New York, pp. 494–8.

St. Louis, M.E., Peck, S.H.S., Bowering, D., Morgan, G.B., Blatherwick, J., Banerjee, S., Kettyls, G.D.M., Black, W.A., Milling, M.E., Hauschild, A.H.W., Tauxe, R.V. and Blake, P.A. (1988). Botulism from chopped garlic: delayed recognition of a major outbreak. *Ann. Intern. Med.*, **108**, 363–8.

Stratton, J.E., Hutkins, R.W. and Taylor, S.L. (1991) Biogenic amines in cheese and other fermented foods: a review. *J. Food Protect.*, **54**, 460–70.

Sugimoto, T., Hayashi, T., Kawashima, K. and Aoki, S. (1986) Reduction of microbial population in spices by gamma-irradiation. *Rep. Natl. Food Res. Inst.*, **48**, 82–5.

Sugiyama, S. (1984) Selection of micro-organisms for use in the fermentation of soy sauce. *Food Microbiol.*, **1**, 339–47.

Sundhagul, M., Piyapongse, S., Munsakul, S. and Bhuntumnomo, K. (1978) Soya sauce industry in Thailand: techno-economic consideration, in *Proc. of ASEAN Workshop on Soya Sauce Manufacturing Techniques*, pp. 37–47.

Surkiewicz, B.F., Johnston, R.W., Elliott, R.P. and Simmons, E.R. (1972) Bacteriological survey of fresh pork sausage produced at establishments under Federal inspection. *Appl. Microbiol.*, **23**, 515–20.
Surkiewicz, B.F., Johnston, R.W. and Carossella, J.M. (1976) Bacteriological survey of frankfurters produced at establishments under federal inspection. *J. Milk Food Technol.*, **39**, 7–9.
Sveum, W.H. and Kraft, A.A. (1981) Recovery of salmonellae from foods using a combined enrichment technique. *J. Food Sci.*, **46**(1), 94–9.
Tainter, D.R. and Grenis, A.T. (2001) *Spices and Seasonings—A Food Technology Handbook*, 2nd edn, John Wiley & Sons Inc., Publications.
Tamagawa, Y., Yamada, K., Takinami, K., Kodama, K. and Suga, T. (1975) *Proceedings of the Annual Meeting on Fermentation Technology*, p. 212.
Thiessen, F.M. (1971) Effect of heat on spices. *Fleischerei*, **22** (11), 15.
Thiessen, F.M. and Hoffmann, K. (1970) Aromatisierung: Veränderung von Gewürzen und Essenzen durch Hitzebehandlung, *Ernährungswirtsch.*, **50**, 317.
Tochikura, T. and Nakadai, T. (1988) *Science and Technology of Soy Sauce* (ed T. Tochikura), Brewing Society of Japan, p. 171.
Tschappe, N., Prager, R., Streckel, W., Fruth, A., Tietze, E. and Bohme, G. (1995) Verotoxigenic *Citrobacter freundii* associated with severe gastroenteritis and cases of haemolytic uraemic syndrome in a nursery school: green butter as the infection source. *Epidemiol. Infect.*, **114**, 441–50.
Tuley, L. (1991) Life after EO, *Food Manufacture*, April, 36–7.
Tuomi, S., Matthews, M.E. and Marth, E.H. (1974) Behaviour of *Clostridium perfringens* in precooled chilled ground beef gravy during cooling, holding, and reheating. *J. Milk Food Technol.*, **37** (10), 494–8.
Tynecka, Z. and Gos, Z. (1973) The inhibitory action of garlic (*Allium sativum* L.) on growth and respiration of some microorganisms. *Acta Microbiol. Pol.*, **5**(l), 51–62.
Uchida, K. (1982) Diversity of lactic acid bacteria in soy sauce brewing and its application. *J. Soc. Brewing* (Japan Nihon Jozo Kyokai Zasshi), **77**, 740.
Ultee, A., Gorris, L.G. and Smid, E.J. (1998) Bactericidal activity of carvacrol towards the food-borne pathogen. *Bacillus cereus. J. Appl. Microbiol.*, **85**, 211–8.
Ultee, A., Kets, E.P. and Smid, E.J. (1999) Mechanisms of action of carvacrol on the food-borne pathogen *Bacillus cereus. Appl. Environ. Microbiol.*, **65**, 4606–10.
Urbain, W.M. (1986) *Food Irradiation*, Academic Press, Orlando, Florida.
USP (1985) United States Patent 4,210,678.
Vajdi, M. and Pereira, R.R. (1973) Comparative effects of ethylene oxide, gamma irradiation and microwave treatments of selected spices. *J. Food Sci.*, **38**, 893–5.
Walkowiak, E., Aleksandrowska, I., Wityk, A. and Walkowicz, I. (1971) Sterilization of spices used in the meat industry by UV irradiation (In Pol.). *Med. Water.*, **27** (11), 694 (Food Sci. Technol. Abstr., **6**, 5S569).
Walsh, S.E., Maillard, J.Y., Russell, A.D., Catrenich, C.E., Charbonneau, D.L. and Bartolo, R.G. (2003) Activity and mechanisms of action of selected biocidal agents on Gram-positive and -negative bacteria. *J. Appl. Microbiol.*, **94**, 240–7.
Werner, H.-P., Klein, H.-J. and Rotter, M. (1970) Die Empfindlichkeit verschiedener Mikroorganismen gegen Äthylenoxyd. *Zentralbl. Bakteriol, Parasitenkd., Infektionskr. Hyg.*, Abt. 1, Orig. 214, 262–271.
WHO. (1973) *Wkly Epidemiol. Rec. World Health Organization Geneva*, (48), 377.
WHO. (1974) *Salmonella* surveillance. *Wkly Epidemiol. Rec. World Health Organization Geneva*, (42) 351.
WHO. (1982) Food-borne disease surveillance: outbreak of *Salmonella oranienburg* infection. *Wkly Epidemiol. Record. World Health Organization Geneva*, 57, 329.
Wilson, C.R. and Andrews, W.H. (1976) Sulfite compounds as neutralizers of spice toxicity for *Salmonella. J. Milk Food Technol.*, **39**, 464–6.
Wistreich, H.E., Thundivil, G.J. and Juhn, H. (1975) Ethanol vapor sterilization of natural spices and other foods. U.S. Patent 3,908,031-1975.
Wood, G.E. (1989) Aflatoxins in domestic and imported foods and feeds. *J. Assoc. Off. Anal. Chem.*, **72**, 543–8.
Yokotsuka, T. (1960) Aroma and flavor of Japanese soy sauce. *Adv. Food Res.*, **10**, 75–134.
Yokotsuka, T. (1986a) Chemical and microbiological stability of shoyu (fermented soy sauce), in *Handbook of Food and Beverage Stability: Chemical, Biochemical, Microbiological, and Nutritional Aspects* (ed G. Charalambous), Academic Press, Orlando, etc., pp. 517–619.
Yokotsuka, T. (1986b) Soy sauce biochemistry, in *Advances in Food Research* (eds C.O. Chichester, E.M. Mrak and B.S. Sweigert), Academic Press, Orlando, pp. 195–329.
Yong, F.M. and Wood, B.J.B. (1976) Microbial succession in experimental soy sauce fermentations. *J. Food Technol.*, **11**, 525–36.
Yoshii, H. (1979) Brewed food and water activity. *J. Soc. Brewing* (Japan Nihon Jozo Kyokai Zasshi), **74**, 213.
Zaika, L.L. (1988) Spices and herbs: their antimicrobial activity and its determination. *J. Food Saf.*, **9**(2), 97–118.
Zehnder, M.J. and Ettel, W. (1982) Zur Keimzahlverminderung in Gewürzen mit Hilfe ionisierender Strahlen. 3. Mitt.: Mikrobiologische, sensorische and physikalisch-chemische Untersuchungen verschiedener Gewürze. *Alimenta*, **20**(4), 95–100.

8 Cereals and cereal products

I Introduction

Cereals are the most efficient human food source, in terms of both energy supply and nutrition. People of all races rely on cereals as their main staple diet, with more than half of the world's population eating rice as their principal food. Therefore, producers, processors, the public and governmental authorities need to be aware of the spoilage, adulteration, and public health problems of these basic foods. This chapter covers only the major grains in commercial production, with less emphasis on small-scale farming and processing (Iizuka, 1957, 1958). For reviews of the microbiology of fermented Oriental and Indian cereal products see Hesseltine (1965, 1979) and Beuchat (1987). Cereals are also the basis for many animal feeds (see Chapter 4), but this chapter covers only human foods.

A *Definitions*

Cereals are the fruiting structure of a variety of grasses. The cereal grains discussed here include wheat, maize, oats, rye, rice, barley, millet, and sorghum. Fresh sweet corn, although technically a cereal, is treated as a vegetable in Chapter 5 because it is not used in dry form. These principal cereal products are utilized in several ways:

Flour is made by grinding cereals and is used in a variety of products.

Batter is a pourable mixture of flour, milk or water, and other ingredients such as sugar, salt, eggs, leavening agents, and fat. Batters are baked or cooked to produce cakes or muffins, or they are used to coat other foods such as meats, fish or vegetables.

Sponge is a batter to which yeast is added. In the sponge process, the yeast is allowed to work in a batter-like mixture before other ingredients are added.

Dough differs from batter in that it is stiff enough to be handled. In addition to the ingredients listed under batter, dough may contain bakers' yeast.

Bread is produced by fermentation of dough with yeast to produce an aerated mix, which is then heated (baked) to produce a rigid, somewhat dried product. Some types of bread are produced without yeast, especially in the Middle Fast.

Pasta is produced from wheat flour, water, semolina, farina, and other ingredients mixed to form a stiff dough of about 30% moisture. The dough is extruded or rolled into a variety of shapes and forms. Some pasta is filled with meat or cheese mixtures. Pasta may be stored frozen or chilled, or dried at about 40°C to a l0–l2% moisture level.

Noodles are a form of pasta containing added egg or egg yolk. Noodles are also made from rice.

Pastries are cakes muffins, donuts, and flaky products made from dough or batter and baked. *Filled products are* dough shells filled with custard, fruit, cream or imitation cream, honey, nuts, meats, spicy fillings or sauces, sometimes topped with sugar, fruit or meringue. Fillings may be fully cooked by baking with the casing (e.g. fruit pies) or cooked separately in bulk and filled into a baked casing (e.g. eclairs or cream pies) or spread onto a baked cake (e.g. cream cakes).

The principal microorganisms of concern are fungi and spore-forming bacteria.

Field fungi are found on or in grains at the time of harvest. Some are pathogenic on the grain and may cause blights, blemishes or discoloration, or produce mycotoxins. Few are able to grow below 0.90 a_w.

Storage fungi invade grains after harvest, causing loss of quality, weight, germinability, and nutrient value. Most are xerophilic and some produce mycotoxins.

Invasive fungi are those that grow within the kernel of grains, and hence are important in causing deterioration. They are usually detected by surface sterilizing and direct plating of grains on suitable agar media.

Contaminant fungi are found superficially on grains, and are of little consequence unless the grain is ground into flour or used as a raw material without heat processing. They are usually detected by dilution plating techniques.

Bacteria of concern are spore-formers, which survive cooking, e.g. *Bacillus cereus*, and non-sporeformers such as salmonellae, which may contaminate grains and flour. Such species may grow in pasta doughs during manufacture, particularly if moisture distribution is uneven, resulting in wet spots.

B *Important properties*

Although cereals and cereal products provide a rich source of nutrients for microbial growth, the reduced water activity of cereals and cereal product prevents the growth of most bacteria. Cereals contain carbohydrates, protein, fat, and fiber (Table 8.1), as well as minerals and vitamins, especially the B group, D and E, and have a near-neutral pH.

Baked cereal products are rarely associated with bacterial foodborne illness. Temperatures required for structural functionality inactivate vegetative bacteria and fungi; therefore interior portions of baked products are essentially free from vegetative pathogens. Bacterial spores may survive cooking procedures, however, the baking process removes water from the product, which reduces the water activity, particularly on the surface. At these reduced water activities, products typically spoil owing to fungal growth before growth of pathogens occurs. Moist cereal products such as boiled rice, raw moist dough, and batters can support bacterial growth if not controlled appropriately. Extending shelf-life of baked products using techniques that inhibit fungal growth must be carefully studied to assure that growth of potential pathogens is controlled.

Fungi present a more significant problem for cereals in all parts of the world. Being more tolerant to reduced water activity than bacteria, fungi can grow and cause spoilage in a wide variety of cereal products. More significant is the ability of fungi to produce mycotoxins before harvest, during drying or during improper drying and storage.

C *Methods of processing*

Kent and Evers (1994) describe general methods used for cereal processing. Harvesting and milling are steps common to most cereal grains. Milled cereals are mixed with other ingredients and converted into a variety of products through baking, boiling, frying, extrusion and fermentation. These processes are described in greater detail in the individual product category sections.

Table 8.1 Proximate analysis of representative food grains

Cereal grain	Carbohydrate (%)	Protein (%)	Fat (%)	Fiber (%)	Water (%)
Wheat (hard)	69.0	14.0	2.2	2.3	13.0
Maize (dry)	72.2	8.9	3.9	2.0	13.8
Rice (brown)	77.4	7.5	1.9	0.9	12.0
Oats (rolled)	68.2	14.2	7.4	1.2	8.3
Rye	73.4	12.1	1.7	2.0	11.0

Harvesting. Harvesting of large areas of cereal production in developed countries is nearly all carried out by mowing and threshing machines, such as combine harvesters. However, in smaller land holdings or in less-developed countries, mowing is carried out with less-specialized machinery or by hand. Threshing often involves animal or human labor and winnowing. Drying may be carried out in the field, in drying yards on the farm, or after transport to a central collecting station, often a cooperative. Wheat, barley, rye, and other temperate climate grains usually dry adequately in the field. Rice, grown in tropical or subtropical regions, is usually harvested wet and then dried on concrete pads in the sun or mechanically. Maize may be field-dried in some favorable climates, but climatic factors often dictate harvesting wet followed by mechanical drying.

Cereal storage varies widely, depending on climate and economic factors. In major grain-producing areas, storage is in bulk, in large concrete or steel silos, with effective insect and rodent control systems. Temperature and/or gas monitoring and control have been installed in many modern facilities. In less-developed countries, storage is mainly in bag stacks, but again may be in warehouses, sometimes with sophisticated control systems also. At local levels, storage may be in bag stacks in sheds, or even in earthen pits.

Milling. The milling process involves removal of debris and extraneous material, optional spraying with water and tempering to adjust moisture levels, removal of the hulls (bran) and the germ from the endosperm, and grinding into flour, meal or grits. The milling process varies with the type of cereal and intended end product.

Further processing. Processing and consumption of cereal grains vary widely. Most cereals are milled to remove outer hulls and to clean or polish the grain. Some staple cereals, particularly rice, and to a lesser extent oats and maize, are then eaten after simple cooking. Others are ground into flour or meal after milling, and then mixed with other ingredients into doughs or batters for baking into breads, biscuits, cakes or pastries or for drying as pasta. Flours also form the basis of a wide variety of dry mixes, sauces, gravies, sausages, meat loaves, canned foods, and confections. Some grains are precooked and partially milled to make breakfast foods, snacks and infant foods. Barley may be fermented to produce beer, and other fermented foods are produced from wheat, rice, sorghum, and maize.

D Types of final products

A wide variety of cereal products are available globally. Flour, starches, and meals, dry mixes, doughs, batters, and coatings, breads, pasta, breakfast cereals, and snack foods, and filled and unfilled pastries are described subsequently. Fermented beverages such as beer are not included.

II Initial microflora

A Fungi

Fungi cause most of the microbiological problems inherent in cereal production. In cereals, they are conveniently divided into two groups, field and storage fungi (Christensen, 1987). Field fungi invade seeds and grains before harvest and are often pathogens or commensals on the grains. For growth, field fungi generally require a_w levels >0.90, equivalent to 20–25% moisture (wet weight basis) (Christensen and Kaufmann, 1974; Magan and Lacey, 1984; Pitt and Hocking, 1997). Damage to seeds from the growth of field fungi includes blemishing, loss of germinability, shriveling, discoloration, and mycotoxin production, occurring before harvest or during drying, but not in storage (Christensen, 1965; Christensen and Kaufmann, 1974; Sauer *et al.*, 1992). The only exception is that where cereals such as maize are stored moist under refrigeration, fungal growth may continue.

Field fungi that survive drying or that recontaminate grain after drying cannot grow. However, they may remain viable for many months during storage. Depending on the precise conditions, field fungi may slowly die during storage, or may survive for long periods. Survival is longest at low temperatures and low-moisture levels (Christensen. 1987; Sauer *et al.*, 1992).

All cereal grains are exposed in the field to a wide variety of microorganisms from dust, water, diseased plants, insects, soil, fertilizer, and animal droppings. The external surfaces of grains at harvest may be contaminated by hundreds of microbial species (Hill and Lacey, 1983). If the quality of grain at harvest is being assessed, contaminant fungi are of no significance. Here, the preferred analytical method is the technique of direct plating following surface sterilization (Samson *et al.*, 1992; Pitt and Hocking, 1997). However, if the grain is to be processed, especially to produce flour, contaminant fungi are important in the assessment of the total fungal load. Many contaminant fungi are capable of growth in products derived from flour. The methodology of choice for assessing grain quality under these circumstances is direct plating without surface sterilization or dilution plating, carried out on the whole grain or on the flour (Samson *et al.*, 1992; Pitt and Hocking, 1997).

Because field fungi grow before harvest, and may have an association with the particular crop, the major fungi occurring in particular commodities are often quite specific (Table 8.2). Many of these field fungi are mycotoxigenic, posing a major threat to world food supplies because they cannot be controlled by post-harvest handling and storage techniques. This is in marked contrast to problems caused by storage fungi.

All cereal crops are susceptible to invasion by field fungi. Infection by particular types of fungi is controlled primarily by climatic factors, especially geographic location and rainfall patterns. Despite the availability of a great deal of information, little attempt has been made to date to provide an integrated picture of fungal infection of cereal crops throughout the world. The most important species invading cereals in the field will be discussed here.

Wheat, barley, and oats. The mycoflora of Scottish wheat, barley, and oats are similar (Flannigan, 1970). The most commonly occurring fungus was *Alternaria alternata* (=*Alt. tenuis*), which was present on >85% of kernels examined. *Cladosporium* species were also very common in barley and oats (85% and 95% of grains, respectively), but rather less so in wheat (77%). Other commonly occurring fungi were *Epicoccum nigrum* and *Penicillium* species.

Table 8.2 The major mycotoxigenic field fungi occurring in cereals

Crop	Fungus	Mycotoxins
Maize	*Fusarium verticillioides*	Fumonisins
	Aspergillus flavus	Aflatoxins
Wheat	*Fus. graminearum*	Deoxynivalenol, nivalenol and zearalenone
	Fus. culmorum	Deoxynivalenol and zearalenone
	Fus. crookwellense	Nivalenol and zearalenone
	Fus. equiseti	Diacetoxyscirpenol
	Fus. avenaceum	Moniliformin
Rye, wheat	*Fus. sporotrichioides*	T-2, HT-2
	Fus. poae	Diacetoxyscirpenol, T-2
Sorghum	*Alternaria alternata*	Tenuazonic acid
Barley	*Penicillium verrucosum*	Ochratoxin A

From Miller, 1994.

Alt. alternata was also the dominant species on English barley, occurring on >75% of a large number of experimental plots sampled over three seasons (Hill and Lacey, 1983). *Cladosporium cladosporioides*, *Aureobasidium pullulans* and *Epi. nigrum* were also very common. Barley in England was reported to be invaded by large numbers of *Penicillium* species, the principal ones being *Penicillium verrucosum*, *Pen. aurantiogriseum*, *Pen. hordei*, *Pen. piceum* and *Pen. roqueforti* (Hill and Lacey, 1984). *Alt. alternata* causes downgrading of cereals owing to gray discoloration and the production of mycotoxins (Watson, 1984). Gray discoloration can result from growth of *Cladosporium* species and *Epi. nigrum* also.

From freshly harvested wheat grains in Egypt, 77 fungal species from 26 genera were isolated (Moubasher *et al.*, 1972). These included 16 species of *Aspergillus* and 21 of *Penicillium*. Other genera of importance were *Alternaria*, *Cladosporium*, and *Fusarium*. No indication was given that any of these species were causing spoilage or unacceptable deterioration. The dominant species were *Aspergillus niger* and *Pen. chrysogenum*.

In a study of freshly harvested barley in Egypt, 37 genera and 109 species were recovered (Abdel-Kader *et al.*, 1979). The dominant genera were *Aspergillus*, represented by 25 species, *Penicillium* (32 species), *Rhizopus*, *Alternaria*, *Fusarium*, and *Drechslera*.

In Japan, an important disease of wheat and barley is termed red mold disease, which is caused by *Fusarium* species (Yoshizawa *et al.*, 1979). They found *Fus. graminearum* to be the dominant species responsible. Grains contained trichothecene mycotoxins.

Korean barley also showed heavy infection with *Fusarium* species (Park and Lee, 1990). Percentage infection in individual samples ranged from 4% to 72%. The average for 23 samples taken from the 1987 crop was about 40%, but was only 19% for 34 samples from the 1989 crop.

In contrast, Thai wheat showed little invasion by *Fusarium* species (Pitt *et al.*, 1994). Principal fungi found were *Alt. alternata* (13% total infection), *Clad. cladosporioides* (19%) and *Pen. aurantiogriseum* (36%). Storage fungi, especially *Eurotium chevalieri* and *Eur. rubrum* were present at much higher levels (each 46% of grains examined) than in other samples reported here, indicating that storage conditions in Thailand are more conducive to growth of such spoilage fungi than exist in temperate zones.

The principal invasive field fungi in US wheat and barley have been reported to be *Alternaria*, *Fusarium*, *Drechslera*, and *Cladosporium*. *Alternaria* was present in nearly 100% of wheat kernels (under the pericarp) and of barley kernels (under the hull) (Christensen, 1965, 1978, 1987).

Almost 100 fungal species were found in wheat samples taken from Australian wheat from the 1992–1994 harvests, US wheat from 1993 and 1994, and European wheat from the 1994 harvest (Pitt *et al.*, 1998b). Frequently occurring species included *Epi. nigrum*, *Clad. cladosporioides*, *Bipolaris sorokiana*, *Drechslera tritici-repentis*, *Nigrospora oryzae*, and in some areas *Fus. graminearum* and other *Fusarium* species. However, the extent of infection with *Alternaria* species exceeded that of all other fungi combined. The two common *Alternaria* species found were *Alt. alternata* and *Alt. infectoria*, though percentage infections varied widely.

Alt. alternata was commonly isolated from samples from both Australian and American sources, with the levels of infection by this species found in individual kernels in a single sample (200 kernels drawn from 500 g samples) sometimes exceeding 80%. Infection levels were much lower in European samples, the highest being 14%. Levels of *Alt. infectoria* were equally variable, reaching up to 100% in Australian samples, 90% in European samples and 70% in American samples.

In Australia, the distribution of infection by *Alt. alternata* and *Alt. infectoria* showed a well defined geographical influence. In the north eastern Australian wheat belt (Queensland and New South Wales), *Alt. alternata* was more common in hard wheats and *Alt. infectoria* in softer general purpose and feed wheats, but overall mean levels were similar across grades. The range of infection across individual samples was very wide, from 0% to 83% for *Alt. alternata* and 0% to 100% for *Alt. infectoria* (Pitt *et al.*, 1998b).

The incidence of *Alt. alternata* was much lower in samples from the southern Australian states (Victoria and South Australia) and very low in those from Western Australia, where mean levels of *Alt. alternata* within a sample were commonly $\leq 2\%$. In contrast, the incidence of *Alt. infectoria* was always high in Southern and Western Australian wheats, where the mean infection levels in the samples from the 1992 and 1993 seasons were always $\geq 45\%$.

Samples from North America showed less obvious geographical trends. Overall, percentages of infection by *Alt. alternata* and *Alt. infectoria* were similar. European wheats were heavily and quite uniformly infected with *Alt. infectoria*, but were virtually free from *Alt. alternata*, irrespective of country of origin and type of wheat (Pitt *et al.*, 1998b).

The significance of these figures lies in the fact that *Alt. alternata* is known to produce mycotoxins, in particular tenuazonic acid, whereas *Alt. infectoria* causes discoloration and downgrading, but is non-toxigenic (Webley *et al.*, 1997).

Maize. Developing ears of maize are encased in strong, protective husks, which greatly reduce invasion by fungi from aerial or dust contamination. The main entry routes for fungi appear to be by systemic infection or insect damage to developing cobs. The following discussion is relevant only to dried maize and not to fresh sweet corn. Because fresh sweet corn is harvested at an early state of maturity, it is not as susceptible to advanced fungal infection as dry maize. *Fusarium* is the predominant pathogenic fungal genus causing spoilage of maize in the ear, the most commonly occurring species being *Fus. verticillioides* (=*Fus. moniliforme*), *Fus. graminearum*, and *Fus. subglutinans* (Burgess *et al.*, 1981; Marasas *et al.*, 1984). *Fus. graminearum* usually causes a generalized rot, with a pronounced reddish discoloration of grains and husk, and with pinkish to red mycelium also visible on the grain surface.

Fus. verticillioides, in earlier literature reported as *Fus. moniliforme*, is endemic in maize in the US (Cole *et al.*, 1973), South Africa (Marasas *et al.*, 1979), Zambia (Marasas *et al.*, 1978), Thailand (Pitt *et al.*, 1993), Australia (Burgess *et al.*, 1994), Indonesia (Pitt *et al.*, 1998a), and probably all other maize-growing areas as well.

The economic importance of these *Fusarium* diseases in maize is exacerbated by the fact that all produce potent mycotoxins of considerable, even devastating, significance to the health of humans and domestic animals. This will be discussed in the following.

Of equal importance to the *Fusarium* diseases is the mycotoxigenic fungus *Asp. flavus*. In the early literature, *Asp. flavus* was regarded only as a storage fungus, but in the mid-1970s, it was realized that freshly harvested maize in the southeastern US sometimes was infected with *Asp. flavus* (Lillehøj *et al.*, 1976a,b; Shotwell, 1977). Maize from the cooler areas in the Midwestern US, however, showed little if any pre-harvest invasion. High growing temperatures, >30°C, favored invasion. Plant stress also appeared to be important, at least under laboratory conditions (Lillehøj, 1983).

Maize from Southeast Asia is heavily invaded by *Asp. flavus* also. It was present in 85% of more than 150 samples of Thai maize, at up to 100% of grains in infected samples and in 17% of all individual grains examined (Pitt *et al.*, 1993). Similar figures were obtained from Indonesia, except that 38% of all individual grains examined were infected (Pitt *et al.*, 1998a).

It has been shown that *Asp. flavus* can grow in young maize plants (Kelly and Wallin, 1986). Systemic growth leading to cob invasion is a possibility. Maize can also be infected pre-harvest by *Penicillium* species. More than 6% of some hundreds of samples of Midwestern US maize were infected with penicillia, the most common species being *Pen. oxalicum* and *Pen. funiculosum* (Mislivec and Tuite, 1970). *Pen. citrinum* was the most common species isolated from 154 samples of maize from Thailand, being found in 67% of samples, up to 60% of kernels in infected samples and 6% of all kernels examined. *Pen. funiculosum* was found in 42% of samples, up to 56% in infected samples and in 4% of kernels overall (Pitt *et al.*, 1993). In Indonesian maize, *Pen. citrinum* was also common, being found in 45% of 82 samples and 4% of kernels overall. *Pen. oxalicum* was found in 10% of samples and 1% of kernels overall, but *Pen. funiculosum* was uncommon (Pitt *et al.*, 1998a). The role of these penicillia in subsequent spoilage is uncertain.

Other fungi commonly isolated from Southeast Asian maize included *Lasiodiplodia theobromae* and *Fus. semitectum*, with significant levels of *Rhizoctonia solani*, *Rhizopus oryzae* and *Trichoderma harzianum* also in Thai maize (Pitt *et al.*, 1993). In addition, Indonesian maize contained high levels of *Asp. niger* and *Eurotium* species, probably indicating long storage (Pitt *et al.*, 1998a).

A much less extensive mycoflora was found in Egyptian maize than in wheat, with numbers of both genera and species being only 50% of those in the latter crop (Moubasher *et al.*, 1972). *Asp. niger* and *Pen. chrysogenum* were the dominant species.

Rice. During cultivation and to the point of husk removal, Southeast Asian rice carries a wide variety of fungi, especially *Trichoconiella padwickii* (=*Alt. padwickii*, *Curvularia* species, *Fus. semitectum*, *Nigrospora oryzae*, *Chaetomium* species, *Phoma* species, and *Diplodia maydis* (Iizuka, 1957, 1958; Majumder, 1974; Kuthubutheen, 1979; Pitt et al., 1998a). The same genera were reported as the dominant fungi on freshly harvested Indian rice (Mallick and Nandi, 1981).

Sorghum. The mycoflora encountered in sorghum from Thailand was rather dissimilar to that found in most other commodities. *Asp. flavus* was present at very high levels, comparable with maize, and in contrast with other small grains. However, *Curvularia lunata*, *Curv. pallescens*, other *Curvularia* species, *Alt. alternata*, *Alt. longissima*, *Fus. moniliforme*, *Fus. semitectum*, *Las. theobromae*, *Nigrospora oryzae*, and *Phoma* species, all present in high numbers, indicated that sorghum is invaded by an unusually wide range of field fungi (Pitt *et al.*, 1994).

Grains are usually contaminated with yeasts as well as molds, the number and variety depending primarily on climatic factors. Most are *Basidiomycetes* requiring high a_w for growth, and normally present no serious problems during storage.

B Bacteria

Bacteria found in cereals include *Bacillus*, *Lactobacillus*, *Pseudomonas*, *Streptococcus*, *Achromobacter*, *Flavobacterium*, *Micrococcus and Alcaligenes* (Fung, 1995). Spore-forming bacilli are the most significant group, as they may survive cooking and subsequently grow if restricting water activity is not achieved. Rope spores are of particular concern for moist bakery products.

Fecal indicators are at a low level in field grains, unless there is considerable animal activity in the fields. The coliform group is an inappropriate indicator of fecal contamination for field grain because many are part of the natural flora of plants. *Bacillus subtilis* and *B. cereus* are present in low numbers.

Weather during the growing season affects bacterial levels, with higher levels persisting through the harvest during humid growing seasons (Legan, 2000). Certain bacterial groups are nearly always present (Deibel and Swanson, 2000), including:

1. Aerobic colony count, from 10^2 to 10^6 cfu/g.
2. Psychrotrophic bacteria, numbering from 10^4 to $>10^5$ cfu/g.
3. Actinomycetes, up to 10^6 cfu/g.
4. Aerobic sporulating bacteria, from 10^0 to 10^5 cfu/g.
5. Coliform group, from 10^2 to 10^4 cfu/g.

In rice, *Pseudomonas*, *Enterobacter*, *Micrococcus*, *Brevibacterium* and *Bacillus* species have been found (Iizuka, 1957, 1958). *B. cereus* is a frequently present pathogen in rice (Johnson, 1984; Kramer and Goepfert, 1989).

III Primary processing

A Effects of processing on microorganisms

Most cereals apart from rice are allowed to dry in the field, or are sometimes dehydrated, and are not further processed before manufacture into retail products. Cereals are almost universally stored, often for long periods, and shipped in the dried state. Storage and shipping can be considered as further processing in the context of this chapter, because major effects on the microflora, especially the mycoflora, can occur.

Unlike other cereals, rice is harvested at 20–24% moisture before full maturity to prevent the grain fissioning. It must then be dried to <14% moisture for microbiologically safe storage. It is usually dehusked before long-term storage. The process of milling to remove the husks strips away many contaminants. Moreover, the process causes substantial heating, so that freshly milled rice has been both dried and rendered effectively sterile (Pitt *et al.*, 1994).

Storage. As noted earlier, field fungi are not capable of growth during storage, and hence have little or no effect on quality; deterioration or mycotoxin production once a_w has fallen <0.90. The field fungi eventually die (Lutey and Christensen, 1963), so that after storage and international transport levels are usually inconsequential—a fortunate fact that reduces the hazard of distributing plant pathogens across international boundaries (Wallace and Sinha, 1975).

Fungi capable of growth at lower water activities rapidly contaminate grains after harvest. Sources include trucks, conveyors, sacks, and particularly storage bins. Dust, generated each time grain is handled, is a major source of fungal spores in the environment of storage facilities and mills. Moreover, grain dust is sometimes collected and returned to the grain. In one study, dust from handling sound US maize contained up to 3×10^4 bacterial spores/g. The number of particles in tailing dust was estimated at 1.7×10^9/g, and the mold spore concentration was up to 2.7×10^6/g. Handling of moderately moldy grain would contribute much higher levels of spores to the dust (Martin and Sauer, 1976).

Stored grains are moribund and lack the defense mechanisms found in fresh grains. Associations between crop and fungus no longer exist. Spoilage fungi are saprophytes, and spoilage of stored grains therefore depends primarily on physical factors (Pitt and Hocking, 1997). The most important of these in the present context are water activity, temperature, and gas composition. The influence of these factors is described in the following section.

Water activity (a_w). Storage fungi have no association with plants, and normally invade grains only after harvest. The exception is *Asp. flavus*, which is capable of invading maize (also peanuts and cottonseed; see Chapter 9) before harvest. Storage fungi by definition are capable of growth at low a_w levels, in contrast to the field fungi. For example, field *Penicillium* species grow down to no lower than 0.86 a_w, whereas the storage species have minima near 0.81–0.83 a_w (Mislivec and Tuite, 1970).

Water activity is the most important parameter governing growth of fungi in grains during storage. The ability of fungi to colonize stored grains is greatly dependent on their water relations, in particular the minimum a_w for growth. Table 8.3 shows the minimum a_w values permitting growth of some common storage fungi.

The most important storage fungi on grains are species of *Aspergillus*, especially species which produce the ascomycete state known as *Eurotium*. *Eurotium* species used to be known as the "*Asp. glaucus* Group" or even as "*Asp. glaucus*", both incorrect terms. Some *Penicillium* species are important spoilage fungi also, especially in cooler climates. Other fungi found in stored cereals include *Wallemia sebi* and a variety of less important genera. All of these fungi fit the definition of xerophile (Pitt, 1975).

Table 8.3 Minimum water activities and approximate moisture contents for growth of some storage fungi in cereals

Fungus	Minimum a_w for growth	Moisture content (%)[a]
Eur. halophilicum	0.68	13.4–14.3
Wallemia sebi	0.69	14.5–15.0
Eur. rubrum	0.70	15.0
Asp. penicillioides	0.73	15.0–15.5
Asp. candidus	0.75	16.5
Asp. ochraceus	0.77	17.0
Asp. flavus	0.80	18.0
Pen. verrucosum	0.78	17.3

From Christensen and Kaufmann (1974, 1977a); Pitt (1975); Pitt and Hocking (1997).
[a]Percentages for wheat. Maize, barley, oats, rye and sorghum are similar.

Table 8.4 Moisture levels and temperatures influencing growth or absence of growth of fungi in grains

	No mold growth			Mold growth			
Grain	Moisture (%)	Temperature (°C)	Time (days)	Moisture (%)	Temperature (°C)	Time (days)	Reference
Wheat	15.0–15.5	5, 10	365	16.0–16.5	5, 10	365	Papavisas and Christensen, 1958
Maize	<17.5	35	32	16.5	25–35	32	López and Christensen, 1967
	15.5	10	365	15.5	25	365	Sauer and Christensen, 1968
	18.5	5	120	18.5	25–30	120	Christensen and Kaufmann, 1977a
Rice	13.5	20–25	120	14.5	20–25	120	Fanse and Christensen, 1966
	14	5–15	465	14.5	30	465	Christensen, 1965
	14	25	150	14.5	30	60	Ito *et al.*, 1971

Data on a_w limits for fungal growth can be applied to stored cereals, to provide minimum moisture contents permitting fungal growth, as shown in Table 8.4. Extensive correlations have been demonstrated between minimum a_w levels for fungal growth obtained under laboratory conditions and growth of fungi in stored cereals. For example, US wheat, stored at 13.8–14.3% will be specifically invaded by *Eur. halophilicum*, one of the most xerophilic *Eurotium* species, which cannot compete against other storage fungi when the a_w-value of the grain is slightly higher (Christensen *et al.*, 1959). Recovery of *Asp. penicillioides*, incorrectly referred to as *Asp. restrictus* in most of the US literature, indicates storage of wheat for some months at 14.0–14.5% moisture (Sauer *et al.*, 1992).

Other *Eurotium* and *Aspergillus* species, and *W. sebi*, are able to grow at a_w >0.75, corresponding to moisture contents >14.5% (Christensen and Kaufmann, 1974).

Temperature. Most fungi grow best at temperate to tropical ambient temperatures, over the range 10–35°C. However, optima vary widely. Many *Aspergillus* and *Eurotium* species have optima in the range 30–40°C (ICMSF, 1996), whereas those of *Penicillium* species are usually lower, 20–30°C. Tropical grain spoilage is normally caused by *Aspergillus* species, whereas in cool temperate climates *Penicillium* is predominant.

Heating may occur in stored grain and then thermophilic fungi are able to grow and may become dominant (see later). Other parameters, especially a_w, significantly influence the temperature range for growth of a fungus. A decrease in a_w restricts the temperature range for growth (Ayerst, 1969), so that the differences in temperature optima noted above are of great importance in grain storage.

Table 8.5 Recommended maximum moisture levels for short- and long-term storage of major grain crops

	Maximum moisture for storage stability (% moisture)	
	Short-term (1 year)	Long-term
Wheat	13–14	11–12
Maize	13	11
Barley	13	11
Oats	11	10
Rice	14	–

From ICMSF, 1980.

Under ambient oxygen tension, fungal growth in cereals depends on the interaction of temperature, water activity, and storage time (Table 8.4). The figures for percent moisture given must be seen as approximate, because various factors, for example, variety, climatic factors, and maturity, affect grain composition, and hence sorption isotherms (ICMSF, 1980). For short storage times, higher moisture levels and higher temperatures can be tolerated. However, such treatment makes the lot a poor risk for further storage (Christensen and Kaufmann, 1964).

On the basis of considerations such as these, recommended maximum moisture levels for grains to be held in short-term (<1 year) and long-term storage have been established (Table 8.5).

Low temperatures are employed for grain storage in some parts of the world, where grains, especially maize, must be harvested wet and cannot for various reasons be dried rapidly. In the US, freshly harvested maize, frequently with a moisture level in excess of 20%, is refrigerated immediately after harvest. However, some fungi, especially *Pen. aurantiogriseum* (=*Pen. cyclopium*, *Pen. martensii*), continue to grow and may cause spoilage and mycotoxin formation during long-term storage (Ciegler and Kurtzman, 1970).

In China, maize in the northern province of Jilin is harvested wet just before the onset of winter, and allowed to freeze. Drying is then carried out from the frozen state throughout the winter months.

Gas composition. Reduced access to oxygen has been somewhat less promising than low water activity and low temperature as a means of inhibiting fungal growth in grains. Although food-borne fungi are primarily aerobes, many can grow slowly in the presence of trace levels of oxygen. Anaerobic conditions, especially with preservatives, have been successfully used in the US (Meiering et al., 1966) and France (Pelhate, 1976) for preserving moist grains for about 2 weeks at 20°C. This type of storage should not be prolonged. *Pen. roqueforti* is capable of growth on grain under low O_2 conditions in the presence of preservatives (Le Bars and Escoule, 1974). Mycotoxin production may result (Häggblom, 1990).

B Spoilage

Fungal damage. Field fungi can cause considerable damage to grains. The main effects are discoloration, which usually results in downgrading, and mycotoxin production. Pink or black tips in wheat are evidence of infection by *Fusarium* or *Alternaria*, respectively. The grain may still be suitable for some purposes, as long as levels of mycotoxins are acceptable. Heavily moldy grain may produce distinctive musty odors. Severe mold growth will be associated with poor gluten quality in flour and breads made from it. Heavily moldy grain will have low germinability and inferior malting properties and hence may be unacceptable to brewers and distillers. The nutritional value of such grain is reduced, but may still be suitable for animal feeds. Trials may be needed to establish whether losses in nutritional value have been severe, and more particularly, whether mycotoxin levels will cause ill thrift, requiring either rejection of the batch or allowance for poorer feed conversions.

Fungi cause deterioration in stored grains in a variety of ways. These include decrease in germinability, discoloration, off-odors, dry matter loss, visible growth, chemical, and nutritional changes, heating, caking, and mycotoxin formation (Sauer *et al.*, 1992). These effects have been reviewed (Christensen and Meronuck, 1986; Sauer, 1988; Mills, 1989).

Spontaneous heating. If not adequately controlled, fungal growth in stored grain can lead to heating and much more serious losses. Temperatures can rise to the point where the ability of seeds to germinate is destroyed, the grain becomes discolored or indeed to the point where spontaneous combustion occurs. Catastrophic losses are less common in the present day, because the causes of heating and the ability to monitor it have improved greatly. Most heating in grains is the direct result of fungal growth and metabolism.

Grain with a moisture content of 10–12% is safe for ambient storage if temperatures remain constant, but is not safe under the influence of temperature gradients within the bulk. This is a problem in geographic areas with wide diurnal or seasonal temperature changes. Temperatures differing within the bulk by only 0.5–1°C may cause moisture to migrate from warm to cool sections of the bin. This may increase the moisture level of the latter areas sufficiently to permit molds to grow (Wallace *et al.*, 1976).

In heating grains a well-recognized succession of fungi occurs. Each fungus raises the temperature to as high a level as it can endure (Christensen and Gordon, 1948), and in so doing produces water as the result of metabolism, which raises the a_w to the point where less xerophilic species can grow. Xerophilic *Eurotium* species can raise the temperature to 35°C, and in so doing may provide enough moisture for *Asp. candidus* and *Asp. flavus* to grow. They in turn can raise the temperature to 50°C and keep it there for weeks. If the heat and water are dissipated, microbial action will cease. However, in some cases thermophilic fungi, such as *Thermoascus aurantiacus*, may take over and raise the temperature to 60–65°C. At this point, the grain darkens and may even look charred. Thermophilic bacteria may then grow, raising the temperature to 70–75°C, producing oxidizable hydrocarbon compounds of low ignition temperature (Milner *et al.*, 1947). In rare instances, chemical oxidation then may ignite these compounds. Spontaneous combustion is more common in oilseeds, such as soybeans, cottonseed, flaxseed and sunflower seeds than in cereal seeds (Milner and Geddes, 1946). Catastrophic losses can sometimes occur (Sauer *et al.*, 1992).

Temperatures of 50–55°C may be sufficient to cause caking and blackening of grain, with a burned appearance, although an ignition temperature has not been approached. Whether fire or microbiological damage has caused this appearance can sometimes be very important for insurance purposes.

Chemical changes. During spoilage, chemical changes take place, the most important being the production of free fatty acids as a result of mold lipase activity on lipid glycerides. Although some evidence points to the formation of free fatty acids in the absence of fungi (Loeb and Mayne, 1952), free fatty acids appear to be an objective measurement of fungal damage in grains.

Increases in free fatty acids during grain storage are proportional to increased moisture content, increased time of storage, increased temperature and increased invasion by fungi. Fatty acid values increase much more rapidly and to a higher final level in fine particles of maize than in broken kernels, and much more rapidly and to a higher final level in broken kernels than in sound maize (Sauer and Christensen, 1969).

The respiration process of fungal growth is similar to that of the grain itself. Therefore, the production of carbon dioxide occurs at a stable, slow rate in clean grains, but increases above this baseline level with the invasion and growth of fungi. The loss in dry matter can be approximated by measuring the carbon dioxide lost to the atmosphere; this correlates well with the amount of fungal activity (Steele, 1969). The grain is generally rejected as food by the time fungal growth has caused 1–2% weight loss, but the point of rejection depends on local customs and availability of alternative food supplies (Saul and Harris, 1979).[1]

C *Pathogens and toxins*

Mycotoxins. Mycotoxins, toxic metabolites produced by common fungi during growth, are the most important of the microbial health hazards in cereals and cereal products. Cereal crops harbor many of the most important mycotoxins. Mycotoxins are low molecular weight compounds which, unlike bacterial toxins, do not cause immediate symptoms of poisoning when ingested. Recent evidence suggests that mycotoxins have played a major role in the health of mankind throughout history (Matossian, 1981, 1989; Austwick, 1984) and are still insidiously involved in human disease and death, particularly in the less-developed regions of the world.

Ergotism. Historically, ergotism was the first mycotoxicosis to be recognized. Ergotism results from ingestion of rye breads on which the fungus *Claviceps purpurea* has grown. The relationship between the disease and the formation of ergots in maturing grain was established by the 10th century, and by 1700 the relationship of the disease with the formation of ergots (fungal sclerotia) in the ovaries of grain, especially rye, had been established. During milling, ergots are not readily separated from sound grain, but become fragmented and dispersed throughout the flour. The toxins are not destroyed during bread making (Fuller, 1968; Matossian, 1989).

The toxins in ergots are a range of alkaloids, derivatives of lysergic acid, which cause constrictions of the peripheral blood vessels, trembling and ultimately sequestration and necrosis of skin. The well-documented decline in mortality in England in the second half of the 18th century coincided with a change from a rye to a wheat-based diet: it has been suggested that this was due to the consequent reduction in poisoning due to ergots (Matossian, 1981). It appears certain that ergotism killed tens of thousands of people in Europe during the Middle Ages.

The last known outbreak of ergotism in Europe occurred in France in 1954 (Fuller, 1968). More than 200 people suffered illness and hallucinations, and several died from consumption of bread baked from ergoty rye. Over 140 people were affected by ergotism from ergoty wild oats in Ethiopia in 1978 (King, 1979).

Poisoning attributed to the consumption of pearl millet infected with *Claviceps* occurred in India in 1956–1957 and 1975 (Patel *et al.*, 1958; Krishnamachari and Bhat, 1976). Some hundreds of people were affected in several outbreaks.

Acute cardiac beri-beri. A second major disease caused by mycotoxins in cereals, known as acute cardiac beri-beri, occurred in Japan, especially in the second half of the 19th century. Acute cardiac beri-beri caused heart distress, then nausea and vomiting, pain and anguish, and in extreme cases, respiratory failure and death. This disease was apparently associated with fungal growth in rice—the "yellow rice syndrome"—probably due to toxins from *Pen. citrinum* and *Pen. islandicum* (Uraguchi, 1969). Acute cardiac beri-beri declined in importance in the early part of the 20th century. This coincided with implementation of a government inspection scheme, which dramatically reduced the sale of moldy rice in Japan (Uraguchi, 1969).

[1] Referenve stopped.

Alimentary toxic aleukia. The third major human disease associated with mycotoxins in grain is alimentary toxic aleukia (ATA), a disease that caused the deaths of many thousands of people in Russia, especially in the Orenburg District north of the Caspian Sea, during 1944–1948. In some localities, mortalities were as high as 60% of those afflicted. ATA is an exceptionally severe disease: symptoms include fever; hemorrhaging from nose, throat, and gums; and necrotic skin lesions—symptoms more similar to those from radiation poisoning than from other mycotoxicoses or bacterial poisoning (Joffe, 1978).

ATA was due to T-2 toxin, a trichothecene produced during growth of *Fusarium* species in rye left over winter in the field owing to acute wartime and post-war labor shortages. It has been shown retrospectively that outbreaks of ATA also occurred in the former Soviet Union in 1932 and in 1913 (Matossian, 1981). No doubt episodes occurred in earlier times too.

None of the toxins just described is causing serious problems at the present time. Current concern over mycotoxins in cereal grains centers on the following (Table 8.2).

1. Aflatoxins, produced in maize by *Asp. flavus* and to a lesser extent *Asp. parasiticus*. Aflatoxins also occur in nuts (see Chapter 9).
2. Trichothecenes and zearalenone, produced by *Fusarium* species in wheat and other small grains, less commonly in maize.
3. Fumonisins, produced by *Fus. verticillioides* (=*Fus. moniliforme*) in maize.
4. Ochratoxin A, formed by *Pen. verrucosum* in barley. Ochratoxin A is sometimes produced in long stored grains by *Asp. ochraceus* and related species, but is seldom of commercial significance in grain intended for human consumption.
5. Toxins, especially tenuazonic acid, produced by *Alternaria* species in small grains.

Each of these problems will be described in the following sections.

Aflatoxins in maize. Aflatoxins are discussed in detail in relation to nuts in Chapter 9. The discussion here will relate specifically to aflatoxins formation in maize. Aflatoxins may be produced by *Asp. flavus* and the closely related *Asp. parasiticus*. In maize, *Asp. flavus* is the dominant fungus, comprising 80–90% of isolates from random samples, and maize usually contains only B aflatoxins, consistent with the dominance of *Asp. flavus* (Davis and Diener, 1983). In the early literature, *Asp. flavus* was regarded only as a storage fungus. However, the discovery of aflatoxins in freshly harvested maize in the southeastern US showed that this fungus could invade pre-harvest (Lillehøj *et al.*, 1976a,b; Shotwell, 1977). Insect damage to cobs (Lillehøj *et al.*, 1980; Hesseltine *et al.*, 1981, Setamou *et al.*, 1998; Hell *et al.*, 2000) is still considered to be the major pre-harvest source; however, *Asp. flavus* can invade maize cobs down the silks without any insect vector (Jones *et al.*, 1980) and can also be systemic in the maize plant (Kelly and Wallin, 1986). Plant stress, such as climatic factors, is also important (Lillehøj, 1983).

Aflatoxin is a major problem in maize in most countries where it is produced (Table 8.6). Several approaches aimed at reducing aflatoxin levels in maize have been advocated in recent years: reduction by effective drying and storage (Siriacha *et al.*, 1991); testing for variation in susceptibility of maize cultivars to *Asp. flavus* invasion, seeking resistance factors (Brown *et al.*, 1993; Windham and Williams, 1998; Brown *et al.*, 1995, 1999, 2001; Tubajika and Damann, 2001); the use of other fungi such as *Asp. niger* as a competitor (Wicklow *et al.*, 1987, 1988); and the use of non-toxigenic *Asp. flavus* in competition with the naturally occurring toxigenic strains (Brown *et al.*, 1991) and the use of *B. subtilis* distributed on dusky sap beetles to reduce *Asp. flavus* infection (Dowd *et al.*, 1998).

Trichothecenes in wheat and other small grains. The principal mycotoxigenic fungi associated with wheat, barley, and other small grain crops are *Fusarium* species, which produce a range of trichothecene

Table 8.6 Levels of total aflatoxin reported in maize and maize meals from various countries

Country	Period	No. of samples	Total aflatoxin: Incidence (%)	Total aflatoxin: Average level (μg/kg)[a]	Reference
Argentina	1983–1994	2271	20	Low	Resnik *et al.*, 1996
Australia, Qld	1978, 1983	979	2	<1	Blaney, 1981; Blaney *et al.*, 1986
Brazil	1985–1986	328	13	9	Sabino *et al.*, 1989
Costa Rica	?	3000	80	70–270[c]	Mora and Lacey, 1997
India	?	2074	26[b]	5–35[c]	Bhat *et al.*, 1997
Indonesia	1988–1992	71	49	65	Pitt *et al.*, 1998a,b
Mexico	1989–1995	?	?	66	Carvajal and Arroyo, 1997
Philippines	1967–1969	98	97	110	Stoloff, 1987
	1988–1992	155	78	53	Pitt, personal communication
Thailand	1967–1969	62	35	93	Stoloff, 1987
	1988–1991	89	64	188	Pitt *et al.*, 1993
Uganda	1966–1967	48	40	53	Stoloff, 1987
US, maize belt	1964–1965, 1967, 1974	1 763	2.5	<1	Stoloff, 1987
US, Southeast	1968–1970, 1974	175	4.1	18	Stoloff, 1987

[a]Average of all samples. Figures for total aflatoxin on average would be 10% higher.
[b]Percentage exceeding 30 μg/kg aflatoxin B_1.
[c]Depending on district.

toxins. The most important are deoxynivalenol (DON) and nivalenol (NIV), and the estrogenic toxin, zearalenone.

As noted earlier, infection of cereals by fungi is greatly affected by climatic factors, especially geographic location and rainfall patterns. *Fus. graminearum* is generally the most common *Fusarium* species in wheat from North America and China (Wang and Miller, 1988), whereas *Fus. culmorum* is dominant in cooler areas such as Finland, France, Poland, and the Netherlands (Snijders and Perkowski, 1990; Saur, 1991). *Fus. avenaceum* is common in wheat from all regions studied (Miller, 1994). *Fus. crookwellense* appears to be rare in Canada and Poland, but common in irrigated wheat in South Africa (Scott *et al.*, 1988). *Fus. poae* and *Fus. sporotrichioides*, together considered to be the most likely cause of ATA (Joffe, 1978), are found more commonly in colder climates (Miller, 1994).

A large literature has built up in recent years concerning the incidence of trichothecene toxins, especially DON and NIV, in wheat from around the world (see review by Wilson and Abramson, 1992). The major problem with DON appears to be in North America, where it was first reported from winter wheats in Ontario and Quebec (Scott *et al.*, 1981; Trenholm *et al.*, 1983). Widespread occurrence was reported later (Abramson *et al.*, 1987; Scott *et al.*, 1989; Trucksess *et al.*, 1995; Trigostockli *et al.*, 1996, 1998). DON has also been reported from wheat in the United Kingdom (Osborne and Willis, 1984), Argentina (Gonzalez *et al.*, 1996; Dalcero *et al.*, 1997), Austria (Adler *et al.*, 1995), Hungary (Fazekas *et al.*, 2000; Rafai *et al.*, 2000), and Scandinavia (Pettersson *et al.*, 1986; Langseth and Elen, 1996; Eskola *et al.*, 2001).

NIV appears to be more common than DON in North Asia and Europe. It has been reported from Japan (Tanaka *et al.*, 1985; Yoshizawa and Jin, 1985), Korea (Lee *et al.*, 1985, 1986), the United Kingdom (Tanaka *et al.*, 1986), Poland (Ueno *et al.*, 1986, Grabarkiewicz-Szczesna *et al.*, 2001), Canada (Scott, 1997), Germany (Müller and Schwadorf, 1993), Sweden (Pettersson *et al.*, 1995) and other countries.

Zearalenone, an estrogenic toxin, is also widespread. It is usually found whenever DON or NIV are present (Miller, 1994; Vrabcheva *et al.*, 1996). It is unclear whether zearalenone is a human carcinogen (Kuiper-Goodman *et al.*, 1987). In Canada, where zearalenone has a relatively high incidence, estimates of ingestion do not exceed tolerable daily intake figures (Kuiper-Goodman, 1994).

Other *Fusarium* species occurring in small grains are of lesser importance.

Fumonisins in maize. *Fus. verticillioides*, previously known as *Fus. moniliforme*, was described from maize more than a century ago. Reports of its possible involvement in human or animal disease date back almost as far, coming from Italy, Russia, and the US by 1904 (Marasas *et al.*, 1984).

The disease of horses and related animals known as equine leukoencephalomalacia (LEM) was known as early as 1850 in the United States maize belts, with epidemics involving hundreds or thousands of horses in 1900, the 1930s, and as recently as 1978–1979. It also occurs in other parts of the world, including Australia, Argentina, China, Egypt, New Caledonia, and South Africa (Marasas *et al.*, 1984). *Fus. moniliforme* was not positively identified as the cause of LEM until 1971 (Wilson, 1971).

The most important toxin produced by *Fus. verticillioides* is fumonisin B (Bezuidenhout *et al.*, 1988), which consists of a 20-carbon aliphatic chain with two ester-linked hydrophilic side chains. It has been established that fumonisins are the toxins responsible for LEM (Marasas *et al.*, 1988).

In animals, fumonisin B has a low acute toxicity, but can induce cancer in rats (Gelderblom *et al.*, 1988; NCTR, 2001), suggesting that this toxin may have a role in human esophageal cancer. Maize is the major staple food in areas of the Transkei where esophageal cancer is endemic, and the most striking difference between areas of low and high incidence was the much greater infection of maize by *Fus. moniliforme* in the high-incidence areas (Marasas *et al.*, 1981). Several isolates of *Fus. moniliforme* from high-incidence areas were acutely toxic to ducklings, but did not produce other known toxins such as moniliformin (Kriek *et al.*, 1981). Fumonisins may also have a role as potentiators of carcinogens such as aflatoxin, but long-term studies on the toxicity of these compounds are not yet complete (Kuiper-Goodman, 1994).

Ochratoxin A in barley. Barley and other grains grown in cool temperate zones are frequently contaminated with ochratoxin A (Krøgh, 1987; Table 8.7). Ochratoxin A is produced by several common fungi, and confusion over the source in that region has persisted until quite recently. The original source of ochratoxin A, *Asp. ochraceus* (sometimes referred to as *Asp. alutaceus*) is rare in cool zones.

The earlier literature reported *Pen. viridicatum* as the producer of ochratoxin A (Scott *et al.*, 1972; Krøgh *et al.*, 1973; Northolt *et al.*, 1979) and this has persisted until recently (e.g. Krøgh, 1987). It

Table 8.7 Natural occurrence of ochratoxin A in foods of plant origin[a]

Food	Country	No. of samples	Positive (%)	Range (μg/kg)	Year
Maize	US	293	1.0	83–166	1971
	France	461	1.3	20–200	1974
	Yugoslavia	542	8.3	6–140	1979
	UK	29	38	50–500	1980
Barley	US	127	14	10–40	1971
	Denmark	50	6	10–190	1974
	Yugoslavia	64	12	14–27	1979
	USSR	48	2	3800	1978
	Poland	296	7	20–470	1981
Wheat	US	577	2	5–115	1976
	Yugoslavia	130	8	14–135	1979
	Denmark	151	1	15–50	1981
Bread	UK	50	2	210	1980
Coffee beans	US	267	7	20–360	1974

Adapted from Krøgh, 1987.

is now well established that *Pen. verrucosum*, a psychrotrophic fungus that appears to have a specific association with barley and other cereals, is the true source (Pitt, 1987; Frisvad, 1989).

Barley is used worldwide for brewing; sometimes ochratoxin is found in European beers (El-Dessouki, 1992; Jiao *et al.*, 1994; Weddeling *et al.*, 1994; Gareis, 1999), though usually at low levels.

Barley is also used in Europe and parts of North America as a major ingredient in pig feeds, and the association of barley, ochratoxin A, and nephritis in pigs was established 25 years ago (Krøgh *et al.*, 1973). Ochratoxin A present in feeds containing barley is transferred to porcine tissues, including kidneys, lean meat, liver, and depot fat (Hald, 1991a), which form important components in the human diet in many European countries.

Studies in Europe and Canada have shown that ochratoxin A is present at readily detectable levels in human blood (Frolich *et al.*, 1991; Hald, 1991b; Petkova-Bocharova and Castegnaro, 1991) and milk (Gareis *et al.*, 1988; Micco *et al.*, 1991, 1995; Breitholtz-Emanuelsson *et al.*, 1993; Hohler, 1998; Skaug *et al.*, 1998, 2001).

More surprising is a report of excessive levels of ochratoxin in human milk in Africa (Jonsyn *et al.*, 1995). The significance of ochratoxin A as a human toxin is far from clear. On the basis that OA is primarily a chronic kidney toxin, and carcinogenic effects are less important, the Joint FAO/WHO Expert Committee on Food Additives (JECFA) has set a tolerable daily intake of 100 ng/kg bw/week, far above the 5 ng/kg bw/week tolerable intake level recommended by the European Commission, who have considered carcinogenicity to be more important (Walker, 2002).

Other grains, including wheat and rye, may also contain ochratoxin A, especially in Europe. Bread made from contaminated grain may contain significant levels of ochratoxin; indeed, bread is probably the major source of this toxin in human blood in Europe (Frank, 1991).

Alternaria toxins in small grains. *Alternaria* species, especially *Alt. alternata*, are widespread in wheat around the world, with reports from Australia (Rees *et al.*, 1984), Iran (Lacey, 1988), Iraq (Sulaiman and Husain, 1985), Korea (Lee *et al.*, 1986) and Spain (Jimenez *et al.*, 1987). *Alt. alternata* produces a range of mycotoxins, the most important being tenuazonic acid. This compound is acutely toxic to a wide range of animals (Visconti and Sibilia, 1994). Little is known about the effects of tenuazonic acid or other *Alternaria* toxins on humans.

Bacterial pathogens. Properly handled cereals are so dry that bacteria cannot grow on them. However, they can mechanically carry the viable cells of many pathogens if the grain is exposed to animal or human contamination. Insects, rodents, birds or human beings may introduce *Salmonella*, *Escherichia*, *Shigella* or *Klebsiella* (Brooks, 1969). Of these, *Salmonella* is the genus of principal concern. The sources may be animal contact in the field; trucks, and railroad cars whose previous cargo was animals, meat scraps, fish meal, poultry or poultry products; insects, mice, rats, and birds in the mill; and human carriers. Once the microorganisms are dried to the low water activity of the cereal, they can remain inactive but viable almost indefinitely, even though their numbers will decrease with time. If the cereal is subsequently exposed to a moist environment, including food, growth of salmonellae may occur.

B. cereus food poisoning has been traced to rice dishes, often because cooked rice has been left at kitchen temperature (Schiemann, 1978). Spores survive the cooking, and germinate and grow if the rice is then held between 10°C and 49°C for several hours (Johnson *et al.*, 1984). Rice should be prepared in small quantities, and refrigerated or held <60°C before consumption (Gilbert *et al.*, 1974). Adding other foods to the rice may speed the growth of *B. cereus* and, therefore, increase the hazard (Morita and Woodburn, 1977).

D CONTROL (cereals)

Summary

Significant hazards	• Trichothecenes, DON and NIV, in wheat outside of Australia. • Ochratoxin A in barley in northern North America and Europe. • Fumonisins, trichothecenes, and aflatoxins in maize.
Control measures	
Initial level (H_0)	• Harvest grain from areas with minimal crop stress if possible. • Dry crops rapidly and completely.
Increase (ΣI)	• Store and transport at <12% moisture. • Avoid temperature fluctuations during storage. • Fumigation and other pest management practices minimize damage that promote subsequent fungal growth.
Reduction (ΣR)	• Usually not possible.
Testing	• Visual observation for fungal growth, product damage, and insect infestation. • Monitoring of temperature changes, or carbon dioxide production during storage as an indicator of fungal growth. • Use of an electronic nose for the same purpose where this is technologically feasible. • After harvest, chemical analysis of mycotoxin concentrations and segregation of lots with higher and lower concentrations can be of value.
Spoilage	• Control measures that prevent mycotoxin production control most spoilage fungi and bacteria.

Comments. The traditional way to avoid microbial growth in grains is to dry them thoroughly and keep them dry. Adequate ventilation in storage bins will remove moisture, prevent condensation, lower and equilibrate temperatures and prevent heating. Bag stacks and manual handling of grain has given way to bulk handling and storage, with great improvements in control of insect and fungal damage, even in tropical areas (Champ and Highley, 1988). Drying technology has shown great advances, with sophisticated computer control of drying rates and temperatures now in use in developed countries.

Monitoring grain storage for increase in temperature or carbon dioxide production provides effective warning systems. Thermocouples placed not >2 m apart are an effective means of detecting hot-spots where spoilage is occurring. Grain is a good heat insulator, so that even a minor rise in temperature at the position of a thermocouple may indicate a nearby hot spot (Christensen and Kaufmann, 1977a,b).

More modern approaches to grain storage rely on fumigation, and sealed storage under controlled atmospheres, especially in tropical and subtropical regions where insect damage is a major problem (Champ *et al.*, 1990). Fumigants are gases added specifically to kill insects, and in some cases they also destroy fungi. Fumigants are usually used as a rapid method for killing insects, and then subsequently removed by ventilation. A variety of gases have been used as fumigants, either singly or in combination, including ethylene dichloride, carbon tetrachloride, carbon disulfide, ethylene dibromide, chloropicrin, hydrogen cyanide, ethylene oxide, methyl chloride, carbon dioxide, methyl bromide, and phosphine. For a variety of reasons, only methyl bromide and phosphine are in widespread use

Table 8.8 Suggested target dosages for gaseous treatments of grain at 25°C

Gas	Time (days)[a]	Concentration[b]	Concentration × time
Carbon dioxide	15	>35%	–
Oxygen	20	<1%	–
Phosphine	7	100 mg/m^3	–
Methyl bromide[c]	1–2	–	150 g h/m^3
Hydrogen cyanide	1	–[d]	–

From Annis, 1990a.
[a] In cases of slow gas introduction or poor gas distribution, increased exposure times may be necessary.
[b] Minimum concentration achieved at end of exposure.
[c] Regulartory prohibition in certain regions owing to safety concerns.
[d] Concentration necessary not well defined.

(Annis, 1990a). Recommended concentrations for use are given in Table 8.8. Environmental considerations are resulting in the phasing out of methyl bromide, and the search for alternative fumigants continues.

Controlled atmospheres may be used for grain storage. This technique relies on continuous application of atmospheres low in O_2 or with high CO_2 concentrations. The recommended approach is to add such gas mixtures to sealed storage and to maintain the grain in totally sealed systems. Where this is not practicable, continuous flow of such gas mixtures may be possible (Annis, 1990a). Recommended O_2 and CO_2 concentrations are given in Table 8.8.

Fumigation and controlled atmospheres help control fungal growth on grains by direct destruction of spores, by inhibition of growth or by killing insects that damage kernels. Fumigants, which merely destroy insects, have no lasting effect (Vandegraft *et al.*, 1973); however, methyl bromide destroys fungi as well as insects (Majumder, 1974), and phosphine has some fungicidal properties (Hocking and Banks, 1993). Use of methyl bromide is restricted in the EU and the US owing to toxicological concerns. Modified atmospheres controlling insects may also have a substantial effect in controlling fungi (Hocking, 1990).

Control of insect and fungal damage in grain stores is of particular importance in the tropics, where most grains are stored in sacks in warehouses unsuitable for sealing and fumigation. Recent approaches to sealing stacks of bags in such stores, fumigating and then maintaining the sealed stacks under controlled atmospheres have shown the potential to reduce grain losses markedly (Annis, 1990b; Graver, 1990).

Many investigators have suggested using heat or chlorine to destroy microorganisms in grains. However, this technology has been little used, with the recognition that such remedial processes do not destroy mycotoxins and are not a substitute for clean grain.

In the same way, ionizing radiation at a level of 2–3 kGy destroys fungi that commonly spoil rice (Iizuka and Ito, 1968; Ito *et al.*, 1971). Such a process is not yet permitted, or is unacceptable to consumers, in many countries.

IV Flours, starches, and meals

A *Effects of processing on microorganisms*

Flour made from wheat is the predominant flour product in international trade; however, flours are also produced from rice, maize, and other cereals. Grains for milling are subjected to a number of cleaning and aspiration steps before tempering. These steps reduce the microbial level of the grains as they enter the milling operation. However, analysis of samples from five Australian flour mills revealed that neither traditional wheat cleaning nor scouring changed the microbial load appreciably, nor did scouring affect the microbial counts of flour (Eyles *et al.*, 1989). Some contamination of wheat occurred during tempering. After tempering for a predetermined time, the grain then passes through a milling and sifting sequence that separates the hulls (bran) and the germ from the endosperm, which is crushed into flour. Maize is sometimes ground to flour, meal or grits without a tempering step.

Sound, clean grains, especially those properly screened and tempered, contain few microorganisms. However, contact with mill machinery introduces contamination of a quantity and variety that is affected by the degree of cleanliness of the mill equipment. A correlation exists between microbial levels in the product and mill sanitation. High mold levels exist in poorly maintained factories, up to 3.4×10^6/g of flour residues on equipment (Christensen and Cohen, 1950) and up to 10^8/g in mill dust (Semeniuk, 1954).

Flour with $\leq 12\%$ water will not support microbial growth (Hesseltine and Graves, 1966). However, water from any source will encourage microbial growth in flour and in residues on mill machinery. Such residues can become damp from high atmospheric humidity, from condensate on cool surfaces, from improper clean-up procedures (Hesseltine and Graves, 1966; Graves *et al.*, 1967) or from insect activity (Thatcher *et al.*, 1953).

Wheat flour may be treated with oxidizing agents such as chlorine dioxide or chlorine, and may also be bleached by treatment with benzoyl peroxide (Thatcher *et al.*, 1953). Bleaching reduces microbial numbers to some degree, but spores are little affected.

The bacterial flora of flour is much more diverse than that of the wheat from which it was made. Both wheat and the flour made from it contain many psychrotrophs, flat sour bacteria, and thermophilic spore-forming bacteria (Hesseltine, 1968). These are of particular interest to processors of canned and chilled foods in which flour is an ingredient. Rope bacilli originate from both insects and insanitary equipment; they are basically soil bacteria.

Aerobic colony counts in flour from different geographical regions range from 10^2 to 10^5/g, (Table 8.9) with most samples having populations in the range of 10^4/g (Eyles *et al.*, 1989; Legan, 1994; Richter *et al.*, 1993). Milling conditions and weather during growth and harvesting influence microbial levels. Although the coliform group is sometimes used as an indicator of sanitary conditions, the use of this group is of questionable value for flour. Coliforms are part of the natural flora of flour and were isolated from 72% to 100% of flours in the UK from the 1988 to 1994 harvest years (Legan, 1994). Interpretation of reported coliform levels in the literature must consider the methodology used, as results generated using most probable number determinations are frequently lower than those obtained using newer methods such as Petrifilm™. Unpublished data suggest that confirmation media (Brilliant Green Bile Broth) may inhibit certain lactose-fermenting, Gram-negative bacteria in flour (Swanson, personal communication). *Escherichia coli* was present in 28–56% of flour from the UK (Legan, 1994) and 13% of flour in the United States (Richter *et al.*, 1993).

Table 8.9 Total aerobic colony count and mold count reported in wheat flour from various countries

Country	Period	No. of samples	Total aerobic colony count/g	Yeast and mold count/g	Reference
Australia	1986	24	4 000	930	Eyles *et al.*, 1989
UK	1988–94	41–66/year	42 000–370 000	1 400–4 900	Legan, 1994
US	1989	>1300	3300–42000	930	Richter *et al.*, 1993

Fungal counts (yeast and mold counts) commonly approach 1 000/g (Table 8.9). The fungi in finished flours are mostly *Penicillium* and *Eurotium* species and *Asp. candidus*. The genera of fungi found in flours differ markedly from those found in the wheat from which the flour was made, demonstrating again the importance of the mill as a source of the microorganisms. Yeasts become important only if flour becomes wet.

Flour and starch form the foundation for many dry mix products. Flour is mixed with other dry ingredients such as powdered egg, dry milk, spices and seasons and leaven agents. These products are sold in the dry state for subsequent use by foodservice establishments and consumers. The dry mixing process has little impact on numbers of microorganisms. Maintenance of dry conditions is important to prevent incidental contamination.

B Saprophytes and spoilage

For flour and maize meal, 12% water is a critical level, below which no microbial growth will occur. Above 12%, some xerophilic fungi can grow; and at 17% some bacteria also can grow (Hesseltine and Graves, 1966). The rate of growth is proportional to the water activity and temperature (Kent-Jones and Amos, 1957). If the moisture level is high, as in a flour and water paste, bacteria will predominate because they grow faster than fungi. Lactic acid bacteria will begin acid fermentation, followed by alcoholic fermentation by yeasts and then oxidation to acetic acid by *Acetobacter*. This sequence is less likely in stored flour because of reductions in viable counts during storage. In the absence of lactic acid bacteria, micrococci may acidify damp flour, and in their absence *Bacillus* spp. may grow, producing lactic acid, gas, alcohol, acetoin, and small amounts of esters and aromatic compounds. It is characteristic of most flour pastes to develop an odor of acetic acid and esters.

Canners often set low specifications for sporulating organisms that can survive canning procedures and spoil the product. Good hygienic practices are necessary to control spore levels.

C Pathogens and toxins

Mycotoxins can be an important health hazard in flours and meals as mycotoxins present in grain will carry through into the flour, surviving heating steps or other procedures designed to kill fungi. In addition, moist flour and maize meal (>14% water) will support fungal growth in the same way that grains do, and mycotoxins can be produced (Seeder *et al.*, 1969; Bullerman *et al.*, 1975).

Of 70 molds isolated from flour and bread, 16 were *Aspergillus*, 48 *Penicillium* and six were other genera. Fifteen of the 48 *Penicillium* species and one strain of *Asp. ochraceus* produced mycotoxins on laboratory media (Bullerman and Hartung, 1973). Mycotoxins do not present a health hazard in dry mixes when mycotoxins are controlled in flour and meal, because fungal growth cannot occur at the low water activities of these products.

Salmonellae also present a health hazard in flours, meals and dry mixes. Salmonellae were present in 0.3–3.0% of wheat flour, with variation due to harvest season and wheat type (Richter *et al.*, 1993). Salmonellae will remain viable in dry flour for several months (Dack, 1961). They are, however, quite sensitive to heat. When maize flour at 10–15% moisture was spray inoculated with 10^5 salmonellae/g and held at 49°C, 99.9% of cells were inactivated in 24 h (van Cauwenberge *et al.*, 1981). Normal cooking of flour-based products inactivates the salmonellae.

B. cereus is common in wheat flour, but is usually present in very low numbers (Eyles *et al.*, 1989). Kaur (1986) reported that *B. cereus* levels in flour were low and only occasionally >10/g. The organism cannot grow in the dry flour and thus does not represent a hazard at this stage.

D CONTROL (flours, starches and meals)

Summary

Significant hazards	• Mycotoxins as for grains. • *Salmonella*.
Control measures	
Initial level (H_0)	• Use grain harvested from areas with minimal crop stress to control mycotoxins. • Test for mycotoxins as appropriate (refer to section in grains).
Increase (ΣI)	• Maintain grain and flour at <12% moisture for mycotoxin control. • Maintain processing facilities dry condition with a minimal use of water. • Use dry cleaning methods for machinery.
Reduction (ΣR)	• Tempering grain in chlorinated water before milling may slightly reduce hazard levels. • Some decrease in mycotoxins occurs during processing.
Testing	• Visual observation for fungal growth, product damage and insect infestation in incoming grain. • Ultraviolet screening of maize. • Test for mycotoxins in incoming grain where appropriate. • Environmental monitoring for salmonellae. • Coliforms are not a useful an indicator. • Spore-forming bacteria when used in canned products.
Spoilage	• Control measures that prevent mycotoxin production control most spoilage fungi and bacteria.

Control measures. The most important control method in flour manufacture is the control of moisture in the grain, in the equipment and in the processing environment. Properly dried grain (<14%) effectively inhibits the growth of bacteria and fungi, which can produce mycotoxins. There are no processing steps that effectively eliminate biological hazards in the production of flour and meal, however, the following procedures should be implemented to reduce the prevalence of potential hazards in this raw commodity.

Tempering of grains for flour manufacture may be carried out in chlorinated water if particularly low levels of microorganisms are desired. This has minimal effect because chlorine is readily inactivated by the organic material present. Milling machinery should be cleaned regularly to avoid accumulation of static material, where fungi and insects thrive. Cleaning of all dry product areas should be done without using water. The flour should contain no more than 15%, and preferably no more than 14% water, in order to reduce fungal growth. Mills should have a program of rodent and insect control and should avoid condensation of moisture where it can drop back into the product or accumulate on the sides of bins, hoppers, conveyors, etc.

After boots and elevators in the flour system have been fumigated, they should be cleaned immediately to remove the insects, which otherwise will decompose and contribute large numbers of microorganisms to the flour.

Testing. Visual observation for fungal growth, product damage and insect infestation in incoming grain can identify conditions that contribute to formation of mycotoxins. Ultraviolet screening of maize is a particularly useful tool for identification of pockets of fungal growth. Mycotoxin tests of incoming grain may be appropriate in regions with persistent mycotoxin problems or in years when conditions are likely to contribute to mycotoxin formation.

Salmonellae are capable of becoming established in wet or moist locations in flour or grain processing facilities. Manufacturing environmental monitoring is useful to prevent establishment of harborage sites for salmonellae. The focus of environmental sampling should be in wet or moist locations, such as tempering areas and condensation, and accumulated static material in equipment.

Use of coliforms as an indicator is of little value because of natural variation in the incoming materials and lack of processing steps to reduce levels.

Starches that are used in canned products require low levels of spores to prevent spoilage after processing.

V Dough

A Effect of processing on microorganism

Dough is a stiff mixture of flour, liquid, and/or fat and other ingredients used to make a wide variety of baked, boiled or fried products. Dough is made primarily from wheat flour, however, flour, and meal from other grains are often used around the world. Rice is a notable example used in Asian cultures. Other starchy, non-cereal commodities such as potato, cassava, and soy may also be used to prepare dough. Subsequent sections on breads, pasta, and pastry describe unique features of specific dough products in greater detail. This section summarizes common features of dough.

The addition of water to dried flour or meal provides a substrate that readily supports microbial growth. Yeasts are used to leaven dough, but the addition of water to flour and meal may also provide an opportunity for spoilage and pathogen growth if controls are not implemented. Temperature, water activity and competitive exclusion are the primary controlling factors.

Processing temperature has a major impact on the handling characteristics of dough, and also on the potential for microbial growth. Soft, pliable, moist doughs, and high fat doughs are frequently handled at cool temperatures (ca. 15°C). Conversely, stiff dry dough may be handled at warmer temperature (30°–40°C) for functional purposes, e.g. yeast dough is held at these warmer temperatures to stimulate yeast activity. Bacterial pathogens are not a hazard of concern in yeast-leavened dough because of competitive exclusion. Although *Staphylococcus aureus* is rarely isolated from wheat flour (Richter *et al.*, 1993), it is a potential concern in dough processed at warmer temperatures without yeasts if the water activity is favorable for growth.

A modern development is the baking of bread in specialist retail bread shops. Bread doughs mixed in a central facility are frozen and distributed to retail outlets, where they may be held frozen or defrosted, proofed, and baked immediately. Such frozen doughs may also be sold direct to consumers, in which case they contain high yeast populations to permit a frozen storage life as long as 18 weeks while maintaining adequate fermentative activity (Kulp, 1991). In many western countries, chilled raw or partially baked doughs designed to be finished by the consumer have also become popular. The products are mostly breads and rolls, cookies, pizza, and baking powder biscuits, primarily packaged in hermetically sealed cardboard, plastic film and metal containers. All chilled doughs are chemically leavened, usually by sodium acid pyrophosphate and sodium bicarbonate. Yeast action may continue during storage and burst the containers (Lannuier and Matz, 1967; Lamprech, 1968). Whereas most doughs are baked immediately, chilled doughs are sometimes held for prolonged periods at temperatures suitable for fungal growth. Water activity of chilled dough ranges from <0.80 in cookie dough to ca. 0.94–0.95 in roll and biscuit dough. Fungal and bacterial counts in the doughs are low unless ingredients are highly contaminated or microbial growth occurs during temperature abuse of higher a_w dough. Sources of contamination can be the flour, dry milk, eggs, sugar, spices, water, flavors and the dough-making equipment (Deibel and Swanson, 2000). Usually the fungal flora of the dough is a reflection of that of the flour (Graves and Hesseltine, 1966). Bacterial levels in fresh dough products vary widely, from 10^3/g in dinner rolls to 10^8/g (lactic acid bacteria) in buttermilk biscuits. All fresh doughs contain wild yeasts (Hesseltine *et al.*, 1969).

B Spoilage

Spoilage is not an issue for dough used immediately after preparation. Doughs in cool storage owe their stability to the following factors (Lannuier and Matz, 1967; Lamprech, 1968).

- Conditions are anaerobic, thus inhibiting the growth of filamentous fungi and aerobic bacteria.
- The leavening agent produces CO_2.
- Formulations are designed to attain a low a_w.
- Refrigeration slows microbial and enzyme activity.
- Manufacturers have strict microbiological specifications for ingredients and maintain faultless sanitation in the manufacturing operation.

If well controlled, the lactic acid bacteria in doughs produce quality sourdough breads. On the other hand, if they are permitted to multiply excessively in refrigerated doughs, packages may split open or burst owing to gas pressure, revealing dough with undesirable odor and flavor (Lannuier and Matz, 1967; Hesseltine *et al.*, 1969). Spoiled doughs contain the kinds of flora listed in Table 8.10.

Fungi are rarely involved in the spoilage of refrigerated doughs, even after 6–7 months of storage. Fungal counts usually remain at similar levels in stored doughs as those in freshly prepared doughs (Graves and Hesseltine, 1966).

C Pathogens and toxins

Refrigerated doughs may contain *Salmonella* species if ingredients such as eggs are contaminated. Dried yeast preparations also sometimes contain salmonellae. *B. cereus* is a fairly common contaminant of refrigerated doughs (Rogers, 1978), but there appear to be no reports of food poisoning attributable to this source. Mycotoxins present in the original flour will probably persist. Temperatures required to set the dough structure of baked dough products inactivate vegetative pathogens. The low levels of *B. cereus* found in flour are also likely to be inactivated during baking (Kaur, 1986).

Table 8.10 Bacteria isolated from spoiled doughs

Species	No. of isolates	Percentage
Leuconostoc mesenteroides	390	35
Leuc. dextranicum	21	2
Lactobacillus spp.	597	53
Streptococcus spp.	45	3
Bacillus spp.	15	1
Micrococcus spp.	51	5
Gram-negative rods	13	1

From Hesseltine *et al.* (1969).

D CONTROL (dough)

Summary

Significant hazards	• *Salmonella*.
Control measures	
Initial level (H_0)	• Use pasteurized ingredients such as eggs and milk. • Specify supplier control programs for other sensitive ingredients.
Increase (ΣI)	• Control water usage, maintaining dry processing conditions.
Reduction (ΣR)	• Subsequent baking to set dough structure destroys *Salmonella* species.
Testing	• Environmental monitoring for *Salmonella*. • Lactic acid bacteria for refrigerated dough. • Routine testing of other dough is not recommended. • Coliforms as indicators is of little value.
Spoilage	• Strict cleaning and disinfection of equipment and maintenance of dry conditions is essential to control potential spoilage organisms in refrigerated dough.

Control measures. The microbial quality of doughs can be maintained by using ingredients (particularly flour) of good quality, and by maintaining good bakery sanitation. Moist environmental niches can readily support the establishment of harborage sites for salmonellae. The subsequent baking process achieves temperatures of ca. 90°C to set dough structure, which greatly exceeds lethality needed to destroy *Salmonella* species.

Refrigerated doughs should have a specified shelf-life so that they can be removed from retail display before they spoil. The quality of sourdough breads can be maintained by careful adherence to traditional procedures or by the use of pure cultures.

Testing. Manufacturing environmental monitoring for salmonellae is useful to prevent establishment of harborage sites, focusing on wet or moist locations, such as condensation, and accumulated static material in equipment. Surveillance for lactic acid bacteria is useful in refrigerated dough products that support lactic growth. Routine testing of other dough is not recommended. Use of coliforms is of little value as an indicator because of the natural variation in the incoming materials and lack of processing steps to reduce levels.

VI Breads

A *Effects of processing on microorganisms*

Standard bread dough. A typical bread contains wheat flour, water or milk, salt, fat, sugar and yeast (usually *Saccharomyces cerevisiae*). Ingredients are mixed; the dough may then be permitted to rise (ferment) for several hours at 24–29°C, before being divided, shaped and finally proofed (fermented) before baking. During this period, which may last as long as 20 h, yeast enzymes ferment sugars and produce carbon dioxide in the dough.

For economic reasons, fermentation time has been reduced to the point where primary fermentation is reduced or eliminated in some products. In conventional bread production, the properties of bakers' yeast are fully utilized, but with high-speed production, some of the rheological functions of the yeast are replaced by high-intensity mixers. As insufficient time elapses for metabolic reactions necessary for flavor development, breads produced by these processes lack the flavor that only full microbial action can contribute (Sugihara, 1977). At the end of fermentation, bread doughs contain very high numbers of yeasts, and a wide variety of other microorganisms acquired from ingredients. Yeasts are the principal contributors to flavor. Bacteria (usually *Lactobacillus* species) contribute flavor in breads with fermentation periods >8 h (Kent-Jones and Amos, 1957). Lactic acid bacteria produce short-chain fatty acids, which probably contribute to good flavor in small amounts, but would give undesirable flavors and aromas in large amounts (Robinson *et al.*, 1958).

During baking, the internal temperature of the loaf closely approaches 100°C. When the crumb reaches 98°C, the optimal baking time has been reached (Stear, 1990). All vegetative microbial cells have been destroyed during this process. Subsequent fungal spoilage problems arise from aerial contamination after baking, from the slicing machine and from cooling and wrapping equipment. The level of fungal spores in the air in bread manufacturing plants has been reported as 100–2500 spores/m^3 (Knight and Menlove, 1961). However, modern facilities have much greater control over spore numbers.

Only spore-forming bacteria survive baking. If the ingredients of the dough are heavily contaminated with spores of the "rope" species, *B. subtilis*, some survival is to be expected during baking. The sources of the spores are multiple, but the highest inoculum may be from the equipment that has been contaminated by previous batches of bread.

Salt-rising bread. This product is unique in that the leavening agent is *Clostridium perfringens* rather than yeast (Robinson, 1967). Although this species can cause foodborne gastroenteritis, it is harmless when used as a leavening agent because it dies during baking. Food poisoning has never been attributed to salt-rising bread (Dack, 1961). *B. cereus* can also be used for the same purpose (Goepfert et al., 1972).

Soda crackers. The formulation for soda crackers is similar to that of bread except that baking soda is added. Yeast (*Sacc. cerevisiae*) is a principal ingredient, but since the industry began in 1840, manufacturers have relied on chance contamination by *Lactobacillus* species for acid formation in a 24 h fermentation period. Later investigations revealed the predominance of *Lactobacillus plantarum* and the secondary role of *Lb. delbruckii* and *Lb. leishmannii* in this function. Pure culture starters with these species reduce the fermentation period to about 6 h, maintain better quality and permit controlled variations in flavor (Sugihara, 1977, 1978a,b).

For crackers and similar biscuits, a continuous fermentation process has been developed using a mixed yeast and *Lactobacillus* culture. The bacteria used are *Lb. plantarum, Lb. fermentans,* and *Lb. casei* (Fox *et al.*, 1989).

Sour dough bread. The traditional method for sourdough bread uses a "starter" built up from the previous batch every 8 h. It requires a high proportion of active starter sponge for each batch of new starter (about 40%) followed by a fermentation period of 7–8 h at 27°C. The initial pH is 4.4–4.5, and the final pH 3.8–3.9. The dough requires a starter that makes up about 11% of the final dough mix. This is followed by a similar 7–8 h period of fermentation with a similar pH drop. The very high proportion of starter sponge contributes a massive inoculum and ensures a highly acid environment that contains a substantial amount of acetic acid. The low pH and the acetic acid prevent growth of spoilage microorganisms and loss of the starter (Kline et al., 1970).

In San Francisco sourdough bread the sourdough yeast is *Sacc. exiguus* (*Torulopsis holmii*), and the heterofermentative bacterium is *Lb. sanfrancisco* (Sugihara, 1977). *Sacc. exiguus* does not ferment maltose, and *Lb. sanfrancisco* ferments only maltose. Thus, the two microorganisms do not compete for the same carbohydrate, a fact that probably contributes to the survival of both species in the starter. The yeast is unusually resistant to acetic acid (Sugihara *et al.*, 1970). Excessive acidification produced by lactobacilli had a deleterious effect on dough rheology and the presence of yeast in the starter improved bread quality (Collar *et al.*, 1994).

Sour rye bread. This was first mentioned in 800 BC, and since the fifth century BC dried starters made up of cakes of fermented whole grain have been available to bakers. The souring microorganisms are *Lb. plantarum*, *Lb. brevis*, and *Lb. fermentans*. *Lb. sanfrancisco*, *Lb. pontis*, *Lb. amylovorus*, *Lb. reuteri*, *Lb. johnsonii*, and *Lb. acidophilus* may also be found in some cultures (Vogel, 1997). Lactic and acetic acids are the principal compounds formed. The technological importance of the starter varies with the pH, acidity, and lactic acid:acetic acid ratio. Only *Lb. brevis* var. *lindneri* was found to give adequate acidity over a wide range of processing conditions (Spicher, 1982). The addition of yeast speeds the fermentation process. Numbers of lactic acid bacteria increase and yeasts decrease with increased fermentation times, with populations in the range of 10^8 and 10^7 cfu/g, respectively (Rosenquist and Hansen, 2000). This bread is generally made with self-perpetuating starter sponges, but pure cultures or the previously mentioned cake, are available commercially (Sugihara, 1977).

Italian panettone. In Italy, panettone dough is the basis for Christmas fruitcake, columba (Easter cake), breakfast rolls and snack cakes. Its production is similarly based on a yeast (*Sacc. exiguus*) and one or more species of bacteria (*Lb. brevis*, *Enterobacter* and *Citrobacter* species). The madre or mother sponge has been perpetuated for centuries in very clean surroundings by a special staff (Sugihara, 1977). The preparation procedure is strikingly like that of sourdough French bread.

Idli. This is an Indian fermented bread, which together with similar products in other Asian and Middle Eastern countries, is prepared from rice and black gram mungo (*Phaseolus mungo*), a legume. The ingredients are soaked in water, combined and permitted to ferment overnight, then steamed and served hot. Leavening and acidification are primarily accomplished by *Leuc. mesenteroides*, with *Streptococcus faecalis*, and *Pediococcus cerevisiae* playing secondary acidifying roles (Mukherjee et al., 1965). Idli is unique in that the leavening action is solely from the activity of a lactic acid bacterium (*Leuc. mesenteroides*).

B Spoilage

Bread, has a short shelf-life owing to staling or mold growth. The interior or crumb has an a_w about 0.94–0.95, whereas the crust has an a_w <0.70. Fungal growth is relatively slow, so that in dry climates, the surface of a slice of bread may dry before fungal growth is evident. In humid conditions or on wrapped bread, however, fungal growth occurs in a few days.

In a 2-year study of German packaged sliced breads made from rye, wheat or blends, 437 samples were stored at 25°C and 70% relative humidity for 7 days, and fungal growth was analyzed (Spicher, 1984a). More than 65% of the samples showed visible growth. Predominant was *Pen. roqueforti* (85% of isolates); other *Penicillium* species and *Aspergillus* species made up the remainder. Breads which had been heat treated showed no spoilage, and only *Pen. roqueforti* was able to grow on breads preserved with propionic or sorbic acid.

A study of 204 samples of Dutch rye bread stored in plastic bags at 24°C for 10 weeks yielded similar results (Hartog and Kuik, 1984). Initial fungal counts were low, with >90% of samples having counts $<10^3$/g. Shelf-lives of samples from smaller bakeries were shorter (42% spoilage within 6 weeks versus 5%), attributed to higher pH and lower levels of preservatives used. *Pen. roqueforti* was again the dominant spoilage species. Its resistance to weak acid preservatives is well documented (Engel and Teuber, 1973; Pitt and Hocking, 1997).

"Chalky mold" is also an important cause of bread spoilage in Europe (Spicher, 1984b). This spoilage is due to *Endomyces fibuliger* (32% of isolates), *Zygosaccharomyces bailii* (24%), *Hyphopichia burtonii* (20%); and *Sacc. cerevisiae* (10%). The lowest growth temperature for these species was 5°C. Calcium propionate (0.3%) delayed spoilage, whereas sorbic acid (0.05%) prevented it.

Similar studies have been carried out on various bakery products (Spicher and Isfort, 1987). Of 150 samples of partially baked bread rolls or baguettes stored for 7 days at 25°C and 70% relative humidity, ~60% showed mold growth, including 92% of those packaged in CO_2, 62% of those without preservatives and 50% of those preserved with propionic acid, and also packaged under CO_2. The predominant spoilage fungi were *Pen. roqueforti* (38%), *Paecilomyces variotii* (14%), *Asp. niger* (8%), *Eur. amstelodami* (6%), and *Moniliella suaveolans* (6%). Packaging under CO_2 may fail because of leaks in packaging material.

The addition of vinegar or dilute acetic acid (5–8%, at the rate of 0.9–1.8 ml/100 cm^2) to bakery goods prevented growth of most fungi during storage (Spicher and Isfort, 1988). However, growth of *Pen. roqueforti* and *Paec. variotii* was prevented only by the addition of 10% vinegar at the higher application rate. Addition of ethanol (1.5% w/w) to the surface of baguettes or packing material also increased mold-free shelf-life (Doulia *et al.*, 2000).

Rope is a bacterial spoilage problem of bread caused by *B. subtilis* (*B. subtilis* var. *mesentericus*) or *B. licheniformis*. The most common source of contamination with these bacteria is flour and equipment that has been in contact with contaminated dough (Stear, 1990). Whole grain breads may also have higher spore counts than other products. The spores survive baking, and germinate and grow within 36–48 h inside the loaf to form a characteristic soft, stringy, brown mass with an odor of ripe cantaloupe. The bacteria are heavily encapsulated, which contributes the mucoid nature of the material. They also produce amylases and proteases that cause the breakdown of the bread structure. Conditions favoring the appearance of rope are: (i) a slow cooling period or storage >25°C; (ii) pH > 5; (iii) high spore level and (iv) moist loaf. The water activity inside a loaf is marginal for *B. subtilis* so that rope may appear in localized areas where the moisture content is high. Calcium propionate, good sanitation and good bakery practice keep it under control (Graves *et al.*, 1967; Pyler, 1973).

Red bread is usually caused by *Serratia marcescens*, which may grow if the moisture level is high enough. Although this is uncommon today, the red color formation has been described as "bleeding bread" (Legan, 2000).

The chapatti, a disc of soft, pliable unleavened bread, is a common wheat food in India. In tropical areas where the temperature is 40–45°C, chapattis that are packaged in polyethylene pouches, where the equilibrium relative humidity reaches 90–95%, spoil after ~7 days owing to the growth of aspergilli (Kameswara Rao *et al.*, 1964, 1966).

The tortilla, a flat maize pancake, is the staple cereal food in all of Central America. Dry maize is cooked with limestone, soaked in water for about 14 h, ground into a wet pasty flour, rolled, and patted into a pancake and then cooked for 4–5 min on a stove top or coals. The uncooked tortilla often has high levels of a general microflora that are largely destroyed during cooking. Tortillas are a very rich, moist medium for microbial spoilage, which will take place within 24 h under tropical conditions (Capparelli and Mata, 1975) and generally <8 days at room temperature. Addition of preservatives, such as propionic acid or fumaric acid, or refrigeration extend shelf-life (Haney, 1989).

C Pathogens and toxins

Some of the molds that grow on bread are capable of producing mycotoxins (Robinson, 1967; Bullerman and Hartung, 1973). The most important of these is *Pen. crustosum*, which has several times caused toxicosis in dogs owing to penitrem A production in moldy bread or hamburger buns (Arp and Richard, 1979; Richard *et al.*, 1981; Hocking *et al.*, 1988). However, this is not a problem for humans who usually will not eat moldy or stale bread. Samples of bread stored at 25°C and 70% relative humidity were also examined for mycotoxin formation (Spicher, 1984c). Of 110 moldy samples studied, 11 contained citrinin ($<5\ \mu g/kg$), four contained ochratoxin ($<8\ \mu g/kg$), and five contained zearalenone ($<5\ \mu g/kg$). These levels are not cause for concern. Aflatoxins, patulin, sterigmatocystin, and penicillic acid were not detected.

In canned non-acid bread, previously inoculated with spores of *Cl. botulinum*, toxin was formed on incubation if the a_w of the bread was ≥ 0.95. This a_w is equivalent to a moisture content of about 40%. As long as the water content did not exceed 36%, no toxin was formed (Wagenaar and Dack, 1954; Denny *et al.*, 1969). There have been no reports of botulism from this product.

Acid breads, whose pH is ≤ 4.8, are safe from botulism, even when the moisture content is 40% (Wagenaar and Dack, 1954; Ingram and Handford, 1957; Weckel *et al.*, 1964).

A substantial trade exists in packaged "part-baked" breads. One concern is with the possible survival of the partial cook by *Cl. botulinum* and other spores. However, retail shelf-life as long as 90 days without refrigeration are common. Growth of spores is prevented by a combination of reduced a_w and reduced pH.

Tortillas are often prepared under very primitive conditions, especially in rural areas of Central America. Fortunately, treating the maize with lime and cooking the tortillas destroys aflatoxins that might be present in the original maize, a fact of major importance to millions of people in Latin America (Ulloa-Sosa and Schroeder, 1969). Cooking also destroys bacterial pathogens; however, tortillas are usually immediately recontaminated, and because of their high moisture content and the prevailing warm temperatures, they can support the growth of *E. coli*, *Staph. aureus*, *B. cereus*, and possibly other bacterial pathogens. If tortillas are not consumed promptly after cooking, they may be responsible for widespread disease outbreaks (Capparelli and Mata, 1975).

Kaur (1986) reported that *B. cereus* levels in flour were low and only occasionally >10/g. The organism was not isolated from baked 400 g loaves made with dough containing ca.10^4/g, but did survive in 800 g loaves. It is not likely that the low levels of *B. cereus* that occur naturally will survive the baking process and grow in baked bread.

D CONTROL (breads)

Summary

Significant hazards	• None for traditional breads.
Control measures	
Initial level (H_0)	• Not applicable.
Increase (ΣI)	• Reduced a_w (<0.86) prevents growth on exterior surfaces after baking. • Interior pH, a_w and preservatives, as appropriate, combined to prevent *Cl. botulinum* growth.

(*continued*)

D CONTROL (Cont.)

Summary

Reduction (ΣR)	• Temperatures required to set dough structure (>90°C) destroy vegetative pathogens.
Testing	• Environmental monitoring for salmonellae to detect harborage sites. • Monitor air quality for fungal spores to reduce contamination after baking. • Routine testing of bread is not recommended.
Spoilage	• Maintain bread and processing environments in a dry condition to reduce mold spoilage. • Cool bread prior to packaging in plastic wrapping to avoid condensation, which promotes spoilage. • Filter air used in cooling and packing areas where possible. • Preservatives may be used to extend shelf-life in some countries. • Maintain clean equipment to minimize development of potential spoilage fungi and bacterial rope spores.

Hazards to be considered. Bacterial pathogen controls are inherent to traditional bread production, therefore concerns are minimal. The baking process inactivates vegetative pathogens at the temperatures required to establish the bread's functional structure. Baking also dehydrates the surface of bread and to a lesser extent the interior portions, which inhibits or slows pathogen growth. Because of this, there are no significant hazards in traditional bread products.

Application of new packaging strategies, such as modified atmosphere or in package heat treatments that inhibit fungal growth, may create an environment that can support the growth of *Cl. botulinum*. It is important to evaluate this potential when evaluating new technologies. Basic controls are listed in the summary mentioned earlier.

Spoilage. The best way to prevent mold spoilage in dry baked goods is to keep them dry. In bread, reduced water activity can be maintained on the loaf surface by vapor-proof packaging, and by avoiding temperature changes that encourage condensation on the inside of the wrapper. For long storage of bread and higher a_w cakes, freezing is best, but refrigeration or even cool storage will increase the shelf-life, particularly if the a_w is relatively low (Seiler, 1964). However, refrigeration of bread accelerates staling.

The number of mold spores on bread has an inverse relation to shelf-life (Seiler, 1964). Mold contamination of baked goods can be reduced by separating all handling and storage facilities for returned baked goods, using filtered air in the bread cooling and wrapping areas, frequent cleaning of the slicing machine and other equipment contacting baked goods before packaging, and using ultraviolet light in critical areas such as wrapping machines (ICMSF, 1980).

The addition of preservatives to breads is permitted in some countries. Propionic acid, sorbic acid or other weak organic acids, or their salts are commonly used. For control of bread mold, calcium propionate is popular at levels of 0.1–0.32% of the weight of flour used. Generally, 0.2% is recommended (Seiler, 1964), but if this amount interferes with yeast fermentation or produces cheese-like flavors, a reduction to 0.1–0.15% is advisable. *Pen. roqueforti*, which is highly resistant to weak acid preservatives (Pitt and Hocking, 1997), remains a problem.

Chapattis generally have a shelf-life of 7 days under tropical conditions. However, with 0.3% sorbic acid and 1.5% salt, they will keep longer than 180 days (Kameswara Rao *et al.*, 1964, 1966). With 0.4% citric acid and 2.5% salt, the sorbic acid level can be reduced to 0.2% (Vijaya Rao *et al.*, 1979).

Control of rope depends on good sanitation and selection of ingredients with low spore levels, quick cooling after baking and before packaging, low pH of the bread and use of preservatives. The dough should be prepared with adequate amount of yeast and fermented vigorously at normal temperature to depress the dough pH sufficiently to inhibit rope development (i.e. pH $\sim$5). An organic acid can also be used for this purpose.

Baking should be thorough enough to reduce the moisture content of the center of the loaf to the point where rope-forming organisms are slow to grow. The bread should be cooled rapidly and should not be wrapped until its temperature reaches $\leq$33°C at the center (Pyler, 1973). Bakery equipment should be kept clean so that it will not contribute spores to the bread.

VII Pasta and noodles

A *Effects of processing on microorganisms*

Pasta is produced from wheat flour, water, semolina, farina, and other ingredients mixed to form a stiff dough that is extruded or rolled into a variety of shapes and forms. Noodles are a form of pasta containing added egg or egg yolk. Oriental noodles are also made from rice, wheat and alkaline wheat.

Traditional dry pasta manufacture includes no cooking step, and bacteria can grow during the mixing and drying operations. Pasta manufacturing equipment is frequently difficult to clean and disinfect, and typically warm temperatures are critical to development of satisfactory rheological properties of the dough. Increases in microbial numbers at the point of extrusion may be a particular problem, and the final dried products may contain very high numbers of microorganisms, as demonstrated in retail pasta surveys conducted in Canada (Rayman *et al.*, 1981), Germany (Spicher, 1985) and Italy (Massa *et al.*, 1986). In the United States, bacterial levels varied from 10^3 to 7×10^7/g. *Asp. flavus* grew from 44 of 47 samples tested, and was the predominant fungus. *Penicillium* species were present in all samples, and several other genera were represented. The counts of fungi were 75–1400 propagules/g (Christensen and Kennedy, 1971). Boiling the pasta kills all significant fungi and vegetative bacterial cells.

Conventionally, the drying temperature for pasta is up to 55°C. It has been raised in some countries to 75°C or even 100°C, resulting in reduced drying times and improved quality (Donnelly, 1991). This has also reduced the potential for microbial growth during drying.

There is a growing retail trade in fresh pasta, both flat and filled (with meat, cheese, etc.). Fresh pasta receives little drying and may require refrigerated storage or modified atmosphere packaging. Water activities are 0.92–0.99. Filled pasta generally receives a pasteurization step to set dough structure. Pasteurizing at 85°C for 10 min effectively eliminated salmonellae, however, *Bacillus* spp. spores survived (Zardetto *et al.*, 1999). Filled pasta had consistently lower microbial levels than flat pasta (Magri *et al.*, 1996).

B *Spoilage*

Spoilage is rare for traditional dry pasta. If drying is too slow, or if it becomes moist or remains moist after manufacture, or is held under refrigeration for long periods, it may spoil from bacterial or fungal growth. For example, *Enterobacter* (*Aerobacter*) *cloacae* has caused gas production in moist macaroni. It is possible, of course, that spoiled ingredients, such as liquid egg, may be incorporated into the pasta mix.

Fresh pasta may spoil owing to fungal growth if modified atmosphere packaging systems fail.

C Pathogens and toxins

Dry pasta and noodles are extruded and dried below pasteurization temperatures, therefore pathogens may survive in the final product. Salmonellae from egg ingredients have been especially troublesome. Although levels of salmonellae are reduced during processing, significant numbers survive if the original contamination is high (Hsieh *et al.*, 1976b). The heat resistance of those *Salmonella* cells that survive increases with the reduction in a_w that occurs during drying (Hsieh *et al.*, 1976a). Once the pasta is fully dry, salmonellae may remain viable for several months (Walsh *et al.*, 1974; Lee *et al.*, 1975). *Salmonella* Infantis and *S.* Typhimurium have been detected in pasta after 360 days storage at room temperature (Rayman *et al.*, 1979). Boiling will destroy the bacteria, but the presence of salmonellae in dried pasta is considered a basis for legal action under the laws of many countries. Salmonellae, *Staph. aureus*, *B. cereus* and *Cl. perfringens* were not commonly isolated in fresh pasta in Italy (Magri *et al.*, 1996, Meloni *et al.*, 1996), and when isolated were more commonly found in filled products. Salmonellae are destroyed using adequate pasteurization temperatures for fresh pasta production (Zardetto *et al.*, 1999). Boiling meat filled pasta for greater than 30 s during consumer preparation also reduced bacterial counts by 6 logs (Zardetto and Fresco, 2000).

Staphylococcal enterotoxins may form as a result of *Staph. aureus* growth in pasta dough and will survive in dried pasta for long periods. *Staph. aureus* was not capable of growth in pasta below a_w 0.86 (Valik and Gorner, 1993). At least one widespread outbreak of staphylococcal food poisoning owing to pasta has occurred (Woolaway *et al.*, 1986). Mixing equipment in pasta factories is often not cleaned more frequently than once a week, and in the early stages of drying, *Staph. aureus* can grow because conditions are nearly ideal (35–40°C, a_w 0.90–0.95, pH ~6). Inadequate or inefficient mixing will add to this hazard. As with *Salmonella*, drying increases the stability of the bacteria, and death during storage may be incomplete (Walsh and Funke, 1975). In a survey of dried pasta products in retail stores in the US, *Staph. aureus* was found in 42 of 1533 packages of macaroni, and in 49 of 1417 packages of noodles; and in related factory investigations, was found in 179 of 350 samples (Walsh and Funke, 1975).

Frozen pasta dishes such as lasagne and ravioli may contain viable *Staph. aureus* and the pasta is one source (Ostovar and Ward, 1976). Staphylococcal enterotoxin can reach a detectable level at about 10^6 cells/g of dough (Ottaviani and Arvati, 1986). Although such levels require abuse, such as holding for 18 h at 25°C or 13 h at 35°C, the resultant toxins are stable in the dried pasta, and would be only partially removed by boiling.

Fresh pasta may also contain *Staph. aureus*, as demonstrated by surveys in Italy (Trovatelli *et al.*, 1988) and Canada (Park *et al.*, 1988). *Staph. aureus* toxin was produced under abusive conditions (4 weeks at 16°C). Glass and Doyle (1991) demonstrated the potential for *Cl. botulinum* toxin formation under extensive abuse conditions (27°C) in a refrigerated, filled pasta with a_w >0.95. Cooking in boiling water inactivated all botulinum toxin formed, however, this should not be viewed as an appropriate safety measure. These products require refrigeration to assure safety. In some countries, retailers are reluctant to refrigerate pasta, especially rice-based pasta, as it hastens retrogradation. However, cooking in boiling water reverses the starch retrogradation process.

Pasta dough, because it contains raw flour, also contains many of the molds from the flour. Some of these species are capable, in appropriate conditions, of producing mycotoxins (Christensen and Kennedy, 1971). However, mycotoxin production poses no threat during processing of pasta as the product would appear spoiled (Stoloff *et al.*, 1978).

Dried spaghetti has been reported to be a source of *Cl. perfringens* spores (Keoseyan, 1971), and this should be taken into consideration in the handling of soups and other prepared foods that contain pasta, and which could act as a favorable growth medium.

Chinese rice noodles, "lo si fan" caused 13 deaths in Malaysia in the year 1988, and aflatoxin was detected (Z. Merican, personal communication).

D CONTROL (pasta and noodles)

Summary

Significant hazards	• *Staph. aureus.* • *Salmonella.* • *Cl. botulinum* (only in refrigerated, fresh pasta packed under modified atmosphere).
Control measures	
Initial level (H_0)	• Use pasteurized egg ingredients to reduce the potential for salmonellae. • Minimize handling of dough with bare hands for *Staph. aureus.*
Increase (ΣI)	• Frequent and thorough cleaning of machinery, especially mixer hubs and extruder heads. • Dry or refrigerate promptly after manufacture. • Chill refrigerated pasta rapidly to <7°C. • Appropriate combinations of a_w and pH can prevent growth of *Cl. botulinum* in refrigerated, modified atmosphere packaged pasta.
Reduction (ΣR)	• Pasteurization can reduce salmonellae and *Staph. aureus* levels. • Boiling pasta dough or dried pasta before consumption inactivates salmonellae. • Reduction of *Cl. botulinum* spores or *Staph. aureus* toxin levels is not possible in pasta production.
Testing	• Periodic in-process testing for *Staph. aureus* is useful to evaluate process control. • Environmental sampling for salmonellae is useful to identify potential harborage sites. • Testing pH and a_w may be appropriate if control of the formulation is used to ensure safety of refrigerated products.

VIII Breakfast cereals and snack foods

A Effects of processing on microorganisms

Manufactured dried breakfast cereal products are usually eaten in Western countries as breakfast foods. They are produced primarily from wheat, maize, oats, and rice. Water is added to the cereal grains, which are processed by flaking, puffing or extrusion. Microbial growth can occur during the moist phase; however, further processing involves heat treatments that reduce the microbial load. Vitamins, minerals, sweeteners, flavorings, and colorings are applied after heating. This provides an opportunity for post-heat treatment contamination before packaging. Breakfast cereals processed in this manner under good hygienic practices typically have aerobic bacterial counts of <1000/g (Deibel and Swanson, 2001).

Whole grain breakfast cereals represent another type of product that may not be exposed to the heat treatments previously described. In a study of the microbial flora of whole grain cereal product in Germany, aerobic bacterial counts were $\sim 10^6$ cfu/g and fungal counts were up to 2.0×10^5/g (Table 8.11) (Spicher, 1979). Although counts of both bacteria and molds were high, pathogen levels were low.

Table 8.11 Average bacterial counts per gram from 184 whole grain cereal products[a]

	Product			
Bacterial type	Whole dehusked grain	Single grain products in husk	Cereal mixtures	Rye and wheat flakes
Mesophiles	0.9×10^6	3.3×10^6	5.5×10^6	1.1×10^4
Coliforms	1	183	697	1
Faecal streptococci	5	50	42	2
Molds	4	1.6×10^5	2.0×10^5	1.4×10^3
Spore-formers	66	116	307	–
Escherichia coli	0	60	3	–
Staphylococcus aureus	0	0	0	–

[a]From Spicher (1979).

The mycoflora of cereal products can be very diverse. One study isolated more than 60 fungal species from cereal flakes (wheat, oats, barley, and rye). *Eurotium* and *Aspergillus* species were dominant (Weidenbörner and Kunz, 1995).

B Spoilage

Microbial spoilage of breakfast cereals is rare because of their extremely low water activity. The water activity must be maintained below the monolayer level to maintain the crunchy character required for processed breakfast cereals, therefore no moisture is available for microbial growth.

C Pathogens and toxins

Outbreaks of foodborne disease related to cereals are uncommon. As with spoilage, the low water activity of the product prevents microbial growth. However, a large multi-state salmonellosis outbreak occurred in the United States related to toasted oat cereal (CDC, 1998). Salmonellae were isolated from the manufacturing environment after the outbreak (Breuer, 1999). Mycotoxin issues discussed previously in the grain section are applicable to cereals if the source of grains is not controlled.

D CONTROL (breakfast cereals, snack foods)

Summary

Significant hazards	• When good hygienic practices are in place, there are no major hazards.
Control measures	
Initial level (H_0)	• Supplier qualification and specification programs.
Increase (ΣI)	• The naturally low a_w of breakfast cereals prevents microbial growth.
Reduction (ΣR)	• Heat treatments required for product functionality effectively destroy pathogens in processed breakfast cereals.
Testing	• Routine manufacturing environmental sampling for salmonellae is useful to identify potential harborage sites.
Spoilage	• There are no spoilage concerns for cereal products due to low water activity.

IX Pastries and filled products

A *Effects of processing on microorganisms*

A wide variety of filled and unfilled sweet products is available around the world, including filled and unfilled cakes, muffins, doughnuts, and cream puffs. Savory filled dough products include Asian egg rolls, wontons, and dimsum; Indian mimosas; Latin American and Spanish empanadas, and tapas; and Western products such as pirogis, meat pies, pasties, pizza, and sausage rolls. All of these products have the potential to combine foods from other food groups with dough or batter, which can influence the a_w and introduce new food safety concerns.

Three basic procedures are used for the manufacture of filled pastries.

1. Filling ingredients are combined, cooked, and dispensed on or into pre-baked pastry tubes, shells or cakes (e.g. chocolate eclairs, Napoleons or imitation cream pies).
2. A preformed, baked or unbaked, pastry shell or wrapper is filled with combined, uncooked filling, the entire pastry is then baked, cooled, and packaged (e.g. coconut custard pies or dim sum).
3. A pre-baked pastry shell is filled with ingredients, some of which have been combined and cooked (as in 1 or 2), but other ingredients are added without cooking and the completed pastry has no final baking (e.g. Nesselrode pies).

Many fillings are excellent microbial growth media. Others are minimal substrates or they may even be inhibitory because of one or more limiting factors, such as low a_w, pH or nutrient content. Particular attention must be given to the interface between filling and cereal product surfaces for products stored and distributed at room temperature. For example, moisture from a filling may increase the a_w of the dough to a level that supports growth, and the dough may buffer pH of an acidic filling to neutral levels. These relationships must be studied carefully to stabilize the products. Preservation is accomplished by alteration of formulation, refrigeration, and/or preservatives.

Cooking fillings to 76–82°C kills all microorganisms except microbial spores, assuming that the entire batch reaches this temperature. However, in Type 1 pastries above, there is considerable opportunity for recontamination of the bulk during cooling, conveying, and dispensing (Silliker, 1969). In Type 3 pastries, there is even greater likelihood of contamination, because some ingredients are not cooked at all. Such a practice can be hazardous (Silliker, 1969; Deibel and Swanson, 2000). Cooking or reheating failures sometimes occur if the product is not all brought to an adequately high temperature.

Browning of meringue over a heat sensitive filling (heating in an oven at ~230°C for about 6 min) offers too little heat to kill bacteria except those in the top layer. Meringue is an excellent insulator because of air bubbles entrapped in the foam. The temperature attained at the meringue–filling interface is often <44°C during browning (Bryan, 1976). Meringue may also be folded into a pastry or dessert with no heat process.

Freezing and frozen storage of pastries may cause some loss of bacterial viability, but rarely eliminates a given population. Aerobic plate counts and coliform group levels on various custard pies dropped to one-half their original level after 76 days at −20°C, except for the very acid lemon-lime pies whose counts dropped to one-half their original level in only 12 days (Kramer and Farquhar, 1977).

B *Spoilage*

As with bread, mold growth is the predominant spoilage problem for pastries. However, the pastry filling or topping may be more susceptible to microbial growth than the cereal product. Many fillings support the growth of spoilage bacteria, especially if they have a high a_w, near to neutral pH, and contain high protein ingredients such as meat, egg or milk. Cooked fillings spoil from spore formers that survive the cooking, other bacteria introduced after the cooking step, or those that survive inadequate cooking. In a German survey of cream cakes, 50% had aerobic plate counts of 1–5 × 10^6 cfu/g, 16% were >10^6 cfu/g, 6.5% were >10^7 cfu/g, with the highest count >10^9 cfu/g (Hartgen, 1983). In Belgium, aerobic plate counts were higher, 12% of samples contained 10^1–10^2 cfu/g of *Staph. aureus*, and 22% contained >10^5 cfu/g of *B. cereus* (Yde, 1982).

Imitation cream pies spoiled in 48 h at room temperature (22°C) with aerobic plate counts of 10^7–>10^8, coliform group levels exceeding 10^6, and *Staph. aureus* up to 10^6 cfu/g (Surkiewicz, 1966). Coliforms and *E. coli* have been found in real and imitation cream cakes and pastries from a variety of manufacturers (Greenwood *et al.*, 1984; Pinegar and Cooke, 1985; Schwab *et al.*, 1985; Michard *et al.*, 1986).

High-sugar icings or low-pH toppings (i.e. fruit) will not support growth of spoilage bacteria but will eventually permit fungi to grow (Silliker, 1969).

C Pathogens

Cream- and custard-filled pastries can cause foodborne disease. In the 5-year period from 1993–1997, 35 of 2751 reported foodborne outbreaks were associated with baked products (CDC, 2000). Salmonellae contributed to 34% of these outbreaks, *Staph. aureus* 9%, 6% viruses and the remainder were of unknown etiology. The report did not specify the type of bakery product. However, a previous report for 1938–1972 identified cream-filled pastries as the primary vehicle with 85.2% from *Staph. aureus* and 12.5% from *Salmonella* (Bryan, 1976).

Staph. aureus, *B. cereus* (Fenton *et al.*, 1984), and salmonellae (Barnes and Edwards, 1992) have been involved in cream filled pastry outbreaks. Temperature abuse and/or inadequate cooking were root causes of the outbreaks. Staphylococcal food poisoning can occur only after the bacteria have multiplied under favorable conditions to reach millions per gram. The minimal temperature for enterotoxin production is 10°C (ICMSF, 1996). It is not surprising, therefore, that foodborne disease outbreaks from cream-filled pastries are attributed primarily to inadequate refrigeration during manufacture or storage (McKinley and Clarke, 1964, Bryan, 1976).

Salmonellae have been reported in many bakery ingredients, such as flour, milk, eggs, butter, cream, cheese, nuts, coconut and dried fruit. It is therefore essential that cooked fillings be protected from direct or indirect contact with such raw ingredients. Children and the elderly or infirm are especially susceptible to infection from small inocula such as one might find in a pastry in which growth had not occurred.

Listeria monocytogenes is sometimes present in the raw materials used in pastry manufacture, including milk and milk products (see Chapter 16), unpasteurized egg and egg products (see Chapter 15), meat (Chapter 1), and poultry (Chapter 2). The surface water activity of unfilled pastries is below the minimum required for *L. monocytogenes* growth and temperatures required to develop the structure of baked products destroy it. However, fillings must be evaluated individually to determine the potential for *L. monocytogenes* growth. For example, *L. monocytogenes* was reported to multiply between 4° and 35°C in whipping cream, although multiplication at 4°C was quite slow (Rosenow and Marth, 1986).

D CONTROL (pastries and filled products)

Summary

Significant hazards	• *Salmonella* spp. • *Staph. aureus*.
Control measures	
Initial level (H_0)	• Use pasteurized eggs and dairy products in uncooked fillings.

(*continued*)

D CONTROL (Cont.)

Summary

Increase (ΣI)	• Wash and sanitize hands before handling cooked product to minimize cross contamination. • Minimize hand contact with cooked product, and keep persons with infections away from the cooked-product area. • Clean and sanitize cooked filling equipment every 4 h unless temperature controlled. • Cool cooked fillings that support pathogen growth rapidly, and maintain at <5°C. • Hold fillings <4 h between 7 and 46°C. • Formulation using combinations of reduced pH, a_w, and preservatives can control pathogens growth. • Separate raw ingredients and processes from cooked products. • Control dusts and aerosols to prevent contamination with potential pathogens and spoilage organisms.
Reduction (ΣR)	• Baking and cooking of fillings destroys vegetative pathogens.
Testing	• Routine testing of pastry products is not recommended. • Environmental monitoring for salmonellae to detect harborage sites. • Monitor air quality for fungal spores to reduce contamination after baking
Spoilage	• Maintain processing environments in a dry condition to reduce mold spoilage. • Cool pastry prior to packaging in plastic wrapping to avoid condensation, which promotes spoilage. • Filter air used in cooling and packing areas where possible. • Preservatives may be used to extend shelf-life in some countries. • Maintain clean equipment to minimize development of potential spoilage fungi.

Hazards to be considered. *Staph. aureus* and *Salmonella* spp. are the primary organisms of concern for pastries on the basis of outbreak history. *B. cereus* and *L. monocytogenes* may be of concern in products with certain fillings (e.g. meat fillings for *L. monocytogenes*), however, outbreak data do not suggest that these organisms are a significant concern for all pastry products.

Control measures

Cooking. An adequate cooking process destroys foodborne pathogens in cream or custard fillings. *L. monocytogenes, Staph. aureus,* and *Salmonella* species are destroyed by bringing custards to a second boil after adding the thickening. Baking custards in a pie shell should also kill these bacteria, but sometimes the cooking is insufficient and some survive. For example, the minimum temperature for thickening egg (78°C) may not destroy all staphylococci, but 91–93°C will do so (Kintner and Mangel, 1953).

In custard, 60°C required 19 min to kill 10^7 salmonellae, or 59 min to kill 10^7 *Staph. aureus*. These times are reduced at 65.7°C, which required 3.5 min to destroy 10^7 salmonellae or 6.6 min to destroy 10^7 staphylococci (Angelotti *et al.*, 1961b).

Egg white should be pasteurized by the supplier and pasteurization is required in most countries, but sometimes salmonellae survive and appear in the meringue used for pastry topping. Adding hot syrup to whipped whites, and whipping again, destroys some salmonellae, but the process cannot be relied upon.

Sanitation. Most contamination occurs while the filling is cooling or during handling. Any direct contact between humans and the filling could introduce *Staph. aureus* as about one-half the human population carries this species on the skin and in the mucous membranes (Surkiewicz, 1966). Bakeries operating under the best sanitary conditions can produce cream-filled pastries containing no *Staph. aureus*, whereas those operating under poor conditions usually cannot (Surkiewicz, 1966). The level of *Staph. aureus* in the cooked, cooled filling can be used as a measure of the degree of human contact.

Only the highest degree of sanitary practice is adequate for handling cooked and cooled cream and custard fillings. They should be protected from dust and droplet contamination from dry foods used as ingredients, and from batters being mixed. Equipment and personnel in the cooked food area should have no contact with raw ingredients. Equipment should be constructed of stainless steel and should be disassembled at each cleaning unless it is specifically designed for a clean in place program (ICMSF, 1980; Chapter 17). Equipment should be cleaned and sanitized every 4 h during production unless other control measures, such as temperature or formulation, are employed to prevent microbial growth. Some equipment (e.g. conveyor belts) may be continuously cleaned and sanitized. Employee contact with product should be minimal and hygienic practices beyond reproach. Contamination with pathogenic bacteria and viruses, such as hepatitis A, as a result of poor personal hygiene by a carrier handling pastries or other baked goods after cooking can have serious consequences (ICMSF, 1980; Chapter 17).

Refrigeration. To prevent potential pathogen growth and/or toxin formation refrigeration at $\leq 5°C$ is essential for most fillings. Most food protection authorities in countries with advanced technology require such refrigeration (Silliker and McHugh, 1967; IFT, 2002).

The first place where good refrigeration is necessary is during the cooling of the cooked filling. Pre-cooling at room temperature is unsatisfactory (Longrée, 1964). Cooling in a large mass is also inefficient, even in a refrigerator, for the center may remain warm for several hours, during which time pathogens may grow. Product should pass through the temperature zone in which *Staph. aureus* and *Salmonella* grow (7–46°C) in ≤ 4 h unless validated to control growth by some other means such as formulation (Hodge, 1960; Angelotti *et al.*, 1961a). To accomplish this, agitation may be necessary, preferably with a refrigerated tube agitator or other device to speed the cooling (Longrée, 1964). Alternatively, the hot filling may be dispensed into pastry shells, which are then refrigerated without delay.

The cooled filling should then be handled expeditiously so that no part of it warms to the danger zone for more than a few minutes. Residual food material on the equipment may remain several hours at ambient temperature unless it is washed away frequently. Finally, transport of the finished pastry and storage at retail should be under refrigeration at $\leq 5°C$.

Formulations. The traditional cream fillings may contain eggs, milk, shortening, sugar, maize starch or flour, salt, vanilla, and other flavorings, and water (Bryan, 1976). Synthetic cream fillings consist mainly of vegetable oils, stabilizers, emulsifiers, and water. They may also contain sugar, coloring, flavoring, fruit purees, dried milk, and starch (Surkiewicz, 1966). If the a_w is high enough, most will support the growth of various bacteria including *Staph. aureus* (Crisley *et al.*, 1964; Surkiewicz, 1966; Silliker, 1969). Adding milk or small amounts of egg markedly increases the growth of *Staph. aureus* (Crisley *et al.*, 1964).

The lowest pH for *Staph. aureus* toxin production is 4.5, and *Salmonella* growth is 3.8, depending on the menstruum (ICMSF, 1996). The pH of many custards used in bakeries is 5.8–6.6 (Bryan, 1976), a range near the optimum for these bacteria. Formulation may be used to reduce the pH of these custards to prevent the growth of pathogens.

Changes in the formulation can alter the a_w or other factors that may permit a pathogen to grow. High-ratio shortenings, for example, permit incorporation of more water in a cake batter than is otherwise possible. If this increase is not accompanied by an increase in sugar content, the increased a_w of the final product may permit the growth of a pathogen (Bryan, 1976).

A a_w <0.87 cannot support *Staph. aureus* toxin production (ICMSF, 1996). The surface of a baked custard pie is markedly dehydrated so that *Staph. aureus* grows only very slowly if at all, whereas it would grow more rapidly in the moist custard (Preonas *et al.*, 1969).

"Butter–cream" fillings have a remarkably good safety record despite the fact that they are stored indefinitely at ambient temperature. Their a_w, sufficiently low to preclude bacterial growth, is adjusted by varying the sugar/water ratio. Growth of *Staph. aureus* sometimes occurs at a ratio of 1.8/1 (or if glucose or invert sugar is used, a ratio of 0.9/1). However, at sucrose/water ratios of 2.1/1 to 3.0/1, no growth generally occurs. Nevertheless, growth has occurred at the interface between a "butter–cream" filling of ratio 2.7/1 and a cake whose a_w was higher than that of the filling (Silliker and McHugh, 1967).

Meat and curry fillings are commonly used in pastry casings worldwide. Examples include egg rolls, tapas, pizza, dim sum, sausage rolls, pirogis, empanadas, etc. Pathogens can grow in these products under temperature abuse. Control measures used for cooked meat products are appropriate to prevent spoilage and pathogenic growth on products of this type (Chapters 1 and 2).

Preservatives. In cake, sorbic acid is effective at a level of 0.03–0.125% of the total batter, although it can affect flavor adversely (Pyler, 1973; Seiler, 1964).

Growth of *Staph. aureus* in fillings that are otherwise excellent microbial media can be prevented by adding preservatives, either incorporated into the filling or shell or sprayed on the surface. The most effective are the weak acids such as propionic, sorbic or benzoic, or their salts, usually added at *ca.* 0.1% (Schmidt *et al.*, 1969; Preonas *et al.*, 1969; Pyler, 1973). Their effectiveness is increased by reducing the pH of the filling to the range 3–5, where a high proportion of the acid is undissociated.

Testing. Routine sampling of the manufacturing environment for salmonellae is useful to identify potential harborage sites. Similar monitoring is useful for refrigerated ready-to-eat products that support *L. monocytogenes* growth. Routine testing for fully baked pastries is of little value as baking destroys vegetative pathogens, and the dehydrated crust prevents growth. Aerobic colony counts or coliforms may provide a useful indicator to monitor process control for products containing filling applied after cooking.

Spoilage. Pastry products are subject to the same spoilage flora as other baked products discussed for bread. Fungal growth is the primary mode of failure.

References

Abdel-Kader, M.I.A., Moubasher, A.H. and Abdel-Hafez, S.I.I. (1979). Survey of the mycoflora of barley grains in Egypt. *Mycopathologia*, **69**, 143–7.

Abramson, D., Clear, R.M. and Nowicki, T.W. (1987). *Fusarium* species and trichothecene mycotoxins in suspect samples of 1985 Manitoba wheat. *Can. J. Plant Sci.*, **67**, 611–9.

Adler, A., Lew, H., Brodacz, W., Edinger, W. and Oberforster, M. (1995) Occurrence of moniliformin, deoxynivalenol, and zearalenone in durum wheat (*Triticum durum* Desf.). *Mycotoxin Res.*, **11**, 9–15.

Angelotti, R., Foter, M.J. and Lewis, K.H. (1961a). Time–temperature effects on salmonellae and staphylococci in foods. I. Behavior in refrigerated foods. II. Behavior at warm holding temperatures. *Am. J. Public Health*, **51**, 76–88.

Angelotti, R., Foter, M.J. and Lewis, K.H. (1961b). Time–temperature effects on salmonellae and staphylococci in foods. III. Thermal death time studies. *App. Microbiol.*, **9**, 308–15.

Annis, P. (1990a) Requirements for fumigation and controlled atmospheres as options for pest and quality control in stored grain, in *Fumigation and Controlled Atmosphere Storage of Grain. Proceedings of an International Conference* (eds B.R. Champ, E. Highley and H.J. Banks), Australian Center for International Agricultural Research, Canberra, A.C.T., pp. 20–8.

Annis, P. (1990b) Sealed storage of bag stacks: status of the technology, in *Fumigation and Controlled Atmosphere Storage of Grain. Proceedings of an International Conference* (eds B.R. Champ, E. Highley and H.J. Banks), Australian Center for International Agricultural Research, Canberra. A.C.T., pp. 203–10.

Arp, L.H. and Richard, J.L. (1979). Intoxication of dogs with the mycotoxin penitrem A. *J. Am. Vet. Med. Assoc.*, **175**, 565–6.

Austwick, P. (1984). Human mycotoxicosis—past, present and future. *Chem. Indy (Lond.)*, **15**, 547–51.

Ayerst, G. (1969). The effects of moisture and temperature on growth and spore germination in some fungi. *J. Stored Products Res.*, **5**, 669–87.

Barnes, G.H. and Edwards, A.T. (1992) An investigation into an outbreak of *Salmonella enteritidis* phage type 4 infection and the consumption of custard slices and trifles. *Epidemiol. Infect.* **109**, 397–403.

Beuchat, L.R. (1987) Traditional fermented food products, in *Food and Beverage Mycology* (ed L.R. Beuchat), Van Nostrand Reinhold, New York, pp. 269–306.

Bezuidenhout, S.C., Gelderblom, W.C.A., Gorst-Allman, C.P., Horak, R.M., Marasas, W.F.O., Spiteller, G. and Vleggaar, R. (1988). Structure elucidation of fumonisins, mycotoxins from *Fusarium moniliforme*. *J. Chem. Soc., Chem. Commun., 1988*, 743–5.

Bhat, R.V., Vasanthi, S., Rao, B.S., Rao, N.R., Rao, V.S., Nagaraja, K.V., Bai, R.G., Prasad, C.A.K., van Chinathan, S., Roy, R., Saha, S., Mukherjee, A., Ghosh, P.K., Toteja, G.S. and Saxena, B.N. (1997) Aflatoxin B_1 contamination in maize samples collected from different geographical regions of India: a multicentre study. *Food Addit. Contam.*, **14**, 151–6.

Blaney, B.J. (1981). Aflatoxin survey of maize from the 1978 crop in the South Burnett region of Queensland. *Queensland J. Agric. Animal Sci.*, **38**, 7–12.

Blaney, B.J., Ramsey, M.D. and Tyler, A.L. (1986). Mycotoxins and toxigenic fungi in insect-damaged maize harvested during 1983 in Far North Queensland. *Aust. J. Agric. Sci.*, **37**, 235–44.

Breitholtz-Emanuelsson, A., Olsen, M., Oskarsson, A., Palminger, I. and Hult, K. (1993) Ochratoxin A in cow's milk and human milk with corresponding human blood samples. *J. Assoc. Off. Anal. Chem. Int.*, **76**, 842–6.

Breuer, T. (1999) CDC investigations: the May 1998 outbreak of *Salmonella agona* linked to cereal. *Cereal Foods World*, **44**, 185–6.

Brooks, M.A. (1969) General relationships between microorganisms and insects, in *Proceedings, Symposium on Biological Contamination of Grain and Animal Byproducts*, University of Minnesota, Minneapolis, MN, pp. 17–20.

Brown, R.L., Cotty, P.J. and Cleveland, T.E. (1991). Reduction in aflatoxin content of maize by atoxigenic strains of *Aspergillus flavus*. *J. Food Protect.*, **54**, 623–6.

Brown, R.L., Cotty, P.J., Cleveland, T.E. and Widstrom, N.W. (1993). Living maize embryo influences accumulation of aflatoxin in maize kernels. *J. Food Protect*,. **56**, 967–71.

Brown, R.L., Cleveland, T.E., Payne, G.A., Woloshuk, C.P., Campbell, K.W. and White, D.G. (1995) Determination of resistance to aflatoxin production in maize kernels and detection of fungal colonization using an *Aspergillus flavus* transformant expressing *Escherichia coli* β-glucuronidase. *Phytopathology*, **85**, 983–9.

Brown, R.L., Chen, Z.Y., Cleveland, T.E. and Russin, J.S. (1999) Advances in the development of host resistance in corn to aflatoxin contamination by *Aspergillus flavus*. *Phytopathology*, **89**, 113–7.

Brown, R.L., Chen, Z.Y., Menkir, A., Cleveland, T.E., Cardwell, K., Kling, J. and White, D.G. (2001) Resistance to aflatoxin accumulation in kernels of maize inbreds selected for ear rot resistance in West and Central Africa. *J. Food Protect.*, **64**, 396–400.

Bryan, F.L. (1976). Public health aspects of cream-filled pastries. A review. *J. Milk Food Technol.*, **39**, 289–96.

Bullerman, L.B. and Hartung, T.E. (1973). Mycotoxin-producing potential of molds isolated from flour and bread. *Cereal Sci. Today*, **18**, 346–7.

Bullerman, L.B., Baca, J.M. and Stott, W.T. (1975). An evaluation of potential mycotoxin-producing molds in corn meal. *Cereal Foods World*, **20**, 248–50, 253.

Burgess, L.W., Dodman, R.L., Pont, W. and Mayers P. (1981). *Fusarium* diseases of wheat, maize and grain sorghum in Eastern Australia, in *Fusarium: Diseases, Biology and Taxonomy* (eds P.E. Nelson, T.A. Toussoun and R.J. Cook), Pennsylvania State University Press, University Park, Pennsylvania, pp. 64–76.

Burgess, L.W., Summerell, B.A., Bullock, S., Gott, K.P. and Backhouse, D. (1994) *Laboratory Manual for Fusarium Research*, 3rd edn, University of Sydney, Sydney.

Carvajal, M. and Arroyo, G. (1997) Management of aflatoxin contaminated maize in Tamaulipas, Mexico. *J. Agric. Food Chem.*, **45**, 1301–5.

Capparelli, E. and Mata, L. (1975) Microflora of maize prepared as tortillas. *Appl. Microbiol.*, **26**, 802–6.

CDC (Centers for Disease Control and Prevention). (1998) Multistate outbreak of *Salmonella* serotype Agona infections linked to toasted oats cereal—United States, April–May, 1998. *Morbid. Mortal. Wkly Rep.*, **47**, 462–4.

CDC. (2000) CDC Surveillance summaries, March 17, 2000. *Morbid. Mortal. Wkly Rep. 2000*, **49** (No. SS-1), 27–45.

Champ, B.R. and Highley, E. (eds) (1988) *Bulk Handling and Storage of Grain*, ACIAR Proceedings No. 22. Australian Center for International Agricultural Research, Canberra, A.C.T.

Champ, B.R., Highley, E. and Banks, H.J. (eds) (1990) *Fumigation and Controlled Atmosphere Storage of Grain*, ACIAR Proceedings No. 25, Australian Center for International Agricultural Research, Canberra, A.C.T.

Christensen. C.M. (1965) Deterioration of stored grains by molds. *Wallerstein Lab. Commun.*, **19**, 31–48.

Christensen, C.M. (1978) Moisture and seed decay, in *Water Deficits and Plant Growth* (ed T.T. Koslowski), *volume 5, Water and Plant Diseases*, Academic Press, New York. pp. 199–219.
Christensen, C.M. (1987) Field and storage fungi, in *Food and Beverage Mycology* (ed L.R. Beuchat), 2nd edn, Van Nostrand Reinhold, New York. pp. 211–32.
Christensen, C.M. and Cohen, M. (1950) Numbers, kinds and sources of molds in flour. *Cereal Chem.*, **27**, 178–85.
Christensen, C.M. and Gordon, DR. (1948) Mold flora of stored wheat and corn and its relation to heating of moist grain. *Cereal Chem.*, **25**, 40–51.
Christensen, C.M. and Kaufmann, H.H. (1964) *Questions and Answers Concerning Spoilage of Stored Grains by Storage Fungi*, Agricultural Extension Service, US Dept of Agriculture, University of Minnesota, St Paul, Minnesota.
Christensen, C.M. and Kaufmann, H.H. (1974) Microflora, in *Storage of Cereal Grains and Their Products*, Monograph Ser. 5, Am. Assoc. Cer. Chem., St. Paul, Minnesota, pp. 158–92.
Christensen, C.M. and Kaufmann, H.H. (1977a) Good grain storage. Agricultural Extension Service, Extension Folder 226. University of Minnesota, St. Paul, Minnesota.
Christensen, C.M. and Kaufmann, H.H. (1977b) Spoilage, heating, binburning and fireburning: their nature, cause and prevention in grain. *Feedstuffs*, **49**, 39, 47.
Christensen, C.M. and Kennedy. B.W. (1971) Filamentous fungi and bacteria in macaroni and spaghetti products. *Appl. Microbiol.*, **21**, 144–6.
Christensen, C.M. and Meronuck, R.A. (1986) *Quality Maintenance in Stored Grains and Seeds*, University of Minnesota Press, Minneapolis, Minnesota.
Christensen, C.M., Papavisas, G.C. and Benjamin, C.R. (1959) A new halophilic species of *Eurotium*. *Mycologia*, **51**, 636–40.
Ciegler, A. and Kurtzman, C.P. (1970) Penicillic acid production by blue-eye fungi on various agricultural commodities. *App. Microbiol.*, **20**, 761–4.
Cole, R.J., Kirksey, J.W., Cutler, H.G., Doupoik, B.L. and Peckham, J.C. (1973) Toxin from *Fusarium moniliforme*: effects on plants and animals. *Science, N.Y.* **179**, 1324–6.
Collar, E.C., Benedito de Barber, C. and Martinez-Anaya, M.A. (1994) Microbial sour doughs influence acidification properties and breadmaking potential of wheat dough. *J. Food Sci.*, **59**, 629–33.
Crisley, F.D., Angelotti, R. and Foter, M.J. (1964) Multiplication of *Staphylococcus aureus* in synthetic cream fillings and pies. *Public Health Rep.*. **79**, 369–76.
Dack, G.M. (1961) Public health significance of flour bacteriology. *Cereal Sci. Today*, **6**, 9–10.
Davis, N.D. and Diener, U.L. (1983) Some characteristics of toxigenic and nontoxigenic isolates of *Aspergillus flavus* and *Aspergillus parasiticus*, in *Aflatoxin and Aspergillus flavus in Corn* (eds U.L. Diener, R.L. Asquith and J.W. Dickens), Auburn University, Auburn, AL, pp. 1–5.
Denny, C.B., Goeke, D.J. and Steinberg, R. (1969) *Inoculation Tests of Clostridium botulinum in Canned Breads with Special Reference to Water Activity*. Research Report 4–69, Washington Research Laboratories, National Canners Association. Washington, DC.
Deibel, K.E. and Swanson, K.M.J. (2001) Cereal and cereal products, in *Compendium of Methods for the Microbiological Examination of Foods* (eds F.P. Downes and K. Ito), 4th edn, Am. Public Health Assoc., Washington, DC, pp. 549–53.
Donnelly, B.J. (1991) Pasta: raw materials and processing, in *Handbook of Cereal Science and Technology* (eds K.J. Lorenz and K. Kulp), Marcel Dekker, New York, pp. 763–92.
Doulia, D., Katsinis, G. and Mougin, B. (2000) Prolongation of the microbial shelf life of wrapped part baked baguettes. *Int. J. Food Prop.*, **3**, 447–57.
Dowd, P.F., Vega, F.E., Nelsen, T.C. and Richard, J.L. (1998) Dusky sap beetle mediated dispersal of *Bacillus subtilis* to inhibit *Aspergillus flavus* and aflatoxin production in maize (*Zea mays* L). *Biocontrol Sci. Technol.*, **8**, 221–35.
Dalcero, A., Torres, A., Etcheverry, M., Chulze, S. and Varsavsky, E. (1997) Occurrence of deoxynivalenol and *Fusarium graminearum* in Argentinian wheat. *Food Additives Contaminants*, **14**, 11–4.
El-Dessouki, S. (1992) Ochratoxin A in beer. *Deutsche Lebensmittel-Rundschau*, **88**, 354–5.
Engel, G. and Teuber, M. (1973) Simple aid for the identification of *Penicillium roqueforti* Thom. *Eur. J. Appl. Microbiol. Biotechnol.*, **6**, 107–11.
Eskola, M., Parikka, P. and Rizzo, A. (2001) Trichothecenes, ochratoxin A and zearalenone contamination and *Fusarium* infection in Finnish cereal samples in 1998. *Food Addit. Contaminants*, **18**, 707–18.
Eyles, M.J., Moss, R. and Hocking, A.D. (1989) The microbiological status of Australian flour and the effects of milling procedures on the microflora of wheat and flour. *Food Aust.*, **41**, 704–8.
Fanse, H.A. and Christensen, C.M. (1966) Invasion of fungi of rice stored at moisture contents of 13.5–15.5%. *Phytopathology*, **56**, 1162–4.
Fazekas, B., Hajdu, E.T., Tar, A.K. and Tanyi, J. (2000) Natural deoxynivalenol (DON) contamination of wheat samples grown in 1998 as determined by high-performance liquid chromatography. *Acta Vet. Hungarica*, **48**, 151–60.
Fenton, P.A., Dobson, K.W., Eyre, A. and McKendrick, M.W. (1984) Unusually severe food poisoning from vanilla slices. *J. Hyg. (Camb.)*, **93**, 377–80.
Flannigan, B. (1970) Comparison of seed-borne mycofloras of barley, oats and wheat. *Trans. Br. Mycol. Soc.*, **55**, 267–76.
Fox, D., Andrade, M., Depalo, M. *et al.* (1989) Arnott's continuous fermentation process. *Aust. J. Biotechnol.*, **3**, 139, 141–2.
Frank, H.K. (1991) Food contamination by ochratoxin A in Germany, in *Mycotoxins, Endemic Nephropathy and Urinary Tract Tumours* (eds M. Castegnaro, R. Pleština, G. Durheimer, I.N. Chenozemsky and H. Bartsch), International Agency for Research on Cancer, Lyon, pp. 77–81.

Frisvad, J.C. (1989) The connection between the Penicillia and Aspergilli and mycotoxins with special emphasis on misidentified isolates. *Arch. Environ. Contamin. Toxicol.*, **18**, 452–67.
Frolich, A.A., Marquardt, R.R. and Omiuski, K.H. (1991) Ochratoxin A as a contaminant in the human food chain: a Canadian perspective, in *Mycotoxins, Endemic Nephropathy and Urinary Tract Tumours* (eds M. Castegnaro, R. Pleština, G. Durheimer, I.N. Chenozemsky and H. Bartsch), International Agency for Research on Cancer, Lyon, pp. 139–43.
Fuller, J.G. (1968) *The Day of St. Anthony's Fire*, McMillan Co., New York.
Fung, D.Y.C. (1995) Microbiological considerations in freezing and refrigeration of bakery foods, in *Frozen and Refrigerated Doughs and Batters* (eds K. Kulp, K. Lorenz and J. Brummer), American Association of Cereal Chemists, pp. 119–254.
Gareis, M., Martlbauer, E., Bauer, J. and Gedek, B. (1988) Bestimmung von Ochratoxin A, in Muttermilch. *Zeitschrift für Lebensmittel-Untersuchung und-Forschung*, **186**, 114–7.
Gareis, M. (1999) [Contamination of German malting barley and of malt-produced from it with the mycotoxins ochratoxin A and B]. *Archiv fur Lebensmittelhygiene*, **50**, 83–7.
Gelderblom, W.C.A., Jaskiewicz, K., Marasas, W.F.O. *et al.* (1988) Fumonisins—novel mycotoxins with cancer-promoting activity produced by *Fusarium moniliforme*. *Appl. Environ. Microbiol.*, **54**, 1806–11.
Gilbert, R.J., Stringer, M.F. and Peace, T.C. (1974) The survival and growth of *Bacillus cereus* in boiled and fried rice in relation to outbreaks of food poisoning. *J. Hyg.*, **73**, 433–44.
Glass, K.H. and Doyle, M.P. (1991) Relationship between water activity of fresh pasta and toxin production by proteolytic *Clostridium botulinum*. *J. Food Prot.*, **54**, 162–5.
Goepfert, J.M., Spira, W.M. and Kim, H.U. (1972) *Bacillus cereus*: food poisoning organism. A review. *J. Milk Food Technol.*, **35**, 213–27.
Gonzalez, H.H.L., Pacin, A., Resnik, S.L. and Martinez, E.J. (1996) Deoxynivalenol and contaminant mycoflora in freshly harvested Argentinian wheat in 1993. *Mycopathologia*, **135**, 129–34.
Grabarkiewicz-Szczesna, J., Kostecki, M., Golinski, P. and Kiecana, I. (2001) Fusariotoxins in kernels of winter wheat cultivars field samples collected during 1993 in Poland. *Nahrung*, **45**, 28–30.
Graver, J.E. van S. (1990) Fumigation and controlled atmospheres as components of integrated commodity management in the tropics, in *Fumigation and Controlled Atmosphere Storage of Grain. Proceedings of an International Conference* (eds B.R. Champ, E. Highley and H.J. Banks), Australian Center for International Agricultural Research, Canberra, A.C.T., pp. 38–52.
Graves, R.R. and Hesseltine, C.W. (1966) Fungi in flour and refrigerated dough products. *Mycopathologia Mycologia Applicata*, **29**, 277–90.
Graves, R.R., Rogers, R.F., Lyons, A.J. and Hesseltine, C.W. (1967) Bacterial and actinocycete flora of Kansas-Nebraska and Pacific Northwest wheat and wheat flour. *Cereal Chem.*, **44**, 288–99.
Greenwood, M.H., Coetzee, E.F.C., Ford, B.M. *et al.* (1984) The microbiology of selected retail food products with an evaluation of viable counting methods. *J. Hyg., Camb.*, **92**, 67–77.
Häggblom, P. (1990) Isolation of roquefortine C from feed grain. *Appl. Environ. Microbiol.*, **56**, 2924–6.
Hald, B. (1991a) Porcine nephropathy in Europe, in *Mycotoxins, Endemic Nephropathy and Urinary Tract Tumours* (eds M. Castegnaro, R. Pleština, G. Durheimer, I.N. Chenozemsky and H. Bartsch), International Agency for Research on Cancer, Lyon, pp. 49–56.
Hald, B. (1991b) Ochratoxin A inhuman blood in European countries, in *Mycotoxins, Endemic Nephropathy and Urinary Tract Tumours* (eds M. Castegnaro, R. Pleština, G. Durheimer, I.N. Chenozemsky and H. Bartsch), International Agency for Research on Cancer, Lyon, pp. 159–64.
Haney, R.L. (1989) Shelf life of corn tortilla extended by preservatives. *Dairy, Food and Environmental Sanitation*, **9**, 552–3.
Hartgen, H. (1983) Mikrobiologische Aspekte zum Keimstatus bei cremehaltigen Konditoreiwaren. *Arch. Lebensm. Hyg.*, **34**, 10–4.
Hartog, B.J. and Kuik, D. (1984) Mycological studies on Dutch rye-bread, in *Microbial Associations and Interactions* (eds I. Kiss, T. Deák and K. Incze), D. Reidel, Dordrecht, Germany, pp. 241–6.
Hell, K., Cardwell, K.F., Setamou, M. and Schulthess, F. (2000) Influence of insect infestation on aflatoxin contamination of stored maize in four agroecological regions in Benin. *African Entomology*, **8**, 169–77.
Hesseltine, C.W. (1965) A millennium of fungi, food and fermentation. *Mycologia*, **57**, 149–97.
Hesseltine, C.W. (1968) Flour and wheat: research on their microbiological flora. *Baker's Digest*, **42**, 40–2, 66.
Hesseltine, C.W. (1979) Some important fermented foods of mid-Asia, the Middle East, and Africa. *J. Am. Oil Chem. Soc.*, **56**, 367–74.
Hesseltine, C.W. and Graves, R.R. (1966) Microbiology of flours. *Econ. Bot.*, **20**, 156–68.
Hesseltine, C.W., Graves, R.R., Rogers, R.F. and Burmeister, H.R. (1969) Aerobic and facultative microflora of fresh and spoiled refrigerated dough products. *App. Microbiol.*, **18**, 848–53.
Hesseltine, C.W., Rogers, R.F. and Shotwell, O.L. (1981) Aflatoxin and mold flora in North Carolina in 1977 corn crop. *Mycologia*, **73**, 216–28.
Hill, R.A. and Lacey, J. (1983) The microflora of ripening barley grain and effects of pre-harvest fungicide application. *Ann. Appl. Biol.*, **102**, 455–65.
Hill, R.A. and Lacey, J. (1984) *Penicillium* species associated with barley grain in the U.K. *Trans. Br. Mycol. Soc.*, **82**, 297–303.
Hocking, A.D. (1990) Responses of fungi to modified atmospheres, in *Fumigation and Controlled Atmosphere Storage of Grain. Proceedings of an International Conference* (eds B.R. Champ, E. Highley and H.J. Banks), Australian Center for International Agricultural Research. Canberra, A.C.T., pp. 70–82.
Hocking, A.D. and Banks, H.J. (1993) The use of phosphine for inhibition of fungal growth in stored grains, in *Controlled*

Atmosphere and Fumigation in Grain Storages, an International Conference (eds S. Navarro and F. Donayahe), Caspit Press, Jerusalem, pp. 173–82.

Hocking, A.D., Holds, K. and Tobin, N.F. (1988) Intoxication by tremorgenic mycotoxin (penitrem A) in a dog. *Aust. Vet. J.*, **65**, 82–5.

Hodge, B.E. (1960) Control of staphylococcal food poisoning. *Public Health Rep.*, **75**, 355–61.

Hohler, D. 1998. Ochratoxin A in food and feed: occurrence, legislation and mode of action. *Zeitschrift fur Ernahrungswissenschaft*, **37**, 2–12.

Hsieh, F., Acott, K. and Labuza, T.P. (1976a) Death kinetics of a pathogen in a pasta product. *J. Food Sci.*, **41**, 516–9.

Hsieh, F., Acott, K. and Labuza, T.P. (1976b) Prediction of microbial death during drying of a macaroni product. *J. Milk Food Technol.* **39**, 619–23.

ICMSF (International Commission on Microbiological Specifications for Foods) (1980) *Microbial Ecology of Foods 1, Factors Affecting Life and Death of Microorganisms*, Academic Press, New York, NY.

ICMSF (1996) *Microorganisms in Foods 5, Characteristics of Microbial Pathogens*, Blackie Academic & Professional, London.

IFT (Institute of Food Technologists) (2001) Evaluation and Definition of Potentially Hazardous Foods. IFT/FDA Contract No. 223-98-2333 Task Order No. 4. www.ift.org/cms/?pid=1000633.

Iizuka, H. (1957) Studies on the microorganisms found in Thai rice and Burma rice. *J. Gen. Appl. Microbiol.*, **3**, 146–61.

Iizuka, H. (1958) Studies on the microorganisms found in Thai rice and Burma rice. Part II. On the microflora of Burma rice. *J. Gen. Appl. Microbiol.*, **4**, 108–19.

Iizuka, H. and Ito, H. (1968) Effect of gamma-irradiation on the microflora of rice. *Cereal Chem.*, **45**, 503–11.

Ingram, M. and Handford, P.M. (1957) The influence of moisture and temperature on the destruction of *Cl. botulinum* in acid bread. *J. Appl. Bacteriol.*, **20**, 442–53.

Ito, H., Shibabe, S. and Iizuka, H. (1971) Effect of storage studies of microorganisms on gamma-irradiated rice. *Cereal Chem.*, **48**, 140–9.

Jiao, Y., Blaas, W., Ruhl, C. and Weber, R. (1994) [Ochratoxin A in foodstuffs (vegetables, cereals, cereal products and beer)]. *Deutsche Lebensmittel-Rundschau*, **90**, 318–21.

Jimenez, M., Santamarina, M.P., Sanchis, V. and Hernandez, E. (1987) Investigation of mycoflora and mycotoxins in stored cereals. *Microbiol. Ailments Nutr.*, **5**, 105–9.

Joffe, A.Z. (1978) *Fusarium poae* and *F. sporotrichioldes* as principal causal agents of Alimentary Toxic Aleukia, in *Mycotoxigenic Fungi, Mycotoxins, Mycotoxicoses—an Encyclopedic Handbook* (eds R.D. Wyllie and L.G. Morehouse), *Volume 3*, Marcel Dekker, New York, pp. 21–86.

Johnson, K.M. (1984) *Bacillus cereus* foodborne illness: an update. *J. Food Protect.*, **47**, 145–53.

Johnson, K.M., Nelson, C.L. and Busta, F.F. (1984) Influence of heating and cooling rates on *Bacillus cereus* spore survival and growth in a broth medium and in rice. *J. Food Sci.*, **49**, 34–9.

Jones, R.H., Duncan, H.E., Payne, G.A. and Leonard, J.L. (1980) Factors influencing infection by *Aspergillus flavus* in silk-inoculated corn. *Plant Dis.*, **64**, 859–63.

Jonsyn, F.E., Maxwell, S.M. and Hendrickse, R.G. (1995) Ochratoxin A and aflatoxins in breast milk samples from Sierra Leone. *Mycopathologia*, **131**, 121–6.

Kameswara Rao, G., Malathi, M.A. and Vijayaraghavan, P.K. (1964) Preservation and packaging of Indian foods. I. Preservation of chapaties. *Food Technol.*, **18**, 108–10.

Kameswara Rao, G., Malathi, M.A. and Vijayaraghavan, P.K. (1966) Preservation and packaging of Indian foods. II. Storage studies on preserved chapaties. *Food Technol.*, **20**, 1070–3.

Kaur, P. (1986) Survival and growth of *Bacillus cereus* in bread, *J. Appl. Bacteriol.*, **60**, 513–6.

Kelly, S.M. and Wallin, J.R. (1986) Systemic infection of maize plants by *Aspergillus flavus*, in *Aflatoxin in Maize, a Proceedings of the Workshop* (eds M.S. Zuber, E.B. Lillehøj and B.L. Renfro), International Maize Improvement and Wheat Improvement Center, Mexico, pp. 187–93.

Kent-Jones, D.W. and Amos, A.J. (1957) *Modern Cereal Chemistry*, 5th edn, Northern Publishers, Liverpool, UK.

Kent, N.L. and Evers, A.D. (1994) *Technology of Cereals*, Pergamon Press, Oxford, New York.

Keoseyan, S.A. (1971) Incidence of *Clostridium perfringens* in dehydrated soup, gravy, and spaghetti mixes. *J. Assoc. Off. Anal. Chem.*, **54**, 106–8.

King, B. (1979) Outbreak of ergotism in Wollo, Ethiopia. Lancet, **i**. 1411.

Kintner, T.C. and Mangel, M. (1953) Survival of staphylococci and salmonellae in puddings and custards prepared with experimentally inoculated dried egg. *Food Res.*, **18**, 492–6.

Kline, L., Sugihara, T.F. and McCready, L.B. (1970) Nature of the San Francisco sour dough French bread process. I. Mechanics of the process. *Baker's Digest*, **44**, 48–50.

Knight, R.A. and Menlove, E.M. (1961) Effect of the bread-baking process on destruction of certain mold spores. *J. Sci. Food Agric.*, **12**, 653–6

Kramer, A. and Farquar, J. (1977) Fate of microorganisms during frozen storage of custard pies. *J. Food Sci.*, **42**, 1138–9.

Kramer, J.M. and Goepfert, R.J. (1989) *Bacillus cereus* and other *Bacillus* species, in *Foodborne Bacterial Pathogens* (ed. M.P. Doyle), Marcel Dekker, New York, pp. 21–70.

Kriek, N.P.J., Marasas, W.F.O. and Thiel, P.G. (1981) Hepato- and cardiotoxicity of *Fusarium verticillioides* (*F. moniliforme*) isolates from southern African maize. *Food and Cosmetics Toxicol.*, **19**, 447–56.

Krishnamachari, K.A.V.R. and Bhat, R.V. (1976) Poisoning by ergoty bajra (pearl millet) in man. *Indian J. Med. Sci.*, **64**, 1624–8.

Krøgh, P. (1987) Ochratoxins in food, in *Mycotoxins in Food* (ed. P. Krøgh), Academic Press, London, pp. 97–121.

Krøgh, P., Hald, B. and Pedersen, E.J. (1973) Occurrence of ochratoxin A and citrinin in cereals associated with mycotoxic porcine nephropathy. *Acta Pathol. Microbiol. Scand. B*, **81**, 689–95.
Kuiper-Goodman, T. (1994) Prevention of human mycotoxicoses through risk assessment and risk management. in *Mycotoxins in Grains: Compounds Other than Aflatoxin* (eds J.D. Miller and H.L. Trenholm), Eagan Press, St Paul, MN, pp. 439–68.
Kuiper-Goodman, T., Scott, PM. and Watanabe, H. (1987) Risk assessment of the mycotoxin zearalenone. *Regulatory Toxicol. Pharmacol.*, **7**, 253–306.
Kulp, K. (1991) Breads and yeast-leavened bakery foods, in *Handbook of Cereal Science and Technology* (eds K.J. Lorenz and K. Kulp), Marcel Dekker, New York, pp. 639–82.
Kuthubutheen, A.J. (1979) Thermophilic fungi associated with freshly harvested rice seeds. *Trans. Br. Mycol. Soc.*, **73**, 357–9.
Lacey, J. (1988) The microbiology of cereal grains from areas of Iran with a high incidence of oesophageal cancer. *J. Stored Products Res.*, **24**, 39–50.
Langseth, W. and Elen, O. (1996) Differences between barley, oats and wheat in the occurrence of deoxynivalenol and other trichothecenes in Norwegian grain. *Phytopathol. Zeitsch.*, **144**, 113–8.
Lamprech, E.D. (1968) Refrigerated dough. *Bulletin of the Association of Food and Drug Officials U.S.*, **32**, 168–73.
Lannuier, G.L. and Matz, S.A. (1967) Refrigerated dough products. *Cereal Sci. Today*, **12**, 478–80.
Le Bars, J. and Escoule, G. (1974) Champignons contaminant les fourrages. Aspects toxicologiques. *Alimentation et in Vie*, **62**, 125–42.
Lee, W.H., Staples, C.L. and Olson, J.C. (1975) *Staphylococcus aureus* growth and survival in macaroni dough and the persistence of enterotoxins in the dried products. *J. Food Sci.*, **40**, 119–20.
Lee, W.Y., Mirocha, C.J., Schroeder, D.J. and Walser, M.M. (1985) TDP-1, a toxic component causing tibial dyschondroplasia in broiler chickens and trichothecenes from *Fusarium roseum* 'Graminearum'. *Appl. Environ. Microbiol.*, **50**, 102–7.
Lee, U.S., Jang, H.S., Tanaka, T., Toyasaki, N. and Sugiura, Y. (1986) Mycological survey of Korean cereals and production of mycotoxins by *Fusarium* isolates. *Appl. Environ. Microbiol.*, **52**, 1258–60.
Legan, J.D. (1994) The microbiological condition of 1994 harvest flours. *Chorleywood Digest*, December, 112–3.
Legan, J.D. (2000) Cereals and cereal products, in *The Microbiological Safety and Quality of Food* (eds B.M. Lund, T.C. Baird-Parker and G.W. Gould), Aspen Publishers, Inc., Gaithersburg, MD.
Lillehøj, E.B. (1983) Effect of environmental and cultural factors on aflatoxin contamination of developing corn kernels, in *Aflatoxin and Aspergillus flavus in Corn* (eds U.L. Diener, R.L. Asquith and J.W. Dickens), Alabama Agricultural Experiment Station, Auburn. AL, pp. 27–34.
Lillehøj, E.B., Fennell, D.I. and Kwolek, W.F. (1976a) *Aspergillus flavus* and aflatoxin in Iowa corn before harvest. *Science, N.Y.*, **193**, 495–6.
Lillehøj, E.B., Kwolek, W.F., Peterson, R.E., Shotwell, O.L. and Hesseltine, C.W. (1976b) Aflatoxin contamination, fluorescence, and insect damage in corn infected a with *Aspergillus flavus* before harvest. *Cereal Chem.*, **53**, 505–12.
Lillehøj, E.B., Kwolek, W.F., Homer, E.S. *et al.* (1980) Aflatoxin contamination of preharvest corn: role of *Aspergillus flavus* inoculum and insect damage. *Cereal Chem.*, **57**, 255–7.
Loeb, J.R. and Mayne, R.Y. (1952) Effect of moisture on the microflora and formation of free fatty acids in rice bran. *Cereal Chem.*, **29**, 163–75.
Longrée, K. (1964) Cooling fluid food under agitation. *J. Am. Dietetic Assoc.*, **31**, 124–32.
López, L.C. and Christensen, C.M. (1967) Effect of moisture content and temperature on invasion of stored corn by *Aspergillus flavus*. *Phytopathology*, **57**, 588–90.
Lutey, R.W. and Christensen, C.M. (1963) Influence of moisture content, temperature and length of storage upon survival of fungi in barley kernels. *Phytopathology*, **53**, 713–17.
Magan, N. and Lacey, J. (1984) Effect of temperature and pH on water relations of field and storage fungi. *Trans. Br. Mycol. Soc.*, **82**, 71–81.
Magri, I., Berveglieri, M., Kumer, E., Contato, E., Sartea, A. and Martinelli, G. (1996) Microbiological quality of alimentary pasta sold in Ferrara. *Igiene Moderna*, **106**, 511–28.
Majumder, S.K. (1974) Control of Microflora and Related Production of Mycotoxins in Stored Sorghum, Rice and Groundnut. Wesley Press, Mysore, India.
Mallick, A.K. and Nandi, B. (198 1) Research: rice. *Rice J.*, **84**, 10–3.
Marasas, W.F.O., Kriek, N.P.J., Steyn, M., van Rensburg, S.J. and van Schalkwyk, D.J. (1978) Mycotoxological investigations on Zambian maize. *Food Cosmet. Toxicol.*, **16**, 39–45.
Marasas, W.F.O., van Rensburg, S.J. and Mirocha, C.J. (1979) Incidence of *Fusarium* species and the mycotoxins, deoxynivalenol and zearalenone, in corn produced in esophageal cancer areas in Transkei. *J. Agric. Food Chem.*, **27**, 1108–12.
Marasas, W.F.O., Wehner, F.C., van Rensburg, S.J. and van Schalkwyk, D.J. (1981) Mycoflora of corn produced in human esophageal cancer areas in Transkei, southern Africa. *Phytopathology*, **71**, 792–6.
Marasas, W.F.O., Nelson, P.E. and Tousson, T.A. (1984) *Toxigenic Fusarium Species: Identity and Mycotoxicology*, Pennsylvania State University Press, University Park, PA.
Marasas, W.F.O., Kellerman, T.S., Gelderblom, W.C.A., Coetzer, J.A.W., Thiel, P.G. and van der Lugt, J.J. (1988) Leukoencephalomalacia in a horse induced by fumonisin B_1 isolated from *Fusarium moniliforme*. *Onderspoort J. Vet. Res.*, **55**, 197–203.
Martin, C.R. and Sauer, D.B. (1976) Physical and biological characteristics of grain dust. *Trans. ASAE*, **19**, 720–3.
Massa, S., Trovatelli, L.D. and Vasta, E. (1986) [Microbiological testing of pasta made with eggs]. *Tecnica Molitoria*, **37**, 267–72.
Matossian, M.K. (1981) Mold poisoning: an unrecognized English health problem, 1550–1800. *Med. His.*, **25**, 73–84.

Matossian, M.K. (1989) *Poisons of the Past: Molds, Epidemics, and History*. Yale University Press, New Haven, CT.
McKinley, T.W. and Clarke, E.J. (1964) Imitation cream filling as a vehicle of staphylococcal food poisoning. *J. Milk Food Technol.* **27**, 302–4.
Meiering, A.G., Bakker-Arkema, F.W. and Bickert, W.G. (1966) Short time scaled storage of high moisture small grains. *Michigan Agricultural Experiment Station, Q. Bull.*, **48**, 465–70.
Meloni, P., Sau, M., Schintu, M. and Contu, A. (1996) Microbial counts in fresh pasta with or without filling. *Igiene Moderna*, **105**, 55–62.
Micco, C., Ambruzzi, M.A., Miraglia, M., Brera, C., Onori, R. and Benelli, L. (1991) Contamination of human milk with ochratoxin A, in *Mycotoxins, Endemic Nephropathy and Urinary Tract Tumours* (eds M. Castegnaro, R. Pleština, G. Dirheimer, I.N. Chernozemsky and H. Bartsch), International Agency for Research on Cancer, Lyon, pp.105–8.
Micco, C., Miraglia, M., Brera, C., Corneli, S. and Ambruzzi, A. 1995. Evaluation of ochratoxin A level in human milk in Italy. *Food Additives and Contaminants*, **12**, 351–354.
Michard, J., Jardy, N., Audiger, M.-T. and Gey, J.L. (1986) Coliforms and fecal coliforms in cream-filled pastries. *Microbial Aliment. Nutrition*, **4**, 205–16.
Miller, J.D. (1994) Epidemiology of *Fusarium* ear diseases of cereals, in *Mycotoxins in Grain: Compounds other than Aflatoxin* (eds J.D. Miller and H.L. Trenholm), Eagan Press, St Paul, MN, pp.19–36.
Mills, J.T. (1989) *Spoilage and Heating of Stored Agricultural Products*, Canadian Government Publishing Center, Ottawa, Ontario.
Milner, M. and Geddes, W.F. (1946) Grain storage studies. IV. Biological and chemical factors involved in the spontaneous heating of soybeans. *Cereal Chem.*, **23**, 449–70.
Milner, M., Christensen, C.M. and Geddes, W.F. (1947) Grain storage studies. VI. Wheat respiration in relation to moisture content, mold growth, chemical deterioration, and heating. *Cereal Chem.*, **24**, 182–9.
Mislivec, P.B. and Tuite, J. (1970) Species of *Penicillium* occurring in freshly-harvested and in stored dent corn kernels. *Mycologia*, **62**, 67–74.
Mora, M. and Lacey, J. (1997) Handling and aflatoxin contamination of white maize in Costa Rica. *Mycopathologia*, **138**, 77–89.
Morita, T.N. and Woodburn, M.J. (1977) Stimulation of *Bacillus cereus* growth by protein in cooked rice combinations. *J. Food Sci.*, **42**, 1232–5.
Moubasher, A.H., Elnaghy, M.A. and Abdel-Hafez, S.I. (1972) Studies on the fungus flora of three grains in Egypt. *Mycopathol. Mycol. Appl.*, **47**, 261–74.
Mukherjee, S.K., Albury, M.N., Pederson, C.S., van Veen, A.G. and Steinkraus, K.H. (1965) Role of *Leuconostoc mesenteroides* in leavening the batter of idli, a fermented food of India. *App. Microbiol.*, **13**, 227–31.
Müller, H.M. and Schwadorf, K. (1993) A survey of the natural occurrence of *Fusarium* toxins in wheat grown in a southwestern area of Germany. *Mycopathologia*, **121**, 115–21.
NCTR (National Center for Toxicological Research) (2001) *TR-496. Toxicology and Carcinogenesis Studies of Fumonisin B_1 (CAS No. 116355-83-0) in F344/N Rats and B6C3F_1 Mice (Feed Studies)*. NIH Publication No. 01-3955, Washington.
Northolt, M.D., van Egmond, H.P. and Paulsch, W.E. (1979) Ochratoxin A production by some fungal species in relation to water activity and temperature. *J. Food Prot.*, **40**, 778–81.
Osborne, B.G. and Willis, K.H. (1984) Studies into the occurrence of some trichothecene mycotoxins in UK home-grown wheat and imported wheat. *J. Sci. Food Agric.*, **35**, 579–83.
Ostovar, K. and Ward, K. (1976) Detection of *Staphylococcus aureus* from frozen and thawed convenience pasta products. *Lebensmittel Wissenschaft und Technologie*, **9**, 218–19.
Ottaviani, F. and Arvati, G. (1986) Experimental evaluation of the possibility of formation of *Staphylococcus aureus* toxins in egg pasta. *Tecnica Molitoria*, **37**, 902–8.
Papavisas, G.C. and Christensen, C.M. (1958) Grain storage studies. XXVI. Fungus invasion and deterioration of wheats stored at low temperatures and moisture contents of 15 to 18 percent. *Cereal Chem.*, **35**, 27–34.
Park, C.E., Szabo, R. and Jean, A. (1988) A survey of wet pasta packaged under a CO_2:N (20:80) mixture for staphylocci and their enterotoxins. *Can. Inst. Food Sci. Technol. J.*, **21**, 109–11.
Park, K.-J. and Lee, Y.-W. (1990) National occurrence of *Fusarium* mycotoxins in Korean barley samples harvested in 1987 and 1989. *Proc. Japn. Assoc. Mycotoxicol.*, **31**, 37–41.
Patel, T.B., Bowman, T.J. and Dallal, U.C. (1958) An epidemic of ergot poisoning through infected bajra (*Pennisetum typhoideum*) in southern parts of Bombay state. *Indian J. Med. Sci.*, **12**, 257–61.
Pelhate, J. (1976) Microflora of moist maize: determination of its development. *Bull. Org. Eur. Mediterr. Prot. Plant. (European and Mediterranean Plant Protection Organization Bulletin)*, **6**, 91–100.
Petkova-Bocharova, T. and Castegnaro, M. (1991) Ochratoxin A in human blood in relation to Balkan Endemic Nephropathy and urinary tract tumours in Bulgaria, in *Mycotoxins, Endemic Nephropathy and Urinary Tract Tumours* (eds M. Castegnaro, R. Pleština, G. Dirheimer, I.N. Chernozemsky and H. Bartsch), International Agency for Research on Cancer, Lyon, pp. 135–7.
Pettersson, H., Hedman, R., Engstrom, B., Elwinger, K. and Fossum, O. (1995) Nivalenol in Swedish cereals: occurrence, production and toxicity towards chickens. *Food Addit. Contaminants*, **12**, 373–6.
Pettersson, H., Kiessling, K.H. and Sandholm, K. (1986) Occurrence of the trichothecene mycotoxin deoxynivalenol (vomitoxin) in Swedish-grown cereals. *Swed. J. Agric. Res.*, **16**, 179–82.
Pinegar, J.A. and Cooke, E.M. (1985) *Escherichia coli* in retail processed food. *J. Hyg., Camb.*, **95**, 39–46.
Pitt, J.I. (1975) Xerophilic fungi and the spoilage of foods of plant origin, in *Water Relations of Foods* (ed. R.B. Duckworth), Academic Press, London, pp. 273–307.

Pitt, J.I. (1987) *Penicillium viridicatum. Penicillium verrucosum* and production of ochratoxin A. *Appl. Environ. Microbiol.*, **53**, 266–9.

Pitt, J.I. and Hocking, A.D. (1997) *Fungi and Food Spoilage*, 2nd edn, Aspen Publishers, Gaithersburg, MD.

Pitt, J.I., Hocking, A.D., Bhudhasamai, K., Miscamble, B.F., Wheeler, K.A. and Tanboon-Ek, P. (1993) The normal mycoflora of commodities from Thailand. 1. Nuts and oilseeds. *Int. J. Food Microbiol.*, **20**, 211–26.

Pitt, J.I., Hocking, A.D., Bhudhasamai, K., Miscamble, B.F., Wheeler, K.A. and Tanboon-Ek, P. (1994) The normal mycoflora of commodities from Thailand. 2. Beans, rice, small grains and other commodities. *Int. J. Food Microbiol.*, **23**, 35–53.

Pitt, J.I., Hocking, A.D., Miscamble, B.F., Dharmaputra, O.S., Kuswanto, K.R., Rahayu, E.S. and Sardjono. (1998a) The mycoflora of food commodities from Indonesia. *J. Food Mycol.*, **1**, 41–60.

Pitt, J.I., Hocking, A.D., Jackson, K.L., Mullins, J.D. and Webley, D.J. (1998b) The occurrence of *Alternaria* species and related mycotoxins in international wheat. *J. Food Mycol.*, **1**, 103–13.

Preonas, D.L., Nelson, A.I., Ordal, Z.J., Steinberg, M.P. and Wei, L.S. (1969) Growth of *Staphylococcus aureus* MF31 on the top and cut surfaces of southern custard pies. *App. Microbiol.*, **18**, 68–75.

Pyler, E.J. (1973) *Baking Science and Technology*, Siebel, Chicago, IL, pp. 210–21.

Rafai, P., Bata, A., Jakab, L. and Vanyi, A. 2000. Evaluation of mycotoxin-contaminated cereals for their use in animal feeds in Hungary. *Food Addit. Contaminants*, **17**, 799–808.

Rayman, M.K., D'Aoust, J.-Y., Aris, B., Maishment, C. and Wasik, R. (1979) Survival of microorganisms in stored pasta. *J. Food Prot.*, **44**, 330–4.

Rayman, M.K., Weiss, K.F. and Reidel, G.W. (1981) Microbiological quality of pasta products sold in Canada. *J. Food Prot.*, **44**, 746–9.

Rees, R.G., Martin. D.J. and Law, D.P. (1984) Black point in bread wheat: effects on quality and germination, and fungal associations. *Aust. J. Exper. Agric. Animal Husb.*, **24**, 601–5.

Resnik, S., Neira, S., Pacin, A., Martinez, E., Apro, N. and Latreite, S.A. (1996) A survey of the natural occurrence of aflatoxins and zearalenone in Argentine field maize – 1983–1994. *Food Addit. Contam*, **13**, 115–20.

Richard, J.L., Bacchetti, P. and Arp, L.H. (1981) Moldy walnut toxicosis in a dog, caused by the mycotoxin, penitrem A. *Mycopathologia*, **76**, 55–8.

Richter, K.S., Dorneanu, E., Eskridge, K.M. and Rao, C.S. (1993) Microbiological quality of flour. *Cereals Foods World*, **38**, 367–9.

Robinson, R.J. (1967) Microbiological problems in baking. *Baker's Digest*, **41**, 80–3.

Robinson, R.J., Lord, T.H., Johnson, I.A. and Miller, B.S. (1958) The aerobic microbiological population of pre-ferments and the use of selected bacteria for flavor production. *Cereal Chem.*, **35**, 295–305.

Rogers, R.F. (1978) *Bacillus* isolates from refrigerated doughs, wheat flour and wheat. *Cereal Chem.*, **55**, 671–4.

Rosenow, E.M. and Marth, E.H. (1986) Growth patterns of *Listeria monocytogenes* in skim, whole and chocolate milk and in whipping cream. *J. Food Prot.*, **49**, 847–8.

Rosenquist, H. and Hansen, A. (2000) The microbial stability of two bakery sourdoughs made from conventionally and organically grown rye. *Food Microbiol.*, **17**, 241–50.

Sabino, M., Prado, G., Inomata, El., Pedroso, M. de O. and Garcia, R.V. (1989) Natural occurrence of aflatoxins and zearalenone in maize in Brazil. Part II. *Food Addit. Contam.*, **6**, 327–31.

Samson, RA., Hocking, A.D., Pitt, J.I. and King, A.D. (eds) (1992) *Modern Methods in Food Mycology*, Elsevier Bioscience, Amsterdam.

Sauer, D.B. (1988) Effects of fungal deterioration on grain: nutritional value, toxicity, germination. *Int. J. Food Microbiol.*, **7**, 267–75.

Sauer, D.B. and Christensen, C.M. (1968) Germination percentage, storage fungi isolated from, and fat acidity values of export corn. *Phytopathology*, **58**, 1356–9.

Sauer, D.B. and Christensen, C.M. (1969) Some factors affecting increase in fat acidity values in corn. *Phytopathology*, **59**, 108–10.

Sauer, D.B., Meronuck, R.A. and Christensen, C.M. (1992) Microflora, in *Storage of Cereal Grains and Their Products* (ed. D.B. Sauer), American Association of Cereal Chemists, St Paul, Minnesota, pp. 313–40.

Saul, R.A. and Harris, K.L. (1979) Losses in grain due to respiration of grain and molds and other organisms, in *Postharvest Grain Loss Assessment Methods* (eds K.L. Harris and C.J. Lindblad), American Oil Chemists' Society, St Paul, MN, pp. 95–9.

Saur, L. (1991) Recherche de geniteurs de resistance à la fusariose de l'épi causée par *Fusarium culmorum* chez le blé et les espèces voisines. *Agronomie*, **11**, 535–41.

Schiemann, DA. (1978) Occurrence of *Bacillus cereus* and the bacteriological quality of Chinese 'take-out' foods. *J. Food Prot.*, **41**, 450–4.

Schmidt, E.W., Gould, W.A. and Weiser, H.H. (1969) Chemical preservatives to inhibit the growth of *Staphylococcus aureus* in synthetic cream pies acidified to pH 4.5 to 5.0. *Food Technol.*, **23**, 1197–9.

Schwab, A.H., Wentz, B.A., Jagow, J.A. *et al.* (1985) Microbiological quality of cream-type pies during processing. *J. Food Prot.*, **48**, 70–5.

Scott, D.B., De Jager, E.J.H. and van Wyk, P.S. (1988) Head blight of irrigated wheat in South Africa. *Phytophylactica*, **20**, 317–19.

Scott, P.M. (1997) Multi-year monitoring of Canadian grains and grain- based foods for trichothecenes and zearalenone. *Food Addit. Contam.*, **14**, 333–9.

Scott, P.M., van Walbeek, W., Kennedy, B. and Anyeti, D. (1972) Mycotoxins (ochratoxin A, citrinin and sterigmatocystin) and toxigenic fungi in grains and other agricultural products. *J. Agric. Food Chem.*, **20**, 1103–9.
Scott, P.M., Lau, P.-Y. and Kanhere, S.R. (1981) Gas chromatography with electron capture and mass spectrometric detection of deoxynivalenol in wheat and other grains. *J. Assoc. Off. Anal. Chem.*, **64**,1364–71.
Scott, P.M., Lombaert, G.A., Pellaers, P. *et al.* (1989) Application of capillary gas chromatography to a survey of wheat for five trichothecenes. *Food Addit. Contam.*, **6**, 489–500.
Seeder, W.A., Mossel, D.A.A. and van Zijl, F.H. (1969) About the growth of molds, especially of *Asp. flavus* on wheat flour with different water content. *Zeitschrift für Lebensmittel-Untersuchung und -Forschung*, **140**, 276–8.
Seiler, D.A.L. (1964) Factors affecting the use of mold inhibitors in bread and cake, in *Microbial Inhibitors in Food* (ed. N. Molin), Almqvist and Wiksell, Stockholm, Sweden, pp. 211–20.
Semeniuk, G. (1954) Microflora, in *Storage of Cereal Grains and Their Products* (eds J.A. Anderson and A.E. Alcock), Monograph Ser. 2, American Society of Cereal Chemists, St Paul, Minnesota, pp. 77–151.
Setamou, M., Cardwell, K.F., Schulthess, F. and Hell, K. (1998) Effect of insect damage to maize ears, with special reference to *Mussidia nigrivenella* (Lepidoptera, Pyralidae), on *Aspergillus flavus* (Deuteromycetes, Moniliales) infection and aflatoxin production in maize before harvest in the Republic of Benin. *J. Econ. Entomol.*, **91**, 433–8.
Shotwell, O.L. (1977) Aflatoxin in corn. *J. Am. Oil Chemists' Soc.*, **54**, 216A–224A.
Silliker, J.H. (1969) Some guidelines for the safe use of fillings, toppings. and icings. *Baker's Digest*, **43**, 51–4.
Silliker, J.H. and McHugh, S.A. (1967) Factors influencing microbial stability of butter-cream type fillings. *Cereal Sci. Tod.*, **12**, 63–5, 73–4.
Siriacha, P., Tanboon-Ek, P. and Buangsuwon, D. (1991) Aflatoxin in maize in Thailand, in Fungi and Mycotoxins, in *Stored Products: Proceedings of an International Conference* (eds B.R. Champ, H. Highley, A.D. Hocking and J.I. Pitt), ACIAR Proceedings No. 36. Canberra, Australian Center for International Agricultural Research, pp. 187–93.
Skaug, M.A., Stormer, F.C. and Saugstad, O.D. 1998. Ochratoxin A: a naturally occurring mycotoxin found in human milk samples from Norway. *Acta Paediatrica*, **87**, 1275–8.
Skaug, M.A., Helland, I., Solvoll, K. and Saugstad, O.D. 2001. Presence of ochratoxin A in human milk in relation to dietary intake. *Food Addit. Contam.*, **18**, 321–7.
Snijders, C.H.A. and Perkowski. J. (1990) Effects of head blight caused by *Fusarium culmorum* on toxin production and weight of wheat kernels. *Phytopathology*, **80**, 566–70.
Spicher, G. (1979) Die mikrobiologische Qualität der derzeit gehandelten Getreidevollkornerzeugnisse. *Getreide, Mehl Brot*, **33**, 290–4.
Spicher, G. (1982) Einige neue Aspekte der Biologic der Sauerteiggärung. *Getreide, Mehl Brot*, **36**, 12–6.
Spicher, G. (1984a) Die Erreger der Schimmelbildung bei Backwaren. I. Die auf verpackten Schnittbroten aufretenden Schimmelpilze. *Getreide, Mehl Brot*, **38**, 77–80.
Spicher, G. (1984b) Die Erreger der Schimmelbildung bei Backwaren. III. Einige Beobachtungen ueber die Biologie der Erreger der 'Kreidekrankheit' des Brotes. *Getreide, Mehl Brot*, **38**, 178–82.
Spicher, G. (1984c) Die Erreger der Schimmelbildung bei Backwaren. II. in verschimmelten Schnittbroten aufretende Mycotoxine. *Deutsche Lebensmittel-Rundschau*, **80**, 35–8.
Spicher, G. (1985) Zur Frage der Hygiene von Teigwaren. 3. Mittelung: Die rnikrobiologisch-hygienische Qualität der derzeit im Handel erhältichen Teigwaren. *Getreide, Mehl Brot*, **39**, 212–5.
Spicher, G. and Isfort, G. (1987) Die Erreger der Schimmelbildung bei Backwaren. IX. Die auf vorgebackenen Broetchen. Toast und Weichbroetchen auftretenden Schimmelpilze. *Deutsche Lebensmittel-Rundschau*, **83**, 246–9.
Spicher, G. and Isfort, G. (1988) Die Erreger der Schimmelbildung bei Backwaren. X. Einfluss von Essigen auf das Wachsturn von Schimmelpilze. *Getreide, Mehl Brot*, **42**, 57–60.
Stear, C.A. (1990) *Handbook of Breadmaking Technology*, Elsevier Science Publishers, Barking, UK.
Steele, J.L. (1969) Deterioration of shelled corn as measured by carbon dioxide production. *Trans. ASAE*, **12**, 685–9.
Stoloff, L. (1987) Aflatoxins – an overview, in *Mycotoxins in Human and Animal Health* (eds J.V. Rodricks, C.W. Hesseltine and M.A. Mehlman), Pathotox Publishers, Park Forest South, IL, pp. 7–28.
Stoloff, L., Trucksess, M., Anderson, P.W., Glabe, E.F. and Aldridge, J.G. (1978) Determination of the potential for mycotoxin contamination of pasta products. *J. Food Sci.*, **43**, 228–30.
Sugihara, T.F. (1977) Non-traditional fermentations in the production of baked goods. Baker's Digest, **51**, 76, 78, 80, 142.
Sugihara, T.F. (1978a) Microbiology of the soda cracker process. I. Isolation and identification of microflora. *J. Food Prot.*, **41**, 977–9.
Sugihara, T.F. (1978b) Microbiology of the soda cracker process. II. Pure culture fermentation studies. *J. Food Prot.*, **41**, 980–2.
Sugihara, T.F., Kline, L. and McCready, L.B. (1970) Nature of the San Francisco sour dough French bread process. II. Microbiological aspects. *Baker's Digest*, **44**, 51–3, 56–7.
Sulaiman, E.D. and Husain, S.S. (1985) Survey of fungi associated with stored food grains in silos in Iraq. *Pakistan J. Sci. Ind. Res.*, **28**, 33–6.
Surkiewicz, B.F. (1966) Bacteriological survey of the frozen prepared foods industry. I. Frozen cream-type pies. *App. Microbiol.*, **14**, 21–6.
Tanaka, T., Hasegawa, A., Matsuki, Y. and Ueno, Y. (1985) A survey of the occurrence of nivalenol, deoxynivalenol and zearalenone in foodstuffs and health foods in Japan. *Food Addit. Contam.*, **2**, 259–65.
Tanaka, T., Hasegawa, A., Matsuki, Y., Lee, U.S. and Ueno, Y. (1986) A limited survey of *Fusarium* mycotoxins nivalenol, deoxynivalenol and zearalenone in 1984 UK-harvested wheat and barley. *Food Addit. Contam.*, **2**, 247–52.

Thatcher, F.S., Coutu, C. and Stevens, F. (1953) The sanitation of Canadian flour mills and its relationship to the microbial content of flour. *Cereal Chem.*, **30**, 71–102.
Trenholm, H.L., Cochrane, W.P., Cohen, H. *et al.* (1983) Survey of vomitoxin contamination of 1980 Ontario, white winter wheat crop: results of survey and feeding trials. *J. Assoc. Official Anal. Chem.*, **66**, 92–7.
Trigostockli, D.M., Deyoe, C.W., Satumbaga, R.F. and Pedersen, J.R. (1996) Distribution of deoxynivalenol and zearalenone in milled fractions of wheat. *Cereal Chem.*, **73**, 388–91.
Trigostockli, D.M., Sanchez-Marinez, R.I., Cortez-Rocha, M.O. and Pedersen, J.R. (1998) Comparison of the distribution and occurrence of *Fusarium graminearum* and deoxynivalenol in hard red winter wheat for 1993–1996. *Cereal Chem.*, **75**, 841–6.
Trovatelli, L.D., Schiesser, A., Massa, S., *et al.* (1988) Microbiological quality of fresh pasta dumplings sold in Bologna and the surrounding district. *Int. J. Food Microbiol.*, **7**, 19–24.
Trucksess, M.W., Thomas, F., Young, K., Stack, M.E., Fulgueras, W.J. and Page, S.W. (1995) Survey of deoxynivalenol in US 1993 wheat and barley crops by enzyme-linked immunosorbent assay. *J. AOAC Int.*, **78**, 631–6.
Tubajika, K.M. and Damann, K.E. (2001) Sources of resistance to aflatoxin production in maize. *J. Agric. Food Chem.*, **49**, 2652–6.
Ueno, Y., Lee, U.S., Tanaka, T., Hasegawa, A. and Strzelecki, F. (1986) Natural occurrence of nivalenol and deoxynivalenol in Polish cereals. *Microbiol. Ailment. Nutr.*, **3**, 321–6.
Ulloa-Sosa, M. and Schroeder, H.W. (1969) Note on aflatoxin decomposition in the process of making tortillas from corn. *Cereal Chem.*, **46**, 397–400.
Uraguchi, K. (1969) Mycotoxic origin of cardiac beriberi. *J. Stored Products Res.*, **5**, 227–36.
Valik, L. and Gorner, F. (1993) Growth of *Staphylococcus aureus* in pasta in relation to its water activity. *Int. J. Food Microbiol.*, **20**, 45–8.
van Cauwenberge, J.E., Bothast. R.J. and Kwolek, W.F. (1981) Thermal inactivation of eight *Salmonella* serotypes on dry corn-flour. *Appl. Environ. Microbiol.*, **42**, 688–91.
Vandegraft, E.E., Shotwell, D.L., Smith, M.L. and Hesseltine, C.W. (1973) Mycotoxin formation affected by fumigation of wheat. *Cereal Sci. Tod.*, **18**, 412–4.
Vijaya Rao, D., Leela, R.K. and Sankaran, R. (1979) Microbial studies on inpack processed chapaties. *J. Food Sci. Technol., India*, **16**, 166–8.
Visconti, A. and Sibilia, A. (1994) *Alternaria* toxins, in *Mycotoxins in Grain: Compounds other than Aflatoxin* (eds J.D. Miller and H.L. Trenholm), Eagan Press. St Paul, MN, pp. 315–36.
Vogel, R.F. (1997) Microbial ecology of cereal fermentations. *Food Technol. Biotechnol.*, **35**, 51–54.
Vrabcheva, T., Gessler, R., Usleber, E. and Martlbauer, E. (1996) First survey on the natural occurrence of *Fusarium* mycotoxins in Bulgarian wheat. *Mycopathologia*, **136**, 47–52.
Wagenaar, R.O. and Dack, G.M. (1954) Further studies on the effect of experimentally inoculating canned bread with spores of *Clostridium botulinum*. *Food Res.*, **19**, 521–9.
Walker, R. 2002. Risk assessment of ochratoxin: current views of the European Scientific Committee on Food, the JECFA and the Codex Committee on Food Additives and Contaminants. *Adv. Exp. Med. Biol.*, **504**, 249–55.
Wallace, H.A.H. and Sinha, R.N. (1975) Microflora of stored grain in international trade. *Mycopathologia*, **57**, 171–6.
Wallace, H.A.H., Sinha, R.N. and Mills, J.T. (1976) Fungi associated with small wheat bulks during prolonged storage in Manitoba. *Can. J. Bot.*, **54**, 1332–43.
Walsh, D.E. and Funke, B.R. (1975) The influence of spaghetti extruding, drying, and storage on survival of *Staphylococcus aureus*. *J. Food Sci.*, **40**, 714–6.
Walsh, D.E., Funke, B.R. and Graalum, K.R. (1974) Influence of spaghetti extruding conditions, drying and storage on the survival of *Salmonella typhimurium*. *J. Food Sci.*, **39**, 1105–6.
Wang, Y.Z. and Miller, J.D. (1988) Screening techniques and sources of resistance to *Fusarium* head blight, in *Wheat Production: Constraints in Tropical Environments* (ed. A.R. Khlatt), Centro Intenacional de Mejoramiento de Maiz Trigo (CIMMYT), Mexico City, pp. 239–50.
Watson, D.H. (1984) An assessment of food contamination by toxic products of *Alternaria*. *J. Food Prot.*, **47**, 485–8.
Webley, D.J., Jackson, K.L., Mullins, J.D., Hocking, A.D. and Pitt, J.I. (1997) *Alternaria* toxins in weather-damaged wheat and sorghum in the 1995–1996 Australian harvest. *Aust. J. Agric. Res.*, **48**, 1249–55.
Weckel, K.G., Hawley, R. and McCoy, E. (1964) Translocation and equilibration of moisture in canned frozen bread. *Food Technol.*, **18**, 1480–2.
Weddeling, K., Bassler, H.M.S., Doerk, H. and Baron, G. (1994) [Orientated tests for the application of the enzyme immunological process for determining deoxynivalenol, ochratoxin A and zearalenone in brewing barley, malt and beer.] *Monatsschrift für Brauwissenschaft*, **47**, 94–8.
Weidenbörner, M. and Kunz, B. (1995) Mycoflora of cereal flakes. *J. Food Prot.*, **58**, 809–12.
Wicklow, D.T., Horn, B.W. and Shotwell, DL. (1987) Aflatoxin formation in pre-harvest maize ears co-inoculated with *Aspergillus flavus* and *Aspergillus niger*. *Mycologia*, **79**, 679–82.
Wicklow, D.T., Horn, B.W., Shotwell, O.L., Hesseltine, C.W. and Caldwell, R.W. (1988) Fungal interference with *Aspergillus flavus* infection and aflatoxin contamination of maize grown in a controlled environment. *Phytopathology*, **78**, 68–74.
Windham, G.L. and Williams, W.P. (1998) *Aspergillus flavus* infection and aflatoxin accumulation in resistant and susceptible maize hybrids. *Plant Dis.*, **82**, 281–4.
Wilson, B.J. (1971). Recently discovered metabolites with unusual toxic manifestations, in *Mycotoxins in Human Health* (ed. I.F.H. Purchase), Macmillan: London, pp. 223–9.

Wilson, D.M. and Abramson, D. (1992) Mycotoxins, in *Storage of Cereal Grains and Their Products* (ed. D.B. Sauer). American Association of Cereal Chemists, St Paul, MN, pp, 341–91.

Woolaway, M.C., Bartlett, C.L.R., Weinhe, A.A., Gilbert, R.J., Murrell, H.C. and Durrell, P. (1986) International outbreak of staphylococcal food poisoning caused by contaminated lasagne. *J. Hyg.*, **96**, 67–73.

Yde, M. (1982) [Microbiological quality of pastries filled with creme patisserie]. *Archives Belges Medicine Sociale, Hygiene, Medicine du Travail et Medecine Legale*, **40**, 455–66.

Yoshizawa, T. and Jin, Y.Z. (1995) Natural occurrence of acetylated derivatives of deoxy-nivalenol and nivalenol in wheat and barley in Japan. *Food Addit. Contam.*, **12**, 689–94.

Yoshizawa, Y., Matsuura, Y., Tsuchiya, Y., Morooka, N., Kitani, K., Ichinoe, M. and Kurata, H. (1979) On the toxigenic Fusaria invading barley and wheat in southern Japan. *J. Food Hyg. Soc. Jpn*, **20**, 216.

Zardetto, S. and Fresco, S. di. (2000) Influence of storage modality on *Salmonella enteritidis* in experimentally inoculated fresh filled pasta. *Tecnica-Molitoria*, **51**, 609–21.

Zardetto, S., Fresco, S. di and Pasqualetto, K. (1999) Heat treatment of fresh filled pasta. II. Influence of heat treatment on normal microbiological content and on inoculated microorganisms. *Tecnica-Molitoria*, **50**, 643–50, 658.

9 Nuts, oilseeds, and dried legumes

I Introduction

A Definitions

Nuts are dry, one seeded fruit, which do not dehisce at maturity, and are usually enclosed by a rigid outer casing or shell. Most nuts grow on large shrubs or trees and are known as *tree nuts*. Tree nuts include almonds (*Prunus amygdalus*), hazelnuts (*Corylus avellana*), pistachios (*Pistachia vera*), Brazil nuts (*Bertholletia excelsa*), pecans (*Carya illinoensis*), coconuts (*Cocos mucifera*), and macadamia nuts (*Macadamia ternifolia*). Although not covered strictly by the botanical definition, walnuts (*Juglans regia*) are usually considered to be nuts. The only major nut not from trees is the peanut (*Arachis hypogaea*), known as "groundnut" in some countries. Botanically, the peanut is a legume, a member of the pea family, but it will be treated as a nut in this chapter. Illipe nuts (*Shorea aptera* and related species) contain 50–70% fat and are exported from Southeast Asia as a cocoa butter substitute.

Peanut butter is the finely ground paste produced from (usually) roasted peanuts. In some countries other edible oils or ingredients are added to peanut butter, but with little effect on microbiology.

Peanut sauces, known as *satay sauces*, are pastes made from dried shelled peanuts, spices and water, usually with other ingredients added. They are widely used in Asian countries, mostly consumed fresh, and are increasingly traded internationally, where they are normally subjected to sterilizing treatments as low-acid products.

Oilseeds are mostly small seeded crops that are grown primarily for oil production. They are drawn from a range of botanical families. Oilseeds include palm nuts (*Elaeis guineensis*, *Ela. oleifera*, and hybrids), rapeseed or canola (*Brassica rapa*, *B. campestris*), sesame (*Sesamum indicum*), sunflowers (*Helianthus annuus*), safflower (*Carthamus tinctorius*), cottonseed (*Gossypium* spp.), and cacao seeds (*Theobroma cacao*). The last named is treated in Chapter 10, along with cocoa butter. Oil from *Zea mays*, known as maize oil or corn oil, is also an important commodity. Dried coconut is known as *copra* when sold in large pieces and as *desiccated coconut* when shredded. Comminuted coconut is known as *coconut cream* and is usually sold canned as a low-acid food. It is sometimes spray dried and sold as a powder.

Dried legumes are the seeds of leguminous plants, members of the family *Leguminosae*. Dried legumes that will be considered in this chapter include soybeans and the many other types of beans that are field dried. Fresh legumes are treated under vegetables in Chapter 5.

Coffee is a beverage made by brewing roasted beans of the coffee tree (*Cafea arabica*, *C. canephora* var. *robusta*, or hybrids). Coffee is consumed almost universally and the coffee growing, and manufacturing industries are amongst the largest in international trade. *Instant coffee* is produced by freeze or spray drying brewed coffee.

B Important properties

Nuts have very high nutritional and calorific values. The pH of all nuts and oilseeds is near neutral, in theory rendering them susceptible to growth of all kinds of microorganisms during development and before natural drying at maturity. In practice, shells provide a highly effective barrier to the entry of bacteria during nut growth. After natural drying, the low a_w of most nuts restricts bacterial spoilage or toxin production. However, contamination of nuts may sometimes occur post-harvest, for example, with salmonellae, leading to concern both directly and with high a_w products to which nuts are added, e.g. dairy products.

In the microbiological context, the most important property of both nuts and oilseeds is their high oil content. This provides a high susceptibility to attack by lipolytic bacteria and by spoilage fungi, with an exceptional potential for mycotoxin production.

Peanuts, with a unique growth habit in soil, are especially vulnerable to fungal invasion before harvest. Many kinds of fungi are found in peanuts, but the presence of *Aspergillus flavus* and the production of aflatoxins are of major concern.

Spoilage and mycotoxigenic fungi sometimes invade other nuts as well, but usually only as the result of insect or mechanical damage (tree nuts) or contamination during drying and processing (pistachios).

Most dried legumes are rich in carbohydrates and low in oils, so they are microbiologically similar to cereals. However, soybeans are up to 20% oil on a moisture free basis (Waggle and Kolar, 1979), so they more resemble oilseeds in their microbiology. Soybeans also contain high levels of protein (40% or more, Waggle and Kolar, 1979), an attractive trait because of the high nutritional quality (Richert and Kolar, 1987).

Fresh coffee beans, known as cherries, are of relatively low nutritional value. The flesh of the fruit must be removed before the beans are fully dried, and this is accomplished by mechanical means or fermentation.

C Methods of processing

Typical primary processing of all of the commodities considered here involves drying, which commonly takes place in the field. Mechanical drying may be used for some crops in particular localities.

Tree nuts. Tree nuts are almost always allowed to dry *in situ*, then harvested mechanically or by hand. Some types are allowed to fall to the ground before collection, resulting in surface contamination by bacteria and fungi. Sometimes dehydration is used in adverse climates or seasons to complete the drying process.

Coconuts are harvested from tall trees either green (for consumption fresh) or at maturity. Nuts are pierced or broken, drained of water, the kernels cut into slices and sun dried to produce copra for oil production. A variety of other coconut products are made using both traditional and modern technologies (Hagenmaier, 1980).

Peanuts. Peanuts are harvested by pulling from the ground while high in moisture content and then drying. In developed countries pulling is mechanical. Pulling requires that the stems (pegs) supporting the nuts must be strong and not yet senescent, so the nuts are usually at least partially dried on the upturned bushes to weaken the pegs before mechanical threshing to remove the nuts from the bushes. Newer types of threshers permit mechanical removal of nuts immediately after pulling, followed by mechanical drying. In damp growing areas or seasons nuts are often mechanically dried after threshing to ensure microbiological stability. Peanuts must be dried to about 8% moisture (below 0.70 a_w) to be microbiologically stable.

The field drying process is usually quite slow: in Australia field drying takes 6–10 days to complete even under good conditions (Pitt, 1989).

In developing countries, harvesting is commonly carried out by hand. Pulling is aided by the use of hoes or forks, then the nuts are usually removed from the bushes by hand and dried in the sun on hessian or plastic sheets. If weather conditions are good, drying may require no more than 2–3 days. Much longer periods may be needed if conditions are adverse.

Oilseeds. All types of oilseeds are usually field dried.

Dried legumes. Generally field drying is used, although in some localities finishing may be by mechanical drying in bins.

Coffee. Coffee cherries are usually picked by hand, or mechanically on large farms, and are dried in the sun or less commonly mechanically. The beans may be dried directly, and separated from the hull afterwards, or mechanically dehulled and dried, or dehulled by fermentation.

D Types of final products

Large quantities of nuts are sold as snack foods, often without further processing or after shell removal. Many nuts are also processed by roasting or frying, and again mostly consumed as snack foods. A proportion of nut production is incorporated into confectionery: chocolates, toffee bars, and muesli bars are three common types of products.

A major end use of peanuts is in peanut butter and related products. Dried peanuts are roasted and ground, usually with the addition of salt and sometimes other oils. In some Southeast Asian countries, peanuts are roasted and ground with added water to produce satay sauces, which may be heat processed as low-acid foods, especially for export. In some peanut producing countries, peanuts are boiled in salty water and consumed as a snack food: this product has a short shelf life. Sometimes boiled peanuts are subsequently roasted in the shell and then vacuum packed to produce a shelf-stable product.

Dried coconut may be sold in large pieces as copra, which is mainly used for oil production, or may be shredded and sold as desiccated coconut for manufacturing or domestic use. Other products from coconut in international trade include canned coconut milk or cream, processed as a low-acid food in cans or pouches; spray-dried coconut cream, a powder made up of oil globules encapsulated in a protein film; desiccated pressed coconut, a by-product of coconut cream manufacture; coconut flour, made from fresh coconut or copra; and coconut oil (Hagenmaier, 1980).

Palm sugars and an alcoholic drink made from the sap of fruiting spath of coconut are popular in Southeast Asia and India. A variety of other coconut products, often made from young coconuts, are sold within the same regions. In Indonesia, a fermented coconut cake called "tempeh bongkrek" is made from coconut or coconut plus soybean fermented with *Rhizopus oligosporus* (Ko *et al.*, 1979).

Oilseeds are normally further processed to provide oil. Residual meals are commonly used as animal feed ingredients.

Dried legumes are usually consumed after boiling to soften them. They may also be used as soup ingredients, either domestically or commercially.

Soybeans have become a very large commercial crop in the United States and Brazil, originally due to demand for soybean oil, a cheap commodity because soybean meal is a high protein, high-value feed ingredient. Since the 1950s, soybeans have been processed into food-grade products including soy flour, soy protein concentrate and isolated soy protein (Waggle and Kolar, 1979). Isolated soy protein, more than 90% pure, with high availability and digestibility, and with an excellent nutritional balance, has become important as a relatively low cost food-grade protein source (Kolar *et al.*, 1985; Richert and Kolar, 1987).

Coffee beans are stored after drying (as "green" coffee) then graded and shipped to manufacturers. Beans are then roasted to varying degrees depending on type and extent of flavor desired. They may be packed and distributed, usually under inert atmospheres, as beans or ground and extracted with hot water, then dried to make instant coffee.

Table 9.1 Fungi commonly invading peanuts before or immediately post harvest

Species	At harvest: average infection (%)	After 2–13 days drying: average infection (%)
Aspergillus flavus	<1	1.6
Lasiodiplodia theobromae	0.9	0.8
Fusarium spp.	13.2	17.0
Macrophomina phaseolina	10.6	24.7
Penicillium funiculosum	1.6	1.4
Rhizoctonia solani	0.9	2.4
Rhizopus spp.	2.1	4.7

From McDonald (1970).

II Initial microflora

A Nuts

Only a few systematic studies have been carried out on the microflora of nuts in the field. Little or no information exists on the microflora of fresh tree nuts. However, the presence of shells provides a very strong protective barrier against infection by both fungi and bacteria. Apart from possible systemic fungal infections from the trees themselves, tree nuts are essentially sterile before harvest. Studies on dried nuts have provided no definitive information about the time of infection of tree nuts by fungi.

The mycoflora of fresh peanuts has been surveyed only rarely. Data from one such study in Nigeria are given in Table 9.1 (McDonald, 1970). Apart from *Asp. flavus*, which was present at low levels, the fungi found were field fungi, with *Fusarium* species, *Lasiodiplodia theobromae* and *Macrophomina phaseolina* dominant.

B Oilseeds

Little information has been published on the pre-harvest microflora of oilseeds. It is to be expected that the bacterial flora will be similar to cereals (Chapter 8), and that some specific field fungi will be present, together with the types commonly found on cereals.

C Legumes

Fresh legumes are readily contaminated externally on the pod with bacteria and fungi. However, the pod provides some protection from entry of microorganisms during the growth phase. Important diseases of fresh legumes include black blight due to *Alternaria alternata*, anthracnoses due to *Colletotrichum* species, dry or soft pod rots due to *Ascochyta* species, and soft rots due to *Pseudomonas* and *Xanthomonas* species (Snowdon, 1991).

D Coffee

Few studies have been carried out on the microflora of fresh coffee cherries. While cherries are fresh and intact few microorganisms are present except as contaminants. *Aureobasidium pullulans*, *Fusarium stilboides*, and *Penicillium brevicompactum* are the most common mould species and *Candida edax* and *Cryptococcus album* the most common yeasts associated with fresh coffee cherries (Frank, 2001). If cherries dry on the tree, to produce what is known as "boia", field fungi such as *Alternaria* and *Cladosporium* are commonly present (M.H. Taniwaki, unpublished).

III Primary processing

A *Effects of processing on microorganisms*

Primary processing of nuts, oilseeds, and legumes usually involves drying under natural conditions. Effective sun drying may reduce the initial microflora. However, if drying occurs under adverse conditions, mycoflora may increase both in kind and in numbers, and mycotoxin production may be initiated. The bacterial flora of these raw commodities reflects the environment in which they are grown and harvested. The numbers and types of bacteria present result from contamination from soil and equipment and to other environmental factors.

Tree nuts

Pistachios. Pistachios were contaminated with up to 14 *Aspergillus* species (Doster and Michailides, 1994). The majority of the isolates came from nuts which had split or been damaged by insects in the orchard. The most commonly encountered species was *Aspergillus niger*, present in 30% of such kernels. *Aspergillus flavus* and *Asp. parasiticus* were also found, with the potential to form aflatoxins, and in addition *Asp. ochraceus* and *Asp. melleus*, potential producers of ochratoxin A (Doster and Michailides, 1994). Mould counts on 143 samples of freshly harvested pistachios in Turkey ranged from 10^3–10^4 per gram and after storage 10^5–10^6 per gram (Heperkan *et al.*, 1994). A significant proportion of kernels (6–16%) was invaded by *Asp. flavus*. Pistachio nuts that split to expose the kernels in the orchard are prone to infection by *Asp. flavus* and hence are likely to contain aflatoxins. The time pistachio nuts split depends largely on cultivar type. Aflatoxin contamination is also more frequent in nuts infected with the navel orange worm (*Amyelois transitella*) (Sommer *et al.*, 1986).

Copra. Coconut meat is probably sterile before the nut is broken. However, the thickness of the meat means that it will dry slowly, leading to growth of bacteria and fungi.

Fungal infection in copra is often very high (Pitt *et al.*, 1993) (Table 9.2). Dominant fungi were *Asp. flavus* (present in 86% of 21 samples and in 20% of all copra pieces examined), *Asp. niger* (present in 43% of samples, and 18% of all pieces), *Rhizopus oryzae* (52% of samples, 25% of all

Table 9.2 Fungal infection in 21 samples of surface disinfected copra from Thailand

Fungus	No. of infected samples (%)	Average infected particles in infected samples (%)	Range of infection in infected samples (%)	Percent of particles infected averaged over all samples
Aspergillus clavatus	3 (14)	23	2–46	3
Asp. flavus	18 (86)	23	2–73	20
Asp. niger	9 (43)	42	3–86	18
Asp. tamarii	5 (24)	8	6–14	2
Endomycopsis fibuliger	4 (19)	11	8–14	3
Eurotium amstelodami	3 (14)	16	12–20	2
Eur. chevalieri	9 (43)	31	2–80	13
Eur. repens	4 (19)	15	5–30	3
Eur. rubrum	8 (38)	29	5–83	11
Mucor spp.	3 (14)	35	10–83	5
Nigrospora oryzae	6 (29)	6	2–14	2
Penicillium citrinum	8 (38)	18	2–48	7
Rhizopus oryzae	11 (52)	48	14–98	25
Sordaria fimicola	8 (38)	29	3–50	11
Samples infected	21 (100)	76	6–100	76

Data of Pitt *et al.* (1993).

Table 9.3 Fungal infection in 45 samples of surface disinfected cashew kernels from Thailand

Fungus	No. of infected samples (%)	Average infected particles in infected samples (%)	Range of infection in infected samples (%)	Particles infected averaged over all samples (%)
Aspergillus flavus	27 (60)	8	2–35	5
Asp. Niger	24 (53)	10	2–66	5
Asp. Sydowii	5 (11)	18	4–70	2
Chaetomium globosum	21 (47)	7	2–30	3
C. fumicola	6 (13)	5	2–10	1
Cladosporium cladosporioides	17 (38)	6	2–22	2
Eurotium amstelodami	7 (16)	20	2–35	3
Eur. chevalieri	18 (40)	8	2–24	3
Eur. rubrum	14 (31)	20	2–90	6
Nigrospora oryzae	26 (58)	8	2–22	5
Penicillium citrinum	13 (29)	7	2–20	2
Pen. olsonii	3 (7)	30	8–40	2
Samples infected	45 (100)	40	6–90	40

Data of Pitt *et al.* (1993).

pieces) and the storage fungi *Eurotium chevalieri* and *Eur. rubrum* (38–43% of samples; 11–13% of all pieces).

Cashews. Dried cashews contain a range of spoilage fungi, but usually at low levels (Table 9.3). This reflects both the arboreal source of the cashew and their very thick shell. Fungal infection can occur during drying under poor conditions, but infection levels are limited. The overall infection level of *Asp. flavus* in 45 samples of cashew nuts from Thailand was only 5% (Pitt *et al.*, 1993).

Others. Thirty-three species of fungi were isolated from 149 hazelnut samples, the most commonly occurring species being *Rhi. stolonifer* and *Penicillium aurantiogriseum* (Senser, 1979). Pecans (37 samples) contained a wide variety of fungi: 119 species from 44 genera. *Aspergillus* species accounted for 48% of the more than 1300 isolates obtained; next came *Penicillium* (19%), *Eurotium* (18%), and *Rhizopus* (8%). The dominant species was *Asp. niger* (293 isolates), followed by *Asp. flavus* (207), *Eur. repens* (132), *Eur. rubrum* (109), *Asp. parasiticus* (100), *Rhi. oryzae* (68), and *Penicillium expansum* (61) (Huang and Hanlin, 1975).

An unusual range of genera was found in pecans that had previously been invaded by weevils in the field (Wells and Payne, 1976). Nearly half of 2300 isolates from several hundred moldy nuts were *Alternaria* or *Epicoccum* species. *Penicillium* species made up 25% of the total, and *Aspergillus* only 1.0%.

Peanuts

Peanuts. Extensive documentation of the mycoflora of dried peanuts was provided by Joffe (1969). During a 5-year period, fungi were isolated from over 400 samples of freshly harvested and stored peanuts from Israel. By far, the most common species encountered was *Asp. niger*, isolated from a low of 8% of kernels in one year to a high of 71%. The other dominant members of the flora were *Asp. flavus* (0–8%); *Penicillium funiculosum* (3–16%); *Pen. purpurogenum* (2–8%); and *Fusarium solani* (0–9%).

Pitt *et al.* (1993) examined more than 100 samples of dried peanuts from Thailand (Table 9.4). Thirty-one fungal species were commonly encountered and a further 26 species occasionally. The dominant fungi were *Asp. flavus* and *Asp. niger*, found in 95% and 86% of all samples respectively. *Asp. tamarii* (31% of samples) and *Asp. wentii* (20% of samples) were also common. Other Aspergilli, except *Asp. candidus* (4% of samples) were rare. *Fusarium semitectum* (19% of samples) and *Fus. equiseti* (10% of

Table 9.4 Major fungal infection in 109 samples of surface disinfected peanut kernels from Thailand

Fungus	No. of infected samples (%)	Average infected particles in infected samples (%)	Range of infection in infected samples (%)	Particles infected averaged over all samples (%)
Aspergillus flavus	103 (95)	44	2–100	41
Asp. niger	94 (86)	38	3–100	33
Asp. tamarii	34 (31)	11	2–40	3
Asp. wentii	22 (20)	22	2–80	4
Chaetomium fumicola	5 (5)	9	2–20	2
Eurotium chevalieri	50 (46)	33	4–100	15
Eur. repens	7 (6)	11	4–20	1
Eur. rubrum	56 (51)	28	2–85	14
Lasiodiplodia theobromae	36 (33)	12	2–40	4
Macrophomina phaseolina	53 (49)	16	2–55	8
Penicillium aurantiogriseum	6 (5)	36	2–100	2
Pen. citrinum	50 (46)	14	2–60	6
Pen. funiculosum	15 (14)	18	2–92	2
Rhizopus oryzae	65 (60)	25	2–95	15
Wallemia sebi	13 (12)	42	18–98	6
Samples infected	109 (100)	84	6–100	84

From Pitt *et al.* (1993).

samples) were the only common *Fusarium* species found; only low percentages of nuts were infected in any sample.

Macrophomina phaseolina and *Lasi. theobromae* (49% and 33% of samples, respectively) were the only field fungi commonly found. *Nigrospora oryzae* was present in 22% of samples, but only low numbers of nuts were infected. Other noteworthy fungi were *Rhizopus oryzae* and *Wallemia sebi* (in 60% and 12% of samples, respectively). *Pen. citrinum* was very common (in 46% of samples) and a wide range of other *Penicillium* species was also detected (Pitt *et al.*, 1993).

Among storage fungi, *Eur. rubrum* and *Eur. chevalieri* (51% and 46% of samples) were very commonly encountered. *Eur. amstelodami* (9% of samples) and *Eur. repens* (6%) were present at much lower frequencies.

Figures for infection in more than 250 peanut samples from Indonesia were similar to those from Thailand (Pitt *et al.*, 1998). Infection with *Asp. flavus* was even higher, with 40% of the 12,500 kernels examined infected by this species.

The incidence of field and storage fungi in peanut samples from farms, middlemen and retailers in Thailand was also compared (Pitt *et al.*, 1993) (Table 9.5). Infection rates by species commonly regarded as field fungi, e.g. *Fus. solani* and *Macrophomina phaseolina*, declined during storage. Some *Penicillium* species usually regarded as storage fungi, notably *Pen. brevicompactum*, *Pen. janthinellum*, and *Pen. pinophilum*, also showed sharp declines during storage. It appears that the traditional view of these species as storage fungi is inappropriate here. In contrast, numbers of *Pen. glabrum* increased during storage, and those of *Pen. citrinum* were unaffected by sampling time.

Oilseeds. There appears to be little published information on the microflora of dried oilseeds.

Legumes. Bacteria on dried legumes are of little consequence when these commodities are consumed after boiling or other heat processing. They may be important if the legumes are incorporated into soups or other high a_w products.

During the drying process, legumes are readily invaded by fungi. However, spoilage is uncommon and mycotoxin production rarely of significance. The mycoflora of various kinds of beans in Thailand

Table 9.5 Comparison of the fungal flora obtained from disinfected peanut kernels collected from farmers and middlemen in Thailand versus those from retail outlets

	Farm and middleman		Retail	
Fungus	No. of infected samples (%)	Average infected particles in infected samples (%)	No. of infected samples (%)	Average infected particles in infected samples (%)
Aspergillus candidus	2 (4)	26	2 (3)	2
Asp. niger	43 (91)	41	51 (82)	35
Asp. Wentii	8 (17)	29	14 (23)	18
Chaetomium fumicola	5 (11)	9	0 (0)	0
Chaetomium spp.	5 (11)	8	5 (8)	3
Cladosporium cladosporioides	6 (13)	6	10 (16)	6
Eurotium amstelodami	3 (6)	21	7 (11)	14
Fusarium solani	3 (6)	6	0 (0)	0
Macrophomina phaseolina	22 (47)	26	32 (52)	11
Penicillium aethiopicum	2 (4)	14	2 (3)	10
Pen. aurantiogriseum	5 (11)	41	1 (2)	2
Pen. brevicompactum	2 (4)	10	0 (0)	0
Pen. citrinum	22 (47)	20	28 (45)	9
Pen. funiculosum	11 (23)	22	4 (6)	8
Pen. glabrum	0 (0)	0	3 (5)	6
Pen. janthinellum	3 (6)	13	0 (0)	0
Pen. pinophilum	4 (9)	17	0 (0)	0
Syncephalastrum racemosum	4 (9)	7	5 (8)	12
Samples infected	47 (100)	90	62 (100)	80

Data of Pitt *et al.* (1993).
Only fungi with notable differences are included. For more complete overall data see Table 9.2.

Table 9.6 Fungi commonly found invading dried legumes in Thailand

	Mung beans		Soybeans		Black beans	
Fungus	Samples infected (%)	Infected particles in infected samples (Av. %)	Samples infected (%)	Infected particles in infected samples (Av. %)	Samples infected (%)	Infected particles in infected samples (Av. %)
Aspergillus flavus	45	4	67	8	61	6
Asp. niger	14	10	12	3	35	7
Asp. penicillioides	0	–	6	20	0	–
Asp. restrictus	0	–	16	12	0	–
Chaetomium globosum	14	5	33	5	26	2
Cladosporium cladosporioides	13	2	49	18	39	4
Eurotium amstelodami	0	–	16	12	13	2
Eur. chevalieri	18	6	33	12	26	4
Eur. rubrum	11	2	51	13	22	3
Fusarium moniliforme	13	7	6	3	0	–
F. semitectum	55	27	29	5	52	11
Lasiodiplodia theobromae	30	6	18	7	22	12
Macrophomina phaseolina	23	9	22	9	1	–
Nigrospora oryzae	11	2	39	5	13	2
Penicillium citrinum	13	5	22	13	17	2

Data from Pitt *et al.* (1994).

were surveyed by Pitt *et al.* (1994) (Table 9.6). Fungal species found were a mixture of field fungi, which presumably had invaded before harvest and during the early stages of drying, and storage fungi, which would have invaded during the late phase of drying or developed in storage. Among the most common field fungi was *Fus. semitectum*, which was present in 55% of mung bean samples, 52% of black bean samples and 29% of those from soybeans. The mycotoxigenic species *Fus. verticillioides*

(=*Fus. moniliforme*) was found in 13% of mung bean samples. It was much less common in soybeans and not found in black beans.

Lasiodiplodia theobromae, a 'universal' field fungus, was present in 30% of mung bean samples, and 18–22% of the other types. *Macrophomina phaseolina*, a common pathogen on beans in the tropics, was found on 22–23% of mung bean and soybean samples, but was seen only once on black beans.

Asp. flavus was common on beans, as on other commodities from Thailand (Pitt *et al.*, 1994). It is unclear whether this species can invade beans before harvest, or is only found as a storage fungus in this commodity. Although the proportion of samples infected was high (45–67%), levels of infection in individual beans in any sample were low (3–6%).

Among storage fungi, three common species of *Eurotium* were present in 11–18% of mung bean samples, 16–51% of soybean samples and 13–26% in those of black beans. The higher levels in soybeans probably reflect longer storage of soybeans than other legumes in Thailand, a probability borne out by the much higher incidence of other storage fungi, *Asp. penicillioides* and *Asp. restrictus*, in soybeans also.

Coffee. Picking cherries and spreading them on drying yards frequently damages coffee cherries, allowing the ingress of microorganisms, particularly fungi. If cherries are picked from the ground, contamination is likely to be high. Drying is often a slow process, in particular because of the environment in which coffee is grown. Coffee tress will not flower above 19EC but require high temperatures to mature, so coffee is commonly grown in upland areas in the tropics. In consequence, drying is often conducted under less than ideal conditions, with morning mists or rain common in some growing areas (Teixera *et al.*, 2001). Fungal growth occurs, sometimes with substantial heating. After drying, most cherries were infected with *Penicillium, Cladosporium, Mucor, Fusarium,* and yeast species (Taniwaki *et al.*, 1999, 2001), though little detail has been published. Most studies have concentrated on determining the source of ochratoxin A, as production of this toxin appears to be initiated during drying.

B Spoilage

Nuts

Tree nuts. Spoilage of tree nuts at harvest is rare. It is usually caused by excessive rain around harvest time, which may result in splits in shells, permitting fungi to enter and cause damage to the nut meats. Spoilage is usually due to discoloration.

Tree nuts contain only low levels of soluble carbohydrates, so slight increases in moisture content lead to sharp rises in a_w. As a result, stored dried nuts are very susceptible to spoilage. Increases in a_w can be due to moisture movement caused by uneven storage temperatures, as may happen in poorly insulated storage facilities or shipping containers. Refrigerated storage is widely used to retard the development of rancidity in stored nuts; if effective dehumidification is not practiced, increases in moisture content may result during storage. Many spoilage fungi grow poorly at low temperature; however, return of the nuts to ambient condition for shipping may result in rapid spoilage.

Marginal increases in moisture content will permit growth of *Eurotium* species. Shipment of nuts in containers across the tropics is a particular hazard, as unsuitable stowage, on decks or near engines, can lead to moisture migration sufficient to cause sporadic spoilage or even total loss. Cases of rampant growth of *Asp. flavus* and high aflatoxin production have been observed under these conditions.

Storage of nuts in tropical countries is of concern. Inadequate facilities may lead to moisture build up from humid air, from damp floors, inadequate ventilation, or moisture migration due to uneven solar heating. Spoilage or mycotoxin production may result.

Peanuts. Spoilage of peanuts due to discoloration is much more common than for tree nuts, due to susceptibility to pre-harvest fungal invasion. Discolored peanuts are usually sorted out before retail packaging, in developed countries by automated color sorting machines, in less developed countries by hand. Color sorting, introduced as a quality control measure, has also proved to be an effective way of controlling aflatoxin levels. Reject nuts may be used in the manufacture of peanut oil, where refining processes remove both fungi and aflatoxins.

Coconut. Coconuts are opened and cut up for drying, and this permits contamination with bacteria and fungi. More than 50 species of fungi were isolated from 25 coconut samples by Zori and Saber (1993). The most common were *Asp. flavus*, *Asp. niger*, *Asp. sydowii*, *Pen. chrysogenum*, *Cladosporium cladosporioides*, *Alt. alternata*, *Rhi. stolonifer*, and *Eur. chevalieri*.

Lipolytic rancidity in coconut is associated with growth of *Micrococcus candidus*, *M. luteus*, *M. flavus*, *Achromobacter lipolyticum*, and *Bacillus subtilis* during the early stages of drying (Minifie, 1989). Rancidity in dried coconut may be caused by xerophilic fungi, usually by *Eurotium* species, which produce off odors and flavors due to ketone formation (Kinderlerer, 1984). An unusual type of spoilage was a cheesy, butyric off flavor due to growth of *Chrysosporium farinicola* (Kinderlerer, 1984).

In the production of coconut milk or cream, the expressed milk typically contains 10^5–10^6 cfu/g of bacteria. Pasteurization at 75°C for 10 min within 2 h of expression is recommended to reduce the microbial load (Hagenmaier, 1980).

Oilseeds. All oilseeds are more or less susceptible to contamination or infection by fungi during growth and drying in the field. However, little information has been published on this topic.

Cottonseed. Cottonseed, a by-product of cotton production, is a valuable source of oil and of residual meal, which is used in stock feeds. Cottonseeds develop inside a tight, effectively impervious boll, which is highly resistant to fungal attack. Nevertheless cottonseed is known to be susceptible to invasion by *Asp. flavus*. This may occur as the result of insect damage, for example, by the cotton bollworm, but the main invasion route has been shown to be entry through the nectaries, glands in the cotton plant near the flowers which attract insects for pollination (Klich *et al.*, 1984).

Legumes. Spoilage of dried legumes is uncommon. Some discoloration often occurs as the result of the growth of storage fungi. However, this is not considered to be such a critical factor as in nuts and does not usually result in downgrading.

Coffee. Spoilage of coffee usually relates to off flavor or aroma development. Little is known of the causes of these problems, but they are most likely to be related to growth of fungi during drying.

C Pathogens

Bacterial pathogens. Bacteria have been reported from tree nuts, but total numbers as estimated by plate counts are usually low. Higher numbers in almonds were related to damaged shells or contamination with soil (King *et al.*, 1970). A variety of genera were isolated, including *Bacillus*, *Brevibacterium*, *Streptococcus*, *Escherichia coli*, and *Xanthomonas*. Total plate count, *Streptococcus* spp. and *E. coli* dropped initially during storage of almonds, then remained constant for more than three months (King *et al.*, 1970).

Contamination in the field or processing plant by pathogens derived from animals can be important. Growth can only occur when favorable conditions of a_w, pH, and temperature permit, but it has been found that very low numbers of bacteria such as *Salmonella* can be a hazard, as oil protects them after ingestion. One outbreak of salmonellosis in peanuts has been reported, in Australia in 1996 (Ng *et al.*, 1996; Oliver, 1996; Scheil *et al.*, 1998) with more than 50 cases of infection and one death (Rouch, 1996; Burnett *et al.*, 2000). *Salmonella* Mbandaka and *S*. Senftenberg were detected in suspect jars of peanut butter at levels as low as 3 cells per gram. This outbreak was probably due to bird or rodent droppings in improperly cleaned equipment in a peanut shelling plant.

Canned peanuts processed by an unlicensed canner resulted in an outbreak of botulism in Taiwan in 1986. Nine people were affected, of whom two died. Only a single batch was apparently contaminated, and the cause was not determined. Lack of distribution records hampered recall, but mass media announcements are credited with a reduction in the severity of the outbreak (Chou *et al.*, 1988). In the United Kingdom, contamination of hazelnuts with *Clostridium botulinum* led to botulism from yoghurt to which inadequately processed hazelnut purée had been added (O'Mahoney *et al.*, 1990).

Aflatoxins. The principal microbial hazard associated with nuts and oilseeds lies with the potential for production of mycotoxins, notably aflatoxins. Aflatoxins are produced by *Asp. flavus* and the closely related species *Asp. parasiticus* during growth in foods or feeds. Crops with high susceptibility to growth of these species and aflatoxin production have a high oil content, though the physiological reasons for this remains unclear. Peanuts, maize, and cottonseed are the three most economically important crops affected by aflatoxins.

Aflatoxins were discovered as the result of the deaths of 100 000 young turkeys in the United Kingdom in 1960 (Sargent *et al.*, 1961). The toxicity was traced to peanut meal originating in Brazil. Subsequent work showed that the cause was the common mould *Asp. flavus*, and the closely related species *Asp. parasiticus*. The use of the new technique of thin layer chromatography soon established that four toxins were implicated, named aflatoxins B_1, B_2, G_1, and G_2, the names being based on the compounds' blue or green fluorescence under ultraviolet light, and their positions on TLC plates (Broadbent *et al.*, 1963).

It was later shown that *Asp. flavus* produces only B aflatoxins, and occurs throughout the tropics and subtropics, while *Asp. parasiticus* produces both B and G toxins, and is much less widespread (Klich and Pitt, 1988; Pitt and Hocking, 1997). Other species capable of producing aflatoxins have been discovered more recently, *Asp. nomius* (Kurtzman *et al.*, 1987), *Asp. ochraceoroseus* (Klich *et al.*, 2000), and *Asp. pseudotamarii* (Ito *et al.*, 2001). However, these three species are very uncommon, have no known food safety significance, and will not be discussed further here. *Aspergillus flavus*, *Asp. parasiticus*, and *Asp. nomius* are closely related and have similar physiology (ICMSF, 1996b; Pitt and Miscamble, 1995).

Soon after their discovery, the acute toxicity of aflatoxins to all domestic animal species was established. Their potential carcinogenicity to animals and, by implication, man, became evident a few years later (Stoloff, 1977). A possible role in human cancer was supported by epidemiological studies (e.g. Peers and Linsell, 1973; Peers *et al.*, 1976). These apparently convincing data were soon confounded by the realization that the hepatitis B virus, endemic in many areas high in liver cancer, was also a liver carcinogen, or at the least a strong potentiator of liver cancer. Indeed, the role of aflatoxins in human liver cancer was dismissed by some authors (Stoloff, 1989; Campbell *et al.*, 1990).

Most recent studies have again supported the position that aflatoxins have a role in human liver cancer. A careful epidemiological study established that liver cancer rates in different areas of Swaziland correlated well with aflatoxin intake, but was independent of hepatitis B, which varied little with geographic region (Peers *et al.*, 1987). A good correlation was also reported between the incidence of liver cancer and the extent and severity of aflatoxin contamination of foodstuffs in Guangxi Province, China (Yeh *et al.*, 1989). In an authoritative report, IARC (1993) considered aflatoxin B_1 to be a Class 1 human carcinogen.

Evidence has been presented that both aflatoxins and hepatitis B virus are involved in the very high incidence of primary liver cancer in some areas of the world, notably parts of Africa, Southeast Asia and China (Wild *et al.*, 1993).

However, the human toxicology of aflatoxins still remains a source of uncertainty. The Joint FAO/WHO Expert Committee on Food Additives (JECFA) summarized this position:

"Risks from specific exposures to aflatoxins [in humans] are difficult to estimate and predict. Questions remain regarding the independence of aflatoxin as a human carcinogen, the extent to which hepatitis B, hepatitis C, and other factors modify the effect of aflatoxin, how findings from countries with high liver cancer rates and high prevalence of hepatitis B may be compared to those from countries with low rates, and how to describe the dose–response curve over the wide range of aflatoxin exposure found worldwide" (JECFA, 1997).

Observations concerning the interaction of hepatitis B and aflatoxins in humans suggest that two separate aflatoxin potencies exist, one in populations in which chronic hepatitis B infections are common and the second where such infections are rare. In consequence, JECFA divided potency estimates for analyses based on toxicological and epidemiological data into two basic groups, applicable to individuals with and without hepatitis B infection (JECFA, 1997). Potency values chosen for hepatitis B positive individuals were 30 times higher than for hepatitis B negative individuals, implying a 30-fold synergy between hepatitis B and aflatoxins.

Very sensitive techniques have now been developed to monitor aflatoxin intake by humans, including DNA–aflatoxin adducts in tissues, aflatoxin adducts in serum and urine (Harrison *et al.*, 1993), and excretion of aflatoxin M_1 in breast milk (Wild *et al.*, 1987).

Detectable aflatoxin levels have been recovered from human serum or milk by these techniques, in developed countries including Australia (El-Nazami *et al.*, 1995), Denmark (Autrup *et al.*, 1991), and the United Kingdom (Harrison *et al.*, 1993), as well as less developed regions, including Kenya (Autrup *et al.*, 1987), China (Ross *et al.*, 1992), Gambia (Allen *et al.*, 1992), Taiwan (Hatch *et al.*, 1993), and Thailand (El-Nazami *et al.*, 1995). In the Taiwan study, a much greater correlation was found between aflatoxin intake and primary liver cancer than with exposure to hepatitis B virus (Hatch *et al.*, 1993).

In Southeast Asia, the major sources of aflatoxins in the human diet are dried peanuts and maize (Pitt and Hocking, 1996). Using their data, and equations relating aflatoxin intake to liver cancer incidence (Kuiper-Goodman, 1991), it has been estimated that aflatoxin causes 12 liver cancer deaths per 100 000 population per annum in Indonesia; the relative contributions of maize and peanuts were estimated to be 7 and 5, respectively. That figure translates to more than 20 000 deaths per annum in Indonesia from aflatoxins (Lubulwa and Davis, 1994).

Aflatoxins are regulated in foods traded throughout the world. Levels permitted by country legislation range from as low as 1 μg/kg total aflatoxins to 50 μg/kg (van Egmond, 1992), though in some developing countries, limits in products for domestic consumption are not enforced. It seems probable that the Codex Alimentarius Commission will set a limit of 15 μg/kg total aflatoxins in peanuts and other commodities traded internationally in the near future. Under that international control, it will still be permissible for individual countries to set lower limits for commodities for local consumption, but not to refuse importation of commodities meeting that international standard. Meeting any lower internal limit will be the responsibility of processors within the country, not international producers or traders.

Levels of aflatoxins that can occur naturally in peanuts and other susceptible commodities are very high. Peanut plants affected by drought can produce kernels without obvious signs of damage but which may contain up to 1 000 μg/kg of aflatoxins (Hill *et al.*, 1983). Kernels affected in this way may be difficult to detect visually or by color sorters, but may cause serious contamination of lots. Much higher levels have been observed in discolored nuts, 20 000 μg/kg or more (Urano *et al.*, 1992).

Aflatoxins are very heat resistant. Heating cottonseed meal of 6.6% moisture content in a jacketed kettle at 100°C for 2 h reduced aflatoxin levels by 50% (Mann *et al.*, 1967); boiling at 100°C for 30 min reduced aflatoxins by 30% (Stoloff and Trucksess, 1981); and 40–50% reduction was achieved in a commercial roasting process (Waltking, 1971; ICMSF, 1996b). However, boiling in a pressure cooker at 116°C with 5% salt, a commercial process in Brazil, reduced aflatoxins by 80–100% (Farah *et al.*, 1983).

Ochratoxin A. The principal public heath hazard associated with coffee is the formation of ochratoxin A. This toxin is discussed more fully in Chapter 8—Cereals.

Nuts

Peanuts. Early work assumed that the presence of *Asp. flavus* in peanuts and subsequent aflatoxin production was primarily a function of inadequate drying or improper storage (Austwick and Ayerst, 1963). While aflatoxin production undoubtedly does occur due to poor storage, it soon became clear that the main time of entry of the fungus was pre-harvest (McDonald and Harkness, 1967; Pettit *et al.*, 1971; Cole *et al.*, 1982). Indeed evidence indicates that, in developed countries such as the United States and Australia, invasion of peanuts by *Asp. flavus* after harvest is rare (Pitt, 1989). Although insect damage or shell cracking may be responsible for nut infection (Graham, 1982), such damage is not a necessary prerequisite for *Asp. flavus* to penetrate the peanut shell. Pre-harvest invasion of peanut seeds most commonly occurs directly through the shell from the surrounding soil.

Asp. flavus may also invade developing peanuts through flowers or the pegs on which the nuts develop (Griffin and Garren, 1976), or systemically (Pitt *et al.*, 1991).

Invasion of peanuts, and hence aflatoxin production, is promoted by pre-harvest drought and/or high temperatures, rather than post-harvest rain (Sanders *et al.*, 1981, 1984, 1985; Cole *et al.*, 1982, 1989). The likely explanation is two-fold. First, in the absence of drought, a vigorously growing peanut plant has a variety of defence mechanisms against invasion, including phytoalexin production (Dorner *et al.*, 1989, 1991). Such defences become ineffective when the plant becomes physiologically weak during drought stress. Secondly, *Asp. flavus* is a xerophile, capable of growth when drought stress reduces competition in soil from its natural enemies—bacteria, amoebae, and soil fungi (Pitt, 1989).

Asp. flavus and *Asp. parasiticus* are capable of growth down to about 0.8 a_w (Pitt and Miscamble, 1995). Toxin production is most abundant at high a_w (greater than 0.95), and is limited below 0.85 a_w. The optimum temperature for growth of both species is between 30°C and 37°C, with a minimum near 8°C and a maximum at about 42°C (ICMSF, 1996b).

Peanuts are readily invaded by both *Asp. flavus* and *Asp. parasiticus*—in contrast to maize, where nearly all aflatoxin is produced by *Asp. flavus*. In the United States and Australia, 50% or more of aflatoxin in peanuts is derived from *Asp. parasiticus*, so that both B and G aflatoxins are normally present (Read, 1989). However, *Asp. parasiticus* appears to be a rare species in some other parts of the world (Pitt *et al.*, 1993; Pitt and Hocking, 2004). Of more than 400 food commodity samples from Thailand, Indonesia and the Philippines, which contained aflatoxin, 95% contained only B aflatoxins (Pitt *et al.*, unpublished).

Almonds. A survey of Californian almonds from the 1972 season showed that 14% of 74 samples were contaminated with aflatoxins, but levels were low, mostly below 20 μg/kg. Commercial sorting was effective in removing almonds containing aflatoxins. Sliced almonds were more uniformly contaminated (Schade *et al.*, 1975). A more comprehensive survey of almonds and walnuts later found that less than one nut in 25 000 was contaminated (Fuller *et al.*, 1977). Of 256 almond samples checked under US Food and Drug Administration compliance programs from 1980–1984, 2.3% contained detectable aflatoxins; only 0.8% exceeded 20 μg/kg (Pohland and Wood, 1987).

Pistachios. Available evidence indicates that pistachio nuts are essentially sterile before harvest, due to thick shells and surrounding tissues. However, in some varieties of pistachio, shells dehisce at maturity, providing an opportunity for microorganisms to enter. Pistachios are also treated by mechanical abrasion and washing to remove the outer nut layers around the shell, providing an opportunity for microbial entry if shell dehiscence has occurred (Doster and Michailides, 1994; Heperkan *et al.*, 1994).

Pistachio nuts have usually been reported to be free of aflatoxin (e.g. Burdaspal and Gorostidi, 1989; Abdel-Gawad and Zohri, 1993; Taguchi *et al.*, 1995). However, of 835 imported pistachio samples checked under US Food and Drug Administration compliance programs from 1980–1984, 2.0% exceeded 20 μg/kg (Pohland and Wood, 1987). In other countries, some reported levels have been very high: up to 400–800 μg/kg (Shah and Hamid, 1989) or even 1300 μg/kg (Tabata *et al.*, 1993).

Pecan nuts. Of 446 pecan nut samples checked under US Food and Drug Administration compliance programs from 1980–1984, only 1.1% contained detectable aflatoxins and only one exceeded 20 μg/kg (Pohland and Wood, 1987). The cause of aflatoxin contamination in pecans has not been determined, as both damaged and undamaged kernels may contain aflatoxin (Pohland and Wood, 1987).

Brazil nuts. Brazil nuts were recognized as a cause for concern before 1980. Analyses of 135 samples in Germany showed 58% aflatoxin free, 22% up to 5 μg/kg and 21 higher, up to 8,000 μg/kg (Woller and Majerus, 1979). In the United States, in 1980, 4.4% of 158 lots examined under Food and Drug Administration compliance programs were found to contain unacceptable levels (>20 μg/kg) of aflatoxins. However, improvements to quality control in producing countries resulted in no unacceptable lots from more than 300 sampled in 1983–1984 (Pohland and Wood, 1987).

Cashews. Spoilage or mycotoxin production by fungi in cashew nuts is very uncommon.

Coconuts. Due to less than ideal drying conditions for coconuts in most producing areas, bacterial contamination and aflatoxin production are major problems in copra and desiccated coconut production.

Salmonellae were frequently been isolated from desiccated coconut (Schaffner *et al.*, 1967; Gilbert, 1982). This was traced to the unhygienic state of collecting centers at farms, where keeping free ranging fowls was the source of the contamination. The *Salmonella* problem has been largely overcome as a result of improved controls.

Bongkrek poisoning, from the Indonesian fermented coconut press cake product "tempeh bongkrek", was first reported in 1895 and 1901, and caused a total of 850 deaths from 1951 to 1975, and 69 in 1977, among unsuspecting Indonesian consumers (Arbianto, 1979). This disease is caused by contamination of the fermenting press cake with *Pseudomonas cocovenenans*, now known as *Burkholderia cocovenenans* (Zhao *et al.*, 1990), and is apparently confined to the Seraryu Valley of Central Java (Cox *et al.*, 1997). *Burkholderia cocovenenans* produces two powerful toxins, bongkrek acid and toxoflavin (Arbianto, 1979; ICMSF, 1996a; Cox *et al.*, 1997). The onset of illness apparently occurs 4–6 h after ingestion, with death following 1–20 h after the onset of symptoms (Cox *et al.*, 1997). Because of the many deaths, the Indonesian Government declared the manufacture of tempeh bongkrek illegal in November 1988, but this is unlikely to eradicate the problem (Buckle and Kartadarma, 1990).

Like other nuts and oilseeds, coconut is susceptible to invasion by *Asp. flavus* during drying. Aflatoxin in copra is a well-recognized hazard, but quantitative data are limited (e.g. Saxena and Mehrotra, 1990; Zohri and Saber, 1993).

Oilseeds. The major microbiological problem in oilseed crops is the growth of *Asp. flavus* and consequent aflatoxin production. High levels of aflatoxins have been found in a variety of oilseeds. In 73 samples of various oilseeds from South Africa, 43% were positive, with aflatoxin levels ranging up to 2 000 μg/kg (Dutton and Westlake, 1985); in linseed oil from India, 44% of 105 samples contained aflatoxins, with levels ranging from 120–810 μg/kg (Sahay *et al.*, 1990); and in sesame seed oil from Pakistan, 55% of 24 samples were positive, ranging up to 440 μg/kg (Dawar and Ghaffar, 1991). Of 45 samples of maize destined for oil manufacture, 87% contained aflatoxins ranging up to 2300 μg/kg (average 250 μg/kg), and 100% of 50 similar peanut samples also contained aflatoxin, ranging up to 22 000 μg/kg, average 1600 μg/kg (Urano *et al.*, 1992).

Dried legumes. As noted above, growth of *Asp. flavus* or other mycotoxigenic fungi in dried legumes (other than peanuts) is rarely sufficient to produce significant mycotoxin levels. Aflatoxins have been reported from a variety of beans including haricot beans and broad beans (Abdalla, 1988; Abdel-Rahim *et al.*, 1989; Mahmoud and Abdalla, 1994), butter beans imported into Japan (Tabata *et al.*, 1993) and soybeans (Pinto *et al.*, 1991; El-Kady and Youssef, 1993; Jacobsen *et al.*, 1995). In general, levels reported were low and of little consequence in trade.

Fusarium semitectum, which appears to have a strong association with beans, is not considered to be an important producer of known mycotoxins (Miller, 1994). The potentially mycotoxigenic species *Fus. verticillioides* (=*Fus. moniliforme*) was found in 13% of mung bean samples in Southeast Asia (Pitt *et al.*, 1994), but it is unclear whether toxin production is a potential problem. Levels of *Asp. flavus* were low in all kinds of Southeast Asian beans, indicating that significant aflatoxin production was unlikely (Pitt *et al.*, 1994, 1998).

Soybeans, with their high oil content, might be expected to be a suitable substrate for aflatoxin production. However, available information indicates that aflatoxin formation in soybeans and soy products is not a commercial problem. Like other beans, soybeans possess factors that are highly inhibitory of the extensive fungal growth needed for toxin production.

Coffee. The principal problem associated with coffee is the production of the mycotoxin ochratoxin A. Although the possibility of significant levels of this toxin being present in coffee beans has been known for some time, only recently has it been established that *Asp. ochraceus* is a major source of ochratoxin A in Brazilian coffee (Pitt *et al.*, 2001) and likely to be the major source world wide. Other known ochratoxin A producers, *Asp. niger* and *Asp. carbonarius*, have also been isolated from coffee, but do not appear to be as important (Frank, 2001; Pitt *et al.*, 2001). Available evidence indicates that the sources of these fungi are environmental, and that entry to cherries is gained during picking and drying (Taniwaki *et al.*, 1999; Pitt *et al.*, 2001). Ochratoxin A is produced during drying (Taniwaki *et al.*, 1999; Bucheli *et al.*, 2001; Teixera *et al.*, 2001). Coffee picked and dried under good agricultural practice appears to contain ochratoxin A only rarely (Taniwaki *et al.*, 1999; Pitt *et al.*, 2001). Nearly all ochratoxin A is produced in the husk rather than in the coffee bean, and immature fruit were less likely to become contaminated than mature or overripe cherries. Removal of defectives (damaged beans) and careful removal of husks reduces ochratoxin A levels in dried green coffee beans (Bucheli *et al.*, 2001). High levels of ochratoxin A in green coffee beans are related to poor manufacturing conditions, i.e. where drying is slow, waste material is admitted to the processing stream, or drying is incomplete and storage inadequate (Teixera *et al.*, 2001).

D CONTROL (primary processing of tree nuts, peanuts, coconut, oilseeds, dried legumes, and coffee)

Summary

Significant hazards[a]	• Aflatoxins in oilseeds, coconut, tree nuts, and especially peanuts. • Ochratoxin A in coffee. • *Salmonella* spp.
Control measures	
Initial level (H_0)	• Initial hazard levels should be very low for tree nuts, coconut, oilseeds, dried legumes, and coffee. • Aflatoxin may already be present at harvest during severe drought. • Segregate immature and damaged nuts from sound nuts. • Harvest tree nuts directly from trees where possible. • Use good agricultural practice.
Increase (ΣI)	• Dry rapidly. Field drying of peanuts can increase aflatoxin levels. • Store at moisture levels <8%. • Control rodent and bird access to equipment and stored commodities.
Reduction (ΣR)	• Sort to remove damaged and immature tree nuts and peanuts. • Color sort for aflatoxins in peanuts.
Testing	• Testing for aflatoxins, using large samples, is essential for control of tree nuts, peanuts, and coconut. • Tests for *Salmonella* spp. is advisable for peanuts, coconut. • Testing for ochratoxin A is essential for coffee.
Spoilage	• Controls for significant hazards will usually control spoilage.

[a]In particular circumstances, other hazards may need to be considered.

Comments. General controls for tree nuts, peanuts, oilseeds, coconut, oilseeds, dried legumes, and coffee are summarized above. Each commodity has specific controls that are summarized below.

Tree nuts and peanuts. Control of aflatoxin formation in susceptible crops is very difficult indeed. Tree nuts should preferably be harvested directly from trees, or collected from the ground at frequent intervals. Mouldy, insect damaged or immature nuts should be segregated. Color sorting is possible for some types of tree nuts.

Peanuts should be dug before or at physiological maturity and should be threshed and dried as efficiently as possible. Aflatoxin testing and segregation of positive loads at sheller intake is recommended (Read, 1989). Storage before shelling should be cool, dry, and free from insects.

After shelling, nuts should be graded to remove immature and damaged kernels; split kernels are often diverted to peanut butter manufacture at this time. Nuts should be color sorted mechanically or by hand or both, then sampled by adequate protocols and tested for aflatoxins (Tiemestra, 1977). Samples testing positive may be color sorted again, and retested. Blanching (skin removal) and roasting may be used to increase the effectiveness of color sorting. It should be kept in mind that roasted peanuts have a short shelf life unless packaged in inert atmospheres.

Shelled nuts should be stored in good conditions: free from moisture, at or below 65% RH, and preferably at about 10°C, to retard development of rancidity. Nuts should be packaged in high quality materials to prevent moisture ingress.

Sampling peanuts for aflatoxin testing is a very important procedure in control. A few highly contaminated kernels may result in unacceptable aflatoxin levels in a large lot of nuts, especially when these are homogenized as in peanut butter or satay manufacture. Sampling procedures have been developed in the United States (Whitaker *et al.*, 1972; Tiemestra, 1977), Australia (Brown, 1984; Read, 1989), the United Kingdom (Coker, 1989) and elsewhere. Two-class sampling plans are in use: samples must be large (8 kg or more) to be effective. It is impossible to guarantee freedom from aflatoxins in processed peanuts or peanut products from a single set of samples, even with the best sampling plans. It is therefore good commercial practice to sample peanuts and test for aflatoxins both at the shelling and processing stages. Only in this way can effective control of aflatoxin in retail peanuts and peanut products be assured.

No effective sorting techniques exist for reducing aflatoxin contamination in peanuts in the shell, pistachios, or cottonseed. Unlike the case of maize, where fluorescence under ultraviolet light can be used as a measure of aflatoxin content, non-destructive chemical testing techniques do not exist for any commodity under discussion here (Steiner *et al.*, 1992).

Mycotoxins other than aflatoxin are rarely reported in tree nuts and peanuts. Control of toxins other than aflatoxins is not recommended at the present state of knowledge.

Coconut. Aflatoxin has been found in coconut, and controls listed above, especially rapid drying, can be used to control the problem. Ochratoxin A has also been reported from coconut (Zohri and Saber, 1993).

Control of bongkrek poisoning in tempeh bongkrek requires process control. Toxin production appears to be related to the concentration of coconut fat in the product (Garcia *et al.*, 1999). Acidifying the coconut press cake to pH 4.5 to 5 with vinegar and incubating at 37°C, the optimum temperature for *Rhi. stolonifer*, is the most appropriate way to suppress toxic bacteria. If *Burk. cocovenenans* contaminates the product, mould growth is inhibited, so product showing poor mould growth should be treated as suspect and not consumed (Arbianto, 1979; Cox *et al.*, 1997).

NaCl (0.5–0.6%) neutralizes the inhibitory effect of *Burk. cocovenenans* on growth of *R. oligosporus*, and greatly reduces production of bongkrek acid (Ko *et al.*, 1979). Higher levels of NaCl inhibited bongkrek acid production even in the presence of 10^7 *Burk. cocovenenans*. Levels of bongkrek acid and toxoflavin were not detectable in tempeh bongkrek made with a high inoculum (10^4–10^5 spores per gram) of *R. oligosporus* in the presence of high NaCl (2%) and pH 4.5–5.5 (Buckle and Kartadarma, 1990).

Oilseeds. Pressing of oilseeds distributes aflatoxin between the oil and the press cake in about equal proportions (Sashidhar, 1993). Aflatoxins are effectively removed from oil by the refining process. Alkali used to remove free fatty acids and other refining processes also detoxify or remove aflatoxins (Parker and Melnick, 1966). The presence of aflatoxins in sesame oil (Dawar and Ghaffar, 1991) and linseed oil (Sahay *et al.*, 1990) indicates that some food oils are unrefined.

Removing aflatoxin from press cakes is not so simple. Physical techniques such as heating are relatively ineffective. Temperatures above 100°C are needed to reduce aflatoxin levels appreciably (see Samarajeewa *et al.*, 1990 for a review).

A variety of chemical techniques have been proposed for detoxification. The earliest effective method was solvent extraction, using acetone–hexane–water, 54:44:2 (Gardner *et al.*, 1968) or 80%

aqueous isopropanol (Rayner and Dollear, 1968), but it was expensive and impractical. Chlorine is effective for decontamination of aflatoxin from contaminated surfaces and glassware (Trager and Stoloff, 1967) but is of little value in foods, where organic material rapidly denatures the chlorine. Concerns over the safety of chlorine decontamination still exist (Samarajeewa *et al.*, 1990); however, reduction in aflatoxin by more than 75% was achieved using chlorine gas (Samarajeewa *et al.*, 1991).

Hydrogen peroxide (0.5%) at pH 4 will detoxify peanut protein isolates (Rhee *et al.*, 1976), but peanut meal required 6% H_2O_2 at pH 9.5 to be detoxified (Sreenivasamurthy *et al.*, 1967). Relatively little change occurred in foods so treated, suggesting H_2O_2 may be a useful treatment under some circumstances.

The most commonly recommended method for aflatoxin detoxification has been ammoniation (see reviews by Park *et al.*, 1988; Samarajeewa *et al.*, 1990). Two techniques have been established: atmospheric pressure ammoniation at ambient temperature, where product is sealed in plastic bags or bins for 2–3 weeks; and ammoniation at high temperature and pressure, where a 1 h treatment is sufficient (Park, 1993).

Ammoniation has been approved for commercial use on contaminated cottonseed meal or maize in certain states of the United States and for some types of animal feeds in other countries (Park, 1993). However, neither ammoniation nor any other procedure has been widely used commercially. The most effective aflatoxin decontamination technique so far found is to feed the contaminated food, press cake or feed to ruminant animals, other than dairy cows. Up to 400 μg/kg of aflatoxin can be tolerated by beef cattle without appreciable reduction in growth rate or contamination of meat (Pohland and Wood, 1987). Also, mixing of feed ingredients to produce feeds with less than 200–300 μg/kg aflatoxin in total is now widely practised.

It is emphasized that feeds contaminated with aflatoxin must not be fed to dairy cows. Aflatoxins are hydroxylated by mammals as a detoxification mechanism and some of the hydroxylated compounds pass into milk. Aflatoxin M_1, the hydroxylated derivative of aflatoxin B_1, is less toxic than B_1, but still sufficiently so to be unacceptable above trace amounts in the diets of young children. Very low limits have been set for aflatoxin M_1 in milk: in Europe limits of 0.1 μg/kg are common (van Egmond, 1992). As about 1% of aflatoxin B_1 in feed is converted to aflatoxin M_1 (Frobish *et al.*, 1986), such a limit requires that dairy rations contain less than 10 μg/kg aflatoxin in the total diet. Mycotoxins other than aflatoxins have been reported only rarely from oilseeds.

Dried legumes. Storage in conditions below 65% relative humidity in conditions free from insects should be adequate for control of microbiological problems with dried legumes. There is no history of significant microbiological safety issues with dried legumes other than peanuts.

Coffee. Control of ochratoxin A production in coffee can be achieved by quality control measures and good practice. As it appears that infection of coffee beans with fungi capable of producing ochratoxin A does not occur on the tree to any real extent, careful control of drying and storage can reduce or eliminate this problem.

Reduce contamination of fresh cherries by toxin producing fungal species by harvesting cherries only from the tree; do not harvest cherries from the ground. Reduce damage to fresh cherries as far as possible.

Segregate cherries that fall from washers or other equipment. Ensure that rapid drying takes place, by mechanical means where necessary. Avoid leaving cherries exposed to moisture ingress, due to either mist or rain, in the drying yard. Turn over cherries frequently during sun drying. Ensure layers of drying cherries are only 1–2 cm deep.

Careful cleaning procedures to eliminate defectives and husk material from dried green coffee can reduce ochratoxin A levels effectively. The use of gravity sorters and color sorters is recommended.

IV Tree nut, peanut, and coconut processing

A *Effects of processing on microorganisms*

Tree nuts and peanuts. The most important step in further processing of nuts from the microbiological viewpoint is roasting. Most kinds of nuts used as snack foods are roasted before packaging for retail sale. Large reductions in numbers and kinds of microorganisms occur during this high temperature process. Roasting is accompanied by drying and by some production of inhibitory compounds due to Maillard browning, so that roasted nuts are an inhospitable environment for microorganisms. This can be an effective method for destroying enteric pathogens such as salmonellae.

Due to the low a_w of the roasted nuts, the heat generated during grinding in peanut butter manufacture provides little change in microbial numbers, but has the effect of distributing cells or toxins throughout a large product bulk. This is particularly important for aflatoxins, where high levels in a few nuts can cause unacceptable contamination in a whole batch.

Coconut. Dried coconuts (copra) are commonly further processed only by shredding, which affords no reduction in bacterial load, and indeed may lead to an increase in bacterial counts. Sterilized coconut milk with or without preservatives was found to support the growth of *Bacillus* spp., *Ps. fluorescens, E. coli, Streptococcus faecalis, Saccharomyces cerevisiae, Clostridium* spp., *Lactobacillus plantarum* and *Salmonella* Typhimurium. Preservatives delayed but did not prevent microbial growth. Some commercial samples of coconut milk were found to contain *Bacillus* and *Clostridium* species (Uboldi-Eiroa *et al.*, 1975).

B *Spoilage*

Microbial spoilage is not a serious problem with most nut products. Inadequate drying before packaging, or poor storage associated with inadequate packaging can lead to visible spoilage from mould growth, or in extreme cases, aflatoxin production.

C *Pathogens*

Problems with mycotoxins in processed nut products are similar to those detailed above for dried unprocessed nuts. No increases in toxins such as aflatoxin are to be expected, but neither do significant reductions occur. High levels of contamination in a few nuts can lead to contamination of a large batch of peanut butter or similar product during manufacture. This problem can result in considerable losses of processed product.

Contamination with pathogens can occur between roasting and packaging from improperly cleaned and/or maintained equipment. Low levels of contamination of peanuts with salmonellae have been known to occur in this way (Oliver, 1996; Scheil *et al.*, 1988). Pathogens on processed nuts can also be important when the nuts are used as ingredients in high a_w products such as yoghurts. Food poisoning due to walnuts incorporated into yoghurt has occurred (O'Mahoney *et al.*, 1990).

Salmonella has been frequently isolated from desiccated coconut (Schaffner *et al.*, 1967; McCoy, 1975; Gilbert, 1982). The source of contamination was traced to animal excreta found on the mill premises, especially in the yard where the coconuts were stacked. Contamination was then passed through the successive stages of manufacture. Two strains of *S*. Senftenberg were found to be the most frequent contaminants of desiccated coconut from Sri Lanka (Coconut Board, 1969). Contamination with levels of coagulase positive *Staphylococcus aureus* exceeding 10^2/g have been reported in commercially produced dehydrated coconut from Brazil (Leitao *et al.*, 1973, 1974).

D CONTROL (treenut, peanut and coconut processing)

Summary

Significant hazards[a]	• Aflatoxin. • *Salmonella*.
Control measures	
Initial level (H_0)	• Use all control methods described under primary processing for most effective control. • Protect incoming materials and equipment from birds and rodents.
Increase (ΣI)	• Store in moisture proof containers at moisture levels <8% (aflatoxin). • Maintain dry processing equipment and environment. • Salmonellae will not increase on dried tree nuts, peanuts, or coconut.
Reduction (ΣR)	• Roasting will reduce salmonellae to some extend, depending on moisture available.
Testing	• Color screening of nuts and peanuts is useful for ingredients. • Testing finished product for aflatoxins presents sampling issues except for homogeneous products such as peanut butter. • Testing for salmonellae is useful. • *E. coli* testing may be a useful indicator.
Spoilage	• Packaging in moisture impermeable films provides adequate protections.

[a]In particular circumstances, other hazards may need to be considered.

Comments. Control of mycotoxins in nut products consists essentially of control at the raw material or primary processing stages. Sorting and diverting discolored, rejected nuts for other uses as previously described is an important, commonly used means to reduce mycotoxins to acceptable levels in finished products.

End product testing is widely practised, both by manufacturers and regulatory agencies. Because the percentage of peanuts containing aflatoxins is usually very low, but the level in individual nuts can be very high, it is impossible to guarantee the absence of aflatoxin in batches for processing. End-product testing provides a further level of assurance of the absence of aflatoxins. Testing of peanut butter, which is a more or less homogeneous product, is a useful check on color sorting and other procedures used on nut batches before processing.

Moisture control in equipment and the environment is necessary to reduce the risk of growth of bacterial pathogens (e.g. salmonellae) in the processing system. Procedures for maintaining, replacing, cleaning, and disinfecting equipment must address this potential hazard. Greater reliance on dry cleaning rather than wet systems is one effective means. Roasting can destroy enteric pathogens. This process should be used as a point of separation between raw nut storage and handling and the process product environment.

The risk of salmonellae in desiccated coconut can be controlled by immersing the raw coconut meat in water at 80°C for 8–10 min (Schaffner *et al.*, 1967) and then preventing recontamination during oven drying and subsequent handling (Simonsen *et al.*, 1987).

V Oilseed products

A Effects of processing on microorganisms

In modern plants, extraction of oils from oilseeds usually involves the use of solvents, often *n*-hexane. As a result, no microbiology is likely.

B Spoilage

Hydrolytic or "soapy" rancidity can occur in confectionery containing fat through the introduction of lipolytic enzymes from moulds present in ingredients such as coconut, milk products, egg albumin, and cocoa. This is especially serious with fats high in lauric acid, i.e. fats in coconut oil, palm kernel oil and butter fat, but not from cocoa butter, illipe fat, palm oil or peanuts, which do not contain lauric acid (Minifie, 1989).

C Pathogens

Problems with pathogens are unlikely in oilseeds products. Aflatoxin is sometimes formed, but is rarely of commercial significance.

D CONTROL (oilseed products)

Summary

Significant hazards[a]	• None identified.
Control measures	
Initial level (H_0)	• Does not apply.
Reduction (ΣR)	• Does not apply.
Increase (ΣI)	• Does not apply.
Testing	• Routing testing is not recommended.
Spoilage	• Dry conditions prevent microbial spoilage.

[a]In particular circumstances, other hazards may need to be considered.

VI Legume products

A Effects of processing on microorganisms

Most processing of dried legumes is unlikely to cause increases in microbial numbers, either of bacteria or fungi.

The first step in soybean processing is oil extraction, using solvents such as n-hexane, which eliminates most microorganisms (Waggle and Kolar, 1979). Further processing to high value products, especially isolated soy protein, involve conditions of high humidity or the addition of water and consequent possibilities for microbial contamination and growth. Further processing normally involves additional heating that is lethal to non-spore-forming bacteria. Little information exists on possible microbiological problems associated with the production of the various soy products that are commonly used as food ingredients.

B Spoilage

Microbial spoilage of legume products has rarely been reported.

C Pathogens

Dried legume products intended as food ingredients have not been associated with food-borne illness. Mycotoxin production, occurring during growth or primary processing, may carry over into products, but levels are usually of little consequence.

The potential exists for recontamination with pathogens, e.g. salmonellae, during wet processing to produce soy protein products. The low levels of mycotoxins potentially present in unprocessed soybeans would be eliminated or extracted during the manufacture of soy protein products.

D CONTROL (legume products)

Summary

Significant hazards[a]	• None identified.
Control measures	
Initial level (H_0)	• Does not apply.
Reduction (ΣR)	• Does not apply.
Increase (ΣI)	• Does not apply.
Testing	• No specific measures for microbial control in legume products have been published.
Spoilage	

[a]In particular circumstances, other hazards may need to be considered.

VII Coffee products

Apart from the potential for processed coffee to contain low levels of ochratoxin A, no potential hazards exist from processed coffee products. Use all control methods described under primary processing for most effective control. Maintain dry conditions to prevent spoilage.

References

Abdalla, M.H. (1988) Isolation of aflatoxin from *Acacia* and the incidence of *Aspergillus flavus* in the Sudan. *Mycopathologia*, **104**, 143–7.

Abdel-Gawad, K.M. and Zohri, A.A. (1993) Fungal flora and mycotoxins of six kinds of nut seeds for human consumption in Saudi Arabia. *Mycopathologia*, **124**, 55–64.

Abdel-Rahim, A.M., Osman, N.A. and Idris, M.O. (1989) Survey of some cereal grains and legume seeds for aflatoxin contamination in the Sudan. *Zentralblatt für Mikrobiologie und Hygiene*, **144**, 115–21.

Allen, S.J., Wild, C.P., Wheeler, J.G., Riley, E.M., Montesano, R., Bennett, S., Whittle, H.C., Hall, A.J. and Greenwood, B.M. (1992) Aflatoxin exposure, malaria and hepatitis B infection in rural Gambian children. *Trans. R. Soc. Trop. Med. Hyg.*, **86**, 426–30.

Arbianto, P. (1979) Bongkrek food poisoning in Java, in *Proceedings of the Fifth International Conference on the Impacts of Appl. Microbiol.* (ed P. Matangkasombut), pp. 371–4.

Austwick, P.K.C. and Ayerst, G. (1963) Toxic products in groundnuts: groundnut microflora and toxicity. *Chem. Ind.*, **1963**, 55–61.

Autrup, H., Seremet, T., Wakhisi, J. and Wasunna, A. (1987) Aflatoxin exposure measured by urinary excretion of aflatoxin B_1-guanine adduct and hepatitis B virus infection in areas with different liver cancer incidence in Kenya. *Cancer Res.*, **47**, 3430–3.

Autrup, J.L., Schmidt, J., Seremet, T. and Autrup, H. (1991) Determination of exposure to aflatoxins among Danish workers in animal-feed production through the analysis of aflatoxin B_1 adducts to serum albumin. *Scand. J. Work, Environ. Health*, **17**, 436–40.

Brown, G.H. (1984) The distribution of total aflatoxin levels in composited samples of peanuts. *Food Technol., Aust.*, **36**, 128–30.

Bucheli, P., Kanchanomai, C., Pittet, A., Goetz, J. and Joosten, H. (2001) Development of ochratoxin A (OTA) during robusta (*Coffea canephora*) coffee cherry drying, and isolation of *Aspergillus carbonarius* strains that produce OTA *in vitro* on coffee cherries, in *Proceedings of 19th Internationa; Science Colloquium on Coffee*, Trieste, Italy, 14–18 May 2001. Published on compact disc, not paginated.

Buckle, K.A. and Kartadarma, E.K. (1990) Inhibition of bongkrek acid and toxoflavin production in tempe bonkrek containing *Pseudomonas cocovenenans*. *J. Appl. Bacteriol.*, **68**, 571–6.

Burdaspal, P.A. and Gorostidi, A. (1989) Aflatoxin contamination of peanuts and other nuts. *Alimentaria* (Madrid), **26**, 51–3.

Burnett, S.L., Gehm, E.R., Weissinger, W.R. and Beuchat, L.R. (2000) Survival of *Salmonella* in peanut butter and peanut butter spread. *Journal of Appl. Microbiol.*, **89**, 472–7.

Campbell, T.C., Chen, J., Liu, C., Li, J. and Parpia, B. (1990) Nonassociation of aflatoxin with primary liver cancer in a cross-sectional ecological survey in the People's Republic of China. *Cancer Res.*, **50**, 6882–93.

Coconut Board. (1969) Investigations into the contamination of Ceylon desiccated coconut. *J. Hyg.*, **67**, 719–29.

Chou, J.H., Hwang, P.H. and Malison, M.D. (1988) An outbreak of type A foodborne botulism in Taiwan due to commercially preserved peanuts. *Int. J. Epidemiol.*, **17**, 899–902.

Coker, R.D. (1989) Control of aflatoxin in groundnut products with emphasis on sampling, analysis, and detoxification, in *Aflatoxins in groundnut: Proceedings of the International Workshop*, 6–9 October, 1987, ICRISAT Center, India, ICRISAT, Patancheru, India, pp. 123–32.

Cole, R.J., Hill, R.A., Blankenship, P.D., Sanders, T.H. and Garren, K.H. (1982) Influence of irrigation and drought stress on invasion by *Aspergillus flavus* of corn kernels and peanut pods. *Dev. Ind. Microbiol.*, **23**, 229–36.

Cole, R.J., Sanders, T.H., Dorner, J.W. and Blankenship, P.D. (1989) Environmental conditions required to induce preharvest aflatoxin contamination of groundnuts: summary of six years' research, in *Aflatoxins in groundnut: Proceedings of the International Workshop*, 6–9 October, 1987, ICRISAT Center, India, ICRISAT, Patancheru, India, pp. 279–87.

Cox, J., Khartadarma, E. and Buckle, K. (1997) *Burkholderia cocovenenans*, in *Foodborne Microorganisms of Public Health Significance* (eds A.D. Hocking, G. Arnold, I. Jenson, K.G. Newton and P. Sutherland), 5th edn, Australian Institute of Food Science and Technology Food Microbiology Group, Sydney, pp. 521–30.

Dawar, S. and Ghaffar, A. (1991) Detection of aflatoxin in sunflower seed. *Pakistan J. Bot.*, **23**, 123–6.

Dorner, J.W., Cole, R.J., Sanders, T.H. and Blankenship, P.D. (1989) Interrelationship of kernel water activity, soil temperature, maturity, and phytoalexin production in preharvest aflatoxin contamination of drought-stressed peanuts. *Mycopathologia* **105**, 117–28.

Dorner, J.W., Cole, R.J., Yagen, B. and Christiansen, B. (1991) Bioregulation of preharvest aflatoxin contamination of peanuts. Role of stilbene phytoalexins, in *Naturally Occurring Pest Bioregulators* (ed P.A. Hedin), *American Chemical Society, Symposium Series, Volume* 449, 352–60.

Doster, M.A. and Michailides, T.J. (1994) *Aspergillus* molds and aflatoxins in pistachio nuts in California. *Phytopathology*, **84**, 583–90.

Dutton, M.F. and Westlake, K. (1985) Occurrence of mycotoxins in cereals and animal feedstuffs in Natal, South Africa. *J. Assoc. Off. Anal. Chem.*, **68**, 839–42.

El-Kady, I.A. and Youssef, M.S. (1993) Survey of mycoflora and mycotoxins in Egyptian soybean seeds. *J. Basic Microbiol.*, **33**, 371–8.

El-Nazami, H.S., Nicoletti, G., Neal, G.E., Donohue, D.C. and Ahokas, J.T. (1995) Aflatoxin M_1 in human breast milk samples from Victoria, Australia and Thailand. *Food Chem.Toxicol.*, **33**, 173–9.

Farah, Z., Martins, M.J.R. and Bachmann, M.R. (1983) Removal of aflatoxin in raw unshelled peanuts by a traditional salt boiling process practiced in the North East of Brazil. *Lebensmittel Wissenschaft und Technologie*, **16**, 122–4.

Frank, J.M. (2001) On the activity of fungi in coffee in relation to ochratoxin A production, in *Proceedings of 19th International Science Colloquium on Coffee*, Trieste, Italy, 14–18 May 2001. Published on compact disc, not paginated.
Frobish. R.A., Bradley, B.D., Wagner, D.D., Long-Bradley, P.E. and Hairston, H. (1986) Aflatoxin residues in milk of dairy cows after ingestion of naturally contaminated grain. *J. Food Prot.*, **49**, 781–5.
Fuller, G, Spooncer, W.W., King, A.D., Schade, J. and Mackey, B. (1977) Survey of aflatoxins in California tree nuts. *J. Am. Oil Chem. Soc.*, **54**, 231A–4A.
Garcia, R.A., Hotchkiss, J.H. and Steinkraus, K.H. (1999) The effect of lipids on bongkrekic (Bonkrek) acid toxin production by *Burkholderia cocovenenans* in coconut media. *Food Addit. Contam.*, **16**, 63–9.
Gardner, H.K., Koltun, S.P. and Vix, H.L.E. (1968) Solvent extraction of aflatoxins from oilseed meals. *Agric. Food Chem.*, **16**, 990–3.
Gilbert, R.J. (1982) The microbiology of some foods imported into England and through the port of London and Heathrow (London) airport, in *Control of the Microbial Contamination of Foods and Feeds in International Trade: Microbial Standards and Specifications* (eds H. Kurata and C.W. Hesseltine), Saikon Publishing Co., Tokyo, pp. 105–119.
Graham, J. (1982) The occurrence of aflatoxin in peanuts in relation to soil type and pod splitting. *Food Technol., Aust.*, **34**, 208–12.
Griffin, G.J. and Garren, K.H. (1976) Colonization of aerial peanut pegs by *Aspergillus flavus* and *A. niger*-group fungi under field conditions. *Phytopathology*, **66**, 1161–2.
Hagenmaier, R.D. (1980) *Coconut Aqueous Processing*, 2nd ed, San Carlos Publishers, University of San Carlos, Cebu City, Philippines.
Harrison, J.C., Carvajal, M. and Garner, R.C. (1993) Does aflatoxin exposure in the United Kingdom constitute a cancer risk? *Environ. Health Perspect.*, **99**, 99–105.
Hatch, M.C., Chen, C.-J., Levin, B., Ji, B.-T., Yang, G.-Y., Hsu, S.-W., Wang, L.-W., Hsieh, L.-L. and Santella, R.M. (1993) Urinary aflatoxin levels, hepatitis-B virus infection and hepatocellular carcinoma in Taiwan. *Int. J. Cancer*, **54**, 931–4.
Heperkan, D., Aran, N. and Ayfer, M. (1994) Mycoflora and mycotoxin contamination in shelled pistachio nuts. *J. Sci. Food Agric.*, **66**, 273–8.
Hill, R.A., P.D. Blankenship, R.J. Cole and T.A. Saunders. (1983) Effects of soil moisture and temperature on preharvest invasion of peanuts by the *Aspergillus flavus* group and subsequent aflatoxin development. *Appl. Environ. Microbiol.*, **45**, 628–33.
Huang, L.H. and Hanlin, R.T. (1975) Fungi occurring in freshly harvested and in-market pecans. *Mycologia*, **67**, 689–700.
IARC (International Agency for Research on Cancer) (1993) Some naturally occurring substances: food items and constituents, heterocyclic aromatic amines and mycotoxins. Monograph 56. Lyon, France: International Agency for Research on Cancer.
ICMSF (International Commission on Microbiological Specifications for Foods) (1996a) *Pseudomnas cocovenenans*, in *Microorganisms in Foods 5. Characteristics of Food Pathogens*, Blackie Academic and Professional, London, pp. 214–6.
ICMSF (International Commission on Microbiological Specifications for Foods) (1996b) Toxigenic fungi: *Aspergillus*, in *Microorganisms in Foods 5. Characteristics of Food Pathogens*, Blackie Academic and Professional, London, pp. 347–81.
Ito, Y., Peterson, S.W., Wicklow, D.T. and Goto, T. (2001) *Aspergillus pseudotamarii*, a new aflatoxin producing species in *Aspergillus* section *Flavi*. *Mycol. Res.*, **105**, 233–9.
Jacobsen, B.J., Harlin, K.S., Swanson, S.P., Lambert, R.J., Beasley, V.R., Sinclair, J.B. and Wei, L.S. (1995) Occurrence of fungi and mycotoxins associated with field mold damaged soybeans in the Midwest. *Plant Dis.*, **79**, 86–9.
JECFA (Joint FAO/WHO Expert Committee on Food Additives). (1997) Aflatoxins B, G and M. 49th Joint FAO/WHO Expert Committee on Food Additives, Rome 17–26 June, 1997. WHO Document PCS/FA/97.17.
Joffe, A.Z. (1969) The mycoflora of fresh and stored groundnut kernels in Israel. *Mycopathol. Mycol. Appl.*, **39**, 255–64.
Kinderlerer, J.L. (1984) Spoilage in desiccated coconut resulting from growth of xerophilic fungi. *Food Microbiol.*, **1**, 23–8.
King, A.D., Miller, M.J. and Eldridge, L.C. (1970) Almond harvesting, processing, and microbial flora. *Appl. Microbiol.*, **20**, 208–14.
Klich, M.A. and Pitt, J.I. (1988) Differentiation of *Aspergillus flavus* from *A. parasiticus* and other closely related species. *Trans. Br. Mycol. Soc.*, **91**, 99–108.
Klich, M.A., Thomas, S.H. and Mellon, J.E. (1984) Field studies on the mode of entry of *Aspergillus flavus* into cotton seeds. *Mycologia*, **76**, 665–9.
Klich, M.A., Mullaney, E.J., Daly, C.B. and Cary, J.W. (2000) Molecular and physiological aspects of aflatoxin and sterigmatocystin biosynthesis by *Aspergillus tamarii* and *A. ochraceoroseus*. *Appl. Microbiol. Biotechnol.*, **53**, 605–9.
Ko, S.D., Kelholt, A.J. and Kampelmacher, E.H. (1979) Inhibition of toxin production in tempe bongkrek, in *Proceedings of the Fifth International Conference on the Impacts of Applied Microbiology* (ed P. Matangkasombut), pp. 375–88.
Kolar, C.W., Richert, S.H., Decker, C.D., Steinke, F.H. and van der Zander, R.J. (1985) Isolated soy protein, in *New Protein Foods, Seed Storage Proteins* (eds A.M. Altschul and H.L. Wilcke), *Voume 5*, Academic Press, New York, pp. 260–99.
Kuiper-Goodman, T. (1991) Approaches to risk assessment for mycotoxins in foods: aflatoxins, in *Mycotoxins, Cancer and Health* (eds G.A. Gray and D.H. Ryan), Louisiana State University Press, Baton Rouge, LA, pp. 65–86.
Kurtzman, C.P., Horn, B.W. and Hesseltine, C.W. (1987) *Aspergillus nomius*, a new aflatoxin-producing species related to *Aspergillus flavus* and *Aspergillus tamarii*. *Antonie van Leeuwenhoek*, **53**, 147–58.
Leitao, M.E. de F., Delazari, I. and Mazzoni, H. (1973–74) Microbiology of dehydrated foods. *Colet. Inst. Tecnol. Aliment*, **5**, 223–41.

Lubulwa, A.S.G. and Davis, J.S. (1994) Estimating the social costs of the impacts of fungi and aflatoxins in maize and peanuts, in *Stored Product Protection: Proceedings of the 6th International Working Conference on Stored-product Protection* (eds E. Highley, E.J. Wright, H.J. Banks and B.R. Champ), CAB International, Wallingford, Oxon, UK, pp. 1017–42.

Mahmoud, A.-L. E. and Abdalla, M.H. (1994) Natural occurrence of mycotoxins in broad bean (*Vicia faba* L.) seeds and their effect on *Rhizobium*-legume symbiosis. *J. Basic Microbiol.*, **34**, 97–103.

McCoy, J.H. (1975) Trends in *Salmonella* food poisoning in England and Wales 1942–72. *J. Hyg.*, **74**, 271–82.

McDonald, D. (1970) Fungal infection of groundnut fruit before harvest. *Trans. Br. Mycol. Soc.*, **54**, 453–60.

McDonald, D. and Harkness, C. (1967) Aflatoxin in the groundnut at harvest in northern Nigeria. *Trop. Sci.*, **9**, 148–61.

Mann, G.E., Codifer, L.P. and Dollear, F.G. (1967) Effect of heat on aflatoxins in oilseed meals. *Agric. Food Chem.*, **15**, 1090–2.

Miller, J.D. (1994) Epidemiology of *Fusarium* ear diseases of cereals, in *Mycotoxins in Grain: Compounds other than Aflatoxin* (eds J.D. Miller and L.H. Trenholm), Eagan Press, St. Paul, MN, pp. 19–36.

Minifie, B.W. (1989) *Chocolate, Cocoa and Confectionery—Science and Technology*, AVI Publishing Co., Westport, CN.

Ng, S., Rouch, G., Dedman, R., Harries, B., Boyden, A., McLennan, L., Beaton, S., Tan, A., Heaton, S., Lightfoot, D., Vulcanis, M., Hogg, G., Scheil, W., Cameron, S., Kirk, M., Feldheim, J., Holland, R., Murray, C., Rose, N. and Eckert, P. (1996) Human salmonellosis and peanut butter. *Commun. Dis. Intell.*, **20**, 326.

Oliver, D. (1996) The *Salmonella* Mbandaka outbreak—an Australian overview. *Commun. Dis. Intell.*, **20**, 326–7.

O'Mahoney, M., Mitchell, E., Gilbert, R.J., Hutchinson, D.N., Begg, N.T., Rodhouse, J.C. and Morris, J.E. (1990) An outbreak of foodborne botulism associated with contaminated hazelnut yoghurt. *Epidemiol. Inf.*, **104**, 389–95.

Park, D.L., Lee, L.S., Price, R.L. and Pohland, A.E. (1988) Review of the decontamination of aflatoxins by ammoniation: current status and regulation. *J. Assoc. Off. Anal. Chem.*, **71**, 685–703.

Park, D.L. (1993) Perspectives on mycotoxin decontamination procedures. *Food Addit. Contam.*, **10**, 49–60.

Parker, W.A. and Melnick, D. (1966) Absence of aflatoxins from refined vegetable oils. *J. Am. Oil Chem. Soc.*, **43**, 635–8.

Peers, F.G. and Linsell, C.A. (1973) Dietary aflatoxins and liver cancer: a population based study in Kenya. *Br. J. Cancer*, **27**, 473–84.

Peers, F.G., Gilman, G.A. and Linsell, C.A. (1976) Dietary aflatoxins and human liver cancer: a study in Swaziland. *Int. J. Cancer*, **17**, 167–76.

Peers, F., Bosch, X., Kaldor, J., Linsell, A. and Pluumen, M. (1987) Aflatoxin exposure, hepatitis B virus infection and liver cancer in Swaziland. *Int. J. Cancer*, **39**, 545–53.

Pettit, R.E, Taber, R.A, Schroeder, H.W. and Harrison, A.L. (1971) Influence of fungicides and irrigation practice on aflatoxin in peanuts before digging. *Appl. Microbiol.*, **22**, 629–34.

Pinto, V.E.F, Vaamonde, G. and Montani, M.L. (1991) Influence of water activity, temperature and incubation time on the accumulation of aflatoxin B_1 in soybeans. *Food Microbiol.*, **8**, 195–201.

Pitt, J.I. (1989) Field studies on *Aspergillus flavus* and aflatoxins in Australian groundnuts, in *Aflatoxins in groundnut: Proceedings of the International Workshop*, 6–9 October, 1987, ICRISAT Center, India, ICRISAT, Patancheru, India, pp. 223–35.

Pitt, J.I. and Hocking, A.D. (1996) Current knowledge of fungi and mycotoxins associated with food commodities in Southeast Asia, in *Mycotoxin contamination in grains* (eds E. Highley and G.I. Johnson), Australian Centre for International Agricultural Research. ACIAR Technical Reports, Canberra, **37**, 5–10.

Pitt, J.I. and Hocking, A.D. (1997) *Fungi and Food Spoilage*, 2nd edn, Aspen Publishers, Gaithersburg, MD.

Pitt, J.I. and Miscamble, B.F. (1995) Water relations of *Aspergillus flavus* and closely related species. *J. Food Prot.*, **58**, 86–90.

Pitt, J.I., Dyer, S.K. and McCammon, S. (1991) Systemic invasion of developing peanut plants by *Aspergillus flavus*. *Lett. Appl. Microbiol.*, **13**, 16–20.

Pitt, J.I., Hocking, A.D., Bhudhasamai, K., Miscamble, B.F., Wheeler, K.A. and Tanboon-Ek, P. (1993) The normal mycoflora of commodities from Thailand. 1. Nuts and oilseeds. *Int. J. Food Microbiol.*, **20**, 211–26.

Pitt, J.I., Hocking, A.D., Bhudhasamai, K., Miscamble, B.F., Wheeler, K.A. and Tanboon-Ek, P. (1994) The normal mycoflora of commodities from Thailand. 2. Beans, rice, small grains and other commodities. *Int. J. Food Microbiol.*, **23**, 35–53.

Pitt, J.I., Hocking, A.D., Miscamble, B.F., Dharmaputra, O.S., Kuswanto, K.R., Rahayu, E.S. and Sardjono. (1998) The mycoflora of food commodities from Indonesia. *J. Food Mycol.*, **1**, 41–60.

Pitt, J.I., Taniwaki, M.H., Teixiera, A.A. and Iamanka, B.T. (2001) Distribution of *Aspergillus ochraceus*, *A. niger* and *A. carbonarius* in coffee in four regions in Brazil, in *Proceedings of the 19th International Science Colloquium on Coffee*, 14–18 May 2001, Trieste, Italy. Published on compact disc, not paginated.

Pitt, J.I. and Hocking, A.D. (2004) Current mycotoxin issues in Australia and Southeast Asia, in *Meeting the Mycotoxin Menace* (eds D. Barug, H. van Egmond, R. Lopez-Garcia, T. van Osenbruggen and A. Visconti), Wageningen Academic Publishers, Wageningen, Netherlands, pp. 67–77.

Pohland, A.E. and Wood, G.E. (1987) Occurrence of mycotoxins in food, in *Mycotoxins in Food* (ed P. Krogh), Academic Press, London, pp. 35–64.

Rayner, E.T. and Dollear, F.G. (1968) Removal of aflatoxins from oilseed meals by extraction with aqueous isopropanol. *J. Am. Oil Chem. Soc.*, **45**, 622–4.

Read, M. (1989) Removal of aflatoxin contamination from the Australian groundnut crop, in *Aflatoxins in groundnut: Proceedings of the International Workshop*, 6–9 October, 1987, ICRISAT Center, India, ICRISAT, Patancheru, India, pp. 133–140.

Rhee, K.C., Natarajan, K.R., Cater, C.M. and Mattil, K.F. (1976) Processing edible peanut protein concentrates and isolates to inactivate aflatoxin. *J. Am. Oil Chem. Soc.*, **42**, 467–71.

Richet, S.H. and Kolar, C.W. (1987) Value of isolated soy protein in food products, in *Cereals and Legumes in the Food Supply* (eds J. Dupont and E.M. Osman), Iowa State University Press, Ames, Iowa, pp. 73–90.
Rouch, G. (1996) Salmonellosis and peanut butter—Australia. Promed—Edr. Http://www.satellife.org/programs/promed-hma/9607/msg00038.htm. As reported in Burnett *et al.* (2000).
Ross, R.K., Yuan, J.-M., Yu, M.C., Wogan, G.N., Qian, G.-S., Tu, J.-T., Groopman, J.D., Gao, Y.-T. and Henderson, B.E. (1992) Urinary aflatoxin biomarkers and risk of hepatocellular carcinoma. *Lancet*, **339**, 943–6.
Sahay, S.S., Prasad, T. and Sinha, K.K. (1990) Post-harvest incidence of aflatoxins in *Linum usitatissimum* seeds. *J. Sci. Food Agric.*, **53**, 169–74.
Samarajeewa, U., Sen, A.C., Cohen, M.D. and Wei, C.I. (1990) Detoxification of aflatoxins in foods and feeds by physical and chemical methods. *J. Food Prot.*, **53**, 489–501.
Samarajeewa, U., Sen, A.C., Fernando, S.Y., Ahmed, E.M. and Wei, C.I. (1991) Inactivation of aflatoxin B_1 in corn meal, copra meal and peanuts by chlorine gas treatment. *Food Cosmetics Toxicol.*, **29**, 41–7.
Sanders, T.H., Hill, R.A., Cole, R.J. and Blankenship, P.D. (1981) Effect of drought on occurrence of *Aspergillus flavus* in maturing peanuts. *J. Am. Oil Chem. Soc.*, **58**, 966A–70A.
Sanders, T.H., Blankenship, P.D., Cole, R.J. and Hill, R.A. (1984) Effect of soil temperature and drought on peanut pod and stem temperatures relative to *Aspergillus flavus* invasion and aflatoxin contamination. *Mycopathologia*, **86**, 51–4.
Sanders, T.H., Cole, R.J., Blankenship, P.D. and Hill, R.A. (1985) Relation of environmental stress duration to *Aspergillus flavus* invasion and aflatoxin production in preharvest peanuts. *Peanut Sci.*, **12**, 90–3.
Sargent, K., Allcroft, R. and Carnaghan, R.B.A. (1961) Groundnut toxicity. *Vet. Rec.*, **73**, 865.
Sashidhar, R.B. (1993) Fate of aflatoxin B_1 during the industrial production of edible defatted peanut protein flour from raw peanuts. *Food Chem.*, **48**, 349–52.
Saxena, J. and Mehrotra, B.S. (1990) The occurrence of mycotoxins in some dry fruits retail marketed in Nainital district of India. *Acta Alimentaria*, **19**, 221–4.
Schade, J.E., McGreevy, K., King, A.D., Mackey, B. and Fuller, G. (1975) Incidence of aflatoxins in California almonds. *Appl. Microbiol.*, **29**, 48–53.
Schaffner, C.P., Mosbach, K., Bibit, V.C. and Watson, C.H. (1967) Coconut and *Salmonella* infection. *Appl. Microbiol.*, **15**, 471–5.
Scheil, W., Cameron, S., Dalton, C., Murray, C. and Wilson, D. (1998) A South Australian *Salmonella* Mbandaka outbreak investigation using a database to select controls. *Aust. NZ J. Public Health*, **22**, 536–9.
Senser, F. (1979) Untersuchungen zum Aflatoxingehalt in Haselnüssen. *Gordian*, **79**, 117–23.
Shah, F.H. and Hamid, A. (1989) Aflatoxins in various foods and feed ingredients. *Pak. J. Sci. Ind. Res.*, **32**, 733–6.
Simonsen, B., Bryan, F.L., Christian, J.H.B., Roberts, T.A., Tompkin, R.B. and Silliker, J.H. (1987) Prevention and control of food-borne salmonellosis through application of Hazard Analysis Critical Control Point (HACCP). *Int. J. Food Microbiol.*, **4**, 227–47.
Snowdon, A.L. (1991) *A Colour Atlas of Post-harvest Diseases and Disorders of Fruits and Vegetables. 2. Vegetables.* Wolfe Scientific, London.
Sommer, N.F., Buchanan, J.R. and Fortlage, R.J. (1986) Relation of early splitting and tattering of pistachio nuts to aflatoxin in the orchard. *Phytopathology*, **76**, 692–4.
Sreenivasamurthy, V.H., Parpia, A.B., Srikanta, S. and Murti, A.S. (1967) Detoxification of aflatoxin in peanut meal by hydrogen peroxide. *J. Assoc. Off. Anal. Chem.*, **50**, 350–4.
Steiner, W.E., Brunschweiler, K., Leimbacher, E. and Schneider, R. (1992) Aflatoxins and fluorescence in Brazil nuts and pistachio nuts. *J. Agric. Food Chem.*, **40**, 2453–7.
Stoloff, L. (1977) Aflatoxins—an overview, in *Mycotoxins in Human and Animal Health* (eds J.V. Rodricks, C.W. Hesseltine and M.A. Mehlman), Pathotox Publishers, Park Forest South, Illinois, pp. 7–28.
Stoloff, L. (1989) Aflatoxin is not a probable human carcinogen: the published evidence is sufficient. *Regul. Toxicol. Pharmacol.*, **10**, 272–83.
Stoloff, L. and Trucksess, M.W. (1981) Effect of boiling, frying, and baking on recovery of aflatoxin from naturally contaminated corn grits or cornmeal. *J. Assoc. Off. Anal. Chem.*, **64**, 678–80.
Tabata, S., Kamimura, H., Ibe, A., Hashimoto, H., Iida, M., Tamura, Y. and Nishima, T. (1993) Aflatoxin contamination in foods and foodstuffs in Tokyo: 1986–1990. *J. Assoc. Off. Anal. Chem. Int.*, **76**, 32–5.
Taguchi, S., Fukushima, S., Sumimoto, T., Yoshida, S and Nishimune, T. (1995) Aflatoxins in foods collected in Osaka, Japan, from 1988 to 1992. *J. Assoc. Off. Anal. Chem. Int.*, **78**, 325–7.
Taniwaki, M.H., Pitt, J.I., Urbano, G.R., Teixeira, A.A. and Leitão, M.F.F. (1999) Fungi producing ochratoxin A in coffee, in *Proceedings of the 18th International Scientific Colloqium on Coffee*, 2–6 August, 1999, Helsinki, Finland, pp. 239–47.
Teixera, A.A., Taniwaki, M.H., Pitt, J.I. and Martins, C.P. (2001) The presence of ochratoxin A in coffee due to local conditions and processing in four regions in Brazil, in *Proceedings of the 19th International Scientific Colloquium on Coffee*, 14–18 May 2001, Trieste, Italy. Published on compact disc, not paginated.
Tiemstra, P.J. (1977) Aflatoxin control during food processing of peanuts, in *Mycotoxins in Human and Animal Health* (eds J.V. Rodricks, C.W. Hesseltine and M.A. Mehlman), Pathotox Publishers, Park Forest South, Illinois, pp. 121–37.
Trager, W.T. and Stoloff, L. (1967) Possible reactions for aflatoxin detoxification. *J. Agric. Food Chem.*, **15**, 679–81.
Uboldi-Eiroa, M.N., Freitas-Leitao, M.F. de, Martin, Z. J. de and Kato, K. (1975) Microbiology of coconut milk. *Colet. Indt. Tecnol. Aliment.*, *6*, 1–10.

Urano, T., Trucksess, M.W., Beaver, R.W., Wilson, D.M., Dorner, J.W. and Dowell, F.E. (1992) Co-occurrence of cyclopiazonic acid and aflatoxins in corn and peanuts. *J. Assoc. Off. Anal. Chem. Int.*, **7**, 838–41.

van Egmond, H.P. (1992) Aflatoxin M_1: occurrence, toxicity, regulation, in *Mycotoxins in Dairy Products* (ed H.P. van Egmond). Elsevier Applied Science, Amsterdam, pp. 11–55.

Waggle, D.H. and Kolar, C.W. (1979) Types of soy protein products, in *Soy Protein and Human Nutrition* (eds H.L. Wilke, D. T. Hopkins and D.H. Waggle), Academic Press, New York, pp. 99–51.

Waltking, A.E. (1971) Fate of aflatoxin during roasting and storage of contaminated peanut products. *J. Assoc. Off. Anal. Chem.*, **54**, 533–9.

Wells, J.M. and Payne, J.A. (1976) Toxigenic species of *Penicillium*, *Fusarium* and *Aspergillus* from weevil-damaged pecans. *Can. J. Microbiol.*, **22**, 281–5.

Whitaker, T.B., Dickens, J.W., Monroe, R.J. and Wiser, E.H. (1972) Comparison of the observed distribution of aflatoxin in shelled peanuts to the negative binomial distribution. *J. Am. Oil Chem. Soc.*, **49**, 590–3.

Wild, C.P., Jansen, L.A.M., Cova, L. and Montesano, R. (1993) Molecular dosimetry of aflatoxin exposure: contribution to understanding the multifactorial etiopathogenesis of primary hepatocellular carcinoma with particular reference to hepatitis B virus. *Environ. Health Perspect.*, **99**, 115–22.

Wild, C.P., Pionneau, F.A., Montesano, R., Mutiro, C.F. and Chetsanga, C.J. (1987) Aflatoxin detected in human breast milk by immunoassay. *Int. J. Cancer*, **40**, 328–33.

Woller, R. and Majerus, P. (1979) Aflatoxine in Paranüssen und Pistazien. *Lebensmittelchemie und Gerichtliche Chemie*, **33**, 115–6.

Yeh, F.-S., Yu, M.C., Mo, C.-C., Luo, S., Tong, M.J. and Henderson, B.E. (1989) Hepatitis B virus, aflatoxins, and hepatocellular carcinoma in Southern Guangxi, China. *Cancer Res.*, **49**, 2506–9.

Zhao, N.X., Ma, M.S., Zhang, Y.P. and Xu, D.C. (1990) Comparative description of *Pseudomonas cocovenenans* (van Damme, Johannes, Cox and Berends 196) NCIB 9450^T and strains isolated from cases of food poisoning caused by consumption of fermented corn flour in China. *Int. J. Syst. Bacteriol.*, **40**, 452–5.

Zohri, A.A. and Saber, M.M. (1993) Filamentous fungi and mycotoxins detected in coconut. *Zentralblatt für Mikrobiologie und Hygiene*, **148**, 325–32.

10 Cocoa, chocolate, and confectionery

I Introduction

A Definitions

Cocoa beans are the seeds of the tree *Theobroma cacao* L. Seeds develop in pods, each containing about 30 beans surrounded by sterile pulp. The pulp consists of parenchymatous cells composed of 80-90% water, 8-13% fermentable sugars (mostly glucose and sucrose), about 0.5% non-volatile acids, mainly citric, and small amounts of amino acids. The pH ranges from 3.6 to 4.0 (Lehrian and Patterson, 1983; Biehl *et al.*, 1989).

The raw beans are composed of two cotyledons, a radicle (germ) and a seed coat (testa). The cotyledons contain about one-third water and one-third fat (cocoa butter), the remainder being starch, sugars, purine bases, phenolic components and non-volatile acids.

Cocoa powders (cocoas) are defined in the Codex Standard 105-1981 (Codex Alimentarius, 1981) as products obtained by mechanical transformation into powder of cocoa press cake produced by partial removal of the fat from cocoa nibs or cocoa by mechanical means. Cocoa butter is also produced during this operation and is defined in the Codex Standard 86-1981 (Codex Alimentarius, 1981).

Chocolate is defined in the Codex Standard 87-1981 (Codex Alimentarius, 1981) as the homogeneous product obtained by an adequate process of manufacture from a mixture of one or more of the following: cocoa nibs, cocoa mass, cocoa press cake, cocoa powder including fat-reduced powder, with or without permitted optional ingredients and/or flavoring agents.

Confectionery is a term for which meaning is different in every country (Minifie, 1989) and may cover a very large number of different products manufactured in various ways. Chocolate confectioneries such as bars, blocks and bonbons, and sugar confectionery such as boiled sweets, toffees, fudge, fondants, jellies and pastilles are discussed here and descriptions and definitions of the most important products can be found in the Codex Standards 142-1983 and 147-1985. A list of all product categories according to CAOBISCO, the Organisation of Confectionery Trade Association within the EC, is given in Nuttall (1999). Flour confectioneries such as cakes and biscuits are considered in Chapter 8.

B Important properties

After preliminary treatments such as fermentation process described subsequently and drying, beans ready for further processing are composed of 87% cotyledon containing only 4-5% water, 12% shell containing 8-10% water and 1% germ.

Cocoa powder contains 9-36% fat and less than 8% moisture, its pH being 5.5-6.2 (natural cocoa) or 7.0-8.0 (alkalized cocoa). Essential composition and quality factors of chocolate are defined in Table 10.1, which corresponds to earlier-mentioned Codex Standard 87-1981.

Confectionery products are a very heterogeneous group of products made with dried milk and other dairy products; cocoa and chocolate products; sugar, honey, syrups or sweeteners; nuts, fruits or jams; starches, gelatin pectin or other thickeners; egg albumen; spices, colors, flavors or acidulants.

Table 10.1 Essential composition and quality factors of chocolates (Codex Alimentarius, 1981, standard 87-1981)

	Constituents (in % milk on the dry matter)						
Product	Cocoa butter	Fat free cocoa solids	Total cocoa solids	Milk fat	Fat free milk	Total fat solids	Sugars
Chocolate	≥18	≥14	≥35	–	–	–	–
Unsweetened chocolate	≥50/ ≥58	–	–	–	–	–	–
Couverture chocolate	≥31	≥2.5	≥35	–	–	–	–
Sweet (plain) chocolate	≥18	≥12	≥30	–	–	–	–
Milk chocolate	–	≥2.5	≥25	≥3.5	≥10.5	≥25	≤55
Milk couverture chocolate	–	≥2.5	≥25	≥3.5	≥10.5	≥31	≤55
Milk chocolate with high milk content	–	≥2.5	≥20	≥5	≥15	≥25	≤55
Skimmed milk chocolate	–	≥2.5	≥25	≤0.5	≥14	≥25	≤55
Skimmed milk couverture chocolate	–	≥2.5	≥25	≤0.5	≥14	≥31	≤55
Cream chocolate	–	≥2.5	≥25	≥7	≥3&≤14	≥25	≤55
Chocolate vermicelli or flakes	≥12	≥14	≥32	–	–	–	–
Milk chocolate vermicelli or flakes	–	≥2.5	≥20	≥3.5	≥10.5	≥12	≤66

II Initial microflora

Cocoa beans are raw agricultural products that are exposed during harvesting and subsequent fermentation to numerous microorganisms, including *Salmonella* spp.

Sound, undamaged beans have few, if any, microorganisms inside the cotyledons (Meursing and Slot, 1968). Immediately after cutting and breaking of the pods and removal of the sterile pulp, beans are contaminated with bacteria and fungi originating mainly not only from soil and air but also from the surface of pods, and the hands and tools of harvesters (Ostovar and Keeney, 1973). Microbial contamination with particular microorganisms occurs from fermentation boxes, baskets or trays, which are constantly reused.

Cocoa fermentation is performed in many ways throughout the world depending on the scale of the plantation and traditions. The principal formats include fermentation in piles, in heaps, in baskets, in wooden boxes or in trays (Shaugnessy, 1992).

A very diverse range of raw materials is used in chocolate and confectioneries. The respective chapters should be consulted for data on the initial flora that are introduced into these products.

III Primary processing

A Effects of processing on microorganisms

The fermentation of the pulp residues surrounding the beans involves external microbial, as well as internal enzymatic, processes and is an important step in the formation of flavor precursors (Schwan *et al.*, 1995). During fermentation, the combination of heat and acids leads to the destruction of the germ, thus avoiding degradation of cocoa fat associated with germination.

More than 60% of the microorganisms isolated do not seem to be essential for successful fermentation (Rombouts, 1952). Although differences in the composition of the microflora can be found due to different sampling methods, inhomogeneities of the fermenting mass, mixing of the beans, etc.; three main stages can be differentiated: (i) production of alcohol; (ii) production of acids; and (iii) utilization of acids. In most cases, this succession is not clear-cut and transitional phases or overlaps occur (Hansen and Welty, 1970; Ostovar and Keeney, 1973).

Yeasts. During the first few days (1–2 days), yeasts form the predominant population, rapidly metabolizing sugars to ethanol, breaking down the mucilaginous pulp by means of excreted pectinolytic enzymes, thus facilitating draining off the sweating fluids (Schwan *et al.*, 1997). Catabolism of citric acid leads to an increase of the pH to about 4.0 (Gauthier *et al.*, 1977; Sanchez *et al.*, 1984, 1985).

By the third day, fermentative yeasts decline and the conditions found, i.e. low oxygen concentration or high concentrations of carbon dioxide, favor the development of lactic acid bacteria (Passos *et al.*, 1984a,b).

Bacteria. In well-aerated portions of the fermenting mass, acetic bacteria become dominant, transforming ethanol to acetic acid (Carr *et al.*, 1980, Passos *et al.*, 1984a,b). Both acetic and lactic acids can be further oxidized to carbon dioxide and water, with a concomitant increase in pH. These exothermic oxidation reactions lead to an increase in temperature of the fermenting mass to 45-50°C, a temperature that is optimal for spore-forming bacilli. Their development is also promoted by the steadily increasing pH value. An important aroma component of cocoa flavor, tetramethyl-pyrazine, was found to be synthesized by *Bacillus subtilis* (Ostovar and Keeney, 1972; Zak *et al.*, 1972).

Microorganisms found in fermenting cacao in Ghana and Malaysia included yeasts *Kloeckera*, *Candida*, *Saccharomyces*, *Hanseniaspora*, *Rhodotorula*, *Debaryomyces*, *Pichia* and *Schizosaccharomyces*; acetic acid bacteria-*Aerobacter rancens*, *Aer. xylinum*, *Aer. ascendens* and *Gluconobacter oxydans*; lactic acid bacteria-*Lactobacillus collinoides*, *Lb. plantarum*, *Lb. fermentum* and *Lb. mali*; and *Bacillus* species-*Bacillus cereus*, *B. licheniformis* and *B. coagulans* (Carr et al., 1979). Fermentation is a complex process and it is not always well understood which microorganisms are essential. Few attempts have been carried out with directed fermentations using cocktails of established strains (Schwan, 1998).

Other lactic acid bacteria present in fermenting cacao beans were *Lb. casei*, *Lb. lactis*, *Lb. bulgaricus*, *Lb. acidophilus* and *Streptococcus lactis*. Other *Acetobacter* species included *Acetobacter lovaniensis* and (from Malaysia) *Aceto. aceti* and *Aceto. roseum*. Other bacteria reported were *B. sphaericus*, *Arthrobacter* spp., *Micrococcus* spp. and *Sarcina* spp. All of these bacteria produced different organic acids associated with cacao fermentation, the major ones being acetic and lactic acids (Jinap, 1994). Malaysian and Brazilian cocoa suffered from excessive acidity and much work has been carried out to overcome this quality problem, thought to be microbiological (Carr *et al.*, 1979, 1980; Chick *et al.*, 1981).

Spoilage of fermenting beans may be due to the development of *Aerobacter* spp. and *Pseudomonas* spp. if the pH rises above 5.0 during fermentation (Ostovar and Keeney, 1973).

Molds. Molds may spoil beans at the outer surfaces of the fermenting heap, in particular if the beans remain unturned for 2-3 days (Roelofsen, 1958). However, after degradation of pulp residues and decrease of temperature, mycelia may penetrate the mass if sufficient oxygen is supplied, causing compositional changes, in particular of fatty acids (Hansen *et al.*, 1973; Hansen, 1975).

Aspergillus fumigatus, the most commonly found mold during fermentation, is particularly harmful in destroying testa and permitting penetration by other molds, such as *Asp. niger*, *Asp. flavus*, *Asp. tamani*, *Eurotium* spp., *Penicillium* spp. and *Mucor* spp. (Chatt, 1953; Roelofsen, 1958). However, the presence of mycotoxins has been reported only rarely (Lenovich, 1979). This is probably due to the presence of inhibitors such as methylxanthines (Buchanan and Fletcher, 1978) or due to the fact that contaminated layers, such as shells, are eliminated during further processing.

During subsequent drying the water content of the beans decreases from about 60% to 6-8%. Artificial drying is very quick and does not permit mold growth, whereas sun drying may take 7 days or more, depending on the atmospheric conditions. Populations of microorganisms normally found on the surface of dried beans consist mainly of mesophilic and thermophilic spores (10^6–10^7/g).

Heat-sensitive bacteria, such as *Enterobacter* spp., *Flavobacterium* spp., *Microbacterium* spp., *Streptococcus* spp., *Micrococcus* spp. and *Streptomyces* spp. (about 10^5cfu/g) as well as yeasts and molds (10^3–10^7cfu/g) have been found (Hansen and Welty, 1970; Barrile *et al.*, 1971; Niles, 1981).The only xerophilic mold species isolated from cocoa beans was *Asp. glaucus*, which may be due to the isolation techniques applied.

Storage in jute bags or silos under unsuitable conditions may cause spoilage of the beans. Molds, in particular xerophilic forms, are able to develop if beans are damaged, improperly dried, or when moisture increases above 8% (Maravalhas, 1966; Hansen and Welty, 1970). Moldy beans are at the origin of off-flavors.

Fungi, especially *Asp. flavus*, commonly found in cacao ferments, were reported to have remarkable lipolytic activity, and to be the main contributors to spoilage of fermented cacao beans (Kavanagh *et al.*, 1970; Hansen *et al.*, 1973).

Aspergillus and *Penicillium* were also reported to cause large increases in carbonyl compounds, methyl ketones, 2-enals and 2,4-dienals in moldy beans (Hansen and Keeney, 1970). Most of the carbonyls were dissolved in the fat phase and remained in butter when the beans were pressed (Hansen and Keeney, 1969).

B Methods of processing

Before processing, beans are cleaned by screening, air currents and magnets to remove extraneous materials. Sound, undamaged beans have few if any microorganisms inside the cotyledons (Meursing and Slot, 1968).

Roasting is an important step in the development of chocolate flavor since basic chemical reactions occur during this process (Zak, 1988; Cros, 1995). Roasting (treatments of 15 min to 2 h at 105–150°C) is the only processing step in the chocolate production allowing for complete destruction of vegetative microorganisms, in particular pathogens such as *Salmonella* spp.

The oldest and most used method starts with cleaned whole beans. Depending on the type of equipment used, and the product characteristics aimed for, roasting of shell-free cocoa nibs or of raw ground (liquid) cocoa mass can also be performed (Heemskerk, 1999). A substantial decrease of the number of spores, i.e. of the total viable count, is achieved mainly by the elimination of shells during winnowing (Lindley, 1972).

Preliminary treatments such as infrared heating (micronizing) or steam-treatment, primarily designed to allow for a better separation of shells and a minimization of fat losses also have a certain bactericidal effect (Minson, 1992, Heemskerk, 1999).

After roasting of the beans, spore-formers such as *B. subtilis, B. coagulans, B. stearothermophilus, B. licheniformis, B. megaterium*, may be found (Barrile *et al.*, 1971; Ostovar and Keeney, 1973).

IV Processed products

A Effects of processing on microorganisms

Chocolate. The subsequent processing steps of the roasted beans, nibs or liquor such as milling and refining, mixing, conching, tempering or molding, have only a small influence on the final flora of chocolate. Even if temperatures of 60-80°C are reached during milling or conching, microorganisms are protected by the low water activity and the high fat content.

The final flora is mainly composed of *Bacillus* spp. (Collins-Thompson *et al.*, 1981), the levels being very much dependent on the original spore load of the raw beans and the type of roasting applied. Slight changes in the distribution of species may be observed after the addition of ingredients such as milk powder or sugar.

The presence of non-sporing bacteria such as faecal indicators or salmonellae is due to recontamination from the environment or from added ingredients.

Cocoa powder. In the production of cocoa powder, alkalization or dutching is a process developed in the early 1800s whereby nibs, less commonly cocoa liquor, cocoa powder or press cake are heated with alkali (usually sodium hydroxide or potassium carbonate) at temperatures of 85–115°C to obtain desired physicochemical changes (flavor, color) as summarized by Kleinert (1988) or Meursing and Zijderveld (1999). This treatment has a strong sterilizing effect due to the combined effect of water, alkali and heat (Minifie, 1989, Meursing and Zijderveld, 1999).

In the case of cocoa powder, the final flora is almost exclusively introduced during further processing of the almost sterile alkalized liquor, i.e. pressing to extract cocoa butter, breaking of the cake and subsequent grinding of the kibbles, cooling and packaging of the powder (Minifie, 1989). The major recontaminants are sporeformers (Gabis *et al.*, 1970; Mossel *et al.*, 1974). During grinding, heat is evolved and cooling air must be dry to prevent growth of molds in ducts and conveyors (Minifie, 1989). Total aerobic count is therefore very appropriate as an indicator of recontamination of cocoa powder. Products with $<10^3$ cfu/g are normal, whereas counts exceeding 10^4cfu/g may indicate poor manufacturing practices (Meursing and Slot, 1968; Collins-Thompson *et al.*, 1978).

Irradiation of cocoa powder has been shown to kill microorganisms effectively, but the organoleptic quality was no longer acceptable (Grünewald and Münzner, 1972).

Confectionery products. Due to the wide range of products, processing may include very weak treatments allowing for little or no killing effect on microorganisms to very strong treatments, such as boiling, allowing for the complete destruction of vegetative bacteria (Slater, 1986; Vendrell *et al.*, 2000).

B Spoilage

Chocolate. Due to its low water activity of 0.4-0.5 (Richardson, 1987), microbial spoilage of chocolate is not possible. Development of molds on the interface of product and packaging material at very high relative humidities and for chocolate prepared with different types of sugars, thus modifying the a_w characteristics of the product, have been reported by Ogunmoyela and Birch (1984). Some xerophilic molds such as *Bettsia alvei*, *Chrysosporium xerophilum* and *Neosartorya glabra* have been isolated from spoilt chocolate and chocolate confectionery. The spoilage of chocolate by *Chrysosporium* species has been studied and described by Kinderlerer (1997).

Soapiness is a chemical defect of unsweetened or white chocolate (Table 10.1) and is most common in products containing coconut and palm oil, which are rich in short- and medium-chain fatty acids. High levels of lipolytic enzymes from *Bacillus* spp. or molds persisting in raw materials such as cocoa liquor or powdered milk may also adversely affect fats used in chocolate and confectionery products (Witlin and Smyth, 1957).

Cocoa powder. Spoilage of cocoa by molds is only observed in case of moisture uptake. Cases of development of off-flavors, i.e. the presence of trichloranisole and derivatives, due to mold contamination of packaging material have been reported (Whitfield *et al.*, 1984).

Confectionery product. Microbial spoilage of confectionery products is best considered within the framework presented in Table 10.2.

Table 10.2 Water activity of various types of confectionery products[a]

Type of product	Water activity
Wafer biscuits	0.15–0.25
Hard candy	0.20–0.35
Roasted nuts	0.4
Caramel	0.4–0.5
Chocolate	0.4–0.5
Raisins	0.5–0.55
Fudge	0.65–0.75
Fondant	0.76
Jellies	0.65–0.75
Nougat	0.4–0.7
Marshmallow	0.6–0.75
Mints	0.75–0.8

[a]From Richardson (1987).

Confectionery products with water activities ranging between 0.5 and 0.8 are prone to spoilage by xerophilic yeasts or molds due to the formation of gas causing fractures or bursting of products, to the formation of slime, color changes or off-odors and off-flavors, or the secretion of enzymes causing liquefaction of products (Mossel and Sand, 1968; Blaschke-Hellmessen and Teuschel, 1970; Windisch, 1977; Pitt and Hocking, 1985; Miller *et al.*, 1986; Jermini *et al.*, 1987). The most important spoilage yeast is *Zygosaccharomyces rouxii*, which is capable of growing in sugar syrups of very high soluble solids content (Pitt and Hocking, 1985).

Spoilage fungi may be introduced through raw materials such as dairy products, flours and starches, sugars, nuts and dried fruits or jams (Zeller, 1963; Mossel and Sand, 1968; Legan and Voysey, 1991; Finoli *et al.*, 1994). They may survive due to processing failures as in the case of preserved fruits (Walker and Ayres, 1970) or from recontamination due to their presence in the environment (Dragoni *et al.*, 1989).

C Pathogens

The only pathogen of concern in chocolate and cocoa powder is *Salmonella* spp., as found in different product surveys (D'Aoust, 1977) and confirmed by epidemiology up to recent years. These products were not recognized as causes of salmonellosis until 1970 and 1973, after two outbreaks. Cocoa powder contaminated with *Salmonella* Durham and used in confectionery products was at the origin of an outbreak affecting 110 people in Sweden (Gästrin *et al.*, 1972). In Canada and in the United States 200 people, mostly children with an average age of 3 years, suffered an intoxication from chocolate contaminated with *S*. Eastbourne (Craven *et al.*, 1975; D'Aoust *et al.*, 1975). Contamination was shown to be due to cross-contamination in the factory due to inadequate separation of clean and unclean zones.

In 1982–1983 an outbreak involving 245 people in the UK, again mostly children, was quickly traced to two types of chocolate bars produced in Italy and contaminated with *S*. Napoli (Gill *et al.*, 1983). Contaminated water penetrating through microleaks of equipment was discussed as a possible source of contamination. Some cases due to *S. nima* reported in Canada were traced back to chocolate coins imported from Belgium (Hockin *et al.*, 1989). Another case involved more than 300 children in an outbreak linked to chocolate contaminated with *S*. Typhimurium (Kapperud *et al.*, 1989a). Epidemiological studies have shown that the strain was identical to strains isolated from birds in the same region (Kapperud *et al.*, 1989b, 1990). Although for a prolonged period of time no outbreak has been reported, a survey in Mexico has shown that *Salmonella* can still be found in finished products (Torres-Vitela *et al.*, 1995). The most recent outbreak has been reported in 2001 involved a few hundreds of consumers and was traced to chocolate manufactured in Germany and contaminated with *S.* Oranienburg (Anonymous, 2002). There are relatively few data on the economical impact of most outbreaks, but available figures compiled by Todd (1985) and Roberts *et al.* (1989) show they can have a fatal outcome for involved companies.

A particular characteristic of salmonellae in chocolate products is its survival over very long periods of time, up to several years in the case of naturally contaminated products (Dockstäder and Groomes, 1971; Rieschel and Schenkel, 1971; Tamminga, 1979). Furthermore, salmonellae show a very high heat resistance in chocolate, which is due to the low water activity and the protective effect of fat. Increased tolerance to heat may also be due to habituation to reduced water activity (Mattick *et al.*, 2000). *S.* Anatum was found to be the most heat-resistant species isolated from chocolate (Barrile *et al.*, 1970).

Temperatures of 70–80°C reached during milling, refining or conching do not provide effective destruction (Goepfert and Biggie, 1968) and even considerable overheating (>100°C) could not achieve complete destruction of small numbers of *S.* Senftenberg (Rieschel and Schenkel, 1971). Addition of 2% of water allowed decontamination at 71°C (Barrile and Cone, 1970).

One remarkable aspect is the very low infective dose reported: an average of 1.6 cells/g was determined for *S.* Napoli (Greenwood and Hooper, 1983); of 0.2–1.0 cells/g for *S.* Eastbourne (D'Aoust and Pivnick, 1976) and as low as 0.005–0.025 cells/g for *S.* Nima (Hockin *et al.*, 1989). The levels of *S.* Oranienburg detected in the contaminated chocolate were of the order of 1 cfu/g or less.

The very low infective doses reported may be a consequence of the short intragastric residence and the protective effect conferred by the fat present in chocolate towards gastric acids (Tamminga *et al.*, 1976; D'Aoust, 1977).

Few outbreaks have been traced back to contaminated confectionery products such as traditional torrone (nougat) in Italy (De Grandi *et al.*, 1987), marshmellows (Lewis et al., 1996). In the first two cases the egg preparation used to manufacture the products were identified as the source of contamination. Recently halva, a popular low moisture confectionery in eastern Mediterranean countries and in Middle East, contaminated with *S.* Typhimurium (Anonymous, 2001) has caused an outbreak. An outbreak of salmonellosis caused by *S.* Mbandaka in 1996 in several states in Australia has been traced back to peanut butter contaminated with as low as 3 cfu/g (Ng *et al.*, 1996; Scheil *et al.*, 1998). Contaminated peanut butter used to flavor extruded snacks had already been at the origin of an outbreak in the UK and the US a few years earlier (Killalea *et al.*, 1996; Shohat *et al.*, 1996). The behavior of the pathogen in peanut butter is similar to that in chocolate (Burnett *et al.*, 2000).

Staph. aureus survives for several months in chocolate, but is not likely to produce significant levels of toxins as it cannot grow (Ostovar, 1973).

Confectionery does not support the growth of disease-causing bacteria and, only rarely, supports that of mycotoxigenic molds. Salmonellae are unable to grow at the a_w of common confections. However, if they enter a confection through the ingredients, they may survive for long periods–up to 8 days in chocolate and egg liqueur stored at 4°C (Warburton *et al.*, 1993). At the low a_w of confections, these bacteria may survive a heating step (Goepfert *et al.*, 1970; Gibson, 1973; De Grandi *et al.*, 1987).

Mycotoxins may be introduced through the use of contaminated ingredients as shown by Bresch *et al.* (2000).

Various confection ingredients may contribute to salmonellae–primarily nuts and coconut (Chapter 9),[1] chocolate (this chapter), milk (Chapter 16), egg albumen (Chapter 15), flour and starches (Chapter 8), spices (Chapter 7) and gelatin (Chapter 1). Moldy nuts could introduce mycotoxins (Chapter 9).

[1] A very recent outbreak due to contaminated almonds has been reported by Chan *et al.* (2002) underlining the risks related with raw materials of agricultural origin, particularly where unprocessed.

D CONTROL (cocoa, chocolate and confectionery)

Summary

Significant hazards[a]	• *Salmonella* spp.
Control measures	
Initial level (H_0)	• Presence in raw cocoa beans unavoidable. Quantitative data are, however, missing. For ingredients used in confectionery products refer to corresponding chapters.
Reduction (ΣR)	• Roasting of raw cocoa beans (chocolate, cocoa powder). • Alkalization or dutching of cocoa liquor (cocoa powder). • Pasteurization, cooking, boiling (confectionery). • Precise figures on the levels of reduction are, however, not published.
Increase (ΣI)	• Selection of added raw materials. • Good hygiene practices.
Testing	• *Salmonella* monitoring program including environmental, line, finished product and critical ingredients to verify effectiveness of preventive measures such as zoning. • Use of indicators such as coliforms/Enterobacteriaceae or total viable counts is recommended.
Spoilage	• For certain confectionery products such as fillings spoilage with osmophilic yeasts or xerophilic moulds known.

[a]In particular circumstances, other hazards may need to be considered.

Hazards to be considered

The only identified significant health hazard of cocoa, chocolate and confectionery products is *Salmonella*.

Control measures

Initial level of hazard (H_0). Raw cocoa beans are a permanent but mostly unavoidable source of salmonellae, a fact confirmed by their regular detection in environmental samples (dust and residues) from raw bean storage and handling areas. However, no quantitative data are available on levels.

Reduction of hazard (ΣR). Roasting of the raw cocoa beans is primarily designed to obtain the desired organoleptic profile. It represents, however, the only killing step (Critical Control Point) for *Salmonella* (Simonsen *et al.*, 1987; Cordier, 1994). It is important to ensure that the design and maintenance of the roasting equipments do not allow further processing of unroasted material. Different processes and processing conditions have been developed but only few microbiological data demonstrating the killing effect are published. This is confirmed by historical and practical experience but data do not allow to calculate precise D-values.

In the case of cocoa powder, alkalization or dutching are primarily applied to obtain the desired organoleptic characteristics of the powder. The conditions applied, however, allow to achieve effects almost equivalent to a sterilization (CCP; Meursing and Zijderveld, 1999). Again here no quantitative data are published to calculate precise D-values.

In the case of confectionery products, more or less severe pasteurization, boiling or cooking conditions are applied during the manufacturing or preparation processes and the majority of them provide appropriate killing (Minifie, 1989). Due to the wide variety of products and processes, they should be examined individually.

Increase of hazard (ΣI). The presence of salmonellae in finished products is due to recontamination during further processing. Survival is likely over prolonged periods of time but no further growth is possible during storage and distribution due to the low water activity of the products.

Raw materials can be potential sources of salmonellae and can be differentiated and classified according to their potential risk (IOCCC, 1991). Processed raw materials such as dairy products, cocoa liquor and powder, crumbs and cocoa butter, egg products, gelatin, flour, lecithin, coconut and starches may need regular checking, but this cannot replace reliance on the supplier's Quality Assurance system and regular audits. Rework is particularly critical and should therefore be handled carefully.

In chocolate factories, water plays an important role in maintaining the temperature of liquid chocolate masses in pipes and storage tanks as well as for tempering and cooling. Microleaks may lead to contamination of the product and it is therefore necessary to guarantee the absence of *Salmonella* by appropriate disinfection methods. The use of water for cleaning should be restricted to a minimum. If wet cleaning is necessary, careful drying is then essential to avoid multiplication of bacteria, possibly pathogens, in wet residues such as milk-powder or sugar.

Recontamination from the processing environment is a further possibility and control can be achieved by an adequate layout of production lines allowing the physical separation of unclean, potentially contaminated zones from clean zones where roasted beans are further processed. Movement of personnel and of vehicles such as fork-lifts must be limited to maintain this separation.

HACCP for cocoa, chocolate and confectionery products is extensively discussed in ICMSF (1988) and Cordier (1994) as well as in the International Office of Cocoa, Chocolate and Confectionery (IOCCC) Code of Hygienic Practice issued by the IOCCC (1991) with a complementary document on Good Manufacturing Practice (IOCCC, 1993).

Testing

At the level of manufacturers, however, the implementation of an environmental sampling program to verify the efficiency of the preventive measures (GMP and HACCP) is more effective. This program can be complemented with checks on critical raw materials, line and finished products samples as appropriate. Hygiene indicators such as coliforms or Enterobacteriaceae and, to a certain extent, of total viable counts can be used to detect deviations and recontamination.

The methods used for the examination of cocoa and cocoa-based products for *Salmonella* require special attention since antibacterial components present in cocoa inhibit growth and hence detection. Addition of skim milk or of casein to pre-enrichment broths is therefore necessary to overcome inhibitory effects (Busta and Speck, 1968; Zapatka *et al.*, 1977). Antibacterial effects are not observed at the higher dilutions used for total counts (Park *et al.*, 1979).

Spoilage

Spoilage with osmophilic yeasts or xerophilic molds is only observed for confectionery products and depends on their water activity. Presence in finished products is mostly due to recontamination through the use of contaminated ingredients or from the environment with airborne molds or contaminated residues on food contact surfaces. Growth is then dependent on the characteristics of the product as well as on the storage and distribution conditions and the shelf-life of the product.

The observation of strict plant hygiene and the testing of residues for fermentative yeasts is important. Visual inspection of certain ingredients such as nuts and dried fruits is frequently sufficient but modern photometric equipment is also used (Finoli *et al.*, 1994) to select raw materials. Additional measures such as reduction of airborne molds (Dragoni *et al.*, 1989), personal hygiene (Kleinert-Zollinger, 1988), separation of raw from processed product, and scheduled examination for microbial content are more or less important, depending on process or product (IOCCC, 1991, 1993).

References

Anonymous (2001) Communicable disease report. *CDR Wkly*, **11**.

Anonymous (2002) International outbreak of *Salmonella oranienburg*, October-December 2001. *WHO Surveill. Newslett.*, March, 3–4.

Barrile, J.C. and Cone. J.F. (1970) Effect of added moisture on the heat resistance of *Salmonella anatum* in milk chocolate. *Appl. Microbiol.*, **19**, 177–8.

Barrile, J.C., Cone, J.F. and Keeney, P.G. (1970) A study of salmonellae survival in milk chocolate. *Manufact.' Conf.*, September.

Barrile, J.C., Ostovar, K. and Keeney, P.G. (1971) Microflora of cocoa beans before and after roasting at 150°C. *J. Milk Food Technol.*, **34**, 369–1.

Biehl, B., Meyer, B., Crone, G., Pallmann, L. and Said, M.B. (1989) Chemical and physical changes in the pulp during ripening and postharvest storage of cocoa pods. *J. Sci. Food Agric.*, **48**, 189–208.

Blaschke-Hellmessen, R. and Teuschel, G. (1970) *Saccharomyces rouxii* Boutroux als Ursache von Gärungserscheinungen in geformten Marzipan- und Persipanartikeln und deren Verhütung im Herstellerbetrieb. *Nahrung*, **18**, 250–67.

Bresch, H., Urbanek, M. and Nusser, M. (2000) Ochratoxin A in food containing liquorice. *Nahrung*, **44**, 276–8.

Busta, F.F. and Speck, M.L. (1968) Antimicrobial effect of cocoa on salmonellae. *Appl. Microbiol.*, **16**, 424–5.

Buchanan, R.L. and Fletcher, A.M. (1978) Methylxanthine inhibition of aflatoxin production. *J. Food Sci.*, **43**, 654–5.

Burnett, S.L., Gehm, E.R., Weissinger, W.R. and Beuchat, L.R. (2000) Survival of *Salmonella* in peanut butter and peanut butter spread. *J. Appl. Microbiol.*, **89**, 472–7.

Carr, J.G., Davies, P.A. and Dougan. J. (1979) Cocoa fermentation in Ghana and Malaysia, in *Proc. 7th Int. Cocoa Res. Conf.*, pp. 573–6.

Carr, J.G., Davies, P.A. and Dougan, J. (1980) Cocoa fermentation in Ghana and Malaysia–further microbial methods and results. University of Bristol, Bristol.

Chan, E.S., Aramini, J., Ciebin, B., Middleton, D., Ahmed, R., Howes, M., Brophy, I., Mentis, I., Jamieson, F., Rodgers, F., Nazarowec-White, M., Pichette, S.C., Farrar, J., Gutierrez, M., Weis, W.J., Lior, L., Ellis, A. and Isaacs, S. (2002) Natural or raw almonds and an outbreak of a rare phage type of *Salmonella enteritidis* infection. *Can. Commun. Dis. Rep.*, **28**, 97–9.

Chatt, E.M. (1953) *Cocoa Cultivation, Processing Analysis*, Wiley Interscience, New York.

Chick, W.H., Mainstone, B.J. and Wai, S.T. (1981) Mitigation of cocoa acidity in Peninsular Malaysia, in *Proc. 8th Int. Cocoa Res. Conf.*, Cartagena, Columbia, October 1981, pp. 759–64.

Codex Alimentarius. (1981) Codex standards for cocoa products and chocolate. Standards 86-1981; 87-1981 and 105-1981.

Collins-Thompson, D.L., Weiss, K.F., Riedel, G.W. and Charbonneau, S. (1978) Sampling plans and guidelines for domestic and imported cocoa from a Canadian national microbiological survey. *Can. Inst. Food Sci. Technol. J.*, **11**, 177–9.

Collins-Thompson, D.L., Weiss, K.F., Riedel, G.W. and Cushing, C.B. (1981) Survey of and microbiological guidelines for chocolate and chocolate products in Canada. *J. Inst. Can. Sci. Technol. Aliment.*, **14**, 203–7.

Cordier, J.L. (1994) HACCP in the chocolate industry. *Food Control*, **5**, 171–5.

Craven, P.C., Mackel, D.C., Baine, W.B., Barker, W.H. and Gangarosa, E.J. (1975) International outbreak of *Salmonella eastbourne* infection traced to contaminated chocolate. *Lancet*, **i**, 788–93.

Cros E. (1995) *Cocoa aroma formation*, in *Cocoa Meetings, Seminar Proceedings*" CIRAD CP Montpellier, pp. 169–79.

D'Aoust, J.Y. (1977) *Salmonella* and the chocolate industry. A review. *J. Food Prot.*, **40**, 718–27.

D'Aoust, J.Y. and Pivnick, H. (1976) Small infection doses of *Salmonella*. *Lancet*, **i**, 866.

D'Aoust, J.Y., Aris, B.J., Thisdele, P., Durante, A., Brisson, N., Dragon, D., Lachapelle, G., Johnston, M. and Laidely, P. (1975) *Salmonella eastbourne* outbreak associated with chocolate. *Can. Inst. Food Sci.Technol. J.*, **6**, 41–4.

De Grandi, D.M., Mistretta, A. and Lelo, S. (1987) Nougat contamination with salmonellae. *Pasticceria Int.*, **58**, 139.

Dockstäder, W.B. and Groomes, R.J. (1971) Detection and survival of salmonellae in milk chocolate. *Bacteriol. Proc.*, **A36**, 7.

Dragoni, J., Balzaretti, C. and Ravaretto, R. (1989) Stagionalita della microflora in ambienti di produzione dolciaria. *Ind. Aliment.*, **28**, 481–6.

Finoli, C., Galli, A., Vecchio, A. and Locatelli, D.P. (1994) Aspetti igienici di prodotti dolciari a base di frutta. *Ind. Aliment.*, **33**, 1201–6.

Gabis, D.A., Langlois, B.E. and Rudnick, W.E. (1970) Microbiological examination of cocoa. *Appl. Microbiol.*, **20**, 644–5.

Gästrin, B., Kaempe, A. and Nystroem, K.G. (1972) *Salmonella durham* epidemi spridd genom kakaopulver. *Laekartidingen.*, **69**, 5335–8.

Gauthier, B., Guiraud, J., Vincent, J.C., Parvais, J.P. and Galzy, P. (1977) Comments on yeast flora from the traditional fermentation of cocoa in the Ivory Coast. *Rev. Ferment. Ind. Aliment.*, **32**, 160–3.

Gibson, B. (1973) The effect of high sugar concentrations on the heat resistance of vegetative micro-organisms. *J. Appl. Bacteriol.*, **36**, 365–76.
Gill, O.N., Sockett, P.N., Bartlett, C.L., Vaile, M.S., Rowe, B., Gilbert, R.J., Dulake, C., Murrell, H.C. and Salmaso, S. (1983) Outbreak of *Salmonella napoli* infection caused by contaminated chocolate bars. *Lancet*, **i**, 574–7.
Goepfert, J.M. and Biggie, R.A. (1968) Heat resistance of *Salmonella typhimurium* and *Salmonella senftenberg* 775W in milk chocolate. *Appl. Microbiol.*, **16**, 1939–40.
Goepfert. J.M., Iskander, J.K. and Amundson, C.H. (1970) Relation of the heat resistance of salmonellae to the water activity of the environment. *Appl. Microbiol.*, **19**, 429–33.
Greenwood,. M.H. and Hooper, W.L. (1983) Chocolate bars contaminated with *Salmonella napoli*: an infectivity study. *Br. Med. J.*, **26**, 139–44.
Grünewald, T. and Münzner, R. (1972) Strahlenbehandlung von Kakaopulver. *Lebensm. Wiss. Technol.*, **5**, 203–6.
Hansen, A.P. (1975) Understanding the microbiological deterioration of cacao. *Candy Snack Ind.*, **140**(Sept), 46–7.
Hansen, A.P. and Keeney, P.G. (1969) Distribution of carbonyls of moldy cacao beans between cocoa butter and cocoa cake fractions. *Int. Chocolate Rev.*, **24**, 2–5.
Hansen. A.P. and Keeney. P.G. (1970) Comparison of carbonyl compounds in moldy and non-moldy cacao beans. *J. Food Sci.*, **35**, 37–40.
Hansen, A.P. and Welty, R.E. (1970) Microflora of raw cocoa beans. *Mycopathol. Mycol. Appl.*, **44**, 309–16.
Hansen, A.P., Welty, R.E. and Shen, R. (1973) Free fatty acid content of cocoa beans infested with storage fungi. *J. Agric. Food Chem.*, **21**, 665–70.
Hansen, A.P., Welty, R.E. and Shen. R. (1973) Free fatty acid content of cacao beans infested with storage fungi. *J. Agric. Food Chem.*, **21**, 665–70.
Heemskerk, R.F.M. (1999) Cleaning, roasting and winnowing, in *Industrial Chocolate Manufacture and Use*, (ed. S.T. Beckett), 3rd edn, Chapter 5, pp. 78–100.
Hockin, J.C., D'Aoust, J.J., Bowering. D., Jessop, J.H., Khama, B., Liar, H. and Milling. M.E. (1989) An international outbreak of *Salmonella nima* from imported chocolate. *J. Food Prot.*, **52**, 51–4.
ICMSF (Intemational Commission on Microbiological Specifications for Foods). (1988) Application of the hazard analysis critical control point (HACCP) system to ensure microbiological safety and quality, *Microorganisms in Foods, Volume 4*, Blackwell Scientific Publications, Oxford.
IOCCC (International Office of Cocoa, Chocolate and Confectionery). (1991) The IOCCC Code of Hygienic Practice based on HACCP for the Prevention of *Salmonella* Contamination in Cocoa, Chocolate and Confectionery Products, IOCCC, Brussels.
IOCCC (International Office of Cocoa. Chocolate and Confectionery). (1993) The IOCCC Code of Good Manufacturing Practice. Specific GMP for the Cocoa, Chocolate and Confectionery Industry, IOCCC, Brussels.
Jermini, M.F., Geiges, O. and Schmidt-Lorenz, W. (1987) Detection. isolation and identification of osmotolerant yeasts from high sugar products. *J. Food Prot.*, **50**, 468–72.
Jinap, S. (1994) Organic acids in cocoa beans–a review. *ASEAN Food J.*,**9**, 3–12.
Kapperud, G., Lassen, J., Aasen. S., Gustavsen, S. and Hellesnes, I. (1989a) Sjokoladeepidemien i 1987. *Tidsshr. Nor Loegeforen*, **109**, 1982–85.
Kapperud, G., Lassen, J., Demmarsnes, K., Kristiansen, B.E., Cougant. D.A., Ask, E. and Jakkola, M. (1989b) Comparison of epidemiological marker methods for identification of *Salmonella typhimurium* isolates from an outbreak caused by contaminated chocolate. *J. Clin. Microbiol.*, **27**, 2019–21.
Kapperud, G., Gustavsen, S., Hellesnes, I., Hansen, A.H., Lassen, J., Him. J., Jakkola, M., Montenegro, M.A. and Helmuth, R. (1990) Outbreak of *Salmonella typhimurium* infection traced to contaminated chocolate and caused by a strain lacking the 60-megadalton virulence plasmid. *J. Clin. Microbiol.*, **28**, 2597–604.
Kavanagh, T.E., Reineccius, G.A., Keeney, P.G. and Weissberger. W. (1970) Mold induced changes in cacao lipids. *J. Am. Oil Chem. Soc.*, **47**, 344–64.
Killalea, D., Ward, L.R., Roberts, D., de Louvois, J., Sufi, F., Stuart, J.M., Wall, P.G., Susman, M., Schweiger, M., Sanderson, P.J., Fisher, I.S.T., Mead, P.S., Gill, O.N., Bartlett, C.L.R. and Rowe, B. (1996) International epidemiological and microbiological study of an outbreak of *Salmonella agona* infection from a ready to eat savoury snack. I. England and Wales and the United States. *Br. Med. J.*, **313**, 1105–7.
Kinderlerer, J.L. (1997) *Chrysosporium* species, potential spoilage organisms of chocolate. *J. Appl. Microbiol.*, **83**, 771–8.
Kleinert, J. (1988) Cocoa mass, cocoa powder, cocoa butter, in *Industrial Chocolate Manufacture and Use* (ed. S. Beckett), Van Nostrand Reinhold Co., New York, pp 58–88.
Kleinert-Zollinger, J. (1988) Hygiene und Qualität in Süsswaren-Betrieben. *Zucker Süsswaren Wirtschaft*, 136–42.
Legan, J.D. and Voysey, P.A. (1991) Yeast spoilage of bakery products and ingredients. *J. Appl. Bacteriol.*, **70**, 361–71.
Lehrian, D.W. and Patterson, G.R. (1983) Cocoa fermentation, in *Biotechnology, a Comprehensive Treatise, Volume 5*, Verlag Chemie, Basel, Switzerland, pp. 529–75.
Lenovich, L.M. (1979) Production of aflatoxin in cocoa beans. *J. Assoc. Anal. Chem.*, **62**, 1076–7.
Lewis, D.A., Paramathasan, R., White, D.G., Neil, L.S., Tanner, A.C., Hill, S.D., Bruce, J.C., Stuart, J.M., Ridley, A.M. and Threlfall, E.J. (1996) Marshmallows cause an outbreak of infection with *Salmonella enteritidis* phage type 4. *Commun. Dis. Rep. CDR Rev.*, **6**, R183–6.
Lindley, P. (1972) Chocolate and sugar confectionery, jams and jellies, in *Quality Control in the Food Industry* (ed. S.M. Herschdoerfer), Volume 3, Academic Press, New York, pp. 259–95.

Maravalhas, N. (1966) Mycological deterioration of cocoa beans during fermentation and storage in Bahia. *Rev. Int. Choc.*, **21**, 375–6.

Mattick, K.L., Joergensen, F., Legan, J.D., Lappin-Scott, H.M. and Humphrey, T.J. (2000) Habituation of *Salmonella* spp. at reduced water activity and its effect on heat tolerance. *Appl. Env. Microbiol.*, **66**, 4921–5.

Meursing, E.H. and Slot. H. (1968) The microbiological condition of cocoa powder, in *The Microbiology of Dried Foods* (eds. E.R. Kampelmacher, M. Ingram and D.A.A. Mossel), Bilthoven, The Netherlands, pp.433–45.

Meursing, E.H. and Zijderveld, J.A. (1999) Cocoa mass, cocoa butter and cocoa powder, in *Industrial Chocolate Manufacture and Use* (ed. by S.T. Beckett), 3rd edn, Chapter 6, pp. 101–14.

Miller, N., Pretorius, H.F. and Van Der, R.W.B. (1986) The effect of storage conditions on mould growth and oil quality of confectionery and high-oil sunflower seeds. *Lebensm. Technol.*, **19**, 101–3.

Minifie, B.W. (1989) *Chocolate, Cocoa and Confectionery–Science and Technology*, AVI Publishing Co., Connecticut.

Minson, E. (1992) Chocolate manufacture–beans through liquor production. *Manuf Conf.*, **72**, 61–7.

Mossel, D.A.A. and Sand, F.E.M.J. (1968) Occurrence and prevention of microbial deterioration of confectionery products. *Conserva*, **17**, 23–33.

Mossel, D.A.A., Meursing, E.H. and Slot, H. (1974) An investigation on the numbers and types of aerobic spores in cocoa powder and whole milk. *Neth. Milk Dairy J.*, **28**, 149–54.

Ng, S., Rouch, G., Dedman, R., Harries, B., Boyden, A., McLennan, L., Beaton, S., Tan, A., Heaton, S., Lightfoot, D., Vulcanis, M., Hogg, G., Scheil, W., Cameron, S., Kirk, M., Feldheim, J., Holland, R., Murray, C., Rose, N. and Eckert, P. (1996) Human salmonellosis and peanut butter. *Commun. Dis. Intell.*, **20**, July.

Niles, E.V. (1981) Microflora of imported cocoa beans. *J. Stored Prod. Res.*, **17**, 147–50.

Nuttall, C. (1999) Chocolate marketing and other aspects of the confectionery industry world-wide, in *Industrial Chocolate Manufacture and Use* (ed. S.T. Beckett), 3rd ed, Chapter 24, pp. 439–59.

Ogunmoyela, G.A. and Birch. G.G. (1984) Effect of sweetener type and lecithin on hygroscopicity and mould growth in dark chocolate. *J. Food Sci.*, **49**, 1088–9, 1142.

Ostovar, K. (1973) A study on survival of *Staphylococcus aureus* in dark and milk chocolate. *J. Food Sci.*, **38**, 663–4.

Ostovar, K. and Keeney, P.G. (1972) Implication of *Bacillus subtilis* in the synthesis of tetramethylpyrazine during fermentation of cocoa beans. *J. Food Sci.*, **37**, 96–7.

Ostovar, K. and Keeney, P.G. (1973) Isolation and characterisation of microorganisms involved in the fermentation of Trinidad's cacao beans. *J. Food Sci.*, **38**, 611–7.

Park, C.E., Stankiewicz, Z.K., Rayman, M.K. and Hauschild, A. (1979) Inhibitory effect of cocoa powder on the growth of a variety of bacteria in different media. *Can. J. Microbiol.*, **25**, 233–5.

Passos. F.M., Lopez, A.S. and Silva, D.O. (1984a) Aeration and its influence on the microbial sequence in cacao fermentations in Bahia with emphasis on lactic acid bacteria. *J. Food Sci.*, **49**, 1470–4.

Passos, F.M., Silva, D.O., Lopez, A., Ferreira, C.L.L.F. and Guimaraes, W.V.G. (1984b) Characterization and distribution of lactic acid bacteria from traditional cocoa bean fermentation in Bahia. *J. Food Sci.*, **49**, 205–8.

Pitt. J.L. and Hocking, A.D. (1985) *Fungi and Food Spoilage*, Academic Press, Sydney.

Richardson, T. (1987) ERH of confectionery food products. *Manuf Conf.*, 65–70.

Rieschel, H. and Schenkel, J. (1971) Das Verhalten von Mikroorganismen, speziell Salmonellen, in Schokoladenwaren. *Alimenta*, **10**, 57–66.

Roberts, J.A., Sockett, P.N. and Gill, O.N. (1989) Economic impact of a nationwide outbreak of salmonellosis: cost–benefit of early intervention. *Br. Med. J.*, **298**, 1227–30.

Roelofsen, P.A. (1958) Fermentation, drying and storage of cacao beans. *Adv. Food Res.*, **8**, 225–96.

Rombouts, JE. (1952) Observation on the microflora of fermenting cacao beans. *Trinidad. Proc. Soc. Appl. Bacteriol.*, **15**, 103–10.

Sanchez. J., Guiraud, J.P. and Galzy. P. (1984) A study of the polygalacturonase activity of several yeast strains isolated from cocoa. *Appl. Microbiol. Biotechnol.*, **20**, 262–7.

Sanchez, J., Daguenet, G., Vincent, J.C. and Galzy, P. (1985) A study of the yeast flora and the effect of pure culture seeding during the fermentation of cocoa beans. *Lebensm. Wiss. Technol.*, **18**, 69–76.

Scheil, W., Carmeron, S., Dalton, C., Murray, C. and Wilson, D. (1998) A South Australian *Salmonella mbandaka* outbreak investigation using a database to select controls. *Aust. NZ J. Public Health*, **22**, 536–9.

Schwan, R.F. (1998) Cocoa fermentation conducted with a defined microbial cocktail inoculum. *Appl. Env. Microbiol.*, **64**, 1477–83.

Schwan, R.F., Rose, A.H. and Board, R.G. (1995) Microbial fermentation of cocoa beans, with emphasis on enzymatic degradation of the pulp. *J. Appl. Bacteriol. Symp. Suppl.*, **79**, 965–1073.

Schwan, R.F., Cooper, R.M. and Wheals, A.E. (1997) Endopolygalacturonidase secretion by *Kluyveromyces marxianus* and other cocoa pulp degrading yeasts. *Enzyme Microb. Technol.*, **21**, 234–44.

Shaughnessy. W.J. (1992) Cocoa beans–planting through fermentation its effect on flavor. *Manuf Conf.*, **72**, 51–8.

Shohat, T., Green, M.S., Merom, D., Gill, O.N., Reisfeld, A., Matas, A., Blau, D., Gal, N. and Slater, P.E. (1996) International epidemiological and microbiological study of an outbreak of *Salmonella agona* infection from a ready to eat savoury snack. II. Israel. *Br. Med. J.*, **313**, 1107–9.

Simonsen, B., Bryan, F.L., Christian. J.H.B., Roberts, T.A., Tompkin, B.R. and Silliker, J.H. (1987) Prevention and control of food-borne salmonellosis through application of HACCP. *Int. J. Food Microbiol.*, **4**, 227–47.

Slater, CA. (1986) Chocolate and sugar confectionery, jams and jellies, in *Quality Control in the Food Industry* (ed. S.M. Herschdoerfer), Volume 3, 2nd ed, Academic Press, London, pp. 139–81.

Tamminga, S.K. (1979) The longevity of *Salmonella* in chocolate. *Antonie van Leeuwenhoek*, **45**, 153–7.

Tamminga, S.K., Beumer, R.R., Kampelmacher, E.R. and van Leusden, F.M. (1976) Survival of *Salmonella eastbourne* and *Salmonella typhimurium* in chocolate. *J. Hyg.*, **76**, 41–7.

Todd, E.D. (1985) Economic loss from foodborne disease and non-illness related recalls because of mishandling by food processors. *J. Food Prot.*, **48**, 621–33.

Torres-Vitela, M.R., Escartin, E.F. and Castillo, A. (1995) Risk of salmonellosis associated with consumption of chocolate in Mexico. *J. Food Prot.*, **58**, 478–81.

Vendrell, M.C., Gallego, A.R., Acosta, F. and Rodriguez, L.A. (2000) Microbiological analysis of confectionery foodstuff: hard candies. *Alimentaria*, **April**, 121–4.

Walker, H.W. and Ayres, J.C. (1970) Yeasts as spoilage organisms, in *The Yeasts* (eds. A.H. Rose and J.S. Harrison), *Volume 3*, Academic Press, New York, pp. 463–527.

Warburton, D.W., Harwig, J. and Bowen, B. (1993) The survival of salmonellae in homemade chocolate and egg liqueur. *Food Microbiol.*, **10**, 4105–10.

Whitfield, F.B., Tindale, C.R., Shaw, K. and Stanley, G. (1984) Contamination of cocoa powder by chlorophenols and chloroanisoles adsorbed from packaging materials. *Chem. Ind.*

Windisch, S. (1977) Nachweis und Wirkung von Hefen in zuckerhaltigen. Lebensmittel. *Alimenta*, 23–9.

Witlin, B. and Smyth. R. D. (1957) "Soapiness" in "white" chocolate candies. *Am. J. Pharm. Sci. Support Public Health*, **129**, 135–42.

Zak, D.L. (1988) The development of chocolate flavor. *Manuf. Conf.*, **68**, 69–74.

Zak, D.L., Ostovar, K. and Keeney. P.G. (1972) Implication of *Bacillus subtilis* in the synthesis of tetramethylpyrazine during fermentation of cocoa beans. *J. Food Sci.*, **37**, 967–8.

Zapatka, F.A., Varney, G.W. and Sioskey, A.J. (1977) Neutralization of the bactericidal effect of cocoa powder on *Salmonella* by casein. *J. Appl Bacteriol.*, **42**, 21–5.

Zeller. M. (1963) Hefeninfizierte Kondensmilch als Ursache von Fehlfabrikaten bei Schokolade. *Arch. Lebensmittelhyg.*, **14**, 6–10.

11 Oil- and fat-based foods

I General introduction

Foods based on oils and fats represent a large proportion of the energy intake in the diet of consumers in most of the world. Nutritional advice is to limit the amount of fat in the overall diet, in particular of saturated fat. As a result, the past decades have shown a reduction in the per capita consumption of oil- and fat-based foods in developed countries and a relative shift to low-fat/low-calorie products. Fats and oils can be attacked by various fat-splitting microorganisms if the conditions for growth are favorable, e.g. temperature, moisture, availability of low-molecular weight nutrients. Enzymes produced by contaminating lipolytic flora can hydrolyze the fat to yield free fatty acids and trigger fatty acid oxidation. At the same time, fats and oils can protect microorganisms so that they may survive for quite some time (Troller and Christian, 1978; Hersom and Hulland, 1980; Gaze, 1985). This would present a hazard in particular if the organisms were infectious pathogens.

Most oil- and fat-based foods contain a certain amount of moisture and non-fat nutrients. Their physical structure is a very important parameter. The products may exist either as a fat-continuous system (i.e. yellow-fat-spreads such as butter and margarine or other dairy and non-dairy spreads and reduced fat spreads) or as a water-continuous system (i.e. mayonnaise, salad dressings, and other water-continuous spreads). This has a strong impact on the microbiological stability of the food. In water-in-oil products, such as margarine, the water is present as well-dispersed fine droplets throughout the fat phase. The inability of microorganisms to move between droplets is a major intrinsic preservation factor. Fat can act as a barrier to microbial growth and for this reason fat-continuous systems are usually much more stable than water-continuous systems. A small category of oil- and fat-based products is characterized by extremely low water contents (e.g. butter oil, ghee, vanaspati, cocoa butter substitutes, and cooking oils) that, generally, limits microbial growth but not necessarily excludes growth under extreme conditions.

Presently, water-in-oil emulsions exist in the market with fat levels ranging from 20% to 80%, whereas products with fat levels as low as 3% have been successfully introduced recently in the USA, UK, and The Netherlands (van Zijl and Klapwijk, 2000). Oil-in-water emulsions do occur also in a considerable range of fat contents. The preservation properties of the wide range of different oil- and fat-based products that are currently safely marketed differ substantially. Product innovations such as inclusion of spices and fresh herbs in water-in-oil products may affect the product structure and, possibly, the microbiological load.

The composition of butter is subject to stringent regulations and for this reason has not changed much over the years, albeit that new types of butter-making processes have been introduced. Reduced fat variants of butter have appeared in the market, in most cases containing 40% butterfat. Products based on mixtures of butter–butter fat and vegetable fat blends (mélanges) have been developed that can be full-fat, medium-fat, or low-fat (Madsen, 1990). Typical butter or margarine manufacturing processes may produce them.

There are no indications that industrially produced oil- and fat-based foods play a significant role in food-borne disease (Delamarre and Batt, 1999; Michels and Koning, 2000; Smittle, 2000; van Zijl and Klapwijk, 2000). However, most of the products are vulnerable to spoilage microorganisms (i.e. acid-tolerant types). The increased attention to hygiene during manufacturing (Mostert and Lelieveld, 2000), the quality of raw materials, and the pasteurization conditions applied have all contributed significantly to product and process designs with a very good safety record. The implementation of HACCP for assuring proper production is essential. Continued attention for safety and quality remains necessary, however, especially with respect to product innovations.

An incident with *Listeria* contaminated butter occurred in Finland due to poor factory hygiene (Lyytikäinen *et al.*, 1999, 2000). In an outbreak due to *Escherichia coli* O157:H7 in the United States in 1993, for which epidemiological evidence implicated bulk mayonnaise as a vehicle of transmission (Anonymous, 1993a), the organism was found to be more tolerant of acidic conditions than either *Salmonella* spp. or *Listeria monocytogenes*, which were the organisms previously considered as the main hazards. Several studies subsequently showed that pathogens such as *Listeria* spp., *Salmonella* spp., and *E. coli* O157:H7 may survive certain conditions under which their acid tolerance is strengthened by exposure to, among other factors, non-lethal acid levels (Leyer and Johnson, 1993; Leyer *et al.*, 1995; Grahan *et al.*, 1996; Duffy *et al.*, 2000; Smith, 2003). Such stress-hardening conditions should be avoided but, in order to do this, knowledge of how ecological conditions in a food product or the production environment impact on the growth–survival of pathogens is needed. The occurrence of stress-hardened pathogens, which may survive in acidic foods and passage through the stomach (pH 1.0–3.0), or which may be less well inactivated by pasteurization, emphasizes the need to assess potential survival of these organisms using acid-adapted cultures when possible.

Keeping up-to-date with information on growth capabilities of pathogenic microorganisms is an essential exercise for food professionals. Changes of manufacturing or marketing practices and launch of product innovations need to be accompanied by a safe product and process design as well as proper practical implementation and control thereof, as is further detailed out in this chapter.

II Mayonnaise and dressings

A *Definitions*

Mayonnaise. Mayonnaise can be a very well-defined product with specific levels of oil (min. 52%), egg yolk (min. 6%), salt (min. 1%), total acid (min. 0.75%), and pH (max. 4.5) fixed in local regulations (Michels and Koning, 2000). The types of oil used are mainly soybean, rapeseed, and sunflower oil and sometimes cottonseed and olive oil. Low-oil mayonnaise products have come on the market since 1980s. This has led to specific labeling of the fat content and a clear indication on the label that the product is a low-fat or reduced-calorie mayonnaise (or similar wording). Although there is legislation to specify the composition of real mayonnaise in many countries, this is less so for low-oil mayonnaise, dressings, and emulsified sauces.

The Codex Alimentarius Regional European Standard for Mayonnaise defines mayonnaise as a condiment sauce obtained by emulsifying edible vegetable oil(s) in an aqueous phase consisting of vinegar, the oil-in-water emulsion being stabilized by hen's egg yolk (FAO/WHO, 1989). This Codex Standard recognizes the following optional ingredients: egg white, egg products, sugar, salt, condiments, herbs, spices, fruits and vegetables including fruit juice and vegetable juice, mustard and dairy products. As acidifying agents, the use of acetic, citric, lactic, malic, and tartaric acids and the salts thereof are allowed. Benzoic acid and sorbic acid and the salts thereof are allowed as preservatives. Other additives may include stabilizers, antioxidants, colors, and flavors (together with a flavor enhancer such as monosodium glutamate). The Standard further stipulates that the total fat content of mayonnaise (from oil and egg yolk) should be 78.5% (w/w) and the technically pure egg yolk content should not be less than 6%. The Association of the Mayonnaise and Condiment Sauce Industry of the EEC adopted a total fat content of minimum 70% (w/w) and a minimal egg yolk content of 5% (w/w) (CIMSCEE, 1991).

According to the US standards of identity for mayonnaise, the vegetable oil content must be at least 65%, the pH may range from about 3.6 to 4.0 with acetic as the predominant acid representing

0.29–0.5% of the total product (US-DHEW, 1975a). The aqueous phase should contain 9–12% salt and 7–10% of sugar.

Low-fat or reduced-calorie mayonnaises, salad dressings, salad creams, and other emulsified products made with oil, emulsifier, and vinegar, typically have a lower fat content than mayonnaise. Salad dressings will have a more fluid consistency to make the product "pourable", while most mayonnaises are "spoonable". They normally contain egg yolk although some are made with dairy-based emulsifiers.

Dressings. Salad dressings are defined in the United States by the FDA as emulsified semi-solid foods prepared from vegetable oils, vinegar, lemon juice, and/or lime juice, egg yolk-containing ingredients, and a cooked or partially cooked starchy paste (US-DHEW, 1975b). The finished product contains no less than 30% of edible vegetable oil and has the equivalent of 4% liquid egg yolk (US-FDA, 1993). The pH is 3.2–3.9 and acetic acid makes up 0.9–1.2% of the total product. The aqueous phase contains 3–4% of salt and 20–30% of sugar (Smittle, 1977).

There are few compositional specifications for dressings or other emulsified sauces (meat and fish sauces). They generally contain less oil than mayonnaise and salad dressings, but all are oil-in-water emulsions in which the emulsifying agent usually is egg yolk. Because of the presence of vinegar or other weak acids in most mayonnaise, dressings, and emulsified sauces, their pH is low, which has a significant influence on the microbiological stability. Low-calorie or low-sodium formulations are more susceptible to spoilage than the traditional products.

B *Important properties*

In Europe, the pH of mayonnaise is typically between 3.0 and 4.2, with 4.5 as the highest value permitted (this is the legal maximum in Denmark). The percentage salt or sugar is not fixed by regulations, but is mostly between 1% and 12% of the aqueous phase. The level of acetic acid in the aqueous phase is typically between 0.8% and 3.0%. Dressings typically have a smaller fat phase than mayonnaise and a starch phase, which helps to give the required consistency. Due to the larger aqueous phase, in which acid and salt are diluted, they are more vulnerable than mayonnaise to microbial spoilage. Typically, the acetic acid content of dressings is 0.5–1.5% of the aqueous phase and the pH 1.0–4.2. Levels of salt (1–4%) and sugar (1–30%) in the aqueous phase contribute little to microbiological stability.

No clear distinction exists worldwide between mayonnaise, low-fat mayonnaise, salad dressings and other emulsified products made with oil, emulsifier and vinegar. They normally contain egg yolk, although some are made with dairy-based emulsifiers or contain no oil and are not emulsified and are two-phase systems like vinaigrettes. This large group of dressings can be "pourable" or "spoonable" (depending on starch content).

Mayonnaise, dressings, and other emulsified sauces can be categorized into inherently stable or unstable products on the basis of their aqueous phase composition and shelf-life.

Stable products. These products are stable at ambient temperatures, not sensitive to spoilage, whether open or closed, because of an aqueous phase composition that inhibits growth of all relevant spoilage organisms, in particular acetic acid-tolerant lactobacilli, yeasts, and molds. Their closed shelf-life will be 6 months to 1 year, limited for organoleptic reasons only. After opening, the sauces are not sensitive to spoilage and they can be kept at ambient temperatures or chilled for as long as their quality remains acceptable.

Unstable products. These products allow (slow to rapid) growth of lactobacilli and/or yeasts. Ingredients used should be selected to minimize the initial contamination; good process design and hygienic practices can eliminate or prevent (re-)contamination during manufacturing. Depending on the specific properties of the preservation system, products may have a relatively long ambient shelf-life in the closed jar (e.g. 6 months to 1 year). However, recontamination during consumer-use restricts the shelf-life of the product after opening to a few weeks (ambient) or months (refrigerated). Where products have a minimal built-in preservation, closed shelf-life may be restricted to a few weeks.

C Methods of processing and preservation

The production of low-acid mayonnaise, salad dressings, and thousand island dressings is similar to that of mayonnaise. However, some of the former products will have a (cooked) starch phase to give the required consistency to the product while others are simply made by incorporating a starch phase into the mayonnaise phase, resulting in a low-oil product. Manufacturing can be by using either batch or continuous processes (Lopez, 1987).

In a batch process (Figure 11.1), liquid (salted) egg yolk is mixed with an acid water phase consisting of vinegar, spices, flavor, salt and/or sugar, and (optional) a (cooked) starch phase. Oil is added under intensive mixing and the resulting coarse emulsion is usually passed through a colloid mill to achieve the small oil droplets (predominantly 5–10 μm) required for a good consistency. Consistency of "spoonable" mayonnaise is expressed in Steven's value, measured by recording the resistance in gram of a measuring gauge pressed into the product at a fixed speed. Steven's values typically are between 50 and 200 g. For liquid products, the Bostwick value is used, which is the run length in cm of the product when released in a measuring tray and measured within 30 s. Typical values are 5–10 cm. The inclusion of (cooked) starch is optional and depends largely on the oil content of the mayonnaise. The freshly made mayonnaise is pumped to a holding vessel, and packaged in glass jars, tubs, buckets, or other containers for consumer/professional use. In a batch manufacturing process of a dressing containing particulates, the particulates can be mixed after the colloid mill with the dressing base. An alternative route to

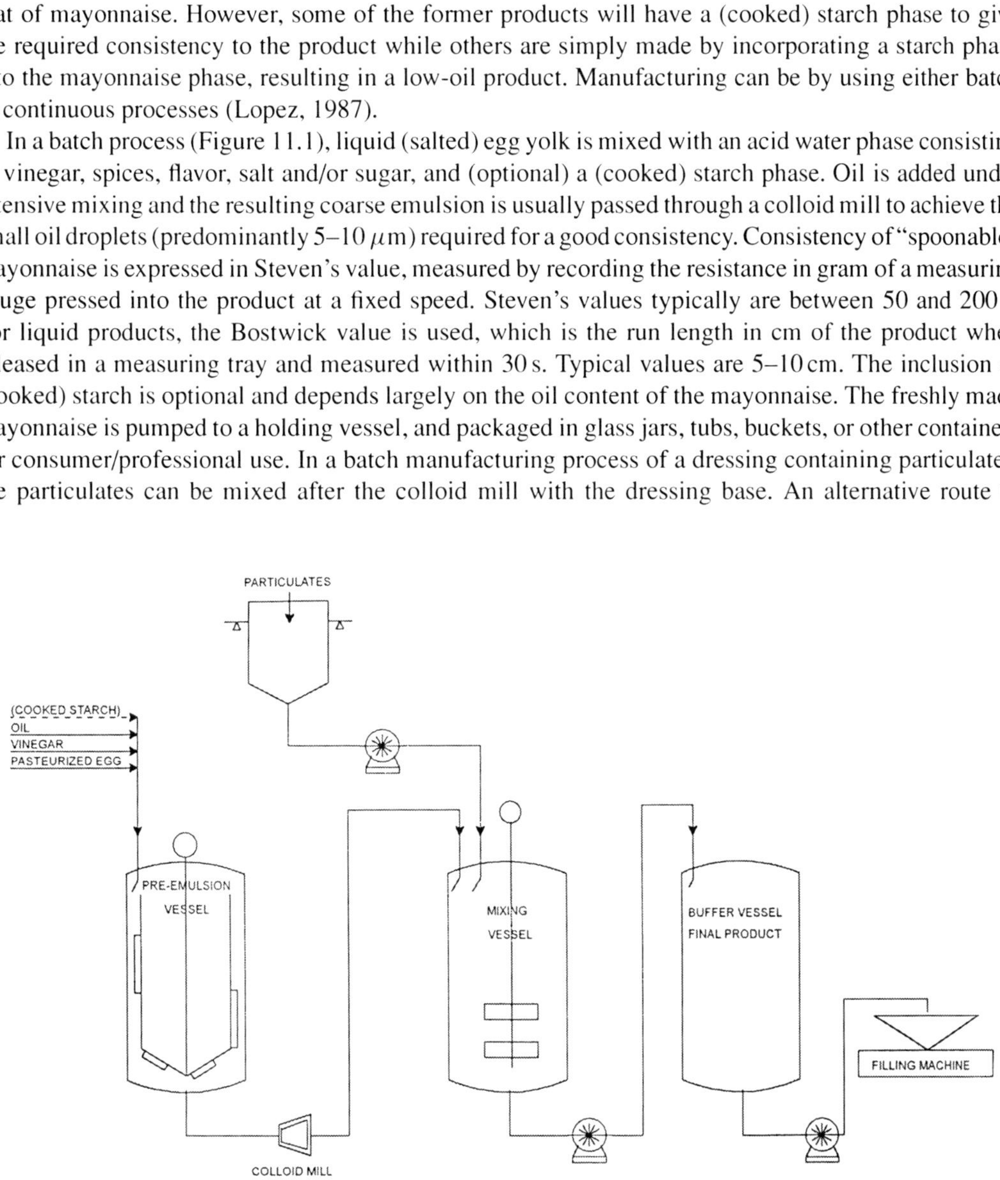

Figure 11.1 Layout of a batch dressings production line.

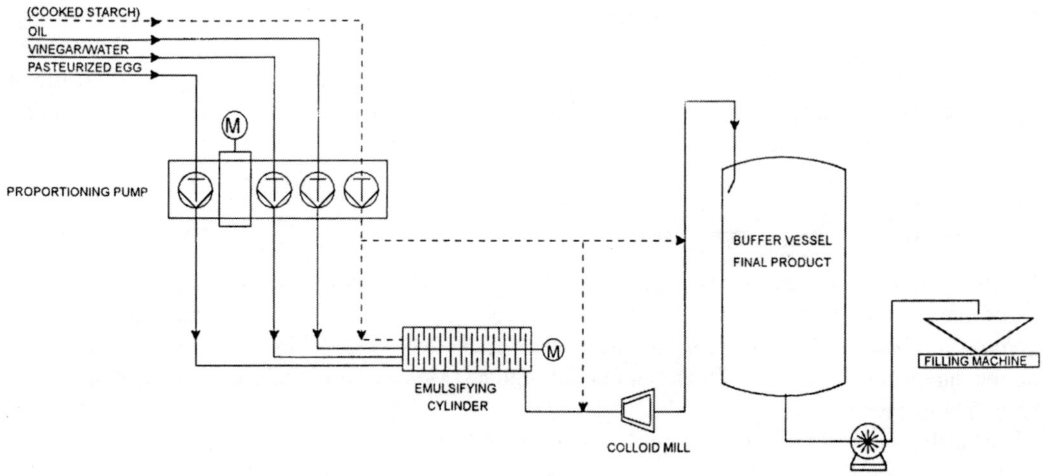

Figure 11.2 Layout of a continuous mayonnaise production line.

preventing product damage is to omit the colloid mill and to introduce the particulates in the first mixer after preparation of the dressing base.

In a continuous process (Figure 11.2), a set of proportioning pumps combines the right amounts of liquid egg, oil, water phase (with vinegar) and cooked starch phase (optional), which are emulsified in an emulsifying cylinder and finally passed through a colloid mill. To ensure the safety of the starch phase, the pH is usually adjusted with vinegar to below pH 4.5 to control any pathogen hazard. When particulates are present, the starch phase can be added (through an in line mixer) directly to the final product buffer vessel.

Whatever process is used, the processing lines should be free of relevant microbial contaminants at the start of the process and the use of hygienically well-designed process and packaging equipment is key to the production of such vulnerable products (EHEDG, 2003; 3-A, 2003). Equipment traditionally used for the manufacture of mayonnaise and dressings has not been easy to clean. However, increasingly more types of hygienic mixers, colloid mills, valves, pumps, and filling machines become available that can be cleaned-in-place (CIP). Their use is recommended as this allows the best control over cleaning and disinfection. Proper manual cleaning is usually still required for mills, fillers, etc., that are difficult to clean by CIP alone (e.g. irregular surfaces). For chilled products hygienic processing, clean (decontaminated) ingredients and proper refrigerated storage are necessary to obtain the desired shelf-life.

In the manufacturing of a microbiologically stable mayonnaise, there is normally no technological need to apply a heat process unless elimination of enzymes is required from ingredients like vinegar or spices. These enzymes could break down the starch, if present. The aqueous phase containing ingredients such as herbs or mustard, which could contain pathogenic or spoilage microorganisms, may be pasteurized before mixing. Also liquid egg yolk or other egg products should be pasteurized (e.g. by the supplier) before use in the manufacturing process.

The overall effect of the manufacturing process on the initial microbial load of the ingredients is negligible when a cold process design is used. The various emulsified products have a continuous water phase and the microbial load is not affected by the presence of oil droplets. The safety and stability characteristics of mayonnaise and dressings are dictated primarily by the low pH (range in use is pH 3.0–4.5) and the preservative effect of acetic acid (added as vinegar) or lactic acid. In mayonnaise, salt and to a much lesser extent sugar can reduce the wateractivity and this helps to inhibit spoilage organisms. Due to the limited water content, the salt-in-brine in the aqueous phase can be as high as 12%, resulting in a water activity of about 0.92. A preservative such as sorbic acid can be used where legally permitted, but a major proportion of the acid will dissolve in the oil phase leaving only 40–60% in the active undissociated form in the aqueous phase to protect against yeast (and lactobacilli). When benzoic acid is used, an even larger proportion of the preservative dissolves in the fat phase, which makes it even less effective.

The paper by Michels and Koning (2000) provides details on product and process design of different product types, including procedures and calculations to evaluate the effect of organic acids–preservatives and overall product stability. Some investigations on the inactivation of infectious pathogens in regular and reduced-calorie ("lite") mayonnaises or yellow-fat spreads, challenging commercially manufactured products, show a protective effect of high fat levels to the survival of pathogens whereas storage at higher temperatures increases inactivation (Hathcox *et al.*, 1995; Holliday *et al.*, 2003). Others have reported increased survival of pathogens at refrigeration temperature as compared to room temperature (Weagant *et al.*, 1994). Good knowledge of the exact chemical and physical composition of the products tested is essential in the interpretation of the results. Whereas challenge tests are commonly performed with rather high numbers of pathogens inoculated onto products, tests evaluating low-level inoculation or natural contamination of a pathogen may be the realistic scenarios. Leuschner and Boughtflower (2001) described a reproducible laboratory-scale procedure for preparation of mayonnaise containing low levels (10–1000 cfu/g in the final product) of *Salmonella* Enteritidis. Systems like these, which simulate a mayonnaise that is naturally contaminated at a low-level, can be used for the validation of the stability and safety of innovative formulations or to test products prepared using new preservation methods. Predictive modeling has been used to simulate pathogen inactivation rates as a function of product formulation and environmental conditions (Membré *et al.*, 1997), as have novel approaches such as neural networks (Xiong *et al.*, 2002).

D Microbial spoilage and pathogens

Mayonnaise and dressings products are often produced using a cold process design, and microbial growth is only controlled by the specific properties of their formulation. These properties (i.e. low pH, presence of acetic acid, etc.) restrict potential problems to certain acid-tolerant microorganisms.

Initial microflora. The microbial load of mayonnaise, dressings, and emulsified sauces comes from the various ingredients and from contamination during processing and packaging. Typical components that can carry spoilage microorganisms are mustard, pickles, dry vegetables and herbs, and blue cheese. Water, refined oil, vinegar, and pasteurized egg are normally free of relevant contamination when they are handled according to Good Manufacturing Practices (GMP) conditions based on the General Principles of Food Hygiene (CAC, 2001a).

- The refined oils used are normally free of microbial contamination as a result of the refining process, which involves steaming at a temperature well above 100°C. The oil has a very low-moisture content of <0.1% and will thus not allow microbial growth.
- When egg is used, it is mainly as pasteurized liquid egg yolk preserved with salt (8–11%) or with salt and potassium sorbate (e.g. 92% egg yolk, 7% salt, and 1% sorbate). With unpasteurized egg preparations, there is a risk of contamination with *S*. Enteritidis and its use is advised against. European Commission (EC) regulations require pasteurized eggs to be distributed and stored at a temperature of $\leq$4°C and *Salmonella* spp. absent in five samples of 25 g. Aerobic Plate Count (APC) should be $\leq 10^5$ cfu/g and coliforms $\leq$100 cfu/g. Commercial pasteurized liquid egg products often have counts

well below these values. The APC values commonly are a few hundred to a few thousand bacteria per g of product. The same holds for coliforms, which often are absent in 0.1 g.

- Vinegar can be obtained through fermentation of different raw materials (e.g. alcohol, wine, malt, cider). The acetic-acid content typically is about 8–11% and thus vinegar is usually not sensitive to spoilage. Artisanal vinegars with low acetic-acid levels occasionally carry spoilage organisms. Vinegar is often pasteurized to eliminate enzyme activity originating from the fermentation process. This also ensures the absence of the rare but extremely acetic acid-tolerant mold *Moniliella acetoabutans* and the acetic acid-resistant *Lactobacillus acetotolerans* (able to multiply in the presence of 9–11% acetic acid at pH 5.0; Entani *et al.*, 1986).
- With mustard, 1.8–2.5% acetic acid typically contained in its formulation prevents survival of infectious pathogens, although high levels of acetic acid-tolerant lactobacilli are known to occur. Yeasts are normally not found because of the antimycotic activity of allylisothiocyanate present in many mustard types.
- Herbs and spices are likely to be contaminated with spoilage organisms or pathogens such as *Salmonella* spp. and *Escherichia coli* O157:H7. Industrial experience (Michels and Koning, 2000) is that the incidence rate of *Salmonella* spp. in raw spices and herbs is about 1%.
- The use of dairy ingredients in emulsified products is quite common, but in most cases these are pasteurized by the producer. Fermented products, e.g. yoghurt and cheese, will have a high load of lactic acid bacteria, and yeasts also may be present. Blue cheeses require special attention due the presence of high levels of acetic acid-tolerant molds and lactobacilli. With soft cheeses in particular, there is a possibility for the presence of *Listeria monocytogenes*.
- Citric acid and acids other than acetic acid or vinegar may be used in emulsified products. Lactic acid typically consists of a 50% or 80% solution of DL-lactic acid with no significant microbial load. Citric acid commonly is used as acid or as concentrated citric juice (30° Brix) with a low pH of ± 3.1. Spoilage is prevented by pasteurization of the concentrate (or preservation with sulfite) to eliminate spoilage yeasts and molds. Other weak (e.g. malic) or strong (e.g. phosphoric) acids that may be used are free of relevant contamination.
- Starch is often used in emulsified products and it can be either "natural", requiring cooking to 85–90°C to set the starch, or "instant", in which case the starch does not require any cooking. The microbial load of natural and instant starches typically is very low, with only a few species of *Bacillus* and *Clostridium* present and no pathogens. However, certain processed starches can contain *Salmonella* spp.
- Common ingredients such as sugar, salt, and preservatives like sorbic and benzoic acid or the salts thereof normally have a very low microbial load. In sugar produced in small industrial operations, osmotolerant yeasts such as *Zygosaccharomyces rouxii* or *Z. bailii* may occur that can cause spoilage of acetic acid-containing emulsified products.
- All products should be made with potable water, which is free of pathogens and acetic acid-tolerant microorganisms.

Spoilage. Microbial spoilage is mainly caused by a small group of acid-tolerant yeasts and lactobacilli. Spoilage by molds is rare because most molds have a limited tolerance to acetic acid (Smittle and Flowers, 1982).

Yeasts. Only yeasts that are resistant to acetic acid are likely to present a spoilage problem to mayonnaise and dressings. Well-known species are *Z. bailii* and *Pichia membranaefaciens*, which can grow in the presence of ±3% acetic acid (Thomas and Davenport, 1985). The name of the latter yeast indicates that it grows as a film on the surface of a medium. In practice growth is only observed when sufficient oxygen is present (Smittle and Flowers, 1982). These two species are probably responsible for the majority of the spoilage incidents caused by yeasts. Other species occasionally observed are *Z. rouxii*,

Saccharomyces cerevisiae, and *Candida magnolia*. Yeasts may cause spoilage by gas formation or by growing as brownish colonies on the surface of mayonnaise, which may appear as small oil droplets. Sometimes growth is limited to $\pm\ 10^4$ cfu/g, probably because of oxygen depletion.

Lactobacilli. *Lb. fructivorans* has been reported as the main lactobacillus species causing spoilage of mayonnaise-type products (Smittle and Flowers, 1982). Others report that *Lb. plantarum* and *Lb. buchneri* are the organisms most frequently isolated from spoiled products in Europe, while *Lb. fructivorans* is less commonly isolated (Michels and Koning, 2000). Lactobacilli occasionally grow to very high numbers without causing evident spoilage. Development of heterofermentative lactobacilli will result in visible spoilage due to gas formation and will cause a decrease in pH.

Molds. The majority of molds cannot grow in the presence of $\geq$0.5% acetic acid. Because jars are steam capped and oxygen is consumed due to oil oxidation, only a little oxygen is generally available which also limits mould growth. Growth of a *Geotrichum* spp. on the surface of mayonnaise has been observed in jars with faulty seals. Although spoilage by molds is very rare, some acid-tolerant types have been reported. Tuynenburg Muys (1971) refers to *M. acetoabutans*, found in acetic acid preserves, which is unique in its ability to multiply in the presence of 8–9% acetic acid. Reference is also made to *Monascus ruber* and *Penicillium glaucumi*, able to grow in the presence of $\geq$1% acetic acid; *Pen. roqueforti*, present in blue cheese, also belongs to this group.

Pathogens. Mayonnaise and dressings with final pH values above 4.1 offer a potential risk of food poisoning by *Salmonella* spp. or other infectious pathogens (*E. coli* O157:H7; *L. monocytogenes*), either because strains of these pathogens can be particularly acid-tolerant or because the acidulant type and its final concentration may not be adequate to kill such pathogens. Even worse, the product properties may be insufficient to prevent multiplication of infectious pathogens or of *Staphylococcus aureus*. This is not a concern with properly formulated industrial products, for which die-off of pathogens of concern has been assured as part of the product design, but it can be for home-made products. Addition of commercial mayonnaise/dressing to salads is considered to retard growth of pathogens and to reduce concern (Smittle, 1977; Doyle *et al.*, 1982). Smittle (2000) reviewed the microbiological safety of mayonnaise, salad dressings and sauces produced in the United States focusing on the death and survival of food-borne pathogens in relation to the product formulation. Through detailed statistical analysis of literature data, this study provided support of the remarkable safety record of these products when commercially produced keeping to safe product formulations, implementation of GMP and HACCP systems. Proper food handling and storage practices were also found to be key in ensuring the safety of mayonnaise-based delicatessen foods in food-service operations. As reported by Bornemeier *et al.* (2003), temperature conditions and product characteristics of typical deli-salads sold in several grocery-stores were conducive of growth of pathogens.

Salmonella. Nearly all incidents involving *Salmonella* spp. were caused by mayonnaise made at home, in restaurants, or in institutional kitchens. The general use of pasteurized egg yolk, an adequate level of vinegar (typically giving more than 1% acetic acid in the water phase) and a pH below 4.5, has prevented food-borne illness from industrially produced mayonnaise during the past 40 years.

A major outbreak with 10000 cases in Denmark in 1955 led to a Danish regulation that the pH of mayonnaise should be below 4.5. Thereafter, there were two more Danish incidents with *Salmonella*-contaminated mayonnaise (*S.* Typhimurium biotype 17 in both cases) from large manufacturers. In the first case, the pH of the mayonnaise was 5.1 and the count of *Salmonella* spp. was 1.8×10^5 cfu/g at the time of analysis, 4 days after the meal (Petersen, 1964). The second outbreak had 41 cases and two fatalities. The mayonnaise was made with raw eggs and had a pH of 6.0; 2 days after the event the *Salmonella* spp. count of the mayonnaise was 6×10^6 cfu/g (Meyer and Oxhøj, 1964).

In 1976, a serious salmonellosis outbreak occurred among passengers on four outgoing and return flights from Las Palmas (Spain). Approximately 500 passengers were ill, with six fatalities. *S*. Typhimurium phage type 96 was isolated from mayonnaise (pH not mentioned) used on the incriminated flights and from a food handler involved in the preparation of the mayonnaise (Davies and Wahba, 1976).

Salmonella Enteritidis has been associated to several cases of food poisoning due to home-made or restaurant-made mayonnaise in the USA, the UK, Argentina, and many other countries (Anonymous, 1988; St Louis *et al*., 1988; Eiguer *et al*., 1990). This serovar of *Salmonella* was responsible for 78% of all outbreaks of known aetiology in Spain (Perales and Audicana, 1988). In a serious US outbreak, 404 of 965 persons at risk (mean age 64.2 year) in a New York City hospital became ill and 9 people died (median age 77.5 years); the source of the incident was hospital-prepared mayonnaise made with raw eggs; contamination with *S*. Enteritidis was traced back to a farm corporation (Telzak *et al*., 1990). The use of raw egg for mayonnaise contributed to an outbreak at a wedding reception in 1992, in which 81 guests and 11 catering staff became ill due to *S*. Enteritidis (Chandrakumer, 1995). In 1995, in Uruguay 600 cases of salmonellosis were traced to sandwiches that were made with small batches of mayonnaise. The source of the contamination was *S*. Enteritidis from unpasteurized eggs (Anonymous, 1995).

Use of vinegar for acidification of mayonnaise to pH 5.0 will prevent multiplication of *S*. Enteritidis in home-made mayonnaise, but the pathogen can survive for a few days at 20°C or 30°C when citric acid or a low level of acetic acid is used as acidulant (Perales and García, 1990; Kurihara *et al*., 1994; Lock and Board, 1994, 1995). In home-made mayonnaise with 0.1% of acetic acid (0.85% acetic acid in the water phase), *S*. Enteritidis survived for 5–6 days at 30°C, but for only 1 day in commercial mayonnaise with 2.26% acetic acid in the water phase. Refrigerated storage of a home-made mayonnaise at 10°C gave less than a factor three reduction in 9 days, while *Salmonella* spp. were eliminated from the commercial mayonnaise (5 $\log_{10}$ reduction) in 3–6 days.

Lock and Board (1994) studied 24 varieties of commercial mayonnaise with pH values of 2.6 to 4.8 and various types of acids and found rapid inactivation of *S. enteritidis* PT 4 (inoculated at 2.5 10^4/g) in all cases; in eight samples, *Salmonella* spp. could not be recovered after 48 h from products stored at 20°C, whereas *Salmonella* spp. were undetectable after 20 min in a fat-free mayonnaise with pH 2.6. At 4°C, *Salmonella* spp. generally survived longer.

Interestingly, not all incidents were caused by *S*. Enteritidis. There is a low incidence of *S*.Typhimurium in eggs. In the UK, 120 of 700 people reported gastrointestinal illness after eating in a large metropolitan building (Mitchell *et al*., 1989). *Salmonella* Typhimurium phage type 49 was isolated from a tartare sauce made from the mayonnaise remaining from the meal. The mayonnaise was made from fresh eggs, oil, and vinegar. The same *Salmonella* type was isolated from the patients and from bird droppings at the farm that supplied the eggs. Further incidents attributed to *S*.Typhimurium are reported in the review by Radford and Board (1993).

The concentrations of acetic acid used in commercially produced mayonnaise cause rapid inactivation of salmonellae and *L. monocytogenes*, thus reducing the risk of any low-level pathogen (re)contamination of pasteurized egg. In typical American reduced-calorie mayonnaise, *Salmonella* spp. and *L. monocytogenes* were inactivated in 3 days in products with a pH below 4.1 and 0.7% acetic acid in the aqueous phase (Glass and Doyle, 1991). In this way, chance introduction of contamination would still result in a product that is safe before it reaches the consumer. Adequate inactivation is less likely for the acid-resistant *E. coli* O157:H7, which survived for 7 days in commercial mayonnaise (Glass *et al*., 1993). Inoculation of 6.5×10^3 *E. coli* O157:H7 in the commercial mayonnaise implicated in the 1993 outbreak showed survival for over 8 days at 20°C and over 34 days at 5°C (Zhao and Doyle, 1994). Rapid inactivation of *Salmonella* spp. was found for salad dressing (pH 3.2–3.3; ± 1% acetic acid; inactivation within 1–6 h) or mayonnaise (pH 3.8-4.0; ± 0.5% acetic acid; inactivation within 1–18 h) kept at ambient temperature or at 37°C.

Another factor that can influence the death rate of salmonellae in mayonnaise is the type of oil used. Extra virgin olive oil, known to contain high levels of phenolic compounds, has been shown to contribute to rapid inactivation in home-made mayonnaise (Radford *et al.*, 1991; ICMSF, 1996, pp. 242–3).

The FDA has established requirements for pH and acetic acid in mayonnaises and dressings containing raw egg preparations (US-FDA, 1990, 1994), meant to ensure inactivation of *Salmonella* spp. by holding the product for 72 h before it is made available to consumers. This regulation considers that raw eggs or raw egg yolk containing ingredients may be used when the final pH is not above 4.1 and the acidity of the aqueous phase, expressed as acetic acid, is not less than 1.4% (US-FDA, 1990). However, due to *Salmonella* outbreaks associated with raw eggs, this last option is rarely used in the USA or Europe.

Staphylococcus aureus. The survival and possible growth of *Staph. aureus* in mayonnaise has been studied extensively (Smittle, 1977). Due to the low pH of these products and the presence of acetic acid, the organism is unable to grow and is normally of no significance for mayonnaise and dressings. In one of the Danish incidents, however, contamination with 1 million of δ-toxin-producing staphylococci was observed in *Salmonella*-contaminated mayonnaise with an elevated pH of 6.0 (Meyer and Oxhøj, 1964). The possible growth of *Staph. aureus* and toxin production in home-made mayonnaise was studied in detail by Gomez-Lucia *et al.* (1987, 1990). Toxin was found only in mayonnaise acidified with vinegar with a pH $\geq$5.0; at pH 4.5, no toxin was formed.

Listeria monocytogenes. Contamination of raw egg yolk with *L. monocytogenes* has been reported and the behavior of the pathogen has been studied in reduced-calorie mayonnaise (Leasor and Foegeding, 1990); with 0.7% of acetic acid in the aqueous phase, *L. monocytogenes* did not grow at 23.9°C and was inactivated by a factor of 10^4 within 3 days (Glass and Doyle, 1991). In four commercial mayonnaise products with pH 3.3–3.9, *Listeria* inactivation at 26.6°C was directly related to aqueous phase acetic acid concentrations; at 2.2% acetic acid, the pathogen was reduced by a factor of 10^8 in 72 h and at 0.67% acetic acid in 192 h (Erickson and Jenkins, 1991).

Escherichia coli O157:H7. Mayonnaise was epidemiologically implicated as the vehicle for transmission of *E. coli* O157:H7 in a 1993 outbreak of food-borne disease in Oregon, USA. In this outbreak, 62 cases were traced to the consumption of contaminated ranch and blue cheese dressings and contaminated seafood salad. The evidence available suggested that the mayonnaise (pH 3.9) was contaminated by tainted meat by the retailer. The bulk mayonnaise as delivered was not suspect (Anonymous, 1993b). Erickson *et al.* (1995) did not find the pathogen in pasteurized egg yolk or the wet processing environment of three plants processing mayonnaise or other emulsified sauces. The primary source for enterohaemorrhagic *E. coli* (EHEC) appears to be cattle. Meat and milk from cattle have been directly linked to many outbreaks, but the Oregon outbreak has shown that recontamination of mayonnaise and dressings with this pathogen may lead to unexpected outbreaks. There is another report of a possible outbreak due to mayonnaise contaminated with a non-motile *E. coli* O101, where 300 people became ill in Eastern Germany in 1988 (Bülte, 1995).

The incidents stimulated research on *E. coli* O157:H7 in mayonnaise, with a focus on its survival in acid environments. *E. coli* O157:H7 has been shown to be unusually resistant to acid pH as suggested by an outbreak associated with consumption of apple cider (Besser *et al.*, 1993) and by apple cider inoculated with *E. coli* O157:H7 (Zhao *et al.*, 1993). *E. coli* O157:H7 survived much better than a control strain of *E. coli* at low pH and numbers remained unchanged after 24 h of incubation in Trypticase Soy Broth at pH 3 and 4 (Miller and Kaspar, 1994). The pathogen may also be more tolerant of some organic acids or other antimicrobials as it has been found to survive well in acid foods and beverages (Duffy *et al.*, 2000; Koodie and Dhople, 2001; Mayerhauser, 2001). In commercial regular mayonnaise and reduced-calorie mayonnaise dressing, *E. coli* O157:H7 survived 7 days at 23.9°C, indicating that it survives better than *Salmonella* spp. and *L. monocytogenes* (Glass *et al.*, 1993).

Weagant *et al.* (1994) studied the survival of three strains of *E. coli* O157:H7 obtained from the mayonnaise implicated in the Oregon outbreak, using an inoculum size of 10^8 cfu/g in a mayonnaise product with pH 3.65, and noted that the pathogens died off quite rapidly at ambient temperature. Survivors were found for a maximum of 72 h at 25°C, whereas they could be recovered for up to 35 days at 7°C. Four different sauces were then made from this mayonnaise and challenge tested. Survival of one EHEC strain in thousand island dressing (pH 3.76) was about 35 days at 5°C; in seafood sauce (pH 4.38) and blue cheese dressing (pH 4.44), the numbers of survivors were about 500 times higher at that point in time. In a fourth, mayonnaise-mustard sauce (pH 3.68), the inactivation at 5°C was very rapid as no survival was observed after 5 days. Similar findings were reported for two strains of *E. coli* O157:H7 inoculated into a commercial mayonnaise (pH 3.91) at a level $>10^6$ cfu/g, where no survival was observed after 96 h at 22°C (Raghubeer *et al.*, 1995). Zhao and Doyle (1994) inoculated 6.5×10^3 *E. coli* O157:H7 in the commercial mayonnaise implicated in the 1993 Oregon outbreak (pH 3.6–3.8 and 0.37% titratable acid) and found survival for 8–21 days at 20°C and 34–55 days at 5°C.

In a ranch salad dressing with pH 4.51 kept at 4°C, *E. coli* O157:H7 survived better (over 17 days) than strains of generic *E. coli* and *Enterobacter aerogenes* that were used as references (no survivors found after 14 days). The reference strains did not survive for 4 days in a mayonnaise product with a composition meeting the FDA requirements for pH and acetic acid in products containing raw egg (US-FDA, 1994). Rapid inactivation of EHEC was observed in commercial mayonnaise with pH $\leq$4.0 by Erickson *et al.* (1995). The authors concluded that intact packages of commercial mayonnaise and mayonnaise dressings pose negligible EHEC contamination and health hazard risks.

In most studies on antimicrobial activity of weak acids done with *E. coli* O157:H7 and other bacterial pathogens, acetic acid proved more inhibitory than lactic acid, with minimal inhibition by citric acid (Smittle, 1977; Conner *et al.*, 1990; Conner and Kotrola, 1995). In Brain Heart Infusion broth at pH 5.0 acidified with lactic acid, visible growth of *E. coli* O157:H7 was evident in 2–3 days at 25°C, but in All Purpose Tween broth acidified with acetic acid to the same pH no growth was observed in 70 days (Davies *et al.*, 1992). Survival and growth at refrigeration temperatures may be dependent on the composition of the bacteriological medium, and consequently on the food formulation (Kauppi *et al.*, 1996).

Other pathogens. Food-borne pathogens like *Clostridium botulinum*, *Cl. perfringens*, and *Bacillus cereus* are unable to grow in mayonnaise and dressings at a pH $\leq$4.5 and are thus of no significance (Michels and Koning, 2000). An outbreak in the UK caused by *B. cereus* (and *Staph. aureus*) has been reported, where the pH of the product involved may have been above 4.6 (Radford and Board, 1993). *Campylobacter jejuni* has been observed in egg yolk but the organism is not heat-resistant, is unable to grow below 30°C and is likely to be inactivated rapidly in the presence of acetic acid.

E CONTROL (mayonnaise and dressings)

Summary

Significant hazards[a]	• *Salmonella* spp. • *E. coli* O157:H7. • *L. monocytogenes*. • *Staph. aureus*.
Control measures	
Initial level (H_0)	• Use pasteurized ingredients (i.e. egg, herbs, spices) or source raw material of appropriate specification from approved suppliers.
Reduction (ΣR)	• Pasteurize starch or water phase. • Use stable formulations (dependent mainly on pH, acetic acid) that cause die-off of infection pathogens.
Increase (ΣI)	• Use a stable product formulation (pH $\leq$4.5; at least 0.2% undissociated acetic acid in the aqueous phase). • Avoid recontamination; physically separate ingredients and processed products. • Use suitable, hygienic equipment and process hygienically (incl. proper cleaning). • Store final product dry; prevent condensation.
Testing	• Ingredients may be contaminated with pathogens of concern at low rates (*Salmonella* spp., for instance, in raw spices and herbs is about 1%); sampling and testing using reasonable sample numbers is not feasible. • When a stable product formulation is used, process control is sufficient to assure consumer safety.
Spoilage	• Spoilage can occur due to acetic-acid resistant microorganisms (i.e. certain yeasts and lactic acid bacteria). The major spoilage problems can be controlled by selecting suitable stable formulations, by preventing contamination via raw materials and the process environment, by hygienic packaging, and chilled storage, and distribution.

[a]In particular circumstances, other hazards may need to be considered.

Control measures. To achieve microbiologically safe and stable products, a process design based on a cold process is generally adequate. With respect to product safety, the formulation should assure that pathogens cannot multiply in case contamination does occur and that viable infectious pathogens are not present in the product at the point of consumption. For microbiologically sensitive products it is most important to prevent contamination originating from raw materials or the processing environment. This means that for each product one should consider whether ingredients used present a potential for contamination. Use of decontaminated ingredients or pasteurization of contaminated ingredients in a starch or water phase will control such a hazard. Chilled storage and distribution could in addition be necessary.

Processing. The production of mayonnaise, dressings, and other emulsified sauces has to be carried out in such a way that (re-)contamination with pathogens is prevented and that spoilage of the final product does not occur. Product and process design and implementation in an actual manufacturing process should follow the principles of HACCP (APHA, 1972; NRC, 1985; ICMSF, 1988). Because of the use of egg and other raw material ingredients in emulsified products, pathogens of concern are in particular *Salmonella* spp., *E. coli* O157:H7, *L. monocytogenes*, and *Staph. aureus*.

Formulation. To prevent growth of pathogens in emulsified sauces, it is recommended that a composition is chosen that does not permit growth of any relevant pathogen. pH alone will be inadequate to control pathogen growth in most emulsified products. Under optimal conditions, for instance, *Salmonella* spp. and *L. monocytogenes* can grow at pH values as low as pH $\pm$ 3.8 and 4.4–4.6, respectively. However, in combination with sufficient acetic acid, multiplication stops at a much higher pH. Stable formulations can be selected when the aqueous phase composition of a product is known by calculation or analysis. The levels of undissociated acetic acid, salt, and sugar can be used to predict stability by using the spoilage prediction chart developed by Tuynenburg Muys (1971). The effects of other acids (e.g. lactic acid), preservatives (e.g. sorbic acid) and natural antimicrobials (e.g. mustard or olive oil) can be considered as well. When a new formulation is considered that is significantly outside the present boundaries of knowledge, its safety and stability should be established by challenge testing, preferably using selected acetic acid-resistant yeasts and lactobacilli.

Michels and Koning (2000) consider that, as a rule, no growth of pathogens will be possible when the emulsified product has a maximum equilibrium pH of 4.5 and at least 0.2% undissociated acetic acid in the aqueous phase. This corresponds to a total acetic acid level of roughly 0.2% at pH 3.0 or to 0.3% at pH 4.5, which is well below the level used in most commercial products.

Decontaminated ingredients. For the industrial manufacture, the absence of infectious pathogens from the final products can best be ensured by the specific use of pasteurized egg preparations and by pasteurization of all other components that may present a microbiological hazard, such as some dairy ingredients or herbs and spices. Pasteurized egg preparations, sourced from approved suppliers (one that controls pathogen hazards by a validated HACCP-based system), are indeed used as the raw material in most industrial productions; sometimes the egg is pasteurized in-house for a second time directly before processing, because a very low incidence of *Salmonella* spp. and *L. monocytogenes* may occur in commercially available pasteurized liquid egg. Recommended pasteurization temperatures for liquid eggs vary from 55.6 to 69°C and the times of exposure vary from 1.5–10 min. In many countries, a specific minimum time–temperature combination is mandatory. The heat treatment required for cooking of a starch phase is generally a few minutes at 85°C, which is well above what is needed for microbiological decontamination. Where ingredients are heated in the acetic acid-containing water phase, most lactobacilli and yeasts are inactivated by a few minutes at 70° and heating to 65°C is frequently adequate.

Hygienic processing. A common source of acetic acid-resistant spoilage organisms is improperly or inadequately cleaned equipment used for manufacturing mayonnaise and dressings. The same may hold for the processing environment. During process design, appropriate hygienic equipment well suited for CIP should be selected. The lay-out of the process line should also be such that the environment is easily cleaned and that cross-contamination from raw ingredients to decontaminated product is prevented. Proper manual cleaning may still be required for traditional colloid mills, fillers, and other equipment that are difficult to clean by CIP alone because of for example crevices in the equipment construction.

Packaging. The glass jars, tubs or other containers used for packaging mayonnaise and dressings are normally free of acetic acid-resistant spoilage organisms and thus there is no need for decontamination or for extremely clean packaging materials.

Distribution. For ambient stable mayonnaise and dressings, distribution is not a problem. For chilled products, temperature of distribution and retail is a point to control and monitoring of temperatures in the chilled chain will help to find and correct deviations.

Consumer use. Most emulsified sauces discussed are only spoiled by acetic acid-resistant organisms, which are not widespread in the consumer's home. Consequently, these products will not easily spoil during consumer use. However, when mild compositions with low acetic acid levels are put on the market, the possibility of spoilage during consumer use should be considered. Such spoilage can be controlled largely, but not completely, by limiting the shelf-life at ambient temperatures of open products or by recommending chilled storage after opening. Labeling instructions can thus be important for vulnerable products.

III Mayonnaise-based salads

A *Definitions*

Mayonnaise-based salads or dressed salads are simply mixtures of a variety of foods described in other chapters of this book with a mayonnaise base. The components may consist of chicken, meat, egg, fish, shellfish, potato, vegetables, herbs, pasta, fruits, or nuts. Apart from the mayonnaise and the main components mentioned, starch, sugar, spices, organic acids, preservatives, and flavors/colors may be present. A specific definition does not exist.

B *Important properties*

Due to the vinegar of the mayonnaise base or the additional acids used, the dressed salads generally have between pH 4.0–5.5; the acetic acid level in the aqueous phase (often between 0.2–0.5%) is typically much lower than that of mayonnaise itself. Most mayonnaise-based salads are kept refrigerated to extend their shelf-life.

C *Methods of processing and preservation*

Mayonnaise-based salads are made by mixing ingredients at ambient or chill temperatures with a special thick mayonnaise or dressing. Some ingredients, like chicken or meat, are cooked but others may not be. Typical ingredients that are used without a heat treatment are raw vegetables. Because they could introduce a very broad spectrum of microorganisms, vegetables are well cleaned and sometimes marinated in special brines to reduce their microbial load. When the complete salad has been mixed, it is packed in consumer-sized tubs or in larger containers for sale by a retailer. When prepared from chilled components in a well-chilled environment, salads can be packed and palletized directly; otherwise quick chilling after packing is required. Acetic acid levels in the aqueous phase of mayonnaise based-salads are still quite low, and therefore these products may be quite vulnerable to spoilage; preservatives (i.e. sorbic acid and benzoic acid) may help to improve stability. The refrigerated shelf-life of a typical mayonnaise-based salad may vary between 2–8 weeks, depending on initial contamination, pH, level of inhibitory acid, and level/type of preservatives (if any) present.

D *Microbial spoilage and pathogens*

The initial microflora of mayonnaise-based salads is made up of the microbial load of the raw materials used. Raw materials used should not introduce pathogens or significant numbers of spoilage organisms. Control of the possible proliferation of the initial microflora depends on the formulation of the product, whether pasteurization is used and whether hygienic conditions are adhered to. A low pH and a suitable level of organic acids in the product formulation, minimize the growth of most spoilage bacteria and pathogens. It is preferred to use pasteurized, or even sterilized, raw materials. Pasteurized materials may be put into an acid brine or salted to render them ambient stable. Several raw materials (e.g. hand-peeled cooked shrimps, vegetables, herbs, and spices) are difficult to obtain with sufficiently low counts and in-house measures (e.g. steam decontamination of herbs and spices) should be taken to reduce their counts upon arrival at the manufacturing site. When it is not possible to effectively eliminate spoilage organisms (e.g. yeasts and lactobacilli in raw herring), the absence of infective pathogens must be ensured by selecting raw materials from approved suppliers. In addition to GAP and other specific control measures for the raw materials, product formulation, processing and expected shelf-life should all be adequate to control the expected level of initial contamination.

Spoilage. Because of the low pH and the presence of a particular level of organic acids, certain yeasts, lactic acid bacteria, and molds are the most frequent cause of spoilage of mayonnaise-based salads. Characteristic spoilage yeasts in, for instance, coleslaw are *Saccharomyces exiguus* and *Sacc. dairensis* (Brocklehurst *et al.*, 1983). Many other yeasts (e.g. *Sacc. cerevisiae*, *Pichia membranaefaciens*, *Z. bailii*, *Z. rouxii*, *Sporobolomyces odorus*, *Trichosporon beigelli*, *Torulaspora delbrueckii*, *Can. sake*, *Can. lambica*, *Can. Vini*, and *Yarrowia lipolytica* have also been isolated from salads (Baumgart *et al.*, 1983; Brocklehurst and Lund, 1984; Bonestroo, 1992). Molds produce visible spoilage, whereas yeasts produce off-flavor, gas formation, or colonies visible as surface growth.

Multiplication of lactobacilli in mayonnaise-based salads may lead to high counts, which are not always accompanied by spoilage. Gas-producing lactobacilli or rope-forming strains produce more evident signs of spoilage. The most frequent lactic acid bacteria that grow in dressed salads are *Lb. plantarum*, *Lb. buchneri*, and *Lb. brevis*. Less common types are *Lb. leichmannii*, *Lb. delbrueckii*, *Lb. casei*, *Lb. fructivorans*, *Lb. confusus*, *Leuconostoc mesenteroides*, and *Pediococcus damnosus* (Baumgart *et al.*, 1983; Erickson *et al.*, 1993). All of these organisms come primarily from vegetables and pickles or the processing environment. Although sorbic and benzoic acids inhibit yeasts, they have little effect on lactobacilli. In the case of benzoic acid, this is because a significant proportion of benzoic acid added partitions into the oil phase of a salad. The rather high pH of salads may decrease the amount of undissociated benzoic acid in the aqueous phase.

A significant consideration for both stability and safety is that large pieces of meat may remain at a relatively high pH within the salad tissues, because the acids diffuse very slowly through solid foods. The equilibrium pH value may be higher than that of the freshly made product, so that a variety of organisms can grow. Marinating pieces of meat or fish (such as herring) to reach a sufficiently low pH before mixing and proper chilling immediately after production may adequately reduce growth of spoilage organisms or pathogens.

Pathogens. Food-borne illness caused by eating contaminated mayonnaise-based salads has occurred when raw materials or ingredients used contain toxins (e.g. potato salad with botulinum toxin, Seals *et al.*, 1981) or that are contaminated with infectious pathogens. An outbreak of typhoid fever in 1974 in Germany, which resulted in 417 cases and five deaths, was associated with contaminated potato salad (Hüpper, 1975). Two related incidents of salmonellosis were caused by *S.* Indiana in a salad product served at a European Union Summit Conference in Maastricht and in a cold buffet eaten at a family dinner

(Beckers *et al.*, 1985). In the first incident, numbers of *S.* Indiana reached levels of 10^7 cfu/g, whereas in the latter incident mainly cross-contamination with *S.* Indiana at a low infective dose without significant multiplication was suspected. Evidently, food-borne bacterial pathogens can survive well in chilled salads. In salads with an elevated pH, held at 22 or 32°C, *Salmonella* and *Staph. aureus* grew well within 24 h in chicken salad (pH 6.1) and in ham salad (pH 5.2), whereas no growth occurred at 4°C (Doyle *et al.*, 1982). Kurihara *et al.* (1994) demonstrated rapid growth of *S.* Enteritidis at 25°C in potato, egg, and crab salads (pH was 5.72, 7.11, and 6.51, respectively) made with 15% of homemade mayonnaise (pH 4.75). This mayonnaise contained 0.1% of acetic acid which, assuming a dry matter level of 30%, results in only 0.02% of total acetic acid in the salad's aqueous phase and thus only a trace of undissociated acetic acid to prevent pathogen multiplication (0.002% at pH 5.7 and even less at the other pH values).

Pathogens other than salmonellae also have been associated with food-borne illness from mayonnaise-based salads. In 1981, enteroinvasive *E. coli* was associated with eating potato salad on a cruise ship (Snyder *et al.*, 1984). *Shigella* is another significant cause of illness and Smith (1987) refers to at least 11 outbreaks in the United States within the period 1975–1981; there were 1500 cases of illness then due to *Shigella flexneri* or *Shig. sonnei* caused by various types of salads of which eight were potato salads. Ten cases of staphylococcal food poisoning occurred in the United States due to potato salad (Bryan, 1988). A total of 41 outbreaks caused by potato salad are listed in Bryan (1988). This high rate of food-borne illness may have been due to the relatively large proportion of neutral potato in these salads and a low level of acetic acid, permitting extended survival and growth of pathogens, particularly when temperature abuse had taken place. In 1981 in Canada, coleslaw was identified as the vehicle of transmission of *L. monocytogenes*, causing 34 peri-natal cases and seven adult cases of listeriosis (Schlech *et al.*, 1983). Erickson *et al.* (1993) studied home-style chicken salad with mayonnaise (pH 5.7) and observed significant growth of *Salmonella* spp. and *L. monocytogenes* at 12.8°C, but not at 4°C; in macaroni salad at pH 4.6, no growth was observed.

E CONTROL (mayonnaise-based salads)

Summary

Significant hazards[a]	• *Salmonella* spp. • *L. monocytogenes.* • *E. coli* O157:H7. • *Staph. aureus.* • *Cl. botulinum* (from some vegetable ingredients).
Control measures	
Initial level (H_0)	• Do not use ingredients that introduce uncontrolled infectious pathogen hazards. • Use pasteurized egg preparations only and use disinfected (cooked, blanched, pasteurized, chemically, or physically disinfected) shrimps, vegetables, herbs, and spices sourced from a reliable supplier (according to agreed specifications). • Disinfect ingredients in-house upon arrival when necessary.
Reduction (ΣR)	• Clean and wash raw vegetables; dice pre-cooked meats and poultry into small pieces, and/or marinate them in acid before mixing them with other ingredients.

(*continued*)

E CONTROL (Cont.)

Summary

Increase (ΣI)	• Prevent growth of any pathogen under the intended (chilled) storage conditions in the mayonnaise-based salad, by using a formulation that will prevent this; consider normal ranges of pH (4.0–5.5), undissociated concentration of acetic acid (AA level in total product often 0.2–0.5%) and salt level) • Keep raw ingredients, packaging material, and other potential sources of contamination separated from processed products; avoid (re-) contamination. • Use strict hygiene in preparing, mixing, packaging, and storing intermediate/finished salads products; use hygienically designed equipment and assure CIP is used. • Restrict the shelf-life to what is safely achievable. • Assure that the chill chain temperature qualifies for the intended shelf-life (e.g. when the intended shelf-life needs storage at 7°C, check that this temperature is achieved throughout the chain).
A Testing	• Ingredient specification and water quality should be verified. Intermediate and finished product should be tested regularly to verify process performance. Routine end product testing is not advised.
Spoilage	• Yeast, lactic acid bacteria, and molds are the most relevant spoilage microorganisms. Generally, control measures for product safety are adequate for product stability. The end product should be kept refrigerated.

[a]In particular circumstances, other hazards may need to be considered.

Comments. *Salmonella* spp., *L. monocytogenes*, *E. coli* O157:H7, and *Staph. aureus* are significant hazards and may be introduced via ingredients or from the environment. Under particular circumstances other pathogens may be relevant. For instance, *Cl. botulinum* toxin can be a hazard in the vegetable ingredients if used in salads.

IV Margarine

A Definitions

Margarine is a water-in-oil emulsion with a high fat content. The US definition of margarine is: "Margarine (or oleomargarine) is the food in plastic form or liquid emulsion containing no less than 80 percent fat" (US-FDA, 1991). In Western Europe, yellow fat products other than butter and melanges are described as: "Products in the form of a solid malleable emulsion, principally of the water-in-oil type, derived from solid and/or liquid vegetable and/or animal fats suitable for human consumption, with a milk fat content of not more than 3% of the fat content"; the category margarines is described as: "The product obtained from vegetable and/or animal fats with a fat content of not less than 80% but less than 90% (EC, 1994).

B Important properties

A wide range of oils and fats may be used, e.g. soya oil, palm oil, and sunflower oil. The type of fat used depends on a whole series of considerations. Within economic constraints such as availability and cost, fat blends are today mixed using linear programming in such a way that they will meet the required functional specifications with regard to cooking performance, spreadability at refrigerator temperature, emulsion stability, flavor release, appearance, and shelf-stability (Madsen, 1989, 1990; Moustafa, 1990). In addition to the fat blend, water (15–20% on product) and emulsifiers are required. The pH of the aqueous phase of a margarine is usually set between pH 3.5 and 6.0, using citric and lactic acid (separate or in combination) as the acidulant. In some countries, such as the UK, products with high salt levels are preferred, but, for instance for dietary reasons, salt may be completely omitted. In effect, generally, the concentration of NaCl may vary from 0 to 2% w/w in the finished product (sometimes up to 2.5% w/w), which corresponds to 0–11% w/w on aqueous phase. Milk or milk products (sweet or cultured skim milk powder, buttermilk powder, sweet or acidified whey powder, milk protein) are generally added as a powder in concentrations of 0.1–1.0% of the product. Other ingredients include vitamins A and D, flavors, and colors such as beta-carotene. In the United States, permissible antimicrobial preservatives include sorbic acid (max. 0.1% of the product) and benzoic acid (max. 0.1%), or a combination thereof. In Western Europe, only sorbic acid or its salts are allowed (acid concentration max. 0.1% of the product; EC, 1995). Because of the neutral taste and favorable partition coefficient of the acid in water-in-oil emulsions, potassium sorbate and sorbic acid are most widely used.

C Methods of processing and preservation

Figure 11.3 illustrates a typical margarine process, in which a proportioning pump mixes an aqueous phase that is maintained at 10–20°C and contains all water-soluble ingredients (generally from brine or stock solutions or from powdered or liquid dairy material such as whey or skim milk) with a fat phase that is held at 45–60°C and contains oil-soluble ingredients (e.g. emulsifiers, flavors, preservatives); the resultant pre-emulsion is a stable fat-continuous mixture throughout which the aqueous phase is dispersed. A high-pressure pump feeds this mixture through a closed system of scraped surface heat exchangers and mixers, which cool it down to 10–20°C and work it into a fine emulsion. Subsequently, the fat phase is allowed to crystallize to a certain extent in a crystallizer (pinstirrer). The sequence, sizes, and configuration of the heat exchangers and mixers are designed to yield the right physical product properties in relation to the crystallization behavior of the fat. The product is then packed into tubs (soft spreadable products) or in wrappers (hard margarines used for baking and cooking). In margarines used for bakery purposes it is not uncommon to omit the dairy material.

The design of the manufacturing process should be such that during production pathogenic microorganisms cannot multiply at all and spoilage microorganisms cannot multiply to unacceptable levels. Inclusion of a pasteurization treatment during processing is optional. When applied, it may be at the stage of the stock solutions, the aqueous phase, and/or the pre-emulsion. With margarines, pasteurization of the aqueous phase is more common than pasteurization of the pre-emulsion. Pasteurization of both the aqueous phase and the pre-emulsion is rarely done. The composition of the solutions, the microbial load, the time and temperature of storage, the hygiene of production and the microbiological stability of the finished product with respect to the intended shelf-life are factors that determine whether pasteurization is necessary at all and what level of inactivation needs to be assured. Margarines with an aqueous phase of only water and salt may not require a pasteurization step. For some mixtures, such as a stock solution of dairy ingredients (which is rather sensitive), the minimum pasteurization requirement applied for milk (15 s at 72°C; IDF, 1986a) may be sufficient. Where appropriate, heat treatments equivalent to

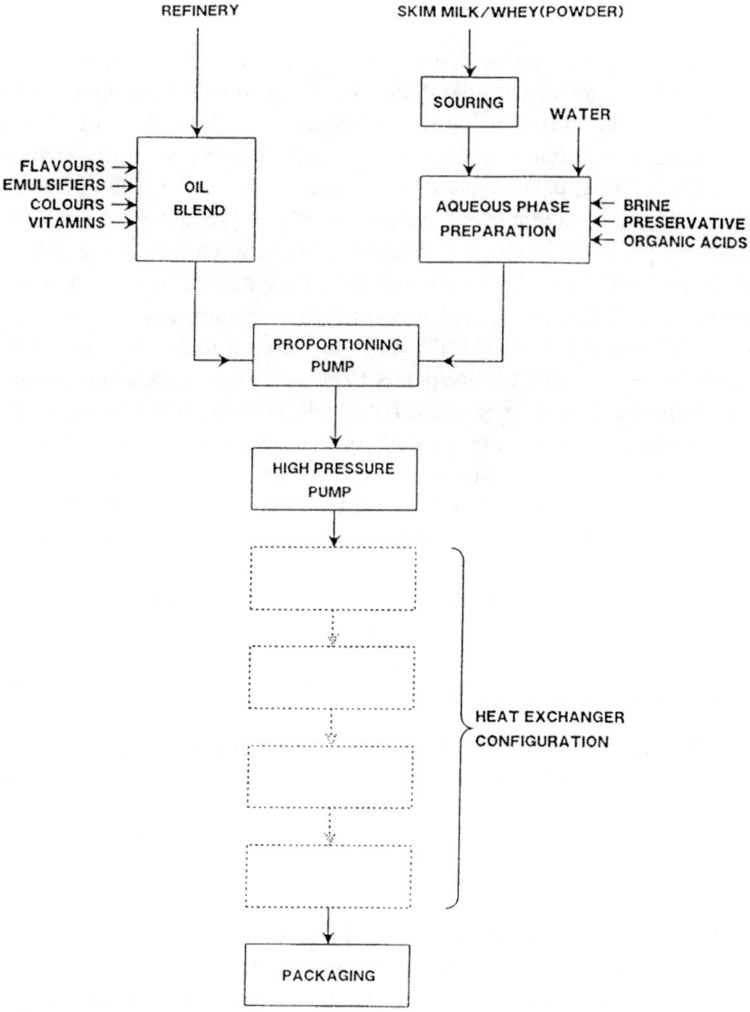

Figure 11.3 Schematic presentation of a margarine process.

6*D* pasteurization should be adequate to control infectious pathogens (i.e. *L. monocytogenes*, *E. coli* O157:H7, *Salmonella* spp., *Staph. aureus*) and most spoilage microorganisms (e.g. yeasts, vegetative bacteria, and molds), but will not inactivate bacterial spores.

Oils and fats can protect microorganisms against heat treatments (Hersom and Hulland, 1980; Troller and Christian, 1978; Gaze, 1985). For margarines, such a protective effect could present a hazard when relevant microorganisms have to be inactivated at the pre-emulsion stage. A protective effect by oils and fats is, however, only found in the absence of water or in the presence of very low levels. Because of the large amount of water present in margarines (15–20% on product), it is not expected that there will be a large protective effect.

The nature of the continuous margarine process requires that any hold-up at the packaging stage will create product that has to be reworked. From the formation of the pre-emulsion onward, the process line will usually be closed, excluding contamination from the factory environment. After secondary packing, the final product is distributed and stored at ambient or chilled temperature, depending on the sensitivity of the product, the contamination level, the type of packing material and/or climatic conditions. From a microbiological point of view, a limitation of the closed shelf-life period is usually not required. However, the shelf-life is generally limited to 3–6 months for reasons of chemical stability. It is advisable to store opened packs and tubs in the refrigerator. Contamination of the product during consumer use is likely to occur but growth of contaminating microorganisms is expected to be limited in view of the fat-continuous nature of the product, the small size of the water droplets, and, when applied, the low storage temperature. In practice, both preserved and non-preserved margarines appear to be very stable products during closed and open shelf-life.

Margarines are true water-in-oil emulsions stabilized by a very special microbiological preservation principle: the aqueous phase is dispersed as small water droplets in a fat-continuous matrix (Tuynenburg Muys, 1969; Gander, 1976; Charteris, 1996; Delamarre and Batt, 1999; van Zijl and Klapwijk, 2000; Holliday and Beuchat, 2003). Due to the fine water dispersion, a contaminating microorganism present in a droplet is restricted in its growth either by space limitations or by exhaustion of the nutrients available in the droplet. This is a feature that margarines have in common with butter, although the latter product may have a coarser emulsion. The continuous fat-phase with a network of fat crystals gives consistency to the product. Emulsifiers that are present at the oil/ water interface play an important role in the physical stability of the structure. Most of the water in a margarine product is normally present as finely dispersed droplets with a diameter $<10\ \mu m$, with 50% of the total water volume in droplets $<3\ \mu m$ (volume weighted geometric mean diameter, $\overline{D}_{3.3}$ in μm). Less than 5% of the water volume will be in droplets with diameters $>10\ \mu m$. The shells of fat crystals that cover the water droplets prevent these from merging together (Juriaanse and Heertje, 1988).

Predictive models have been developed (Verrips and Zaalberg, 1980; Verrips *et al.*, 1980; Verrips, 1989), which quantitatively predict microbiological stability. Such models can be used effectively during product design to define the growth–multiplication factors for relevant microorganisms, when microbiological growth characteristics, concentrations of nutrients, emulsion characteristics and antimicrobial factors present are known (Verrips, 1989; Klapwijk, 1992; ter Steeg *et al.*, 1995, 2001; Guerzoni *et al.*, 1997). Parameters that describe the emulsion characteristics are the volume weighted geometric mean diameter ($\overline{D}_{3.3}$ in μm) and the geometric standard deviation (e^{σ}) of the droplet size distribution (Alderliesten, 1990; 1991). These parameters can be determined using pulsed-field-gradient NMR (Packer and Rees, 1972; van de Enden *et al.*, 1990; Mooren *et al.*, 1995). Alternatively, droplet sizes can be estimated by microscopic observation, although this technique is not sufficiently accurate to describe the population of droplet sizes in quantitative terms.

Figure 11.4 illustrates how a margarine emulsion with a lower $\overline{D}_{3.3}$ but a higher e^{σ} can be made up of a larger number of larger volume droplets, which adds to the mold sensitivity of the product. The estimated D_{min} (minimal droplet diameter to allow onset of mold germination and outgrowth) adds to the visualization how the droplet-size distribution contributes to mold stability. The contribution of e^{σ} to mold stability will be reduced, when $\overline{D}_{3.3}$ increases. At higher $\overline{D}_{3.3}$ the volume fraction of droplets prone to mold spoilage will hardly increase, when e^{σ} goes up.

Once the multiplication of relevant microorganisms can be predicted at the expected storage and transportation temperatures, it is possible to specify the stringency of process hygiene requirements and the pasteurization requirements, in order to ensure that microbiological limits for such products will not be exceeded during the intended shelf-life. The next stage is to assess the ability of the process to meet these requirements. If the existing process equipment is not sufficiently hygienic, microbiological problems are to be expected, and corrective measures are necessary. This can be either using other raw materials, hygienic improvement of process equipment, introduction of additional pasteurization steps, product reformulation, or a combination of any of those measures.

The preparation of the aqueous phase and the stock solutions are microbiologically vulnerable steps. There may be contamination from the environment and human sources. When temperatures over

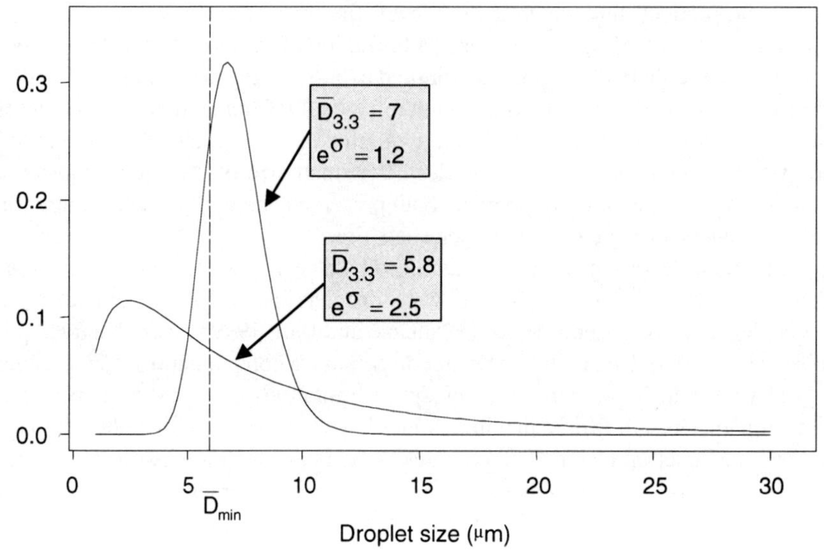

Figure 11.4 Droplet diameter, distribution width (e^{σ}) and the minimal diameter ($\overline{D}_{min}$) determine the volume of mold vulnerable droplets of an emulsion (ter Steeg *et al.*, 2001).

40°C are used for extended periods, growth of thermophilic microorganisms may occur. Sensitive raw materials, i.e. ingredients such as dairy materials and proteins that may harbor spoilage and pathogenic microorganisms, must always be heat treated at some stage in the process. Occasionally the aqueous phase or the complete emulsion (pre-emulsion or premix) is pasteurized, but the latter approach is more commonly used for the manufacture of low-fat and very low-fat spreads, which are generally more vulnerable to microbial contamination. Evidently, recontamination of the pasteurized product must be prevented since there is no further heat treatment in the manufacturing process.

The above considerations about the contribution of the emulsion to the stability of margarines are only relevant for those situations where a microbiologically unstable aqueous phase is used. Often this is not the case, because of a high salt content, low pH, lack of nutrients or a combination of these. Low-calorie spreads often have unstable aqueous phases in combination with coarse emulsions. The development of low-fat spreads necessitated the establishment of models allowing predictions to be made on the stability of aqueous phases and the stabilizing contribution of the emulsion.

D Microbial spoilage and pathogens

Spoilage. Mold growth is the major stability problem associated with margarines. Many mold species (*Trichoderma viride*, *Tr. harzianum*, *Aspergillus* spp., *Alternaria* spp., *Pen. roqueforti*, *Pen. expansum*, *Paecylomyces variotti*, *Geotrichum candidum*, and *Cladosporium* spp.) have been associated with spoilage (Beerens, 1980; Hocking, 1994). In contrast to other food spoilage microorganisms, molds grow through the fat matrix of the product. Hence, very low populations of mold spores (sometimes below the detection limit) can cause a product to fully spoil. The aqueous phase composition (i.e. salt level, pH, preservatives) affects mold growth and fine emulsions seem less likely to spoil. Conditions favoring mold development include high ambient temperatures (> 10°C), poor packaging hygiene, and development of free moisture on the outer surfaces of the product. Differences between product and air temperature can induce condensation of water, hence temperature cycling can cause deterioration of product. Parchment is often used for wrappers and in some instances as cover leaves in tub products. If the parchment absorbs moisture, cellulolytic fungi (notably *Tr. viride*) can grow. The use of parchment impregnated with potassium sorbate can reduce the incidence of mold growth.

Molds in margarine are easily detected by using media like oxytetracycline glucose yeast extract agar, with the exception of *Cladosporium* species which can be difficult to culture (Tuynenburg Muys *et al.*, 1966; Tuynenburg Muys, 1971). In some instances, the problem of mold growth is signaled by visible presence of mold colonies, but usually only detected by microbiological investigations after receiving off-flavor complaints. Occasionally mold growth is indicated by the presence of small watery droplets on the surface of the product, secreted by fungal mycelium.

During growth, molds produce off-flavor, generate free fatty acids, break down the emulsion and degrade any preservatives present. Liewen and Marth (1984, 1985) found that *Pen. roqueforti* and *Tr. harzianum* were capable to adapt and grow in the presence of levels of undissociated sorbic acid as high as >300 ppm. *Penicillium* spp. have been shown to metabolize sorbic acid into *trans* 1,3-pentadiene in a decarboxylation step (Finol *et al.*, 1982). *Paecilomyces variotii* has been isolated from a pentadiene-containing margarine and was demonstrated to convert sorbic acid into pentadiene (Sensidoni *et al.*, 1994). Also *Trichoderma* and *Aspergillus* species posses this trait. Because of the strong smell of pentadiene, which is detectable before mold growth is visible, it is very unlikely that the presence of this compound would not be detected by the consumer. Degradation of benzoate to benzaldehyde may also occur, but is less frequently observed.

Lipolytic yeasts (i.e. *Can. lipolytica*) occasionally cause spoilage problems. In contrast to the filamentous fungi, yeasts do not grow through the fat matrix and thus their growth is restricted in fine emulsions. Yeasts may grow at extreme values of salt concentration and acidity, potentially causing off-flavor defects and hydrolysis of fat; spoilage will be less dramatic than with molds. High yeast populations in properly emulsified products indicate insufficient hygienic precautions, such as poor process line hygiene. There are certain aqueous phase compositions that allow the growth of yeasts but not of bacteria. When both bacteria and yeast are able to grow, it is more likely that bacterial problems would occur because bacteria grow much faster than yeasts.

Lipolytic bacteria such as Micrococcaceae, Pseudomonadaceae (*Pseudomonas*, *Flavobacterium*), and bacilli may grow in coarse or unstable emulsions. For this reason, in case of suspected spoilage problems, margarine should be tested for lipolytic microorganisms on media like the Eijkman-plate (Tuynenburg Muys and Willemse, 1965). Occasionally, spoilage may be caused by Enterobacteriaceae, such as *Enterobacter* spp., generally as a result of a post-pasteurization contamination. These organisms generally do not cause spoilage because growth, if it can occur, is restricted by the continuous fat phase. Enterobacteriaceae, together with yeasts, serve in practice as good indicator microorganisms for the hygiene of a production line.

Pathogens. There are no genuine reported cases of food-borne illness associated with consumption of margarine (Beerens, 1980). The possibility of staphylococcal enterotoxin formation in unstable high-salt margarine is a hazard, which may require consideration. A recall due to contamination with such enterotoxin from blended margarine and butter products has been reported (Anonymous, 1992a). The food poisoning outbreak involved more than 265 cases of illness in two states in southwestern United States. Among 15 outbreak isolates from patients and food sources, a single, enterotoxin A-producing strain of *Staph. intermedius* was identified as the etiologic agent. *Staph. intermedius* was isolated from a margarine, a vegetable margarine, and different samples of butter blend (Khambaty *et al.*, 1994). It is likely, however, that the toxin originated from the butter (Charteris, 1996), as butter is known for its potential as a medium for staphylococcal food poisoning. In the UK, the death of an elderly woman was alleged to be associated with margarine contaminated with *L. monocytogenes* and the product was recalled from the market. Analysis of the original samples and samples taken from the incriminated lot, however, revealed that these all were negative for *L. monocytogenes* when tested by public health authorities. The conclusion was that the batch of margarine was free of *Listeria* spp. (Barnes, 1989).

E CONTROL (margarine)

Summary

Significant hazards[a]	Although there are no reported cases of food-borne illness associated with consumption of margarine, significant hazards to which adequate control measures need to be established are: • *Salmonella* spp. • *E. coli* O157:H7 • *L. monocytogenes*.
Control measures	
Initial level (H_0)	• Use appropriate ingredients (source from approved supplier on basis of agreed specifications). • Pasteurise ingredients in-house before use, when contamination with pathogens of concern cannot be excluded. • Use potable water from a reliable source.
Reduction (ΣR)	• Use preservatives where appropriate. • Pasteurise water or pre-emulsion phase.
Increase (ΣI)	• Use the appropriate product structure: i.e. fine water-in-oil dispersion to prevent or limit growth of microorganisms. • Avoid (re-)contamination (keep to GMP; physically separate ingredients and processed products). • Use suitable, hygienic equipment and process hygienically (incl. proper cleaning). • Store final product dry; prevent condensation.
Testing	• Ingredient specification and water quality should be verified. • End product testing is not advised when a stable product formulation/ structure is used and process control is adequate.
Spoilage	• Generally, spoilage problems are controlled by using a stable formulation and an appropriate product composition, i.e. correct pH, clean ingredients and appropriate (fine) water dispersion. It is advisable to keep product refrigerated during open shelf-life.

[a]In particular circumstances, other hazards may need to be considered.

Control measures. A process design based on a cold process is generally adequate to achieve microbiologically safe and stable products, provided that suitable raw materials are employed. Correct product composition and emulsion characteristics should assure safety of margarine products at the point of consumption (Charteris, 1995; van Zijl and Klapwijk, 2000). For microbiologically sensitive products, it is most important to prevent contamination originating from raw materials or the processing environment. This means that for each product one should consider whether ingredients used present a potential for contamination. Use of decontaminated ingredients or pasteurization of contaminated ingredients in a starch or water phase will control such a hazard. In the case of very sensitive products, refrigerated distribution should be considered.

Raw materials. The microbiological quality of ingredients should be ensured by purchasing on the basis of microbiological specifications (van Zijl and Klapwijk, 2000) from audited suppliers. The quality of a raw material, as obtained from the supplier, can be a CCP in an HACCP plan when that raw material is at risk of being contaminated with relevant pathogens and does not undergo a decontamination step before it is used in the production of a margarine.

Refined oil is virtually sterile due to the high temperatures applied during refining. Microbiological problems can be prevented by keeping the oil storage and transport system free from water. Oil should therefore be monitored for the absence of water. As water is heavier than oil, water condensate will accumulate at the bottom of tanks or in low insufficiently flushed parts of process equipment, creating unexpected and undesirable problems with molds. This happens only when the oil system is improperly designed.

Some ingredients may be added to the oil (e.g. flavors, colors, vitamins, and emulsifiers). Spices and herbs added for flavoring purposes may be contaminated with pathogens of concern (i.e. *Salmonella* spp., *L. monocytogenes*, *E. coli* O157:H7) and may also contain high numbers of spoilage microorganisms (Chapter 7). Spices and herbs should be decontaminated either by the supplier or, directly prior to processing, by the margarine manufacturer. Special attention should be given to lecithin from, for instance, soybean oil, since this can be contaminated with lipolytic microorganisms as well as *Salmonella* spp.

Dairy ingredients should always be made from pasteurized milk. Therefore, tight microbiological criteria apply for dairy ingredients. Example criteria (taken from a typical manufacturer—not legal requirements) for powders: e.g. aerobic plate count <10000 cfu/g, Enterobacteriaceae (as indicator for hygiene) <100 cfu/g, yeasts and molds <100 cfu/g, *Bacillus cereus* <1000-cfu/g, and absence of infectious pathogens. Examples of criteria for liquids dairy ingredients, e.g. aerobic plate count <1000 cfu/g, Enterobacteriaceae (as indicator for hygiene) <10 cfu/g, thermophilic bacteria <100 cfu/g, yeasts and molds <100 cfu/g, *B. cereus* <100 cfu/g, and absence of infectious pathogens. Buttermilk powder may contain high counts of thermophilic spore-formers, which is important when either stock solution, the aqueous phase, or the pre-mix is stored at a temperature allowing growth. Because the presence of bacterial spores in the finished product cannot be prevented, the composition and the physical structure of the finished product should limit their outgrowth.

Other ingredients such as salt, preservatives, and acidulants are generally free of microbial contamination. Water must always be of potable quality and drawn from a reliable source. The microbial content should be regularly monitored.

Composition. The stability of a margarine is influenced by the product composition (i.e. correct pH, the appropriate level of ingredients and fine water dispersion). Therefore, the process must be controlled to obtain these desired properties. The pH value of the product after equilibration with the fat phase may be critical, particularly when a preservative is being used. Critical product properties should always be incorporated into monitoring and verification programs and the results subjected to trend analysis.

Processing. Manufacture requires a well-cleaned and disinfected line. It is important to pay proper attention to cleaning and disinfection procedures; equipment should preferably be cleanable in place. The effectiveness of cleaning and disinfection should be verified by analyzing start-up samples. Contamination from the environment and human sources should be avoided throughout the production process. The equipment can be a source of contamination when it is improperly designed (e.g. dead spaces occur), improperly maintained, allows build-up of product residues or remains wet during idle periods. Also, reciprocating or rotary shafts (in pumps) may carry external contaminants to the product stream. Therefore, the state of the equipment is an important factor in determining the time of a production run until the next cleaning cycle (Lelieveld and Mostert, 1992).

The preparation of the aqueous phase and the stock solutions are microbiologically vulnerable steps. Sensitive raw materials should be heat-treated at some stage in the manufacturing process; pasteurized materials should be used when they enter the process at a stage when there is no heat-decontamination step further downstream. Permissible time and temperatures should be specified and controlled at this stage. The microbiological status at this point should be verified by taking samples at regular intervals. When temperatures over 40°C are used for extended periods, samples should also be checked for growth of thermophilic microorganisms.

A stable end product may be obtained when a suitable formulation is used and the emulsion characteristics are properly controlled. In a stable product, multiplication of contaminating bacteria or yeasts is not possible and, more likely, death may occur. In this case, more lenient microbiological specifications and sampling schemes may apply. However, when emulsion characteristics are not well controlled and a too coarse emulsion allows microbial growth, then the specifications for fresh product and the sampling scheme should be more stringent. The stability of a product can be checked by comparing microbial growth in stored (e.g. 2 weeks at 20°C) samples with fresh samples.

Packaging. Packaging equipment usually is not easily cleaned. Manual cleaning needs to be supervised and closely monitored. Since end product may be exposed to the environment at packaging, the hygiene needs to be controlled carefully: air quality should be adequate (when necessary, air filtration should be used to limit contamination of the product with mold spores; in exceptional cases a laminar air flow cabinet is needed) and packaging materials should be of good microbiological quality (complying to specifications regarding mold counts; stored, transported and handled appropriately for a cold-filled food product). Secondary packaging materials, especially recycled cardboard (Scholte, 1995), may be a specific source of mold spores and should best not be handled in the packaging area. Packaging, storage, and distribution conditions should not permit free moisture to develop either on the product surface or on the packaging material.

Occasionally, product is made that is not packaged immediately after production. This may be at the start-up of the process when the packaging machine is mechanically out of order, or when the product has not been packed correctly. The handling of product flow at this point is critical with respect to microbiological product quality. If unpacked product is reworked through a closed return pipeline and hygiene is properly controlled, the product can re-enter the main line without a decontamination step. Packaged product must be returned to the refinery to regain the oil. This is microbiologically sound because this will involve a heat step. Rework that is exposed to the environment is subject to environmental contaminants and should be hygienically handled.

V Reduced-fat spread

A Definitions

Much of the above information about margarine (with over 80 fat) is equally applicable to many of the spreads with a reduced fat content. The difference is relative rather than absolute, as long as the spreads are true water-in-oil emulsions. Generally, water-in-oil spreads vary in fat content between 20% and 80%; below 20% fat spreads are oil-in-water emulsions. Products with a fat content between 41% and 62% are commonly referred to as reduced-fat spreads, whereas products with a fat content between 10% and 41% are called low-fat or light spreads (EC, 1994). Spreads made exclusively from milk fat are commonly referred to as dairy spreads (EC, 1994; CAC, 2001b). Spreads made from vegetable and animal fats, with a milk fat content of 10–80% of the total fat content, are called "blends" (EC, 1994). Other spreads are called "fat spreads" (EC, 1994; CAC, 2003) or "non-dairy spreads", if they contain ≤3% milk fat of the total fat content. The following is applicable for a typical reduced fat-spread (water-in-oil; about 40% fat content); other spreads are not discussed separately.

B Important properties

There are several differences between an 80% fat and a 40% fat water-in-oil emulsion (Keogh *et al.*, 1988):

- The moisture content of the reduced-fat product is higher and salt and other water-soluble preservatives (e.g. organic acids) are diluted to such an extent that they lose their effect on microbiological stability. This dilution factor cannot be compensated for by adding more salt or acid because an unacceptable taste results.
- In order to structure the aqueous phase of the reduced-fat product, there is a tendency to add biopolymers (vegetable or animal protein, thickeners), which may increase the microbiological vulnerability of the aqueous phase.
- The water droplet size distribution of the reduced-fat spread often has a higher mean diameter and a larger distribution width compared to margarine products because of physical effects.

As a consequence, reduced-fat spreads are more vulnerable to microbiological problems than margarines. This increased vulnerability may partly be counteracted by introduction of in-line pasteurization, use of a preservative or increased attention to process line hygiene and equipment design. Preservatives generally allowed are sorbic acid (EC, 1995) and benzoic acid or their salts; permissible concentrations in, for example, 0.1% in $\geq$ 60% fat spreads and 0.2% in $<$60% fat spreads. To ensure a sufficient level of the undissociated acid, the pH should be sufficiently low (i.e. $<$5.3; best $<$5.0). A pH of $<$4.5 is generally not possible because of precipitation of dairy proteins. The salt level in the aqueous phase of a reduced-fat spread is generally lower than in margarine or butter, except for spreads containing 70–80% fat. The salt level of the product ranges in general from 0% to 1% and rarely contributes to the microbiological stability of the spread.

The structure of reduced-fat spreads in principle is similar to that of butter or margarine, although the aqueous phase may be structured by thickeners and larger water droplets may occur that make the water dispersion coarser (Keogh *et al.*, 1988; Madsen, 1990; van Zijl and Klapwijk, 2000). Reasonable emulsion characteristics for a 40% fat spread containing protein are, for instance, 50% of the water volume in droplets with diameters $<15\,\mu$m and less than 5% in droplets with diameters $>90\,\mu$m ($\overline{D}_{3.3} = 15\,\mu$m, $e^{\sigma} = 3.0$). Reduced-fat spreads lacking protein are easier to produce and often have a finer water dispersion, e.g. 50% of droplets have ø $<10\,\mu$m and $<$5% of droplets have ø $>45\,\mu$m ($\overline{D}_{3.3} = 10\,\mu$m, $e^{\sigma} = 2.5$). Fat-continuous spreads containing 20% fat often have a very coarse water dispersion, e.g. 50% of droplets have ø $<50\,\mu$m and $<$5% of droplets have ø $>300\,\mu$m ($\overline{D}_{3.3} = 50\,\mu$m, $e^{\sigma} = 30$). Next to water dispersion, the stability of the emulsion is important for microbiological stability. Coalescence of droplets can release water at the product surface allowing, for instance, mold growth.

C Methods of processing and preservation

The manufacturing process for reduced-fat spreads is comparable to that of margarine. Strict hygiene is necessary where the aqueous phase forms a water continuum (pre-emulsion phase). Also the microbiological stability of the complete formulation will determine the hygienic requirements of the production process (Klapwijk, 1992). With vulnerable formulations, in-line pasteurization of the complete emulsion is applied. Pasteurization aims at a 5–6 log reduction of vegetative microorganisms, for instance by applying 70°C for 2 min. Growth of microorganisms before pasteurization should be prevented, because metabolic products (e.g. enzymes, off-flavors, toxins) might remain and affect the stability, quality, or the safety of the spread. It is advisable to decontaminate potentially suspect raw material before addition to the aqueous phase or pre-emulsion. Interim and finished product should be tested regularly to verify that microbial growth (including that of thermophiles) does not occur during processing. Equipment should be designed to prevent contamination during processing, especially after the last heat-decontamination step, and it should be possible to properly clean and disinfect the process line. If the downstream equipment is a closed system preventing contamination from outside, microbiological stability will be obtained (Lelieveld and Mostert, 1992). The pasteurize and the configuration of accompanying pipe-work should safeguard against under-pasteurization and should prevent leakage from the cooling system to the product and, thus, recontamination after pasteurization. Proper hygiene during packing is also important, especially with regard to contamination by molds. The most vulnerable products may require special measures such as a laminar flow cabinet with sterile air and decontamination of packing materials (comparable to aseptic packing of liquid products).

Post-processing temperature abuse may be another factor to account for, as it may cause condensation of moisture on the surface of the spread and destabilization of the physical product structure. Both factors can compromise the preservative effects of the compartmentalized structure providing opportunity for incidental (cross-)contaminants to proliferate. This may be especially relevant for the large-size, multiple-use containers commonly used in food-service establishments, which also get more popular for household in parts of the world. Challenge tests can help evaluate whether product formulations are sufficiently robust under abuse conditions. Cirigliano and Keller (2001), surface inoculated different commercial margarine and reduced-fat spreads with *L. monocytogenes* and found no growth for 7 days during storage between 5 and 23°C. Holliday *et al.* (2003) challenged two reduced-fat spreads (pH 4.05 and 49% fat; pH 5.37 and 61% fat) and a light-margarine (pH 5.34 and 31% fat), all containing preservatives, after temperature abuse (keeping products for 1 h at 37°C, 85% relative humidity) and physical abuse and stored the products at 4.4 or 21°C for up to 21 days. These products were found to be sufficiently robust, as products did not support growth of any of the pathogens at either temperature for the duration of storage. In a related study, Holliday and Beuchat (2003) showed that the inactivation rate of individual pathogens in commercial yellow-fat spreads products varies according to product formulation (i.e. pH, emulsion characteristics, salt, fat, and preservatives content) and storage temperature.

Closed shelf-life of reduced-fat spreads is generally limited to 3–6 months, often for non-microbiological reasons. The more sensitive product formulations may require refrigerated storage, especially during consumer use. During consumer use, the product may become contaminated with a variety of microorganisms originating from the air, bread crumbs, or from other foods. With very vulnerable products, e.g. non-preserved low-fat spreads containing a dairy ingredient or water-continuous spreads, the closed shelf-life duration may be quite limited (e.g. 2–3 weeks) due to microbiological reasons. For most spreads, the open shelf-life duration is, however, not limited.

D *Microbial spoilage and pathogens*

Spoilage. Many reduced-fat spreads, in principle, may allow growth of yeasts, molds, spoilage bacteria (e.g. Enterobacteriaceae, pseudomonads, aerobic spore-formers), and even pathogenic bacteria (discussed below) when present or introduced during production/open shelf-life. Compared with margarine, the aqueous phase is often more vulnerable, whereas the final emulsion exerts less protection. Most frequently, microbiological problems are caused by molds of the same types as those causing spoilage of butter and margarine. The more coarse and less stable emulsions of reduced-fat spreads may favor mold-spore germination, mycelial development, and sporulation. Preservatives such as benzoate and sorbate greatly reduce mold problems during distribution and consumer use, provided they applied at the appropriate concentration and pH.

Pathogens. Although there are no reports of food-borne illnesses caused by the consumption of reduced-fat spreads, the microbiological vulnerability of this type of product requires use of suitable raw materials, pasteurized when necessary to assure absence of vegetative infectious pathogens, and attention for contamination by such pathogens during the manufacturing processing.

E CONTROL (reduced-fat spread)

Summary

Significant hazards[a]	Although there are no reported cases of food-borne illness associated with consumption of reduced-fat spreads, significant hazards to which adequate control measures need to be established are: • *Salmonella* spp. • *E. coli* O157:H7. • *L. monocytogenes*.
Control measures	
Initial level (H_0)	• Use appropriate ingredients (source from approved supplier and/or pasteurize ingredients in-house before use). • Use potable water from a reliable source.
Reduction (ΣR)	• Use preservatives where appropriate. • Pasteurise water or pre-emulsion phase; vulnerable formulations may require in-line pasteurization of the complete emulsion.
Increase (ΣI)	• Use appropriate product structure: i.e. an appropriately fine water-in-oil dispersion to prevent or limit growth of microorganisms. • Avoid (re-)contamination; very vulnerable formulations may require specific facilities for aseptic handling. • Use suitable, hygienic equipment and process hygienically (incl. proper cleaning). • Keep end-product refrigerated (closed and open shelf-life); store dry and prevent condensation.
Testing	• Ingredient specification and water quality should be verified. • Interim and finished product should be tested regularly to verify process performance. • Routine end product testing is not advised when a stable product formulation/structure is used and process control is adequate.
Spoilage	• Generally, spoilage problems are controlled by using a stable formulation and an appropriate product composition, i.e. correct pH, clean ingredients, and appropriate water dispersion. Special attention needs to be given to the quality (specifications) of the stabilizing agents.

[a]In particular circumstances, other hazards may need to be considered.

Control measures. Considering the relative vulnerable nature of a reduced-fat spread formulation, a process design based on a cold manufacturing process alone may not be sufficient to achieve adequate microbiological product safety and stability. The use of suitable (pasteurized) raw materials is essential, in-line pasteurization is commonly employed and recontamination during processing (incl. packaging) should be avoided. In the case of very sensitive products, refrigerated distribution should be considered.

Raw materials. To meet the right microbiological standards, raw materials must be selected as in the case of margarine. Special attention must be given to materials that are used to stabilize the aqueous phase such as thickeners, proteins and gelatin (van Zijl and Klapwijk, 2000). In general, gelatin will contain only low numbers of thermophilic spore-formers because of the way it is manufactured, but these can grow rapidly, for instance in a stock solution of gelatin in water at 50–55°C. A low amount of potassium sorbate may help preserve the gelatine stock solution. Starches may contain *Salmonella* spp. and occasionally contain high numbers of bacterial spore-formers, including *Bacillus* (particularly *B. cereus* in, for instance, rice starch). Typical microbiological specifications for starches and gums are aerobic plate count $<10^4$ cfu/g, Enterobacteriaceae <100 cfu/g, yeasts and molds <500 cfu/g, *B. cereus* <1000 cfu/g, absence of infectious pathogens. Because the finished product may include bacterial spores, its composition and the physical structure should not allow their outgrowth.

Processing. Although in-line pasteurization is often applied, the microbiological status of the raw materials and of the processing facilities before pasteurization needs to be appropriate. Although vegetative microorganisms may be killed by pasteurization, products of their metabolism (which might not be destroyed) remain. Intermediate product should be tested to verify that microbial (including thermophiles) growth does not occur in the aqueous phase. It is advisable to decontaminate any potentially suspect raw material, apply heat pasteurization to the aqueous phase or the complete emulsion and avoid recontamination by physical measures to effectively control infective pathogen hazards.

Composition. Reduced-fat spreads are mildly preserved products. Therefore, it is important to monitor all aspects of product composition that will affect their vulnerability toward microbiological spoilage. The level of preservative, the equilibrium pH of the aqueous phase in the finished product, organic acids, and emulsion characteristics are all important properties that need to be verified.

Packaging. Because reduced-fat spreads are particularly mold-sensitive, adequate measures will specifically control mold contamination after the pasteurization stage and at packaging (incl. control of air and packaging material quality). Actually, proper hygiene during packaging will control spoilage in general, but the efficacy of the control measures (incl. cleaning and disinfection of the packaging machine) must be monitored.

Distribution. Chilled distribution reduces the chance of mold problems during closed shelf-life and, in most cases, prevents the growth of other microbial contaminants as well. Refrigeration is certainly beneficial during open shelf-life, where consumer use may cause contamination with for instance spoilage microorganisms. Low temperatures may also be required to maintain the desired physical properties of the product.

VI Butter

A *Definitions*

Butter consists of at least 80% milk fat, a small amount of non-fat milk solids, water and, most often, salt (sodium chloride). The fat phase is the continuous phase. A lactic starter culture, colors, and neutralizing agents may be used during manufacture. In the United States, Western-Europe and many other countries the composition of butter is strictly regulated: e.g. "the product with a milk fat content of not less than 80% but less than 90%, a maximum water content of 16%, and a maximum dry non-fat milk-material content of 2%" (EC, 1994). It is normally not permitted to add preservatives other than salt or anti-oxidants to butter (Murphy, 1990).

Consumers in Europe favor cultured butter, which is unsalted, or at most salted to a level not exceeding 0.5% (w/w). Sweet cream butter, often with 1.5–2% (w/w) salt, is sold almost exclusively in countries like UK, US, Australia, Japan, and India. Occasionally, consumers prefer unsalted, sweet butter. Cultured butter contains lactic acid instead of lactose; it may be of neutral pH, but usually has a pH of 4.6–5.3. The primary ingredient for butter is cream, and the initial microflora of butter is therefore to a large extent derived from the cream. In microbiological terms, unsalted, sweet butter is the most unstable type of butter.

B *Important properties*

Like margarine, butter is a water-in-oil emulsion in which the water is finely dispersed (Keogh *et al.*, 1988; Madsen, 1990; van Zijl and Klapwijk, 2000). Different processing regimes may result in different microstructures. Microscopic methods have been developed to characterize the water dispersion in butter. Electron-microscopic examinations demonstrated that, in properly worked butter, the aqueous phase is almost exclusively present as isolated, globular, or elongated droplets with diameters $<30\,\mu m$, covered with high-melting butter fat crystals (Buchheim and Dejmek, 1990). When the water in butter is compartmentalized in sufficiently small droplets that are well dispersed throughout the product, outgrowth of microorganisms will be limited and die-off may occur during storage. When the droplets are too coarse or when water channels have been formed, the aqueous phase composition (notably pH and salt-in-water concentration) becomes important for the keeping quality (Jensen *et al.*, 1983).

C *Methods of processing and preservation*

The processes for making butter have shown a shift away from batch to continuous processing (IDF, 1986b, 1987; Murphy, 1990, van Zijl and Klapwijk, 2000). The technology is quite advanced and cleanable-in-place, stainless steel equipment is common (IDF, 1996). Neutralization of cream is accomplished in some countries by the addition of alkaline salts, like calcium oxide, sodium carbonate, or sodium hydroxide. Neutralization prevents excessive loss of fats during churning and eliminates objectionable oily or fishy flavors. This treatment has no bactericidal effect on the microbiological population of cream.

The butter process can be roughly divided into two stages: (1) a churning stage, in which the fat in the cream is concentrated to about 80% after which the so-called buttermilk is drained off, and (2) a working stage, in which the concentrated cream is worked into butter. In between, ingredients such as salt, colors, and concentrated starter cultures may be added.

The butter process starts with pasteurized cream. Pasteurization is designed to destroy relevant vegetative microorganisms, especially pathogens, but will not eradicate bacterial spore-formers and some of the more heat-resistant vegetative spoilage flora. Pasteurization also aims at the destruction of enzymes in the raw cream that may reduce the organoleptic acceptability of the product and will completely liquefy the fat for subsequent control of crystallization (IDF, 1986b; Varnam and Sutherland, 1994). It is usual to pasteurize the cream after separation, even when the original milk is pasteurized (Murphy, 1990). Heat treatments (plate heat exchanger) involve temperatures in the range of 85–112°C (IDF, 1986b); 85–95°C for 10–30 s is most commonly used (Varnam and Sutherland, 1994). Batch pasteurization at longer holding times is obsolete or confined to low technology plants in some areas of the world (van den Berg, 1988). Since there are no further decontamination treatments after the pasteurization step, subsequent processing should be designed to prevent the multiplication of those microorganisms surviving pasteurization and to minimize recontamination.

Churning of the cream involves intensive agitation at 5–7°C, due to which the fat globules in the cream are converted to fat granules. The granules are then separated from the other constituents of the cream, leaving buttermilk. This results in a two-fold concentration of the fat in the cream. Most microorganisms are retained in the buttermilk, the aqueous phase in which the bacterial count is greater than that of the cream or the butter produced from it. Salt will be added, when at all, after the last point from which buttermilk drainage can occur. A concentration of 2% salt in product with 16% water will result in 11% salt in the aqueous phase [salt/(salt + water)]. This increased salt concentration may inhibit microorganisms, although microbial deterioration may occur even in highly salted butter when the salt is not uniformly distributed over the compartmentalized aqueous phase.

In mainland Europe, most butter traditionally has been made from soured cream using starter cultures and an aqueous phase pH of about 4.6. Since the resulting sour butter milk has limited market value, increasingly more use is being made of the so-called NIZO process which starts with sweet cream and yields sweet buttermilk. In this process, a (concentrated) lactic starter permeate (containing a high concentration of lactic acid) and, separately, an aromatic starter culture is worked directly into the butter after churning (Veringa *et al.*, 1976; IDF, 1986b). These natural concentrates are commercially available. The resulting sweet buttermilk, after drying, has a higher market value than dried acid buttermilk. Butter resulting from this process has a desirable sour taste and aroma. It also has a reduced copper content and is therefore less sensitive to oxidation.

The butter is worked mechanically after churning to obtain the right physical properties, with the water being dispersed into minute droplets in a fat-continuous matrix. As with margarine, this greatly increases the shelf-stability of the product. A major process aim is to obtain small water droplets. Both under-working and over-working produce a final butter with too coarse water droplet size distributions or even free moisture. As droplet size distributions are skewed with a tail of larger droplets, this means that about 50% of the total water volume should be in droplets with diameters $<3\ \mu m$, while at the same time less than 5% of the water volume should be in droplets with diameters $>10\ \mu m$ (van Zijl and Klapwijk, 2000).

Butter may be packed in bulk (usually up to 25-kg units) or in consumer-size portions (10–500g). Parchment has traditionally been used as the primary packing material, but other materials have become available (IDF, 1987). Bulk-packed butter may be stored for long periods at very low temperatures, e.g. for 6 months at $-15°C$ or a year at $-30°C$. Consumer-size portions should always be stored at chill temperatures ($\leq 10°C$) and even then have a storage time in general limited to 6–12 weeks. Re-packing of butter may result in enlargement of the water droplets, making the product microbiologically less stable.

D Microbial spoilage and pathogens

Initial microflora. The microflora of butter is derived mainly from the cream used. The nature, sources, and control of microorganisms likely to be present in industrial fresh milk and cream are discussed in Chapter 16. In the modern dairy situation, milk is collected on farms in refrigerated bulk tanks and transported by road tankers to the dairy factory. The microbiological quality of farm-separated cream differs considerably from that of fresh cream separated at a dairy plant, but farm-separation nowadays only occurs in areas with less advanced dairy industries. For instance, when cream is held on a farm for a week under poor hygienic conditions and without refrigeration, souring (e.g. by growth of *Lactococcus lactis* and other undesirable microorganisms) may have occurred, growth of yeasts and molds (e.g. *Geotrichum candidum*) may be abundant and Gram-negative aerobic bacteria (e.g. members of the genera *Pseudomonas*, *Alcaligenes*, *Acinetobacter/Moraxella* and *Flavobacterium*) may have proliferated, resulting in proteolytic and lipolytic changes (Foster *et al.*, 1957).

Spoilage. Butter manufacture is a fairly sensitive process from a microbiological point of view and, therefore, microbial spoilage needs to receive due attention. Although farm-separated cream may be badly deteriorated, subsequent processing such as neutralization, vacuum treatment and the use of butter cultures may eliminate and/or mask off-flavors. The butter from this cream is acceptable, although much inferior to that made from good-quality factory-separated cream.

Yeast and molds are important spoilage microorganisms of butter and can result in surface discoloration and off-flavor (references in Varnam and Sutherland, 1994; van Zijl and Klapwijk, 2000). A variety of mold genera (*Penicillium*, *Oospora*, *Mucor*, *Geotrichum*, *Aspergillus* and *Cladosporium*) have been implicated. Some yeasts are lipolytic and can grow in the presence of high concentrations of salt, at low pH, and at low temperatures. In retail samples of butter, *Rhodotorula* spp., *Saccharomycopsis lipolytica*, *Cryptococcus laurentii*, and *Can. diffluens* were reported as the predominant yeasts (Fleet and Mian, 1987).

Psychrotrophic Gram-negative bacteria such as *Pseudomonas* spp. and *Flavobacterium* spp. may develop and cause off-odor formation and rancidity (Driessen, 1983; Jooste *et al.*, 1986; Champagne *et al.*, 1993). Growth of *Alteromonas putrefaciens* or *Flavobacterium malodoris* may lead to surface taints (Foster *et al.*, 1957; Jooste *et al.*, 1986) very quickly affecting the mass of the product and accompanied by the development of a putrid, decomposed, or cheesy flavor (apparently from isovaleric acid or a closely related compound). Certain *Pseudomonas* spp. are associated with the formation fruity odors or black discolorations of the butter (Foster *et al.*, 1957). A variant of *Lactococcus lactis* (formerly *Streptococcus lactis* var. *maltigenes*) may cause the so-called malty-flavor defect related to the formation of 3-methylbutanal (Jackson and Morgan, 1954). When the organism grows extensively in cream before pasteurization, the flavor can carry over to the pasteurized cream.

Rancid flavor comes mainly from free butyric acid that arises from the hydrolysis of butter fat. This reaction may be catalyzed by naturally occurring lipase in the milk or by enzymes secreted by microorganisms. Milk lipase is destroyed by pasteurization (Driessen, 1983). Most, but not necessarily all, lipases of microbial origin will be fully inactivated by pasteurization. The carry-over of residual enzyme activity, bacterial toxins, and off-flavor to the pasteurized cream are the major reasons to strictly limit microbial growth in non-pasteurized cream, even though the previously mentioned microorganisms will all be destroyed by proper pasteurization.

Pathogens. Since the milk used for the production of butter may carry vegetative pathogens (e.g. *L. monocytogenes*, *E. coli* O157:H7; Chapter 16), pasteurization of the milk upon arrival at the dairy factory and measures to control of post-pasteurization recontamination are important for microbiological safety (Murphy, 1990; van Zijl and Klapwijk, 2000). Essentially all commercial butter is made from pasteurized cream and since the physical–chemical characteristics of butter (notably the fine water dispersion and fat continuity) are more inhibitory than those of pasteurized milk, butter should present even less of a food-borne disease problem than pasteurized milk. Sims *et al.* (1969) reported that butter made from inoculated cream supported growth of *Salmonella* at 25°C, while populations decreased during storage below 4.4°C. El-Gazzar and Marth (1992) reported on large decreases in *Salmonella* for unsalted butter stored at −17.8°C and 23.3°C. Lanciotti *et al.* (1992) observed growth of *L. monocytogenes* in light butter and concluded that the pathogen might be less affected by space and the nutritional limitations of the compartmentalized structure compared with butter. Holliday *et al.* (2003) investigated the survival and growth characteristics of mixtures of 5 serotypes of *Salmonella*, five strains of *E. coli* O157:*H*7, and 6 strains of *L. monocytogenes* in 3 types of commercial butter: sweet cream whipped salted butter (pH 6.4; 78% fat), sweet cream whipped unsalted butter (pH 4.51; 78% fat), and salted light butter (pH 4.58; 43% fat; with preservatives). The products were subjected to temperature abuse, by holding at 37°C under high relative humidity (85%) for 1 h to induce condensation of water on the surface, before storing at 4.4°C or 21°C for up to 21 days. Sweat cream whipped salted butter supported surface growth of all three pathogens and of only *L. monocytogenes* during storage at 21°C and at 4.4°C, respectively. The other two products did not support growth of any of the three pathogens at either temperature for the duration of storage. All pathogens tested were inactivated more rapidly in products stored at 21°C than at 4.4°C, and in products containing preservatives and acidulants. It is advisable to conduct challenge studies to determine pathogen survival and growth in butter, especially under abuse conditions, and assure that product characteristics include additional hurdles to pathogen growth. A known antimicrobial such as garlic might be considered to provide an additional defence to growth of certain pathogens (Zhao *et al.*, 1990), although an outbreak of *Campylobacter enteritidis* involving garlic butter has been reported (Anonymous, 1996). Adler and Beuchat (2002) investigated the viability of mixtures of different sero-types and strains of three pathogens (5 serotypes of *Salmonella*, five strains of *E. coli* O157:H7 and 6 strains of *L. monocytogenes*) inoculated onto unsalted butter with or without garlic in dependence of the storage temperature. None of the pathogens was able to grow, although they all retained viability at 4.4°C regardless of the presence of garlic; addition of garlic speeded-up inactivation at 21°C and 37°C.

A number of cases concerned contamination with *Staph. aureus* or its toxin, but over the years a variety of vegetative pathogenic bacteria were implicated as well. Nevertheless, considering the extremely large volumes that butter is marketed on, there are relatively few reported cases of butter associated food-borne illness to date.

In 1970, butter was voluntarily recalled by a manufacturer in the United States because of excessively high numbers of bacteria and the suspicion of staphylococcal contamination (Anonymous, 1970). In the same year, a case of food poisoning occurred in a restaurant implicating whipped butter prepared from butter contaminated by *Staph. aureus* (US-DHEW, 1970). Notably, butter from the same brand was implicated as the cause of a single case of typical staphylococcal food poisoning and was shown to contain staphylococcal enterotoxin A. (US-DHEW, 1970). In 1977, whipped butter produced by a single manufacturing plant in the United States was implicated in a multi-state outbreak of presumed staphylococcal food poisoning, which involved over 100 people; up to 10^7 cfu/g *Staph. aureus* were isolated from lots of the whipped butter involved (US-DHEW, 1977). Whipping of butter increases the risk of *Staph. aureus* growth, because water or milk may be added to the butter before whipping and because the salt level in the final product is reduced. Therefore, the product should be subjected to strict hygiene, refrigerated storage ($\leq$7°C) and a restricted shelf-life. The toxin of *Staph. aureus* is very heat stable (Brunner *et al.*, 1991) and, once formed as a result of poor sanitary conditions in the pre-pasteurization stage, might carry over to the pasteurized product.

A food poisoning caused by consumption of blended margarine and butter products contaminated by *Staph. intermedius* occurred in the United States in 1991 and involved over 265 people; since a single strain was identified as the causative agent, a common source of contamination rather than post-process contamination was suggested (Anonymous 1992b; Khambaty *et al.*, 1994; Bennet, 1996). From Germany, an incident with green butter contaminated with *Citrobacter freundii* was reported; the butter was made by adding organically grown parsley and a genetically identical *Cit. freundii* was isolated from most patients and from the parsley of the organic garden (Tschäpe *et al.*, 1995). An outbreak in 1995 of *C. enteritis* involved about 30 people who had eaten at a restaurant in Louisiana and appeared to be associated with garlic butter prepared by the restaurant (Anonymous, 1996). In several cases, *L. monocytogenes* (Marth and Ryser, 1990; Massa *et al.*, 1990; Harvey and Gilmore, 1992) or *Salmonella* spp. (Cavalcante dos Santos *et al.*, 1995) contamination was suspected, although these bacteria were not detected in the samples studied. In a case-control study of a cluster of peri-natal listeriosis cases in California carried out over a period of 6 months in 1987–1988, butter was identified as a possible vehicle for infection (Chun *et al.*, 1990) although no direct evidence was obtained. A recall in the United States in 1994 involved unsalted and lightly salted butter contaminated with *L. monocytogenes* (Anonymous, 1994; Proctor *et al.*, 1995), but no cases of food poisoning were linked to this contamination.

A serious outbreak of food poisoning occurred in 1999 and implicated butter from a Finnish dairy plant. Butter in small-sized packages (7 and 10 g) were found to be contaminated with *L. monocytogenes* serotype 3A (Lyytikäinen *et al.*, 1999, 2000). Poor sanitation within the plant led to contamination of the butter packages. Eighteen people developed listeriosis, four of whom died. The mean age of patients was 57 years (range 18–85); all had serious underlying illnesses and were undergoing treatment in several hospitals. Most of the disease-causing strains (14/18) were of a rare serotype 3a which was isolated from samples taken from different places in the butter production facility, including environmental and butter samples from the dairy plant and butter samples from the cold store of the dairy. Isolates were also obtained from the kitchen of a hospital. The isolates from butter and epidemic isolates were indistinguishable by pulsed-field gel electrophoresis (PFGE) and their PFGE-type had not been isolated previously from any other foods in Finland. *L. monocytogenes* was detected at low levels ($<10^2$ cfu/g) in most of the small butter packs examined. One pack contained $>10^4$ cfu/g, a level in food well documented to cause outbreaks of listeriosis. Estimates of the doses of the pathogen actually ingested range from 14–2200 cfu/day and 2.2×10^4 and 3.1×10^5, based on hospital kitchen data and contamination found in retail samples, respectively (Maijala *et al.*, 2001). *L. monocytogenes* may survive for months in butter when temperatures are very low (−18°C) and will grow slowly if temperatures are 4°C or 13°C (Olsen *et al.*, 1988). European Union Directives specify that *L. monocytogenes* should not be detectable in 1 g of butter.

E CONTROL (butter)

Summary

Significant hazards[a]	• *Salmonella* spp. • *E. coli* O157:H7. • *L. monocytogenes.* • *Staph. aureus.*
Control measures	
Initial level (H_0)	• Limit growth of microorganisms in cream/milk raw material by controlling time/temperature of storage before us. • Assure that the milk or cream used to start butter manufacture with is adequately pasteurized. • Use appropriate ingredients (source from approved supplier or pasteurize ingredients in-house before use).
Reduction (ΣR)	• Use preservatives where appropriate. • Pasteurise before churning.
Increase (ΣI)	• Minimize possible growth in milk/cream before pasteurization (keep refrigerated). • Use appropriate product structure to prevent or limit growth of microorganisms; assure proper distribution of moisture and salt throughout the product. • Avoid (re-)contamination after pasteurization. • Use of hygienic equipment and process hygiene (incl. proper cleaning) is critical. • Store final product dry; prevent condensation.
Testing	• Ingredient specification and water quality should be verified. • Intermediate and finished product should be tested regularly to verify process performance. • Routine end product testing is not advised when a stable product formulation/structure is used and process control is adequate.
Spoilage	• Yeast and molds are the most relevant spoilage microorganisms. The end product should be kept refrigerated during open shelf-life. Freezing may allow very long shelf-life.

[a]In particular circumstances, other hazards may need to be considered.

Control measures. To achieve microbiologically safe and stable products, a process design based on a cold process may be completely adequate, but dairy raw materials and ingredients should be of suitable quality; pasteurization of the cream is commonly applied. As butter is a rather vulnerable product, strict processing-line hygiene, microbiological quality of the processing environment, and air humidity—all need to be carefully controlled. Appropriateness of the product composition and emulsion characteristics of the final product should be ascertained, especially with types of butter containing less than 80% fat and considering post-processing temperature abuse. Refrigeration is necessary during open shelf-life.

Raw materials. Cream should be of suitable microbiological quality. Suggested microbiological criteria for cream are: aerobic plate count <1000 cfu/mL; yeast, mold, and coliform counts <1 cfu/mL (Murphy, 1990). *Bacillus cereus*, which may be relevant for sweet, unsalted, or low-salted products with coarse water droplets, might be considered for monitoring purposes in view of its increasing occurrence in milk and the ability of some strains to germinate and grow slowly at refrigeration temperature (te Giffel *et al.*, 1996). To date no *B. cereus* related outbreaks have occurred. Growth of microorganisms in the milk or cream before use for butter should be kept to a minimum (CCP), by controlling time and temperature of storage.

Starter cultures used for souring of cream are responsible for lactic acid formation and a concomitant decrease in pH. Pure or mixed concentrated cultures of lactococci and leuconostocs from commercial suppliers will provide a great degree of control over the acidification process. The starter culture should not become a source of contamination. Therefore, the number of subcultures should be limited.

Ingredients such as salt, coloring agents (e.g. beta-carotene), and neutralizers (e.g. sodium carbonate) are generally free of microbial contamination because of the way they are manufactured; the chemicals should be of food-grade quality. When water is used in butter manufacture after pasteurization, e.g. for washing, the water should be of potable quality.

Use of a preservative in butter may be considered. A concentration of 0.033% undissociated sorbic acid on aqueous phase, for example, may be sufficient to control mold spoilage of a protein-containing 40%-fat spread (0.12% potassium sorbate on product, pH 4.9) during closed and open shelf-life (van Zijl and Klapwijk, 2000). To limit mold contamination, a laminar flow cabinet at the packaging stage may be necessary. Product (mold) spoilage may be limited further by storage at refrigeration temperature (≤7°C).

Processing. Pasteurization of the cream before churning is commonly done. When high numbers of microorganisms are present before pasteurization, not all hazardous microorganisms or the products of their metabolism may be inactivated by the usual pasteurization conditions. Bacterial spores are not inactivated by pasteurization and can germinate and grow when milk or cream of neutral pH is not sufficiently cooled. Any failure in pasteurization could lead to the survival of pathogens. Pasteurization is a critical control point when there are no further decontamination treatments down-stream in the butter manufacture process. Subsequent process steps should be designed such that multiplication of those microorganisms surviving pasteurization is prevented and recontamination is minimized.

Moisture and salt distribution is important with respect to the microbiological stability of butter. The pH is an important product parameter of sour cream butter. The verification programme should incorporate measurements of these factors and include trend analysis. If butter is used for making whipped butter or spiced butter, care should be taken that pathogens are not introduced due to poor hygiene or contaminated ingredients. After whipping or cold mixing, the time before using the product should be strictly limited even under refrigeration.

Recontamination of butter with microorganisms usually results from inadequate post-pasteurization hygiene (van Zijl and Klapwijk, 2000). Care should be taken that the equipment is properly cleaned and disinfected before start-up of the process. Preventing recontamination is also a function of equipment design, maintenance, cleaning and disinfection procedures, and proper supervision. Process line hygiene is important for the microbiological verification program. Samples from the start-up of the process as well as from the end of the run should be analyzed.

Packaging. Product may be exposed to the environment during packaging and recontamination should be prevented. High humidity in the room and lack of proper ventilation may allow molds to grow on walls and ceilings, thus producing foci of contaminating organisms from which air currents may carry mold spores to the product. The packing material should be of a good microbiological quality. Of particular importance is the level of contamination with molds (i.e. their spores). Cardboard cartons used as shipping containers may be an important source of mold spores, especially when recycled cardboard is used. The cardboard packing material should therefore not be handled in the packing environment but in a separate room. When butter is moved from a cool area to an area of high humidity, e.g. at the time it is being printed and wrapped, moisture is apt to condense on the surface possibly triggering growth of mold spores. Careful handling is necessary to prevent damage to the wrappers. Repackaging butter (e.g. from large blocks to portion packs) may cause the spread in localized areas of microbial growth to the entire production lot. Hence, butter that is repackaged should be fresh or of extremely good microbiological quality. Frozen butter must be thawed in a dry and well-ventilated room to prevent moisture condensation on the butter. Butter should be kept free from moisture during distribution. Cool, dry storage without exposure to high humidities during transport is necessary to prevent mold spoilage problems.

VII Water-continuous spreads

Water-continuous products with fat contents as low as 3% have been introduced successfully on the market. These oil-in-water products are developed for spreading on bread as low-calorie alternatives. Water-continuous spreads lack the protective effect of the fat barrier, but this can partly be compensated for by adding more preservative. Nevertheless, the products are still likely to be more sensitive to microbial growth. Although they are used as if they were a fat-continuous yellow fat spread, it should be realized that their preservation relies partly on a different principle. Water-continuous spreads may actually be so vulnerable to mold spoilage that special measures such as a laminar flow cabinet with sterile air and decontamination of packing materials may be necessary, comparable to aseptic packing. Water-continuous spreads will always require refrigerated storage. The shelf-life typical of these products may be limited, for microbiological reasons, to 2–3 weeks.

The physical structure is critical for product stability and this is based on the physical stability of the emulsions discussed in the previous sections of this chapter. Physical stability can be ensured by a proper product design, appropriate combinations of temperature–time for storage and distribution, and proper handling during open and closed shelf-life.

Other intrinsic factors contributing to stability, either alone or in combination, are salt-on-water, pH, limited availability of nutrients and presence of preservatives (e.g. sorbic acid; benzoic acid; bacteriocins or diacetyl resulting from lactic acid bacteria involved in milk fermentation). Although 0.033% undissociated sorbic acid on aqueous phase may be sufficient to control mold spoilage during refrigerated shelf-life in a protein-containing, aseptically packaged 40%-fat spread (0.12% potassium sorbate on product, pH 4.9), water-continuous spreads may require higher levels of undissociated sorbic acid (i.e. >0.04%) to prevent mold spoilage under these conditions. Higher levels of undissociated sorbic acid can be reached by increasing the sorbic acid level or by decreasing the pH. The pH level of water-continuous spreads, however, is not much lower than 4.9 because lower values cause precipitation of dairy proteins.

Up to now, water-continuous spreads have not been associated with any serious microbiological problems, probably because considerable attention is paid to their production (including careful product design, high quality starting materials, and proper process hygiene) and maybe also because they have not been marketed extensively. However, considering the trend for major product innovations in this field and the appearance of new, sensitive types of products on the market-place, it should be emphasized that microbiological knowledge and examinations need to be kept up-to-date for such potentially problematic products as water-continuous spreads. Additionally, it is advisable to validate new product and processing concepts using appropriate challenge tests.

VIII Miscellaneous products

This group includes butter oil, ghee, vanaspati, cocoa butter substitutes, and cooking oils. All these products have an extremely low-water content (<0.5%) and, therefore, generally do not allow microbial growth. However, when stored under moist conditions, mold spoilage may occur on the product surface. Also survival of infectious pathogens, in principle, is possible although there is no epidemiological evidence to indicate that this occurs in practice.

Some specific properties are:

- Butter oil is milk fat separated from cream or butter after heat destabilization; it is purely milk fat without non-fat milk solids.
- Ghee is prepared by cooking (up to 120°C) butter or concentrated cream under atmospheric pressure to evaporate the water; as a result, ghee differs from butter oil by taste and the presence of non-fat milk solids.
- Vanaspati is a blend of butter oil and vegetable fat; the butter oil from ghee or fermented cream is used to give the product the correct flavor.
- Cocoa butter substitute is a purely vegetable fat product used to replace expensive cocoa butter in some applications; cooking oils consist mostly, but not exclusively, of vegetable oil.

References

3-A (3-A Sanitary Standards Inc.). (2003) http://www.3-a.org/standards/standards.htm.

Adler, B.B. and Beuchat, L.R. (2002). Death of *Salmonella*, *Escherichia coli* 0157:H7, and *Listeria monocytogenes* in garlic butter as affected by storage temperature. *J. Food Prot.*, **65,** 1976–80.

Alderliesten, M. (1990) Mean particle diameters. Part I: Evaluation of definition systems. *Particle Particle Syst. Charact.*, **7**, 233–41.

Alderliesten, M. (1991) Mean particle diameters. Part II: Standardization of nomenclature. *Particle Particle Syst. Charact.*, **8**, 237–41.

Anonymous. (1970) Firm voluntarily recalls butter because of contamination. *Food Chem. News*, **12** (13), 33.

Anonymous. (1988) *Salmonella enteritidis* phage type 4: chicken and egg. *Lancet*, **ii**, 720–2.

Anonymous. (1992a) FDA Enforcement Report, February, **26**, 3.

Anonymous. (1992c) Outbreak that didn't seem right spread like butter. *Food Prot. Rep.*, **8**, (7–8), 5–6.

Anonymous. (1993a) Cross-contamination/different strain in Oregon *E. coli* case. *Food Chem. News*, **35** (8), 32–3.

Anonymous. (1993b) Tainted mayo blamed in Sizzler *E. coli* cases. *Nation's-Restaurant-News*, April, **19**, 3.

Anonymous. (1994) Two class 1 recalls caused by *Listeria*. *Food Chem. News*, **35** (48), 14.

Anonymous. (1995) El emporio de los sandwiches ofreció sus garantías para erradicar la "Salmonella". *El Pais*, 9 May.

Anonymous. (1996) *Campylobacter* retains viability in butter. *At a Glance* (Center for Food Safety and Quality Enhancement of the University of Georgia), **5** (2), 2.

APHA (American Public Health Association). (1972) *Proceedings of the 1971 National Conference on Food Protection*. Food and Drug Administration, USA.

Barnes, P.J. (1989) *Listeria*. A threat to margarine? *Lipid Technol.*, **1** (2), 46–7.

Baumgart, J., Weber, B. and Hanekamp, B. (1983) Mikrobiologische Stabilität von Feinkosterzeugnisse. *Fleischwirtschaft*, **63**, 93–4.

Beckers, H.J., Daniels-Bosman, M.S.M., Ament, A., Daenen, J., Hanekamp, A.W.J., Knipschild, R, Schuurman, A.H.H. and

Bijkerk, H. (1985) Two outbreaks of salmonellosis caused by *Salmonella indiana*. A survey of the European Summit outbreak and its consequences. *Int. J. Food Microbiol.*, **2**, 185–95.
Beerens, H. (1980) Hygiène des fabrications et propriétés bactériologiques des margarines. *Revue Française des Corps Gras*, **27**, 221–3.
Bennett, R. W., 1996. Atypical toxigenic Staphplococcus and *non-Staphylococcus aureus* species on the horizon? An update. *J. Food Protect.*, **59**, 1123–6.
van den Berg, J.C.T. (1988) *Dairy technology in the tropics and subtropics*. PUDOC, Wageningen.
Besser, R.E., Lett, S.M., Weber, J.T., Doyle, M.P., Barren, T.J., Wells, J.G. and Griffin, P.M. (1993) An outbreak of diarrhea and hemolytic uremic syndrome from *Escherichia coli* O157:H7 in fresh-pressed apple cider. *J. Am. Med. Assoc.*, **269**, 2217–20.
Bonestroo, M.H. (1992) Development of fermented sauce-based salads—assessment of safety and stability, *Doctoral Thesis*, Agricultural University, Wageningen, The Netherlands.
Bornemeier, V.L., Albrecht, J.A., and Sumner, S.S. (2003) Survey of mayonnaise-based salads for microbial safety and quality. *Food Prot. Trends*, **23** (5), 387–92.
Brocklehurst, T.F. and Lund, B.M. (1984) Microbiological changes in mayonnaise-based salads during storage. *Food Microbiol.*, **1**, 5–12.
Brocklehurst, T.F., White, C.A. and Dennis, C. (1983) The microflora of stored coleslaw and factors affecting the growth of spoilage yeasts in coleslaw. *J. Appl. Bacteriol.*, **55**, 57–63.
Brunner, K., Rodriguez, R. and Wang, A. (1991) *Staphylococcus enterotoxin production and thermal stability in mushrooms*. Food Research Institute, Wisconsin, Annual Report, 58–9.
Bryan, F.L. (1988) Risks associated with vehicles of foodborne pathogens and toxins. *J. Food Prot.*, **51**, 498–508.
Buchheim, W. and Dejmek, P. (1990) Milk and dairy-type emulsions, in *Food Emulsions* (eds K. Larsson and S.E. Friberg), Marcel Dekker, New York, pp. 203–46.
Bülte, M. (1995) Enterohämorrhagische *E. coli*-Stämme (EHEC)-Aktuell in der Bundesrepublik Deutschland? 1. Pathogenitätspotential von EHEC-Stämmen-Bedeutung als Lebensmittelinfektionserreger. *Fleischwirtschaft*, **75**, 1430–2.
CAC (Codex Alimentarius Commission). (2001a) Codex Alimentarius—Joint FAO/WHO Food Standards Programme. Food Hygiene—Basic Texts, 2nd ed (revised 2001), ISBN 9251046190.
CAC (Codex Alimentarius Commission). (2001b) *Fats, Oils and Related Products*, 2nd ed (Revised 2001), Joint FAO/WHO Food Standards Programme, *Volume 8*, ISBN 9251046824.
CAC (Codex Alimentarius Commission). (2003) Proposed draft standard for fat spreads and blended spreads (at step 6 of the procedure). Report of the eighteenth session of the codex committee on fats and oils, London, United Kingdom, 3—7 February 2003. Alinorm 3/17. Appendix IV. Available: ftp://ftp.fao.org/codex/alinorm03/Al03_17e.pdf.
Cavalcante dos Santos, E.G., da Costa Raimondo, S.M. and Guimaraes Robbs, P. (1995) Microbiological evaluation of butter purchased from the market of Rio de Janeiro. I. Indicator and pathogenic microorganisms. *Rev. Microbiol. Sao Paulo*, **26** (3), 224–9.
Champagne, P.C., Laing, R.R., Roy, D., Mafu, A.A. and Griffiths, M.W. (1993) Psychrotrophs in dairy products: their effects and their control. *Crit. Rev. Food Sci. Nutr.*, **34**, 1–30.
Chandrakumer, M. (1995) From outbreak to prosecution. *Int. Food Hyg.*, **5**, 27, 29.
Charteris, W.P. (1996) Microbiological quality assurance of edible table spreads in new product development. *Intl. J. Dairy Technol.* **49,** 87–98.
Chateris, W. P. (1995) Physicochemical aspects of the microbiology of edible table spreads. *Intl. J. Dairy Technol.* **48,** 87–96.
CIMSCEE (Comité des Industries des Mayonnaises et Sauces Condimentaires de la Communauté Économique Européenne. (1991) Code of Practice Mayonnaise, CIMSCEE.
Chun, L., Mascola, L., Thomas, J.C., Bibb, W.F., Schwartz, B., Salminen, C. and Heseltine, P. (1990) A case-control study of a cluster of perinatal listeriosis identified by an active surveillance system in Los Angeles County, California, September 1987–February 1988, in *Foodborne Listeriosis* (eds A.J. Miller, J.L. Smith and G.A. Somkuti), Elsevier, Amsterdam, 75–9.
Cirigliano, M.C. and Keller, A.M. (2001) Death kinetics of *Listeria monocytogenes* in margarine, yellow fat spreads, and toppings. *Program Abstracts Book, 88th Annual Meeting International Association of Food Protection*. p. 102.
Conner, D.E. and Kotrola, J.S. (1995) Growth and survival of *Escherichia coli* O157:H7 under acidic conditions. *Appl. Environ. Microbiol.*, **61**, 382–5.
Conner, D.E., Scott, V.N. and Bernard, D.T. (1990) Growth, inhibition, and survival of *Listeria monocytogenes* as affected by acidic conditions. *J. Food Prot.*, **53**, 652–5.
Davies, A.R., Slade, A., Blood, R. and Gibbs, P.A. (1992) *Effect of temperature and pH value on the growth of verotoxigenic E. coli*. Leatherhead Food Research Association, Report No. 691.
Davies, R.F. and Wahba, A.H. (1976) *Salmonella* infections of charter flight passengers. Report on a visit to Spain (Canary Islands) 26 February-2 March 1976. WHO Regional Office for Europe, Copenhagen.
Delamarre, S. and Batt, C.A. (1999) The microbiology and historical safety of margarine. *Food Microbiol.*, **16**, 327–33.
Doyle, M.P., Bains, N.J., Schoeni, J.L. and Foster, E.M. (1982) Fate of *Salmonella typhimurium* and *Staphylococcus aureus* in meat salads prepared with mayonnaise. *J. Food Prot.*, **45**, 152–6.
Driessen, F.M. (1983) Lipases and proteases in milk, *Doctoral Thesis*, Agricultural University, Wageningen, The Netherlands.
Duffy, G., Riordan, D.C.R, Sheridan, J.J., Call, J.E., Whiting, R.E., Blair, I.S. and McDowell, D.A. (2000) Effect of pH on survival, thermotolerance, and verotoxin production of *Escherichia coli* O157:H7 during simulated fermentation and storage. *J. Food Prot.*, **63**, 12–8.

EC (European Commission). (1994) European parliament and council regulation No 2991/94/EC laying down standards for spreadable fats. *Off. J. Eur. Commun.*, **L 316**, 2–7.

EC (European Commission). (1995) European parliament and council directive 95/2/EC on food additives other than colours and sweeteners. *Off. J. Eur. Commun.*, **L 61**, 1–40.

EHEDG (European Hygienic Engineering and Design Group). (2003) http://www.ehedg.org/guidelines.htm.

Eiguer, T., Caffer, M.I. and Fronchkowsky, G.B. (1990) Importancia de la *Salmonella enteritidis* en brotes de enfermedades transmitidas por alimentos en Argentina, años 1986–1988. *Rev. Argen. Microbiol.*, **22**, 31–6.

El-Gazzar, F.E. and Marth, E.H. (1992) Salmonellae, Salmonellosis, and dairy foods: a review. *J. Dairy Sci.*, **75**, 2327–43.

van de Enden, J.C., Waddington, D., van Aalst, H., van Kralingen C.G. and Packer, K.J. (1990) Rapid determination of water droplet size distributions by pulse field gradient-NMR. *J. Colloid Interf. Sci.*, **140**, 105–13.

Entani, E., Masai, H. and Suzuki, K.-I., (1986) *Lactobacillus acetotolerans*, a new species from fermented vinegar broth. *Int. J. Syst. Bacteriol.*, **36**, 544–9.

Erickson, J.P. and Jenkins, P. (1991) Comparative *Salmonella* spp. and *Listeria monocytogenes* inactivation rates in four commercial mayonnaise products. *J. Food Prot.*, **54**, 913–6.

Erickson, J.P., McKenna, D.N., Woodruff, M.A. and Bloom, J.S. (1993) Fate of *Salmonella* spp., *Listeria monocytogenes*, and indigenous spoilage microorganisms in home-style salads prepared with commercial real mayonnaise or reduced calorie mayonnaise dressings. *J. Food Prot.*, **56**, 1015–21.

Erickson, J.P., Stamer J.W., Hayes, M., McKenna, D.N. and Van Alstine, L.A. (1995) An assessment of *Escherichia coli* O157:H7 contamination risks in commercial mayonnaise from pasteurized eggs and environmental sources, and behavior in low-pH dressings. *J. Food Prot.*, **58**, 1059–64.

Finol, M.L., Marth, E.H. and Lindsay, R.C. (1982) Depletion of sorbate from different media during growth of *Penicillium* species. *J. Food Prot.*, **45**, 398–404.

Fleet, G.H. and Mian, M.A. (1987) The occurrence and growth of yeasts in dairy products. *Int. J. Food Microbiol.*, **4**, 145–55.

FAO/WHO (Food and Agriculture Organisation/World Health Organization). (1989) CODEX standard for mayonnaise (Regional European Standard) CODEX STAN 168-1989. ftp://ftp.fao.org/codex/standard/en/CXS_168e.pdf.

Foster, E.M., Nelson, F.E., Speck, M.L., Doetsch, R.N. and Olson, J.C. (1957) *Dairy Microbiology*, Prentice-Hall, Englewood Cliffs, New Jersey, USA.

Gander, K.-F. (1976) Margarine oils, shortenings and vanaspati. *J. Am. Oil Chem. Soc.*, **53**, 417–20.

Gaze, J.E. (1985) The effect of oil on the heat resistance of *Staphylococcus aureus*. *Food Microbiol.*, **2**, 277–83.

te Giffel, M.C., Beumer, R.R., Bonestroo, M.H. and Rombouts, F.M. (1996) Incidence and characterization of *Bacillus cereus* in two dairy processing plants. *Neth. Milk Dairy J.*, **50**, 479–92.

Glass, K.A. and Doyle, M.P. (1991) Fate of *Salmonella* and *Listeria monocytogenes* in commercial, reduced-calorie mayonnaise. *J. Food Prot.*, **54**, 691–5.

Glass, K.A., Loeffelholz, J., Harried, M. and Nelson, J.H. (1993) Survival of *Escherichia coli* O157:H7 in mayonnaise and mayonnaise dressing. Food Research Institute Annual Meeting, May 13.

Gomez-Lucia, E., Goyache, J.H., Blanco, J.L., Garayzabal, J.F.F., Orden, J.A. and Suárez, G. (1987) Growth of *Staphylococcus aureus* and enterotoxin production in homemade mayonnaise prepared with different pH values. *J. Food Prot.*, **50**, 872–5.

Gomez-Lucia, E., Goyache, J.H., Orden, J.A., Doménech, A., Hernández, F.J., Ruiz-Santa-Quiteria, J.A. and Suárez, G. (1990) Influence of temperature of incubation on *Staphylococcus aureus* growth and enterotoxin production in homemade mayonnaise. *J. Food Prot.*, **53**, 386–90.

Grahan, C.G.M., O'Driscoll, B. and Hill, C. (1996) Acid adaptation of *Listeria monocytogenes* can enhance survival in acid foods and during milk fermentation. *Appl. Environ. Microbiol.*, **62**, 3128–32.

Guerzoni, M.E., Lanciotti R., Westall, F. and Pittia, P. 1997. Interrelationship between chemico-physical variables, microstructure and growth of *Listeria monocytogenes* and *Yarrowia lipolytica* in food model system. *Sci. Alim.*, **17**, 507–22.

Harvey, J. and Gilmour, A. (1992) Occurrence of *Listeria* species in raw milk and dairy products produced in Northern Ireland. *J. Appl. Bacteriol.*, **72**, 119–25.

Hathcox, A.K., Beuchat, L.R. and Doyle, M.P. (1995) Death of enterohemorrhagic *Escherichia coli* O157:H7 in real mayonnaise and reduced-calorie mayonnaise dressing as influenced by initial population and storage temperature. *Appl Environ. Microbiol.*, **61**, 4172–7.

Hersom, A.C. and Hulland, ED. (1980) *Canned Foods. Thermal Processing and Microbiology*, Churchill Livingstone, London.

Hocking, A. D. (1994) Fungal spoilage of high-fat foods. *Food Australia*, **46**, 3–33.

Holliday, S.L. and Beuchat, L.R. (2003) Viability of *Salmonella*, *Escherichia coli* O157:H7, and *Listeria monocytogenes* in yellow fat spreads as affected by storage temperature. *J. Food Prot.*, **66**, 549–58.

Holliday, S.L., Adler, B.B., Beuchat, L.R. (2003) Viability of *Salmonella*, *Escherichia coli* O157 : H7, and *Listeria monocytogenes* in butter, yellow fat spreads, and margarine as affected by temperature and physical abuse. *Food Microbiol.*, **20**, 159–168.

Hüpper, H. (1975) Typhusepidemie in Baden-Württenberg 1974. *Bundesgesundheidsblatt*, **18** (9), 142–5.

ICMSF (International Commission on Microbiological Specifications for Foods). (1988) *Microorganisms in Foods 4: Application of the hazard analysis critical control point (HACCP) system to ensure microbiological safety and quality*, Blackwell, Oxford.

ICMSF (International Commission on Microbiological Specifications for Foods). (1996) *Microorganisms in Foods 5: Characteristics of Microbial Pathogens*, Blackie Academic & Professional, London.

IDF (International Dairy Federation). (1986a) Monograph on pasteurized milk. *Int. Dairy Fed. Bull.*, **200**.

IDF (International Dairy Federation). (1986b) Continuous Butter Manufacture. *Int. Dairy Fed. Bull.*, **204**, 1–36.
IDF (International Dairy Federation). (1987) Packaging of butter, soft cheese, fresh cheese. *Int. Dairy Fed. Bull.*, **214**, 3–11.
IDF, 1996. Codex standards in the context of world trade agreements—IDF General Recommandations for the Hygienic Design of Dairy Equipment. IDF Bulletin 310.
Jackson, H.W. and Morgan, M.E. (1954) Identity and origin of the malty aroma substance from milk cultures of *Streptococcis lactis*, var. *maltigenes*. *J. Dairy Sci.*, **37**, 1316–24.
Jensen, H., Danmark, H. and Mogensen, G. (1983) Effect of storage temperature on microbiological changes in different types of butter. *Milchwissenschaft*, **38**, 482–4.
Jooste, P.J., Butz, T.J. and Lategan, P.M. (1986) The prevalence and significance of *Flavobacterium* strains in commercial salted butter. *Milchwissenschaft*, **41**, 69–73.
Juriaanse, A.C. and Heertje, I. (1988) Microstructure of shortenings, margarine and butter—a review. *Food Microstruct.*, **7**, 181–8.
Kauppi, K.L., Tatini, S.R., Harrell, F. and Feng, P. (1996) Influence of substrate and low temperature on growth and survival of verotoxigenic *Escherichia coli*. *Food Microbiol.*, **13**, 397–405.
Keogh, M.K., Quigley, T., Connelly, J.F. and Phelan, J.A. (1988) Anhydrous milk fat. 4. Low-fat spreads. *Irish J. Food Sci. Technol.*, **12**, 53–75.
Khambaty, F.M., Bennett, R.W. and Shah, D.B. (1994) Application of pulsed-field gel electrophoresis to the epidemiological characterization of *Staphylococcus intermedius* implicated in a food-related outbreak. *Epidemiol. Infect.*, **113**, 75–81.
Klapwijk, P.M. (1992) Hygienic production of low-fat spreads and the application of HACCP during their development. *Food Control*, **3**, 183–9.
Koodie, L. and Dhople, A.M. (2001) Acid tolerance of *Escherichia coli* O157:H7 and its survival in apple juice. *Microbios*, **104**, 167–175.
Kurihara, K., Mizutani, H., Nomura, H. *et al.* (1994) Behavior of *Salmonella enteritidis* in home-made mayonnaise and salads. *Jpn. J. Food Microbiol.*, **11**, 35–41.
Lanciotti, R., Massa, S., Guerzoni, M.E. and DiFabio, G. (1992) Light butter: natural microbial population and potential growth of *Listeria monocytogenes* and *Yersinia enterocolitica*. *Lett. Appl. Microbiol.*, **15**, 256–8.
Leasor S.B. and Foegeding, P.M. (1990) *Listeria* species in commercially broken raw liquid whole egg. *J. Food Prot.*, **52**, 777–80.
Lelieveld, H.L.M. and Mostert, M.A. (1992) Hygienic aspects of the design of food plants, in *Food Production, Preservation and Safety* (ed P. Palet), Ellis Horwood Ltd., Chichester, UK.
Leuschner, R.G.K. and Broughtflower M.P. (2001) Standardized laboratory-scale preparation of mayonnaise containing low levels of *Salmonella enterica* serovar *enteritidis*. *J. Food Prot.*, **64**, 623–9.
Leyer, G.E. and Johnson, E.A. (1993) Acid adaptation induces cross-protection against environmental stresses in *Salmonella typhimurium*. *Appl. Environ. Microbiol.*, **59**, 1842–7.
Leyer, G.E., Wang, L.L. and Johnson, E.A. (1995) Acid adaptation of *Escherichia coli* O157:H7 increases survival in acid foods. *Appl. Environ. Microbiol.*, **61**, 3752–5.
Liewen, M.B. and Marth, E.H. (1984) Inhibition of penicillia and aspergilli by potassium sorbate. *J. Food Prot.*, **47**, 554–6.
Liewen, M.B. and Marth, E.H. (1985) Growth of sorbate-resistent and -sensitive strains of *Penicillium roqueforti* in the presence of sorbate. *J. Food Prot.*, **48**, 525–9.
Lock, J.L. and Board, R.G. (1994) The fate of *Salmonella enteritidis* PT4 in deliberately infected commercial mayonnaise. *Food Microbiol.*, **11**, 499–504.
Lock, J.L. and Board, R.G. (1995) The fate of *Salmonella enteritidis* PT4 in home-made mayonnaise prepared from artificially inoculated eggs. *Food Microbiol.*, **12**, 181–6.
Lopez, A. (1987) Mayonnaise and salad dressing products, in *A complete course in canning*, 12th ed, *Book III, Processing Procedures for Canned Food Products*, Canning Trade Inc, Baltimore, MD, 420–36.
Lyytikäinen, O., Ruutu, P., Mikkola, J., Siitonen, A., Maijala, R., Hatakka, M. and Autio, T. (1999) An out-break of listeriosis due to *Listeria monocytogenes* serotype 3a from butter in Finland. *Eurosurveill. Wkly*, **3** (11), 1–2.
Lyytikäinen, O., Autio, T., Maijala, R., Ruutu, P., Honkanen-Buzalski, T., Miettinen, M., Hatakka, M., Mikkola, J., Anttila, V.J., Johansson, T., Rantala, L., Aalto, T., Korkeala, H. and Siitonen, A. (2000) An outbreak of *Listeria monocytogenes* serotype 3a infections from butter in Finland. *J. Infect. Dis.*, **181**(5), 1838–41.
Madsen, J. (1989) Technological problems in margarine and low-calorie spreads, in *Food Colloids; Second International Symposium* (eds R.D. Bee, P. Richmond and J. Mingins), Royal Society of Chemistry, Cambridge, UK, pp. 267–71.
Madsen, J. (1990) Low-calorie spread and melange production in Europe, in *Edible Fats and Oils Processing. Basic Principles and Modern Practices* (ed D.R. Erickson), American Oil Chemists' Society, Champaign, Illinois, pp. 221–7.
Maijala, R., Lyytikäinen, O., Autio, T., Aalto, T., Haavisto, L. and Honkanen-Buzalski. (2001) Exposure to *Listeria monocytogenes* within an epidemic caused by butter in Finland. *Journal of Food Microbiol.*, **70**, 97–107.
Marth, E.H. and Ryser, E.T. (1990) Occurrence of *Listeria* in foods: milk and dairy foods, in *Foodborne Listeriosis* (eds A.J. Miller, J.L. Smith and G.A. Somkuti), Elsevier, Amsterdam, 151–64.
Massa, S., Cesaroni, D., Poda, G. and Trovatelli, L.D. (1990) The incidence of *Listeria* spp. in soft cheeses, butter and raw milk in the province of Bologna. *J. Appl. Bacteriol.*, **68**, 153–6.
Mayerhauser, C.M. (2001) Survival of enterohemorrhagic *Escherichia coli* O157:H7 in retail mustard. *J. Food Prot.*, **64**, 783–7.
Meyer, M. and Oxhøj, P. (1964) En musetyfusepidemi. *Medlemsblad Danske Dyrlaegeforening*, **47**, 810–9.
Membré, J.-M., Majchrzak H.M. and Jolly, I. (1997) Effects of temperature, pH, glucose, and citric acid on the inactivation of *Salmonella typhimurium* in reduced-calorie mayonnaise. *J. Food Prot.*, **60**, 1497–1501.

Michels, M.J.M. and Koning, W. (2000) Mayonnaise, dressings, mustard, mayonnaise-based salads, and acid sauces, in *The Microbiological Safety and Quality of Food* (eds B.M. Lund, T.C. Baird-Parker and G.W. Gould), , *Volume I*, Aspen Publishers Inc, Gaithersburg, Maryland, pp. 807–35.

Miller, L.G. and Kaspar, W. (1994) *Escherichia coli* O157:H7. Acid tolerance and survival in apple cider. *J. Food Prot.*, **57**, 460–4.

Mitchell, E., O'Mahoney, M., Lynch, D., Ward, L.R., Rowe, B., Uttley, A., Rogers, T., Cunningham, D.G. and Watson, R. (1989) Large outbreak of food poisoning caused by *Salmonella typhimurium* definitive type 49 in mayonnaise. *Br. Med. J.*, **298**, 99–101.

Mooren, M.M.W., Gribnau, M.C.M. and Voorbach, M.A. (1995) Determination of droplet size distributions in emulsions by pulsed field gradient NMR, in *Characterization of Food: Emerging Methods* (ed A.G. Gaonkar), Elsevier Science B.V., Amsterdam, pp. 151–62.

Mostert, M.A. and Lelieveld, H.L.M. (2000) Overall approach to hygienic processing in *Encycl. Food Microbiol.* (eds R.K. Robinson, C.A. Batt and P.D. Patel), Academic Press, London, pp. 1802–5.

Moustafa, A. (1990) Margarines and spreads in the United States, in *Edible Fats and Oils Processing. Basic Principles and Modern Practices* (ed D.R. Erickson), American Oil Chemists' Society, Champaign, Illinois, pp. 214–20.

Murphy, M.F. (1990) Microbiology of butter, in *Dairy Microbiology, Volume 2, The Microbiology of Milk Products* (ed R.K. Robinson), 2nd edn, Elsevier, London, pp. 109–30.

NRC (US National Research Council) *Food Protection Committee. Subcommittee on Microbiological Criteria (1985) An evaluation of the role of microbiological criteria for foods and food ingredients.* National Academy Press, Washington, DC.

Olsen, J.A., Yousef, A.E. and Marth, E.H. (1988) Growth and survival of *Listeria monocytogenes* during making and storage of butter. *Milchwissenschaft*, **43**, 487–9.

Packer, K.J. and Rees, C.J. (1972) Pulsed NMR studies of restricted diffusion I. Droplet size distributions in emulsions. *J. Colloid Interf. Sci.*, **40**, 206–18.

Perales, I. and Audicana, A. (1988) Salmonella enteritidis and eggs. *Lancet*, ii, 1133z.

Perales, I. and García, M.I. (1990) The influence of pH and temperature on the behaviour of *Salmonella enteritidis* phage type 4 in home-made mayonnaise. *Lett. Appl. Microbiol.*, **10**, 19–22.

Petersen, P.J. (1964) Et udbrud af levnedsmidelinfektion fremkaldt af mayonnaise inficeret med *Salmonella typhimurium. Medlemsblad Danske Dyrlaegeforening*, **47**, 284–7.

Proctor, M.E., Brosch, R., Mellen, J.W., Garrett, L.A., Kaspar C.W. and Luchansky, J.B. (1995) Use of pulsed-field gel electrophoresis to link sporadic cases of invasive listeriosis with recalled chocolate milk. *Appl. Environ. Microbiol.*, **61**, 3177–9.

Radford, S.A., Tassou, C.C., Nychas, G.J.E. and Board, R.G. (1991) The influence of different oils on the death rate of *Salmonella enteritidis* in homemade mayonnaise. *Lett. Appl. Microbiol.*, **12**, 125–8.

Radford, S.A. and Board, R.G. (1993) Review: fate of pathogens in home-made mayonnaise and related products. *Food Microbiol.*, **10**, 269–78.

Raghubeer, E.V., Ke, J.S., Campbell, M.L. and Meyer, R.S. (1995) Fate of *Escherichia coli* O157:H7 and other coliforms in commercial mayonnaise and refrigerated salad dressing. *J. Food Prot.*, **58**, 13–8.

Schlech,W.F. III, Lavigne, P.M., Bortolussi, R.A., Allen, A.C., Haldane, E.V., Wort, A.J., Hightower, A.W., Johnson, S.E., King, S.H., Nicholls, E.S. and Broome C. (1983) Epidemic listeriosis—evidence for transmission by food. *N. Eng. J. Med.*, **308**, 203–6.

Scholte, R.P.M. (1995) Spoilage fungi in the industrial processing of food, in *Introduction to Food-borne Fungi* (eds R.A. Samson, E.S. Hoekstra, J.C. Frisvad and O. Filtenborg), Ponsen & Looyen, Wageningen, pp. 275–88.

Seals, J.E., Snyder, J.D., Edell, T.A., Hatheway, C.L., Johnson, C.J., Swanson, R.C. and Hughes, J.H. (1981) Restaurant-associated type A botulism: transmission by potato salad. *Am. J. Epidemiol.*, **113**, 436–45.

Sensidoni, A., Rondinini, G., Peressini, D., Maifreni, M. and Bortolomeazzi, R. (1994) Presence of an off-flavour associated with the use of sorbates in cheese and margarine. *Italian J. Food Sci.*, **2**, 237–42.

Sims, J.E., Kelley, D.C. and Foltz, V.D. (1969) Effect of time and temperature on salmonellae in inoculated butter. *J. Milk Food Technol.*, **32**, 485–8.

Smith, J.L. (1987) *Shigella* as a foodborne pathogen. *J. Food Prot.*, **50**, 788–801.

Smith, J.L. (2003) The role of gastric acid in preventing foodborne disease and how bacteria overcome acid conditions. *J. Food Prot.*, **66**, 1292–1303.

Smittle, R.B. (1977) Microbiology of mayonnaise and salad dressing: a review. *J. Food Prot.*, **40**, 415–22.

Smittle, R.B. (2000) Microbiological safety of mayonnaise, salad dressings and sauces produced in the United States: a review. *J. Food Prot.*,**63**, 1144–53.

Smittle, R.B. and Flowers R.S. (1982) Acid tolerant microorganisms involved in the spoilage of salad dressings. *J. Food Prot.*, **45**, 977–83.

Snyder, J.D., Wells, J.G., Yashuk, J., Puhr, N. and Blake P.A. (1984) Outbreak of invasive *Escherichia coli* gastroenteritis on a cruise ship. *Am. J. Trop. Med.Hyg.*, **33**, 281–4.

St. Louis, M.E., Morse, D.L., Potter, M.E., Demelfi, T.M., Guzewich, J.J., Tauxe, R.V. and Blake, P.A. (1988) The emergence of grade A eggs as a major source of *Salmonella enteritidis* infections. New implications for the control of salmonellosis. *J. Am. Med. Assoc.*, **259**, 2103–7.

ter Steeg, P.F., Otten, G.D., Alderliesten, M., de Weijer, R., Naaktgeboren, G., Bijl, J., Kershof, I. and van Duijvendijk, A.M. (2001) Modelling the effects of (green) antifungals, droplet size distribution and temperature on mold outgrowth in water-in-oil emulsions. *Int. J. Food Microbiol.*, **67**, 227–39.

ter Steeg, P.F., Pieterman, F.H. and Hellemons, J.C. (1995) Effects of air/nitrogen, temperature and pH on energy-dependent growth and survival of *Listeria innocua* in continuous culture and water-in-oil emulsions. *Food Microbiol.*, **12**, 471–85.

Telzak, E.E., Budnick, L.D., Zweig Greenberg, M.S., Blum, S., Shayegani, M., Benson, C.E. and Schultz, S. (1990) A nosocomial outbreak of *Salmonella enteritidis* infection due to the consumption of raw eggs. *N. Eng. J. Med.*, **323**, 394–7.

Thomas, D.S. and Davenport R.R. (1985) *Zygosaccharomyces bailii*—a profile of characteristics and spoilage activities. *Food Microbiol.*, **2**, 157–69.

Troller, J.A. and Christian, J.H.B. (1978) *Water Activity and Food*, Academic Press, New York.

Tschäpe, H., Prager, R., Streckel, W., Fruth, A., Tietze, E. and Bohme, G. (1995) Verotoxigenic *Citrobacter freundii* associated with severe gastroenteritis and cases of haemolytic uraemic syndrome in a nursery school: green butter as the infection source. *Epidemiol. Inf.*, **114**, 441–50.

Tuynenburg Muys, G. (1969) Microbiology of margarine. *Process Biochem.*, **4**, 31–4.

Tuynenburg Muys, G. (1971) Microbial safety in emulsions. *Process Biochem.*, **6**, 25–8.

Tuynenburg Muys, G. and Willemse, R. (1965) The detection and enumeration of lipolytic microorganisms by means of a modified Eykman-plate method. *Antonie van Leeuwenhoek*, **31**, 103–12.

Tuynenburg Muys, G., van Gils, H.W. and de Vogel, P. (1966) The determination and enumeration of associate microflora of edible emulsions. Part II: the microbiological investigation of margarine. *Lab. Practice*, **15**, 975–84.

US-D HEW (US Department of Health, Education and Welfare). (1970) Staphylococcal food poisoning traced to butter-Alabama. *Morb. Mort. Wkly Rep.*, **19**, 271.

US-DHEW (US Department of Health, Education and Welfare). (1975a) *Dressings for food. Mayonnaise*, 21 CFR 25.1, US Government Printers Office, Washington, DC.

US-DHEW (US Department of Health, Education and Welfare). (1975b) *Dressings for Food. Salad. Dressing*, 21 CFR 25.3, US Government Printers Office, Washington, DC.

US-DHEW (US Department of Health, Education and Welfare). (1977) Presumed staphylococcal food poisoning associated with whipped butter. Morb. Mort. Wkly Rep., **26**, 268.

US-FDA (US Food and Drug Administration). (1990) *Code of Federal Regulations Title 21*, Parts 101.100 and 169.140, US Government. Printing Office, Washington, DC.

US-FDA (US Food and Drug Administration). (1991) *Code of Federal Regulations. Margarine. Title 21, Chapter 1, Part 166 revised as of April 1, 1991 of the Code of Federal Regulations*, US Government Printing Office, Washington, DC.

US-FDA (U.S. Food and Drug Administration). (1993) *Code of Federal Regulations, Title 21, Part 169. Food Dressings and Flavorings*. U.S. Government Printing Office, Washington, DC.

US-FDA (U.S. Food and Drug Administration). (1994) *Code of Federal Regulations, Title 21, 1. Subpart G. Exemptions from food labelling requirements § 101.100*. U.S. Government Printing Office, Washington, DC.

Varnam, A.H. and Sutherland, J.P. (1994) Butter, margarine and spreads, in *Milk and Milk Products. Technology, Chemistry and Microbiology* (eds A.H. Varnam and J.P. Sutherland), Chapman & Hall, London, 224–73.

Veringa, H.A., van den Berg, G. and Stadhouders, J. (1976) An alternative method for the production of cultured butter. *Milchwissenschaft*, **31**, 658–62.

Verrips, C.T. (1989) Growth of microorganisms in compartmentalized products, in *Mechanisms of Action of Food Preservation Procedures* (ed G.W. Gould), Elsevier, London, New York, pp. 363–99.

Verrips, C.T. and Zaalberg, J. (1980) The intrinsic microbiological stability of water-in-oil emulsions. I. Theory. *Eur. J. Appl. Microbiol. Biotechnol.*, **10** (3), 187–96.

Verrips, C.T., Smid, D. and Kerkhof, A. (1980) The intrinsic microbiological stability of water-in-oil emulsions. II. Experimental. *Eur. J. Appl. Microbiol. Biotechnol.*, **10** (1–2), 73–85.

Weagant, S.D., Bryant, J.L. and Bark, D.H (1994) Survival of *Escherichia coli* O157:H7 in mayonnaise and mayonnaise-based sauces at room and refrigerated temperatures. *J. Food Prot.*, **57**, 629–31.

Xiong, R, Xie, G., Edmondson, A.S. and Meullenet, J.-F. (2002) Neural network modelling of the fate of *Salmonella enterica* serovar Enteritidis PT4 in home-made mayonnaise prepared with citric acid. *Food Control*, **13**, 525–33.

Zhao, T., Doyle, M.P. and Berg, D.E. (1990) Fate of *Campylobacter jejuni* in butter. *J. Food Prot.*, **63**, 120–2.

Zhao, T. and Doyle, M.P. (1994) Fate of enterohemorrhagic *Escherichia coli* O157:H7 in commercial mayonnaise. *J. Food Prot.*, **57**, 780–3.

Zhao, T., Doyle, M.P. and Besser, R.E. (1993) Fate of enterohemorrhagic *Escherichia coli* O157:H7 in apple cider with and without preservatives. *Appl. Environ. Microbiol.*, **59**, 2526–30.

van Zijl, M.M. and Klapwijk, P.M. (2000) Yellow fat products (butter, margarine, dairy and nondairy spreads). *The Microbiological Safety and Quality of Food* (eds B.M. Lund, T.C. Baird-Parker and G.W. Gould), *Volume I*, Aspen Publishers Inc, Gaithersburg, Maryland, pp. 784–806.

12 Sugar, syrups, and honey

I Introduction

Sucrose, in commercial practice commonly referred to as sugar, is the most widely distributed sugar in nature and therefore easily manufactured in large quantities. Other simple sugars, such as dextrose (glucose), fructose, and lactose, or complex ones such as mannitol, sorbitol, or xylitol play also an important economic role.

Ninety nine percent of sucrose worldwide is derived from two main sources, sugar cane (*Saccharum officinarum*) and sugar beets (*Beta vulgaris*). Depending on the region, sucrose is also obtained from several types of palms (~1% of the worldwide production) such as the wild date palm (*Phoenix sylvestris*), coconut palm (*Cocos nucifera*), palmyra (*Borassus flabellifera*), and others; from sweet sorghum, *Sorghum bicolor* (L.) Moench (~0.05%); or from maple trees, *Acer saccharum* and *A. rubrum* (~0.01%) (FAO, 1998).

Fructose is obtained by ion-exchange separation from invert sugar, which is a hydrolysate of sucrose containing dextrose and fructose, or purified from high fructose maize (corn) syrup. Dextrose is manufactured from starch after complete hydrolysis, whereas lactose is commercially produced from whey. Polyols are usually manufactured by hydrogenation of easily available carbohydrates such as glucose syrup, fructose, or xylose derived from hemicellulose. Specifications for sucrose, dextrose, lactose, and fructose are given in the Codex Standards 4-1981, 7-1981 or 8-1981, 11-1981, and 102-1981 (Codex, 1994).

Sugar syrups are concentrated aqueous solutions manufactured from sucrose, from the sap of maple trees or obtained after hydrolysis of starches from potato, maize, or wheat. Specifications for glucose syrup are given in the Codex Standard 9-1981 (Codex, 1994). Honey is naturally produced by honey bees predominantly from nectars of flowering plants. Its sugar composition varies greatly according to the type of plant; specifications are provided in the Codex Standard 12-1981 (Codex, 1994).

II Cane sugar

A *Initial microflora*

The numbers of microorganisms found on the surface of the green cane stalks and in exudates depend on the climatic conditions but *Bacillus* spp., *Enterobacter* spp., *Flavobacterium spp.*, *Pseudomonas* spp., *Xanthomonas* spp., *Lactobacillus* spp., yeasts, and molds at levels ranging between 10^3 and 10^9 have been described (Nuñez and Colmer, 1968; Tilbury, 1970; Klaushofer *et al.*, 1998). Detailed studies showed that the composition of the populations were related to the sugar content and pH of the exudates (Duncan and Colmer, 1964; Bevan and Bond, 1971). The pink sugar cane mealybug (*Saccharococcus sacchari*), commonly found in sugar cane fields, excretes an acidic honeydew (pH ca. 3) which favors the development of acidophilic bacteria and yeasts (Ashbolt and Inkerman, 1990). After the mealybugs die, these microorganisms are replaced by less tolerant genera such as *Erwinia* spp. and *Leuconostoc* spp.. Damage to the cane caused by insects, frost, or other causes lead to the internal development of microorganisms. This development and in particular that of *Leuconostoc mesenteroides* which deposits dextran in the tissues are detrimental to sugar yields (Chen and Chung, 1993).

B Effects of processing on microorganisms

The major stages and their implications for the microbiology of sugar cane manufacture are summarized in Table 12.1. Detailed reviews on sugar processing including recent developments have been published by Belotti *et al.* (2002) and Blackwell (2002).

Harvesting. Mature cane may be harvested green or after burning to remove the leaves, either manually with machetes (cane knives) or mechanically. Sometimes, cane that is cut mechanically is chopped into pieces (billets) of ~30 cm in length directly on the field, thus exposing many additional cut surfaces to contamination.

Leuconostoc mesenteroides is the microorganism of major concern since growth to high levels will lead to losses in the yield of sucrose of up to 1.5% due to formation of acids, dextran, and slime (Tilbury, 1975; Salunkhe and Desai, 1988; Cerutti de Guglielmone *et al.*, 2000). Tilbury (1970) found ~25% of the swab samples from cane stalks to contain *Leuc. mesenteroides* with levels ranging between <1 000 and 50 000. Levels can increase rapidly after burning, harvesting (Bevan and Bond, 1971), and growth. Thus, losses are minimized by rapid further processing. Trials with biocides to minimize losses have not been very successful (Desai and Salunkhe, 1991). In warm and dry conditions, yeasts may become the major microbial population (Lionnet and Pillay, 1988).

Extraction and processing. Cane is processed to raw sugar in a sequence of operations, which is outlined in detail by Desai and Salunkhe (1991) and van der Poel *et al.* (1998). These operations are: (i) cutting; (ii) crushing and milling to separate the raw juice from the bagasse; (iii) liming; (iv) heating; (v) clarification to separate the mud; (vi) evaporation to concentrate the juice; (vii) crystallization to obtain massecuite; and finally (viii) centrifugation. Most of these steps have an effect on the microflora.

Table 12.1 Characteristics of the different steps in cane sugar manufacture

Process step	Temperature (°C)	pH	% Dry matter	Predominant microflora	Microorganisms	Results/problems
Post-harvesting	25–30	5.7–7.7	19	Mesophilic	*Leuconostoc*	Souring, Sugar loss/ dextran formation
Crushing and extraction	25–30	5.0–5.6	10–18	Mesophilic	*Leuconostoc*, Enterobacter Yeasts	Souring, Sugar-loss, Alcohol production
Clarification	80–100	8.0	10–18	None	None	NA
Evaporation/ crystallization/ centrifugation	60–100	5.0–6.0	NA	None	Survival of thermophilic spores	NA
Microorganisms from Processing equipment						
Recontamination of raw sugar	25–30	5.6–6.0	70–90+	Osmophiles if $a_w > 0.65$	*Z. rouxii*, Xerophilic molds	Sugar loss, Invert sugars produced, Acid produced, a_w, rises
Refining	70–90	5.0–8.0	70–90+	None	Surviving thermophilic spores	NA
Refined sugar	25–30	N/A	99+	None	Surviving thermophilic spores	Introduction of spores into final products

Data from various sources.
N/A, not available.

During crushing and milling, the cane is shredded and passed sequentially through a series of rollers with extraction water flowing in the opposite direction. Juice from the first roller may contain up to 19% sugar, and from the last roller, <5%. The juice from the different rollers is combined to raw juice, and the extracted fibrous residue of cane is called bagasse. The raw juice is an ideal medium for growth of many microorganisms, but only a few compete successfully. Raw juice has a Brix (percent sucrose w/w, or equivalent in soluble solids) of 10-18, a pH of 5.0-5.6, is rich in inorganic and organic salts, amino acids, and other nutrients, and a temperature usually about 25-30°C. The bacterial count of the first expressed juice ranges between 10^5 and 10^7 per mL for normal cane and about 10^8 per mL for sour cane in which *Leuconostoc* spp. and other acid-producing microorganisms have multiplied, thus lowering the pH. The changes in the microbial flora during extraction and milling has been investigated by Lillehoj *et al.* (1984) with a special focus on *Leuconostoc* spp.

Recontamination with *Leuconostoc* spp. (which produce slime) is observed in mills where insufficient attention is given to hygiene. In particular, accumulation of bagasse produced during milling allows microbial development (Klaushofer *et al.*, 1998). In some mills, *Enterobacter* spp. predominate; in others, yeasts compete well.

The clarification process follows readily after extraction and permits minimalization of sugar losses unless it is delayed. It involves addition of lime to increase the pH to about 8.0 and rapid heating to 80-100°C to destroy vegetative microorganisms. Maintaining a high temperature in diffusion plants is an effective means of microbiological control. Sedimentation and filtration is then applied to remove scums, precipitates, and suspended solids as filter cake mud. Clarification decreases the microbial count by 99.999% (Table 12.2), but dextran and mesophilic or thermophilic spores remain (Chen and Chung, 1993). The clarified juice is then submitted to evaporation and crystallization at about 60°C to obtain a concentrated sugar suspension, the so-called massecuite. This is cleaned by centrifugation to remove the residual liquid phase, known as mollasses, which consists of sucrose, inverted sugar, organic acids, amino acids, nitrogenous compounds, minerals, and polysaccharides, and to obtain raw cane sugar crystals.

Raw cane sugar is the end-product of the cane mill and the raw material for the refinery (Desai and Salunkhe, 1991; van der Poel *et al.*, 1998). It has a pH of 5.0-6.0, a water activity of about 0.65, and a sucrose concentration of 95-99%, with about 0.5% of residual molasses, which surround the crystals.

Table 12.2 Bacterial content of sugar cane during processing[a]

Product	Mesophiles/mL (range)	Thermophiles/mL (range)
Raw juice (early)[b]	8×10^6–1.6×10^7	1×10^1–1×10^2
Raw juice (late)[b]	6×10^8–8×10^8	–
Clarified effluent	0–11	0–8
Press juice	0–5×10^4	3×10^3–2×10^5
Evaporator	2×10^2–3×10^4	2×10^2–2×10^3
Storage tank	1×10^3–7×10^3	2×10^4–4×10^4
Crystallizer	2×10^3–4×10^4	3×10^2–2×10^4
Massecuite[c]	1×10^3–1×10^4[d]	2×10^3–2×10^4
Raw sugar	3×10^2–5×10^3[d]	2×10^2–2×10^3
Molasses	3×10^3–3×10^5	1×10^3–2×10^4

[a]Adapted from Owen (1977).
[b]Early in season and late in season.
[c]Mixture of sugar crystals and molasses.
[d]Per gram.

The microbial flora of raw sugar consists of bacterial spores, which will survive the thermal processes (Chen and Chung, 1993). Xerophilic yeasts are often present due to recontamination in storage tanks or by wet sugar residues which support their multiplication in the processing lines. This is often due to the poor hygienic design of equipment: long periods between cleaning and wet sugar residues favor growth to levels as high as 10^6 yeasts per gram (Tilbury, 1980). Several of these yeasts are also thermotolerant under these concentrated conditions (Bärwald and Hamad, 1984; Anderson *et al.*, 1988) and are often not killed during vacuum-pan crystallization. Air-borne molds such as *Aspergillus* and *Penicillium* spp. may contaminate products during crystallization, centrifugation, or drying.

Growth of yeasts in the molasses film is favored in inadequately centrifuged raw cane sugar or in case of moisture uptake. This may cause important economic losses due to the formation of invert sugar. Mold spoilage is, however, not of major concern, probably due to the absence of oxygen in silos as opposed to jute bags frequently used in the past.

Refining. The refining of raw cane sugar to food grade crystalline sugar is designed to remove impurities and produce crystals of sucrose >99.9% pure (Desai and Salunkhe, 1991; van der Poel *et al.*, 1998). The refining of raw cane sugar includes the following procedures which have an impact on the microbiological quality of the final product. During affination, raw sugar and molasses are separated by centrifugation and at the same time washed with water under high pressure. During this step, the residual molasses including microorganisms are eliminated. The washed sugar is then dissolved in hot water of about 70°C to obtain a 66° Brix syrup. This syrup is then mixed with lime and carbon dioxide, or phosphoric acid, to precipitate impurities, including bacteria, which are then removed by filtration.

Deionization is performed to remove ash followed by decolorizing through charcoal beds and ion exchange resins. The final steps, evaporation, crystallization, and drying, produce a crystalline sugar with levels of microorganisms ranging between <100 and <1000 cfu/g (Müller *et al.*, 1988).

C Spoilage

Sour cane results when *Leuconostoc* spp. and other acid-forming bacteria grow in harvested cane (Tilbury, 1968; McCowage and Atkins, 1984). These bacteria produce invert sugar, lactic acid and acetic acid, and frequently dextran, which is used as an indicator of stale or old cane. Losses of sucrose may be substantial unless the time between harvest and crushing is minimized. In hot, humid climates, up to 15% of total sugar may be lost for each day between harvesting and crushing (Tilbury, 1975), whereas in warm but dry climates, the loss is usually much less and under such conditions ethanol is the preferred indicator (Lionnet and Pillay, 1988).

Dextran is a polysaccharide that causes significant processing problems for both raw sugar factories and refineries. Dextran increases the viscosity of the process liquid, necessitating slower processing. Dextran can damage pumps and can necessitate an increased frequency of cleaning of equipment such as vacuum pans. It also inhibits growth of sugar crystals and hence slows crystallization rates.

Dextran has an impact on yield and quality of sugar as well as on processing rates. Several publications have reviewed the causes and effects of this polysaccharide throughout the process (McCowage and Atkins, 1984; Clarke *et al.*, 1996; Clarke, 1997).

In cane diffusers, thermophiles ferment invert sugar to lactic acid (95%) and small quantities of formic acid, acetic acid, and glycollic acid. This is an important problem in cane processing (Oldfield *et al.*, 1974a; McMaster, 1975). The main acid producers are *Bacillus stearothermophilus* and *B. coagulans* (McMaster and Ravnö, 1977). In some cases, gas (carbon dioxide and hydrogen)

Table 12.3 Minimum water activities permitting growth of osmophilic yeasts from raw sugars[a,b]

Yeast	Water activity
Zygosaccharomyces rouxii	0.65
Saccharomyces bisporus var. *mellis*	0.70
Torulopsis candida (1)	0.65
Torulopsis candida (2)	0.70
Torulopsis etchellsii	0.70
Torulopsis versatilis	0.70
Hansenula anomala	0.75

[a]From Tilbury (1967).
[b]Tested in sucrose/glycerol syrups for 12 weeks at 27°C.

may be produced; and in battery diffusers, gas pressure may increase to levels affecting the sugar extraction.

Raw sugar is frequently stored for many months before shipping over long distances. Unless precautions are taken, spoilage may cause important economic losses.

Xerophilic yeasts appear to be most active but xerophilic molds may also cause, or contribute to, spoilage. The reasons for spoilage and the methods to prevent it follow.

Yeasts found in raw sugar are mainly xerophilic. Their natural reservoirs appear to be sugar cane, bagasse, filter cake mud, and wet sugary materials in the mill. *Zygosaccharomyces rouxii* is most common, but other species of *Zygosaccharomyces* and species of *Pichia*, *Candida*, *Dekkeromyces*, and *Endomycopsis* have been found (Tilbury, 1968; Skole *et al.*, 1977). The minimum a_w permitting growth of some yeasts commonly found in raw sugar is given in Table 12.3.

Spoilage is caused by the growth of these xerophilic yeasts in the molasses film of the raw sugar and the rate of growth depends primarily on its water activity. The a_w of raw sugar varies widely, from 0.575 to 0.825. Spoilage does not occur <0.65 but may progress rapidly at values >0.7. During storage, the increase of reducing (mainly invert) sugar and of the relative humidity of the atmosphere will cause an increase of a_w (Klaushofer *et al.*, 1998). During growth in the molasses film, the fructose component of the invert sugars is metabolized, and water and organic acids are produced. The increase of a_w and the decrease the pH favor further growth of xerophilic yeasts, and the decreased pH causes hydrolysis of sucrose to produce more invert sugar. Inversion may also result from the activity of invertase produced by a few xerophilic species of yeasts (Klaushofer *et al.*, 1998).

Under favorable conditions, growth of yeasts may continue during bulk storage and transport, and populations may reach 10^7–10^8 cfu/g, levels affecting the organoleptic qualities of the final product. Within a few months after reaching the maximum, the population of viable yeasts may decline by $>99.99\%$ (Tilbury, 1968).

The effect of different conditions and the interaction of different yeasts have been investigated under laboratory conditions and results published by Tilbury (1968).

The only refined product that has a history of spoilage is liquid sugar; granulated sugar rarely has been reported to be spoiled, and then only following accidental wetting (Müller, 1989). The main factor influencing its deterioration in storage is the temperature.

D Pathogens

Cane sugar has never been associated with food-borne poisoning outbreaks, and reference to *Salmonella* spp. in connection with microbiological criteria (Chen and Chung, 1993) is to be seen as part of the verification applied to raw materials used to manufacture other products. The processing and refining of sugar eliminate or destroy vegetative and probably also pathogenic microorganisms present in the raw materials. Isolation of *Clostridium botulinum* from raw sugar and molasses, brown sugar lumps, and sugar used as bee-feed has been reported by Nakano *et al.* (1992). Spores of this pathogen were not detected in refined sugar or in samples taken during production, however.

E CONTROL (cane sugar)

Summary

Significant hazards[a]	• No significant hazard.
Control measures	
Initial level (H_0)	• Does not apply.
Reduction (ΣR)	• Does not apply.
Increase (ΣI)	• Does not apply.
Testing	• In most situations, testing for *Salmonella* or hygiene indicators such as coliforms or Enterobacteriaceae is only performed as a verification for the adherence of good hygiene practices during processing and handling. • Testing for specific parameters is perfomed in special cases, for example, sporeformers in sugar used for canning.
Spoilage	• Growth of xerophilic yeasts possible if a_w>0.65.

[a]In particular circumstances, such as the use of sugar as ingredient for specific products or processes, other hazards may need to be considered.

Comments

Raw materials. In the field, decreasing the interval between harvesting and processing decreases the opportunity for microbial growth. This interval should be not longer than 24-36 h for whole stalk cane and 8-12 h for chopped cane in hot humid weather or 18 h in cool dry weather. Cutting with sharp knives decreases ragged cuts and minimizes entry of bacteria through the cut surfaces. In mechanized harvesters, cleanliness and sanitizing of chopping boxes are desirable, but opportunities for such care are minimal.

Measures applied during processing. Application of formaldehyde to billets decreases spoilage but is not economical (Egan, 1971). Formaldehyde is also not an acceptable food process additive under US Food and Drug Administration regulations. During processing, it is essential to maintain good hygiene and to give sufficient attention to the cleanliness of equipment to avoid build-up of slime-forming bacteria (Klaushofer *et al.*, 1998). Accumulation of bagasse residues needs to be avoided and ideally mills should be steamed at regular intervals and cleaned during stops or shut-down. Growth and thus sugar losses are minimized by maintaining temperature in diffusers well above 70°C, preferably around 85°C. This is much more efficient than the use of expensive bacteriostats.

Raw sugar. The prevention of spoilage of raw sugar depends on obtaining and maintaining an a_w of <0.65. If this is achieved, the numbers and types of contaminants are of little consequence because they are unable to grow. Thus, procedures to control the water activity are important and are complemented by adherence to good hygiene practice. These steps are as follows.

(i) Centrifugation sufficient to eliminate as much wash water as possible, to reduce any increase in a_w.

(ii) Artificial drying of the raw sugar to an $a_w < 0.65$, if necessary.
(iii) Storage in sealed silos at an RH below 65%.

Storage stability of final products. The overriding requirement is to avoid water uptake by the use of suitable packaging or storage under conditions of temperature and relative humidity that prevent the a_w rising >0.65.

III Beet sugar

A Initial microflora

The microflora comes from the soil adhering to the beets. The genera identified from beet tissues are *Pseudomonas*, *Arthrobacter*, *Erwinia*, *Flavobacterium*, *Streptomyces*, and yeasts as well as mesophilic or thermophilic *Bacillus* spp. and *Clostridium* spp. (Bugbee *et al.*, 1976). *Bacillus* spp. are especially capable of causing spoilage during processing.

B Effects of storage and processing on microorganisms

The major processing steps and their impact to the microbiology of beet sugar are summarized in Table 12.4. Details on the processing steps are provided by Desai and Salunkhe (1991) and van der Poel *et al.* (1998).

Storage, fluming and washing. Beets are harvested before the onset of winter. After topping and cleaning, they are often stored in piles on the field for several days to months. Normally, a properly covered and ventilated pile will maintain a temperature of about 1.5-5°C even if the external temperature drops as low as −35°C.

If piles are not well prepared, sugar beets may spoil due to damages caused by freezing or overheating to temperatures ~50°C. Beets stored without adequate air circulation, mainly due to pockets of trash or earth, may overheat to ~50°C within 2 days and thereafter show evidence of microbial spoilage

Table 12.4 Characteristics of the different steps in beet sugar manufacture

Process step	Temperature (°C)	pH	%Dry matter	Predominant/ microflora	Microorganisms	Results/problems
Beet storage	0–15	7.5	25	Psychrophilic/ mesophillic	*Bacillus* spp., *Leuconostoc* spp.	Slime formation, dextrans, levans
Beet fluming	0–15	7.5–9.0	N/A	Psychrophilic	*Pseudomonas* spp., *Flavobacterium* spp.	Acid production, corrosion
Beet washing	0–15	7.0	N/A	Psychrophilic	As above	As above
Extraction system	70	6.0–6.5	0.5–15	Thermophilic	*B.stearothermophilus*, *Clostridium* spp., thermophilic cocci	Acid production, sugar loss, hydrogen sulfide production
Raw juice system	35–40	6.0–6.5	14–15	Mesophillic		None
Raw juice cistern	25 or 55	6.0–6.5	14–15	Meso/thermophilic		None
Preliminary	60	6.0–11.0	14–15	Thermophilic	*B.licheniformis*	Acid production, sugar loss
Mainlining	80	12.5	14–15	None		
Thin juice system	70–128	9.2	14–15	None		
Thick juice system	70	8.6–8.8	70	None		
Boiling crystallization	70–80	8.8	70–92	None		
Crystal/syrup	40–50	7.2	99/90	None	Survival of thermophilic spores	Introduction of spores into final products

Adapted from Nystrand (1984).
N/A, not available

(Bugbee and Cole, 1976; Cole and Bugbee, 1976). Spoilage leads to the production of dextran, levan, or inverted sugars (Oldfield et al., 1971; Cole and Bugbee, 1976).

At the factory, beets are flumed in water to washers at 30-40°C, where residual soil is removed. Microbial numbers can be reduced by using fresh or chlorinated recirculated water (Carruthers and Oldfield, 1955; Moroz, 1963). Residual soil and leaching of sugar allow rapid growth to levels of 10^6–10^7 bacteria/mL. In general, the higher the bacterial population in the flume water, the higher the population in the juice from the beets.

Extraction. Beets are cut into cossettes (thin V-shaped strips) and extracted with hot water in counter-current diffusers.

Diffusers normally have two ancillary systems, one for the extraction water and one for the juice, that pump, transport, filter, strain, heat, and store liquids generated during the process. Details of various beet processing lines are provided by Salunkhe and Desai (1988).

In the diffusers, gradients of sucrose (from 0.5% at the tail to 15.0% at the head) invert sugars, minerals and nitrogenous compounds, the temperature (25-75°C), dissolved oxygen, or the pH (5.0-8.0) occur.

The microorganisms that enter the head end of the diffuser reflect the microflora of the soil in which the beets were grown. However, in these gradients only a few groups of microorganisms will proliferate. As shown by Hollaus *et al.* (1997), the majority of the flora is composed of mesophilic microorganisms, mainly *Lactobacillus* and *Leuconostoc* spp., although coliforms may be present also. Yeasts may grow well in some places but, in general, molds and yeasts do not grow sufficiently to cause losses of economic consequence.

The few thermophilic microorganisms are of particular significance. In the part of the extraction plant where temperatures reach 65-75°C, growth of *Bacillus stearothermophilus* or *B. coagulans* is favored (Klaushofer *et al.*, 1971). Counts may reach levels of 10^6-10^7 cfu/mL within few hours and produce sufficient acid to reduce the pH from 6.5 or 7.0 to 5.2-5.4 (Oldfield *et al.*, 1974a). However, in more anaerobic extraction systems such as batch diffusers, thermophilic *Clostridium* spp., some producing hydrogen sulfide, may grow (Belamri *et al.*, 1991, 1993; Pollach *et al.*, 2002). If lower temperatures are used, the losses of sucrose during extraction can be high.

Avoiding decrease in pH is important. For example, a decrease of 0.5 pH units below 6.5 may mean that *Bacillus* spp. become dominant, and a sugar loss of 0.16-0.19% may occur. The corresponding figure for *Clostridium* spp. is 0.10-0.13% (Hollaus, 1977).

The water used in the diffusers is either fresh or recycled from the last extraction step. This recycled water still contains 0.5-1.0% recoverable sucrose allowing for growth of microorganisms present in pipes, strainers, filters, tanks, or diffusion cells. Development can be avoided by heating the water to >80°C or by using biocides (Brigidi *et al.*, 1985; Franchi and Bocchi, 1994).

Liming, carbonation, evaporation, filtration, and crystallization. The raw juice is heated to 80-90°C and lime added as an aqueous suspension (milk) or as a slurry of calcium saccharates to precipitate colloidal material. Carbon dioxide is introduced (as bubbles) in two steps, first to improve removal of sludge and second to precipitate residual lime. Precipitates are eliminated during filtration and the filtrate (thin juice) is treated with sulfur dioxide to lower the pH and destroy colors. The juice is then submitted to ion exchange before being evaporated to obtain thick juice with a water activity of 0.88. The juice is then filtered and finally the standard liquor obtained is crystallized to raw beet sugar or massecuite. The microbiology of the concentrated extract has been reviewed by Hein *et al.* (2002). Details of different industrial processes are provided by Desai and Salunkhe (1991) and van der Poel *et al.* (1998).

Refining. There are three main steps during which microbial growth may occur: in deionization beds, charcoal beds, and sweet waters. Deionization of clarified liquor is carried out at low density (55°Brix) and at a temperature (50°C) that permits growth of thermophiles. In charcoal beds, a_w and temperature preclude growth, but elution of sugar before regenerating the charcoal beds by heating at high temperatures may allow growth of thermophilic *C. thermosaccharolyticum* that are able to produce large quantities of gums and slimes (Belamri *et al.*, 1991).

Sweet waters are sugar-containing waters from several sources including bag washers, dust collectors, spillage, and wash water from filters, charcoal decolorizing beds, and deionizers. The pH ranges from 4.5 to 7.5 (usually about 5.5), the Brix from 0° to 60°, and the temperature from 15°C to 75°C. The temperature determines the type of microorganisms able to grow and mesophilic or thermophilic bacteria including *Leuconostoc*, *Lactobacillus*, *Streptococcus*, and *Bacillus* spp. at levels of 10^4-10^7 cfu/mL have been described (Tilbury, 1975; Tilbury *et al.*, 1976). The yeasts most frequently present are *Candida*, *Zygosaccharomyces*, and *Pichia* spp., but others may be present also. Many are xerophilic and some actively produce invertase. At suitable temperatures, there may be 10^5-10^6 yeasts/mL. Because sweet waters are used during refining, for example, to melt sugar, or to dilute high Brix solutions, the microbial quality of sweet waters is important.

Spore-forming bacteria compete well in conditions that are unsuitable for asporogenous lactic acid bacteria. They survive high temperatures and are usually the only microorganisms present when a_w and temperature become suitable for microbial growth. This is reflected in the flora of the refined sugar which is mainly formed by mesophilic or thermophilic, aerobic or anaerobic *Bacillus* spp. or *Clostridium* spp. (Hollaus, 1977; de Lucca *et al.*, 1992; Hollaus *et al.*, 1997). Other sources of recontamination include the air, contaminated dust, or packaging material (Pollach *et al.*, 1998).

C Spoilage

During the extraction process, microbial activities create the following problems.

Formation of acids and sugar losses. The pH of raw juice ranges between 6.0 and 6.7. Microbial growth and acid formation during extraction are the main causes of sugar losses. Species vary widely in their ability to degrade sucrose (Table 12.5). Thermophilic bacteria, for example, growing at 70°C in diffusion juice may reach viable populations of 10^6-10^7 cfu/mL and reduce the pH to 5.2-5.4. Numerous studies on the metabolism of microorganisms growing in diffusers and their impact on sugar losses have been recorded. Results are summarized and discussed by Klaushofer *et al.* (1998).

Table 12.5 Rate of destruction of sucrose by microorganisms from beet juice

Organism	Temperature (°C)	Sucrose destroyed (mg/10^9 cells/h)
Desulfotomaculum (Clostridium) nigrificans[a]	55	0
Enterobacter (Aerobacter) aerogenes	35	0.1–0.4
Flavobacterium, Micrococcus,[b] *Streptococcus*[b]		
Leuconostoc mesenteroides[b]	35	2–8
Lactobacillus[b]		
Clostridium thermohydrosulfuricum[a,c]	66	2–3
Clostridium thermohydrosulfuricum[a,c]	70	7–8
Clostridium thermosaccharolyticum[a]	66	2–3
Bacillus thermophilus[b,c]	55	10–40
Bacillus stearothermophilus[a]	65	108–160
Bacillus subtilis[b]	55	20–60
Saccharomyces[b]	35	1500–3000

[a]From Klaushofer and Parkkinen (1966).
[b]Devillers (1955).
[c]Not listed in the eighth edition of Bergey's Manual (Buchanan and Gibbons, 1974).

Corrosion. Steel in diffusers and ancillary systems corrodes from reaction with lactic acid. The rate of corrosion at 70°C is about twice that at 20°C and increases ~4-fold for each decrease of 1 pH unit in the range 6.2-4.2 (Allen *et al.*, 1948; Carruthers and Oldfield, 1955). Using lime to increase the pH of the diffusion water in the supply tank, decreases the overall corrosion rate but increases the depth of pitting. Inhibition of microbial growth by chlorination of the diffusion water inhibits corrosion in the recirculation system but not in the diffusers.

Formation of slime. Growth of mesophilic microorganisms such as *Leuconostoc mesenteroides* is only possible in cases where the temperature falls <40°C, e.g. in the countercurrent juice/cosette heat exchanger, at the exhausted pulp outlet or in the pulp presses. This is, however, well known now and only occurs exceptionally (Chen and Chung, 1993, Tallgren *et al.*, 1999).

Formation of nitrite. Nitrate is present in sugar beets, usually at 20-200 ppm of nitrate nitrogen; nitrite is absent. In raw juice, the nitrite level is usually 2–15 ppm, occasionally up to 75 ppm. In continuous diffusers, *B. stearothermophilus* is the prevalent thermophile, which reduces nitrate to nitrite or to nitrogen gas depending on the strain. Reduction of nitrate to nitrite at different steps of the process by *Thermus*, a highly thermophilic non-sporing Gram-negative bacterial genus, has been described by Hollaus *et al.* (1997).

Nitrite formed by these microorganisms may combine with other chemicals such as bisulfite formed from sulfur dioxide added during processing (Carruthers *et al.*, 1958; Oldfield *et al.*, 1974b). Nitrite combines with the bisulfite, reducing its efficacy and also forms imidodisulfonate which co-crystallizes with sucrose, thus increasing the ash content and causing malformed crystals that impede centrifugation of the massecuite.

D Pathogens

As for section II.

E CONTROL (beet sugar)

Summary

Significant hazards[a]	• No significant hazard.
Control measures	
Initial level (H_0)	• Does not apply.
Reduction (ΣR)	• Does not apply.
Increase (ΣI)	• Does not apply.
Testing	• In most situations, testing for *Salmonella* or hygiene indicators such as coliforms or Enterobacteriaceae is only performed as a verification for the adherence of good hygiene practices during processing and handling. • Testing for specific parameters is perfomed in special cases, for example, spore-formers in sugar used for canning.
Spoilage	• Growth of xerophilic yeasts possible if $a_w > 0.65$.

[a]In particular circumstances, such as the use of sugar as ingredient for specific products or processes, other hazards may need to be considered.

Comments

Raw materials. The microbial content of flume water may be reduced by using fresh instead of recirculated water; however, this is not usually practical. Heavy chlorination may be useful in flumes carrying damaged beets (Moroz, 1963), but chlorination does not destroy spore-forming thermophiles (Carruthers and Oldfield, 1955).

Measures applied during processing. Ideally, the temperature in diffusers and ancillary systems should be at 75°C throughout to minimize microbial growth. At 70°C, abundant growth of *B. stearothermophilus* may occur; whereas at 80°C, excessive extraction of pectin may interfere with clarification. If extraction is performed at temperatures >70°C, stagnant regions in the pipelines and extraction plant are eliminated and minimal sugar losses are accepted, then operation without formalin is possible (Hollaus and Pollach, 1993). However, when extraction is performed at temperatures between 60°C and 72°C, then addition of bactericides is an essential part of the control measures. Several preservatives are effective, e.g. quaternary ammonium compounds, benzoate, formaldehyde, and metabisulfite. Of the legally permissible compounds, metabisulfite appears the most economical. Formaldehyde, although effective, is permitted for use in diffusers but not during the refining process.

In diffusers, formalin (30-50% aqueous solution of formaldehyde) is still the most effective agent used to eliminate bacterial populations (Klaushofer *et al.*, 1998). It is added to those cells that are most susceptible to microbial growth as evidenced by low pH. Discontinuous high dosage application of formalin is much more effective than continuous dosing. A beet processing factory will use about 0.25 kg formalin (40% formaldehyde) per tonne of beets processed (Guerin *et al.*, 1972).

Numerous other biocides have been investigated, including sulfur dioxide (Chen and Rauh, 1990), cationic substances (Franchi and Bocchi, 1994), glutaraldehyde (Accorsi, 1994), and hydrogen peroxide (Duffaut and Godshall, 2002). Only the latter seems to provide acceptable results. Mixtures of surfactants with an oxidizing agent such as hydrogen peroxide or peracetic acid have been proposed as well (Bowler *et al.*, 1996).

One aspect of control is related to the choice between microbial production of lactic acid and addition of mineral acid, preferably sulfuric, to diffusion water. Make-up water for continuous diffusers has a pH of 7.0–9.0, and cossettes leaving the tail cell must be at pH 6.0 or less for efficient removal of water by pressing; dried cossettes are used as cattle feed. In some operations, sufficient lactic acid is generated by fermentation to obtain the desired low pH, but lactic acid production during diffusion is not sufficiently controllable or predictable to ensure that the pH of spent cossettes will be <6.0 (Oldfield *et al.*, 1974a).

Control measures to minimize microbial growth and losses in sugar processing are also discussed by Pollach *et al.* (1998), Day (2000), and Trost and Steele (2002).

Storage stability of final product. The key requirement is avoidance of moisture uptake, as for cane sugar. In addition, the low ambient temperatures in areas where beets are grown and processed may be responsible for moisture migration or condensation on the product.

F Microorganisms in refined sugar capable of spoiling other food

Certain microorganisms that grow during extraction and refining can survive the process or gain entrance after processing. Usually there are fewer than 10^2 per g, but if present in sufficient numbers in the refined sugar they may result in serious spoilage in foods that have sugar as an ingredient (Chen and Chung, 1993).

The bacteria most commonly present in sugar are *Bacillus* spp., which do not grow in the raw sugar but may grow in dilute sugar solutions during the refining process. *Desulfotomaculum* (formerly

Clostridium) *nigrificans*, *Cl. butyricum*, and thermophilic *Bacillus* spp. may also be present and are of concern if they persist through the refining process and are present in the refined sugar used in canned food.

The bacteria of most concern are the following (Goldoin *et al.*, 1982).

1. *Bacillus stearothermophilus* and *B. coagulans*, which may grow in canned food, producing acid without gas. As the can of container is not distended, the condition is described as "flat sour", and the two species are designated flat sour organisms. *Bacillus stearothermophilus* is a particular nuisance because it forms spores that are very heat resistant and grows at temperatures up to 75°C. However, growth will not occur at pH values <5.2. In contrast, *B. coagulans* does not grow at temperatures above 65°C and is less heat resistant than *B. stearothermophilus*, but can grow at pH 4.2.
2. *Clostridium thermosaccharolyticum*, which grows well at 72°C, but less well at 75°C. In canned food, it may produce sufficient acid to cause hydrogen swells.
3. *Desulfotomaculum nigrificans*, which grows optimally at 55°C. It may cause sulfide stinker spoilage in canned food.
4. Mesophilic bacteria, yeasts, and molds that can grow at the pH of soft drinks. The most common molds are *Aspergillus* and *Penicillium* spp., whereas species in several other genera are found less frequently (Tilbury, 1968).

IV Palm sugar

A *Initial microflora*

The microflora originates mainly from the inflorescences and spathes of the palm or from slits made in the trunk. Microorganisms have been identified as *Acetobacter aceti*, *Aceto. rancens*, *Aceto. suboxydans*, *Leuconostoc dextranicum*, *Lactobacillus* spp., *Micrococcus* spp., *Pediococcus* spp., and *Bacillus* spp. as well as yeasts such as *Saccharomyces cerevisiae* and *Schizosaccharomyces pombe* (Faparusi and Bassir, 1971; Shamala and Skreekantiah, 1988). The pH of the sap is close to neutrality and if concentration is delayed, growth of lactobacilli may rapidly reduce the pH to 4, making conditions favorable for alcoholic yeast fermentation (Faparusi and Bassir, 1971).

B *Effects of processing on microorganisms*

The palms that provide the sap and the processing methods used vary greatly around the world. Generally, the inflorescences are tenderized by bruising, and the sap extracted or sap is collected from slits made in the trunk. Lime is added to the sap to prevent fermentation. After precipitation of the calcium carbonate, the sap is strained and concentrated by boiling, often in open pans. Crystallization occurs and the syrup is poured into molds where it solidifies rapidly (Naim and Husin, 1986; Hamilton and Murphy, 1988). Prolonged boiling destroys vegetative microorganisms, and concentration for crystallization reduces the water activity to 0.80-0.83, preventing the growth of bacteria including spore formers. The product is hygroscopic and molds may grow if it is held inadequately packaged in humid conditions (Naim and Husin, 1986).

C *Spoilage*

Contamination may occur at all points in the process, often leading to spoilage of the sap.

D *Pathogens*

As for section II.

E CONTROL (palm sugar)

Summary

Significant hazards[a]	•	No significant hazard.
Control measures		
Initial level (H_0)	•	Does not apply.
Reduction (ΣR)	•	Does not apply.
Increase (ΣI)	•	Does not apply.
Testing	•	In most situations, testing for *Salmonella* or hygiene indicators such as coliforms or Enterobacteriaceae is only performed as a verification for the adherence of good hygiene practices during processing and handling.
	•	Testing for specific parameters is perfomed in special cases, for example, spore-formers in sugar used for canning.
Spoilage	•	Growth of xerophilic yeasts possible if a_w >0.65.

[a]In particular circumstances, such as the use of sugar as ingredient for specific products or processes other hazards may need to be considered.

Comments. Lime water may be used to control fermentation during extraction, unless the sap is used to prepare a fermented beverage using yeasts. The control of spoilage microorganisms is achieved through the boiling process. Sap collected in the evening may be boiled for preservation and held overnight for further processing with overnight sap. Boiling is also necessary to inactivate invertase, which will invert the sucrose and prevent the setting of the final sugar concentrate in the molds (Naim *et al.*, 1985). Packaging that retards or prevents moisture absorption will prevent the growth of molds on the surface of the crystalline sugar.

V Syrups

A *Initial microflora*

Three major categories of sugar syrups are in use: (i) syrups made by dissolving refined sugar in water; (ii) syrups made by hydrolyzing starches (potato, maize, or wheat) chemically or enzymically; and (iii) naturally occurring tree saps, such as maple syrup.

Sugar syrups or liquid sugars are raw materials manufactured for a wide variety of end users and products, and therefore fulfill different requirements such as concentration, composition and ratio of different sugars, technological properties, and presence of additives (Blanchard and Katz, 1995; Kearsley and Dziedzic, 1995). The initial microflora of liquid sugar is the same as those described in Sections II and III for cane and beet sugar.

Maple syrup is mainly produced in Canada and the North-Eastern United States. Sap drawn directly from maple trees (*Acer. saccharum* or *Acer. rubrum*) is sterile. But under the normal conditions of collection, yeasts and bacteria can be found. Most of the bacteria were identified as *Pseudomonas fluorescens* (Morselli and Feldheim, 1988), which was attributed to poor hygiene during collection. Late season raw sap contained *Pseudomonas*, *Aerobacter*, *Leuconostoc*, and *Bacillus* spp. (Kissinger, 1974). Other studies have shown that the microorganisms in the maple sap were related to those found on the trees and were *Bacillus* spp. and actinomycetes (Parker *et al.*, 1994). Harvesting conditions and temperatures may be conducive to rapid growth and to spoilage such as green sap due to the development of fluorescent pseudomonads, red sap to yeasts and some bacteria, milky sap to bacilli, and ropy sap to *Enterobacter agglomerans* excreting exopolysaccharides (Britteen and Morin, 1995).

B Effect of processing on microorganisms

Liquid sugar is refined sugar concentrated after the decolorizing step or made by dissolving crystalline refined sugar in water. It has a sugar content of 66–76° Brix.

Invert syrup is manufactured using microbial invertase, hydrochloric acid, or ion exchange on very acidic cationic resins (Pancoast and Junk, 1980). These processes are performed at temperatures ranging between 65°C (enzymatic) and 90°C (chemical). Starches from maize, wheat, potato, or tapioca are used to manufacture glucose, fructose, or maltose syrups. Details of the processes are provided by Blanchard and Katz (1995), Olsen (1995), and Le Bot and Gouy (1995).

Precise microbiological data on the different types of processes are not available. In view of the processing conditions, no major changes are however likely to occur, except the destruction of vegetative cells and of some spores.

For both liquid sugar and sugar syrups, recontamination with xerophilic yeasts may occur during intermediate storage in tanks, during transport in pipes, pumps, or trucks.

Maple sap is an aqueous solution containing 1-9% sucrose (Morselli and Feldheim, 1988), which is concentrated to obtain syrup and sugar. This is usually done in special pans and molds as described in detail by Salunkhe and Desai (1988).

C Spoilage

Depending on their sugar content, syrups have a water activity ranging between 0.70 and 0.85, and are therefore prone to spoilage by xerophilic yeasts (Vindelov and Arneborg, 2002).

Different xerophilic yeasts and molds may grow in sugar syrups. *Zygosaccharomyces rouxii*, in particular, is able to grow at a_w values as low as 0.65 (Tokuoka, 1993). Growth is usually slow because lag times and mean generation times generally are inversely proportional to the a_w. Bacteria do not grow. Three factors determine growth of yeasts and molds:

- Size of inoculum.
- Availability of nutrients other than sucrose.
- Gradients with increasing a_w.

Liquid sugars with gradients of a_w are most susceptible to spoilage because strains of yeast with moderate xerophilic characteristics can grow quickly in solutions with high a_w (low concentration of sugar); by adaptation or selection they grow across the gradient to the lowest a_w (highest concentration of sugar). Thus, large populations of xerophilic yeasts may become established in concentrated solutions of sugar. Gradients of a_w occur because water does not readily mix with highly concentrated solutions of sugar unless agitated. Thus, pockets of water in improperly dried equipment, and condensate that forms on the ceilings and walls of tanks and runs down to the surface of sugar concentrates, provide gradients. Also, improperly washed pipes and valves containing diluted sugar solutions may provide suitable sites for growth of yeasts to large populations.

The presence of xerophilic yeasts is detrimental to the syrup itself and also in products manufactured from these syrups. Development of yeasts may lead to changes in the organoleptic or textural characteristics of the syrup, while contamination of confectionery products may cause spoilage due to fermentation and gas production (Pitt and Hocking, 1997).

Quality failure of packaged maple syrup through visible mold growth or blown containers from yeast growth may result from improper packaging and storage.

D Pathogens

As for cane and beet sugar, this group of products has never been linked to outbreaks of food-borne disease.

The presence of *Clostridium botulinum* was reported in 13 of 1 010 samples of maize syrup (Kautter *et al.*, 1982) at levels of about 50 spores/g. No spores were found in two additional surveys of different types of syrups and products manufactured with this ingredient (Hauschild *et al.*, 1988; Lilly *et al.*, 1991). No cases of botulism have been associated with sugar syrups.

The method of production of maple syrup is not conducive to the survival of pathogens, and at an a_w of 0.83-0.86 for packaged syrups (Troller and Christian, 1978) the growth of bacterial pathogens is unlikely.

E CONTROL (syrups)

Summary

Significant hazards[a]	• No significant hazard.
Control measures	
Initial level (H_0)	• Does not apply.
Reduction (ΣR)	• Does not apply.
Increase (ΣI)	• Does not apply.
Testing	• In most situations, testing for *Salmonella* or hygiene indicators such as coliforms or Enterobacteriaceae is only performed as a verification for the adherence of good hygiene practices during processing and handling. • Testing for specific parameters is perfomed in special cases, for example, sporeformers in sugar used for canning.
Spoilage	• Growth of xerophilic yeasts possible if a_w >0.65.

[a]In particular circumstances, such as the use of sugar as ingredient for specific products or processes, other hazards may need to be considered.

Comments. The basis for the prevention of spoilage is to prevent recontamination. This is achieved by the application of good manufacturing practice and good hygienic practice including an appropriate hygienic design of the equipment and processing lines. Avoiding the presence of condensation and thus of spots with increased a_w is of particular importance. Destruction can be achieved using biocides and by steam sterilization of the processing equipment. Recontamination during storage can be reduced by appropriate protection and the use of tanks equipped with air filters and UV lamps (Fiedler, 1994).

In the case of maple, syrup control is essential to achieve the best commercial grade of maple syrups and to prevent deterioration once packaged. Traditionally, maple syrup is packed at the draw-off temperature of 99-103°C directly from the evaporator or finishing pan. Filled containers are laid on their sides to allow syrup to sterilize the head-space and closure. Destruction of spoilage microorganisms can also be achieved by sterilizing the product before packaging (Dumont *et al.*, 1993) or by UV irradiation of surfaces of syrups during storage (Dumont *et al.*, 1991), ozone, however, has been shown to be of limited use (Labbe *et al.*, 2001).

VI Honey

The honey bee or hive bee, *Apis mellifera*, makes the bulk of the world's honey. An additional important hive bee in South and East Asia is *Apis cerana* (Crane, 1979). The bees harvest nectar from the nectaries of flowering plants. The sugar content of nectars from different plants varies from about 5% to about 80%, with great differences in the sugars present and their proportions. Evaporation of water in the hive leads to fully ripened honey, with a water content of ≤20%. Chemical alterations occur, particularly the inversion of sucrose to glucose and fructose. The wax cells in the hive are then sealed, preventing the absorption of water and avoiding the risk of fermentation.

Honey is traded in different forms, liquid or crystalline products or a mixture of both, solid crystallized or granulated honey due to glucose crystallization, spreadable creamed honey or comb and chunk honey. The internationally accepted norms for honey are given in the Codex Standard 12-1981 (Codex, 1994).

Important properties. The composition of honeys varies widely and depends predominantly upon the composition of the nectar; climatic conditions and extraction procedures have minor influences. The composition also varies greatly between the producing countries, but in general terms the glucose content (ca. 30–35%) is usually lower than the fructose content (ca. 35-45%). Moisture content is usually between 15% and 21%, sucrose about 1-3%, ash between 0.09-0.33%, and pH falls in the range 3.2-4.5 (White, 1978, 1987). The main physical attributes of honey are largely determined by the types and concentrations of sugars. Of major importance with respect to spoilage by fermentation is a_w.

Table 12.6 shows the relationship between water content and ERH for a typical clover honey (Martin, 1958). Over the temperature range 4-43°C, honey maintained an ERH of 59% ± 4% at a water content of 18%.

Methods of processing and preservation. Before extraction, the thin wax coverings or caps that seal the ripe honey in the cells are removed with a heated knife. The cappings still contain a considerable amount of honey, which is recovered by straining, centrifuging, or melting the cappings.

Honey is usually extracted from the uncapped combs by centrifugation. Heather honey, which is thixotropic, cannot be extracted in this way. The honey flows into a tank where coarse wax material is removed by baffles, by skimming, or by settling. It is then strained and piped into containers. Heating

Table 12.6 Approximate equilibrium between relative humidity of air and the water content of a clover honey[a]

ERH (%)	Water content (%)
50	15.9
55	16.8
60	18.3
65	20.9
70	24.2
75	28.3
80	33.1

[a]From the data of Martin (1958).

is essential at various stages in the extraction and handling of honey, but excessive heat is deleterious, leading particularly to the production of hydroxymethylfurfural which causes darkening and lowers quality (Townsend, 1975). Temperatures of 71–77°C for short periods serve to destroy most xerophilic yeasts present, have little effect on color, and reduce the tendency to crystallization. Details on processing are given by Crane (1979).

A *Initial microflora*

The microbiology of honey has been reviewed thoroughly by Snowdown and Cliver (1996). These authors have underlined the fact that microorganisms of interest to the honey processing industry are those adapted to the characteristics of honey, i.e. high sugar contents, acidity, and the presence of antimicrobials.

The microbial content is, as shown in different surveys, generally low with counts <100 cfu/g, exceptionally up to 1 000 or 10 000 (Snowdown and Cliver, 1996). In most studies, *Bacillus* spp. have been identified as the main microflora, originating from the pollen, nectar, and bee, as well as from external sources such as sugar solutions used to feed bees. *Bacillus* spp. form also the predominant flora in feces from bee larvae and adults, followed by Gram variable pleomorphic bacteria. Molds, actinomyces, Gram-negative rods and yeasts have been isolated as well but not *Clostridium* spp. (Gilliam and Valentine, 1976; Gilliam and Prest, 1987; Gilliam *et al.*, 1988).

Changes occur during ripening of the honey, and vegetative bacteria such as *Gluconobacter* spp. and *Lactobacillus* spp. present at the beginning disappear due to reduced a_w (Ruiz-Argüeso and Rodriguez-Navarro, 1975; Snowdown and Cliver, 1996).

The microflora of commercial importance are the xerophilic yeasts, which may cause fermentation if the a_w is sufficiently high, and the spores of bacteria or fungi that are pathogenic to bees or toxigenic to humans.

The most frequently reported yeasts are *Zygosaccharomyces* spp. but a wide range of other genera also occur in unprocessed honey (Tysset and Rousseau, 1981; Snowdown and Cliver , 1996). Molds are found in honey usually at low levels of up to a few hundreds cfu/g, the most frequently isolated being *Aspergillus* and *Penicillium* spp. (Tysset *et al.*, 1970; Gilliam and Prest, 1987). Xerophilic fungi have not been reported. Two spore-forming bacteria, *B. larvae* and *Cl. botulinum*, are uncommon. However, *B. larvae*, the causal agent of "foul brood" or "American plague" of bees, is of high economic importance. Spores of *Cl. botulinum* have been isolated from 7% to 16% of honey samples of various origins at levels ranging between 140 and 80 000 spores/kg (Sugiyama *et al.*, 1978; Huhtanen *et al.*, 1981; Hauschild *et al.*, 1988; Criseo *et al.*, 1993; Lund and Peck, 2000; Nevas *et al.*, 2002). Spores of *Cl. botulinum* appear to survive for long periods in honey (Nakano *et al.*, 1992) and an increased incidence seems to be linked to growth and sporulation in diseased bees in hives (Nakano *et al.*, 1994).

B *Effect of processing on microorganisms*

As extracted from the honeycomb, honey usually has a water content near 18%, corresponding to an a_w of about 0.60. A strain reported as *Zygosaccharomyces bailii* isolated from honey grew at a_w 0.65 (Leveau and Bouix, 1979), but this was probably more correctly *Z. rouxii*, the only *Zygosaccharomyces* species capable of growth at such low a_w (Pitt and Hocking, 1997). The minimum a_w for growth of a group of xerophilic yeasts inoculated into honey was above 0.68 (Esteban-Quilez and Marcos-Barrado, 1976). The heating given to honey after extraction, to control crystallization, provides a pasteurization process in spite of the increased heat resistance provided by the reduced a_w (Gibson, 1973). Spores of *Cl. botulinum* are not inactivated by such treatments, however.

C Spoilage

The number of yeasts in honey is usually dependent on the moisture content, increasing as the moisture content increases. Counts up to 10^6 cfu/g have been reported (Graham, 1992). *Zygosaccharomyces* spp. are common, particularly the xerophilic species, *Z. rouxii* (Jermini *et al.*, 1987), probably the most common cause of spoilage of honey (Pitt and Hocking, 1997). Another important species of this genus in honey is *Z. bisporus* (Hocking, 1988). Growth of yeasts causes fermentation and leads to unacceptable organoleptic changes.

D Pathogens

Clostridium botulinum in honey has been implicated in several cases of infant botulism (Dodds, 1993; Aureli *et al.*, 2002; Tanzi and Gabay, 2002). Infant botulism has recently been reviewed by Midura (1996) and Cox and Hinkel (2002). As a source of the bacteria causing infant botulism, honey appears to be less important than some other environmental sources (Long *et al.*, 1985). Outbreaks have, in general, been associated with very heavily contaminated honey. For further details, see Kautter *et al.* (1982), Guilfoyle and Yager (1983), Hauschild *et al.* (1988), and Lilly *et al.* (1991).

E CONTROL (honey)

Summary

Significant hazards[a]	• *Clostridium botulinum.*
Control measures	
Initial level (H_0)	• Presence in honey seems unavoidable. An incidence of 7–16 % and maximal levels of 80 000 cfu/kg have been reported.
Reduction (ΣR)	• So far no possibilities to achieve consistant reductions in raw honey.
Increase (ΣI)	• No increase due to the low a_w.
Testing	• Testing for *Clostridium botulinum* is not recommended. • Testing for sulfite-reducing spores provides information on the general hygiene of honey but low levels or absence will not necessarily indicate absence of *Cl. botulinum.*
Spoilage	• Growth of xerophilic yeasts possible if $a_w > 0.65$.

[a]In particular circumstances, other hazards may need to be considered.

Comments. No practical procedures exist that can prevent the occasional contamination of honey in the hive by spores of *Cl. botulinum* (Hazzard and Murrell, 1989), or that can ensure their destruction in normal processing. Should their elimination be essential, an effective method is heating to autoclaving temperatures, but dilution before processing and reconcentration after processing are probably necessary. Killing effect should in any case be validated.

As recommended by the American Dietetic Association, for prevention of infant botulism, it is important not to use honey as a sweetener in preparations for infants <9-12 months of age (Anonymous, 2003; CDC, 2004).

Spoilage. The heating that honey receives during processing should inactivate contaminating xerophilic yeasts, which are the spoilage agents of concern. However, recontamination from equipment and from the air in the processing establishment is common. This can be minimized by adherence to good hygienic practices. Control of any yeasts present after the extraction process depends on the maintenance of an a_w of 0.65 or below. To achieve this, absorption of moisture must be prevented by appropriate packaging.

References

Accorsi, C.A. (1994) Glutaraldehyd als Desinfektionsmittel in Extraktionsanlagen. *Zuckerind.*, **119**, 124–8.

Allen, L.A., Cairns, A., Eden, G.E., Wheatland. A.B., Wormwell, F. and Nurse, T.J. (1948) Microbiological problems in the manufacture of sugar from beet. Part I. Corrosion in the diffusion battery and in the recirculation system. *J. Soc. Chem. Ind. London.*, **67**, 70–7.

Anderson, P.J., McNeil, K.E. and Watson, K. (1988) Isolation and identification of thermotolerant yeasts from Australian sugar cane mills. *J. Gen. Microbiol.*, **134**, 1691–8.

Anonymous (2003) Position of the American Dietetic Association: Food and water safety. *J. Am. Diet. Assoc.*, **103**, 1203–18.

Ashbolt, N.J. and Inkerman, P.A. (1990) Acetic acid bacterial biota of the pink sugar cane mealybug, *Saccharococcus sacchari*, and its environs. *Appl. Environ. Microbiol.*, **56**, 707–12.

Aureli, P., Franciosa, G. and Fenicia, L. (2002) Infant botulism and honey in Europe: a commentary. *Pediatr. Infect. Dis. J.*, **21**, 866–8.

Bärwald, G. and Hamad S.H. (1984) The presence of yeasts in Sudanese cane sugar factories. *Zuckerind*, **109**, 1014–6.

Belamri, M., Mekkaoui, A.K. and Tantaoui-Elaeaki, A. (1991) Saccharolytic bacteria in beet juices. *Int. Sugar J.*, **93**, 210–5.

Belamri, M., Douiri, K., Fakhereddine, L. and Tantaoui-Elakari, A. (1993) Preliminary study on the saccharolytic activity of thermophilic bacteria from extraction beet juice. *Int. Sugar J.*, **95**, 17–22.

Belotti, A., Journet, G., Neve, H. and Urbaniack, J. (2002) Some recent sugar manufacturing equipment design innovations. *Intern. Sugar J.*, **104**, 214–20.

Bevan, D. and Bond, J. (1971) Microorganisms in field and mill–a preliminary survey, in *Proceedings 38th Conference of the Queensland Society of Sugar Cane Technologists*, pp. 137–43.

Blackwell, J. (2002) Recent developments in sugar processing. *Int. Sugar J.*, **104**, 28–42.

Blanchard, P.H. and Katz, F.R. (1995) Starch hydrolysates, in *Food Polysaccharides and their Applications* (ed. A.M. Stephen), Marcel Dekker, Inc., New York, 99–122.

Bowler, G., Malone, J.W.G. and Pehrson, R. (1996) Recent advances in the application of peracetic acid formulations in the European beet sugar industry. *Zuckerind*, **121**, 414–6.

Brigidi, P., Marzola, M.G. and Trotta, F. (1985) Inhibition of thermophilic aerobic sporeformers from diffusion juices by antiseptic substances based on quaternary ammonium compounds. *Zuckerind*, **110**, 302–4.

Britteen, M. and Morin, A. (1995) Functional characterization of the exopolysaccharide from *Enterobacter agglomerans* grown on low-grade maple sap. *Lebensm. Wiss. Technol.*, **28**, 264–71.

Buchanan, R.E. and Gibbons, N.E. (eds.) (1974) *Bergey's Manual of Determinative Bacteriology*, 8th edn., Williams & Wilkins, Baltimore Maryland.

Bugbee, W.M. and Cole, D.F. (1976) Sugarbeet storage rot in the Red River Valley 1974-75. *J. Am. Soc. Sugar Beet Technol.*, **19**, 19–24.

Bugbee, W.M., Cole. D.F. and Nielsen, G. (1975) Microflora and invert sugars in juice from healthy tissue of stored sugar beets. *Appl. Microbiol.*, **29**, 780–1.

Carruthers, A. and Oldfield, J.F.T. (1955) The activity of thermophilic bacteria in sugar-beet diffusion systems. *8th Ann. Tech. Conf. Br.*, Sugar Corp., Nottingham, England.

Carruthers, A., Gallagher, P.J. and Oldfield, J.F.T. (1958) *Nitrate reduction by thermophilic bacteria in sugar beet diffusion systems*. Report of British Sugar Corporation, Nottingham, England.

CDC (2004) Botulism–General Information. www.cdc.gov/ncidod/dbmd/diseaseinfo/botulism_g.htm.

Cerutti de Guglielmone, G., Diez, O., Cardenas, G. and Oliver, G. (2000) Sucrose utilization and dextran production by *Leuconostoc mesenteroides* isolated from the sugar industry. *Sugar J.*, **62**, 36–40.

Chen, J.C.P. and Chung, C.C. (1993) *Cane Sugar Handbook. A Manual for Cane Sugar Manufacturers and their Chemists*, 12th edn, John Wiley & Sons Ltd., Chichester, UK.

Chen, J.C.P. and Rauh, J.S. (1990) Technical and economic justification for the use of sugar process chemicals, in 49th Annual Conference of the Hawaiian Sugar Technologists. F48–57.

Clarke, M.A. (1997) Dextran in sugar factories: causes and control part II. *Sugar Azucar, Nov*, 22–34.

Clarke, M.A., Roberts, E.J and Garegg, P.J. (1996) Sugarbeet and sugarcane polysaccharides: a brief review. *Proc. Sugar Process. Conf. Res.*, 368–88.

Codex (Codex Alimentarius Commission) (1994) *Sugars, Cocoa Products and Chocolate and Miscellaneous Products, volume 11*, 2nd edn, Joint FAO/WHO Food Standards Programme, Codex Alimentarius Commission, Rome.
Cole, D.F. and Bugbee, W.M. (1976) Changes in resident bacteria, pH, sucrose and invert sugar levels in sugarbeet roots during storage. *Appl. Environ. Microbiol.*, **31**, 754–7.
Cox, N. and Hinkle, R. (2002) Infant botulism. *Am. Fam. Physician*, **65**, 1388–92.
Crane, E. (1979) *Honey. A Comprehensive Survey*, Heinemann, London.
Criseo, G., Bolignano, M.S. and de Leo, F. (1993) Isolazione di *Clostridium botulinum* tipo B da campioni di miele di origine Siciliane. *Riv. Sci. Aliment.*, **22**, 175–81.
Day, D.F. (2000) Microbiological control in sugar manufacturing and refining, in *Handbook of Sugar Refining: A Manual for the Design and Operation of Sugar Refining Facilities*.
De Lucca, A.J., II, Kitchen, R.A., Clarke, M.A. and Goynes, W.R. (1992) Mesophilic and thermophilic bacteria in a cane sugar refinery. *Zuckerind*, **117**, 237–40.
Desai, B.B. and Salunkhe, D.K. (1991) Sugar crops, in *Foods of Plant Origin: Production, Technology, and Human Nutrition* (eds. by D.J.C. Salunkhe and S.S. Deshpande), Van Nostrand Reinhold, New York, pp. 413–89.
Duffaut, E. and Godshall, M.A. (2002) Hydrogen peroxide as a processing aid in the cane factory, in Proceedings of the 2002 Sugar Processing Research Conference, New Orleans. pp. 189–202.
Dumont, J., Lessard, D. and Allard, G.B. (1991) Treatment of spring maple sap using ultraviolet radiation. *Can. Inst. Food Sci. Technol. J.*, **24**, 259–63.
Dumont, J., Saucier, L., Allard, G.B. and Aurouze, B. (1993) Microbiological, physocochemical and sensory quality of maple syrup aseptically packaged in paper-based laminate. *Int. J. Food Sci. Technol.*, **28**, 83–93.
Duncan, C.L. and Colmer, A.R. (1964) Coliforms associated with sugarcane plants and juices. *Appl. Microbiol.*, **12**, 173–7.
Dodds, K.L. (1993) Worldwide incidence and ecology of infant botulism, in *Clostridium botulinum*. Ecology and control in foods (eds. A.H.W. Hauschild and K.L. Dodds), Marcel Dekker, New York, pp. 105–17.
Egan, B.T. (1971) Post harvest deterioration losses in sugar cane. *Sugar J.* **33**, 9–13.
Esteban-Quilez, M.A. and Marcos-Barrado. A. (1976) Actividad de agua de miel y desarollo de levaduaras osmotolerantes. *Anal. Bromotol.*, **28**, 33–44.
FAO (Food and Agriculture Organisation of the United Nations) (1998) *Production Yearbook, volume 51*, Food and Agriculture Organisation, Rome.
Faparusi, S.I. and Bassir, O. (1971) Microflora of fermenting palm sap. *J. Food Sci. Technol.* (Mysore), **8**, 206.
Fiedler, B. (1994) Effekt von Desinfektionsmitteln auf osmophile Hefen während der Herstellung und Verarbeitung von Zucker. *Zuckerind*, **119**, 130–3.
Franchi, F. and Bocchi, A. (1994) Control of diffusion juice and press water. *Int. Sugar J.*, **96**, 80–3.
Gibson, B. (1973) The effect of high sugar concentrations on the heat resistance of vegetative microorganisms. *J. Appl. Bacteriol.*, **36**, 365–76.
Gilliam, M. and Prest, D.B. (1987) Microbiology of feces of the larval honey bee, *Apis mellifera*. *J. Invert. Pathol.*, **49**, 70–5.
Gilliam, M. and Valentine, D.K. (1976) Bacteria isolated from the intestinal contents of foraging worker honey bees, *Apis mellifera*: the genus *Bacillus*. *J. Invert. Pathol.*, **28**, 275–6.
Gilliam, M., Lorenz, B.J. and Richardson, G.V. (1988) Digestive enzymes and micro-organisms in honey bees, *Apis mellifera*: influence of streptomycin, age, season and pollen. *Microbios*, **55**, 95–114.
Goldoin, D.S., Souza, L.G., da Costa, S.M. and da Silva, A.A. (1982) Microbiology of crystal sugar distributed in commerce (in Portuguese). *Brasil Açucareiro* **100**, 331–35. (Abstract 83-2-12-10866-FSTA).
Graham, J.M. (1992) *The Hive and the Honey Bee*, Dadant and Sons, Hamilton, IL.
Guerin, B., Guerin, M.-S. and Loilier, M. (1972) Emploi en sucrerie d'un nouvel inhibiteur de développements microbiens. *Sucr. Fr.*, **113**, 203–11.
Guilfoyle, D.E. and Yager, J.F. (1983) Survey of infant foods for *Clostridium botulinum* spores. *J. Assoc. Off. Anal. Chem.*, **66**, 1302–4.
Hamilton, L.S. and Murphy, D.H. (1988) Use and management of nipa palm (*Nypa fruticans* Arecaceae): a review. *Econ. Bot.*, **42**, 206–13.
Hauschild, A.H.W., Hilsheimer, R., Weiss, K.F. and Burke, R.B. (1988) *Clostridium botulinum* in honey, syrups and dry infant cereals. *J. Food Prot.*, **51**, 892–4.
Hazzard, A.R. and Murrell, W.G. (1989) *Clostridium botulinum*, in *Foodborne Microorganisms of Public Health Significance* (eds. K.A. Buckle *et al.*), 4th edn, Australian Institute of Food Science and Technology (NSW Branch) Food Microbiology Group, Pymble, NSW, Australia, pp. 177–208.
Hein, W., Pollach, G. and Rösner, G. (2002) Studien zu mikrobiologischen Aktivitäten bei der Dicksaftlagerung. *Zuckerind*, **127**, 243–57.
Hocking, A.D. (1988) Moulds and yeasts associated with foods of reduced water activity: ecological considerations, in *Food Preservation by Moisture Control* (ed. C.C. Seow), Proc. Conference Penang, Malaysia 21–24 September 1987, Elsevier Applied Science, Barking, UK, pp. 57–72.
Hollaus, F. (1977) Die Mikrobiologie bei der Rübenzuckergewinnung: Praxis der Betriebskontrolle und Massnahmen gegen Mikroorganismen. *Zschrft. Zuckerind.*, **27**, 722–6.
Hollaus, F. and Pollach, G. (1993) Untersuchungen über den Monosaccharid-Abbau während der Rübenextraktion. *Zuckerind*, **118**, 169–79.

Hollaus, F., Hein, W., Pollach, G., Scheberl, A. and Messner, P. (1997) Nitritbildung im Dünnsaftbereich durch *Thermus*-Arten. *Zuckerind*, **122**, 365–8.
Huhtanen, C.N., Knox, D. and Shimanuki, H. (1981) Incidence and origin of *Clostridium botulinum* spores in honey. *J. Food Prot.*, **44**, 812–4.
Jermini, M.F.G., Geiges, O. and Schmidt-Lorenz, W. (1987) Detection, isolation and identification of osmotolerant yeasts from high-sugar products. *J. Food Prot.*, **50**, 468–72, 478.
Kautter, D.A., Lilly, T., Jr., Solomon, H.M. and Lynt, R.K. (1982) *Clostridium botulinum* spores in infant foods: a survey. *J. Food Prot.*, **45**, 1028–9.
Kearsley, M.W. and Dziedzic, S.Z. (1995) *Handbook of Starch Hydrolysis Products and their Derivatives*, Blackie Academic and Professional, Glasgow.
Kissinger, J.C. (1974) Collaborative study of a modified resazurin test for estimating bacterial count in maple sap. *J. Assoc. Off. Anal. Chem.*, **57**, 544–7.
Klaushofer, H. and Parkkinen, E. (1966) Concerning taxonomy of highly thermophilic aerobic sporeformers found in juices from sugar factories. *Z. Zuckerind.*, **16**, 125–30.
Klaushofer, H., Hollaus, F. and Pollach, G. (1971) Microbiology of beet sugar manufacture. *Process Biochem.*, **6**, 39–41.
Klaushofer, H., Clarke, M.A. Rein, P.W. and Mauch, W. (1998) Microbiology, in *Sugar Technology–Beet and Cane Sugar Manufacture*. (eds. P.W. Van der Poel, H. Schiweck and T. Schwartz), Verlag Dr. Albert Bartens KG, Berlin.
Labbe, R.G., Kinsley, M. and Wu, J. (2001) Limitations in the use of ozone to disinfect maple sap. *J. Food Prot.*, **64**, 104–7.
Le Bot, Y. and Gouy, P.A. (1995) Polyols from starch, in *Handbook of Starch Hydrolysis Products and their Derivatives*, Blackie Academic & Professional, Glasgow.
Leveau, J.Y. and Bouix, M. (1979) Etude des conditions extrèmes de croissance de levures osmophiles. *Ind. Alim. Agric.*, **96**, 1147–50.
Lillehoj, E.B., Clarke, M.A. and Tsang, W.S.C. (1984) *Leuconostoc* spp. in sugarcane processing samples. *Proc. Sugar Process. Res. Conf.*, 141–51.
Lilly, T., Jr., Rhodehamel, E.J., Kautter, D.A. and Solomon, H.M. (1991) *Clostridium botulinum* spores in corn syrup and other syrups. *J. Food Prot.*, **54**, 585–7.
Lionnet, G.R.E. and Pillay, J.V. (1988) Ethanol as an indicator of cane delays under industrial conditions *Proc. Annu. Congr. S. Afr. Sugar Technol. Assoc.*, **62**, 6–8.
Long, S.S., Gajewski, J.L., Brown, L.W. and Gilligan, P.H. (1985) Clinical, laboratory and environmental features of infant botulism in Southeastern Pennsylvania. *Pediatrics*, **75**, 935–41.
Lund, B.M. and Peck, M.W. (2000) *Clostridium botulinum*, in *The Microbiological Safety and Quality of Food* (eds. B.M. Lund, T.C. Baird-Parker and G.W. Gould), *volume 2*, Aspen Publishers, Maryland.
Martin, E.C. (1958) Some aspects of hygroscopic properties and fermentation of honey. *Bee World*, **39**, 165–78.
McCowage, R.J. and Atkins, P.C. (1984) Dextran–an overview. The Australian experience *Proc. Res. Int. Dextran Workshop.*, 7–39.
McMaster, L. (1975) Thermophilic bacteria associated with the cane sugar diffusion process. *M.Sc. Thesis*, University of Natal, Durban, South Africa.
McMaster, L. and Ravnö, A.B. (1977) The occurrence of lactic acid and associated microorganisms in cane sugar processing. *Proc. Int. Soc. Sugar Cane Technol.*, **16**, 1–15.
Midura, T.F. (1996) Update: infant botulism (review). *Clin. Microbiol. Rev.*, **9**, 119–25.
Moroz, R. (1963) Methods and procedures for the analyses of microorganisms in sugar, in *Principles of Sugar Technology* (ed. P. Honig), *volume 3*, Elsevier, Amsterdam, pp. 373–449.
Morselli, M.F. and Fedheim, W. (1988) Ahornsirup–eine Übersicht. (Maple syrup–a review.). *Zschr. Lebensmitt. Unters. Forschung*, **186**, 6–10.
Naim, S.H. and Husin, A. (1986) Coconut palm sugar. Cocoa and coconut: progress and outlook pp. 943–94. Rajaratnam and Chew Poh Soon eds.
Müller, G. (1989) Microbial counts in sugar (sucrose) and the influence of humid storage conditions. *Lebensmittelind*, **36**, 253–5.
Müller,G., Gertknecht, E.and Strubel, S. (1988) Microbiological and physicochemical analyses of cane sugar and manufactured affinated sugars. *Lebensmittelind*, **35**, 169–71.
Naim, S.H. and Husim, A. (1986) Coconut palm sugar. Cocoa and coconut: progress and outlook pp. 943–94. Rajaratnam and Chew Poh Soon eds.
Nakano, H., Yoshikuni, Y., Hashimoto, H. and Sakaguchi, G. (1992) Detection of *Clostridium botulinum* in natural sweetening. *Int. J. Food Microbiol.*, **16**, 117–21.
Nakano, H., Kizaki, H. and Sakaguchi, G. (1994) Multiplication of *Clostridium botulinum* in dead honey-bees and bee pupae, a likely source of heavy contamination of honey. *Int. J. Food Microciol.*, **16**, 117–21.
Nevas, M., Hielm, S., Lindström, M., Horn, H., Koivuletho, K. and Korkeala, H. (2002) High prevalence of *Clostridium botulinum* type A and B in honey samples detected by polymerase chain reaction. *Int. J. Food Microbiol.*, **72**, 45–52.
Nuñez, W.J. and Colmer. A.R. (1968). Differentiation of *Aerobacter-Klebsiella* isolated from sugarcane. *Appl. Microbiol.*, **16**, 1875–8.
Nystrand, R. (1984) Microflora in beet sugar extraction. PhD Thesis, University of Lund, Sweden.

Oldfield, J.F.T., Dutton, J.V. and Teague, H.I. (1971) The significance of invert and gum formation in deteriorated beet. *Int. Sugar J.*, **73**, 3–8, 35–40, 66–8.
Oldfield, J.F.T., Dutton, J.V. and Shore, M. (1974a) Effects of thermophilic activity in diffusion on sugar beet processing. Part I. *Int. Sugar J.*, **76**, 260–3.
Oldfield, J.F.T., Dutton, J.V. and Shore, M. (1974b) Effects of thermophilic activity in diffusion on sugar beet processing. Part II. *Int. Sugar J.*, **76**, 301–5.
Olsen, H.S. (1995) Enzymatic production of glucose syrups, in *Handbook of Starch Hydrolysis Products and their Derivatives* (eds. M.W. Kearsley, and S.Z. Dziedzic), Blackie Academic & professional, Glasgow, 26–64.
Owen, W.L. (1977) Microbiology of sugar manufacture and refining, in *Cane Sugar Handbook* (eds. Meade, G.P. and Chen, J.C.P.), 10th edn., Wiley, New York pp. 405–22.
Pancoast, H.M. and Junk, W. (1980) Handbook of Sugars. 2nd ed. Westport, CT, AVI Publishing Co., Inc.
Parker, S., Shortle, W.C. and Smith, K.T. (1994) Identification of Gram-positive bacteria isolated from initial stages of wound-initiated discoloration of red maple. *Eur. J. Forest Pathol.*, **24**, 48–54.
Pitt, J.I. and Hocking, A.D. (1997) *Fungi and Food Spoilage*, 2nd edn, Blackie Academic and Professional, London.
Pollach, G., Hein, W. and Rösner, G. (1998) New findings towards solving microbial problems in sugar factories. *Zuckerind*, **124**, 622–37.
Pollach, G., Hein, W., Leitner, A. and Zöllner, P. (2002) Detection and control of strictly anaerobic sporeforming bacteria in sugar beet tower extractors. *Zuckerind*, **7**, 530–7.
Ruiz-Argüeso, T. and Rodriguez-Navarro. A. (1975) Microbiology of ripening honey. *Appl. Microbiol.*, **30**, 893–6.
Salunkhe, D.K. and Desai, B.B. (1988) *Postharvest Biotechnology of Sugar Crop*. CRC Press, Boca Raton, FL.
Scarr, M.P. and Rose, D. (1966) Study of osmophilic yeasts producing invertase. *J. Gen. Microbiol.*, **45**, 9–16.
Shamala, T.R. and Skreekantiah, K.R. (1988) Microbiological and biochemical studies on traditional Indian palm wine fermentation. *Food Microbiol.*, **5**, 157–62.
Skole, R.D., Hogu, J.N. and Rizzuto, A.B. (1977) Microbiology of sugar: a taxonomic study. *Tech. Sess. Cane Sugar Refin. Res.*, New Orleans.
Snowdown, J.A. and Cliver, D.O. (1996) Microorganisms in honey. *Int. J. Food Microbiol.*, **31**, 1–26.
Sugiyama, H., Mills, D.C., and Kuo, L.J.C. (1978) Number of *Clostridium botulinum* in honey. *J. Food Prot.*, **41**, 848–50.
Tallgren, A.H., Airaksinen, U., von Weissenberg U., Ojamo, H., Kuresito, J. and Leisola, M. (1999) Exopolysaccharide-producing bacteria from sugar beets. *Appl. Environ. Microbiol.*, **65**, 862–4.
Tilbury, R.H. (1967) Studies on the microbiological deterioration of raw cane sugar, with special reference to osmophilic yeasts and the preferential utilisation of laevulose in invert. MSc Thesis, University of Bristol, UK.
Tilbury, R.H. (1968) Biodeterioration of harvested sugar cane, in *Biodeterioration of Materials. Microbiological and Allied Aspects* (eds. A.H. Walters and J.J. Elphick), Elsevier, Amsterdam, pp. 717–30.
Tilbury, R.H. (1970) Biodeterioration of harvested sugar cane in Jamaica. *Ph.D. Thesis*, Aston University, Birmingham, UK.
Tilbury, R.H. (1975) Occurrence and effects of lactic acid bacteria in the sugar industry, in *Lactic Acid Bacteria in Beverages and Food* (eds. J.G. Carr, C.V. Cutting and G.C. Whiting), Academic Press, London, pp. 177–91.
Tilbury, R.H. (1980) Xerotolerant yeasts at high sugar concentrations, in *Microbial Growth and Survival in Extremes of Environment* (eds. G.W. Gould and J.E.L. Corry), Academic Press, London, 103–28.
Tilbury, R.H., Orbell, C.J., Owen, J.W. and Hutchinson, M. (1976) Biodeterioration of sweetwaters in sugar refining. *Proc. 3rd Int. Biodegrad. Symp., Applied Science*, London, pp. 533–43.
Tokuoka, K. (1993) Sugar- and salt-tolerant yeasts. *J. Appl. Bacteriol.*, **74**, 101–10.
Townsend, G.F. (1975) Processing and storing liquid honey, in Honey: a comprehensive survey. (ed. Crane, E.), Heinemann, London, UK. pp. 269–92.
Troller, J.A. and Christian, J.H.B. (1978) *Water Activity and Food*, Academic Press, London.
Trost, L.W. and Steele, M. (2002) Control of microbiological losses prior to cane delivery, and during sugar processing. *Int. Sugar J.*, **104**, 118, 120–3.
Tysset, C. and Rousseau, M. (1981) Problem of microbes and hygiene of commercial honey. *Rev. Med. Vet.*, **132**, 591–600.
Tysset, C., Brisou, J., Durand, C. and Malaussene, J. (1970) Contribution to the study of the microbial infection of healthy honey bees (*Apis mellifera*): inventory of bacterial populations by negative Gram. *Assoc. Diplom. Microbiol. Fac. Pharm. Univ. Nancy Bull.*, **116**, 41–53.
Van der Poel, P.W., Schiweck, H. and Schwartz, T. (1998) *Sugar Technology–Beet and Cane Sugar Manufacture*, Verlag Dr. Albert Bartens KG, Berlin.
Vindelov, J. and Arneborg, N. (2002) Effects of temperature, water activity, and syrup film composition on the growth of *Wallemia sebi*: development and assessment of a model predicting growth lags in syrup agar and crystalline sugar.
White. J.W., Jr. (1978) Honey. *Adv. Food Res.*, **24**, 287–374.
White, J.W., Jr. (1987) Wiley led the way: a century of federal honey research. *J. Assoc. Off Anal. Chem.*, **70**, 181–9.

13 Soft drinks, fruit juices, concentrates, and fruit preserves

I Introduction

Soft drinks, fruit juices, and fruit preserves represent well-defined and unique ecosystems because of their particular combination of physical and chemical characteristics. Microbiological stability is largely determined by low pH, low oxygen content, pasteurization, and preservatives. This chapter surveys the compositional characteristics of these products and their influence on the microflora of raw materials and end products. It also outlines and discusses the hygienic, preservation, and control requirements for the different product types.

A Foods covered

Carbonated soft drinks. They account for about 50% of the soft-drink market and include colas, sparkling fruit drinks, tonics, ginger ales, shandy, and carbonated teas. They are non-alcoholic beverages made by absorbing carbon dioxide. They may also contain fruit juices, pulp, or peel extracts. The amount of carbon dioxide should not be less than that which will be absorbed by the beverage at a pressure of 1 bar (1 atmosphere of pressure; ~100 kPa) and a temperature of 15°C (typically 1–4 v/v). Many countries have defined the ingredients that may be used additionally in carbonated soft-drinks, such as nutritive sweeteners, flavoring, coloring, and acidification agents, foaming and emulsifying agents, stabilizing or viscosity-producing agents, caffeine, quinine, and chemical preservatives. Variants of soft drinks containing alcohol have been put on the market, typically containing 4–6% alcohol. Their microbiology is similar to that of regular soft drinks.

Non-carbonated (still) soft drinks. Traditionally, they are predominantly fruit based, e.g. fruit drinks, fruit juice drinks, and fruit squashes, but do not contain carbon dioxide.

Fruit juices. Fruit juices are the unfermented, but fermentable, liquids obtained from the edible part of sound, appropriately mature and fresh fruits or fruits maintained in fresh condition by physical means or other suitable treatments. Juices can be obtained by mechanical extraction processes or by reconstitution of concentrated fruit juice with potable water. Juices can be cloudy or clear and must have the essential characteristics typical of the juice of the fruit from which it comes. Diluting and or blending is a common practice as many fruit juices are either too acid or too strongly flavored to be pleasant for consumption. Fruit juice drinks typically contain no less than 20% fruit juice.

Concentrated fruit juice. It is the product that complies with the definition of fruit juice, except that water has been physically removed, e.g. by evaporation. In the production of juice that is to be concentrated, appropriate technological processes are used and may include simultaneous diffusion of the pulp cells or fruit pulp by water, provided that the water extracted juice is added in-line to the primary juice, before the concentration process. Two or more kinds of fruit juice concentrates can be mixed together. Fruit juice concentrates may have added or restored aromatic substances, volatile flavor components, pulp and cells, all of which must be recovered from the same kinds of fruits and be obtained by physical means. The microbiology of fruit juice concentrates is greatly influenced by their reduced water activity.

Fruit nectars and cordials. These are the unfermented but fermentable pulpy-liquid drinks prepared from one or more fruits, plus one or more sweeteners, and other ingredients. Fruit nectars can also be obtained by adding water with or without the addition of sugars, other carbohydrate sweeteners such as honey and/or other sweeteners to concentrated fruit juices. Fruit nectars typically contain no less than 50% pure fruit juice. Cordials are prepared from fruit (as fruit juice, comminuted fruit, or orange peel extract), water, and sugars.

Fruit purées. Fruit purees are the unfermented but fermentable products obtained by appropriate processing (e.g. sieving, grinding, milling) the edible part of the whole or peeled fruit without removing the juice. The fruit must be sound, appropriately mature, and fresh or preserved by physical means or by treatment applied in accordance with the provisions of the Codex Alimentarius Commission. Concentrated fruit purée may be obtained by the physical removal of water of the fruit purée.

Ready-to-drink tea based beverages. These range from relatively unformulated, still products formed from direct leaf extraction, which may be slightly sweetened and flavored with lemon or other fruits, to carbonated soft drinks made from instant tea solids and lemon juice which may have a lower pH value and be preserved with weak acids. Because of their diversity, these products have a wide range of microbiological susceptibilities.

Fruit preserves or jams. They are viscous or semi-solid products containing one or more fruits, together with permitted sweetening agents and gelling ingredients such as pectin, carrageenin, agar, guar gum, alginate, or methylcellulose.

Coconut milks and coconut water. These are products derived from separated endosperm (kernel) of coconut palm (*Cocos nucifera* L.). Coconut milk is the dilute emulsion of comminuted coconut endosperm in water. The Draft standard for aqueous coconut products (Anonymous, 2002) describes standards for different types of coconut products (regular, light and skim milks, and creams) and, amongst others prescribes that coconut milks shall be treated with heat pasteurization, sterilization, or ultrahigh temperature (UHT) process. Coconut water is the albumen of the coconut. It is a white milky liquid that will change into flesh as the fruit matures.

Standards of identity for several of the above products have been or are developed by the Codex Alimentarius Commission (CAC, 1991a,b; FAO/WHO, 2003).

B Important properties

Acidity. Fruits, the main raw material used in fruit juices, soft drinks, and preserves, usually have pH values between 2.0 and 4.5. This results from the high level of organic acids present, e.g. 0.2% in pears to 8.5% in limes. The highest pH values are found in tomatoes (average 4.3, range 3.5–5.0; Powers, 1976). There are some exceptions, like prickly pears, which have pH 6.4–6.5 (Gurrieri *et al.*, 2000), and melons (e.g. cantaloupe 6.2–6.7; Banwart, 1979). Banana's would be around pH 5.0. Besides citric acid (which may account for 95% of the total acid content in citrus fruit), ascorbic acid, malic acid, tartaric acid, and quinic acid may also be present (Gurrieri *et al.*, 2000; Stratford *et al.*, 2000). Fruit concentrates, which frequently serve as raw materials in soft drinks and fruit preserve manufacture, have the same pH range. Cola-based drinks also contain phosphoric acid. Tables 13.1–13.3 give examples of the composition and typical pH values of commercial soft drinks, concentrated drinks, and fruit preserves.

Water activity. Water activity (a_w) plays an important role in the preservation of soft drinks and fruit preserves. Many of them, especially jams, marmalades and concentrated soft drinks, have a high sugar

Table 13.1 Examples of the composition, the optional additives, and pH values of single strength soft drinks

Orange drink (with fruit)	g/kg	Cola drink(without fruit)	g/kg
Orange concentrate (60°Brix)	13.20	Cola concentrate	6.4
Citric acid	1.4	Cola essence	0.3
Sucrose syrup (67°Brix)	141.0	Sucrose (powder)	84.1
Benzoic acid	0.07	Tartaric acid	2.2
Water	844	Caffeine	0.131
Refraction: 10°Brix[a]		Orthophosphoric acid	0.652
Carbonated (2.5 v/v of CO_2)[b]: pH = 2.95		Water	906
Noncarbonated: pH = 3.2		Refraction: 8.6°Brix	
		Carbonated (3.5 v/v of CO_2): pH = 2.6	
Other examples, using lemon, pineapple, cherry, raspberry, etc. as the fruit, show a mean pH of 3.0 when carbonated and 3.3 when not		Other examples like tonic, lemon-lime, etc. show a mean pH of 2.8–2.9	

[a]The dry matter content of a particular product is often expressed as "°Brix"; this refers to that sucrose concentration which has the same refraction as the formula concerned: 15EBrix means, for example, that the product has the same refraction as a 15% (*w*/*w*) sucrose solution.
[b]The CO_2 content is often expressed in "volumes of CO_2 dissolved in one volume of water": when the CO_2 content is 2.5 *v*/*v*, this means that 2.5 volumes of CO_2 are dissolved in 1 volume of water at the specified temperature.

Table 13.2 Examples of the composition, the optional additives, and the pH values of concentrated soft drinks[a]

Orange squash (with fruit)	g/kg	Grenadine drink (without fruit)	g/kg
Orange concentrate (61°Brix)	55	Grenadine aroma	4
Citric acid	15	Sucrose syrup (67°Brix)	955
Sucrose (powder)	509	Coloring agents	0.4
Sorbic acid	0.5	Tartaric acid	2.5
Water	434	Sorbic acid	0.300
		Water	42
Refraction: 59°Brix; pH 2.5		Refraction: 64°Brix; pH 2.4	
Other examples, with different types of fruit concentrates (apple, cherry, black currant, grapefruit, lemon, pineapple, etc.) have a mean pH of 2.7		Other examples, like peppermint syrup and other artificially flavored concentrated drinks have a mean pH of 2.4	

[a]W. Kooiman and W. Baggerman (unpublished data).

Table 13.3 Examples of a composition, the optional additives, and the pH values of fruit preserves[a]

Strawberry jam	g/kg	Low-calorie strawberry spread	g/kg
Strawberries	400	Strawberries	422
Sucrose	600	Sucrose	331
Pectin	4	Pectin	7.5
Citric acid	5	Citric acid	2.5
Sorbic acid	0.3	Coloring agent	0.24
Refraction: 65°Brix; pH 3.3		Sorbic acid	0.5
		Refraction: 33°Brix; pH 3.5	
Other types of jam mostly have pH 3.2–3.8		Most low-calorie spreads have pH 3.4–4.0	

[a]W. Kooiman and W. Baggerman (unpublished data).

content. In jams, sugar levels of 55–65% *w*/*w* are common. Concentrated fruit drinks, which from a microbiological point of view can be considered as jams without gelling agents, have sugar levels ranging from 40–65% *w*/*w*. As a consequence, a_w of these products is frequently below 0.90. Sugar levels in soft drinks and fruit juices range from 5% to 15% *w*/*w*, i.e. 5 to 15°Brix. Actual data on a_w of various types of sugar solutions, i.e. sucrose, glucose, inverted sugar, and glucose syrups are given in ICMSF (1980) Chapter 4, Table 4.2. On the reasonable assumption that there are no solute–solute

Table 13.4 Example of a_w calculation for a model formulation of a citrus drink

Ingredient	Weight in formulation (g/100g)	Water content[a] (g/100g)	Concentration of individual components g/kg of total water	Molality	a_w
Citrus concentrate (60°Brix)	12.0	4.8	(12/42.8)1000 = 280		0.976[b]
Sucrose	40.0		(40/42.8)1000 = 935	2.73	0.940[c]
Glucose	10.0		(10/42.8)1000 = 234	1.30	0.975[c]
Water	38.0	38.0			

[a]The water content of the final mixture is 38.0 g plus the 4.8 g, which comes from the citrus concentrate, i.e., 42.8/100 g.
[b]The $a_{w.1}$ is calculated, using the experimental datum that the concentrate has an a_w value of 0.9000. Because this concentrate is diluted with water over a factor of 12/50 (i.e. 0.24×), the interpolated a_w value equals 0.976.
[c]Calculated by interpolation in Table 4.2. (ICMSF, 1980).

interactions, a_w of mixed solutions containing two or more solutes may be calculated using the following equation (Ross, 1975):

$$a_w = a_{w.1} \times a_{w.2} \times a_{w.3}, \ldots, \text{etc.}$$

where $a_{w.1}$ represents the water activity of the first solution, $a_{w.2}$ that of the second solution, and so forth. By using this equation and the data of Table 4.2 in ICMSF (1980), the water activity of a particular mixture containing 40% *w/w* sucrose ($a_{w.1} = 0.959$), and in addition, 40% *w/w* glucose ($a_{w.2} = 0.933$) can be calculated as follows:

$$a_w = 0.959 \times 0.933 = 0.895.$$

Table 13.4 presents a calculation of a_w in a model concentrated-citrus drink.

Nutrients. The carbohydrate content of soft drinks and fruit preserves is often high and consists mostly of easily metabolized hexoses (e.g. glucose and fructose), pentoses, and pectins. The organic acid content is usually high but variable and has a strong influence on the pH; many fungi can use organic acids as carbon sources under aerobic conditions (see ICMSF (1980) Chapter 7). Organic nitrogen is frequently present but only at very low levels, i.e. 0.05–0.15% and about 60% of it is in the form of free amino acids. In contrast to popular belief, most soft drinks and fruit preserves contain only very low levels of vitamins. Exceptions are black currant juice and rosehips, which contain naturally high levels of vitamin C. Orange juice is high in Vitamine C, folate and potassium. Other fruits or fruit concentrates may be supplemented in some countries with ascorbic acid (vitamin C) as a contributor to flavor and color stability as well as for nutritional reasons. Group B vitamins are practically absent in these products.

Oxygen and redox potential. Dissolved oxygen is readily consumed by fruit particles or sulfite, is bound by ascorbic acid, or is removed when the product is pasteurized. As a result, the redox potential is often very low. Replacement of oxygen by CO_2 in a carbonated beverage creates a specific ecosystem. Aseptic filling into oxygen-permeable packaging materials, or any packaging system that permits a large head space, may drastically increase the level of dissolved oxygen or the redox potential. This again has a strong influence upon the ecosystem.

Natural antimicrobial substances. Many fruits, e.g. citrus fruits and cranberries, naturally contain antimicrobial substances such as essential oils, benzoic acid, or sorbic acid. Tea contains alkaloids (the methylated derivatives of purine, caffeine, theobromine, and theophylline) and catechins (flavonols) with bacteriostatic effects that have been particularly characterized against pathogens (Vanos *et al.*, 1987). Even anti-botulinum properties of some teas, or particular varieties of tea, have been reported (Hara *et al.*, 1989; Horiba *et al.*, 1991). However, care must be taken not to extrapolate these findings to all tea based drinks as tea containing milk and sugar will support the growth of *Cl. botulinum* and allow toxin formation (M.B. Cole, personal communication). Although individual compounds are rarely available in a sufficiently high concentration to assure complete inhibition of microbial growth, they may serve as additional preservative factors.

Some types of soft drinks are stable without heat treatment. For example, some carbonated cola drinks are stable because of low pH (2.6), high CO_2 level (3.5 *v/v*) and the antimicrobial activity of orthophosphoric acid and components of the cola oils, particularly terpenes.

C Initial microflora

Many genera of microorganisms occur in/on fresh fruits (see Chapter 6) that are the raw materials used in the manufacture of soft drinks, fruit juices, and fruit preserves. Although this group of products presents low public health risks, the various ingredients used in their manufacture may contain small numbers of pathogens or adventitious contaminants (See Chapter 6, Section IIIC). Fungi, and in particular yeasts, form the main flora of fruits before processing because of the acidic pH.

Yeasts. More than 110 species of yeasts have been listed as associated with foods, of which a large proportion occur on fruits, and more than 40 with soft drinks (Barnett *et al.*, 2000). The major genera include *Candida*, *Dekkera* (asexual stage *Brettanomyces*), *Hanseniaspora* (asexual stage *Kloeckera*), *Pichia*, *Saccharomyces*, and *Zygosaccharomyces*. Many of these species are adventitious and are sensitive to heat processing or preservatives, so are uncommonly isolated from fruit products. The major yeast species of concern are discussed under Spoilage (Section III).

The major part of the microflora of fruits is removed by pasteurization or the use of adequate levels of preservatives. Only heat-resistant fungi, preservative resistant yeasts and the acid dependant, thermotolerant bacterium *Alicyclobacillus* may survive these preservation techniques. These exceptional microorganisms will be discussed below.

Filamentous fungi (molds). The filamentous fungi most commonly isolated from fresh fruits and fruit juices belong to the following genera *Penicillium*, *Byssochlamys*, *Aspergillus*, *Paecilomyces*, *Mucor*, *Cladosporium*, *Fusarium*, *Botrytis*, *Talaromyces*, and *Neosartorya* (see also Chapter 6). Like the yeasts, many kinds do not survive processing or are sensitive to preservatives, so relatively few species are important in spoilage of fruit products (see Sections II and III).

Bacteria. Many bacterial genera have been isolated from fresh fruits and their juices. Lactic acid bacteria (*Lactobacillus* and *Leuconostoc* spp.), acetic acid bacteria (*Gluconobacter* and *Acetobacter* spp.), and some spore-forming bacteria such as *Bacillus coagulans*, *Clostridium butyricum*, and *Cl. pasteurianum* (the latter especially in tomato-based products) have been implicated in the spoilage of soft drinks. *Streptococcus* and *Pediococcus* species are found in lower frequency in processed fruit juice. Strictly anaerobic microorganisms, such as *Propionibacterium cyclohexanicum*, a recently described bacterium isolated from deteriorated orange juice, may also deteriorate orange juice (Kusano *et al.*, 1997).

A thermoacidophilic spore-forming bacterium, *Alicyclobacillus acidoterrestris*, can also cause spoilage of juices (See Section III, C).

II Potential food safety hazards

Any microorganism present on or below the fruit surface may potentially contaminate fruit juices, concentrates, and fruit preserves. Fruit juices, especially when unpasteurized, are considered to pose a risk to human health (NACMCF, 1997). Several outbreaks have been reported due to consumption of contaminated fruit juices (Parish, 1997, 2000; Burnett and Beuchat, 2000; FDA, 2000; Orlandi *et al.*, 2002; Dawson, 2003). While spoilage yeasts are generally not considered to pose a risk to healthy consumers, recent developments in the medical field have indicated that a number of yeasts previously thought to be innocuous are capable of damaging the human body (Stratford *et al.*, 2000).

A Mycotoxins

The growth of fungi in fruit-based products may occasionally be accompanied by the formation of mycotoxins. Use of raw material of appropriate quality should minimize introduction of mycotoxins in the processed product. The only mycotoxins currently considered to be important in fruit products are patulin, which is produced by the growth of *Pen. expansum* in apples and pears, and ochratoxin A, resulting from growth of *Asp. carbonarius* in grapes. Patulin is now considered to be of only low toxicity to humans. However the presence of patulin in juices is considered to be evidence of the use of unsound fruit in juice manufacture. WHO has recommended a limit of 50 μg/kg for patulin in juices and juice products.

The formation of ochratoxin A by *Asp. carbonarius* (and a few isolates of the closely related species *Asp. niger*) was discovered only recently (Abarca *et al.*, 1994; Téren *et al.*, 1996; Heenan *et al.*, 1998). The principal fruit affected is the grape; ochratoxin A has been detected in red grape juice (Zimmerli and Dick, 1996) as well as dried vine fruits and wines (see Chapter 6).

B Bacterial pathogens

The presence of salmonellae in fruit juices and ciders has been reported on several occasions. In 1975, the Centers for Disease Control and Prevention reported a *Salmonella* Typhimurium outbreak traced back to a commercial apple cider (CDC, 1975). A total of 300 illnesses were reported and it was established that grounder apples from an orchard fertilized with manure were used to make the cider (FDA, 2000). In 1991, 23 people in Fall River, Massachusetts, USA, became ill from apple juice (16 developing bloody diarrhoea, 4 progressing to HUS); a local farm selling unpasteurized cider at a roadside used dropped apples and inadequate washing systems (FDA, 2000). In 1993, *Salmonella* Typhi caused 69 illnesses in the USA, as a consequence of consumption of contaminated reconstituted orange juice although an infected food handler seemed to be the root-cause (Birkhead *et al.*, 1993). D'Aoust (1994) described an outbreak caused by apple cider containing *S.* Typhimurium, which resulted in 286 illnesses. In 1995, an outbreak of *S. enterica* serotype Hartford (*S.* Hartford) infections affected visitors of a theme park in Orlando, Florida, USA, who drank locally produced unpasteurized orange juice. The contamination was a consequence of inadequately sanitized processing equipment (CDC, 1995, Parish, 1997, 1998b; Cook *et al.*, 1998). Amphibians were suspected to be the source of contamination (Parish, 2000). In 1999, two outbreaks caused by *S.* Muenchen contaminated unpasteurized orange juice produced by a single manufacturer resulted in 300 cases of illness and one fatality (CDC, 1999). The same year, an outbreak occurred in Adelaide, Australia, where some 500 laboratory confirmed cases of *Salmonella* infection were associated to consumption of fresh, chilled, unpasteurized orange juice (Parish, 2000). In 2000, 47 confirmed cases of salmonellosis occurred in seven USA states as a result of commercial unpasteurized orange juice contaminated with *S.* Enteritidis (Anonymous, 2000).

Escherichia coli O157:H7 is another pathogen of concern in fruit juices. This acid-tolerant pathogen is known to be associated with feces of a range of animals (including cattle). In 1980 in Toronto, Canada, fourteen children became ill after drinking fresh apple juice and 13 developed bloody diarrhea and hemolytic uraemic syndrome (HUS), with one case resulting in death. The source of contamination is unknown, although it has been postulated that fecal contamination of the apples occurred (Parish, 2000). Several US states reported cases of *E. coli* O157:H7 infection relating back to juices expressed from apples contaminated with fecal material (McLellan and Splittstoesser, 1996). In 1996 in Connecticut,

USA, 14 cases of illness (with 3 people developing HUS) resulted from drinking apple cider from a small cider mill, which apparently had used dropped apples contaminated with *E. coli* O157:H7 (FDA, 2000). Two outbreaks of illness associated to *E. coli* O157:H7 were described in the Western United States and British Columbia, Canada, epidemiologically linked to a particular brand of unpasteurized apple juice. This outbreak caused illness in 70 people, including a child who died from HUS. The contamination of the juice originated from an orchard frequented by deer that were subsequently shown to carry *E. coli* O157:H7 (Besser *et al.*, 1993; Anonymous, 1996a; Mshar *et al.*, 1997; Cody *et al.*, 1999).

Salmonella Typhimurium DT104 and *Escherichia coli* O157:H7 have been shown to survive for up to 21 days in preservative-free apple cider (pH 3.3–3.5) stored at 4°C and 10°C (Zhao *et al.*, 1993; Parish *et al.*, 1997; Roering *et al.*, 1999). It has also been shown that *E. coli* O157:H7 is able to survive on apples fallen on the ground in orchards and used for cider production during storage at 4°C, 10°C, or 25°C due to pH increase associated with mold growth (Fisher and Golden, 1998).

Sliced and diced tomatoes, tomato juices, and tomato concentrates support the growth of pathogens and toxin formation because of their relatively high pH. For this reason, acidification of tomato products to pH 4.1 or 4.2 is common practice. The minimum pH permitting the growth of salmonellae is 3.8 to 4.0, depending on the nature of the acid and the temperature (Ferreira and Lund, 1987). In moderately acidic products such as apple juice (US Department of Health, Education and Welfare, 1975), salmonellae do not grow, but may survive under adverse conditions of temperature, redox potential, or a_w to cause salmonellosis (Chung and Goepfert, 1970). Survival is longer when the pH is high, the storage temperature is low, or the sugar concentration is high (Mossel, 1963).

Considering the significance of the highly infectious *E. coli* O157:H7 and that of several *Salmonella* spp. as food-borne diseases associated with fruit juices and ciders, it is important to apply adequate pasteurization or equivalent (combination) treatments (Uljas and Ingham, 1999; FDA, 2001).

Little information is available on coconut milk or coconut water as vehicles for Food-borne pathogens but salmonellae and *E. coli* O157:H7 should be considered as significant hazards, especially in case of no or inadequate pasteurization. Notably, fresh frozen coconut milk was implicated in an incident in 1991 involving *Vibrio cholerae* O1 (Taylor, 1993).

C *Viruses*

Viruses have been implicated in outbreaks due to consumption of unpasteurized or inadequately heat-treated juice or preparation of reconstituted juice using contaminated water. For instance, more than 3000 infections with Norwalk-like caliciviruses (NLV) occurred in Australia due to unpasteurized orange juice as a result of use of potable water contaminated by waste-water and sewage containing NLV (Fleet *et al.*, 2000). A variety of fruits and fruit juices have transmitted hepatitis A (Cliver, 1983) or viral gastroenteritis caused by small round structured viruses, including NLV (Caul, 1993).

In general, most documented Food-borne viral outbreaks can be traced to food that has been manually handled by an infected food-handler and that is not heated or otherwise treated afterwards, rather than to industrially processed foods (Koopmans and Duizer, 2002). It is key that sufficient attention be given to good agriculture practice (GAP) and good manufacturing practice (GMP) to avoid introduction of viruses onto the raw material and into the food-manufacturing environment. If viruses are present in food *before* processing, residual viral infectivity may be present after some industrial processes. If viruses are present in foods *after* processing, they remain infectious in most circumstances and in most foods for several days or weeks, especially if kept cooled (at 4°C). The HACCP systems should adequately consider possible presence and persistence of viruses as well stringent personal hygiene during preparation (Koopmans and Duizer, 2002).

The destruction of viruses in juices is variable. For example, grape juice and apple juice inactivated polio virus type 1, but pineapple, tomato, grapefruit, and orange juices did not. Ascorbic acid solutions destroyed the virus, but not when added to fruit juices (Konowalchuk and Speirs, 1978a,b). Acidification of juices has been reported not to completely inactivate NLV or hepatitis A viruses (Koopmans and Duizer, 2002).

D Parasites

The relationship between enteric parasitic protozoa, the environment, contamination of food, and human illness is extremely complex (Orlandi *et al.*, 2002). Environmental factors play a significant role in the transmission of most Food-borne parasitic diseases. This impact is particularly apparent with protozoa, which are readily transported to food by contaminated water (Slifko *et al.*, 2000a). Fecal contamination of water sources used in crop irrigation, food processing, and meal preparation all are important sources of human infection. In this regard, contamination of fresh fruits and vegetables is of great concern. With raw or under-pasteurized fruits, the long-surviving encysted forms of Food-borne parasites should specifically be considered as potential hazards.

Juices and ciders have been implicated in Food-borne outbreaks of cryptosporidiosis (Orlandi *et al.*, 2002; Dawson, 2003). While parasites have caused outbreaks associated to fresh fruits in several cases (Orlandi *et al.*, 2002), and thus control of fruit contamination is important, most of the problems seen with juices were caused by environmental contamination or inadequate personnel hygiene combined to insufficient sanitation or no or inadequate pasteurization. *Cryptosporidium* spp. and *Cryp. parvum* were the etiologic agents in outbreaks in the USA due to consumption of unpasteurized apple juice and apple cider (Millard *et al.*, 1994; CDC, 1997; Mshar *et al.*, 1997). In one of these outbreaks there were at least 191 cases of illnesses; the apples used for the cider were probably contaminated by feces when they fell to the ground in a cow pasture (Millard *et al.*, 1994; FDA, 2000). In an outbreak in Maine, USA, children who drank contaminated apple cider developed cryptosporidiosis; the apples used for the cider were likely to have become contaminated when they were washed using well water (CDC, 1997; FDA, 2000).

Food-borne parasites are significant hazards in fruit juices and ciders and they need to be brought under control through appropriate GHP, personnel hygiene, and processing. Orlandi *et al.* (2002) and Dawson (2003) provide overviews of the effect of food processing technologies on food-borne parasites.

III Spoilage

As soft drinks and fruit juices are acidic products that usually contain substantial amounts of fermentable sugars, they are vulnerable for a number of spoilage yeasts and molds and a few acid-tolerant bacteria (Stratford *et al.*, 2000). The range of spoilage microorganisms is restricted due to inhibitory factors such as low pH, low water activity, and the presence of weak-acid preservatives. Among the yeasts, a significant spoilage organism is *Zygosaccharomyces bailii*, which is quite resistant to chemical preservatives. Several fungi are able to grow at the low pH and reduced a_w of soft drinks, juices, and fruit preserves. Only lactic and acetic acid bacteria are able to grow. In fruit juices, also thermoacidophilic sporeforming *Alicyclobacillus acidoterrestris* is able to grow.

A Preservative resistant yeasts

Yeasts predominate in spoilage of acid fruit products because of high acid tolerance, frequent ability to grow anaerobically and, in certain species, preservative resistance. Compared to most filamentous fungi, yeasts possess limited biochemical pathways and have quite fastidious nutritional requirements. Fruit juices are generally rich in simple carbohydrates and complex nitrogen sources, and hence are an

Table 13.5 Thermal resistance of sporing and non-sporing yeasts[a]

Temperature of heating (°C)	Percentage of non-sporing (A) and sporing (S) strains[b] showing survival after heating for			
	10 min		20 min	
	A	S	A	S
65°	None	16	None	None
62.5°	25	60	3	30
60°	75	~100	40	~80

[a]From Put *et al.* (1976).
[b]The non-sporogenous yeasts belonged to the genera *Brettanomyces*, *Candida*, *Kloeckera*, and *Torulopsis* (35 strains). The sporogenous yeasts belonged to the genera *Debaryomyces*, *Hansenula*, *Kluyveromyces*, *Lodderomyces*; *Pichia* and *Saccharomyces* (85 strains).

ideal substrate for yeasts. However, purely synthetic soft drinks usually lack nitrogen sources suitable for yeasts, and rarely spoil.

Yeasts that form ascospores are more heat resistant than those that do not (Table 13.5). Decimal inactivation (D) values of ascospores of *Sacc. cerevisiae* and *Sacc. chevalieri*, common spoilage yeasts in canned fruit products, were 10 times those of vegetative cells (Put *et al.*, 1976). However, these ascospores are much less heat resistant than those of filamentous fungi or bacterial spores. For example $D_{60°C}$ values of ascospores of both the above species were approximately 10 min and z values of approximately 5°C.

Growth of yeasts is usually accompanied by formation of CO_2 and alcohol. Yeasts may also produce turbidity, flocculation, pellicles, and clumping. If pectinesterases are produced, pectin can also be degraded causing spoilage: the natural pectin cloud can be destroyed. Organic acids and acetaldehyde, which contribute for a "fermented flavor", may also be formed.

The following are the most important yeasts causing spoilage of fruit juices and soft drinks (Fleet, 1992; Deak and Beuchat, 1996; Pitt and Hocking, 1997, Barnett *et al.*, 2000):

Sporogenous yeasts	Asexual name for the same species
Debaryomyces hansenii	–
Dekkera anomala, *D. bruxellensis*	*Brettanomyces anomalus*, *Bret. bruxellensis*
Henseniaspora uvarum	*Kloeckera apiculatum*
Issatchenkia orientalis	*Candida krusei*
Kluyveromyces thermotolerans	*Candida dattila*
Lodderomyces elongisporus	–
Pichia anomala, *P. fermentans*, *P. guilliermondii*	*Candida pelliculosa*, *Cand. lambica*, *Cand. guilliermondii*
Saccharomyces cerevisiae, *Sacc. kluyveri*	*Candida robusta*,
Torulaspora delbrueckii	*Candida colliculosa*
Zygosaccharomyces bailii, *Z. fermentati*, *Z. Microellipsoides*, *Z. rouxii*	–
No sexual stage known	*Candida boidinii*, *Cand. etchellsii*, *Cand. inconspicua*, *Cand. sake*, *Cand. stellata*, *Cand. tropicalis*

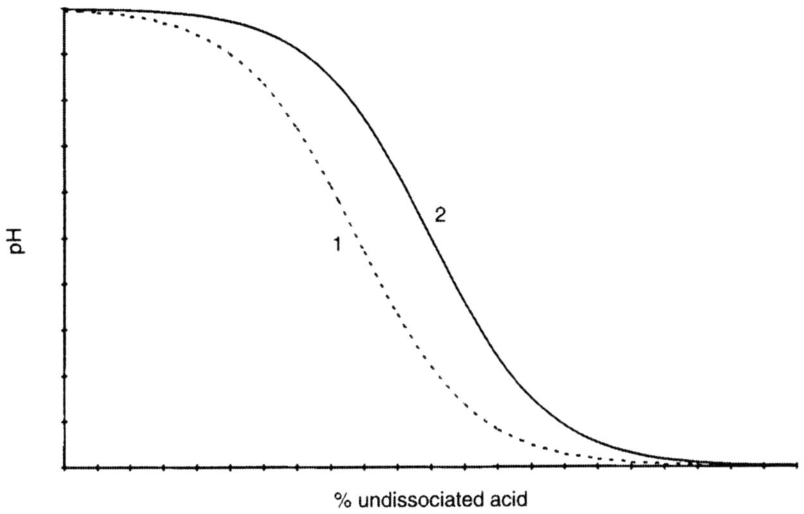

Figure 13.1 Influence of pH on the percentage of undissociated benzoic acid (1) and sorbic acid (2); y = pH, x = % undissociated acid.

Spoilage of fruit juices and fruit juice products is greatly influenced by the presence of preservatives. Many acid, liquid foods, including those under consideration here, are preserved by the addition of sorbic acid, benzoic acid, or sulfur dioxide, alone or in combination. All these products are susceptible to spoilage by preservative resistant yeasts. By far the most significant of these is *Zygosaccharomyces bailii*, which is able to vigorously ferment glucose and fructose solutions, producing CO_2 even under gas over pressures up to 500 kPa (5 bar) or more. Distortion or leakage of cans or plastic bottles may result; glass containers may even shatter. This yeast is acid tolerant, xerophilic, and extremely resistant to weak-acid preservatives (Pitt and Richardson, 1973; Berry, 1979; Thomas and Davenport, 1985; Cole and Keenan, 1986, 1987a,b; Cole *et al.*, 1987; Pitt and Hocking, 1997).

Inhibition of microbial growth by weak acids is strongly pH dependent, being greater at low pH (Freese *et al.*, 1973) (see Figure 13.1). Efficacy presumably depends on undissociated acid molecules, which are believed to pass freely into the microbial cell (Macris, 1975) due to high solubility in the phospholipid portion of the plasma membrane (Cramer and Presteguard, 1977). The major antifungal effect of benzoic acid is to decrease cellular pH, which inactivates phosphofructokinase and to a lesser extent hexokinase (Krebs *et al.*, 1983). An extrapolation cannot be made, however, that only the undissociated form is active. Indeed, specific inhibitory effects of anionic species have been demonstrated for both sorbic acid (Eklund, 1983) and benzoic acid (Cole and Keenan, 1986).

The exceptional resistance of *Z. bailii* to weak-acid preservatives was originally attributed to an inducible, energy requiring system, which pumps preservatives from the cell (Warth, 1977). This hypothesis has been shown to be incorrect (Cole and Keenan, 1987b) as in exponentially growing cells the concentration of weak acid is exactly that which is predictable from the cell intracellular pH and the pK_a value of the acid. Resistance instead appears to be initially due to the ability of cells to tolerate chronic intracellular pH drops, by means of a pH resistant phosphofructokinase enzyme. Subsequently, cells compensate for the decline in pH by re-establishing a 'normal' pH value through a reduced protoplast volume and increased acid efflux rate (Cole and Keenan, 1987b).

The degree of weak-acid resistance is such that if present in a final product, *Z. bailii* will cause spoilage of most preserved foods. This yeast can cause spoilage from very low inocula (J.I. Pitt, Personal communication), which makes detection in the processing plant very difficult. It is therefore essential that effective control measures be aimed at this yeast.

Some products are less susceptible to spoilage by *Z. bailii*, including synthetic products such as soft drinks and water ices that lack a nitrogen source or are made with sucrose, which *Z. bailii* usually cannot assimilate (Pitt and Hocking, 1997). However, at low pH (<4.0) sucrose slowly inverts to fructose and glucose, permitting *Z. bailii* to grow.

B Filamentous fungi (molds)

Most molds are not nutritionally demanding, but unlike yeasts, are, with a few exceptions, strict aerobes. The low redox potential and/or low oxygen tension in fruit juices and soft drinks restricts the development of molds. Molds cause spoilage by forming colonies on the surface, flocculation, or floating mycelia within the product, or clarification of the pectin cloud by breakdown.

Many molds are also xerophilic. Spoilage of jams or preserves is usually caused by *Eurotium* species, though xerophilic Penicillia, especially *Penicillium corylophilum*, occur from time to time (Pitt and Hocking, 1997).

Spoilage of pasteurized fruit based products is usually due to heat-resistant fungi. The most important species are *Byssochlamys fulva*, *Byss. nivea*, *Neosartorya fischeri*, and *Talaromyces* species (Pitt and Hocking, 1997). Techniques for detecting and identifying such species have been outlined (Murdock and Hatcher, 1978; Hocking and Pitt, 1984; Beuchat and Pitt, 2001).

Ascospores of heat-resistant fungi come from soil and do not develop in food factories. Thus the types of fruit juices and products that need to be screened for such fungi are those which may have contact with the soil before or at harvest, principally this applies to grapes, passionfruits, pineapples, mangoes, strawberries, and other berries (Pitt and Hocking, 1997).

C Bacteria

Acetic acid bacteria such as *Gluconobacter (Acetomonas)* spp. and *Acetobacter* spp. are among the main spoilage bacteria because of their ability to grow at relatively low pH (i.e. 3.0–3.5) and in low nutrient levels. They are strict aerobes, and any developments in packaging techniques and materials that increase the amount of oxygen present in the product will increase the incidence of spoilage by these bacteria. *Acetomonas* spp. can be detected in large quantities in processing plants, but should readily be contained by normal plant hygiene (Stratford *et al.*, 2000).

Lactobacillus and *Leuconostoc* spp. have been frequently isolated from fruits and spoiled soft drinks. These lactic acid bacteria can also cause spoilage as recontaminants where adapted strains are allowed to build up in the factory environment. Most lactobacilli isolated are heterofermentative, i.e. they produce CO_2, acetic acid, and ethanol from sugar. The properties that favor the growth of lactic acid bacteria in soft drinks and fruit concentrates are as follows:

1. Ability to grow very well at low redox potential.
2. Acid tolerance, as most species can grow at pH 3.5, and 5–10% of them at pH 3.5–3.0. Despite this, spoilage rarely occurs in drinks below pH 4.0.
3. High CO_2-tolerance.
4. Ability to grow at temperatures used for fruit and fruit-juice processing.
5. Ability to grow in the presence of more than 30% *w/w* sucrose.

The growth of lactic acid bacteria leads to turbidity and opalescence of soft drinks and concentrates and, sometimes, visible bubbles of gas or increased pressure and bursting of containers. In addition,

Table 13.6 Heat resistance of some lactic acid bacteria when heated in single-strength or concentrated orange juice[a]

	Heat resistance when heated in			
	Single-strength juice[b]		Concentrated juice[b]	
Organism bacterium	D ($^{65.5}$ min)	z (°C)	D ($^{65.5}$ min)	z (°C)
Leuconostoc: 13 strains, composite suspension	0.04	3.89	0.23	5.56
Lactobacillus: 2 strains, composite suspension	0.28	3.89	1.20	10.0

[a]Adapted from Murdock *et al.* (1953).
[b]The single strength and the concentrated orange juice had a dry matter content of 98° and 42.0° Brix, respectively, and pH values of 3.7 and 3.45, respectively.

Leuconostoc and some *Lactobacillus* spp. produce dextrans and levans, which form a gummy slime or "ropiness". The "buttermilk" off-flavor frequently associated with the growth of *Lactobacillus* species in fruit products is due to the formation of diacetyl. Product flavors are changed by the formation of acids such as acetic and gluconic.

Lactic acid bacteria are relatively heat sensitive (Table 13.6). For example the $D_{65.5^\circ C}$ values of isolates of *Leuconostoc* and *Lactobacillus* spp. heated in a single strength orange juice concentrate (42°Brix, pH 3.45) were 0.04 min and 0.28 min, respectively (Murdock *et al.*, 1953). Higher D and z values were obtained in concentrates.

The principal bacterium of concern in tomato juice canning is *Bacillus coagulans*, which can produce "flat sour" spoilage. High-Temperature Short-Time (HTST) treatment is designed to reduce the population of *B. coagulans*, which requires relatively high temperatures for germination but can grow over a rather wide temperature range (minimum 5–20°C, maximum 35–50°C (Holt *et al.*, 1994).

The thermoacidophilic spore-forming bacterium *Alicyclobacillus acidoterrestris* has rapidly gained importance in spoilage of shelf-stable fruit juices, fruit juice blends, and lemonades where it produces off-flavors and visible growth (Baumgart *et al.* 1997, ABECitrus, 1999; Stratford *et al.*, 2000). First isolated from deteriorated orange juice by Cerny *et al.* (1984), this microorganism was classified as *Bacillus acidocaldarius*, then as *Bacillus acidoterrestris* (Deinhard *et al.*, 1987) and then assigned to the new genus, *Alicyclobacillus* (Wisotzkey *et al.*, 1992). The bacterium has been isolated from a variety of commercial juices (apple, orange, grapefruit, lime) or juice blends (lemonade fruit juice blend; fruit/carrot juice blend) (Splittstoesser *et al.*, 1994; Yamazaki *et al.*, 1996, 1997; Baumgart *et al.*, 1997; Borlinghaus and Engel, 1997; Parish, 1997; Pinhatti *et al.*, 1997; Splittstoesser *et al.*, 1998a,b; Wisse and Parish, 1998; ABECitrus, 1999). *Alicyclobacillus acidoterrestris* was also detected in deteriorated acidic and isotonic beverages in Japan (Yamazaki *et al.*, 1996; 1997). Detection of *Alicyclobacillus* in apparently sound fruit juices suggests that deterioration may be incidental, requiring adequate conditions for development (Prevedi *et al.*, 1995).

Alicyclobacillus acidoterrestris is strictly aerobic and requires high temperature (optimum between 45°C and 70°C) and low pH (optimum between 2.5 and 4.5) for growth (ABECitrus, 1999). The reported D values range from 60.8 to 94.5 min at 85°C, 10–20.6 min at 90°C and 2.5–8.7 min at 95°C. Z values are between 7.2°C and 11.3°C (Eiroa *et al.*, 1999; ABECitrus, 1999). The pH, acid, and temperature of processing of the juice influence the thermal resistance of spores of *A. acidoterrestris* (Murakami *et al.*, 1998, Pontius *et al.*, 1998). The outgrowth ability of the spores is influenced by various factors such as oxygen availability, type of product, and residual spore level (Prevedi *et al.*, 1997).

This type of spoilage is easily detected due to a distinct medicinal or antiseptic off-odor attributed to guaiacol, a metabolic by-product of the bacterium (Pettipher *et al.*, 1997, Orr *et al.*, 2000). Off-flavor due to 2,6-dibromophenol can also occur (Baumgart *et al.*, 1997).

Spoilage caused by *A. acidoterrestris* in acidic beverages can be prevented by the addition of lysozyme (Yamasaki *et al.*, 1997). It was demonstrated that polylysine, protamine, acetic acid, sucrose ester, and polyglycerol esters inhibited the outgrowth of spores in Trypticase Soy broth at pH 4.0 (Yamasaki *et al.*, 1997).

IV Processing

The method for processing of a particular soft drink, fruit juice, or fruit preserve is dictated basically by balancing organoleptic quality (flavor, color, texture, and cloud-stability), distribution of fruit particles (shreds in marmalade, fruit in jams, etc), ingredient and formulation characteristics (pH, a_w, preservatives, etc.) and processing technology. In order to avoid contamination by bacteria, yeast and molds, ready-to-drink teas and fruit beverages must be processed and filled at high temperatures (Bakka, 1995; Zhao *et al.*, 1997). While processing typically involves a thermal treatment, more recently products (juices and preserves) stabilized using non-thermal treatments such as high hydrostatic pressure have been developed.

Outbreaks associated with fruit juices have caused the US regulators to alter labeling laws, advise on suitable process control options, and require manufacturers to use food safety management systems based on HACCP principles (FDA, 2001). Under this regulation, juice processors must attain at minimum a 5-$\log_{10}$ reduction in the pertinent pathogen(s) using appropriate thermal pasteurization or equivalent measures and document meeting this performance criterion in their systems (NACMCF, 1997). Overviews of the effect of food processing technologies on food-borne parasites have been provided (Orlandi *et al.* 2002; Dawson, 2003). Combinations of processes achieving a 5-$\log_{10}$ reduction have been investigated (Uljas and Ingham, 1999). Process sanitation and fruit surface treatments help to reduce the microbial load entering the production process (Bakka, 1995; Pao and Davis, 2001; Pao *et al.*, 2001).

A model system may serve as a guide for estimating the expected intrinsic stability of various soft drinks and concentrates. The data used to generate the predictive model in Figure 13.2 were obtained using a model system of yeast nitrogen base and fructose buffered to different pH values inoculated

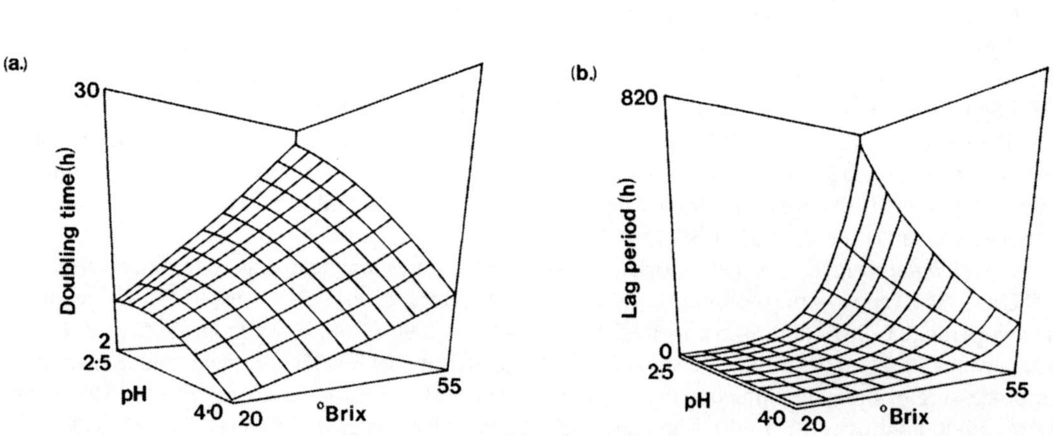

Figure 13.2 The effect of pH and sugar concentration (Brix) on (a) the predicted doubling time and (b) the lag period of *Zygosaccharomyces bailii*.

with *Z. bailii* (1000 cells/mL). Fructose was used as it is preferentially fermented by *Z. bailii*, unlike most other yeasts (Emmerich and Radler, 1983). The following equations can be used to predict the effect of °Brix [F] and hydrogen-ion concentration $[H^+]$ (mM) on doubling time Dt (h) and lag period (h) of *Z. bailii* NCYC 563 (Figure 13.2).

$$Dt = -2.47 + 8.30 \times 10^{-2}[F] + 2.07 \times 10^{-2}[H^+] - 1.80 \times 10^{-3}[F]^2 - 8.40 \times 10^{-1}[H^+]^2 + 1.29 \times 10^{-1}[F][H^+]$$

Lag $= e^x$, where

$$x = -2.14 - 1.67 \times 10^{-2}[F] - 2.90 \times 10^{-2}[H^+] + 1.25 \times 10^{-3}[F]^2 - 4.70 \times 10^{-2}[H^+]^2 + 1.29 \times 10^{-1}[F][H^+].$$

By substitution into these equations, the strong inhibitory effect of both factors acting in combination can be predicted. For example, at Brix 55° and pH 2.5, an initial contamination level of 1–2 cells/mL would take longer than 1 month to reach detectable spoilage levels.

Likewise, the interaction between two weak acids, benzoic acid, and sorbic acid, on the doubling time of *Z. baillii* has been modeled and a significant interaction between the two acids at pH 2.5 found (Cole and Keenan, 1986):

$$\text{Dt} = 6.46 + 1.57 \times 10^{-3}[\text{Benz}] + 1.14 \times 10^{-3}[\text{Sorb}] + 8.7 \times 10^{-6}[\text{Benz}][\text{Sorb}]$$

where [Benz] and [Sorb] represent the concentration (in μM) of benzoic acid and sorbic acid, respectively. These examples show that predictive modeling can be a useful tool to quantify the effects of several factors on the growth of *Z. bailii*, not only to assess existing preservative systems but also during product formulation. Notably, Andres *et al.* (2002) established predictive models for the prediction of the shelf-life of orange juices as a function of fruit washing pre-processing, the use of different preservatives, and the use of different packaging films for the final product. Automated techniques for CO_2 measurement (Guerzoni *et al.*, 1990) and indirect conductimetry (Deak and Beuchat, 1994) have been used to produce predictive models that describe the effect of preservatives and other factors on the growth of yeasts. To control the growth of fungi when the intrinsic stability of the product is insufficient the required stability may be obtained by (thermal or non-thermal) pasteurization, chilled storage, addition of preservatives or combination(s) thereof.

A Heat processing

Heat treatments aim to destroy microorganisms capable of growth in a product during normal conditions of storage. In the case of soft drinks, fruit juices and fruit preserves, pasteurization will generally meet this objective. Most fruit-juice manufacturers aim for the inactivation of enzymes and spoilage microorganisms that are more resistant than the relevant pathogens. Pasteurization causes flavor loss and alteration so that tight sanitation, HACCP, or combinations of alternative technologies are preferred to achieve the requisite reduction in counts. Appropriate pasteurization conditions can readily be achieved by either industrial or artisanal fruit juice manufacturers, to meet specified product criteria, such as a 5-$\log_{10}$ reduction of infectious pathogens (FDA, 2001).

Pasteurization can be achieved by hot filling, in-pack tunnel pasteurization, or in-line HTST pasteurization (e.g. 4–10 s at 90–95°C) followed by aseptic packaging. Table 13.7 lists typical pasteurization times and temperatures applied in practice. However, some filamentous fungi produce ascospores that can withstand pasteurization.

Most of the normal flora of fruits, bacteria, and fungi is removed by pasteurization or the use of adequate levels of preservatives. However, heat-resistant fungi, preservative-resistant yeasts and the

Table 13.7 Examples of pasteurization conditions for soft drinks and fruit preserves[a]

	Pasteurization treatment	
Product	Time (min)	Temperature (°C)
Carbonated orange drink (2.5 *v/v* CO_2)	10–20	65–70
Orange juice, non-carbonated	10–15	70–75
Concentrated drink (orange syrup) pH = 3.0, 50°Brix	1–5	85–90
Strawberry jam (pH = 3.3, 58°Brix)	5–10	80
Low-calorie strawberry jam (pH = 3.5, 30°Brix)	5–15	80–85

[a]These times and temperatures, taken from various sources, should be viewed as examples. They are not recommendations.

acid dependent, thermotolerant bacterium *Alicyclobacillus* may survive these preservation techniques to various extents. Also, acid adapted microorganisms that have increased heat resistance may survive pasteurization (Mazzotta, 2001).

Tomato juice, because of its higher pH (average 4.3), must receive a higher treatment heat to achieve microbiological stability, e.g. an HTST treatment of 118°C for 0.5–1.5 min ($F_0 = 0.23$ and 0.7, respectively). For juice of pH 4.1 or below, a lower heat, e.g. 100°C for 1.5 min, may suffice, especially if an edible acid is added to achieve this.

The treatment to be applied depends on a_w, pH, type of product, and presence of anti-microbial substances (Corry, 1976; Juven *et al.*, 1978; Beuchat, 1981a,b; Beuchat, 1983; Jermini and Schmidt-Lorenz, 1987). The effect of a_w on the heat resistance of *Z. bailii* is shown in Table 13.8, which compares the estimated times for 6-fold decimal reduction at 0.86 a_w and 0.96 a_w at a range of temperatures.

Pasteurization values (*P* values) have been traditionally used (Shapton *et al.*, 1971) to compare the equivalency of time–temperature treatments for optimum process setting. The use of *D* and *z* values is based on the assumption of logarithmic death but death curves are usually not logarithmic over the whole range. The use of thermal death times was suggested as an alternative (Put *et al.*, 1976) although these are difficult to apply to the calculation of the equivalence of heat treatments. The development of alternative mathematical models, e.g. Cole *et al.* (1993) to accurately describe the death kinetics of vegetative microorganisms under a wide range of environmental conditions may be of value here.

Table 13.8 Influence of temperature and water activity on thermal death (6*D* values) of 10^6 cells of *Zygosaccharomyces bailii* at pH 4.5[a]

	Time (min)	
Temperature (°C)	a_w 0.963	a_w 0.858
58	56.4	300
59	30.0	165
60	15.6	89.4
61	8.4	41.4
62	4.2	20.4
64	1.2	5.4

[a]From Jermini and Schmidt-Lorenz (1987).

Pasteurization is a quite effective method of preservation as 2 min at 52°C will kill most yeasts, while 10 min at 60°C kills most vegetative mold cells (Stratford *et al.*, 2000). It has already been pointed out that heat-resistant fungal ascospores cannot be destroyed by pasteurization. Heat treatments above 90°C are necessary to inactivate *Bysschloamys* ascospores (Beuchat and Rice, 1979).

Heat-resistance studies have shown that *Alicyclobacillus* spores can survive the usual hot fill processes that are given to commercial juices (Splittstoesser *et al.*, 1998b). While sugar concentrations above 18° Brix are inhibitory (as are ethanol levels above 6%), increasing the Brix makes spores more heat resistant. The spores are less resistant as the pH is decreased and the type of organic acid can be important (Pontius *et al.*, 1998). A successful commercial process for *Alicyclobacillus* control might require a 5-$\log_{10}$ spore kill (Splittstoesser *et al.*, 1998b).

As mycotoxins such as patulin are relatively heat stable, it should be noted that while pasteurization may inactivate most molds effectively, it may not inactivate preformed mycotoxins in the product.

B Chilled storage

Soft drinks and fruit juices may be preserved for a limited length of time by chilled storage. The chilled shelf-life of unpasteurized or raw juice should be limited (below 7 days). Chilled pasteurized juices may have shelf-lives up to 45–60 days. Chilled storage is particularly advised for the period of open shelf-life. A number of spoilage fungi are able to grow at 5°C, although relatively slowly. Bacteria gradually lose viability in chilled juice (Stratford *et al.*, 2000).

C Preservatives

Many fruit juices, preserves, and soft drinks cannot be stabilized by pasteurization, because the heat treatment damages their physical and organoleptic properties, or by pasteurization alone, because of the presence of heat-resistant microorganisms. In those cases where the required intrinsic stability cannot be achieved by reformulation, chemical preservatives are then usually added. These preservatives should be used judiciously and only when there is a clear need to increase shelf-life, prevent spoilage, or minimize the food-poisoning risk. It should also be recognized that the effectiveness of a preservative, added at a certain concentration to a particular soft drink or preserve, not only depends on the pH and water activity but also on the presence of other growth inhibitors. Such mixtures of preservatives may reinforce or reduce their separate growth-inhibitory activities, i.e. may work synergistically, additively, or antagonistically.

Many compounds, including antibiotics and various flavoring agents (Chichester and Tanner, 1972) are described in the literature as antimicrobial in soft drinks and fruit preserves. However, not all of these compounds may be used legally. Because of the great variation in types and levels of preservatives permitted in foods, data, and recommendations of the Food and Agriculture Organisation/World Health Organisation (FAO, 2001b) may be used as a reference if a particular country does not list permitted preservatives. An example of the permitted levels of weak-acid preservatives in different countries is given in Table 13.9. For relevant data on physical, chemical, and toxicological properties of several preservatives, see Table 7.1 in ICMSF (1980). The effectiveness of preservatives against a variety of microorganisms is described in Table 7.2 in ICMSF (1980). In most countries, only benzoic acid, sorbic acid, sulfites, *p*-hydroxybenzoic acid, and its esters (parabens) and carbon dioxide are permitted as preservatives in fruit based products. In general, sulfite and parabens are slightly more effective against molds than yeasts. Differences between benzoic acid and sorbic acid are probably due to the difference in pK_a (Figure 22.1). Sorbic acid, sulfite, and parabens are reasonably effective inhibitors of bacterial growth at the range of pH values of fruit juices and concentrates.

Table 13.9 Permitted concentrations of preservatives in soft drinks in various countries[a] (mg/kg = ppm)

Country	Ba	Sa	PHB	SO_2	Other
Australia	400	400		115	
Austria	30	30			
Benlux	100	100			
Brazil	500	100			
Canada	1000	1000		100	
China	200	200			
Denmark	200	500	200		100 Formic Acid
Eire	160	300	160	70	
Europe	150	300			150Ba + 200Sa
Finland	500	1000	300		
France	160				
Germany	1000	1000			4000 Formic Acid
Greece	1000	1000	500		
Hongkong	160	400	160		
India	120		70		
Indonesia	400	400	100	70	
Italy	160			20	
Japan	600		100		
Malaysia	350	350		140	
Pakistan	600		350		
Philippines	100			100 m-PHB	
Portugal	160	300			300Ba + Sa
Singapore	160	300	160	70	
Spain	600	600			600Ba + Sa
Sweden	1000	1000	1000	50	
Switzerland	100	100			
Taiwan	250				
Thailand	200	200		70	
Turkey	300	700			
UK	160	300	160	70	
USA	1000		1000		

Key Ba = Benzoic acid, Sa = Sorbic acid, PHB = Parahydroxybenzoic acid, SO_2 = Sulfur dioxide.
[a]EC (1995) and other sources.

Strains of the spoilage yeast *Z. bailii* are tolerant to preservative concentrations in excess of those permitted legally (Pitt, 1974; Splittstoesser *et al.*, 1978; Thomas and Davenport, 1985). The increasing awareness of the potential toxicological effects of weak acids means that there is a pressure for industry to deploy preservative systems that do not rely on high concentrations of individual weak acids. Systems of preservation therefore should increasingly exploit the factors that improve preservative efficacy, such as low pH (Cole and Keenan, 1986) and high sugar concentration (Brix) (Tuynenburg Muys, 1971). Also, increased use is made of suitable combinations of preservatives, exploiting synergisms that exist between weak acids (Moon, 1983; Cole and Keenan, 1986, 1987a,b). The combined effects of Brix, pH, and the preservatives sulfur dioxide, sorbic acid, and benzoic acid on the probability of growth of *Z. bailii* is shown in Figure 13.3. The experiment was carried out in a buffered model fruit drink using a miniaturized system for growth and by recording time to visible growth. The results were analyzed using predictive modeling techniques (Cole *et al.*, 1987). For example, Figure 13.3a shows that at pH 4.0, the probability of growth of *Z. bailii* is not decreased by increasing Brix or sorbic acid at the highest concentrations used in this experiment. Even at the lowest pH value, pH 2.5 (Figure 13.3c) neither sorbic acid nor Brix alone will decrease the probability of growth very much. However, in combination they have a dramatic effect, the beginning of which can be seen at pH 2.79 (Figure 13.3b). Therefore, there is a synergy between Brix and sorbic acid and this synergy is pH dependent.

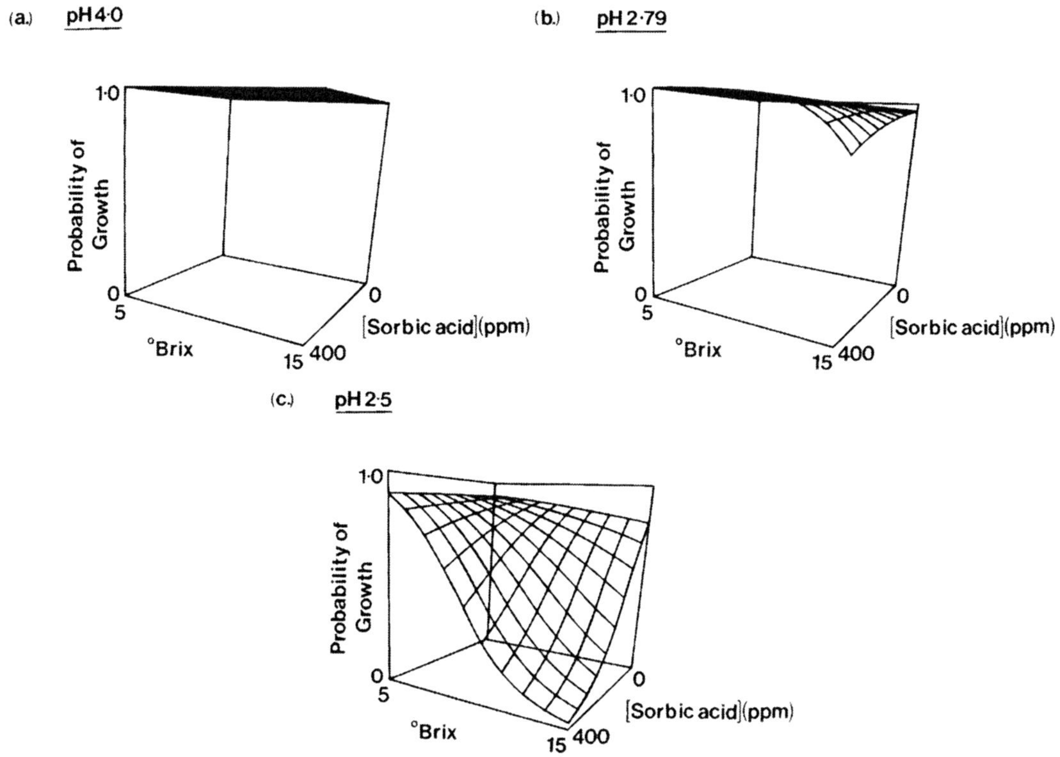

Figure 13.3 The combined effects of sorbic acid and sugar concentration (Brix) on the probability of growth of *Zygosaccharomyces bailii* at (a) pH 4.0, (b) 2.79, and (c) 2.5.

Many microbial species can adapt to unfavorable growth conditions, i.e. to low pH, low a_w, or low preservative concentration. The selection of yeasts resistant to benzoic acid under industrial conditions is well recognized. For example, *Z. bailii*, at pH 3.0 and in the presence of 40% (*w/w*) sucrose, has a benzoic acid tolerance of approximately 700 mg/kg, but the organisms can adapt to benzoic acid concentrations of up to 1200 mg/kg in the same circumstances (Pitt and Hocking, 1997).

The risk of adaption/selection of resistant types should be reduced as far as possible by the following methods:

- Good manufacturing practice, i.e. use of hygienic equipment and strict attention to factory hygiene.
- Avoidance of raw materials that have been treated with preservatives (such as benzoic or sorbic aid) because these may introduce adapted organisms into the factory.
- Checks on end products and stored batches for the occurrence of adapted yeasts (not routine testing)
- The use of mixtures of preservatives and/or preservative treatments with different mechanisms of inhibition.

An additional antimicrobial factor in carbonated beverages is the elevated concentration of dissolved carbon dioxide brought about by using pressures in excess of 100 kPa (1 bar). Yeasts, however, are relatively resistant to inhibition by carbon dioxide (Thom and Marquis, 1984; Ison and Gutteridge, 1987). The CO_2 content is usually expressed as the volume of CO_2 dissolved in a unit volume of water (*v/v*) at a specified temperature.

Table 13.10 Tolerance of yeasts to dissolved carbon dioxide[a]

	Volumes of CO_2				
	Control	1.11	2.23	3.34	4.45
Brettanomyces bruxellensis	+	+	+	+	+
Bret. intermedius	+	+	+	+	+
Bret. naardenensis	+	+	+	+	+
Candida intermedia	+	+	+	+	−
Debaryomyces hansenii	+	+	−	−	−
Dekkera anomala	+	+	+	+	+
Hansenula anomala	+	+	+	+	−
Klyveromyces lactis	+	+	+	−	−
Pichia membranefaciens	+	+	−	−	−
Saccharomyces cerevisiae	+	+	+	−	−
Schizosaccharomyces pombe	+	+	+	−	−
Zygosaccharomyces bailii	+	+	+	+	−
Z. microellipsoides	+	+	+	−	−
Z. rouxii	+	+	−	−	−

+ = visible growth within 40 d.
[a]From Ison and Gutteridge (1987).

The tolerances of a number of yeast strains to dissolved CO_2 are shown in Table 13.10.

The antimicrobial activity of CO_2 is strongly affected by the following:

- The sugar concentration—sugar protects against inactivation by CO_2.
- The pH value and the presence of organic or inorganic acid—microbial inactivation increases in proportion to the decrease in pH and/or the increase of undissociated acid.
- The initial number of contaminants—although CO_2 is an effective agent when yeast numbers are low, it will not always prevent spoilage when the initial numbers are high.

Carbon dioxide inhibits the development of molds in soft drinks mainly because of the resulting anaerobic atmosphere. Carbonation is frequently applied in combination with benzoic acid (to about 75 mg/kg of the soft drink) and/or a heat treatment. For further information on CO_2, see Chapter 10 in ICMSF (1980).

Kniel *et al.* (2003) reported on the ability of organic acids (malic acid, citric acid, and tartaric acid) and hydrogen peroxide (H_2O_2) added to fruit juices (apple, orange, and grape juices) to inhibit survival of *C. parvum* oocysts; for example, they found that addition of 0.025% of H_2O_2 resulted in a $>5\text{-log}_{10}$ reduction of the infectivity of the parasite in a bioassay.

Dimethyl dicarbonate (DMDC) is an alkyl ester of pyrocarbonic acid that hydrolyzes rapidly in water to form methanol and carbon dioxide. DMDC is extremely effective as a cold sterilant, being effective against yeast and bacteria at concentrations between 150 and 250 ppm. The mechanism of inactivation seems strongly related to enzyme inactivation (Ough, 1993). In the US, DMDC was previously only permitted as a yeast inhibitor in wine and wine substitutes, but following an amendment to the Food Additive Order it is now permitted in ready-to-drink teas at a concentration of 250 ppm or less (Anonymous, 1996b). In the EC, it is permitted as a preservative (E242) at the same concentration for non-alcoholic flavored drinks, alcohol-free wine and liquid-tea concentrate (EC, 1995).

D Stabilization of concentrated fruit products

Concentrated fruit juices and soft drinks are non-carbonated. Three basic systems are used to achieve stability as follows:

1. By combination of low a_w (65°–70°Brix) and low pH (2–2.5).
2. By hot filling and/or pasteurization. To compensate for increased suger, the heat treatment should be more severe, e.g. 1–5 min at 85–90°C (see Table 13.7). Preservatives may be added to achieve the shelf-life required after opening.
3. By addition of chemical preservatives. In many cases, this is the only option. The amount of sorbic acid or benzoic acid used depends on the other intrinsic variables and need to be in line with requirements of relevant regulatory agencies, but will usually be between 300 and 1500 mg/kg. Concentrates should be stored at constant temperature to prevent condensation and possible fungal growth on the surface. This can also be prevented by minimizing the available oxygen in the head-space or, where permitted, dusting the surface of the concentrate with sorbate.

E Stabilization of fruit preserves

Nearly all fruit preserves, including jams and marmalades require heating to 80–85°C to denature fruit enzymes and must be hot filled before they gel. Heating after sealing or inverting the container to decontaminate the lid provide added protection, because the low pH prevents germination. Only bacterial spores survive the heat treatment, so the product is stable. However, product may be recontaminated after opening, so sorbic acid or benzoic acid (200–500 mg/kg) may be added to inhibit fungi. The preservation of jams can also be achieved by the use of high-pressure treatment (400–600 MPa) producing products that retain their original flavor and color (Horie *et al.*, 1991).

F Combination of pasteurization and preservatives

Combination of pasteurization with the addition of preservatives allows lower heat treatment and lower preservative concentration. An example of such a combination is a low-calorie fruit spread (W.J. Kooiman, unpublished: see ICMSF, 1980, pp. 660–661). Such a product with a pH of 3.8–4.0 and containing 30% w/w sucrose can be made by one of the following methods:

1. Pasteurization, i.e. heating for 1–5 min at 80°C.
2. Addition of 800–1200 mg/kg sorbic acid if not pasteurized.
3. Application of a combination of a milder heat treatment (i.e. hot filling at 65°C with 600 mg/kg of sorbic acid.

Of six genera of yeasts studied by Beuchat (1981b) without exception the presence of 500 ppm potassium sorbate or sodium benzoate in the heating medium caused more rapid inactivation rates of yeasts compared to heating in the absence of preservatives. Heat-damaged cells also display an increased sensitivity to potassium sorbate in recovery medium (Shibasaki and Tsuchido, 1973). Interestingly, benzoic acid appears to show a greater interaction with heat than sorbic acid (Beuchat, 1983).

Tropical spices may prove useful in preservation of fruits or fruit juices by hurdle technology. For instance, cinnamaldehyde was found to be suitable for surface disinfection of tomatoes (Smid *et al.*, 1996) while the microbial stability of mango juice heated at 55°C for 15 min can be markedly increased by supplementation with extracts of ginger and nutmeg (Ejechi *et al.*, 1998).

The exact conditions for use and the suitability of combinations of natural preserves depend strongly on the product concerned. It should be realized that many natural preservatives may have a marked impact on the organoleptic properties of the fruit products (Smid and Gorris, 1999). With some preservatives, such as cinnamaldehyde, the impact on the organoleptic properties may add a beneficial taste aspect to the original characteristics of products such whole fruits or fruit preserves.

G Alternative non-thermal methods

New technologies not relying on temperature effects for inactivation of microorganisms thermal, such as high hydrostatic pressure (HHP) and pulsed electric fields (PEF) but also a relatively old technology such as irradiation may be used to destroy spoilage microorganisms and pathogens in fruit juices and related products (IFT, 2000).

High hydrostatic pressure (HHP). Inactivation of yeasts and vegetative bacteria in fruit products by HPP is typically effective because of the inherent low pH of these foods. Limiting parameters are usually the presence of enzymes that affect quality, as some enzymes may not be inactivated as quickly or sufficiently by HPP compared with thermal treatment. The application of HPP for pasteurization of orange juice (Ogawa *et al.*, 1990, 1991) and grapefruit juice (Yuge *et al.*, 1993) has been reported. Citrus juices pressure treated at 400 MPa for 10 min at 40°C showed no spoilage after 2–3 months storage at room temperature, while 300 MPa or higher for 10 min was sufficient to cause a 3-$\log_{10}$ reduction in yeast and mold counts (Ogawa *et al.*, 1992). Parish (1998a) reported the following $\log_{10}$ reduction (D) values using HPP of Hamlin orange juice:

- For *Sacc. cerevisiae* ascospores: D-values of 4 and 76 s at 500 and 350 MPa, respectively.
- For *Sacc. cerevisiae* vegetative cells: D-values of 1 and 38 s at 500 and 350 MPa, respectively.
- For native flora: D values of 3 and 74 s at 500 and 350 MPa, respectively.

HPP at 500 MPa of orange and apple juices inoculated with *Sacc. cerevisiae* resulted in D values of 0.18 min and 0.15 min, respectively (Zook *et al.*, 1999). HPP was shown to eliminate *Sacc. cerevisiae, Z. rouxii, Staphylococcus* spp., and *Salmonella* spp., with a starting inoculum of 10^5–10^6cfu/g in jams and preserves (Horie *et al.*, 1991). High hydrostatic pressure (80 000 ψ; >60 s) inactivated *C. parvum* oocysts in apple and orange juice by at least 3 $\log_{10}$ (Slifko *et al.*, 2000b). This technology also inactivated *Byssochlamys nivea* ascospores suspended in apple and cranberry juice concentrates (Palou *et al.*, 1998). HPP has also been reported as being a non-thermal process that can allow the production of juices in compliance with the FDA performance criterion for fruit juices where a 5-$\log_{10}$ inactivation step is required (FDA, 2001). Indeed, inactivation (5-$\log_{10}$ reduction) of *L. monocytogenes* and *E. coli* O157 in apple, tomato and orange juices was possible with pressures of 500 MPa for 5 min (Jordan *et al.*, 2001).

Pulsed electric fields (PEF). PEF treatments have been reported to extend the shelf life of orange juice (Jia *et al.*, 1999; Yeom *et al.*, 2000) and apple juice (Ortegea-Rivas *et al.*, 1998). High voltage bipolar pulsed electric fields (30, 26, 22 and 18 kV/cm and treatment times of 172, 144, 115, and 86 microseconds) achieved a 5-$\log_{10}$ reduction for pathogens, including *E. coli* O157:H7 (Evrendilek *et al.*, 1999).

Ultraviolet light. UV light treatment has been reported to be effective in the reduction of pathogens and spoilage microorganisms in juices. Effectiveness varied with flow rate and exposure. A sound HACCP program in conjunction with UV light treatment is recommended for juice producers (Morris, 2000; Sastry *et al.*, 2000).

Irradiation. Irradiation of apple juice with 1.8 kGy is sufficient to achieve the 5-$\log_{10}$ ($5D$) inactivation of *E. coli* O157:H7 recommended by the FDA (2001) Reported D values for non-acid-adapted *E. coli* O157:H7 cells ranged from 0.12 to 0.21 kGy but were somewhat higher for acid-adapted cells: 0.22–0.31 kGy (Buchanan *et al.*, 1998).

CONTROL (Soft drinks, carbonated and non-carbonated)

Significant hazards[a]	• None.
Control measures	• Good Manufacturing Practices (GMP).
Testing	• No routine testing is recommended.
Spoilage	• Yeasts are controlled by adequate processing and preservatives.

[a]Under particular circumstances, hazards may need to be considered.

Hazards to be considered. No significant microbiological hazards are associated with soft drinks because of the nature of the product and processing methods used for production.

Control measures. Good manufacturing practices control significant health hazards and spoilage concerns. Stabilization strategies are discussed in Section IV.

Water is an important ingredient in soft-drinks, tea, reconstituted juices, sport-drinks, etc. Water of the appropriate quality can be obtained by a number of treatments, including ion exchange or reverse osmosis, filtration, or decontamination (e.g. chlorine or UV treatment).

Packaging can be a source of microbiological contamination. Packaging material thus needs to be properly cleaned and decontaminated, especially when recycled or re-used (e.g. return bottles).

Testing. Testing for pathogens or their indicators is not recommended for soft drinks. Since control of the processing conditions applied is essential for proper stabilization, the following conditions should be monitored as appropriate:

- Temperature of heat treatment/pasteurization (or equivalent alternative non-thermal method).
- Temperature during storage of raw materials and filling.
- Controls on carbonation by routine measurements of CO_2 pressures in packed products.
- Measurements of pH, solids concentration, filling volume, product viscosity, preservative, acid, etc. to ensure that product has been correctly formulated.
- Controls of closure integrity of bottles, cans, glass jars, or other materials.

Spoilage. Carbonated and non-carbonated soft drinks spoil due to growth of acid-tolerant and preservative-resistant yeasts that cause fermentative spoilage (*Z. bailii* and others). Proper cleaning and sanitizing of equipment (GMPs) and good hygiene practices should be followed.

CONTROL (Fruit juice and related products)

Significant hazards[a]	• Salmonellae. • *E. coli* O157:H7. • *Cryptosporidium parvum.* • Patulin (apples and pears). • Ochratoxin A (grapes).

(*continued*)

E CONTROL (Cont.)

Control measures	
Initial level (H_0)	• Good agricultural practices. • Avoid use of dropped ("windfalls") or damaged fruit. • Good hygienic practices.
Reduction (ΣR)	• Clean and wash fruit. • 5-$\log_{10}$ reduction through pasteurization or other treatment (see IV. Processing).
Increase (ΣI)	• Refrigerated storage ($\leq 8^{\circ}$C) to prevent mycotoxin production.
Testing	• Equipment hygiene monitoring for pasteurized and unpasteurized juice. • No product testing is recommended for canned or concentrated fruit juices, purees, and nectars. • Formulation and process control monitoring.
Spoilage	• Yeast controlled by pasteurization, equipment hygiene, temperature, and preservatives. • *Alicyclobacillus* control requires equipment hygiene control for shelf stable products.

[a]Under particular circumstances, other hazards may need to be considered.

Hazards to be considered. In principle, significant hazards in fruit juices and related products are the same as those of concern for the fresh fruits from which they are prepared (See Chapter 6). The hazards of most concern are the acid-tolerant bacterial pathogen *E. coli* O157:H7 and salmonellae; the mycotoxins patulin, produced by the growth of *Pen. expansum* in apples and pears, and ochratoxin A, from growth of *Asp. carbonarius* in grapes; and parasites such as *Cryptosporidium parvum*. Other hazards that sometimes need to be controlled are small round structured viruses and Hepatitis A, particularly if fruit is extensively handled and water is of questionable quality.

Control measures

Initial level of hazard (H_0). Pre-harvest and post-harvest Good Agricultural Practices (GAP) are necessary to keep the contamination level of fruits as low as possible. Some pathogens are present in the soil so they easily contaminate the surface of fruits that are harvested. Use of dropped ("windfalls") and/or damaged fruits must be avoided. Handlers also need to observe good hygiene practices during preparation of these products. Proper transportation, distribution, and packaging of the fruits are important measures to control the hazards. The control measures recommended for raw fruits and vegetables (see Chapters 5 and 6) apply to those used for production of non-pasteurized juices.

As is the case with with soft drinks, water is an important ingredient in reconstituted juices. Water of the appropriate quality can be obtained by a number of treatments, including ion exchange or reverse osmosis, filtration or decontamination (e.g. chlorine or UV treatment).

Packaging can also be a source of microbiological contamination and should be properly cleaned and decontaminated, especially when recycled or re-used (e.g. return bottles).

Reduction of hazard (ΣR). Proper cleaning, washing, and sanitizing of fresh fruits are measures that reduce the level of contamination. Information about sanitizers applicable to fruits can be found in Chapter 6. It must be noted that legal use of disinfectants differs from country to country. Cross-contamination between sanitized and non-sanitized fruits needs to be avoided.

A minimal standard of 5-$\log_{10}$ reduction is advocated (e.g. FDA, 2001; Al-Taher and Knutson, 2004), although the initial contamination level of the fresh fruits should be suitably low. See Section IV for a discussion of various strategies to achieve this reduction.

Increase of hazard (ΣI). Factory hygiene is a major factor in the control of product stability, as most cases of yeast spoilage can be attributed to bad factory hygiene. Inadequate factory hygiene has also been linked to illness outbreaks involving fruit juices (Al-Taher and Knutson, 2004). Therefore, scrupulous attention to cleaning of the lines, fillers, and cooling meter (if used) downstream from the pasteurizer is essential to prevent recontamination of the product. This should include thermal as well as chemical sanitation. Such processes are essential in products without preservatives, as any kind of adventitious fermentative yeast contamination will lead to spoilage. Aerial contamination with yeast and molds is another important factor to control. Major vectors such as dust and insects may import the microbes into the factory environment from nearby sources (orchards, wineries, breweries, fruits/vegetables storage facilities).

Proper storage under refrigeration (≤8°C) is required to prevent fungal growth and subsequent potential mycotoxin formation. Refrigeration is also necessary for non-pasteurized fruit juices with lower acid concentrations such as tomato and melon to prevent growth of bacterial pathogens.

Stabilization though use of preservatives may also prevent an increase of hazards (See Section IV).

Testing. Process control monitoring indicated for carbonated beverages is also appropriate for fruit juice and related products. Because of the extensive heat treatment received, no product testing is recommended for canned or concentrated fruit juices, purees, and nectars.

Only raw materials (and not the final product) should be analyzed microbiologically on a routine basis because the ingredients, after being combined, will generally be subjected to conditions that would tend to destroy the microorganisms. End-product sampling and inspection does not deliver reliable control, but microbiological enumeration can be used for verification purposes for non-pasteurized and pasteurized juice. Several traditional methods, e.g. standard plate counts, counts of yeasts and molds or direct microscopic examination, are available for this (FDA, 1998; Downes and Ito, 2001). *E. coli* may be used as an indicator of enteric pathogens in non-pasteurized juice.

Automated impedance monitoring is a useful tool for rapid detection of yeasts in fruit juices (Zindulis, 1984, Deak and Beuchat, 1994) and heat-resistant microorganisms in a range of processes (Nielsen, 1992).

The simplest and most effective way to screen for preservative resistant yeasts is to spread or streak product onto plates of malt acetic acid agar (MAA) (Pitt and Richardson, 1973). MAA is a suitable medium for monitoring raw materials, process lines and products containing preservatives for resistant yeasts.

The diacetyl test maybe a tool for checking the sanitary conditions during the manufacture of raw materials, especially citrus and apple juices and concentrates (Murdock, 1967, 1968). This test detects diacetyl and acetylcarbinol, which are end products of bacterial growth and is specifically limited to lactic acid bacteria.

ATP photometry may be useful for rapid hygiene testing results.

Spoilage. The shelf-life of non-pasteurized fruit juices is short due to enzymatic activity and presence of high number of microorganisms. Product should be chilled (≤8°C) during storage. See Section III—Spoilage to refer to relevant spoilage agents in these products. Stabilization strategies using preservatives are discussed in Section IV—Processing.

The efficacy of pasteurization is affected by the composition of the product, with high levels of sugars being protective, but low pH and presence of preservatives increasing the lethality of the heat process. It should be considered that pasteurization will not always destroy heat-resistant lactic acid bacteria, spores of thermotolerant *Alicyclobacillus* spp. and heat-resistant fungi (*Byssochlamys fulva*, *Byss. nivea*, *Talaromyces macrosporus*, and *Neosartorya fischeri*). The most common manifestation of spoilage are fermentation, when containers swell due production of CO_2, and formation of cloudiness. *Alicyclobacillus* spp. does not produce gas, but may cause medicinal taints. The growth of *Alicyclobacillus* spp. is minimized when the product is kept chilled; however, psychrotrophic yeasts, and some lactic acid bacteria may spoil chilled products.

Due to low pH and a_W, the growth of spoilage microorganisms in concentrated fruit juice is minimal. *Alicyclobacillus* spp. can grow at low pH, but needs oxygen and warm temperatures. Preservatives (CO_2, sorbic acid, or benzoic acid) are commonly added to concentrated fruit juices to inhibit the growth of spoilage microorganisms and extend the shelf-life of the product. Post-processing contamination should be avoided. Often aseptic/hot filling is used.

Preservative resistant yeasts (*Z. bailii*, *Z. rouxii*, *Sacc. cerevisiae*, and *Sacc. pombe*) cause fermentative spoilage of fruit purees, nectars, cordials, preserves, and jams. They grow slowly but produce large amounts of CO_2, causing distortion or explosion of containers. Because of its ability to adapt, it is important that populations of *Z. bailii* are not allowed to build up in the factory environment, where exposure to preservative-containing residues of product can lead to major problems. Post-processing contamination should be avoided (often aseptic/hot filling is used).

H Tea-based beverages

The diversity of tea-based beverages does not allow for a generic summary of significant hazards and controls that would be appropriate for all products. For simple tea-based beverages, adequate pasteurization and avoidance of post-process recontamination prevent significant safety and spoilage concerns. Addition of fruit juices may require use of controls discussed under fruit juices above. Addition of significant protein sources, such as milk and soy protein, requires adequate validation of the control of potential hazards, including *Cl. botulinum*.

References

Abarca, M.L., Bragulat, M.R., Castella, G. and Cabanes, F.J. (1994) Ochratoxin A production by strains of *Aspergillus niger* var. *niger*. *Appl. Environ. Microbiol.*, **60**, 2650–2.

ABECitrus, (1999) Acidothermophilic sporeformimg bacteria (ATSB) in orange juices: detection methods, ecology and involvement in the deterioration of fruit juices. http://www.abecitrus.com.br/pesq_us.html.

Al-Taher, F. and Knutson, K. (2004) Overview of the FDA juice HACCP rule. *Food Prot. Trends*, **24**, 222–38.

Andres, S.C., Giannuzzi, L. and N.E. Zaritsky (2002) Mathematical modelling of microbial growth in packaged refrigerated orange juice treated with chemical preservatives. *J. Food Sci.*, **55**, 724–8.

Anonymous (1996a) An outbreak of *Escherichia coli* O157:H7 infections associated with drinking unpasteurized commercial apple juice: British Columbia, California, Colorado and Washington. *Morb. Mort. Wkly Rep.*, **45**(44), 975.

Anonymous (1996b) Additive cleared as yeast inhibitor for use in sport and fruit drinks. *Food Chem. News*, **38**(15), 19.

Anonymous (2000) *Juice Recall*, Associated Press, April 21, 2000.

Anonymous (2002) Draft standard for Aqueous Coconut products (Codex CL 2002/19-PFV).

Bakka, R. (1995) Sanitation challenges of hot-fill. *Beverage World*, **114**(1603), 98, 100–2.

Banwart, G.J. (1979) *Basic Food Microbiology*, AVI Publishing Company, Westport.

Barnett, J.A., Payne, R.W. and Yarrow, D. (2000) *The Yeasts: Characteristics and Identification*, 3rd edn, Cambridge University Press, Cambridge, UK.

Baumgart, J., Husemann, M. and Schmidt, C. (1997) *Alicyclobacillus acidoterrestris*: occurrence, importance, and detection in beverages and raw materials for beverages. *Flussiges Obst.*, **64**(4), 178–80.

Berry, J.M. (1979) Yeast problems in the food and beverage industry, in *Food Mycology* (ed M.E. Rhodes), G.K. Hall and Co., Boston, M.A., pp. 82–90.

Besser, R.E., Lett, S.M., Weber, J.T., Doyle, M.P., Barrett, T.J., Wells, J.G. and Griffin, P.M. (1993) An outbreak of diarrhoea and hemolytic uremic syndrome from *Escherichia coli* O157:H7 in fresh-pressed apple cider. *J. Am. Med. Assoc.*, **269**(17), 2217–20.

Beuchat, L.R. (1981a) Effects of potassium sorbate and sodium benzoate on inactivating yeasts heated in broths containing sodium chloride and sucrose. *J. Food Prot.*, **44**, 10, 765–9.

Beuchat, L.R. (1981b) Synergistic effects of potassium sorbate and sodium benzoate on the thermal inactivation of yeasts. *J. Food Sci.*, **46**, 771–7.

Beuchat, L.R. (1983) influence of water activity on growth, metabolic activities and survival of yeasts and molds. *J. Food Prot.*, **46**, 135–41.

Beuchat, L.R. and Pitt, J.I. (2001) Detection and enumeration of heat resistant molds, in *Compendium of Methods for the Microbiological Examination of Foods* (eds F.P. Downes and K. Ito), 4th edn, American Public Health Assn., Washington, DC, pp. 217–22.

Beuchat, L.R. and Rice, S.L. (1979) *Byssochlamys* spp. and their importance in processed fruits. *Adv. Food Res.*, **25**, 237–88.

Birkhead, G.S., Morse, D.L., Levine, W.C., Fudual, J.K., Kondraki, S.F., Chang, H.G., Shayegani, M., Novick, L. and Blake, P.A. (1993) Typhoid fever at a resort hotel in New York: a large outbreak with an unusual vehicle. *J. Infect. Dis.*, **167**, 1228–32.

Borlinghaus, A. and Engel, R. (1997) Incidence of *Alicyclobacillus* sp. in commercial apple juice concentrates—development and validation of test method. *Flussiges Obst.*, **64**(6), 306–9.

Buchanan, R.L., Edelson, S.G., Snipes, K. and Boyd, G. (1998) Inactivation of *Escherichia coli* O157:H7 in apple juice by irradiation. *Appl. Environ. Microbiol.*, **64**, 4533–5.

Burnett, S.L. and Beuchat, L.R. (2000) Human pathogens associated with raw produce and unpasteurized juices, and difficulties in decontamination. *J. Ind. Microbiol. Biotechnol.*, **25**, 281–7.

Caul, E.O. (1993) Outbreaks of gastroenteritis associated with SRSV's. *Public Health Lab. Ser. Microbiol. Digest*, **10**(1), 2–8.

CAC (Codex Alimentarius Commission). (1991a) Guidelines for mixed fruit juices CAC/GL 11. ftp://ftp.fao.org/codex/standard/en/CXG_011e.pdf).

CAC (Codex Alimentarius Commission). (1991b) Guidelines for mixed fruit nectars CAC/GL 12. (ftp://ftp.fao.org/codex/standard/en/CXG_012e.pdf).

CDC (Centers for Disease Control and Prevention). (1975) *Salmonella* Typhimurium outbreak traced to a commercial apple cider—New Jersey. *Morb. Mort. Wkly rep.*, **28**(44), 522–3.

CDC (Centers for Disease Control and Prevention). (1995) Outbreak of *Salmonella* Hartford infections among travellers to Orlando, Florida. EPI-AID Trip Rpt.95–62.

CDC (Centers for Disease Control and Prevention). (1997) Outbreaks of *Escherichia* coli O157:H7 infection and cryptosporidiosis associated with drinking unpasteurised apple cider—Connecticut and New York, October 1996. *Mort. Morb. Wkly Rep. 1997*, **46**, 4–8.

CDC (Centers for Disease Control and Prevention). (1999) Outbreak of *Salmonella* serotype Muenchen infections associated with unpasteurized orange juice. United States and Canada. June 1999. *Morb. Mort. Wkly Rep.*, **48**, 582–5.

Cerny, G., Hennlich, W. and Poralla, K. (1984) Fruchtsaftverderb durch Bacillen: isolierung und charakterisierung des verderserrgers. *Z. Lebens. Unters. Forsh.*, **179**, 224–7.

Chichester, D.V. and Tanner, F.W. (1972) Antimicrobial food additives, in *Handbook of Food Additives* (ed T.E. Furia), 2nd edn, CRC Press, Cleveland, Ohio, pp. 115–84.

Chung, K.C. and Goepfert, J.M. (1970) Growth of *Salmonella* at low pH. *J. Food Sci.*, **35**, 326–8.

Cliver, D.O. (1983) *Manual of Food Virology.*,World Health Organization, Geneva.

Cody, S.H., Glynn, M.K., Farrar, J.A., Cairns, K.L., Griffin, P.M., Kobayashi, J., Fyfe, M., Hoffman, R., King, A.S., Lewis, J.H., Swaminathan B., Bryant, R.G. and Vugia, D.J. (1999) An outbreak of *Escherichia coli* O157:H7 infection from unpasteurised commercial apple juice. *Ann. Int. Med.*, **130**(3), 202–9.

Cole, M.B. and Keenan, M.H.J. (1986) Synergistic effects of weak-acid preservatives and pH on the growth of *Zyosaccharomyces bailii. Yeast*, **2**, 93–100.

Cole, M.B., Davies, K.W., Munro, G., Holyoak, C.D. and Kilsby, D.C. (1993) A vitalistic model to describe the thermal inactivation of *Listeria monocytogenes. J. Ind. Microbiol.*, **12**, 232–9.

Cole, M.B., Franklin, J.G. and Keenan, M.H.J. (1987) Probability of growth of the spoilage yeast *Zygosaccharomyces bailii* in a model fruit drink system. *Food Microbiol.*, **4**, 115–9.

Cole, M.B. and Keenan, M.H.J. (1987a) Effects of weak acids and external pH on the intracellular pH of *Zygosaccharomyces bailii*, and it implications in weak-acid resistance. *Yeast*, **3**, 23–32.

Cole, M.B. and Keenan, M.H.J. (1987b) A quantitative method for predicting shelf life of soft drinks using a model system. *J. Ind. Microbiol.*, **2**, 59–62.

Cook, K.A., Dobbs, T.E., Hlady, W.G., Wells, J.G., Barrett, T.J., Puhr, N.D., Lancette, G.A., Bodager, D.W., Toth, B.L., Genese, C.A., Highsmith, A.K., Pilot, K.E., Finelli, L. and Swerdlow, D.L. (1998) Outbreak of *Salmonella* serotype Hartford infections associated with unpasteurized orange juice. *J. Am. Med. Assoc.*, **280**(17), 1504–9.

Corry, J.E.L. (1976) The effect of sugars and polyols on the heat resistance and morphology of osmophilic yeasts. *J. App. Bacteriol.*, **40**, 269–76.

Cramer, J.A. and Presteguard, J.H. (1977) NMR studies of pH induced transport of carboxylic acids across phospholipid vesicle membranes. *Biochem. Biophys. Res. Commun.*, **75**, 295–301.

Dawson, D. (2003) Food-borne Protozoan Parasites. ILSI Europe Report Series. ILSI Europe, Brussels. Belgium. ISBN 1-57881-159-7. Available at: http://europe.ilsi.org.

D'Aoust, J.Y. (1994) *Salmonella* and the international food trade. *Int. J. Food Microbiol.*, **24**, 11–31.
Deak, T. and Beuchat, L.R. (1994) Use of indirect conductimetry to predict the growth of spoilage yeasts with special consideration of *Zygosaccharomyces bailii*. *Int. J. Food Microbiol.*, **23**, 405–17.
Deak, T. and Beuchat, L.R. (1996) *Handbook of food spoilage yeasts*, CRC Press, Boca Raton, FL.
Deinhard, G., Blanz, P., Poralla, K. and Altan, E. (1987) *Bacillus acidoterrestris* sp. nov., a new thermotolerant acidophile isolated from different soils. *Syst. Appl. Microbiol.*, **10**, 47–53.
Downes, F.P. and Ito, K. (2001) *Compendium of Methods for the Microbiological Examination of Foods*, 4th edn, American Public Health Association, Washington, DC.
EC (European Commission). (1995) Food additives other than colours and sweetners. European Parliament and Council Directive No. 95/2/EC. *Off. J. Eur. Commun.*, **L61**, 1–40.
Eiroa, M.N.U., Junqueira, V.C.A. and Schmidt, F.L. (1999) *Alicyclobacillus* in orange juice: occurrence and heat resistance of spores. *J. Food Prot.*, **62**, 883–6.
Ejechi, B.O., Souzey, J.A. and Akpomedaye, D.E. (1998) Microbial stability of mango (*Mangifera indica* L.) juice preserved by combined application of mild heat and extracts of two tropical spices. *J. Food Prot.*, **61**, 725–7.
Eklund, T. (1983) The antimicrobial effect of dissociated and undissociated sorbic at different pH levels. *J. Appl. Bacteriol.*, **54**, 383–9.
Emmerich, W. and Radler, F. (1983) The anaerobic metabolism of glucose and fructose by *Saccharomyces bailii*. *J. Gen. Microbiol.*, **129**, 3311–8.
Evrendilek, G.A., Zhang, Q.H. and Richter, E.R. (1999) Inactivation of *Escherichia coli* O157:H7 and *Escherichia coli* 8739 in apple juice by pulsed electric fields. *J. Food Prot.*, **62**, 793–6.
FAO/WHO (Food and Agriculture Organization/World Health Organization). (2001b). Evaluation of certain food additives and contaminants (Fifty-seventh report of the Joint FAO/WHO Expert Committee on Food Additives). WHO Technical Report Series, No. 909, 2002. Available at: http://www.who.int/pcs/jecfa/trs909.pdf.
FAO/WHO (Food and Agriculture Organization/World Health Organization). (2003) Proposed Draft Codex General Standard for Fruit Juices and Nectars (At Step 5/8). FAO/WHO joint FAO/WHO Food Standards Programme ad hoc Codex Intergovernmental Task Force on Fruit and Vegetable Juices. Appendix II. Alinorm 03/39A. Available: ftp://ftp.fao.org/codex/alinorm03/al0339Ae.pdf.
FDA (US Food and Drug Administration). (1998) *Bacteriological Analytical Manual*, 8th edn, Revision A, AOAC International, Gaithersburg, MD, USA.
FDA (US Food and Drug Administration). (2000) Department of Health and Human Services. Available: http://vm.cfsan.fda.gov/~acrobat/fr98424b.pdf.
FDA (US Food and Drug Administration). (2001) Hazard Analysis and Critical Control Point (HAACP); Procedures for the Safe and Sanitary Processing and Importing of Juice; Final Rule. *Fed. Reg.*, **66**, FR 6137–202.
Ferreira, M.A.S.S. and Lund, B.M. (1987) The influence of pH and temperature on initiation of growth of *Salmonella* spp. *Lett. App. Microbiol.*, **5**, 67–70.
Fisher, T.L. and Golden, D.A. (1998) Fate of *Escherichia coli* O157:H7 in ground apples used for cider production. *J. Food Prot.*, **61**, 1372–4.
Fleet, G.H. (1992) Spoilage yeasts. *Crit. Rev. Biotechnol.*, **12**, 1–44.
Fleet, G.H., Heiskanen, P., Reid, I. and Buckle, K.A. (2000) Food-borne viral illness—status in Australia. *Int. J. Food. Microbiol.*, **59**, 127–36.
Freese, E., Sheu, C.W., Galliers, E. (1973) Function of lipophilic acids as antimicrobial food additives. *Nature*, **241**, 321–5.
Guerzoni, M.E., Gardini, F. and Duan, J. (1990) Interactions between inhibitory factors on microbial stability of fruit based systems. *Int. J. Food Microbiol.*, **10**, 1–18.
Gurrieri, S., Miceli, L., Lanza, C.M., Tomaselli, F., Bonomo, R.P. and Rizzarelli, E. (2000) Chemical characterization of sicilian prickly pear (*Opuntia ficus indica*) and perspectives for the storage of its juice. *J. Agric. Food Chem.*, **48**, 5424–31.
Hara, Y., Watanabe, M. and Sakaguchi, G. (1989) Studies of antibacterial effects of tea polyphenols. I. The fate of *Clostridium botulinum* spores inoculated into tea drinks. *J. Jpn. Soc. Food Sci. Technol. [Nippon Shokuhin Kogyo Gakkaishi]*, **36**(5), 375–379. [Food Science & Technology Abstract 90_03_H0047]
Heenan C.N., Shaw K.J. and Pitt J.I. (1998) Ochratoxin A production by *Aspergillus carbonarius* and *A. niger* isolates and detection using coconut cream agar. *J. Food Mycol.*, **1**, 67–72.
Hocking, A.D. and Pitt, J.I. (1984) Food spoilage fungi. II. Heat resistant fungi. *CSIRO Food Res.Q.* **44**, 73–82.
Holt, J.G., Krieg, N.R., Sneath, P.H.A., Staley, J.T. and Williams, S.T. (1994) *Bergey's Manual of Determinative Bacteriology*, 9th edn, Lippincott Williams & Wilkins, Baltimore, Maryland, pp. 787.
Horiba, N. Maekawa, Y., Ito, M., Matsumoto, T. and Nakamura, H. (1991) A pilot study of Japanese green tea as a medicament: Antibacterial and bactericidal effects. *J. Endodont.*, **17**(3), 122–4.
Horie, Y., Kimura, K., Ida, M., Yosida, Y. and Ohki, K. (1991) Jam preparation by pressurization. *Nippon Nogeikagaku Kaishi*, **65**, 975–80.
ICMSF (International Commission on Microbiological Specifications for Foods) (1980) *Microbial Ecology of Foods, Volume 1, Factors Affecting Growth and Death of Microorganisms*, Academic Press, New York.
IFT, 2000, Kinetics of Microbial Inactivation for Alternative Food Processing Technologies. IFT/FDA Contract No. 223-98-2333. Available through: http://vm.cfsan.fda.gov/~comm/ift-hpp.html
Ison, R.W. and Gutteridge, C.S. (1987) Determination of the carbonation tolerance of yeasts. *Lett. Appl. Microbiol.*, **5**, 11–3.

Jermini, M.F.G. and Schmidt-Lorenz, W. (1987) Growth of osmotolerant yeasts at different water activity values. *J. Food Prot.*, **50**, 404–10.

Jia, M., Zhang, Q.H. and Min, D.B. (1999) Pulsed electric field processing effects on flavor compounds and microorganisms of orange juice. *Food Chem.*, **65**, 445–51.

Jordan, S.L., Pascual, C., Bracey, E. and Mackey, B.E. (2001) Inactivation and injury of pressure-resistant strains of *Escherichia coli* O157 and *Listeria monocytogenes* in fruit juices. *J. Appl. Microbiol.*, **91**, 463–9.

Juven, B.J., Kanner, J. and Weisslowicz, H. (1978) Influence of orange juice composition on the thermal resistance of spoilage yeasts. *J. Food Sci.*, **43**, 1074–76.

Kniel, K.E., Sumner, S.S., Lindsay, D.S., Hackney, C.R., Pierson, M.D., Zajac, A.M., Golden, D.A. and Fayer, R. (2003) Effect of organic acids and hydrogen peroxide on *Cryptosporidium parvum* viability in fruit juices. *J. Food Prot.*, **66**, 1650–7.

Konowalchuk, J. and Speirs, J.I. (1978a) Antiviral effects of apple beverages. *Appl. Environ. Microbiol.*, **36**, 798–801.

Konowalchuk, J. and Speirs, J.I (1978b) Antiviral effect of commercial juices and beverages. *Appl. Environ. Microbiol.*, **35**, 1219–20.

Koopmans, M. and Duizer, E. (2002) *Food-borne Viruses: an Emerging Problem, ILSI Europe Report Series*, ILSI Europe, Brussels, Belgium, ISBN 1-57881-130-9. Website: http://europe.ilsi.org.

Krebs, H.A., Wiggins, D., Stubbs, M., Sols, A. and Bedoya, F. (1983) Studies on the mechanism of the antifungal action of benzoate. *Biochem.J.*, **214**, 657–63.

Kusano, H., Hideko, Y., Niwa, M. and Yamasato, K. (1997) *Propionobacterium cyclohexanicum* sp. nov., a new acid-tolerant ω-cyclohexyl fatty acid-containing propionibacterium isolated from spoiled orange juice. *Int. J. Syst. Bacteriol.*, **47**(3), 825–31.

Macris, B.J. (1975) Mechanism of benzoic acid uptake by *Saccharomyces cerevisiae*. Appl. Microbiol., **30**, 503–6.

Mazzotta, A. (2001) Thermal inactivation of stationary-phase and acid-adapted *Escherichia coli* O157:H7, *Salmonella* and *Listeria monocytogenes* in fruit juices. *J. Food Prot.*, **64**, 315–20.

McLellan, M.R. and Splittstoesser, D.F. (1996) Reducing the risk of *E. coli* in apple cider. *Food Technol.*, **50**, 174.

Millard, P.S., Gensheimer, K.F., Addiss, D.G., Sosin, D.M., Beckett, G.A., Houck-Jankoski, A. and Hudson, A. (1994) An outbreak of cryptosporidiosis from fresh-pressed apple cider. *J. Am. Med. Assoc.*, **272**, 1592–6.

Moon, N.J. (1983) Inhibition of the growth of acid tolerant yeasts by acetate, lactate and propionate and their synergistic mixtures. *J. Appl. Bacteriol.*, **55**, 453–60.

Morris, C.E. (2000) US developments in non-thermal juice processing. *Food Eng. Ingredients*, **2000**, 26–7.

Mossel, D.A.A. (1963) La survie des salmonellae dans les differents produits alimentaires. *Ann. Inst. Pasteur, Paris*, **104**, 551–69.

Mshar, P.A., Dembek, Z.F., Cartter, M.L., Hadler, J.L., Fiorentino, T.R., Marcus, R.A., McGuire, J., Shiffrin, M.A., Lewis, A., Feuss, J. Kyke, J. van, Toly, M., Cambridge, M., Gruzewich, J., Keithly, J., Dziewulsky, D., Braun-Howland, E., Ackman, D., Smith, P., Coates, J. and Ferrara, J. (1997) Outbreaks of *Escherichia coli* O157:H7 infection and cryptosporidiosis associated with drinking unpasteurized apple cider—Connecticut and New York, October 1996. *Morb. Mort. Wkly Rep.*, **46**, 4–8.

Murakami, M., Tedzuka, H. and Yamazaki, K. (1998) Thermal resistance of *Alicyclobacillus acidoterrestris* spores in different buffers and pH. *Food Microbiol.*, **15**(6), 577–82.

Murdock, D.I. (1967) Methods employed by the citrus concentrate industry for detecting diacetyl and acetylmethylcarbinol. *Food Technol.*, **21**, 643–7.

Murdock, D.I. (1968) Diacetyl test as a quality control tool in processing frozen concentrated orange juice. *Food Technol.*, **22**, 90–4.

Murdock, D.I. and Hatcher, W.S. (1978) A simple method to screen fruit juices and concentrates for heat-resistant mold. *J. Food Prot.*, **41**, 254–6.

Murdock, D.I., Troy, V.S. and Folinazzo, J.F. (1953) Thermal resistance of lactic acid bacteria and yeast in orange juice and concentrate. *Food Res.*, **18**, 85–9.

NACMCF (National Advisory Committee on Microbiological Criteria for Foods) (1997) Recommendations on fresh juice. Available at http://vm.cfsan.fda.gov/~mow/nacmcf.html.

Nielsen, P.V. (1992) Rapid detection of heat resistant fungi in fruit juices by an impedimetric method, in *Modern methods in Food Mycology* (eds R.A. Samson, A.D. Hocking, J.I. Pitt and A.D. King), Elsevier, Amersterdam, pp. 311–9.

Ogawa, H., Fukuhisa, K., Sugawara, K., Kubo, Y. and Fukumoto, H. (1990) Effect of hydrostatic pressure on sterilization of citrus juice, in *Pressure Processed Food* (ed R. Hayashi), Sanei Press, Kyoto, pp. 179–91.

Ogawa, H., Fukuhisa, K. Sugawara, K., Kubo, Y. and Fukumoto, H. (1991) Effect of hydrostatic pressure on sterilization of citrus juice, in *High Pressure Science for Food* (ed R. Hayashi), Sanei Press, Kyoto, pp. 353–60.

Ogawa, H., Fukuhisa, K. and Fukumoto, H. (1992) Effect of hydrostatic pressure on sterilization and preservation of citrus juice, in *High Pressure and Biotechnology* (eds C. Balny, R. Hayashi, K. Heremans and P. Masson), *Volume 224*, Colloque INSERM/John Libbey Eurotext Ltd., London, pp. 269–78.

Orlandi, P.A., Chu, D.-M.T., Bier, J.W. and Jackson, G.J. (2002) Parasites and the food supply. Food Technology, **56**(4), 72–81.

Orr, R.V., Shewfelt, R.L., Huang, C.J., Tefera, S. and Beuchat, L. (2000) Detection of guaiacol produced by *Alicyclobacillus acidoterrestris* in apple juice by sensory and chromatographic analyses and comparison with spore and vegetative cell populations. *J. Food Prot.*, **63**, 1517–22.

Ortega-Rivas, E., Zarate-Rodriguez, E. and Barbosa-Canovas, G.V. (1998) Apple juice pasteurisation using ultrafiltration and pulsed electric fields. *Trans. Inst. Chem. Eng.*, **76**, C, 193–7.

Ough, C.S. (1993) Dimethyl dicarbonate and diethyl dicarbonate, in *Antimicrobials in Foods* (eds P.M. Davidson and A.L. Branen), Marcel Dekker Inc., New York, pp. 343–68.

Palou, E., Lopez-Malo, A., Barbosa-Canovas, G.V., Welti-Chanes, J., Davidson, P.M. and Swanson, B.G. (1998) Effect of oscillatory high hydrostatic pressure treatments on *Byssochlamys nivea* ascospores suspended in fruit juice concentrates. *Lett. Appl. Microbiol.*, **27**, 375–8.

Pao, S. and Davis, C.L. (2001) Maximizing microbiological quality of fresh orange juice by processing sanitation and fruit surface treatments. *Dairy, Food Environ. Sanit.*, **21**, 287–91.

Pao, S., C.L. Davis and M.E. Parish (2001) Microscopic observation and processing validation of fruit sanitizing treatments for the enhanced microbiological safety of fresh orange juice. *J. Food Prot.*, **64**, 310–4.

Parish, M.E. (1997) Public health and non-pasteurized fruit juices. *Crit. Rev. Microbiol.*, **23**, 109–19.

Parish, M.E. (1998a) High pressure inactivation of *Saccharomyces cerevisiae*, the endogenous microflora and pectinmethylesterase in orange juice. *J. Food Saf.*, **18**, 57–65.

Parish, M.E. (1998b). Coliforms, *Escherichia coli* and *Salmonella* serovars associated with a citrus-processing facility implicated in a Salmonellosis outbreak. *J. Food Prot.*, **61**, 280–4.

Parish, M.E. (2000) Relevancy of *Salmonella* and pathogenic *E. coli* to fruit juices. *Fruit Processing*, **7**, 246–50.

Parish, M.E., Narciso, J.A. and Friedrich, L.M. (1997) Survival of *Salmonella* in orange juice. *J. Food Saf.*, **17**, 273–81.

Pettipher, G.L., Osmundson, M.E. and Murphy, J.M. (1997) Methods for the detection and enumeration of *Alicyclobacillus acidoterrestris* and investigation of growth and production of taint in fruit juice and fruit juice containing drinks. *Lett. Appl. Microbiol.*, **24**(3), 185–9.

Pinhatti, M.E.M.C., Variane, S., Eguchi, S.Y. and Manfio, G.P. (1997) Detection of acidothermophilic bacilli in industrialized fruit juices. *Fruit Processing*, **7**(9), 350–3.

Pitt, J.I. (1974) Resistance of some food spoilage yeasts to preservatives. *Food Technol. Aust.*, **26**, 238–41.

Pitt, J.I. and Hocking, A.D. (1997) *Fungi and Food Spoilage*, 2nd edn, Aspen Publishers, Gaithersburg, MD.

Pitt, J.I. and Richardson, K.C. (1973) Spoilage by preservative-resistant yeasts. *CSIRO Food Res. Q.*, **33**, 80–5.

Pontius, A.J., Rushing, J.E. and Foegeding, P.M. (1998) Heat resistance of *Alicyclobacillus acidoterrestris* spores as affected by various pH values and organic acids. *J. Food Prot.*, **61**(1), 41–6.

Powers, J.J. (1976) Effect of acidification of canned tomatoes on quality and shelf life. *CRC Crit. Rev. Food Sci. Nutr.*, **7**, 371–95.

Prevedi, P., Colla, F. and Vicini, E. (1995) Characterization of *Alicyclobacillus*, a sporeforming thermophilic acidophilic bacterium. *Industria Conserve*, **70**, 128–32.

Prevedi, P., Quintavalla, S., Lusardi, C. and Vicini, E. (1997) Heat resistance of *Alicyclobacillus* spores in fruit juices. *Industria Conserve*, **72**(4), 353–8.

Put, H.M., De Jong, J. and Sand, F.E.M.J. and van Grinsven, A.M. (1976) Heat resistance studies on yeast spp. causing spoilage in soft drinks. *J. Appl. Bacteriol.*, **50**, 135–52.

Roering, A.M., Luchansky, J.B., Ihnot, A.M., Ansay, S.E., Kaspar, C.W. and Ingham, S.C. (1999) Comparative survival of *Salmonella typhimurium* DT104, *Listeria monocytogenes* and *Escherichia coli* O157:H7 in preservative-free apple cider and simulated gastric juice. *Int. J. Food Microbiol.*, **46**, 262–9.

Ross, K.D. (1975) Estimation of water activity in intermediate moisture foods. *Food Technol.*, **29**(3), 26–34.

Sastry, S.K., Datta, A.K. and Worobo, R.W. (2000) Ultraviolet light. *J. Food Sci.*, (**Supplement**), 90–2.

Shapton, D.A., Lovelock, D.W. and Laurita-Longo, R. (1971) The evaluation of sterilisation and pasteurisation processes from temperature measurements in degrees Celsius (°C). *J. Appl. Bacteriol.*, **34**, 491–500.

Shibasaki, I. and Tsuchido, T. (1973) Enhancing effect of chemicals on the thermal injury of microorganisms. *Acta. Aliment. Acad. Sci. Hung.*, **2**(3), 327–49.

Slifko, T.R., Smith, H.V. and Rose, J.B. (2000a) Emerging parasite zoonoses associated with water and food. *Intl. J. Parasitol*, 12–3, 1379–93.

Slifko, T.R., Raghubeer, E. and Rose, J.B. (2000b) Effect of high hydrostatic pressure on *Cryptosporidium parvum* infectivity. *J. Food. Prot.*, **63**, 1261–7.

Smid, E.J. and Gorris, L.G.M. (1999) *Natural Antimicrobials for Food Preservation. Handbook of Food Preservation* (ed By M.S. Rahman. Marcel Dekker), New York, 1999, pp. 285–308.

Smid, E.J., Hendriks, L., Boerrigter, H.A.M. and Gorris, L.G.M. (1996) Surface disinfection of tomatoes using the natural plant compound *trans*-cinnamaldehyde. *Postharvest Biol. Technol.*, **9**, 343–50.

Splittstoesser, D.E., Queale, D.T. and Mattich, L.P. (1978) Growth of *Saccharomyces bisporus* var *bisporus*, a yeast resistant to sorbic acid. *Am. J. Enol. Viticult.*, **29**, 272–6.

Splittstoesser, D.F., Churey, J.J. and Lee, C.Y. (1994) Growth characteristics of aciduric spore-forming bacilli isolated from fruit juices. *J. Food Prot.*, **57**, 1080–3.

Splittstoesser, D.F., Lee, C.Y. and Churey, JJ. (1998a) Control of *Alicyclobacillus* in the juice industry. *Dairy Food Environ. Sanit.*, **18**, 585–7.

Splittstoesser, D.F., Worobo, R.W. and J.J. Churey (1998b). Food Safety and You: Alicyclobacillus: An Emerging Problem for New York's Processors of Fruit Juices. Venture (The Newsletter of the New York State Venture Center) Summer 1998 Vol. 1 No.3. Available at http://www.nysaes.cornell.edu/fst/fvc/Venture/venture3_safety.html

Stratford M., Hofman, P.D. and Cole, M.B. (2000) Fruit juices, fruit drinks and soft drinks, in *The Microbiological Safety and Quality of Food* (eds B.M. Lund, T.C. Baird-Parker and G.W. Gould), Aspen Publishers, Gaithersburg, MD, USA.

Taylor, J.L., Tutle J., Praukul, T., O'Brien, K., Barrett, T.J., Jolbarito, B., Lim, Y.L., Vugia, D.J., Morris, J.G. Jr., Tauxe, R.V. and

Dwyer, M. (1993) An outbreak of cholera in Maryland associated with imported commercial frozen coconut milk. *J. Infect. Dis.*, **167**, 1330–5.

Téren, J., Varga, J., Hamari, Z., Rinyu, E. and Kevei, F. (1996) Immunochemical detection of ochratoxin A in black *Aspergillus* strains. *Mycopathologia*, **134**, 171–6.

Thom, S.R. and Marquis, R.E. (1984) Microbial growth modification by compressed gases and hydrostatic pressures. *Appl. Environ. Microbiol.*, **74**, 780–7.

Thomas, D.S. and Davenport, R. (1985) *Zygosaccharomyces bailii*—a profile of characteristics and spoilage activities. *Food Microbiol.*, **2**, 157–69.

Tuynenburg Muys, G. (1971) Microbial safety of emulsions. *Process Biochem.*, June, 25–8.

U.S. Department of Health, Education and Welfare (1975) *Salmonella typhimurium* outbreak traced to a commercial apple cider—New Jersey. *Morb. Mort. Wkly Rep.*, 24, 87–8.

Uljas, H.H. and Ingham, S.C. (1999) Combinations of intervention treatments resulting in 5-$\log_{10}$-unit reductions in numbers of *Escherichia coli* O157:H7 and *Salmonella typhimurium* DT104 organisms in apple cider. *Appl. Environm. Microbiol.*, **65** (5), 1924–9.

Vanos, V., Hofstaetter, S. and Cox, L. (1987) The microbiology of instant tea. *Food Microbiol.*, **4**, 19–33.

Warth, A.D. (1977) Mechanism of resistance of *Saccharomyces bailii* to benzoic, sorbic and other weak acids used as food preservatives. *J. Appl. Bacteriol.*, **43**, 215–30.

Wisse, C.A. and Parish, M.E. (1998) Isolation and enumeration of sporeforming, thermoacidophilic, rod-shaped bacteria from citrus processing environments. *Dairy, Food Environ. Sanit.*, **18**(8), 504–9.

Wisotzkey, J.D., Jurtshuk, P., Fox, G., Deinard, G. and Poralla, K. (1992) Comparative sequence analyses on the 16S rRNA (rDNA) of *Bacillus acidocaldarius*, *Bacillus acidoterrestris* and *Bacillus cycloheptanicus* and proposal of creation of a new genus *Alicyclobacillus* gen. Nov. *Int. J. Syst. Bacteriol.*, **42**, 263–9.

Yamazaki, K., Teduka, H. and Shinano, H. (1996) Isolation and identification of *Alicyclobacillus acidoterrestris* from acidic beverages. *Biosci., Biotechnol. Biochem.*, **60**, 543–5.

Yamazaki, K., Isoda, C., Tedzuka, H., Kawai, Y. and Shinano, H. (1997) Thermal resistance and prevention of spoilage bacterium, *Alicyclobacillus acidoterrestris*, in acidic beverages. *Nippon Shokuhin Kagaku Kaishi*, **44**(12), 905–11.

Yeom, H.W., Streaker, C.B., Zhang, Q.H. and Min, D.B. (2000) Effects of pulsed electric fields on the activities of microorganisms and pectin methyl esterase in orange juice. *J. Food Sci.*, **65**, 1359–63.

Yuge, N., Mieda, H., Mutsushika, O. and Tamaki, T. (1993) Bitterness inhibition in grapefruit juice by high pressure treatment, in *High Pressure Bioscience and Food Science* (ed R. Hayashi), Sanei Press, Kyoto, pp. 350–4.

Zhao, T., Doyle, M.P. and Besser, R.E. (1993) Fate of enterohemorrhagic *Escherichia coli* O157:H7 in apple cider with and without preservatives. *Appl. Environm. Microbiol.*, **59**, 2526–30.

Zhao, T., Clavero, M.R.S., Doyle, M.P., Beuchat, L.R. (1997) Health relevance of the presence of fecal coliforms in iced tea and in leaf tea. *J. Food Prot.*, **60**, 215–8.

Zimmerli, B. & Dick, R. (1996) Ochratoxin A in table wine and grape-juice: occurrence and risk assessment. *Food Addit. Contam.*, **13**, 655–68.

Zindulis, J. (1984) A medium for the impedimetric detection of yeasts in foods. *Food Microbiol.*, **1**, 159–67.

Zook, C.D., Parish, M.E., Braddock, R.J. and Balaban, M.O. (1999) High pressure inactivation kinetics of *Saccharomyces cerevisiae* ascospores in orange and apple juices. *J. Food Sci.*, **64**, 533–5.

14 Water

I Introduction

A Important properties

Water is the quantitatively most important inorganic constituent of living cells and the one on which all life processes depend. Water is also one of the most important elements on our planet, a large proportion being bound as ice. It plays an important role in climate, transport, and agriculture.

Water is an essential part of our nutrition, both directly as drinking water or indirectly as a constituent of food. Water is not only essential for life, it also remains a most important vector of illness and infant mortality in many developing countries and even in technologically more advanced countries (Ford, 1999). It is also a key parameter influencing survival and growth of microorganisms in foods and other microbial environments.

B Methods of processing and preservation

Numerous studies on the effect of different types of treatments on water quality have been performed and published. Results depend on many factors including whether the environment is natural or artificial, physicochemical parameters, and analytical techniques applied (WHO, 1993).

One of the major objectives of the physicochemical treatments applied to raw water is to eliminate pathogens in order to supply users with safe drinking and processing water. Both national and international rely on the determination of total viable counts and/or coliform counts to determine if the treatments applied have been sufficient.

Producing water of appropriate quality is becoming increasingly difficult because of the ever-increasing demand. Treatments applied need to be adapted to the initial quality of the raw water. This often requires the use of combinations of different treatments (Olson and Nagy, 1984; Payment *et al.*, 1985) such as storage in open reservoirs or ponds, coagulation, filtration, and treatment with activated carbon or disinfectants (see Section II.C).

In some countries for certain types of bottled water, e.g. natural mineral water, treatments are limited to filtration in order to keep the original characteristic flora of the water.

C Types of final products

The following types of water will be discussed:

- Drinking water;
- Process or product water and
- Bottled waters including mineral water.

II Drinking water

A Definitions

It is difficult to establish an universal standard for drinking water due to different sociological conditions, varying climates, and other specific circumstances found all over the world. To facilitate the development of national standards, WHO began in 1979 to prepare Guidelines for Drinking Water Quality (WHO, 1993, 1996). In volume I of the guidelines, values and rationales for all parameters and constituents governing the quality of water are given. Other aspects such as application of guidelines, sampling plans, handling of samples, compliance and surveillance, actions in case of deviations are also covered in this document. In volume 2, health criteria for single vectors and parameters are discussed. The WHO guidelines and the list of parameters considered are undergoing constant revisions and updates and the latest version can be consulted on the website of the organization.

The quality of drinking water is further defined in numerous national or supranational documents such as the USA Food Code (FDA, 1993) and the European Union's Directive on drinking water (EC, 1980a, 1998).

B Initial microflora

Raw water for the make-up of drinking water originates from two sources, either surface or ground waters. The choice of a source of water depends on several factors including the availability of adequate quantities throughout the year and their amenability to treatment. Protection of the water sources from domestic, industrial or agricultural pollution is essential as well.

Surface waters from rivers, lakes and reservoirs are normally easily accessible. These sources, in particular the lowland ones are more exposed to contamination with suspended material and microorganisms. The levels and type of contamination depends very much on the environment and the quantities of waste, waste water or effluents from sewage treatment plants directed to them. Changes in microbial loading can be quite rapid depending on the climatic conditions. Heavy rains, for examples, can lead to rapid increases due to runoff from fields.

Surface water may contain a wide variety of harmless heterotrophic microorganisms such as *Flavobacterium* spp., *Pseudomonas* spp., *Acinetobacter* spp., *Moraxella* spp. (formerly *Achromobacter* spp.) and *Chromobacterium* spp., as well as numerous unidentified or unidentifiable bacteria (Geldreich, 1983; Olson and Nagy, 1984). In tropical areas, the microbial flora may be dominated by mesophilic and thermotolerant bacteria and the diversity is greater (Hazen and Toranzos, 1990).

A wide range of potentially infectious agents including bacteria, viruses, protozoa, and helminths may be introduced to water sources (WHO, 1993). Pathogenic bacteria linked to waterborne outbreaks include: *Campylobacter jejuni*, *Escherichia coli*, *Salmonella* spp., *Shigella* spp., *Vibrio cholerae*, *Yersinia enterocolitica*, *Aeromonas hydrophila* and *Escherichia coli* O157 (Jones and Watkins, 1985, Szewzyk *et al.*, 2000, Frost, 2001, Theron and Cloete, 2002 and Leclerc *et al.*, 2002).

Pollution by sewage of human or animal origin, after overflow or infiltration from septic tanks or from land applications of sewage and sludge, is the major source of enteric viruses such as enteroviruses, reovirus, adenovirus, hepatitis A virus, rotavirus and Norwalk agent (Gerba, 1987; Gerba and Rose, 1990; Fleet *et al.*, 2000; Schaub and Oshiro, 2000).

Protozoa including *Entamoeba histolytica*, *Giardia intestinalis* and *Cryptosporidium parvum*, and helminths such as *Ascaris lumbricoides*, are frequently found in water in tropical zones and are considered as one of the most serious threats to human health. Outbreaks of *Cryptosporidium*, *Cyclospora* and *Giardia* have also been reported from temperate climates (Hibler and Hancock, 1990; Rose, 1990; MacKenzie *et al.*, 1994, Marshall *et al.*, 1997, Wright and Collins, 1997; Steiner *et al.*, 1997; Orlandi *et al.*, 2002, Rose *et al.*, 2002). The number of reports on the occurrence of toxins due to development of cyanobacteria is also increasing (Hunter, 1991 and Szewzyk *et al.*, 2000).

Ground waters comprise water from springs or from wells and boreholes used to catch water from the aquifers by means of pumps. Deep wells or boreholes provide usually water of excellent bacteriological quality. Ground water is therefore often used without any treatment, except physicochemical ones to reduce hardness or eliminate off-flavors and odors. The water pumped from shallow wells or boreholes, however, is more exposed to pollution and contamination (Wilson *et al.*, 1983).

Depending on the type of aquifer, the type of soil and its protective effect against pollution, the physicochemical characteristics of the water, and the levels of the microbial flora may be very low (Bischofberger *et al.*, 1990). Levels as high as 10^5–10^7 cfu/mL have, however, been reported. In such cases, microbial and chemical contaminants may reach ground water sources through wells, due to infiltration, leakage of solids at the surface, leaks in pipelines, effects of agricultural treatments, cross-contamination between aquifers, water fluxes due to rainfalls, etc. (Wilson *et al.*, 1983, Ghiorse and Wilson, 1988; Kolbel-Boelke *et al.*, 1988).

C Primary processing of raw water

Different treatments are applied depending on the origin of the water. They have evolved over the years to address new issues such as formation of halogenated compounds after chlorination or the presence of pesticides or nitrates in the raw water.

Due to its quality, ground water from deep aquifers is usually only submitted to a disinfection step, in some instances removal of metallic cations is necessary. In the case of surface water several steps are necessary to obtain water suitable for drinking. Depending on its origin, physicochemical treatments of the water may be necessary to remove color, turbidity, or metallic cations. Different steps, alone or in combination, are used to obtain a microbiological quality corresponding to the standards in force.

Pre-treatment (impoundment)

Surface water cleared of large debris and foliage by mechanical filters can be directed to storage tanks and ponds. These systems represent important water resources in case of drought and can be used as buffers in case of sporadic pollution.

Such ponds contribute appreciably the improvement of the water quality both chemically and microbiologically. This is due to the natural degradation of unwanted compounds, sedimentation, and the effect of sunlight (Geldreich *et al.*, 1980). Up to 99% of the viruses and pathogens present in raw water can be eliminated during impoundment (Payment *et al.*, 1985).

The efficiency of the exposure to sunlight depends on factors such as holding time, temperature, and the extent of recontamination due to aquatic fauna or recreational activities (Geldreich, 1972; Fennel *et al.*, 1974). Exposure of water to solar radiation to prepare safe drinking water on a small scale (e.g. households) is proposed as simple and cheap alternative for developing countries (McGuigan *et al.*, 1999). For example, it has been shown to reduce the risk of cholera in young children (Conroy *et al.*, 2001).

Coagulation, flocculation and clarification

Natural sedimentation can be accelerated by coagulation or flocculation. For this purpose techniques are applied using aluminum or iron salts or, less frequently, plant extracts or clays, which co-precipitate with particles in the water. Quality and efficiency of the procedure, however, depend upon factors such as pH, temperature, or the chemical quality of the water. Different types of clarifiers such as upward flow clarifiers are then used to separate the flocs from the water. Removal of 50-95% of microorganisms or viruses and up to 90% of parasites can be achieved (Payment *et al.*, 1985; Logsdon, 1990).

Filtration

Natural filtration. Rapid gravity filtration (20 m/h, particles of 2.5-15 mm diameter) using sand and/or anthracite on a gravel bed allows to eliminate flocs carried over from the clarification. Water may then be treated by slow filtration through man-made sand beds or natural systems such as dunes or riverbanks. The combination of physical factors, such as sedimentation, adsorption, or electrostatic binding; and of biological factors, such as metabolic activities, allow for 10^3 to 10^4-fold reductions of the microbial populations as well as of the reduction or elimination of chemical compounds, can be achieved (Slezak and Sims, 1984; Logsdon and Rice, 1985).

The efficiency of the slow filtration (0.1-0.4 m/h for particles of 0.15-0.4 mm diameter) is strongly dependent on the quality of the raw water, its flow rate and temperature, the depth of the bed, the size of sand particles and/or the stability and age of the filter (Bellamy *et al.*, 1985). The use of slow sand filters requires, however, large land areas per volume of water treated and hence cannot be applied generally. In certain regions, appropriate natural topographical configurations, such as dunes or riverbanks, represent inexpensive systems for water filtration (0.2 m/day).

Filtration with activated carbon. Filtration of surface water with activated carbon may be necessary to improve its quality by adsorbing or degrading compounds causing off-flavors, off-odors or undesirable color. However, the use of activated carbon filters may have an adverse effect on microbiological quality through the release of carbon particles loaded with bacteria or bacterial aggregates (Camper *et al.*, 1986, 1987).

Other filtration techniques. Different other filtration techniques such as micro-, ultra- and nanofiltration are also frequently used in water processing plants.

Disinfection

This step is intended to complement primary treatments such as those outlined above by inactivating remaining microorganisms, particularly pathogens, and maintaining a residual bactericidal effect for an extended period. Recommendations on the evaluation of the effect of disinfectants used for the water treatment have been made by NSF International (1999).

For disinfection to be effective, the raw water, in particular surface water, must be of relatively high purity and quality and low in substances that protect microorganisms or inactivate disinfecting agents. Also, the types and state of the microorganisms, pH, and temperature greatly affects the efficiency of the disinfecting procedure. Detailed information on individual disinfectants, their chemical characteristics, mechanism of action, biocidal activity and analytical methods for their detection can be found in different compilations (Safe Drinking Water Committee, 1980; Denyer and Stewart, 1998; Russell *et al.*, 1999; Maillard, 2002). Disinfection procedures as described below may yield different by-products some of which can have adverse health effects (Boorman, 1999).

Chlorine and chloramines. Chlorine is the most widely used agent for disinfecting water, although it is not always the most effective. Disinfection can be achieved with chlorine gas (Cl_2) or sodium or calcium hypochlorite (NaOCl and $Ca(OCl)_2$), which react with water to form hypochlorous acid (HOCl) or hypochlorite ions, depending on the pH (White, 1992). Levels of at least 0.5 ppm free chlorine and a contact time of at least 30 min at pH <8.0 are recommended by the WHO to achieve optimal results (WHO, 1996). During chlorination, chlorine will react with ammonia or organic amines to form chloramines. Although less oxidizing, these compounds are more stable and will persist for a longer time. Depending on the predominant chemical species (i.e. chlorine, hypochlorous acid, and/or hypochlorite ions), 99.9% or more of the bacteria can be destroyed. The efficacy of the process is influenced by dose, contact time, pH, and presence of organic compounds.

Although a relatively good viricidal activity can be obtained with chlorine, inactivation is very limited with chloramines. Protozoan cysts are, in general, more resistant to oxidizing disinfectants than bacteria

Table 14.1 Inactivation of microorganisms by free chlorine (Huss *et al.*, 2003).

Organism	Water	Cl_2 Residues mg/L	Temp. (°C)	pH	Time (min)	Reduction %	Ct[a]
E. coli	BDF[b]	0.2	25	7.0	15	99.997	ND[c]
E. coli	CDF[d]	1.5	4	?	60	99.9	2.5
E. coli + GAC[e]	CDF	1.5	4	?	60	≪10	≫60
L. pneumophila (water grown)	Tap	0.25	20	7.7	58	99	15
L. pneumophila (media grown)	Tap	0.25	20	7.7	4	99	1.1
Acid-fast							
Mycobacterium chelonei	BDF	0.3	25	7.0	60	40	≫60
Virus							
Hepatitis A	BDF	0.5	5	10.0	49.6	99.99	12.3
Hepatitis A	BDF	0.5	5	6.0	6.5	99.99	1.8
Parasites							
G. lamblia	BDF	0.2–0.3	5	6.0	–	99	54–87
G. lamblia	BDF	0.2–0.3	5	7.0	–	99	83–133
G. lamblia	BDF	0.2–0.3	5	8.0	–	99	119–192

[a] *Ct* product of disinfectant concentration (*C*) in mg/L and contact time (*t*) in minutes for 99% inactivation (modified after Sobsey (1989)).
[b] BDF = buffered demand free.
[c] ND = no data.
[d] CDF = chlorine demand free.
[e] GAC = granular activated carbon.

and viruses. Thus, parasites such as *Cryptosporidium* can survive the concentrations and exposure times normally applied (Oppenheimer and Aieta, 1997). A killing effect of more than 3 log-units can only be achieved using very high levels of free chlorine (80 ppm) for 2–3 h of exposure (Anonymous, 1997). Survival of bacteria within protozoa during chlorination has also been reported (King et al., 1988) and the subject reviewed by Barker and Brown (1994).

The effect of free chlorine on different microorganisms is shown in Table 14.1.

Chlorine dioxide. Chlorine dioxide is very reactive and is therefore generated at the point of dosage, usually by mixing sodium chlorite, hydrochloric acid, and hypochlorite. Details on the different chemical processes are provided in (Holah, 1997). Chlorine dioxide is effective over a wide pH range (5-9) and target levels of 0.3 ppm for a contact time of at least 20 min are usually recommended. Chlorine dioxide is an effective bactericidal and viricidal agent, and several studies have shown that it was more effective against parasites such as *Cryptosporiudium* but that significantly higher concentrations and contact times were necessary to achieve a killing effect of 1.5-3 log units (Liyanage *et al.*, 1997).

Bromine. This is an active disinfectant over a wider pH range than chlorine and has been used more widely over the last years. It is accepted by several agencies in the USA and Europe as an alternative in cooling and retort water, for the drinking water treatment and other applications. Bromine is generated by the dissolution and hydrolysis of a solid bromine carrier in water. Bromine retains its biocidal activity at higher pH-values than chlorine (Holah, 1997). Bromine, however, is more expensive and monitoring is not quite as simple as chlorine. In addition off-taint compounds formed after reaction with natural phenolic compounds have a much lower threshold level than similar compounds formed from chlorine.

Ozone. This is a highly toxic and unstable gas, which has been widely used for the disinfection of drinking water. The disinfection effect is due to its direct strong oxidizing potency, as well as to the formation of radicals after breakdown in the water. Ozone has good bactericidal and viricidal action and may have applications as an anti-parasitic agent, thus ozone is more efficient than chlorine in eliminating oocysts of *Cryptosporidium* (Clark *et al.*, 2002). Concentrations of 0.4 ppm are typically used with contact times of 4 min. The efficacy of ozone disinfection depends on the quality of the water, since the molecules are sensitive to inactivation by organic matter. Ozone has several other effects such as the oxidation of organic compounds, which may cause off-tastes and off-flavors, the removal of colors and to some extent the oxidation of pesticide residues. However, unlike chlorine, residual activity is not maintained over a prolonged period of time and recontamination or growth within the distribution system must therefore be prevented by other means. The degradation of organic compounds leads to an increase in available organic carbon, which may favor the development of such microorganisms and in particular, the formation of biofilms.

Ultra-violet light. Ultra-violet light with wavelength in the range of 200-310 nm (optimum 254 nm) is effective for the destruction of microorganisms. Spores, pigmented microorganisms as well as viruses show higher resistance and higher doses are necessary. UV is not effective against protozoa, but several alternatives using different approaches have been discussed by Clancy and Fricker (1998).

The efficacy of UV light is strongly dependent on water quality, particularly the suspended particles, which protect the microbes. The intensity of the lamps, flow rates, and thickness of the treated water films are important factors, whereas pH and temperature do not influence efficacy. As with ozone, no residual activity occurs, so no protection exists against the growth of injured bacterial cells that recover in the presence of light (photo-reactivation).

Other agents. Iodine is as effective as chlorine as a disinfectant, with optimal activity in a slightly alkaline environment. Chemically, iodine is less reactive than chlorine, and the long-term health effects are not known. Water treatment with iodine is more expensive than with other available biocides. Other disinfection methods using ferrate, high pH conditions, hydrogen peroxide, ionizing radiation, potassium permanganate, and silver have been described. Most have been used only for specific applications and only very infrequently for routine treatment. Trials have also been performed using combination of different disinfectants in sequence and obtaining a synergistic effect for lower operating costs (Clancy and Fricker, 1998).

Distribution of drinking water

The sole purpose of disinfection is to destroy pathogens. The presence of viable bacteria in drinking water is common and includes *Pseudomonas* spp., *Flavobacterium spp.*, *Aeromonas* spp., *Micrococcus* spp., and *Bacillus* spp. (McFeters *et al.*, 1986). While levels of 100-300 cfu/mL have been considered acceptable in national or international standards (EC, 1980a) in the past, precise limits have been replaced by statements such as "No significant increase over that normally observed" (UK; Anonymous, 1988) or "No abnormal changes at 22°C" (e.g. EC, 1998).

Increased numbers of microorganisms at the point of consumption or use reflect growth and/or recontamination during storage or transport through the distribution system to the final user. Length and design of the distribution network is determined by the age of the system, the distance, and location of consumers and the topography of the zone. Long and complex distribution systems increase the risk of contamination (Geldreich *et al.*, 1980; Maul *et al.*, 1992).

Growth in the distribution system may be due to regrowth or aftergrowth, i.e. recovery of injured cells or development of native microorganisms, including some opportunistic pathogens including the genera *Legionella*, *Aeromonas* and *Mycobacterium*. Aftergrowth may also promote development of higher organisms such as protozoa, crustaceans, nematodes, etc., leading to aesthetic or technical problems such as clogging of filters.

Bacteria developing in distribution systems normally have low temperature optima and grow between 15-20°C. They usually belong to the genera *Flavobacterium*, *Pseudomonas*, *Arthrobacter*, *Aeromonas*, etc. Deterioration of the microbiological quality of the water will depend on the presence of metabolizable organic compounds as well as greatly reduced levels or the absence of residual disinfectant (Hutchinson and Ridgway, 1977; Walker and Percival, 2000). Most major organic components such as humic acids originating from the source of the water or the treatment plant will not support significant growth of surface slimes or formation of off-flavours or off-tastes (Whitfield, 1998). Soil, which permits growth, may be introduced directly through crevices in walls of reservoirs, breaks in pipes, or during maintenance work. Impurities may be the result of cross-contaminations with dirty water or air or through release from established biofilms. Some organic substances may also originate from construction materials, coatings or sealants of reservoirs, pipes or fittings of the network through leaching or exposure at surface water interfaces (Colbourne, 1985; Schoenen and Scholer, 1985).

Development of bacteria may occur in the free water phase, on the surface of suspended particles or in biofilms accumulating throughout the distribution system (LeChevallier *et al.*, 1987). Attachment and biofilm development is dependent on materials used to build the different parts of the system as mentioned above. A biofilm is a community of microorganisms embedded in a matrix of polymers excreted by bacteria and adhering to a solid surface. These communities are often heterogenous, e.g. with aerobic and anaerobic zones and are constantly replenished. Nutrients come from the water phase or are generated within the biofilm. Cells in biofilms are protected from disinfectants. The biofilm formation may be minimized in drinking water systems (and disinfection efficiency improved) by limiting the amount of organic carbon (Chandy and Angles, 2001). Although chloramines show a better penetration of biofilms than chlorine, they are less reactive (Costerton, 1984; McCoy *et al.*, 1991; Ganesh-Kumar and Anand, 1998).

Recontamination is also an important factor influencing quality of the drinking water. The presence of excessive numbers of microorganisms may be due to single episodes such as breakdowns in the treatment plant or contamination during building or maintenance of the distribution system. Secondary recontamination by heterotrophic or pathogenic bacteria may of course also occur from soil introduced through breakage or leaks, by contamination with non-potable water through cross-connections, back-siphoning, air-intake, or release of bacteria from biofilms. Typical examples are several outbreaks due to *Cryptosporidium* and presence of this parasite in drinking water, when detected through monitoring, triggers advices to boil water as a control measure to protect consumers (Anonymous, 1995).

In recent years, a particular attention has been given to the development of risk assessments for *Cryptosporidium parvum* (Gale, 2001; AFSSA, 2002), but also for other pathogens such as rotavirus and bovine spongiform encephalopathy (BSE) (Gale, 1996; Gale *et al.*, 1998).

D Pathogens

Infective agents transmitted by water cover a wide range from viruses through bacteria, fungi, protozoa, and parasites to helminths. Waterborne outbreaks have been reviewed by several authors such as Levine and Craun (1990), Furtado *et al.* (1998), Tillett, *et al.* (1998) and more recently by Szewzyk *et al.* (2000), Craun *et al.*, (2002), and Leclerc *et al.* (2002).

E Spoilage

Spoilage is mainly due to organoleptic deterioration of water quality may also occur due to the release of compounds with earthy, musty, or chemical odors (Gerber, 1983; Montiel *et al.*, 1987; Cabral and Fernandez, 2002).

F CONTROL (drinking water)

Summary

Significant hazards	• Bacteria: *Campylobacter jejuni*, *E. coli* O157 and other pathogenic *E. coli*, *Salmonella* spp., *Shigella* spp., *Vibrio cholerae*, *Yersinia enterocolitica*. • Viruses: Hepatitis A, Small Round Structured Virus (SRSV). • Parasites: *Entamoeba histolitica*, *Giardia intestinalis*, *Cyclospora cayatenensis*, *Cryptosporidum parvum*, *Ascaris lumbricoides*.
Control measures	
Initial level (H_0)	• Varies depending on source of water.
Increase (ΣI)	• Good Hygienic Practices to prevent biofilm formation. • Inspection, surveillance and preventive maintenance of the distribution systems to prevent environmental contamination.
Reduction (ΣR)	• Impoundment (surface water)–elimination of up to 99% of pathogens. • Coagulation, flocculation, clarification–elimination of up to 50–90% of pathogens. • Filtration (natural)–varying degrees of reduction. • Filtration (filters)–elimination of up to 50–90% of pathogens. • Disinfection–can reduce up to 99.9% of bacteria, effect on parasites is much more limited. • When there is reason to suspect a water supply is contaminated the population affected should be advised to boil the water before drinking.
Testing	• *E. coli*, coliforms, fecal coliforms as indicators. • Analysis for residual disinfectants, where appropriate.
Spoilage	• Microbial spoilage of potable drinking water is not a concern.

Hazards to be considered

Numerous microbial pathogens need to be considered when designing appropriate controls for drinking water. These are listed in the table above.

Control measures

Initial level (H_0). The presence of significant hazards in raw water depends on its origin. The type and initial levels will vary widely: the risk of pathogens will usually be the lowest in water from deep aquifers and highest in surface waters.

Increase (ΣI). Increase is usually due to either microbial growth within the distribution systems (e.g. release of bacteria from biofilms) and/or recontamination from the environment (e.g. cross-connections with non-potable systems or leaks that lead to contamination with bacteria, viruses, and parasites). Unacceptable increases in microbial content are best controlled by the application of Good Hygienic Practices (particularly preventive maintenance of the distribution systems), inspection, surveillance, and testing.

Reduction (ΣR). Reduction in levels of pathogens occurs during the combined reductions that occur through the primary processing steps described above. The selection and integration of reduction steps depends on the quality of the water source with adjustments being made as necessary.

Testing

It is the responsibility of water authorities or suppliers (in case of private sources) to ensure the microbiological safety of potable water supplies. Regular monitoring of the water for pathogens and indicators of recontamination, including faecal contamination performed by water authorities and/or water companies provides important information to users as to possible microbiological deviations. Guidelines on microbiological criteria are provided by WHO or by local legislations.

Microbiological testing by users of the water can only be regarded as verification for the adherence to Good Hygiene Practices during distribution up to the point of use. Analysis for residual disinfectants, where applicable, is much more useful and is therefore recommended. These results will provide rapid and continuous information on the status of the water and on the potential for deviations, allowing for immediate application of corrective actions.

Normally, water samples are tested at regular intervals for the presence of indicator bacteria to verify potability. *Escherichia coli* is widely used as an indicator of faecal pollution, both in temperate and tropical zones (Barbaras, 1986). Coliforms, faecal streptococci, and gas-producing clostridia (sulfite-reducing clostridia) commonly found in the feces of humans and animals are also traditionally used to indicate recent fecal pollution (Hutchinson and Ridgway, 1977). More recently, Leclerc *et al.* (2001) have discussed the limitations of such marker organisms and proposed to further investigate the suitability of novel genomic types of Enterobacteriaceae as monitoring tools. For indicators to indicate potential presence of pathogenic organisms, they must be present in greater numbers more resistant to disinfectants that may be used for treating the water and easily detectable and identifiable. These criteria are not always fulfilled. The role of bacteriophages as indicator for fecal contamination and thus of enteric viruses has been proposed by several authors and is discussed in details by Leclerc *et al.* (2000). The WHO and EU guidelines are provided in Tables 14.2 and 14.3.

Table 14.2 Bacteriological quality of drinking water (WHO, 1996)

Organisms	Guideline value
All water intended for drinking	
E. coli or thermotolerant coliform bacteria[a]	Not detectable in any 100 mL sample
Treated water entering the distribution system	
E. coli or thermotolerant coliform bacteria[b]	Not detectable in any 100 mL sample
Total coliform bacteria	Not detectable in any 100 mL sample
Treated water in the distribution system	
E. coli or thermotolerant coliform bacteria[c]	Not detectable in any 100 mL sample
Total coliform bacteria	Not detectable in any 100 mL sample. In the case of large supplies, where sufficient samples are examined: Not detectable in 95% of samples taken during any 12-months period

[a] Immediate investigative action must be taken if either *E. coli* or total coliform bacteria are detected. The minimum action in the case of total coliform bacteria is repeat sampling; if these bacteria are detected in the repeat sample, the cause must be determined by immediate further investigation.

[b] Although *E. coli* is the more precise indicator of fecal pollution, the count of thermotolerant coliform bacteria is an acceptable alternative. If necessary, proper confirmatory tests must be carried out. Total coliform bacteria are not acceptable indicators of the sanitary quality of rural water supplies, particularly in tropical areas where many bacteria of no sanitary significance occur in almost all untreated supplies.

[c] It is recognized that in the great majority of rural water supplies in developing countries, fecal contamination is widespread. Under these conditions, the national surveillance agency should set medium-term targets for the progressive improvement of water supplies, as recommended in Volume 3 of *Guidelines for Drinking-water quality*.

Table 14.3 Microbiological criteria for drinking water (EC, 1998)

Parameter	Parametric value	Method of examination
E. coli	0/100 mL	ISO, 9308-1
Enterococci	0/100 mL	ISO, 7899-2
Indicator colony count, 22°C	[No abnormal change][a]	Pr EN ISO 6222
Coliform bacteria	0/100 mL	ISO, 9308-1

[a]Former directive 80/778/EC (EC, 1980a) used 100 cfu/mL as guidelines.

Taking into account these issues and acknowledging the unreliability of end-product testing as a means of ensuring the safety of water, emphasis should be on application of HACCP principles. Points for control are dependent on the factors controlling growth, survival, and contamination as outlined in the earlier sections. A detailed discussion of the preventive measures to ensure safe drinking water supply is given by Havelaar (1994). The implementation of more global water management systems combining a geographical information system, information on the water supply structures as well as on incidents and outbreaks would allow to respond much more rapidly to the increased requirements for public health protection (Kisteman *et al.*, 2001).

In the case of ground water, protection of the aquifer is essential, a complementary disinfection being possible if necessary. For surface water, the situation is more complex and the different treatments applied such as sedimentation after coagulation and flocculation, the different types of filtration and disinfection need to be strictly controlled, as well as the storage system of treated water.

During distribution, recontamination can be controlled or minimized by maintaining the integrity of the distribution system.

Testing for residual biocidal activity of a disinfectant (e.g. free available chlorine) seems to be an appropriate complementary method, which can be applied in-line or more frequently than microbial testing. Simple dip-stick kits are available for such determinations. Payment (1999), however, discusses the limitations of this approach, in particular with respect to the early detection of sporadic cases of waterborne diseases. Extensive chemical analysis of water for pollutants is performed at more infrequent intervals. Issues related to sampling plans, the sample storage and handling, the use of indicators as well as with analytical methods for water and which may influence reliability of results and conclusions have been discussed by Pipes (1990).

Testing for microbiological parameters is frequently performed applying standardized methods (Schmidt, 2003) and new developments using DNA microarrays may allow for rapid analysis of several pathogens and indicators at the time in the near future (Straub and Chandler, 2003).

III Process or product water

A *Definition*

Water plays a major role in food production. The food processing industry is in fact the fifth largest water consuming industry in the world, just after industries transforming or manufacturing metals, chemicals, petroleum, and paper (Emery, 1989). Water used in processing may be purchased from Water Companies or Authorities or, depending on the location, made up from surface or groundwater at the factory itself.

Apart from its use as drinking water (see Section II) or as bottled water (see Section IV), water is used in food factories for a number of purposes, most of which are directly linked with processing. Water usage may be subdivided into different classes depending on its contact with the products (Terplan and Bierl, 1980).

Use of non-potable water, such as from ponds or rivers, may be permitted for fire-fighting, cooling of motors, etc. The impact on the safety of products needs, however, to be assessed within the HACCP study.

In view of the reduced availability of drinking water as well as of the increasing costs for both the purchase and the discharge of water, reuse of water is attracting more attention from the food industry. Reuse is, however, associated with microbiological risks that need to be addressed through appropriate preventive measures including HACCP (Casari and Knøchel, 2002).

Ingredient. Water is an important ingredient in different types of products such as drinks and beverages, culinary preparations, meats or canned products. This water, which could also be added in the form of ice, should therefore not contain microbial or chemical contaminants that may cause the products to become hazardous to consumers. Ingredient water should meet drinking water standards.

Direct contact. Water is used as a technological aid during food production. Because of uptake, absorption or penetration into packaging material, inclusion of a certain quantity of water in the product is often unavoidable. Thus, contaminants in the water could represent a hazard in the finished product or lead to its spoilage. Examples of such uses of water are washing, conveying, and blanching of vegetables or fruits; scalding, cleaning and chilling of poultry or other slaughtered animals; storage of fish and meat in ice; washing to remove certain components during cheese or butter manufacturing; cutting of products, or lubricating conveyor belts, etc.

Indirect contact. Water is used as an aid for cleaning utensils and equipment. Inadequate draining of equipment of water may lead to contamination of the product depending on the techniques used. Precautions must be taken to permit adequate drainage of residual water, to eliminate dead spots in cleaning in place systems, to dry utensils, etc.

Accidental contact. Accidental contact occurs when water is not intended to contact food, but where it occurs due to engineering defaults, wrong manipulations by operators or defects in equipment. Examples are cooling of cans, water circulating in closed heat-exchange systems, aerosols formed and dispersed during uncontrolled cleaning or hosing with high-pressure systems, condensation due to uncontrolled temperature differences.

B *Initial microflora*

The initial microflora of raw water made up and of the water supplied and used as processing water will largely correspond to the flora described in Section II.

C *Primary processes of processing or product water*

The requirements for process or operations water are the same as for drinking water (Jacob, 1988; Anonymous, 1993; FDA, 1993). The same procedures as those described in Section II are applicable to treat raw water used as processing or product water in food manufacturing sites. Water purchased from water authorities or water companies is usually stored in tanks or cisterns before use. Treatments with biocides or UV are increasingly been applied to take into account possible contamination during distribution or growth during storage before use. The same problems may occur during distribution in a food plant even though it is on a more limited scale than throughout a city water and controls must be in place to prevent recontamination.

In processing plants, however, use of water of a different quality, i.e. with higher bacterial counts, may be permitted for certain applications. This is particularly true for processes where the quality of the water will not affect the wholesomeness of the final product. The use of recycled water provides substantial savings of drinking water and reduces the quantity of discharged wastewater. An example is the transport of vegetables or fruits directly after harvesting and before cleaning and blanching. This could be performed by means of recycled water without influencing the quality of processed vegetables and fruits.

Water fulfilling specific physico-chemical requirements is needed in certain cases and for specific types of products. Denitrification is done by means of ion exchange, electrodialysis, or biological means; organics or colors are eliminated by passing the water through activated carbon columns; different types of filtrations may be used to obtain ultra-pure water; dealkalization is needed to prepare water for soft drinks; ion exchange, reverse osmosis, or oxidation are applied to eliminate salts or iron and manganese. The impact of these technologies on the microbiological quality of the water has been studied by different authors such as for softeners (Stamm *et al.*, 1969; Parsons 2000); for activated carbon and electrostatic filters (Tobin *et al.*, 1981); for point of use devices (Geldreich *et al.*, 1985); for reverse osmosis (Payment *et al.*, 1989) and for activated carbon columns (Camper *et al.*, 1986).

While certain treatments will not affect the microflora, others may increase or decrease levels depending on the technology applied. Increases are mainly due to multiplication in poorly maintained systems such as ion exchange columns or filtration units. The levels reached depend on the initial levels in the water, the growth conditions (nutrients, pH, temperature) on technological parameters such as flow and turbulence as well as on the maintenance schedules.

Although *Legionella pneumophila* is not a water- or food-borne pathogen, water authorities, water companies, and food manufacturer need to take it into consideration for reason of public and personal safety. A review of the situation, monitoring and control measures is provided by the ASHRAE (2000) and EHEDG (2002).

D CONTROL (process or product water)

Summary

Significant hazards	• Similar to drinking water, depending on the source.
Control measures	
Initial level (H_0)	• Varies depending on source of water.
Increase (ΣI)	• Similar to those for drinking water if produced from raw water. • Maintain systems to prevent cross-connections and biofilm formation.
Reduction (ΣR)	• May require filtration and/or disinfection for recycled process water.
Testing	• Testing described for drinking water is appropriate for ingredient and cleaning applications. • Higher levels of organisms may be permissible in situations where recycled water will not have a detrimental impact on food quality and safety.
Spoilage	• Process or product water must be of high quality and not subject to excessive microbial growth.

Hazards to be considered

Processing water has significant hazards similar to those for drinking water. If city water is used, then hazard analysis has to be performed taking into account possibilities of recontamination during distribution to and within the facility. A thorough knowledge of the local situation is important.

If the water is from the user's private water supply, e.g. at the factory level, then the same situation applies as for drinking water supplied by water authorities.

Control measures

Initial levels (H_0). Initial microbial population levels may differ between city water and recycle water. If water from water authorities or companies is used, then lower levels may be expected. It is essential that knowledge of the local situation is considered through hazard analysis.

Increase (ΣI). Control measures for water used in food processing facilities are similar to those described in the previous section on drinking water. Care must be taken that non-potable water does not contaminate the water system carrying potable or drinking water.

Bacterial growth in distribution systems, release of bacterial biofilms and/or recontamination from environment through cross-connections, leaks, etc., can increase levels of significant hazards. The control measures discussed for drinking water are applicable to process water. It is important to recognize that higher levels of disinfectant agents may be required to compensate for organic compounds in recycled water.

Reduction (ΣR). Reduction methods described for drinking water should be used if process water is made from raw water by the user. If water from water authorities or companies is used, then an additional treatment with a biocide may be necessary control growth and recontamination that may occur during distribution up to the point of use within the plant. Ultraviolet light treatment may be useful considering water supply and anticipated hazards. Normally, the may decide to use additional biocides for quality reasons (e.g. reduce microorganisms in the water supply that may cause unacceptable product spoilage). Additional biocides should not be needed to ensure the safety of the water supply if the supplier is reliable.

Testing

It is the responsibility of water authorities or suppliers (in case of private sources) to ensure the microbiological safety of water supplies. Regular monitoring of the water for pathogens and indicators of recontamination, including fecal contamination performed by water authorities and/or water companies provides important information to users as to possible microbiological deviations. Guidelines on microbiological criteria are provided by WHO or by local legislations.

Microbiological testing by users of the water can only be regarded as verification for the adherence to Good Hygiene Practices during distribution up to the point of use. Analysis for residual disinfectants, where applicable, is much more useful and is therefore recommended. These results will provide rapid and continuous information on the status of the water and on the potential for deviations, allowing for immediate application of corrective actions. Normally, water samples are tested at regular intervals for the presence of indicator bacteria to verify potability.

Recycled processing water requires testing to ensure the system is adequate to control the microbiological concerns that are relevant to the food being produced. Once the system for reprocessing the water has been validated to be controllable and effective, a monitoring program should be established. The routine testing program for monitoring should be based on the data accumulated during validation and experience gained with the system.

Spoilage

Multiplication of microorganisms, in particular in biofilms, may lead to the formation of off-flavors and off-odors in the water supply. Routine maintenance, including thorough cleaning and sanitizing is necessary to prevent biofilm formation.

IV Bottled water

A Definitions

Two categories of bottled water are be considered:

- spring or mineral water; and
- other bottled water.

Spring and mineral (natural) water. Spring and mineral water are drawn from underground sources such as a bore holes or a springs. They differ according to their composition and content in minerals. Depending on the type they have to fulfill the requirements for dissolved solids as defined by the Codex Alimentarius Commission (1993, 1994) or local legislations. In Europe labeling as "natural" implies the absence of any biocidal treatments before bottling.

According to the European Union's Directive (EC, 1980b), natural mineral water can be clearly distinguished from ordinary drinking water by its nature, i.e. its mineral content, its trace elements, and other constituents certain declared (health) effects or its origin.

Other bottled water. Bottled water is water commercialized in different types of containers such as bottles, cans, bags in different sizes. Bottled water can be either water from springs and wells or drinking water from the distribution system. Water can be submitted to a number of treatments such as carbonation, distillation, ionization or purification as well as bactericidal treatments such as filtration or ozonation. Further details on different types of bottled waters and allowed treatments are summarized by Warburton and Austin (2000). Standards are provided by the International Bottled Water Association (IBWA, 1995) or by local Authorities (FDA, 1995), by the EU (EC, 1998) and are under preparation at Codex level.

B Initial microflora

Mineral water. Low numbers of bacteria of the order of 10-100 cfu/mL may be present as the water emerges from natural sources or drilled from underground sources. This microflora known as the autochthonous or indigenous flora, is composed mainly of non-fermentative Gram-negative bacteria, i.e. *Pseudomonas*, *Moraxella*, *Acinetobacter*, *Flavobacterium*, and *Xanthomonas* (Guillot and Leclerc, 1993a,b; Elomari *et al.*, 1995; Leclerc and da Costa, 1998). Some Gram-positive bacteria such as *Micrococcus* and *Arthrobacter* have also been isolated (Hunter, 1993). These autochthonous bacteria are as a rule very small in cell size are usually aerobic and able to grow at low temperatures. Their requirements for nitrogen and organic nutrients are low. They are, nevertheless, chemo-organotrophic rather than autotrophic.

Other bottled water. The initial microflora of water tapped from sources or from the distribution system for drinking water is the same as described in earlier sections.

C Primary processing

Mineral water. Natural mineral water should originate from an underground water table as a spring and be tapped at one or more natural or bore exits. These underground sources must be protected from all risks of pollution and contamination.

The composition, temperature, and other specific characteristics must remain stable within the limits of natural fluctuation and be preserved during catchment and bottling. According to the EU directive, for example, natural mineral water may not be subjected to any treatment or additions other than to separate unstable elements by filtration or decanting, if necessary preceded by oxygenation. It is also

Table 14.4 Sampling plans for testing microbiological quality of natural mineral water according to CAC (1993, 1994, 1996) and EC (1980, 1998)

Agency	Parameter	Volume (mL)	*n*	*c*	*m*	*M*
CAC	Coliforms	250	5	1	0	2
	Escherichia coli	250	5	0	0	
	Fecal streptococci	250	5	1	0	2
	Sulfite reducing clostridia	250	5	1	0	2
	Pseudomonas aeruginosa	250	5	0		
EC	ACC[a] (20–22°C; at source)	1	(*)[b]	0	20	
	ACC (37°C; at source)	1	(*)	0	5	
	ACC (20–22°C; 12 h after bottling)	1	(*)	0	100	
	ACC (37°C; 12 h after bottling)	1	(*)	0	20	
	Coliforms	250	(*)	0	0	
	Escherichia coli	250	(*)	0	0	
	Fecal streptococci	250	(*)	1	0	
	Sulfite reducing clostridia	250	(*)	1	0	
	Pseudomonas aeruginosa	250	(*)	0		

[a] ACC = Aerobic Colony Count
[b] (*) not specified.

possible to extract the natural gas with subsequent re-introduction during bottling to avoid corrosion of the equipment. These treatments are only allowed as long as they do not alter the essential constituents and properties of the water.

Information concerning the particular properties of a natural mineral water may appear on the label. The microbiological quality must comply with international, regional or national regulations. Examples for are provided in Table 14.4.

Conditions for collecting and distributing natural mineral water are also specified in regulations. In particular, transport of the water in containers other than those authorized for distribution to the ultimate consumer is prohibited. The same principles have been incorporated in the revised Codex Standard for Natural Mineral Water (CAC, 1996), especially concerning the origin, the particular characteristics of the water, and the authorized treatments. This standard does, however, not recognize possible health benefits, but does include both microbiological and chemical criteria. Recommendations for collecting, processing and marketing of such waters are provided in the Codex Code of Practice for Mineral Water (CAC, 1994).

Similar descriptions for mineral water have been adopted by other countries, but in other legislation. The concept of a natural mineral water may, however, differ from that adopted by the European countries.

Other bottled water. The requirements for other bottled water are in general less strict than those for natural mineral water. For example, the constancy of the constituents is not required, the origin of the water can be a surface water and a bactericidal treatment can be permitted or even required. This category of water cannot be linked to health claims in the EU (EC, 1996). Moreover, in this document "spring water" is defined as a potable water of underground origin, protected from pollution, microbiologically safe and not subjected to a bactericidal treatment.

Consumers expect a very high standard of quality and safety of bottled water and in practice these bottled waters must comply with the same strict hygiene requirements as for piped drinking water. Examples of regulations are provided in Table 14.5.

C *Effects of processing on microorganisms*

Mineral water. In the case of mineral water, including natural mineral water, no processing steps are allowed. Therefore modifications of the natural flora present in the aquifer will only be due to recontamination in intermediate storage tanks or piping before bottling due to poor hygienic conditions.

Table 14.5 Regulations for bottled mineral water as outlined by EC (1980a, 1998) and US (FDA, 1995)

Agency	Parameter	Volume (mL)	*n*	*c*	*m*	*M*
EC	ACC[a] (22°C)	1	(*)[b]	0	100	
	ACC (37°C)	1	(*)	0	20	
	Coliforms	250	(*)	0	0	
	Fecal coliforms	250	(*)	0	0	
	Fecal streptococci	250	(*)	1	0	
	Sulfite reducing clostridia	250	(*)	1	0	
	Pseudomonas aeruginosa	250	(*)	0		
United States	Coliforms (MPN)	100	10	1	2.2	9.2
	Coliforms (membrane filter)	100	10	1	1	4

[a]ACC = Aerobic Colony Count
[b](*) not specified.

Further changes during shelf-life of the water are due to the growth of the natural flora or of contaminants in the bottle.

After bottling, when the open system (the source) is changed for a closed system (the filled bottle), multiplication of the autochthonous bacteria begins. This growth is characterized by an alternating increase and decrease in the number of bacteria. The characteristic initial increase in bacterial counts, just after bottling, is supposed to be caused by: (i) the adsorption of organic compounds to the bottle surface, thus increasing their concentration sufficiently to permit growth; and (ii) the increase of dissolved oxygen during the filling operation. When tested within 12 h after bottling, aerobic (heterotrophic) plate counts at 20-22°C and 37°C should be less than 10 times that of the water taken directly from the source. During the distribution and storage of non-carbonated mineral water, the autochthonous microflora may grow, and counts of up to 10^4–10^5/mL (20-22°C for 72 h) are commonly found. Such growth is more common in waters bottled in plastic bottles and is believed to be due to the fact that plastic provides a surface suitable for microbial growth (Jones *et al.*, 1999). Such growth is completely normal and is characteristic of all types of low-nutrient water. It, in no way, affects the suitability or safety of the water for human consumption (Hunter, 1993, Leclerc and Moreau, 2002). For this reason, specifications for total aerobic counts cannot be used in such cases as an indicator of good hygienic manufacturing practice.

Contamination of the source or recontamination before bottling may account for the presence of non-autochthonous microorganisms or enteric microorganisms, including pathogens, viruses, or parasites. Recontamination of mineral water by transitory or permanent contaminants, as defined by Schmidt-Lorenz (1976) should therefore be prevented at different stages. Growth of these microorganisms is usually inhibited by the natural microflora and the low levels of nutrients. Slow but definite decreases after several weeks is frequently observed as demonstrated in different inoculation studies. Numerous inoculation studies have been performed for *E. coli*, *Pseudomonas aeruginosa*, salmonellae, staphylococci, and *Bacillus* spp. or *E. coli* O157 and several of them have been reviewed and compiled by Warburton and Austin (2000). The evaluation of the results and in particular the comparison between published studies is often difficult, due to the different experimental design chosen by different authors. In a number of cases information concerning, for example, the presence of a competitive flora is missing and the relevance of the inoculation levels can be questioned.

Other bottled waters. The effects of processing on the primary flora are outlined in Section II. At the bottling plant, as in the case of mineral water, recontamination may occur during storage or in the pipe system used to transport the water. The microbiology of treated bottled waters was reviewed in depth by Edberg (1998).

Some of the treatments applied, such as filtration through activated carbon, chemical softening, or reverse osmosis, may have a detrimental effect on the microbiological quality, i.e. an increase in coliforms and heterotrophic plate counts (Geldreich *et al.*, 1985; Camper *et al.*, 1987).

For water originating from unprotected source or from surface water a bactericidal treatment is normally required to assure its safety. Carbonation of water to a level of three to four volumes of CO_2 has a strong bactericidal effect on most bacteria. The reduction in pH (four volumes of CO_2 will reduce the pH to approximately 3.3) and the effect of carbon dioxide will cause salmonellae, staphylococci, vibrios, and many other bacteria to die within a few hours (Burge and Hunter, 1990; Hunter, 1993). The effect of carbonation is reflected by surveys on the quality of bottled water showing much less rejection than of carbonated ones (Ruskin *et al.*, 1991; Warburton *et al.*, 1998). However, although carbonation is clearly effective in destroying enteric pathogens, it should not be used as a substitute for maintaining high standards of hygiene during the processing of carbonated bottled waters.

One element needing consideration is the recontamination of the water during handling at the consumer level. This is the case, for example, in water coolers where the greatest potential for recontamination occurs during the change of the bottles, either during handling or due to contact with elements of the water coolers (Eckner, 1992; Levesque *et al.*, 1994; Hunter and Barrell, 1999).

C Pathogens

Surveys performed in different countries on the quality of commercialized bottled water have shown deviations, including the presence of potentially pathogenic microorganisms (Edberg, 1998). In numerous publications it is difficult, however, to figure out whether mineral water and other bottled water are discussed, which have, as outlined above, a totally different ecology.

The two categories of products have, however, only been associated to outbreaks in two occasions, in a cholera (Blake *et al.*, 1977a,b) and a typhoid outbreak (Buttiaux and Boudier, 1960). In the cholera epidemic in Portugal in 1974, in which about 3000 persons were affected, bottled, non-carbonated mineral water was discussed as one of the possible vehicles based on epidemiological investigations. The pathogen itself, however, has never been isolated from the water. The absence of reported outbreaks in recent years (1985–1997) has been confirmed in a survey performed by Bohmer and Resch (2000) on published data from Central and Northwestern Europe, United States and Canada.

D Spoilage

Spoilage of bottled waters (of all type) has only rarely been published. The few reported cases are linked with the visible growth of moulds or streptomyces and have been detected due to visual or organoleptic deviations of the products (Fujikawa *et al.*, 1997, 1999; Cabral and Fernandez, 2002).

E CONTROL (Natural mineral water)

Summary

Significant hazards	• Similar to drinking water, depending on the source.
Control measures	
Initial level (H_0)	• Catchment for natural mineral water must be free from contamination. • Absent or very low depending on the protection of the catchment or the quality of the drinking water used for other waters.
Increase (ΣI)	• Protect catchment from contamination for natural mineral water. • Good Hygiene Practices, including routine maintenance of lines.

(*continued*)

E CONTROL (Cont.)

Summary

Reduction (ΣR)	• None allowable for natural mineral water. • For other bottled waters biocidal treatments—chlorination, ozone or UV treatment, filtration or others, depending on local legislation.
Testing	• Use methods specific to water analysis, including filtration steps. • Coliforms are useful indicators of Good Manufacturing practices for treated bottled waters. • *E. coli* testing of the water source may be useful. • Legislation may require other testing.
Spoilage	• Must protect from recontamination to prevent unacceptable microbial growth.

Control measures

The quality and safety of bottled water must be controlled at several points. Details apply for both types of bottled water. Biocidal treatments do not apply to natural mineral water.

Source. Prevention of contamination of the underground water by environmental pollution such as faecal material is very important, particularly for waters that do not receive purification and bactericidal treatments.

Pumps, pipes, and reservoirs. These may represent sources of contamination if poorly maintained or where flow rates are low or holding times long. Significant increases in counts can be due to growth in water, e.g. in dead ends or stagnant water, or to the release of organisms from biofilms. Secondary contaminations during manual handling, e.g. during breakdowns of the line, must be prevented.

Containers. Plastic bottles and caps are normally formed at high temperatures and bacterial loads, mainly spores, are very low and not of significance. The use of recycled glass bottles represents the main hazard if cleaning is insufficient or unsatisfactory with regard to the risk of microbiological, physical, or chemical contamination.

Air. Air used to flush bottles should be filtered and the filters maintained to avoid contamination.

Bottling plant. The plant should apply Good Hygienic Practices and HACCP as part of a general process control system. When the water receives a bactericidal treatment, the effectiveness of this treatment must be controlled and monitored. Carbonated waters should have a minimum content of 250 mg/L free CO_2.

If not treated with biocides, lubricating oils may be at the origin of secondary contamination. Maintenance work performed on the lines can also be a source of contamination when necessary precautions are not taken.

Testing

Microbiological tests of bottled waters are of limited value, particularly when they are already in the distribution chain, because the results are often difficult to interpret. When bacteriological criteria have to be established, it is necessary to differentiate between waters that receive a bactericidal treatment or are carbonated from those that are not thus treated. The origin of water (e.g. water from a protected source or surface water) and the point of examination (e.g. at the source) have to be considered. Finally, the point of sampling is important (i.e. during distribution and storage, before or after bactericidal treatment).

The Code of Hygienic practice for Mineral Water (CAC, 1994) provides microbiological criteria for *E. coli* and other coliforms and fecal streptococci. The presence of such bacteria at the source or during marketing would indicate either a contamination of the water at the source or a recontamination caused by insanitary practices during bottling.

Testing non-treated bottled water for "total aerobic counts" is relevant only if the water is examined at the source and within the 12 h following bottling and kept refrigerated at 4°C. Since these aerobic microorganisms can grow in the bottle, such tests should not be applied to bottles in storage and distribution.

References

AFSSA (Agence Française de Sécurité Sanitaire des Aliments). (2002) Rapport sur les infections à protozoaires liées aux aliments et à l'eau: évaluation scientifique des risques associés à *Cryptosporidium* sp.

Anonymous (1988) Operational guidelines for the protection of drinking water supplies. Water Authorities Association.

Anonymous (1993) Council Directive on the Hygiene of Foodstuffs (93/43/EEC). Official Journal of the European Communities no. L 175, Brussels.

Anonymous (1997) Modelling the disinfection of *Cryptosporidium*. *Cryptosporidium Capsule*, **3**, 13–4.

ASHRAE (American Society of Heating, Refrigerating and Air-conditioning Engineers, Inc.). (2000) ASHRAE Standard–Minimizing the risk of legionellosis associated with building water systems. *ASHRAE Guideline* 12-2000.

Barbaras, S. (1986) Monitoring natural waters for drinking water quality. *WHO Statist. Q.*, **39**, 32–45.

Barker, J. and Brown, M.R.W. (1994) Trojan horses of the microbial world: protozoa and the survival of bacterial pathogens in the environment. *Microbiology*, **140**, 1253–9.

Bellamy, W.D., Silverman, G.P., Hendricks, D.W. and Logsdon, G.S. (1985) Removing *Giardia* cysts of the anaerobic dysentery organisms from water. *J. Am. Water Works Assoc.*, **77**, 52–60.

Bischofberger, T., Cha, S.K., Schmitt, R. and Schmidt-Lorenz, W. (1990) The bacterial flora of non-carbonated, natural mineral water from springs to reservoir and glass and plastic bottles. *Int. J. Food Microbiol.*, **11**, 51–72.

Blake, P.A., Rosenberg, M.L., Costa, J.B., *et al.* (1997a) Cholera in Portugal, 1974. I. Modes of transmission. *Am. J. Epidemiol.*, **105**, 337–43.

Blake, P.A., Rosenberg, M.L., Florencia, J., Costa, J.B., Quintino, L.D.P. and Gangarosa, E.J. (1977b) Cholera in Portugal, 1974 II. Transmission by bottled mineral water. *Am. J. Epidemiol.*, **105**, 344–8.

Bohmer, H. and Resch, K.L. (2000) Mineralwasser oder Leitungswasser? Eine systematische Literaturanalyse zur Frage der mikrobiellen Sicherheit. *Forsch. Komplementarmed. Klass. Naturheilkd.*, **7**, 5–11.

Boorman, G.A. (1999) Drinking water disinfection byproducts: review and approach to toxicity evaluation. *Environ. Health Perspect.*, **107**, 207–17.

Burge, S.H. and Hunter, P.R. (1990) The survival of enteropathogenic bacteria in bottled mineral water. *Riv. Italiana Ig.*, **50**, 401–6.

Buttiaux. R. and Boudier, A. (1960) Comportement des bactéries autotrophes dans les eaux minérales conservées en récipients hermétiquement clos. *Ann. Inst. Pasteur, Lille*, **11**, 43–54.

Cabral, D. and Fernandez, P (2002) Fungal spoilage of bottled mineral water. *Int. J. food Microbiol.*, **72**, 73–6.

Camper. A.K., LeChevallier, M.W., Broadaway, S.C. and McFeters, G.A. (1986) Bacteria associated with granular activated carbon particles in drinking water. *Appl. Env. Microbiol.*, **52**, 434–8.

Camper. A.K., LeChevallier, M.W., Broadaway, S.C. and McFeters, G.A. (1987) Operational variables and the release of colonized granular activated carbon particles in drinking water. *J. Am. Water Works Assoc.*, **79**, 70–4.

CAC (Codex Alimentarius Commission). (1993) Conversion of the Codex European regional standard for natural mineral waters to a world-wide Codex standard. CL 1993/4-NMW, FAO, Rome.

CAC (Codex Alimentarius Commission). (1994) Code of Hygienic Practice for Mineral waters, CAC/RCP 33-1985, Vol. 17-1994, FAO, Rome.

CAC (Codex Alimentarius Commission). (1996) Draft revised Standard for Natural Mineral Water. CL 1996/3-NMW,. FAO, Rome.

Casari, S. and Knøchel, S. (2002) Application of HACCP to water reuse in the food industry. *Food Control*, **13**, 315–27.

Chandy, J.P. and Angles, M.L. (2001) Determination of nutrients limiting biofilm formation and the subsequent impact on disinfectant decay. *Water Res.*, **35**, 2677–82.

Clark, R.M., Sivagenesan, M., Rice, E.W. and Chen, J. (2002) Development of a Ct equation for the inactivation of *Cryptosporidium* oocysts with ozone. *Water Res.*, **36**, 3141–9.
Colbourne, J.S. (1985) Materials usage and their effects on the microbiological quality of water supplies. *J Appl. Bacteriol., (Symp. Suppl.)*, **14**, S47–S59.
Conroy, R.M., Meegan, M.E., Joyce, T., McGuigan, K. and Barnes, J. (2001) Solar disinfection of drinking water protects against cholera in children under 6 years of age. *Arch. Dis. Child*, **85**, 293–5.
Costerton, J.W. (1984) The formation of biocide-resistant biofilms in industrial, natural and medical systems. *Dev. Ind. Microbiol.*, **25**, 363–72.
Clancy, J.L. and Fricker, C. (1998) Control of *Cryposporidium*: how effective is drinking water treatment? *Water Quality Int.*, **July/August** 1998.
Craun, G.F., Nwachuku, N., Calderon, R.L. and Craun, M.F. (2002) Outbreaks in drinking-water systems, 1991–1998. *J. Environ. Health*, **65**, 16–23, 28, 31–32.
Denyer, S.P. and Stewart, G.S.A.B. (1998) Mechanism of action of disinfectants. *Int. Biodet. Biodegrad.*, **41**, 261–8.
EC (European Commission). (1980a) Council Directive relating to the quality of water intended for human consumption (80/778/EEC). *Off. J. Eur. Commun.*, **L 229**, 11–28, Brussels.
EC (European Commission). (1980b) Council Directive of 15 July 1980 on the approximation of the laws of Member States relating to the exploitation and marketing of natural mineral waters (80/777/EEC). *Off. J. Eur. Commun.*, **L 229**, 1–10, Brussels.
EC (European Commission). (1996) Common Position (EC) No.7/96, Amending Directive 80/777/EEC on the approximation of the laws of Member States relating to the exploitation and marketing of natural mineral waters. *Off. J. Eur. Commun.*, **C 50**, 44, Brussels.
EC (European Commission). (1998) Proposal for a Council Directive concerning the quality of water intended for Human consumption. *Off. J. Eur. Commun.*, **C91**, 1, Brussels.
Eckner, K.F. (1992) Comparison of resistance to microbial contamination of conventional and modified water dispensers. *J. Food Prot.*, **55**, 627–31.
Edberg, S.C. (1998) Microbiology of treated bottled water, in *Technology of Bottled Water* (eds D.A.G. Senior and P.R. Ashurst), CRC Press, Shefield, UK.
EHEDG (European Hygienic Equipment Design Group). (2002) *The Prevention and Control of Legionella spp. (incl. Legionnaire's disease) in food factories.* Document 24. Published by CCFRA Technology Ltd., Campden, UK.
Elomari. M., Coroler, L., Izard, D., and Leclerc, H. (1995) A numerical taxonomic study of fluorescent *Pseudomonas* strains isolated from natural mineral waters. *J. Appl. Bacteriol.*, **78**, 71–81.
Emery, D.F. (1989) Water quality: problems with an essential resource. *Cereal Foods World*, **34**, 483–6.
FDA (Food and Drug Administration). (1993) *Food Code.* US Public Health Service, Food and Drug Administration, Washington, DC, 20204, USA.
FDA (Food and Drug Administration) (1995) Beverages: Bottled water: Final rule. *Federal Register,*. 21 CFR Part 103, **60**, 57075–130.
Fennel, H., James, D.B. and Morris, J. (1974) Pollution of a storage reservoir by roosting gulls. *J. Soc. Water Treat. Exam.*, **23**, 5–24.
Fleet, G.H., Heiskanen, P., Reid, I. and Buckle, K.A. (2000) Foodborne viral illness–status in Australia. *Int. J. Food Microbiol.*, **59**, 127–36.
Ford, T.E. (1999) Microbiological safety of drinking water: United States and global perspectives. *Environ. Health Perspect.*, **107** (Suppl. 1), 191–206.
Frost, J.A. (2001) Current epidemiological issues in human campylobacteriosis. *J. Appl. Microbiol. (Symp. Supplement)*, **90**, 85–95.
Fujikawa, H., Wauke, T., Kusunoki, J., Noguchi, G., Takahashi, Y., Ohta, K. and Itoh, T. (1997) Contamination of microbial foreign bodies in bottled mineral water in Tokyo, Japan. *J. Appl. Microbiol.*, **82**, 287–91.
Fujikawa, H., Aketagawa, J., Nakazato, M., Wauke, T., Tanura, H., Morozumi, S. and Itoh, T. (1999) Growth of moulds inoculated into commercial mineral water. *Lett. Appl. Microbiol.*, **28**, 211–5.
Furtado, C., Adak, G.K., Stuart, J.M., Wall, P.G., Evans, H.S. and Casemore, D.P. (1998) Outbreaks of waterborne infectious intestinal disease in England and Wales, 1992–1995. *Epidemiol. Inf.*, **121**, 109–19.
Gale, P. (1996) Developments in microbiological risk assessment models for drinking water–a short review. *J. Appl. Bacteriol.*, **81**, 403–10.
Gale, P. (2001) Developments in microbiological risk assessment for drinking water. *J. Appl. Microbiol.*, **91**, 191–205.
Gale, P., Young, C., Stanfield, G. and Oakes, D. (1998) Development of a risk assessment for BSE in the aquatic environment. *J. Appl. Microbiol.*, **84**, 467–77.
Ganesh-Kumar, C. and Anand, S.K. (1998) Significance of microbial biofilms in food industry: a review. *Int. J. Food Microbiol.*, **42**, 9–27.
Geldreich, E.E. (1972) Buffalo Lake recreational water quality: a study of bacteriological data interpretation. *Water Res.*, **6**, 913–24.
Geldreich, E.E. (1983) Microbiology of water. *J. Wat. Pollut. Control Fed.*, **55**, 869–81.
Geldreich, E.E., Nash. H.D., Spurio, D.F. and Reasoner, D.J. (1980) Bacterial dynamics in a water supply reservoir: a case study. *J. Am. Wat. Works Assoc.*, **72**, 31–40.

Geldreich, E.E., Taylor. R.H., Blamon, J.C. and Reasoner, D.J. (1985) Bacterial colonization of point-of-use water treatment devices. *J. Am. Water Works Assoc.*, **77**, 72–80.

Gerba, C.P. (1987) Transport and fate of viruses in soils: field studies, in *Human Viruses in Sediments, Sludges and Soils* (eds V.C. Rao and J.L. Melnick), CRC Press, Boca Raton. FL, pp. 141–54.

Gerba, C.P. and Rose, J.B. (1990) Viruses in source and drinking water, in *Drinking Water Microbiology* (ed G.A. McFeters), Springer-Verlag, Berlin, pp. 380–96.

Gerber, N.N. (1983) Volatile substances from *Actinomycetes*: their role in the odor pollution of water. *Wat. Sci. Technol.*, **1**, 115–25.

Ghiorse, W.C. and Wilson, J.T. (1988) Microbial ecology of the terrestrial subsurface. *Adv. Appl. Microbiol.*, **33**, 107–72.

Guillot, F. and Leclerc, H. (1993a) Biological specificity of bottled natural mineral waters: characterization by ribosomal ribonucleic acid gene restriction patterns. *J. Appl. Bacteriol.*, **75**, 292–8.

Guillot, F. and Leclerc, H. (1993b) Bacterial flora in natural mineral waters: characterization by ribosomal ribonucleic acid gene restriction patterns. *Syst. Appl. Microbiol.*, **16**, 483–93.

Havelaar, A.H. (1994) Application of HACCP to drinking water supply. *Food Control*, **5**, 145–52.

Hazen, T.C. and Toranzos, G.A. (1990) Tropical source water, in *Drinking Water Microbiology* (ed G.A. McFeters), Springer Verlag, pp. 32–53.

Hibler, C.P. and Hancock, C.M. (1990) Waterborne giardiasis, in *Drinking Water Microbiology* (ed G.A. McFeters), Springer Verlag, pp. 32–53.

Holah, J.T. (1997) Microbiological control of food industry process waters: Guidelines on the use of chlorine dioxide and bromine as alternatives to chlorine. Guideline no. 15. Campden and Chorleywood Food Research Association.

Hunter, P.R. (1991) An introduction to the biology, ecology and potential public health significance of the blue-green algae. *PHLS Microbiol. Digest*, **8**, 13–5.

Hunter, P.R. (1993) The microbiology of bottled natural mineral waters. *J. Appl. Microbiol.*, **74**, 345–52.

Hunter, P.R. and Barrell, R.A. (1999) Microbiological quality of drinking water from office water dispensers. *Commun. Dis. Public Health*, **2**, 67–8.

Huss, H.H., Ababouch, L. and Gram, L. (2003) Assessment and Management of Seafood Safety. FAO Fish Techn. Pap. 444.

Hutchinson, M. and Ridgway, J. (1977) Microbiological aspects of drinking water supplies. *Soc. Appl. Bacteriol., Symp. Ser.*, **6**, 179–218.

IBWA (International Bottled Water Association). (1995) Model Bottled Water Regulation. IBWA, Alexandria, VA.

Jacob, M. (1988) Regulation of water quality for the food industry. *Br. Food J.*, **90**, 114–6.

Jones, F. and Watkins, J. (1985) The water cycle as a source of pathogens. *J. Appl Bacteriol.* (*Symp. Suppl.*), **14**, S27–36.

Jones, C.R., Adams, M.R., Zhdan, P.A. and Chamberlain, A.H. (1999) The role of surface physicochemical properties in determining the distribution of the autochthonous microflora in mineral water bottles. *J. Appl. Microbiol.*, **86**, 917–27.

King, C.H., Shotts, E.B., Jr., Wooley, R.E. and Porter, K.G. (1988) Survival of coliform and bacterial pathogens within protozoa during chlorination. *Appl. Environ. Microbiol.*, **5**, 3023–33.

Kistemann, T., Herbst, S., Dangendorf, F. and Exner, M. (2001) GIS-based analysis of drinking-water supply structures: a module for microbial risk assessment. *Int. J. Hyg. Environ. Health*, **203**, 301–10.

Kolbel-Boelke, J., Anders, E.M. and Nehrkorn, A. (1988) Microbial communities in the saturated groundwater environment. II: Diversity of bacterial communities in a pleistocene sand aquifer and their in vitro activities. *Microbiol. Ecol.*, **16**, 31–48.

LeChevallier, M.W., Babcock, T.M. and Lee, R.G. (1987) Examination and characterization of distribution system biofilms. *Appl. Environ. Microbiol.*, **53**, 2714–24.

Leclerc, H. and da Costa, M.S. (1998) The microbiology of natural mineral waters, in *Technology of Bottled Water* (eds D.A.G. Senior and P.R. Ashurst), CRC Press, Shefield, UK.

Leclerc, H. and Moreau, A. (2002) Microbiological safety of natural mineral water. *FEMS Microbiol. Rev.*, **26**, 207–22.

Leclerc, H., Edberg, S., Pierzo, V. and Delattre, J.M. (2000) Bacteriophages as indicators of enteric viruses and public health risk in groundwaters. *J. Appl. Microbiol.*, **88**, 5–21.

Leclerc, H., Mossel, D.A., Edberg, S.C. and Struijk, C.B. (2001) Advances in the bacteriology of the coliform group: their suitability as markers of microbial water safety. *Annu. Rev. Microbiol.*, **55**, 201–34.

Leclerc, H., Schwartzbrod, L. and Dei-Cas, E. (2002) Microbial agents associated with waterborne diseases. *Crit. Rev. Microbiol.*, **28**, 371–409.

Levesque, B., Simard, P., Gauvin, D. Gingras, S., Dewailly, E. and Letarte, R. (1994) Comparison of the microbiological quality of water coolers and that of municipal water systems. *Appl. Env. Microbiol.*, **60**, 1174–8.

Levine, W.C. and Craun, G.F. (1990) Waterborne disease outbreaks, 1986–1988. *Morb. Mort. Wkly. Rep.*, **39**, 1–13.

Liyanage, L.R.J., Finch, G.R. and Belosevic, M. (1997) Effect of aqueous chlorine and oxychlorine compounds on *Cryptosporidium parvum* oocysts. *Env. Sci. Technol.*, **31**, 1992–4.

Logsdon, G.S. (1990) Microbiology and drinking water filtration, in *Drinking Water Microbiology* (ed GA. McFeters), Springer Verlag, pp. 120–46.

Logsdon, G.S. and Rice, E.W. (1985) Evaluation of sedimentation and filtration for micro-organism removal, in *Proceedings of the* 1985 *American Water Works Association, Annual Conference*, American Water Works Association, Denver, Colorado, pp. 1177–97.

MacKenzie, W.R., Hoxie, N.J., Proctor, M.E. *et al.* (1994) A massive outbreak in Milwaukee of *Cryptosporidium* infection transmitted through the public water supply. *N. Engl. J. Med.*, **331**, 161–7.

Maillard, J.Y. (2002) Bacterial target sites for biocide action. *J. Appl. Microbiol. (Symp. Suppl.)*, **92**, 16S–27S.

Marshall, M.M., Naumovitz, D., Ortega, Y. and Sterling, C.R. (1997) Waterborne protozoan pathogens. *Clin. Microbiol. Rev.*, **10**, 67–85.

Maul, A., El-Shaarawi, A.H. and Block, J.C. (1992) Bacterial distribution and sampling strategies for drinking water networks, in *Principles and Practice of Disinfection, Preservation and Sterilization* (eds A.D. Russell, W.B. Hugo and G.A.J. Ayliffe), 2nd edn, Blackwell Scientific Publications.

McCoy, W.E., Bryers, J.D., Robbins, J. and Costerton, J.W. (1991) Observations of fouling biofllm formation. *Can. J. Microbiol.*, **27**, 910–7.

McFeters, G.A., LeChevallier, M.W., Singh, A. and Kippin, J.S. (1986) Health significance and occurrence of injured bacteria in drinking water. *Water Sci. Technol.*, **18**, 227–31.

McGuigan, K.G., Joyce, T.M. and Conroy, R.M. (1999) Solar disinfection: use of sunlight to decontaminate drinking water in developping countries. *J. Med. Microbiol.*, **48**, 785–7.

Montiel, A., Ourard, J., Rigal. S. and Bousquet, G. (1987) Etude de l'origine et du mécanisme de formation de composés sapides responsables de goûts de moisi dans les eaux distribuées. *L'eau*, **82**, 3–83.

NSF International (1999) *Protocol for Equipment Verification Testing for Incativation of Microbioloogical Contaminants*, Ann Arbor, MI, USA.

Olson, B.H. and Nagy. L.A. (1984) Microbiology of potable water. *Adv. Appl. Microbiol.*, **30**, 73–132.

Oppenheimer, J. and Aieta, M. (1997) Evaluating disinfectant requirements to inactivate *Cryptosporidium parvum* in potable water supplies. *Cryptosporidum Capsule*, **2**, 7–10.

Orlandi, P.A., Chu, D.M. T., Bier, J.W. and Jackson, G.J. (2002) Parasites and the food supply. *Food Technol.*, **56**, 72–81.

Parsons, S.A. (2000) The effect of domestic ion-exchanger water softeners on the microbiological quality of drinking water. *Water Res.*, **34**, 2369–75.

Payment, P., Trudel, M. and Plante, R. (1985) Elimination of viruses and indicator bacteria at each step of treatment during preparation of drinking water at seven water treatment plants. *Appl. Env. Microbiol.*, **49**, 1418–28.

Payment, P., Gamache, F. and Paquette, G. (1989) Comparison of microbiological data from two water filtration plants and their distribution system. *Water Sci. Technol.*, **21**, 287–9.

Payment, P. (1999) Poor efficacy of residual chlorine disinfectant in drinking water to inactivate waterborne pathogens in distribution systems. *Can. J. Microbiol.*, **45**, 709–15.

Pipes, W.O. (1990) Microbiological methods and monitoring of drinking water. in *Drinking Water Microbiology* (ed GA. McFeters), Springer Verlag, pp. 428–51.

Rose, J.B. (1990) Occurrence and control of *Cryptosporidium* in drinking water, in *Drinking Water Microbiology* (ed G.A. McFeters), Springer Verlag, pp. 294–321.

Rose, J.B., Huffman, D.E. and Gennaccaru, A. (2002) Risk and control of waterborne cryptosporidiosis. *FEMS Microbiol. Rev.*, **26**, 113–23.

Ruskin, R.H., Krishna, J.H. and Beretta, G.A. (1991) Microbiological quality of selected bottled water brands in the US Virgin Islands. In Abstracts of the 91st General Meeting of the American Society for Microbiology, May 5–9, 1991, Dallas, TX, p. 317.

Russell, A.D., Hugo, W.B. and Ayliffe, G.A.J. (eds) (1999) *Principles and Practice of Disinfection, Preservation and Sterilization*, 3rd edn, Blackwell Science.

Safe Drinking Water Committee (eds) (1980) *Drinking Water and Health, Volume 2*, National Academy Press.

Schaub, S.A. and Oshiro, R.K. (2000) Public health concerns about calicivirus as water borne contaminants. *J. Infect. Dis.*, **181**, S374–80.

Schmidt, S. (2003) International standardization of water analysis: basis for comparative assessment of water quality. *Environ. Sci. Pollut. Res. Int.*, **10**, 183–7.

Schmidt-Lorenz, W. (1976) Microbiological characteristics of natural mineral water. *Ann. Inst. Super. Sanitá*, **12**, 93–112.

Schoenen, D. and Schöler, M. (1985) *Drinking Water Materials: Field Observations and Methods of Investigation*, Ellis Horwood Ltd., Chichester, UK.

Slezak, L.A. and Sims, R.C. (1984) The application and effectiveness of slow sand filtration in the United States. *J. Am. Water Works Assoc.*, **76**, 38–43.

Sobsey, M.D. 1989. Inactivation of health-related microorganisms in water by disinfection processes. *Wat. Sci. Technol.*, **21**, 179–95.

Stamm, J.M., Engelhard, W.E. and Parsons, J.E. (1969) Microbiological study of water-softener resins. *Appl. Microbiol.*, **18**, 376–86.

Steiner, T.S., Thielman, N.M. and Guerrant, R.L. (1997) Protozoal agents: what are the dangers for the public water supply? *Annu. Rev. Med.*, **48**, 329–40.

Straub, T.M. and Chandler, D.P. (2003) Towards a unified system for detecting waterborne pathogens. *J. Microbiol. Methods*, **53**, 185–97.

Szewzyk, U., Szewzyk, R., Manz, W. and Schleifer, K.H. (2000) Microbiological safety of drinking water. *Annu. Rev. Microbiol.*, **54**, 81–127.

Terplan, G. and Bierl, J. (1980) Technologisch-mikrobiologische Anforderungen an Trink- und Brauchwasser in Lebensmittelbetrieben. *Swiss Food*, **2**, 28–32.

Theron, J. and Cloete, T.E. (2002) Emerging waterborne infections: contributing factors, agents and detection tools. *Crit. Rev. Microbiol.*, **28**, 1–26.

Tillett, H.E., de Louvois, J. and Wall, P.G. (1998) Surveillance of outbreaks of waterborne infectious disease: categorising levels of evidence. *Epidemiol. Inf.*, **120**, 37–42.
Tobin, R.S., Smith, D.K. and Lindsay, J.A. (1981) Effects of activated carbon and bacteriostatic filters on microbiological quality of drinking water. *Appl. Env. Microbiol.*, **41**, 646–51.
Walker, J.T. and Percival, S.L. (2000) Control of biofouling indrinking water systems, in *Industrial Biofouling* (eds J. Walker, S. Surman and J. Jass), John Wiley & Son Ltd.
Warburton, D.W. and Austin, J.W. (2000) Bottled water, in *The Microbiological Safety and Quality of Food* (eds B.M. Lund, T.C. Baird-Parker and G.W. Gould), *Volume 1*, An Aspen Publication.
Warburton, D., Harrison, B., Crawford, C., Foster, R., Fox, C., Gour, L. and Krol, P. (1998) A further review of the microbiological quality of bottled water sold in Canada: 1992–1997 survey results. *Int. J. Food Microbiol.*, **39**, 221–6.
White, G.C. (1992) *Handbook of Chlorination and Alternative Disinfectants*, 3rd edn, Van Nostrand Reinhold, New York.
Whitfield, F.B. (1998) Microbiology of food taints. *Int. J. Food Sci. Technol.*, **33**, 31–51.
Wilson, J.T., McNabb, J.F., Balkwill, D.L. and Ghiorse, W:C. (1983) Enumeration and characterization of bacteria indigenous to shallow water table aquifer. *Ground Water*, **21**, 134–142.
WHO (World Health Organization). (1993) Recommendations, *Guidelines for Drinking Water Quality, Volume 1*, WHO, Geneva.
WHO (World Health Organization). (1996) Health criteria and other supporting information, *Guidelines for Drinking Water Quality, Volume 2*, WHO, Geneva.
Wright, M.S. and Collins, P.A. (1997) Waterborne transmission of *Cryptosporidium*, *Cyclospora* and *Giardia*. *Clin. Lab. Sci.*, **10**, 287–90.

15 Eggs and egg products

I Introduction

A Definitions

This chapter encompasses the microbiology of avian eggs and egg products, primarily from the domestic chicken. Eggs from other birds such as ducks, turkeys, geese, guinea fowl, quail, and ostrich appear in international commerce at a lower tonnage.

B Important properties

Biologically, the function of the egg is reproduction. It must nourish the embryo until it develops into a chick and hatches (21 days for the domestic chicken), and must protect it against microbial entry from the environment. The different components provide multiple protective barriers (Board *et al.*, 1994).

The egg consists of (1) the cuticle, a largely proteinaceous coating on the exterior of the shell; (2) the shell, which is mostly calcium carbonate; (3) the outer coarse membrane; (4) the inner fine membrane; (5) the outer thin white (albumen layer #1); (6) the thick white (albumen layer #2); (7) the inner thin white (albumen layer #3); (8) the chaliziferous layer (albumen layer #4), which culminates in the chalaza cords which anchor the yolk in the centre of the egg; (9) the vitelline membrane that encloses the yolk; and (10) the yolk (Figure 15.1). Shortly after the egg is laid, evaporation of water reduces the volume of the contents, and an air chamber (sac) forms between the inner and outer membranes at the blunt end of the egg. The total time needed for the formation of an egg from the ovary to laying is ~24 h. Eggs from different fowl vary with respect to the proportions of the different components.

Tables 15.1 and 15.2 give the proximate analysis of hens' eggs and the nutrient content of different raw egg products. Fat-soluble vitamins (A, D, E, and K) are present in the yolk; water-soluble vitamins (B complex) are present either in the white, the yolk or both. Approximately equal amounts of trace minerals are present in the yolk and white, sometimes in combination with proteins and lipids.

The egg shell is a remarkable natural package. The yolk is as rapidly perishable as milk, yet the fragile shell, if undamaged and dry, will usually keep an uncontaminated egg edible for many months, even when stored at room temperature. Before contamination and spoilage can occur, the responsible microorganisms encounter several efficient barriers: (1) the physical barrier to penetration provided by the shell and its membranes and (2) the antimicrobial components of the egg white that make it an antagonistic medium for microbial growth. However, the transovarian transmission of *Salmonella* spp. allows the organism to avoid the barriers of the shell and associated membranes. For additional information on the structure and composition of the eggshell, see reviews by Solomon *et al.* (1994) and Sparks (1994).

Resistance to penetration. The relative importance of the egg structures contributing to retarding penetration are, in decreasing order of importance: cuticle>inner membrane>shell>outer membrane (Lifshitz *et al.*, 1964). Obviously, cracks that penetrate the inner membrane allow microorganisms to bypass these barriers and permit easy entry by spoilage and pathogenic bacteria. Therefore, most countries reject as inedible eggs whose whites are leaking to the outside surface. Similarly, if the shell is very dirty, the microbial challenge is greater and microorganisms are likely to penetrate sooner and in greater numbers.

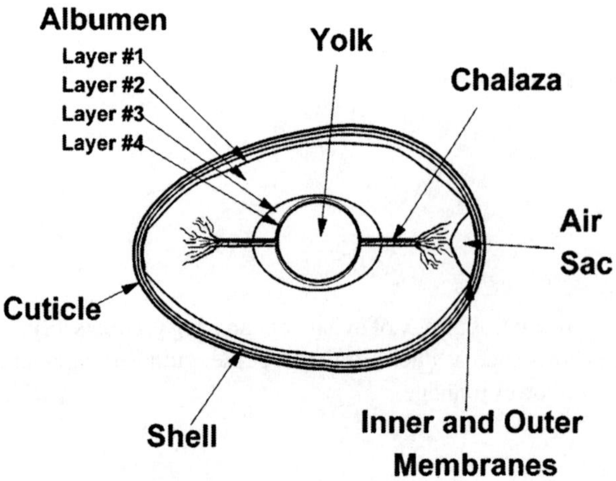

Figure 15.1 Structure of an avian egg.

Table 15.1 Composition of hen's egg

Egg component	Percentage
Entire egg	
Water	73.6
Solids	26.4
Organic matter	25.6
Proteins	12.8
Lipids	11.8
Carbohydrates	1.0
Inorganic matter	0.8
Fractions in the yolk	
Water	48.0
Solids	51.3
Organic matter	50.2
Proteins	16.6
Lipids	32.6
Carbohydrates	1.0
Inorganic matter	1.1
Fractions in the albumen (white)	
Water	87.9
Solids	12.2
Organic matter	11.6
Proteins	10.6
Lipids	Trace
Carbohydrates	0.9
Inorganic matter	0.6
Proteins in the albumen[a]	
Ovalbumin	5.4
Conalbumin	13.0
Ovomucoid	11.0
Lysozyme	3.5
Ovomycin	1.5
Flavoprotein–Apoprotein	0.8
Ovoinhibitor	0.1
Avidin	0.05
Globulin and others	8.0

Adapted from Board (1969).
[a]Values are expressed as percentage of egg white solids.

Table 15.2 The nutrient composition of the edible portion of fresh raw hen's egg and egg components (USDA, 1991)

	Whole egg	Yolk	White	Sugared yolk	Salted yolk
Proximate (g/100 g)					
Water	75.33	48.81	87.81	49.51	49.52
Protein	12.49	16.76	10.52	14.19	14.19
Lipid, total	10.02	30.87	0.00	23.41	23.40
Carbohydrate, total	1.22	1.78	1.03	11.47	1.48
Ash	0.94	1.77	0.64	1.42	11.40
Minerals (mg/100 g)					
Calcium	49	137	6	105	109
Iron	1.4	3.5	0.03	2.7	2.8
Magnesium	10	9	11	8	8
Phosphorous	178	488	13	374	374
Potassium	121	94	143	90	91
Sodium	126	43	164	56	3932
Zinc	1.1	3.1	<0.1	2.4	2.4
Copper	<0.1	<0.1	<0.1	<0.1	<0.1
Manganese	<0.1	<0.1	<0.1	<0.1	0.1
Vitamins					
Ascorbic acid (mg/100 g)	0.00	0.00	0.00	0.00	0.00
Thiamin (mg/100 g)	0.06	0.17	0.01	0.13	0.13
Riboflavin (mg/100 g)	0.51	0.64	0.45	0.55	0.55
Niacin (mg/100 g)	0.07	0.02	0.09	0.03	0.03
Pantothenic acid (mg/100 g)	1.26	3.81	0.12	2.89	2.89
B_6 (mg/100 g)	0.14	0.39	<0.01	0.30	0.30
Folacin (μg/100 g)	47	146	3	110	110
B_{12} (μg/100 g)	1.0	3.1	0.2	2.4	2.4
Vitamin A (IU)	635	1945	–	1475	1475

The outer surface of the shell is covered with the cuticle, a thin stratum of minute glycoprotein spheres, which makes the shell resistant to the entry of water. The cuticle extends far down into the pores of guinea fowl eggs, but only a short distance into the pores of chicken eggs. With duck eggs, the cuticle tends only to cap the pore canals. If the cuticle is damaged, there is a greater susceptibility to microbial entry (Board and Halls, 1973; Seviour and Board, 1972; Bruce and Drysdale, 1994). The amount of infiltration due to damage is related directly to the extent to which the pores are no longer plugged with cuticle. If dirty eggs are cleaned with abrasive substances the cuticle is damaged, but it is fairly resistant to water, detergents, or gentle rubbing with a cloth. Even localized damage may permit bacterial entry through a few pores. The protection offered by the undamaged cuticle generally lasts at least 4 days, after which it begins to fail, presumably owing to cracks that develop as the cuticle dries (Baker, 1974). Eggs without cuticle, or eggs treated experimentally with chemicals to remove cuticle, spoil much faster than normal eggs (Vadehra *et al.*, 1970). The incidence of eggs contaminated with *Pseudomonas* spp. and Enterobacteriaceae tends to increase with flock age (Bruce and Johnson, 1978), which may partially reflect the increased incidence of eggs with poor cuticles with flock age (Bruce and Drysdale, 1994).

The egg shell has numerous pores; a hen's eggs has 6000–10000 (Bruce and Drysdale, 1994). The number of pores per egg tends to increase with the age of the flock (Rahn *et al.*, 1981). Shells with high specific gravity (i.e. fewer pores) have been reported to offer greater resistance to bacterial penetration. Table 15.3 provides an example, showing that spoilage began in 3 days when shell specific gravity was low, but not until 10–12 days when the specific gravity was high. The percentages of eggs spoiled after 8 weeks at 10°C showed similar differences (Table 15.4). The penetration by *Salmonella* was also more rapid for low specific gravity shells (Table 15.5). However, other investigations found that the penetration

Table 15.3 The effect of eggshell specific gravity and time of bacterial challenge on the time of first fluorescent spoilage of eggs (Sauter and Peterson, 1969.)

	Specific gravity of the shell		
Challenge time (min)	1.070	1.077	1.085
1	8[a]	10	12
3	4	10	12
5	3	11	12

[a]Time in days.

Table 15.4 Effect of eggshell specific gravity and time of bacterial challenge on the total incidence of *Pseudomonas* contamination of eggs after 8 weeks of storage (Sauter and Peterson, 1969.)

	Specific gravity		
Challenge time (min)	1.070	1.077	1.085
1	69.2[a]	43.3	21.5
3	77.5	54.2	26.7
5	84.2	75.8	36.7

[a]Percent infected eggs.

Table 15.5 Proportion of eggs of different shell qualities penetrated by various salmonellae in 24 h (Sauter and Peterson, 1974.)

	Specific gravity of the shell		
Salmonella spp.	1.070	1.080	1.090
S. Anatum	19.4[a]	7.5	3.8
S. Brandenburg	68.1	17.1	7.2
S. Typhimurium	82.1	48.7	21.2
Average of 12 *Salmonella* spp.	47.5	21.4	10.0

[a]Percent infected eggs.

by *Pseudomonas fluorescens* was not dependent on shell porosity, but was affected by the age of the egg and the number of bacteria on the shell surface (Brooks, 1960; Hartung and Stadelman, 1963; Sparks and Board, 1984). Stress in the laying hen can cause oviduct damage, leading to ultrastructural defects in the eggshell and increased susceptibility to bacterial invasion (Nascimento and Solomon, 1991).

The outermost of the two shell membranes is porous and does not provide a barrier to bacterial entry. The inner membrane usually delays entry for a few days due to its fine structure (Elliott, 1954; Garibaldi and Stokes, 1958; Gillespie and Scott, 1950; Board, 1965a). The superior protection provided by the inner membrane is not due to its thickness and weight, as it is one-sixth as heavy and one-third as thick as the outer membrane (Lifshitz and Baker, 1964), but to the lack of pores in the inner membrane. Although some motile bacteria can apparently penetrate by wriggling through the membranes tightly overlapping fibers (Baker, 1974), electron microscopic studies indicate that the majority of bacteria penetrate the membrane through the albuminous cementing matrix. The keratin core of the membrane and its polysaccharide mantle are unaffected. Zones of hydrolysis surrounding bacteria observed in the membranes support the hypothesis that penetration of the membrane is enzyme-mediated (Stokes *et al.*, 1956; Brown *et al.*, 1965). Using eggshell membrane removed from broiler hatching eggs at approximately monthly intervals through the productive life of a commercial flock in a laboratory set-up. Berrang *et al.* (1999) studied membrane penetration by *Salmonella* Typhimurium. Although penetration by *S.* Typhimurium cells was evident, there was no clear correlation with specific ultrastructural elements of the membrane.

Wet and dirty shells combined with a fall in temperature facilitate entry of bacteria. Temperature reduction causes the air sac to contract, resulting in a negative pressure. The more rapid the temperature fall, the greater the pressure differential between the interior and exterior of the egg. As the pressure differential across the shell equalizes, water and bacteria are aspirated through the shell and trapped at the surface of the inner membrane. This becomes more pronounced as the egg ages and the air sac increases in size and is one of several factors that make stored eggs more susceptible to penetration.

Table 15.6 Antimicrobial factors in the albumen of the hen's egg

Component	Activity
Lysozyme (muramidase)	Lysis of cell walls of Gram-positive bacteria; flocculation of bacterial cells; hydrolysis of β-1,4-glycosidic bonds
Conalbumin	Chelation of iron, copper, and zinc, especially at high pH; chelation of cations
pH 9.1–9.6	Provides unsuitable environment for growth of many microorganisms; enhances chelating activity of conalbumin
Avidin	Binds biotin, making it unavailable for bacteria that require it.
Low non-protein nitrogen	Nutritionally fastidious organisms cannot grow
Ovoinhibitor	Inhibits fungal proteases
Ovomucoid	Inhibits trypsin, but does not affect growth of Gram-negative bacteria
Uncharacterized proteins	Inhibit trypsin and chymotrypsin; combine with vitamin B_6; chelate calcium; inhibit ficin and papain

Adapted from Garabaldi (1960) and Board (1969).

Antimicrobial factors in the albumen (white). The albumen kills or prevents the growth of a wide variety of microorganisms, whereas the yolk or a mixture of yolk and albumen does not (Haines, 1939; Brooks, 1960). Table 15.6 summarizes the principal factors in egg white that help to control the growth of bacteria. Lysozyme, conalbumin, and the alkaline pH are the most important. All the adverse factors listed apply to the thick white; only the high pH applies to the two layers of thin white (Baker, 1974).

Lysozyme, first named in 1909 because it "lysed" bacterial cells, is a muramidase and attacks the murein layer or the murein sacculus of bacterial cells. Gram-positive bacteria are especially sensitive to lysozyme; Gram-negative are less so, partly because the murein layer is protected by the outer cell membrane. The alkaline conditions in the egg sensitize cells, making them more susceptible to lysis. The activity of lysozyme is reduced when yolk is mixed with the white (Galyean *et al.*, 1972).

Conalbumin is important because it sequesters (chelates) metal ions, particularly iron, copper, and zinc, making them unavailable to bacteria. Many bacteria are unable to grow in the presence of conalbumin. Gram-positive bacteria are generally more susceptible to conalbumin than Gram-negative. Bacteria able to grow in the presence of conalbumin usually have an extended lag phase and a decreased growth rate. The ability to grow appears to be related to the bacterium having an active system for acquiring essential trace minerals (i.e. siderophores). For example, pseudomonads growing in egg white often produce a mixture of green fluorescent chelators, collectively termed "pyoverdine." This material has a high affinity for metal ions essential for growth of *Pseudomonas* spp., and competes successfully with conalbumin. Unlike conalbumin, pyoverdine releases the metals to the bacterial cell (Elliott, 1954; Elliott *et al.*, 1964; Garibaldi, 1960, 1970). Pyoverdine is associated with, or may be identical to, the fluorescent hydroxymate transport compounds that neutralize the effect of conalbumin (Garibaldi, 1970). Salmonellae produce phenolate compounds that act similarly, permitting them to penetrate and multiply (Garibaldi, 1970). Experimentally added metal salts of iron, aluminum, copper, manganese, and zinc will saturate the binding capacity of conalbumin, the excess being available for microbial growth (Sauter and Peterson, 1969).

When freshly laid, hens' eggs have a pH in the range of 7.6–7.8. The loss of carbon dioxide from the egg to the atmosphere results in the pH of the egg white rising to 9.1–9.6 after 1–3 days of storage at room temperature. Most bacteria do not grow well at this alkaline pH value, which also enhances the chelating activity of conalbumin.

C Types of products

Most eggs are marketed and consumed as shell eggs. For example, in the United States ~70% of eggs reach the consumers as shell eggs (AEB, 2003). However, other egg products are an increasingly important part of commerce. For commercial use in various food and foods service operations, eggs are broken from their shells and may then be mixed whole or separated into whites and yolks. They are sometimes sugared, salted, or mixed with other ingredients when manufacturing various foods. Commercial liquid egg products are normally pasteurized. They can also be de-sugared by enzymatic or microbial action and dried for use as a food ingredient. Liquid egg products are used as ingredients of a broad range of processed food products, particularly bakery products, confectionery, drinks, special diets, infant foods, sauces, mayonnaise, and noodles. Co-extruded, cooked egg products (e.g. "long eggs") are an example of the numerous advanced food technologies used to manufacture egg products.

A variety of traditional specialty egg products are consumed in the Far East, for example pidan, the "thousand-year egg" (Su and Lin, 1993). These traditional alkalized eggs are produced using eggs from various birds and are preserved by immersion in NaOH and NaCl.

II Initial microflora

Eggs become contaminated by two primary means, transovarian or trans-shell infection (Bruce and Drysdale, 1994). Freshly laid eggs may be contaminated through the oviduct, and the presence of certain bacterial species can be indicative of an infected bird. Trans-shell infection involves the initial contamination of the egg surface, followed by the subsequent penetration by the microorganism into the albumen or in some cases directly into the yolk. The surface of the newly formed egg is contaminated with a variety of enteric microorganisms because of the anatomy of the bird; the intestinal, urinary, and reproductive tracts of fowl share a common orifice. The surface of eggs also become contaminated by microorganisms from the environment where it is laid.

A Transovarian transmission

Common spoilage microorganisms are not associated with transmission via the oviduct (Miller and Crawford, 1953; Jordan, 1956; Philbrook *et al.*, 1960; Board *et al.*, 1964). However, the transovarian transmission of salmonellae can occur and has significant consequences for animal health (e.g. *Salmonella enterica* serovar Pullorum, *S. enterica* serovar Gallinarum) or public health (e.g. *Salmonella enterica* serovar Enteritidis, shortened to *S.* Enteritidis). It is worth noting that these three species are phylogenetically closely related (Stanley and Baquar, 1994). *Salmonella enterica* serovar Typhimurium, *S. enterica* serovar Heidelberg, *S.* Typhimurium var. *copenhagen*, and other serovars have also been isolated from the ovaries of laying hens (Snoeyenbos *et al.*, 1969; Barnhart *et al.*, 1991). Between 1930 and 1946, *S.* Enteritidis in duck eggs caused epidemics of human salmonellosis in Europe, however, it was not proven that this involved transovarian transmission (Scott, 1930; Humphrey, 1994b).

Beginning in the mid-1980s, *S.* Enteritidis was increasingly isolated from poultry and poultry eggs (Dreesen *et al.*, 1992; Cogan and Humphrey, 2003). This was accompanied by a sharp increase in human salmonellosis in several European, North American, and South American countries that was epidemiologically associated to the consumption of foods, particularly those containing raw or undercooked egg (Hopper and Mawer, 1988; St. Louis *et al.*, 1988; Humphrey, 1990a; Rodriguez *et al.*, 1990; Duguid and North, 1991; CDC, 1992; ACMSF, 1993; Binkin *et al.*, 1993; Caffer and Eiguer, 1994; Fantasia and Filetici, 1994; Glosnicka and Kunikowska, 1994; Mishu *et al.*, 1994; Morse *et al.*, 1994). Although there was a simultaneous increase in *S.* Enteritidis cases on both continents, the European cases were largely associated with phage type 4 (PT4), whereas the North American cases involved other phage types (Cowden *et al.*, 1989; Khakhria *et al.*, 1991; Humphrey, 1994b; Angulo and Swerdlow, 1999; Cogan and Humphrey, 2003). Investigation of the contamination of chicken farms with salmonellae by sampling feed and eggs from commercial layer farms in eastern Japan between

1993 and 1998, revealed *S*. Enteritidis amongst the isolates recovered from eggs that were linked to the occurrence of salmonellae in feed (Shirota *et al.*, 2001). Following the sharp increase in *S*. Enteritidis infections noted in the USA between 1976 and 1996, increasing from 0.6 to 3.6 *S*. Enteritidis isolates per 100 000 inhabitants per annum, respectively, a marked decrease was noted by 1998, when isolation frequency had fallen again to 2.2 *S*. Enteritidis isolates per 100 000 inhabitants per annum (CDC, 2000). For the USA, costs associated with human salmonellosis due to *S*. Enteritidis were estimated to range from $150 million to $870 million annually (FSIS, 1998a).

Salmonella Typhimurium has traditionally been the predominant species of salmonellae implicated in egg-associated outbreaks. The prevalence of *S*. Typhimurium in hens' eggs is low (Philbrook *et al.*, 1960; Chapman *et al.*, 1988). Since the late 1970s, *S*. Enteritidis has emerged as the major cause of salmonellosis in the United States, Europe, and South America. A significant increase in the incidence of *S*. Enteritidis infection has also been reported in Yugoslavia, Finland, Sweden, Norway, and the UK. For example, from 1981 and 1988, the annual number of human isolates of *S*. Enteritidis in England increased from 392 to 12 522. In 1987, none of the six egg-associated outbreaks of salmonellosis investigated in England was caused by *S*. Enteritidis PT4. In 1988, 19 of 34 egg-associated outbreaks were caused by that serovar. In 1990, it was reported that in the northeastern United States the incidence of *S*. Enteritidis had increased >6-fold since 1976, particularly in the summer months (Rodriquez *et al.*, 1990). Isolation rates for the same serovar also increased in the mid-Atlantic and south-Atlantic regions, and *S*. Enteritidis was the second most common serovar reported.

Hen eggs became a principal source of the pathogen. The emergence of *S*. Enteritidis as the leading cause of human salmonellosis in many countries is attributed to this serovar's unusual ability to colonize the ovarian tissue of hens and to be present in the contents of intact shell eggs (Cogan and Humphrey, 2003). Most food-borne *S*. Enteritidis infection is associated with the consumption of raw eggs and foods containing raw eggs such as homemade egg nog, cookie batter, homemade ice cream, mayonnaise, Caesar salad dressing, and Hollandaise sauce. In fact, 77–82% of *S*. Enteritidis outbreaks were associated with Grade A shell eggs (Mishu *et al.*, 1994) or egg-containing foods (St. Louis *et al.*, 1988). Undercooked eggs and products containing undercooked eggs such as soft custards, French toast, soft fried eggs, and poached eggs, are also significant sources of *S*. Enteritidis. According to a FDA report (1997), between 128 000 and 640 000 *Salmonella* infections are annually associated with the consumption of *S*. Enteritidis-contaminated eggs. The CDC estimated that 75% of all outbreaks of salmonellosis were due to consumption of raw or inadequately cooked Grade A whole shell eggs (FDA, 2003).

These findings aroused large scale investigations in both the United Kingdom and United States. In the UK, *S*. Enteritidis infections decreased significantly, since the 1990s, with the introduction of measures to control the disease in laying hens, in particular programs of vaccination (Cogan and Humphrey, 2003). The decline noted in *S*. Enteritidis related human illness between 1996 and 1998 in the US has been tentatively attributed to the implementation of egg quality assurance programs on farms and improved egg handling (CDC, 2000). Invasion and infection of the ovaries can give rise to the transfer of *S*. Enteritidis to the follicle (yolk), whereas infection of the oviduct leads to the deposition of microorganisms in the albumen. Experimental infections of hens with *S*. Enteritidis resulted in the production of contaminated eggs (Humphrey *et al.*, 1989b, 1989c; Gast and Beard, 1990a; Gast, 1994). Ovarian infection of hens appears to have become more common, leading to eggs infected either in the ovary or during transit of the ovum from the ovary through the infundibulum and oviduct and before the shell and shell membranes are formed (Gast and Beard, 1990b; Barnhart *et al.*, 1991; Clay and Board, 1991; Baskerville *et al.*, 1992).

The principal site for in-shell contamination appears to be the albumen. Depending on the site of the oviduct infection, *S*. Enteritidis can be found at specific sites within the egg white. The closer the site of the initial infection to the ovaries, the closer the salmonellae will be located to the yolk. Although it might be assumed that ovarian infection would cause deformities or dysfunction in the ovaries, and that such infections would be self-limiting in that the animal would quickly cease laying, this has generally not been the case. Acute *S*. Enteritidis infections are typically limited to chicks less than 1-month-old, but in chicks of that age can cause high rates of mortality (up to 20%) (Lister, 1988; O'Brien, 1988; Suzuki, 1994). However, while certain strains reduce productivity (Shivaprasad *et al*., 1990; Humphrey *et al*., 1991b, 1991c; Gast, 1994), experience in the United Kingdom and the United States indicates that egg production in flocks infected with *S*. Enteritidis may not be adversely affected (Hopper and Mawer, 1988; Cooper *et al*., 1989; ACMSF, 1993; Gast, 1994). This is despite the fact that *S*. Enteritidis strains isolated from European eggs are predominately phage type 4 (de Louvois, 1993b), which had previously been associated with avian disease. Sub-clinical infections of laying birds do not cause significant reductions in fertility.

In artificially contaminated shell eggs, the rate and extent of infection were influenced by the size of inoculum, the site of contamination relative to yolk movement, and the presence of iron in the inoculum (Clay and Board, 1991). Iron, which is normally scarce in the albumen (Table 15.2), is retained on the mantles on the fibers of the shell membrane and only trace amounts are required to induce microbial growth and contamination of the underlying albumen. This can occur when unsatisfactory methods of washing are used. It has also been shown that extracts of hens' feces have sufficient iron to promote the growth in albumen *in vitro* (Humphrey, 1994a). Migration of *S*. Enteritidis from the albumen to the yolk occurred in artificially contaminated eggs within a few days of inoculation (Braun and Fehlhaber, 1995). The rate of migration was positively related to the level of contamination, the storage temperature, and the age of the eggs. Another investigation found that *S*. Enteritidis grew more slowly when inoculated into the yolks of eggs from sero-positive hens compared with eggs from uninfected hens (Bradshaw *et al*., 1990), and it was hypothesized that the secretion of antibodies into the yolk may be one of the egg's multiple antimicrobial barriers (Rose *et al*., 1974).

Salmonella Enteritidis infection in a layer flock does not result in 100% of the birds being positive for the pathogen. The percentage of excreters may vary from under 0.6% to 30%, and the prevalence of egg-shell contamination varies widely in flocks infected with *S*. Enteritidis (Table 15.7). Examining 2 525 eggs produced by 16 free-range hens over a period of 242 days, Muller *et al*. (1994) detected only 19 eggs that were positive for *S*. Enteritidis. Following the total egg production of 35 hens, Humphrey *et al*. (1989b) found 1% of 1 119 eggs were positive for *S*. Enteritidis, with the positive eggs being laid by only 10 of the 35 birds. When both shells and egg contents were examined, 1.1% of shells and 1.9% of egg contents were positive. Where yolk and albumen were examined separately, the microorganism was isolated only from albumen in 12 of 16 eggs. This is in agreement with studies of artificially

Table 15.7 The rate of detection of *Salmonella* Enteritidis positive eggs for flocks known to be infected with the pathogen

No. of eggs examined	% of *S*. Enteritidis positive eggs	Reference
372	1.1	Perales and Audicana (1989)
998	0.5	Perales and Audicana (1989)
68	7.4	Humphrey *et al*. (1989a)
1119	1.0	Humphrey *et al*. (1989b)
32	19.0	Humphrey *et al*. (1989b)
70	2.8	Buchner *et al*. (1992)
349	1.4	Buchner *et al*. (1992)
630	0.0	Buchner *et al*. (1992)
1070	0.3	Buchner *et al*. (1992)
309	1.3	Buchner *et al*. (1992)
30	3.3	Buchner *et al*. (1992)
16560	0.06	Poppe *et al*. (1992)

infected flocks (Gast, 1994; Humphrey, 1994a). The pathogen was isolated with high frequency from albumen, while all yolk content samples were negative. When tissues from 580 birds from 7 flocks were examined, 4.5% from two flocks were *S.* Enteritidis positive (Poppe *et al.*, 1992). Subsequent examination of eggs from these flocks indicated that the rate of contamination was <0.06% (Poppe, 1994). Comparison of *S.* Enteritidis phage types isolated from poultry and clinical samples in Canada indicated that both were predominately PT8, PT13, and PT13a (Poppe, 1994), and that 97% of the PT8 and PT13 isolates contained a 36 MDa plasmid that hybridized with the 60 MDa virulence from *S.* Typhimurium (Poppe *et al.*, 1989). However, it is not clear what significance this plasmid has as a virulence factor that affects the pathogenicity of *S. enteritidis* isolates in humans and birds (Suzuki, 1994).

The rate of *S.* Enteritidis contamination in shell eggs at retail sale is very low. For example, a survey of eggs destined to British retail markets indicated *S.* Enteritidis contamination of 0.04–0.11%, and of contamination with *S.* Enteritidis PT4 of 0.03–0.08% (de Louvois, 1993a). The overall contamination for all salmonellae was 0.15–0.27%. One reason for the low *S.* Enteritidis contamination in retail eggs is the co-mingling of the few positive eggs with large numbers of eggs from non-infected flocks. In the United States, it is estimated that one in 20 000 eggs is contaminated with *S.* Enteritidis; and that ~70% of these contaminated eggs would go to the retail market (FSIS, 1998a; Whiting *et al.*, 2000).

While there have been advances in detection methods (Barrow, 1994; Helmuth and Schroeter, 1994; McClelland and Pinder, 1994; Thorns *et al.*, 1994; van der Zee, 1994; Gast and Holt, 1995; Holt *et al.*, 1995; McElroy *et al.*, 1995; Wang *et al.*, 1995), the low rates of pathogen positive eggs from infected flocks make routine microbiological examination of eggs impractical. There have been instances where *S.* Enteritidis has not been isolated from the shells of eggs whose contents contained *S.* Enteritidis PT4 (Mawer *et al.*, 1989). In the majority of instances, the initial number of salmonellae in contaminated eggs is low, e.g. 10–20 cfu/egg according to Humphrey *et al.* (1991a), although higher levels have been reported (Humphrey, 1994b; Muller *et al.*, 1994). An exception is a report of a clean, intact egg that had 10^7 cfu/g (Salvat *et al.*, 1991). Antimicrobial factors in the albumen may complicate the detection of salmonellae, leading to false negative results (Humphrey, 1994b).

Feeds do not appear to be a major source for *S.* Enteritidis infections of flocks (ACMSF, 1993; 1996 at Annex E, pp. 131–135), although historically some feed components were associated with other serovars (Williams, 1981). Recommendations for breaking the *Salmonella* cycle include the acquisition of birds from parent flocks demonstrably free from salmonellae (Cox *et al.*, 1990), fumigation or sanitation of hatchery eggs and nesting materials, use of competitive exclusion (Seuna and Nurmi, 1979), the use of *Salmonella*-free feeds (Marthedal, 1973), and use of vaccines. Factors that impact the degree of stress encountered by the birds also appears to influence the proportion of eggs that are contaminated with *S.* Enteritidis (Suzuki, 1994).

According to the ACMSF (2001), competitive exclusion techniques and vaccination can quite successfully be employed to reduce *Salmonella* infection of chicks, as well as shedding and transmission of the pathogen. Competitive exclusion aims to increase the resistance of chicks to colonization of the gut by salmonellae, thereby helping to limit the organism multiplying in chicks and spreading and persisting in flocks. Competitive exclusion is best applied as a part of a wider *Salmonella* control program. Vaccination substantially reduces shedding of *S.* Enteritidis in feces of vaccinated birds, which is likely to reduce shell surface contamination and internal egg contamination. Vaccination could effectively and rapidly eliminate persistent salmonellae from broiler breeding sites, and, from UK experience, appear effective in egg production sites (vaccinating layer hens). Though not always providing total elimination of salmonellae, vaccination reduced prevalence of infection in flocks and the level of salmonellae in the environment, complementing the effect of other pathogen control measures (disinfection, rodent control, etc). In the United Kingdom, regulations mandate testing of breeder layer and broiler flocks (ACMSF, 2001).

A quantitative risk assessment carried out under the auspices of FAO and WHO (FAO/WHO, 2002) evaluated strategies for risk reduction associated with *S.* Enteritidis in eggs. Based on the data and assumptions used to develop the risk model, control of prevalence, either proportion of flocks infected or proportion of hens within flocks that are infected, has a direct effect in reducing probability of illness per egg serving. However, reduction of shell egg storage times (to 7 days or less) and reduction of storage temperatures (to 7°C or less) had greater effects in reducing risk of illness. On the whole, it was predicted that egg storage times and temperatures can disproportionately influence the risk of illness per serving. Numbers of organisms initially present in eggs at the time of lay seems less important, presumably because the risk model predicted a larger impact associated with the potential multiplication of *S.* Enteritidis in eggs and egg products, underlying the need to ensure adequate refrigeration at all stages to minimize any growth. Testing flocks, combined with diversion of eggs from positive flocks to pasteurized egg product rather than the shell egg market, was predicted to reduce public health risk substantially. Although this introduced more contaminated eggs to liquid egg product, by controlling the length of time and adequate refrigeration before pasteurization, diversion of eggs from test positive flocks also reduced the apparent risk from egg products. Vaccination may reduce risk of illness by ~75%, but is typically less effective because producers would only vaccinate test-positive flock.

Using data from a National Salmonella Control Program carried out in Finland, a probabilistic model was build to study the transmission of salmonellae in the primary broiler production chain and used to quantify the effect of interventions used in the program such as eliminating *Salmonella* positive flocks (Ranta and Maijala, 2002).

B Contamination in the cloacae

The surface of egg-shells can become contaminated with virtually any microorganism excreted by the birds. For salmonellae other than *S.* Enteritidis, this is the most important source of contamination. Many different serovars of the genus *Salmonella* can be isolated from laying flocks. Ebel *et al.* (1992) examined 23 431 pooled caecal samples from spent hens collected from 406 layer houses to assess the prevalence of total salmonellae and *S.* Enteritidis. Salmonellae were recovered from 24% of the pooled samples, and *S.* Enteritidis from 3%. The overall prevalence of *Salmonella* positive flocks was 86%. Salmonellae on the surface of eggshells generally die rapidly (Baker, 1990), but survival can be extended substantially when the eggs are stored at high relative humidities (Lancaster and Crabb, 1953) or low temperatures (Lancaster and Crabb, 1953; Rizk *et al.*, 1966; Baker, 1990).

An investigation of 42 flocks of the ovaries (intestines were not investigated) of spent laying hens found 76% were positive for *Salmonella*, with *S.* Heidelberg the most frequent (56.5%) of the 14 serovars detected (Barnhart *et al.*, 1991), while only 2.4% of sampled birds from those houses had *S.* Enteritidis contaminated ovaries. Consumption of undercooked eggs has been identified as a risk factor for *S.* Heidelberg infections in the United States (CDC, 2003).

Flocks are frequently infected with *Campylobacter jejuni*, and it would be expected that this pathogen would also be found on egg surfaces. However, of 226 eggs from hens experimentally colonized and actively excreting *C. jejuni*, only two were positive (Doyle, 1984). The microorganism appears to have little ability to penetrate the albumen of hen or turkey eggs (Acuff *et al.*, 1982; Doyle, 1984; Neill *et al.*, 1985; Shane *et al.*, 1986; Shanker *et al.*, 1986). Sahin *et al.* (2003) showed that *C. jejuni* indeed has limited ability to penetrate the eggshell, but that the pathogen was able to survive for up to 14 days when inoculated into the egg yolk and the eggs. Stored eggs are consistently negative for *Campylobacter* (Shane *et al.*, 1986). On the egg surface, the pathogen dies off rapidly due to the humidity and temperature conditions during storage (Kollowa and Kollowa, 1989). While there is one reported outbreak of campylobacteriosis linked to consumption of undercooked eggs (Finch and Blake, 1985), it is generally considered that transmission of this pathogen via eggs is unlikely (Bruce and Drysdale, 1994; Sahin *et al.*, 2003).

C Contamination in the production environment

The production environment is the other primary source of microorganisms on the exterior surface of eggs, including contact with feces, nesting material, dust, feedstuffs, shipping and storage containers, human beings, rodents, and invertebrates. A small survey of shell egg processing plan sanitation programs in the United States found high levels of total aerobic counts on egg contact surfaces on several types of equipment (Jones *et al.*, 2003). Spoilage rates are substantially greater in eggs from heavily contaminated environments (Harry, 1963; Smeltzer *et al.*, 1979). Some eggs come from small family farms, where collection is by hand, but the vast majority is produced in large, semi-automated facilities where the hens are individually caged. In such facilities, the eggs roll under gravity from the cages to troughs, a system reported to produce eggs with lower contamination rates than eggs laid into nests (Harry, 1963; Quarles *et al.*, 1970; Carter *et al.*, 1973). The eggs are then transferred by hand to pressed paper or polystyrene trays, and transported in cases to be candled and graded. The nature and amount of shell contamination vary according to the storage system and the time until the eggs are collected. The earlier the eggs are collected after laying, the lower is the contamination of the shell, even under unfavorable conditions (North, 1984).

Nests must be clean and dry, and surfaces in contact with the shells should be dry and free from visible feces and other soil (Joyce and Chaplin, 1978; Smeltzer *et al.*, 1979; Tullet, 1990). Because eggs emerge from the cloaca and pass by the anus of the chicken, complete freedom from faecal matter is impossible. Wet feces-smeared nests, wet hands of collectors, the laying of eggs on dirty floors, and wet equipment all foster penetration of the surface bacteria into the shell, especially when the egg is cooling from its initial temperature of 40–42°C.

The laying environment can be a potentially important source of salmonellae for the external surface of the egg. For example, Jones *et al.* (1995) isolated salmonellae from 73%, 64%, 100%, and 100% of samples from egg belts, egg collectors, ventilation fans, and wash water, respectively. In comparison, only 8% of the eggshells and none of the internal contents of eggs prior to collection yielded salmonellae. A diverse group of *Salmonella* serovars has been isolated from the environment of egg production facilities and the surface of the eggs, including *S.* Agona, *S.* Typhimurium, *S.* Infantis, *S.* Derby, *S.* Heidelberg, *S.* California, *S.* Montevideo, and *S.* Mbandaka. Earlier studies cited by Cantor and McFarlane (1948) included *S.* Thompson, *S.* Typhimurium, *S.* Bareilly, *S.* Oranienburg, *S.* Montevideo, *S.* Tennessee, *S.* Derby, *S.* Essen, and *S.* Worthington among the salmonellae isolated from egg shells. Poppe *et al.* (1991) found that 53% of randomly selected Canadian egg production houses had *Salmonella*-positive environments, with the level specifically for *S.* Enteritidis 3%. Dust generated during the hatching process has been strongly implicated in *Salmonella* transmission, which complicates the cleaning and disinfecting processes for hatchers. Electrostatic charging of particles in enclosed spaces can be an effective means of reducing air-borne dust, and thus to reduce air-borne bacteria within hatching cabinets thereby reducing the presence of *Salmonella* in cecal contents of chicks (Mitchell *et al.*, 2002).

Effective disinfection of production facilities, control of rodents and other vermin, chlorination of drinking water, and prevention of cross-contamination are among the factors that have been identified as important for disrupting the transmission of *S.* Enteritidis to future flocks in a production facility (McIlroy *et al.*, 1989; O'Brien, 1990; Dawson, 1992; Edel, 1994; Giessen *et al.*, 1994; Mason, 1994; Davies and Wray, 1995; Wierup *et al.*, 1995). Periodic evaluation of the laying environment for *S.* Enteritidis has been used in a number of these control programs to identify infected flocks.

L. monocytogenes can be isolated from both poultry flocks and the birds' immediate environment (Gray, 1958; Bailey *et al.*, 1989). Consequently, *Listeria* spp. are to be expected on the shells of freshly laid eggs.

III Shell eggs

A *Effect of initial processing*

Transport and storage. After eggs have been collected and packed in cases or small packages, they are shipped within a few days to the final purchaser (direct marketing), a packing station, or (where permitted) a washing station. Davies and Breslin (2003) conclude that contamination of eggs in farm egg-packing plants may be a significant contributory factor to surface contamination of shell eggs and advocate keeping the packing environment as dry as possible, effective cleaning and disinfection followed by full drying in order to control salmonellae. In many instances, countries have established regulatory or legislative standards relating to quality, weight, packaging, handling, labeling, transport, and dating of shell eggs. For example, according to EC Egg Marketing Standards Regulation 1907/90 and Commission Regulation 1274/91, eggs have to be shipped to the licensed packing station at least every third working day or once a week when the intervening storage temperature does not exceed 18°C. Such standards can vary substantially between countries.

Eggs should be stored with the blunt end up. This helps keep the yolk, which has a lower specific gravity than the white, from drifting toward the inner membrane. If the yolk comes in contact with the inner membrane, microorganisms that penetrate the membrane at that site can directly contaminate the yolk, bypassing the protective barriers in the white. When this occurs, spoilage proceeds rapidly (Board, 1964; Brown *et al.*, 1970).

Storage temperatures <8°C inhibit the growth of salmonellae and related mesophiles and slow the loss of internal quality. At temperatures up to 18°C, the egg's natural barriers degrade only slowly with time. As the effectiveness of these barriers decline, the egg becomes increasingly susceptible to bacterial penetration and growth (Elliot, 1954; Brown *et al.*, 1970; Humphrey, 1994b). When storage temperatures are increased >18°C, the degradation of the antimicrobial barriers accelerates (Humphrey, 1994b). In the United States, shell eggs in distribution and at retail are required to be held under refrigeration at ≤7.2°C (FSIS, 1998b; FDA, 1999).

Cold-stored eggs brought into a warm, moist atmosphere can become wet from condensation (sweating). If the eggs are returned to the cold room while still wet, surface bacteria may be aspirated through the shell because of the pressure differential as the air sac contracts (Forsythe *et al.*, 1953). The relative importance of this as a means of increasing bacterial penetration of the egg has been debated since some investigators have not observed increased penetration except after several periods of sweating due to alternating storage temperatures (Vadehra and Baker, 1973). Sweating increases the extent of spoilage to a greater degree when the shell is dirty (Forsythe *et al.*, 1953). This could account for the discrepancies in reported results. It may also reflect the sensitivity of the testing. For example, shell eggs showed no internal contamination with *Yersinia enterocolitica* immediately after being inoculated by immersion in water containing 10^6 cfu/mL, followed by exposure to pressure or temperature differentials (Amin and Draughon, 1990). However, after 14 days of incubation at 10°C, the population of *Y. enterocolitica* within the eggs exceeded 10^6 cfu/mL, with all eggs being positive. Rapid air-cooling by forced convection makes eggs more prone to penetration by *S.* Enteritidis (Fajardo *et al.*, 1995). While the shells of cooled and uncooled eggs both had microscopic cracks, those of the rapidly cooled eggs were larger and more numerous. The use of cryogenic gases has been investigated as a means of reducing damage while enhancing rapid cooling of shell eggs (Curtis *et al.*, 1995).

Penetration increases with duration of contact with contaminated material, especially during storage at high relative humidities. This is true both for spoilage bacteria and salmonellae (Simmons *et al.*,

Table 15.8 Proportion of slightly dirty eggs penetrated by spoilage bacteria during storage for 9 months at 1.7–4.4°C and 65–80% relative humidity, as affected by cleaning method

Cleaning method	Number of eggs tested	Percentage of eggs penetrated
Dry cleaned with mechanical sander	577	3.5
Washed in detergent, rinsed in water	276	7.3
Washed in detergent, no rinse	286	7.0
Washed in detergent-sanitizer, rinsed in water	278	13.3
Washed in detergent-sanitizer, no rinse	284	4.2

From Miller (1959).

1970). The relative humidity during storage should be between 70% and 85% (Henderson and Lorenz, 1951). Below 70%, there is a rapid loss of weight by evaporation that adversely affects quality. Above 85%, microbial penetration is enhanced, and molds may grow, particularly in the air sac.

Cleaning. Countries give different emphasis to the cleaning of shell eggs, reflecting the continuing debate on its efficacy. The European Union forbids the cleaning of Grade A eggs intended for human consumption, although it does not forbid cleaning (washing) of eggs prior to their use for egg products (EC, 1991, 2003). On the contrary, cleaning of eggs is required in the United States and Canada (see CFIS, 2003; USDA, 2003) Those require it do so, in part, because removing fecal material from the egg surface is assumed to reduce the risk of pathogenic bacteria penetrating into the egg. However, others have indicated that, particularly for eggs that are to be stored for extended periods, washing the surface increases spoilage rates due to increased penetration by bacteria (Sparks, 1994). Manufacturers of egg products typically prefer clean eggs, and often stipulate that shell eggs are visibly clean, intact, and free of physical deterioration.

Eggs can either be dry cleaned or washed. Dry cleaning is usually with a stiff brush, sandpaper, or steel wool. Mechanical dry cleaners are often difficult to clean, and require frequent changes of brushes. Dry cleaning also removes the cuticle, making eggs more susceptible to microbial penetration and spoilage should they subsequently become wet (Brown *et al.*, 1965). During dry cleaning, microorganisms on the shell surface can be forced into the pores of the egg shell, encouraging penetration. However, if eggs are stored under proper humidity control, dry cleaning can be at least as effective as washing the eggs (Table 15.8).

Eggs can be washed in a water bath or combined with mechanical cleaning by brushes or steel wool. However, most modern egg washing machines employ spray-wash systems. The typical continuous egg washer consists of three stages: a wash chamber where the eggs are washed with warm water and detergent using high pressure jets, a rinse chamber which usually includes a sanitizing agent, and a drying chamber (Sparks, 1994). Most processors in North America wash all eggs as received to avoid the labor of sorting (Forsythe, 1970), but in some countries only dirty eggs are washed. Many attempts have been made to enhance the keeping quality of eggs through improved washing and/or disinfecting of the shell.

Temperature of the wash-water has a marked influence on the microbial count (Lucore *et al.*, 1997) and the survival of infectious pathogens such as salmonellae (Meckes *et al.*, 2003). A number of factors related to egg washing are known to affect microbial penetration and hence spoilage, listed below and reviewed by Stadelman (1994).

- Washing eggs in a liquid that is at a lower temperature than the eggs results in the liquid (plus bacteria in the liquid) being drawn through the pores (Haines, 1938; Haines and Moran, 1940; Brant and Starr, 1962). The temperature of the solution should be at least 12°C higher than the eggs' temperature.
- Visibly dirty eggs tend to have a higher spoilage rate than those that appear clean.
- Any process that wets the shell tends to increase spoilage, e.g. sweating, cleaning with a wet cloth.

Table 15.9 Effect of washing eggs on spoilage during and after storage

Original condition	Washed	Percent spoiled
Clean	No	0.6
Dirty	No	12.7
Clean	Yes	5.8
Dirty	Yes	19.9

Adapted from Lorenz and Starr (1952).

Presumably water enters pores by capillary action even when there is no temperature/pressure differential.

- Damage to the cuticle results in increased microbial penetration. Investigations have indicated that continuous spray washers do not damage the cuticle (Kuhl, 1987). However, Sparks (1994) found that such washers decreased the ability of the cuticle to resist water uptake by the egg.
- Wash water containing iron will increase the iron level in the albumen, neutralizing the antimicrobial effect of conalbumin. Wash water should have <1–2 ppm Fe(III), and levels above ~5 ppm have been associated with greatly accelerated spoilage (Garibaldi and Bayne, 1962; Board *et al.*, 1968) and growth of pathogenic bacteria (Becirevic *et al.*, 1988).
- Minimizing of the levels of microorganisms present in the wash water through the use of high quality water and disinfectants or alkaline detergents reduces the microbiological impact of washing.

The recommended requirements for commercial egg washing include the following.

Quality of the eggs. Only fresh, intact eggs that ideally have been cooled to 10°–14°C should be washed. By holding eggs at 10°–14°C, an adequate temperature differential between the egg and the wash water can be achieved readily. Badly soiled and damaged eggs spoil more frequently and rapidly than clean eggs, and washing can increase spoilage during subsequent storage, regardless of whether the eggs were dirty before washing or not (Table 15.9). Washing should take place as soon as feasible after collection; the microorganisms that have had time to penetrate to the inner membrane are not readily removed or destroyed. The eggs should be carefully handled at all times to prevent physical damage and contamination.

Washing requirements (machine). The eggs should be conveyed in such a manner that jets of wash water and/or brushes have complete access to each egg. The machine must use potable water, low in metal salts, with <2 ppm iron. Eggs washed in natural water containing 4.8 ppm iron showed 6.2% spoilage by pseudomonads on storage, whereas those washed with water containing 0.2 ppm iron showed only 0.8% spoilage (Garibaldi and Bayne, 1962). Whether hard water, high in calcium or magnesium, encourages spoilage has not been reported. The washing temperature should be 40°–42°C, or at least 12°C warmer than the eggs; high enough to ensure adequate washing but not risk cuticle damage. The wash water must be clean, in general purified or filtered to ensure that the organic matter and the microbiological load are maintained at low levels. Meckes *et al.* (2003) found enumeration of *E. coli* or total coliforms was a good indicator of the potential contamination of wash water by salmonellae. An alkaline, low foam detergent can be used to raise the wash water to pH 10–11 and improving the dirt-removing efficiency of the water.

Alkaline detergents are used because acid detergents attack the shell. Washing eggs experimentally with 1–3% acetic acid destroyed many microorganisms and cleaned the surface of the shell, but reduced shell thickness and egg quality (Heath and Wallace, 1978). Simple alkaline compounds such as trisodium phosphate or sodium metasilicate are satisfactory for this purpose, as are the more complicated proprietary mixes (Swanson, 1959). A good detergent will physically remove up to 92% of the bacteria on the shell surface (Forsythe *et al.*, 1953; Bierer *et al.*, 1961a,b). The effectiveness of washing for eliminating *S.* Enteritidis and other salmonellae is dependent on wash water pH and temperature (Holley and Proulx, 1986; Catalano and Knabel, 1994a,b; Meckes *et al.*, 2003). Pathogen survival was more likely with low washing temperature (32°–35°C) and low pH (9–10) compared with high temperature (38°–43°C) and high pH (11–12.5)(Catalano and Knabel, 1994a). The rate of destruction of *S.* Enteritidis in wash water increased at high detergent concentration and low egg solids. Cross-contamination by *S.* Enteritidis was observed with washwater at pH 9 but not at pH 11 (Catalano and Knabel, 1994b). *Yersinia enterocolitica* (Southam *et al.*, 1987), *S.* Typhimurium (Meckes *et al.*, 2003), and *L. monocytogenes* (Brackett, 1988; Laird *et al.*, 1991) all persist in egg wash water. The inactivation of *L. monocytogenes* and *S.* Typhimurium in egg wash water as a function of temperature, pH, chlorine concentration, and egg solids content was described by a linear equation that can be used to predict the pathogens' survival (Leclair *et al.*, 1994).

A final rinse with clean water containing a sanitizer should be applied (USDA, 1975a). Commonly used sanitizes include 100–200 ppm of chlorine, quaternary ammonium compounds or calcium hypochlorite, or 12–25 ppm of iodine. A potable water final rinse is required with iodine. Alternative sanitizers have been studied (Knape *et al.*, 1999; Kuo *et al.*, 1997; McKee *et al.*, 1998) as well as combined use with ultraviolet radiation (Favier *et al.*, 2001). The use of sanitizers destroys many of the remaining bacteria. In addition to treating the surface of the egg, these compounds also sanitize the recirculating conveyors. The temperature of the rinse-water, 43°–45°C, at the surface of the egg should always be slightly higher than the wash water. Some investigators have shown that iodophors, quaternary ammonium compounds (Sauter *et al.*, 1962), or chlorine-bromine compounds (Forsythe, 1970) are more effective, particularly if permitted to remain on the egg-shells without a subsequent clean water rinse. Some prefer detergent-sanitizers, because they clean and sanitize in one step. Organic matter in the wash water, however, will destroy much of the sanitizer's effectiveness, so a two-step (wash-sanitize) procedure is typically more effective.

Almost all commercially available machines recirculate the hot, detergent/sanitizer treated water. It passes through a series of filters, which remove most of the organic material. The water slowly overflows and is "topped off" with new solution. After about 4 h, the machine is emptied and the cycle started again. Providing the refill process is at an adequate rate and detergent levels are maintained, the microbiological load can be held at an acceptable level. Processing water in egg product plants is being increasingly reused as water acquisition and disposal become more expensive. Adequate water processing procedures should be used to treat all recirculated processing water. There is currently no general agreement upon water reuse guidelines; requirements can vary substantially in different municipalities, states, and countries.

Post-washing. Immediately after rinsing, the eggs should be quickly and completely dried. Quick drying reduces the risk that bacteria remaining on the surface of the egg are aspirated into the egg as it cools to ambient temperature. Careful handling of the eggs after washing is important to avoid recontamination. Following complete drying, the eggs should be carefully candled using clean conveyors and equipment. The candled eggs should be placed into new, clean packages, and then stored and distributed in a manner that assures that the egg surface remains dry and is not recontaminated. Condensation on the eggs should be avoided.

All cracked eggs must be removed as these, in most countries, are considered hazardous because their contents may have been exposed to bacterial pathogens and because they could be used in products that are not thoroughly cooked. In a risk assessment of use of cracked eggs in Canada (Todd, 1996), it was found that cracked eggs are 3–9 times more likely to cause outbreaks of salmonellosis that uncracked shell eggs when marketed for direct consumption or use in commercial food production and food service.

Shell coatings. Most countries do not allow shell coatings for commercially produced fresh eggs. Mineral oil (paraffin oil) sprays have been used to protect clean shell eggs from loss of water and the associated increase in air cell volume during cold storage. The oil does not protect aged eggs from their high susceptibility to penetration and growth by spoilage bacteria (Elliott, 1954). Experiments have shown that coatings of alginates, polymethacrylic acid, and certain butyl rubbers help to maintain egg quality as well as oil (Rutherford and Murray, 1963). Corn prolamine, polyvinyldene chloride, epolene emulsion, and hydrolyzed sugar derivative plus shellac have all been reported to greatly retard penetration by *Ps. fluorescens* and *Salmonella* Typhimurium (Tryhnew *et al.*, 1973).

Water glass (sodium silicate) interacts with the silicate in the shell to produce an impervious calcium silicate barrier that improves retention of functional qualities during storage. It has been used in times of acute shortage and when no other possibilities for long-term storage of eggs existed.

In-shell pasteurization (thermostabilization) of intact eggs. From 1867 to the present, there have been many reports in the scientific and patent literature on the efficacy of heat to kill bacteria on and near the surface of shell and membranes. The heat establishes a nearly impervious layer of coagulated protein immediately beneath the shell membranes. This reduces evaporation in subsequent storage. There have been reports of loss to functional properties resulting from such processing (Goresline *et al.*, 1950; Knowles, 1956). The heat should be applied within 24 hr of collection because microorganisms that have penetrated to the white already will not be destroyed by the minimal heat treatment applied (Feeney *et al.*, 1954). While this process has not been used commercially for many years, recent *S.* Enteritidis outbreaks have stimulated renewed interest into the development of techniques for pasteurizing intact shell eggs (Hou *et al.*, 1996). Intact pasteurized shell eggs are now commercially available in the United States and the United Kingdom. Ionizing radiation in doses up to 3.0 kGy may be applied to shell eggs in the USA (FDA, 2000).

B Spoilage

The spoilage of eggs is related to the ability of microorganisms to penetrate the egg and overcome the egg's multiple antimicrobial barriers. The microorganisms most commonly present on the egg surface are not necessarily ones most often associated with spoilage (Mayes and Takeballi, 1983) (Table 15.10). While the microflora of the eggshell varies qualitatively and quantitatively between geographical regions and bird types (Table 15.11), the microorganisms associated with spoilage tend to be the same. This is generally interpreted as indicating that it is the intrinsic defense mechanisms of the egg that select organisms that are capable of growth in that environment (Bruce and Drysdale, 1994). A major cause of spoilage during and immediately after removal of eggs from storage is fluorescent pseudomonads (Lorenz and Starr, 1952; Ayres, 1960). This reflects the fact that fluorescent pseudomonads, which are ubiquitous in soil and water, are frequently the first to penetrate and grow because they are motile, produce a fluorescent pigment that competes for metal ions with the conalbumin of the white, and are resistant to other protective mechanisms of the white. An egg showing bright fluorescence in most or all of the white when examined with a candler emitting long-wave ultraviolet light (black light) always contains a high number of bacteria. Such eggs are not readily detected using a white-light candler, and the odors of decomposition in the early stages are mild, often detectable only after soft cooking (Elliott, 1954). *Pseudomonas* spp. may colonize the external side of the inner membrane, sometimes permitting the fluorescent pigments to diffuse into the white before bacterial cells actually penetrate (Elliott, 1954). Although *Pseudomonas* spp. can grow on the membrane when it is isolated from other parts of the egg

Table 15.10 Microflora on the eggshell and within spoiled eggs

Type of microorganism	Frequency of occurrence[a]	
	On the shell	In rotten eggs
Micrococcus	+++	+
Achromobacter	++	+
Enterobacter	++	–
Alcaligenes	++	+++
Arthrobacter	++	+
Bacillus	++	+
Cytophaga	++	+
Escherichia	++	+++
Flavobacterium	++	+
Pseudomonas	++	+++
Staphylococcus	++	–
Aeromonas	+	++
Proteus	+	+++
Sarcina	+	–
Serratia	+	–
Streptococcus	+	+

[a]Number of plus signs indicates relative frequency of occurrence.
From Mayes and Takeballi, (1983) as adapted from Bruce and Drysdale (1994).

Table 15.11 Microflora of eggs from different bird types (adapted from Bruce and Drysdale, 1994)

Bacterium	Duck[a]	Duck[b]	Waterfowl[a]	Hen[a]	Hen[b]	Turkey[c]
Enterobacteriaceae	65.4[d]	40	66.0	11.8	31.5	71.4
Staphylococcus	2.5	4	11.4	23.0	9.2	7.7
Micrococcus	1.2	0	21.3	63.8	34.6	0
Streptococcus/*Enterococcus*	0	0	0	1.2	15.3	8.5
Pseudomonas	16.0	56	0	0	2.5	1.5
Acinetobacter	6.2	0	0	0	0	0
Bacillus	8.6	0	0.9	0	1.2	3.9
Molds	0	0	0	0	0.2	1.6
Unidentifited	0	0	0	0	5.5	5.4

[a]Seviour and Board (1972).
[b]Bruce and Johnson (1978).
[c]Bruce and Drysdale (1983).
[d]% of isolates.

and immersed in saline (Elliott and Brant, 1957; Board, 1965a), antibacterial activity associated with the membrane has been reported several times. The presence of lysozyme in both membranes (Vadehra *et al.*, 1972) could partially explain this phenomenon. Pyoverdine-producing pseudomonads typically penetrate and grow in shell eggs more quickly than any other group of bacteria. They are frequently the only microorganism present in stored eggs (Lorenz *et al.*, 1952).

Besides pseudomonads, a limited number of other bacteria are capable of acting as primary invaders of shell eggs. Examples include strains of the genera *Alcaligenes*, *Proteus*, *Flavobacterium*, and *Citrobacter*. In addition, a number of other genera, such as *Acinetobacter*, *Moraxella*, *Alcaligenes*, *Proteus*, *Escherichia*, *Flavobacterium*, and *Enterobacter* are able to grow in eggs once the eggs' defences have been broached by a primary invader (Florian and Trussell, 1957; Elliott, 1958; Ayres, 1960). Presumably, these secondary invaders are able to use the metal ions sequestered by the siderophores produced by the primary invaders.

Table 15.12 Bacterial genera isolated from various types of rotten shell eggs

Type of rot in decreasing order of frequency	Bacterial genera isolated
Green	*Pseudomonas*
Colorless	*Acinetobacter–Moraxella*
Black	*Pseudomonas*
	Proteus
	Aeromonas
	Alcaligenes
	Enterobacter
Pink	*Pseudomonas*
Red	*Pseudomonas*
	Serratia

From Alford *et al.* (1950); Florian and Trussell (1957); Mayes and Takeballi (1983).

The nature of bacterial rots associated with eggs depends on the bacterial species/strain or mixture of species/strains present (Table 15.12). For example, non-proteolytic *Ps. putida* produces fluorescence in the white, whereas lecithinase-producing *Ps. fluorescens* breaks down the diffusion barrier at the surface of the yolk and turns the white pink. This is probably due to Fe^{3+} ovotransferrin chromogen. *Ps. maltophilia* produces a characteristic "nutty" flavor and causes a slight crusting of the yolk with streaks of ferric sulfide on the surface. Spoilage by pseudomonads is favored in cold-stored eggs (Lorenz and Starr, 1952; Ayres and Taylor, 1956). Strongly proteolytic microorganisms digest the albumen, blackening the yolk. The bacteria mostly commonly associated with black rot are *Alcaligenes, Escherichia, Aeromonas*, and *Proteus* (Stadelman, 1994). Other microorganisms do not cause macroscopic changes but can form populations as large as those of the "rot producers" (Board, 1965b). These include *Alcaligenes faecalis, Enterobacter* spp. (*cloaca*), and some *Ps. fluorescens* strains. These organisms may not be detected at candling or when the eggs are broken out, and thus contaminate egg products (Johns and Berard, 1945, 1946).

Mold growth on eggs from free-range flocks from small farms has been occasionally seen when egg collection is unduly delayed. Molds can also cause spoilage during refrigerated storage when the humidity is too high. Mold growth on the surface of eggs is referred to as "whiskers", and is most often associated with *Cladosporium herbarum* (Board *et al.*, 1994). The hyphae can penetrate the shell's pores and membrane and spread throughout the interior of the egg.

C Pathogens

Salmonellae have long been associated with eggs and are the most important human pathogen carried by this food product (Cogan and Humphrey, 2003). Some species of salmonellae can enter the egg via transovarian transmission. Others penetrate from the surface of the egg in the same way as spoilage bacteria.

On rich media, and under otherwise optimal conditions, the maximum temperature for growth of salmonellae ranges from 43°C to 46°C (Elliott and Heiniger, 1965). A few strains have been reported to grow at 5–7°C (Matches and Liston, 1968) and occasionally at lower temperatures (d'Aoust, 1991). However, the minimum growth temperature for most salmonellae is ±7°C (Mackey *et al.*, 1980). Under the less than optimal conditions that are encountered on and in the egg, salmonellae cannot penetrate and grow below 10°C (Stokes *et al.*, 1956; Ayres and Taylor, 1956; Simmons *et al.*, 1970; Ruzickova, 1994). At temperatures between that of a traditional "cold room" for eggs (15°C) and the newly laid egg (40°C), salmonellae can penetrate and grow (Stokes *et al.*, 1956; Licciardello *et al.*, 1965). Salmonellae do not usually alter the odor or appearance of eggs (Vadehra and Baker, 1973), so that an egg containing millions of salmonellae could pass visual inspection and become a severe health hazard.

Experimental contamination of the yolk revealed that *S.* Enteritidis multiplies rapidly at elevated storage temperatures, reaching populations of 10^9 cfu/egg in 24 h (Braun and Fehlhaber, 1995). *S.* Enteritidis has a generation time of ~30 min at 37°C (Humphrey *et al.*, 1995). At 15.5°C, the generation time increased to 3.5 h, and no multiplication was observed at 7–8°C after 94 days (Ruzickova, 1994). Similarly, another investigation indicated that *S.* Enteritidis grew in eggs at 13°C, but failed to grow at 7°C (Agger, as cited by Stadelman, 1994). The temperature range supporting growth of *S.* Enteritidis is more restricted for egg white than yolk (Ruzickova, 1994). The albumen of freshly laid eggs has an initial detrimental effect on *S.* Enteritidis resulting in a 50% reduction in viability in 4 h at 37°C (Bradshaw *et al.*, 1990), but surviving cells are resistant to the antimicrobial properties of the albumen. There seemed to be little differences between the potential of *S.* Enteritidis isolates of different phage types (i.e. 4, 8, 13a, and 14b) to multiply rapidly in egg yolk and to survive for several days in egg albumen (Gast and Holt, 2001). There is some evidence that *S.* Enteritidis competes poorly with other organisms within the albumen of eggs stored at ambient temperatures (Dolman and Board, 1992).

The growth of *S.* Enteritidis PT4 in naturally infected eggs is largely governed by the initial location of the microorganism in relation to the egg yolk, the age of the egg, and the storage temperature. Storage studies with naturally infected eggs indicate that at 20°C growth will probably not occur before the 21st day (Humphrey, 1994). In the majority of inoculated eggs in which the microorganism was placed at different sites in the albumen and then stored at 20°C, salmonellae did not grow rapidly until the end of approximately 3 weeks. Storage at room temperature had no appreciable effect on the prevalence of *Salmonella*-positive eggs, but those held for more than 21 days were more likely to be heavily contaminated (Humphrey, 1994a). It should be noted that the size of the inoculum and the composition of the medium used to suspend the pathogen cells in prior to inoculation may have a marked effect on its growth (Cogan *et al.*, 2001).

The highest incidence of outbreaks of salmonellosis occurs in the summer months when eggs, if unrefrigerated, are subject to higher ambient temperatures. In the area of the albumen closest to the yolk, growth-promoting factors reach sufficiently high concentrations to permit growth and to support large populations of the organism. It was observed that migration of yolks towards the air sacs contaminated with *S.* Enteritidis enhanced its multiplication. Changes in the albumen and the vitelline membrane during storage slowly alter conditions so that they are increasingly favorable for *S.* Enteritidis growth (Humphrey *et al.*, 1991c). Higher storage temperatures speed up the break down of the egg's defense mechanisms (Humphrey and Whithead, 1993). The ACMSF (1993) concluded that the normal antimicrobial barriers in shell eggs are sufficient to control *S.* Enteritidis growth as long as the eggs are less than 21 days old and the temperature has not exceeded 20°C. However, if either condition is exceeded, then the eggs should be stored at temperatures not exceeding 8°C. Increased assurance of salmonellae control in shell eggs can be achieved by storing all eggs at the lower temperature.

Eggs boiled long enough to solidify the yolk (for example, 10 min) are sufficiently hot to kill all salmonellae, but other types of cooking that leave the yolk liquid (e.g. soft-boiled eggs) are not always sufficient to inactivate *S.* Enteritidis and other salmonellae (Stafseth *et al.*, 1952; Licciardello *et al.*, 1965; Baker *et al.*, 1983; Humphrey *et al.*, 1989a; Baker, 1990; Humphrey, 1994b). Various methods of boiling an egg in households may lead to different temperature curves in the interior. Prior exposure of *S.* Enteritidis PT4 to 4°C or 8°C significantly increases the sensitivity of *S.* Enteritidis PT4 (and presumably other salmonellae) to heat treatments (Humphrey, 1990b).

D CONTROL (shell eggs)

Summary

Significant hazards[a]	• Salmonellae, especially *Salmonella* Enteritidis.
Control measures	
Initial level (H_0)	• Keep to appropriate farm measures (rearing of flock controls, farm hygiene, elimination of contaminated flock, etc.). • Remove cracked eggs.
Increase (ΣI)	• Keep eggs refrigerated (best below 8°C). • Avoid free water on the eggs: dry well if washed, store under suitable RH, avoid condensation due to changes in temperature.
Reduction (ΣR)	• Wash the eggs, where possible; chlorinated water can be used. Note: washing will reduce eggs from Grade A to Grade B in Europe.
Testing	• *Salmonella* monitoring program including environmental, line, finished product, and critical ingredients to verify effectiveness of preventive measures such as zoning.
Spoilage	• Keep shell eggs dry and chilled to prevent spoilage.

Control measures. Control of bacteria in shell eggs requires an integrated effort that begins at the egg production facility and ends with the consumer. Laying hens should be raised under conditions that minimize stress and environmental contamination of the egg after it is laid. Cages, litter, and nesting materials should be clean and kept as free of feces as possible. Eggs should be collected at least daily; as often as every 4 h is ideal. Eggs should be kept dry at this and all other stages of handling, transport and sale. All eggs should be stored with the blunt end upward to prevent migration of the yolk. Measures taken in egg-production were reviewed by Humphrey (1994b).

While there continues to be debate over the need to refrigerate shell eggs, prompt chilling after collection to below 10°C retards growth of many spoilage and pathogenic bacteria. Cooling should be done to minimize damage to the cuticle and the shell, and only when the egg surface is dry to prevent aspiration of bacteria into the egg.

Eggs should be candled, using white-light and black-light candles, or in some other way inspected to remove as inedible spoiled, leaking, or otherwise unacceptable eggs. These techniques help segregation of eggs with punctured yolks, shell cracks, and have practical quality control applications.

If the eggs are washed, the water should be at 42°C or higher, so that the wash water is at least 12°C warmer than the eggs. The water should be potable and low in iron content. It should contain an alkaline detergent, such as sodium metasilicate or trisodium phosphate, and should be continuously replenished to allow an overflow. The cleaned eggs should be rinsed in a spray of fresh water containing a suitable disinfectant such as chlorine at 100–200 ppm, with this final rinse being done at a temperature 1–2°C warmer than the wash water. The washing machine should be emptied, cleaned, and refilled with clean detergent solution at least daily. Washing should be performed in a manner that minimizes damage to the cuticle.

The shells should be dried immediately after washing, and recooled to below 15°C (preferably below 10°C). Shell eggs should not be frozen, since freezing can damage the shell. Movements in and out of storage should be in a manner that prevents condensation on the shell surface (sweating). All surfaces in contact with the shells should be clean and dry. The humidity of storage facilities should be maintained between 70% and 80% RH, avoiding changes in temperature leading to condensation, to ensure that the surface of the egg remains dry without accelerating the loss of moisture from the egg and the concomitant loss of quality.

Special considerations. *Salmonella* is the primary pathogen of concern, and its control requires interventions that disrupt both transovarian and trans-shell contamination of the egg. The general considerations outlined above have a beneficial impact on the control of salmonellae. There are a number of other potential means for reducing or controlling the incidence of *Salmonella*, including *S.* Enteritidis, particularly at the production level. The acquisition of birds from parent flocks that are salmonella-free is an important means of reducing the number of contaminated flocks. Other control measures include chlorination of drinking water, competitive exclusion (Seuna and Nurmi, 1979), cleaning and disinfection of poultry houses between occupation by laying flocks (Schlosser *et al.*, 1999), and vermin control programs. A number of control programs have included microbiological testing of the laying environment as a means of identifying infected flocks (ACMSF, 2001), but the efficacy of this approach and the actions that should be taken once a positive flock has been identified are not agreed internationally. The multiplication of *S.* Enteritidis and other salmonellae in contaminated eggs can readily be controlled by refrigeration. Routine microbiological testing of shell eggs for *S.* Enteritidis, other *Salmonella* serovars, or other pathogens is not recommended due to the low frequency of contamination. However, occasional testing as part of a HACCP verification program can provide useful information concerning the adequacy of control programs over time. Control programs in Canada, Netherlands, Sweden, the United Kingdom, and the United States are summarized by Altekruse *et al.* (1993) and more recently in the United States by Schlosser *et al.* (1999).

Adequate cooking of shell eggs so that the yolk is no longer soft is a means by which the final user can be assured that any salmonellae present are inactivated.

IV Liquid eggs

Shell eggs fit for human consumption may be separated from their shells to produce liquid, concentrated, dried, crystallized, frozen, quick frozen, coagulated, or reduced cholesterol products. Such products have been produced from hens', ducks', turkeys', guinea fowls', or quails' eggs, but not a mixture of eggs from different species. The liquid egg may be homogenized as whole egg or separated into white and yolk. Salt, sugar, or acidulants may be added to liquid eggs destined for further processing. All liquid egg products should be pasteurized, chilled, filled into containers or tanks, and shipped refrigerated or frozen.

A Effects of processing on microorganisms

Breaking, separating, and homogenizing. The initial microflora of liquid egg consists of a diverse mixture of Gram-positive and Gram-negative bacteria that come from (i) the shell which is often contaminated with fecal and other matter; (ii) an occasional contaminated egg content, (iii) processing equipment (such as breaking utensils, pipes, pumps, filters, pails, churns, and holding tanks) and the processing environment, and (iv) food handlers. Unless carefully designed, equipment used continuously for long periods of time may be difficult to clean, and there can be pockets of liquid and semi-stagnant films where bacteria will accumulate and multiply.

Unless eggs are already clean, they should be washed immediately before the breaking operation. This operation should be done in a separate room in order to prevent cross-contamination. They need not be dried after washing so long as they drain enough that water from the shells does not run directly into the liquid egg product (USDA, 1975b; EC, 1989). Washing dirty eggs before breaking can

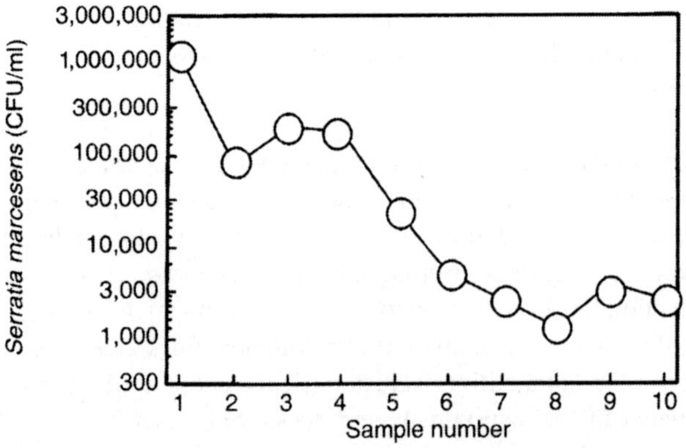

Figure 15.2 Contamination of liquid egg by the machine-breaking of eggs infected with *Serratia marcescens* (Kraft, 1967b). Each sample represents the contents of 102 eggs. Sample 1–4 were from eggs infected with *S. marcescens*. Samples 5–10 were from fresh uninfected eggs broken on the same machine after samples 1–4, without cleanup.

reduce the aerobic plate counts of liquid egg by several orders of magnitude (Penniston and Hedrick, 1947).

A single spoiled egg can contaminate breaking equipment and add millions of bacteria to the liquid egg. Candling before the breaking operation can detect most spoiled eggs, but some types of spoilage are hard to detect. For example, fluorescent rots by *Pseudomonas* spp. are difficult to see using only a white-light candler. Their mild odors are also deceiving (Johns and Berard, 1945; Elliott, 1954; Mercuri *et al.*, 1957). Similarly, the *Acinetobacter–Moraxella* group can produce colorless rots that can enter the liquid egg undetected. With automatic breaking machines, examination for rots is even more difficult. Efficient use of such machines depends on a uniform supply of sound, clean, unspoiled eggs (Forsythe, 1970). The use of cracked eggs increases the rate of contamination by salmonellae and spoilage organisms in egg mixes (Baker, 1974). The use of incubator rejects eggs, meaning eggs that have been subjected to incubation but removed from it as infertile or unhatchable, is not allowed in some countries (EU, the United States, and Canada). A spoiled egg can contaminate breaking equipment and subsequent liquid egg product (Figure 15.2).

Crushing the entire egg, followed by separation of the shell by centrifugation is a technique that has been used to produce liquid egg, although this procedure is not allowed in the EU. The mixing of the melange with shell can lead to heavy contamination, even when the eggs are surface disinfected, rinsed, and dried immediately before crushing. Several countries have regulations that disallow the use of crushing technology.

Homogenization of liquid eggs is usually accomplished in a large mixing vat called a churn or in continuous homogenizer systems. Here the individual eggs, white, or yolks are thoroughly mixed. During this process, microbial contamination is uniformly distributed throughout the batch. Egg mixes should be further processed immediately. If this is not possible, the product should be immediately put into short-term storage at temperatures of not >4°C.

Pasteurization. *Salmonella* is the target pathogen for which egg pasteurization treatments were designed. Fortunately, salmonellae are not particularly heat resistant. However, in egg products, surrounded by proteins and fats, their heat resistance increases. The times and temperatures that kill salmonellae are at or near temperatures that adversely affect the physical and functional properties of egg products. The albumen is the most sensitive; it is denatured in a very few minutes at or above 60°C. Homogenized whole egg and yolk are reasonably stable at this temperature.

Table 15.13 Pasteurization temperatures and times for liquid whole egg required by regulations in various countries (Cunningham, 1990)

Country	Time (s)[a]	Temperature (°C)
Australia	150	62
China	150	63
Denmark	90–180	65–69
England	150	64
Poland	180	68
United States	210	60

[a]Times for average particle. With higher temperatures turbulence within the mix must be increased to minimize heat damage to the eggs.

Recommended pasteurization temperatures for liquid eggs that have received no chemical additives vary from 55.6°C to 69°C, and the times of exposure vary from 10 to 1.5 min. Lower temperatures and shorter times increase the risk of survival of salmonellae, whereas higher temperatures and longer times increase the damage to functional properties (whipping, emulsifying, binding, coagulation, flavor, texture, color, and nutrition) (Forsythe, 1970). The minimum pasteurization temperatures and times required by various countries vary substantially (Table 15.13).

The thermal resistance of salmonellae, including *S.* Enteritidis, in liquid egg products is dependent on the physical and chemical characteristics of the individual products. Furthermore, thermal resistance varies among *Salmonella* serovars and strains. Examining 17 strains of *S.* Enteritidis, Shah *et al.* (1991) reported that the $D_{57.2^\circ C}$ and $D_{60^\circ C}$ values in liquid whole egg ranged from 1.21 to 2.81 min and 0.20 to 0.52 min, respectively. There was no increased resistance in strains from egg-associated outbreaks of gastroenteritis. However, Humphrey *et al.* (1993, 1995) reported that among *S.* Enteritidis phage type 4 isolates, stationery phase cultures of clinical isolates had substantially greater resistance to heat, acid, and hydrogen peroxide than chicken or egg isolates. Stationary phase cells were ~10-fold more heat resistant than log phase cells. Garibaldi *et al.* (1969b) reported that the $D_{60^\circ C}$ value for *S.* Typhimurium in liquid whole egg and yolk was 0.27 and 0.40 min, respectively. There are differences in thermal resistance among isolates and phage types of *Salmonella* Enteritidis. Palumbo *et al.* (1995) found that the $D_{60^\circ C}$ value for four strains of *S.* Enteritidis in yolk ranged from 0.5 to 0.75 min, with *z*-values ranging from 4°C to 6°C. Single strains of *S.* Senftenberg and *S.* Typhimurium had $D_{60^\circ C}$ values of 0.73 and 0.67, respectively, with *z*-values of 4.1°C and 3.2°C. *Salmonella* Enteritidis PT4 was somewhat more heat resistant than some poultry associated *Salmonella* isolates, but not to an extent that would impact pasteurization effectiveness (Humphrey *et al.*, 1990). Heat resistance and acid tolerance of *S.* Enteritidis increase if cells are pre-exposed to elevated temperatures (37–48°C) (Shah *et al.*, 1991; Humphrey *et al.*, 1993). Garibaldi *et al.* (1969b) reported that in liquid egg white *S.* Typhimurium had $D_{54.8^\circ C}$ and $D_{56.7^\circ C}$ values of 0.64 and 0.25 min, respectively. Palumbo *et al.* (1996) used a mixture of the six *Salmonella* isolates to evaluate thermal resistance in liquid egg white. The $D_{56.6^\circ C}$ value was 1.44 min with a *z*-value of 4.0°C. Palumbo *et al.* (1995, 1996) found reasonable agreement between the thermal resistance values from sealed tube studies and plate pasteurizer studies. When evaluating the heat resistance of six strains of *Salmonella* (including Enteritidis, Heidelberg, and Typhimurium) in liquid whole egg and shell eggs, Brackett *et al.* (2001) found that $D_{57.6}$-values ranged from 3.05 to 4.09 min, with significant differences between the strains (alpha = 0.05). When ~7$\log_{10}$ cfu/g of a six-strain cocktail was inoculated into the geometric center of raw shell eggs and the eggs was heated at 57.2°C, the *D*-values of the pooled salmonellae ranged from 5.49 to 6.12; a heating period of 70 min or more resulted in no surviving salmonellae being detected (i.e. an 8.7-log reduction per egg).

An atypical strain that is not destroyed by current egg pasteurization practices is *S.* Senftenberg 775W. Originally isolated from eggs in 1946, this strain has a heat resistance that is 10–20 times greater

Table 15.14 Reduction of numbers of various bacteria during pasteurization of liquid egg white

Pasteurization scheme			Log_{10} reduction of bacterial counts			
°C	Holding time (min)	pH	APC	Salmonellae	Coliforms	Reference
57.2	1	9.0	4	–	>3	Ayres and Slosberg (1949)
57.2	10	9.0	5	–	>3	Ayres and Slosberg (1949)
60	"flash"	9.0	2	–	>3	Ayres and Slosberg (1949)
55.6	4	–	2.6	>4	–	Kline *et al.* (1966)
56.7	2	–	2.6	>4	–	Kline *et al.* (1966)
57.2	2.5	9.2	2.0	–	>5.4	Barnes and Corry (1969)
		9.3	2.7	–	>5.9	

Table 15.15 Reduction of numbers of various bacteria during pasteurization of liquid whole egg

Pasteurization scheme		Log_{10} reduction of bacterial counts			
°C	Holding time (min)	Aerobic plate count	Salmonella	Coliforms	Reference
60	10	–	4.3	–	Gibbons *et al.*(1946)
	20	–	5.6	–	
	25	–	>7.3	–	
	30	2	–	–	
62.8	"flash"	0.6–1.3	+	>4.8	Goresline *et al.* (1951)
63.9	2.5	1.3–3.3	–	>4	Murdock *et al.* (1960)
64.4	3	3.5	–	–	Mulder and van der Hulst (1973)
65.2	9	2.2–2.3	–	>2	Murdock *et al.* (1960)
66.1–62.8	2.5	>5.1	>6	>3	Heller *et al.* (1962)
67–68	1.75	1.7–2.8	–	>2	Murdock *et al.* (1960)

than other salmonellae (Osborne *et al.*, 1954). In one study with albumen at pH 9.1, *S.* Senftenberg 775W had a $D_{57.8^\circ C}$ value at of 2.1–2.4 min, while that for *S.* Typhimurium was 0.125 min (Corry and Barnes, 1968). Among hundreds of isolates tested over a period of more than 30 years, no one has re-isolated this strain or found another strain with comparable heat resistance. Therefore, pasteurization times and temperatures have been designed to destroy the typical, less-resistant strains.

The degree of protection offered by pasteurization is related to the numbers of salmonellae originally present. Most recommendations for egg pasteurization reduce the levels of salmonellae in inoculated eggs by 1000–10000-fold (Tables 15.14 and 15.15). Generally, adequate pasteurization (adhering to recommended pasteurization times and temperatures) would give an adequate safety margin as it eliminate virtually all salmonellae present in unpasteurized liquid egg under hygienic processing operation conditions. Considering that a single contaminated egg would be mixed with a large number of *Salmonella*-free eggs, it is unlikely that a batch of egg mix with high levels of salmonellae would be encountered with a hygienic processing operation. However, this does emphasize the importance of using only raw eggs with an incidence and prevalence of salmonellae that is as low as possible.

Pasteurization reduces the aerobic plate count in liquid eggs by 100–1000-fold, usually to about 100 cfu/g (Tables 15.14 and 15.15). Survivors are mostly *Micrococcus, Staphylococcus, Bacillus* spp., and a few Gram-negative rods (Shafi *et al.*, 1970). Payne *et al.* (1979) found that the major organisms surviving after heating at 65°C for 3 min was *Microbacterium lacticum* and *Bacillus* spp. None of the isolates were capable of growth at 5°C, but several were capable of relatively rapid growth at 10°C and 15°C.

Although the transmission of *L. monocytogenes* to humans via pasteurized egg products has not been documented as yet, the pathogens and other *Listeria* spp. have been isolated from liquid whole egg (Leasor and Foegeding, 1989; Moore and Madden, 1993) Characterizations of the heat resistance of

Table 15.16 The heat resistance characteristics of *Listeria monocytogenes* in liquid egg products

Product	$D_{°C}$ value (min)	z-value(°C)	System/strains	Reference
Whole egg	D_{51} 14.3–22.6; $D_{55.5}$ 5.3–8.0; D_{60} 1.3–1.7; D_{66} 0.06–0.20	5.9–7.2	Sealed capillary tube, one strain	Foegeding and Leasor (1990)
Yolk	$D_{61.1}$ 0.7–2.3; $D_{63.3}$ 0.35–1.28; $D_{64.4}$ 0.19–0.82	5.1–11.5	Sealed tubes, five individual strains of *L. monocytogenes* and one of *L. innocua*	Palumbo *et al.* (1995)
White	$D_{55.5}$ 13.0; $D_{56.6}$ 12.0; $D_{57.7}$ 8.3	11.3	Sealed tubes, mixture of five strains of *L. monocytogenes* and one *L. innocua*	Palumbo *et al.* (1996)

L. monocytogenes (Table 15.16) indicate that current U.S. minimum pasteurization requirement (60°C for 3.5 min) would achieve a 2.1–2.7$\log_{10}$ reduction in the pathogen in liquid whole egg (Foegeding and Leasor, 1990). A similar estimate (2.5$\log_{10}$ reduction) was observed for the current minimum pasteurization requirement for liquid yolk (Palumbo *et al.*, 1995). The extent of inactivation of *L. innocua* was <10-fold in 3.5 min when egg white was heated using both sealed ampoules and a plate pasteurizer at 56.6°C and 57.7°C (Palumbo *et al.*, 1996). It has been concluded that current minimum pasteurization requirements would be sufficient to control *L. monocytogenes* in extended shelf life egg products only if the initial levels of the pathogen were low (Foegeding and Leasor, 1990; Foegeding and Stanley, 1990; Palumbo *et al.*, 1995, 1996). Moore and Madden (1993) considered that current pasteurization practices were adequate based on the absence of *Listeria* spp. in 500 daily samples of the pasteurized product.

Continuous pasteurization systems heat the liquid egg to the target temperature and then maintain it at that temperature for a specified amount of time through the use of holding tubes of appropriate lengths. Such pasteurizers require adequate control systems that can assure a constant rate of flow. This includes automatic equipment for temperature monitoring and recording, automatic control devices to prevent insufficient heating, and a safety system (including appropriate recording devices) that diverts inadequately heated product and prevents it from mixing with fully pasteurized product.

Humectants added to liquid egg products to decrease the water activity also increase the thermal resistance of salmonellae and other pathogens. For example, Palumbo *et al.* (1995) reported the $D_{63.3°C}$ values for salmonellae in yolk, yolk + 10% sucrose, and yolk + 10% NaCl (a_w s 0.989, 0.978, and 0.965) to be 0.21, 0.72, and 11.50 min. Inactivation of *L. monocytogenes* in these products gave $D_{63.3°C}$ values for yolk and yolk + 10% sucrose of 0.81 and 1.05 min, while that for yolk + 10% NaCl was 10.5 min after an initial lag period of 14.8 min during which time there was no decline in pathogen levels.

If salted yolks are destined for use in high acid salad dressing or high acid mayonnaise production, no pasteurization is necessary. However, greater care is needed to prevent post-processing contamination. The yolks can be pasteurized before the salt is added, taking precautions to avoid contamination. Acetic acid or other organic acids can be added to lower the pH to ≤4.6, where salmonellae die more rapidly. *Salmonella* Enteritidis may be more acid tolerant than *S.* Typhimurium (Humphrey *et al.*, 1993). Acidified salted yolks can be pasteurized in 1 min at 60°C (Garibaldi, 1968).

Pasteurizer temperatures often cause egg material to coagulate on the hot surfaces of the heating plates (Ling and Lund, 1978), which adversely affects both the functional quality of the product and the effectiveness of the microbial inactivation. Therefore, research has centered on means to:

- Repair damaged quality by adding chemicals such as whipping aids at point of use.
- Increase the sensitivity of salmonellae to heat by adding chemicals or altering pH
- Prevent adverse effects on quality by adding chemicals before pasteurization.

Egg white is the most heat sensitive egg component. Heating unaltered egg white at 62°C for 3.5 min alters 3–5 % of the ovomucin, 90–100% of the lysozyme, and >50% of the conalbumin (Lineweaver *et al.*, 1967). The US minimum heat treatment of 56.7°C for 3.5 min increases markedly the whipping time to prepare meringue. Even a minimal heat treatment of 3 min at 54.4°C doubled the whipping time, but whipping aids such as triethyl citrate or triacetin restored this function to near normal (Kline *et al.*, 1965). In this temperature range, a 2°C rise in temperature increases the damage 2.5–3-fold, whereas it increases the destruction rate of salmonellae only 2-fold. The kinetics of quality loss vs. destruction of salmonellae, in combination with the build-up of coagulated material on pasteurizer plates at higher temperatures, makes it unlikely that unaltered egg white would be pasteurized at temperatures >60°C (Kline *et al.*, 1965; Lineweaver *et al.*, 1967). Authorities in the United Kingdom have recommended 57°C for 2.5–3 min (Corry and Barnes, 1968; Hobbs and Gilbert, 1978).

Unaltered homogenized whole egg is less sensitive to damage, in part because iron from the yolk satisfies the chelating capacity of the conalbumin, and in doing so, stabilizes it (Cunningham, 1966). Egg yolk is relatively stable, but is difficult to handle because of its high viscosity.

Lower heating temperatures can be used if H_2O_2 is added to eggs at ~0.1–1.0% when it increases the heat sensitivity of salmonellae. For example, Palumbo *et al.* (1996) reported that the $D_{56.6°C}$ value in liquid egg white was 1.44 min, while $D_{53.2°C}$ value in liquid egg white treated with 0.875% H_2O_2 was 1.54 min. Treatment with catalase after pasteurization breaks down excess H_2O_2 (Lloyd and Harriman, 1957; Rogers *et al.*, 1966). Palumbo *et al.* (1996) reported that the addition of H_2O_2 did not enhance destruction of *L. monocytogenes* in liquid egg white. "Electroheating" has been used commercially as an alternative means of heating liquid egg products (Reznik and Knipper, 1994).

The thermal resistance of bacteria is also affected by the pH of the heating menstruum. Typically, bacteria are most resistant near their optimal pH for growth, and become increasingly less resistant as the pH deviates from this optimum. A freshly laid hen's egg has a pH in the range 7.6–7.8, rising as the egg ages to 9.1–9.6. Adjusting the pH of egg white from its normal level near 9 to between 6.5–6.7 increases the stability of both the egg white and salmonellae to heat, but salmonellae less so than the egg white (Lineweaver *et al.*, 1967). At pH values below 7, salmonellae become more susceptible to heating, especially in the presence of organic acids. The heat required to kill salmonellae in eggs can be reduced substantially by adjusting the pH to 5.5 or 6 with citric, lactic, acetic, formic, or propionic acids (Lategan and Vaughn, 1964). When the pH of egg white is ≥9, salmonellae do not grow, but if the pH is adjusted to pH 6.8, they grow well (Banwart and Ayres, 1957). Care must be taken to ensure that pasteurized, acidified egg white is not recontaminated with salmonellae and then temperature abused.

As more eggs are produced specifically for liquid egg products, an increasing percentage of eggs are entering the breaking facility before the pH of the egg white has reached pH 9.1–9.6. Cotterill (1968) reported that the temperature to achieve a 99.99% (4-D) reduction of *S.* Oranienburg in egg white within a specified time increased from 55.0°C at pH 9.4 to 58.6°C at pH 8.5. Garibaldi *et al.* (1969a) found that the *D*-value for *S.* Typhimurium in egg white was 4.6 times greater at pH 7 than at pH 9. Palumbo *et al.* (1996) reported that reducing the pH of egg white to 7.8 approximately triples the thermal resistance of *S.* Enteritidis (Figure 15.3). Interestingly, the opposite relationship was observed with *L. monocytogenes*, which Palumbo *et al.* (1996) attributed to the effect of pH on the inactivation of lysozyme.

Altering the pH of egg white to increase the thermal destruction of bacteria has adverse effects on the egg white proteins, primarily conalbumin. The adjustment of the pH to 7 with lactic acid enhances the stability of ovalbumin, lysozyme, and ovomucoid (Cunningham and Lineweaver, 1965). The addition

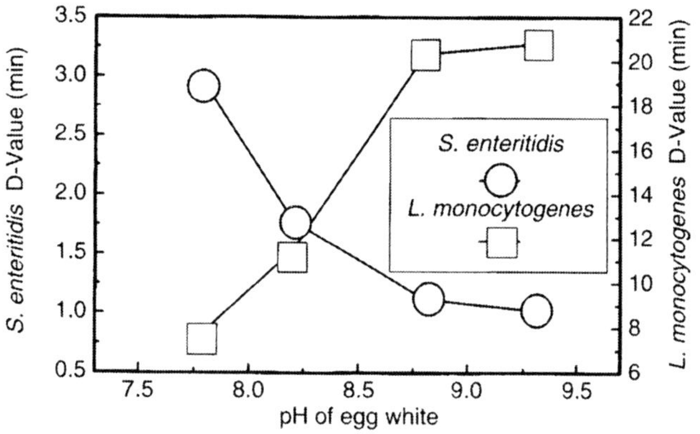

Figure 15.3 Effect of pH on the $D_{56.6^\circ C}$ value for *Salmonella* Enteritidis and *Listeria monocytogenes* heated in liquid egg white (adapted from Palumbo *et al.*, 1996).

of a metal salt satisfies the chelating activity of the conalbumin, and makes it relatively stable to heat. Either Fe^{3+} or Al^{3+} will work, but because Fe^{3+} turns the albumen pink, Al^{3+} as aluminum sulfate is the metal of choice (Cunningham, 1966). With the adjustment of pH to 7 and the addition of Al^{3+} egg white can be pasteurized at 60°–62°C for 3.5–4 min. The total amount of protein altered is <1%; however, whipping time is still increased and the white requires the addition of a whipping aid (Lineweaver *et al.*, 1967). Aluminum sulfate is not permitted as a food additive in some countries.

Other compounds also stabilize conalbumin or decrease *Salmonella* resistance. The addition of 0.5–0.75% sodium polyphosphate in egg white permits effective destruction of salmonellae at 52.2–55°C for 3.5 min, without damage to functional properties (Chang *et al.*, 1970; Kohl, 1971). Disodium ethylenediamine tetra-acetic acid (EDTA) at 7 mg/mL of egg white killed 10^6 salmonellae in >24 h at 28°C, and 70 mg of sodium polyphosphate per mL of egg white killed 10^6 salmonellae in 60 h at 28°C. When the egg white was adjusted with lactic acid to pH 5.3, and EDTA added at 7 mg/mL, *Salmonella* heat resistance decreased 100-fold. These sequestrants make Ca^{2+} and Mg^{2+} unavailable to the microorganisms. Microorganisms then become susceptible to attack by lysozyme (Garibaldi *et al.*, 1969a).

Filling and chilling. Once an egg product has been pasteurized, it must be handled with care to prevent recontamination from unpasteurized egg, insanitary equipment, containers, dust, or human or animal sources. It must also be chilled quickly, preferably using a heat exchanger, or if not, then filled into cans and cooled within 1.5 h to 7°C or below to prevent growth of any surviving microorganisms. The temperature of the egg product should pass quickly through the temperature range that supports rapid microbial growth (50–7°C).

Salt. The addition of 10% salt to yolks decreases the water activity (a_w) of the egg yolk to about 0.90 (i.e. 20.3 g salt in 100 g of water phase). Salmonellae will not grow at this a_w regardless of temperature, and die off within in a matter of weeks (Banwart, 1964; Cotterill and Glauert, 1972; Ijichi *et al.*, 1973). However, when unpasteurized 10% salted yolks are shipped and used immediately after manufacture, die-off may not be complete and the product contain salmonellae at point of use.

Freezing and thawing. Containers of liquid eggs to be frozen should be placed in a freezer at −23 to −40°C immediately after chilling and should be stacked in such a manner that they will freeze

Table 15.17 Effect of freezing on the microflora of pasteurized and unpasteurized liquid whole egg (Wrinkle *et al.*, 1950)

	Unpasteurized		Pasteurized	
Genera	Before freezing	After freezing	Before freezing	After freezing
Acinetobacter–Moraxella	0[a]	2	–	–
Enterobacter	3	0	–	–
Alcaligenes	25	20	4	8
Bacillus	7	2	83	84
Chromobacterium	2	2	–	–
Escherichia	6	7	3	0
Flavobacterium	29	27	4	0
Gram^{+} cocci	5	4	3	0
Proteus	16	18	4	8
Pseudomonas	7	16	–	–
Salmonella	2	0	–	–
Streptothrix	0	2	–	–

[a]Percent of isolates.

promptly. Temperatures for prolonged storage should be at or below −18°C; a few microorganisms can grow slowly at or above −10°C (Michener and Elliott, 1964).

Freezing and frozen storage reduce the numbers of microorganisms but usually not to the point of extinction of a given strain, particularly not in the protective proteinaceous egg matrix. While freezing and frozen storage reduce salmonellae, these cannot be relied on to eliminate the pathogen. *L. monocytogenes* levels remained unchanged in frozen (−18°C) liquid whole egg during 6 months of storage (Brackett and Beuchat, 1991). An example of the effects of freezing on pasteurized and unpasteurized whole egg is depicted in Table 15.17.

Improper thawing can lead to unacceptable increases in microbial content (Forsythe, 1970). For example, when a can of liquid eggs is allowed to thaw completely in a warm room, the temperature of the outside portion will have been in the temperature range that supports bacterial growth for many hours. Frozen products should be thawed under conditions that permit the temperature of the thawed material to rise to >4°C for only short periods. Some manufacturers thaw in a refrigerator at about 4°C; others immerse the cans in cold running water; others have installed crushers to break up the partly frozen mass; still others remove by centrifuge or otherwise, the liquid egg material as it thaws (Lawler, 1965).

Special treatment with alcohol. When used for the production of egg containing liqueurs, egg products receive less than the recommended heat treatment. The preservative effect of alcohol kills salmonellae within 6 days of storage when the alcohol content is 13% and above (Bolder *et al.*, 1987; Warburton *et al.*, 1993).

Irradiation. It is technically feasible to control salmonellae by using ionizing radiation (Comer *et al.*, 1963; Schaffner *et al.*, 1989; Slater and Sanderson, 1989; Kijowski *et al.*, 1994). In the United States, ionizing radiation in doses up to 3.0 kGy may be used on shell eggs. A dose of 3.0 kGy is sufficient to eliminate salmonellae and Enterobacteriaceae from liquid egg white, egg yolk, and whole egg with and without added 50% sugar or 11% salt. *Salmonella* Enteritidis appears to be somewhat more radiation resistant than *S.* Typhimurium (Thayer *et al.*, 1990). The combination of irradiation and thermal processing has been shown to reduce the requirements of both treatments to eliminate *S.* Enteritidis in liquid whole egg (Schaffner *et al.*, 1989).

B Spoilage

The contaminating organisms at time of breaking are primarily those on the egg-shell (Table 15.10) and within the occasional spoiled egg. If not pasteurized immediately, the broken out eggs must be cooled promptly to 7°C or below, especially if cracked or dirty eggs were used.

After the liquid eggs are pasteurized, they should again be cooled promptly. Although organisms that spoil eggs under refrigeration have largely been destroyed (Speck and Tarver, 1967), a delay in cooling will permit mesophilic bacteria to grow and will permit the few psychrotrophs present to build up rapidly. In one study, unpasteurized whole eggs held at 4°C for 8–10 days had bacterial levels of 3×10^6 cfu/g (Steele *et al.*, 1967). Pasteurized eggs, frozen immediately and then stored at 4°C for 45 days, had fewer than 100 cfu/g. Pasteurized eggs held at 13°C for 24 h before freezing, followed by storage at 4°C, developed 3×10^6 cfu/g in 24 days.

The refrigerated shelf life of pasteurized egg products (i.e. time until spoilage is evident) is remarkably long. Clean eggs that are broken, cooled, mixed, pasteurized, and cooled under ideal sanitary conditions, remain edible for 20–22 days in the refrigerator (Wilkin and Winter, 1947; Kraft *et al.*, 1967a). If made from dirty eggs, even though the egg product is pasteurized, the refrigerated shelf life may be only 2 or 3 days; more microorganisms will survive because of the high original level of contamination (Baker, 1974). Most samples of pasteurized eggs from commercial operations in the United States had a shelf life of 12–15 days at 2°C (Vadehra *et al.*, 1969; York and Dawson, 1973). Before pasteurization came into general use in Europe and North America during the 1960s and early 1970s, the shelf life of refrigerated whole egg was only 5–7 days (Wilkin and Winter, 1947; Wrinkle *et al.*, 1950). Ultrapasteurization systems (i.e. heating over 60°C for <3.5 min) in combination with aseptic packaging systems can produce whole egg products with significantly extended shelf lives, e.g. 3–6 months at 4°C (Ball *et al.*, 1987). It was concluded that ultrapasteurization could be used to effectively produce liquid whole egg that is free from *L. monocytogenes* (Foegeding and Stanley, 1990).

Salted yolks, particularly when batch pasteurized and filled hot into cans, have a long shelf life, even without rapid cooling, and stored at room temperature, because the bacterial vegetative cells capable of growth in 10% salt have been killed by the heat. Eventually a few spores may germinate and grow (Cotterill *et al.*, 1974).

Pasteurization destroys microorganisms such as *Pseudomonas*, *Acinetobacter*, and *Enterobacter* spp.,which grow in raw albumen. This leaves mesophilic organisms like micrococci, staphylococci, *Bacillus* spp., enterococci, and catalase-negative rods able to grow if the product is temperature abused (Barnes and Corry, 1969; Shafi *et al.*, 1970). In one investigation, the primary microorganisms surviving ultrapasteurization were *Bacillus circulans*, a *Pseudomonas* isolate, and *Enterococcus faecalis* (Foegeding and Stanley, 1987). The pseudomonad and the enterococcus were capable of growth at 4°C and 10°C, respectively.

Table 15.18 lists the changes produced by different genera when growing in pure culture in liquid whole egg. Most of the genera would not be expected to survive pasteurization, and their presence would indicate post-pasteurization contamination. The odors of spoilage are much more intense in yolk or whole egg than in white. In white, isolated strains caused no putrid odor or H_2S (Imai, 1976). Most spoilage organisms reduced the pH of the white slightly and produce a small amount of trimethylamine. On the other hand, the yolk develops fishy, moldy, and ammoniacal odors, with high levels of H_2S and trimethylamine and a high total volatile base level.

C Pathogens

Salmonellae, the most important pathogens in liquid egg, are discussed in detail in sections describing the effect of processing procedures on microorganisms and methods of control. Since the heat

Table 15.18 Changes produced by different genera of bacteria originally isolated from liquid egg and then inoculated in pure culture into sterile egg (Wrinkle *et al.*, 1950)

Genus	Number of strains inoculated	Changes produced
Acinetobacter–Moraxella	1	No change in 72 h
Enterobacter	4	Slight acid odor after 72 h
Alcaligenes	17	12 very sour odor in 60 h, 1 musty odor in 48 h, 4 no detectable change
Bacillus	8	6 coagulation within 18 h, 1 very sour odor in 24 h, 1 no detectable change
Chromobacterium	2	No change in 72 h
Escherichia	15	4 slight acid odor in 60 h, 8 sour odor after 60 h, 3 no detectable change
Flavobacterium	11	2 coagulation in 120 h, 1 fecal odor after 60 h, 4 slight odor in 120 h, 4 no change in 72 h
Proteus	10	3 very flat-sour odor in 60 h, 2 coagulation in 18 h, 4 very sour odor in 60 h, 1 no detectable change
Pseudomonas	7	3 very sour odour in 60 h, 2 gas within 60 h,1 sour odor in 60 h, 1 no detectable change
$Gram^{+}$ cocci	3	2 produced gas in 18 h, 1 no detectable change

treatments used for liquid egg products do not produce shelf stable products, they should be held refrigerated (Gibbons *et al.*, 1944). Salmonellae grow rapidly in yolks either aerobically or anaerobically (Lineweaver *et al.*, 1969; Lawler, 1965).

Staphylococcus aureus grows well in liquid whole egg held above 15.6°C. Staphylococci are a potential hazard in salted yolks because they grow readily at the reduced a_w (0.90) of the product. If they survive pasteurization or are re-introduced post-pasteurization, they must reach levels of 10^5 cfu/mL, or more, before toxin forms. This would require severe temperature abuse, e.g. storage of the egg product for several days at ambient temperatures (Ijichi *et al.*, 1973). It would also require the presence of sufficient oxygen since staphylococcal enterotoxins synthesis would not be expected at this a_w under anaerobic conditions.

While pasteurized liquid egg products have not been associated with cases of human listeriosis to date, the presence of *L. monocytogenes* in unpasteurized processed eggs (Leasor and Foegeding, 1989; Nitcheva *et al.*, 1990; Moore and Madden, 1993) has become a significant concern, particularly in products with extended refrigerated shelf life. A survey of commercially broken raw eggs from 11 establishments in the United States detected *Listeria* spp. in 36% of the samples (Leasor and Foegeding, 1989). *Listeria innocua* was found in each of the 15 positive samples, whereas *L. monocytogenes* was present in only 2 (5%). However, it is often difficult to isolate low levels of *L. monocytogenes* when present with *L. innocua*. In another survey, *Listeria* spp. were isolated from 72% of in-line filters from the processing of raw blended whole egg, with *L. innocua* isolated from 62% and *L. monocytogenes* from 38% (Moore and Madden, 1993). The mean level of *Listeria* spp. in the raw liquid whole egg was one organism per mL.

L. monocytogenes has been observed to grow in raw and pasteurized liquid whole egg and yolk at temperatures ranging from 5°C to 30°C, but was inactivated in unpasteurized liquid albumen at pH 7.0–8.9, presumably due to the action of lysozyme (Khan *et al.*, 1975; Foegeding and Leasor, 1990; Sionkowski and Shelef, 1990). Schuman and Sheldon (2003) investigated the use of nisin in pH-adjusted liquid whole egg, and found that the bacteriocin could delay or prevent the growth of *L. monocytogenes*. The generation times among different *L. monocytogenes* strains in liquid whole egg ranged from 24–51 h at 4°C and 8–31 h at 10°C (Foegeding and Leasor, 1990). The pathogen survived for extended periods in liquid whole egg stored at 0°C (Brackett and Beuchat, 1991). Liquid egg products appear to have a low potential for becoming a source of enterohemmorrhagic *Escherichia coli*, since its prevalence in eggs appears low and current pasteurization temperatures are sufficient to inactivate the microorganism (Erickson *et al.*, 1995).

D CONTROL (liquid eggs)

Summary

Significant hazards[a]	• Salmonellae, especially *Salmonella* Enteritidis. • *L. monocytogenes* for products with extended refrigerated shelf life.
Control measures	
Initial level (H_0)	• Same control measures as mentioned for shell-eggs apply. • Somewhat higher H_0 may be expected, due to common practice to use eggs of lesser quality for production of liquid egg products. H_0 up to 100 cfu/g *Salmonella* has been reported.
Increase (ΣI)	• Limit time between lay and processing. • Use high quality, intact eggs. • Chill quickly to below 7°C after break-out and pasteurization. • Keep product at as low a temperature as possible, ideally below 3°C to minimise pathogen proliferation. • Sanitize equipment; use proper process hygiene. • Freezing promptly can be used for long shelf life products.
Reduction (ΣR)	• Pasteurization aiming at a 4 to 5D reduction for salmonellae. • Pasteurization; 60°C for 3.5 min will achieve 2.1–2.7 D reduction (Foegeding & Leasor, 1990). • Lower temperatures need to be used with liquid egg white products, compared with liquid whole egg (yolk and egg white not separated) product.
Testing	• Periodic verification testing for salmonellae. • Coliform, *E. coli*, or Enterobacteriaceae may be useful indicators for process control. • Environmental monitoring for salmonellae would be useful in post pasteurization areas. • α−Amylase activity testing may be useful in certain situations; see subsequent discussion.
Spoilage	• Freeze liquid eggs for prolonged shelf life. • Good Hygienic Practices are essential to control spoilage.

Hazards to be considered. Salmonellae are well recognized as a significant hazard in egg products. *L. monocytogenes* needs to be considered in products with extended refrigerated shelf life, although there is no epidemiological evidence of human listeriosis associated to liquid egg products.

Control measures. To prevent heavy bacterial contamination of liquid egg, eggs for breaking should be washed, candled to remove rots, and examined for odor and appearance at breakout. Equipment that contacts contaminated eggs at breakout should be washed and sanitized before reuse (Forsythe, 1970). Ideally, automated equipment should be designed so that it is cleaned and sanitized on a continuous basis. All product-contact equipment such as pipes, churns, tanks, and pails should be thoroughly cleaned and sanitized at least daily. Equipment should be designed and installed for easy cleaning, as for instance described in E-3-A Sanitary Standards (IAMFES, 1976a–d).

To slow microbial growth, liquid egg products should be chilled to 7°C or below promptly after breakout and after pasteurizing. If the product is to be frozen, cans of eggs should be placed promptly into a freezer at −23 to −40°C and stored at or below −18°C.

Pasteurizers should have the following characteristics (Murdock *et al.*, 1960; Lineweaver *et al.*, 1969; Forsythe, 1970; Kaufman, 1969, 1972):

- Automatic flow control, including flow diversion valves to divert inadequately heated egg.
- Automatic control of temperature.
- Higher pressure on the pasteurized side than on the unpasteurized to prevent raw melange from leaking into the pasteurized material.
- Recording thermometers at entrance and exit of the pasteurizer and the cooler.
- Leak detection devices.

Because of the marginal nature of the thermal processing used to pasteurize liquid egg products, it important that the levels of *Salmonella* and *L. monocytogenes* are as low as possible in the raw product. Excessively high levels of these pathogens will not be completely inactivated. This requires that eggs used have both a low numbers and prevalence of the two pathogens. Particular care is needed for products intended for extended refrigerated storage due to the inherent heat resistance of *L. monocytogenes*, its ubiquitous presence in processing environments, and its ability to grow at refrigeration temperatures. In addition to adequate heat treatment, such products should be kept as cold as possible, preferably close to freezing to avoid or minimise pathogen proliferation. Temperature abuse substantially increases the risk of growth of both pathogens.

Testing. The efficacy of integrated processes used to produce and pasteurize liquid egg products should be validated and periodically verified by appropriate laboratory testing to assure that they are capable of achieving the degree of microbial inactivation needed. The egg products industry has traditionally relied heavily on microbiological testing programs that have focused on testing for salmonellae and either coliforms, *E. coli*, or Enterobacteriaceae. These organisms should not be found in the pasteurized material. The rationale for testing for salmonellae is obvious, i.e. the pasteurization is designed to eliminate this organism. However, most of the egg entering the pasteurizer does not contain salmonellae, so coliforms, *E. coli*, or Enterobacteriaceae have been used as process integrity indicators (i.e. heating sufficiency). These indicator organisms are virtually always present in raw egg melange and their heat resistance is similar to that of the salmonellae. *Escherichia coli* is also an indicator of temperature abuse since it would also not be expected to reach substantial numbers unless the product was temperature abused. However, commercially produced egg products sometimes (though rarely) contain salmonellae in 25 g portions when other Enterobacteriaceae are not detected in 0.1 or 1 g samples (van Schothorst and van Leusden, 1977). A positive finding for Enterobacteriaceae or coliforms could mean either that pasteurization was inadequate or that post-pasteurization contamination occurred. While microbiological testing of this type helps verify process adequacy, the level of testing is insufficient to assure safety on a lot-by-lot basis. The resources that have traditionally been expended in testing of this type would be better used to develop, validate, and implement better process controls.

In some countries (e.g. the United Kingdom), an enzyme assay for α-amylase is used to verify the efficacy of pasteurization (Brooks and Shrimpton, 1962; Murdock *et al.*, 1960; Shrimpton *et al.*, 1962). At the temperatures and times preferred in the United Kingdom (64.4°C for 2.5 min for liquid whole egg), α-amylase is destroyed. The α-amylase test is rapid, accurate, convenient, and inexpensive, whereas tests for microorganisms are slow, expensive, and often give variable results among workers and among laboratories. However, α-amylase is not destroyed by the lower heating temperatures preferred in the United States (60°C for 3.5 min; Lineweaver *et al.*, 1969) and other countries, and is not applicable in such instances. The α-amylase test cannot be used with salted or sugared eggs, and does not detect post-pasteurization contamination.

Two methods are employed to detect the use of incubator reject eggs (Robinson *et al.*, 1975):

- An enzyme assay comparable to the α-amylase test;
- A test for 3-hydroxy butyric acid, formed upon inhibition of embryonic growth within the egg.

V Dried eggs

A Effects of processing on microorganisms

Three methods are widely used for the dehydration of liquid egg products.

- Spray drying, where the liquid is atomized by squirting through a nozzle.
- Drying on a heated surface (pan or drum drying).
- Freeze drying.

Water may be removed by ultra filtration or reverse osmosis before final drying.

The microbiology of all three is essentially the same. Drying kills many of the bacteria initially present in the liquid egg. However, once the egg material is dry, the microbial population is stabilized, and further declines occur only slowly with extended storage, even at ambient temperatures. Only rarely is there a complete extinction of surviving strains. The predominant microorganisms in the dried product are enterococci and aerobic spore-forming bacilli, the most resistant members of the original microflora. The number of salmonellae may be reduced as much as 4 logs during drying (Gibbons and Moore, 1944). Fermented albumen or temperature-abused whole egg can have high initial bacterial levels, such that survival of some bacteria is likely (Ayres and Slosberg, 1949; Gibbons *et al.*, 1944). Salmonellae are the principal microbial problem in dried eggs, and the problem becomes more serious if they grow during fermentation for glucose removal.

Glucose removal (fermentation). Dried whites normally contain about 0.6% free carbohydrate, primarily as glucose. On storage, particularly above 15°C, the aldehyde group of the glucose combines with the amino groups of the proteins, reducing their solubility, causing off flavors, and forming insoluble brown compounds (Maillard reaction products). Removal of glucose from the liquid egg white before drying prevents these reactions. Glucose removal also improves to a lesser extent the stability of dried whole egg and egg yolk (Stewart and Kline, 1941; Paul *et al.*, 1957; Forsythe, 1970; Kilara and Shahini, 1973).

The earliest method used to remove glucose was simply to permit the natural egg microflora to grow at temperatures of 21–29°C for 2–7 days. The length of time was judged by observations of bubbling, consistency, and clarity of samples. Scums and sediments were discarded and ammonia added to clear the liquid. The reactions were not well controlled, and often led to objectionable odors and proteolysis. This procedure allowed the growth of enterococci, *Enterobacter aerogenes*, and other bacteria. In addition, salmonellae could grow, presenting a health hazard (Ayres, 1958).

Some manufacturers add bacterial starter cultures to rapidly ferment egg whites. The temperature is raised to 35°C so that the glucose is metabolized within 12–24 h. Enterococci are favored; *Enterobacter* spp. and lactic organisms are not as competitive at the natural pH of egg white (>9.0), although all three organisms are used by different manufacturers. Enterococci do not cause proteolysis or off-odors, but do acidify the whites to approximately pH 6. This fermentation method may permit the rapid growth of salmonellae after the pH drops below 8.

Table 15.19 Commercial and experimental methods for the removal of glucose from liquid eggs

Microorganism or agent	Comment	Reference
Natural flora	*Enterobacter*, enterococci, and other bacteria	Ayres (1958)
Coliforms	Early pure culture studies	Stuart and Goresline (1942a, 1942b)
Saccharomyces apiculatis	1% inoculum gave yeasty flavor	Hawthorne and Brooks (1944)
Saccharomyces cerevisiae	Yeasty odor from large inocula can be eliminated by 0.1% yeast extract which stimulates activity of small inocula. Can centrifuge to remove yeast.	Ayes and Stewart (1947), Hawthorne (1950), Carlin and Ayres (1953), Kline and Sonoda (1951), Ayres (1958)
Streptococcus lactis and *Streptococcus faecalis* subsp. *liquefaciens*	Resting cells, 37°C for 3 h, yeast extract at 0.1% inhibits acid production.	Kaplan *et al.* (1950), Ayres (1958), Galuzzo *et al.* (1994)
Glucose oxidase and catalase	Glucose oxidized to gluconic acid, catalase destroys the H_2O_2 that is formed.	Baldwin *et al.* (1953), Carlin and Ayres (1953), Scott (1953), Paul *et al.* (1957), Ayres (1958)
Enterobacter aerogenes	Acetyl methyl carbinol production can be minimized by using small inocula with 0.1% yeast extract for stimulation.	Ayres (1958)
Cell-free yeast extracts	Desugars in 4–5 h at 5°C.	Niewiarowicz *et al.* (1967)
Escherichia coli	Shows antagonism for *Salmonella*.	Mickelson and Flippin (1960), Flippin and Mickelson (1960)
Lactobacillus brevis, *Lactobacillus casei*, *Lactobacillus fermenti*, *Lactobacillus plantarum*	25°C optimal temperature, lactobacilli eliminated by pH adjustment prior to hot room treatment	Mulder and Bolder (1988)

It is generally recommended that manufacturers employ pure culture starters. The first step is to reduce the pH of the egg from its normal 9.0 to 7.0–7.5 with an organic acid such as lactic acid. The starter culture is then added, and allowed to ferment the available carbohydrate for 12–24 h at 30–33°C. It has been claimed that bacterial cultures are best because the finished product has high whipping quality, good odor, and solubility (Forsythe, 1970). However, others prefer alternative yeast or enzyme treatments (Table 15.19). Glucose oxidase treatment has been investigated as a potential means of inactivating *Salmonella* Enteritidis and other bacteria in liquid whole egg (Dobbenie *et al.*, 1995).

In many of the fermentation procedures listed in Table 15.19, salmonellae can grow if the pH range of the egg white is reduced to 6–8, but not if the pH is maintained at ≥ 9 (Banwart and Ayres, 1957). After fermentation, pasteurization is essential to kill salmonellae (Kline and Sonoda, 1951; Ayres and Stewart, 1947), so that any present are not carried into the dried product. In the United States, whites are usually fermented, dried, and then pasteurized.

Destruction of salmonellae by hot storage. Despite the rather efficient procedures developed to pasteurize liquid eggs prior to drying, salmonellae are occasionally present in the final dried packaged product. This can be due to insufficient pasteurization, excessively high initial numbers, or post-pasteurization contamination. After the product has been dried, microorganisms can still be destroyed by hot storage (hot room treatment).

Microorganisms are most heat sensitive when wet; their heat resistance increasing as the environment becomes dry. While the pasteurization times for a wet product are typically in the range of 2–5 min, those for dried egg products are several days. Examples of times and temperatures for "pasteurization" of dried egg white are listed in Table 15.20. These exposures have no serious effect on functional qualities.

Table 15.20 Times and temperatures of hot room storage to destroy salmonellae in dried egg albumen

Pretreatment	Temperature (°C)	Time (days)	Reference
Fermented, pan dried	48.9	20	Ayres and Slosberg (1949)
	54.4	8	
	57.2	4	
3% Moisture	50	9	Banwart and Ayres (1956)
6% Moisture	50	6	Banwart and Ayres (1956)
Spray dried	54.4	7	USDA (1975b)
Pan dried	51.7	5	USDA (1975b)
Adjusted to pH 9.8 with ammonia, pan dried	49	14	Northolt *et al.* (1978)
Treated with citric acid	55	14	Northolt *et al.* (1978)
Spray dried	49	14	Northolt *et al.* (1978)

The advantage of hot storage over the standard wet pasteurization is that there is little possibility of recontamination as long as the containers remain closed during and after the treatment period. The bactericidal effect of the hot room treatment depends on the moisture content, the temperature, and the nature of treatments preceding the heat treatment (e.g. method of fermentation, use of ammonia or citric acid, method of drying) (Banwart and Ayres, 1956; Carlson and Snoeyenbos, 1970; Northolt *et al.*, 1978). As the time required for inactivation of all the organisms depends upon the number originally present, potentially a shorter treatment could be applied to materials with low levels of contamination.

Investigators have demonstrated that salmonellae in egg powders can also be inactivated by irradiation (Matic *et al.*, 1990; Narvaiz *et al.*, 1992). The irradiation resistance of a mixture of *S.* Enteritidis, *S.* Typhimurium, and *S.*Lille was 0.8 kGy; a reduction of 10^3 required 2.4 kGy. A similar degree of inactivation was achieved by irradiating egg powder at 1.0 kGy, and holding the product for 3 weeks.

B Spoilage

Potentially, microorganisms could remain alive indefinitely, but not grow. However, most bacteria die slowly over time depending on a variety of factors such as species, temperature, pH, water activity, and atmosphere. On reconstitution with water or on accidental wetting during storage, surviving bacteria will grow resuscitate and spoil the product.

C Pathogens

Salmonella-contaminated dried egg has caused numerous outbreaks of food-borne illness by direct ingestion, and additional outbreaks by the way of other foods cross-contaminated by dried egg in the kitchen. Before 1965, dried eggs were often contaminated. Subsequently, low levels (0.1–0.01 cfu/g) of *S.* Enteritidis, *S.* Typhimurium, and *S.* Lille were isolated from whole egg powder produced in Yugoslavia (Matic *et al.*, 1990).

Since that time, pasteurization, either wet or dry, has come into general use. National governments have also established and enforced regulations requiring pasteurization. The incidence of salmonellae in dried egg products is now typically at a low, but largely undetermined level.

L. monocytogenes can survive essentially unchanged for extended periods in dried powdered whole egg and egg yolk stored at refrigeration temperatures, but declines over time when stored at 20°C (Brackett and Beuchat, 1991).

D CONTROL (dried eggs)

Summary

Significant hazards[a]	• Salmonellae, especially *Salmonella* Enteritidis.
Control measures	
Initial level (H_0)	• Same control measures as mentioned for shell-eggs apply.
Increase (ΣI)	• Use proper equipment (no cracks etc., integrity). • Sanitize equipment, use proper process hygiene. • Avoid recontamination during processing. • Maintain dry product and production environment.
Reduction (ΣR)	• Hot storage may reduce levels of salmonellae, with reductions influenced by moisture levels, temperature, and other factors.
Testing	• *Salmonella* monitoring program including environmental and finished product.
Spoilage	• Dry conditions prevent spoilage.

Control measures. Drying may reduce contamination, but no quantitative data are available; it is probably not a robust process for reducing contamination and should not be included in the product and process design.

The procedures described in earlier sections apply also to dried eggs. In addition, the drier should be made of impervious materials without cracks, crevices, and dead-end pockets where product can remain wet and warm. Dried eggs, in common with other dried foods, must be protected against condensation dripping into the dried product. Frequently tank lids, drier hoods, and conveyor covers are cooler than the stream of dried egg, so that condensate forms on their undersides. Dust from the product contributes nutrients so that bacteria, including salmonellae, can grow in the droplets. When the droplets or clumps of damp egg fall back into the main mass they can contribute localized areas of *Salmonella* contamination. Sampling and analysis are a difficult means to detect such infrequent and localized contamination. The best way to control the problem is to prevent condensation. Warming the surfaces of equipment where condensation occurs is one answer.

Cleaning with water of an area used for a dried product may introduce additional hazards. Instead, a vacuum cleaner, dry brushes, and cloths may be used, followed by 95% ethyl alcohol as a disinfectant. If it seems advisable to use water, the equipment must be thoroughly and rapidly dried before reuse.

VI Further processed egg products

There is a wide variety of processed products containing egg. Products such as meringue pie, mousses, marzipan, egg nog, or dry diet mixes, when not or insufficiently cooked, remain potential hazards from salmonellae that might have survived or were re-introduced after pasteurization. Cooked foods such as custard, cream cakes, and angel cake could be cross-contaminated in the preparation area from a contaminated egg ingredient, particularly from a dried egg product. Baked goods that reach $\geq 71°C$ are safe from salmonellae.

References

ACMSF (Advisory Committee on the Microbiological Safety of Food). (1993) *Salmonella in Eggs*, Her Majesty's Stationery Office, London (ISBN 0 11 321568 1).

ACMSF (Advisory Committee on the Microbiological Safety of Food) (1996) *Report on Poultry Meat*, Her Majesty's Stationery Office, London (ISBN 0 11 321969 5).

ACMSF (Advisory Committee on the Microbiological Safety of Food). (2001) *Second Report on Salmonella in Eggs*. Her Majesty's Stationery Office, London (ISBN 0 11 322466 4).

Acuff, G.R.C., Vanderzant, C., Gardner, F.A. and Golan, F.A. (1982) Examination of turkey eggs, poults and brooder house facilities for *Campylobacter jejuni*. *J. Food Prot.*, **45**, 1279–81.

AEB (American Egg Board). (2003) www.aeb.org.

Alford, L.R., Holmes, N.E., Scott, W.J. and Vickery, J.R. (1950) Studies in the preservation of shell eggs. I. The nature of wastage in Australian export eggs. *Aust. J. Appl. Sci.*, **1**, 208–14.

Altekruse, S.F., Tollefson, L.K. and Bögel, K. (1993) Control strategies for *Salmonella enteritidis* in five countries. *Food Control*, **4**(1), 10–6.

Amin, M.K. and Draughon, F.A. (1990) Infection of shell eggs with *Yersinia enterocolitica*. *J. Food Prot.*, **53**, 826–30.

Angulo, F.J. and Swerdlow, D.L. (1999) Epidemiology of human *Salmonella enterica* serovar *enteritidis* in the United States, in *Salmonella enterica Serovar enteritidis in Humans and Animals* (ed A.M. Saeed), Iowa State University Press, Ames, Iowa, pp. 33–42.

Ayres, J.C. (1958) Methods for depleting glucose from egg albumen before drying. *Food Technol.*, **12**, 186–9.

Ayres, J.C. (1960) The relationship of organisms of the genus *Pseudomonas* to the spoilage of meat, poultry, and eggs. *J. Appl. Bacteriol.*, **23**, 471–86.

Ayres, J.C. and Slosberg, H.M. (1949) Destruction of *Salmonella* in egg albumen. *Food Technol.*, **3**, 180–3.

Ayres, J.C. and Stewart, G.F. (1947) Removal of sugar from raw egg white by yeast before drying. *Food Technol.*, **1**, 519–26.

Ayres, J.C. and Taylor, B. (1956) Effect of temperature on microbial proliferation in shell eggs. *Appl. Microbiol.*, **4**, 355–9.

Bailey, J.S., Fletcher, D.L. and Cox, N.A. (1989) Recovery and serotype distribution of *Listeria monocytogenes* from broiler chickens in the southeastern United States. *J. Food Prot.*, **52**, 148–50.

Baker, R.C. (1974) Microbiology of eggs. *J. Milk Food Technol.*, **37**, 265–8.

Baker, R.C. (1990) Survival of *Salmonella enteritidis* on and in shelled eggs, liquid eggs and cooked egg products. *Dairy, Food Environ. San.*, **10**, 273–5.

Baker, R.C., Hogarty, S., Poon, W. and Vadhera, D.V. (1983) Survival of *Salmonella typhimurium* and *Staphylococcus aureus* in eggs cooked by different methods. *Poult. Sci.*, **62**, 1211–6.

Baldwin, R.R., Campbell, H.A., Thiessen, R., Jr. and Lorant, G.J. (1953) The use of glucose oxidase in the processing of foods with special emphasis on the desugaring of egg white. *Food Technol.*, **7**, 275–82.

Ball, H.R., Jr., Hamid-Samimi, M., Foegeding, P.M. and Swartzel, K.R. (1987) Fuctionality and microbial stability of ultrapasteurized, aseptically packaged refrigerated whole egg. *J. Food Sci.*, **52**, 1212–8.

Banwart, G.J. (1964) Effect of sodium chloride and storage temperature on the growth of *Salmonella oranienburg* in egg yolk. *Poult. Sci.*, **43**, 973–6.

Banwart, G.J. and Ayres, J.C. (1956) The effect of high temperature storage on the content of *Salmonella* and on the functional properties of dried egg white. *Food Technol.*, **10**, 68–73.

Banwart, G.J. and Ayres, J.C. (1957) The effect of pH on the growth of *Salmonella* and functional properties of liquid egg white. *Food Technol.*, **11**, 244–6.

Barnes, E.M. and Corry, J.E.L. (1969) Microbial flora of raw and pasteurized egg albumen. *J. Appl. Bacteriol.*, **31**, 97–107.

Barnhart, H.M., Dreesen, D.W., Bastien, R. and Pancorbo, O.C. (1991) Prevalence of *Salmonella enteritidis* and other serovars in ovaries of layer hens at time of slaughter. *J. Food Prot.*, **54**, 488–91.

Barrow, P.A. (1994) Serological diagnosis of *Salmonella* serotype *enteritidis* infections in poultry by ELISA and other tests. *Int. J. Food Microbiol.*, **21**, 55–68.

Baskerville, A., Humphrey, T.J., Fitzgeorge, R.B., Cook, R.W., Chart, H., Rowe, B. and Whitehead, A. (1992) Airborne infection of laying hens with *Salmonella enteritidis* phage type 4. *Vet. Rec.*, **130**(18), 395–8.

Becirevic, M., Popovic, M. and Becirevic, N. (1988) Influence of iron and water on the growth of *S. typhimurium* and enteropathogenic *E. coli* in eggs and albumen. *Vet Yugoslav.*, **37**, 41–9.

Berrang, M.E., Frank, J.F., Buhr, R.J., Bailey, J.S. and Cox, N.A. (1999) Eggshell membrane structure and penetration by *Salmonella typhimurium*. *J. Food Prot.*, **62**, 73–6.

Bierer, B.W., Valentine, H.D., Barnett, B.D. and Rhodes, W.H. (1961a) Germicidal efficiency of egg washing compounds on eggs artificially contaminated with *Salmonella typhimurium*. *Poult. Sci.*, **40**, 148–52.

Bierer, B.W., Barnett, B.D. and Valentine, H.D. (1961b) Experimentally killing *Salmonella typhimurium* on egg shells by washing. *Poult. Sci.*, **40**, 1009–14.

Binkin, N., Scuderi, G., Novaco, F., Giovanardi, G.L., Paganelli, G., Ferrari, G., Cappelli, O., Ravaglia, L., Zilioli, F., Amadei, V., Magliani, W., Viani, I., Ricco, D., Borrini, B., Magri, M., Alessandrini, A., Bursi, G., Barigazzi, G., Fantasia, M., Filetici, E. and Salmaso, S. (1993) Egg-related *Salmonella enteritidis*, Italy, 1991. *Epidemiol. Infect.*, **110**(2), 227–37.

Board, P.A., Hendon, L.P. and Board, R.G. (1968) The influence of iron on the course of bacterial infection of the hen's egg. *Br. Poult. Sci.*, **9**, 211–5.

Board, R.G. (1964) The growth of gram-negative bacteria in the hen's egg. *J. Appl. Bacteriol.*, **27**, 350–64.

Board, R.G. (1965a) Bacterial growth on and penetration of the shell membranes of the hen's egg. *J. Appl. Bacteriol.*, **28**, 197–205.

Board, R.G. (1965b) The properties and classification of the predominant bacteria in rotten eggs. *J. Appl. Bacteriol.*, **28**, 437–53.

Board, R.G. (1969). Microbiology of the hen's egg. *Adv. Appl. Microbiol.*, **11**, 245–81.

Board, R.G. and Halls, N.A. (1973) The cuticle: a barrier to liquid and particle uptake by the shell of the hen's egg. *Br. Poult. Sci.*, **14**, 69–97.

Board, R.G., Ayres, J.C., Kraft, A.A. and Forsythe, R.H. (1964) The microbiological contamination of egg shells and egg packing materials. *Poult. Sci.*, **43**, 584–95.

Board, R.G., Clay, C., Lock, J. and Dolman, J. (1994) The egg: A compartmentalized aseptically packaged food, in *Microbiology of the Avian Egg* (eds R.G. Board and R. Fuller), Chapman and Hall, London, pp. 43–61.

Bolder, N.M., van der Hulst, M.C. and Mulder, R.W.A.W. (1987) Survival of spoilage and potentially pathogenic microorganisms in egg nog. *Lebensmittel Wissenschaft Technol.*, **20**, 151–4.

Brackett, R.E. (1988) Presence and persistence of *Listeria monocytogenes* in food and water. *Food Technol.*, **42**, 162–4.

Brackett, R.E. and Beuchat, L.R. (1991) Survival of *Listeria monocytogenes* in whole egg and egg yolk powders and in liquid whole eggs. *Food Microbiol.*, **8**, 331–7.

Brackett, R.E., Schuman, J.D., Ball, H.R. and Scouten, A.J. (2001) Thermal inactivation kinetics of *Salmonella* spp. within intact eggs heated using humidity-controlled air. *J. Food Prot.*, **64**, 934–8.

Bradshaw, J.D., Shak, D.B., Forney, E. and Madden, J.M. (1990) Growth of *Salmonella enteritidis* in yolk of shell eggs from normal and seropositive hens. *J. Food Prot.*, **53**, 1033–6.

Brant, A.W. and Starr, P.B. (1962) Some physical factors related to egg spoilage. *Poult. Sci.*, **41**, 1468–73.

Braun, P. and Fehlhaber, K. (1995) Migration of *Salmonella enteritidis* from the albumen into the egg yolk. *Int. J. Food Microbiol.*, **25**, 95–9.

Brooks, J. (1960) Mechanism of the multiplication of *Pseudomonas* in the hen's egg. *J. Appl. Bacteriol.*, **23**, 499–503.

Brooks, J. and Shrimpton, D.H. (1962) α-amylase in whole egg and its sensitivity to pasteurization temperatures. *J. Hyg.*, **60**, 145–51.

Brown, W.E., Baker, R.C. and Naylor, H.B. (1965) The role of the inner shell membrane in bacterial penetration of chicken eggs. *Poult. Sci.*, **44**, 1323–7.

Brown, W.E., Baker, R.C. and Naylor, H.B. (1970) The effect of egg position in storage on susceptibility to bacterial spoilage. *Can. Inst. Food Technol. J.*, **3**, 29–32.

Bruce, J. and Drysdale, E.M. (1983) The bacterial flora of candling reject and dead-in-shell turkey eggs. *Br. Poult. Sci.*, **24**, 391–5.

Bruce, J. and Drysdale, E.M. (1994) Trans-shell transmission, in *Microbiology of the Avian Egg* (eds R.G. Board and R. Fuller), Chapman and Hall, London, pp. 63–91.

Bruce, J. and Johnson, A.L. (1978) The bacterial flora of unhatched eggs. *Br. Poult. Sci.*, **19**, 681–9.

Buchner, L., Wermter, R., Henkel, S. and Ahne, B. (1992) Aum Nachweis von *Salmonellen* in Huehnererern unter Berueck-sichligung eines Stichprobenplanes in Jahr 1991 (Detection of *Salmonella* in eggs based on a 1991 sampling survey). *Arch. Lebensmittelhyg.*, **43**, 99–100.

Caffer, M.I. and Eiguer, T. (1994) *Salmonella enteritidis* in Argentina. *Int. J. Food Microbiol.*, **21**, 15–9.

Cantor, A. and McFarlane, V.H. (1948) *Salmonella* organisms on and in chicken eggs. *Poult. Sci.*, **27**, 350–5.

Carlin, A.F. and Ayres, J.C. (1953) Effect of the removal of glucose by enzyme treatment on the whipping properties of dried albumen. *Food Technol.*, **7**, 268–70.

Carlson, V.L. and Snoeyenbos, G.H. (1970) Effect of moisture on salmonellae populations in animal feeds. *Poult. Sci.*, **49**, 717–25.

Carter, T.A., Gentry, R.F. and Bressler, G.O. (1973) Bacterial contamination of hatching eggs and chicks produced by broiler systems. *Poult. Sci.*, **52**, 2226–36.

Catalano, C.R. and Knabel, S.J. (1994a) Incidence of *Salmonella* in Pennsylvania egg processing plants and destruction by high pH. *J. Food Prot.*, **57**, 587–91.

Catalano, C.R. and Knabel, S.J. (1994b) Destruction of *Salmonella enteritidis* by high pH and rapid chilling during simulated commercial egg processing. *J. Food Prot.*, **57**, 592–5.

CDC (Centers for Disease Control). (1992) Outbreak of *Salmonella enteritidis* infection associated with consumption of raw shell eggs, 1991. *Morb. Mort. Wkly Rep.*, **41**, 369–72.

CDC (Centers for Disease Control). (2000) Outbreaks of *Salmonella* serotype Enteritidis infection associated with eating raw or undercooked shell eggs—United States, 1996-1998. *Morb. Mort. Wkly Rep.*, **49**(4), 73–9.

CDC (Centers for Disease Control). (2003) Data provided via FoodNet on www.cdc.gov/foodnet//default/htm.

CFIS (Canadian Food Inspection Service) (2003). www.inspection.gc.ca/english/anima/eggoeu/eggoeue.shtml.

Chang, P.K., Powrie, W.D. and Fennema, O. (1970) Sodium hexametaphosphate effect on the foam performance of heat-treated and yolk-contaminated albumen. *Food Technol.*, **24**, 63–7.

Chapman, P.A., Rhodes, P. and Rylands, W. (1988) *Salmonella typhimurium* phage type 141 infections in Sheffield during 1984 and 1985: Association with hens' eggs. *Epidemiol. Infect.*, **101**, 75–82.

Clay, C.E. and Board, R.G. (1991) Growth of *Salmonella enteritidis* in artificially contaminated hens' shell eggs. *Epidemiol. Infect.*, **106**, 271–81.

Cogan, T.A. and Humphrey T.J. (2003) The rise and fall of *Salmonella* Entertidis in the UK. *J. Appl. Microbiol.*, **94**, 114S–119S (supplement).

Cogan, T.A., Domingue, G., Lappin-Scott, H.M., Benson, C.E., Woodward, M.J. and Humphrey T.J. (2001) Growth of Salmonella enteritidis in artificially contaminated eggs: the effects of inoculum size and suspending media. *Int. J. Food Microbiol.*, **70**, 131–41.

Comer, A.G., Anderson, G.W. and Garrard, E.H. (1963) Gamma irradiation of *Salmonella* species in frozen whole egg. *Can. J. Microbiol.*, **9**, 321.

Cooper, G.L., Nicholas, R.A.J. and Bracewell, C.D. (1989) Serological and bacteriological investigations of chickens from flocks infected with *Salmonella enteritidis*. *Vet. Rec.*, **125**, 567–72.

Corry, J.E.L. and Barnes, E.M. (1968) The heat resistance of salmonellae in egg albumen. *Br. Poult. Sci.*, **9**, 253–60.

Cotterill, O.J. (1968) Equivalent pasteurization temperatures to kill salmonellae in liquid egg products at various pH levels. *Poult. Sci.*, **47**, 354–65.

Cotterill, O.J. and Glauert, J. (1972) Destruction of *Salmonella oranienburg* in egg yolk containing various concentrations of salt at low temperatures. *Poult. Sci.*, **51**, 1060–1.

Cotterill, O.J., Glauert, J., Steinhoff, S.E. and Baldwin, R.E. (1974) Hot-pack pasteurization of salted egg products. *Poult. Sci.*, **53**, 636–45.

Cox, N.A., Bailey, J.S., Maudlin, J.M. and Blakenship, L.C. (1990) Presence and impact of salmonella contamination in commercial broiler hatcheries. *Poult. Sci.*, **69**, 1606–9.

Cowden, J.M., Lynch. D., Joseph, C.A., O'Mahony, M., Mawer, S.L., Rowe, B. and Bartlett, C.L. (1989) Case-control study of infections with *Salmonella enteritidis* phage type 4 in England. *Br. Med. J.*, **299**, 771–3.

Cunningham, F.E. (1966) Process for pasteurizing liquid egg white, in *The Destruction of Salmonellae*, ARS 74-37, USDA Agric. Res. Serv., Albany, CA, pp. 61–5.

Cunningham, F.E. (1990) Egg production pasteurization, in *Egg Science and Technology* (eds W.J. Stadelman and O.J. Cotteril), 3rd edn, Haworth Press, Bringhamton, NY.

Cunningham, F.E. and Lineweaver, H. (1965) Stabilization of egg-white proteins to pasteurizing temperatures above 60°C. *Food Technol.*, **19**, 136–41.

Curtis, P.A., Anderson, K.E. and Jones, F.T. (1995) Cryogenic gas for rapid cooling of shelleggs before packaging. *J. Food Prot.*, **58**, 389–94.

D'Aoust, J.-Y. (1991) Psychrotrophy and foodborne *Salmonella*. *Int. J. Food Microbiol.*, **13**, 207–16.

Davies, R.H. and Wray, C. (1995) Observations on disinfection regimens used on *Salmonella enteritidis* infected poultry units. *Poult. Sci.*, **74**, 638–47.

Davies, R.H. and Breslin, M. (2003) Investigation of *Salmonella* contamination and disinfection in farm egg-packing plants. *J. Appl. Microbiol.*, **94**, 191–6.

Dawson, P.S. (1992) Control of *Salmonella* in poultry in Great Britain. *Int. J. Food Microbiol.*, **15**, 215–7.

Dobbenie, D., Uyttendaele, M. and Debevere, J. (1995) Antibacterial activity of the glucose oxidase/glucose system in liquid whole egg. *J. Food Prot.*, **58**, 273–9.

Dolman, J. and Board, R.G. (1992) The influence of temperature on the behaviour of mixed bacterial contamination of the shell membrane of the hen's egg. *Epidemiol. Infec.*, **108**, 115–21.

Doyle, M.P. (1984) Association of *Campylobacter jejuni* with laying hens and eggs. *Appl. Environ. Microbiol.*, **47**, 533–6.

Dreesen, D.W., Barnhart, H.M., Burke, J.L., Chen, T. and Johnson, D.C. (1992) Frequency of *Salmonella enteritidis* and other salmonellae in the ceca of spent hens at time of slaughter. *Avian Dis.*, **36**, 247–50.

Duguid, J.P. and North, R.A.E. (1991) Egg and *Salmonella* food-poisoning: An evaluation. *J. Med. Microbiol.*, **34**, 65–72.

Ebel, E.D., David, M.J. and Mason, J. (1992) Occurrence of *Salmonella enteritidis* in the U.S. commercial egg industry: Report of a national spent hen survey. *Avian Dis.*, **36**, 646–54.

Edel, W. (1994) *Salmonella enteritidis* eradication programme in poultry breeder flocks in The Netherlands. *Int. J. Food Microbiol.*, **21**, 171–8.

EC (European Commission) (1991) OJ L 121, 16.5.1991, p. 11. Article 5(2) of Commission Regulation (EEC) No 1274/91 of 15 May 1991 introducing detailed rules for implementing Regulation (EEC) No 1907/90.

EC (European Commission) (1989) Council Directive 89/437/EEC on hygiene and health problems affecting the production and placing on the market of egg products (OJ L211, p87, 2/7/1989).

EC (European Commission) (2003) final report from the commission to the council with regard to developments in consumption, washing and marking of eggs. COM(2003) 479. Brussels 6.8.2003.

Elliott, L.E. and Brant, A.W. (1957) Effect of saline and egg shell membrane on bacterial growth. *Food Res.*, **22**, 241–50.

Elliott, R.P. (1954) Spoilage of shell eggs by pseudomonads. *Appl. Microbiol.*, **2**, 158–64.

Elliott, R.P. (1958) Determination of pyoverdine, the fluorescent pigment of pseudomonads in frozen whole egg. *Appl. Microbiol.*, **6**, 247–51.

Elliott, R.P. and Heiniger, P.K. (1965) Improved temperature-gradient incubator and the maximal growth temperature and heat resistance of *Salmonella*. *Appl. Microbiol.*, **13**, 73–6.

Elliott, R.P., Straka, R.P. and Garibaldi, J.A. (1964) Polyphosphate inhibition of growth of pseudomonads from poultry meat. *Appl. Microbiol.*, **12**, 517–22.

Erickson, J.P., Stamer, J.W., Hayes, M., McKenna, D.N. and VanAlstine, L.A. (1995) An assessment of *Escherichia coli* O157:H7 contamination risks in commercial mayonnaise from pasteurized eggs and environmental sources, and behavior low-pH dressing. *J. Food Prot.*, **58**, 1059–64.

Fajardo, T.A., Anantheswaran, R.C., Puri, V.M. and Knabel, S.J. (1995) Penetration of *Salmonella enteritidis* into eggs subjected to rapid cooling. *J. Food Prot.*, **58**, 473–7.

Fantasia, M. and Filetici, E. (1994) *Salmonella enteritidis* in Italy. *Int. J. Food Microbiol.*, **21**, 7–13.

FAO/WHO (2002) Risk Assessment on Salmonella in eggs and broiler chickens. Interpretive summary, *Microbiological Risk Assessment Series 1*, FAO/WHO, ISBN 92-5-104873-8 (full report available at: ftp://ftp.fao.org/es/esn/food/RA_Salmonella_report.pdf).

Favier, G.L., Escudero, M.E. and de Guzman, A.M. (2001) Effect of chlorine, sodium chloride, trisodium phosphate, and ultraviolet radiation on the reduction of *Yersinia enterocolitica* and mesophilic aerobic bacteria from eggshell surface. *J. Food Prot.*, **64**, 1621–3.

FDA (U.S. Food and Drug Administration). (1997) Food safety from farm to table: a national food safety initiative. Report to the president. Available at http://vm.cfsan.fda.gov/~mow/chap1.html

FDA (Food and Drug Administration). (1999) Safe handling label statement and refrigeration at retail requirements for shell eggs [Proposed rule]. *Fed. Reg.*, **64**, 36491–516.

FDA (Food and Drug Administration) (2000) Irradiation in the production, processing, and handling of food [Final rule]. *Fed. Reg.*, **65**, 45280–2.

FDA (U.S. Food and Drug Administration) (2003) Foodborne Pathogenic Microorganisms and natural toxins handbook (Bad Bug Book), updated since Jan 1992. *Salmonella* spp. Available at http://vm.cfsan.fda.gov/~mow/chap1.html.

Feeney, R.E., MacDonnell, L.R. and Lorenz, F.W. (1954) High temperature treatment of shell eggs. *Food Technol.*, **8**, 242–5.

Finch, M.J. and Blake, P.A. (1985) Foodborne outbreaks of campylobacteriosis: The U.S. experience, 1980–1982. *Am. J. Epidemiol.*, **122**, 262–8.

Flippin, R.S. and Mickelson, M.N. (1960) Use of salmonellae antagonists in fermenting egg white. I. Microbial antagonists of salmonellae. *Appl. Microbiol.*, **8**, 366–70.

Florian, M.L.E. and Trussell, P.C. (1957) Bacterial spoilage of shell eggs IV. Identification of spoilage organisms. *Food Technol.*, **11**, 56–60.

Foegeding, P.M. and Leasor, S.B. (1990) Heat resistance and growth of *Listeria monocytogenes* in liquid whole egg. *J. Food Prot.*, **53**, 9–14.

Foegeding, P.M. and Stanley, N.W. (1987) Growth and inactivation of microorganisms isolated from ultrapasteurized egg. *J. Food Sci.*, **52**, 1219–23, 27.

Foegeding, P.M. and Stanley, N.W. (1990) *Listeria monocytogenes* F5069 thermal death times in liquid whole egg. *J. Food Prot.*, **53**, 6–8.

Forsythe, R.H. (1970) Egg processing technology—Progress and sanitation programs. *J. Milk Food Technol.*, **33**, 64–73.

Forsythe, R.H., Ayres, J.C. and Radlo, J.L. (1953) Factors affecting the microbiological populations of shell eggs. *Food Technol.*, **7**, 49–56.

FSIS (Food Safety and Inspection Service). (1998a) *Salmonella enteritidis* risk assessment: shell eggs and egg products. Available at: www.fsis.usda.gov/ophs/risk/index.htm.

FSIS (Food Safety Inspection Service). (1998b) Refrigeration and labeling requirements for shell eggs [Final rule]. *Fed. Reg.*, **63**, 45663–75.

Galuzzo, S.J., Cotterill, O.J. and Marshall, R.T. (1994) Fermentation of whole egg by heterofermentative streptococci. *Poult. Sci.*, **53**, 1575–84.

Galyean, R.D., Cotterill, O.J. and Cunningham, F.E. (1972) Yolk inhibition of lysozyme activity in egg white. *Poult. Sci.*, **51**, 1346–53.

Garibaldi, J.A. (1960) Factors in egg white which control growth of bacteria. *Food Res.*, **25**, 337–44.

Garibaldi, J.A. (1968) Acetic acid as a means of lowering the heat resistance of *Salmonella* in yolk products. *Food Technol.*, **22**, 1031–3.

Garibaldi, J.A. (1970) Role of microbial iron transport compounds in the bacterial spoilage of eggs. *Appl. Microbiol.*, **20**, 558–60.

Garibaldi, J.A. and Bayne, H.G. (1962) The effect of iron on the *Pseudomonas* spoilage of farm washed eggs. *Poult. Sci.*, **41**, 850–3.

Garibaldi, J.A. and Stokes, J.L. (1958) Protective role of shell membranes in bacterial spoilage of eggs. *Food Res.*, **23**, 282–90.

Garibaldi, J.A., Ijichi, K. and Bayne, H.G. (1969a) Effect of pH and chelating agents on the heat resistance and viability of *Salmonella typhimurium* Tm-1 and *Salmonella senftenberg* 775W in egg white. *Appl. Microbiol.*, **17**, 491–6.

Garibaldi, J.A., Straka, R.P. and Ijichi, K. (1969b) Heat resistance of *Salmonella* in various egg products. *Appl. Microbiol.*, **17**, 491–6.

Gast, R.K. (1994) Understanding *Salmonella enteritidis* inlaying chickens: The contributions of experimental infections. *Int. J. Food Microbiol.*, **21**, 107–16.

Gast, R.K. and Beard, C.W. (1990a) Production of *Salmonella enteritidis* contaminated eggs by experimentally infected hens. *Avian Dis.*, **34**, 438–46.

Gast, R.K. and Beard, C.W. (1990b) Isolation of *Salmonella enteritidis* from internal organs of experimentally infected hens. *Avian Dis.*, **34**, 991–3.

Gast, R.K. and Holt, P.S. (1995) Iron supplementation to enhance the recovery of *Salmonella enteritidis* from pools of egg contents. *J. Food Prot.*, **58**, 268–72.

Gast, R.K. and Holt, P.S. (2001) Multiplication in egg yolk and survival in egg albumen of *Salmonella enterica* serotype Enteritidis strains of phage types 4, 8, 13a, and 14b. *J. Food Prot.*, **64**, 865–8.

Gibbons, N.E. and Moore, R.L. (1944) Dried whole egg powder. XII. The effect of drying, storage, and cooking on the *Salmonella* content. *Can J. Res.*, Sect. F, **22**, 58–63.

Gibbons, N.E., Moore, R.L. and Fulton, C.O. (1944) Dried whole egg powder. XV. The growth of *Salmonella* and other organisms in liquid and reconstituted egg. *Can. J. Res.*, Sect F, **22**, 169–73.

Gibbons, N.E., Fulton, C.O. and Reid, M. (1946) Dried whole egg powder. XXI. Pasteurization of liquid egg and its effect on quality of the powder. *Can. J. Res.*, Sect. F, **24**, 327–37.

Giessen, A.W., Ament, A.J.H.A and Notermans, S.H.W. (1994) Intervention strategies for *Salmonella enteritidis* in poultry flocks: A basic approach. *Int. J. Food Microbiol.*, **21**, 145–54.
Gillespie, J.M. and Scott, W.J. (1950) Studies in the preservation of shell eggs. IV. Experiments in the mode of infection by bacteria. *Aust. J. Appl. Sci.*, **1**, 514–30.
Glosnicka, R. and Kunikowska, D. (1994) The epidemiological situation of *Salmonella enteritidis* in Poland. *Int. J. Food Microbiol.*, **21**, 21–30.
Goresline, H.E., Moser, R.E. and Hayes, K.M. (1950) A pilot scale study of shell egg thermostabilization. *Food Technol.*, **4**, 426–30.
Goresline, H.E., Hayes, K.M., Moser, R.E., Howe, M.E. and Drewniak, E.E. (1951) Pasteurization of liquid egg under commercial conditions to eliminate *Salmonella*. *US Dept. Agric. Circ.* No. 897.
Gray, M.L. (1958) Listeriosis in fowls—a review. *Avian Dis.*, **2**, 296–314.
Haines, R.B. (1938) Observations on the bacterial flora of the hen's egg, with a description of new species of *Proteus* and *Pseudomonas* causing rots in eggs. *J. Hyg.*, **38**, 338–55.
Haines, R.B. (1939) Microbiology in the preservation of the hen's egg. *G. B. Dep. Sci. Ind. Res. Food Invest. Board, Spec. Rep.*, 47.
Haines, R.B. and Moran, T. (1940) Porosity of, and bacterial invasion through the shell of the hen's egg. *J. Hyg.*, **40**, 453–61.
Harry, E.G. (1963) The relationship between egg spoilage and the environment of the egg when laid. *Br. Poult. Sci.*, **4**, 91–100.
Hartung, T.E. and Stadelman, W.J. (1963) *Pseudomonas fluorescens* penetration of egg shell membranes as influenced by shell porosity, age of egg, and degree of bacterial challenge. *Poult. Sci.*, **42**, 147–50.
Hawthorne, J.R. (1950) Dried albumen: Removal of sugar by yeast before drying. *J. Sci. Food Agric.*, **1**, 199–201.
Hawthorne, J.R. and Brooks, J. (1944) Dried egg. VIII. Removal of the sugar of egg pulp before drying. A method of improving the storage life of spray-dried whole egg. *J. Soc. Chem. Ind.*, **63**, 232–4.
Heath, J.L. and Wallace, J. (1978) Dilute acid immersion as a method of cleaning shell eggs. *Poult. Sci.*, **57**, 149–55.
Heller, C.L., Roberts, B.C., Amos, A.J., Smith, M.E. and Hobbs, B.C. (1962) The pasteurization of liquid whole egg and the evaluation of the baking properties of frozen whole egg. *J. Hyg.*, **60**, 135–43.
Helmuth, R. and Schroeter, A. (1994) Molecular typing methods for *S. enteritidis*. *Int. J. Food Microbiol.*, **21**, 69–77.
Henderson, S.M. and Lorenz, F.W. (1951) Cooling and holding eggs on the ranch. Calif. Agric. Exp. Stn., Circ. No. 405. Univ. California, Davis.
Hobbs, B.C. and Gilbert, R.J. (1978) The vehicle of infection, in *Food Poisoning and Food Hygiene*, 4th edn, Arnold, London, pp. 51–76.
Holley, R.A. and Proulx, M. (1986) Use of egg washwater pH to prevent survival of *Salmonella* at moderate temperatures. *Poult. Sci.*, **65**, 922–8.
Holt, P.S., Gast, R.K. and Greene, C.R. (1995) Rapid detection of *Salmonella enteritidis* in pooled liquid egg samples using a magnetic bead-ELISA system. *J. Food Prot.*, **58**, 967–72.
Hopper, S.A. and Mawer, S.L. (1988) *Salmonella enteritidis* in a commercial layer flock. *Vet. Rec.*, **123**, 351.
Hou, H., Singh, R.K., Muriana, P.M. and Stadelman, W.J. (1996) Pasteurization of intact shell eggs. *Food Microbiol.*, **13**, 93–101.
Humphrey, T.J. (1990a) Public health implications of the infection of egg-laying hens with *Salmonella enteritidis* phage type 4. *World's Poult. Sci.*, **46**, 5–13.
Humphrey, T.J. (1990b) Heat resistance in *Salmonella enteritidis* phage type 4: The influence of storage temperature before heating. *J. Appl. Bacteriol.*, **69**, 493–7.
Humphrey, T.J. (1994a) Contamination of egg shell and contents with *Salmonella enteritidis*: A review. *Int. J. Food Microbiol.*, **21**, 31–40.
Humphrey, T.J. (1994b) Contamination of eggs with potential human pathogens, in *Microbiology of the Avian Egg* (eds R.G. Board and R. Fuller), Chapman & Hall, London, pp. 93–116.
Humphrey, T.J. and Whitehead, A. (1993) Egg age and the growth of *Salmonella enteritidis* PT4 in egg contents. *Epidemiol. Infec.*, **111**, 209–19.
Humphrey, T.J., Greenwood, M., Gilbert, R.J., Rowe, B. and Chapman, P.A. (1989a) The survival of salmonellas in shell eggs cooked under simulated domestic conditions. *Epidemiol. Infec.*, **103**, 35–45.
Humphrey, T.J., Baskerville, A., Mawer, S., Rowe, B. and Hopper, S. (1989b) *Salmonella enteritidis* phage type 4 from the contents of intact eggs. A study involving naturally infected hens. *Epidemiol. Infec.*, **103**, 415–23.
Humphrey, T.J., Baskerville, A., Chart, H. and Rowe, B. (1989c) Infection of egg-laying hens with *Salmonella enteritidis* PT4 by oral inoculation. *Vet. Rec.*, **125**, 531–2.
Humphrey, T.J., Chapman, P.A., Rowe, B. and Gilbert, R.J. (1990) A comparative study of the heat resistance of salmonellas in homogenized whole egg, egg yolk or albumen. *Epidemiol. Infect.*, **104**, 237–41.
Humphrey, T.J., Whitehead, A., Gawler, A.H.L., Henley, A. and Rowe, B. (1991a) Numbers of *Salmonella enteritidis* in the contents of naturally contaminated hens eggs. *Epidemiol. Infec.*, **106**, 489–96.
Humphrey, T.J., Baskerville, A., Chart, H., Rowe, B. and Whitehead, A. (1991b) *Salmonella enteritidis* PT4 infection in specific pathogen free hens: Influence of infecting dose. *Vet. Rec.*, **129**, 482–5.
Humphrey, T.J., Chart, H., Baskerville, A. and Rowe, B. (1991c) The influence of age on the response of SPF hens to infection with *Salmonella enteritidis* PT4. *Epidemiol. Infect.*, **106**, 33–43.
Humphrey, T.J., Richardson, N.P., Stutton, K.M. and Rowbury, R.J. (1993) Effects of temperature shift on acid and heat tolerance in *Salmonella enteritidis* phase type 4. *Appl. Environ. Microbiol.*, **59**, 3120–2.

Humphrey, T.J., Slater, E., McAlpine, K., Rowbury, R.J. and Gilbert, R.J. (1995) *Salmonella enteritidis* phage type 4 isolates more tolerant of heat, acid, or hydrogen peroxide also survive longer on surfaces. *Appl. Environ. Microbiol.*, **61**, 3161–4.
IAMFES (International Association of Milk, Food, and Environmental Sanitarians). (1976a) E-3-A sanitary standards for liquid egg products cooling and holding tanks. No. E-1300. *J. Milk Food Technol.*, **39**, 568–75.
IAMFES (International Association of Milk, Food, and Environmental Sanitarians). (1976b) E-3-A sanitary standards for fillers and sealers of single service containers for liquid egg products. No. E-1700. *J. Milk Food Technol.*, **39**, 576–9.
IAMFES (International Association of Milk, Food, and Environmental Sanitarians). (1976c) E-3-A sanitary standards for egg breaking and separating machines. No. E-0600. *J. Milk Food Technol.*, **39**, 651–3.
IAMFES (International Association of Milk, Food, and Environmental Sanitarians). (1976d) E-3-A sanitary standards for shell egg washers. *J. Milk Food Technol.*, **39**, 654–6.
Ijichi, K., Garibaldi, J.A., Kaufman, V.F., Hudson, C.A and Lineweaver, H. (1973) Microbiology of a modified procedure for cooling pasteurized salt yolk. *J. Food Sci.*, **38**, 1241–3.
Imai, C. (1976) Some characteristics of psychrophilic bacteria isolated from green rotten eggs. *Poult. Sci.*, **55**, 606–10.
Johns, C.K. and Berard, H.L. (1945) Further bacteriological studies relating to egg drying. *Sci. Agric.*, **25**, 551–65.
Johns, C.K. and Berard, H.L. (1946) Effect of certain methods of handling upon the bacterial content of dirty eggs. *Sci. Agric.*, **26**, 11–5.
Jones, F.T., Rives, D.V. and Carey, J.B. (1995) *Salmonella* contamination in commercial eggs and an egg production facility. *Poult. Sci.*, **74**, 753–7.
Jones, D.R., Northcutt, J.K., Musgrove, M.T., Curtis, P.A., Anderson, K.E., Fletcher, D.L. and Cox, N.A. (2003) Survey of shell egg processing plant sanitation programs: effects on egg contact surfaces. *J. Food Prot.*, **66**, 1486–9.
Jordan, F.T.W. (1956) The transmission of *Salmonella gallinarum* through the egg. *Poult. Sci.*, **35**, 1019–25.
Joyce, D.A. and Chaplin, N.R.C. (1978) Hygiene and hatchability of duck eggs—a field study. *Vet. Rec.*, **103**, 9–12.
Kaplan, A.M., Solowey, M., Osborne, W.W. and Tubiash, H. (1950) Resting cell fermentation of egg white by streptococci. *Food Technol.*, **4**, 474–7.
Kaufman, V.F. (1969) Detection of leaks in the regeneration section of egg pasteurizers. *J. Milk Food Technol.*, **32**, 94–8.
Kaufman, V.F. (1972) Locating leaks in egg pasteurizers. *J. Milk Food Technol.*, **35**, 461–3.
Khakhria, R., Duck, D. and Lior, H. (1991) Distribution of *Salmonella enteritidis* phage types in Canada. *Epidemiol. Infect.*, **106**, 25–32.
Khan, M.A., Newton, J.A., Seaman, A. and Woodbine, M. (1975) The survival of *Listeria monocytogenes* inside and outside its host, in *Problems of Listeriosis* (ed M. Woodbine), Leicester University Press, Leicester, England, p. 75.
Kijowski, J., Lesnierowski, G., Zabielski, J., Fiszer, W. and Magnuski, T. (1994) Radiation pasteurization of frozen whole egg, in *Egg Uses and Processing Technologies* (eds J.S. Sim and S. Nakai), CAB International, Wallingford, UK, pp. 340–8.
Kilara, A. and Shahani, K.M. (1973) Removal of glucose from eggs: A review. *J. Milk Food Technol.*, **36**, 509–13.
Kline, L. and Sonoda, T.T. (1951) Role of glucose in the storage deterioration of whole egg powder. I. Removal of glucose from whole egg melange by yeast fermentation before drying. *Food Technol.*, **5**, 90–4.
Kline, L., Sugihara, T.F., Bean, M.L. and Ijichi, K. (1965) Heat pasteurization of raw liquid egg white. *Food Technol.*, **19**, 1709–18.
Kline, L., Sugihara, T.F. and Ijichi, K. (1966) Further studies on heat pasteurization of raw liquid egg white. *Food Technol.*, **20**, 1604–6.
Knape, K.D., Carey, J.B., Burgess, R.P., Kwon, Y.M. and Ricke, S.C. (1999) Comparison of chlorine with an iodine-based compound on eggshell surface microbial populations in a commercial egg washer. *J. Food Saf.*, **19**, 185–94.
Knowles, N.R. (1956) The prevention of microbial spoilage in whole shell eggs by heat treatment methods. *J. Appl. Bacteriol.*, **16**, 107–18.
Kollowa, J. and Kollowa, C. (1989) Occurrence and survival of *Campylobacter jejuni* on the shell surface of hen eggs. *Monatshefte für Veterinarmedizin*, **44**, 63–5.
Kohl, W.F. (1971) A new process for pasteurizing egg whites. *Food Technol.*, **25**, 1176–84.
Kraft, A.A., Ayres, J.C., Forsythe, R.H. and Schultz, J.R. (1967a) Keeping quality of pasteurized liquid egg yolk. *Poult. Sci.*, **46**, 1282.
Kraft, A.A., Torrey, G.S., Ayres, J.C. and Forsythe, R.H. (1967b) Factors influencing bacterial contamination of commercially produced liquid egg. *Poult. Sci.*, **46**, 1204–10.
Kuhl, H. (1987) Washing and sanitizing hatching eggs. *Int. Hatch. Prac.*, **2**(3), 20–1.
Kuo, F.-L., Kwon, Y.M., Carey, J.B., Hargis, B.M., Krieg, D.P. and Ricke, S.C. (1997) Reduction of salmonella contamination on chicken egg shells by a peroxidase-catalyzed sanitizer. *J. Food Sci.*, **62**, 873–84.
Laird, J.M., Bartlett, F.M. and McKellar, R.C. (1991) Survival of *Listeria monocytogenes* in egg washwater. *Int. J. Food Microbiol.*, **12**, 115–22.
Lancaster, J.E. and Crabb, W.E. (1953) Studies on disinfection of eggs and incubators. *Br. Vet. J.*, **109**, 139–48.
Lategan, P.M. and Vaughn, R.H. (1964) The influence of chemical additives on the heat resistance of *Salmonella typhimurium* in liquid whole egg. *J. Food Sci.*, **29**, 339–44.
Lawler, F.K. (1965) Thaw frozen eggs fast. *Food Eng.*, **37**(8), 72, 75–6.
Leasor, S.B. and Foegeding, P.M. (1989) *Listeria* species in commercially broken raw liquid whole egg. *J. Food Prot.*, **52**, 777–80.
Leclair, K., Heggart, H., Oggel, M., Bartlett, F.M. and McKellar, R.C. (1994) Modelling the inactivation of *Listeria monocytogenes* and *Salmonella typhimurium* in simulated egg wash water. *Food Microbiol.*, **11**, 345–53.
Licciardello, J.J., Nickerson, J.T.R. and Goldblith, S.A. (1965) Destruction of salmonellae in hard boiled eggs. *Am. J. Public Health*, **55**, 1622–8.

Lifshitz, A. and Baker, R.C. (1964) Some physical properties of the egg shell membranes in relation to their resistance to bacterial penetration. *Poult. Sci.*, **43**, 527–8.
Lifshitz, A., Baker, R.C. and Naylor, H.B. (1964) The relative importance of chicken egg exterior structures in resisting bacterial penetration. *J. Food Sci.*, **29**, 94–9.
Lineweaver, H., Cunningham, H.E., Garibaldi, J.A. and Ijichi, K. (1967) *Heat Stability of Egg White Proteins under Minimal Conditions that kill Salmonellae, ARS 74-39*, USDA Agric. Res. Serv. Albany, California.
Lineweaver, H., Palmer, H.H., Putnam, G.W., Garibaldi, J.A. and Kaufman, V.F. (1969) *Egg Pasteurization Manual. ARS 74-48*, USDA Agric. Res. Serv., Albany, California.
Ling, A.C. and Lund, D.B. (1978) Fouling of heat transfer surfaces by solutions of egg albumen. *J. Food Prot.*, **41**, 187–94.
Lister, S.A. (1988) *Salmonella enteritidis* infection in broilers and broiler breeders. *Vet. Rec.*, **123**, 350.
Lloyd, W.E. and Harriman, L.A. (1957) Method of treating egg whites. U.S. Patent 2,776,214.
Lorenz, F.W. and Starr, P.B. (1952) Spoilage of washed eggs. I. Effect of sprayed versus static water under different washing temperatures. *Poult. Sci.*, **31**, 204–14.
Lorenz, F.W., Starr, P.B., Starr, M.P. and Ogasawara, F.X. (1952) The development of *Pseudomonas* spoilage in shell eggs. I. Penetration through the shell. *Food Res.*, **17**, 351–60.
de Louvois, J. (1993a) *Salmonella* contamination of eggs: A potential source of human salmonellosis. *PHLS Microbiol. Digest*, **10**, 158–62.
de Louvois, J. (1993b) *Salmonella* contamination of eggs. *Lancet*, **342**, 366–7.
Lucore, L.A., Jones, F.T., Anderson, K.E. and Curtis, P.A. (1997) Internal and external bacterial counts from shells of eggs washed in a commercial-type processor at various wash-water temperatures. *J. Food Prot.*, **60**, 1324–8.
Mackey, B.M., Roberts, T.A., Mansfield, J. and Farkas, G. (1980) Growth of *Salmonella* on chilled meat. *J. Hyg., Camb.*, **85**, 115–24.
Mason, J. (1994) *Salmonella enteritidis* control programs in the United States. *Int. J. Food Microbiol.*, **21**, 155–69.
Matches, J.R. and Liston, J. (1968) Low temperature growth of *Salmonella*. *J. Food Sci.*, **33**, 641–5.
Matic, S., Mihokovic, V., Katusin-Razem, B. and Razem, D. (1990) The eradication of *Salmonella* in egg powder by gamma irradiation. *J. Food Prot.*, **53**, 111–4.
Marthedal, H.E. (1973) The occurrence of salmonellosis in poultry in Denmark, 1935—1971 and the eradication programme established, in *The Microbiological Safety of Foods* (eds B.C. Hobbs and J.H.B. Christian), Academic Press, New York, pp. 211–27.
Mawer, S.L., Spain, G.E. and Rowe, B. (1989) *Salmonella enteritidis* phage type 4 and hens' eggs. *Lancet*, **i**, 280–1.
Mayes, F.J. and Takeballi, M.A. (1983) Microbial contamination of the hen's egg: A review. *J. Food Prot.*, **46**, 1092–8.
McClelland, R.G. and Pinder, A.C. (1994) Detection of *Salmonella typhimurium* in dairy products with flow cytometry and monoclonal antibodies. *Appl. Environ. Microbiol.*, **60**, 4255–62.
McElroy, A.P., Cohen, N.D. and Hargis, B.M. (1995) Evaluation of a centrifugation method for the detection of *Salmonella enteritidis* in experimentally contaminated chicken eggs. *J. Food Prot.*, **58**, 931–3.
McIloy, S.G., McCracken, R.M., Neill, S.D. and O'Brien, J.J. (1989) Control, prevention and eradication of *Salmonella enteritidis* infection in broiler and broiler breeder flocks. *Vet. Rec.*, **125**, 545–8.
McKee, S.R., Kwon, Y.M., Carey, J.B., Sams, A.R. and Ricke, S.C. (1998) Comparison of a peroxide-catalysed sanitizer with other egg sanitizers using a laboratory-scale sprayer. *J. Food Saf.*, **18**, 173–83.
Meckes, M.C., Johnson, C.H. and Rice, E.W. (2003) Survival of *Salmonella* in waste egg wash water. *J. Food Prot.*, **66**, 233–6.
Mercuri, A.J., Thompson, E., Rown, J.D. and Norris, K.H. (1957) Use of the automatic green-rot detector to improve the quality of liquid egg. *Food Technol.*, **11**, 374–7.
Michener, H.D. and Elliott, R.P. (1964) Minimum growth temperatures for food-poisoning, fecal-indicator and psychrophilic microorganisms. *Adv. Food Res.*, **13**, 349–96.
Mickelson, M.N. and Flippin, R.S. (1960) Use of salmonellae antagonists in fermenting egg white. II. Microbiological methods for the elimination of salmonellae from egg white. *Appl. Microbiol.*, **8**, 371–7.
Miller, W.A. (1959) Dry cleaning slightly soiled eggs versus washing to prevent penetration of spoilage bacteria. *Poult. Sci.*, **38**, 906–10.
Miller, W.A. and Crawford, L.B. (1953) Some factors influencing bacterial penetration of eggs. *Poult. Sci.*, **32**, 303–9.
Mishu, B., Koehler, J., Lee, L.A., Rodrigue, D., Brenner, F.H., Blake, P. and Tauxe, R.V. (1994) Outbreaks of Salmonella enteritidis infections in the United States, 1985–1991. *J. Infect. Dis.*, **169**(3), 547–52.
Mitchell, B.W., Buhr, R.J., Berrang, M.E., Bailey, J.S. and Cox, N.A.(2002) Reducing airborne pathogens, dust and *Salmonella* transmission in experimental hatching cabinets using an electrostatic space charge system. *Poult. Sci.*, **81**, 49–55.
Moore, J. and Madden, R.H. (1993) Detection and incidence of *Listeria* species in blended raw egg. *J. Food Prot.*, **56**, 652–4, 60.
Morse, D.L., Birkhead, G.S., Guardino, J., Kondracki, S.F. and Guzewich, J.J. (1994) Outbreak and sporadic egg-associated cases of *Salmonella enteritidis*: New York's experience. *Am. J. Public Health*, **84**, 859–60.
Mulder, R.W.A.W. and Bolder, N.M. (1988) Removal of glucose from egg white products by *Lactobacillus* strains. *Archiv für Geflügelkunde*, **52**, 251–4.
Mulder, R.W.A.W. and van der Hulst, M.C. (1973) The microflora of liquid whole egg made from incubator reject eggs. *J. Appl. Bacteriol.*, **36**, 157–63.
Muller, C., Haberthur, F. and Hoop, R.K. (1994) Monitoring of *Salmonella enteritidis* in eggs from a naturally infected free-range laying hen flock. *Mitteilungen aus dem Gebiete der Lebensmitteluntersuchung und Hygiene*, **85**, 235–44.

Murdock, C.R., Crossley, E.L., Robb, R., Smith, M.E. and Hobbs, B.C. (1960) The pasteurization of liquid whole egg. *Mon. Bull. Minist. Health Public Health Lab. Serv.*, **19**, 134–52.
Narvaiz, P., Lescano, G. and Kairiyama, E. (1992) Physiochemical and sensory analyses on eggpowder irradiated to inactivate *Salmonella* and reduce microbial load. *J. Food Saf.*, **12**, 263–82.
Nascimento, V.P. and Solomon, S.E. (1991) The transfer of bacteria (*Salmonella enteritidis*) across the eggshell wall of eggs classified as poor quality. *Animal Technol.*, **42**, 157–65.
Neill, S.D., Campbell, J.N. and O'Brien, J.J. (1985) Egg penetration by *Salmonella enteritidis. Avian Pathol.*, **14**, 313–20.
Niewiarowicz, A., Trojan, M. and Zielinska, T. (1967) Removal of glucose from raw egg white with the aid of enzyme containing yeast extract. *Przem. Spozyw.*, **21**, 15–7.
Nitcheva, L., Yonkova, V., Popov, V. and Manev, C. (1990) *Listeria* isolation from foods of animal origin. *Acta Microbiol. Hung.*, **37**, 223–5.
North, M.O. (1984) Maintaining hatching egg quality, in *Commercial Chicken Production Manual*, 3rd edn, AVI Pub. Co, Inc., Westport, CN, pp. 71–84.
Northolt, M.D., Wiegersma, N. and van Schothorst, M. (1978) Pasteurization of dried egg white by high temperature storage. *J. Food Technol.*, **13**, 25–30.
O'Brien, J.D.P. (1988) *Salmonella enteritidis* infection in broiler chickens. *Vet. Rec.*, **122**, 214.
O'Brien, J.D.P. (1990) Aspects of *Salmonella enteritidis* control in poultry. *World's Poult. Sci. J.*, **46**, 119–24.
Osborne, W.W., Straka, R.P. and Lineweaver, H. (1954) Heat resistance of strains of *Salmonella* in liquid whole egg, egg yolk, and egg white. *Food Res.*, **19**, 451–65.
Palumbo, M.S., Beers, S.M., Bhaduri, S. and Palumbo, S.A. (1995) Thermal resistance of *Salmonella* spp. and *Listeria monocytogenes* in liquid egg yolk and egg yolk products. *J. Food Prot.*, **58**, 960–6.
Palumbo, M.S., Beers, S.M., Bhaduri, S. and Palumbo, S.A. (1996) Thermal resistance of *Listeria monocytogenes* and *Salmonella* spp. in liquid egg white. *J. Food Prot.*, **59**, 1182–6.
Paul, P., Symonds, H., Varozza, A. and Stewart, G.F. (1957) Effect of glucose removal on storage stability of egg yolk solids. *Food Technol.*, **11**, 494–8.
Payne, J., Gooch, J.E.T. and Barnes, E.M. (1979) Heat-resistant bacteria in pasteurized whole egg. *J. Appl. Bacteriol.*, **46**, 601–23.
Penniston, V.A. and Hedrick, L.R. (1947) The reduction of bacterial count in egg pulp by use of germicides in washing dirty eggs. *Food Technol.*, **1**, 240–4.
Perales, I. and Audicana, A. (1989) The role of hens' eggs in outbreaks of salmonellosis in north Spain. *Int. J. Food Microbiol.*, **8**, 175–80.
Philbrook, F.R., MacCready, R., van Roekel, H., Anderson, E.S., Smyser, C.F., Sanen, F.J. and Groton, W.M. (1960) Salmonellosis spread by a dietary supplement of avian source. *N. Engl. J. Med.*, **263**, 713–8.
Poppe, C. (1994) *Salmonella enteritidis* in Canada. *Int. J. Food Microbiol.*, **21**, 1–5.
Poppe, C., Curtiss, R., Gulig, P.A. and Gyles, C.L. (1989) Hybridization studies with a DNA probe derived from the virulence region of the 60 MDa plasmid of *Salmonella typhimurium*. *Can J. Vet. Res.*, **53**, 378–84.
Poppe, C., Irwin, R.J., Forsberg, C.M., Clarke, R.C. and Oggel, J. (1991) The prevalence of *Salmonella enteritidis* and other *Salmonella* spp. among Canadian registered commercial layer flocks. *Epidemiol. Infect.*, **106**, 259–76.
Poppe, C., Johnson, R.P., Forsberg, C.M. and Irwin, R.J. (1992) *Salmonella enteritidis* and other *Salmonella* in laying hens and eggs from flocks with *Salm*onella in their environment. *Can. J. Vet. Res.*, **56**, 226–32.
Quarles, C.L., Gentry, R.F. and Bressler, G.O. (1970) Bacterial contamination in poultry houses and its relationship to egg hatchability. *Poult. Sci.*, **49**, 60–6.
Rahn, H., Christensen, V.L. and Edens, F.W. (1981) Changes in shell conductance, pores and physical dimensions of egg and shell during the first breeding cycle of turkey hens. *Poult. Sci.*, **60**, 2536–41.
Ranta, J. and Maijala, R. (2002) A probabilistic transmission model of *Salmonella* in the primary broiler production chain. *Risk Analysis*, **22**, 47–58.
Reznik, D. and Knipper, A. (1994) Method of eletroheating liquid egg and product thereof. U. S. Patent 5,290,583. March 1, 1994.
Rizk, S.S., Ayres, J.C. and Kraft, A.A. (1966) Effect of holding condition on the development of salmonellae in artificially inoculated hens' eggs. *Poult. Sci.*, **45**, 823–9.
Robinson, D.S., Barnes, E.M. and Taylor, J. (1975) Occurrence of β-hydroxybutyric acid in incubator reject eggs. *J. Sci. Food Agric.*, **26**, 91–8.
Rodriguez, D.C., Tauxe, R.V. and Rowe, B. (1990) International increase in *Salmonella enteritidis*: A new pandemic. *Epidemiol. Infec.*, **105**, 21–7.
Rogers, A.B., Sebring, M. and Kline, R.W. (1966) Hydrogen peroxide pasteurization process for egg white, in *The Destruction of Salmonellae*, ARS 74-37, USDA Agric. Res. Serv., Albany, California, pp68–72.
Rose, M., Orlans, E. and Buttress, N. (1974) Immunoglobulin classes in the hen's egg; their secretion in yolk and white. *Eur. J. Immunol.*, **4**, 521–3.
Rutherford, P.P. and Murray, W.W. (1963) The effect of selected polymers upon the albumen quality of eggs after storage for short periods. *Poult. Sci.*, **42**, 499–505.
Ruzickova, V. (1994) Growth and survival of *Salmonella enteritidis* in selected egg foods. *Vet. Med.*, **39**(4), 187–95.
Sahin, O, Kobalka, P. and Zhang, Q. (2003) Detection and survival of *Campylobacter* in chicken eggs. *J. Appl. Microbiol.*, **95**, 1070–9.

Salvat, G., Protais, J., Lahellec, C. and Colin, P. (1991) Excretion rate of *Salmonella enteritidis* in laying hens following a production of *Salm. enteritidis* contaminated eggs responsible for foodborne disease, in *Quality of Poultry Products III, Safety and Marketing Aspects* (eds R.W.A.W. Mulder and A.W. De Vries), Spelderholt Centre for Poultry Research and Information Serveces, Beekbergen, The Netherlands, pp. 35–42.

Sauter, E.A. and Peterson, C.F. (1969) The effect of egg shell quality on penetration by *Pseudomonas fluorescens*. *Poult. Sci.*, **48**, 1525–8.

Sauter, E.A. and Peterson, C.F. (1974) The effect of egg shell quality on penetration by various salmonellae. *Poult. Sci.*, **53**, 2159–62.

Sauter, E.A., Peterson, C.F. and Lampman, C.E. (1962) The effectiveness of various sanitizing agents in the reduction of green rot spoilage in washed eggs. *Poult. Sci.*, **41**, 468–73.

Schaffner, D.F., Hamdy, M.K., Toledo, R.T. and Tift, M.L. (1989) *Salmonella* inactivation in liquid egg by thermoradiation. *J. Food Sci.*, **54**, 902–5.

Schlosser, W.D, Henzler, D.J., Mason J., Kradel D., Shipman L., Trock S.,Hurd S.H., Hogue A.T., Sischo W., and Ebel E.D. (1999) The *Salmonella enterica* Serovar Enteritidis Pilot Project, Chapter 32, in *Salmonella enterica Serovar Enteritidis in Humans and Animals: Epidemiology, Pathogenesis, and Control* (ed A.M. Saeed), Iowa State University Press, Ames, IA, USA.

Schuman, J.D. and B.W. Sheldon (2003) Inhibition of *Listeria monocytogenes* in pH-adjusted pasteurized liquid whole egg. *J. Food Prot.*, **66**, 999–1006.

Scott, D. (1953) Glucose conversion in the preparation of albumen solids by glucose oxidase-catalase system. *J. Sci. Food Agric.*, **1**, 727–30.

Scott, W.M. (1930) Food poisoning due to eggs. *Br. Med. J.*, **11**, 56–8.

Seuna, E. and Nurmi. E. (1979) Therapeutic trials with antimicrobial agents and cultured caecal microflora in *Salmonella infantis* infections in chickens. *Poult. Sci.*, **58**, 1171–4.

Seviour, E.M. and Board, R.G. (1972) The behaviour of mixed bacterial infections in the shell membranes of the hen's egg. *Br. Poult. Sci.*, **13**, 33–43.

Shafi, R., Cotterill, O.J. and Nichols, M.L. (1970) Microbial flora of commercially pasteurized egg products. *Poult. Sci.*, **49**, 578–85.

Shah, D.B., Bradshaw, J.G. and Peeler, J.T. (1991) Thermal resistance of egg-associated epidemic strains of *Salmonella enteritidis*. *J. Food Sci.*, **56**, 391–3.

Shane, M., Gifford, D.H. and Yogasundrum, Y. (1986) *Campylobacter jejuni* contamination of eggs. *Vet. Res. Commun.*, **10**, 487–92.

Shanker, S., Lee, A. and Sorrell, T.C. (1986) *Campylobacter jejuni* in broilers: the role of vertical transmission. *J. Hyg.*, **96**, 153–9.

Shirota. K., Katoh, H., Murase, T., Ito, T. and Otsuki, K. (2001) Monitoring of layer feed and eggs for Salmonella in eastern Japan between 1993 and 1998. *J. Food Protection*, **64**, 734–7.

Shivaprasad, H.L., Timoney, J.F., Morales, S., Lucio, B. and Baker, R.C. (1990) Pathogenesis of *Salmonella enteritidis* infection in laying chickens. I. Studies on egg transmission, clinical signs, fecal shedding, and serologic responses. *Avian Dis.*, **34**, 548–57.

Shrimpton, D.H., Monsey, J.B., Hobbs, B.C. and Smith, M.E. (1962) A laboratory determination of the destruction of α-amylase and salmonellae in whole egg by heat pasteurization. *J. Hyg.*, **60**, 153–62.

Simmons, E.R., Ayres, J.C. and Kraft, A.A. (1970) Effect of moisture and temperature on ability of salmonellae to infect shell eggs. *Poult. Sci.*, **49**, 761–8.

Sionkowski, P.J. and Shelef, L.A. (1990) Viability of *Listeria monocytogenes* in egg wash water. *J. Food Prot.*, **50**, 103–7.

Slater, C. and Sanderson, D.C.W. (1989) Salmonellosis and eggs. *Br. Med. J.*, **298**, 322.

Smeltzer, T.I., Orange, K., Peel, B. and Range, G.I. (1979) Bacterial penetration in floor and nest-box eggs from meat and layer birds. *Aust. Vet. J.*, **55**, 592–3.

Snoeyenbos, G.H., Smyser, C.F. and van Roekel, H. (1969) *Salmonella* infections of the ovary and peritoneum of chickens. *Avian Dis.*, **13**, 668–70.

Solomon, S.E., Bain, M.M, Cranstoun, S. and Nascimento, V. (1994) Hen's egg structure and function, in *Microbiology of the Avian Egg* (eds R.G. Board and R. Fuller), Chapman and Hall, London, pp. 1–24.

Southam, G., Pearson, J. and Holley, R.A. (1987) Survival and growth of *Yersinia enterocolitica* in egg wash water. *J. Food Prot.*, **50**, 103–7.

Sparks, H.C. (1994) Shell accessory materials: Structure and function, in *Microbiology of the Avian Egg* (eds R.G. Board and R. Fuller), Chapman and Hall, London, pp. 25–42.

Sparks, N.H.C. and Board, R.G. (1984) Cuticle, shell porosity, and water uptake through hens' eggshells. *Br. Poult. Sci.*, **25**, 267–76.

Speck, M.L. and Tarver, F.R. Jr. (1967) Microbiological populations in blended eggs before and after commercial pasteurization. *Poult. Sci.*, **46**, 1321.

Stadelman, W.J. (1994) Contaminants of liquid egg products, in *Microbiology of the Avian Egg* (eds R.G. Board and R. Fuller), Chapman and Hall, Ltd., London, pp. 139–51.

Stafseth, H.J., Cooper, M.M. and Wallbank, A.M. (1952) Survival of *Salmonella pullorum* on the skin of human beings and in eggs during storage and various methods of cooking. *J. Milk Food Technol.*, **15**, 70–3.

St. Louis, M.E., Morse, D.L., Potter, M.E., DeMelfi, T.M., Guzewich, J.J., Tauxe, R.V. and Blake, P.A. and the *Salmonella enteritidis* Working Group. (1988) The emergence of Grade A eggs as a major source of *Salmonella enteritidis* infections. *J. Am. Med. Assoc.*, **259**, 2103–7.

Stanley, J. and Baquar, N. (1994) Phylogenetics of *Salmonella enteritidis*. *Int. J. Food Microbiol.*, **21**, 79–87.
Steele, F.R. Jr., Vadehra, D.V. and Baker, R.C. (1967) Recovery of bacteria following pasteurization of liquid whole egg. *Poult. Sci.*, **46**, 1322.
Stewart, G.R. and Kline, R.W. (1941) Dried egg albumen. I. Solubility and color denaturation. Proc. *Inst. Food Technol.*, pp. 48–56.
Stokes, J.L., Osborne, W.W. and Bayne, H.G. (1956) Penetration and growth of *Salmonella* in shell eggs. *Food Res.*, **21**, 510–8.
Stuart, L.S. and Goresline, H.E. (1942a) Bacteriological studies on the "natural" fermentation process of preparing egg white for drying. *J. Bacteriol.*, **44**, 541–9.
Stuart, L.S. and Goresline, H.E. (1942b) Studies of bacteria from fermenting egg white and the production of pure culture fermentations. *J. Bacteriol.*, **44**, 625–32.
Su, H.P. and Lin, C.W. (1993) Manufacture of pidan from various poultry eggs and their physico-chemical properties, in *Quality of Poultry Products* (ed Y. Nys.), (World Poultry Science Assoc.), Blanche Francaise de la W.P.S.A., pp. 314–20.
Suzuki, S. (1994) Pathogenicity of *Salmonella enteritidis* in poultry. *Int. J. Food Microbiol.*, **21**, 89–105.
Swanson, M. (1959) Shell egg preservation in the midwest: Progress in shell treatments, in *Conference on Eggs and Poultry*, ARS 74-12, USDA Agric. Res. Serv. Albany, California, pp. 41–2.
Thayer, D.W., Boyd, G., Muller, W.S., Lipson, C.A., Hayne, W.C. and Baer, S.H. (1990) Radiation resistance of *Salmonella*. *J. Indus. Microbiol.*, **5**, 383–90.
Thorns, C.J., McLaren, I.M. and Sojka, M.G. (1994) The use of latex particle agglutination to specifically detect *Salmonella enteritidis*. *Int. J. Food Microbiol.*, **21**, 47–53.
Todd E.C.D. (1996) Risk assessment of use of cracked eggs in Canada. *Int. J. Food Microbiol.*, **30**, 125–43.
Tryhnew, L.J., Gunaratne, K.W.B. and Spencer, J.V. (1973) Effect of selected coating materials on the bacterial penetration of the avian egg shell. *J. Milk Food Technol.*, **6**, 272–5.
Tullet, S.G. (1990) Science and the art of incubation. *Poult. Sci.*, **69**, 1–15.
USDA (U.S. Department of Agriculture) (1975a) *Regulations Governing the Grading of Shell Eggs and United States Standards, Grades, and Weight Classes of Shell Eggs*. 7 CFR Part 56, U.S. Govt. Print. Off., Washington, DC.
USDA (U.S. Department of Agriculture) (1975b) *Regulations Governing the Inspection of Eggs and Egg Products*, 7 CFR Part 59, U.S. Govt. Print. Off., Washington, DC.
USDA (U.S. Department of Agriculture) (1991) *Composition of Foods*, Agricultural Handbook Number 8, U.S. Govt. Print. Off., Washington, DC.
USDA (U.S. Department of Agriculture) (2003) *Code of Federal Regulation §590.515 Egg cleaning operations and §590.516 Sanitizing and drying of shell eggs prior to breaking*, 9 CFR part 590, U.S. Govt. Print. Off., Washington, DC.
Vadehra, D.V. and Baker, R.C. (1973) Effect of egg shell sweating on microbial spoilage of chicken eggs. *J. Milk Food Technol.*, **36**, 321–2.
Vadehra, D.V., Steele, F.R. Jr. and Baker, R.C. (1969) Shelf life and culinary properties of thawed frozen pasteurized whole egg. *J. Milk Food Technol.* **32**, 362–4.
Vadehra, D.V., Baker, R.C. and Naylor, H.B. (1970) Infection routes of bacteria into chicken eggs. *J. Food Sci.*, **35**, 61–2.
Vadehra, D.V., Baker, R.C. and Naylor, H.B. (1972) Distribution of lysozyme activity in the exteriors of eggs from *Gallus gallus*. *Comp. Biochem. Physiol. B*, **43**, 503–8.
van Schothorst, M. and van Leusden, F.M. (1977) Microbiologische specificaties van eiprodukten. (Microbiological specifications for egg products.) *Voedingsmiddelentechnologie.*, **10**(22), 16–9.
Wang, H., Blais, B.W. and Yamazaki, H. (1995) Raid and economical detection of *Salmonella enteritidis* in eggs by the polymyxin-cloth enzyme immunoassay. *Int. J. Food Microbiol.*, **24**, 397–406.
Warburton, D.W., Harwig, J. and Bowen, B. (1993) The survival of *Salmonella* in homemade chocolate and egg liquer. *Food Microbiol.*, **10**, 405–10.
Whiting, R.C., Hogue, A., Schlosser, W.D., Ebel, E.D. Morales, R.A., Baker, A. and Mcdowell, A.R.M. (2000) A quantitative process model for *Salmonella* Enteritidis in shell eggs. *J. Food Sci.*, **65**, 864–9.
Wierup, M.B., Engstrom, A. Engvall and Wahlstrom, H. (1995) Control of *Salmonella enteritidis* in Sweden. *Int. J. Food Microbiol.*, **25**, 219–26.
Wilkin, M. and Winter, A.R. (1947) Pasteurization of egg yolk and white. *Poult. Sci.*, **26**, 136–42.
Williams, J.E. (1981) Salmonellas in poultry foods—a worldwide review. Part 1. *Worlds Poult. Sci. J.*, **37**, 16–9.
Wrinkle, C., Weiser, H.H. and Winter, A.R. (1950) Bacterial flora of frozen egg products. *Food Res.*, **15**, 91–8.
York, L.R. and Dawson, L.E. (1973) Shelf life of pasteurized liquid whole egg. *Poult. Sci.*, **52**, 1657–8.
van der Zee, H. (1994) Conventional methods for the detection and isolation of *Salmonella enteritidis*. *Int. J. Food Microbiol.*, **21**, 41–6.

16 Milk and dairy products

I Introduction

The purpose of this chapter is to give the reader an appreciation of the complex relationship between microorganisms and dairy products. Much of the technology of dairy processing is long established and is reviewed in detail elsewhere (Varnam and Sutherland, 1994; Spreer, 1998; Robinson, 2002; Tetra Pak, 2003). The microbiology of butter is discussed in Chapter 11. This chapter discusses milk (raw or heat-treated) for human consumption, concentrated and dried milks, and fermented milks.

A *Definitions*

"Milk" is the product of normal secretion of the mammary gland of mammals. This chapter focuses on milk obtained from cows, with milk obtained from other animals, including sheep, goats, buffaloes, camels, or horses, mentioned where appropriate.

"Milk for direct consumption" is intended for sale directly to the consumer. This includes raw milk, where legally permitted, and processed milks. For microbiological considerations, however, any milk that has not been heated to pasteurization temperatures or higher is considered raw. Other fluid milks for direct consumption are pasteurized, sterilized, ultra-high-temperature (UHT)-treated and include whole milk, low-fat milk, skim milk, and flavored milks.

"Cream" is the fat-rich part of milk that is separated by skimming or by other techniques. According to their fat content ($\sim$10–55%), different types of cream can be differentiated by classification depending on local regulations.

"Concentrated milks" are those from which part of the water has been removed, e.g. concentrated milk, evaporated milk, or sweetened condensed milk. These products may be reconstituted or used in the condensed form. Standards for this category of products have been established by Codex Alimentarius (1999a,b).

"Dried dairy products" normally contain <5% residual moisture and include dried whole milk, skim milk or non-fat dry milk (NFDM), cream, buttermilk, cheese, and whey. Low heat and instantized milks are special forms of dried milks. Standards for this category of products have been established by Codex Alimentarius (1999c).

"Cultured or fermented milks" are milk products intended for consumption after fermentation by lactic acid bacteria or by fungi and lactic acid bacteria.

"Cheese" is the product of casein coagulation in the milk, followed by separation and removal of the whey from the curd. Casein coagulation in the milk is achieved by addition of rennet (the majority of cheeses), or by addition of acids (Harzer cheese), or by a combination of both coagulation techniques (e.g. Speisequark). Standards for this category of products have been established by Codex Alimentarius (2001, 2003).

"Whey cheeses" are made by coagulation of whey proteins through heat treatment of whey and subsequent ripening by fermenting starter cultures. The formation of curd and subsequent coagulation are brought about by acidification typically using fermenting starter cultures or by adding acids directly to the milk. Apart from certain fresh cheese, curd is then textured, salted, formed, pressed, and finally ripened. Cheese varieties included fresh, soft, semi-soft, and hard, as well as processed and blended cheeses. Standards for this category of products have been established by Codex Alimentarius (1999d).

"Ice cream and ice milk" are formulated milk products intended for consumption in the frozen or partially frozen state.

A detailed listing of all definitions used for dairy products can be found in the "General Standard for the Use of Dairy Terms" issued by Codex Alimentarius (1999e).

B Importance of microorganisms and other important properties

Milk as synthesized in the milk-producing glands of various mammals is designed by nature to meet specifically the nutritional needs of the suckling calves. "Average" cow milk is composed of approximately 87.3% water, 4.2% fat, 4.6% lactose, 3.25% protein, and 0.65% mineral substances (Walstra and Jenness, 1984; Nickerson, 1995; Schlimme and Buchheim 1999; Walstra *et al.*, 1999). However, the composition of milk varies widely among, and even within, breeds of cows. Milk composition also depends on feeding practices, age of the animals, and phase of lactation (Toppino *et al.*, 2001). The secretion of the first (usually five) day post-parturition is called colostrum, which is extremely rich in protein (up to 27%) and particularly in immunoglobulins, enabling the newborn ruminant to resist various infections (Barrington and Parish, 2001). Colostrum is rarely used as human food and then only in certain regions. Milk from other animals, for example buffaloes, sheep, horses, and camels, differ considerably from cows' milk with respect to composition and physical properties.

Metabolic disorders in milk-producing animals, and errors in handling during milking, handling, transport, and storage may cause defects in the sensory, chemical, and physical qualities of raw milk. Inflammatory changes of the mammary gland including sub-clinical mastitis lead to changes in milk composition and technological suitability (e.g. lower heat stability of whey protein, reduced yield in cheese production, off-flavor of milk products). Residues of veterinary drugs used in animal husbandry are undesirable and can impact the production of fermented dairy products and cheese. They may also be illegal and some may even pose potential health hazards to the consumer.

Microorganisms are important in milk and dairy products for three principal reasons:

- Pathogens or their toxins may constitute health hazards.
- Spoilage microorganisms or their metabolites may cause spoilage.
- Lactic acid bacteria and others may contribute in the preservation of milk and in the production of desirable flavor and physical characteristics.

To minimize milk-borne food poisoning and spoilage problems, routes of contamination of detrimental microorganisms and factors influencing their destruction and proliferation must be known and understood. Furthermore, new products and new processing technologies must be evaluated to ensure their safety and effectiveness.

Statistics of food-borne disease outbreaks show that milk and milk products contribute to a lesser extent to food-borne illness than other foods of animal origin. A survey of US food-borne diseases between 1990 and 2001, in which both the food and the etiological agent were identified, listed a total of 1 589 outbreaks involving 73 425 cases. Of these, 65 outbreaks (2 866 cases) were linked to dairy products (De Buyser *et al.*, 2001). The European Surveillance Programe for Control of Food-borne Infections and Intoxications reports that 7.8% of investigated outbreaks where the food was known were traced to milk and milk products and 0.6% to cheeses. The major cause for milk-borne disease is consumption of raw milk and recontaminated processed and mixed dairy products such as cream fillings (FAO/WHO Collaborating Centre for Research and Training in Food Hygiene and Zoonoses, Berlin, 2001). Gillepsie *et al.* (2003) reviewed 27 milk-borne outbreaks between 1992 and 2000 in England and Wales representing 2% of all food-borne outbreaks. They highlighted the importance of VTEC O157 and the continued role of unpasteurized milk in human disease.

C Methods of processing and preservation

Processing of raw milk in dairy factories is intended

- To produce microbiologically safe products with acceptable shelf life.
- To develop or maintain desired sensory qualities (appearance, flavor, and texture).
- To isolate particular constituents of milk that are used directly, or as part of other foods, or for non-food purposes.

Processing technologies applied involve:

- Application of heat or cold, or combination of both, such as batch or HTST (high-temperature short time) pasteurization, UHT treatment, sterilization, chilling/refrigeration, and freezing;
- Mechanical treatments such as separation, centrifugation, homogenization, filtration, micro- and nano-filtration, high-pressure treatment, and reverse osmosis.
- Removal of water by, for example, concentration, dehydration, ultrafiltration, curing/ageing, e.g. condensed or dried products, ripening of cheese.
- Microbiological fermentation or chemical acidification.
- Combinations of these processing technologies.

D Types of final products

The different procedures and processes applied in the dairy industry result in a vast number of widely differing products. The examples given in the respective sub-sections below are those in which milk, or milk constituents, form a major part and determine the characteristics of the end product. The discussion of the types of product categories follows technology and processes in as much as they have an impact on the microbiological hazards and spoilage microorganisms to be controlled.

II Raw milk–initial microflora

Raw milk has a mixed microflora derived from the interior of the udder, from the farm or milking barn environment, and from persons and equipment in contact with the milk. Milk is an excellent substrate for growth of all microorganisms present, including pathogens and spoilage organisms. Levels of microorganisms provide information on the hygiene level during milking and subsequent steps. Programs to improve milk hygiene have led to a decrease in the total bacterial count of fresh milk. In many countries, counts at the farm level that were initially $\geq 10^6$ cfu/mL were reduced to $<2 \times 10^4$. Germany reported in 1977 an average count of 500 000 cfu/mL, and in 2002 an average count of 20 000 (Suhren and Reichmuth, 2003).

Sources of the initial microflora include:

- The interior of the udder.
- Udder and teat surfaces.
- Milking equipment, milk transport line and storage tank.
- The environment, such as air and water.
- Persons handling milk.

The levels and composition of the initial microflora may be directly affected by natural microbicidal systems in the milk and inhibitory substances/veterinary drugs used to maintain udder health and to treat diseased animals.

A Interior of the udder

Milk is held in the udder by the capillary forces of the lactiferous duct network and the sphincter muscle at the lower end of the teat canal. During milking, hormonal influences and intermittent pressure (hand milking) or vacuum (machine milking) applied on the teat forces milk through the teat canal, which may serve as a portal for microbial access to the udder. In healthy animals the secretory tissue of the udder is free of microorganisms, as shown by Tolle and Heeschen (1975) with aseptically drawn milk. However, the mucosal membrane of the streak canal has a microflora that includes streptococci, staphylococci, and micrococci (normally >50%), and *Corynebacterium* spp., coliforms, lactic acid bacteria, and other bacteria. The level of contamination through the streak canal may vary between 10^2 and 10^4 cfu/mL milk.

Infection of an udder quarter affects the microflora of milk. Milk from quarters with clinical mastitis is usually altered in its appearance (clots, changes of color, or viscosity). Changes are routinely detected by inspection of foremilk. Milk from such animals must be separated and should not be used for human consumption.

Many animals suffer from sub-clinical mastitis, often a chronic inflammatory disease of the mammary tissue (with simultaneous reduction in milk yield in the affected quarters). It is one reason for shedding high numbers of microorganisms and increase of somatic cell counts in milk. The number of microorganisms shed into milk varies with the stage of mastitis. Several detailed discussions of mastitis in bovines have been published (Tolle, 1980; O'Shea, 1987; Watts, 1988; Harmon, 1995; IDF, 2001; Bradley, 2002). Mastitis in dairy small ruminants has been reviewed by Bergonier *et al.* (2003). The organisms most frequently involved in sub-clinical mastitis include *Staphylococcus aureus*, *Streptococcus agalactiae*, *Strep. dysgalactiae*, and *Strep. uberis*, as well as coagulase-negative staphylococci, and *Mycoplasma* (Gonzalez and Wilson, 2003). *Escherichia coli* and coliforms, *Corynebacterium bovis*, *Arcanobacterium pyogenes*, *Listeria monocytogenes*, *Pseudomonas aeruginosa*, and yeasts also cause mastitis. The role of viruses in bovine mastitis was reviewed by Wellenberg *et al.* (2002).

Microorganisms causing mastitis are important for a number of reasons:

- They change the composition of the milk.
- They can be pathogenic for man (e.g. *Brucella* spp., *Staph. aureus*).
- Residues of antibiotics used for mastitis treatment may be found in milk.

Inflammatory processes in the udder change the milk composition substantially, particularly the content of casein, but also of enzymes and other technologically relevant components (Le Roux *et al.*, 2003; Pyorala, 2003).

Microscopic examinations of stained smears play an important role in the detection of mastitis milk. Bacteria are often engulfed by polymorphonuclear leukocytes and macrophages. Few polymorphonuclear leukocytes are found in normal milk but soon after infection their number increases dramatically, often exceeding several million per milliliter. The somatic cell count of milk consisting of macrophages, lymphocytes, polymorphonuclear leukocytes, and epithelial cells of the mammary gland is commonly used as a screening test to identify infected animals (Le Roux *et al.*, 2003; Schukken *et al.*, 2003; de Haas *et al.*, 2004). In several countries, the somatic cell count in bulk tank milk is used as a means of determining the payment for milk to farmers and helps to maintain healthy herds. The normal level of somatic cells in healthy uninfected quarters is considered to be <100 000 cells/mL of milk (Hamann and Reichmuth, 1990; Peeler *et al.*, 2003).

Ruminants may also be infected with zoonotic agents, which are then shed in the milk, e.g. *Mycobacterium bovis*, *Brucella abortus*, *Br. melitensis*, *Br. suis*, *L. monocytogenes*, salmonellae, or *Coxiella burnetii*. Tuberculosis and brucellosis in milk animals are systematically targeted because of their high economic importance for the animal stock. Many countries are already declared to be free from these diseases according to the criteria established by the Office International des Epizooties (OIE, 2003).

Among viruses shed into the milk, Foot and Mouth Disease virus is of primary importance, since it can be easily spread countrywide to other livestock unless appropriate precautions are taken. Preventive management measures are discussed by Schagemann (1994), Crispin *et al.* (2002), James and Rushton, (2002), Bouma *et al.* (2003), and OIE (2003). Although Foot and Mouth Disease virus is classified as a zoonotic agent, human health issues are mostly restricted to occupational health (Lopez-Sanchez *et al.*, 2003).

B Udder and teat surfaces

Contamination of milk during milking is possible if there is insufficient cleaning and disinfection of the exterior of the udder. The contribution of microorganisms from these sources in freshly drawn milk may range from less than 100 to several thousand cfu per milliliter depending on the care taken in cleaning and disinfecting the udder surfaces prior to milking.

Depending on the mechanical milking technique applied, milk may come into contact with the external surface of the teat. The microflora of the skin is then washed-off and contaminates the milk. This primary contamination is usually lower when cows are kept on pastures rather than in cow sheds. Keeping animals in-house can result in higher counts of anaerobic bacteria in the milk. Cleaning and disinfection of teats with approved disinfectants reduces microbial numbers, mainly staphylococci and streptococci.

Soil, bedding, feed residues, and manure are also present on the surface of the udder, teats, and adjacent body surfaces (Kotimaa *et al.*, 1991). Pathogens and spoilage microorganisms are associated with these materials. Spore-forming bacteria such as *Bacillus* spp., and clostridia from the soil, feeding stuff, and from manure readily find their way into milk (Driehuis and Oude Elferink, 2000; te Giffel *et al.*, 2002). Although mastitis may also be caused by several pathogens such as *Salmonella* spp., *Listeria* spp., *Campylobacter* spp., *Yersinia* spp., or enteropathogenic *E. coli*, the primary source of their presence in milk is post-secretory contamination with fecal material from infected cows or from animals harboring and shedding the organism without clinical signs. Fecal contamination of milk with *Mycobacterium avium* subsp. *paratuberculosis* is the main pathway for the infection of suckling calves (Kennedy *et al.*, 2001). Such occasional contamination at low levels is difficult to eliminate, even under good hygienic conditions of milking.

Some spoilage microorganisms are particularly important to further processing of milk. These include thermoduric bacteria such as *Microbacterium* spp. or *Enterococcus* spp., which are important for products made from pasteurized milk, *Bacillus sporothermodurans*, *Clostridium tyrobutyricum*, and *Cl. butyricum* from silage and manure for UHT-milk and hard-cheese production, respectively (Ingham *et al.*, 1998, Vaerewijck *et al.*, 2001; Scheldemann *et al.*, 2002).

C Milk handling equipment

Milk-handling equipment includes components of milking machines such as teatcup inflations, milk tubes and airline hoses, pails, strainers, milk cans, churns, coolers, bulk-milk refrigerated cooling tanks, milk transport pipelines, tank trucks, and other equipment and accessories. It is well recognized that this equipment may contribute substantially to the raw-milk microflora (Druce and Thomas, 1972; Thomas and Thomas, 1977; Palmer, 1980; McKinnon *et al.*, 1988; Rasmussen *et al.*, 2002; Schreiner and Ruegg, 2002; IDF, 2004a).

Cleaning and disinfection of such equipment plays an essential role in maintaining the quality of the raw milk. Adequate cleaning effectively removes residues and microorganisms from equipment surfaces. Residual microorganisms are then readily killed by either disinfectants or heat (steam or hot water). However, if such surfaces remain wet for long periods, then extensive growth of the few surviving microorganisms is possible. Milk residues left on equipment surfaces after inadequate cleaning provide ample nutrients and ambient temperatures favor microbial growth. The effectiveness of cleaning and disinfection is further reduced by milk-stone, a high mineral deposit, which can gradually build-up on equipment surfaces that are inadequately cleaned. Very hard water or use of highly alkaline cleaners enhances development of milk-stone and acid-type cleaners must be used periodically to reduce or eliminate it.

During subsequent use of equipment contamination of the raw milk takes place. The type and number of microorganisms introduced in the raw milk from contaminated surfaces depend largely on the efficiency of prior cleaning and disinfection (Mackenzie, 1973; Faille *et al.*, 2003). Gram-negative psychrotrophic microorganisms such as *Pseudomonas*, *Alcaligenes*, *Flavobacterium*, and *Chromobacterium* spp., or thermoduric bacteria that readily adhere to solid surfaces, such as steel, enter the milk from inadequately cleaned handling equipment (Lewis and Gilmour, 1987; Barnes *et al.*, 1999; Jayarao and Wang, 1999).

Prolonged inadequate cleaning of equipment may also lead to the formation of biofilms. Microorganisms such as micrococci, enterococci, aerobic spore-formers, and certain lactobacilli become embedded in the biofilm, multiply in it, and are protected from the effects of detergent and disinfectant solutions. The total count of microorganisms in such biofilms commonly ranges from 10^3 to 10^{11} cfu/g. The microflora in deposits typically comprises streptococci (5–20%), micrococci (20–50%), corynebacteria (10–16%), coliforms (0.8–30%), other Gram-negative bacteria (11–27%), and aerobic spore-formers (0.5–3.6%) (Austin and Bergeron, 1995; de Jong *et al.*, 2002). Many of these bacteria are thermoduric or thermoresistant and can lead to problems during further processing of the milk (Thomas and Thomas, 1977; Meers *et al.*, 1991; Wong, 1998; Flint *et al.*, 2002; Yoo and Chen, 2002).

D Environment

The air in the milking environment contributes a negligible number of organisms to milk, unless it is extremely dusty. More important than the number of bacteria introduced *via* air-borne contamination is the type. Microorganisms commonly found in air include micrococci, yeasts, and spores of *Bacillus* and *Clostridium* spp. and fungi (Palmer, 1980), all of which may survive heat processes and cause flavor or physical defects in processed products. Ventilation fans, drains, dust, and worker activity contribute to air-borne contamination (Cousin, 1982). Water supplies in farms often contain coliforms and psychrotrophic organisms (Palmer, 1980; Cousin, 1982), and when used to rinse dairy equipment may be a source of contamination. For this purpose, water should be of drinking water quality.

E Persons handling milk

Milking performed by hand under poor hygiene conditions and by sick farmers was, in the past, a source of transmission of diphtheria and scarlet fever through contaminated milk. Even in modern milk production, persons can, through handling the udder, contribute to the spread of mastitis agents within the herd.

F Antimicrobial factors naturally present in milk

Milk contains several inherent antimicrobial factors, which may inhibit or minimize development of microorganisms during transport and storage under difficult conditions, in particular in sub-tropical and tropical countries (Ekstrand, 1989; IDF, 1986a, 1991a). The occurrence, functions, and properties of antimicrobial components of milk have been reviewed in several recent publications (Isaacs, 2001; Clare *et al.*, 2003; Florisa *et al.*, 2003; Kilara and Panyam, 2003; Pellegrini, 2003). Lysozyme is an enzyme that cleaves bonds in peptidoglycan, a major component of bacterial cell walls and therefore

Gram-positives are the most susceptible microorganisms. Gram-negative bacteria contain little peptidoglycans and are also protected by an outer layer of lipopolysaccharides (Masschalck and Michiels, 2003). Most bacterial endospores are resistant although some are germinated by lysozyme (e.g. non-proteolytic strains of *Cl. botulinum* and some strains of *Cl. perfringens*). Lactoferrin is an iron-binding protein able to chelate iron and to deprive bacteria essential mineral. Other milk proteins that bind vitamin B_{12} and folic acid may inhibit certain microorganisms. Lactoperoxidase has no direct antibacterial effect. In its presence, H_2O_2 oxidizes naturally present thiocyanate to yield highly reactive, short-lived antimicrobial substances, mainly hypothiocyanite. These substances, in turn, oxidize essential protein sulfydryl groups in microbial membranes, resulting in altered cellular system functions, and subsequent inhibition of growth and/or death of a wide range of bacteria (Wolfson and Sumner, 1993; Haddadin *et al.*, 1996; Kussendrager and van Hooijdonk, 2000). Bacterial spores and fungi are not affected. Maternal immunoglobulins may be present in milk, and inactivate microorganisms, or their toxins through agglutination (Korhonen *et al.*, 2000; Tizard, 2001).

G *Inhibitory substances and veterinary drug residues*

Antibiotics have been used for two main veterinary purposes: to treat animal diseases and to promote growth of young animals. However, a general concern on the development of antibiotic resistance in pathogens has resulted in reduced use of antibiotics in animal feed, with new regulations on feed additives under development.

The presence of antibiotic residues in milk can have different adverse effects.

- Influence on microbiological methods for the determination of the microbiological quality of milk.
- Impact on the fermentation and production of fermented dairy products.
- Potential problems for the public health.

Numerous methods have been developed for the detection of residues of antibiotics in milk. They have been reviewed by Suhren and Heeschen (1996) and IDF (2004b).

Starter cultures used in the production of fermented milk products and cheese are usually susceptible to antimicrobial residues. The minimum inhibitory concentration (MIC) is dependent on the type of residue and the species/strains of starter cultures used. In general, mesophilic starter cultures are less susceptible than thermophilic. Fermentation failures due to the presence of antibiotic residues and the inhibition of starters can cause, for example, off-flavors in Emmental cheese or an insufficient pH decrease during yoghurt production. Similar effects are observed in natural fermentation of raw milk causing deviations in the organoleptic profile of raw-milk cheeses (Mäkinen, 1995; Suhren, 1996).

Milk containing such antimicrobial residues, and which might cause technological problems during further use, is penalized in many countries by a price reduction governed by milk quality payment schemes.

For the protection of the consumers, "safe levels" or maximum residue limits (MRLs) for residues of veterinary drugs in food have been fixed on the basis of the "no effect level" (NEL) or Acceptable Daily Intake (ADI)-concept (Codex Alimentarius, 1993, 1995; WHO, 2002b) taking into account the potential health risks for consumers. These include:

- Microbiological aspects (selection pressure in the intestinal flora, growth of resistant microorganisms, development of resistance in pathogenic microorganisms).
- Pharmacological–toxicological aspects.
- Immunopathological (allergic) aspects.

According to the available knowledge it can be concluded that no single method is able to detect all anti-infectives in milk for which maximum residue limits have been fixed. Reviews on the impact of antimicrobial resistant microorganisms, in particular of pathogens, and the management of the treatments of infectious diseases in cattle, have been reviewed (van den Bogaard and Stobberingh, 1999, 2000; Lathers, 2001; McDermott *et al.*, 2002; Robertson, 2003; Makovec and Ruegg, 2003).

III Raw milk for direct consumption

In raw milk for direct consumption there is no opportunity to reduce pathogens or spoilage microorganisms from the primary or secondary contamination. Shelf life is limited by the growth of the microflora, including spoilage organisms. The only means of prolonging shelf life while maintaining the original characteristics of the raw milk is to chill it (see Section IV).

In some regions or countries, raw milk is sold directly from the farm, or, when produced from certified and regularly monitored herds, distributed regionally at retail. However, because of the risks for public health, many countries have stringent regulations that restrict, or completely prohibit, the retail sale of raw milk (see Section IV).

A Effects of handling of raw milk on microorganisms

Collection and transport. In smallholder dairy farms, raw milk is frequently processed in the household or distributed in the immediate neighborhood. In regions with an extended milk-handling infrastructure, milk is first stored at the farm or cooperative group level, collected by milk tankers at defined intervals, and transported directly to a dairy plant for further processing.

The composition of milk makes it an excellent growth medium for many microorganisms. Milk should be cooled to below 4°C within two hours after milking unless it is heat processed (exact regulations vary with the country). The closer milk is kept to 0°C prior to processing, the better the retention of the microbiological quality. At temperatures between 0°C and 5°C, growth of bacteria is slowed, but undesirable changes in milk may nevertheless take place within a few days. The extent of such changes depends on the type and number of microorganisms present and the enzymes they produce. In inadequately cooled milk, lactic acid producing bacteria, in particular *Lactococcus* spp., quickly suppress the initial multiplication of Gram-negative bacteria, which grow rapidly and contribute to the proteolytic and lipolytic degradation of milk components. There are many interactions, synergisms, and antagonisms that influence the sensory properties of the milk before it is spoiled. Main representatives of this flora are species of *Pseudomonas, Aeromonas, Flavobacterium, Alcaligenes*, and *Acinetobacter* (e.g. Dogan and Boor, 2003).

Long distances between farms, collection centres and dairy plants can result in raw milk being transported for many hours, increasing the probability of bacterial growth, and spoilage.

Filtration, separation, and clarification. Straining raw milk through filter cloths was common a practice, and is still used in some places today. While straining removes visible fecal and other soil particles, it does little to lower the microbial load of the milk. If not replaced between milkings, the filter cloth can contribute to cross-contamination of clean milk. Modern milking equipment may include a filtration step.

When raw milk reaches the processing plant, the usual practice is to purify, cool, and store it at a temperature not exceeding 6°C until processed. The separation process produces three fractions: skimmed milk, cream, and sediment (separator or clarifier "slime"). A large proportion of microorganisms are physically removed from the milk and concentrated with other particulate matter in the slime, which contains millions of bacteria and bacterial spores per gram. Because of its high population and the presence of undesirable microorganisms such as *Listeria* spp. and psychrotrophs, slime must be isolated and eliminated without contaminating other areas (Varnam and Sutherland, 1994).

Clarification removes suspended particulate matter and adhering microorganisms by centrifugation. Special clarifiers (e.g. Bactofuge®) have been designed to reduce bacterial populations in raw milks, thus reducing the potential for subsequent spoilage defects (Lembke *et al.*, 1984; Kirschenmann, 1989; Spreer, 1998) such as gas production in cheese due to *Cl. tyrobutyricum* or *Cl. butyricum*. Agitation during separation and clarification may break up bacterial clumps, increasing the apparent number of bacteria capable of forming colonies on agar media or observed in a microscopic field.

B Spoilage

A wide range of undesirable organoleptic and physical changes in raw milk are caused by microorganisms (Cousin, 1982; Suhren, 1988; Meers *et al.*, 1991; Muir, 1996a,b,c; Sorhaug and Stepaniak, 1997). While refrigerated storage of raw milk prevents growth of most types of mesophilic microorganisms such as lactic bacteria psychrotrophic microorganisms such as species of *Pseudomonas*, *Flavobacterium*, *Micrococcus*, *Bacillus*, *Enterobacter*, *Aeromonas*, and *Alcaligenes* are able to grow. Although present initially in low numbers, dependent on hygiene conditions, extended refrigerated storage allows their growth to high numbers. Psychrotrophic growth is accompanied by production of enzymes such as heat-stable proteinases and lipases. Defects such as malty, rancid, yeasty, bitter, fruity, or putrid off-flavors and appearance of purple (*Chromobacterium* spp.) and reddish (*Serratia* spp.) pigments have been associated with growth of psychrotrophic microorganisms. Ropiness may result from exopolysaccharide production by *Alcaligenes viscolactis* (Champagne *et al.*, 1994). Levels of 10^6 to 10^7 cfu/mL are usually required before organoleptic defects become evident.

C Pathogens

Every pathogen that may be associated with milk-providing animals, handlers, equipment, and the environment may be an accidental contaminant of milk.

The literature is replete with accounts of outbreaks due to the consumption of raw milk. Before the widespread use of pasteurization in the 1930s, milk and dairy products were a major vehicle for transmission of human disease such as typhoid fever, diphtheria, septic sore throat, tuberculosis, and brucellosis (Bryan, 1983). However, since the enforcement of pasteurization in combination with a better health status of the animal stock, the number of such outbreaks has declined dramatically. Of 19 milk-related outbreaks of food-borne illness in the US between 1990 and 2001, 11 were caused by raw milk (CSPI, 2004). Most cases of milk-borne diseases in England and Wales were, until the late 1990s, associated with raw-milk consumption. In Scotland, regulatory changes banning the sale of raw milk dramatically decreased the number of cases of salmonellosis from this source (Sharp *et al.*, 1988).

Mycobacterium bovis and *M. tuberculosis*, as well as *Br. abortus* and *Br. melitensis*, have contributed much to milk-borne diseases. In some countries, bovine tuberculosis and brucellosis in ruminants have been eradicated by enforcement of strict slaughter policies when the diseases are detected. Some countries use vaccination for the eradication of brucellosis from the livestock. The OIE (Office International des Epizooties) reports cases of these diseases in various countries according to their prevalence in the livestock.

Raw milk is an important vehicle for salmonellae causing human infection (Vogt *et al.*, 1981; Bryan, 1983; Potter *et al.*, 1984; Anonymous, 2003). *Salmonella* Dublin and *S.* Typhimurium are the serovars most frequently associated with cattle in the United Kingdom, the United States and France (McManus and Lanier, 1987; D'Aoust, 1989a,b) and those two serovars have been at the origin of several outbreaks (Vlaemynck, 1994). *S.* Dublin is a relatively rare, but particularly virulent, serovar that primarily attacks consumers over 40 years of age with underlying disorders. In California, the risk of acquiring a *S.* Dublin infection was calculated to be 84 times higher for persons who consumed raw milk compared with those who did not (Potter *et al.*, 1984; Richwald *et al.*, 1988). More than 80% of affected individuals are usually hospitalized and the overall mortality rate is 25% (Blaser and Feldman, 1981; Potter *et al.*, 1984). Other serovars, such as *S.* Anatum, *S.* Thompson, *S.* Heidelberg, *S.* Enteritidis, and *S.* Newport, have also been frequently recovered from raw milk (Marth, 1987; D'Aoust, 1989a,b; Fitzgerald *et al.*, 2003), the last displaying multiple resistance to antibiotics used clinically.

The link between campylobacteriosis and the consumption of raw milk seemed obvious, although evidence was not demonstrated until 1978 (Potter *et al.*, 1983). This was perhaps due to the rapid die-off of *C. jejuni* in raw milk, leading to the epidemiological link being established indirectly (Hahn, 1994). Of all campylobacteriosis outbreaks reported to the Centers for Disease Control (CDC) between 1980 and 1982, 61% involved raw-milk consumption (Potter *et al.*, 1983). Outbreaks are often characterized by the high number of affected consumers (500–3 500), probably due to the extended distribution zones of milk. Numerous cases and outbreaks of campylobacteriosis have been reported from many countries (Jones *et al.*, 1981; Robinson and Jones, 1981; Stalder *et al.*, 1983; Kornblatt *et al.*, 1985; Thurm and Dinger, 1998; Thurm *et al.*, 1999, Kalman *et al.*, 2000, Anonymous, 2002; Neimann *et al.*, 2003; Peterson, 2003).

Different groups of pathogenic *Escherichia coli*, enteropathogenic (EPEC), enterotoxigenic (ETEC), enteroinvasive (EIEC), and enterohaemorrhagic or verotoxigenic (EHEC/VTEC), cause food-borne outbreaks (Kornacki and Marth, 1982; Doyle and Cliver, 1990; Olsvik *et al.*, 1991; Molenda, 1994; Thielman, 1994; Willshaw *et al.*, 2000). *Escherichia coli* O157:H7 is of particular concern due to its low infective dose and severe clinical consequences (Karmali, 1989; Tarr, 1994; Burnens, 1996; Su and Brandt, 1996), but involvement of other serovars is increasingly reported. Although these pathogens have been isolated from lamb, pork, and venison (Doyle, 1991), cattle have been identified as the major reservoir (Martin *et al.*, 1986; Orskov *et al.*, 1987; Blanco *et al.*, 1995; Chapman, 1995; Willshaw *et al.*, 2000). Raw milk becomes contaminated with pathogenic *E. coli* through exposure to fecal material. A clinical case of mastitis due to *E. coli* is easily identified and should not be a problem provided good milking practices are applied (Bryan, 1983; Chapman *et al.*, 1993; Bleem, 1994; Kuntze *et al.*, 1996). Raw milk has been implicated in several outbreaks of illness due to pathogenic *E. coli*, in most cases O157:H7 (de Buyser *et al.*, 2001; CSPI, 2001; Gillepsie *et al.*, 2003).Cases have been reported from the United States, UK, Scotland, Germany, and the former Soviet Union (Bryan, 1983; Martin *et al.*, 1986; Beutin, 1995, 1996; Keene *et al.*, 1997; Sharp *et al.*, 1994; Gillepsie *et al.*, 2003). Raw goats milk was the cause of four cases of hemolytic uremic syndrome in children in the Czech Republic in 1995 (Bielaszewska *et al.*, 1997) and in Canada (McIntyre *et al.*, 2002). Klinger and Rosenthal (1997) reviewed the safety of milk and milk products from sheep and goats.

The most probable source of *Staph. aureus* found in raw milk is the infected bovine udder. The prevalence of *Staph. aureus* in quarter milk samples ranges from 5% to 22.4% (IDF, 1980a), values confirmed by other studies and surveys (Gilmour and Harvey, 1990). In cases of sub-clinical illness levels of 10^4–10^5 cfu/mL are reported, and up to 10^8 cfu/mL in case of clinical mastitis. Different authors have shown that up to 30% of that population is able to produce enterotoxins, mainly the types C and D (Jarvis and Lawrence, 1970; Olson *et al.*, 1970; Garcia *et al.*, 1980). Although the pathogen is a poor competitor and is often overgrown by other microorganisms, raw milk has been implicated in outbreaks of staphylococcal intoxication before rapid cooling of milk and pasteurization became accepted practices (Bryan, 1983). In 1960, for example, an outbreak resulted after milk was held in milk churns at 15.6°C (60°F) prior to consumption. Pasteurization kills *Staph. aureus*, but preformed enterotoxins are not destroyed by heating (Bennett, 1991).

Listeria monocytogenes is often recovered from raw milk although reported prevalence varies. Hayes *et al.* (1986) recovered the pathogen from 15 (12%) of 121 raw-milk samples examined. Lovett *et al.* (1987) surveyed the prevalence of *L. monocytogenes* in raw milk from three different geographic areas of the USA. An overall prevalence of 4.2% was determined, with extremes of 7% in Massachusetts and absence in samples from California, but usually with less than one cell per mL. Dominguez-Rodriguez *et al.* (1985) isolated *L. monocytogenes* from 45.3% of raw-milk samples from Spanish dairies and Beumer *et al.* (1988) in only 4.5% in bulk milk in the Netherlands. Mastitis plays only a minor role and different studies indicate the prevalence of *L. monocytogenes* in milk samples obtained aseptically from mastitic cows is very low and ranges between 0.02% and 0.2% depending on the studies (Prentice, 1994). In a summary of several surveys covering more than 7 000 samples, Farber and Peterkin (2000) reported positives ranging from 0% to 81% of the samples of raw milk. The major sources for *Listeria* spp. on the farm are poor silage conditions, fecal contamination, and improperly cleaned equipment, especially if not dried, on which *Listeria* spp. can grow.

Potel (1951) provided the first epidemiological link between the consumption of raw milk and listeriosis when *L. monocytogenes* was isolated from stillborn twins and the mother who had consumed contaminated raw milk. Other outbreaks of listeriosis in humans have since then been epidemiologically associated with raw-milk consumption (Seeliger, 1961; Gray and Killinger, 1966). The vast majority of dairy related listeriosis outbreaks have, however, been linked to cheese products (see subsequetnt paragraphs).

Surveys to determine the prevalence of *Yersinia enterocolitica* in raw milk have been conducted in many countries, including Australia, Canada, Brazil, Czechoslovakia, France, Japan, and the United States (Schiemann and Toma, 1978; Hughes, 1979; Vidon and Delmas, 1981; Delmas, 1983; Moustafa *et al.*, 1983). Prevalence varied between 10% and 81.4%. High prevalence was also reported for goat milk (Jensen and Hughes, 1980; Walker and Gilmour, 1986). However, strains of *Y. enterocolitica* possessing virulent characteristics, or biotypes and serovars associated with human illness, are rarely isolated from raw milk (Moustafa *et al.*, 1983; Jayarao and Henning, 2001), although non-virulent serotypes are.

Other microbial pathogens implicated in outbreaks of milk-borne illness include *Strep. agalactiae*, other haemolytic streptococci and *Coxiella burnetii* and have been thoroughly reviewed (IDF, 1994).

Raw milk has also been implicated as a source of viral infection in human illness outbreaks. Examples of transmission of viruses causing hepatitis A, poliomyelitis, encephalitis and gastroenteritis were reviewed by Bryan (1983) and Schagemann (1994).

Raw milk may contain mycotoxins, in particular aflatoxin M_1 (AFM_1), but other mycotoxins have also been reported including aflatoxin B_1, G_1, sterigmatocystin, ochratoxin, and others (Galvano *et al.*, 1996). The presence of AFM_1 has been reported in cow's milk and also in milk from sheep, goat, buffalo, and camel (Saad *et al.*, 1989; Vandana *et al.*, 1991). The presence of AFM_1 is the result of a metabolic hydroxylation of aflatoxin B_1 by mammals with subsequent excretion in the milk, primarily in the protein fraction (Piva *et al.*, 1995). The prevalence depends on several factors including seasonal effects, type of feed, type of agricultural system, etc., and is reviewed by Galvano *et al.* (1996). Normally AFM_1 disappears from milk approximately 3 to 4 days following removal of contaminated feed. The Codex Alimentarius (2001a) has defined maximal AFM_1 limits in milk and most countries have established limits which are, however, extremely heterogeneous ranging between "zero" and 1.0 μg/kg milk. The exclusion of aflatoxin-contaminated feed has led in many countries to a decrease in milk contamination, e.g. Germany reported in 1976 on the average 35 ng/mL whereas in 2000, 95% of ex-farm milk samples were practically aflatoxin free (Blüthgen and Ubben, 2000).

Similar results were obtained in surveys in different European countries and published by, for example, Gareis and Wolff (2000), Martins and Martins (2000), Roussi *et al.* (2002), and Rodriguez Velasco *et al.* (2003). On the other hand, studies on the occurrence of AFM_1 in milk in the southern and western regions of India indicated levels in the range of 0.05 to 3.0 μg/L, the high levels being related to the use of contaminated feed (Vasanthi and Bath, 1998). These results indicate that the problem may still be present in some parts of the world.

D CONTROL (raw milk for direct consumption)

Summary

Significant hazards[a]	• Depending on the geographical region and farming and hygienic practices, one or more of the pathogens listed in Table 16.1 can be found in raw milk.
Control measures	• Prohibition or special requirements for direct sale of raw milk. • Risk communication.
Initial level (H_0)	• Health status of milk-producing animals through national/international eradication programs for tuberculosis/brucellosis and FMD virus and associated animal health monitoring programs, including for clinical and sub-clinical mastitis. • Application of Good Hygiene Practices during milking. • Adequate water supply (drinking water quality). • Use of aflatoxin-free feed. • Prudent use of antimicrobial drugs observing recommendations of veterinary authorities.
Reduction (ΣR)	• No reduction process available. • Use of natural antimicrobial systems (e.g. lactoperoxidase) where possible.
Increase (ΣI)	• Cooling of milk $\leq$4-6°C.
Testing	• Surveillance activities related to pathogens as indicated above. • Mastitis diagnosis through somatic cell counts and microbiological analyses. • Determination of microbial counts in bulk tank milk for verification of hygiene (cleaning, disinfection, cooling). • Testing for veterinary drugs and aflatoxins.
Spoilage	• Determination of total viable counts using traditional or automated methods (IDF, 1990a; ISO, 2002). Determination of psychrotrophic (IDF, 1991d) or thermoduric microorganisms (Frank *et al.*, 1985).

[a]In particular circumstances, other hazards may need to be considered.

Hazards to be considered

Drinking raw milk containing bovine tubercle bacilli was a primary cause of alimentary tuberculosis in infants in central Europe in the 19th and 20th century. In regions where consumption of raw milk still continues, multiple outbreaks of salmonellosis, campylobacteriosis, and infections with entero-haemorrhagic *E. coli* have been reported since 1980, including milk from certified herds (Sharpe, 1987; D'Aoust, 1989b). Consumption of raw milk therefore poses a significant threat to human health. A WHO consultation recommended that risk groups should be strongly discouraged from consuming raw milk (WHO, 1995).

All the microbial hazards that may initially be present in the raw milk (see Table 16.1) are of serious concern. Their possible presence in raw milk will depend on the health status of the animals, milking hygiene, and distribution/sales systems and will therefore vary from region to region. Experience

Table 16.1 Diseases transmitted through milk and their most important sources

Causative agent	Disease	Man	Milking animal	Environment
Ba. anthracis	Anthrax		×	×
Cl. botulinum	Botulism			×
Br. melitensis, B. abortus	Brucellosis		×	
Campylobacter jejuni	Campylobacteriosis		×	×
Vibrio cholerae	Cholera	×		×
E. coli spp.	Pathogenic *E. coli* infections	×	×	×
Cl. perfringens	*Cl. perfringens* infections			×
Corynebacterium diphteriae	Diphtheria	×		
Listeria monocytogenes	Listeriosis		×	×
Salmonella Paratyphi	Paratyphus	×	×	×
Salmonella Enterica serovars	Salmonellosis (exclusively typhus and paratyphus)	×	×	×
Shigella spp.	Shigellosis	×		×
Staph. aureus	*Staph. aureus* intoxication		×	
Streptococcus spp.	*Streptococcus* infections	×	×	×
Mycobacterium bovis, M. tuberculosis	Tuberculosis	×	×	
Adenoviruses	Adenovirus-infection	×		
Various enteric viruses	Enterovirus infection incl. Poliomyelitis and Coxsackie Virus	×		
Food and Mouth Disease Virus	Food and Mouth Disease		×	
Hepatitis A virus	Hepatitis	×?[1]		
Tick-encephalitis virus	Tick-encephalitis		×	
Coxiella burnetti	Q-fever		×	
Entamoeba hystolytica	Amoebiasis	×		×
Cryptosporidiae spp.	Cryptosporidiosis	×	×	×
Toxoplasma gondii	Toxoplasmosis	×		×

Modified from Kaplan *et al.* (1962).
[1] Not fully demonstrated.

indicates that raw milk will often be contaminated with low levels of pathogens. Therefore measures short of pasteurization cannot be relied upon in large-scale production to provide safe raw milk for human consumption.

Control measures

Initial level of hazard (H_0). International recommendations and statistics on animal tuberculosis and brucellosis are found in publications of the Office International des Epizooties (OIE). This organization defines the absence of the disease agent in a country and provides the status to be "free of". In general national programs follow specific recommendations of this organization.

To maintain the initial level of other pathogens as low as possible, several measures have to be applied. Generalized hygiene programs to control raw-milk production are provided on an international level by IDF (1994), FAO/WHO Codex Alimentarius, or national organizations such as the National Mastitis Council in the USA (1987).

Animals must be maintained in a healthy condition and provided with feed of good hygienic quality, e.g. free of aflatoxins. Implementation of adequate mastitis control programs will reduce the incidence of contaminated raw milk. Application of good hygiene practices during milking is a further important element. Personnel should be well versed in hygienic milk production practices and proper use of equipment. If milking is not performed in a separate milking parlor, air-borne and particulate contamination must be minimized during milking by restricting movement of bedding, silage, and forage materials. Walls, ceilings, and floors should be kept free from loose materials. Appropriate cleaning of the teat surfaces and adjacent parts of the animal, followed by disinfection where possible. Drying of the udder is very important to avoid subsequent multiplication. Equipment, and in particular, surfaces in direct contact with the raw milk, need to be cleaned and disinfected carefully after each milking.

Reduction of hazard (ΣR). There are no processes that will reduce the bacterial load of raw milk for direct consumption.

Filtration of raw milk only eliminates larger particles such as soil, hay, but will not significantly change the bacterial load. The activation of the lactoperoxidase system should be restricted to situations where cooling systems are not available, mainly in sub-tropical or tropical countries.

Increase of hazard (ΣI). Immediately after milking, microbial growth will initiate and freshly drawn milk should therefore be cooled immediately to 4-6°C, or preferably lower (even if some regulations are less stringent), to slow microbial growth. However, freezing should be avoided because it can induce undesirable physicochemical changes.

Testing

A number of countries have established criteria relevant to quality and hygiene, such as microbiological counts, somatic cell count, and sediment content. Published legal criteria for different types of raw milk were reviewed by Milner (1995) and Otte-Südi (1996a). Traditional microbiological methods are standardized internationally (ISO, 2002) or automated techniques such as the Bactoscan® are used.

Values for raw milk for the production of heat-processed dairy products in the European Union and the United States are as follows:

	European Union	United States
Total viable count	$<1 \times 10^5$/mL at delivery and 3×10^5/mL before further processing	$<1 \times 10^5$/mL at delivery and 3×10^5/mL before further processing
Somatic cells	$<4.0 \times 10^5$/mL	7.5×10^5/mL

With respect to inhibitory substances/veterinary drugs, an integrated system has been proposed to ensure a high technological quality of milk and its safety for consumers (Suhren, 1996). It is a "hurdle" system that combines different methods and the responsibility of all parties involved. It uses elements of the HACCP concept to minimize risks and to ensure safe products and comprises two different elements:

1. Application of different detection methods with different targets to detect a range of inhibitory substances.
2. Shared responsibilities of veterinarians, dairy farmers, processing establishments, and food inspection.

This integrated system has proved to be effective in many countries. For example, in Germany the percentage of "inhibitor-positive" samples raw milk at the farm is generally <0.1%, while it is a little higher when tanker milk is tested. The dominating compounds have been β-lactams and sulfonamides.

Spoilage

Depending on the intended use of the raw milk, certain groups of microorganisms may be used to identify specific bacterial contaminations relevant to further processing of the milk. Thermoduric bacteria, for example, may indicate problems with cleaning and disinfection, as they are frequently found in biofilms on the milking equipment. Bacteriological methods are standardized internationally (IDF, 1990a), for psychrotrophic bacteria (IDF, 1991d) or used according to relevant publications, such as for thermoduric bacteria (Frank *et al.*, 1985).

IV Processed fluid milk

A *Introduction*

Milk products are an essential part of a balanced diet and fluid milks for direct consumption represent a major proportion. The composition of various market milks (Table 16.2) is usually expressed in terms of fat content and percent non-fat milk solids (NFMS). This is commonly specified in laws and regulations of national or supranational bodies (Staal, 1981, 1986).

Table 16.2 Typical composition of fluid milks

Product	Milk fat (%)	NFMS (%)
Whole milk	3.25 minimum	8.25
Low-fat milk	1.5–2.0	8.25
Skim milk	0.5 maximum	8.25
Flavored low-fat milks	0.5–2.0	8.25

B *Initial processing steps*

Homogenization. This processing step may take place before or after further treatments of the milk. Milk fat mostly (98%) consists of triglycerides with a minor fraction of mono- and diglycerides, phospholipids, and other fats. The fat forms globules surrounded by a phospholipid membrane. In unhomogenized raw milk, globules may coalesce to form a compact layer of cream. The sweeping action of the rising fat globule clusters can carry microorganisms upward and concentrates them in the cream layer. The homogenizer is a pump that forces milk through small orifices under high pressure. Fat globules are therefore reduced in size and remain in suspension throughout the milk for long periods of time. However, homogenization of raw milk accelerates hydrolysis of milk fat by lipases, which may result in development of rancid flavors and may be confused with bacterial spoilage. Homogenization has little effect on the microbiology of fluid milk products, except that it breaks up clumps of bacteria (Lanciotti *et al.*, 1994).

Heat-treated, or otherwise processed, milk may be homogenized in sterile equipment after the thermal processing prior to aseptic packaging. Inadequately cleaned and disinfected or incompletely sterilized homogenization equipment can adversely affect the quality and safety of the product and lead to recontamination.

C *Basic procedures to reduce the initial microflora*

Definitions of heat treatments as applied to milk and milk products have been discussed by IDF (1985).

Thermization. Growth of psychrotrophic microorganisms, such as *Pseudomonas* spp., to high numbers will result in the production of heat-stable proteases and lipases. High levels of these enzymes can lead to enzymatic deterioration of finished products manufactured using such milk. To increase its keeping quality during storage at low temperatures, raw milk can be submitted to a mild heat treatment or "thermization" in a continuous flow system, i.e. heating at 57–68°C for a maximum of 30 s, followed by rapid cooling to <6°C. In general, a reduction of 3–4 $\log_{10}$ cycles can be expected, but thermization does not control vegetative pathogens fully. It should be specifically noted that *L. monocytogenes* is able to survive thermization (Mackey and Bratchell, 1989), and will multiply during subsequent chilled storage. Thermization may also heat-shock endospores of *Bacillus* spp. Subsequent storage may then permit their germination and possibly allow their inactivation when milk is subsequently pasteurized (van den Berg, 1984).

Thermization is mild enough to have minimal effects on the physical properties of raw milk. Heat inactivation of enzymes present in the milk varies depending on the enzyme and its source. Alkaline phosphatase activity is reduced by thermization but remains present.

Pasteurization. According to the IDF/WHO/FAO definition of pasteurization (IDF, 1995), milk products are pasteurized to ensure their safety by minimizing numbers of vegetative pathogens to a level considered safe for public health. Many spoilage bacteria and yeasts are also controlled. Pasteurization is not only applied to ensure microbiological safety and to extend shelf life during refrigerated distribution (IDF, 1986), but also to meet criteria for suitability of the milk for further processing.

Pasteurization can be carried out as batch process, also called Low Temperature/Long Time (LTLT)-pasteurization, where the product is heated in an enclosed tank. In industrial milk processing, LTLT treatments between 62°C and 65°C for 30–32 min are normally applied. Pasteurization can also be performed as a dynamic process, a High Temperature/Short Time (HTST) pasteurization. In this instance, the product is heated in a heat exchanger, then maintained at the desired temperature under turbulent flow conditions in a holding tube to ensure the desired killing effect. Currently, the most common HTST conditions are temperatures ranging between 71°C and 78°C for at least 15 s (often 30-40 s) or very brief heating for a few seconds at 85°C to 127°C (Kessler, 1987). Time–temperature requirements vary from country to country and are frequently under regulatory control (Staal, 1986) as are the construction and operation of heating equipment.

The most severe treatments are normally applied to products with higher fat or solids, the mildest for standard liquid milk. The actual treatments applied to the milk are often more severe to ensure that heating requirements are met.

Sterilization and UHT treatment. Treatments at higher temperatures are performed either in batches in closed containers such as cans or bottles, or continuously with subsequent aseptic packaging. Sterilization of milk usually involves treatments of 119.5–120°C for 10–30 min, the lower range being applied when a UHT treatment is performed before filling and sterilizing in closed containers. Single UHT treatment are performed at temperatures not less than 135°C for at least 1 s; usually combinations of 135–150°C and holding times from 1 to 5 s are applied.

Miscellaneous methods. Modern food processing techniques such as microfiltration (e.g. Trouve *et al.*, 1991; Grandison and Glover, 1994; Corredig *et al.*, 2003; Papadatos *et al.*, 2003); high pressure treatment (e.g. Garcia-Graells *et al.*, 2003; Hayes and Kelly, 2003; Harte *et al.*, 2003), ultrasonication (e.g. Vercet *et al.*, 2002), electromagnetic treatment (e.g. Bendicho *et al.*, 2002) or addition of carbon dioxide (e.g. Ma *et al.*, 2003) are currently being explored for their use in the dairy industry. One advanced technology, which is already being applied is microfiltration, where the milk is processed and sterilized by filtration, which also allows the separation of various milk components. The technique is not yet widely used, however. A combination of filtration and heat treatment can result in "fresh" milk with an extremely extended shelf life. In this procedure, separate heat treatments are applied to skimmed milk (mild) and cream, and the skimmed milk fraction is further filtered before recombination allowing the elimination of heat-resistant microorganisms, predominantly responsible for spoilage.

D Cleaning and disinfection

Food contact surfaces are an important source of contamination, but unclean non-contact surfaces can also contribute by generating contaminated dust particles and aerosols. Manual cleaning using detergents, acids, or alkali, followed by rinsing and disinfecting, is routinely used in milk and dairy products facilities.

To enhance the efficiency of cleaning and disinfecting procedures, automated cleaning-in-place (CIP) systems are frequently used in the dairy industry. Such systems are designed to provide high and reproducible standards of cleanliness with the minimum of manual efforts. Disassembly and manual cleaning may be required for sensitive parts such as filters or complex elements that cannot be cleaned effectively by CIP. Automation may also facilitate coordination between production and cleaning schedules. To ensure proper performance, equipment design, the type of soil to be removed, water quality, and detergent and disinfectant usage must be considered when installing a CIP system. The design, and use, of CIP systems in the dairy industry was reviewed by the International Dairy Federation (IDF, 1979). The conclusions are still valid. Even if not mandatory, strict separation of CIP systems for the raw and heat-treated milk is strongly recommended.

Proper operation of cleaning and disinfection programs must be documented. Documentation may be manual or by using recording charts for time, temperature, flow rate, disinfectant concentration, pH, etc.

E Effects of processing on microorganisms

Pasteurization aims to reduce numbers of potentially pathogenic vegetative bacteria present to levels that do not constitute a public health concern. Additional treatments may include microfiltration and bactofugation of the separated skimmed milk to increase the effect of subsequent heat treatment (IDF, 1995.) Such treatments are applied for products with extended shelf life. The risk of recontamination during subsequent handling and filling becomes even more crucial.

Freshly pasteurized milk usually contains less than 1000 cfu/mL, but higher levels are observed if the initial levels in the raw milk are much higher than 10^6 cfu/mL. In the case of ultra-pasteurized milk (extended shelf life, ESL) the initial levels are much lower. Hence, minimal LTLT and HTST treatments may permit the survival of thermoduric and spore-forming organisms, particularly if they are present at high levels, making it difficult to meet norms for pasteurized products. Thermoduric organisms include vegetative microorganisms showing an increased heat-resistance such as *Micrococcus* spp., *Enterococcus faecalis*, and *Ent. faecium*, and some lactobacilli (Deibel and Hartman, 1984). Endospores of *Bacillus* and *Clostridium* spp. display a range of heat-resistance.

An important source of microorganisms in pasteurized milk is post-process contamination. Pseudomonads are typical post-process contaminants (Eneroth *et al.*, 2000) but *Bacillus* spp. may also be reintroduced after the heat treatment (Schraft *et al.*, 1996). Since surviving thermoduric bacteria may adhere and multiply in the cooling section of the pasteurizer, care must be taken that plants are regularly evaluated for the prevalence of biofilms, and HTST systems not operated too long without cleaning (Bouman *et al.*, 1982; Sharma and Anand, 2002). Cooling, filling, and low-temperature storage of pasteurized milk do not inactivate microorganisms, but must be controlled to minimize growth of those surviving, to minimize recontamination and limit growth of possible post-process contaminants.

F Spoilage

Pasteurized milk from modern well-operated plants can have a shelf life of well over 10 days under refrigeration (Otte-Südi, 1996b). Commercially manufactured ultra-pasteurized milk is much more stable and remains stable for periods as long as 10 weeks under refrigeration (Boor, 2001).

As in the case of raw milk, pasteurized and ultra-pasteurized fluid milk products will support abundant growth of microorganisms if contaminated.

Spoilage may involve:

- Growth of surviving spore-forming bacteria (*Bacillus* and *Clostridium* spp.);
- Growth of thermoduric bacteria;
- Growth of contaminating psychrotrophic (Gram-negative) bacteria; and
- Activity of heat-stable enzymes produced pre-pasteurization.

Microorganisms causing spoilage are usually spore-forming or thermoduric bacteria surviving the pasteurization process (Meers *et al.*, 1991), from the contamination of equipment, e.g. as biofilms (Bouman *et al.*, 1982; Carpentier and Cerf, 1993; Wong, 1997) or from post-process contamination by environmental microorganisms such as Enterobacteriaceae (Varnam and Sutherland, 1994). Further details on spoilage flora such as species of *Pseudomonas*, *Flavobacterium*, *Chromobacterium*, *Alcaligenes*, *Bacillus*, and coliforms commonly found have been reviewed (Cousin, 1982; Meers *et al.*, 1991; Ternstrom *et al.*, 1993; Deeth *et al.*, 2002). Psychrotrophic strains of both thermoduric or spore-forming organisms grow at temperatures as low as 5°C and can cause spoilage or be hazardous to health (Crielly *et al.*, 1994). If initially present in high numbers, sufficient growth may occur to cause spoilage within 10–14 days of refrigerated storage.

Microbial spoilage of refrigerated market milk products is recognized primarily by development of off-flavors, often described as unclean, putrid, and fruity, while physical changes such as ropiness and partial coagulation are less common defects. The intensity of flavor defects depends on the extent of microbial enzymatic decomposition of milk proteins, fat, and to some extent lactose. A particular aspect of spoilage is that due to preformed enzymes, such as heat-resistant lipases or proteases of psychrotrophic microorganisms, principally *Pseudomonas* spp. (Champagne *et al.*, 1994; Muir, 1996a,b,c; Stevenson *et al.*, 2003), commonly causing bitterness of the product. *Bacillus cereus* is particularly troublesome because it produces lecithinase that acts on the phospholipids of the milk fat globule, forming small proteinaceous fat particles that adhere to the surfaces of glasses (IDF, 1992). The defect due to the development of *B. cereus* is referred to as "sweet curdling" and is particularly a problem in the warm summer months (Christiansson *et al.*, 1999). The time required for changes to occur depends on the initial numbers and type of microorganisms present, the pasteurization conditions and the storage temperature (Schröder *et al.*, 1982; Mourgues *et al.*, 1983; Schröder and Bland, 1984). For example, the average time required for spoilage of HTST pasteurized milk stored at 1.7, 5.6, and 10°C was 17, 12, and 6.9 days, respectively (Hankin *et al.*, 1977). Pasteurized milks from modern well-operated plants have >10 days of shelf life under refrigeration, and even longer if the milk is aseptically packed (Otte-Südi, 1996b).

G Pathogens

Minimum pasteurization treatments specified by law or regulation generally allow destruction of pathogens likely to be present initially in raw milk with a sufficient margin of safety. Although pasteurized market milk products have a remarkable safety record and present little health hazard, several outbreaks (campylobacteriosis, salmonellosis, yersiniosis, etc.) have been linked to pasteurized milk. Such outbreaks are usually due to inadequate pasteurization, post-pasteurization contamination, or temperature abuse during use (Snyder *et al.*, 1978; Sharpe, 1987; Doyle, 1989).

Salmonellae do not survive pasteurization and their presence in pasteurized milk is the consequence of either improper heat treatments or post-pasteurization contamination. The outbreaks in the United Kingdom (*Salmonella* Braenderup; Rampling *et al.*, 1987) and in the United States (Adams *et al.*, 1984) are examples of inappropriate heat treatments. In the latter example, temperatures as low as 54.5°C for 30 min were recorded.

The largest outbreak of salmonellosis in the history of the United States occurred in Northern Illinois. Pasteurized low-fat milk (2%) contaminated with *S.* Typhimurium was the cause of more than 16 000 cases (Anonymous, 1985a,b; Ryan *et al.*, 1987). Inspections of the incriminated dairy plant revealed no evidence of improper pasteurization and Bradshaw *et al.* (1987) confirmed that the outbreak strain possessed no abnormal heat resistance. Although the cause was never completely elucidated, a thorough investigation of the plant and production lines revealed a cross-connection between the pipes used for raw and pasteurized milk that may have been at the origin of the contamination (Lecos, 1986).

It is generally accepted that current minimum pasteurization standards (71.7°C, 15 s or 62.8°C, 30 min) are sufficient to inactivate *L. monocytogenes* in milk (Donnelly *et al.*, 1987; Bunning *et al.*, 1988). Nevertheless, viable *Listeria* spp. including *L. monocytogenes* have been isolated from pasteurized milk in different countries at frequencies ranging from 0.9% to about 5% (Harvey and Gilmour, 1992; Moura *et al.*, 1993; Ahrabi *et al.*, 1998). Fernandez-Garayzabal *et al.* (1986) attributed the presence of *L. monocytogenes* in 21.4% of the samples of pasteurized milk treated at 78°C for 15 s to the protective effect provided by type A leukocytes. A similar explanation was given for the reasons of the outbreak of *L. monocytogenes* 4b in Massachusetts affecting 49 consumers (Fleming *et al.*, 1985). These explanations were, however, questioned by other researchers (Donnelly, 1990; Lovett *et al.*, 1990).

L. monocytogenes is frequently found in the wet environments of milk processing plants (Jung and Busse, 1988; Jeong and Frank, 1994a,b; Pritchard *et al.*, 1995) and may cause post-processing contamination.

Campylobacter jejuni has been the cause of an outbreak involving 110 patients after the consumption of inadequately pasteurized milk (Fahey *et al.*, 1995). Additional sporadic cases of campylobacteriosis in the United Kingdom were traced to birds pecking through the caps of milk bottles. *Campylobacter* was isolated from the beaks of jackdaws and magpies as well as from the contaminated milk (Hudson *et al.*, 1990; Southern *et al.*, 1990; Stuart *et al.*, 1997).

Although swine are recognized as the major reservoir of *Y. enterocolitica* in nature (Doyle *et al.*, 1981), several studies have demonstrated the presence of this species in pasteurized milk and outbreaks have been reported (CSPI, 2004). Schiemann and Toma (1978) recovered *Y. enterocolitica* only from 1 of 165 samples of pasteurized dairy products produced in Ontario, Canada. Similarly, Moustafa *et al.* (1983) recovered *Y. enterocolitica* from 1% of pasteurized milk samples examined, while Tibana *et al.* (1987) recovered the pathogen from 13.7% of pasteurized milk samples produced in Rio de Janeiro, Brazil. Twenty-two of the 41 isolates were capable of producing heat-stable toxin in culture media but not in sterile whole milk.

Since *Y. enterocolitica* is rapidly inactivated at pasteurization temperatures (Francis *et al.*, 1980; Lovett *et al.*, 1982), its presence in finished product is most likely to be the result of post-pasteurization contamination. This is certainly true for reported outbreaks (Ackers *et al.*, 2000). In 1976, consumption of contaminated chocolate milk at a school resulted in the illness of 36 children from *Y. enterocolitica* serotype O8. The same serotype was isolated from the milk and was probably introduced during manual mixing of pasteurized milk with chocolate syrup without subsequent heat treatment (Black *et al.*, 1978). In another outbreak, pasteurized milk was implicated as the vehicle of infection of 148 persons in three states (Tacket *et al.*, 1984) and of 19 people in Nevada in 1996 (CSPI, 2004). A very large outbreak was probably caused by the use of packaging material contaminated with pig faeces, although the implicated serovar could never be isolated from swine (Schiemann, 1989). Greenwood *et al.*, (1990) indicated in the investigation of the incriminated dairy bottling plant the possibility of a contaminated filler valve causing the outbreak.

Escherichia coli O157:H7 was responsible for a large outbreak (more than 100 persons) due to the consumption of pasteurized milk. During inspection of the processing plant, the same phage-type was isolated from pipes to the bottling machine and from rubber seals from the same machine, suggesting post-process contamination (Upton and Coia, 1994). The reasons for another outbreak involving 114 persons are not known and could have been either a failure in the heat treatment or post-process contamination, since the milk was processed at the farm (Goh *et al.*, 2002).

Overall, the risk of *B. cereus* enterotoxin caused gastroenteritis from pasteurized milk is very low, although it has been implicated in an outbreak in the Netherlands involving more than 250 consumers (van Netten *et al.*, 1990). Taking into account the quantity of pasteurized milk sold, the number of cases due to *B. cereus* worldwide is very low, perhaps because spoilage of the products deters consumers, and the quantity of toxin required to cause illness seems high (IDF, 1992; Langeveld *et al.*, 1996). The prevalence of *B. cereus* in milk varies greatly. In some countries, a prevalence of 2% has been reported (Wong *et al.*, 1988), whilst in other areas every sample is positive (Notermans *et al.*, 1997). Some strains of *B. cereus* strains are capable of growing and producing enterotoxin at chill temperatures (<7°C) (Dufrenne *et al.*, 1995; Notermans *et al.*, 1997). When stored at 6°C, spoilage occurred before high levels of *B. cereus* were present, however, significant growth to levels above 10^5 cfu/mL was seen at 8–12°C (Notermans *et al.*, 1997). In studies with human volunteers, only very weak symptoms were noticed when milk containing $>10^7$ cfu *B. cereus* per mL was ingested (Langeveld *et al.*, 1996). Investigations to determine the origin of this spore-forming species have shown that post-process contamination is a frequent cause of its presence in the finished products (Lin *et al.*, 1998; Eneroth *et al.*, 2001).

In the mid 90s, PCR-based detection demonstrated the presence of *Mycobacterium avium* subsp. *paratuberculosis* (MAP) in retail pasteurized milk in the UK (Millar *et al.*, 1996; Grant *et al.*, 2002a). This has caused much debate since MAP in cattle and sheep causes an inflammatory chronic infection of the gut called Johne's disease. In humans, a disease with similar symptoms, Crohn's Disease (CD), is becoming more common. No firm link has been made between occurrence of MAP and CD, and several factors (genetic predisposition, abnormal immune response) appear to influence development of the disease. However, studies have failed to demonstrate that MAP is not involved in CD (European Commission, 2000; Harris and Lammerding, 2001; Chamberlin *et al.*, 2001; Bernstein *et al.*, 2004). *Mycobacterium avium* subsp. *paratuberculosis* is very difficult to culture because it grows very slowly, with incubation for up to 1 year recommended. Studies have demonstrated that despite a 4-6 $\log_{10}$ reduction during commercial pasteurization, some individual cells of MAP may survive the milk pasteurization process (Grant *et al.*, 2002b; Hammer *et al.*, 2002). Increasing the pasteurization time, e.g. from 15 to 25 s at 73°C, or temperature up to 90°C, appear to have no increased lethal effect (Lund *et al.*, 2002; Hammer *et al.*, 2002).

Although there is the potential for other pathogens to be present in pasteurized milk as a result of post-processing contamination, there are very few documented cases of illness resulting from such contamination. *Staph. aureus* intoxications were reported from pasteurized milk, and recontamination during filling of containers was assumed to be the reason (Geringer, 1983). A large outbreak of staphylococcal enterotoxin poisoning was caused by chocolate milk containing between 94 and 184 ng of type A toxin per carton (Evenson *et al.*, 1988). If staphylococcal enterotoxins or aflatoxin M_1 are present in raw milk, they also will be present in pasteurized milk.

H CONTROL (processed fluid milk)

Summary

Significant hazards[a]	• Hazards in raw milk are summarized in Table 16.1 and several of them are also significant in pasteurized fluid milk. • Pathogens may also be present in the environment of milk processing lines.

(*continued*)

H CONTROL (Cont.)

Summary

Control measures	
Initial level (H_0)	• Low levels of pathogens are likely in raw milk; use milk from healthy animals, milked according to good milking practices. • Cool milk during intermediate storage on the farm and during transportation to dairy plants ($\leq 6°C$ recommended). • Consider the use of the activated lactoperoxidase system in regions where cooling is not available only as an interim solution.
Reduction (ΣR)	• Pasteurization at 72-75°C for at least 15 s reduces vegetative bacteria by at least 5 $\log_{10}$ cycles. • Ultra-pasteurization treatments will eliminate vegetative bacterial pathogens and a large fraction of spores. • Packaging materials used to fill the pasteurized milk should be treated e.g. with hydrogen peroxide.
Increase (ΣI)	• Strict cleaning and disinfection regimes applied to the equipment after processing. • Post-process contamination during filling and packing or during distribution due to damaged packaging material (e.g. microleaks). • Growth of surviving psychrotrophic thermoduric or spore-forming microorganisms during storage and distribution. • Growth is enhanced at higher storage temperatures.
Testing	• Routine testing of pasteurized milks for pathogens is not recommended. Monitoring of incubated packs for verification according to defined sampling plans and trend analyses can be performed.
Spoilage	• Surviving psychrotrophic spore-formers or microorganisms from post-process contamination.

[a]In particular circumstances, other hazards may need to be considered.

Hazards to be considered

Low numbers of vegetative pathogens may survive pasteurization if they are present in raw milk exceeding numbers of 10^5 cfu/mL. Post-pasteurization contamination by pathogens (e.g. *L. monocytogenes*) from the processing environment and insufficiently cleaned processing lines can occur.

Control measures

Initial level of hazard (H_0). High quality raw milk should be used in production of pasteurized liquid milk.

Reduction of hazard (ΣR). Pasteurization is the critical processing step to ensure the absence of vegetative cells of pathogens and to reduce the numbers of spoilage microorganisms.

Increase of hazard (ΣI). Temperature and flow rate (holding time) are the critical parameters in the heat treatments applied to eliminate sensitive microorganisms. Properly operating monitoring devices such as thermometers and flow meters are crucial, as well as flow diversion valves in case the temperature drops below a set critical limit. In the case of plate heat exchangers, the heat-treated milk must be protected against recontamination. In counterflow systems, this is achieved by maintaining an overpressure on the side of the treated milk thus avoiding ingress from the non-sterile side, either raw milk or cooling fluid which can be highly contaminated (Strantz *et al.*, 1989). Post-processing contamination of pasteurized milk is a very common source of Gram-negative psychrotrophic bacteria such as pseudomonads, and is possibly the most common reason for spoilage of pasteurized milk (see subsequent paragraphs) (Eneroth *et al.*, 2000). All equipment surfaces that might come into contact with heat-treated milk such as pumps, pipes, balance tanks, intermediate storage tanks or fillers, need to be cleaned and disinfected after use or before resuming processing. Rinsing with water of high microbiological quality, if not mandatory, is strongly recommended to remove chemical residues. Maintenance of adequate refrigeration is required during storage and distribution of packed pasteurized milk (Mottar and Waes, 1986). Details of a generalized HACCP program to control pasteurized milk production were presented by ICMSF (1988).

Testing

Routine testing for pathogens in pasteurized milk is not recommended. Due to the short shelf life of the products no testing of the product is usually performed before release. Incubation of packed units is not useful due to the presence of surviving spore-formers. Testing for indicators such as aerobic mesophilic counts or coliforms can be used to verify the effectiveness of pasteurization and the absence of post-processing contamination during further handling and filling.

Regulations often require that aerobic plate counts remain below 10 000 or 20 000 cfu/mL and coliforms below 1 or 10 cfu/mL after pasteurization (USPHS FDA, 1978; Heeschen, 1992; Milner, 1995; Otte-Südi, 1996b). Absence of coliforms does not necessarily reflect the quality of pasteurized milk during sale and the determination of Gram-negative psychrotrophs is often useful to predict shelf-life stability (Bishop and White, 1986; Otte-Südi, 1996a,b).

Pathogens, such as *Salmonella* spp. and *L. monocytogenes* are likely to be present in raw milk and can thus gain entrance to a dairy-processing environment. It is important that steps are taken to control environmental contamination and to verify the effectiveness of the measures through an environmental monitoring program to detect these pathogens in the post-pasteurization area of the processing plant (Flowers *et al.*, 1992; Stahl *et al.*, 1996).

Spoilage

Spoilage is usually due to the growth of surviving or contaminating psychrotrophic microorganisms, in particular *Pseudomonas* spp. Spoilage is characterized by the development of off-tastes and off-odors and physical changes of the product (curdling or coagulation). Spoilage during distribution may occur when the raw milk used was highly contaminated causing excessive survival or the presence of enzymes surviving the pasteurization process. In such cases the shelf life of the product may be shortened to avoid consumer complaints.

Common criteria for pasteurized milk after a 6 days storage at 6°C are $n = 5$, $c = 1$, $m = 50\,000$, and $M = 500\,000$ for aerobic mesophilic counts in 1 mL at 21°C for 25 h, and $n = 5$, $c = 1$, $m = <0.3$, $M = 5$ for coliforms in 1 mL.

I Shelf-stable milk

Effects of processing on microorganisms. The purpose of sterilization is to achieve a 9-$\log_{10}$ reduction of *B. stearothermophilus* and similar thermophilic spore-forming bacteria. Those heat treatments are sufficient to ensure a 12-$\log_{10}$ reduction of *Cl. botulinum* and thus to ensure the safety of sterilized products (Westhoff, 1981). Time and temperature parameters are calculated according to the heat resistance of selected types of spores (*Cl. botulinum*, *Bacillus* spp.) using recognized mathematical formulae (Burton, 1988). Effectiveness of treatments depends on the initial total spore load, which is governed by the quality of the raw materials. Characteristics of spores and basic parameters influencing their destruction have been reviewed (Russell, 1982; Russell *et al.*, 1992; Holdsworth, 1992).

Sterilization of fluid milks can be performed in closed containers such as bottles or cans. More commonly, however, milk is UHT heated continuously either by a direct treatment with steam injection or by an indirect treatment by means of heat exchangers, followed by aseptic packaging (Burton, 1988; Robinson, 2002). The aim is to manufacture a product that remains stable during distribution and storage up to its consumption. However, from a statistical point of view, and despite the severe heat treatments applied in UHT processes, absolute sterility is never reached. The terms "commercial sterility" or "biological stability" are used, describing a product free from pathogens and stable under the anticipated storage and distribution conditions, which may vary greatly from temperate to tropical regions.

Spoilage. With modern equipment running under optimal conditions, spoilage rates of UHT products as low as1/5 000-1/10 000 (Cerf, 1989) can normally be achieved.

Spoilage of UHT-products is usually due to post-processing contamination rather than to survival of very heat-resistant microorganisms, provided the spore load of raw materials is not abnormally high. Cases of spoilage are commonly linked with raw materials such as cocoa powder used for chocolate drinks (Antoine and Donawa, 1990). Most failures are traceable to difficult to clean points in the processing line after heating, to leaks due to sealing deficiencies, package pinholes, or recontamination before, or during, closure. Under these circumstances, any type of spoilage caused by organisms from the environment can be expected.

Psychrotrophs, such as *Bacillus* spp. and *Pseudomonas* spp., are readily destroyed by UHT treatments. Their lipolytic and proteolytic enzymes formed before heat treatment, however, may survive in sufficient quantities to cause product deterioration during storage (Law, 1979). Gelling of the milk or development of bitter flavors in UHT milks is, at some storage temperatures, linked to the degree of proteolysis of the product (Collins *et al.*, 1993).

Over the last 10 years, an increasing number of spoilage cases of UHT-milk involving highly heat-resistant bacterial spores have been reported. A particular characteristic of these cases is the growth of the mesophilic spore-former to levels of only about 10^5 cfu/mL without further multiplication and without any visible modification of the product (Kessler *et al.*, 1994; Hammer *et al.*, 1996). The new species involved, *Bacillus sporothermodurans*, produces spores able to survive classical UHT treatment but is not pathogenic (Hammer and Walte, 1996; Petterson *et al.*, 1996). Modern molecular detection and identification methods has revealed the diversity of the strains (Guillaume-Gentil *et al.*, 2002) and their sources, from the feed and the environment of dairy processing plants, where reworking of processed milk seems to contribute to selecting these highly resistant forms (Scheldeman *et al.*, 2002).

Pathogens. Proper heat treatment assures the absence of pathogens in shelf-stable milk. The possibility of post-processing contamination of UHT-treated milk with a pathogen cannot be completely excluded, but in correctly designed and controlled processes it is extremely rare. UHT-products have not been linked to a major outbreak or to repeated smaller numbers of cases of food-borne disease (Bryan, 1983, WHO–Surveillance Programme for Europe.). In a recent survey, de Buyser *et al.* (2001) indicated that UHT products have been linked with 1.5% of the outbreaks related to dairy products, but the causes and origin of the outbreaks were not clear.

CONTROL (shelf-stable milk)

Summary

Significant hazards[a]	• *Cl. botulinum* and toxigenic bacilli.
Control measures	
Initial level (H_0)	• Low levels of pathogens in raw milk are likely (see section on raw milk). Use milk from healthy animals, milked according to good milking practices; cool milk during intermediate storage on the farm and during transportation to dairy plants (≤6°C recommended).
Reduction (ΣR)	• Heat treatments applied are sufficient to destroy vegetative microorganisms and to reduce *Cl. botulinum* by approx. 12 $\log_{10}$ units. • Process milk according to HACCP principles with special attention to the control of heating and aseptic filling. • Cleaning and disinfection after processing are essential.
Increase (ΣI)	• Post-process contamination may occur during aseptic filling and packing and during distribution if packaging materials damaged. • Growth of contaminating microorganisms during storage and distribution.
Testing	• Routine testing for pathogens is not recommended. • Incubation tests to detect deviations from Good Hygienic Practice. In cases of sensitive products (e.g. baby foods), 100% can be tested by non-destructive methods (e.g. vacuum test).
Spoilage	• Spoilage may occur due to the growth of contaminating microorganisms. Parameters are off-taste, off-odor, curdling/coagulation.

[a]In particular circumstances, other hazards may need to be considered.

Hazards to be considered

Heating technology is applied so that microbial hazards arising from raw milk are controlled. Problems arising from contaminated raw milk indicate inappropriate heating technology. Recontamination may occur during filling the heated product into containers such as bottles and cartons.

Control measures

Initial level of hazard (H_0). Initial level of the processed milk will depend on the previous processing steps, starting with the raw milk up to the UHT treatment.

Reduction of hazard (ΣR). Sterilization processes may be characterized by the number of decimal reductions of viable spores achieved during processing. As time-temperature parameters for UHT purposes are almost constant and defined, the initial spore load is critical relative to the potential for survivors. Spore loads in raw milk may fluctuate according to the season (Phillips and Griffiths, 1986). The type of spores, and hence their resistance to heating, can also vary depending on the origin of raw materials, e.g. milk, cocoa powder, malt extract, or carrageenan. This should be taken into account when developing new products or when modifying existing recipes. Consequently, modification of process parameters is sometimes necessary (Brown, 2000).

Cleanliness and sterility of all equipment that will come into contact with milk is a prerequisite to starting production. Heat exchangers should be carefully monitored during maintenance and dismantling. Defective heat exchangers may be a source of cross-contamination. Cooling water, or other cooling fluids, may be a source of contamination, including when cans are cooled, and should be disinfected.

Increase of hazard (ΣI). Design and maintenance of the process equipment is of utmost importance and material should be chosen carefully. Particular attention should be paid to aspects such as connections of pipes, presence of bends or dead ends, cleanability of valves, pumps, etc. These technical aspects are important to guarantee efficient cleaning, in particular cleaning in place (CIP) where physicochemical parameters such as flow rates, temperature of fluids, and concentration of chemicals can be critical to effective cleaning.

The fillers can be considered as "open and unprotected parts" of UHT-lines. Installation in locations separate from processing areas help minimize the risk of contamination. Sterile airlocks, steam seals, and other protection permit their installation in processing areas. Good environmental hygiene to avoid build-up of contaminants in this zone, quality of air, and hygiene of operators are further important aspects of product protection.

Packaging material can be classified as a raw material and its microbiological quality depends on its manufacturing process (Reuter, 1987). Further contamination by dust particles carrying microorganisms should be avoided during transport, handling, and storage. Sterilization of packaging is important and should be carefully controlled. An efficient treatment is possible only if the key parameters of each treatment such as time, temperature, dose, or concentration or sterilant are carefully monitored and any deviations corrected.

When milk is UHT-treated and then filled into bottles under non-aseptic conditions prior to final in-bottle sterilization, filling can be done at 50-70°C. In such processes, sporulation of thermophilic bacilli may occur in the filler and spores will not be inactivated in the subsequent sterilization. Cleaning and disinfection every 6-8 h is needed to limit build-up of *Bacillus* spp. and avoid the onset of the sporulation cycle.

Testing

Incubation tests alone are not sufficient to guarantee the quality and safety of a product and should be integrated into a preventive system based on HACCP (ICMSF, 1988; Cordier, 1990). Checking the final products is not feasible, for technical and practical reasons, and even analysis of large numbers of samples would not assure complete safety. Analysis of 100% of the manufactured units is normally only performed during the commissioning of new equipment or the start-up of new lines. As experience increases, however, sampling frequency is gradually reduced to for example 0.1% or less or to a fixed number of units per lot. For this verification, routine samples withdrawn at regular intervals and event samples taken after start or end of production, after a line stops, roll, or strip changes, etc. are analyzed.

Quality and safety of the final product depends on adequate sealing and complete integrity of seals. Visual inspection is an effective means of monitoring. Maintenance of this integrity throughout distribution from the manufacturer to the consumer is essential.

Spoilage

Spoilage is characterized by the development of off-tastes and off-odors and physical changes of the product (curdling or coagulation) similar to the one observed for pasteurized products. Incubation tests for five days at 30°C should result in total counts of ≤ 10 per 0.1 mL after incubation of the agar plates for 72 h at 37°C. This test is also applicable when checking for *B. sporothermodurans*. Incubation tests of a fixed number of samples up to 100% of the products (for sensitive products) are performed as verification of the process. Microbiological tests performed after incubation allow spoilage microorganisms to be identified and provide useful information for trouble shooting. Indicator tests include the total count of mesophilic bacteria.

V Cream

Cream is the fat rich fraction of the milk, which is separated by skimming or other means. Classification depends on legal aspects and is normally performed according to the fat-content: from half-cream (12%) to double cream or clotted cream (48% and 53%). Reviews on cream processing include Davis and Wilbey (1990), Bøgh-Sørensen (1992), Jöckel (1996) and Walstra *et al.* (1999).

A Effect of processing on microorganisms

Fat separation is performed by means of centrifuges and separators that lead to concentration of microorganisms in the fat phase. Separation is normally performed at 45–50°C or higher to inhibit possible growth of microorganisms. Separation at low temperature (5°C) allows cream with better physicochemical properties to be obtained. After separation cream is standardized to the desired fat content. Cream is submitted to treatments ranging from pasteurization to sterilization or UHT treatment with basically the same effects as those described for milk. Homogenization is applied either before or after heat treatment and conditions adapted to the type of cream manufactured. After filling the products are cooled to improve their physical characteristics. Driessen and van den Berg (1992) reviewed the microbiology of cream, in particular of pasteurized creams

B Spoilage

The microbiological quality of raw cream depends on the quality of the milk used to prepare it. Cream represents a more sensitive product than milk and this is due to differences in consumption habits. While milk is consumed regularly and rapidly once opened cream is used for special occasion and packs may remain open for longer periods.

Behavior of spore-forming bacteria is similar to that in milk, sweet curdling or bitter cream being caused by the multiplication of *B. cereus*, but other spore-formers have also been linked to organoleptic defects (Davis and Wilbey, 1990).

Recontamination after heat processing may occur as for milk, and extent of spoilage will depend on the product and storage conditions. Due to the high fat content, lipolytic spoilage organisms such as *Pseudomonas* spp. or yeasts will impact strongly on product quality.

C Pathogens

Because of the higher fat content, and its protective effect on microorganisms, heat treatments are stronger than those applied for fluid milks. Most cases and outbreaks of salmonellosis are linked with desserts or dishes prepared with cream, indicating contamination during preparation. This observation is true for other pathogens such as *Staph. aureus* (Jöckel, 1996). Outbreaks attributed to salmonellae were reviewed by Becker and Terplan (1986). Notable sources of recontamination of whipped cream are whipping machines used in restaurants and bakeries. Prevention of outbreaks is only possible through improved hygienic design and strict adherence to hygiene practices (Jöckel, 1996).

D Control

The measures and recommendations for fluid milk also apply to cream. Common sources of recontamination of whipped cream are "whipping machines" used in restaurants and bakeries. To prevent such recontamination, these machines must meet underlying hygienic design requirements and maintenance recommendations.

VI Concentrated milks

Concentrated products can be sub-divided into three main groups: (i) condensed and evaporated milks, (ii) sweetened condensed milk, and (iii) products (retentates) obtained by reverse osmosis, microfiltration, or ultrafiltration. There are numerous general reviews on the production, technology, and microbiological aspects of these milk products, e.g. Nelson (1990), Caric (1994), Spreer (1998), Bylund (2001), and De Jong and Verdurmen (2001).

The products represented in the three groups above are characterized by a reduced water content achieved by evaporation. They are stabilized microbiologically by a combination of "hurdles", usually heat treatments followed by the addition of sugar, or by benefitting from the low pH of the initial product (e.g. buttermilk, acid whey). They are commercialized either as small units for retail, or in bulk as ingredients for the manufacture of ice cream, confectionery, bakery, and other types of foods, allowing for lower costs of transport and storage.

A Effects of processing on microorganisms

For all types of concentrated or condensed products, the initial processing steps such as clarification, storage, or standardization are performed as under Section III-A and have the same effect on the microbial flora.

Concentrated milk was an important product in 1950s, but has subsequently lost importance. Product for retail is concentrated from Grade A milk about 1:3 to reach concentrations of 10–12% fat and about 36% total milk solids, usually consumed after dilution in water. Acid products such as sour milk, acid whey, or buttermilk may also be concentrated. Final pasteurization of the products is performed at slightly higher temperature than fluid milk to overcome the protective effect of increased solids, e.g. 80°C/25 s (U.S. Department. of Health Education and Welfare, 1994b). The keeping quality at low temperature is similar to that of pasteurized milk. Bulk concentrated milk is usually made from manufacturing-grade milk and is concentrated according to the final use as ingredient (2.5:1 to 4:1), preheated to temperatures between 65 and 90°C depending on the desired organoleptical characteristics, then concentrated. Thermophilic microorganisms may develop to high levels if the hygiene is not optimal. The concentrate is then submitted to a heat treatment, which is lower than a sterilization to provide a product with a limited shelf life, in particular at ambient temperature.

Unsweetened condensed or evaporated milk contains at least 7.5–9% milk fat and 25.9–31% total milk solids and is prepared from fresh milk or recombined milk using dry milk (EEC, 1992; U.S. Department. of Health Education and Welfare, 1995b; Codex Alimentarius, 1999a). After standardization depending on the type of final product, the milk is submitted to preheating (100–120°C for 1–3 min) followed by chilling to 70°C to stabilize the proteins. It is then concentrated on free-falling type evaporators at temperatures of ca 45–70°C (Milner, 1995; Tetra Pak, 2003). After concentration, milk is homogenized, supplemented with authorized stabilizers, cooled to 14°C for immediate packing or to 5–8°C for intermediate storage. The milk is then canned and the cans sterilized in batch or continuous autoclaves (110–120°C for 15–20 min). The application of UHT treatments (140°C for about 3 s) is also known. All these heat treatments are sufficient to destroy spore-forming bacteria.

Preheating is required to stabilize milk components, but leads also to inactivation of vegetative microorganisms and of spores with low-heat resistance. The concentration process cannot be relied upon to achieve a lethal effect. In fact, peculiar to this group of products are evaporators operating at about 40–50°C, increasing the risk of build-up of the population of mesophilic and thermophilic spore-forming bacteria unless operated under hygienic conditions (Milner, 1995). Even temperatures of around 70°C, reached in certain installations, are not sufficient to destroy spores and may actually permit growth of certain thermophilic spore-forming bacteria or vegetative forms such as *Thermus thermophilus*, which has been increasingly reported (Langeveld, 1995).

Sweetened condensed milk normally contains 8% milk fat, 28% total milk solids and usually sugar (sucrose) to achieve a sugar ratio of 42–64%, depending on the product, and is prepared from fresh or recombined milk (U.S. Department of Health Education and Welfare, 1995b). Sugar is added before concentration or as syrup late during this operation. Afterwards, the concentrate is allowed to cool, seeded with lactose to promote crystallization of lactose as very small crystals, and then filled. Products at water activities around 0.85, although not sterile, can be stored at room temperature.

Products containing lower levels of sugar, e.g. semi-finished products for confectionery or bakery, may require refrigeration to avoid growth of spore-formers, thermoduric organisms or other contaminants.

Retentates from reverse osmosis, microfiltration, and ultrafiltration are produced to concentrate desirable components from raw materials such as milk, buttermilk, or whey by means of selective membranes. These retentates can also be used as sources of solids for confectionery or bakery products, ice cream, yogurts, or cheese. Reverse osmosis is applied for separation of components of low molecular weight (cut-off around 500 Da) from their solvents, normally water. Ultrafiltration is a sieving process during which molecules such as proteins are retained while water and solutes are eliminated. From a microbiological point of view increases in numbers of most groups of microorganisms can be observed in retentates (Veillet-Poncet *et al.*, 1980); the extent is dependent on residence time, temperatures and quality of raw materials used. Conditions for growth or inhibition of microorganisms differ from those of untreated raw materials, linked to changes in levels of components such as ions or proteins. Reviews of applications include Glover (1986), El-Gazzar and Marth (1991), Caric (1994), and Spreer (1995).

B Spoilage

Concentrated milk is a favorable medium for microbial growth and the spoilage problems are of the same type as those observed in pasteurized milk (see Section IV-F). Shelf life varies from a few days to weeks, depending on the number of heat-resistant survivors and post-process contaminants. Concentrated acid products are more stable due to the low pH. Fungal development may possibly result from surviving spores, but is more likely to be due to post-processing contamination.

Problems linked to shelf-stable evaporated milk are similar to those observed for sterilized or UHT-products (see Section IV-I). *Bacillus stearothermophilus*, *B. coagulans*, and *B. licheniformis* are the most frequently isolated spoilage organisms (Kalogridou-Vassiliadou *et al.*, 1989; Kalogridou-Vassiliadou, 1992). Obligate thermophiles are a problem only if the product is stored at elevated temperatures.

In the case of sweetened condensed milk ($a_w \sim 0.85$), only halophilic and xerophilic microorganisms, e.g. micrococci and fungi are able to grow due the low water activity (Tudor and Board, 1993). Growth of xerophilic yeasts may cause blowing of cans and development of off-flavors. Occasionally, the brown mold *Wallemia sebi* forms small brownish "buttons" of mycelium and coagulated casein on the product surface if there is too much headspace and cans have not been gassed after filling.

C Pathogens

In concentrated and condensed milk, concentrated acid products and evaporated milk, the same comments apply as for pasteurized or sterilized milk (see Section IV-G). The primary concern is control of post-process contamination.

No case of food-borne disease caused by sweetened condensed milk has been reported. The water activity of around 0.85 has an inhibitory effect on all pathogens. This a_w is only borderline for *Staph.*

aureus, which will grow at an $a_w \sim 0.85$. However, in the anaerobic environment existing in sweetened condensed milk, growth and enterotoxin formation are inhibited.

Sweetened condensed milk may be produced with lower sugar content, i.e. higher water activity, and used for example for pastry and confectionery. In such products, growth of pathogens such as *Staph. aureus* is possible if the temperature is favorable and the product has been abused. Spores of *Clostridium* and *Bacillus* spp. may be present in sweetened condensed milk (Bhale *et al.*, 1989) but outgrowth and multiplication is prevented by the low a_w.

Growth and survival of pathogens in ultrafiltered milk were studied by Haggerty and Potter (1986) and shown to be similar to that in milk, except that *L. monocytogenes* grew faster and achieved higher populations (El-Gazzar *et al.*, 1991).

D CONTROL (concentrated milks)

Summary

Significant hazards[a]	• *Cl. botulinum* and bacilli.
Control measures	
Initial level (H_0)	• Low levels of pathogens in unprocessed milk are likely (see section on initial microflora of raw milk).
Reduction (ΣR)	• Heat treatments applied are sufficient to destroy vegetative microorganisms. • Process milk according to HACCP principles with special attention to the control of heating and aseptic filling. • Cleaning and disinfection after processing are essential.
Increase (ΣI)	• Post-process contamination may occur during filling and packing and during distribution if packaging materials are damaged. • Growth of contaminating microorganisms during storage and distribution.
Testing	• Routine testing for pathogens is not recommended. • Incubation tests to detect of deviations from Good Hygienic Practice.
Spoilage	• Spoilage may occur due to the growth of contaminating microorganisms (mostly mesophilic and facultative thermophilic bacilli) during evaporation, reverse osmosis or ultrafiltration. Parameters are off-taste, off-odor, curdling/coagulation.

[a]In particular circumstances, other hazards may need to be considered.

Hazards to be considered

Apart from the theoretical presence of *Cl. botulinum*, surviving bacilli in the product are of concern.

Control measures

Initial level of hazard (H_0). Low levels are likely, however quantitative data are missing.

Reduction of hazard (ΣR). The reduction effect on microorganisms is very similar to the UHT-milk procedure.

Increase of hazard (ΣI). Evaporators running at temperatures around 40–50°C, if not run under hygienic conditions, may lead to increases in numbers of spore-forming bacteria, which may affect stability of products. The sticky nature of sweetened condensed milk increases the difficulty of cleaning and residues may become a problem throughout the entire line, and in particular in the crystallization tanks or in the complex fillers, which are inherently difficult to clean adequately. Cans and lids, tubes, and other packaging material should be cleaned and sterilized before filling, as no heat treatment is applied to the product after filling. Critical to control is maintenance of the water activity below that allowing growth of surviving spores and post-processing contamination.

In the case of reverse osmosis and ultrafiltration, an important aspect in terms of microbial growth are the large surfaces of membranes offering an ideal support for the development of biofilms. Increases in cell numbers in the liquid phase may be due to the constant release of organisms or of fragments of the biofilm reflected in sudden increases in populations of microorganisms. Cleaning and disinfection are thus very important and must be adapted to the type of membranes used, bearing in mind that bacteria embedded in biofilms are more resistant to cleaning and more tolerant of sanitizing agents (Defrise and Gekas, 1988; Carpentier and Cerf, 1993).

Testing

Routine testing for pathogenic bacteria is not recommended. In case of suspicion (e.g. incorrect sealing of containers), the detection of vegetative bacteria may be useful to identify the defect. Integrity testing of containers may be part of the testing procedure.

Spoilage

Microbiological spoilage concerns for concentrated and condensed milks, and for sweetened condensed milk with low sugar content are similar to pasteurized or UHT milk. The control measures in particular the integrity control of the containers are also similar.

VII Dried dairy products

Many milk products, including whole milk, skim milk, whey, buttermilk, cheese, and cream, may be dried by the application of heat. The manufacturing processes of dried dairy products are described in detail in Masters (1985), Knipschildt (1986), Caric (1994), Anonymous (1995a), Spreer (1998), and Bylund (2001).

Dried milk may be rehydrated and consumed directly, but more commonly, dried dairy products are used as ingredients in a number of products such as bakery, chocolate and confectionery, baby foods, ice cream, animal feeds, and culinary products.

A Effects of processing on microorganisms

Before drying, milk is submitted to preliminary treatments such as clarification, standardization, and heating (see Section III-A, IV-B-E). Two types of dried milk powder are manufactured by spray drying or roller (drum) drying. The spray-dried product has the better solubility. A further differentiation can be made according to the intensity of the heat treatment applied. In the case of low-heat milk powders, the heat treatments correspond to a pasteurization; in the case of medium-heat powders, temperatures of 85–95°C for 20–30 s are applied and temperatures above 120°C for up to 30 s are used to obtain high-heat powders.

Drying does not provide a controlled killing effect. The extent of microbial destruction during drying depends on the types of microorganisms present, and on the drying temperature of the exit air in spray drying or on the drum temperature and retention time for drum drying. Various vegetative bacteria, including Gram-negative Enterobacteriaceae, survive the drying process (Daemen and van der Stege, 1982). Doyle *et al.* (1985) determined that *L. monocytogenes* also survives a typical spray-drying process. Therefore, dairy products for drying must be given a heat treatment equal to or greater than pasteurization, and the product must be protected against contamination between the pasteurizer, the dryer, and the packaging operations. After dehydration, the products will not support microbial growth.

Instant dry milk evolved from the desire for an easily soluble dried milk for use in beverages. Dried milks do not wet or disperse easily when added to water. The instantizing process rewets surfaces of dried particles in steam or atomized water droplets, causing them to agglomerate in clusters. The product is then redried to 5% moisture or less. The principal microbiological problems associated with instant dry milk occur upon accidental contamination during rewetting or after it is reconstituted, since the water activity of the dried product is too low to permit growth.

Subsequent processing steps such as cooling of the powder, intermediate storage, mixing, and packaging may influence the microbiological quality by recontamination from the line or the environment. During storage of dry milks, some vegetative microorganisms present may slowly die-off (Thompson *et al.*, 1978) but others may survive over prolonged periods. Spore-formers, being the most resistant, retain viability indefinitely.

B Spoilage

Due to the extremely low water activity (0.3–0.4) of dried dairy products, development of spoilage organisms is not possible. However, presence of water derived from condensation on wet packaging material, may allow development of molds or bacteria and must be avoided. Spoilage may occur in reconstituted products similar to pasteurized milks (see Section IV-F).

C Pathogens

Salmonella. Several outbreaks of salmonellosis have been traced to dry milk products. One of the first nationwide outbreaks occurred in the United States during 1964–1965 (Collins *et al.*, 1968; ICMSF, 1980) and was associated with non-fat dry milk that was produced in one plant and instantized in other factories. After an investigation in 156 plants located in 23 states, milk and environmental samples were found to contain salmonellae, initiating a number of improvements in processing and sanitation measures. Non-fat dry milk contaminated with *S.* Typhimurium and *S.* Agona occurred in Oregon (USA) in 1979 (Anonymous, 1979). Recommendations concerning preventive measures from different authors (e.g. IDF, 1991b; Burgess *et al.*, 1994) and the implementation of HACCP have greatly contributed to improvements. Nevertheless, outbreaks in different countries are reported from time to time, including dairy based infant formulae, which are manufactured in a similar way (Becker and Terplan, 1986; Rowe *et al.*, 1987; Gelosa, 1994; Usera *et al.*, 1996; Anonymous, 1997a; Threlfall *et al.*, 1998; Bornemann *et al.*, 2002; Forsyth *et al.*, 2003). These outbreaks demonstrated that failures in preventive systems, such as presence of water allowing multiplication of salmonellae, or zones that are difficult to maintain and to clean (isolation from a drying tower), were the origin of contamination, and that salmonellae with particular characteristics (lactose positive) were involved. Another small outbreak occurred in Canada in 1992 from infant formulae, produced in the United States, and resulted in various recall of products containing several lots of dry milk manufactured by the same company (Anonymous, 1993). Other outbreaks were reviewed by Mettler (1994).

Listeria monocytogenes. There have been no outbreaks of listeriosis linked to dry dairy products. Doyle *et al.* (1985) examined the fate of *L. monocytogenes* during spray drying and ambient storage of dry product and observed a reduction of about 1.0- to 1.5-$\log_{10}$ cells per gram during the drying process and noted that, although the organism progressively died during storage, some samples tested positive for *L. monocytogenes* for up to 12 weeks. *L. monocytogenes* is, however, readily killed by the heat treatments applied before drying and presence in finished products is unlikely.

Proper heat treatments and prevention should preclude the presence of *Listeria* spp. in dry zones of powder processing plants. The difference in the incidence of *L. monocytogenes* or *Listeria* spp. can be seen in the investigation of Gabis *et al.* (1989) in 18 plants manufacturing dry dairy products, all samples from the dry zones showed negative results. Surveys of the incidence of *L. monocytogenes* over more than 10 years have not detected it in dry milk products (Pak *et al.*, 2002).

Staphylococcus aureus. A very large outbreak causing more than 13 000 cases has recently been reported in Japan (Asao *et al.*, 2003) due to preformed staphylococcal enterotoxin in the powder. This was traced back to poor hygienic and manufacturing practices during processing of liquid milk, in particular the storage conditions. Other cases of illness have been due to contamination and abuse of reconstituted products (Umoh *et al.*, 1985; El Dairouty, 1989).

Bacillus cereus. The presence of *B. cereus* at low levels in dried milk has been reported (Becker *et al.*, 1989, 1994), with over 60% of milk powder supplied in the U.S. positive for *B. cereus*. Although outbreaks of food poisoning due to *B. cereus* have not been directly attributed to dry dairy products, temperature abuse of the reconstituted product is of major concern. A review and assessment of the risks related to the presence of *B. cereus* in infant formulae has been published by the Food Standards Australia and New Zealand (2004).

Enterobacter sakazakii. *Enterobacter sakazakii* has been implicated in sporadic outbreaks causing neonatal meningitis and several of these cases have been associated with the consumption of contaminated infant formulae. In several case studies, environmental contamination in the hospital and temperature–time abuse has been underlined as one of the main factors contributing to the outbreaks. Outbreaks caused by *E. sakazakii* have been reviewed by Nazarowec-White and Farber (1997), Lai (2001), and Iversen and Forsythe (2003).

The presence in milk powder of the potentially pathogenic *Pandoraea norimbergensis* was reported by Moore *et al.* (2001) but no outbreaks have been reported.

Mycotoxins. Occasionally, dried milk has been found to contain aflatoxin M_1 (Galvano *et al.*, 1996). Although the amount of toxin present in fluid milk is reduced somewhat by the drying process, a significant percentage of it appears to survive the process and will survive for extended periods in the dry product (Marth, 1987). Other mycotoxins have not been found at any significant level in dried milk products and appear unlikely to occur.

D CONTROL (dried dairy products)

Summary

Significant hazards[a]	• Several pathogens may be present in raw milk (see Table 16.1). • Pathogens may also be present in the environment of milk processing.
Control measures	
Initial level (H_0)	• Low levels of pathogens may be present in raw milk (see Section on raw milk).

(*continued*)

D CONTROL (Cont.)

Summary

	• Use milk from healthy animals, milked according to good milking practices. • Cool milk during intermediate storage on the farm and during transportation to dairy plants ($\leq 6^{\circ}$C recommended).
Reduction (ΣR)	• Various heat treatments during processing such as pasteurization or sterilization destroy vegetative microorganisms and ensure the reduction of pathogenic spore-formers such as *B. cereus* to very low levels.
Increase (ΣI)	• During cooling of milk powder, intermediate storage, filling and packing, post-process contamination with salmonellae is possible if it is present in the closed processing environment. However, multiplication of salmonellae in the product is not possible due to the low water activity.
Testing	• Finished products are tested by manufacturers as verification, along with testing of raw materials (used in dry mixing operations), line samples to assess the hygiene status of the line, and the processing environment to assess the absence of pathogens such as salmonellae. The hygienic status of line and environment is also assessed using hygiene indicators such as coliforms or Enterobacteriaceae.
Spoilage	• Spoilage of milk powders and similar products is not possible due to the low water activity.

[a]In particular circumstances, other hazards may need to be considered.

Hazards to be considered

Depending on the processing parameters, bacterial spores may survive and be present in the powder. Contamination during or post process by pathogens from the environment (salmonellae) is to be considered.

Control measures

Initial level of hazard (H_0). The quality of the raw milk is important and the microbial load, in particular spore-forming and thermoduric bacteria, influences the final quality, in particular for low- and medium-heated products, where only partial destruction of microbes may be achieved (Muir *et al.*, 1986).

Reduction of hazard (ΣR). Production of dried milk includes various heating steps sufficient to eliminate pathogenic bacteria. Heating procedures vary but usually start with at least a pasteurization, followed by concentration steps between 40°C and 90°C and final drying at elevated temperatures on heated drums (150–160°C, 1 s) or by spray drying (180–220°C).

The final quality of dried dairy powders depends on the quality of raw materials, processing parameters, and environmental factors. Detailed recommendations for the hygienic manufacture of spray-dried milk, casein, and whey products have been published (IDF, 1991b; Mettler, 1994).

Increase of hazard (ΣI). Salmonellae and other pathogens may be introduced into zones where raw milk is handled. Spreading to dry zones must be avoided by strict separation of raw milk areas from spray drying or roller drying and subsequent processing steps. This is achieved by appropriate layout of production lines, maintenance of buildings and equipment, control of movements of personnel and materials, filtration of air, and application of dry cleaning.

Personal hygiene, dry cleaning, wet cleaning, and pest control must be monitored: relative humidity and efficiency of air filtration can be checked visually or by physicochemical methods. Any accumulation of moisture due to condensation or other causes is a potential source of growth and contamination.

Testing

Analysis for salmonellae in line samples and environmental samples (air filters, ventilation systems, room and equipment surfaces, sewer systems, etc.) are used to verify the application of the system and its effectiveness. Trend analysis for Enterobacteriaceae or coliforms can provide further information on the performance of the system although there is not a direct correlation to the presence of salmonellae.

Spoilage

Spoilage due to bacteria or fungi is not possible, due to the low water activity of the dried product. Attention must be given to the storage of the final product to maintain this condition.

VIII Ice cream and frozen dairy desserts

Many different types of ice cream and frozen desserts are manufactured and descriptions vary from country to country.

Ice cream can be sub-divided into four main categories according to the main ingredients used: (i) ice cream made exclusively from milk products, (ii) ice cream containing vegetable fat, (iii) sherbet ice cream containing fruit juice, milk and milk solids non fat, and (iv) water ice manufactured from water, sugar, and fruit juices or concentrates. The first two categories account for more than 80% of the products manufactured. Other classifications can be found, e.g. Papademas and Bintsis (2000).

The composition of different products is regulated by international or national legislations (Pappas, 1988b; U.S. Department Health, Education and Welfare, 1995c; FSA, 1995). Minimum fat standards vary from country to country. In the U.S., ice cream must contain a minimum of 10% milk fat. Frozen desserts are shaped and flavored forms of these products, and may be retailed "on a stick," such as chocolate-coated ice cream bars and "pops" (Table 16.3). Parfaits, custards, and "French" ice cream contain approximately 1.5% added egg yolk solids. Mousse is whipped cream that has been flavored, colored, stabilized, sweetened, and frozen.

The microflora of ice cream before pasteurization is that of the individual mix ingredients. For further details, see Sections in this chapter on milk, cream, condensed milk, dried milk, and buttermilk. The

Table 16.3 Typical composition (%) of frozen dairy products

	Ice cream	Ice milk	Sherbet	Frozen yogurt
Fat	8–18	3–6	1–2	2.5
NFMS[a]	9–12	11–12	2–5	9
Sugar	13–16	15–16	26	25
Stabilizer/Emulsifier	0.3–0.5	0.5	0.3	0.3

US Department of Health Education and Welfare (1995c); Walstra *et al.* (1999); Bylund (2000).
[a]Non-fat milk solids.

microflora of additional ingredients is discussed in other chapters: butter, Chapter 11; sugar, Chapter 12; chocolate, Chapter 10; fruits, Chapter 6; nuts, Chapter 9; eggs, Chapter 15.

A Effects of processing on microorganisms

Modern technology used in the manufacture of ice cream and other frozen desserts is well documented (Rothwell, 1985; Arbuckle, 1986; Marshall and Arbuckle, 1996; Robinson, 2002). Similar steps are used in the production of different frozen products. Milk fat sources include fresh cream, butter oil, sweet butter, whole milk, and other dairy products. Non-fat milk solids are derived from whole milk, skim milk, condensed milk, dried whey, buttermilk, and non-fat dry milk. Sugar is added as sucrose or a blend of sucrose and corn syrup solids. Small amounts of emulsifier improve whipping properties, while stabilizers improve body and texture and prevent ice crystal formation. Authorized ingredients vary from country to country. Some countries, such as Italy and Ireland, traditionally did not mandate ingredient usage (Pappas, 1988b).

Mix ingredients are combined and may be held for a short time to allow the stabilizer to hydrate. Time-temperature regulations vary from country to country. In the United States, parameters for pasteurization are based on, but higher than, time-temperature combinations used for milk. Heat treatments must usually be 3°C higher than those used for simple milk for each minimum holding time. A large range of legislative requirements for the pasteurization of ice cream mix exists, and some have been summarized by Papademas and Bintsis (2002). Ice mixes, because of their low pH (4.5 or less), are not pasteurized.

Ice cream mix is pre-heated and homogenized to improve body and texture in the frozen product and then pasteurized. The homogenizer often functions as the metering pump for the HTST pasteurizer. Flavors may be added before or after pasteurization and homogenization; color, fruits, candies, chocolate, and nuts are generally added after pasteurization. Pasteurized mix is cooled and transferred to an ageing tank, where the mix is stored for at least 4 h at temperatures ranging between 2°C and 5°C. This phase is necessary to allow for stabilization of the mix and crystallization of the fat and adsorption of protein onto the fat globules, to occur. After ageing, ice cream mix is promptly frozen, unless it is to be used for soft-serve ice cream. Hardened ice cream is frozen in a two-step process. First, partial freezing to −5°C to −8°C occurs while air is beaten into the mix. Overrun, or the increase in volume of the ice cream due to incorporation of air, ranges from 30% to 100% of the mix volume. Partially frozen mix is packaged and immediately placed in a hardening room or freezing tunnel where it is frozen to −25°C to −30°C. The low temperature of frozen ice cream completely prevents microbial growth. Soft-serve ice cream is usually drawn from the freezer at about −6°C to −7°C with an overrun of about 40%.

B Spoilage

Heat treatments applied to ice cream mixes are frequently more severe than the minimal requirements. Microbial destruction is extensive, and survivors are primarily spores. Post-pasteurization aerobic plate counts are usually a few hundred per milliliter or lower. Fruits, nuts, flavors, colors, and other ingredients added after pasteurization may contribute significant contamination. If ice cream mix is frozen promptly, microbial spoilage does not occur. If there is an uncontrolled storage delay between pasteurizing and freezing, microorganisms can grow and spoilage may occur.

Mix for soft-serve ice cream must be transported to retail soft-serve outlets where it is stored until soft-frozen and dispensed to consumers. Both contamination and temperature abuse of the mix may easily occur. Facilities for cleaning and sanitizing the freezer and associated equipment often are inadequate (Holm *et al.*, 2002). Improper refrigeration may permit bacterial growth to levels above legal limits and cause spoilage. Martin and Blackwood (1971) demonstrated that aerobic plate counts on mix immediately after UHT-pasteurization at 140.6°C were below 80 cfu per mL. Shelf life of the mix at 4.4°C, 10°C, and 15°C was 3–4 weeks, 2–3 weeks, and 1–2 weeks, respectively. Aerobic plate counts on mix stored at 4.4°C for only two weeks exceeded the bacteriological criteria for ice cream. In a further trial, UHT pasteurized mix was aseptically sampled at the pasteurizer to eliminate post-pasteurization contamination. No microorganisms were evident when counts were made immediately after pasteurization and fewer than 20 cfu per mL were detected after storage at 4.4°C, 10°C, or 15°C for 8 weeks. When the mix was inoculated with about 10^6 *B. cereus* spores and UHT treated at 137.7°C, approximately 10^4 spores per mL survived and that population remained constant for 6 weeks at 4.4°C, then gradually declined. When stored at 10°C and 15°C, populations increased in one week to over 2×10^7 and 10^8 per mL, respectively. Curdling and proteolysis were visibly evident upon further storage. The microbiological quality of the mix after pasteurization may be maintained by chilled storage (e.g. 3-5°C) for a limited time.

C Pathogens

When ice cream made from raw milk and cream is consumed, the same dangers inherent to consumption of raw milk may result. With few exceptions, most outbreaks have involved home-made ice cream (Taylor *et al.*, 1984; Barrett, 1986) and have been due to the use of raw milk, cream, or eggs, inadequate heat treatment, or contamination by infected handlers. Similar problems have sometimes resulted in outbreaks associated with the consumption of commercially manufactured ice cream (Snyder *et al.*, 1978; U.S. Department of Health, Education and Welfare, 1994a). Raw, or improperly pasteurized, eggs containing salmonellae were involved in one outbreak. In another, the mix was contaminated with *Staph. aureus*, and subsequent temperature abuse permitted growth and enterotoxin production. Pathogens, if present, may survive in ice cream for many months (Wallace, 1938). *Salmonella* survived for 7 years in ice cream (Georgala and Hurst, 1963). Several recalls of frozen dairy products including frozen novelties, ice cream, ice milk, and sherbet due to contamination by *L. monocytogenes* have occurred in the U.S. since 1985 (Ryser and Marth, 1991). Although millions of gallons of product were recalled, no direct link to listeriosis was documented. As shown by the risk assessment on ready-to-eat foods performed by the WHO (2002a), however, ice cream does not play a role as a cause of outbreaks attributed to this pathogen (Pak *et al.*, 2002; FDA, 2003). Because *L. monocytogenes* does not survive pasteurization, post-pasteurization contamination from the processing environment is the source of this contaminant (Miettinen *et al.*, 1999). The inability to grow at freezing temperatures minimizes the risk associated with this, and other pathogens (Kozak *et al.*, 1996).

An outbreak of *S.* Enteritidis food-borne illness was associated with nationally distributed ice cream products in the United States in October 1994 (Oemichen, 1995). Eighty confirmed cases were reported to the Minnesota Department of Health over 20 days prior to that report, but the final number of cases was estimated to be 225,000 (O'Ryan *et al.*, 1996; Vought and Tatini, 1998). Contamination probably resulted during transport of pasteurized ice cream mix in tankers that were also used to transport unpasteurized raw egg. The mix was not subsequently re-pasteurized (Hennessy *et al.*, 1996; O'Ryan *et al.*, 1996).

D CONTROL (ice cream and frozen dairy desserts)

Summary

Significant hazards[a]	•	Pathogens present in ingredients used to manufacture ice cream, e.g. milk, cream, milk derivatives, fruits, egg as described in the corresponding chapters.

(continued)

D CONTROL (Cont.)

Summary

Control measures	
Initial level (H_0)	• Low levels of pathogens in the raw ingredients are likely–see appropriate Sections.
Reduction (ΣR)	• Heat treatments, usually pasteurization applied are sufficient to destroy vegetative microorganisms including pathogens such as salmonellae and *L. monocytogenes*.
Increase (ΣI)	• Post-process contamination during different steps such as cooling, ageing, freezing, or filling of the ice cream with salmonellae or *L. monocytogenes* is possible if present in the closed processing environment. Growth in ice-cream is not possible due to the low temperature.
Testing	• Testing of finished products by manufacturers is done as verification along with testing of raw materials (used without previous heat treatment). Line samples to assess the hygiene status of the line and the processing environment to assess the absence of pathogens such as *L. monocytogenes* is relevant for these products. The hygiene status of line and environment is also assessed using hygiene indicators such as coliforms or Enterobacteriaceae. • Typical criteria (e.g. EEC, 1992) include mandatory criteria: *L. monocytogenes* and salmonellae, as well as hygiene indicators such as total viable counts and coliforms.
Spoilage	• Microbial spoilage of ice cream and similar products is not possible due to low storage temperature.

[a]In particular circumstances, other hazards may need to be considered.

Hazards to be considered

Salmonellae are certainly the pathogen of most concern in ice cream. *Listeria monocytogenes*, although frequently included in legislation, has not been linked to outbreaks.

Control measures

Initial level of hazard (H_0). Low levels of pathogens in unprocessed ingredients are likely.

Reduction of hazard (ΣR). Control of salmonellae is achieved through the application of appropriate heat treatments of the ice cream mix and ingredients, as far as possible. Careful selection of suppliers of other critical ingredients added after the pasteurization is an important control measure.

Increase of hazard (ΣI). Increases are usually due to post-heat treatment contamination during handling and further processing of the ice cream mix, either due to poorly cleaned equipment or from the processing environment. Slow growth of contaminating microorganisms is only possible during steps where the ice cream mix is exposed to higher temperatures (0-5°C).

Testing

Several factors are important in the production of high-quality ice cream (Bigalke and Chappel, 1984). Adherence to Good Hygiene Practices can be monitored through hygiene indicators such as total viable counts, coliforms or Enterobacteriaceae, which are readily killed by heat treatments.

Environmental monitoring (*Listeria* spp. and *L. monocytogenes*) provides useful information on the effectiveness of the implemented preventive measures such as zoning (IDF, 1994; U.S. Dept. of Health Education and Welfare, 1995c).

Spoilage

Due to the storage and distribution conditions used, spoilage is not possible.

IX Fermented milks

Fermented or acidified milks are the oldest dairy products and for centuries a significant portion of milk has been consumed in this form. Around 400 names of fermented milk products exist worldwide (Kurmann *et al.*, 1992). Numerous products, however, are very similar and have only a regional importance and only a few have any commercial importance.

Based on the metabolism of the flora, fermented milks are best sub-divided into three groups: lactic, yeast-lactic, and mold-lactic fermentations (Robinson and Tamime, 1990). The first group can be further sub-divided according to the characteristics of the lactic flora into mesophilic, thermophilic fermentations, and include as well therapeutic products.

All products are the result of the transformation of lactose into mainly lactic acid, and this is responsible for the typical taste of fermented milks. The identity of different products is the result of the metabolism of the fermenting microorganisms and the flavor compounds formed. Acetate, acetaldehyde, lactate, fatty acids, peptides, ethanol, CO_2 or diacetyl are typical products of their lactose and nitrogen metabolism (Desmazeaud, 1990; Marshall and Tamime, 1997). Product consistency is also an important parameter. The desired characteristics are obtained by using starters producing polymers (ropy strains), by the fortification with milk solids, or in certain cases and depending on the legislation (Pappas, 1988a), of thickeners or sugar.

The initial steps of the manufacture of fermented milks are the same as those outlined under Section III. A particular aspect is the fortification with non-fat milk solids to reach about 14% in finished products, about 5% being proteins and <1.5% fat. This can be achieved by addition of dairy powders, in particular skimmed milk powder (Robinson and Tamime, 1993; Tamime and Marshall, 1997a,b). After recombination, filtration and de-aeration are recommended procedures to eliminate undissolved particles and to provide optimal growth conditions for the starter cultures used.

Heat treatments ranging from thermization to UHT-processes have been described (Tamime and Robinson, 1985) and may have an impact on the organoleptic properties of the products. After cooling to the desired optimal fermentation temperature, starter cultures are inoculated at rates that may vary from 2% to 3% or 10% to 30% depending on the products manufactured (Marshall, 1987; Tamime and Robinson, 1988a; Tamime *et al.*, 1995; Robinson *et al.*, 2002). During fermentation the desired texture is obtained and the complex mechanisms involved are described in details in Tamime and Marshall (1997a,b). For set-type products, gel-formation takes place in the container while stirred-type products are bulk products where the gel is broken before further processing.

For further discussions of the technological and microbiological aspects of fermented milks refer to Driessen and Puhan (1988), IDF (1988), Kurmann and Rasic (1988), Robinson and Tamime (1990, 1993), Spreer (1995), and Teuber (2000).

A Effect of processing on microorganisms

As discussed in the previous paragraphs, milk used to manufacture fermented milks is submitted to processing steps such as clarification, homogenization and heat treatments. The effects of these different technological steps have been described in previous paragraphs (Sections III-A and IV-B,C).

Pasteurization is sufficient to kill vegetative cells and the rapid development of the starter cultures is sufficient to inhibit outgrowth and development of spore-formers. This is due to the production of organic acids, in particular lactic acid, and for certain strains of bacteriocins (Hoover, 1993; Klaenhammer, 1993; de Vuyst and Vandamme, 1994).

In certain cases, and in particular when slow fermenting strains are used, development of spore-forming bacteria is more rapid, thus affecting the quality of the products and causing losses (Chandan, 1982; Alm, 1983). In such cases sterilization of the milk (e.g. UHT) is necessary.

Heat treatments also destroy phages (bacteriophages) that may be present. This processing step as well as de-aeration of the milk will also have a beneficial effect on the activity of starter cultures: (Tamime and Marshall, 1997a,b).

Products obtained by mesophilic lactic fermentation

Cultured buttermilk and cream. These products are based on fermented skimmed milk (about 9.5% solids-not-fat). Fortification with additional skimmed milk powder and fat is frequently used to obtain a desirable final viscosity. Milk is pre-heated, homogenized and heated to either 85°C for 30 min, or 95°C for up to 5 min. After cooling to 22°C, starters are added (1%). Combinations of mesophilic lactic acid producers, such as *Lactococcus lactis* subsp . *lactis* or *Lc. lactis* subsp. *cremoris*, and aroma/flavor producers, such as *Lc. lactis* biovar. *diacetylactis*, and *Leuconostoc mesenteroides* subsp. *cremoris*, are used (Vedamuthu, 1985; Marshall, 1987). Flavor enhancement by diacetyl is achieved by increasing the oxygen content of the milk. Incubation at 22°C is then carried out for 12–14 h until the milk reaches a pH of 4.6-4.7. The product is then cooled, filled, and distributed refrigerated, usually below 10°C depending on local regulations. The cultured cream, also known as sour cream, is a fat-rich product (10-40%) with the flavor and aroma of buttermilk. Fermented milks can also be produced from goat's or buffalo's milk. Reviews include IDF (1986b), Anifantakis (1990), and Abrahamsen and Rysstad (1991).

Products obtained by thermophilic lactic fermentation. Products prepared with starter cultures showing a growth optimum between 37°C and 45°C are termed "thermophilic".

Yoghurt. A broad variety of yoghurts exist, which are sub-divided according to different criteria such as legislation (full, medium, light); type of gel (set, stirred, drinkable); plain or flavored or type of post-fermentation processing.

Yoghurt may be manufactured from full milk, skimmed, or partially skimmed milk or non-fat dry milk after recombination. Non-fat skimmed milk is frequently added to reach 11-15%, to improve the rheological characteristics of the final product (IDF, 1988; U.S. Department of Health Education and Welfare, 1995a).

Milk is processed as described under Sections III-A and IV B, C, the type of heat treatment being adapted to the type of yoghurt to be manufactured. After cooling to 40–45°C (certain varieties to 30-35°C) starter cultures are added (1-3%). Normally well-defined mixtures of *Lactobacillus delbrueckii* subsp. *bulgaricus* and *Strep. thermophilus* are used, but other strains are known depending on the countries (Weber, 1996a,b; Tamime and Marshall, 1997a,b). Selection and ratio of the strains is chosen according to their capacities to produce flavors and exopolymers (Rohm *et al.*, 1994; Rohm and Kovac, 1994, 1995). Fermentation is continued until a desired pH, between 4.0 and 4.5, is reached. Growth and further acidification is minimized or inhibited by cooling before further processing, to 12-15°C in the case of stirred yoghurt, to 2-5°C for set yoghurt.

Newer processes applying technologies such as ultrafiltration and reverse osmosis as well as semi-continuous and continuous procedures, the production of "low-acid" yoghurt or the use of milk replacers such as soya milk or whey and the preparation of fruit based products are described in Wegner (1996) and Tamime and Marshall (1997a,b).

Products based on probiotic strains. Claims regarding health benefits associated with the consumption of fermented milks were first published at the beginning of the 20th century (Metchnikoff, 1907, 1908). Numerous publications have since associated metabolic products of lactic acid bacteria or the bacteria themselves with beneficial effects on human health (Hitchins and McDonough, 1989; O'Sullivan *et al.*, 1992; Fuller, 1994). Although most claims need to be considered with caution (Tamime and Marshall, 1997a,b) effects such as establishment of a fermentable microbial balance in the gut flora, establishment of probiotic strains, inhibition of pathogens, improvement of resorption of food components such as lactose, have been reported. The concept of prebiotics and possible dietary modulation of the colonic microbiota is discussed by Gibson and Roberfroid (1995).

The term probiotic (Greek "for life") in humans and animals has been defined by Fuller (1989) as "a live microbial feed supplement, which beneficially affects the host animal by improving its intestinal microbial balance". The normal human intestinal microflora is extremely diverse and forms a complex ecosystem, reviewed Tamime *et al.* (1995). To fulfil their task strains have to belong to the normal intestinal flora, must survive in the fermented product, survive the acidic pH in the stomach, and survive and if possible establish and multiply in the intestine. The mostly used strains are therefore *Bifidobacterium* spp., *Lb. acidophilus*, and *Lb. casei*, either as pure cultures or in combination with other species.

A number of products have been commercialized containing different species of *Bifidobacterium*, in particular *Bif. bifidum* and *Bif. longum*, often with other lactic acid bacteria as well. Processing steps are similar to those outlined for yoghurt. Frequently a specific starter medium is used (Klaver *et al.*, 1993) to grow the fastidious organisms. Mixed cultures may be utilized to overcome slow acid production as well as the production of acetic acid, which causes an undesirable taste.

Products based on lactobacillus include *Lb. paracasei* subsp. *paracasei*, *Lb. paracasei* subsp. *rhamnosus*, *Lb. acidophilus*, and *Lactobacillus* strain GG. Numerous products are sold under the label Yakult, a therapeutic product distributed worldwide. Acidophilus milk with a titrable acidity of <0.8% is obtained after slow fermentation with *Lb. acidophilus* and is claimed to alleviate intestinal disorders (Salji, 1992). Due to the poor competitiveness of the starter, a high heat treatment of 95°C for 60 min, or even UHT, may first be necessary to eliminate spore-forming microorganisms. After cooling to 37°C milk is inoculated with the starter (2-5%) and incubated for up to 24 h, cooled to 5°C and then packed (Chandan, 1982; Alm, 1983). Survival of the strain depends on the acid content and temperature (about 0.65% lactic acid, less than 5°C) sometimes under a modified atmosphere (Tamime and Robinson, 1988b). Shelf life is generally a few days. A detailed compilation of numerous different products is given in Tamime and Marshall (1997a,b).

Other fermented milks. Products based on a combined fermentation of lactic acid bacteria and yeasts are characterized by the presence of ethanol up to 2% and CO_2. Typical products are kefir, koumiss (horse milk), shubat (camel milk), and acidophilus-yeast milk, their origin being the regions encompassing the Caucasus and Mongolia.

In the case of kefir, the starter culture is an irregular kefir grain of a diameter of 1-6 mm or more depending on the growth conditions (Koroleva, 1991). The flora is complex and variable but is normally formed of bacteria such as *Lc. lactis* subsp. *lactis*, *Lb. delbrueckii* subsp. *bulgaricus*, and yeasts such as *Torulaspora delbrueckii*, *Candida kefir*, *Saccharomyces cerevisiae*, *Kluyveromyces marxianus*, etc.

(IDF, 1991a; Tamime and Marshall, 1997a,b) and large-scale production is therefore cumbersome. Manufacture of kefir has been described in details by Koroleva (1988, 1991). Other similar products, such as koumiss, acidophilus-yeast milk or acidophilin, are described by Koroleva (1991) and Lozovich (1995).

The combination of the mold *Geotrichum candidum* and lactic acid bacteria is particular to the Finnish product Viili (Meriläinen, 1984). Concentrated or strained fermented milk products called Ymer (Denmark), Skyr (Iceland), Labneh (Lebanon), etc. are popular in several countries. For these products, after fermentation the whey is drained off by using either cloth bags (traditional), mechanical devices (nozzles, membranes) or by recombination and formulation. Details on composition and processes can be found in Kurmann *et al.* (1992), Akin and Rice (1994), and Tamime and Marshall (1997a,b).

B Spoilage

The manufacture of fermented milks requires milk of good quality. High levels of contaminating microorganisms will cause metabolic changes such as increased excretion of enzymes or formation of peptides and amino acids, which may influence the activity of the starter cultures. Both enhancement (overacidification) or inhibition can be observed as a result of such metabolites. Furthermore the presence of heat-resistant enzymes will affect the organoleptic characteristics of the products during their storage and distribution (Riber, 1989).

As with other products, fermented milks must be manufactured under hygienic conditions. Due to their low pH (3.9-4.6) and storage and distribution at low temperatures, bacterial contaminants are not likely to grow. However, growth of spore-forming bacteria and other recontaminants may occur in case of slow acidification.

Phages are undesired contaminants, which, in case of recontamination after the heat treatment, can lead to the destruction of the starter culture and thus to significant losses (IDF, 1991c; Smaczny and Krämer, 1984; Batt *et al.*, 1995). In certain cases rapid growth of contaminating wild strains of lactic acid bacteria may lead to a strong undesired post-acidification of products.

Acid-tolerant fungi, especially yeasts able to grow at low pH and temperatures are the most frequent spoilage microorganisms. Yeast growth leads to blowing (production of CO_2), off-flavors and off-odors, while mold growth is usually visual. Contamination may occur through air, but is mainly due to ingredients such as fruit concentrates, cereals, honey, chocolate and cocoa, nuts or spices, stabilizing agents such as thickeners, contaminated packaging material (dust), or poor hygiene on the processing lines (Spillmann and Geiges, 1983; Foschino and Ottogalli, 1988; IDF, 1988; Maimer and Busse, 1990; Seiler and Wendt, 1992; Deak and Beuchat, 1995; Filtenborg *et al.*, 1996).

C Pathogens

The low pH, the presence of lactic acid, of other organic acids as well as in certain cases of inhibitory compounds such as bacteriocins generate an unfavorable environment for pathogenic microorganisms. Fermented milks have thus only been associated with outbreaks in very few cases.

Factors that may interfere with the normal fermentation can lead to health hazards. In the case of staphylococcal food poisoning, which occurred in France (Mocquot and Hurel, 1970), the high sugar content favored the development of *Staph. aureus* and toxin formation while inhibiting the lactic acid bacteria.

Consumption of hazelnut/acorn yoghurt caused an outbreak of botulism in the UK. Investigation revealed that the added hazelnut/acorn purée had been under-processed and prepared with artificial sweeteners instead of sugar. This allowed *Cl. botulinum* spores to germinate, grow, and to produce toxin (O'Mahony *et al.*, 1990).

If post-fermentation contamination occurs, pathogens may survive for short periods of time. Minor and Marth (1972) inoculated *Staph. aureus* into cultured buttermilk, sour cream, and yogurt. When the initial population was 10^2 per gram, staphylococci were not recovered after 24 h. For a population of 10^5 per gram, however, viable staphylococci were recovered for up to 1 week, although growth did not occur. Survival was longest in sour cream, followed in descending order by buttermilk and yogurt. Choi *et al.* (1988) showed that several strains of *L. monocytogenes* inoculated into cultured buttermilk could be recovered after storage for up to 3 weeks. Ryser and Marth (1991) reviewed the behavior of *L. monocytogenes* in fermented milks. This organism has survived in milk fermented with mesophilic and thermophilic starter cultures. Length of survival in stored buttermilk and yoghurt was related to starter culture and final pH. The lower the pH, the shorter the survival time.

When fermented product was prepared with milk containing *S.* Typhimurium, recovery varied depending on the starter culture, the level of inoculum and the incubation temperature (Park and Marth, 1972a). The lower the acid production, the longer recovery was possible. The length of time recovery was possible from cultured milks stored at 11°C also varied with the starter culture used (Park and Marth, 1972b). *Yersinia enterocolitica* survived yoghurt making and could be recovered from yoghurt for 1 week when stored at 5°C (Ahmed *et al.*, 1986). *Enterobacter aerogenes* and *E. coli* were rapidly inactivated in 4 days at 7.2°C when added individually to samples of yoghurt (Goel *et al.*, 1971). Frank and Marth (1977a,b) demonstrated that *E. coli* was inhibited by lactic fermentation. However, in 1991, an outbreak of *E. coli* O157:H7 was associated with consumption of yoghurt. The causative organism could not be isolated from the milk or yoghurt, but the product was epidemiologically associated with infection (Morgan *et al.*, 1993).

D CONTROL (fermented milks)

Summary

Significant hazards[a]	• Several pathogens may be present in raw milk (see Table 16.1). • Pathogens may be present in the environment of milk processing, e.g. *Salmonella, Listeria* spp. • Pathogens may be introduced by ingredients (see respective chapters).
Control measures	
Initial level (H_0)	• See pasteurized milk and chapters on specific ingredients.
Reduction (ΣR)	• The major controls for pasteurized milk products are also appropriate for fermented milks.
Increase (ΣI)	• Contamination and subsequent growth after storage.
Testing	• Routine testing for pathogens is not recommended.
Spoilage	• By fungi, especially yeasts as a consequence of recontamination.

[a]In particular circumstances, other hazards may need to be considered.

Hazards to be considered

Pathogens established in the processing environment, e.g. salmonellae, *Listeria spp., B. cereus* may contaminate fermented products during or post processing.

Control measures

Initial level of hazard (H_0). All aspects valid for pasteurized milk have to be taken into account (see Section IV-H). The quality of the raw milk, as expressed by total viable count, somatic cell count, and absence of inhibitory substances, is important. Although the last has no direct influence on the occurrence of pathogens in milk, their presence can greatly affect the fermentation process. Transportation, storage, and appropriate heat treatment in the milk plant are also prerequisites for the production of a reliable product. Ingredients added after pasteurization must be free of vegetative pathogenic bacteria.

Reduction of hazard (ΣR). Pasteurization of the raw milk and quality of ingredients, especially fruits and flavors added after pasteurization, are of primary importance. Control of starter activity through proper starter selection, protection from phage infection and monitoring acid development are also critical.

Increase of hazard (ΣI). Prevention of contamination by air and from containers during packaging, along with proper storage temperature during distribution, minimizes potential quality problems.

Testing

Microbiological criteria have been recommended for finished fermented milks and may be useful to verify control (Robinson, 2002).

Spoilage

Ingredients added after pasteurization should contain only very low levels of potential spoilage fungi, especially yeasts.

The primary concern in fermented milks is post process contamination with fungi from the process environment, or from added ingredients. The processing environment should be routinely monitored for yeasts and molds.

X Cheese

According to different authors, from 400 to 1 000 varieties of cheese are produced throughout the world (Burkhalter, 1981; Kalantzopoulos, 1993; Fox, 2004). Cheese is believed to have evolved in Iraq some 8 000 years ago, probably after attempts to store milk over prolonged periods. Production and types of cheese manufactured evolved and diversified over time, but for many years cheese making remained an art rather than a scientific process. Differences between varieties are the result of modifications made in one or more basic steps of cheese making. Standardization of production steps to obtain varieties with stable characteristics started in the ninth century (Scott, 1986). However, even today variations are observed for the same type of cheese, depending on the manufacturer, the origin and type of milk used. Representative cheese varieties are listed in Table 16.4.

Table 16.4 Classification of selected cheeses[a]

Category	Cheese/Country of origin
Fresh (unripened)	Cottage (UK); Quarg; Cream (UK); Ricotta (I); Petit Suisse (F)
Soft (ripened)	Brie (F); Camembert (F); Bel Paese (I); Neufchatel (F)
Semisoft	Munster (F); Limburger (B); Roquefort (F); Stilton (UK); Port Salut (F); Gorgonzola (I); Tilsit (CH); Brick (USA); Vacherin Mont d'Or (CH)
Hard	Edam (NL); Gouda (NL); Cheddar (UK); Grana (I); Emmental (CH); Gruyère (CH); Provolone (I); Fontina (I)

Adapted from IDF (1981) and Fox (2004), Processed cheese.
[a]Further varieties such as heavily salted Feta or Domiati may be classified differently.

Production of cheese converts highly perishable milk to a less perishable product. Several characteristics contribute to the preservation of cheese. Manufacture can be sub-divided into different steps:

- Transformation of the milk, either raw or heat treated, into fresh cheese curd by acidification,
- coagulation,
- dehydration, and
- shaping and salting.

The degree of dehydration during which fat and casein are concentrated from 6- to 12-fold is achieved by combining cutting of the coagulum, cooking, stirring, pressing, and salting. The freshly molded and pressed curd is then ripened. During this phase regulation of the microflora, of moisture, pH and salt occur which will determine the characteristic flavor, aroma, and texture of different cheese varieties. Technologies, as well as microbiological, biochemical, and physicochemical changes occurring during cheese making, are complex and have been reviewed by Eckhoff-Stork (1976), Kosikowski (1977), Eck (1987), Spreer (1995), Zickrick (1996), and Fox (2004).

Compositional standards for many cheeses are often specified by international and national regulatory agencies (Codex Alimentarius, 2001b; IDF, 1981; U.S. Department of Health, Education and Welfare, 1984). Classification of cheeses can be based on different criteria such as moisture content, concentration of calcium, rheological properties, cooking temperature, secondary microflora or type of ripening and the different approaches are discussed by Fox (2004).

A Effects of processing on microorganisms

The use of high quality milk, both chemically and microbiologically, is important for the manufacture of cheese of good quality. Aspects concerning the microflora of raw milk are discussed under Section 16.II. The milk must be free of inhibitory substances and certain specific requirements may be applied, such as the absence of spores of clostridia in milk used for the production of certain hard cheeses (Zangerl and Ginzinger, 1993). Cooled raw milk is normally submitted to the same steps as outlined under 16.III and 16.IV.

The effects of pre-treatments such as filtration, clarification, or bactofugation on the microbial flora are described under Section III. Bactofugation is of particular interest in cheese making as tool to reduce the level of spores, which survive heat treatments and may germinate and multiply during further processing of the cheese (Zickrick, 1996).

Heating. Raw milk is still used in various countries to manufacture cheese, but in others the milk is always heat-treated before further processing. Carefully controlled heating is needed and overheating must be avoided, so as not to hamper coagulation and curd-forming properties of the milk. Treatments such as thermization, sub-pasteurization, or pasteurization are applied (see Section 16III.C.). Sub-pasteurization may range from 64°C to 70°C for 15 to 20 s, depending upon the type of cheese being made. The treatment has been reviewed by Johnson *et al.* (1990a,b,c) and reduces populations of vegetative microorganisms, in particular of pathogens (Zottola and Jezeski, 1969; D'Aoust *et al.*, 1987; Farber *et al.*, 1988). However, because this process is less severe than pasteurization, many enzymes remain active and the flavor of cheese made from this type of milk is similar to that made from raw milk.

Acidification and starter cultures. Acidification throughout the initial phases up to the early stage of ripening is a basic and key step in cheese making, normally achieved by production of lactic acid by starter cultures. However, acids are added directly for certain cheeses, such as Mozzarella, UF Feta, Mascarponi, or Cottage cheese.

For many years, a natural indigenous flora was allowed to develop. However, due to inconsistent results, and to avoid undesirable effects such as production of off-flavors and/or gas, today selected and specific starter cultures are used. The amount and type of starter and the form in which they are added, either as liquid, frozen, lyophilized or dried cultures, depends on the type of cheese manufactured. The literature on starter cultures is abundant and comprehensive reviews are available (Cogan and Hill, 1993; Zickrick, 1996).

One critical factor that influences the type of starter used is the temperature at which the curd is cooked. Mesophilic starters are used when the cooking temperature is less than 40°C. *Lc. lactis* subsp. *lactis*, *Lc. lactis* subsp. *cremoris*, *Lc. lactis* subsp. *lactis* biovar *diacetylactis*, and *Leuconostoc* spp. are frequently used singly or in combination as mesophilic starters. All of them produce L(+) lactic acid from lactose. In addition, *Lc. lactis* subsp. *lactis* biovar *diacetylactis* uses citric acid to produce flavor compounds, principally diacetyl and CO_2. Thermophilic starters are used when curd-cooking temperatures are high (45-54°C). Single or mixed strains of *Strep. salivarius* subsp. *thermophilus* and *Lb. delbrueckii* subsp. *bulgaricus* are used, depending upon the type of cheese. Mixed cultures using mesophilic and thermophilic cultures at intermediate temperatures are also possible.

Additional microorganisms may be used to impart specific characteristics to the cheeses produced. *Propionibacterium freudenreichii* subsp. *shermanii* and other species are used in the manufacture of Emmentaler cheeses to produce the gas required for eye formation and propionic acid, an important flavor compound (Britz and Riedel, 1991).

Non-starter lactic acid bacteria including lactobacilli, leuconostoc, and pediococci and contribute to flavor development in some ripened cheeses (Law, 1984; Thomas, 1986; Khalid and Marth, 1990). Particular species of enterococci, e.g. *Ent. faecalis* and *Ent. faecium*, are considered by some cheese makers to be important contributors to the organoleptic properties of particular cheeses (Jensen *et al.*, 1975; Asperger, 1992). Another bacterium, the orange-pigmented *Brevibacterium linens*, is involved in the ripening of the so-called smear surface cheeses, e.g. Limburger (Seiler, 1988).

"Lactic acid bacteria" are difficult to identify using biochemical characteristics. Aguirre and Collins (1993) used 16S rRNA sequencing to identify lactic acid bacteria from human clinical infections, including endocarditis, bacteraemia, and urinary tract infections, many of which were resistant to vancomycin. Their conclusion that many isolates were indistinguishable from species used in the food industry was initially greeted with hostility. Subsequent to that report, much more attention has been paid to the identity of starter cultures and non-starter lactic acid bacteria, their role in cheese making, their virulence factors, and possible health implications (Franz *et al.*, 1999; Giraffa, 2003; Klein, 2003).

It was long believed that vegetative pathogens initially present in the raw milk would lose viability during storage and maturation of the cheese but research indicated that some pathogens, including salmonellae and *L. monocytogenes*, may survive this storage requirement (D'Aoust *et al.*, 1985; Ryser and Marth, 1991; Spahr and Url, 1993). Another benefit of pasteurization is the inactivation of spoilage microorganisms and enzymes capable of producing flavor and texture defects. Control of the cheese making process is easier, and more uniform product results, when non-starter microorganisms are eliminated by pasteurization, although some desired flavor development may be lost.

Secondary starter cultures such as yeasts and molds are used to impart unique characteristics to several cheeses.

Yeasts are involved in the degradation of lactic acid in the surface of certain cheeses such as Camembert or Romadur and increase the pH thus favoring the development of corynebacteria. They are also involved in the formation of aromatic compounds (Siewert, 1979). Lipolytic and proteolytic activities of *Geotrichum candidum* contribute to aroma formation.

In addition to lactic starter cultures, filamentous fungi (molds) are used to impart unique characteristics to a number of types of cheeses (Eck, 1987). *Penicillium roqueforti* is used in the production of blue veined cheeses such as Roquefort, Gorgonzola, and Stilton. Camembert and Brie cheeses are produced by the use of *Pen. camemberti*. Other fungi such as *Geotrichum* and *Mucor* spp. also may be involved. Aqueous or powdered spore and mycelium preparations may be added to the milk in the vat or can be applied to the surface of the cut-curd particles or formed cheeses (Molimard and Spinnler, 1996).

Curd formation. Coagulation of the casein fraction of the milk produces an aqueous gel formed by either a weak proteolysis, or acidification to about pH 4.6 or >4.6 in combination with heating. During this process fat is entrapped in a gel. Enzymatic coagulation usually follows by means of rennet, coagulants from plants, chymosin from stomachs from young ruminants or substitutes such as pepsin (bovine or porcine), acid proteinases of fungal origin or chymosin produced by genetically modified microorganisms (Dalgleish, 1993; Foltmann, 1993; Fox and McSweeney, 1997). The curd is allowed to set for a time that varies with the type of cheese made–as long as 16 h in the case of long-set cottage cheese, or less than one hour for most semi-hard and hard cheeses, or 5 min for Swiss cheese.

Whey removal. After the gel is cut or broken into small pieces (1-2 cm^3), syneresis occurs and whey is expelled. The extent of this process depends on the milk composition, pH of the whey, cooking temperature, stirring speed, and time. This step is important for the characteristics of the final products and can be influenced by the cheese maker (Walstra, 1993). When heated, and the curd particles have become sufficiently firm and the proper level of acidity has been reached, the curd is shaped. The shape of the form, and amount of pressure used to press the cheese into the form, depends on the cheese variety being made.

Fresh cheeses are produced by washing, draining, dressing, and packaging the heated curd. Consequently, those effects described for fermented milks would apply here.

Salting. The last operation is salting (before or after shaping and pressing). Addition of sodium chloride has several effects including control of microbial growth and metabolism, control of enzymatic activities and impact on texture (Guinee and Fox, 1993).

Ripening. While some cheeses are consumed fresh and constitute an important part of total consumption (IDF, 1990c), most varieties require ripening. During this phase water is lost and complex biochemical reactions take place as a result of interaction of the coagulant, indigenous milk enzymes, starter bacteria and secondary microorganisms and their enzymes. These biochemical changes favor development of flavor and texture characteristics (Fox, 1989; Fox *et al.*, 1993).

Ripening periods vary in duration depending on the type of cheese. Generally, softer high-moisture cheeses undergo relatively short ripening periods, while well-aged sharp-flavored cheeses may ripen for a year or more. Ripening temperature and humidity depend on cheese type. High humidity encourages the growth of the microorganisms involved in surface ripening of cheese, and low humidity is necessary for maturation of most hard cheeses where internal enzymatic activity is encouraged and surface microbial growth is discouraged.

Some cheeses undergo high-temperature manufacturing procedures such as the stretching process in Mozzarella cheese. Elevated temperatures inactivate many spoilage and pathogenic bacteria that could contaminate the curd. Placing cheeses in brines with high NaCl concentrations inactivates some bacteria, selects for others, and possibly adds some to the cheese. Production of metabolic end products such as propionic acid by starter cultures also contributes to inhibition of some microorganisms. Blue-veined cheeses are made to have an open internal texture to permit penetration of oxygen required for mold growth. "Spiking", by passing metal rods about 3 mm diameter and about 20 mm apart are passed through the cheese wheels at or near the beginning of the ripening period, helps such gaseous exchange.

Processed cheeses. Cold-pack cheese is made by grinding and mixing natural cheeses. Since no heat is applied, these cheeses used must be made from pasteurized milk or must have been aged for at least 60 days at not less than 1.7°C (see Section X.C. regarding survival of pathogens during aging.) Cold-pack cheeses may contain specific dairy ingredients added in either liquid or dry form, e.g. cream, milk, whey or buttermilk as well as acidifying agents, salt, coloring, flavoring, and mold-inhibiting agents.

Processed cheeses are prepared by grinding and blending one or more natural cheeses into a plastic mass. The composition of a product reflects the composition of the cheese(s) used in its production.

The ingredient cheese is cleaned, trimmed, and ground or shredded by machine, mixed with melting salts, such as polyphosphates, emulsifying agents and, depending on the type, with cream, milk, skim milk, buttermilk, non-fat dry milk or whey.

The ingredients are melted and the mass submitted to heat treatments ranging from 85 to 95°C or under pressure to 110°C or more for several minutes. Hot, semi-liquid cheese is then poured into moisture-proof containers or is chilled over rollers before packaging as pre-sliced processed cheese. Although fat exerts a protective effect on microorganisms during heating, good destructive effects are obtained at temperatures above 115°C. Enzymes involved in ripening are also destroyed. Stability of processed cheeses under refrigeration or at room temperature is dependent upon heat treatment and additions such as salt, nitrite, potassium sorbate, lactic acid and phosphate, pH, method of filling and packaging.

Processed cheese spreads may contain sufficient water so that the product is spreadable at room temperature. Edible stabilizers such as gums, gelatin and alginate, along with sugar, dextrose, and corn syrup may also be added. As in processed cheeses, the cheeses used must contribute at least 51% of the product weight of processed cheese spreads.

B Spoilage

Spoilage of the different cheese varieties may be either bacterial or fungal, caused by microorganisms surviving the different processing steps or introduced as recontamination. The many manipulations of the coagulated milk and curd offer numerous opportunities for entry of contaminants, especially psychrotrophs. For example, washing cottage cheese curd with water that has not been properly disinfected commonly adds psychrotrophic Gram-negative bacteria which can readily spoil the product, sometimes accompanied by visual and physical defects such as green or yellow slime and fruity or putrid off odors (ICMSF, 1988). The types of spoilage microorganisms, and their effects, depend on the characteristics of the cheese, leading to visual or organoleptic defects, either at the surface or in the interior.

For a number of cheeses fungal growth is part of the process, to achieve the desired organoleptic characteristics. However, growth of other species of fungi, especially from the genera *Penicillium*, *Mucor*, *Monilia*, *Aspergillus*, *Cladosporium*, etc. leads to undesirable changes affecting the quality of products. Moreover, many common cheese spoilage fungi can produce mycotoxins. Development of visible molds at the surface of cheeses is often the first sign of spoilage, followed by the appearance of musty off-taints and odors. Yeast spoilage of fresh cheeses is characterized by the formation of gas, off-odors, or visual defects.

Bacterial spoilage can be sub-divided into "early blowing" and "late blowing" of cheeses. Early blowing is observed in fresh cheese or after few days of ripening. Yeasts, but more frequently bacteria such as coliforms or *Bacillus subtilis* able to ferment lactose are the cause of these defects. Late blowing occurs during storage and ripening in hard cheeses and can be observed after 10 days in Gouda or Edamer, and after up to five months in Emmental. Late blowing is due the formation of butyric acid leading to the formation of gas and off-flavors. Clostridia, mainly *Cl. tyrobutyricum*, but also *Cl. butyricum*, are involved (IDF, 1990b). *Clostridium tyrobutyricum* is more frequent in winter and is present in milk from cows fed silage, while *Cl. butyricum* is more frequent in summer. Even 10 spores per litre of milk may result in the late blowing.

Paradoxically, gasiness in Swiss and Emmentaler is frequently caused by the propionic acid bacteria used to produce the typical eye structure of these cheeses. Nisin, nitrite, or lysozyme have been used to some extent to control the late gas or "blown" defect caused by clostridia in Swiss, Edam, and related cheeses. Defects of Swiss and related cheeses have been discussed extensively (Langsrud and Reinbold, 1974; Hettinga and Reinbold, 1975; IDF, 1990b). High hydrostatic pressure plus a germinant, such as nisin or lysozyme, is being explored as a means of reducing the counts of *Bacillus* spp. (Lopez-Pedemonte *et al.*, 2003).

The cold-pack cheese class of products do not undergo any form of heat process. As a consequence, microorganisms, principally fungi, from the raw materials can play a major role in spoilage (Marth, 1987). Preservatives such as potassium sorbate and lactic acid are often used to stabilize the formulation. As a consequence the major problem is growth and gas formation by the relatively sorbate resistant lactobacilli originally from the cheese. Due the composition and strict chilled distribution, clostridia and other vegetative bacteria are not usually a problem in these products. Air-borne contamination or contaminants from improperly disinfected packaging equipment can contribute spoilage microorganisms to cold pack products, which can cause defects under aerobic conditions, or in the absence of inhibitors, or if the species is preservative resistant.

C Pathogens

Several factors influence the presence and survival of pathogens in cheese: characteristics of the pathogen, such as heat, acid, and salt tolerances, the number initially present, and their physiological condition affect their ability to survive the cheese-making process. The steps employed in the cheese-making process also affect pathogen survival. The temperatures used in storage and processing, acid production by starter organisms, addition of salt, and other inhibitors and the curing process are important parameters (Bachmann and Spahr, 1995).

Considering the huge quantities produced worldwide, cheese has an admirable history with respect to microbiological safety. However, it has been the vehicle for several food-borne disease outbreaks (Johnson *et al.*, 1990a,b,c; Zottola and Smith, 1991, 1993; Flowers *et al.*, 1992; Kerr *et al.*, 1996 and Rampling, 1996). Johnson *et al.* (1990a,b,c) assigned pathogenic bacteria to three risk groups based on epidemiological data, incidence in milk and characteristics of the individual species. *Salmonella* spp., *L. monocytogenes*, and *E. coli* O157:H7 are considered high risk microorganisms, all three having been involved in disease outbreaks due to the consumption of contaminated cheese. Medium risk species include Group A and C streptococci, *Y. enterocolitica*, *Br. abortus*, *M. bovis*, *Ps. aeruginosa*, *Cox. burnetii*, and *Aeromonas hydrophila*. Group A streptococci were involved in one outbreak implicating home-made cheese from raw milk. The commercial significance of this species is low. *Staphylococcus aureus* and *Cl. botulinum* were categorized as low risk species. Although *Staph. aureus* was the causative agent of several outbreaks in 1950s and 1960s, improvements in the technology of cheese production has minimized this risk. Although most of the bacteria in the medium- and low-risk categories have not been involved in disease outbreaks, they are of concern because they can either grow in, or have been isolated from, milk. Control of *Cl. botulinum* in dairy products, and especially cheeses, was reviewed by Collins-Thompson and Wood (1993). There is no indication that *M. bovis*, *Br. abortus*, *C. jejuni*, *Y. enterocolitica*, *Cl. perfringens* or *Cl. botulinum* can grow during the actual cheese fermentation barring complications in starter activity or other manufacturing parameters (Northolt, 1984). Laboratory systems are being developed to study the possible survival of *M. avium* subsp. *paratuberculosis* (Donaghy *et al.* 2003).

Brucella melitensis is a concern in white/fresh cheeses made from unpasteurized goat and sheep's milk, most commonly in Mediterranean countries (Kaufmann and Martone, 1980). Outbreaks have occurred in raw milk goat cheese contaminated with this species in Spain (Mendez Martinez *et al.*, 2003) and Malta (Anonymous, 1995b), and in Mexican cheese made from raw milk (Eckman, 1975, Thapar and Young, 1986).

Salmonella. Although analysis of commercially made cheeses rarely results in isolation of salmonellae, these species may grow during cheese manufacture (Hargrove *et al.*, 1969) and can survive in various cheeses for more than 60 days (Goepfert *et al.*, 1968; White and Custer, 1976; D'Aoust *et al.*, 1985). Several outbreaks of salmonellosis due to consumption of contaminated cheese have been attributed to faulty control of the cheese-making process (Fontaine *et al.*, 1980; D'Aoust *et al.*, 1985) or use of contaminated raw milk (Wood *et al.*, 1984; Sharpe, 1987; D'Aoust, 1989b).

In 1976, seven lots of pasteurized Cheddar cheese contaminated with *S.* Heidelberg were incriminated in an outbreak in USA involving 28 000 to 36 000 people (Fontaine *et al.*, 1980). Another outbreak associated with Cheddar cheese contaminated with *S.* Muenster occurred in Canada in 1982 (Wood *et al.*, 1984). Two years later, another outbreak of salmonellosis associated with Cheddar cheese occurred in Canada with more than 1 500 confirmed cases due to *S.* Typhimurium. These two Canadian outbreaks support earlier research reports on the ability of *Salmonella* spp. to grow and survive more than 60 days of refrigerated storage (Goepfert *et al.*, 1968; Hargrove *et al.*, 1969; White and Custer, 1976). Seven outbreaks involving fresh soft cheese made from unpasteurized milk contaminated with *S.* Enteritidis, *S.* Typhimurium DT104 (three separate outbreaks), *Salmonella* sp., and *S.* Newport, respectively, were implicated in outbreaks in several countries (Altekruse *et al.*, 1998; Cody *et al.*, 1999; Villar *et al.*, 1999; de Valk *et al.*, 2000). In Europe, outbreaks have occurred involving Irish soft cheese in England and Wales contaminated with *S.* Dublin (Maguire *et al.*, 1992), goat's milk cheese in France contaminated with *S.* Paratyphi B (Desenclos *et al.*, 1996), cheese from the Doubs region in France contaminated with *S.* Dublin (Vaillant *et al.*, 1996), and Cantal cheese in France contaminated with *S.* Enteritidis PT 8 (two separate outbreaks) (Haeghebaert *et al.*, 2003). The cheeses involved in these outbreaks were made from unpasteurized milk.

Listeria monocytogenes. Contaminated Mexican style cheese was responsible for an outbreak of listeriosis in 1985 in California (James *et al.*, 1985; Linnan *et al.*, 1988). Equipment and the plant environment were heavily contaminated with *L. monocytogenes*. Improper pasteurization of milk was alleged, but not confirmed (Johnson, 1990a,b,c). Further outbreaks caused by contaminated soft cheese such as Vacherin Mont d'Or were reported during the following years (Bille and Glauser, 1988; Bannister, 1987). The higher incidence of *L. monocytogenes* in soft, mold-ripened cheeses than in hard cheeses was confirmed in subsequent surveys in the USA, France, Italy, Denmark, Cyprus, Spain, Switzerland and West Germany (Ryser and Marth, 1991). Mold-ripened cheeses have high moisture levels, high pH due to lactate metabolism by molds, and are extremely susceptible to surface contamination during the ripening process. Particularly repeated "smearing" of cheese during the ripening contributes to the spread of *Listeria* spp. over the whole batch and into the whole cheese-making environment. Several thorough reviews of the significance of *L. monocytogenes* to the dairy industry have been published (Ryser and Marth, 1991; Pearson and Marth, 1990; Gellin and Broome, 1989; Griffiths, 1989). *Listeria monocytogenes* multiplies during the manufacturing processes used in the production of Camembert, Brie, blue, and Feta cheese. Increases in numbers of listeriae were associated with increases in the pH of cheese during ripening. *Listeria monocytogenes* was isolated from the brine used in the production of Brie and Feta cheeses and could be detected during extended periods of storage from Colby, Cheddar, and cold-pack cheeses. Slicing and repacking of cheese can also contribute to serious cross-contamination. Depending on the type of cheese, multiplication during subsequent storage may occur.

Enteropathogenic Escherichia coli. Enteropathogenic *E. coli* (EPEC) is a cause of gastroenteritis in humans and other animals. Willshaw *et al.* (2000) reviewed food-borne illness caused by *E. coli*. In general, *E. coli* does not grow well during the cheesemaking process. Low pH and salt are inhibitory but if starter activity is impaired, *E. coli* can grow and survive the cheese-making process (Park *et al.*, 1973; Frank and Marth, 1977a,b; Frank and Marth, 1978). Outbreaks of EPEC gastroenteritis due to the consumption of contaminated cheese have been reported with post-pasteurization contamination (Marier *et al.*, 1973) and mishandling during shipping or distribution the causes of the problems (MacDonald *et al.*, 1985).

Enterohaemorrhagic Escherichia coli (EHEC). Unlike other types, *E. coli* O157:H7 is relatively acid tolerant and this was confirmed for the cheese-making process by Reitsema and Henning (1996). This pathogen was found during surveys (Djuretic *et al.*, 1997; Quinto and Capeda, 1997) and an outbreak was reported involving the consumption of contaminated soft cheese (Deschenes *et al.*, 1996), Cheddar cheese curds in Wisconsin (Anonymous, 2000), Fromage frais in France (Anonymous, 1994a), farm produced goat's milk cheese in Scotland (Anonymous, 1994b) and Lancashire-type cheese in England (Anonymous, 1997b).

Staphylococcus aureus. *Staphylococcus aureus* is frequently found in milk at low levels. Milk from cows suffering mastitis milk is a significant source of enterotoxigenic strains of *Staph. aureus* in milk (Olson *et al.*, 1970) but it is inactivated by pasteurization and sub-pasteurization of milk (Zottola *et al.*, 1970), and inhibited by lactic acid fermentation (Auclair *et al.*, 1981). The small number of outbreaks caused by *Staph. aureus* and reported since 1965 have been attributed to use of contaminated starter cultures or faulty control of the cheese-making process (Zehren and Zehren, 1968a,b; Johnson *et al.*, 1990a,b,c). Slow acid production by starter cultures could permit growth of staphylococci to numbers high enough for significant enterotoxin production. High *Staph. aureus* populations ($>10^6$ per g of cheese or mL of whey) are associated with the presence of toxin (Tatini *et al.*, 1971a,b). Cheese vats with low acidity should be analyzed for *Staph. aureus* population and the presence of enterotoxins. However, because *Staph. aureus* populations decline rapidly during ripening, the thermonuclease assay (TNA) may be used to assess staphylococcal growth (Stadhouders *et al.*, 1978; Park *et al.*, 1978). Attempts have been made to estimate the risk of illness from *Staph. aureus* in unripened cheese and the most important risk factors (Lindqvist *et al.*, 2002).

Clostridium botulinum. *Clostridium botulinum* spores may be present in milk and may survive pasteurization, but conditions in the cheese prevent germination and/or growth which is necessary for toxin production. Interactions between water activity, salt, pH, and antimicrobial agents produced by starter culture organisms prevent outgrowth of the spores. Outbreaks involving cheese were listed by Collins-Thompson and Wood (1993) and Lund and Peck (2000). An outbreak involving commercial cheese spread was reported in Argentina by de Lagarde (1974) showing that spores of the pathogen were introduced with onions used as ingredients. Several cases of botulism occurred between 1973 and 1978 in Switzerland and France, after consumption of ripened Brie. The epidemiological study showed that straw on which the cheeses were stored during ripening was probably at the origin of the contamination with spores (Billon *et al.*, 1980). In Italy, temperature-abused Mascarpone was at the origin of an outbreak (Aureli *et al.*, 1996; Simini, 1996).

Survival, growth, and toxin production in processed cheeses depend on the processing conditions and the parameters such as pH, $a_{w,}$ and the presence of preservatives. Several recent studies have investigated the interaction of different parameters (Tanaka *et al.*, 1979; Tanaka, 1982; Tanaka *et al.*, 1986a,b; Ter Steeg and Cuppers, 1995; Ter Steeg *et al.*, 1995).

Other bacteria. Sporadic outbreaks of cheese-borne illness associated with other pathogenic bacteria have occurred. Gastroenteritis caused by *Shigella* spp. transmitted through cheese is not common, but careless milk-handling practices by infected persons can result in direct contamination of milk and cheese (Rubinstein-Szturn *et al.*, 1964). An outbreak of shigellosis was reported in Scandinavia in 1982 and Brie cheese purchased in France was contaminated with *Shigella sonnei.*

Biogenic amines. The occurrence, mechanism of formation, and catabolism of biologically active amines in foods have been reviewed (Joosten, 1988; Stratton *et al.*, 1991; Petridis and Steinhart, 1996), and their importance reviewed by Santos (1996). Tyramine and histamine are often found in cheese but present little hazard, except to mono- and diamine oxidase deficient persons. Deficiencies in these enzyme may be genetic or the result of therapy using amine oxidase inhibitors. Tyramine and histamine are vasoactive. High levels of tyramine can cause critical increases in blood pressure. Histamine is a strong capillary dilator and has hypotensive effects. Symptoms of histamine poisoning mimic the symptoms of food allergy reactions, including flushing, rapid pulse, fall in blood pressure, and headache.

Tyramine and histamine are formed during cheese ripening from enzymatic decarboxylation of parent amino acids, tyrosine, and histamine. Causative agents are often mesophilic lactobacilli and Enterobacteriaceae. However, precursors (histidine and tyrosine) are present only in ripened cheese in sufficient quantities to allow for significant amine build-up. High levels can be found in particular in cheese from raw milk and mold ripened cheese.

Mycotoxins. Two basic mechanisms can cause mycotoxin contamination of cheese: their presence in the milk used to manufacture the cheese, or the consequence of growth of contaminant fungi in the cheese (Morris and Tatini, 1987). For example, aflatoxin M_1 in milk results from the presence of detectable levels of aflatoxin B_1 in feed. In particular, dairy cows should not be fed rations, which may contain aflatoxin, such as cottonseed.

Growth of toxigenic fungi can be controlled by careful cleaning and disinfection, especially in curing rooms, and use of filtered air. Spoilage fungi, notably *Penicillium commune*, but also many other *Penicillium* species, may produce mycotoxins in cheese. Badly moldy cheese is not suitable for retail sale or reworking. Treatment of cheese with natamycin inhibits surface mold growth. Some starter cultures have been shown to produce mycotoxins. *Penicillium camemberti* isolates produce cyclopiazonic acid (Still *et al.*, 1978), but not usually in cheese (Schoch *et al.*, 1984). *Penicillium roqueforti* produces PR toxin in pure culture (Lafont *et al.*, 1976), but only very low levels in cheese (Finoli *et al.*, 2001). It also produces roquefortine C and mycophenolic acid, but these compounds have very low toxicity (Finoli *et al.*, 2001). Mycotoxins have rarely been isolated from cheese and those cheeses from which isolation was possible were so severely molded that consumption was unlikely.

D CONTROL (fresh and ripened cheese)

Processed cheese concerns and control measures differ from those for fresh and ripened cheese, and will be discussed separately.

Summary

Significant hazards[a]	• Pathogens which can be present in raw milk (see Table 16.1). • Pathogens which can be present in the environment of milk processing, cheese production including storage for ripening.
Control measures	
Initial level (H_0)	• Low levels of pathogens in raw milk are likely–see section on raw milk. Use milk from healthy animals, milked according to good milking practices; cool milk during intermediate storage on the farm and during transportation to dairy plants ($\leq 6°C$ recommended). • Animal health status and milking hygiene is especially important when the milk is used for the production of ripened soft cheeses.
Reduction (ΣR)	• Pasteurization treatments when applied to the milk for cheese manufacture are sufficient to eliminate vegetative microorganisms. • Milk for raw-milk cheeses does not undergo an initial treatment to significantly reduce microorganisms.

(*continued*)

D CONTROL (Cont.)

Summary

Increase (ΣI)	• Post process contamination during different steps such as whey removal, salting, or ripening with pathogens such as salmonellae, *Listeria monocytogenes*, and pathogenic *E. coli* is possible. • Use of equipment designed for cleanability and easy disinfection.
Testing	• Particular surveillance for zoonotic agents in herds delivering milk for the production of raw-milk cheeses. • Microbiological criteria are enforced e.g., by the EU (EEC, 1992), USA (US FDA CFSAN, 1998), Australia and New Zealand (ANZFA, 2002). • Hygiene status of the line and the environment is assessed using sampling plans and tests for salmonellae, *Listeria monocytogenes* and VTEC. • With the exception of mold-ripened cheeses environmental and periodic testing of the finished product is recommended.
Spoilage	• Microorganisms introduced post-process, such as yeast, coliforms, or undesirable molds, or from psychrotrophic organisms introduced during curd-washing steps. Surface contamination with yeast and molds are of concern. • "Late blowing" spoilage caused by *Clostridium butyricum* and *Cl. tyrobutyricum* or heat-resistant lactic acid bacteria surviving heat treatment of milk.

[a]In particular circumstances, other hazards may need to be considered.

Hazards to be considered

The type of pathogen depends on the type of cheese but *L. monocytogenes, Salmonella* spp. and pathogeneic *E. coli.* are the most frequently involved. In special cases *S. aureus* and *Cl. botulinum* are also of concern.

Control measures

Initial level of hazard (H_0) and *reduction of hazard* (ΣR): As already indicated for most milk products, control of cheese quality begins with use of high-quality milk. Adequate heat treatment of that milk reduces potential hazards. When cheeses are made from pasteurized milk, all regular preventative measures are appropriate. Vegetative pathogens present in the raw milk were expected to lose viability during storage of the cheese, but some pathogens may survive this storage requirement (D'Aoust *et al.*, 1985; Ryser and Marth, 1991; Spahr and Url, 1993). Pasteurization also inactivates spoilage organisms and enzymes capable of producing flavor and texture defects. When non-starter organisms are eliminated by pasteurization, control of the cheesemaking process is easier and more uniform product results, although some desired flavor development may be lost.

Starter culture activity is a critical factor. When starter is added, the temperature of the milk must be appropriate for starter growth, and acid development monitored, either as development of titratable acidity or as decrease in pH. Inadequate development of acidity by the starter culture can be due to several factors including residual antibiotics in milk from cows undergoing antibiotic therapy for mastitis (Cogan and Hill, 1993). Low residual levels of disinfecting agents may also be inhibitory to starter cultures (IDF, 1980). The lactoperoxidase system, high oxygen levels, and immunoglobulins present in milk can also contribute to starter failure (IDF, 1991a). A very serious cause of slow acid development is the infection of starter cultures with bacteriophages, with *Lc. lactis* subsp. *lactis*, and *Lc. lactis* subsp. *cremoris* particularly susceptible to infection. Knowledge of the sensitivities of each strain used is essential in developing starter handling systems. Control measures have been outlined by Cogan and Hill (1993). Strains used in multiple-strain starter cultures should have different phage-sensivity patterns because if one strain becomes infected by bacteriophage, the other strains present will continue to produce acid. Starter cultures used should not harbor lysogenic phages, as conversion to the lytic form of bacteriophage will reduce acid production. A defined strain culture programme should be used. Chlorine or peracetic acid-based disinfectants should be used in equipment cleaning and sanitizing programmes. Culture preparation rooms should be isolated and specially constructed to prevent air-borne contamination. Finally, phage-inhibitory media should be used for the preparation of bulk starter. Lactic phages survive minimum pasteurization treatments (Zottola and Marth, 1966), and bulk starter milk must be heated to assure destruction of the phage. Time and temperature for phage inactivation varies with the phage type, but a typical process is 82-88°C for at least 1 h. Concentrated frozen or freeze-dried starter cultures are available commercially. Direct vat setting (DVS) culture products are also available.

Production of cheese from raw milk with less than 60 days ripening requires control of pathogens and other unwanted microorganisms in the cow, during milking and in the milk, during storage, and transfer from farm to manufacturing facility. With exception of effects from lactic acid bacteria and fungal cultures, no control treatments are present. Acid production to pH 5 or below should slow down or limit the growth of pathogens (Mossel, 1983). Unfortunately many soft cheeses do not develop such acidity, and during ripening the pH rises. In the laboratory, lactic acid bacteria, enterococci, brevibacteria, *Geotrichum* spp., and other cheese cultures may exhibit significant inhibitory properties against pathogens including *L. monocytogenes* (Barnby-Smith, 1992). Lactic acid bacteria may limit pathogen development during ripening (Sulzer and Busse, 1991). Antibodies present in the milk and cheese produced *in vivo* against pathogens (e.g. *Brucella*) also may have some effects on other unwanted microorganisms (Plommet *et al.*, 1988).

Legislation in some countries (e.g. USA) requires pasteurization of milk intended for production of soft cheese unless the cheese is ripened for at least 60 days. This requirement is based on microbiological safety considerations (Johnson *et al.*, 1990a,b,c). In other countries, including some highly industrialized countries in Europe, production, and consumption of soft cheeses made from raw milk (not heated over 40°C) are common (EEC, 1992). Unfortunately, when some soft cheese are ripened for 60 days they develop unwanted organoleptic properties and lose much of their market value. Thermization (63-65°C for 15-20 s) has been recommended in conjunction with 60 days ripening (Johnson *et al.*, 1990a,b,c) but quality problems similar to ripening raw-milk cheese develop. Thus, production of cheese from raw milk continues in many parts of the world, and in some cases, does not include 60 day ripening.

Increase of hazard (ΣI). Strict plant sanitation, including adequate cleaning and disinfection, will minimize post-pasteurization and post-fermentation contamination during whey removal, ripening, cutting, and packaging. Equipment and water that may contact the product also must be free of spoilage and pathogenic microorganisms. Wash water should be chlorinated at 5-10 ppm and acidified to pH 5.0 with food grade phosphoric acid. Sanitation in ripening rooms is especially important. Waxing or plastic-coating the cheese, clean storage shelves, and careful control of humidity will help to minimize contamination. Preservative sprays such as natamycin and sorbate also reduce surface mold growth.

Testing

Traditional controls for soft cheeses made from raw milk depend on final product analysis and elimination of infected batches of product. Microbiological analysis after abusive challenges in accelerated shelf life testing have also been used as verification that cheese making effectively controls pathogens. Large manufacturers may be able to employ these techniques but this approach is not feasible for small plants or farms. Where raw-milk cheeses are being produced, some national surveillance networks for public health exist and indicate low numbers of outbreaks and cases (Huchot *et al.*, 1993).

Spoilage

Processed cheeses are very susceptible to mold growth, and most types are normally kept under refrigeration. Many retail packs are vacuum packed or gas flushed. Thus spoilage is usually confined to molds that are psychrotrophic or able to grow in low-oxygen atmospheres.

The most common spoilage species in retail cheeses are *Penicillium commune* and *Pen. roqueforti*, but other species from *Penicillium* subgenus *Penicillium* are also common. Yeasts and yeast-like fungi are often present on cheeses, and may be used for flavor development especially in "smeared" cheeses. However these same yeasts when present on Cheddar or Gurda type cheeses can cause taints and off flavors due to lipolytic and proteolytic activity (Pitt and Hocking, 1997). Heat-resistant molds occasionally cause spoilage of heat processed soft cheeses. Ascospores of *Byssochlamys fulva* and other heat-resistant species may be present in raw milk (from dust contamination) and easily survive pasteurization. Prolonged storage and inadequate cooling can permit growth (Pitt and Hocking, 1997).

Some types of cheese are permitted to contain preservatives such as sorbate in some countries, and a number of fungal species including *Pen. roqueforti* are capable of decarboxylating sorbate to 1-3-pentadiene and other compounds which cause a "kerosene" off-flavor in cheese and other food products (Pitt and Hocking, 1997).

Moldy cheese is unsuitable for sale or remanufacture. Prevention of *Penicillium* growth and spoilage relies on low-temperature storage, low-oxygen atmospheres, integrity of packaging materials, intact rinds, preservative impregnated wrappers, and rapid turnover of stock.

CONTROL (processed cheese)

Summary

Significant hazards[a]	•	Pathogens present in the cheeses to be processed, such as *Salmonella, Listeria monocytogenes, E. coli* O157:H7, *Cl. botulinum*. Pathogens present in ingredients used to manufacture processed cheese such as milk, cream milk derivates, gums, and others are described in the corresponding chapters.
Control measures		
Initial level (H_0)	•	Low levels of pathogens or toxins in the cheese to be processed are less likely than in ingredients. Raw material should be purchased from approved suppliers and tested.
Reduction (ΣR)	•	Heat treatments are usually sufficient to eliminate vegetative bacteria. In cold-packed processed cheese there is no process to reduce microorganisms during the processing.

(*continued*)

D CONTROL (Cont.)

Summary

Increase (ΣI)	• Improperly formulated processed cheeses can be able to support growth of *Clostridium botulinum*. Strict adherence to formula specifications is critical. If present in the process environment post process contamination with *Salmonella* or *Listeria monocytogenes* is possible. Use of equipment designed for cleanability and easy disinfection.
Testing	• Use of microbiological criteria for incoming raw material, the ingredients and final products at risk as appropriate.
Spoilage	• Bacterial spores from non-botulinal clostridia can survive heat treatment and grow out during ambient storage. Spore loads of critical raw materials should be monitored.

[a]In particular circumstances, other hazards may need to be considered.

Hazards to be considered

The potential hazard of *Cl. botulinum* growth is of primary concern.

Control measures

Initial level of hazard (H_0). Control of *Cl. botulinum* is accomplished by selecting the appropriate product composition (Tanaka, 1986a,b). Cheese to be processed incoming ingredients may contain other pathogens such as *Salmonella, Listeria monocytogenes, E. coli* O157:H7.

Reduction of hazard (ΣR). Due to heat decontamination and hot filling, microbiological problems result either from bacterial spore-formers (clostridia) surviving from the cheese raw materials or from post-process contamination (Collins-Thompson and Wood, 1993). Bacterial spores will survive the heat step and therefore the product formulation must prevent the outgrowth of such spores under the relevant storage and distribution conditions. Some benefit may be gained by using cheese from bactofugated milk to reduce the risk of spoilage by clostridia. However, the potential hazard of *Cl. botulinum* growth is of primary concern. Control of *Cl. botulinum* is accomplished by selecting the appropriate product composition (Tanaka, 1986a,b). The main factors determining stability and safety are pH, a_w, non-fat dry matter, presence of preservatives, and temperature. It has been amply demonstrated that nisin will contribute to the stability of the formulation *vis-à-vis* anaerobic spore-formers, including *Cl. botulinum*.

Increase of hazard (ΣI). Monitoring of the environment is recommended in order to avoid post process contamination with pathogens such as salmonellae and *L. monocytogenes*. The main factors determining stability and safety of the finished product are pH, a_w, non-fat dry matter, present of preservatives, and temperature.

Testing

Microbiological criteria have been proposed for cheese by the European Union (EEC, 1992, US FDA/CFSAN, 1998), including *L. monocytogenes*, *Salmonella*, *Staph. aureus*, Type I *E. coli*, and coliforms (30°C). Although no criteria have been recommended for yeasts and molds, surface contamination is of primary concern. Thus, routine environmental and periodic testing of finished product for yeasts and molds is recommended, with the exception of mold-ripened products.

Spoilage

Preservatives such as potassium sorbate and lactic acid are often used to stabilize processed cheese formulations. Another potential spoilage hazard is growth and gas formation by the relatively sorbet-resistant lactobacilli originally from the cheese.

Therefore processed cheese formulations can be categorized as follows:

- ambient stable, if pasteurized and filled properly, and
- products which must be stored and distributed chilled.

Some formulations may be so unstable that even brief temperature abuse may trigger clostridial growth. If that cannot be prevented, risks may be the unacceptable and the products should be sterilized.

Post-process contamination with vegetative spoilage organisms (molds, Enterobacteriaceae and lactobacilli) could still lead to spoilage.

The main areas for control for processed cheese therefore include

- Spore load of raw materials: for critical raw materials (such as the Emmenthaler cheese) strict control should be applied to prevent high spore loads.
- Composition control of pH, moisture, NaCl concentration, and phosphate concentration.
- Heating regime: time and temperature should be recorded (start-up is usually a weak point).
- Filling operation: time–temperature should be controlled and recorded. If this is not applicable equipment, factory hall and personnel hygiene is of utmost importance. Usually such products are packed under modified atmosphere conditions, implying that proper sealing is crucial. At any time, great care should be taken to avoid post-process contamination. Storage and distribution time and temperature has to be recorded.

References

Abrahamsen, R.K. and Rysstad, G. (1991) Fermentation of goat's milk with yoghurt starter culture bacteria: a review. *Cult. Dairy Products J.*, **26**, 20–6.

Ackers, M.L., Schoenfeld, S., Markman, J., Smith, M.G., Nicholson, M.A., De Witt, W., Cameron, D.N., Griffin, P.M. and Slutsker, L. (2000) An outbreak of *Yersinia enterocolitica* 0:8 infection associated with pasteurized milk. *J. Infect. Dis.*, **181**, 1834–7.

Adams, D., Well, S., Brown, R.F., Gregorio, S., Townsend, L., Scags, J.W. and Hinds, M.W. (1984) Salmonellosis from inadequately pasteurised milk: Kentucky. *Morb. Mortal. Wkly Rep.*, **33**, 504–5.

Aguirre, M. and Collins, M.D. (1993) Lactic acid bacteria and clinical infection. *J. Appl. Bacteriol.*, **75** (1), 95–107.

Ahmed, A.-H.A., Moustafa, M.K. and El-Bassiony, T.A. (1986) Growth and survival of *Yersinia enterocolitica* in yogurt. *J. Food Prot.*, **49**, 983–5.

Ahrabi, S.S., Erguven, S. and Gunalp, A. (1998) Detection of *Listeria* in raw and pasteurized milk. *Cent. Eur. J. Public Health*, **6**, 254–5.

Akin, N. and Rice, P. (1994) Main yoghurt and related products in Turkey. *Cult. Dairy Products J.*, **29**, 23–9.

Alm, L. (1983) Arla acidophilus—an updated product with a promising future. *Nordisk Mejerindustri*, **10**, 395–7.

Altekruse, S.F., Timbo, B.B., Mowbray, J.C., Bean, N.H. and Potter, M.E. (1998) Cheese-associated outbreaks of human illness in the United States, 1973 to 1992: sanitary manufacturing practices protect consumers. *J. Food Prot.*, **61**, 1405–7.

Anifantakis, E.M. (1990) Manufacture of sheep's milk products, in *Proceedings, XXIII International Dairy Congress, volume 1*, Mutual Press, Ottawa, pp. 420–32.

Anonymous. (1979) Salmonellosis associated with consumption of nonfat powdered milk: Oregon. *Morb. Mortal. Wkly Rep.*, **28**, 129–30.

Anonymous. (1985a) Milk-borne salmonellosis: Illinois. *Morb. Mortal. Wkly Rep.*, **34**, 200.

Anonymous. (1985b) Update: milk-borne salmonellosis: Illinois. *Morb. Mortal. Wkly Rep.*, **34**, 215–6.

Anonymous. (1993) *Salmonella* serotype Tennessee in powdered milk products and infant formulae–Canada and United States, 1993. *Morb. Mortal. Wkly Rep.*, **42**, 516–7.
Anonymous. (1994a) Two clusters of haemolytic uraemic syndrome in France. *Commun. Dis. Rep.*, **4**, 29.
Anonymous. (1994b) *Escherichia coli* O157 phage type 28 infections in Grampian. Communicable Diseases and Environmental Health, Scotland, *Wkly Rep.*, **28**, 1.
Anonymous. (1995b) Brucellosis associated with unpasteurised milk products abroad. *Commun. Dis. Rep.*, **5**, 1.
Anonymous. (1997a) *Salmonella anatum* infection in infants linked to dried milk. *Commun. Dis. Rep.*, **7**, 33 and 36.
Anonymous. (1997b) Verocytotoxin producing *Escherichia coli* O157. *Commun. Dis. Rep., Rev.*, **7**, 409 and 412.
Anonymous. (2000) Outbreak of *Escherichia coli* O157:H7 infection associated with eating fresh cheese curds–Wisconsin, June 1998. *J. Am. Med. Assoc.*, **284**, 2991–2.
Anonymous. (2002) Outbreak of *Campylobacter jejuni* infections associated with drinking unpasteurized milk procured through a cow-leasing program–Wisconsin 2001. *Morb. Mortal. Wkly. Rep.*, **51**, 548–9.
Anonymous. (2003) Multistate outbreak of *Salmonella* serotype *typhimurium* infections associated with drinking unpasteurized milk–Illinois, Indiana, Ohio, and Tennessee, 2002-2003. *Morb. Mortal. Wkly Rep.*, **52**, 613–5.
Antoine, J.C. and Donawa, A.L. (1990) The spoilage of UHT-treated chocolate milk by thermoduric bacteria. *J. Food Prot.*, **53**, 1050–1.
ANZFA (Australia New Zealand Food Standards). (2002) Code Standard 1.6.1 Microbiological limits for food.
Arbuckle, W.S. (1986) *Ice Cream*, AVI Publishing Company, Westport, CT.
Asao, T., Kumeda, Y., Kawai, T., Shibata, T., Oda, H., Haruki, K., Nakazawa, H. and Kozaki, S. (2003) An extensive outbreak of staphylococcal food poisoning due to low-fat milk in Japan: estimation of enterotoxin A in the incriminated milk and powdered skim milk. *Epidemiol. Infect.*, **130**, 33–40.
Asperger, H. (1992) Zur Bedeutung des mikrobiologischen Kriteriums Enterokokken für fermentierte Milchprodukte. *Lebensmittelindustrie und Milchwirtschaft*, **113**, 900–5.
Auclair, J., Accolas, J.-P., Vassal, L. and Mocquot, G. (1981) Microbiological problems in cheese manufacture. *Bull. Int. Dairy Fed.*, **136**, 20–6.
Aureli, P., Franciosa, G. and Pourshaban, M. (1996) Foodborne botulism in Italy, *Lancet*, **348**, 1594.
Austin, J.W. and Bergeron, G. (1995) Development of bacterial biofilms in dairy processing lines. *J. Dairy Res.*, **62**, 509–19.
Bachmann, H.P. and Spahr, U. (1995) The fate of potentially pathogenic bacteria in Swiss hard and semihard cheeses made from raw milk. *J. Dairy Sci.*, **78**, 476–83.
Bannister, B.A. (1987) *Listeria monocytogenes* meningitis associated with eating soft cheese. *J. Infect.*, **15**, 165–8.
Barnby-Smith, F.M. (1992) Bacteriocins: applications in food preservation. *Trends Food Sci. Technol.*, **3**, 133–7.
Barnes, L.M., Lo, M.F., Adams, M.R. and Chamberlain, A.H. (1999) Effect of milk proteins on adhesion of bacteria to stainless steel surfaces. *Appl. Environ. Microbiol.*, **65**, 4543–8.
Barrett, N.J. (1986) Communicable disease associated with milk and dairy products in England and Wales: 1983–1984. *J. Infect.*, **12**, 265–72.
Barrington, G.M. and Parish, S.M. (2001) Bovine neonatal immunology. *Vet. Clin. North Am. Food Anim. Pract.*, **17**, 463–76.
Batt, C.A., Erlandson, K. and Bsat, N. (1995) Design and implementation of a strategy to reduce bacteriophage infection of dairy starter cultures. *Int. Dairy J.*, **5**, 949–62.
Becker, H. and Terplan, G. (1986) Salmonellen in Milchtrockenprodukten. *Deutsche Molkerei Zeitung*, **42**, 1398–403.
Becker, H., El-Bassiony, T.A. and Terplan, G. (1989) Incidence of *Bacillus cereus* and other pathogenic microorganisms in infant food. *Zentralblatt der Bakteriologie und Hygiene, I. Abteilung.*, **179**, 198–216.
Becker, H., Shaller, G., von Wiese, W. and Terplan, G. (1994) *Bacillus cereus* in infant foods and dried milk powders. *Int. J. Food Microbiol.*, **23**, 1–15.
Bendicho, S., Espachs, A., Arantegui, J. and Martin, O. (2002) Effect of high intensity pulsed electric fields and heat treatments on vitamins of milk. *J. Dairy Res.*, **69**, 113–23.
Bennett, R.W. (1991) Effects of thermal processing on *Staphylococcus aureus* enterotoxins, in *Proceedings,10th International Congress of Canned Foods* (ed L. Mermaz), Paris, pp. 441–52.
Bergonier, D., de Cremoux, R., Rupp, R., Lagriffoul, G. and Berthelot, X. (2003) Mastitis of dairy small ruminants. *Vet. Res.*, **34**, 689–716.
Bernstein, C.N., Blanchard, J.F., Rawsthorne, P. and Collins, M.T. (2004) Population-based case control study of seroprevalence of *Mycobacterium paratuberculosis* in patients with Crohn's disease and ulcerative colitis. *J. Clin. Microbiol.*, **42**, 1129–35.
Beumer, R.R., Cruysen, J.J.M. and Birtantie, I.R.K. (1988) The occurrence of *Campylobacter jejuni* in raw cow's milk. *J. Appl. Bacteriol.*, **65**, 93–6.
Beutin, L. (1995) Zur Epidemiologie von Infektionen durch enterohaemorrhagische *E. coli* (EHEC) in der BRD. *Bundesgesundheitsblatt*, **38**, 428–9.
Beutin, L. (1996) Infektionen mit enterohaemorrhagischen *Escherichia coli* (EHEC). *Bundesgesundheitsblatt*, **39**, 426–9.
Bhale, P., Sharma, S. and Smika, R.N. (1989) Clostridia in sweetened condensed milk and other associated deteriorative changes. *J. Food Science Technol.*, **26**, 46–8.
Bielaszewska, M., Janda, J., Blahova, K., Minarikova, H., Jikova, E., Karmali, M.A., Laubova, J., Sikulova, J., Preston, M.A., Khakhria, R., Karch, H., Klazarova, H. and Nyc, O. (1997) Human *Escherichia coli* O157:H7 infection associated with the consumption of unpasteurized goat's milk. *Epidemiol. Infect.*, **119**, 299–305.

Bigalke, D. and Chapple, R. (1984) Ice cream microbiological quality. Part I. Controlling coliform and other microbial contamination in ice cream. *Dairy Food Sanit.*, **4**, 318–9.

Bille, J. and Glauser, M.P. (1988) Zur Listeriose-Situation in der Schweiz. *Bulletin des Bundesamtes für Gesundheitswesens*, **3**, 28–9.

Billon, J., Guérin, J. and Sebald, M. (1980) Etude de la toxinogénèse de *Clostridium botulinum* au cours de la maturation de fromages à pâte molle. *Le Lait*, **60**, 392–6.

Bishop, J.R. and White, C.H. (1986) Assessment of dairy product quality and potential shelf-life–a review. *J. Food Prot.*, **49**, 739–53.

Black, R.E., Jackson, R.J., Tsai, T., Medvesky, M., Shayegani, M., Feeley, J.C., MacLeod, K.I.E. and Wakelee, A.M. (1978) Epidemic *Yersinia enterocolitica* infection due to contaminated chocolate milk. *N. Engl. J. Med.*, **298**, 76–9.

Blanco, J.E., Blanco, M. and Blanco, J. (1995) Enterotoxigenic, verotoxigenic and necrotoxigenic *Escherichia coli* in foods and clinical samples. Role of animals as reservoir of pathogenic strains for humans. *Microbiologia*, **11**, 97–110.

Blaser, M.J. and Feldman, R.A. (1981) *Salmonella* bacteremia: reports to the Center of Disease Control. 1968–1979. *J. Infect. Dis.*, **143**, 743–6.

Bleem, A. (1994) *Escherichia coli* O157:H7 in raw milk. A review. *Animal Health Insight*, **Spring/Summer**, 1–8.

Blüthgen, A. and Ubben, E.-H. (2000) Zur Kontamination von Futtermitteln und Tankwagensammelmilch mit den Aflatoxinen B1 and M1 in Schleswig-Holstein–ein aktueller Überblick. *Kieler Milchwirtschaftliche Forschungsberichte*, **52**, 335–54.

Bøgh-Sørensen, T. (1992) Cream pasteurization technology. *Bulletin of the IDF*, **271**, Chapter 7.

Boor, K.J. (2001) Fluid dairy product quality and safety: looking to the future. *J. Dairy Sci.*, **84**, 1–11.

Bornemann, R., Zerr, D.M., Heath, J., Koehler, J., Grandjean, M., Pallipamu, R. and Duchin, J. (2002) An outbreak of *Salmonella* serotype Saintpaul in a children's hospital. *Infect. Control Hosp. Epidemiol.*, **23**, 671–6.

Bouma, A., Elbers, A.R., Dekker, A., de Koeijer, A., Bartels, C., Vellema, P., van der Wal, P., van Rooij, E.M., Pluimers, F.H. and de Jong, M.C. (2003) The foot-and-mouth disease epidemic in The Netherlands in 2001. *Prevent. Vet. Med.*, **57**, 155–66.

Bouman, S., Lund, D., Driessen, F.M. and Schmidt, D.G. (1982) Growth of thermoresistant streptococci and deposition of milk constituents on plates of heat exchangers during long operating times. *J. Food Prot.*, **45**, 806–12, 815.

Bradley, A. (2002) Bovine mastitis: an evolving disease. *Vet. J.*, **164**, 116–28.

Bradshaw, J.G., Peeler, J.T., Corwin, J.J., Barnett, J.E. and Twedt, R.M. (1987) Thermal resistance of disease-associated *Salmonella typhimurium* in milk. *J. Food Prot.*, **50**, 95–6.

Britz, T.J. and Riedel, K.H.J. (1991) A numerical taxonomic study of *Propionibacterium* strains from dairy sources. *J. Appl. Bacteriol.*, **71**, 407–16.

Brown, K.L. (2000) Control of bacterial spores. *Br. Med. Bull.*, **56**, 158–71.

Bryan, F.L. (1983) Epidemiology of milk-borne diseases. *J. Food Prot.*, **46**, 637–49.

Bunning, V.K., Donnelly, C.W., Peeler, J.T., Briggs, E.H., Bradshaw, J.G., Crawford, R.G., Beliveau, C.M. and Tierney, J.T. (1988) Thermal inactivation of *Listeria monocytogenes* within bovine milk phagocytes. *Appl. Environ. Microbiol.*, **54**, 364–70.

Burkhalter, G. (1981) Catalogue of Cheese. International Dairy Federation, Bulletin no. 141, Brussels.

Burgess, K., Heggum, C., Walker, S. and van Schothorst, M. (1994) Recommendations for the hygienic manufacture of milk and milk-based products. Bulletin of the International Dairy Federation no. 292, Brussels.

Burnens, A.P. (1996) Bedeutung von *Escherichia coli* O157 und anderen Verotoxin bildenden *E. coli*. *Mitteilungen auf dem Gebiete der Lebensmitteluntersuchung und Hygiene*, **87**, 73–83.

Burton, H. (1988) *Ultra-high-temperature Processing of Milk and Milk Products*. Elsevier Applied Science, New York, USA.

Bylund, G. (2001) *Dairy Processing Handbook*, 2nd Edn, Tetra Pak Processing Systems AB, S-221 86 Lund Sweden.

Caric, M. (ed) (1994) *Concentrated and Dried Dairy Products*. VCH Publishers, Inc., New York.

Carpentier, B. and Cerf, O. (1993) Biofilms and their consequences with particular reference to hygiene in the food industry. *J. Appl. Bacteriol.*, **75**, 499–511.

Center for Science in the Public Interest (CSPI). (2004) Outbreak alert–closing the gap in our Federal Food Safety Net. 24pp.

Cerf, O. (1989) Statistical Control of UHT Milk, in *Aseptic Packaging of Food* (ed H. Reuter), Technomic Pub. Corp., Lancaster (Pennsylvania), pp. 244—57.

Chamberlin, W., Graham, D.Y., Hutten, K., El-Zimaity, H.H., Schwartz, M.R., Naser, S., Shafran, I. and El-Zaatari, F.A. (2001) Review article: *Mycobacterium avium* subsp. *Paratuberculosis* as one cause of Crohn's disease. *Aliment. Pharmacol. Ther.*, **15**, 337–46.

Champagne, C.P., Laing, R.R., Roy, D., Mafu, A.A. and Griffiths, M.W. (1994) Psychrotrophs in dairy products: their effects and their control. *CRC Crit. Rev. Food Sci. Nutr.*, **34**, 1–30.

Chandan, R.C. (1982) Other fermented dairy products, in *Prescott and Dunn's Industrial Microbiology* (ed G. Reed), 4th edn, AVI, Wesport, pp. 113–84.

Chapman, P.A. (1995) Verocytotoxin-producing *Escherichia coli*: an overview with emphasis on the epidemiology and prospects for control of *E. coli* O157. *Food Control*, **6**, 187–93.

Chapman, P.A., Wright, D.J. and Higgins, R. (1993) Untreated milk as a source of verotoxigenic *Escherichia coli* O157. *Vet. Rec.*, **133**, 171–2.

Choi, H.K., Schaak, M.M. and Marth, E.H. (1988) Survival of *Listeria monocytogenes* in cultured buttermilk and yogurt. *Milchwissenschaft*, **43**, 790–2.

Clare, D.A., Catignani, G.L. and Swaisgood, H.E. (2003) Biodefense properties of milk: the role of antimicrobial proteins and peptides. *Current Pharm. Des.*, **9**, 1239–55.

Codex Alimentarius. (1993) *Guidelines for the Establishment of a Regulatory Programme for Control of Veterinary Drug Residues in Foods*. Codex Standard 16–1993.
Codex Alimentarius. (1995) *Residues of Veterinary Drugs in Foods, volume 3*, 2nd edn, revised.
Codex Alimentarius. (1999a) *Evaporated milks*. Codex Standard A-3, 1971, Rev. 1-1999.
Codex Alimentarius. (1999b) *Sweetened condensed milks*. Codex Standard A-4, 1971, Rev. 1, 1999.
Codex Alimentarius. (1999c) *Milk powders and cream powders*. Codex Standard 207.
Codex Alimentarius. (1999d) *Whey Cheeses*. Codex Standard A-7, Rev. 1, 1999.
Codex Alimentarius. (1999e) *General standard for the use of dairy terms*. Codex Standard 206.
Codex Alimentarius. (2001a) Maximum level for aflatoxin M_1 in milk. Codex Standard 232.
Codex Alimentarius. (2001b) Group standard for unripened cheese including fresh cheese. Codex Standard 221.
Codex Alimentarius. (2003) General standard for cheese. A6, 1978, Rev-1, 1999, Amended 2003.
Cody, S.H., Abbott, S.L., Marfin, A.A., Schulz, B., Wagner, P., Robbins, K., Mohle-Boetani, J.C. and Vugia, D.J. (1999) Two outbreaks of multidrug-resistant *Salmonella* serotype *typhimurium* DT104 infections linked to raw-milk cheese in Northern California. *J. Am. Med. Assoc.*, **281**, 1805–10.
Cogan, T.M. and Hill, C. (1993) Cheese starter cultures, in *Cheese: Chemistry, Physics and Microbiology* (ed P.F. Fox), *volume 1, General Aspects*, Chapman and Hall, London, Chapter 6, pp. 193–255.
Collins, E.B., Traeger, M.D., Goldsby, J.B., Borig, J.R., III, Cohoon, D.B. and Barr, R.N. (1968) Interstate outbreak of *Salmonella newbrunswick* infection traced to powdered milk. *J. Am. Med. Assoc.*, **203**, 838–44.
Collins, S.J., Bester, B.H. and McGill, A.E.J. (1993) Influence of psychrotrophic bacterial growth in raw milk on the sensory acceptance of UHT skim milk. *J. Food Prot.*, **56**, 418–25.
Collins-Thompson, D.L. and Wood, D.S. (1993) Control in Dairy Products, in *Clostridium botulinum: Ecology and Control in Foods* (eds. A.H.W. Hauschild and K.L. Dodds), Marcel Dekker, N.Y., pp. 261–77.
Cordier, J.L. (1990) Quality assurance and quality monitoring of UHT-products. *J. Soc. Dairy Technol.*, **43**, 42–5.
Corredig, M., Roesch, R.R. and Dalgleish, D.G. (2003) Production of a novel ingredient from buttermilk. *J. Dairy Sci.*, **86**, 2744–50.
Cousin, M.A. (1982) Presence and activity of psychrotrophic microorganisms in milk and dairy products: a review. *J. Food Prot.*, **45**, 172–207.
Crielly, E.M., Logan, N.A. and Anderton, A. (1994) Studies on the *Bacillus* flora of milk and milk products. *J. Appl. Bacteriol.*, **77**, 256–63.
Crispin, S.M., Roger, P.A., O'Hare, H. and Binns, S.H. (2002) The 2001 foot and mouth disease epidemic in the United Kingdom: animal welfare perspectives. *Rev. Sci. Technol.*, **21**, 877–83.
D'Aoust, J.Y. (1989a) Salmonella, in *Foodborne Bacterial Pathogens* (ed M.P. Doyle), Marcel Dekker Inc., New York, pp. 321–445.
D'Aoust, J.Y. (1989b) Manufacture of dairy products from unpasteurized milk: a safety assessment. *J. Food Prot.*, **52**, 906–14.
D'Aoust, J.Y., Warburton, D.W. and Sewell, A.M. (1985) *Salmonella typhimurium* phage-type 10 from Cheddar cheese implicated in a major Canadian foodborne outbreak. *J. Food Prot.*, **48**, 1062–6.
D'Aoust, J.Y., Emmons, D.B., McKellar, R., Timbers, G.E., Todd, E.C.D., Sewell, A.M. and Warburton, D.W. (1987) Thermal inactivation of *Salmonella* species in fluid milk. *J. Food Prot.*, **50**, 494–501.
Daemen, A.L.M. and van der Stege, H.J. (1982) The destruction of enzymes and bacteria during the spray drying of milk and whey. 2. The effect of the drying conditions. *Netherlands Milk Dairy J.*, **36**, 211–29.
Dalgleish, D.G. (1993) The enzymatic coagulation of milk, in *Cheese: Chemistry, Physics and Microbiology* (ed P.F. Fox), *volume 1, General Aspects*, Chapman and Hall, London, Chapter 3, pp. 69–100.
Davis, J.G. and Wilbey, R.A. (1990) Microbiology of cream and dairy desserts, in *Dairy Microbiology* (ed R.K. Robinson), *volume 2, The Microbiology of Milk Products*, Elsevier Applied Science Publishers, pp. 41–108.
De Buyser, M.L., Dufour, B., Maire, M. and Lafarge, V. (2001) Implication of milk and milk products in food-borne diseases in France and in different industrialised countries. *Int. J. Food Microbiol.*, **20**, 1–17.
De Jong, P. and Verdurmen, R.E.M. (2001) Concentrated and dried dairy products, in *Mechanisation and Automation in Dairy Technology* (eds. Y.T. Adnan and B.A. Law), CRC Press, Sheffield, Chapter 4.
de Jong, P., te Giffel, M.C. and Kiezebrink, E.A. (2002) Prediction of the adherence, growth and release of microorganisms in production chains. *Int. J. Food Microbiol.*, **74**, 13–25.
de Haas, Y., Veerkamp, R.F., Barkema, H.W., Grohn, Y.T. and Schukken, Y.H. (2004) Associations between pathogen-specific cases of clinical mastitis and somatic cell count patterns. *J. Dairy Sci.*, **87**, 95–105.
de Lagarde, E.A. (1974) *Boletin Informativo del Centro Panamericano de Zoonosis, volume 1*, Centro Panamericano de Zoonosis, Buenos Aires, Argentina.
De Valk, H., Delarocque-Astagneau, E., Colomb, G., Ple, S., Godard, E., Vaillant, V., Haeghebaert, S., Bouvet, P.H., Grimont, F., Grimont, P. and Desenclos, J.C. (2000) A community-wide outbreak of *Salmonella enterica* serotype *typhimurium* infection associated with eating a raw milk soft cheese in France. *Epidemiol. Infect.*, **124**, 1–7.
de Vuyst, L. and Vandamme, E.J. (eds) (1994) *Bacteriocins of Lactic Acid Bacteria*, Blackie Academic and Professional, London.
Deak, T. and Beuchat, L.R. (eds) (1995) *Handbook of Food Spoilage Yeasts*. CRC Press, Boca Raton, FL.
Deeth, H.C., Khusniati, T., Datta, N. and Wallace, R.B. (2002) Spoilage patterns of skim and whole milk. *J. Dairy Res.*, **69**, 227–41.
Defrise, D. and Gekas, V. (1988) Microfiltration membranes and the problem of microbial adhesion. *Process. Biochem.*, **August**, 105–16.

Deibel, R.H. and Hartman, P.A. (1984) The enterococci, in *Compendium of Methods for the Microbiological Examination of Foods* (ed M.L. Speck), 2nd ed, American Public Health Association, Washington, DC, pp. 405–7.

Delmas, C. (1983) La contamination du lait par *Yersinia enterocolitica*. *Med. Nutr.*, **19**, 208–10.

Deschenes, G., Casenave, C., Grimont, F., Desenclos, J.C., Benoit, S., Collin, M., Baron, S., Mariani, P., Grimont, P.A. and Nivet, H. (1996) Cluster of cases of haemolytic uraemic syndrome due to unpasteurised cheese. *Pediatr. Nephrol.*, **10**, 203–5.

Desenclos, J.C., Bouvet, P., Benz-Lemoine, E., Grimont, F., Desqueyroux, H., Rebierem, I. and Grimont, P.A. (1996) Large outbreak of *Salmonella enterica* serotype *paratyphi* B infection caused by a goats' milk cheese, France, 1993: a case finding and epidemiological study. *Br. Med. J.*, **312**, 91–4.

Desmazeaud, M.J. (1990) Rôle des cultures de micro-organismes dans la flaveur et la texture des produits laitiers fermentés, in *Proceedings, XXIII International Dairy Congress, volume 2*, Huhest Press, Ottawa, pp. 1155–77.

Djuretic, T., Wall, P.G. and Nichols, G. (1997) General outbreaks of infectious intestinal disease associated with milk and dairy products in England and Wales: 1992 to 1996. *Commun. Dis. Rep.*, **7**, R41–R45.

Dogan, B. and Boor, K.J. (2003) Genetic diversity and spoilage potentials among *Pseudomonas* spp. isolated from fluid milk products and dairy processing plants. *Appl. Environ. Microbiol.*, **69**, 130–8.

Dominguez-Rodriguez, L., Garayzabal, J.F.F., Boland, J.A.V., Ferri, E.R. and Fernandez, G.S. (1985) Isolation de microorganismes du genre *Listeria* à partir de lait crû destiné à la consommation humaine. *Can. J. Microbiol.*, **31**, 938–41.

Donaghy, J.A., Totton, N.L. and Rowe, M.T. (2003) Iodixanol development of a laboratory-scale technique to monitor the persistence of *Mycobacterium avium* subsp. *paratuberculosis* in Cheddar cheese. *Sci. World J.*, **3**, 1241–8.

Donnelly, C.W. (1990) Concerns of microbial pathogens in association with dairy foods. *J. Dairy Sci.*, **73**, 1656–61.

Donnelly, C.W., Briggs, E.J. and Donnelly, L.S. (1987) Comparison of heat resistance of *Listeria monocytogenes* in milk as determined by two methods. *J. Food Protect.*, **50**, 14–7.

Doyle, M.P. (ed) (1989) *Foodborne Bacterial Pathogens*, Marcel Dekker, Inc., New York.

Doyle, M.P. (1991) *Escherichia coli* O157:H7 and its significance in foods. *Int. J. Food Microbiol.*, **12**, 289–302.

Doyle, M.P. and Cliver, D.O. (1990) *Escherichia coli*, in *Foodborne-Diseases* (ed D.O. Cliver), Academic Press, London, pp. 210–5.

Doyle, M.P., Hugdahl, M.B. and Taylor, S.L. (1981) Isolation of virulent *Y. enterocolitica* from porcine tongues. *Appl. Environ. Microbiol.*, **42**, 661–6.

Doyle, M.P., Meske, L.M. and Marth, E.H. (1985) Survival of *Listeria monocytogenes* during the manufacture and storage of nonfat dry milk. *J. Food Prot.*, **48**, 740–2.

Driehuis, F. and Oude Elferink, S.J. (2000) The impact of the quality of silage on animal health and food safety: a review. *Vet. Q.*, **22**, 212–6.

Driessen, F.M. and Puhan, Z. (1988) Technology of mesophilic fermented milk. *Bull. Int. Dairy Fed.*, **227**, 75–81.

Driessen, F.M. and van den Berg, M.G. (1992) Microbiological aspects of pasteurized cream. Bulletin of the International Dairy Federation. no. 271, Chapter 2.

Druce, R.G. and Thomas, S.B. (1972) Bacteriological studies on bulk milk collection: pipeline milking plants and bulk milk tanks as sources of bacterial contamination of milk–a review. *J. Appl. Bacteriol.*, **35**, 253–70.

Dufrenne, J., Bijwaard, M., te Giffel, M., Beumer, R. and Notermans, S. (1995) Characteristics of some psychotrophic *Bacillus cereus* isolates. *Int. J. Food Microbiol.*, **27**, 175–83.

Eck, A. (1987) *Cheesemaking—Science and Technology*. Lavoisier Publishing Inc., New York/Paris.

Eckhoff-Stork, N.M. (1976) *The World Atlas of Cheese* (A. Bailey, transl.). Paddington Press, London.

Eckman, M.R. (1975) Brucellosis linked to Mexican cheese. *J. Am. Med. Assoc.*, **232**, 636–7.

EEC (1992) Sanitary rules for the production and the market of raw milk, heat treated milk and milk based products. Directive 92/46/EEC, 16 June 1992, OJ L268 of 14 September 1992.

Ekstrand, B. (1989) Antimicrobial factors in milk–a review. *Food Biotechnol.*, **3**, 105–26.

El Dairouty, K.R. (1989) Staphylococcal intoxication traced to non-fat dried milk. *J. Food Protect.*, **52**, 901–2.

El-Gazzar, F.E. and Marth, E.H. (1991) Ultrafiltration and reverse osmosis in dairy technology: a review. *J. Food Prot.*, **54**, 801–9.

El-Gazzar, F.E., Bohner, H.F. and Marth, E.H. (1991) Growth of *Listeria monocytogenes* at pH 4.32 and 40°C in skim milk and in retentate and permeate from ultrafiltered skim milk. *J. Food Prot.*, **54**, 338–42.

Eneroth, A., Ahrne, S. and Molin, G. (2000) Contamination routes of Gram-negative spoilage bacteria in the production of pasteurized milk, evaluated by randomly amplified polymorphic DNA (RAPD). *Int. Dairy J.*, **10**, 325–31.

Eneroth, A., Svensson, B., Molin, G. and Christiansson, A. (2001) Contamination of pasteurized milk by *Bacillus cereus* in the filling machine. *J. Dairy Res.*, **68**, 189–96.

European Commission (2000) Possible links between Crohn's disease and Paratuberculosis. *SANCO/B3/R16/2000*, Brussels.

Evenson, M.L., Hinds, M.W., Bernstein, R.S. and Bergdoll, M.S. (1988) Estimation of human dose of staphylococcal food poisoning involving chocolate milk. *Int. J. Food Microbiol.*, **31**, 311–6.

Fahey, T., Morgan, D., Gunneburg, C., Adak, G.K. and Kaczmarski, E. (1995) An outbreak of *Campylobacter jejuni* enteritis associated with failed milk pasteurization. *J. Infect.*, **31**, 137–43.

Faille, C., Fontaine, F., Lelievre, C. and Benezech, T. (2003) Adhesion of *Escherichia coli*, *Citrobacter freundii* and *Klebsiella pneumoniae* isolated from milk: consequence on the efficiency of sanitation procedures. *Water Sci. Technol.*, **47**, 225–31.

FAO/WHO Collaborating Centre for Research and Training in Food Hygiene and Zoonoses (2001) WHO Surveillance Programme for control of foodborne infections and intoxications in Europe, 17th report 1993–1998 (eds. Schmidt K. and Tirado C.), Berlin, ISSN0948-0307 ISBN 3-9311675-70-X.

FDA (Food and Drug Administration) (2003) Quantitative assessment of relative risk to Public Health from foodborne

Listeria monocytogenes among selected categories of ready-to-eat foods. FDA/Center for Food Safety and Applied Nutrition. September 2003.

Farber, J. and Peterkin (2000) Listeria monocytogenes, in *The microbiological Safety and Quality of Food* (eds. B. M. Lund, T. C. Baird-Parker and G. W. Gould), Aspen Publishers Inc., Maryland, United States, Chapter 44.

Farber, J.M., Sanders, G.W., Speirs, J.I., D'Aoust, J.Y., Emmons, D.B. and McKellar, R. (1988) Thermal resistance of *Listeria monocytogenes* in inoculated and naturally contaminated raw milk. *Int. J. Food Microbiol.*, **7**, 227–36.

Fernandez-Garayzabal, J.F., Dominguez-Rodriguez, L., Vazquez-Boland, J.A., Blanco-Cancelo, J.L. and Suarez-Fernandez, G. (1986) *Listeria monocytogenes* dans le lait pasteurisé. *Can. J. Microbiol.*, **32**, 149–50.

Filtenborg, O., Frisvad, J.C. and Thrane, U. (1996) Moulds in food spoilage. *Int. J. Food Microbiol.*, **33**, 85–102.

Finoli, C., Vecchio, A., Galli, A. and Dragoni, I. (2001) Roquefortine C occurence in blue cheese. *J. Food Prot.*, **64**, 246–51.

Fitzgerald, A. C., Edrington, T. S., Looper, M. L., Callaway, T. R., Genovese, K. J., Bischoff, K. M., McReynolds, J. L., Thomas, J. D., Anderson, R. C., and Nisbet, D. J. (2003) Antimicrobial susceptibility and factors affecting the shedding of *E. coli* O157:H7 and *Salmonella* in dairy cattle. *Lett. Appl. Microbiol.*, 37, 392–98.

Fleming, D.W., Cochi, S.L., MacDonald, K.L., Brondum, J., Hayes, P.S., Plikaytis, B.D., Holmes, M.B., Audurier, A., Broom, C.V. and Reingold, A.L. (1985) Pasteurized milk as a vehicle of infection in an outbreak of listeriosis. *N. Engl. J. Med.*, **312**, 404–7.

Flint, S., Brooks, J., Bremer, P., Walker, K. and Hausman, E. (2002) The resistance to heat of thermo-resistant streptococci attached to stainless steel in the presence of milk. *J. Ind. Microbiol. Biotechnol.*, **28**, 134–6.

Florisa, R., Recio, I., Berkhout, B. and Visser, S. (2003) Antibacterial and antiviral effects of milk proteins and derivatives thereof. *Curr. Pharm. Des.*, **9**, 1257–75.

Flowers, R.S., Andrews, W., Donnelly, C.W. and Koenig, E. (1992) Pathogens in milk and milk products, in *Standard Methods for the Examination of Dairy Products*, 16th edn, American Public Health Association.

Foltmann, B. (1993) General and molecular aspects of rennets, in *Cheese: Chemistry, Physics and Microbiology* (ed P.F. Fox), *volume 1, General Aspects*, Chapman and Hall, London, Chapter 2, pp 37–68.

Fontaine, R.E., Cohen, M.L., Matrin, W.T. and Vernon, M.T. (1980) Epidemic salmonellosis from Cheddar cheese: surveillance and prevention. *Am. J. Epidemiol.*, **111**, 247–53.

Forsyth, J.R., Bennett, N.M., Hogben, S., Hutchinson, E.M., Rouch, G., Tan, A. and Taplin, J. (2003) The year of the *Salmonella* seekers–1977. *Austr. NZ J. Public Health*, **27**, 385–9.

Foschino, R. and Ottogalli, G. (1988) Episodio di bombaggio in yogurt ai cereali causato da muffe del genere *Mucor*. *Annali. Microbiologia ed. Enzymologia*, **38**, 147–53.

Fox, P.F. (1989) Proteolysis during cheese manufacture and ripening. *J. Dairy Sci.*, **72**, 1379–400.

Fox, P.F. (2004) Cheese: an overview, in *Cheese: Chemistry, Physics and Microbiology* (ed P.F. Fox), *volume 1*, Chapman and Hall, London, Chapter 1, pp. 1–36.

Fox, P.F. and McSweeney, P.L.H. (1997) Rennets: their role in milk coagulation and cheese ripening, in *Microbiology and Biochemistry of Cheese and Fermented Milk* (ed B.A. Law), Blackie Academic and Professional, Chapter 1, pp. 1–49.

Fox, P.F., Law, J., McSweeney, P.L.H. and Wallace, J. (1993) Biochemistry of cheese ripening, in *Cheese: Chemistry, Physics and Microbiology* (ed P.F. Fox),*volume 1*, Chapman and Hall, Chapter 10, pp. 389–438.

Francis, D.W., Spaulding, P.L. and Lovett, J. (1980) Enterotoxin production and thermal resistance of *Yersinia enterocolitica* in milk. *Appl. Environ. Microbiol.*, **40**, 174–6.

Frank, F.J. and Marth, E.H. (1977a) Inhibition of enteropathogenic *Escherichia coli* by homofermentative lactic acid bacteria in skim milk. I. Comparison of strains of *Escherichia coli*. *J. Food Prot.*, **40**, 749–53.

Frank, F.J. and Marth, E.H. (1977b) Inhibition of enteropathogenic *Escherichia coli* by homofermentative lactic acid bacteria in skim milk. I. Comparison of strains of lactic acid bacteria and enumeration of methods. *J. Food Prot.*, **40**, 754–9.

Frank, F.J. and Marth, E.H. (1978) Survey of soft and semisoft cheese for the presence of fecal coliforms and serotypes of enteropathogenic *E. coli*. *J. Food Prot.*, **41**, 198–200.

Frank, J.F., Hankin, L., Koburger, J.A. and Marth, E.H. (1985) Tests for groups of microorganisms, in *Standard Methods for the Examination of Dairy Products* (ed G.H. Richardson), American Public Health Association, Washington, DC.

Franz, C.M., Holzapfel, W.H. and Stiles, M.E. (1999) Enterococci at the crossroads of food safety? *Int. J. Food Microbiol.*, **47**(1–2), 1–24.

FSA (Food Standards Agency) (1995) The Food Safety (General Food Hygiene) Regulations 1995.

Fuller, R. (1989) A review–probiotics in man and animals. *J. Appl. Bacteriol.*, **66**, 365–78.

Fuller, R. (1994) Probiotics: an overview, in *Human Health: the Contribution of Microorganisms* (ed S.A.W. Gibson), Springer Verlag, London, pp. 63–73.

Gabis, D.A., Flowers, R.S., Evanson, D. and Faust, R.E. (1989) A survey of 18 dry dairy product processing plant environments for *Salmonella*, *Listeria* and *Yersinia*. *J. Food Prot.*, **52**, 122–4.

Galvano, F., Galofaro, V. and Galvano, G. (1996) Occurrence and stability of aflatoxin M_1 in milk and milk products: a worldwide review. *J. Food Prot.*, **59**, 1079–90.

Garcia, M.L., Moreno, B. and Bergdoll, M.S. (1980) Characterization of staphylococci isolated from mastitic cows in Spain. *Appl. Environ. Microbiol.*, **39**, 548–53.

Garcia-Graells, C., Van Opstalm, I., Vanmuysen, S.C. and Michiels, CW. (2003) The lactoperoxidase system increases efficacy of high-pressure inactivation of foodborne bacteria. *Int. J. Food Microbiol.*, **81**, 211–21.

Gareis, M. and Wolff, J. (2000) Relevance of mycotoxin contaminated feed for farm animals and carryover of mycotoxins to food of animal origin. *Mycoses*, **43**, 79–83.

Gellin, B.G. and Broome, C.V. (1989) Listeriosis. *J. Am. Med. Assoc.*, **261**, 1313–20.

Gelosa, L. (1994) Latte in polvere per la prima infanzia contaminato da *Salmonella bovis morbificans*. *Ind. Alimen.*, **33**, 20–4.

Georgala, D.L. and Hurst, A. (1963) The survival of food poisoning bacteria in frozen foods. *J. Appl. Bacteriol.*, **26**, 346–58.

Geringer, M. (1983) Lebensmittelvergiftungen durch enterotoxinbildende *Staphylococcus aureus* Stämme in H-Milch und einem UHT-Milchmischgetränk. *Tieraerztliche Umschau*, **38**(2), 98–109.

Gibson, G.R. and Roberfroid, M.B. (1995) Dietary modulation of the human colonic microbiota: introducing the concept of prebiotics. *J. Nutr.*, **125**, 1401–12.

Giraffa, G. (2003) Functionality of enterococci in dairy products. *Int. J. Food Microbiol.*, **88**, 215–22.

Gillepsie, I.A., Adak, G.K., O. Brien, S.J. and Bolton, F.J. (2003) Milkborne general outbreaks of infectious intestinal disease, England and Wales, 1992–2000. *Epidemiol. Infect.*, **30**, 461–8.

Gilmour, A. and Harvey, J. (1990) Staphylococci in milk and milk products. *J. Appl. Bacteriol.*, **Symposium Supplement**, 147S–166S.

Glover, F.A. (1986) Modifications in the composition of milk, in *Modern Dairy Technology* (ed R.K. Robinson), *volume 1, Advances in Milk Processing*, Elsevier Applied Science Publishers, New York, pp. 235–71.

Goel, M.C., Kulshrestha, D.C., Marth, E.H., Francis, D.W., Bradshaw, J.G. and Read, R.B., Jr. (1971) Fate of coliforms in yogurt, buttermilk, sour cream and cottage cheese during refrigerated storage. *J. Milk Food Technol.*, **34**, 54–8.

Goepfert, J.M., Olson, N.F. and Marth, E.H. (1968) Behavior of *Salmonella typhimurium* during the manufacture and curing of Cheddar cheese. *Appl. Microbiol.*, **16**, 862–6.

Goh, S., Newman, C., Knowles, M., Bolton, F.J., Hollyoak, V., Richards, S., Daley, P., Counter, D., Smith, H.R. and Keppie, N. (2002) *E. coli* O157 phage type 21/28 outbreak in North Cumbria associated with pasteurized milk. *Epidemiol. Infect.*, **129**, 421–57.

Gonzalez, R.N. and Wilson, D.J. (2003) Mycoplasmal mastitis in dairy herds. *Vet. Clin. North Am. Food Anim. Pract.*, **19**, 199–221.

Grandison, A.S. and Glover, F.A. (1994) Membrane processing of milk, in *Modern Dairy Technology* (ed R.K. Robinson), *volume 1*, Chapman & Hall, London, pp. 273–312.

Grant, I.R., Ball, H.J. and Rowe, M.T. (2002a) Incidence of *Mycobacterium paratuberculosis* in bulk raw milk and commercially pasteurized cow's milk from approved dairy processing establishments in the United Kingdom. *Appl. Environ. Microbiol.*, **68**, 2428–35.

Grant, I.R., Hitchings, E.I., McCartney, A., Ferguson, F. and Rowe, M.T. (2002b) Effect of commercial-scale high-temperature, short-time pasteurisation on the viability of *Mycobacterium paratuberculosis* in naturally infected cow's milk. *Appl. Environ. Microbiol.*, **68**, 602–7.

Gray, M.L. and Killinger, A.H. (1966) *Listeria monocytogenes* infections. *Bacteriol. Rev.*, **30**, 309–82.

Greenwood, M.H., Hooper, W.L. and Rodhouse, J.C. (1990) The source of *Yersinia* spp. in pasteurized milk and investigation at a dairy. *Epidemiol. Infect.*, **104**, 351–60.

Griffiths, M.W. (1989) *Listeria monocytogenes*: its importance in the dairy industry. *J. Sci. Food Agric.*, **47**, 133–57.

Guillaume-Gentil, O., Scheldeman, P., Marugg, J., Herman, L., Joosten, H., and Heyndrickx, M. (2002) Genetic heterogeneity in *Bacillus sporothermodurans* documented by ribotyping and repetitive extragenic palindromic PCR fingerprinting. *Appl. Environ. Microbiol.*, **68**, 4216–24.

Guinee, T.P. and Fox, P.F. (1993) Salt in cheese: physical, chemical and biological aspects, in *Cheese: Chemistry, Physics and Microbiology* (ed P.F. Fox), *volume 1, General Aspects*, Chapman and Hall, London, Chapter 7, pp. 257–302.

Haddadin, M.S., Ibrahim, S.A. and Robinson, R.K. (1996) Preservation of raw milk by activation of the natural lactoperoxidase system. *Food Control*, **7**, 149–52.

Haeghebaert, S., Sulem, P., Deroudille, L., Vanneroy-Adenot, E., Bagnis, O., Bouvet, P., Grimont, F., Brisabois, A., Le Querrec, F., Hervy, C., Espie, E., de Valk, H. and Vaillant, V. (2003) Two outbreaks of *Salmonella enteritidis* phage type 8 linked to the consumption of Cantal cheese made with raw milk, France, 2001. *Euro Surveill.*, **8**, 151–6.

Haggerty, P. and Potter, N.N. (1986) Growth and death of selected microorganisms in ultrafiltered milk. *J. Food Prot.*, **49**, 233–5.

Hahn, G. (1994) Campylobacter jejuni, in *The Significance of Pathogenic Microorganisms in Raw Milk*, Monograph of International Dairy Federation, Brussels.

Hamann, J. and Reichmuth, J. (1990) Exogene Einflüsse auf den Zellgehalt der Milch unter Berücksichtigung des Gesundheitszustandes der Milchdrüse. *Milchwissenschaft*, **45**, 286–90.

Hammer, P. and Walte, H.G. (1996) Zur Pathogenität hitzeresistenter mesophiler Sporenbildner aus UHT-Milch. *Kieler Milchwirtschaftliche Forschungsberichte*, **48**, 151–61.

Hammer, P., Lembke, F., Suhren, G. and Heeschen, W. (1996) Characterization of a heat-resistant mesophilic *Bacillus* species affecting quality of UHT-milk. *Kieler Milchwirtschaftliche Forschungsberichte*, **47**, 303–11.

Hammer, P., Kiesner, C., Walte, H.-G., Knappstein, K. and Teufel, P. (2002) Heat resistance of *Mycobacterium avium* ssp. *paratuberculosis* in raw milk tested in a pilot plant pasteuriser. *Kieler Milchwirtschaftliche Forschungsberichte*, **54**, 275–303.

Hankin, L., Dillman, W.F. and Stephens, G.R. (1977) Keeping quality of pasteurized milk for retail sale related to code date, storage temperature and microbial counts. *J. Food Prot.*, **40**, 848–53.

Hargrove, R.E., McDonough, F.E. and Mattingly, J.A. (1969) Factors affecting survival of *Salmonella* in cheddar and Colby cheese. *J. Milk Food Technol.*, **32**, 480–4.

Harmon, R. (1995) Mastitis and milk quality, in *Milk Quality* (ed F. Harding), Blackie Academic and Professional, London. pp. 25–39.

Harris, J.E. and Lammerding, A.M. (2001) Crohn's Disease and *Mycobacterium avium* subsp. *paratuberculosis*: current issues. *J. Food Prot.*, **64**, 2103–10.
Harte, F., Luedecke, L., Swanson, B. and Barbosa-Canovas, G.V. (2003) Low-fat set yogurt made from milk subjected to combinations of high hydrostatic pressure and thermal processing. *J. Dairy Sci.*, **86**, 1074–82.
Harvey, J. and Gilmour, A. (1992) Occurrence of *Listeria* species in raw milk and dairy products produced in Northern Ireland. *J. Appl. Bacteriol.*, **72**, 119–25.
Hayes, M.G. and Kelly, A.L. (2003) High pressure homogenisation of milk (b) effects on indigenous enzymatic activity. *J. Dairy Res.*, **70**, 307–13.
Hayes, P.S., Feeley, J.C., Graves, L.M., Ajello, G.W., Fleming, D.W. (1986) Isolation of *Listeria monocytogenes* from raw milk. *Appl. Environ. Microbiol.*, **51**, 438–40.
Heeschen, W.H. (1992) Part 2. End-product criteria for milk and milk products: 2.1 The European Community, *Int. Dairy Fed. Bull.*, **276**, 20–8.
Hennessy, T.W., Hedberg, C.W., Slutsker, L., White, K.E., Besser-Wie, J.M., Moen, M.E., Coleman, W.W., Edmonson, L.M., MacDonald, K.L. and Osterjholm, M.T. (1996) A national outbreak of *Salmonella enteritidis* infections from ice cream. The investigation team. *N. Engl. J. Med.*, **334**, 1281–6.
Hettinga, D.H. and Reinbold, G.W. (1975) Split defect of Swiss cheese. II. Effect of low temperatures on the metabolic activity of *Propionibacterium*. *J. Milk Food Technol.*, **38**, 31–5.
Hitchins, A.D. and McDonough, F.E. (1989) Prophylactic and therapeutic aspects of fermented milk. *Am. J. Clin. Nutr.*, **49**, 675–84.
Holdsworth, S.D. (1992) *Aseptic Processing and Packaging of Food Products*, Elsevier Applied Science, London.
Holm, S., Toma, R.B., Reiboldt, W., Newcomer, C. and Calicchia, M. (2002) Cleaning frequency and the microbial load in ice cream. *Int. J. Food Sci. Nutr.*, **53**, 337–42.
Hoover, D.G. (1993) Bacteriocins with potential for use in foods, in *Antimicrobials in Foods* (eds. P.M. Davidson and A.L. Branen), Marcel Dekker, New York, pp. 181–90.
Huchot, B., Bohnert, M., Cerf, O., Farrokh, C. and Lahellec, C. (1993) Does cheese made from raw milk pose a health problem? HF-Doc 223 Supplement, International Dairy Federation: a review of international epidemiological data.
Hudson, S.J., Sobo, A.O., Russell, K. and Lightfoot, N.F. (1990) Jackdaws as potential source of milk-borne *Campylobacter jejuni* infection. *Lancet*, **335**, 1160.
Hughes, D. (1979) Isolation of *Yersinia enterocolitica* from milk and a dairy farm in Australia. *J. Appl. Bacteriol.*, **46**, 125–30.
ICMSF (International Commission on Microbiological Specifications for Foods). (1980) *Microbial Ecology of Foods, volume 2, Food Commodities*, Academic Press, New York, NY.
ICMSF (International Commission on Microbiological Specifications for Foods). (1988) *Microorganisms in Foods 4. Application of the Hazard Analysis Critical Control Point (HACCP) System to Ensure Microbiological Safety and Quality*, Blackwell Scientific Publications, Ltd., Oxford.
IDF (International Dairy Federation). (1979) Design and use of CIP systems in the dairy industry. *Bull. Int. Dairy Fed.*, no. 117.
IDF (International Dairy Federation). (1980) Starters in the manufacture of cheese. *Bull. Int. Dairy Fed.*, no. 120, Brussels, Belgium.
IDF (International Dairy Federation). (1981) IDF-Catalogue of cheeses. *Bull. Int. Dairy Fed.*, no. 141, Brussels, Belgium.
IDF (International Dairy Federation). (1986) Pasteurized milk. *Bull. Int. Dairy Fed.*, no. 200.
IDF (International Dairy Federation). (1986a) Protective proteins in milk-Biological significance and exploitation. *Bull. Int. Daily Fed.*, no. 191.
IDF (International Dairy Federation). (1986b) Ewe's and goat's milk and milk products. *Bull. Int. Daily Fed.*, no. 202.
IDF (International Dairy Federation). (1988) Fermented milks-Science and Technology. *Bull. Int. Daily Fed.*, no. 227, Brussels, Belgium.
IDF (International Dairy Federation). (1990a) Methods for assessing the bacteriological quality of raw milk from the farm. *Bull. Int. Daily Fed.*, 256.
IDF (International Dairy Federation). (1990b) Methods of detection and prevention of anaerobic sporeformers in relation to the quality of cheese. *Bull. Int. Daily Fed.*, no. 251.
IDF (International Dairy Federation). (1990c) Consumption statisitcs for milk and milk products (1988). *Bull. Int. Daily Fed.*, no. 246, Brussels, Belgium.
IDF (International Dairy Federation). (1991a) Significance of the indigenous antimicrobial agents of milk to the dairy industry. *Bull. Int. Daily Fed.*, no. 264, Brussels, Belgium.
IDF (International Dairy Federation). (1991b) IDF recommendations for the hygienic manufacture of spray dried milk powders. *Bull. Int. Daily Fed.*, no. 267, Brussels, Belgium.
IDF (International Dairy Federation). (1991c) Practical phage control. *Bull. Int. Daily Fed.*, no. 263, Brussels, Belgium.
IDF (International Dairy Federation). (1991d) Milk—Enumeration of psychotrophic micro-organisms—Colony count technique at 6.5°C. International IDF Standard 101A:1991.
IDF (International Dairy Federation). (1992) *Bacillus cereus* in milk and milk products. *Bull. Int. Daily Fed.*, no. 275, Brussels, Belgium.
IDF (International Dairy Federation). (1994) Recommendations for the hygienic manufacture of milk and milk based products. *Bull. Int. Daily Fed.*, no. 292, Brussels, Belgium.
IDF (International Dairy Federation). (1995) Heat induced changes in milk. Second edition–Revision of bulletin 238/1989.
IDF (International Dairy Federation). (2001) Mastitis Newsletter no. 24. *Bull. Int. Daily Fed.*, no. 367, Brussels, Belgium.

IDF (International Dairy Federation). (2004a) Quality management at farm level—code of good hygienic practices for milking with automatic milking systems. *Bull. Int. Daily Fed.*, no. 386, Brussels, Belgium.

IDF (International Dairy Federation). (2004b) Suitability and application of available test kits for the detection of residues of antimicrobials in milk from species other than the cow–a review. *Bull. Int. Daily Fed.*, no. 390, Brussels, Belgium.

Ingham, S.C., Hassler, J.R., Tsai, Y.W. and Ingham, B.H. (1998) Differentiation of lactate-fermenting, gas-producing *Clostridium* spp. isolated from milk. *Int. J. Food Microbiol.*, **43**, 173–183.

Isaacs, C.E. (2001) The antimicrobial function of milk lipids. *Adv. Nutr. Res.*, **10**, 271–285.

ISO. (2002) Microbiology—horizontal method for the enumeration of micro-organisms—colony count technique at 30°C, ISO 4833.

Iversen, C. and Forsythe, S. (2003) Risk profile of *Enterobacter sakazakii*, an emergent pathogen associated with infant milk formula. *Trends Food Sci. Technol.*, **14**, 443–54.

James, A.D. and Rushton, J. (2002) The economics of foot and mouth disease. *Rev. Sci. Technol.*, **21**, 637–44.

James, S.M., Fannon, S.L., Agree, B.A., Hall, B., Parker, E., Vogt, J., Run, G., Williams, J., Lieb, L., Prendergast, T., Werner, S.B. and Chin, J. (1985) Listeriosis associated with Mexican-style cheese: California. *Morb. Mortal. Wkly Rep.*, **34**, 357–9.

Jarvis, A.W. and Lawrence, R.C. (1970) Production of high titers of enterotoxins for the routine testing of staphylococci. *Appl. Microbiol.*, **19**, 698–9.

Jayarao, B.M. and Wang, L. (1999) A study on the prevalence of Gram-negative bacteria in bulk tank milk.*J. Dairy Sci.*, **82**, 2620–4.

Jayarao, B.M. and Henning, D.R. (2001) Prevalence of foodborne pathogens in bulk tank milk. *J. Dairy Sci.*, **84**, 2157–62.

Jensen, J.P., Reinbold, G.W., Washam, C.J. and Vedamuthu, E.R. (1975) Role of enterococci in Cheddar cheese: organoleptic considerations. *J. Milk Food Technol.*, **38**, 142–5.

Jensen, N. and Hughes, D. (1980) Public health aspects of raw goat's milk production throughout New South Wales. *Food Technol. Austr.*, **32**, 336–8.

Jeong, D.K. and Frank, J.F. (1994a) Growth of *Listeria monocytogenes* at 10°C in biofilms with microorganisms isolated from meat and dairy processing environments. *J. Food Prot.*, **57**, 576–86.

Jeong, D.K. and Frank, J.F. (1994b) Growth of *Listeria monocytogenes* at 21°C in biofilms with microorganisms isolated from meat and dairy processing environments. *Lebensmittel Wissenschaft und Technologie*, **27**, 415–24.

Jöckel, J. (1996) Mikrobiologie der Sahnererzeugnisse, in *Milch und Milchprodukte* (ed H. Weber), Behr's Verlag Hamburg, Chapter 3, pp. 69–104.

Johnson, E.A., Nelson, J.H. and Johnson, M . (1990a) Microbial safety of cheese made from heat-treated milk, Part I. Executive summary, introduction and history. *J. Food Prot.*, **53**, 441–52.

Johnson, E.A., Nelson, J.H. and Johnson, M. (1990b) Microbial safety of cheese made from heat-treated milk, Part II. Microbiology. *J. Food Prot.*, **53**, 519–40.

Johnson, E.A., Nelson, J.H. and Johnson, M. (1990c) Microbial safety of cheese made from heat-treated milk, Part III. Technology, discussion, recommendations, bibliography. *J. Food Prot.*, **53**, 610–23.

Jones, P.H., Willis, A.T., Robinson, D.A., Skirrow, M.B. and Josephs, D.S. (1981) *Campylobacter enteritis* associated with the consumption of free school milk. *J. Hyg. (Camb.)*, **87**, 155–62.

Joosten, H.M.L.J. (1988) The biogenic amine contents of Dutch cheese and their toxicological significance. *Netherlands Milk Dairy J.*, **42**, 25–42.

Jung, W. and Busse, M. (1988) Bekämpfung von Listerien im Molkereibetrieb. *Deutsche Milchwirtschaft* **12**, 393.

Kalantzopoulos, G.C. (1993) Cheeses from ewes' and goats' milk, in *Cheese: Chemistry, Physics and Microbiology* (ed P.F. Fox), *volume 2*, Chapman and Hall, London, pp. 507–53.

Kalman, M., Szollosi, E., Czermann, B., Zimanyi, M., Szekeres, S. and Kalman, M. (2000) Milkborne campylobacter infection in Hungary. *J. Food Prot.*, **63**, 1426–9.

Kalogridou-Vassiliadou, D. (1992) Biochemical activities of *Bacillus* species isolated from flat sour evaporated milk. *J. Dairy Sci.*, **75**, 2681–6.

Kalogridou-Vassiliadou, D., Tzanetakis, N. and Manolkidis, K. (1989) *Bacillus* species isolated from flat sour evaporated milk. *Lebensmittel Wissenschaften und Technologie*, **22**, 287–91.

Kaplan, M.M., Abdussalam, M. and Bijlenga, G. (1962) Diseased transmitted through milk. *WHO*, Monograph Series, **48**, 11–74.

Karmali, M.A. (1989) Infection by verotoxin-producing *Escherichia coli*. *Clin. Microbiol. Rev.*, **2**, 15–38.

Kaufmann, A.F. and Martone, W.J. (1980) Brucellosis, in *Maxcy-Rosenau Public Health and Preventive Medicine* (ed J.M. Last), 11th edn, Appleton-Century Crofts, New York, pp. 419–22.

Keene, W.E., Hedberg, K., Herriott, D.E., Hancock, D.D., McKay, R.W., Barrett, T.J. and Fleming, D.W. (1997) A prolonged outbreak of *Escherichia coli* O157:H7 infections caused by commercially distributed raw milk. *J. Infect. Dis.*, **176**, 815–8.

Kennedy, D., Holmstroem, A., Plym Forshell, K., Vindel, E. and Suarez Fernandez, G. (2001) On-farm management of paratuberculosis in dairy herds. *Bull. IDF*, no. 362, 18–31.

Kerr, K.G., Nice, C.S. and Lacey, R.W. (1996) Cheese and salmonella infection. *Br. Med. J.*, **312**, 1099–100.

Kessler, H.G. (1987) Pasteurisieren und Thermisieren von Milch - eine kritische Analyse der Erhitzungsbedingungen. *Deutsche Molkerei Zeitung*, **6**, 146–53.

Kessler, H.G., Pfeifer, C. and Schwöppe, C. (1994) Untersuchungen über hitzeresistente mesophile *Bacillus*-Sporen in UHT-Milch. Konsequenzen für Erhitzungsbedingungen. *Deutsche Milchwirtschaft*, **13**, 588–92.

Khalid, N.M. and Marth, E.H. (1990) Lactobacilli–their enzymes and role in ripening and spoilage of cheese. *J. Dairy Sci.*, **73**, 2669–84.

Kilara, A. and Panyam, D. (2003) Peptides from milk proteins and their properties. *Crit. Rev. Food Sci. Nutr.*, **43**, 607–33.
Kirschenmann, B. (1989) Bactofugation. *Deutsche Molkerei Zeitung*, **21**, 654–7.
Klaenhammer, T.R. (1993) Genetics of bacteriocins produced by lactic acid bacteria. *FEMS Microbiol. Rev.*, **12**, 39–85.
Klaver, F.A.M., Kingma, F., Martin, J., Timmer, K. and Weerkamp, A.H. (1993) Growth and survival of bifidobacteria in milk. *Netherlands Milk Dairy J.*, **47**, 151–64.
Klein, G. (2003) Taxonomy, ecology and antibiotic resistance of enterococci from food and the gastro-intestinal tract. *Int. J. Food Microbiol.*, **88**(2–3), 123–31.
Klinger, I. and Rosenthal, I. (1997) Public health and the safety of milk and milk products from sheep and goats. *Rev. Sci. Tech.*, **16**, 482–8.
Knipschildt, M.E. (1986) Drying of milk and milk products, in *Modern Dairy Technology* (ed R.K. Robinson), *volume 1, Advances in Milk Processing*, Elsevier Applied Science Publishers, New York, pp. 131–234.
Korhonen, H., Marnila, P. and Gill, H.S. (2000) Milk immunoglobulins and complement factors. *British J. Nutr.*, **84**, S75–S80.
Kornacki, J.L. and Marth, E.H. (1982) Foodborne illness caused by *Escherichia coli*: a review. *J. Food Prot.*, **45**, 1051–67.
Kornblatt, A.N., Barett, T., Morris, G.K. and Tosh, F.E. (1985) Epidemiologic and laboratory investigation of an outbreak of *Campylobacter enteritis* in a rural area. *Am. J. Epidemiol.*, **122**, 844–9.
Koroleva, N.S. (1988) Technology of kefir and kumys. *Bull. Int. Dairy Fed.*, no. 227, 96–100.
Koroleva, N.S. (1991) Products prepared with lactic acid bacteria and yeasts, in *Therapeutic Properties of Fermented Milks* (ed. R.K. Robinson), Elsevier Applied Science, London, pp. 159–79.
Kosikowski, F.V. (1977) *Cheese and Fermented Milk Foods*, 2nd edn, Edwards, Ann Arbor, Michigan.
Kotimaa, M.H., Oksanen, L. and Koskela, P. (1991) Feeding and bedding materials as sources of microbial exposure on dairy farms. *Scand. J. Work Environ. Health*, **17**, 117–22.
Kozak, J., Balmer, T. Byrne, R. and Fisher, K. (1996) Prevalence of *Listeria monocytogenes* in foods-Incidence in dairy products. *Food Control*, **7**, 215–21.
Kuntze, U., Becker, H., Maertlbauer, E., Baumann, C. and Burow, H. (1996) Nachweis von verotoxinbildenden *E. coli*–Stämmen in Rohmilch und Rohmilchkäse. *Archiv für Lebensmittelhygiene*, **47**, 141–4.
Kurmann, J., Rasic, A.J.L. and Kroger, M. (1992) *Encyclopedia of Fermented Fresh Milk Products*, Van Nostrand Reinhold, New York.
Kurmann, J.A. and Rasic, J.L. (1988) Technology of fermented special products. *Bull. Int. Dairy Fed.*, no. 227, 101–9.
Kussendrager, K.D. and van Hooijdonk, A.C. (2000) Lactoperoxidase: physico-chemical properties, occurrence, mechanism of action and applications. *Br. J. Nutr.*, **84**, S19–S25.
Lafont, P., Lafont, J., Payen, J., Chany, E., Bertin, G. and Frayssinet, C. (1976) Toxin production by 50 strains of *Penicillium* used in the cheese industry. *Food Cosmet. Toxicol.*, **14**, 137–9.
Lai, K.K. (2001) *Enterobacter sakazakii* infections among neonates, infants, children and adults. Case reports and a review of the literature. *Medicine*, **80**, 113–22.
Lanciotti, R., Sinigaglia, M., Angelini, P. and Guerzoni, M.E. (1994) Effects of homogenization pressure on the survival and growth of some food spoilage and pathogenic micro-organisms. *Lett. Appl. Microbiol.*, **18**, 319–22.
Langeveld, L.P. (1995) Adherence, growth and release of bacteria in a tube heat exchanger for milk. *Netherland Milk Dairy J.*, **49**, 207–20.
Langeveld, L.P.M., van Spronsen, W.A., van Berestejin, C.H. and Notermans, S.H.W. (1996) Consumption by healthy adults of pasteurized milk with a high concentration of *Bacillus cereus*: a double-blind study. *J. Food Prot.*, **59**, 723–6.
Langsrud, T. and Reinbold, G.W. (1974) Flavor development and microbiology of Swiss cheese–a review. IV. Defects. *J. Milk Food Technol.*, **37**, 26–41.
Lathers, C.M. (2001) Role of veterinary medicine in public health: antibiotic use in food animals and humans and the effect on evolution of antibacterial resistance. *J. Clin. Pharmacol.*, **41**, 595–9.
Law, B.A. (1979) Reviews of the progress of dairy science: enzymes of psychrotrophic bacteria and their effects on milk and milk products. *J. Dairy Res.*, **46**, 573–88.
Law, B.A. (1984) Microorganisms and their enzymes in the maturation of cheeses. *Prog. Ind. Microbiol.*, **19**, 245–83.
Lecos, C. (1986) Of microbes and milk: probing America's worst *Salmonella* outbreak. *Dairy Food Sanit.*, **6**, 136–40.
Lembke, F., Krusch, U., Prokopek, D., Rathjen, G. and Teuber, M. (1984) Use of centrifuges to reduce the added nitrite content in semi-hard cheese. *Kieler Milchwirtch. Forschungsber.*, **36**, 3–64.
Le Roux, Y., Laurent, F. and Moussaoui, F. (2003) Polymorphonuclear proteolytic activity and milk composition change. *Vet. Res.*, **34**, 629–45.
Lewis, S.J. and Gilmour, A. (1987) Microflora associated with the internal surfaces of rubber and stainless steel milk transfer pipelines. *J. Appl. Bacteriol.*, **62**, 327–33.
Lin, S., Schraft, H., Odumeru, J.A. and Griffiths, M.W. (1998) Identification of contamination sources of *Bacillus cereus* in pasteurized milk. *Int. J. Food Microbiol.*, **43**, 159–71.
Lindqvist, R., Sylven, S. and Vagsholm, I. (2002) Quantitative microbial risk assessment exemplified by *Staphylococcus aureus* in unripened cheese made from raw milk. *Int. J. Food Microbiol.*, **78**, 155–70.
Linnan, M.F., Mascola, L., Lou, X.O., Goulet, V., May, S., Salminen, C., Hird, D.W., Yonekura, L., Hayes, P., Weaver, R., Andurier, A., Plikaytis, B.D., Fannin, S.L., Kleks, A. and Broome, C.V. (1988) Epidemic listeriosis associated with Mexican-style cheese. *N. Engl. J. Med.*, **319**, 823–8.
Lopez-Pedemonte, T.J., Roig-Sagues, A.X., Trujillo, A.J., Capellas, M. and Guamis, B. (2003) Inactivation of spores of *Bacillus cereus* in cheese by high hydrostatic pressure with the addition of nisin or lysozyme. *J. Dairy Sci.*, **86**, 3075–81.

Lopez-Sanchez, A., Guijarro Guijarro, B. and Hernandez Vallejo, G. (2003) Human repercussions of foot and mouth disease and other similar viral diseases. *Med. Oral*, **8**, 26–32.

Lovett, J., Bradshaw, J.G. and Peeler, J.T. (1982) Thermal inactivation of *Yersinia enterocolitica* in milk. *Appl. Environ. Microbiol.*, **44**, 517–9.

Lovett, J., Francis, D.W. and Hunt, J.M. (1987) *Listeria monocytogenes* in raw milk: detection, incidence and pathogenicity. *J. Food Prot.*, **50**, 188–92.

Lovett, J., Wesley, I.V., Vandermaaten, M.J. Bradshaw, J.G., Francis, D.W., Crawford, R.G., Donnelly, C.W. and Messer, J.W. (1990) High-temperature short-time pasteurization inactivates *Listeria monocytogenes*. *J. Food Prot.*, **53**, 743–8.

Lozovich, A. (1995) Medical use of whole and fermented mare milk in Russia. *Cult. Dairy Products J.*, **30**, 18–21.

Lund, B.M., Gould, G.W. and Rampling, A.M. (2002) Pasteurisation of milk and the heat resistance of *Mycobacterium avium* subsp. *paratuberculosis*: a critical review of the data. *Int. J. Food Microbiol.*, **77**, 135–45.

Lund, B.M. and Peck, M.W (2000) Clostridium botulinum, in *The Microbiological Safety and Quality of Food* (eds. B.M. Lund, T.C. Baird-Parker and G.W. Gould), *volume II*, Aspen Publishers Inc., Gaithersburg, MD, Chapter 41, pp. 1057–109.

Ma, Y., Barbano, D.M. and Santos, M. (2003) Effect of CO_2 addition to raw milk on proteolysis and lipolysis at 4 degrees C. *J. Dairy Sci.*, **86**, 1616–31.

MacDonald, K.L., Edison, M., Strohmeyer, C., Levy, M.E., Wells, J.G., Puhr, N.D., Wachsmuth, K., Nargett, N.T. and Cohen, M.L. (1985) A multistate outbreak of gastrointestinal illness caused by enterotoxigenic *Escherichia coli* in imported semisoft cheese. *J. Infect. Dis.*, **151**, 716–20.

Mackenzie, E. (1973) Thermoduric and psychrotrophic organisms on poorly cleansed milking plants and farm bulk milk tanks. *J. Appl. Bacteriol.*, **36**, 457–63.

Mackey, B.M. and Bratchell, N. (1989) The heat resistance of *Listeria monocytogenes*–A review. *Lett. Appl. Microbiol.*, **9**, 89–94.

Maimer, E. and Busse, M. (1990) Die Hefen von Fruchtzubereitungen. *Deutsche Milchwirtschaft*, **41**, 847, 850–1.

Maguire, H., Cowden, J., Jacob, M., Rowe, B., Roberts, D., Bruce, J. and Mitchell, E. (1992) An outbreak of *Salmonella dublin* infection in England and Wales associated with a soft unpasteurized cows' milk cheese. *Epidemiol. Infect.*, **109**, 389–96.

Mäkinen, M. (1995) Technological significance of residues for the dairy industry, in *Proceedings of the symposium: Residue of antimicrobial drugs and other inhibitors in milk*, Kiel 1995, 136-143.IDF, Brussels, ISBN 92 9098 021 4.

Makovec, J.A. and Ruegg, P.L. (2003) Antimicrobial resistance of bacteria isolated from dairy cow milk samples submitted for bacterial culture: 8 905 samples (1994–2001). *J. Am. Vet. Med. Assoc.*, **222**, 1582–9.

Marier, R., Wells, J.G., Swanson, R.C., Callahan, W. and Mehlman, I.J. (1973) An outbreak of enteropathogenic *Escherichia coli* foodborne disease traced to imported French cheese. *Lancet*, **ii**, 1376–8.

Marshall, R.T. and Arbuckle, W.S. (1996) *Ice Cream*, Chapman and Hall, New York.

Marshall, V.M.E. (1987) Fermented milks and their future trends: I. Microbiological aspects. *J. Dairy Res.*, **54**, 559–74.

Marshall, V.M.E. and Tamime, A.Y. (1997) Physiology and biochemistry of fermented milk, in *Microbiology and Biochemistry of Cheese and Fermented Milk* (ed B.A. Law), Blackie Academic and Professional, Chapter 4, pp. 153–92.

Marth, E.H. (1987) Dairy Products, in *Food and Beverage Mycology* (ed L.R. Beuchat), Van Nostrand Reinhold, New York.

Martin, J.H. and Blackwood, P.W. (1971) Effect of pasteurization conditions, type of bacteria, and storage temperature on the keeping quality of UHT-processed soft-serve frozen dessert mixes. *J. Milk Food Technol.*, **34**, 256–9.

Martin, M.L., Shipman, L.D., Wells, J.G., Potter, M.E., Hedberg, K., Wachsmuth, I.K., Tauxe, R.V., Davis, J.P., Arnolai, J. and Tilleli, J. (1986) Isolation of *Escherichia coli* O157:H7 from dairy cattle associated with two cases of haemolytic-uremic syndrome. *Lancet*, **ii**, 1043.

Martins, M.L. and Martins, H.M. (2000) Aflatoxin M_1 in raw and ultra high temperature-treated milk commercialized in Portugal. *Food Addit. Contam.*, **17**, 871–4.

Masters, K. (1985) Spray drying, in *Evaporation, Membrane Filtration and Spray Drying in Milk Powder and Cheese Production* (ed R. Hansen), North European Dairy Journal, Copenhagen, Denmark, pp. 299–346.

Masschalck, B. and Michiels, C.W. (2003) Antimicrobial properties of lysozyme in relation to foodborne vegetative bacteria. *Crit. Rev. Microbiol.*, **29**, 191–214.

McDermott, P.F., Zhao, S., Wagner, D.D., Simjee, S., Walker, R.D. and White, D.G. (2002) The food safety perspective of antibiotic resistance. *Anim. Biotechnol.*, **13**, 71–84.

McKinnon, C.H., Bramley, A.J. and Morant, S.V. (1988) An in-line sampling technique to measure the bacterial contamination of milk during milking. *J. Dairy Res.*, **55**, 33–40.

McIntyre, L., Fung, J., Paccagnella, A., Isaac-Renton, J., Rockwell, F., Emerson, B. and Preston, T. (2002) *Escherichia coli* O157 outbreak associated with the ingestion of unpasteurized goat's milk in British Columbia, 2001. *Can. Commun. Dis. Rep.*, **28**, 6–8.

McManus, C. and Lanier, J.M. (1987) *Salmonella, Campylobacter jejuni* and *Yersinia enterocolitica* in raw milk. *J. Food Prot.*, **50**, 51–5.

Meers, R.R., Baker, J., Bodyfelt, F.W. and Griffiths, M.W. (1991) Psychrotrophic *Bacillus* spp. in fluid milk products: a review. *J. Food Prot.*, **54**, 969–78.

Mendez-Martinez, C., Paez-Jimenez, A., Cortes-Blanco, M., Salmoral-Chamizo, E., Mohedano-Mohedano, E., Plata, C., Varo Baena, A. and Martinez-Navarro, F. (2003) Brucellosis outbreak due to unpasteurized raw goat cheese in Andalucia (Spain), January–March 2002. *Euro Surveill.*, **8**, 164–8.

Meriläinen, V.T. (1984) Yoghurt and cultured buttermilk, in *Milk–The Vital Force XXI. International Dairy Congress*, D. Reidel Publishing Company, Dordrecht, pp. 661–72.

Metchnikoff, E. (1907) Quelques remarques sur le lait aigri (Scientifically Soured Milk and its Influence in Arresting Intestinal Putrefaction), Putnam, New York.

Metchnikoff, E. (1908) *The Prolongation of Life.*, Putnam, New York.
Mettler, A.E. (1994) Present day requirements for effective pathogen control in spray dried milk powder production. *J. Soc. Dairy Technol.*, **47**, 95–107.
Miettinen, M.K., Bjørkroth, K.J. and Korkeala, H.J. (1999) Characterization of *Listeria monocytogenes* from an ice cream plant by serotyping and pulsed field gel electrophoresis. *Int. J. Food Microbiol.*, **46**, 187–92.
Millar, D., Ford, J., Sanderson, J., Withey, S., Tizard, M., Doran, T. and Hermon-Taylor, J. (1996) IS900 PCR to detect *Mycobacterium paratuberculosis* in retail supplies of whole pasteurised cow's milk in England and Wales. *Appl. Environ. Microbiol.*, **62**, 3446–52.
Milner, J. (1995) *LFRA Microbiology Handbook. 1. Dairy Products*, Leatherhead Food Research Association, Leatherhead, UK.
Minor, T.E. and Marth, E.H. (1972) Fate of *Staphylococcus aureus* in cultured buttermilk, sour cream, and yogurt during storage. *J. Milk Food Technol.*, **35**, 302–6.
Mocquot, G. and Hurel, C. (1970) The selection and use of some microorganisms for the manufacture of fermented and acidified milk products. *J. Soc. Dairy Technol.*, **23**, 130–6.
Molenda, J.R. (1994) *Escherichia coli* (including O157:H7): an environmental health perspective. *Dairy Food Environ. Sanit.*, **14**, 742–7.
Molimard, P. and Spinnler, H.E. (1996) Review: compounds involved in the flavor of surface mold-ripened cheeses: origins and properties. *J. Dairy Sci*, **79**, 169–84.
Moore, J.E., Coenye, T., Vandamme, P. and Elboren, J.S. (2001) First report of *Pandoraea norimbergensis* isolated from food–potential clinical significance. *Food Microbiol.*, **18**, 113–4.
Morgan, D., Newman, C.P., Hutchinson, D.N., Walker, A.M., Rowe, B. and Majid, F. (1993) Verotoxin producing *Escherichia coli* infections associated with yogurt. *Epidemiol. Infect.*, **111**, 181–7.
Morris, H.A. and Tatini, S.R. (1987) Progress in cheese technology - safety aspects with microbiological emphasis, in *Milk–The Vital Force* (ed Organizing Committee of the XXII International Dairy Congress), Reidel Publishing Co., pp. 187–94 D.
Mossel, D.A.A. (1983) Seventy-five years of longitudinally microbial safety assurance in the dairy industry in the Netherlands. *Netherlands Milk Dairy J.*, **37**, 240–5.
Mottar, J. and Waes, G. (1986) Quality control of pasteurized milks. *Bull. Int. Dairy Fed.*, no. 200, 66–70.
Moura, S.M., Destro, M.T. and Franco, B.D. (1993) Incidence of *Listeria* spp. in raw and pasteurized milk produced in Sao Paulo, Brazil. *Int. J. Food Microbiol.*, **19**, 229–37.
Mourgues, R., Deschamps, N. and Auclair, J. (1983) Influence de la flore thrmorésistante du lait cru sur la qualité de conservation du lait pasteurisé exempt de recontaminations post-pasteurisation. *Le Lait.*, **63**, 391–404.
Moustafa, M.K., Ahmed, A.A.-H. and Marth, E.H. (1983) Occurrence of *Yersinia enterocolitica* in raw and pasteurized milk. *J. Food Prot.*, **46**, 276–8.
Muir, D.D. (1996a) The shelf-life of dairy products. 1. Factors influencing raw milk and fresh products. *J. Soc. Dairy Technol.*, **49**, 24–32.
Muir, D.D. (1996b) The shelf-life of dairy products. 2. Raw milk and fresh products. *J. Soc. Dairy Technol.*, **49**, 44–8.
Muir, D.D. (1996c) The shelf life of dairy products. 3. Factors influencing intermediate and long life dairy product. *J. Soc. Dairy Technol.*, **49**, 67–72.
Muir, D.D., Griffiths, M.W., Philips, J.D., Sweetsur, A.W.H. and West, J.G. (1986) Effect of the bacterial quality of raw milk on the bacterial quality and some other properties of low heat and high heat dried milk. *J. Soc. Dairy Technol.*, **39**, 115–8.
National Mastitis Council (1987) *Laboratory and Field Handbook on Bovine Mastitis*, Arlington, VA.
Nazarowec-White, M. and Farber, J.M. (1997) *Enterobacter sakazakii*: a review. *Int. J. Food Microbiol.*, **34**, 103–13.
Neimann, J., Engberg, J., Molback, K. and Wegener, H.C. (2003) A case-control study of risk-factors for sporadic campylobacter infections in Denmark. *Epidemiol. Infect.*, **130**, 353–66.
Nelson, F.E. (1990) The microbiology of concentrated milks, in *Dairy Microbiology* (ed. R.K. Robinson), *volume 1, The Microbiology of Milk Products*, Elsevier Applied Science Publishers, London, pp. 271–88.
Nickerson, S.C. (1995) Milk production: factors affecting milk composition in *Milk Quality* (ed. F. Harding), Blackie Academic and Professional, London, pp. 3–24.
Northolt, M.D. (1984) Growth and inactivation of pathogenic micro-organisms during manufacture and storage of fermented dairy products. A review. *Netherlands Milk Dairy J.*, **38**, 135–50.
Notermans, S., Dufrenne, J., Teunis, P., Beumer, R., Te Giffel, M. and Weem, P. (1997) A risk assessment study of *Bacillus cereus* present in pasteurized milk. *Food Microbiol.*, **14**, 143–151.
O'Ryan, M.D., Djuretic, T., Wall, P.G. and Nichols, G. (1996) An outbreak of *Salmonella* infection from ice cream. *N. Engl. J. Med.*, **33**, 824–5.
O'Sullivan, M.G., Thornton, G., O'Sullivan, G.C. and Collins, J.K. (1992) Probiotic bacteria: myth or reality. *Trends Food Sci. Technol.*, **3**, 309–14.
Oemichen, W.L. (1995) The Schwan's *Salmonella enteritidis* experience. *J. Assoc. Food Drug Off.*, **59**, 48–68.
OIE (Office International des Epizooties) (2003) *Terrestrial Animal Health Code*, 12th edn, ISBN 92–9044–583–1.
Olson, J.C., Jr., Casman, E.P., Baer, E.F. and Stone, J.E. (1970) Enterotoxigenicity of *Staphylococcus* cultures isolated from acute cases of bovine mastitis. *Appl. Microbiol.*, **20**, 605–7.
Olsvik, O., Wateson, Y., Lund, A. and Hornes, E. (1991) Pathogenic *Escherichia coli* found in food. *Int. J. Food Microbiol*, **12**, 103–14.
O'Mahony, M., Mitchell, E., Gilbert, R.J., Hutchinson, D.N., Begg, N.T., Rodhouse, J.C. and Morris, J.E. (1990) An outbreak of foodborne botulism associated with contaminated hazelnut yoghurt. *Epidemiol. Infect.*, **104**, 389–95.

Orskov, F., Orskov, I. and Villar, J.A. (1987) Cattle as reservoir of verotoxin-producing *Escherichia coli* O157:H7. *Lancet*, **11**, 276.

O'Shea, J. (1987) Machine milking factors affecting mastitis-A literature review. *Bull. Int. Dairy Fed.*, no. 215, 5–32.

Otte-Südi, I. (1996a) Mikrobiologie der Rohmilch, in *Mikrobiologie der Lebensmittel, Milch und Milchprodukte* (ed. H. Weber), Behr's Verlag, Hamburg, pp. 1–35.

Otte-Südi, I. (1996b) Mikrobiologie der pasteurisierten Trinkmilch, *in Mikrobiologie der Lebensmittel, Milch und Milchprodukte* (ed. H. Weber), Behr's Verlag, Hamburg, pp. 39–65.

Pak, S.J., Spahr, U., Jemmi, T. and Salman, M.D. (2002) Risk factors for *L. monocytogenes* contamination of dairy products in Switzerland 1990–1999. *Prev. Vet. Med.*, **53**, 55–65.

Palmer, J. (1980) Contamination of milk from the milking environment. *Bull. Int. Dairy Fed.*, no. 120, 16–21.

Papadatos, A., Neocleous, M., Berger, A.M. and Barbano, D.M. (2003) Economic feasibility evaluation of microfiltration of milk prior to cheesemaking. *J. Dairy Sci*, **86**, 1564–77.

Papademas, P. and Bintsis, T. (2002) *Dairy Microbiol. Handbook*, 3rd edn.

Pappas, C.P. (1988a) A comparative study of laws and regulations on compositional requirements for yogurt in EC member states. *Br. Food J.*, **90**, 195–8.

Pappas, C.P. (1988b) Comparative laws and regulations on compositional requirements for ice-cream in the EC. *Br. Food J.*, **90**, 250–4.

Park, C.E., El Derea, H.B. and Rayman, M.K. (1978) Evaluation of staphylococcal thermonuclease (TNA) assay as a means of screening foods for growth of staphylococci and possible enterotoxin production. *Can. J. Microbiol.*, **24**, 1135–9.

Park, H.S. and Marth, E.H. (1972a) Behavior of *Salmonella typhimurium* in skimmilk during fermentation by lactic acid bacteria. *J. Milk Food Technol.*, **35**, 482–8.

Park, H.S. and Marth, E.H. (1972b) Survival of *Salmonella typhimurium* in refrigerated cultured milks. *J. Milk Food Technol.*, **35**, 489–95.

Park, H.S., Marth, E.H. and Olson, N.F. (1973) Fate of enteropathogenic strains of *Escherichia coli* during the manufacture and ripening of Camembert cheese. *J. Milk Food Technol.*, **36**, 543–6.

Pearson, L.J. and Marth, E.H. (1990) *Listeria monocytogenes*–threat to a safe food supply: a review. *J. Dairy Sci.*, **73**, 912–28.

Peeler, E.J., Green, M.J., Fitzpatrick, J.L. and Green, L.E. (2003) The association between quarter somatic-cell counts and clinical mastitis in three British dairy herds. *Prev. Vet. Med.*, **59**, 169–80.

Pellegrini, A. (2003) Antimicrobial peptides from food proteins. *Curr. Pharm. Des.*, **9**, 1225–38.

Peterson, M.C. (2003) *Campylobacter jejuni* enteritis associated with consumption of raw milk. *J. Environ. Health*, **65**, 20–1, 24, 26.

Petridis, K.D. and Steinhart, H. (1996) Biogenic amines in hard cheese production II. Control points study in standardised Swiss cheese production. *Deutsche Lebensm. Rundsch.*, **92**, 142–6.

Petterson, B., Lembke, F., Hammer, P., Stackebrandt, E., Priest, F.G. (1996) *Bacillus sporothermodurans*, a new species producing highly heat resistant endospores. *Int. J. Syst. Bacteriol.*, **46**, 759–64.

Phillips, J.D. and Griffiths, M.W. (1986) Factors contributing to the seasonal variation of *Bacillus* spp. in pasteurized dairy products. *J. Appl. Bacteriol.*, **61**, 275–85.

Pitt, J.I. and Hocking, A.D. (1997) *Fungi and Food Spoilage*, 2nd edn, Aspen Publishers, Gaithersburg, MD.

Piva, G., Galvano, F. and Carini, E. (1995) Detoxification methods of aflatoxin. A review. *Nutr. Res.*, **15**, 689–715.

Plommet, M., Fensterbank, R., Vassal, L., Auclair, J. and Mocquot, G. (1988) Survival of *Brucella abortus* in ripened soft cheese made from a naturally infected cow's milk. *Lait*, **68**, 115–20.

Potel, J. (1951) Die Morphologie, Kultur und Tierpathogenität des *Corynebacterium infantiseptum. Zeitschrift der Bakteriologie Parasitenkunde und Infektionskrankheiten. Hygiene Abteilung. 1 Originale*, **156**, 490–3.

Potter, M.E., Blaser, M.J., Sikes, R.K. and Kaufmann, A.F. (1983) Human *Campylobacter* infection associated with certified raw milk. *Am. J. Epidemiol.*, **117**, 475–83.

Potter, M.E., Kaufman, A.F., Blake, P.A., and Feldman, R.A. (1984) Unpasteurized milk: the hazards of a health fetish. *J. Am. Med. Assoc.*, **252**, 2048–52.

Prentice, G.A. (1994) Listeria monocytogenes, in *The Significance of Pathogenic Microorganisms in Raw Milk*, Monogr Int. Dairy Fed., Brussels.

Pritchard, T.J., Flanders, K.J. and Donnelly, C.W. (1995) Comparison of the incidence of *Listeria* on equipment versus environmental sites within dairy processing plants. *Int. J. Food Microbiol.*, **26**, 375–84.

Pyorala, S. (2003) Indicators of inflammation in the diagnosis of mastitis. *Vet. Res.*, **34**, 565–78.

Quinto, E.J. and Capeda, A. (1997) Incidence of toxigenic *Escherichia coli* in soft cheese made with raw or pasteurised milk. *Lett. Appl. Microbiol.*, **24**, 291–5.

Rampling, A. (1996) Raw milk: cheese and *Salmonella*. *Br. Med. J.* **312**, 67–8.

Rampling, A., Taylor, C.E.D. and Warren, J.E. (1987) Safety of pasteurised milk. *Lancet*, **ii**, 1209.

Rasmussen, M.D., Bjerring, M., Justesen, P. and Jepsen, L. (2002) Milk quality on Danish farms with automatic milking systems. *J. Dairy Sci.*, **85**, 2869–978.

Reitsema, C.J. and Henning, D.R. (1996) Survival of enterohemorrhagic *Escherichia coli* O157:H7 during the manufacture and curing of cheese. *J. Food Prot.*, **59**, 460–4.

Reuter, H. (1987) Kriterien zur Beurteilung von aseptischen Abfüll-und Verpackungssystemen, in *Asepticher Verpacken von Lebensmitteln* (ed. Reuter, H.), Hamburg, Behr's Verlag, pp. 121–33.

Riber, R.F. (1989) Three major areas that causes defects in cultured dairy products. *Cult. Dairy Products J.*, **24**, 4, 6, 7–9.

Richwald, G.A., Greenland, S., Johnson, B.J., Friedland, J.M., Goldsteink, E.J. and Plichta, D.T. (1988) Assessment of the excess risk of *Salmonella dublin* infection associated with the use of certified raw milk. *Public Health Rep.*, **103**, 489–93.

Robertson, J.R. (2003) Establishing treatment protocols for clinical mastitis. *Vet. Clin. North Am. Food Anim. Pract.*, **19**, 223–34.

Robinson, D.A. and Jones, D.M. (1981) Milk-borne *Campylobacter* infection. *Br. Med. J.*, **282**, 1374–6.

Robinson, R.K. (ed) (2002) The microbiology of milk and milk products, in *Dairy Microbiology Handbook*, 3rd edn, John Wiley & Sons, Inc., Publication, New York.

Robinson, R.K. and Tamime, A.Y. (1990) Microbiology of fermented milks, *in Dairy Microbiology–The Microbiology of Milk Products* (ed. R.K. Robinson), *volume 2*, 2nd edn, Elsevier Applied Science Publisher, London, pp. 291–343.

Robinson, R.K. and Tamime, A.Y. (1993) Manufacture of yoghurt and other fermented milks *in Modern Dairy Technology–Advances in Milk Products* (ed. R.K. Robinson), *volume 2*, 2nd edn, Elsevier Applied Science Publishers, London, pp. 1–48.

Robinson, R.K., Tamime, A.Y. and Wsolek, M. (2002) Microbiology of fermented milks, in *Dairy Microbiology Handbook* (ed. R.K. Robinson), John Wiley & Sons, Inc., Publication, New York.

Rodriguez Velasco, M.L., Calonge Delso, M.M. and Ordonez Escudero, D. (2003) ELISA and HPLC determination of the occurrence of aflatoxin M(1) in raw cow's milk. *Food Addit. Contam.*, **20**, 276–280.

Rohm, H. and Kovac, A. (1994) Effects of starter cultures on linear viscoelastic and physical properties of yogurt gels. *J. Texture Stud.*, **25**, 311–29.

Rohm, H. and Kovac, A. (1995) Effects of starter cultures on small deformation rheology of stirred yoghurt: 1. Evaluation of flow curves. *Lebensmittel-Wisenschaft und -Technologie*, **28**, 319–22.

Rohm, H., Kovac, A. and Kneifel, W. (1994) Effects of starter cultures on sensory properties of set-style yoghurt determined by quantitative descriptive analysis. *J. Sens. Stud.*, **9**, 171–86.

Rothwell, J. (1985) Microbiology of frozen dairy products, in *Microbiology of Frozen Foods* (ed. R.K. Robinson), Elsevier Applied Science Publishers, New York, pp. 209–231.

Roussi, V., Govaris, A., Varagouli, A. and Botsoglou, N.A. (2002) Occurrence of aflatoxin M(1) in raw and market milk commercialized in Greece. *Food Addit. Contam.*, **19**, 863–8.

Rowe, B., Hutchinson, D.N., Gilbert, R.J., Hales, B.H., Begg, N.T., Dawkins, H.C., Jacob, M. and Rae, F.A. (1987) *Salmonella ealing* infections associated with consumption of infant dried milk. *Lancet*, **i**, 900–3.

Rubinstein-Szturn, S., Courterier, A.L. and Maka, G. (1964) A cheese contaminated with *Shigella sonnei* as a cause of food poisoning. *Bull. Acad. Natl. Med. Paris*, **148**, 480–2.

Russell, A.D. (1982) *The Destruction of Bacterial Spores*, Academic Press, London.

Russell, A.D., Hugo, W.B. and Ayliffe, G.A.J. (1992) *Principles and Practice of Disinfection, Preservation and Sterilization*, Blackwell Scientific Publications, Oxford.

Ryan, C.A., Nickles, M.K., Hargett-Bean, N.T., Potter, M.E., Endo, T., Mayer, L., Langkop, C.W., Gibson, C., McDonald, R.C., Kenney, R.T., Puhr, N.D., McDonnell, P.J., Martin, R.J., Cohen, M.L. and Blake, P.A. (1987) Massive outbreak of antimicrobial resistant salmonellosis traced to pasteurized milk. *J. Am. Med. Assoc.*, **258**, 3269–74.

Ryser, E.T. and Marth, E.H. (1991) *Listeria, listeriosis, and food safety*, Marcel Dekker, Inc., New York.

Saad, A.M., Abdelgadir, A.M. and Moss, M.O. (1989) Aflatoxin in human and camel milk in Abu Dhabi, United Arab Emirates. *Mycotoxins Res.*, **5**, 57–60.

Salji, J. (1992) Acidophilus milk products: food with a third dimension. *Food Sci. Technol. Today*, **6**, 142–7.

Santos, M.H.S. (1996) Biogenic amines–their importance in foods. *Int. J. Food Microbiol.*, **29**, 213–31.

Schagemann, G. (1994) Viruses, in *The Significance of Pathogenic Microorganisms in Raw Milk*, Monograph of International Dairy Federation, Brussels.

Scheldeman, P., Herman, L., Goris, J., De Vos, P. and Heyndrickx, M. (2002) Polymerase chain reaction identification of *Bacillus sporothermodurans* from dairy sources. *J. Appl. Microbiol.*, **92**, 983–91.

Schiemann, D.A. (1989) Yersinia enterocolitica and Yersinia pseudotuberculosis, in *Foodborne Bacterial Pathogens* (ed. M.P. Doyle), Marcel Dekker, Inc., New York, pp. 601–72.

Schiemann, D.A. and Toma, S. (1978) Isolation of *Yersinia enterocolitica* from raw milk. *Appl. Environ. Microbiol.*, **35**, 54–58.

Schlimme, E. and Buchheim, W. (1999) *Milch und ihre Inhaltsstoffe*, Th. Mann Verlag, Gelsenkirchen.

Schoch, U., Lüthy, J. and Schlatter, C. (1984) Mutagenitätsprüfung industriell verwendeter *Penicillium camemberti-* und *P. roqueforti* - Stämme. *Z. Lebensm.-Unters. Forsch.*, **178**, 351–5.

Schraft, H., Steele, M., McNab, B., Odumeru, J., and Griffiths, M.W. (1996) Epidemiological typing of *Bacillus* spp. isolated from food. *Appl. Environ. Microbiol.*, **62**, 4229–32.

Schreiner, D.A. and Ruegg, P.L. (2002) Effects of tail docking on milk quality and cow cleanliness. *J. Dairy Sci.*, **85**, 2503–11.

Schröder, M.J.A. and Bland, M.A. (1984) Effect of pasteurization temperature on the keeping quality of whole milk. *J. Dairy Res.*, **51**, 569–78.

Schröder, M.J.A., Cousins, C.M. and McKinnon, C.H. (1982) Effect of psychrotrophic post-pasteurization contamination on the keeping quality at 11°C and 5°C of HTST-pasteurized milk in the UK. *J. Dairy Res.*, **49**, 619–30.

Schukken, Y.H., Wilson, D.J., Welcome, F., Garrison-Tikofsky, L. and Gonzalez, R.N. (2003) Monitoring udder health and milk quality using somatic cell counts. *Vet. Res.*, **34**, 579–96.

Scott, R. (ed) (1986) *Cheesemaking Practice*, Elsevier Applied Science Publisher, London.

Seeliger, H.P.R. (1961) *Listeriosis*, Hafner Publishing Co., New York, NY.

Seiler, G. (1988) Identification of cheese-smear coryneform bacteria. *J. Dairy Res.*, **53**, 439–49.

Seiler, H. and Wendt, A. (1992) Die CO_2—Messung in Fruchtcontainern. *Deutsche Milchwirtschaft*, **43**, 158, 159–62.
Sharma, M. and Anand, S.K. (2002) Bacterial biofilm on food contact surfaces: a review. *J. Food Sci. Technol.*, **39,** 573–93.
Sharp, J.C., Collier, P.W., Forbes, G.I. and Hill, T.W. (1988) Surveillance programme for the control of foodborne infections and intoxications in Europe: the first 6 year's experience in Scotland, 1980–1985. *Bull. World Health Organ.*, **66,** 471–476.
Sharp, J.C.M., Cola, J.E., Curnow, J. and. Reilly, W.J. (1994) *Escherichia Coli* O157 infections in Scotland. *J. Clin. Microbiol.*, **40**, 3–9.
Sharpe, J.C.M. (1987) Infections associated with milk and dairy products in Europe and North America, 1980-856. *Bull. WHO*, **65**, 397–406.
Siewert, R. (1979) Zur Bedeutung von Hefen bei der Reifung von Camembert und Brie. *Deutsche Molkerei Zeitung*, **107**, 1134–8.
Simini, B. (1996) Outbreak of foodborne botulism continues in Italy. *Lancet*, **348**, 813.
Smaczny, T. and Krämer, J. (1984) Säuerungsstörungen in der Joghurt-, Bioghurt- und Biogarde-Produktion, bedingt durch Bakteriocine und Bakteriophagen von *Streptococcus thermophilus*. II. Verbreitung und Charakterisierung der Bakteriophagen. *Deutsche Molkerei Zeitung*, **105**, 614–8.
Snyder, I.S., Johnson, W. and Zottola, E.A. (1978) Significant pathogens in dairy products. in *Standard Methods for the Examination of Dairy Products* (ed. E.H. Marth), American Public Health Association, Washington, DC, pp. 11–32.
Sorhaug, T. and Stepaniak, L. (1997) Psychrotrophs and their enzymes in milk and dairy products: quality aspects. *Trends Food Sci. Technol.*, **8**, 35–41.
Southern, J.P., Smith, R.M.M. and Palmer, S.R. (1990) Bird attack on milk bottles: possible mode of transmission of *Campylobacter jejuni* to man. *Lancet*, **336**, 1425–7.
Spahr, U. and Url, B. (1993) Behavior of pathogenic bacteria in cheese. *Int. Dairy Fed.*, no. 223, **Supplement**, Brussels, Belgium.
Spillmann, H. and Geiges, O. (1983) Identifikation von Hefen und Schimmelpilzen aus bombierten Joghurt-Packungen. *Milchwissenschaft*, **38**, 129–32.
Spreer, E. (ed) (1995) *Technologie der Milchverarbeitung*, Behr's Verlag, Hamburg.
Spreer, E. (1998) *Milk and Dairy Technology*, Marcel Dekker.
Staal, P.F.J. (1981) Legislative aspects. *Bull. Int. Dairy Fed.*, no. 133, 122–8.
Staal, P.F.J. (1986) Legislation/statutory regulations applicable to pasteurized fluid milk in a selected number of countries. *Bull. Int. Dairy Fed.*, no. 200, 71–9.
Stadhouders, J., Cordes, M.M. and van Schouwenberg-van Foeken, A.W.J. (1978) The effect of manufacturing conditions on the development of staphylococci in cheese. Their inhibition by starter bacteria. *Netherlands Milk Dairy J.*, **32**, 193–203.
Stahl, V., Garcia, E., Hezard, B. and Fassel, C. (1996) Prevention of *Listeria monocytogenes* in dairy farms and dairy processing plants. *Pathol. Biol.*, **44**, 816–24.
Stalder, H., Isler, R., Stutz, W., Salfinger, M., Lauwers, S. and Vischer, W. (1983) Beitrag zur Epidemiologie von *Campylobacter jejuni*. *Schweiz. Med. Wochenschr.*, **113**, 245–9.
Stevenson, R.G., Rowe, M.T., Wisdom, G.B. and Kilpatrick, D. (2003) Growth kinetics and hydrolytic enzyme production of *Pseudomonas* spp. isolated from pasteurized milk. *J. Dairy Sci.*, **70**, 293–6.
Still, P., C. Eckardt and L. Leistner. (1978) Bildung von Cyclopazonsäure durch *Penicillium camembertii*-isolate von Käse. *Fleischwirtschaft*, **58**, 876–8.
Strantz, A.A., Zottola, E.A., Petran, R.L., Overdahl, B.J. and Smith, L.B. (1989) The microbiology of sweet water and glycol cooling systems used in HTST pasteurizers in fluid milk processing plants in the United States. *J. Food Prot.*, **52**, 799–804.
Stratton, J.E., Hutkins, R.W. and Taylor, S.L. (1991) Biogenic amines in cheese and other fermented foods: a review. *J. Food Prot.*, **54**, 460–70.
Stuart, J., Sufi, F., McNulty, C. and Park, P. (1997) Outbreak of *Campylobacter enteritis* in a residential school associated with bird pecked bottle tops. *Commun. Dis. Rep CDC Rev.*, **7**, R38–R40.
Su, C. and Brandt, L.J. (1996) *Escherichia coli* O157:H7 infection in humans. *Ann. Int. Med.*, **123**, 698–714.
Suhren, G. (1988) Producer microorganisms, in *Enzymes of Psychrotrophs in Raw Food* (ed. R.C. Mekeller), Elsevier Applied Science, New York, pp. 3–34.
Suhren, G. (1996) Untersuchungen zum Einfluss von Rückständen von antimikrobiell wirksamen Substanzen in Mich auf kommerziell eingesetzte Starterkulturen in Modellversuchen (Influence of residues of antimicrobials in milk on commercially applied starter cultures–model trials). *Kieler Milchwirtschaftliche Forschungsberichte*, **96**, 131–49.
Suhren, G. and Heeschen, W. (1996) Detection of inhibitors in milk by microbial tests. A review. *Nahrung*, **40**, 1–7.
Suhren, G. and Reichmuth, J. (2003) Measurability and development of the hygienic value of the raw material milk. *Kieler Milchwirtschaftliche Forschungsberichte*, **55**, 5–36.
Sulzer, G. and Busse, M. (1991) Die Entwicklung von Listerien auf Camembert und deren Beeinflussung durch Keime mit einer Hemmwirkung auf Listerien. *DMZ-Lebensmittelindustrie und Michwirtschaft* (112), 82–4.
Tacket, C.O., Narain, J.P., Sattin, R., Lofgren, J.P., Konigsberg, C., Renøtorff, R.C., Rausa, A., Davis, B.R. and Cohen, M.L. (1984) A multistate outbreak of infections caused by *Yersinia enterocolitica* transmitted by pasteurized milk. *J. Am. Med. Assoc.*, **251**, 483–6.
Tamime, A.Y. and Marshall, V.M.E. (1997a) Microbiology and technology of fermented milks, in *Microbiology and Biochemistry of Cheese and Fermented Milk* (ed. B.A. Law), Blackie Academic and Professional, Chapter 1, pp. 1–49.
Tamime, A.Y. and Marshall, V.M.E. (1997b) Microbiology and technology of fermented milks, in *Microbiology and Biochemistry of Cheese and Fermented Milk* (ed. B.A. Law), Blackie Academic and Professional, Chapter 3, pp. 57–72.
Tamime, A.Y. and Robinson, R.K. (1985) *Yoghurt–Science and Technology*, Pergamon Press, Oxford.
Tamine, A.Y. and Robinson, R.K . (1988a) Technology of thermophilic fermented milk. *Bull. Int. Dairy Fed.*, no. 227, 82–95.

Tamime, A.Y. and Robinson, R.K. (1988b) Fermented milks and their future trends: II. Technological aspects. *J. Dairy Res.*, **55**, 281–307.
Tamime, A.Y., Marshall, V.M. and Robinson, R.K. (1995) Microbiology and technological aspects of milk fermented by bifidobacteria. *J. Dairy Res.*, **62**, 151–87.
Tanaka, N. (1982) Challenge of pasteurized process cheese spreads with *Clostridium botulinum* using in-process and post-process inoculation. *J. Food Prot.*, **45**, 1044–50.
Tanaka, N., Goepfert, J.M., Traisman, E. and Hoffbeck, W.M. (1979) A challenge of pasteurized process cheese spread with *Clostridium botulinum* spores. *J. Food Prot.*, **42**, 787–9.
Tanaka, N., Traisman, E., Plantinga, P., Finn, L., Flom, W., Meske, L. and Guggisberg, J. (1986a) Evaluation of factors involved in antibotulinal properties of pasteurized process cheese spreads. *J. Food Prot.*, **49**, 526–31.
Tanaka, N., Traisman, E., Plantinga, P., Finn, L., Flom, W., Meske, L. and Guggisberg, J. (1986b) Erratum: evaluation of factors involved in antibotulinal properties of pasteurized process cheese spreads. *J. Food Prot.*, **49**, 754.
Tarr, P.I. (1994) *Escherichia coli* O157:H7: overview of clinical and epidemiological issues. *J. Food Prot.*, **57**, 632–6.
Tatini, S.R., Jezeski, J.J., Olson, J.J., Jr. and Casman, E.P. (1971a) Factors influencing the production of staphylococcal enterotoxin A in milk. *J. Dairy Sci.*, **54**, 312–20.
Tatini, S.R., Jezeski, J.J., Morris, H.A., Olson, J.J., Jr. and Casman, E.P. (1971b) Production of staphylococcal enterotoxin A in Cheddar and Colby cheeses. *J. Dairy Sci.*, **54**, 815–25.
Taylor, D.N., Bopp, C., Birkness, K. and Cohen, M.L. (1984) An outbreak of salmonellosis associated with a fatality in a healthy child: a large dose and severe illness. *Am. J. Epidemiol.*, **119**, 907–12.
te Giffel, M.C., Wagendorp, A., Herrewegh, A. and Driehuis, F. (2002) Bacterial spores in silage and raw milk. *Antonie Van Leeuwenhoek*, **81**, 625–30.
Ter Steeg, P.F. and Cuppers, H.G.A.M. (1995) Growth of proteolytic *Clostridium botulinum* in process cheese products. II. Predictive modelling. *J. Food Prot.*, **58**, 1100–8.
Ter Steeg, P.F., Cuppers, H.G.A.M., Hellemons, J.C. and Rijke, G. (1995) Growth of proteolytic *Clostridium botulinum* in process cheese products. I. Data acquisition for modeling the influence of pH, sodium chloride, emulsifying salts, fat dry basis, and temperature. *J. Food Prot.*, **58**, 1091–9.
Ternstrom, A., Lindberg, A.M. and Molin, G. (1993) Classification of the spoilage flora of raw and pasteurized bovine milk, with special reference to *Pseudomonas* and *Bacillus*. *J. Appl. Bacteriol.*, **75**, 25–34.
Tetra Pak (2003) *Dairy Processing Handbook*, Tetra Pak, Lund.
Teuber, M. (2000) Fermented Milk Products, in *The Microbiological Safety and Quality of Food* (eds. B.M. Lund, T.C. Baird-Parker and G.W. Gould), Aspen Publishers Inc., Gaithersburg, MD., Chapter 23, pp. 535–89.
Thapar, M.K. and Young, E.J. (1986) Urban outbreak of goat cheese brucellosis. *Pediatr. Infect. Dis.*, **5**, 640–3.
Thielman, N.M. (1994) Enteric *Escherichia coli* infections. *Infect. Dis.*, **7**, 582–91.
Thomas, S.B. and Thomas, B.F. (1977) The bacterial content of milking machines and pipeline milking plants. Part II of a review. *Dairy Ind. Int.*, **42**, 16–23.
Thomas, T.D. (1986) Oxidative activity of bacteria from Cheddar cheese. *NZ J. Dairy Sci. Technol.*, **21**, 37–47.
Thompson, S.S., Harmon, L.G. and Stine, C.M. (1978) Survival of selected organisms during the spray drying of skimmilk and storage of nonfat dry milk. *J. Food Prot.*, **41**, 16–9.
Threlfall, E.J., Ward, L.R., Hampton, M.D., Ridley, A.M., Rowe, B., Roberts, D., Gilbert, R.J., Van Someren, P., Wall, P.G. and Grimont, P. (1998) Molecular fingerprinting defines a strain of *Salmonella* enterica serotype Anatum responsible for an international outbreak associated with formula dried milk. *Epidemiol. Infect.*, **121**, 289–93.
Thurm, V. and Dinger, E. (1998) Subtyping of outbreak related strains as a useful method in the surveillance of *Campylobacter* infections, in *Proceedings, 4th World Congress Foodborne Infections and Intoxications*, 7–12 June 1998 (eds. K. Noeckler, P. Teufel, K. Schmidt and E. Weise), ISBN 3-931675-34-3.
Thurm, V., Dinger, E. Lyytikäinen O., Petersen L., Wiebelitz A., Lange D., Fischer R., Oppermann H. and Mäde D. (1999) Infektionsepidemiologie lebensmittelbedingter *Campylobacter*-Infektionen–Untersuchung eines Ausbruchs in Sachsen-Anhalt mittels epidemiologischer, mikrobiologischer und molekularbiologischer Methoden. *Bundesgesundheitsblatt*, **42**, 206–11.
Tibana, A., Warnken, M.B., Nunes, M.P., Ricciardi, I.D. and Noleto, A.L.S. (1987) Occurrence of *Listeria* species in raw and pasteurized milk in Rio de Janeiro, Brazil. *J. Food Prot.*, **50**, 580–3.
Tizard, I. (2001) The protective properties of milk and colostrum in non-human species. *Advan. Nutr. Res.*, **10**, 139–66.
Tolle, A. (1980) The microflora of the udder. *Bull. Int. Dairy Fed.*, no. 120, 4–10.
Tolle, A. and Heeschen, W. (1975) Der aseptische Milchentzug über implantierte Dauerkatheter als Modell einer pulsierungsfreien Melktechnik und als Grundlage zum Studium der spezifischen und unspezifischen Infektionsabwehr. *Berichte über Landwirtschaft* **190**, Special Issue, 60–92.
Toppino, P.M., Degano, L., Itabashi, H., Boevre, L., Tamminga, S., Kennelly, J.J., Erasmus, L.J., Hermansen, J.B.H. and Rulquin, H. (2001) Influence of feed in major components of milk. *Bull. Int. Dairy Fed.*, no. 366, Brussels, Belgium.
Trouve, E., Maubois, J.L., Piot, M., Madec, M.N., Fauquant, J., Rouault, A., Tabard, J. and Brinkman, G. (1991) Rétention de différentes espèces microbiennes lors de l'épuration du lait par microfiltration en flux tangentiel. *Lait*, **71**, 1–13.
Tudor, E.A. and Board, R.G. (1993) Food spoilage yeasts, in *The yeasts* (eds. A.H. Rose and J.S. Harrison), *volume 5, Yeast Technology*, Academic Press, London, pp. 435–516.
US Department of Health, Education and Welfare. (1984) Cheeses and related cheese products, Code of Federal Regulations, Title 21, Chapter 1, Part 133. US Gov. Print. Off., Washington, D.C.

US Department of Health Education and Welfare. (1994a) Outbreak of *Salmonella enteritidis* associated with nationally distributed ice cream products–Minnesota, South Dakota, and Wisconsin 1994. *Mort. Morb. Wkly Rep.*, **43**(40), 740–1.

US Department of Health Education and Welfare. (1994b) General Specifications for Dairy Plants Approved for USDA Inspection and Grading Service, Quality Specifications for raw milk. Code of Federal Regulations, Title 7, Subtitle B, Chapter 1, Part 58. 132–141. US Gov. Print. Off., Washington, D.C.

US Department of Health Education and Welfare. (1995a) Requirements for specific standardized milk and cream, Code of Federal Regulations, Title 21, Chapter 1, Part 131, Subpart A.

US Department of Health Education and Welfare. (1995b) Requirements for specific standardized milk and cream, Code of Federal Regulations, Title 21, Chapter 1, Part 131, Subpart B.

US FDA CFSAN (1998) Food Compliance Program. Domestic and imported cheese and cheese products, Chapter 3.

US Department of Health Education and Welfare. (1995c) Frozen Desserts, Requirements for Specific Standardized Frozen Desserts, Code of Federal Regulations, Title 21, Chapter 1, Part 135, Subpart B.

Umoh, V.J., Obawede, K.S. and Umoh, J.U. (1985) Contamination of infant powdered milk in use with enterotoxigenic *Staphylococcus aureus*. *Food Microbiol.*, **2**, 255–61.

Upton, P. and Coia, J.E. (1994) Outbreak of *Escherichia coli* O157:H7 infection associated with pasteurised milk supply. *Lancet*, **344**, 1015.

Usera, M.A., Echeita, A., Aladuena, A., Blanco, M.C., Reymundo, R., Prieto, M.I., Tello, O., Cano, R., Herrera, D. and Martinez-Navarro, F. (1996) Interregional foodborne salmonellosis outbreak due to powdered infant formula contaminated with lactose-fermenting *Salmonella virchow*. *Eur. J. Epidemiol.*, **12**, 377–81.

USPHS, FDA (United States Public Health Service, Food and Drugs Administration). (1978) Grade A Pasteurized Milk Ordinance, 1978. Recommendations of the US Public Health Service/Food and Drug Administration. US Government Printing Office, Washington, D.C.

Vaerewijck, M.J., De Vos, P., Lebbe, L., Scheidemann, P., Hoste, B. and Heyndrickx, M. (2001) Occurrence of *Bacillus sporothermodurans* and other aerobic spore-forming species in feed concentrates for dairy cattle. *J. Appl. Microbiol.*, **91**, 1074–84.

Vaillant, V., Haeghebaert, S., Desenclos, J.C., Bouvet, P., Grimont, F., Grimont, P.A. and Burnens, A.P. (1996) Outbreak of *Salmonella dublin* infection in France, November-December 1995. *Eur. Surveillance*, **1**, 9–10.

van den Berg., L. (1984) The thermization of milk. *Bull. Int. Dairy Fed.*, no. 182, 3–12.

van den Bogaard, A.E. and Stobberingh, E.E. (1999) Antibiotic usage in animals: impact on bacterial resistance and public health. *Drugs*, **58**, 589–607.

van den Bogaard, A.E. and Stobberingh, E.E. (2000) Epidemiology of resistance to antibiotics. Links between animals and humans. *Int. J. Antmicrobial Agents*, **14**, 327–35.

van Netten, P., van de Moosdijk, A., van Itoensel, P., Mossel, D.A.A. and Perales, I. (1990) Psychrotrophic strains of *Bacillus cereus* producing enterotoxin. *J. Appl. Bacteriol.*, **69**, 73–9.

Vandana, T., Chauhan, R.K.S. and Tiwari, V. (1991) Aflatoxin detection in milk samples of cattle . *Natl. Acad. Sci. Lett.*, **14**, 391–2.

Varnam, A.H. and Sutherland, J.P. (1994) *Milk and Milk Products. Technology, Chemistry, and Microbiology*, Chapman & Hall, London.

Vasanthi, S. and Bhat, R.V. (1998) Mycotoxins in foods–Occurrence, health and economic significance and food control measures. *Indian J. Med. Res.*, **108**, 212–24.

Vedamuthu, E.R. (1985) What is wrong with cultured buttermilk? *Dairy Food Sanit.*, **5**, 8–13.

Veillet-Poncet, L., Tayfour, A. and Millière, J.B. (1980) Etude bactériologique de l'ultrafiltration du lait et du stockage au froid du rétentat. *Lait*, **60**, 351–74.

Vercet, A., Oria, R., Marquina, P., Crelier, S. and Lopez-Buesa, P. (2002) Rheological properties of yoghurt made with milk submitted to manothermosonication. *J. Agric. Food Chem.*, **50**, 6165–71.

Vidon, D.J.M. and Delmas, C.L. (1981) Incidence of *Yersinia enterocolitica* in raw milk in Eastern France. *Appl. Environ. Microbiol.*, **41**, 355–9.

Villar, R.G., Macek, M.D., Simons, S., Hayes, P.S., Goldoft, M.J., Lewis, J.H., Rowan, L.L., Hursh, D., Patnode, M. and Mead, P.S. (1999) Investigation of multidrug-resistant *Salmonella* serotype *typhimurium* DT104 infections linked to raw-milk cheese in Washington state. *J. Am. Med. Assoc.*, **281**, 1811–6.

Vlaemynck, G. (1994) Salmonella, in *The Significance of Pathogenic Microorganisms in Raw Milk*, Monograph of International Dairy Federation, Brussels.

Vogt, R.L., Hackey, A. and Allen, J. (1981) *Salmonella enteritidis* serotype derby and consumption of raw milk. *J. Infect. Dis.*, **144**, 608.

Vought, K.J. and Tatini, S.R: (1998) *Salmonella enteritidis* contamination of ice cream associated with a 1994 multistate outbreak. *J. Food Prot.*, **61**, 5–10.

Walker, S.J. and Gilmour, A. (1986) The incidence of *Yersinia enterocolitica* and *Yersinia enterocolitica*-like bacteria in goats milk in Northern Ireland. *Lett. Appl. Microbiol.*, **3**, 49–52.

Wallace, G.I. (1938) The survival of pathogenic microorganisms in ice cream. *J. Dairy Sci.*, **21**, 35–6.

Walstra, P. (1993) The Synthesis of Curd, in *Cheese: Chemistry, Physics and Microbiology* (ed P.F. Fox), *volume 1, General Aspects*, Chapman and Hall, London, Chapter 5, pp. 141–91.

Walstra, P. and Jenness, R. (1984) *Dairy Chemistry and Physics*. John Wiley and Sons. New York.

Walstra, P., Geurts, T.J., Noomen, A., Jellema and van Boekel, M.A.J.S. (1999) *Dairy Technology–Principles of Milk Properties and Processes*. Marcel Dekker, Inc., Basel.

Watts, J.L. (1988) Etiological agents of bovine mastitis. *Vet. Microbiol.*, **16**, 41–66.
Weber, H. (ed) (1996a) *Mikrobiologie der Lebensmittel - Milch und Milchprodukte*, Behr's Verlag, Hamburg.
Weber, H. (1996b) Starterkulturen in der milchverarbeitenden Industrie, in *Milch und Milchprodukte* (ed. H. Weber), Behr's Verlag, Hamburg, Chapter 4, pp. 105–52.
Wegner, K. (1996) Mikrobiologie der Sauermilcherzeugnisse, in *Milch und Milchprodukte* (ed. H. Weber), Behr's Verlag, Chapter 5, pp. 153–230.
Wellenberg, G.J., van der Poel, W.H. and Van Oirschot, J.T. (2002) Viral infections and bovine mastitis: a review. *Vet. Microbiol.*, **88**, 27–45.
Westhoff, D.D. (1981) Microbiology of ultrahigh temperature milk. *J. Dairy Sci.*, **64**, 167–73.
White, C.H. and Custer, E.W. (1976) Survival of *Salmonella* in Cheddar cheese. *J. Milk Food Technol.*, **39**, 328–31.
Wilshaw, G.A., Cheasty, T. and Smith, G. (2000) Escherichia coli., in *The Microbiological Safety and Quality of Food* (eds. B.M. Lund, T.C. Baird-Parker and G.W. Gould), *volume II*, Aspen Publishers Inc., Gaithersburg, MD, Chapter 43, pp. 1136–77.
Wolfson, L.M. and Sumner, S.S. (1993) Antibacterial activity of the lactoperoxidase system: a review. *J. Food Prot.*, **56**, 887–92.
Wong, A.C. (1998) Biofilms in food processing environments. *J. Dairy Sci.*, **81**, 2765–70.
Wong, H.C., Chang, M.H. and Fan, J.Y. (1988) Incidence and characterization of *Bacillus cereus* isolates contaminating dairy products. *Appl. Environ. Microbiol.*, **54**, 699–702.
Wood, D.S., Collins-Thompson, D.L., Irvine, D.M. and Myhr, A.N. (1984) Source and persistence of *Salmonella muenster* in naturally contaminated Cheddar cheese. *J. Food Prot.*, **47**, 20–2.
World Health Organisation (1995) Report of a WHO consultation on public health implications of consumption of raw milk and meat and their products. Kiel, Germany, 17–20 December 1995.
World Health Organisation (2002a) Exposure assessment of *Listeria monocytogenes* in ready-to eat foods (ed. T. Ross, E. Todd and M. Smith), WHO MRA-00/02, Geneva.
World Health Organisation (2002b) Evaluation of certain veterinary drug residues in food. Joint FAO/WHO Expert Committee on Food Additives. World Health Organisation Technical Reports Series, 911, pp. 66.
Yoo, J.A. and Chen, X.D. (2002) An emission pattern of a thermophilic bacteria attached to or imbedded in porous supports. *Int. J. Food Microbiol.*, **73**, 11–21.
Zangerl, P. and Ginzinger, W. (1993) Ein HACCP-Konzept für Hartkäsereien. *Milchwirtschaftliche Berichte*, **115**, 99–102.
Zehren, V.L. and Zehren, V.F. (1968a) Examination of large quantities of cheese for staphylococcal enterotoxin A. *J. Dairy Sci.*, **51**, 635–44.
Zehren, V.L. and Zehren, V.F. (1968b) Relation of acid development during cheese making to development of staphylococcal enterotoxin A. *J. Dairy Sci.*, **51**, 645–9.
Zickrick, K. (1996) Mikrobiologie der Käse, in *Milch und Milchprodukte* (ed. H. Weber), Chapter 7, pp. 255–351.
Zottola, E.A. and Jezeski, J.J. (1969) Comparisons of short-time holding procedures to determine thermal resistance of *Staphylococcus aureus*. *J. Dairy Sci.*, **52**, 1855–7.
Zottola, E.A. and Marth, E.H. (1966) Thermal inactivation of bacteriophages active against lactic streptococci. *J. Dairy Sci.*, **49**, 1388–1342.
Zottola, E.A. and Smith, L.B. (1991) Pathogens in cheese. *Food Microbiol.*, **8**, 171–82.
Zottola, E.A. and Smith, L.B. (1993) Growth and survival of undesirable bacteria in cheese, in *Cheese: Chemistry, Physics and Microbiology* (ed. P.F. Fox), *volume 1*, Chapman and Hall, Chapter 12, pp. 471–92.
Zottola, E.A., Schmeltz, D.L. and Jezeski, J.J. (1970) Effect of short-time subpasteurization treatments on the destruction of *Staphylococcus aureus* in milk for cheese manufacture. *J. Dairy Sci.*, **52**, 1707–14.

17 Fermented beverages

I Introduction

A *Definitions*

Fermentation is any process during which the enzymes of a microorganism convert an energy source to one or more chemical products, which may be for foods or for industrial or pharmaceutical use. Yeasts are by far the most important microorganisms used for liquid fermentations, though bacteria are used in some milk-based products such as yoghurts.

The most important yeast fermentation product is alcohol, and this chapter will discuss the major product of such fermentations, beer, and wine.

Solid fermentations, i.e. of fruit and vegetables, are treated under other chapters.

B *Important properties*

Beer, wine, and other alcoholic beverages usually contain >4% alcohol, which gives the products their character. Products from grapes (wines) have an acidic pH. Desirable pH values for white wine are 3.0–3.4 and for red wines 3.3–3.7 (Rankine, 1989). Together with the alcohol content, these low pH levels provide stability against most microorganisms. Products from apples (cider) are also of acid pH. Products from cereals (beers) are generally of higher pH, 4.0–4.5, and here the addition of carbon dioxide provides additional microbial stability. Products from honey (mead) are also of higher pH.

C *Methods of processing*

Beer. Barley is the main substrate for production of beer, though wheat is also used. Barley is treated to produce malt by "steeping", in which a sufficient amount of water is added to induce germination of the barley. Enzymes produced during germination break down the starch in the kernels into small polysaccharides, making it fermentable by yeasts. After this process is complete, the barley is kiln dried to produce malt. Wort is produced from the dry malt by the addition of hops and sometimes adjuncts (starch in the form of rice, maize, etc.), and boiling in water.

The wort is then cooled, oxygenated, and fermented using specific yeast species and strains, usually *Saccharomyces cerevisiae* or *Sacc. pasteurianus*, which produce lager beer or ale. An example of a fermentation chart for a lager beer showing the temperature and the decrease in pH and fermentable extract (°Plato) over time is given in Figure 17.1 [professional brewers often use the Plato (°P) scale, instead of specific gravity, as a measure of the sugar levels in wort and beer]. A specific gravity (SG) of 1.004 is equivalent to 1° Plato (1% sucrose) and 1.040 to 10° Plato (10% sucrose). Hence, each Plato degree accounts for 0.004 SG. To convert SG to Plato, divide the digits to the right of the decimal point by 4. For example, 1.044 is 11° Plato and 1.054 is 13.5° Plato). After fermentation and maturation (low temperature and secondary fermentation), the beer is sterile filtered or plate pasteurized before bottling or canning, or pasteurized in tunnels after filling and sealing.

Wines. The grape juice ("must") is extracted from the grapes by crushing and often maceration of the grapes. A starter culture, usually *Sacc. cerevisiae*, may be added, or fermentation may rely on naturally

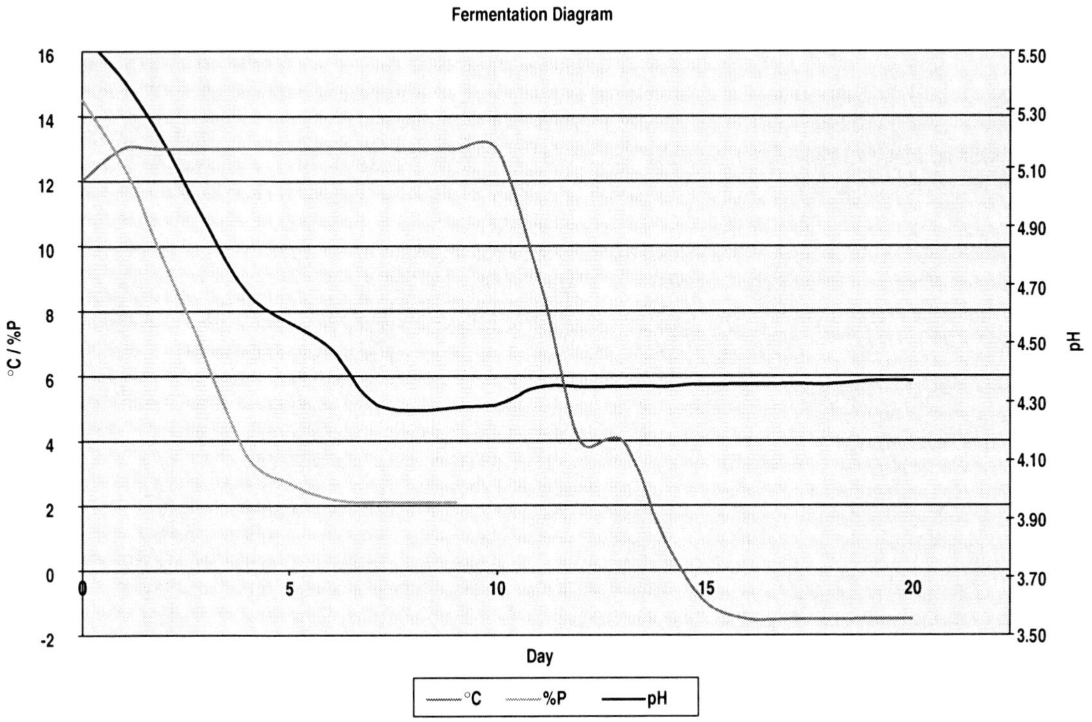

Figure 17.1 Fermentation chart for lager beer: changes in temperature (°C), pH, and fermentable extract (%P.)

occurring yeasts present in the grapes. Greater control of the fermentation process is possible if starter cultures are used. In the production of white wines, the skin of the grapes is removed from the must before fermentation; while red wines are produced by fermentation with the skins present. Controlled temperatures are sometimes used, especially for white wine production.

Traditionally, SO_2 has been used to control undesirable microorganisms as well as unwanted color deterioration (browning) due to enzyme activity. However, the increasing focus on health issues has lead to a reduction in the use of SO_2. Fermentation may involve microorganisms from the grapes or winery environment as well as starter cultures, usually *Saccharomyces* spp. The purpose of using a starter culture is to establish a controlled dominance of the culture yeast over the native microflora.

To produce wines of lower acidity, the primary fermentation is often followed by malolactic fermentation, where L-malic acid is converted into L-lactic acid. The microorganisms responsible for the malolactic fermentation are lactic acid bacteria (LAB) dominated by several species of *Lactobacillus* spp. (*Lactobacillus casei, Lb. plantarum, Lb. sake, Lb. brevis, Lb. fructivorans,* etc.), *Pediococcus* spp. *(Pediococcus parvulus, Ped. pentosaceus,* and *Ped. damnosus)* and *Leuconostoc* species. Because of the detrimental effect on shelf life of pasteurization, sterile filtration is usually the preferred method for stabilizing the wine microbiologically before bottling.

Fortified wines, sherries, and distilled products. Sherries are produced by using special alcohol-tolerant yeast strains. The introduction of oxygen during fermentation also assists the formation of higher levels of alcohol than is produced in wines. Fortified wines are produced by the addition of distilled alcohol to wines. Distilled products are made from a variety of substrates with the addition of alcohol. Such products have no microbiological issues and are not considered further.

D Types of final products

Ale types of beer are mainly produced in Great Britain and differ from the conventional European lager beers on a number of parameters, e.g. malt type, yeast type, and fermentation temperature. The ale yeast *(Sacc. cerevisiae)* is always a top fermenting strain and fermentation is normally at temperatures of 20–23°C. The lager strains *(Sacc. pasteurianus)* are usually bottom fermenters and used at 12–17°C.

Beer is classified according to alcohol content, typically ranging from 0% to 6% alcohol by volume, but there are no international standards for this classification.

Generally speaking, wines are classified according to color (red, white, and rosé) and alcohol content. Table wines have an alcohol content of 7–14% v/v. Fortified wines such as port, sherry, and madeira usually have an alcohol content of 14–21% v/v. Wines may also be classified according to the grape variety, taste (dry, semidry, semisweet, and sweet), and content of CO_2(still or sparkling).

II Initial microflora

A Grains

The initial microflora of malt is dominated by microorganisms indigenous to the barley (Flannigan, 1969, 1996; see also Chapter 8). The initial microflora of adjuncts like maize and rice that are used in some breweries is also described in Chapter 8.

A variety of fungi (especially species of *Fusarium* and *Aspergillus*) may invade barley before malting. *Fusarium* plays a role in cold, wet climates, invading the kernels during the growing season. Kernels infected with *Fusarium* species may be responsible for gushing (over-foaming) of the final beer (Flannigan, 1996).

During steeping, some microorganisms are capable of growth. Increases in bacterial numbers are variable, depending on temperature and a_w, with *Aureobacterium, Alcaligenes, Clavibacterium, Flavobacterium, Erwinia*, and *Pseudomonas* being the most common genera (Petters *et al.*, 1988). Fungi that commonly increase in numbers include *Fusarium, Mucor, Eurotium, Alternaria*, and *Rhizopus* (Flannigan, 1996). *Aspergillus clavatus* is of particular concern, as this species can build to unacceptably high levels if malting temperatures are elevated or spontaneous heating occurs (Flannigan *et al.*, 1984; Flannigan, 1986). In extreme cases, blue green mats of *Asp. clavatus* may form on grain during malting (Shlosberg *et al.*, 1991). Under these conditions *Asp. clavatus* can be allergenic and is reported to be the cause of "malt workers' lung" (Riddle *et al.*, 1968; Flannigan, 1986).

Kilning reduces the microbial population substantially, but does not usually eliminate the flora (Flannigan, 1996).

The microflora of finished malt commonly consists of 10^6–10^8 cfu/g, consisting largely of *Erwinia, Pseudomonas*, and *Bacillus* species. *Lactobacilli* can be of the order of 10^4 per gram (Flannigan, 1996). Fungal counts are commonly 10^3–10^4 cfu/g, with great variability reported in the genera present (Flannigan, 1996).

Malt may be stored or transported before use, during which time it may pick up moisture, often with resultant increases in storage fungi including *Asp. candidus* and *Asp. versicolor* (Flannigan, 1996).

B Grapes

Many studies on the microflora of grapes have been published, but sound quantitative data are still needed (Fleet and Heard, 1992). In general terms, the microflora on grapes is dominated by yeasts (*Kloeckera, Hanseniaspora, Candida, Pichia*, and *Kluyveromyces* spp.), LAB, acetic acid bacteria (*Gluconobacter* and *Acetobacter* spp.), and fungi (*Botrytis, Penicillium, Aspergillus, Mucor, Rhizopus, Alternaria, Ucinula*, and *Cladosporium* spp.) (Walker, 2000).

III Primary processing

A *Effects of processing on microorganisms*

Beer manufacture. The heat treatment of the malt and adjuncts (i.e. boiling to produce wort) eliminates the initial microflora except for spore-forming bacteria and fungi. These microorganisms are usually of little relevance but may cause problems with low acid wines.

The yeast cultures used for the fermentation of beer are produced from pure cultures of selected strains of *Sacc. pastorianus* (lager beer) or *Sacc. cerevisiae* (ale beer).

Growth of microorganisms in beer and wine is restricted by the alcohol content, the low pH, the presence of yeast metabolites (organic acids, fatty acids, and acetaldehyde), and the low redox potential created by the production of CO_2. In beer, the content of oxygen is typically <0.5 mg/L. In beer, the microflora is also controlled by the presence of CO_2(~0.5% w/v). Further, the hops contain a number antimicrobial compounds, including α- and β-acids, iso-humulone, *trans*-isohumulone, *trans*-humulinic acid, and colupolone. Most lactic acid bacteria are inhibited by the levels of hop bitters used (15–55 mg/L of iso-alpha-acids). *Leuconostoc* and *Lactococcus* spp. are sensitive, as are most *Lactobacillus* and *Pediococcus* spp. However, some *Lactobacillus* and *Pediococcus* spp. isolated from spoiled beer can resist concentrations >65 mg/L of these acids (Simpson, 1993).

Winemaking. Aseptic crushing of sound, mature grapes will yield a total yeast population of 10^3–10^5 cfu/mL. The apiculate yeast *Kloeckera apiculata* and *Hanseniaspora* species dominant, accounting for 50–70% of the population. Lesser populations of *Candida, Cryptococcus, Rhodotorula, Pichia,* and *Kluyveromyces* species will also be present. Fermentative species, i.e. *Saccharomyces* species, occur only at low levels (Martini and Martini, 1990). Numerous factors affect this population, including climate and weather at the time of picking, use of fungicides in the vineyard, and physical damage to the grapes (Fleet and Heard, 1992). The surfaces of winery equipment are also important sources of yeasts and other contaminants. Importantly, the winery equipment provides an important source of *Sacc. cerevisiae*, which also may or may not be added as a starter culture.

During fermentation, the viable populations of yeasts increase to 10^8–10^9 cfu/mL, with growth following the typical growth curve for microorganisms in batch culture. In the absence of a starter culture, a range of yeast species become established, including various species of *Kloeckera, Hanseniaspora, Candida*, and *Pichia*. These genera die off within 2–3 days of the start of fermentation, and are replaced by *Saccharomyces* species, which are able to withstand the increasing concentrations of alcohol produced (Fleet and Heard, 1992).

The fermentation process is influenced by the use of sulfur dioxide in the fermentation, traditionally considered to limit growth of yeasts other than *Saccharomyces* species. Recently, this view has been challenged (Heard and Fleet, 1988a). Fermentation temperature affects both the rate of fermentation and, probably more important, the ecology of the process (Heard and Fleet, 1988b; Fleet and Heard, 1992).

Lactic acid bacteria are also an integral component in the ecology of winemaking. The principal species associated with grape fermentation are *Leuconostoc oenos, Pediococcus parvulus, Ped. pentosaceus*, and various *Lactobacillus* species (Fleet and Heard, 1992).

B *Spoilage*

Spoilage of beer. In the wort and during the first days of fermentation, oxygen is sufficient and the pH is relatively high (5.0–5.5), consequently enterobacteria (e.g. *Obesumbacterium proteus, Hafnia protea, Rahnella aquatilis, Enterobacter agglomerans*, and *Klebsiella terrigena*) may proliferate, affecting the fermentation process and causing sulfury off-flavors in the finished beer. Further, aerobic yeasts

Table 17.1 *Lactobacillus* and *Pediococcus* species ranked according to importance in beer spoilage[a]

Lactobacillus			
Group 1	Group 2	Group 3	Pediococcus
Lb. brevis	*Lb. brevisimilis*	*Lb. delbrueckii*	*Ped. damnosus*
Lb. lindneri	*Lb. malefermentans*	*Lb. fermentum*	*Ped. inopinatus*
Lb. curvatus	*Lb. parabuchneri*	*Lb. fructivorans*	*Ped. dextrinicus*[b]
Lb. casei			*Ped. pentosaceus*[b]
Lb. buchneri			
Lb. coryneformis			
Lb. plantarum			

[a]From Back (1987), Farrow *et al.* (1988), and Priest (1996). It should be noted that not all of the *Lactobacillus* species are now recognized as valid (Hammes and Vogel, 1995).
[b]Usually, no growth occurs in beer due to sensitivity to low pH, but may occur in pitching yeast and fermenting wort.

including *Pichia, Brettanomyces, Dekkera, Debaromyces, Filobasidium,* and *Candida* spp. may grow and produce acetic acid and esters, changing the flavor of the finished beer.

The lactic acid bacteria are considered the most hazardous of all beer spoilage microorganisms, especially *Ped. damnosus.* Whether introduced during fermentation or at bottling, LAB may grow in the beer, producing diacetyl (e.g. *Ped. damnosus* and *Lb. brevis*) or fruity (*Micrococcus kristinae*) off-flavors in the beer. The most important spoilage species in *Lactobacillus* are listed in Table 17.1.

Wild yeasts, defined as "yeasts not deliberately used and not under full control" (Priest, 1981) can grow in the fermenting beer, affecting the fermentation and producing off-flavors, such as phenolics. The group of wild yeasts able to spoil beer is very diverse and is normally divided into *Saccharomyces* and non-*Saccharomyces* wild yeasts. The *Saccharomyces* wild yeasts, dominated by *Sacc. cerevisiae* and *Sacc. pastorianus* (Jespersen *et al.*, 2000), are often regarded as the most hazardous. In a study of yeast samples from 45 lager breweries, >50% of the wild yeast strains isolated belonged to *Sacc. cerevisiae* (van der Kühle and Jespersen, 1998). Wild yeasts other than *Saccharomyces* include *Pichia, Candida, Kluyveromyces, Torulaspora, Brettanomyces*, and *Zygosaccharomyces* species (Campell and Msongo, 1991; Campell, 1996; van der Kühle and Jespersen, 1998).

Over the years, brewers have had great success reducing the levels of oxygen in the beer to achieve better taste stability. The production of non-pasteurized or flash pasteurized beer is also increasing. As a result of these developments, strictly anaerobic Gram-negative bacteria including *Pectinatus cerevisiiphilius* and *Megashaera* spp. are becoming more and more important as beer spoilers, especially in beers with pH >4.1–4.3 and ethanol <5% (w/v). Growth of these bacteria may occur during maturation, but they are most often seen as contaminants during bottling. The anaerobic bacteria result in strong off-flavors of the beer due to their production of fatty acids (propionic, acetic, succinic, and butyric acids), mercaptans, dimethyl sulfide, and hydrogen sulfide (Seidel-Rüfer, 1990).

Spoilage of wines. Similarly, growth of unwanted microorganisms on the grapes, in the must, during fermentation and maturation of wine can result in off-flavors and may affect the primary and secondary fermentation. As with beer, wild yeasts including *Zygosaccharomyces, Brettanomyces,* and *Schizosaccharomyces* play an important role as spoilers causing estery, mousy, or phenolic off-flavors, haze or turbidity. Some yeasts can also produce low levels of a variety of sulfur compounds, adversely affecting aroma and flavor (Rauhut, 1992).

The oxygen levels of fermenting grapes and wines are higher than that of beer allowing acetic acid bacteria, including *Gluconobacter* and *Acetobacter* spp., to produce acetic acid or even ropiness. Lactic acid bacteria (*Lactobacillus, Pediococcus,* and *Leuconostoc* spp.) are to a large extent controlled by the alcohol but may cause acidity, diacetyl off-flavors, mousiness, etc., and ropiness. Spore-forming *Clostridium* and *Bacillus* spp. may also produce off-flavors, especially in wines of low acidity.

Table 17.2 Occurrence of ochratoxin A in beer

Origin of sample	Incidence (%)	Detection limit (μg/kg)	Positives (μg/kg) Mean	Range
Germany	111/358 (31)	0.1	NS	0.1–1.5
Canada (11 were imports)	26/41 (63)	0.01	0.06	0.01–0.2
United Kingdom (14 were imports)	14/16 (88)	0.002	0.014	0.002–0.052
Switzerland	7/7 (100)	0.01	0.012 (median)	0.01–0.033
Total	158/422 (37)			

Adapted from Scott (1996).

A major problem in wines is the growth of the spoilage yeast *Zygosaccharomyces bailii*. This yeast has a high tolerance to both ethanol and preservatives, as well as the ability to grow in high sugar concentrations (Sponholz, 1992). It can cause spoilage of white wines that contain residual sugars (glucose) by carbon dioxide production in the bottle, or by growth, which produces an undesirable haze. In warmer climates, white wines are often filter sterilized through membrane filters just before bottling to reduce the hazard from this yeast. *Brettanomyces* species can also cause wine spoilage by the production of haze or volatile acidic off-flavors (Sponholz, 1992). Certain other yeasts (*Candida, Metschnikowia,* and *Pichia* spp.) can form films on bottled wines and produce off-flavors due to acetaldehyde production (Sponholz, 1992).

Acetic acid and lactic acid bacteria occasionally cause wine spoilage also (Sponholz, 1992).

A variety of moulds may cause earthy and corky off-flavors mainly from growth on corks, wooden barrels, etc. (Lee and Simpson, 1993; Chattonet *et al.*, 1994). The principal species are *Penicillium glabrum* and *Pen. spinulosum* (Lee and Simpson, 1992). Several possible mechanisms for cork taint exist, and these are revieweded by Lee and Simpson (1992).

C *Pathogens*

There have been no reports of illness due to enteropathogenic microorganisms associated with beer or wine.

Mycotoxins in beer. Because beer is made from barley, it might be expected occasionally to contain ochratoxin A. Data from 422 samples of beer are summarized in Table 17.2. Although 158 (37%) of the samples listed here were positive for ochratoxin A, the maximum level observed was only 1.5 μg/kg. Reported means ranged from 0.06 to 0.8 μg/kg. The latter figure was exceptional, and the true mean for European beers is probably near 0.1 μg/kg (Table 17.2). The beer fermentation process resulted in a 2–13% reduction in ochratoxin A contamination (Scott *et al.*, 1995; Scott, 1996).

Mycotoxins in wines. Attention was drawn to the possibility of ochratoxin A contamination of wines by Zimmerli and Dick (1995, 1996), who developed sensitive methodology for ochratoxin detection. Extensive data on ochratoxin A in wines from Europe and worldwide is now available and is summarized in Tables 17.3 and 17.4. Although the results indicate clearly that a high proportion of wines contain ochratoxin A, levels reported have almost always been low.

The highest ochratoxin A level observed in more than 200 samples reported in Table 17.3 was 7 μg/L in an Italian red wine. Methods of reporting make quantitative data difficult to extract, but it appears that figure was exceptional. Few of these or many other reported samples from Europe and elsewhere contained more than 1 μg/L ochratoxin A. Höhler (1998) reported that wines from Germany, France, and Spain had median concentrations of ochratoxin A <0.15 μg/L, samples from Italy, Portugal, and Macedonia contained median levels a little higher, i.e. 0.3–0.4 μg/L, whereas the median for five

Table 17.3 Occurrence of ochratoxin A (ng/L) in major types of European wines

Type	Number of samples	Number (%) positive	90th percentile	Maximum
White	58	14 (24)	400	1400
Rosé	51	18 (35)	200	2400
Red	172	79 (46)	500	7000
Total	281	112 (40)	400	7000

From Majerus *et al.* (2000). Limit of detection 10 ng/L.

Table 17.4 Occurrence of ochratoxin A (ng/L) in wines worldwide

Country	No. of samples	Positive (%)	Median	Maximum
Germany	11	3 (27)	30	240
Switzerland	1	1 (100)	–	70
France	20	6 (30)	130	780
Italy	10	7 (70)	300	7000
Spain	6	3 (50)	130	190
Greece	2	1 (50)	–	110
Portugal	2	2 (100)	320	340
Macedonia	6	4 (67)	430	890
Tunisia	5	5 (100)	1630	1850
South Africa	2	1 (50)	–	50
USA	2	1 (50)	–	80
Chile	9	2 (20)	210	210
Australia	5	1 (20)		220
Total	81	37 (46)		7000

From Höhler (1998).

samples from Tunisia was 1.6 μg/kg (Table 17.4). In a recent study of 600 wines from Australia, only one contained in excess of 0.5 μg/kg (Hocking *et al.*, 2003).

However, because wine consumption is high in some populations, wine cannot be ignored as a source of ochratoxin A, especially in regions where this toxin is present in other foods as well (Pitt and Tomaska, 2002).

D CONTROL (fermented beverages)

Summary

Significant hazards	• Ochratoxin A (wine).
Control measures	
Initial level (H_0)	• Use sound fruit.
Increase (ΣI)	• None.
Reduction (ΣR)	• Fermentation may reduce ochratoxin levels.
Testing	• Routine testing is not recommended under normal circumstances. • Ochratoxin A testing may be necessary if the condition of the fruit used for production is not known.
Spoilage	• Yeast starter culture control. • Pasteurization and filtration can reduce the potential for spoilage.

Comments

Beer. Although ochratoxin A has been reported in beer, reported levels are low and stringent controls are therefore not considered necessary in routine beer production. Use of high-quality grain ingredients will easily control the hazard. Microbiological considerations in the control of spoilage of beer follow.

Culture yeast. The culture yeast must be free from bacteria or wild yeast contaminants. The yeast produced during the fermentation is harvested and reused for several generations. Each cycle increases the potential for contamination of the yeast. The analytical quality control of the yeast culture is challenging, as it is necessary to detect low levels of contaminating bacteria and wild yeasts, including *Saccharomyces* spp., in the presence of up to 10^8 cfu/mL of yeast culture. One of the preferred methods is to use media containing levels of $CuSO_4$ sufficient to inhibit the culture yeast, but allowing the contaminants to grow.

Contamination of wort, yeast, fermenting wort, or beer from equipment. A high sanitary status of all process equipment in the brewery after wort boiling is essential to avoid microbiological contamination. Aseptic bottling of sterile filtered or plate pasteurized beer requires tightened hygienic conditions in the bottling area. Cleaning of the lines often includes thermal and chemical sanitation. Lactic acid and anaerobic Gram-negative bacteria (*Pectinatus* and *Megashaera* species) are known to be introduced during bottling, later establishing themselves in lubricants, drains, etc.

Pasteurizing equipment must be properly maintained to ensure effective processing. Pasteurization failure may occur with plate pasteurizers when cracks or pin holes in the plates allow contact between pasteurized and non-pasteurized beer or water. Non-uniform temperature distribution can also result in pasteurization failures in tunnel pasteurizers due to poor cleaning and maintenance of the pasteurizer.

Wine. In production of wines, controlled use of $SO_2$and controlled temperatures during fermentation plays an important role in controlling the native lactic acid bacteria and acetic acid bacteria and to ensure the numerical superiority of the culture yeast to the native yeasts (Fugelsang, 1997). As in the breweries, a high sanitary status of all equipment and facilities is essential to avoid microbial spoilage. To a certain extent, preservatives like sorbic acid, potassium sorbate, $SO_2$and to a lower extent, dimethyldicarbonate are used to inhibit the development of microorganisms in wines, especially alcohol-free wines.

Filter sterilization of white wine is practiced in some countries especially to prevent growth of *Zygosacc. bailii.*

References

Back, W. (1987) Neubeschreibung einer bierschaedlichen Laktobazillen—Art. *Lactobacillus brevisimilis* spec. nov. Monatsschr. Brau, **17**, 484–8.

Campell, I. (1996) Wild yeast in brewing and distilling, in *Brewing Microbiology* (eds F.G. Priest, I Campell), 2nd ed, Chapman and Hall, London, pp. 193–208.

Campell, I. and Msongo, H.S. (1991) Growth of aerobic wild yeast. *J. Inst. Brew.*, **97**, 279–82.

Chattonet, P., Guimberteau, D., Dubourdieu, D. and Boidron, J.N. (1994) Nature et origine des odeurs de 'moisi' dans les caves. Incidences sur la contamination des vins. *J. Int. Sci. Vigne Vin*, **28**(2), 131–51.

Farrow, J.A.E., Phillips, B.A. and Collins, M.D. (1988). Nucleic acid studies on some heterofermentative lactobacilli: description of *Lactobacillus malefermentans* sp. nov. and *Lactobacillus parabuchneri* sp. nov. *FEMS Microbiol. Lett.*, **55**, 163–8.

Flannigan, B. (1969) Microflora of dried barley grain. *Trans. Br. Mycol. Soc.*, **53**, 371–9.

Flannigan, B. (1986) *Aspergillus clavatus*—an allergenic, toxigenic deteriogen of cereals and cereal products. *Int. Biodeterior.*, **22**, 79–89.

Flannigan, B. (1996) The microflora of barley and malt, in *Brewing Microbiology* (eds F.G. Priest and I. Campbell), Chapman and Hall, London, pp. 83–125.

Flannigan, B., Day, S.W., Douglas, P.E. and McFarlane, G.B. (1984) Growth of mycotoxin-producing fungi associated with malting of barley, in *Toxigenic Fungi—their Toxins and Health Hazard* (eds H. Kurata and Y. Ueno), Elsevier, Amsterdam, pp. 52–60.

Fleet, G.H. (ed). (1992) *Wine Microbiology and Biotechnology*, Harwood Academic Publishers, Chur, Switzerland.

Fleet, G.H. and Heard, G.M. (1992) Yeasts—growth during fermentation, in *Wine Microbiology and Biotechnology* (ed. G.H. Fleet), Harwood Academic Publishers, Chur, Switzerland, pp. 27–54.

Fugelsang, K.L. (1997) *Wine microbiology*, Chapman and Hall, London, pp. 117–42.

Hammes, W.P. and Vogel, R.F. (1995) The genus *Lactobacillus*, in *The Lactic Acid Bacteria. 2. The Genera of Lactic Acid Bacteria* (eds B.J.B. Wood and W.H.Holzapfel), Blackie Academic and Professional, London, pp. 19–54.

Heard, G.M. and Fleet, G.H. (1988a) The effect of sulfur dioxide on yeast growth during natural and inoculated wine fermentation. *Aust. NZ Wine Ind. J.*, **3**, 57–60.

Heard, G.M. and Fleet, G.H. (1988b) The effects of temperature and pH on the growth of yeast species during the fermentation of grape juice. *J. Appl. Bacteriol.*, **65**, 23–8.

Höhler, D. (1998) Ochratoxin A in food and feed: occurrence, legislation and mode of action. *Z. Ernähringswiss.*, **37**, 2–12.

Hocking, A.D., Varelis, P., Pitt, J.I., Cameron, S.F. and Leong, S.-L.L. (2003) Occurrence of ochratoxin A in Australian wine. *Aust. J. Grape Wine Res.*, **9**, 72–8.

Jespersen, L., van der Kühle, A. and Petersen, K.M. (2000) Phenotypic and genetic diversity of *Saccharomyces* contaminants isolated from lager breweries and their phylogenetic relationship with brewing yeasts. *Int. J. Food Microbiol.*, **60**, 43–53.

Lee, T.H. and Simpson, R.F. (1992) Microbiology and chemistry of cork taints in wine, in *Wine Microbiology and Biotechnology* (ed G.H. Fleet), Harwood Academic Publishers, Chur, Switzerland, pp. 353–72.

Majerus, P., Bresch, H. and Otteneder, H. (2000) Ochratoxin A in wines, fruit juices and seasonings. *Arch. Lebensmittelhyg.*, **51**, 95–7.

Martini, A. and Martini, A.V. (1990) Grape must fermentation past and present, in *Yeast Technology* (eds J.F.T. Spencer and D.M. Spencer), Springer Verlag, Berlin, pp. 105–23.

Petters, H.I., Flannigan, B. and Austin, B. (1988) Quantitative and qualitative studies of the microflora of barley malt production. *J. Appl. Bacteriol.*, **65**, 279–97.

Priest, F.G. (1981) in *An Introduction to Brewing Science and Technology, Part II*. The Institute of Brewing, London, pp. 23–31.

Priest, F.G. (1996) Gram-positive brewery bacteria, in *Brewing Microbiology*. (eds F.G. Priest and I. Campell), 2nd edn, Chapman and London, pp. 127–61.

Rankine, B. (1989) Making Good Wine, in *A Manual of Winemaking Practice for Australia and New Zealand*, Sun Books, Sydney.

Rauhut, D. (1992). Yeasts—production of sulfur compounds, in *Wine Microbiology and Biotechnology* (ed G.H. Fleet), Harwood Academic Publishers, Chur, Switzerland, pp. 183–203.

Riddle, H.F.V., Channell, S., Blyth, W., Weir, D.M., Lloyd, M., Amos, W.M.G. and Grant, I.W.B. (1968) Allergic alveolitis in a maltworker. *Thorax*, **23**, 271–80.

Scott, P. (1996) Mycotoxins transmitted into beer from contaminated grains during brewing. *J. AOAC Int.*, **79**, 875–82.

Scott, P.M., Kanhere, S.R., Lawrence, G.A., Daley, E.F. and Farber, J.M. (1995) Fermentation of wort containing added ochratoxin A and fumonisins B_1 and B_2. *Food Addit. Contam.*, **12**, 31–40.

Seidel-Rüfer, H. (1990) Pectinatus und andere morphologisch ähnliche Gram-negative, anaerobe Stäbchen aud dem Brauereibereich. *Monatsschr. Brau.*, **43**, 101–5.

Shlosberg, A., Zadikov, I., Perl, S., Yakobson, B., Varod, Y., Elad, D., Rapoport, E. and Handji, V. (1991) *Aspergillus clavatus* as the probable cause of a lethal mass mycotoxicosis in sheep. *Mycopathologia*, **114**, 35–9.

Simpson, W.J. (1993) Studies on the sensitivity of lactic acid bacteria to hop bitter acids. *J. Inst. Brew.*, **99**, 405–11.

Van der Kühle, A. and Jespersen, L. (1998) Detection and identification of wild yeasts in lager breweries. *Int. J. Food Microbiol.*, **43**, 205–13.

Walker, G.M. (2000) Microbiology of wine making in *Encyclopoedia of Food Microbiology* (eds R.K.Robinson, C.A.Batt and P.D. Patel), Academic Press, London.

Zimmerli, B. and Dick, R. (1995) Determination of ochratoxin A at the ppt level in human blood, serum, milk and some foodstuffs by high-performance liquid chromatography with enhanced fluorescence detection and immunoaffinity column cleanup methodology and Swiss data. *J. Chromatogr. B*, **666**, 85–99.

Zimmerli, B. and Dick, R. (1996) Ochratoxin A in table wine and grape-juice: occurrence and risk assessment. *Food Addit. Contam.*, **13**, 655–68.

Appendix I Objectives and accomplishments of the ICMSF

History and purpose

The International Commission on Microbiological Specifications for Foods (ICMSF, the Commission) was formed in 1962 through the action of the International Committee on Food Microbiology and Hygiene, a committee of the International Union of Microbiological Societies (IUMS). Through the IUMS, the ICMSF is linked to the International Union of Biological Societies (IUBS) and to the World Health Organization (WHO) of the United Nations.

In the 1960s, there was growing recognition of food-borne disease and greatly increased microbiological testing of foods. This, in turn, created unforeseen problems in international trade in foods. Different analytical methods and sampling plans of doubtful statistical validity were being used. Furthermore, analytical results were interpreted using different concepts of biological significance and acceptance criteria, creating confusion and frustration for both the food industry and the regulatory agencies.

In this environment, the ICMSF was founded to (i) assemble, correlate, and evaluate evidence about the microbiological safety and quality of foods; (ii) consider whether microbiological criteria would improve and assure the microbiological safety of particular foods; (iii) propose, where appropriate, such criteria, and (iv) recommend methods of sampling and examination.

Forty years later, the primary role of the Commission remains to give guidance on: (i) appraising and controlling the microbiological safety of foods and (ii) microbiological quality, since this influences consumer acceptance and the losses due to spoilage. Meeting those objectives assists international trade, national control agencies, the food industry, international agencies concerned with humanitarian food distribution and consumer interests.

Functions and membership

The ICMSF provides basic scientific information through extensive study, and makes recommendations without prejudice on the basis of that information. Results of the studies are published as books, discussion documents, or refereed papers. Major publications of the Commission are listed in Appendix III.

The ICMSF functions as a Working Party, not as a forum reading papers. Meetings consist of discussions within subcommittees, debating to achieve consensus, editing draft materials, and planning. Most work is done between meetings by the Editorial Committee and members, sometimes with the help of non-member consultants.

Since 1962, 33 meetings have been held in 20 countries (Australia, Brazil, Canada, Chile, Denmark, Dominican Republic, Egypt, England, France, Germany, Italy, Mexico, South Africa, Spain, Switzerland, The Netherlands, USA, the former USSR, Venezuela, and the former Yugoslavia). During its meetings, Commission members frequently participate in symposia organized by microbiologists or public health officials of the host country.

Currently, the membership consists of 16 food microbiologists from 11 countries, with combined professional interests in research, public health, official food control, education, product and process development, and quality control, from government laboratories in public health, agriculture, and food technology; from universities; and from the food industry (see Appendix II). The ICMSF is also assisted by consultants, specialists in particular areas of microbiology, who are critical to the success of the

Commission (see Appendix II for lists of the consultants, contributors, and reviewers). New members and consultants are selected for their expertise, not as national delegates. All work is voluntary without fees or honoraria.

Two sub-commissions (Latin American and South-East Asian) promote activities of the ICMSF among food microbiologists in their regions and facilitate communication worldwide (see Appendix II).

The ICMSF raises its own funds to support its meetings. Support has been obtained from government agencies, WHO, IUMS, IUBS, and the food industry (over 80 food companies and agencies in 13 countries). Grants for specific projects and seminars/conferences have been provided by a variety of sources. Some funds are received from the sale of its books.

Recent projects

Microorganisms in Foods 5. Characteristics of Microbial Pathogens (1996) is a thorough, but concise, review of the literature on growth, survival, and death responses of food-borne pathogens. It is intended as a quick reference manual to assist in making decisions in support of HACCP plans and to improve food safety.

Microorganisms in Foods 6. Microbial Ecology of Food Commodities (1998) updates and extends ICMSF (1980b). For 16 commodity areas, it describes the initial microbial flora and the prevalence of pathogens, the microbiological consequences of processing, typical spoilage patterns, episodes implicating those commodities with food-borne illness, and measures to control pathogens.

Microorganisms in Foods 7. Microbiological Testing in Food Safety Management (2002) introduces the concept of food safety objectives (FSO) and their use for the establishment of HACCP plans and microbiological criteria. The book gives an update of the statistical aspects of sampling and the choice of the "cases" which determine the stringency of sampling plans. It replaces as such the first part of *Microorganisms in Foods 2:* Sampling for Microbiological analysis: Principles and Specific Applications (1986). It illustrates how systems such as HACCP and GHP provide greater assurance of safety than microbiological testing, but also identifies circumstances in which microbiological testing still plays a useful role.

Microorganisms in Foods 6. Microbial Ecology of Food Commodities 2nd edition (2005) keeps the overall structure of each chapter, brings up-to-date consideration of the pathogens of concern and, particularly, treats the means by which those pathogens can be controlled systematically.

Discussion documents prepared for the Joint Food and Agriculture Organization (FAO) and World Health Organization (WHO) Food Standards Program, and Codex Alimentarius Commission.

1. Establishment of sampling plans for microbiological safety criteria for foods in international trade.
2. Discussion of sampling plans for *L. monocytogenes*, *Salmonella*, *Campylobacter*, and verocytotoxin-producing *E. coli* in foods in international trade.
3. Recommendations for the future management of microbiological hazards for foods in international trade.
4. Principles for the establishment of FSO and related control measures.

The recommendations of ICMSF for sampling foods and acceptance criteria for *Listeria monocytogenes* were subsequently published as "Sampling plans for *L. monocytogenes*" (*Int. J. Food Microbiol.*, 1994, **22**, 89–96), as was "Establishment of microbiological safety criteria for foods in international trade" (*World Health Stat. Q.*, 1997, **50**, 119–23).

At the request of the Secretariat of Codex, the ICMSF developed recommendations for revision of Principles for the Establishment and Application of Microbiological Criteria for Foods, published in the Procedural Manual of Codex.

Addressing the need for a scientific basis in risk assessment, a Working Group of the ICMSF published "Potential application of risk assessment techniques to microbiological issues related to international trade in food and food products" (*J. Food Protect.*, 1998, **61** (8), 1075–86).

Past and future

For almost 25 years, the major efforts of the ICMSF were devoted to methodology. This resulted in improved comparisons of microbiological methods and better standardization (17 refereed publications). Among many significant findings it was established that, when analyzing for salmonellae, analytical samples could be bulked (composited) into a single test with no loss of sensitivity. This made practical the collection and analysis of the large number of samples recommended in some sampling plans.

With the rapid development of alternative methods and rapid test kits, and the ever expanding list of biological agents involved in food-borne illness, the Commission reluctantly discontinued its program of comparison and evaluation of methods, recognizing that issues of methodology were being addressed effectively by other organizations.

A long-term objective of the Commission has been to enhance the microbiological safety of foods in international commerce. This was initially addressed through two books that recommended uniform analytical methods (ICMSF, 1978), and sound sampling plans and criteria (ICMSF, 1974, 1978, 2nd edn, 1986). The Commission then developed a book on the microbial ecology of foods (ICMSF 1980a,b) intended to familiarize analysts with processes used in the food industry and microbiological aspects of foods submitted to the laboratory. Knowledge of the microbiology of the major food commodities, and the factors affecting the microbial content of these foods, helps the analyst to interpret analytical results.

At an early stage, the Commission recognized that no sampling plan can ensure the absence of a pathogen in food. Testing foods at ports of entry, or elsewhere in the food chain, cannot guarantee food safety. This led the Commission to explore the potential value of HACCP for enhancing food safety. A meeting in 1980 with the WHO led to a report on the use of HACCP for controlling microbiological hazards in food, particularly in developing countries (ICMSF, 1982). The Commission then developed a book on the principles of HACCP and procedures for developing HACCP plans (ICMSF, 1988), covering the importance of controlling the conditions of producing/harvesting, preparing, and handling foods. Recommendations are given for the application of HACCP from production/harvest to consumption, together with examples of how HACCP can be applied at each step in the food chain.

The Commission next recognized that a major weakness in the development of HACCP plans is the process of hazard analysis. It has become more difficult to be knowledgeable about the many biological agents recognized as responsible for food-borne illness. ICMSF (1996) summarizes important information about the properties of biological agents commonly involved in food-borne illness, and serves as a quick reference manual when making judgments on the growth, survival, or death of pathogens.

Subsequently, the Commission updated its volume on the microbial ecology of food commodities (ICMSF, 1998).

Microorganisms in Foods 7. The Role of Microbiological Testing in Systems Managing Food Safety (2002) illustrated how systems such as HACCP and GHP provide greater assurance of safety than microbiological testing, but also identified circumstances where microbiological testing still plays a useful role. It also introduced the concept of FSO as a public health goal to be met to provide the appropriate level of health protection.

We believe that the original objectives of the Commission are still relevant today. The European Union, the many other political changes occurring throughout the world, the growth of developing countries seeking export markets, and the increased trade in foods worldwide, as evidenced by the passage of GATT and NAFTA, all point to the continuing need for the independent recommendations, such as those of the Commission. It is essential that import/export policies be established as uniformly as possible and on a sound scientific basis. The overall goal of the Commission will continue to be to enhance the safety of foods moving in international commerce. The Commission will continue to strive to meet this goal through a combination of educational materials, promoting the use of food safety management systems using microbiological FSO, HACCP and GHP, and recommending sampling plans and microbiological criteria where they have been developed according to Codex principles and offer increased assurance of microbiological safety. The future success of the ICMSF will continue to depend upon the efforts of members, support from consultants who generously volunteer their time, and those who provide the financial support so essential to the activities of the Commission.

Appendix II ICMSF participants

Officers

Dr. M. B. Cole (from 2000), National Center for Food Safety and Technology (NCFST), 6502 S. Archer Road, Summit-Argo, Illinois 60501, USA.

Secretary

Prof. Lone Gram (from 2003), Danish Institute for Fisheries Research, Department of Seafood Research, Soltofts Plads, c/o Danish Technical University Bldg 221, DK-2800 Kgs. Lyngby, Denmark.
Prof. Mike van Schothorst, Food Safety Consultant, Ch. du Grammont 20, La Tour-de-Peilz, CH-1814, Switzerland (retired 2003).

Treasurer

Dr. Jeffrey M. Farber (from 2000), Health Canada, Food Directorate, Microbiology Research Division, Banting Research Centre, Tunney's Pasture, Ottawa, Ontario K1A OL2, Canada.

Members

Dr. Robert L. Buchanan, U.S. Food and Drug Administration, Center for Food Safety and Applied Nutrition, 5100 Paint Branch Parkway, College Park, MD 20740, USA.
Dr. Jean-Louis Cordier, Quality Management, Nestec. SA, Av. Nestlé, CH-1800, Vevey, Switzerland.
Dr. Susanne Dahms, COE Biometrics Europe, Schering AG, D-13342 Berlin, Germany.
Dr. R.S. Flowers, Silliker Laboratories, 900 Maple Road, Homewood, Illinois 80430, USA.
Prof. Bernadette D.G.M. Franco, Departamento de Alimentos e Nutricao Experimental, Faculdade de Ciencias Farmaceuticas, Universidade de São Paulo, Av. Prof. Lineu Prestes 580, 05508-900, São Paulo, SP, Brazil.
Prof. Leon Gorris, Quantitative Hazard Assessment, Unilever, Colworth House, Sharnbrook (Bedford) MK44 1LQ, England.
Dr. Fumiko Kasuga, Division of Safety Information on Drugs, Food and Chemicals, National Institute of Health Sciences, 1-18-1 Kamiyoga, Setagaya-ku, Tokyo 158-8501, Japan.
Prof. Jean-Louis Jouve, Food Quality and Standard Service, Food and Nutrition Division, Food and Agriculture Organization of the United Nations, Via delle Terme di Caracalla, 0100 Rome, Italy.
Dr. Fumiko Kasuga, Section Chief, Division of Safety Information on Drugs, Food and Chemicals, National Institute of Health Sciences, 1-18-1 Kamiyoga, Setagaya-ku, Tokyo 158-8501, Japan.
Dr. Anna M. Lammerding, Food Safety Risk Assessment Unit, Laboratory for Zoonosis, Health Canada, 160 Research Lane, Guelph, Ontario N1G 5BZ, Canada.
Ms. Zahara Merican, ZM Consultancy, 56 B Jalan TR 2/2 Tropicana G&C Resort, 47410 Petaling Jaya, Selangor, Malaysia.
Dr. John I. Pitt, Honorary Research Fellow, Food Science Australia, P.O. Box 52, North Ryde NSW 1670, Australia (retired 2002).

Dr. Morris Potter, Center for Food Safety and Applied Nutrition, FDA, 60 Eighth Street, NE, Atlanta, GA 30309, USA.

Dr. Terry A. Roberts, Food Safety Consultant, 59 Edenham Crescent, Reading RG1 6HU, UK (retired 2000).

Dr. Katherine M.J. Swanson, Mendota Heights, MN 55120, USA.

Dr. Paul Teufel, Institute for Hygiene and Food Safety, Federal Dairy Research Centre, Hermann-Weigmann Strasse 1, D-24103 Kiel, Germany.

Dr. R. Bruce Tompkin, Food Safety Consultant, 1319 West 54th Street, La Grange, IL 60525, USA (retired 2002).

Past members of the ICMSF

Dr. A.C. Baird-Parker	UK	1974–1999
Dr. M.T. Bartram	USA	1967–1968
Dr. H.E. Bauman	USA	196–1977
Dr. F.L. Bryan	USA	1974–1996[a]
Dr. L.Buchbinder*	USA	1962–1965
Prof. F.F. Busta	USA	1985–2000[b]
Dr. R. Buttiaux	France	1962–1967
Dr. J.H.B. Christian	Australia	1971–1991[c]
Dr. D.S.Clark	Canada	1963–1985[d]
Dr. C. Cominazzini	Italy	1962–1983
Dr. C.E. Dolman*	Canada	1962–1973
Dr. M.P. Doyle	USA	1989–1999
Dr. R.P. Elliott*	USA	1962–1977
Dr. Otto Emberger	Czechoslovakia	1971–1986
Dr. M. Eyles	Australia	1996–1999
Dr. J.Farkas	Hungary	1991–1998
Mrs. Mildred Galton*	USA	1962–1968
Dr. E.J. Gangarosa	USA	1969–1970
Dr. F. Grau	Australia	1985–1999
Dr. J.M. Goepfert	Canada	1985–1989[e]
Dr. H.E. Goresline*	USA/Austria	1962–1970
Dr. Betty C. Hobbs*	UK	1962–1996
Dr. A. Hurst	UK/Canada	1963–1969
Dr. H. Iida	Japan	1966–1977
Dr. M. Ingram*	UK	1962–1974[f]
Dr. M. Kalember-Radosavljevic	Former Yugoslavia	1983–1992
Dr. K. Lewis*	USA	1962–1982
Dr. John Liston	USA	1978–1991
Dr. Holger Lundbeck*	Sweden	1962–1983[g]
Dr. S. Mendoza	Venezuela	1992–1998
Dr. G. Mocquot	France	1964–1980
Dr. G.K. Morris	USA	1971–1974
Dr. D.A.A. Mossel*	The Netherlands	1962–1975
Dr. N.P. Nefedjeva	USSR	1964–1979
Dr. C.F. Niven, Jr.	USA	1974–1981
Dr. P.M. Nottingham	New Zealand	1974–1986
Dr. J.C. Olson, Jr.	USA	1968–1982
Dr. John I. Pitt	Australia	1987–2002

Dr. H. Pivnick	Canada	1974–1983
Dr. T.A. Roberts	UK	1978–2000[h]
Dr. F. Quevedo	Peru	1965–1998
Dr. A.N. Sharpe	Canada	1985–1998[i]
Dr. J. Silliker	USA	1974–1987[j]
Bent Simonsen	Denmark	1963–1987
Dr. H.J. Sinell	Germany	1971–1992
Dr. G.G. Slocum*	USA	1962–1968
Dr. F.S. Thatcher*	Canada	1962–1973[k]
Dr. R.B. Tompkin	USA	1982–2002
Prof. M. van Schothorst	Switzerland	1973–2003

*Founding member.
[a]Secretary, 1981–1991.
[b]Treasurer, 1989–1998.
[c]Chairman, 1980–1991.
[d]Secretary–Treasurer, 1963–1981.
[e]Treasurer, 1987–1989.
[f]Ex-offimember, 1962–1968.
[g]Chairman, 1973–1980.
[h]Chairman, 1991–2000.
[i]Treasurer, 1989–1998.
[j]Treasurer, 1981–1987.
[k]Chairman, 1962–1973.
[l]Secretary, 1991–2003.

Members of the Latin American Subcommission

Chairperson

Dra. Maria Alina Ratto, General Manager, Microbiol S.A., Joaquin Capello 222, Lima 18, Peru. E-mail:microbl@terra.com.pe.

Secretary/Treasurer

Lic. Ricardo A, Sobol, Director Tecnico, Food Control S.A., Santiago del Estero 1154, 1075 Buenos Aires, Argentina. E-mail: 50601@foodcontrol.com.

Honorary members

Prof. Fernando Quevedo, Food Quality and Safety Assurance International, Buenos Aires 188, Miraflores, Lima 18, Peru. E-mail: fquevedo@amauta.rcp.net.pe.

Prof. Sebastião Timo Iaria, Av. Angelica 2206, apto 141, 01228-200, São Paulo, SP, Brazil. E-mail: stiaria@aol.com.br.

Prof. Silvia Mendoza, Conjunto Residencial E1, Av. Washington Torre 1A, piso 12 apto 123, Caracas, Venezuela. E-mail: silmendoza@cantr.net.

Prof. Nenufar Sosa de Caruso, Alimentarius, Tomas de Tezanos 1323, Montevideo, Uruguay. E-mail: alimenta@adinet.com.uy.

Members

Prof. Bernadette D.G.M. Franco, Departamento de Alimentos e Nutricao Experimental, Faculdade de Ciencias Farmaceuticas, Universidade de São Paulo, Av. Prof. Lineu Prestes 580, 05508-900, São Paulo, SP, Brazil. E-mail: bfranco@usp.br.

Dra. Eliana Marambio, Coventry 1046, Depto 405, Ñuñoa, Santiago, Chile. E-mail: emarambio@entelchile.net.
Profa. Janeth Luna Cortéz, Universidad de Bogota, Carrera 4 No. 22-61 Of 436, Santafé de Bogotá, DC, Colombia. E-mail: ingeneria.alimentos@utadeo.edu.co.
Dra. Dora Martha González, Sarmiento 2323, Montevideo, Uruguay. E-mail: dmgonzal@adinet.com.uy.
Profa. Pilar Hernandez S., Universidad Central de Venezuela, Apartado 40109, Caracas 1040-A, Venezuela. E-mail: hernands@camelot.rect.ucv.ve.

Former members of the Latin American Subcommission

Dra. Ethel G.V. Amato de Lagarde	Argentina
Dr. Rafael Camperchioli	Paraguay
Dr. Cesar Davila Saa	Ecuador
Dr. Mauro Faber de Freitas Leitao	Brazil
Dra. Josefina Gomez-Ruiz*	Venezuela
Dra. Yolanda Ortega de Gutierrez	Mexico
Dr. Hernan Puerta Cardona	Colombia
Dra. Elvira Regus de Pons	Dominican Republic

*Former Chairperson.

Members of the South-East Asian Subcommission

Chairperson

Ms. Zahara Merican, ZM Consultancy, 56 B Jalan TR 2/2 Tropicana G&C Resort, 47410 Petaling Jaya, Selangor, Malaysia.

Secretary

Ms. Quee Lan Yeoh, Biotechnology Research Centre, Malaysian Agricultural Research and Development Institute, P.O. Box 12301, GPO 50774 Kuala Lumpur, Malaysia.

Treasurer

Dr. Lay Koon Pho, School of Chemical and Life Sciences, Singapore Polytechnic, 500 Dover Road, Singapore 13951.

Members

Dr. Ir. Ratih Dewanti-Hariyadi, Department of Food Technology and Human Nutrition, Faculty of Agricultural Technology, Bogor Agricultural University (IBP), P.O. Box 220, Bogor, Indonesia.
Dr. Kim Loon Lor, Senior Manager, Food Research and Development, SATS Catering Pte Ltd., SATS Inflight Catering Centre, P.O. Box 3, Singapore Changi Airport, Singapore 918141
Dr. Reynaldo C. Mabesa, Assoc. Professor, Institute of Food Science and Technology, University of the Philippines at Los Banos, Los Banos, Laguna 4031, Philippines.
Ms. Wongkhalaung Chakamas, Deputy Director, Institute of Food Research and Product Development (IFRPD), Kasetsart University, P.O. Box 1043, Kasetsart, Bangkok 10903, Thailand.

Appendix III Publications of the ICMSF

Books

Food and Agriculture Organization and International Atomic Energy Agency/ICMSF (1970) Microbiological specifications and testing methods for irradiated foods. Technical Report Series No. 104, Vienna: Atomic Energy Commission.

ICMSF. (1978) *Microorganisms in Foods 1. Their Significance and Methods of Enumeration*, 2nd edn, University of Toronto Press, Toronto (ISBN 0-8020-2293-6, reprinted 1982, 1988 with revisions).

ICMSF. (1980a) *Microbial Ecology of Foods. Volume 1. Factors Affecting Life and Death of Microorganisms*,Academic Press, New York (IBSN 0-12-363501-2).

ICMSF (1980b) *Microbial Ecology of Foods. Volume 2. Food Commodities*, Academic Press: New York (IBSN 0-12-363502-0).

ICMSF (1986) *Microorganisms in Foods 2. Sampling for Microbiological Analysis: Principles and Specific Applications*, 2nd edn, University of Toronto Press, Toronto (ISBN 0-8020-5693-8). (Available outside North America from Blackwell Scientific Publications, Ltd., Osney Mead, Oxford OX2 0EL, UK, first edition: 1974; revised with corrections, 1978.)

ICMSF (1988) *Microorganisms in Foods 4. Application of the Hazard Analysis Critical Control Point (HACCP) System to Ensure Microbiological Safety and Quality*, Blackwell Scientific Publications, Oxford (ISBN 0-632-02181-0). (Also published in paperback under the title HACCP in Microbiological Safety and Quality 1988, ISBN 0 632 02181 0.)

ICMSF (1996) *Microorganisms in Foods 5. Characteristics of Microbial Pathogens*, Blackie Academic & Professional, London (ISBN 0 412 47350 X).*

ICMSF (1998) *Microorganisms in Foods 6. Microbial Ecology of Food Commodities*, Blackie Academic & Professional: London (ISBN 0 412 47350 X).*

ICMSF (2002) *Microorganisms in Foods 7. Microbial Testing in Food Safety Management*, Kluwer Academic/Plenum Publishers, New York (ISBN 0 306 47262 7).

*Available from Springer at http://www.springeronline.com.

WHO publications

1. ICMSF (Authors: Silliker, J.H., Baird-Parker, A.C., Bryan, F.L., Olson, J.C., Jr., Simonsen, B. and van Schothorst, M.)/WHO. (1982) Report of the WHO/ICMSF meeting on Hazard Analysis: Critical Control Point System in Food Hygiene, WHO/VPH/82.37, World Health Organization, Geneva (also available in French).
2. ICMSF (Authors: Simonsen, B., Bryan, F.L., Christian, J.H.B., Roberts, T.A., Silliker, J.H. and Tompkin, R.B.). (1986) Prevention and control of foodborne salmonellosis through application of the hazard analysis critical control point system. Report, International Commission on Microbiological Specifications for Foods (ICMSF), WHO/CDS/VPH/86.65, World Health Organization, Geneva.
3. Christian, J.H.B. (1983) *Microbiological Criteria for Foods* (Summary of recommendations of FAO/WHO expert consultations and working groups 1975-1981), WHO/VPH/83.54, World Health Organization, Geneva.

Other ICMSF technical papers

1. Thatcher, F.S. (1963) The microbiology of specific frozen foods in relation to public health: report of an international committee. *J. Appl. Bacteriol.*, **26**, 266–85.
2. Simonsen, B., Bryan, F.L., Christian, J.H.B., Roberts, T.A., Tompkin, R.B. and Silliker, J.H. (1987) Report from the International Commission on Microbiological Specifications for Foods (ICMSF). Prevention and control of foodborne salmonellosis through application of hazard analysis critical control point (HACCP). *Int. J. Food Microbiol.*, **4**, 227–47.
3. International Commission on Microbiological Specifications for Foods (ICMSF). (1994) Choice of sampling plan and criteria for *Listeria monocytogenes*. *Int. J. Food Microbiol.*, **22**, 89–96.
4. International Commission on Microbiological Specifications for Foods (ICMSF). (1997) Establishment of microbiological safety criteria for foods in international trade. *World Health Stat. Q.*, **50**, 119–23.
5. International Commission on Microbiological Specifications for Foods (ICMSF). (1998) Potential application of risk assessment techniques to microbiological issues related to international trade in food and food products. *J. Food Protect.*, **61** (8): 1075–86.
6. International Commission on Microbiological Specifications for Foods (ICMSF) [M van Schothorst, Secretary]. (1998) Principles for the establishment of microbiological food safety objectives and related control measures. *Food Control*, **9** (6), 379–84.

Translations

Thatcher, F.S. and Clark, D.S. (1973) *Microorganisms in Foods 1. Their Significance and Methods of Enumeration* [in Spanish: Garcia, B. (translator)], Editorial Acribia, Zaragoza, Spain.

ICMSF (1981) Microorganismos de los Alimentos 2. Métodos de Muestreo para Análisis Microbiológicos: Principios y Aplicaciones Especificas, Ordonez Pereda, J.A. and Diaz Hernandez, M.A. (translators), Editorial Acribia , Zaragoza, Spain.

ICMSF (1983) Ecología Microbiana de los Alimentos 1. Factores que Afectan a la Supervivencia de los Microorganismos en los Alimentos, Burgos Gonzalez, J. *et al.* (translators), Editorial Acribia, Zaragoza, Spain.

ICMSF (1984) Ecología Microbiana de los Alimentos 2. Productos Alimenticios, Sanz Perez, B. *et al.* (translators), Editorial Acribia, Zaragoza, Spain.

ICMSF (1988) El sistema de análisis de riesgos y puntos críticos. Su aplicación a las industrias de alimentos, Malmenda, P.D. and Garcia, B.M. (translators), Editorial Acribia, Zaragoza, Spain.

ICMSF (1996) Microorganismos de los Alimentos: Caraterísticas de los patógenos microbianos. Manuel Ramis Vergés (translator), Editorial Acribia, SA, Zaragoza, Spain.

ICMSF (1998) Microorganismos de los Alimentos: Ecología microbiana de los productos alimentarios. Bernabé Sanz Pérez, José Fernandez Salguero, Manuel Ramis Vergés, Francisco León Crespo, Juan Antonio Ordoñez Pereda (translators), Editorial Acribia, SA, Zaragoza, Spain (ISBN 84 200 0934 2).

About the ICMSF

Bartram, M.T. (1967) International microbiological standards for foods. *J. Milk Food Technol.*, **30**, 349–51.

Saa, C.C. (1968) The Latin American Subcommittee on microbiological standards and specifications for foods. *Rev. Facultad Quím. Farm.*, **7**, 8.

Cominazzini, C. (1969) The International Committee on microbiological specifications for foods and its contribution to the maintenance of food hygiene (in Italian). *Croniche Chimico*, **25**, 16.

Saa, C.C. (1969) El Comité Internacional de Especificaciones Microbiológicas de los Alimentos de la IAMS. Rev. *Facultad Quím. Farm.*, **8**, 6.

Mendoza, S. and Quevedo, F. (1971) Comisión Internacional de Especificaciones Microbiológicas de los Alimentos. *Bol. Inst. Bacteriol. Chile*, **13**, 45.

Thatcher, F.S. (1971) The International Committee on microbiological specifications for foods. Its purposes and accomplishments. *J. Assoc. Off. Anal. Chem.*, 54, 814–36.

Clark, D.S. (1977) The International Commission on Microbiological Specifications for Foods. *Food Technol.*, **32**, 51–4, 67.

Clark, D.S. (1982) International perspectives for microbiological sampling and testing of foods. *J. Food Protect.*, **45**, 667–71.

Anonymous (1984) International Commission on Microbiological Specifications for Foods. *Food Lab. Newslett.*, **1** (1), 23–25 (Box 622, S-751 26 Uppsala, Sweden).

Quevedo, F. (1985) Normalización de alimentos y salud para América Latina y el Caribe. 3. Importancia de los criterios microbiológicos. *Boletín de la Oficina Sanitaria Panamericana*, **99**, 632–40.

Bryan, F.L. and Tompkin, B.T. (1991) The International Commission on Microbiological Specifications for Foods (ICMSF). *Dairy Food Environ. Sanit.*, **11**, 66–8.

Anonymous (1996) The International Commission on Microbiological Specifications for Foods (ICMSF): update. *Food Control*, **7**, 99–101.

Index

Lightning Source UK Ltd.
Milton Keynes UK
UKOW041422040212

186627UK00008B/6/P

9 780306 486753